Structure of Materials

Blending rigorous presentation with ease of reading, this is a self-contained textbook on the fundamentals of crystallography, symmetry, and diffraction. Emphasis is placed on combining visual illustrations of crystal structures with the mathematical theory of crystallography to understand the complexity of a broad range of materials. The first half of the book describes the basics of crystallography, discussing bonding, crystal systems, symmetry, and concepts of diffraction. The second half is more advanced, focusing on different classes of materials, and building on an understanding of the simpler to more complex atomic structures. Geometric principles and computational techniques are introduced, allowing the reader to gain a full appreciation of material structure, including metallic, ceramic, amorphous, molecular solids, and nanomaterials. With over 430 illustrations, 400 homework problems, and structure files available to allow the reader to reconstruct many of the crystal structures shown throughout the text, this is suitable for a one-semester advanced undergraduate or graduate course within materials science and engineering, physics, chemistry, and geology.

Additional resources for this title, including solutions for instructors, data files for crystal structures, and appendices are available at www.cambridge.org/9780521651516.

All crystal structure illustrations in this book were made using CrystalMaker®: a crystal and molecular visualization program for Mac and Windows computers (http://www.crystalmaker.com).

Marc De Graef is a Professor in the Department of Materials Science and Engineering at the Carnegie Mellon University in Pittsburgh, USA, where he is also Co-director of the J. Earle and Mary Roberts Materials Characterization Laboratory. He received his Ph.D. in Physics in 1989 from the Catholic University of Leuven. An accomplished writer in the field, he is on the Board of Directors for the Minerals, Metals and Materials Society (TMS).

Michael E. McHenry is Professor of Materials Science and Engineering, with an appointment in Physics, at the Carnegie Mellon University in Pittsburgh, USA. He received his Ph.D. in Materials Science and Engineering in 1988 from MIT, before which he spent 3 years working in industry as a Process Engineer. Also an accomplished writer, he is Publication Chair for the Magnetism and Magnetic Materials (MMM) Conference.

Structure of Materials: An Introduction to Crystallography, Diffraction, and Symmetry

Marc De Graef
Carnegie Mellon University, Pittsburgh

Michael E. McHenry
Carnegie Mellon University, Pittsburgh

CAMBRIDGE UNIVERSITY PRESS
Cambridge, New York, Melbourne, Madrid, Cape Town, Singapore, São Paulo, Delhi, Dubai, Tokyo, Mexico City

Cambridge University Press
The Edinburgh Building, Cambridge CB2 8RU, UK

Published in the United States of America by Cambridge University Press, New York

www.cambridge.org
Information on this title: www.cambridge.org/9780521651516

First published 2007
3rd printing 2010

Printed in the United Kingdom at the University Press, Cambridge

A catalog record for this publication is available from the British Library

ISBN 978-0-521-65151-6 Hardback

in memory of Mary Ann (McHenry) Bialosky (1962–99), a devoted teacher, student, wife and mother, who was taken from us much too soon

M. E. M.

for Marie, Pieter, and Erika

M. D. G.

Contents

Preface

In the movie *Shadowlands*,[1] Anthony Hopkins plays the role of the famous writer and educator, C. S. Lewis. In one scene, Lewis asks a probing question of a student: *"Why do we read?"* (Which could very well be rephrased: *Why do we study?* or *Why do we learn?*) The answer given is simple and provocative: *"We read to know that we are not alone."* It is comforting to view education in this light. In our search to know that we are not alone, we connect our thoughts, ideas, and struggles to the thoughts, ideas, and struggles of those who preceded us. We leave our own thoughts for those who will follow us, so that they, too, will know that they are not alone. In developing the subject matter covered in this book, we (MEM and MDG) were both humbled and inspired by the achievements of the great philosophers, mathematicians, and scientists who have contributed to this field. It is our fervent hope that this text will, in some measure, inspire new students to connect their own thoughts and ideas with those of the great thinkers who have struggled before them and leave new and improved ideas for those who will struggle afterwards.

The title of this book (*The Structure of Materials*) reflects our attempt to examine the atomic structure of solids in a broader realm than just traditional crystallography, as has been suggested by Alan Mackay, 1975. By combining visual illustrations of crystal structures with the mathematical constructs of crystallography, we find ourselves in a position to *understand* the complex structures of many modern engineering materials, as well as the structures of naturally occurring crystals and crystalline biological and organic materials. That all important materials are not crystalline is reflected in the discussion of amorphous metals, ceramics, and polymers. The inclusion of quasicrystals conveys the recent understanding that materials possessing long-range orientational order without 3-D translational periodicity must be included in a modern discussion of the structure of materials. The discovery of quasicrystals

[1] MEM is grateful to his good friend Joanne Bassilious for recommending this inspirational movie.

has caused the *International Union of Crystallographers* to redefine the term *crystal* as "any solid having an essentially discrete diffraction pattern." This emphasizes the importance of diffraction theory and diffraction experiments in determining the structure of matter. It also means that extensions of the crystallographic theory to higher dimensional spaces are necessary for the correct interpretation of the structure of quasicrystals.

Modern crystallography education has benefitted tremendously from the availability of fast desktop computers; this book would not have been possible without the availability of wonderful free and commercial software for the visualization of crystal and molecular structures, for the computation of powder and single crystal diffraction patterns, and a host of other operations that would be nearly impossible to carry out by hand. We believe that the reader of this book will have an advantage over students of just a generation ago; he/she will be able to directly visualize all the crystal structures described in this text, simply by entering them into one of these visualization programs. The impact of visual aids should not be underestimated, and we have tried our best to include clear illustrations for more than 100 crystal structures. The structure files, available from the book's web site, will be useful to the reader who wishes to look at these structures interactively.

About the structure of this book

The first half of the book, Chapters 1 through 13, deals with the basics of crystallography. It covers those aspects of crystallography that are mostly independent of any actual material, although we make frequent use of actual materials as examples, to clarify certain concepts and as illustrations. In these chapters, we define the seven crystal systems and illustrate how lattice geometry computations (bond distances and angles) can be performed using the metric tensor concept. We introduce the reciprocal space description and associated geometrical considerations. Symmetry operations are an essential ingredient for the description of a crystal structure, and we enumerate all the important symmetry elements. We show how sets of symmetry elements, called point groups and space groups, can be used to succinctly describe crystal structures. We introduce several concepts of diffraction, in particular the structure factor, and illustrate how the International Tables for Crystallography can be used effectively.

In the second half of the book, Chapters 15 through 25, we look at the structures of broad classes of materials. In these chapters, we consider, among others, metals, oxides, and molecular solids. The subject matter is presented so as to build an understanding of simple to more complex atomic structures, as well as to illustrate technologically important materials. In these later chapters, we introduce many geometrical principles that can be used to understand the structure of materials. These geometrical principles, which enrich the material presented in Chapters 1 through 13, also allow us to gain insight into the structure of quasicrystalline and amorphous materials, discussed in advanced chapters in the latter part of the text.

In the later chapters, we give examples of crystallographic computations that make use of the material presented in the earlier chapters. We illustrate the relationship between structures and phases of matter, allowing us to make elementary contact with the concept of a *phase diagram*. Phase relations and phase diagrams combine knowledge of structure with concepts from thermodynamics; typically, a thermodynamics course is a concurrent or subsequent part of the curriculum of a materials scientist or engineer, so that the inclusion of simple phase diagrams in this text strengthens the link to thermodynamics. Prominent among the tools of a materials scientist are those that allow the examination of structures on the nanoscale. Chapters in the latter half of the book have numerous illustrations of interesting nanostructures, presented as extensions to the topical discussions.

Chapter 14 forms the connection between the two halves of the book: it illustrates how to use the techniques of the first half to study the structures of the second half. We describe this connection by means of four different materials, which are introduced at the end of the first Chapter. Chapter 14 also reproduces one of the very first scientific papers on the determination of crystal structures, the 1913 paper by W. H. Bragg and W. L. Bragg on *The Structure of the Diamond*. This seminal paper serves as an illustration of the long path that scientists have traveled in nearly a century of crystal structure determinations.

Some topics in this book are more advanced than others, and we have indicated these sections with an asterisk at the start of the section title. The subjects covered in each chapter are further amplified by 400 end-of-chapter reader exercises. At the end of each chapter, we have included a short historical note, highlighting how a given topic evolved, listing who did what in a particular subfield of crystallography, or giving biographical information on important crystallographers. Important contributors to the field form the main focus of these historical notes. The selection of contributors is not chronological and reflects mostly our own interests.

We have used the text of this book (in course-note form) for the past 13 years for a sophomore-level course on the structure of materials. This course has been the main inspiration for the book; many of the students have been eager to provide us with feedback on a variety of topics, ranging from "This figure doesn't work" to "Now I understand!" Developing the chapters of the book has also affected other aspects of the Materials Science and Engineering curriculum at CMU, including undergraduate laboratory experiments on amorphous metals, magnetic oxides, and high temperature superconductors. Beginning in June, 1995, in conjunction with the CMU Courseware Development Program, multimedia modules for undergraduate students studying crystallography were created. The first module, "Minerals and Gemstones," coupled photographic slides generously donated by Marc Wilson, curator of the Carnegie Museum of Natural History's Hillman Hall of Minerals and Gems (in Pittsburgh, PA), with crystal shapes and atomic arrangements. This and subsequent software modules were made available on a CD in the Fall of 1996; as updated versions become available, they will be downloadable through the book's web site. This software development work was heavily supported by our undergraduate students, and helped to shape the focus of the text. A module on the "History of Crystallography" served as a draft for the *Historical notes* sections of this book.

The text can be used for a one-semester graduate or undergraduate course on crystallography; assuming a 14-week semester, with two 90-minute sessions per week, it should be possible to cover Chapters 1 through 14 in the first 11–12 weeks, followed by selected sections from the later chapters in the remainder of the semester. The second half of the book is not necessarily meant to be taught "as is"; instead, sections or illustrations can be pulled from the second half and used at various places in the first half of the book. Many of the reader exercises in the second half deal with the concepts of the first half.

Software used in the preparation of this book

Some readers might find it interesting to know which software packages were used for this book. The following list provides the name of the software package and the vendor (for commercial packages) or author web site. Weblinks to all companies are provided through the book's web site.

- **Commercial packages**:
  - Adobe Illustrator [http://www.adobe.com/]
  - Adobe Photoshop [http://www.adobe.com/]
  - CrystalMaker and CrystalDiffract [http://www.crystalmaker.com/]
- **Shareware packages**:
  - QuasiTiler [http://www.geom.uiuc.edu/apps/quasitiler/]
  - Kaleidotile (Version 1.5) [http://geometrygames.org/]
- **Free packages**:
  - teTEX [http://www.tug.org/]
  - TeXShop [http://www.texshop.org/]
  - POVray [http://www.povray.org/]

The web site for this book runs on a dedicated Linux workstation located in MDG's office. The site can be reached through the publisher's web site, or, directly, at the following Uniform Resource Locator:

http://som.web.cmu.edu/

Acknowledgements

Many people have (knowingly or unknowingly) contributed to this book. We would like to thank as many of them as we can remember and apologize to anyone that we have inadvertently forgotten. First of all, we would like to express our sincere gratitude to the many teachers that first instructed us in the field of the Structure of Materials. Michael McHenry's work on the subject of quasicrystals and icosahedral group theory dates back to his Massachusetts Institute of Technology (MIT) thesis research (McHenry, 1988). Michael McHenry acknowledges Professor Linn Hobbs, formerly of Case Western Reserve University and now at MIT, for his 1979 course *Diffraction Principles and Materials Applications* and the excellent course notes which have served to shape several of the topics presented in this text. Michael McHenry also acknowledges Professor Bernard Wuensch of MIT for his 1983 course *Structure of Materials*, which also served as the foundation for much of the discussion as well as the title of the book. The course notes from Professor Mildred Dresselhaus' 1984 MIT course *Applications of Group Theory to the Physics of Solids* also continues to inspire. Michael McHenry's course project for this course involved examining icosahedral group theory, and was suggested to him by his thesis supervisor, Robert C. O'Handley; this project also has had a profound impact on his future work and the choice of topics in this book.

Marc De Graef's first exposure to crystallography and diffraction took place in his second year of undergraduate studies in physics, at the University of Antwerp (Belgium), in a course on basic crystallography, taught by Professor J. Van Landuyt and Professor G. Van Tendeloo, and in an advanced diffraction course, also taught by Van Landuyt. Marc De Graef would also like to acknowledge the late Professor R. Gevers, whose course on analytical mechanics and tensor calculus proved to be quite useful for crystallographic computations as well. After completing a Ph.D. thesis at the Catholic University of Leuven (Belgium), MDG moved to the Materials Department at UCSB, where the first drafts of several chapters for this book were written. In 1993, he moved to the Materials Science and Engineering

Department at Carnegie Mellon University, Pittsburgh, where the bulk of this book was written.

We are especially grateful to Professor Jose Lima-de-Faria for providing us with many of the photographs of crystallographers that appear in the Historical notes sections of the book, as well as many others cited below. His unselfish love for the field gave the writers an incentive to try to emulate his wonderful work.

We would like to acknowledge the original students who contributed their time and skills to the Multimedia courseware project: M. L. Storch, D. Schmidt, K. Gallagher and J. Cheney. We offer our sincere thanks to those who have proofread chapters of the text. In particular, we thank Nicole Hayward for critically reading many chapters and for making significant suggestions to improve grammar, sentence structure, and so on. In addition, we would like to thank Matthew Willard, Raja Swaminathan, Shannon Willoughby and Dan Schmidt for reading multiple chapters; and Sirisha Kuchimanchi, Julia Hess, Paul Ohodnicki, Roberta Sutton, Frank Johnson, and Vince Harris for critical reading and commenting on selected chapters. We also thank our colleague Professor David Laughlin for critical input on several subjects and his contribution to a Special tutorial at the 2000 Fall Meeting of The Minerals, Metals & Materials Society (TMS), "A Crystallography and Diffraction Tutorial Sponsored by the ASM-MSCTS Structures Committee."

There is a large amount of literature on the subject of structure, diffraction, and crystallography. We have attempted to cite a manageable number of representative papers in the field. Because of personal familiarity with many of the works cited, our choices may have overlooked important works and included topics without full citations of *all* seminal books and papers in that particular area. We would like to apologize to those readers who have contributed to the knowledge in this field, but do not find their work cited. The omissions do not reflect on the quality of their work, but are a simple consequence of the human limitations of the authors.

The authors would like to acknowledge the National Science Foundation (NSF), Los Alamos National Laboratory (LANL), the Air Force Office of Scientific Research (AFOSR), and Carnegie Mellon University for providing financial support during the writing of this book.

We would also like to thank several of our colleagues, currently or formerly at CMU, for their support during the years it has taken to complete the text: Greg Rohrer, Tresa Pollock, David Laughlin, and Alan Cramb. In particular, we would like to thank Jason Wolf, supervisor of the X-ray Diffraction facility; Tom Nuhfer, supervisor of the Electron Optics facility; and Bill Pingitore, MSE undergraduate laboratory technician at CMU.

We would like to thank our editors at Cambridge University Press, Tim Fishlock, Simon Capelin, Michelle Carey, and Anna Littlewood for their patience. This book has taken quite a bit longer to complete than we had originally anticipated, and there was no pressure to hurry up and finish it off. In this time of deadlines and fast responses, it was actually refreshing to be able to take the time needed to write and re-write (and, often, re-write again) the various sections of this book.

Michael McHenry would like to acknowledge the support and encouragement of his wife, Theresa, during the many years he has been preoccupied with this text. Her patience and encouragement, in addition to her contributions to keeping hardware and software working in his household during this process, were instrumental in its completion. Marc De Graef would like to thank his wife, Marie, for her patience and understanding during the many years of evening and weekend work; without her continued support (and sporadic interest as a geologist) this book would not have been possible. Last but not least, the authors acknowledge their children. Michael McHenry's daughter Meghan and son Michael lived through all of the travails of writing this book. Meghan's friendship while a student at CMU has helped to further kindle the author's interest in undergraduate education. Her friends represent the best of the intellectual curiosity that can be found in the undergraduates at CMU. Michael McHenry's son Michael has developed an interest in computer networking and helped to solve many of a middle-aged (old!) man's problems that only an adept young mind can grasp. We hope that he finds the joy in continued education that his sister has.

Both of Marc De Graef's children, Pieter and Erika, were born during the writing of this book, so they have lived their entire lives surrounded by crystallographic paraphernalia; indeed, many of their childhood drawings, to this day, are made on the back of sheets containing chapter drafts and trial figures. Hopefully, at some point in the future, they will turn those pages and become interested in the front as well.

Figure reproductions

This book on the structure of materials has been enriched by the courtesy of other scientists in the field. A number of figures were taken from other authors' published or unpublished work, and the following acknowledgements must be made:

The following figures were obtained from J. Lima-de-Faria and are reproduced with his permission: 1.8(a),(b); 3.15(a); 4.4(a),(b); 5.11(a),(b); 6.4(a),(b); 7.12(a),(b); 8.20(a),(b); 9.15(b); 10.13(a),(b); 15.15(a); 16.18(a),(b); 19.25(a); 20.19(b); 21.18(a),(b); 22.23(a); 24.23(a),(b).

The following figures were obtained from the Nobel museum and are reproduced with permission: 2.10(a),(b); 3.15(b); 11.25(a),(b); 12.9(a),(b); 13.18(a),(b); 15.15(b); 22.23(b); 23.19(b); 25.28(a),(b);

The 1913 article by W. L. and W. H. Bragg on the structure determination of diamond (historical notes in Chapter 14, W. H. Bragg and W. L. Bragg (The Structure of the Diamond) *Proc. R. Soc. A*, **89**, pp. 277–291 (1913)) was reproduced with permission from The Royal Society.

The following figures were reproduced from the book *Introduction to Conventional Transmission Electron Microscopy* by M. De Graef (2003) with permission from Cambridge University Press: 3.3; 5.7; 7.1; 7.7; 7.8; 7.10; 8.15; 11.16; 13.5; 13.6; 13.8(a); 13.10; 13.11; 13.12.

Insets in Fig. 1.2 courtesy of D. Wilson, R. Rohrer, and R. Swaminathan; Fig. 1.5 courtesy of P. Ohodnicki; Fig. 11.8 courtesy of the Institute for Chemical Education; Fig. 13.13 courtesy of ANL; Fig. 13.14(a) photo courtesy of ANL, (b) picture courtesy of BNL; Fig. 13.16(b) courtesy of ANL; Fig. 13.17(a) courtesy of A. Hsiao and (b) courtesy of M. Willard; Figure in Box 16.6 courtesy of M. Skowronski; Figure in Box 17.6 courtesy of M. Tanase, D. E. Laughlin and J.-G. Zhu; Figure in Box 17.9 courtesy of K. Barmak; Fig. 17.29(a) courtesy of Department of Materials, University of Oxford; Fig. 17.29(b) courtesy of T. Massalski; Figure in Box 18.4 courtesy of E. Shevshenko and Chris Murray, IBM; Fig. 18.29(a) courtesy of the Materials Research Society, Warrendale, PA; Fig. 18.29(b) courtesy of A. L. Mackay; Figure in Box 19.1 courtesy of E. Shevshenko and Chris Murray, IBM; Fig. 19.25(b) courtesy of C. Shoemaker; Fig. 20.10: Tilings were produced using QuasiTiler from the Geometry Center at the University of Minnesota – simulated diffraction patterns courtesy of S. Weber; Fig. 20.7 courtesy of J. L. Woods; Fig. 20.14, R. A. Dunlap, M. E. McHenry, R. Chaterjee, and R. C. O'Handley, Phys. Rev. B **37**, 8484–7, 1988, Copyright (1988) by the American Physical Society; Fig. 20.17 courtesy of F. Gayle, NIST Gaithersburg; Fig. 20.18 courtesy of W. Ohashi and F. Spaepen; (a) and (b) were originally published in Nature (Ohashi and Spaepen, 1987) and (c) appears in the Harvard Ph.D. thesis of W. Ohashi; Fig. 20.19(a) courtesy of the Materials Research Society, Warrendale, PA; Figure in Box 21.1 courtesy of M. Willard; Fig. 21.6(a) and (b) courtesy of J. Hess and (c) N. Hayward; Fig. 21.16 courtesy of R. Swaminathan; Figure in Box 22.7 courtesy of R. Swaminathan; Figure in Box 23.4 courtesy of M. Hawley, LANL; Fig. 23.8(a) courtesy of S. Chu; Fig. 23.19(a) courtesy of B. Raveau; Fig. 25.1(b) L. Bosio, G. P. Johari, and J. Teixeira, Phys. Rev. Lett., **56**, 460–3, 1986, Copyright (1986) by the American Physical Society; Figure in Box 25.5 courtesy of M. Bockstaller.

Atomic coordinates of known higher fullerenes have been graciously made available at the website of Dr. M. Yoshida; http://www.cochem2.tutkie.tut.ac.jp/Fuller/Fuller.html.

Symbols

Roman letters

(H, K, L)	*Quasicrystal Miller indices*
$(n_1 n_2 n_3 n_4)$	*Penrose vertex configuration*
(u, v, w)	*Lattice node coordinates*
(x, y, z)	*Cartesian coordinates*
ΔE	*Energy difference*
Δp_x	*Momentum uncertainty*
ΔS	*Entropy change*
ΔT	*Temperature difference*
Δx	*Position uncertainty*
$\hbar$	*Normalized Planck constant*
$\mathbf{A}_i^*, \mathbf{C}^*$	*Hexagonal reciprocal basis vectors*
c	*Velocity of light in vacuum*
$\mathbf{D}_i(\theta)$	*Rotation matrix in i-dimensional space*
ν	*Frequency of an electromagnetic wave*
$\overline{M_\mathrm{n}}$	*Number average molecular weight*
$\overline{M_\mathrm{w}}$	*Weight average molecular weight*
$\overline{M}$	*Average molecular weight*
$\overline{r^2}$	*Radius of gyration*
$\overline{X_n}$	*Degree of polymerization*
$\mathcal{T}$	*Plane tiling*
$\mathbf{A}, \mathbf{B}, \mathbf{C}$	*Face centering vectors*
$\mathbf{a}, \mathbf{b}, \mathbf{c}$	*Bravais lattice basis vectors*
$\mathbf{a}^*, \mathbf{b}^*, \mathbf{c}^*$	*Reciprocal basis vectors*
$\mathbf{a}_i^*$	*Reciprocal basis vectors*
$\mathbf{a}_i$	*Bravais lattice basis vectors*
$\mathbf{C}_h$	*Chiral vector*
$\mathbf{E}$	*Electrical field vector*
$\mathbf{e}_i$	*Cartesian basis vectors*
$\mathbf{e}_r$	*Radial unit vector*
$\mathbf{F}$	*Interatomic force vector*
$\mathbf{g}$	*Reciprocal lattice vector*
$\mathbf{g}_{hkl}$	*Reciprocal lattice vector*
$\mathbf{I}$	*Body centering vector*
$\mathbf{j}$	*Electrical current density vector*
$\mathbf{k}$	*Wave vector*
$\mathbf{M}$	*Magnetization vector*
$\mathbf{n}$	*Unit normal vector*
$\mathbf{P}$	*General material property*
$\mathbf{Q}$	*Higher-dimensional scattering vector*
$\mathbf{r}$	*General position vector*
$\mathbf{S}$	*Poynting vector*
$\mathbf{t}$	*Lattice translation vector*
$\mathcal{F}$	*General field*
$\mathcal{G}_m^n$	*m-D symmetry group in n-D space*
$\mathcal{P}$	*Percentage ionic character*
$\mathcal{P}$	*Probability*
$\mathcal{R}$	*General material response*
$\mathcal{S}(k)$	*k-th order Fibonacci matrix*
$\mathcal{T}$	*Bravais lattice*
$\mathcal{W}$	4×4 *symmetry matrix*

$\mathcal{O}$	*General symmetry operator*
σ	*Lennard-Jones distance parameter*
RDF(r)	*Radial distribution function*
$\tilde{x}_j$	*Normal coordinates*
$\{a, b, \gamma\}$	*Net parameters*
$\{a, b, c, \alpha, \beta, \gamma\}$	*Lattice parameters*
A	*Absorption correction factor*
A	*Atomic weight*
A	*Electron affinity*
a_R	*Quasicrystal lattice constant*
a_{ij}	*Direct structure matrix*
b	*Neutron scattering length*
$B(T)$	*Debye–Waller factor*
B_i	*Magnetic induction components*
b_{M}	*Neutron magnetic scattering length*
b_{ij}	*Reciprocal structure matrix*
D	*Detector*
D	*Distance between two points*
D_i	*Electric displacement components*
d_{hkl}	*Interplanar spacing*
E	*Electric field strength*
E	*Electronegativity*
E	*Number of polygon edges*
E	*Photon energy*
e	*Electron charge*
E_i	*Electric field components*
E_n	*Energy levels*
E_{p}	*Potential energy*
E_{kin}	*Kinetic energy*
F	*Number of polygon faces*
$f(s)$	*Atomic scattering factor*
f^{el}	*Electron scattering factor*
F_k	*Fibonacci numbers*
F_{hkl}	*Structure factor*
G	*Optical gyration constant*
$g(r)$	*Pair correlation function*
g^*_{ij}	*Reciprocal metric tensor*
g^*_i	*Reciprocal lattice vector components*
g_{ij}	*Direct space metric tensor*
h	*Planck's constant*
H_i	*Magnetic field components*
h_i	*Heat flux components*
$H_{\mathrm{c1}}(T)$	*Lower critical field*
$H_{\mathrm{c2}}(T)$	*Upper critical field*
I	*Intensity*
I	*Ionization potential*
$i(k)$	*Reduced intensity function*
I_0	*Incident beam intensity*
I_{hkl}	*Diffracted beam intensity*
j	*Electrical current density*
J_{c}	*Critical current density*
K	*Normalization constant*
K, L, M, . . .	*Spectroscopic principal quantum numbers*
k_{B}	*Boltzmann constant*
L	*Potential range*
l	*Angular momentum quantum number*
$L(x, y)$	*2-D lattice density*
L, S	*Fibonacci segment lengths*
l_i	*Direction cosines*
L_n	*Lucas numbers*
$L_p(\theta)$	*Lorentz polarization factor*
M	*Debye–Waller factor*
m	*Magnetic quantum number*
m	*Particle mass*
m_0	*Electron rest mass*
m_i	*Mass flux components*
m_n	*Neutron rest mass*
M_W	*Molecular weight*
n	*Principal quantum number*
n, l, m	*Atomic quantum numbers*
N_{e}	*Number of free electrons*
P	*Synchrotron total power*
p	*Subgroup index*
$P(\mathbf{r})$	*Patterson function*
$P(\theta)$	*Polarisation factor*

$p_i, q_i, \ldots$	*General position vector components*
p_{hkl}	*Multiplicity of the plane* (hkl)
r	*Radial distance*
r_N	*Nuclear radius*
R_p	*Profile agreement index*
r_{ws}	*Wigner–Seitz radius*
$R_{nl}(r)$	*Radial atomic wave function*
R_{wp}	*Weighted profile agreement index*
S	*Sample*
s	*Scattering parameter*
s	*Spin quantum number*
$s, p, d, f, g, \ldots$	*Spectroscopic angular momentum quantum numbers*
s_i	*Planar intercepts*
T	*Absolute temperature*
T	*Target*
T	*Triangulation number*
t	*Grain size*
T_0	*Equal free-energy temperature*
T_c	*Superconductor critical temperature*
T_g	*Glass transition temperature*
T_L	*Liquidus temperature*
T_N	*Nëel temperature*
T_{rg}	*Reduced glass transition temperature*
T_{x1}	*Primary recrystallization temperature*
T_{x2}	*Secondary recrystallization temperature*
u_i	*Lattice translation vector components*
V	*Accelerating voltage*
V	*Electrostatic potential drop*
V	*Number of polygon vertices*
V	*Unit cell volume*
$V(r)$	*Radial electrostatic potential*
$V_c(r)$	*Coulomb interaction potential*
$V_r(r)$	*Repulsive interaction potential*
$Y_{lm}(\theta, \phi)$	*Angular atomic wave function*
Z	*Atomic number*
a	*Anorthic*
c	*Cubic*
h	*Hexagonal*
m	*Monoclinic*
o	*Orthorhombic*
R	*Rhombohedral*
t	*Tetragonal*

Greek letters

(r, θ, ϕ)	*Spherical coordinates*
α	*Madelung constant*
α_{ij}	*General coordinate transformation matrix*
χ	*Mulliken electronegativity*
$\chi(k)$	*Absorption function (EXAFS)*
$\Delta\beta_{ij}$	*Change of impermeability tensor*
δ_{ij}	*Identity matrix*
δ_{ij}	*Kronecker delta*
ϵ	*Lennard-Jones energy scale parameter*
ϵ^*_{ijk}	*Reciprocal permutation symbol*
ϵ_0	*Permittivity of vacuum*
ϵ_F	*Fermi energy level*
ϵ_{ijk}	*Permutation symbol*
ϵ_{ij}	*Strain tensor*
λ	*Photon/electron/neutron wave length*
λ	*radiation wave length*
μ	*Linear absorption coefficient*

μ/ρ	*Mass absorption coefficient*
ν	*Photon frequency*
ν_0	*Zero-point motion frequency*
Ω	*Atomic volume*
ϕ	*Chiral angle*
ϕ	*Phase of a wave*
$\Psi(\mathbf{r})$	*General wave function*
ρ	*Density*
$\rho(\mathbf{r})$	*Charge density*
$\rho_{\text{atom}}(r)$	*Spatially dependent atomic density*
σ	*Electrical conductivity*
σ	*Scattering cross section*
σ_{ij}	*Electrical conductivity tensor*
σ_{ij}	*Stress tensor*
τ	*Golden mean*
θ_{hkl}	*Bragg angle*
e^*_{ijk}	*Normalized reciprocal permutation symbol*
e_{ijk}	*Normalized permutation symbol*

Special symbols

(ϕ, ρ)	*Stereographical projection coordinates*
$(\mathrm{D}\vert\mathbf{t})$	*Seitz symbol*
$(hkil)$	*Hexagonal Miller–Bravais indices*
(hkl)	*Miller indices of a plane*
$[uvtw]$	*Hexagonal Miller–Bravais direction indices*
$[uvw]$	*Direction symbol*
$\square$	*Vacancy*
$\cdot$	*Vector dot product operator*
det	*Determinant operator*
$\exists$	*"there exists"*
$\forall$	*"for all, for each"*
$\in$	*"belongs to, in"*
$\langle uvw\rangle$	*Family of directions*
$\leftrightarrow$	*Isomorphism*
$\oplus$	*Direct product operator*
$\mathcal{F}$	*Fourier transform operator*
$\rightarrow$	*Homomorphism*
$\subset$	*group–subgroup relation symbol*
$\times$	*Vector cross product operator*
$\vert\ \vert$	*Norm of a vector*
$\{hkl\}$	*Family of planes*

CHAPTER

1 Materials and materials properties

> *"We proceed to distribute the figures [solids] we have described between fire, earth, water, and air . . . Let us assign the cube to earth, for it is the most immobile of the four bodies and most retentive of shape; the least mobile of the remaining figures (icosahedron) to water; the most mobile (tetrahedron) to fire; the intermediate (octahedron) to air. There still remained a fifth construction (dodecahedron), which the god used for embroidering the constellations on the whole heaven."*
>
> Plato, *Timaeus*, 427–347 BC

1.1 Materials and structure

The practice of using organic and inorganic materials is many millennia old. Oxide pigments were used in early cave paintings, flint tools were used in the *Stone Age* and precious metal smelting was prevalent in the Nile Valley as early as 5000 years ago (Klein and Hurlbut, 1985). Extractive metallurgy led to the use of metals in the *Bronze Age* and *Iron Age*. The extraordinary advances made possible by electronic materials have led some to suggest that we are in the midst of the *Silicon Age*. It is clear that the prior materials ages evolved slowly through the accumulation of empirical knowledge. The present materials age is evolving at a more rapid pace through the development of synthesis, structure, properties and performance relationships, the *materials paradigm*.

In this book, we will introduce many concepts, some of them rather abstract, that are used to describe solids. Since most materials are ultimately used in some kind of application, it seems logical to investigate the link between the atomic structure of a solid, and the resulting macroscopic properties. After all, that is what the materials scientist or engineer is really interested in: how can we make a material useful for a certain task? What type of material do we need for a given application? And why can some materials not be used for particular applications? All these questions must be answered when a material is considered as part of a design. The main focus of the book is on the fundamental description of the positions and types of the atoms, the ultimate building blocks of solids, and on the experimental techniques used to determine how these atoms are arranged.

We now know that many of the materials we use every day are *crystalline*. The concept of crystalline solids and the development of experimental techniques to characterize crystals are recent developments, although certain kernels of thought on the basic building blocks of solids can be traced to much earlier times. For example, the quote beginning this chapter is attributed to the Greek philosopher Plato (427–348/347 BC); in his dialogue *Timaeus*, he discussed his theory of the structure of matter. He postulated that the basic particles of earth, air, fire, and water had the form of the regular *Platonic solids* (Fig. 1.1). Plato believed that it was possible to group these basic particles into crystal shapes that filled space. In our current understanding of the structure of solids, the shapes that are combined to fill space are known as *unit cells*, and we distinguish seven major shapes, more formally known as the seven *crystal systems*.

For crystalline solids we will define a standardized way to describe crystal structures. We will also describe experimental methods to determine where the atoms are in a given crystal structure. We will rely on mathematical techniques to develop a clear and unambiguous description of crystal structures, including rules and tools to perform crystallographic computations (e.g., what is the distance between two atoms, or the bond angle between two bonds, etc.). We will introduce the concept of *symmetry*, a unifying theme that will allow us to create classifications for crystal structures.

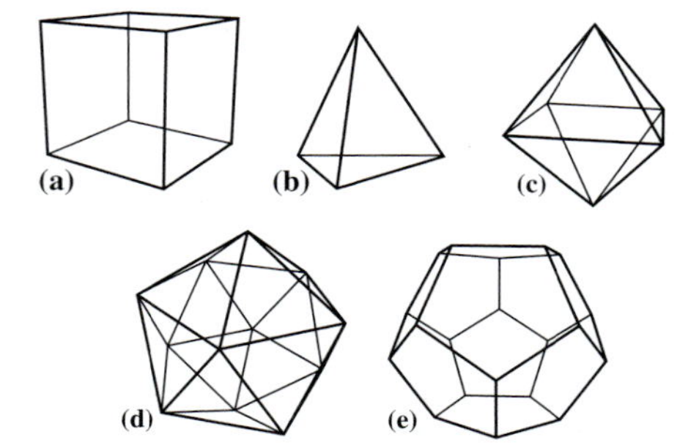

Fig. 1.1. The five Platonic solids: (a) cube, (b) tetrahedron, (c) octahedron, (d) icosahedron, and (e) pentagonal dodecahedron.

1.2 Organization of the book

The first half of the book, Chapters 1 through 13, deals with the basics of crystallography. It covers those aspects of crystallography that are mostly independent of any actual material, although we will frequently use actual materials as examples to clarify certain concepts and as illustrations. The second half of the book, Chapters 15 through 25, looks at the structure of broad classes of materials. In these chapters, we consider metals, oxides, and molecular solids. This subject matter helps the reader build an understanding of atomic structures, from simple to complex. Where possible, we also illustrate technologically important materials. In these later chapters, we will introduce many geometrical principles that can be used to understand the structure of materials. Such principles enrich the material presented in Chapters 1 through 13, and allow us to gain insight into the structure of quasicrystalline and amorphous materials discussed in advanced chapters in the second half of the text.

Chapter 14 forms the connection between the two halves of the book: it illustrates how techniques of the first half are used to study the structures of the second half. We will discuss this connection by means of four different materials, which will be introduced later in this first chapter. Some topics are more advanced than others, and we have indicated these sections with an asterisk at the start of the section title. Each chapter has an extensive problem set, dealing with the concepts introduced in that chapter. At the end of each chapter, we have included a short historical note, highlighting how a given topic evolved, listing who did what in a particular subfield of crystallography, or giving biographical information on important crystallographers.

In the later chapters, we give examples of crystallographic computations that make use of the material presented in the earlier chapters. We illustrate the relationship between structures and phases of matter, allowing us to make elementary contact with the concept of a *phase diagram*. Phase relations and phase diagrams combine knowledge of structures and thermodynamics.[1] Prominent among the tools of a materials scientist are those that allow examination of structures on a nanoscale. Chapters in the latter half of the book will have further illustrations of interesting nanostructures.

We begin, in this chapter, with a short discussion of length scales in materials. Then we introduce the concepts of *homogeneity* and *heterogeneity*. We will talk about material properties and propose a general definition for a material property. We continue with a discussion of the directional dependence of certain properties and introduce the concepts of *isotropy* and *symmetry*. We conclude the chapter with a preview of some of the things this book has to offer.

[1] In a materials science or materials engineering curriculum, phase relations and diagrams are typically the subject of the course following a structures course.

1.3 About length scales

When we talk about crystals, most of us will think about the beautiful crystalline shapes that can be found in nature. Quartz crystals are ubiquitous, and we can recognize them by their shape and color. Many naturally occurring crystals have sizes in the range from a few millimeters to a few centimeters. These are objects that we can typically hold in our hands. When it comes to describing the structure of a crystal at the atomic level, we must reduce the length of our measuring stick by many orders of magnitude, so it might be useful to take a brief look at the relevant length scale. In addition, when we wish to study, say, the distance between a pair of atoms, we must use an experimental measuring stick that is capable of measuring such tiny distances. The human eye is obviously not capable of "seeing atoms," but there are several alternative observation methods that *are* capable of operating at the atomic length scale.

The size of an atom is of the order of 10^{-10} meters. This particular distance is known as the Ångström, i.e., $1\,\text{Å} \equiv 10^{-10}\,\text{m}$. It is convenient to stick to the so-called *metric system*, and the closest standard metric unit is the *length unit, nanometer* (nm), which is defined as $1\,\text{nm} \equiv 10^{-9}\,\text{m}$, so that $1\,\text{Å} = 0.1\,\text{nm}$. In this book, we will use the nanometer as the standard unit of length, so that we can express all other distances in terms of this unit. For instance, 1 micrometer (μm) equals 10^3 nm, and one millimeter (mm) is equal to 10^6 nm. An illustration of the range of object sizes from the atomistic to the "human" length scale is shown in Fig. 1.2.[2] The central vertical axis represents the size range on a logarithmic scale; going up one tick mark means a factor of 10 larger. To the right of the figure, there are a few examples of objects for each size range. In the scientific community, we distinguish between a few standard size ranges:

- *macroscopic*: objects that can be seen by the unaided eye belong to the class of the macroscopic objects. An example is the quartz crystal shown in the top circle to the right of Fig. 1.2.
- *microscopic*: objects that can be observed by means of optical microscopy. The second circle from the top in Fig. 1.2 shows individual grains in a $SrTiO_3$ polycrystalline material. The lines represent the boundaries between grains, the darker spots are pores in the ceramic material.
- *nanoscale*: objects with sizes between 1 nm and 100 nm. The third circle shows a set of nano-size particles of a MnZn ferrite with composition $Mn_{0.5}Zn_{0.5}Fe_2O_4$.

[2] With "human length scale" we mean objects that can be found in our societies: chairs, houses, vehicles, and so on, i.e., objects with sizes typically less than 10^2 m or 10^{11} nm. In this book, we will have no need for larger sizes.

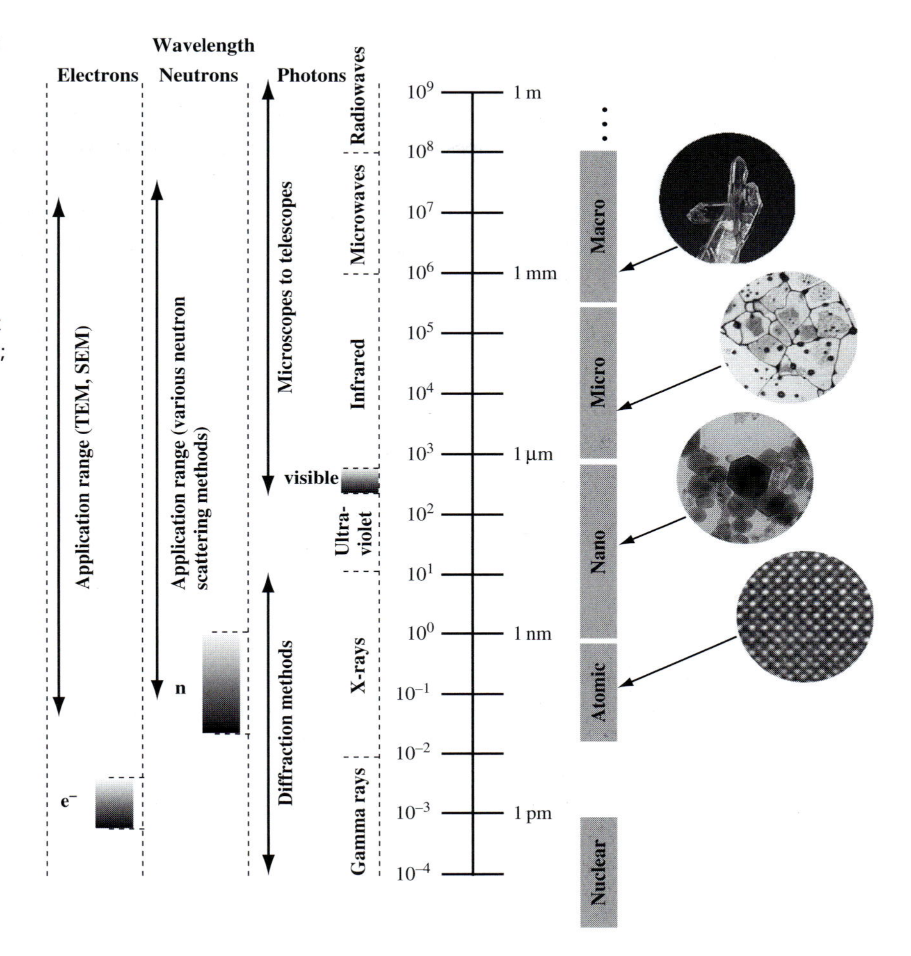

Fig. 1.2. Schematic illustration of the various length scales, from macroscopic, to microscopic, to nanoscale, to the subatomic. The left hand side of the figure shows the experimental techniques that are used to cover the various length scales. The images in the circles on the right are (from the top down): a quartz crystal (courtesy of D. Wilson); grains in a $SrTiO_3$ ceramic (courtesy of G. Rohrer); nano-crystalline particles of $Mn_{0.5}Zn_{0.5}Fe_2O_4$ (courtesy of R. Swaminathan); atomic resolution image of $BaTiO_3$.

- *atomistic*: the bright dots in this image correspond to the Ba and Ti ions in a tetragonal $BaTiO_3$ crystallite. This is a so-called high resolution image, obtained by transmission electron microscopy, where one can distinguish individual columns of atoms with a distance between the atoms of around 0.2 nm.

The ability to observe an object of a certain size is closely linked to the *wave length* of the radiation used for the observation. Consider circular waves that travel on a large pond after you toss a rock into the water. If an object, much smaller than the distance between the crests of the waves (the *wave length*), floats on the water nearby, then the waves will pass by the object without being perturbed by the object; the object will move up and down with the passing waves. If the object is large compared to the wave length, say, a concrete pillar or a wall, then the waves will be perturbed, since they have to travel around the object; often, part of the waves will be reflected by the object. If waves are not perturbed by an object, then this object is essentially *invisible* to those waves. If we use visible light, with a wave length of around 500 nm, to look at viruses (with a typical size between 3 and 300 nm), then the light waves will not be perturbed significantly by the viruses, and, therefore, we will not be able to observe viruses using optical microscopy methods.

To determine the smallest thing the human eye can see, we must understand the structure of the eye. The human eye is a sphere with an approximate diameter of 25 mm. It has a lens with an opening (pupil) of about 3.5 mm. The inside back surface of the eye is covered with two types of light receptors: rods and cones. The cones are concentrated in a small area, 0.3 mm diameter, directly opposite the lens. This area is known as the *fovea*. There are about 15 000 cones in the fovea, leading to a cone density of about 200 000 per mm^2. Each cone is about 1.5 μm in diameter, and the average spacing between cones is 2.5 μm. For convenience, we can imagine the cones to be packed in a hexagonal array, as shown by the small gray disks in Fig. 1.3.

If we consider an object at a distance of 250 mm from the eye, then this object will be imaged by the lens onto the fovea with a magnification factor of $M = 0.068$ (Walker, 1995). Consider a set of narrowly spaced lines, with a line density of ρ lines per millimeter (lpm). The eye lens will image this grid of lines onto the fovea, so that the line density at the fovea becomes $\rho_f = \rho/M$ (since the eye demagnifies the object size, the line density will become larger). The highest line density that can be "seen" by the fovea corresponds to each line being projected onto a row of cones, and the next row does not have a line projected on it. Since the average spacing of the cones is 2.5 μm, the smallest possible distance at which the lines can still be resolved by individual rows of cones is $2.5\sqrt{3} = 4.3$ μm. A line spacing of 4.3 μm leads to a line density at the fovea of $\rho_f = 230$ lpm, which corresponds to a line density at the object of 15.6 lpm. So, at a distance of 250 mm, the human eye, in the best possible conditions, can see the individual lines in a grid with about 16 lines per mm.

The discussion in the preceding paragraph describes an idealized case; in reality, the highest resolvable line density at a distance of 250 mm

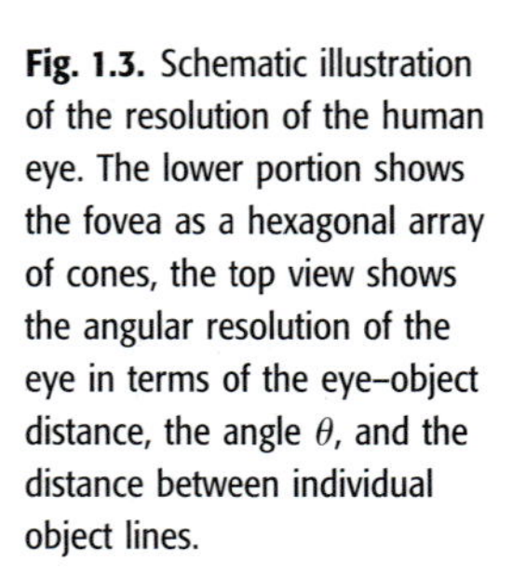

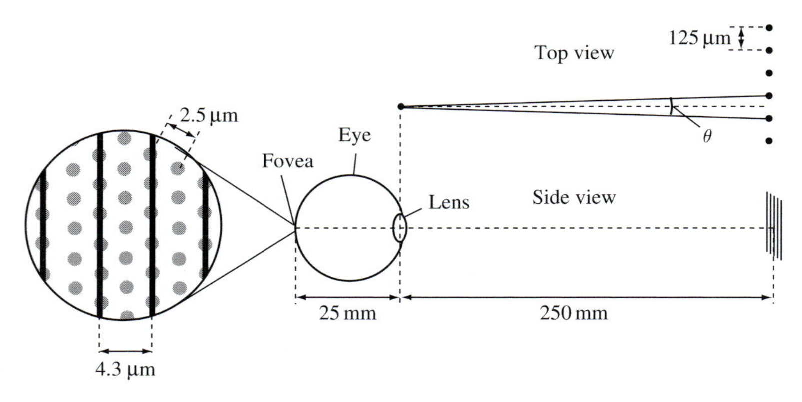

Fig. 1.3. Schematic illustration of the resolution of the human eye. The lower portion shows the fovea as a hexagonal array of cones, the top view shows the angular resolution of the eye in terms of the eye–object distance, the angle θ, and the distance between individual object lines.

is about 8 lpm. It is more convenient to express these numbers in terms of the angle between two lines leaving the eye and reaching two neighboring lines of the grid, as shown in the top view of Fig. 1.3. Simple trigonometry shows that $\theta = 0.125/250$, so that $\theta = 5 \times 10^{-4}$ radians, or 0.029°, which is equivalent to 1.7 arc minutes. Rounding up, we can say that the human eye has a visual resolution limit of 2 arc minutes per line pair. Similar numbers are obtained when the object consists of dots or other shapes. It should be clear to the reader that the human eye is not capable of resolving microscopic objects, let alone the distance between atoms.[3]

Returning to Fig. 1.2, the columns on the left hand side of the figure indicate the range of applicability of a number of important materials characterization methods. Each method relies on the use of a particular type of particle: electrons, neutrons, and photons (or electromagnetic radiation). All three of these methods are capable of producing information about the atomic structure of matter. In this book, we will discuss mostly the use of X-rays for structure determination, but we will also describe briefly how neutrons and electrons can be used to obtain similar information.

1.4 Wave–particle duality and the de Broglie relationship

By the early part of the 20th century it had been established that electromagnetic (EM) radiation (light) has both a wave and particle (photon) character, the *wave–particle duality*. James Clerk Maxwell proposed a *theory of electromagnetism* (Maxwell's equations) which put the wave nature of light on a formal mathematical basis by the late nineteenth century. However, the *photoelectric effect* was explainable only in terms of the particle aspects of light (Albert Einstein, 1905). Einstein's formula relates the energy of a photon, E, to the frequency of radiation, ν:

$$\boxed{E = \mathrm{h}\nu,} \qquad (1.1)$$

The wave length of electromagnetic radiation is related to the frequency, ν, of the wave (the number of cycles per second) by means of the following relation:

$$\boxed{\lambda = \frac{\mathrm{c}}{\nu},} \qquad (1.2)$$

[3] It is noteworthy that the human eye, as compared to a digital camera, has a remarkably high pixel density. While the cone and rod densities vary across the eye, on average, the eye corresponds to a 600 megapixel camera for a 120° field of view! Of course, the eye produces a continuous stream of images, more like a movie than a still image. Nevertheless, the amount of information processed by the eye and brain is truly remarkable.

where c = 299 792 458 m/s is the velocity of light in vacuum. A consequence of Einstein's explanation of the photoelectric effect was that EM waves could be thought of as having particle-like momentum.

The wave length, λ, of electromagnetic radiation spans many orders of magnitude, from the long wave length radio waves (see Fig. 1.2) to visible light to X-rays and gamma rays. Visible light cannot be used to observe the atomistic length scale, but X-rays have a wave length that is comparable to the distance between atoms. Hence, X-ray waves will be perturbed by atoms, and we can make good use of this perturbation, as we will see in later chapters.

Similar relations can be derived for electrons and neutrons. Louis de Broglie, using formulae from Einstein's special theory of relativity, argued that if electromagnetic waves also have a particle nature, should not particles such as the electron also have a wave nature? For a particle with mass, m, moving with velocity v, he proposed an associated wave characterized by the wave length:

$$\lambda = \frac{\mathrm{h}}{mv}. \tag{1.3}$$

For electrons accelerated by a voltage, V, the electron wave length is given by:

$$\lambda = \frac{\mathrm{h}}{\sqrt{2m_0eV}}, \tag{1.4}$$

where $\mathrm{h} = 6.626075 \times 10^{-34}$ J s is Plank's constant, $e = 1.602177 \times 10^{-19}$ C is the electron charge, and $m_0 = 9.109389 \times 10^{-31}$ kg is the rest mass of the electron. In 1927, Davisson and Germer showed experimentally that electrons do indeed have wave character by causing them to undergo diffraction like X-rays, through a crystal lattice (Davisson and Germer, 1927a). This experiment laid the basis for measuring crystal structures by the method of low energy electron diffraction (LEED), and, later on, for the invention of the transmission electron microscope.

In the case of high energy electrons where the accelerating voltage, V, is large, a relativistic correction is made and the electron wave length is given by:

$$\lambda = \frac{\mathrm{h}}{\sqrt{2m_0eV(1 + \frac{e}{2m_0c^2}V)}}. \tag{1.5}$$

Table 1.1 lists a few representative electron wave lengths used in scanning and transmission electron microscopes.

The wave length of neutrons also follows from the de Broglie relation:

$$\lambda = \frac{\mathrm{h}}{m_n v}, \tag{1.6}$$

Table 1.1. Electron wave lengths (in pm) for selected acceleration voltages V for scanning electron microscopes (left two columns) and transmission electron microscopes (right two columns).

V (Volt)	λ (pm)	V (Volt)	λ (pm)
1 000	38.76	100 000	3.701
5 000	17.30	200 000	2.508
10 000	12.20	300 000	1.969
20 000	8.59	400 000	1.644
		1 000 000	0.872

where $m_n = 1.674\,929 \times 10^{-27}$ kg is the neutron rest mass and v is its velocity. However, neutrons are not charged particles and, therefore, they are not accelerated by a voltage. Neutrons are created in nuclear fission processes inside nuclear reactors, as described in more detail in Chapter 13. Typically, a wide range of neutron velocities emerges from the reactor, and by selecting only neutrons within a certain narrow velocity window, one can select a particular wave length. For instance, to obtain a neutron with a wave length of 0.1 nm, one would have to select a velocity window at $v = 3.96 \times 10^3$ m/s, or approximately 4 km/s. It is also possible to have neutrons reach thermal equilibrium, so that their kinetic energy is:

$$E_{\text{kin}} = \frac{1}{2} m_n v^2 = \frac{3}{2} \text{k}_\text{B} T, \tag{1.7}$$

where $\text{k}_\text{B} = 1.38 \times 10^{-23}$ J/molecule/K is the Boltzmann constant. The de Broglie relation then becomes:

$$\lambda = \frac{\text{h}}{\sqrt{3 m_n \text{k}_\text{B} T}}. \tag{1.8}$$

1.5 What is a material property?

1.5.1 Definition of a material property

We choose materials to perform well in certain applications. For instance, we use steel beams and cables in bridges, because they provide the strength and load-bearing capacity needed. We use plastics in toys because they can be molded into virtually any shape and they are strong and light weight. When we use a material in a certain application, we know that it will be subjected to particular external conditions, e.g., a constant load, or a high temperature, or perhaps an electrical current running through the material. In all these cases, we must make sure that the material responds in the desired way. For a bridge deck held up by steel cables, we want the cables to retain their strength all year round, regardless of the weather and temperature, and regardless of the number of cars and trucks crossing the bridge. For a computer chip, we want the semiconductor material to behave predictably for the lifetime of the computer.

In general, we want a material to have a particular response to a given external influence. This basic statement can be cast in more formal, mathematical terms. We will represent the external influence by the symbol $\mathcal{F}$, which stands for *Field*. This could be an electrical or magnetic field, a temperature field, the earth's gravitational field, etc. The material will respond to this field, and the *Response* is described by the symbol $\mathcal{R}$. For instance, the response of a steel beam to an external load (i.e., a weight at the end of the beam) will be a deflection of the beam. The response of a conductor to an electrical field applied between its two ends will be an electrical current running through the conductor. In the most general sense, the relation between field and response is described by:

$$\mathcal{R} = \mathcal{R}(\mathcal{F}), \tag{1.9}$$

i.e., the material response is a function of the externally applied field. It is one of the tasks of a materials scientist to figure out what that function looks like.

Once we recognize that the behavior of a material under certain external conditions can be expressed in mathematical terms, we can employ mathematical tools to further describe and analyze the response of this material. We know from calculus that, for "well-behaved" functions, we can always expand the function into powers of its argument, i.e., construct a Taylor expansion.[4] For equation 1.9 above, the Taylor expansion around $\mathcal{F} = 0$ is given by:

$$\mathcal{R} = \mathcal{R}_0 + \frac{1}{1!}\left.\frac{\partial \mathcal{R}}{\partial \mathcal{F}}\right|_{\mathcal{F}=0} \mathcal{F} + \frac{1}{2!}\left.\frac{\partial^2 \mathcal{R}}{\partial \mathcal{F}^2}\right|_{\mathcal{F}=0} \mathcal{F}^2 + \frac{1}{3!}\left.\frac{\partial^3 \mathcal{R}}{\partial \mathcal{F}^3}\right|_{\mathcal{F}=0} \mathcal{F}^3 + \ldots \tag{1.10}$$

where $\mathcal{R}_0$ describes the "state" of the material at zero field. There are two possibilities for $\mathcal{R}_0$:

(i) $\mathcal{R}_0 = 0$: in the absence of an external field ($\mathcal{F} = 0$), there is no permanent (or remanent) material response. For example, if the external field is an applied stress, and the material response is a strain, then at zero stress there is no strain (assuming linear elasticity).

[4] Recall that a Taylor expansion of a function $f(x)$ around $x = 0$ is given by

$$f(x) = f(0) + \sum_{n=1}^{\infty} \frac{1}{n!}\left.\frac{\mathrm{d}^n f}{\mathrm{d}x^n}\right|_{x=0} x^n$$

where $n! = 1 \times 2 \times 3 \times \ldots \times (n-1) \times n$ is the factorial of n. If the function f depends on other variables in addition to x, then the derivatives $\mathrm{d}^n/\mathrm{d}x^n$ must be replaced by partial derivatives $\partial^n/\partial x^n$.

(ii) $\mathcal{R}_0 \neq 0$: in the absence of an external field ($\mathcal{F} = 0$), there is a permanent material response. For example, in a ferromagnetic material, the net magnetization is in general different from zero, even at zero applied field.

If we truncate the series after the second term (i.e., we ignore all derivatives of $\mathcal{R}$ except for the first one), then the expression for $\mathcal{R}$ is simplified dramatically:

$$\mathcal{R} = \mathcal{R}_0 + \left.\frac{\partial \mathcal{R}}{\partial \mathcal{F}}\right|_{\mathcal{F}=0} \mathcal{F} = \mathcal{R}_0 + \mathbf{P}\mathcal{F} \quad \text{with} \quad \mathbf{P} = \left.\frac{\partial \mathcal{R}}{\partial \mathcal{F}}\right|_{\mathcal{F}=0.} \tag{1.11}$$

This is a *linear* equation between the applied field and the response. The quantity $\mathbf{P}$ is a *material property*. Ignoring the higher order derivatives of $\mathcal{R}$ is generally known as *linear response theory*. This approximation simplifies things considerably and, for many purposes, it is a useful and accurate approximation.

Let us consider an example. An electrical conductor, say, a copper wire, is placed between the terminals of a battery. If the wire is 3 meters long, and the battery is capable of producing a 9 V voltage drop, then there is an electric field, E, of 9 volts per 3 meter, or $E = 3\,\text{V/m}$. In response to this field, a current will flow through the wire. The amount of current depends on the cross section of the wire, so it is convenient to work in terms of current density (current per unit area, or A/m^2), j. For most conductors, the relation between current density and electric field is linear, i.e.,

$$j = \sigma E,$$

where σ is known as the *electrical conductivity*, and has units of A/Vm or $1/\Omega$m, where Ω stands for ohm ($1\,\text{ohm} = 1\,\text{V/A}$). Let us compare this equation with the Taylor expansion in Eq. 1.10. The external field $\mathcal{F}$ is equal to E, and the response $\mathcal{R}$ is equal to j. First of all, when there is no voltage, there will be no current, so that $\mathcal{R}_0 = j_0 = 0$. There is no dependence on powers of E, so there is only one term in the series, namely:

$$j = \left.\frac{\partial j}{\partial E}\right|_{E=0} E \quad \text{and hence} \quad \sigma \equiv \left.\frac{\partial j}{\partial E}\right|_{E=0}.$$

We conclude that σ is equal to the first derivative of the current density with respect to the electric field. This proportionality factor does not depend on j or E, therefore we call σ a material property. In more general terms, *a linear material property is the proportionality factor between an applied field and the resulting material response.*

1.5.2 Directional dependence of properties

In the previous section, we saw that the current density, j, in a conductor is proportional to the applied electric field, E. The proportionality factor is the conductivity σ. All three quantities in the previous relation were scalar quantities. However, we can imagine taking a rectangular block of a conducting material, and applying an electric field between the top and bottom surfaces, or between the front and back surfaces, or between opposite corners. This means that the electric field has both a magnitude and a direction, hence it can be represented by a vector, $\mathbf{E}$.[5] The same thing can be said of the current density, since the current has a magnitude and it runs in a particular direction. Hence, we have a vector $\mathbf{j}$. The relation between electric field and current density then reads:

$$\mathbf{j} = \sigma \mathbf{E}.$$

Since σ is a scalar (i.e., a number), this means that the current density vector is always parallel to the electric field vector. Well, not quite. When we defined the conductivity, we started from the relation:

$$\sigma = \left. \frac{\partial j}{\partial E} \right|_{E=0}.$$

But this relation is only valid for scalar j and E. We must incorporate the fact that both $\mathbf{j}$ and $\mathbf{E}$ are vectors into this equation. Both vectors have components with respect to a standard Cartesian reference frame: $\mathbf{j} = (j_x, j_y, j_z)$ and $\mathbf{E} = (E_x, E_y, E_z)$. So, instead of having only one single value for σ, now we have a total of nine values! Here's how that works. Consider the following expression:

$$\left. \frac{\partial j_x}{\partial E_x} \right|_{E=0}.$$

In other words, this is the derivative of the x-component of the current density with respect to the x-component of the electric field. This derivative will have a particular value (a scalar value) which we will represent by σ_{xx}. Similarly, we can define

$$\sigma_{xy} = \left. \frac{\partial j_x}{\partial E_y} \right|_{E=0},$$

and so on. There are nine such relations, which can be summarized by writing:

$$\sigma_{ij} = \left. \frac{\partial j_i}{\partial E_j} \right|_{E=0},$$

[5] In this book, we will always use bold characters to represent vectors.

where the subscripts or indices i and j take on the values x, y, and z. The relation between the current density vector and the electric field vector is then given by:

$$j_x = \sigma_{xx}E_x + \sigma_{xy}E_y + \sigma_{xz}E_z;$$
$$j_y = \sigma_{yx}E_x + \sigma_{yy}E_y + \sigma_{yz}E_z;$$
$$j_z = \sigma_{zx}E_x + \sigma_{zy}E_y + \sigma_{zz}E_z.$$

This relation expresses the fact that the current density, in response to an electric field, need not be parallel to this electric field. Each component of the current density is written as a linear combination of *all* the components of the electric field.

What we learn from the above example is that a material property is not always represented by a simple scalar. If the property connects a vector field to a vector response, then the material property has nine elements, which can be written as a 3×3 matrix. Mathematicians call such a matrix a *tensor*.[6] The question then arises: Do we need nine numbers for the electrical conductivity of every material, or is it possible that some materials need fewer numbers? The answer to this question will become clear in the next section, where we introduce the concept of symmetry. Before we do so, let us first consider the possibility that a material property varies with location in the material.

It is intuitively clear that an external field can depend on location. For instance, the temperature at one end of a material can be different from the temperature at the other end. In mathematical terms, this means that the *gradient* of the temperature does not vanish. It is possible for a material property to show a similar dependence on position within the material. Consider, for instance, a cube of pure silicon. It is clear that the chemical composition of this cube is the same everywhere, since there is only one chemical element present. We say that the composition is *homogeneous*, i.e., the composition does not depend on position. Similarly, the electrical conductivity of pure silicon is the same everywhere, so that the electrical conductivity is homogeneous. Imagine, next, that we implant phosphorus atoms on one side of the cube, to a depth of a few hundred microns. Since the phosphorus concentration is not a constant throughout the cube, we say that the composition is *heterogeneous*, i.e., the concentration depends on the location in the material. Since phosphorus has five electrons in its outer shell, whereas silicon has only four, we see intuitively that the electrical conductivity in the regions that contain P must be different from that of the other regions. In other words, the electrical conductivity of P-doped silicon is heterogeneous if the P is not distributed in a homogeneous way.

[6] The definition and properties of tensors need not concern us here. It is sufficient that the reader understands that material properties often consist of multiple scalars, arranged in a particular form (in this case, a 3×3 matrix).

1.5.3 A first encounter with symmetry

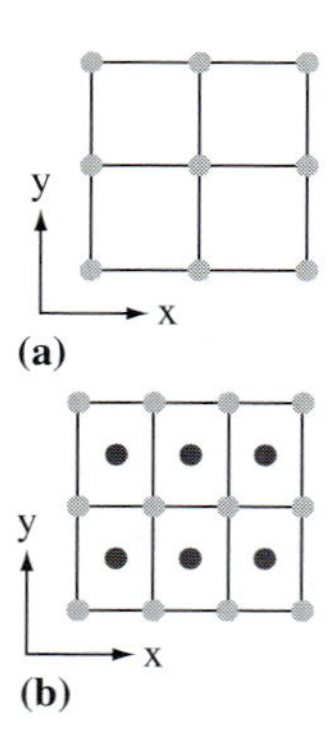

Fig. 1.4. Illustration of two simple 2-D crystal structures: (a) is based on a square grid with one type of atoms, while (b) is a rectangular grid with two different kinds of atoms. The electrical conductivities in the x and y directions for (a) are expected to be the same, whereas they are most likely different for the second structure.

One might ask what the previous section has to do with crystallography. That's a very good question, and we will attempt to answer it superficially in this section. Consider a 2-D material in which the atoms are arranged as shown schematically in Fig. 1.4(a). All atoms are identical, and they are located on the nodes of a square grid. If we apply an electric field along the x-axis, we will generate a certain current density (assuming that our 2-D material is a conductor). If we apply the same field strength in the y direction, then there is no reason why the current density along y should be any different from that along x. After all, the structure looks exactly the same along the x and y directions. The relation between electric field and current density in this 2-D material can be written as:

$$\begin{pmatrix} j_x \\ j_y \end{pmatrix} = \begin{pmatrix} \sigma_{xx} & \sigma_{xy} \\ \sigma_{yx} & \sigma_{yy} \end{pmatrix} \begin{pmatrix} E_x \\ E_y \end{pmatrix}. \tag{1.12}$$

Since the x and y directions in the crystal are equivalent, we can interchange them. In other words, we interchange the subscripts in the material property matrix:

$$\begin{pmatrix} \sigma_{xx} & \sigma_{xy} \\ \sigma_{yx} & \sigma_{yy} \end{pmatrix} \rightarrow \begin{pmatrix} \sigma_{yy} & \sigma_{yx} \\ \sigma_{xy} & \sigma_{xx} \end{pmatrix}.$$

If we apply the electric field along the same direction as before, we obtain:

$$\begin{pmatrix} j_x \\ j_y \end{pmatrix} = \begin{pmatrix} \sigma_{yy} & \sigma_{yx} \\ \sigma_{xy} & \sigma_{xx} \end{pmatrix} \begin{pmatrix} E_x \\ E_y \end{pmatrix}. \tag{1.13}$$

The response in this case must be equal to the response in (1.12), so that we must have:[7]

$$\begin{pmatrix} \sigma_{xx} & \sigma_{xy} \\ \sigma_{yx} & \sigma_{yy} \end{pmatrix} = \begin{pmatrix} \sigma_{yy} & \sigma_{yx} \\ \sigma_{xy} & \sigma_{xx} \end{pmatrix}, \tag{1.14}$$

which means that

$$\sigma_{xx} = \sigma_{yy} \qquad \text{and} \qquad \sigma_{xy} = \sigma_{yx}. \tag{1.15}$$

[7] This procedure is mathematically not entirely rigorous. An exact derivation requires the use of the transformation formula for a second rank tensor, which is beyond the scope of this textbook. The exact derivation for the case illustrated above would result in the following equalities:

$$\sigma_{xx} = \sigma_{yy} \qquad \text{and} \qquad \sigma_{xy} = -\sigma_{yx}.$$

So, we have established that, for a crystal with the structure shown in Fig. 1.4(a), the components of the conductivity matrix are related to one another by the above relations. The fact that the current densities in the x and y directions must be equal to each other is a reflection of the *symmetry* of the underlying crystal structure.[8] The square character of the grid directly leads to relations 1.15.

This is an example of how the symmetry of a structure imposes constraints on the physical (or material) properties of the structure. This simple observation provides an immediate motivation for a textbook on crystal structures: many material properties are directly determined by the underlying structure of the material, i.e., the precise distribution of the atoms. To understand material properties, and to design materials with new properties, we must, therefore, understand how the atoms are arranged. This consists of two parts: first, we must learn the proper language to describe crystal structures; then, we must learn how to determine where the atoms are located. We will learn both of these aspects in the first half of the book. Then, we will apply what we have learned to a large variety of crystal structures in the second half.

Before we provide a further illustration of what this book is all about, we must conclude the example that we started at the beginning of this section. There is more to material properties than just the underlying crystal structure. Material properties must also satisfy additional laws of physics, in particular, the laws of *thermodynamics*. In the case of electrical conductivity, one can show that the matrix representing the conductivity must always be a symmetric matrix, i.e., $\sigma_{ij} = \sigma_{ji}$. If we apply this to Equations 1.15, taking into account the footnote on page 14, we find that $\sigma_{xy} = -\sigma_{xy}$, and this can only be true if $\sigma_{xy} = 0$. Hence, thermodynamics and symmetry combine to predict that for the crystal structure shown in Fig. 1.4(a), the relation between current density and electric field must be

$$\begin{pmatrix} j_x \\ j_y \end{pmatrix} = \begin{pmatrix} \sigma & 0 \\ 0 & \sigma \end{pmatrix} \begin{pmatrix} E_x \\ E_y \end{pmatrix} = \sigma \begin{pmatrix} E_x \\ E_y \end{pmatrix}.$$

However, if the crystal structure is based on a rectangular grid rather than a square grid, it can be shown (reader exercise) that $\sigma_{xx} \neq \sigma_{yy}$ so that the relation becomes:

$$\begin{pmatrix} j_x \\ j_y \end{pmatrix} = \begin{pmatrix} \sigma_{xx} & 0 \\ 0 & \sigma_{yy} \end{pmatrix} \begin{pmatrix} E_x \\ E_y \end{pmatrix}.$$

For the crystal structure shown in Fig. 1.4(b), it is intuitively clear that the conductivity along the x and y directions must be different, since the sequence of atoms in each direction is different.

[8] Note that we will define what a crystal structure is in Chapter 3.

When a material property does not depend on the direction of the applied field, then that property is known as an *isotropic* property. Properties that do depend on the direction of the field are *anisotropic* properties. The electrical conductivity in the crystal structure of Fig. 1.4(a) is isotropic, but in Fig. 1.4(b) the conductivity is anisotropic. Note that it is possible to have anisotropic properties that are homogeneous or heterogeneous across a crystal; if a property is heterogeneous, it means that the value of the material constants (e.g., the value of the electrical conductivity σ) varies with location in the crystal, perhaps due to chemical inhomogeneities.

Note that the above arguments do not say anything at all about the magnitude of the conductivity parameters. Instead, symmetry and thermodynamics only state which parameters must vanish, and how each parameter is related to the others. The magnitude of the parameters must follow from a different branch of physics, known as *solid state physics*, which would use quantum mechanics and other tools to express the conductivity in terms of more fundamental parameters (i.e., the charge distribution in the material). Once again, knowledge of the underlying crystal structure is essential for these kinds of computations.

There are many material properties. The most important ones are linear properties, meaning that there is a direct proportionality between the field and the response. Others are quadratic in the field, or even higher order. Each material property is represented mathematically by a tensor. Tensors of rank zero are scalars, rank one results in a vector, rank two in a 3×3 matrix, and so on. Table 1.2 shows some of the more important material properties that are represented by tensors. The tensors are grouped by rank, and are also labeled (in the last column) by E (equilibrium property) or T (transport property). The number following this letter indicates the maximum number of independent, non-zero elements in the tensor, taking into account symmetries imposed by thermodynamics. The *Field* and *Response* columns contain the following symbols: ΔT = temperature difference, ΔS = entropy change, E_i = electric field components, H_i = magnetic field components, ϵ_{ij} = mechanical strain, D_i = electric displacement, B_i = magnetic induction, σ_{ij} = mechanical stress, $\Delta\beta_{ij}$ = change of the impermeability tensor, j_i = electrical current density, $\nabla_j T$ = temperature gradient, h_i = heat flux, $\nabla_j c$ = concentration gradient, m_i = mass flux, ρ_i^a = anti-symmetric part of resistivity tensor, ρ_i^s = symmetric part of resistivity tensor, $\Delta\rho_{ij}$ = change in the component ij of the resistivity tensor, l_i = direction cosines of electromagnetic wave direction in crystal, and G = optical gyration constant.

It is clear from this table that there are quite a few important material properties. While the details of this table go far beyond this textbook, it is instructive to see that the symmetry of the underlying crystal structure of a material has an influence on *all* these properties.

Table 1.2. Materials property and transport tensors (adapted from Nowick (Nowick, 1995)).

Property	Symbol	Field	Response	Type/#
	Tensors of Rank 0 (Scalars)			
Specific heat	C	ΔT	$T\Delta S$	E/1
	Tensors of Rank 1 (Vectors)			
Electrocaloric	p_i	E_i	ΔS	E/3
Magnetocaloric	q_i	H_i	ΔS	E/3
Pyroelectric	p'_i	ΔT	D_i	E/3
Pyromagnetic	q'_i	ΔT	B_i	E/3
	Tensors of Rank 2			
Thermal expansion	α_{ij}	ΔT	ϵ_{ij}	E/6
Piezocaloric effect	α'_{ij}	σ_{ij}	ΔS	E/6
Dielectric permittivity	κ_{ij}	E_j	D_i	E/6
Magnetic permeability	μ_{ij}	H_j	B_i	E/6
Optical activity	g_{ij}	$l_i l_j$	G	E/6
Magnetoelectric polarization	λ_{ij}	H_j	D_i	E/9
Converse magnetoelectric polarization	λ'_{ij}	E_j	B_i	E/9
Electrical conductivity (resistivity)	σ_{ij} (ρ_{ij})	E_j (j_j)	j_i (E_i)	T/6
Thermal conductivity	K_{ij}	$\nabla_j T$	h_i	T/6
Diffusivity	D_{ij}	$\nabla_j c$	m_i	T/6
Thermoelectric power	Σ_{ij}	$\nabla_j T$	E_i	T/9
Hall effect	R_{ij}	B_j	ρ_i^a	T/9
	Tensors of Rank 3			
Piezoelectricity	d_{ijk}	σ_{jk}	D_i	E/18
Converse piezoelectricity	d'_{ijk}	E_k	ϵ_{ij}	E/18
Piezomagnetism	Q_{ijk}	σ_{jk}	B_i	E/18
Converse piezomagnetism	Q'_{ijk}	H_k	ϵ_{ij}	E/18
Electro-optic effect	r_{ijk}	E_k	$\Delta\beta_{ij}$	E/18
Nernst tensor	Σ_{ijk}	$\nabla_j T B_k$	E_i	T/27
	Tensors of Rank 4			
Elasticity	s_{ijkl} (c_{ijkl})	σ_{kl} (ϵ_{kl})	ϵ_{ij} (σ_{ij})	E/21
Electrostriction	γ_{ijkl}	$E_k E_l$	ϵ_{ij}	E/36
Photoelasticity	q_{ijkl}	σ_{kl}	$\Delta\beta_{ij}$	E/36
Kerr effect	p_{ijkl}	$E_k E_l$	$\Delta\beta_{ij}$	E/36
Magnetoresistance	ξ_{ijkl}	$B_k B_l$	ρ_{ij}^s	T/36
Piezoresistance	Π_{ijkl}	σ_{kl}	$\Delta\rho_{ij}$	T/36
Magnetothermoelectric power	Σ_{ijkl}	$\nabla_j T B_k B_l$	E_i	T/54
Second order Hall effect	ρ_{ijkl}	$B_j B_k B_l$	ρ_i^2	T/30
	Tensors of Rank 6			
Third order elasticity	c_{ijklmn}	$\epsilon_{kl}\epsilon_{mn}$	σ_{ij}	E/56

1.5.4 A second encounter with symmetry

The previous section illustrated how the symmetry of an arrangement of atoms affects a particular material's property tensor, in this case the electrical conductivity tensor. In the relation between the current density and the applied electric field, there is no "zeroth-order" term j_0, i.e., there is no current density when there is no applied field. In this section we consider another, pictorial, representation of the role of symmetry in determining the internal energy of a material. The *magnetocrystalline anisotropy energy density* is the internal energy per unit volume associated with the orientation of the *magnetization vector*, $\mathbf{M}$, in a crystal. The *magnetization* is defined as the magnetic dipole moment per unit volume. In a *ferromagnetic material*, like Fe, these dipole moments are aligned to give rise to a permanent magnetization.[9]

Magnetocrystalline anisotropy refers to the fact that the magnetization of a single crystalline material prefers to be oriented parallel to certain easy internal crystallographic directions. For example, a cubic Fe single crystal has a lower internal magnetocrystalline anisotropy energy density, if the magnetization vector points along a direction with 4-fold as opposed to 3-fold or 2-fold symmetry. A cubic Co single crystal has a lower internal magnetocrystalline anisotropy energy density, if the magnetization vector points along a direction with 3-fold as opposed to 4-fold or 2-fold symmetry. Figure 1.5 shows a surface of constant magnetocrystalline anisotropy energy density in a Cartesian coordinate system for each of these materials. Note that both of these surfaces reflect all of the symmetries of

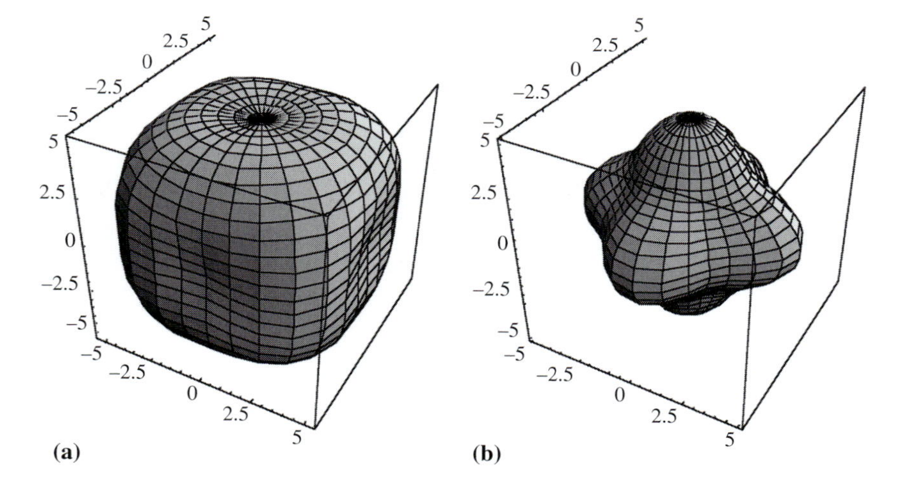

Fig. 1.5. Surface of constant magnetocrystalline anisotropy energy density in Cartesian coordinates for (a) a cubic Fe (a) and Co (b) single crystal (courtesy of P. Ohodnicki).

[9] Magnetization in a ferromagnet is an example of a material response in which there is a remanent response: even in the absence of a magnetic field there is a remanent magnetization vector which is oriented in such a way as to minimize the internal energy density.

the cubic crystal. However, the minimum energy for the Fe single crystal corresponds to directions pointing to the center of the faces (4-fold axes), while for the Co single crystal, the minimum corresponds to directions pointing to the cube corners (3-fold axes). This example illustrates that the symmetry of a crystal lattice not only determines what properties this material can exhibit, it also governs the energetics of various physical phenomena.

1.6 So, what is this book all about?

In the last section of this introductory chapter, we take a brief look at some of the topics covered in more detail in later chapters. To focus our attention on a few concrete examples, we have selected four different materials that are easy to find, in case the reader would like to repeat some of the observations presented in this section and in Chapter 14:

- sugar: regular sugar (sucrose) that you can get in a grocery store;
- salt: standard table salt (essentially pure sodium chloride);
- nickel: the 5 cent coin in the USA is made of a Cu–Ni alloy;
- glass: a simple glass slide for an optical microscope.

Let us assume that we would like to find out what the atomic structure is of these four materials. Of course, these are rather basic materials, and their structures have been understood for a long time. Nevertheless, we will "pretend" that we do not know what the crystal structures of these materials look like.

A simple observation with an optical microscope reveals that sugar and salt crystallites show clearly developed facets, indicative of the underlying crystalline character of the material. We know already, from the discussions in Section 1.3, that we cannot use optical microscopy to determine the crystal structure. We must use X-rays with a wave length similar to the interatomic spacing. The most straightforward way of obtaining the experimental data needed to identify the crystal structure is to obtain a *powder diffraction pattern*. Here's how this is done.

Consider the experimental set up in Fig. 1.6. A beam of electrons is accelerated by a potential drop in the range of 10–50 kV. The beam reaches a metal target, T, typically made of Cu, Mo, W, Co, or Cr, where the kinetic energy of the electrons is converted into X-rays and heat. The X-rays are then collimated through a narrow slit and projected onto the material. The sample, S, is mounted on a stage that can rotate around an axis normal to the plane of the drawing. The angle between the sample plane and the X-ray beam direction is usually represented by the symbol θ. Finally, a detector, D, is placed at some distance, r, from the sample. The detector also rotates around the same axis, but at twice the angular velocity. In other words,

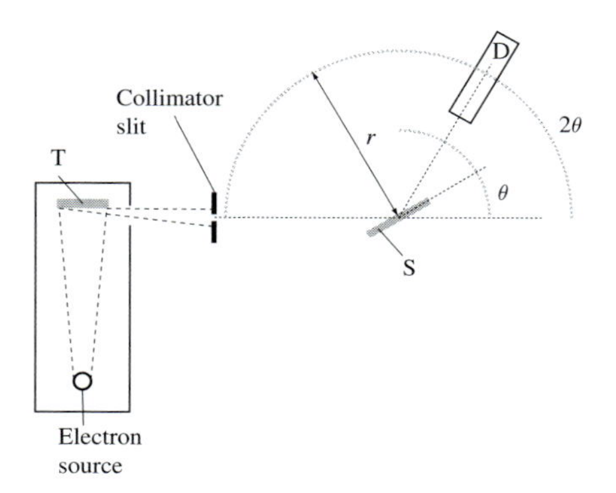

Fig. 1.6. Schematic illustration of the experimental set up for a powder diffraction experiment.

when the sample rotates by 10°, the detector rotates 20°. This experimental configuration is known as a θ–2θ diffractometer, and it is one of the basic tools of modern crystallography.

The detector, D, measures the number of X-ray photons, I, that are scattered by the sample over an angle 2θ with respect to the incident beam direction. When we plot the scattered intensity I as a function of the angle 2θ, we obtain a *powder diffraction pattern*. We have carried out precisely this observation on our four basic materials. The resulting powder diffraction patterns are shown in Fig. 1.7.

We note that the four patterns in Fig. 1.7 are distinctly different from each other. Sugar has a large number of peaks, whereas table salt only has a few in the angular range shown ($10° \leq 2\theta \leq 60°$). Glass has only one very broad peak, and for the nickel coin, only two peaks fall in the selected angular

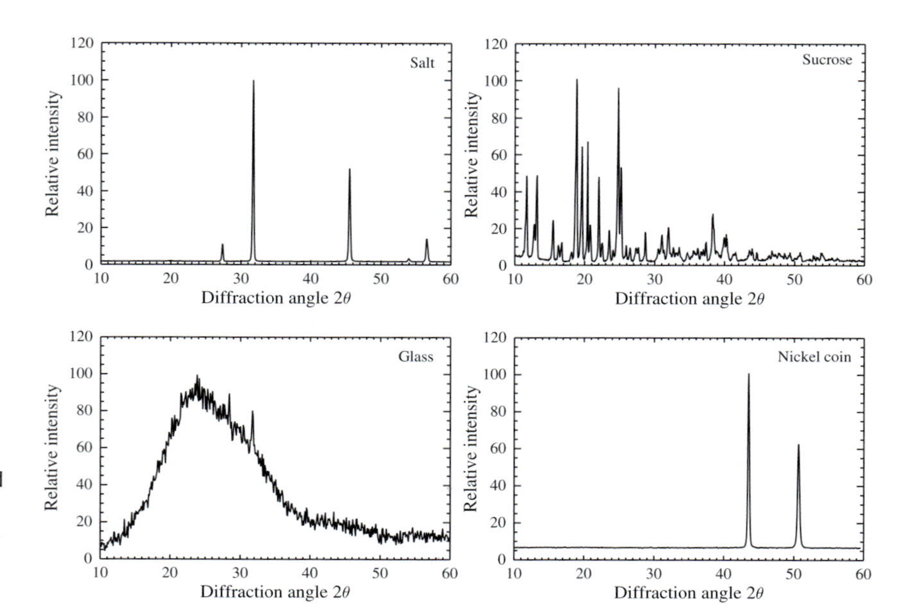

Fig. 1.7. Powder diffraction patterns for the angular range $10° \leq 2\theta \leq 60°$ for the four materials (salt, sugar, glass and nickel coin). The intensities have been scaled so that the most intense peak of each pattern corresponds to 100.

range. Note that there are two characteristic values associated with each peak: the location of the peak, measured in degrees, and its relative intensity. We will see in the first part of this book that the position of the peaks is related to the shape of the unit cell of the material, whereas the relative intensities are determined by the positions of the atoms inside that unit cell. In other words, a given crystal structure produces a particular set of peaks, and these peaks uniquely identify the structure. This means that we can use powder diffraction patterns such as the ones shown in Fig. 1.7 as *fingerprints* for crystal structures. It is one of the purposes of this book to teach the reader how the powder diffraction pattern of a crystal structure is related to that structure.

1.7 Historical notes

The final section of each chapter is entitled "*Historical notes*", and contains a few paragraphs describing important events in the history of crystallography and/or important scientists who have made a significant impact in the field. The authors have attempted to present a wide range of characters, from the early beginnings of crystallography all the way to the present day. These historical sections do not provide a complete history of crystallography, and they are not even in chronological order. Instead, each section reflects on a particular episode or person, usually related to one of the topics treated in the corresponding chapter. The main purpose of these sections is to provide the reader with background information that is often not as well known as the rest of the chapter. Usually, a few citations are provided as well, so that the interested reader may find out more about a particular person or event. Obviously, in this day of internet search engines and on-line encyclopedias, it should not be too difficult for the reader to locate additional information; the names and events in the historical sections will provide the search terms for such explorations.

(a)

(b)

Fig. 1.8. (a) G. Agricola (1494–1555), and (b) C. Huyghens (1629–1695) (pictures courtesy of J. Lima-de-Faria).

Georgius Agricola (1494–1555) was a German physician and metallurgist. The science of mineralogy is widely viewed as originating with the works he authored on mining, geology, and mineralogy. Crystallography is closely related to the field of mineralogy. Similarly, the roots of modern materials science are intertwined with early mineralogy. The book *De Natura Fossilium* (Agricola, 1546) contains a classification of minerals based on physical properties such as color, density, transparency, lustre, taste, odor, shape, and texture. The *powers* of minerals were attributed to their natural properties instead of a divine origin. The importance of the geometrical shape of crystals, a precursor to ideas of geometrical crystallography, was emphasized in this work.

Agricola is considered to be the father of the experimental approach to science. He published an influential book *De Re Metallica* (Agricola, 1556) describing metallic minerals. This book was translated into English in 1912 by **Herbert Clark Hoover**, a mining engineer who went on to become the United States' 31st president.

Christian Huyghens (1629–95) was a Dutch astronomer, mathematician, philosopher, and physicist. Huyghens studied and mastered geometrical optics and developed techniques for lens grinding which aided his career in astronomy. In his *Traité de la Lumière* ("Treatise on Light," Huyghens, 1690) Huyghens described his research in physical optics. This work included many notable achievements in the field of crystallography. Huyghens was the originator of the wave theory of light and the *Huyghen's principle* is a useful tool to understand *diffraction*. Huyghens was also an accomplished clock designer and builder, and he published a book on probability theory.

1.8 Problems

(i) *Electron diffraction and the de Broglie wave length*: Consider electrons with kinetic energies of 1 eV; 100 eV; and 10 keV.

  (a) Find the de Broglie wave length in each case, and consider whether the electron would be appropriate for use in electron diffraction determinations of crystal structures. (Ignore relativistic corrections.)
  (b) Calculate m_0c^2 for an electron and the size of the relativistic correction for the case of 100 keV electrons.

(ii) *Thermal neutron de Broglie wave length*: Consider thermal neutrons at a temperature, T, of 300 K. i.e., neutrons that are in thermal equilibrium with their surroundings.

  (a) Calculate the neutron thermal velocity. Compare this with the thermal velocity of an electron at 300 K.
  (b) Calculate the de Broglie wave lengths for the neutron and electron of (a).
  (c) Would either of these be useful in resolving atomic positions?

(iii) *X-ray diffraction*: Assuming that the reader has access to a $\theta - \theta$ or $\theta - 2\theta$ diffractometer, collect all four study materials discussed in this chapter, and obtain powder diffraction patterns for the range $10° \leq 2\theta \leq 120°$. For salt, sugar, and nickel, create a spreadsheet file containing (in columns) the values for 2θ, the raw (experimental) intensity, and the scaled intensity (scaled such that the highest intensity equals 100). This data file can be used later in this book (in particular in Chapter 14), to compare the experimental measurements with theoretical predictions.

(iv) *Diffusivity tensor*: The defining relationship between field and response in diffusion is Fick's first law:

$$\mathbf{J_a} = -D\nabla C.$$

The response vector, $\mathbf{J_a}$, the atomic flux, is given by:

$$\mathbf{J_a} = ((J_a)_x, (J_a)_y, (J_a)_z).$$

It has units of # atoms/m^2/s. The field vector is the concentration gradient which is given by:

$$\nabla C = \left(\frac{\partial C}{\partial x}, \frac{\partial C}{\partial y}, \frac{\partial C}{\partial z} \right).$$

Here C is the atomic concentration, in units of # atoms/m^3. The material property associated with diffusion is the diffusivity, or diffusion coefficient, D. For an orthorhombic crystal, Fick's first law can be written as:

$$\begin{pmatrix} (J_a)_x \\ (J_a)_y \\ (J_a)_z \end{pmatrix} = \begin{pmatrix} D_{xx} & 0 & 0 \\ 0 & D_{yy} & 0 \\ 0 & 0 & D_{zz} \end{pmatrix} \begin{pmatrix} \frac{\partial C}{\partial x} \\ \frac{\partial C}{\partial y} \\ \frac{\partial C}{\partial z} \end{pmatrix}$$

(a) What are the units of the diffusion coefficient?

(b) Perform the matrix multiplication to express three equations relating the three components of the flux vector to the three components of the concentration gradient.

(c) What do they imply about atomic fluxes in different directions?

CHAPTER

2 The periodic table of the elements and interatomic bonds

> *"Where the telescope ends, the microscope begins. Which of these two has the grander view?"*
>
> Victor Hugo, *les Miserables*. Cosette, bk III, ch. 3

We begin this chapter with a description of the building blocks of matter, the atoms. We will discuss the periodic table of the elements, and describe several trends across the table. Next, we introduce a number of concepts related to interatomic bonds. We enumerate the most important types of bonds, and how one can describe the interaction between atoms in terms of interaction potentials. We conclude this chapter with a brief discussion of the influence of symmetry on binding energy.

2.1 About atoms

2.1.1 The electronic structure of the atom

The structure of the periodic table of the elements can be understood readily in terms of the structure of the individual atoms. It is, therefore, ironic that the table of the elements was established long before the discovery of quantum theory and the structure of the atom by Bohr in 1913 (Bohr, 1913a,b,c). Bohr introduced his atomic model for the hydrogen atom, consisting of a negatively charged electron orbiting a positively charged nucleus. Nowadays, we take it for granted that the atomic nucleus consists of protons and neutrons, and that a cloud of electrons surrounds the nucleus, but in the nineteenth century and the early part of the twentieth century this was not at all obvious. Let us briefly review the history of the constituents of the atom:

e^- The electron was discovered by J. J. Thomson in 1897; the name *electron* was coined a few years earlier (1891) by J. Stoney, who first used it to indicate the unit of electric charge.[1] The discovery of the electron marks the beginning of a new era, and it is hard to imagine what our societies would look like if the electron had not yet been discovered.

p The proton was first identified by Wien in 1898 and Thomson in 1919, and subsequently named by Rutherford in 1920. It is a positively charged particle of mass 1.6726×10^{-27} kg, which is about 1836 times heavier than the electron. The word *proton* stems from the Greek and means "the first one."

n The neutron was discovered by Sir James Chadwick in 1932. It is a neutral particle with mass nearly equal to that of the proton. It is the strong interaction force between neutrons and protons that keeps the nuclei of atoms from flying apart under the intense electrostatic repulsion between the positively charged protons.

2.1.2 The hydrogenic model

Let us assume that a nucleus of charge Ze, with Z the number of protons, is fixed in the origin of the reference frame and that the position of the single electron is described by a position vector $\mathbf{r}$. It is convenient to work in the so-called *spherical coordinate system*, where the coordinates of a point are given by (r, θ, ϕ) (see Fig. 2.1). It is easy to show that they are related to the cartesian coordinates (x, y, z) in the following way:

$$\begin{aligned} x &= r\sin\theta\cos\phi & r &= \sqrt{x^2+y^2+z^2} \\ y &= r\sin\theta\sin\phi & \theta &= \arccos z/r \\ z &= r\cos\theta & \phi &= \arctan y/x \end{aligned}$$

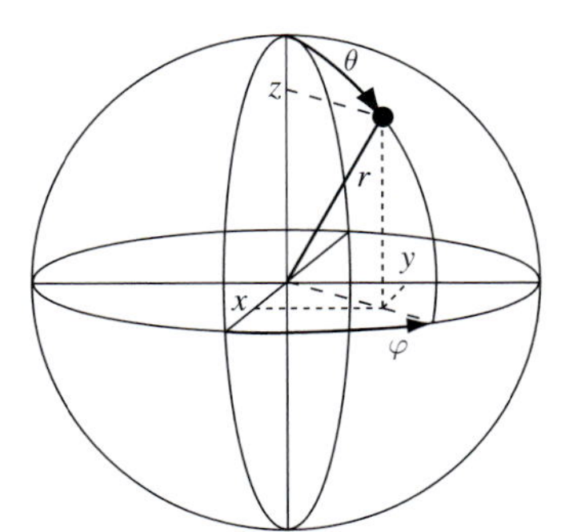

Fig. 2.1. Graphical representation of spherical coordinates.

[1] For a review of electron related research during the first century after its discovery we refer to the book *Electron: a centenary volume* (Springford, 1997). For a detailed account of the early history of elementary particle research we recommend the book *Inward Bound: of Matter and Forces in the Physical World* by A. Pais (Pais, 1986). Both books provide the reader with an in-depth view of how scientists' thoughts were forced to fundamentally change when new discoveries were made.

These coordinates are similar to the commonly used two-dimensional coordinates to locate a position on the globe (longitude and latitude); in that case r would be equal to the radius of the earth.

For the hydrogenic atom (hydrogenic means that the nuclear charge can be a multiple of e) the only relevant force is the Coulomb force between the electron and the nucleus. The electrostatic potential energy V is:

$$V(r) = \frac{-Ze^2}{4\pi\epsilon_0 r},$$

with $\epsilon_0 = 8.85419 \times 10^{-12}$ F/m the permittivity of vacuum. For a hydrogen atom $Z = 1$.

In the first quarter of the twentieth century it was established that the atomistic world obeys the laws of *quantum theory*. Quantum theory or quantum mechanics describes the behavior of particles in terms of a complex-valued function, $\Psi(\mathbf{r})$, known as the *wave function*. The probability, $\mathcal{P}$, to find a particle at a certain location $\mathbf{r}$ is expressed by the modulus-squared of the wave function, evaluated at that position:

$$\mathcal{P}(\mathbf{r}) = \Psi(\mathbf{r})\Psi^*(\mathbf{r}) = |\Psi(\mathbf{r})|^2,$$

where the asterisk * denotes complex conjugation. The wave function Ψ is the solution to a differential equation, known as the *Schrödinger equation* (Schrödinger, 1926). While the details of the mathematics for the hydrogen atom are beyond what we need in this book, it is instructive to consider the formal expression for the wave function of the hydrogen atom. Quantum theory predicts that the general wave function of this system (see historical note on page 51) can be written as the product of two functions; one function depends only on the distance to the origin and is known as the *radial* part, the other function depends only on the angular coordinates and is hence known as the *angular* part. The standard notation for the hydrogenic wave function is:

$$\Psi_{nlm}(r, \theta, \phi) = R_{nl}(r)Y_{lm}(\theta, \phi).$$

The numbers n, l, and m are integer numbers, known as *quantum numbers*; they follow from the detailed mathematical theory.

The integer numbers n, l, and m describe the type of "orbit" the electron will occupy. Each orbit has an associated energy which, in the simplest approximation, depends only on the number n. The energy E corresponding to the different wave functions is:

$$E_n = \frac{-m_0 e^4 Z^2}{8\epsilon_0^2 \mathrm{h}^2 n^2} = -13.6\frac{Z^2}{n^2}\mathrm{eV}.$$

The different energy levels are determined by the quantum number n, which is generally known as the *principal quantum number*.

Table 2.1. Correspondence of the principal and angular momentum quantum numbers and the spectroscopic letter notations.

n	l
$1 \to$ K	$0 \to$ s
$2 \to$ L	$1 \to$ p
$3 \to$ M	$2 \to$ d
$4 \to$ N	$3 \to$ f
$5 \to$ O	$4 \to$ g
...	...

One can show that the radial functions $R_{nl}(r)$ are defined for values of l for which

$$l \leq n - 1.$$

In other words, if $n = 2$ then $l = 0, 1$, if $n = 4$ then $l = 0, 1, 2, 3$, and so on. The number l is known as the *azimuthal* or *angular momentum quantum number*.

Finally, the functions $Y_{lm}(\theta, \phi)$, called spherical harmonics, are only defined for values of m for which

$$|m| \leq l.$$

Hence, for $l = 1$ we have $m = -1, 0, 1$, and for $l = 3$ we have $m = -3, -2, \ldots, 3$. The number m is known as the *magnetic quantum number*. For every value of l there are $2l + 1$ possible values of m. Each triplet of numbers (nlm) defines a possible *state* for the electron. States with identical energy (defined by n) but different values of (lm) are called *degenerate states*.

It is convenient to employ the historical notation for the angular momentum quantum number; the different angular momentum states are referred to by lowercase letters and the principal quantum numbers are referred to by uppercase letters (see Table 2.1).

2.2 The periodic table

In the previous section, we have described how quantum theory gives rise to orbitals and quantum numbers for the hydrogenic atom. It turns out that hydrogen is the *only* atom for which such an analytical solution can be obtained; all other atoms are so complex that only numerical methods can be used to solve their Schrödinger equations! Atoms with atomic number larger than $Z = 1$ are more complex because an additional interaction arises which

Table 2.2. Wavelengths and colors for the most important spectral lines of hydrogen. The experimentally measured wave length is shown for comparison.

Transition	ΔE_{ji} (eV)	λ (nm)	Exp. (nm)	Color
$2 \rightarrow 1$	10.20	121.552	121.566	UV
$3 \rightarrow 1$	12.09	102.559	102.572	UV
$4 \rightarrow 1$	12.75	97.2414	97.253	UV
$5 \rightarrow 1$	13.05	94.9623	94.974	UV
$3 \rightarrow 2$	1.89	656.380	656.280	red
$4 \rightarrow 2$	2.55	486.207	486.133	blue-green
$5 \rightarrow 2$	2.86	434.113	434.047	violet
$4 \rightarrow 3$	0.66	1875.37	1875.1	IR
$5 \rightarrow 3$	0.97	1281.99	1281.8	IR
$5 \rightarrow 4$	0.31	4051.73	4050.0	IR

is not present in hydrogen: electron–electron repulsive Coulomb interaction. One of the consequences of this interaction is that the energy of an orbital depends not only on the principal quantum number n, but also on the azimuthal quantum number l.

Using the energy expression for the hydrogen atom,

$$E_n = \frac{-13.6}{n^2}\text{eV},$$

we can readily explain the *spectral lines* of hydrogen, measured by a spectroscopy experiment. The electron is normally in the lowest energy state or *ground state*, which is the state with $n = 1$. When the electron is excited to a higher energy shell, e.g., by absorption of thermal or electromagnetic energy, then it can fall back to a lower energy orbital, emitting a quantum of energy. According to Planck, the wave length of the emitted photon is given by

$$\lambda = \frac{\text{hc}}{\Delta E}, \tag{2.1}$$

where ΔE is the difference in energy between the initial and final levels and c the velocity of light in vacuum. If we denote by ΔE_{ji} the energy difference between levels j and i, then we can construct a table that lists the wave lengths of the corresponding emitted radiation quanta (Table 2.2). The third column indicates the theoretical value for the wave length, using Eq. 2.1, and the fourth column shows the experimentally measured wave length, along with the part of the electromagnetic spectrum that the wave length belongs to.

The successful prediction of the wave lengths of these spectral lines provided the first indication that quantum mechanics was indeed the correct theory, and soon the theory was applied to a large variety of different systems.

It remains an extremely successful theory and its accuracy is surpassed by only very few other theories.[2]

In 1925, George Eugene Uhlenbeck and Samuel Goudsmit postulated that, in addition to spinning around the nucleus, an electron might also have an "internal state of motion," similar to spinning around its own axis. This is called the *spin*, and it is represented by the *spin quantum number s*. Like the energy, spin is *quantized* and the electron can be in only two possible spin states, $s = +\hbar/2$ or $s = -\hbar/2$. These are often referred to as *spin-up* and *spin-down*, respectively. Each electron in an atom thus has four quantum numbers:

(i) n – principal quantum number, $n = 1, 2, 3, \ldots$;
(ii) l – angular momentum quantum number, $l \leq n - 1$;
(iii) m – magnetic quantum number, $|m| \leq l$;
(iv) s – spin quantum number, $s = \pm\frac{1}{2}$ (in units of $\hbar \equiv h/2\pi$).

The principal quantum number n indicates the *shell*, the azimuthal quantum number l indicates the *subshell*. The magnetic quantum number is used to specify the *orbital*.

How are the electrons in atoms with $Z > 1$ arranged? To answer this question we must introduce the *Pauli exclusion principle*. This principle states that no two electrons in a quantum mechanical system can have all identical quantum numbers. It thus follows that there can only be *two* electrons per orbital, one with spin-up and one with spin-down. In addition we find that the maximum number of electrons in a shell is given by $N_n = 2n^2$, or $N_K = 2$, $N_L = 8$, $N_M = 18$, $N_N = 32$, $N_O = 50$, ...

The order in which the shells must be filled can be remembered easily by means of the drawing in Fig. 2.2: write down the principal and azimuthal quantum number combinations, with the principal quantum number changing between rows. Then draw a vertical line from top to bottom and thread the line through the list. This defines the order in which the subshells are being filled. Table 2.3 shows the filling of the orbitals for all atoms in the periodic table.

```
     Start
n
1  1s
2  2s  2p
3      3s  3p  3d
4          4s  4p  4d  4f
5              5s  5p  5d  5f  5g
6                  6s  6p  6d  6f  6g  6h
7                      7s  7p  7d  7f  7g
8                          8s  8p
```

Fig. 2.2. Graphical determination of the filling order for subshells.

[2] In fact, the most accurate theory currently known is *Quantum Electrodynamics*: QED has been used to predict properties of elementary particles and the agreement between theory and experiment is often better than 12 significant digits!

Table 2.3. Filling of the orbitals for all atoms in the periodic table.

Shell		K	L		M			N				O			
Subshell		1s	2s	2p	3s	3p	3d	4s	4p	4d	4f	5s	5p	5d	5f
1	H	1													
2	He	2													
3	Li	2	1												
4	Be	2	2												
5	B	2	2	1											
6	C	2	2	2											
7	N	2	2	3											
8	O	2	2	4											
9	F	2	2	5											
10	Ne	2	2	6											
11	Na	2	2	6	1										
12	Mg	2	2	6	2										
13	Al	2	2	6	2	1									
14	Si	2	2	6	2	2									
15	P	2	2	6	2	3									
16	S	2	2	6	2	4									
17	Cl	2	2	6	2	5									
18	Ar	2	2	6	2	6									
19	K	2	2	6	2	6		1							
20	Ca	2	2	6	2	6		2							
21	Sc	2	2	6	2	6	1	2							
22	Ti	2	2	6	2	6	2	2							
23	V	2	2	6	2	6	3	2							
24	Cr	2	2	6	2	6	5	1							
25	Mn	2	2	6	2	6	5	2							
26	Fe	2	2	6	2	6	6	2							
27	Co	2	2	6	2	6	7	2							
28	Ni	2	2	6	2	6	8	2							
29	Cu	2	2	6	2	6	10	1							
30	Zn	2	2	6	2	6	10	2							
31	Ga	2	2	6	2	6	10	2	1						
32	Ge	2	2	6	2	6	10	2	2						
33	As	2	2	6	2	6	10	2	3						
34	Se	2	2	6	2	6	10	2	4						
35	Br	2	2	6	2	6	10	2	5						
36	Kr	2	2	6	2	6	10	2	6						
37	Rb	2	2	6	2	6	20	2	6			1			
38	Sr	2	2	6	2	6	10	2	6			2			
39	Y	2	2	6	2	6	10	2	6	1		2			
40	Zr	2	2	6	2	6	10	2	6	2		2			
41	Nb	2	2	6	2	6	10	2	6	4		1			
42	Mo	2	2	6	2	6	10	2	6	5		1			
43	Tc	2	2	6	2	6	10	2	6	6		1			
44	Ru	2	2	6	2	6	10	2	6	7		1			
45	Rh	2	2	6	2	6	10	2	6	8		1			
46	Pd	2	2	6	2	6	10	2	6	10					
47	Ag	2	2	6	2	6	10	2	6	10		1			
48	Cd	2	2	6	2	6	10	2	6	10		2			
49	In	2	2	6	2	6	10	2	6	10		2	1		
50	Sn	2	2	6	2	6	10	2	6	10		2	2		

Table 2.3. Filling of the orbitals for all atoms in the periodic table (continued; the K, L, and M shells are completely filled and are not shown in this portion of the table).

Shell		N				O				P				Q	
Subshell		4s	4p	4d	4f	5s	5p	5d	5f	6s	6p	6d	6f	7s	7p
51	Sb	2	6	10		2	3								
52	Te	2	6	10		2	4								
53	I	2	6	10		2	5								
54	Xe	2	6	10		2	6								
55	Cs	2	6	10		2	6			1					
56	Ba	2	6	10		2	6			2					
57	La	2	6	10		2	6	1		2					
58	Ce	2	6	10	2	2	6			2					
59	Pr	2	6	10	3	2	6			2					
60	Nd	2	6	10	4	2	6			2					
61	Pm	2	6	10	5	2	6			2					
62	Sm	2	6	10	6	2	6			2					
63	Eu	2	6	10	7	2	6			2					
64	Gd	2	6	10	7	2	6	1		2					
65	Tb	2	6	10	9	2	6			2					
66	Dy	2	6	10	10	2	6			2					
67	Ho	2	6	10	11	2	6			2					
68	Er	2	6	10	12	2	6			2					
69	Tm	2	6	10	13	2	6			2					
70	Yb	2	6	10	14	2	6			2					
71	Lu	2	6	10	14	2	6	1		2					
72	Hf	2	6	10	14	2	6	2		2					
73	Ta	2	6	10	14	2	6	3		2					
74	W	2	6	10	14	2	6	4		2					
75	Re	2	6	10	14	2	6	5		2					
76	Os	2	6	10	14	2	6	6		2					
77	Ir	2	6	10	14	2	6	9							
78	Pt	2	6	10	14	2	6	9		1					
79	Au	2	6	10	14	2	6	10		1					
80	Hg	2	6	10	14	2	6	10		2					
81	Tl	2	6	10	14	2	6	10		2	1				
82	Pb	2	6	10	14	2	6	10		2	2				
83	Bi	2	6	10	14	2	6	10		2	3				
84	Po	2	6	10	14	2	6	10		2	4				
85	At	2	6	10	14	2	6	10		2	5				
86	Rn	2	6	10	14	2	6	10		2	6				
87	Fr	2	6	10	14	2	6	10		2	6			1	
88	Ra	2	6	10	14	2	6	10		2	6			2	
89	Ac	2	6	10	14	2	6	10		2	6	1		2	
90	Th	2	6	10	14	2	6	10	1	2	6	1		2	
91	Pa	2	6	10	14	2	6	10	2	2	6	1		2	
92	U	2	6	10	14	2	6	10	3	2	6	1		2	
93	Np	2	6	10	14	2	6	10	4	2	6	1		2	
94	Pu	2	6	10	14	2	6	10	5	2	6	1		2	
95	Am	2	6	10	14	2	6	10	7	2	6			2	
96	Cm	2	6	10	14	2	6	10	7	2	6	1		2	
97	Bk	2	6	10	14	2	6	10	8	2	6	1		2	
98	Cf	2	6	10	14	2	6	10	10	2	6			2	
99	Es	2	6	10	14	2	6	10	11	2	6			2	
100	Fm	2	6	10	14	2	6	10	12	2	6			2	
101	Md	2	6	10	14	2	6	10	13	2	6			2	
102	No	2	6	10	14	2	6	10	14	2	6			2	
103	Lw	2	6	10	14	2	6	10	14	2	6			2	

There are a few elements for which the order defined in Fig. 2.2 is not followed; this is mostly due to Coulomb interactions between the electrons in the various shells. A complete numerical analysis of the electronic structure of all elements does yield the correct answer for all elements, but this is beyond the scope of this book.

2.2.1 Layout of the periodic table

We can now use our understanding of the electronic structure of the atom to construct the periodic table of the elements. Each horizontal row in the table corresponds to a different principal quantum number n. From Table 2.3 we find that the K-shell is completely filled after only two elements, H (hydrogen) and He (helium). This is a consequence of the fact that the $1s$ orbital can only accommodate two electrons with opposite spin. The electronic structure is commonly denoted by a symbol of the type $1s^1$, where the first symbol indicates the principal quantum number, the second the subshell, and the superscript how many electrons occupy that particular level. For the He atom, the symbol is $1s^2$.

A completely filled shell is a particularly stable configuration. Elements with a completely filled shell are non-reactive and are generally known as the *inert gases*. We can now construct the periodic table (see Fig. 2.3):

- The last inert gas element in every row has a completely filled shell. The inert gases are Helium (He), Neon (Ne), Argon (Ar), Krypton (Kr), Xenon (Xe), and Radon (Rn).
- In the leftmost column we begin to fill a new shell; these elements, Hydrogen (H), Lithium (Li), Sodium (Na), Potassium (K), Rubidium (Rb), Cesium (Cs), and Francium (Fr), all have an electronic structure of the type $[IG]ns^1$, where the principal quantum number n numbers the rows in the table. The symbol [IG] stands for the electronic structure of the preceding inert element.

Periodic table of the elements

n	1	2	1	2	3	4	5	6	7	8	9	10	1	2	3	4	5	6
1	H																	He
2	Li	Be											B	C	N	O	F	Ne
3	Na	Mg											Al	Si	P	S	Cl	Ar
4	K	Ca	Sc	Ti	V	Cr	Mn	Fe	Co	Ni	Cu	Zn	Ga	Ge	As	Se	Br	Kr
5	Rb	Sr	Y	Zr	Nb	Mo	Tc	Ru	Rh	Pd	Ag	Cd	In	Sn	Sb	Te	I	Xe
6	Cs	Ba	La	Hf	Ta	W	Re	Os	Ir	Pt	Au	Hg	Tl	Pb	Bi	Po	At	Rn
7	Fr	Ra	Ac	*Rf*	*Db*	*Sg*	*Bh*	*Hs*	*Mt*	*Ds*	*Rg*							

Zintl line

	1	2	3	4	5	6	7	8	9	10	11	12	13	14
Lanthanides	Ce	Pr	Nd	Pm	Sm	Eu	Gd	Tb	Dy	Ho	Er	Tm	Yb	Lu
Actinides	Th	Pa	U	*Np*	*Pu*	*Am*	*Cm*	*Bk*	*Cf*	*Es*	*Fm*	*Md*	*No*	*Lw*

Fig. 2.3. The periodic table of the elements. The transuranic (unstable) elements are indicated in italics. The numbers above each column refer to the filling of the different electron orbitals; these are two columns for the *s*-orbitals, six for the *p*-orbitals, 10 for the *d*-orbitals, and 14 for the *f*-orbitals.

- The elements in between are inserted according to the filling order shown in Fig. 2.2. Since the first shell can only contain two electrons, there are only two elements on the first row, H and He. The configuration ns^2np^6 is the configuration of the outer orbitals for all inert elements, except He which is represented by $1s^2$. Since the radius (or more precisely, the expectation value of the radial distance) for the d and f orbitals is smaller than that of the s-and p-orbitals, the d and f electron levels actually lie inside the p orbitals; they are known as the *inner shells*.
- In the second row, we fill the $2s$ level (Li and Berillium (Be)) and then the $2p$ subshell (Boron (B), Carbon (C), Nitrogen (N), Oxygen (O), Fluorine (F), and Ne).
- The filling order diagram then imposes the order $3s$, $3p$, so we fill in another eight elements (Na, Magnesium (Mg), Aluminum (Al), Silicon (Si), Phosphorus (P), Sulfur (S), Chlorine (Cl), and Ar).
- In the fourth row, the filling order is $4s$, $3d$, and $4p$, so there are 18 elements in this row. The row starts with K and Calcium (Ca). Elements with incompletely filled *inner* shells are known as transition elements. The fourth row transition elements (known as the *first transition series*) go from Scandium (Sc, $[Ar]4s^23d^1$) to Zinc (Zn, $[Ar]4s^23d^{10}$): Sc, Titanium (Ti), Vanadium (V), Chromium (Cr), Manganese (Mn), Iron (Fe), Cobalt (Co), Nickel (Ni), Copper (Cu), and Zn. These elements all have the same outer shell structure (except for Cr and Cu), and they all look very similar from a chemical point of view. The electron distribution is spherical (because of the outer s level). The fourth row is completed with the elements Gallium (Ga), Germanium (Ge), Arsenic (As), Selenium (Se), Bromium (Br), and Kr.
- In the fifth row, the filling order is $5s$, $4d$, and $5p$, so again there are 18 elements in the row. The row starts with Rb and Strontium (Sr). The overall structure is similar to that of the fourth row, with the *second transition series* going from Yttrium (Y, $[Kr]5s^24d^1$) to Cadmium (Cd, $[Kr]5s^24d^{10}$): Y, Zirconium (Zr), Niobium (Nb), Molybdenum (Mo), Technetium (Tc), Ruthenium (Ru), Palladium (Pd), Silver (Ag), and Cd. The fifth row is concluded by the six elements Indium (In), Tin (Sn), Antimony (Sb), Tellurium (Te), Iodine (I), and Xe.
- On the sixth row, the filling order is $6s$, $4f$, $5d$, and $6p$. There are 32 elements in the row. To avoid drawing a very wide diagram, the filling of the $4f$ subshell is drawn in a separate row below the main diagram. The row starts with Cs and Barium (Ba). The 14 extra elements after Lanthanum (La) are known as the *lanthanide series* or the *rare earth* elements. The radius of the $4f$ subshell is smaller than that of the $4d$ shell, so the Lanthanides all have similar chemical properties. The rare earth elements are Cerium (Ce), Praesodymium (Pr), Neodymium (Nd), Promethium (Pm), Samarium (Sm), Europium (Eu), Gadolinium (Gd), Terbium (Tb), Dysprosium (Dy), Holmium (Ho), Erbium (Er), Thulium (Tm), Ytterbium (Yb), and Lutetium (Lu). The elements La ($[Xe]6s^25d^1$)

through Mercury (Hg, $[Xe]6s^25d^{10}$) form the *third transition series*: La, Hafnium (Hf), Tantalum (Ta), Tungsten (W), Rhenium (Re), Osmium (Os), Irridium (Ir), Platinum (Pt), Gold (Au), and Hg. The sixth row is completed with the elements: Thallium (Tl), Lead (Pb), Bismuth (Bi), Pollonium (Po), Attinium (At), and Rn.

- On the seventh row, the filling order is $7s$, $5f$, $6d$, and $7p$. Again there are 32 elements, with the 14-member *actinide series* drawn at the bottom of the diagram. The actinides also have similar chemical properties. The row starts with Fr and Radium (Ra). The series from Actinium (Ac, $[Rn]7s^26d^1$) until the end of the table is known as the *fourth transition series*. The actinides are Thorium (Th), Proactinium (Pa), Uranium (U), Neptunium (Np), Plutonium (Pu), Americium (Am), Curium (Cm), Berkelium (Bk), Californium (Cf), Einsteinium (Es), Fermium (Fm), Mendelevium (Md), Nobelium (No), and Lawrencium (Lr). Elements beyond U (the *transuranic elements*) have unstable nuclei and short lifetimes; they are indicated in italic font in Fig. 2.3. The remaining elements in the fourth transition series are Rutherfordium (Rf), Dubnium (Db), Seaborgium (Sg), Bohrium (Bh), Hassium (Hs), Meitnerium (Mt), Darmstadtium (Ds), and Roentgenium (Rg). The elements with atomic number larger than 111 have not (yet) been named.[3]
- One can draw an imaginary diagonal line to the left of B, Si, As, Te, and At. This is the so-called *Zintl line*. The Zintl line separates metals on its left from non-metals on its right.
- Finally, some of the vertical columns in the table have specific names:
  - Column 1: *Alkali metals* (s^1)
  - Column 2: *Alkaline earth metals* (s^2)
  - Column 3: *Rare earth-like metals* (ds^2)
  - Column 11: *Noble metals* (closed d-shell)
  - Column 17: *Halogens* (s^2p^5)
  - Column 18: *Inert gases* (outer shell filled)

This concludes the construction of the periodic table. The familiar appearance of the table finds its roots in the underlying quantum mechanical nature of the interaction between electrons and nucleonic particles.

2.2.2 Trends across the table

2.2.2.1 Atom size

We have seen in Chapter 1 that the nucleus of an atom has a diameter of about 10^{-14} m; the electron cloud around the nucleus has a diameter of

[3] It took a research group at the Gesellshaft für Schwerionenforschung in Darmstadt, Germany, 11 days to produce one atom of element 109! And then it disintegrated after only a few milliseconds.

about 10^{-10} m, which means that the atom is relatively empty. Despite this emptiness, it is the outer shell of the electron cloud that determines almost all chemical properties of the atom. To a first approximation, an atom can be regarded as a sphere, with a relatively constant radius; as we will see later on, this radius may change, depending on the environment.

The atomic radius increases whenever a shell is added to the atom. Thus, the radius increases from top to bottom across the periodic table. For the alkali metals, the radii are $r_{\mathrm{Li}} = 0.157\,\mathrm{nm}$, $r_{\mathrm{Na}} = 0.192\,\mathrm{nm}$, $r_{\mathrm{K}} = 0.238\,\mathrm{nm}$, $r_{\mathrm{Rb}} = 0.251\,\mathrm{nm}$, and $r_{\mathrm{Cs}} = 0.270\,\mathrm{nm}$. For the inert gases we have $r_{\mathrm{He}} = 0.05\,\mathrm{nm}$, $r_{\mathrm{Ne}} = 0.160\,\mathrm{nm}$, $r_{\mathrm{Ar}} = 0.192\,\mathrm{nm}$, $r_{\mathrm{Kr}} = 0.197$, and $r_{\mathrm{Xe}} = 0.218\,\mathrm{nm}$.

Across a row, more electrons are added and an identical number of protons is added to the nucleus. The forces between the nucleus and the electron cloud thus increase from left to right across the table, which means that the atomic radii on average decrease from left to right. For row 4 we find $r_{\mathrm{K}} = 0.238\,\mathrm{nm}$, $r_{\mathrm{Ca}} = 0.197\,\mathrm{nm}$, $r_{\mathrm{Sc}} = 0.160\,\mathrm{nm}$, $r_{\mathrm{Ti}} = 0.147\,\mathrm{nm}$, $r_{\mathrm{V}} = 0.136$, and so on. Irregularities in this sequence are usually due to irregularities in the filling order of the orbitals.

The atomic size is an important parameter when considering the stability of crystal structures. If an atom is too large to fit into a crystal lattice, then the lattice will either be distorted or the crystal structure will change to accommodate the larger atom.

2.2.2.2 Electronegativity and related quantities

Electronegativity scales measure the relative abilities of atomic species to (1) attract more electrons to themselves or (2) give up electrons to neighboring atoms in molecules or solids. If we have two atoms, A and B, then their relative electronegativities should predict whether A is more likely to rob an electron from B or vice versa. With this description in mind, the larger the electronegativity, the "hungrier" is the atom for more electrons. Atoms like Cl, F, etc. are more electronegative than Li, Na, etc. On the other hand, electronegativity is not defined so that it has a precise quantitative meaning. There are many electronegativity scales (Pauling (Pauling, 1932), Mulliken (Mulliken, 1949), . . .) and they are largely empirical. The absolute values are not important in comparing electronegativity scales, only the relative values.

Perhaps one of the most precise and physically significant electronegativity scales is the so-called *Mulliken electronegativity scale* named after Robert S. Mulliken. The Mulliken electronegativity, represented by the symbol χ, is defined as:

$$\chi = \frac{1}{2}(I + A), \tag{2.2}$$

where the *ionization potential*, I, is a measure of an atom's proclivity for giving up an electron, and the *electron affinity*, A, is a measure of an atom's ability to attract an additional electron.

The ionization potential, I, is the energy required to pull away the most loosely bound electron from the atom so as to leave behind a singly charged positive ion. It is the energy required to drive the reaction:

$$C \rightarrow C^{+} + e^{-},$$

where C represents the atom. The lowest ionization potentials are displayed by the alkali metals. This can readily be understood since removal of one electron from an alkali metal leaves an ion with a completely filled shell, a very stable state. When we move across a row in the periodic table, I increases in a somewhat irregular fashion until we reach the inert gas column. Some examples: $I(\mathrm{Na}) = 0.82 \times 10^{-18}$ J (or 0.82 aJ, where a stands for atto $= 10^{-18}$), $I(\mathrm{Ar}) = 2.53$ aJ. The ionization potential decreases going down a column because the average diameter of the atom increases and the outer electron is more loosely bound.

The electron affinity, A, is the energy gained when an electron is brought from infinity up to a neutral atom; in other words, it is the energy required to drive the reaction:

$$C + e^{-} \rightarrow C^{-}.$$

It is hence a measure for the stability of a negatively charged ion. There is once again a systematic variation across the periodic table.

The Mulliken electronegativity is the average of I and A. The true meaning of the Mulliken electronegativity can be made clear by the following thought experiment (Fig. 2.4): imagine that the total electron charge Q of an atom is not a discrete quantity, measured in multiples of the electron charge, but instead can take on any real value. The atom in its neutral state has n electrons, with a total charge of $Q = -ne$ and a total energy $E(n)$. One can then imagine a smoothly varying atomic energy $E(n)$ as a function of the electronic occupation n. Positively ionized states have less electronic charge and fall to the left of the neutral atom in Fig. 2.4, with the ionic state C^{+} corresponding to $Q = -(n-1)e$. Negatively charged states fall to the right of the neutral atom, and correspond to $Q = -(n+1)e$ for the ionic state C^{-}. The four points on the curve in Fig. 2.4 correspond to actual quantum mechanical calculations for the sulfur atom S. From this figure we can now derive the significance of the Mulliken electronegativity. The cord connecting $E(n-1)$ with $E(n+1)$ obviously has a slope $(I+A)/2$, given the prior definitions of I and A, and their graphical representation on the figure. This cord is also seen to be parallel to a line tangent to $E(n)$ at the neutral atom position. It is, therefore, clear that:

$$\chi = \frac{1}{2}(I + A) \approx \left.\frac{\partial E}{\partial n}\right|_{n}.$$

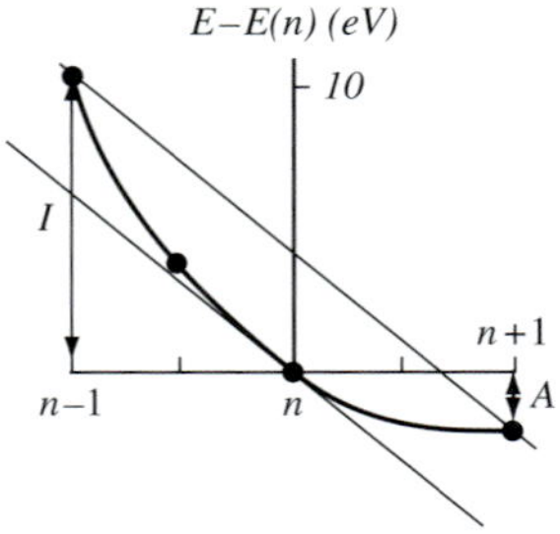

Fig. 2.4. Schematic illustration of the energy function used to define the Mulliken electronegativity scale (McHenry *et al.*, 1987)

Table 2.4. Experimentally determined Mulliken electronegativities (in eV).

Atom	I_{expt} [a]	A_{expt} [b]	χ
B	8.296	0.28	4.238
C	11.254	1.263	6.264
O	13.614	1.46	7.537
F	17.420	3.40	10.410
Al	5.984	0.44	3.212
Si	8.149	1.39	4.769
S	10.357	2.077	6.217
Cl	13.01	3.62	8.315
Li	5.39	0.618	3.004
N	14.54	0 [c]	7.27
Na	5.138	0.548	2.843
P	10.55	0.746	5.648

[a] (Moore, 1970)
[b] (Radzig and Smirnov, 1985)
[c] (Ebbing, 1984)

The derivative of total energy with respect to the electron occupation number is generally known as the *chemical potential* for electrons. Thus, we conclude that the Mulliken electronegativity represents a chemical potential for electrons. Table 2.4 summarizes experimental ionization potentials and electron affinities for some light atoms and values of the Mulliken electronegativity derived from these values.

The electronegativity also allows us to define elements as metals and non-metals.

- *metals*

  (i) have few electrons in the outer shell (three or less)
  (ii) form cations by losing electrons
  (iii) have low electronegativity

- *non-metals*

  (i) have four or more electrons in the outer shell
  (ii) form anions by gaining electrons
  (iii) have high electronegativity

As mentioned before, the *Zintl line* delineates the boundary between metals and non-metals.

2.2.2.3 Valence

Valence is a measure of the number of *other* atoms with which the atom of a given element tends to combine. It is also known as the *oxidation number*. Let us consider two examples:

(i) Sodium chloride (NaCl): The ionization energies and electron affinities of both elements are:

$$I(\text{Na}) = 0.82\,\text{aJ} \qquad I(\text{Cl}) = 2.08\,\text{aJ}$$
$$A(\text{Na}) = 0.11\,\text{aJ} \qquad A(\text{Cl}) = 0.58\,\text{aJ}$$

From these numbers we see that Na is more likely than Cl to give up an electron, whereas Cl is more likely than Na to accept one. Upon doing so, both elements acquire an inert gas configuration: Na^+ has the [Ne] configuration and Cl^- has the [Ar] configuration. The resulting molecule of NaCl is hence very stable. We conclude that sodium likes to bond with another atom and form a positive ion; the *valence* of sodium is hence +1, that of chlorine is −1.

(ii) Magnesium chloride ($MgCl_2$): Magnesium has the [Ne]$3s^2$ electronic structure and thus likes to donate two electrons to obtain the [Ne] inert gas configuration. Chlorine can only accept one single electron since its valence is −1, as derived above. Therefore, Mg prefers to bond to two negative ions and hence has a valence of +2.

This procedure can be continued until valences are self-consistently assigned to all elements. Many elements can have multiple valence states. In many compounds, the valence of one or more of the elements helps determine the resulting crystal structure.

2.3 Interatomic bonds

2.3.1 Quantum chemistry

The wave model of the atom proposed by Schrödinger provided the mathematical basis for *quantum chemistry*. Quantum chemistry is the branch of chemistry that applies quantum mechanics to problems such as chemical energy and the reactivity of atoms and molecules. While solutions to the Schrödinger equation are required to address these problems in quantitative detail, it is possible to illustrate the key results of quantum chemistry in a simpler, more intuitive manner, as will be done in the following sections.

The first calculation in quantum chemistry was that of the chemical binding energy in the H_2 molecule performed by Walter Heitler and Fritz London (Heitler and London, 1927). The results of this model are often used (as below) to motivate the description of a covalent chemical bond. The Heitler–London theory was followed by the *valence bond theory* proposed by American scientists John C. Slater and Linus Carl Pauling and the *molecular orbital theory* proposed by Mulliken. In recent years these techniques have been supplanted by *density functional theory*. While the details of quantum chemistry are beyond the scope of this book, we do illustrate relevant examples of its results throughout the text.

2.3.2 Interactions between atoms

A completely filled shell is a particularly stable electronic configuration. Atoms that are only a few electrons short of such a configuration have a strong tendency to form bonds with other atoms, such that the shells of both atoms become completely filled. There is thus a *driving force* for atoms to form bonds. The energy of the pair of atoms after they have formed a bond is lower than the energy of the atoms separated from each other at infinity. This energy difference is generally known as the *binding energy*.

If we take the infinite separation state as the zero energy state – we are always free to choose the zero of the energy scale – then we can plot the interaction energy (or potential energy) of the two atoms versus their internuclear spacing, r. Figure 2.5 shows a typical plot of the potential energy $V(r)$ versus the distance r. As the atoms approach each other, the energy is lowered, indicating that bond formation will occur. The curve then goes through a minimum for a particular spacing r_m and begins to rise again. For much smaller spacings, the curve rises steeply since the atoms repel each other. This reflects the Pauli exclusion principle: no two electrons can occupy the same volume in space and have all identical quantum numbers. This is valid regardless of the origin of those electrons, i.e., regardless of whether the electrons belong to atom A or atom B.

The potential energy curve, $V(r)$, is often known as the *interatomic potential*. Regardless of the types of atoms involved in the bond, the interaction potential always has a shape similar to that shown in Fig. 2.5; repulsive at short distances, and attractive at larger distance, with an equilibrium spacing r_m somewhere in between. The exact form of the potential must be determined either from first principles quantum mechanical computations, or it can be fitted to experimental measurements of quantities like the heat of formation,

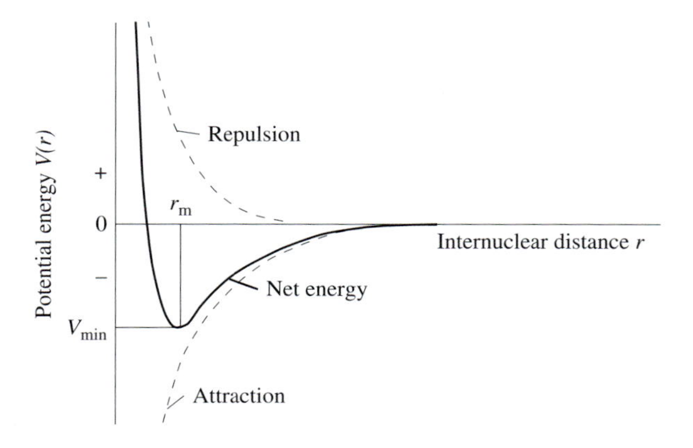

Fig. 2.5. Attractive and repulsive components of the potential energy curve for the interaction between two atoms.

the spacing between the atoms, and so on. The interatomic potential is not only assumed to be valid for a single pair of atoms, each pair in a solid interacts according to the same potential function; the strength of the interaction depends on the distance between the atoms.

The simplest type of interatomic potential depends only on the distance between atoms, and not on the direction. It is thus a *radial* or *central* potential. Only two atoms interact with each other and therefore it is a *pair* or *two-body* potential. Many-body interactions can be taken into account in more advanced potential models, as can directional effects. For instance, we will see later on that in the diamond crystal structure the bonds are strongly directional, i.e., they point in certain directions in space. Therefore, the interaction between two neighboring carbon atoms can not be described by a simple radial pair potential, and direction dependent terms must be included. There are several ways in which many-body interactions can be included and one of the more popular formalisms is known as the *embedded atom method*.

From classical mechanics, we know that there is a standard relation between a potential energy function and the resulting force. The force experienced at a distance $r = r_1$ is given by the negative gradient of the potential energy, evaluated at that distance r_1:

$$\mathbf{F}(r_1) = -\left.\nabla V(r)\right|_{r=r_1} = -\left.\frac{\partial V}{\partial r}\right|_{r=r_1} \mathbf{e}_r,$$

where the vector $\mathbf{e}_r$ is a unit vector along the line connecting the two atoms. An atom is said to be in equilibrium in a crystal when the total sum of all forces exerted on the atom averages to zero over a sufficiently long time interval. Since atoms vibrate around their lattice positions, the instantaneous force is nearly always different from zero, but, on average, over a large number of vibrations (or equivalently, a sufficiently long time), the total force vanishes and the average position of the atom remains constant.

2.3.3 The ionic bond

Electronegativity differences between two elements are important in determining the bond character between them. Let us consider the alkali-halide LiF. It consists of a strongly electronegative element (fluorine), and a strongly "electropositive" element (lithium). The electronic structure and electronegativities are:

Element	config.	Pauling	Mulliken
Li	$[He]2s^1$	1.0	3.004
F	$[He]2s^22p^5$	3.98	10.410

The E_N-difference between these elements is about 3 on the Pauling scale, and 7.4 on the Mulliken scale. Lithium will, as an electropositive element, give up an electron easily to acquire the [He] configuration, and F will accept one to acquire the [Ne] configuration. Both of these configurations are spherical and stable and the two ions will be bonded by a strong *Coulombic* or electrostatic force. This is known as an *ionic bond*.

We know from the discussion of the variation of atomic radii across the periodic table that the addition or removal of an electron changes the radius of the atom. In particular, since the nuclear charge does not change upon formation of an ion, the addition of an electron will increase the radius of the atom because of the increased repulsive interaction between the orbital electrons. Removal of the outer electron will decrease the size, because now the attraction of the nucleus becomes stronger. We can illustrate this with the NaCl compound:

$$\begin{array}{cc|cc} & \overset{\textit{electron transfer}}{\rightarrow} & & \\ \text{Na} & \text{Cl} & \text{Na}^+ & \text{Cl}^- \\ [\text{Ne}]3s^1 & [\text{Ne}]3s^23p^5 & [\text{Ne}] & [\text{Ar}] \\ r_{\text{Na}} = & 0.192\,\text{nm} & r_{\text{Na}^+} = & 0.095\,\text{nm} \\ r_{\text{Cl}} = & 0.099\,\text{nm} & r_{\text{Cl}^-} = & 0.181\,\text{nm} \end{array}$$

The interaction energy between the two ions is expressed by the sum of two terms:

(i) *Coulombic attraction*: the electrostatic potential energy of two charged particles at a distance r from each other is given by

$$V_c(r) = \frac{Z_1 Z_2 e^2}{4\pi\epsilon_0 r},$$

where Z_i is the number of electrons moved for the formation of ion i. Note that Z_i is positive if the electrons were added to the ion, and negative if they were removed. In the case of LiF, the product Z_1Z_2 is equal to -1.

(ii) *Repulsion due to exclusion principle*: if there were no repulsive interaction between the two charged particles, their distance would decrease to zero, since the force between oppositely charged particles is attractive. The repulsive force due to the Pauli exclusion principle is described by an empirical expression:

$$V_r(r) = \frac{b}{r^n},$$

where $b > 0$ and n are adjustable parameters; n usually ranges between 7 and 9. For sufficiently small distances, this term dominates the interaction energy and the slope of the curve becomes negative, indicating a repulsive force.

From the potential energy expression, we can derive an expression for the force by taking the negative radial gradient:

$$\mathbf{F}(r) = \left(\frac{Z_1 Z_2 e^2}{4\pi\epsilon_0 r^2} + \frac{nb}{r^{n+1}} \right) \mathbf{e}_r. \tag{2.3}$$

Schematic energy versus distance curves are shown in Fig. 2.5. The upper dashed curve is the repulsive potential energy, the lower dashed curve represents the attractive potential energy. The sum of the two energies has a minimum at the distance r_{m} and a value of V_{min}. This distance is equal to the sum of the radii of the two ions. One can compute analytically the value for r_{m} by requiring that the force must be zero for $r = r_{\mathrm{m}}$. It is left as an exercise for the reader to show that for the particles of Fig. 2.5, using Eq. 2.3 as the expression for the interaction force, the minimum energy occurs for a separation of

$$r_{\mathrm{m}} = \left(\frac{e^2}{4\pi\epsilon_0 nb} \right)^{\frac{1}{1-n}}.$$

The equilibrium distance can also be derived graphically by drawing the total force versus distance and determining where it becomes equal to zero. This is shown schematically in Fig. 2.6, which shows the energy and force for the so-called *Lennard-Jones potential*

$$\frac{V(r)}{\epsilon} = \left[\left(\frac{\sigma}{r}\right)^{12} - 2\left(\frac{\sigma}{r}\right)^{6} \right].$$

The crystal structures formed by ionic compounds are determined by the relative sizes of the participating ions. One could intuitively compare this to

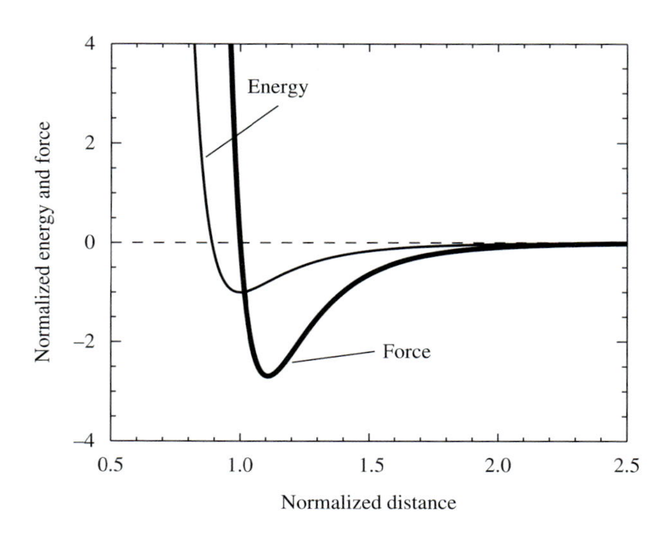

Fig. 2.6. Typical potential energy curve for the interaction between two atoms; the force is shown in a thicker line. The potential is the Lennard-Jones potential $V(r)/\epsilon = \left[(\sigma/r)^{12} - 2\,(\sigma/r)^{6}\right]$, with σ the equilibrium interatomic spacing and ϵ the depth of the potential well.

the problem of packing two sets of marbles with different diameters in such a way that each type is surrounded by the largest possible number of the other type. In Chapter 22 we will return to this type of bond and discuss its implications for the crystal structure.

2.3.4 The covalent bond

When the electronegativity difference between two atoms is small, then there is no reason for either of the atoms to become an ion. In that case, atoms will attempt to acquire a noble gas electron configuration by *sharing* electrons with neighboring atoms, rather than *transferring* them completely. Consider the hydrogen molecule H_2 as an example. Each hydrogen atom has a configuration of $1s^1$. If we bring the two atoms together from infinity, then there are two possibilities for the combination of electron spins, as shown in Fig. 2.7: the spins are anti-parallel, in which case the electrons will share a *bonding state* with lower energy than the separated atoms. If the spins are parallel, then according to the Pauli exclusion principle there will be a repulsive interaction preventing the two electrons from occupying the same location in space. This leads to an increase in energy and the corresponding state is called an *anti-bonding* state. The covalent bond is also a *directional* bond, i.e., it is located between the atomic nuclei. Since there is an electrostatic repulsion between two electrons with opposite spin, and also an attractive force because of the tendency to form an inert gas configuration, there is again an equilibrium distance between the atom cores. This distance can be used to define the *covalent radius* of the atom.

There is another factor that stabilizes this type of bond: electrons are identical particles and if we label the electron on H_A by the number 1 and the electron on H_B by the number 2, then we cannot distinguish between the following configurations:

$$H^1_A + H^2_B \qquad H^2_A + H^1_B$$

Both of these states have identical energy, and, therefore, the molecule can *resonate* between them. The fact that the molecule can choose between two

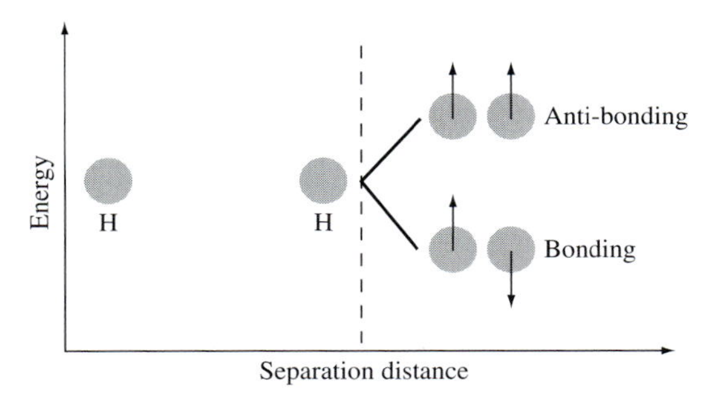

Fig. 2.7. Formation of bonding and anti-bonding states in the hydrogen H_2 molecule. This simple illustration lies at the basis of the Heitler–London model for the covalent bond.

states with identical energy is called *resonance*. Resonance will in general lower the energy of a system of interacting particles, and it accounts for the very existence of the covalent bond.

Resonance also leads to the concept of *electron promotion* or *hybridization*. Consider the carbon atom, with electronic structure [He]$2s^2 2p^2$. One would normally expect the electrons to occupy the following orbitals: $2s^2\, 2p_x^1\, 2p_y^1$. However, by *promoting* an electron from the $2s$ level to the $2p_z$ orbital, the atom can create four new energy states *with identical energy*: $2s^1\, 2p_x^1\, 2p_y^1\, 2p_z^1$. There are then many ways (24 to be exact) in which the four electrons can be distributed into the four available states and the atom can resonate between them. In general, the more *degenerate* states are available to a system, the more likely it becomes that the system will actually be in one of those states.[4] This leads in the case of the carbon atom to the formation of the sp^3 hybridized orbital, which can accommodate four electrons with parallel spins. The four orbitals extend into space with mutual angles of 109.5° and form a *tetrahedral* shape. Each electron can form a covalent bond with another atom, and in the case of carbon this leads to the formation of *diamond*. The mathematical description of covalent bonds is complicated by the fact that the bonds have a directional character; this usually leads to rather complex expressions for the potential energy and forces. Covalent bonds are very important in organic chemistry, as they lie at the basis of life itself.

2.3.5 The metallic bond

Electropositive atoms will readily give up their weakly bound outer electron(s) and become positive ions in a "sea" of electrons. The valence electrons are not closely associated with any particular atom and they can move freely throughout the structure, rendering metals excellent conductors. Other properties which distinguish metals from other forms of solid matter are the high thermal conductivity and the optical opacity. About three-quarters of all elements are metallic.

The bonds from any one atom must be regarded as spherically distributed around that atom. The crystal structures formed rely on the concept of *closest packing*. This can be understood intuitively by considering a set of identical size marbles; if one throws the marbles in a box and shakes it, then they will tend to occupy positions in which every marble is surrounded by 12 neighbors. This type of structure is a *close-packed* structure.

In a close-packed structure, each atom is surrounded by six other atoms in the same plane, as shown in Fig. 2.8(a). There are also three neighbors

[4] We will see in later chapters that the higher the *symmetry* of a system, the more equivalent point positions are available to it. The same holds for the number of possible energy states: the more degenerate states that are available to a system, the lower the energy of the system. Nature tends to prefer states with high symmetry.

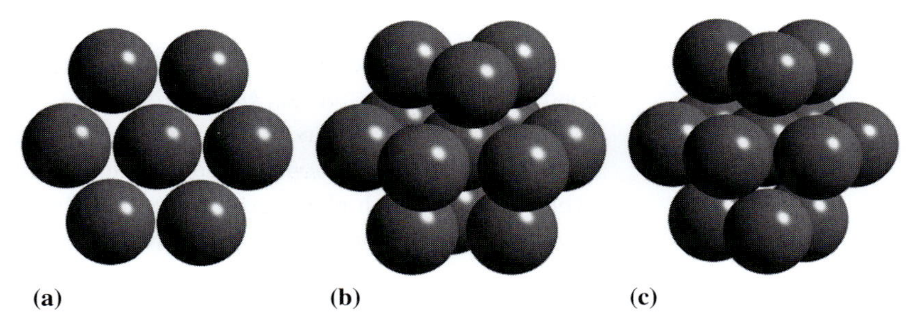

Fig. 2.8. (a) A single close-packed plane, (b) cubic close-packed stacking, and (c) hexagonal close-packed stacking. Note how the bottom three atoms in (b) and (c) have a different position with respect to the center plane. The atoms have been separated from each other to more clearly show their arrangement; in a close-packed structure, the central atom would touch all 12 nearest neighbors.

above and three below the plane, bringing the total to twelve. There are two possibilities for the atoms above and below the plane. In Fig. 2.8(b), the three atoms above the central plane occupy positions that are different from the atoms below the plane. This is known as the *cubic closest packing* or *ccp*. When the atoms above and below the central plane occupy identical positions (Fig. 2.8(c)), then the environment is known as the *hexagonal closest packing* or *hcp*. Both of these arrangements are important in the study of the structure of metals, and we will come back to them in detail in Chapter 17.

2.3.6 The van der Waals bond

When large, neutral molecules come close together, there is a tendency for them to attract each other. This is due to small instantaneous charge redistributions which cause an effective *polarization* of the molecule. Polarization refers to the fact that the centers of gravity of positive and negative charges do not coincide. One can show that the energy associated with this type of interaction can be written as:

$$V(r) = -\frac{3}{4}\frac{\mathrm{h}\nu_0\alpha^2}{r^6},$$

where ν_0 is the frequency of the zero-point motion of the atom, α is the polarizability, and r the distance between the centers of the molecules or atoms. The important thing about this interaction is that it is attractive, and that it falls off as $1/r^6$.

There is thus an attractive force between any two molecules or atoms, even in the absence of ionic, covalent, or metallic bonding between them. This force is caused by instantaneous polarization and is known as the *van der Waals force*. It is mostly spherical in nature (for spherical molecules) and does not have a strong directional component. It is a weak force, which is why it is only observed clearly in materials where none of the other bonding types are present. A typical example of such a material would be a solid inert element, say solid He or Ar. Another more recent example of van der Waals bonding is solid C_{60}, or *Buckminsterfullerene*. The "molecules" bonded by van der Waals forces can also be two-dimensional, and *graphite* is a prime example of such a situation. The carbon atoms within the layers are in an sp^2 hybridized state with planar covalent bonds; the remaining electron occupies an orbital perpendicular to the plane. Neighboring planes are weakly bonded together through van der Waals forces between those electron orbitals. The van der Waals bond is hence responsible for the softness of graphite, and for the fact that it is a very good lubricant.

2.3.7 Mixed bonding

Real bonds are rarely of one particular pure type; usually, there is some mixing between ionic and covalent bond character. Linus Carl Pauling suggested the following formula to compute the percentage ionic character, $\mathcal{P}$, of a bond from the electronegativities (according to the Pauling scale):

$$\%\text{ ionic character } \mathcal{P} = \left(1 - e^{-\frac{1}{4}(E_N^{\mathrm{A}} - E_N^{\mathrm{B}})^2}\right)(100\%)$$

The percentage ionic character hence depends in an exponential way on the electronegativity difference between the elements. As an example, let us determine which of the two compounds ZnSe and GaAs is the more ionic. The electronegativities for the elements are given by:

$$E_N^{\mathrm{Zn}} = 1.6 \qquad E_N^{\mathrm{Se}} = 2.4$$

$$E_N^{\mathrm{Ga}} = 1.6 \qquad E_N^{\mathrm{As}} = 2.0$$

The percentage ionic character $\mathcal{P}$ is given by:

$$\mathcal{P}_{\mathrm{Zn\,Se}} = \left(1 - e^{-\frac{1}{4}(0.8)^2}\right)(100\%) = 14.78\%$$

$$\mathcal{P}_{\mathrm{Ga\,As}} = \left(1 - e^{-\frac{1}{4}(0.4)^2}\right)(100\%) = 3.92\%$$

and we find that ZnSe is more ionic than GaAs. A similar scale can be defined for the Mulliken electronegativity.

2.3.8 Electronic states and symmetry

The symmetry of *crystalline electric fields* can be important in determining properties such as the optical and magnetic properties in a variety of systems, most importantly in ionic solids with rare earth or transition metal cations. Although a complete discussion of crystal field theory is beyond the scope of this book, we can use a simple illustration of this theory to show how the symmetry of an interaction potential influences the energy levels of a quantum mechanical system.

Much of the energetics of the electron(s) localized on an ion in an ionic solid depends on the Coulomb interaction between nearest neighbors. Coulomb interactions between electrons on the site of interest and charges external to the ion are described by the electrostatic potential. For transition metals, the outermost d electrons are most strongly perturbed by this crystalline potential. For rare earth elements, the crystalline potential influences the energy levels of the f electrons but not as noticeably because of shielding by the outermost s and p electrons.

The perturbing potential of the central ion partially lifts the $(2l+1)$ degeneracy of the ground state energies of the ion. The symmetry derived splittings are illustrated for a transition metal in Fig. 2.9, which illustrates the orientation of various d orbitals with respect to ions in a cubic environment. Symmetry dictates that the d_{xy}, d_{xz} and d_{yz} orbitals have the same energy (hence, they are equivalent) but that this energy is different from that of the $d_{x^2-y^2}$ and d_{z^2} pair of orbitals. Note that the energies must be different because the first three orbitals are oriented towards nearest neighbor ion sites and the second set towards second nearest neighbor sites. In the octahedral arrangement of the nearest neighbors, the triplet of orbitals acquires a lower energy than the doublet, whereas for cubic symmetry the situation is reversed. The notations E and T_2 refer to group theoretical labels for the various energy levels.

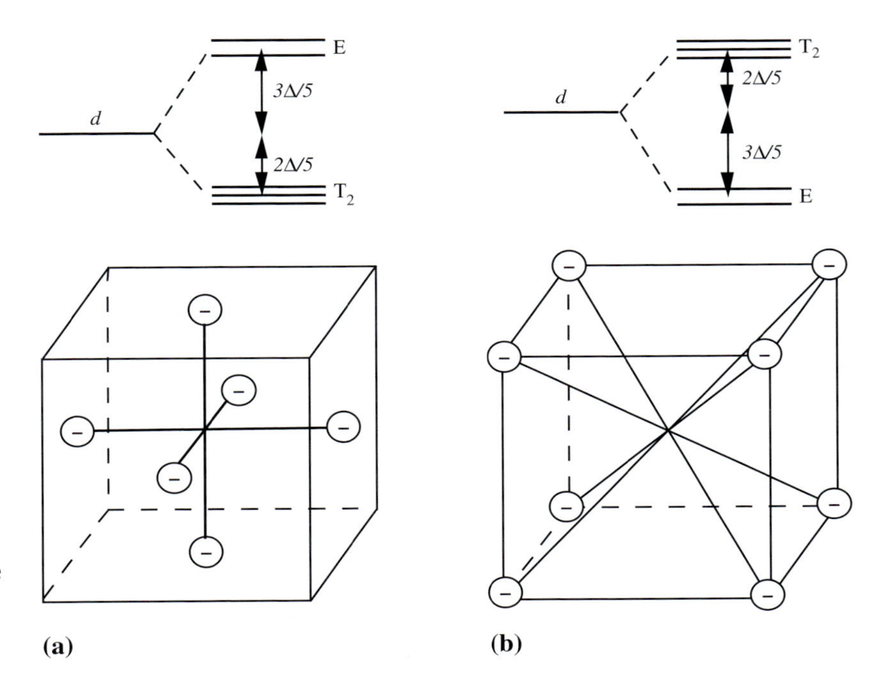

Fig. 2.9. Crystal field splitting of d orbitals due to the charge on neighboring ions in (a) an octahedral and (b) a cubic environment.

Table 2.5. Physical and structural properties associated with the four interatomic bond types (based on Table 6.03 in (Evans, 1966)).

Property	Ionic	Covalent
Structural	Non-directed, giving structures of high coordination	Spatially directed and numerically limited, giving structures of low coordination and low density
Mechanical	Strong, giving hard crystals	Strong, giving hard crystals
Thermal	Fairly high melting point, low expansion coefficient, ions in melt	High melting point, low expansion coefficient, molecules in melt
Electrical	Moderate insulators, conduction by ion transport in melt	Insulators in solid and melt
Optical and magnetic	Absorption and other properties primarily those of individual ions	High refractive index. Absorption profoundly different in solution or gas

Property	Metallic	van der Waals
Structural	Non-directed, giving structures of very high coordination and high density	Formally analogous to metallic bond
Mechanical	Variable strength, gliding common	Weak, giving soft crystals
Thermal	Variable melting point, long liquid interval	Low melting point, large expansion coefficient
Electrical	Conduction by electron transport	Insulators
Optical and magnetic	Opaque, properties similar in liquid	Properties those of individual molecules

2.3.9 Overview of bond types and material properties

We conclude this chapter with a brief account of the relation between bonding types and material properties. The different bond types are idealized bonds, and "real" materials usually show some form of mixing between two or more types. Semiconductors are mostly covalent, metals are mostly metallic, ceramics have both covalent and ionic character, and most polymers lie between covalent and "secondary" bond types (which includes the weaker van der Waals bonds). The bond types can also be correlated to properties which span many different materials, as shown in Table 2.5 (Evans, 1966).

2.4 Historical notes

The atom is the building unit of all matter, in the solid, liquid, and gaseous state. The Greek philosopher **Leucippus** and his student **Democritos** are among the first to speculate about the structure of matter; around 400 BC, they

Table 2.6. Comparison between Mendeleev's predictions for the element *Ge* and the currently accepted values.

Property	Prediction	Current value
color	dark gray	grayish-white
atomic weight	72	72.59
density (g/cm^3)	5.5	5.35
atomic volume (cm^3/g-atom)	13	13.5
specific heat (cal/g/°C)	0.073	0.074
oxide stoichiometry	XO_2	GeO_2
oxide density (g/cm^3)	4.7	4.703
chloride stoichiometry	XCl_4	$GeCl_4$
chloride boiling point	<100°C	86°C
chloride density (g/cm^3)	1.9	1.844

proposed that all matter is built from *indivisible* minute particles which they called *atoms*. It took nearly 2000 years before substantial further advances were made. **Robert Boyle** coined the concept of *elements* in 1661; to classify the various elements, or pure substances, he determined their color when burnt in a flame, a technique that is now known as *spectroscopy*. Based on his own experiments and those of his colleagues, **Antoine Lavoisier** and **Joseph Priestley**, **John Dalton** put forward the idea that the defining characteristic of an element is the weight (actually, the *mass*) of one of its atoms.

By the year 1870, researchers had isolated about 65 different elements and began to notice certain trends and similarities between the chemical properties of certain elements. **John Newlands** discovered in 1865 that, if the elements were ranked according to their atomic weight, certain properties repeated themselves with a periodicity of eight elements. His ideas were not accepted in the scientific community until **Dmitri Mendeleev** in 1869 suggested a similar arrangement of the elements, which is now known as the *periodic table of the elements*. Mendeleev noticed that there were gaps in his table, presumably with undiscovered elements. The importance of his classification scheme becomes clear when we consider the fact that he *predicted* the existence of *eka-silicon* (now known as Germanium, Ge). He predicted several of the properties of Ge, as listed in Table 2.6; note how close his predictions are to the currently accepted values for pure Ge!

It wasn't until the famous experiments of **Joseph John Thomson** that the existence of the internal structure of the atom was shown. Thomson discovered negatively charged particles, called *electrons*. **Robert Millikan** determined the ratio of charge to mass for the electron and showed that the mass was about 2000 times smaller than that of the hydrogen atom. **Henry Moseley** then showed that the number of electrons in a particular atom is equal to the atomic number of that element. The presence of negative particles in a neutral atom requires the existence of other, positively charged particles. In 1904,

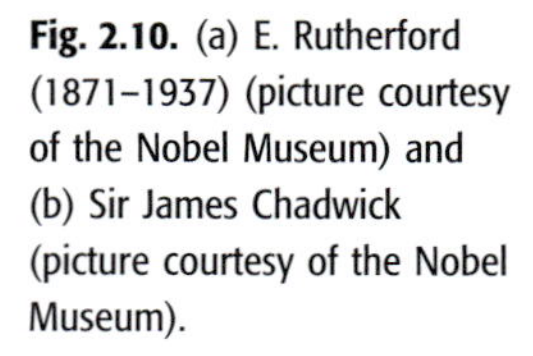

Fig. 2.10. (a) E. Rutherford (1871–1937) (picture courtesy of the Nobel Museum) and (b) Sir James Chadwick (picture courtesy of the Nobel Museum).

Hantaro Nagaoka conjectured that the atom must look like a miniature solar system. This implies a relatively open structure, which was experimentally confirmed by **Ernest Rutherford** (Fig. 2.10a) and his co-workers.

Rutherford fired so-called alpha particles (essentially the nucleus of a helium atom) onto a sheet of gold and found that most particles went right through the foil. Only a few were bounced back. This means that (1) the atom is fairly empty, and (2) that the mass of the atom must be located in a dense, small nucleus. The positive charge (two units) of the alpha particles showed that the nucleus must consist of positively charged particles, *protons*, which are held together by "glue", made up of neutral particles or *neutrons*. Protons were first proposed by Rutherford in 1913 (Rutherford coined the name "proton" in 1920). **Sir James Chadwick** discovered the neutron in 1932. Table 2.7 lists several important events related to our understanding of the structure of the atom. For the interested reader we recommend the book by Abraham Pais (Pais, 1986) listed in the References. Pais describes the evolution of our understanding of the structure of the atom since 1815.

To make a long story short, it was soon recognized that the structure of the atom is fundamentally *different* from that of the solar system. In fact, classical mechanics (or Newtonian mechanics) predicts that the atom must be unstable.[5]

A radically new theory was needed to explain the structure of the atom. Such a theory was constructed in the first part of the twentieth century and

[5] From electrodynamics we know that an electric charge must radiate energy when it is accelerated or decelerated. Since the electron is on a curved orbit around the nucleus, it is subjected to a constant acceleration and must hence continuously radiate energy. Therefore, the electron must spiral down towards the nucleus and the atom must cease to exist (at least, that is what classical mechanics predicts).

Table 2.7. Brief list of some of the more important discoveries and events around the end of the nineteenth and early twentieth century.

Year	Event
1853	First observation of hydrogen spectrum
1864	Maxwell's theory of electromagnetism
1869	Mendeleev's periodic table of the elements
1874	First estimate of fundamental charge e
1895	Discovery of X-rays (Röntgen)
1896	Discovery of radioactivity (Bequerel)
1897	First speculations about existence of electrons
1898	Identification of α and β-rays
1899	Measurement of e; discovery of electron
1900	Discovery of γ-rays
1900	Planck discovers the quantum theory
1905	Einstein postulates the photon
1906	Rutherford discovers α-particle scattering
1911	Rutherford coins the term "nucleus"
1913	Emergence of the proton–electron model
1921	Discovery of strong nuclear force
1925	Foundation of quantum theory
1932	Discovery of neutron

is known as *quantum theory*. This theory is based upon two fundamental concepts.

First, **Max Planck** (1858–1947) stated in 1900 that particles moving in an atomistic world can only occupy certain *energy states* or *energy levels*. They can only change state by jumping up or down to another energy level. Since energy is conserved, this must happen by *emission* or *absorption* of a well defined amount of energy. Such an amount is called a *quantum*, and we say that the energy levels of an atom are *quantized*.

Planck also determined the magnitude of such an energy quantum (we now call these emitted or absorbed quanta *photons*) and established that it was equal to the frequency of the radiation multiplied by a new universal constant, *Planck's constant* $\mathrm{h} = 6.626 \times 10^{-34}$ Js, i.e., $E = \mathrm{h}\nu$

Second, **Louis de Broglie** postulated in 1923 that every particle must have a *dual character*. Depending on the externally imposed conditions, a particle may behave as a particle or as a wave (hence the concept of *particle–wave duality*). de Broglie also described how to compute the wave length corresponding to a particle: $\lambda = \mathrm{h}/p$, where h is Planck's constant and p the particle momentum.

If matter can be considered to have both particle and wave properties, then one can ask the question: what is the size of the particle, or, what is the wave length of the wave? To measure the wave length accurately, one needs to measure the distance between many consecutive wavecrests and then divide by the number of crests. If a particle is very small, then the number of wave crests that will "fit" in the particle is rather small, hence the accuracy with which the wave length can be determined is rather poor. On the other hand, if a particle is described by a wave with many crests, then it becomes difficult to determine the exact location of the particle. There is, hence, an intrinsic uncertainty related to the wave – particle aspect of matter. **Werner Heisenberg** was the first to recognize the importance of this uncertainty and he stated what is now known as the *Heisenberg Uncertainty Principle*: it is impossible to simultaneously determine, with arbitrary accuracy, the position and the wave length (and hence the momentum, through de Broglie's relation) of a particle. In mathematical terms this is usually stated as:

$$\Delta x \Delta p_x \geq \hbar \equiv \frac{\mathrm{h}}{2\pi}$$

where the symbol Δ refers to the uncertainty in the measurement of either the position component x, or the x component of the momentum, p_x.

Max Born then suggested in the late 1920s that the wave representation of a particle tells us what the *probability* is of finding the particle at a certain position: if the wave amplitude is high, then it is highly probable that the particle will be located at or near the maximum. If the wave amplitude is low, then it is unlikely that the particle will be found there. This brings us to an important observation: in the world of atoms and subatomic particles, the particle waves represent probabilities of finding the particles at a certain location. We cannot say anything with complete certainty, but we *can* make statements such as: there is a 40% probability of finding the particle at this location. During the 1920s and 1930s, both **Werner Heisenberg** and **Erwin Schrödinger** derived mathematical techniques to compute these probabilities. Since the wave nature of particles is essential to these descriptions, the central quantity in quantum theory is the *wave function*; it is usually represented by the symbol $\Psi(\mathbf{r})$. The wave function is the solution of a differential equation, commonly known as the Schrödinger equation.

2.5 Problems

(i) *Ionization energy*: Calculate the energy required to ionize a hydrogen atom, i.e., the energy required to remove the electron and convert the atom into a positive ion H^+.

(ii) *Quantum numbers I*: Write out explicitly the 4 quantum numbers for *each* of the 17 electrons in a chlorine (Cl) atom.

(iii) *Quantum numbers II*: Write out explicitly the 4 quantum numbers for *each* of the 14 electrons in a silicon (Si) atom.

(iv) *Multi-electron atoms and X-rays*: In Hydrogenic atoms, each shell is labeled by the spectroscopic symbols: K, L, M, N, O, etc. corresponding to the $n = 1, 2, 3, 4, 5$, etc. quantum states, with energy levels predicted by the Bohr model. The Bohr model applies to a single electron bound to a nucleus of charge $+Ze$ (Z is the atomic number). This model is modified in Hartree theory for multi-electron atoms. An approximate result of the Hartree theory is that electrons in different shells experience a screened nuclear potential which can be parameterized in terms of an effective charge $+Z_n e$ where for the K-shell $Z_1 = Z–2$; for the L-shell $Z_2 = Z–10$, and so on.

Calculate for a Cu atom ($Z = 29$):

(a) the energy required to pull an electron from the K-shell;
(b) the energy required to pull an electron from the L-shell;
(c) what would be the minimum accelerating voltage required for an X-ray tube to strip a K electron from a ^{29}Cu anode?
(d) what would be the wavelength of an X-ray emitted for a K_α transition in which an electron from the L-shell fell into the empty state in the K-shell after stripping the K electron? How does this wavelength compare with a typical atomic radius?

(v) *Systematic property variation*: Consider the series of elements in the row of the periodic table running from rubidium (Z = 37) to indium (Z = 49). Is there any evidence of a systematic variation of the melting temperature with electronic structure ? If so, explain this variation.
(vi) *Atomic computations I*: Calculate the approximate number of atoms in a sphere with diameter 1 cm. Assume that the sphere is made of pure titanium and that there are no gaps between the spherical atoms.
(vii) *Atomic computations II*: Knowing that the mass of one mole of SiC (silicon carbide) is equal to the combined atomic mass of the elements Si and C, compute the density of SiC if one mole occupies a cubic volume with an edge length of 2.32 cm.
(viii) *Atomic computations III*: A steel nail with a total surface area of 5 cm^2 is coated with a Zinc layer to prevent corrosion. If the Zinc layer is 50 μm in thickness, calculate the total number of atoms in the zinc layer, and the total mass of the coating.
(ix) *Mulliken electronegativity*: Using data from the literature (e.g., Radzig and Smirnov (1985), McHenry *et al.* (1987)), calculate the Mulliken electronegativities for the third row transition metal elements.
(x) *Interatomic bonds*: Compute the equilibrium distance, r_m, for the interaction force given in Eq. 2.3 on page 42. Then, repeat the computation for the Lennard-Jones potential given on page 42.
(xi) *Bonding types*: Consider the following three materials as prototypes of various classes of electronic materials:

(a) Al metal
(b) Si semiconductor
(c) SiO_2 insulator

Sketch or otherwise depict the bonding in each of these three materials.

(xii) *Pair potential*: Consider an interatomic pair potential of the form:

$$\frac{V(r)}{\epsilon} = \left[\left(\frac{d}{r} \right)^n - \frac{n}{m} \left(\frac{d}{r} \right)^m \right]$$

where m and n are exponents with $m < n$.

(a) Identify the attractive and repulsive terms in the expression.
(b) Differentiate the potential to determine the equilibrium spacing of the atoms. What is the equilibrium spacing for a Lennard-Jones potential?
(c) Determine the value of the minimum of the $V(r)$ curve for the general potential and for the specific Lennard-Jones potential.

CHAPTER

3 What is a crystal structure?

"In mathematics, if a pattern occurs, we can go on to ask, Why does it occur? What does it signify? And we can find answers to these questions. In fact, for every pattern that appears, a mathematician feels he ought to know why it appears."

W. W. Sawyer, mathematician

3.1 Introduction

In this chapter, we will analyze the various components that make up a *crystal structure*. We will proceed in a rather pragmatic way, and begin with a loose "definition" of a crystal structure that most of us could agree on:

> A crystal structure is a regular arrangement of atoms or molecules.

We have some idea of what atoms and molecules are – at least, we think we do . . . And we also have some understanding of the words "regular arrangement." The word "regular" could imply the existence of something that repeats itself, whereas "arrangement" would imply the presence of a *pattern*. But, there are many possible patterns: the words on this page form a pattern of lines; migrating birds often fly in V-shaped formations; musicians in a marching band walk in an orderly way; the kernels on a piece of corn are arranged in neatly parallel rows; and so on. All of these words, *regular arrangement, pattern, orderly, repeats itself*, are commonly used words in our everyday language, but they are not sufficiently precise for a scientific description of what a crystal structure really is.

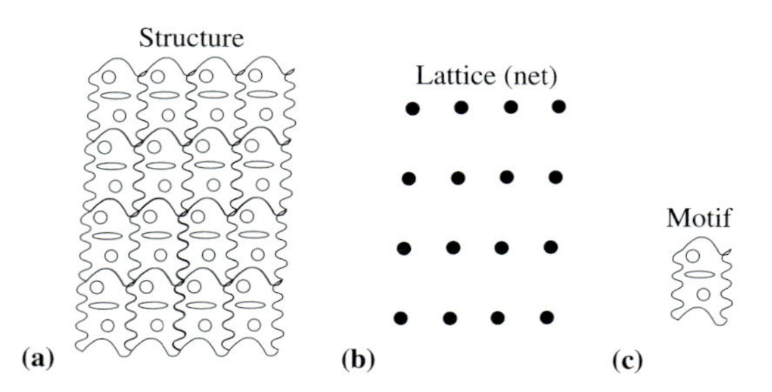

Fig. 3.1. (a) A periodic pattern consisting of (b) a 2-D net and (c) a motif. The motif is repeated at each point of the 2-D net, to create the pattern in (a).

So, we will need to define more rigorously what we mean by a regular arrangement. We can understand this concept intuitively by considering the drawing in Fig. 3.1. Figure 3.1(a) shows a periodic drawing; although this is clearly not a drawing of a crystal structure, the drawing does illustrate some of the more fundamental aspects of crystals. The drawing consists of a *motif*, shown in Fig. 3.1(c), which is repeated by translating it from one point, chosen as the origin in Fig. 3.1(b), to other points arranged in a two-dimensional pattern. The set of points constitutes what we will call a *net* in two dimensions (2-D) and a *lattice* in three dimensions (3-D). The motif represents the *decoration* of that net/lattice. In exactly the same way, a crystal structure can be described as a 3-D lattice, decorated with atoms or molecules. Hence, our "regular arrangement" is now restricted to be a "lattice." In the next section, we will describe in a more rigorous way what a lattice is.

Before we do so, let us return to one of the examples of patterns given in the previous paragraph: the *marching band*. Consider a marching band in which the members occupy positions on 10 rows of 3 musicians each. When the band assembles itself into this formation, the rows and columns are well defined, and all musicians are nicely lined up, with the nearest musicians in front, behind, and to left and right at, say, 1.5 meters from each other. Once the band starts marching, however, it becomes much harder for the musicians to maintain this formation with great accuracy; as a spectator, we expect them to keep their formation as best they can, and, not infrequently, the band which does this best may also end up being more popular (assuming their music sounds good, too!). Depending on the discipline and/or motivation of the band members, the formation may remain nearly perfect throughout the march (as would be expected for a military marching band), or it may be more loosely related to the original formation, with each musician staying within, say, half a meter of his/her supposed position. At any moment in time, only a few of the musicians will be precisely at their nominal position, but on average, over the duration of the march, all of them will have been where they were supposed to be. This is illustrated in Fig. 3.2: (a) shows the initial positions on a regular square grid. At an instant of time, each musician may deviate somewhat from these positions. In (b), the trajectory of each musician during

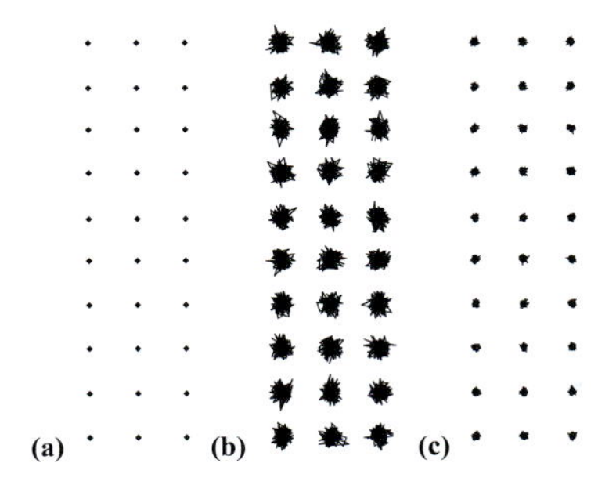

Fig. 3.2. (a) Initial positions of the musicians of a marching band (atoms on a crystal lattice); (b) trajectories of all musicians relative to their initial position for the length of the march for a loosely structured band (atoms at elevated temperature), and (c) for a highly disciplined band (atoms at low temperature).

the entire march is drawn with respect to that musician's initial position. It is clear that, on average, the musicians did keep to the initial formation. The size of the "trajectory cloud" around each site is an indication of how much each musician deviated on average from the formation. For a military marching band, we would expect the diameter of these clouds to be very small, as shown in (c).

Now we can abandon the marching band and replace each band member by an atom.[1] The atoms are positioned on a grid and, as a function of time, they move around their own grid site in a somewhat random way. The surrounding atoms prevent them from moving too far from their initial positions, so that, on average, over a relatively long time, each atom appears to occupy the perfect grid position. The magnitude of the instantaneous deviations is determined, not by motivation or discipline, but by the *temperature* of the atom assembly. A high temperature means that the atoms have a high kinetic energy, so their excursions from the average position can become quite large; whereas at a low temperature, there is insufficient kinetic energy available for large excursions, and the *vibration amplitude* will remain small. This kind of atom motion is known as *thermal motion* or *thermal vibration*. It is present in every crystal structure and it is convenient to ignore it in a structural description of crystals.[2] The thermal motion of atoms only becomes important in the determination of the crystal structure by means of a suitable form of radiation (X-rays, electrons, neutrons) and can be adequately described by means of the so-called *Debye–Waller factor*, which will be introduced in Chapter 11. From here on, we will always consider the average position to be the "real" position of the atom; this is an approximation, but it turns out to be a very convenient

[1] The reader who also happens to be a member of a marching band may rest assured: there will be no further verbal abuse of marching bands in this book!

[2] Thermal vibration is not limited to materials with a crystalline structure; it also occurs in liquids and gases, where there is no periodic structure. The vibrations are related to the curvature of the interatomic interaction potential introduced in the previous chapter. A small curvature around the equilibrium distance indicates a small restoring force for excursion away from this position; hence, the vibration frequency will be low. For a large curvature, the restoring force is large, and therefore the vibration frequency is high.

one because most mathematical relations to be derived in the remainder of this book become independent of time.

The average position of an atom in a crystal structure does not change with time, so we can slightly revise our initial loose definition of a crystal structure to:

> A crystal structure is a time-invariant, three-dimensional arrangement of atoms or molecules on a lattice.

We will take this statement as a starting point for this chapter. First, we need to define more precisely what we mean by the term "lattice."

3.2 The space lattice

3.2.1 Basis vectors and translation vectors

The historical comments in Box 3.1 show how René-Just Haüy built models of crystals by stacking rectangular blocks in such a way that the assembly resembled the external shape (or *form*) of macroscopic crystals. By assuming the existence of a single shape, he was able to construct many different forms, thereby explaining the large variety of crystal forms (or shapes) observed in nature. We will take Haüy's block model as the starting point for the introduction of the space lattice. First of all, we consider the most general block shape, an outline of which is shown in Fig. 3.3. If we take one of the corners of the block as the origin, then we can define three vectors along the three edges of the block. We shall call them **a**, **b**, and **c**. Note that the angles between these vectors need not be 90°, and that the lengths of these vectors need not be the same.

The main advantage of defining these basis vectors is that we can easily identify the coordinates of all of the corners of the block. For instance, the corner opposite the origin has position vector $\mathbf{a}+\mathbf{b}+\mathbf{c}$. Alternatively, we can write the *coordinates* of this point as $(1, 1, 1)$, since the position vector corresponds to $1\times\mathbf{a}+1\times\mathbf{b}+1\times\mathbf{c}$. Note that we will always write coordinates between parentheses, with commas separating the individual components.

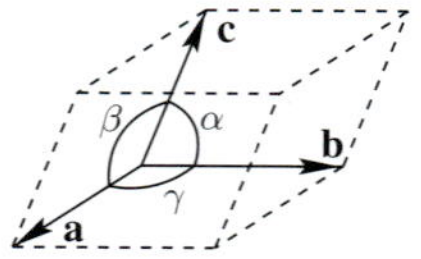

Fig. 3.3. Illustration of a general building block (a unit cell), with the three basis vectors **a**, **b**, and **c** (figure reproduced from Fig. 1.1 in *Introduction to Conventional Transmission Electron Microscopy*, M. De Graef, 2003, Cambridge University Press).

Next, we consider a stack of blocks, as in Haüy's models. Since each block is identical to every other block, and they are stacked edge to edge and face to face, it is easy to see that we can jump from the origin to *any* corner of any block in the stack, by taking *integer linear combinations* of the three basis vectors. The coordinates of each block corner can therefore be written as triplets of integers, which we will denote by (u, v, w). Note that these integers can take on all possible values, including negative ones, since we can take the origin at any point in the stack.

Box 3.1 Haüy's crystal models

René-Just Haüy (1743–1822) was a French priest and mineralogist. His building block theory of crystal structures led directly to the lattice model (Haüy, 1784, 1801, 1822). He suggested that crystals are composed of arrays of subdivisible blocks, called *integral molecules*, with shapes specific to the crystal. Haüy showed how, replicating the same blocks in different ways, he could construct different external shapes. This was taken as an explanation as to why the same substance could have crystals with different external forms. Haüy showed further that the building block theory implied that the overall symmetry of a crystal must be the same as that of its constituent parts. Nowadays, we no longer talk about "integral molecules," but, instead, we use the name "unit cell." The figure below, taken from Vol. 5 of Haüy (1801), shows how the rhombic dodecahedron shape (on the left) can be obtained by starting from a cubic crystal shape (on the right), and adding layers of cubic building blocks, with each new layer one unit cell smaller on all sides than the previous one.

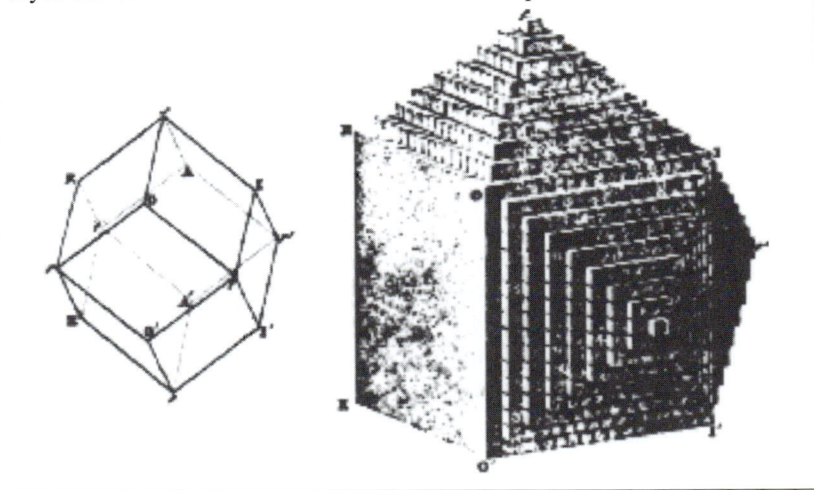

Instead of considering blocks, we will forget about the outline of the blocks, and only consider the corner points. We can then jump from the origin to the point with coordinates (u, v, w) by using the *translation vector* (or *lattice vector*) $\mathbf{t}$, defined as

$$\mathbf{t} = u\mathbf{a} + v\mathbf{b} + w\mathbf{c}. \tag{3.1}$$

All corner points can be reached by integer linear combinations of the three basis vectors. We shall call the collection of corner points the *space lattice*, and each individual corner a *node* or *lattice point*. A space lattice is thus a set of nodes, related to one another by the *translation vectors* $\mathbf{t}$. In 2-D, there are only two basis vectors, **a** and **b**, and the integer linear combinations of these vectors make up the nodes of the net.

A space lattice (net) is the geometrical image of the operation of the translation operators on the node at the origin.

3.2.2 Some remarks about notation

At this point, it is useful to introduce a shorthand notation for the translation vector **t**. In addition to writing its components as the integer triplet (u, v, w), we will also write them as (u_1, u_2, u_3), or as u_i $(i = 1 \ldots 3)$. Similarly, we will often write $\mathbf{a}_i$ for the basis vectors, where $\mathbf{a}_1 \equiv \mathbf{a}$, $\mathbf{a}_2 \equiv \mathbf{b}$, and $\mathbf{a}_3 \equiv \mathbf{c}$. This appears to be a superfluous complication of the notation, but it will turn out to be extremely useful for all kinds of crystallographic computations, as we will see in the following chapters.

The *set* (or collection) of translation vectors of any space lattice necessarily contains an infinite number of elements; we will denote this set by the calligraphic symbol $\mathcal{T}$:

$$\mathcal{T} = \{\mathbf{t} \mid \mathbf{t} = u\mathbf{a} + v\mathbf{b} + w\mathbf{c},\ (u, v, w) \text{ integers}\}. \tag{3.2}$$

This expression reads as follows: $\mathcal{T}$ is the set of all vectors **t** that can be written as linear combinations of the type $u\mathbf{a} + v\mathbf{b} + w\mathbf{c}$, with u, v, and w restricted to be integers.

Before we continue with a description of space lattices, it is worthwhile taking a brief "notational excursion." In crystallographic computations, it is often useful to be as economic as possible with symbols: the fewer symbols needed to describe a concept, the less likely that errors will be made. So, at this point we will introduce a device which will allow us to shorten all the expressions that we have discussed so far. This device is commonly known as the *Einstein summation convention*, and it is stated as follows:

A summation is implied over every subscript which appears twice on the same side of an equation.

Here is how it works. We start from the expression for the translation vector, and rewrite it in a few different ways, using various notations that we are already familiar with:

$$\begin{aligned}\mathbf{t} &= u\mathbf{a} + v\mathbf{b} + w\mathbf{c};\\ &= u_1\mathbf{a}_1 + u_2\mathbf{a}_2 + u_3\mathbf{a}_3;\\ &= \sum_{i=1}^{3} u_i\mathbf{a}_i.\end{aligned}$$

This last expression uses the summation sign $\sum$. It is obvious that this last expression is shorter in length than the other two, but now it has grown in the vertical direction... We all know that we are living and working in a three-dimensional (3-D) space, so it is rather clear that the sum goes from $i = 1$ to $i = 3$. So, why don't we simply drop the summation sign altogether and write:

$$\boxed{\mathbf{t} \equiv u_i \mathbf{a}_i,} \tag{3.3}$$

and we remind ourselves that there is an *implied summation* over the index i. We know that there is a summation, *since the index i appears twice on the same side of the equation*, once on the u_i and once on the $\mathbf{a}_i$. This notation convention (dropping the summation signs) is the Einstein summation convention, which we will use profusely throughout this text.[3] Since this looks a little confusing, let's practice this convention on a few examples.

First of all, consider the expression

$$\mathbf{t} = u_j \mathbf{a}_j.$$

Is there an implied summation? Yes, there is, since the subscript j is repeated twice on the same side of the equation! So, this equation really reads as:

$$\mathbf{t} = \sum_{j=1}^{3} u_j \mathbf{a}_j.$$

This also illustrates an important point: it does not matter which letter of the alphabet we use for the subscript, as long as we use the same letter twice. The subscript is therefore known as a *dummy* subscript or a dummy index.

Let's look at a slightly more complicated expression:

$$??? = b_j u_i \mathbf{a}_i.$$

First we deal with the right hand side. The index i occurs twice, so there is a summation implied over i. The index j occurs only once, so there is no summation over j. So, the equation really reads as:

$$??? = b_j \sum_{i=1}^{3} u_i \mathbf{a}_i.$$

But what would we have on the left hand side? That's a good question! If we use the relation $\mathbf{t} = u_i \mathbf{a}_i$, then we would have

$$b_j \mathbf{t} = b_j u_i \mathbf{a}_i.$$

[3] The reader may find a comment on the notation used in this book in Box 3.2.

Box 3.2 Alternative notation

There exists an alternative notation, frequently used in the physics literature. This notation employs both subscripts and superscripts. The components of a vector are denoted with superscripts, as in:

$$\mathbf{t} = u^i \mathbf{a}_i = \sum_{i=1}^{3} u^i \mathbf{a}_i.$$

The Einstein summation convention then reads: *A summation is implied over every index which appears twice on the same side of an equation, once as a subscript and once as a superscript.* While there are some advantages of this notation over the one used in this book, in particular when we start describing *reciprocal space* in Chapter 6, the authors decided to simplify the notation, and to only consider subscripts for both vector components and basis vectors.

This illustrates another important rule when working with subscripts: *subscripts must be balanced on both sides of the equation.* This means that, if a subscript is present on one side of the equation, and no summation is implied over this subscript, then it must also be present on the other side of the equation.

Finally, let's look at a more complicated example, which we will encounter in a later chapter:

$$F = \epsilon_{ijk} p_i q_j r_k.$$

Leaving aside for now the exact meaning of the symbol ϵ_{ijk}, simply note that it is possible for symbols to have more than one subscript.[4] We see that there are three different subscripts, and each of them occurs twice, so there must be three summations:

$$F = \sum_{i=1}^{3} \sum_{j=1}^{3} \sum_{k=1}^{3} \epsilon_{ijk} p_i q_j r_k.$$

Since all indices are used up in the summations, there can be no index on the left-hand side of the equation. This concludes some simple examples. We will make extensive use of the summation convention in this text, so it is

[4] Think about matrices! A 3×3 matrix A has three rows and three columns, and each entry of the matrix is labeled by two subscripts, as in A_{ij}. This stands for the entry on row i and column j. The symbol ϵ_{ijk} is actually a $3 \times 3 \times 3$ matrix, so we need three indices to describe each of its entries.

important for the reader to be familiar with this notation. There are a few more exercises at the end of this chapter.

3.2.3 More about lattices

Having defined what a lattice is, we can take a closer look at the consequences of this definition. If we translate the lattice by any of the lattice vectors **t**, then we obtain the same lattice again. In other words, if you were to look at an infinite lattice, then look away while someone else translates this lattice by **t**, then you would not be able to see the difference between the lattices before and after translation; they would coincide. If the translation vector was not a lattice vector, then you *would* be able to see the difference, since the translated lattice would not coincide with the original one. This means that the lattice is *invariant* under any translation by a lattice vector **t**. As a consequence of this invariance, *all lattice points are identical*. This is illustrated in Fig. 3.4: we can choose any lattice point as the origin, and the surroundings of all lattice points are identical, as indicated by the thin lines around points 0, 1, and 2.

The space lattice is a purely mathematical abstraction and *does not contain any atoms or molecules at all*. However, we can take a molecule and attach it to each lattice point to obtain a crystal structure. We thus find that

A crystal structure consists of a 3-D space lattice which is decorated with one or more atoms.

The lattice is a 3-D assembly of mathematical points, which reflect the *translational symmetry* of the complete crystal. In general, any 3-D lattice can be fully described by stating the lengths of the 3 basis vectors and their mutual angles. According to the *International Tables for Crystallography* (Hahn, 1989) the following notation should be used to describe the dimensions of a 3-D lattice:

$$\boxed{\begin{aligned} a &= \text{length of } \mathbf{a}; \\ b &= \text{length of } \mathbf{b}; \\ c &= \text{length of } \mathbf{c}; \end{aligned}} \tag{3.4}$$

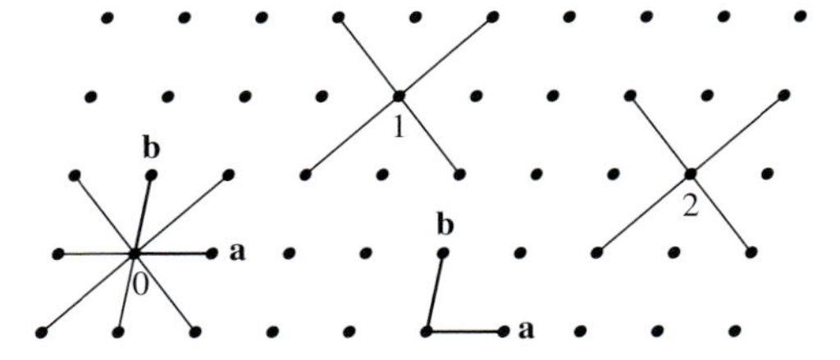

Fig. 3.4. All lattice points have identical surroundings and every point can be selected as the origin of the lattice.

$$\boxed{\begin{aligned}\alpha &= \text{angle between } \mathbf{b} \text{ and } \mathbf{c};\\ \beta &= \text{angle between } \mathbf{a} \text{ and } \mathbf{c};\\ \gamma &= \text{angle between } \mathbf{a} \text{ and } \mathbf{b}.\end{aligned}} \quad (3.5)$$

It is easy to remember the angle designations: for any pair of vectors, say, **a** and **c**, take the missing letter (in this case b) and turn it into a greek letter (in this case β). These six quantities fully specify the space lattice (see Fig. 3.3). The choice of the shortest lattice vector as either **a**, **b**, or **c** will depend on the symmetry of the lattice. We will often write the six numbers as $\{a, b, c, \alpha, \beta, \gamma\}$; they are known as the *lattice parameters*. For a 2-D net, the *net parameters* are usually written as $\{a, b, \gamma\}$.

The volume defined by the three basis vectors (shown by the dotted lines in Fig. 3.3) is known as the *unit cell* of the space lattice. It is customary to define the vectors in such a way that the reference frame is right-handed. If the mixed vector product $(\mathbf{a} \times \mathbf{b}) \cdot \mathbf{c}$ is positive, then the reference frame is right-handed; if the product is negative, then the reference frame is left-handed. We will define the dot and cross products in the following chapters. Next, we will attempt to answer the question: *how many different space lattices/nets are there?* We will consider 2-D nets before describing the 3-D lattices.

3.3 The four 2-D crystal systems

Consider the net parameters $\{a, b, \gamma\}$. If we take arbitrary values for all three parameters, then we end up with a net similar to that shown in Fig. 3.5. This is known as an *oblique net*. There are no special conditions on any of the net parameters. The oblique net has a low symmetry;[5] if we place a line normal to the drawing in Fig. 3.5(a), through one of the nodes of the net, then it is easy to see that, if we rotate the net by 180° around this line, all the nodes of the rotated net will coincide with the original nodes. A node at a position $u\mathbf{a} + v\mathbf{b}$ will end up at position $-u\mathbf{a} - v\mathbf{b}$ after the rotation, and this is again an integer linear combination of the basis vectors. This means that the new rotated node coincides with one of the original nodes, so that the original and rotated nets are indistinguishable.

There is one special value for the angle γ. When $\gamma = 90°$, the unit cell of the net becomes a rectangle, and the resulting net is the *rectangular*

[5] We will define and discuss the concept of symmetry extensively in Chapter 8. For now it is sufficient for the reader to understand what a rotation is.

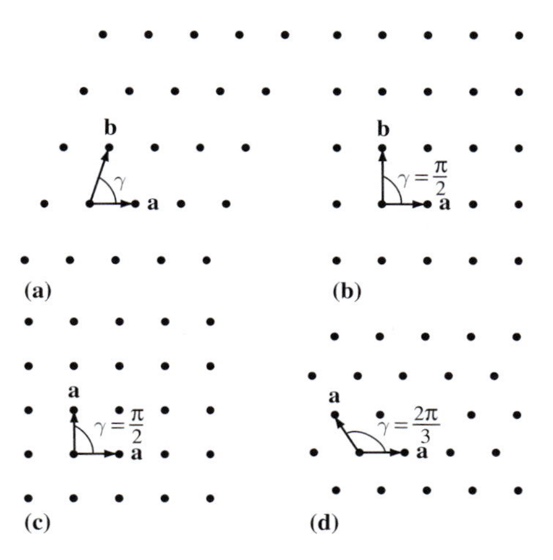

Fig. 3.5. Examples of the oblique (a), rectangular (b), square (c), and hexagonal (d) 2-D nets.

net, illustrated in Fig. 3.5(b). The rectangular net has the same rotational symmetry (a 180° rotation axis normal to the plane of the drawing). In addition, it also has mirror symmetry. This is easily verified by holding this book in front of a mirror and looking at Fig. 3.5(b); there is no difference between the original and its mirror image. For the oblique net, we see that the image in the mirror leans over to the left, whereas the original leans over towards the right. Therefore, the oblique net does not have mirror symmetry, and we say that the rectangular net has a higher symmetry than the oblique net. The net parameters of the rectangular net are usually written as $\{a, b, \pi/2\}$.

Next, we again start from the oblique net, but this time we take the two basis vectors to have equal length, so that the net parameters are $\{a, a, \gamma\}$. In this case, there are two special angles γ, for which the resulting net has a higher symmetry than the oblique net. If $\gamma = 90°$, then the net is based on a square unit cell, and is called the *square net*, as shown in Fig. 3.5(c). The higher symmetry is easy to spot, since a rotation of 90° around any axis going through a node (perpendicular to the plane of the drawing) leaves the net invariant.

Finally, the last 2-D net is obtained by setting $\gamma = 120°$. This is the *hexagonal net*; it is easy to see that this net is invariant under a rotation of 60°, hence the name *hexa*gonal. Note that we could also have selected $\gamma = 60°$; the resulting net would have been indistinguishable from the one shown in Fig. 3.5(d). The international convention is to select $\gamma = 120°$ for the hexagonal net. These four nets are the only possible nets that can be generated with only two basis vectors **a** and **b**. We say that, in 2-D, there are only four possible *crystal systems*: *oblique*, *rectangular*, *hexagonal*, and *square*. Table 3.1 summarizes the net parameter symbols, the crystal system name and an example of a unit cell for each of the 2-D crystal systems.

Table 3.1. The four 2-D crystal systems.

Condition/symbol	Crystal system	Drawing
no condition, $\{a, b, \gamma\}$	**OBLIQUE**	
$\gamma = 90°$, $\{a, b, 90°\}$	**RECTANGULAR**	
$a = b, \gamma = 120°$, $\{a, a, 120°\}$	**HEXAGONAL**	
$a = b, \gamma = 90°$, $\{a, a, 90°\}$	**SQUARE**	

3.4 The seven 3-D crystal systems

There are seven fundamentally different combinations of basis vectors in 3-D. In the most general case, we select arbitrary numbers for the set of six lattice parameters $\{a, b, c, \alpha, \beta, \gamma\}$. This generates the *triclinic* or *anorthic* lattice.[6] Figure 3.3 shows an example of a triclinic unit cell. When we translate this unit cell by integer linear combinations of its basis vectors, we obtain the triclinic lattice. No matter how we rotate this lattice, there are no rotation axes for which the lattice is invariant.

Next, we can assign special values to some or all of the lattice parameters, as we did for the 2-D case. We look for combinations of lattice parameters for which we can identify rotational symmetry in the resulting lattice. It turns out that we can have a single 180° rotation axis when two of the

[6] The name *triclinic* can be split into two parts: *tri* which stands for "three," and *clinic*, which comes from the Greek word *klinein* for "to bend or slope." In other words, we need three angles to describe this unit cell. The second name, *anorthic*, is a combination of *an*, which means "not," and *ortho*, which stands for "perpendicular," meaning that none of the three angles is a right angle.

three angles α, β, and γ are equal to 90°.[7] It is customary to select β to be the angle that is not equal to 90°, so that we arrive at the lattice parameters $\{a, b, c, \pi/2, \beta, \pi/2\}$. This is known as the *monoclinic* lattice.[8] Table 3.2 shows the lattice parameters and conditions for each of the 3-D crystal systems, along with simple sketches of the corresponding unit cells.

If we select two of the lattice parameters, a and b, to be equal to each other then we can create another 3-D lattice by putting the angles equal to 90°, 90°, 120°, or $\{a, a, c, \pi/2, \pi/2, 2\pi/3\}$. This is similar to the hexagonal 2-D net $\{a, a, 2\pi/3\}$, but now there is a third dimension to the unit cell, perpendicular to the 2-D drawing of Fig. 3.5(d). This is known as the 3-D *hexagonal* lattice. Similar to its 2-D analogue, the 3-D hexagonal lattice has a 60° rotation axis along the $\mathbf{c}$ direction.

If all three lengths a, b, and c are equal to each other, then we find that there is in general no new lattice unless the three angles are also equal to each other: $\{a, a, a, \alpha, \alpha, \alpha\}$. The resulting lattice is known as the *rhombohedral* lattice. Along the direction corresponding to the body diagonal of this unit cell, a rotation of 120° leaves the lattice invariant. An alternative name for this system is *trigonal*, indicating that the three angles are equal to each other.[9]

A special case of the rhombohedral lattice is found when the angle α is set equal to 90°. In that case, $\{a, a, a, \pi/2, \pi/2, \pi/2\}$, we have a *cubic* lattice. Note that there are now several rotation axes that will leave the unit cell invariant. We can rotate the cube by 90° around any axis normal through one of the faces and going through the center of the cube, by 120° around the body diagonals, and by 180° around any axis going through the centers of two edges of opposite sides of the cube. It is also clear that, when we look at a cube in a mirror, we will see the same cube, so that the cube also has mirror symmetry. We will describe all these symmetry properties in a much more systematic way later on in this book. For now, it suffices that the reader obtain just a simple intuitive understanding of what symmetry means.

Starting from the monoclinic unit cell, we can put the angle β equal to 90°, so that we obtain a lattice for which all three angles are equal, but the lengths of the basis vectors are not equal: $\{a, b, c, \pi/2, \pi/2, \pi/2\}$. This is the *orthorhombic* lattice, with a unit cell which is shaped like a right-angled rhombus. This shape will be familiar to the reader, since most packaging boxes have this shape. It is easy to convince yourself that this shape has three 180° rotation axes, going through the centers of opposite faces.

Finally, we can put two of the three parameters of the orthorhombic lattice equal to each other, as in $\{a, a, c, \pi/2, \pi/2, \pi/2\}$. This is the *tetragonal*

[7] We postpone a more rigorous proof of the existence of the seven 3-D crystal systems until Chapter 8, where we will define all symmetry operators.

[8] *Mono* means "one," indicating that one of the three angles is not a right-angle.

[9] The Greek word *gonia* means "angle."

Table 3.2. The 7 three-dimensional crystal systems.

Condition/symbol	Crystal system	Drawing
no conditions $\{a, b, c, \alpha, \beta, \gamma\}$	**TRICLINIC (ANORTHIC)**	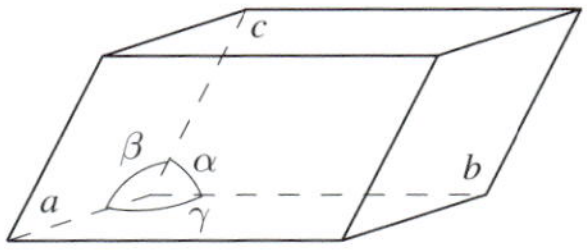
$\alpha = \gamma = 90°$[†] $\{a, b, c, 90, \beta, 90\}$	**MONOCLINIC**	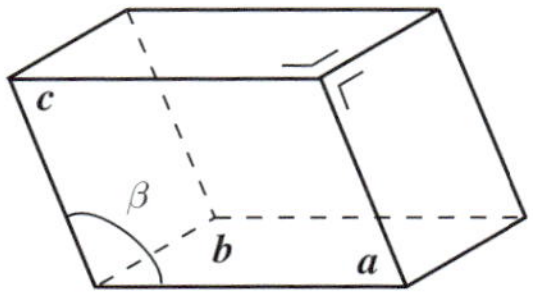
$a = b$, $\alpha = \beta = 90°$, $\gamma = 120°$ $\{a, a, c, 90, 90, 120\}$	**HEXAGONAL**	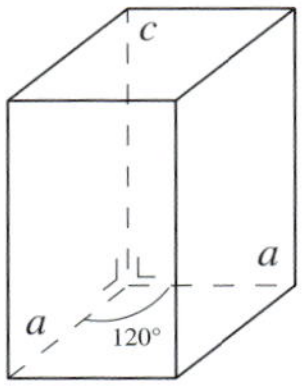
$a = b = c$, $\alpha = \beta = \gamma$ $\{a, a, a, \alpha, \alpha, \alpha\}$	**RHOMBOHEDRAL (TRIGONAL)**	
$\alpha = \beta = \gamma = 90°$ $\{a, b, c, 90, 90, 90\}$	**ORTHORHOMBIC**	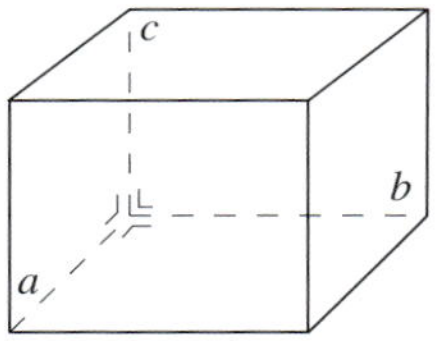
$a = b$, $\alpha = \beta = \gamma = 90°$ $\{a, a, c, 90, 90, 90\}$	**TETRAGONAL**	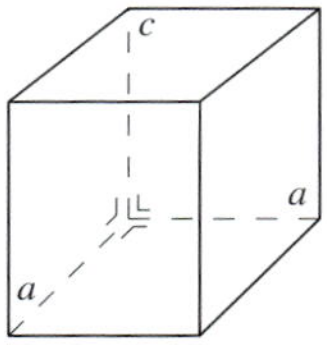
$a = b = c$, $\alpha = \beta = \gamma = 90°$ $\{a, a, a, 90, 90, 90\}$	**CUBIC**	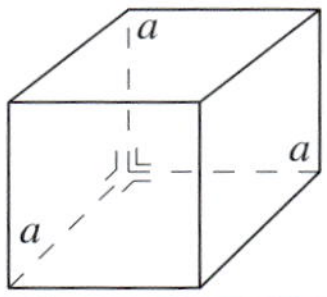

[†] The angle β is usually chosen to be larger than 90°

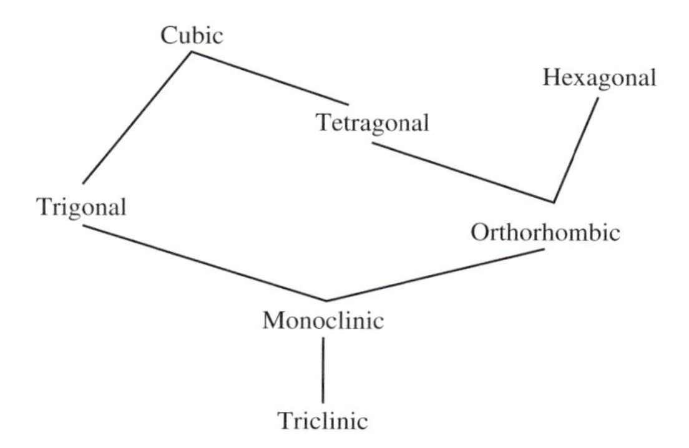

Fig. 3.6. The seven crystal systems ranked according to symmetry. The closer the system is to the top of the drawing, the higher its symmetry.

lattice, which has a single rotation axis of 90° going through the centers of the opposite square faces, and 180° rotation angles through the center of the opposite non-square faces. This concludes the enumeration of the seven 3-D crystal systems (Table 3.2).

The 3-D crystal systems can be ranked by their symmetry (for a more complete description of symmetry, see Chapter 8). This ranking is shown in Fig. 3.6. Starting from the cubic symmetry, we can, by successive distortions, create a triclinic lattice.

3.5 The five 2-D Bravais nets and fourteen 3-D Bravais lattices

Consider the 2-D lattice in Fig. 3.7. We can define a unit cell for this lattice in an infinite number of ways; a few possibilities are shown in the figure. The unit cells numbered 1, 2, and 3 are so-called *primitive* unit cells, because they contain only one lattice point. The number of nodes in a cell can be computed in two different ways:

(i) Displace the outline of the unit cell, so that the corners of the cell no longer coincide with lattice sites. Now count the number of sites inside the displaced unit cell. This is illustrated by the dashed cell outlines in Fig. 3.7.

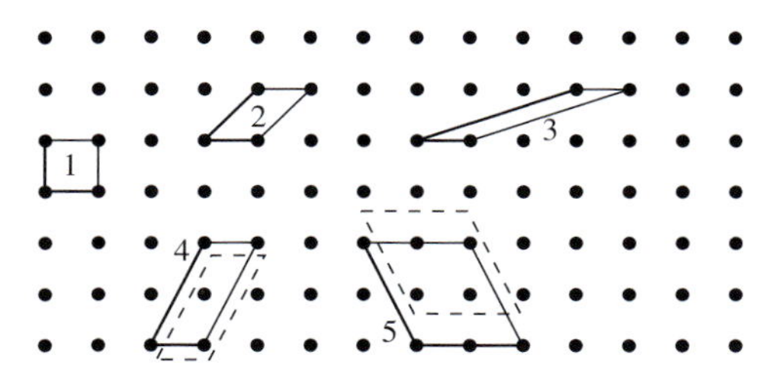

Fig. 3.7. A few possible unit cells in a 2-D square net.

(ii) In 2-D, count the number of sites inside the unit cell (N_{in}); add to that 1/2 of the number of sites on the unit cell boundaries (N_{edge}), and add to that 1/4 of the sites on the unit cell corners (N_{corner}). In other words:

$$N_{2D} = N_{\text{in}} + \frac{1}{2}N_{\text{edge}} + \frac{1}{4}N_{\text{corner}}.$$

In 3-D, the total number of sites in a unit cell is given by:

$$N_{3D} = N_{\text{in}} + \frac{1}{2}N_{\text{face}} + \frac{1}{4}N_{\text{edge}} + \frac{1}{8}N_{\text{corner}},$$

where N_{face} is the number of sites in the faces of the unit cell.

For all three of the cells 1, 2, and 3, we find that there are 4 sites located at the corners and none inside or on the edges, hence the number of sites in the unit cell is 1.

The unit cells numbered 4 and 5 in Fig. 3.7 are *non-primitive* unit cells, because they contain more than one lattice site. From the displaced unit cell outlines (indicated with a dashed line) we find that there are $N_{2D} = 2$ for cell 4, and $N_{2D} = 4$ for cell 5. Although these cells could be used to describe this 2-D net, they are not as convenient as cell 1. In general, one describes a lattice with the simplest (not necessarily the smallest) possible unit cell, in this case cell 1. Note also that cell 1 is the only cell of the five shown that reflects the squareness of the net. It is useful to select as unit cell, the cell that reflects the *symmetry* of the net. This is also true for 3-D lattices.

From the definitions of the seven 3-D crystal systems, we know that there are seven primitive unit cells. They are denoted by a two-letter symbol: the first letter (lowercase) indicates the crystal system (*a* for anorthic or triclinic, *m* for monoclinic, *o* for orthorhombic, *t* for tetragonal, *h* for hexagonal, *c* for cubic, and, strangely enough, *no letter* for trigonal or rhombohedral). The second letter (uppercase) indicates the type of cell, which in this case is *primitive* or *P*. The exception to this rule is the rhombohedral or trigonal system, which is indicated by the symbol *R*. The primitive cubic unit cell is hence represented by the symbol *cP*, the primitive tetragonal cell by *tP*, etc.

We can then ask: can we add additional lattice points to the primitive lattices or nets, in such a way that we still have a lattice (net) belonging to the same crystal system? We will first illustrate this for the 2-D nets. We know that, in order for a collection of nodes to form a net, the surroundings of each node must be identical. If we consider a rectangular net with lattice parameters $\{a, b, \pi/2\}$, and add a node at the position $\mathbf{a}/3 + \mathbf{b}/2$, as shown in Fig. 3.8(a), then it is clear that the surroundings of the point A are not the same as those of the point B. While A has as a neighbor the point located at $\mathbf{r}_B = \mathbf{a}/3 - \mathbf{b}/2$ from A, this point B does not have a point located at $\mathbf{a}/3 - \mathbf{b}/2$ from itself (this location is indicated by a gray circle). Therefore, the surroundings of A and B are not identical, so this is not a net. There

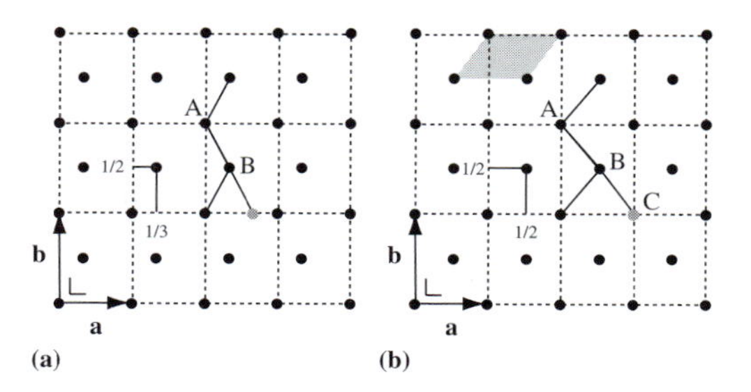

Fig. 3.8. (a) Adding the point $\mathbf{r}_B = \mathbf{a}/3+\mathbf{b}/2$ to each unit cell of a rectangular net does not produce a new net, since all points are no longer identical; (b) adding the point $\mathbf{r}_B = (\mathbf{a}+\mathbf{b})/2$, produces the centered rectangular net. A primitive cell for this net is shown in gray.

is, however, a special position B inside the rectangular unit cell, for which the surroundings are identical to those of A. This is the point at the center of the cell, as shown in Fig. 3.8(b). If $\mathbf{r}_B = (\mathbf{a}+\mathbf{b})/2$, then there is a point located at this position relative to B, namely the point C. Hence, the surroundings of A, B, and C are identical – in fact, all of the nodes have identical surroundings – so that this is a new net.

The attentive reader might say: "*Wait a minute! This is not a new net, because I can select a smaller, primitive unit cell (in gray in Fig. 3.8(b)) which fully defines this net. Furthermore, this primitive cell indicates that this net is an oblique net, not a rectangular one!*" This is absolutely correct. We could indeed use the primitive cell to describe the complete net. However, this primitive oblique cell does not reveal that the net actually has a higher symmetry! Indeed, looking at the primitive unit cell in a mirror, we see that the mirror image is not the same as the original cell. The mirror image of the rectangular cell with a node at its center *is* the same as the original, so it makes sense to use this non-primitive cell to describe the net. This simple example illustrates two important ideas:

- It is always possible to define a primitive unit cell, for every possible net (this is also true for 3-D lattices).
- If a non-primitive cell can be found, that describes the symmetry of the net (lattice), then that cell should be used to describe the net (lattice). Since the surroundings of every node must be identical, we can only add new nodes at locations that are *centered* in the middle between the original lattice sites.

The 2-D net shown in Fig. 3.8(b) is, therefore, a new net, known as the *centered rectangular net*. If we try to do the same thing with the other 2-D nets, we find (this is left as an exercise for the reader) that there are no new nets to be found. We conclude that in 2-D, there are only five possible nets: four of them are primitive (oblique, rectangular, hexagonal, and square) and one is centered (centered rectangular). We call these five nets the *2-D Bravais nets*. The five 2-D Bravais nets are shown in Fig 3.9.

We can repeat this procedure in three dimensions. In this case, there are three possible ways to add nodes at the center in between existing nodes.

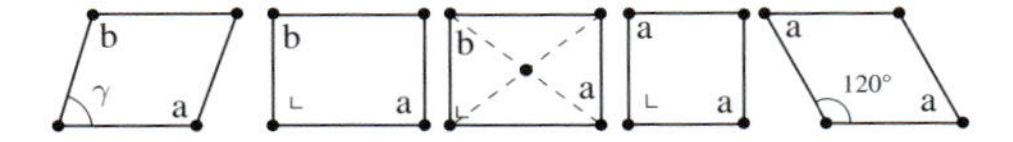

Fig. 3.9. The five two-dimensional Bravais lattices.

- *Body centering*: we add a lattice site in the center of the unit cell, at the location (1/2, 1/2, 1/2). For every site **t**, there is then an additional site $\mathbf{t}+(\mathbf{a}+\mathbf{b}+\mathbf{c})/2$. The vector $\mathbf{I} \equiv (\mathbf{a}+\mathbf{b}+\mathbf{c})/2$ is known as the *body centering vector*. Note that this vector is *not* a translation vector of the lattice since its components are not integer numbers. The symbol for a body centered lattice is I, from the German word for body centered: "Innenzentriert."
- *Face centering*: we add a lattice site to the center of all faces of the unit cell, at the locations (1/2, 1/2, 0), (1/2, 0, 1/2), and (0, 1/2, 1/2). For every site **t** there are then three additional sites $\mathbf{t}+(\mathbf{a}+\mathbf{b})/2$, $\mathbf{t}+(\mathbf{a}+\mathbf{c})/2$, and $\mathbf{t}+(\mathbf{b}+\mathbf{c})/2$. The vectors $\mathbf{C}=(\mathbf{a}+\mathbf{b})/2$, $\mathbf{B}=(\mathbf{a}+\mathbf{c})/2$, and $\mathbf{A}=(\mathbf{b}+\mathbf{c})/2$ are known as the *face centering vectors*. The symbol for a face centered lattice is F.
- *Base centering*: we add a lattice site to the center of only one face of the unit cell, at the location (1/2, 1/2, 0) *or* (1/2, 0, 1/2) *or* (0, 1/2, 1/2). The base centering vectors are identical to the face centering vectors, except that only one of them is present. If the plane formed by the basis vectors **a** and **b** is centered, then the lattice is known as a C-centered lattice. If the **a**–**c** plane is centered, the lattice is B-centered and if the **b**–**c** plane is centered then the lattice is A-centered.

One can show that for two-face centering not all lattice points have the same surroundings, and hence two-face centering cannot give rise to a new lattice.

We can now apply these five forms of centering (A, B, C, I, and F) to all seven primitive unit cells. In several cases we do generate a new lattice, in other cases we can redefine the unit cell and reduce the cell to another type. Consider the following example. The primitive tetragonal unit cell *tP* shown in Figure 3.10(a) is C-centered in Fig. 3.10(b). This is not a new cell, however, since we can redefine the unit cell by the thick lines in Fig. 3.10(c), which form a new, smaller *primitive tetragonal* unit cell with lattice parameter $a_t = a\sqrt{2}/2$. We find that a C-centered tetragonal cell *tC* is equivalent to tP and hence does not form a new lattice. Repeating this exercise for all possible

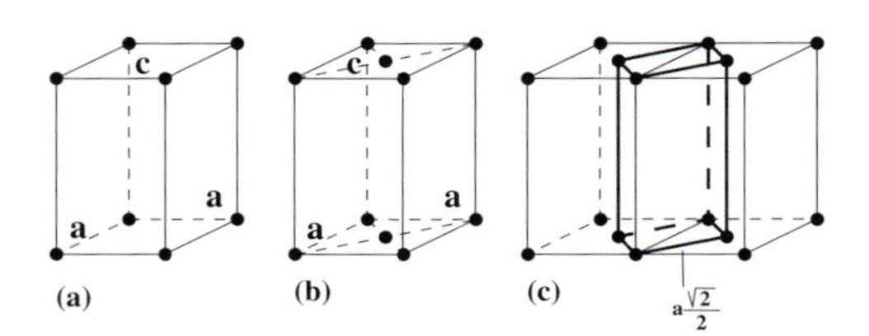

Fig. 3.10. (a) *tP* lattice, (b) *tC* lattice, (c) equivalence of *tC* and *tP* lattices.

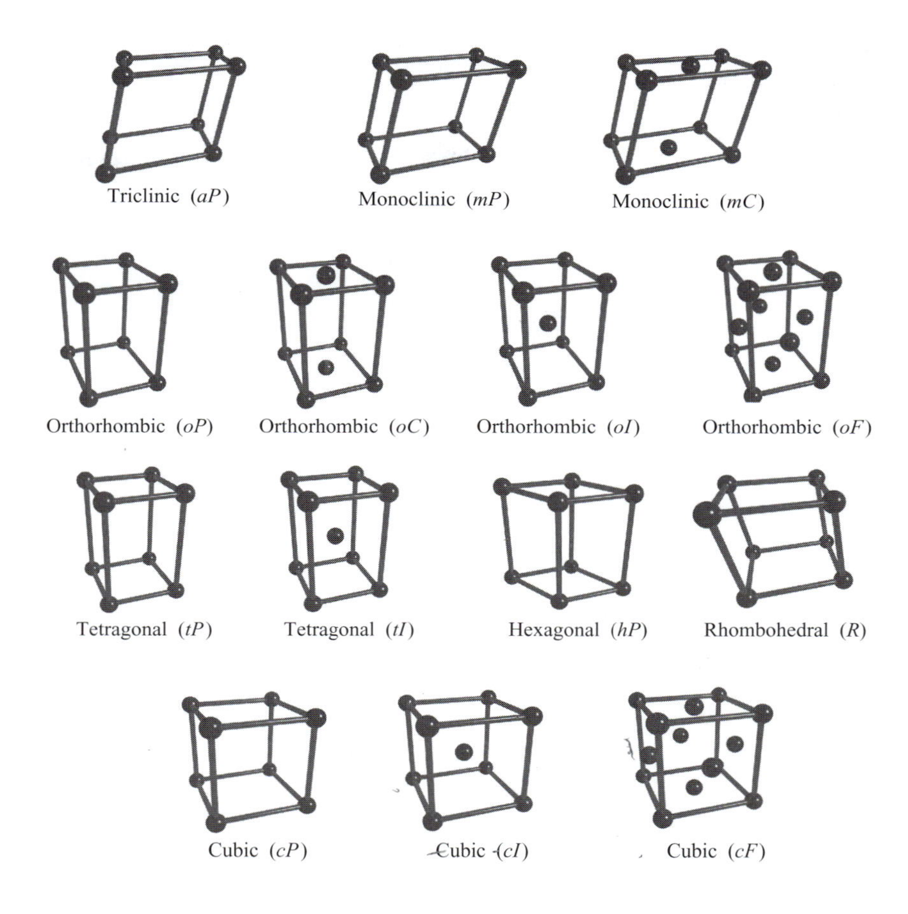

Fig. 3.11. The fourteen 3-D Bravais lattices.

types of lattice centering (there are $5 \times 7 = 35$ possibilities!) we end up with seven additional lattice types that cannot be reduced to primitive ones of the same crystal system: *mC, oC, oI, oF, tI, cI,* and *cF*. All fourteen 3-D Bravais lattices are shown in Fig. 3.11.

3.6 Other ways to define a unit cell

It is always possible to describe a lattice with a primitive unit cell. Hence, all 14 Bravais lattices can be described by primitive cells, even when they are centered. As an example, consider the *cF* lattice in Fig. 3.12a. By selecting shorter vectors $\mathbf{a}_p$, $\mathbf{b}_p$, and $\mathbf{c}_p$ we can define a primitive rhombohedral unit cell with angle $\alpha = 60°$. This cell does not reflect the cubic symmetry of the *cF* lattice, but is has the advantage that it contains only one lattice site. In solid state physics, it is often convenient to work with the primitive unit cells of all the Bravais lattices, rather than with their non-primitive (and higher symmetry) versions.

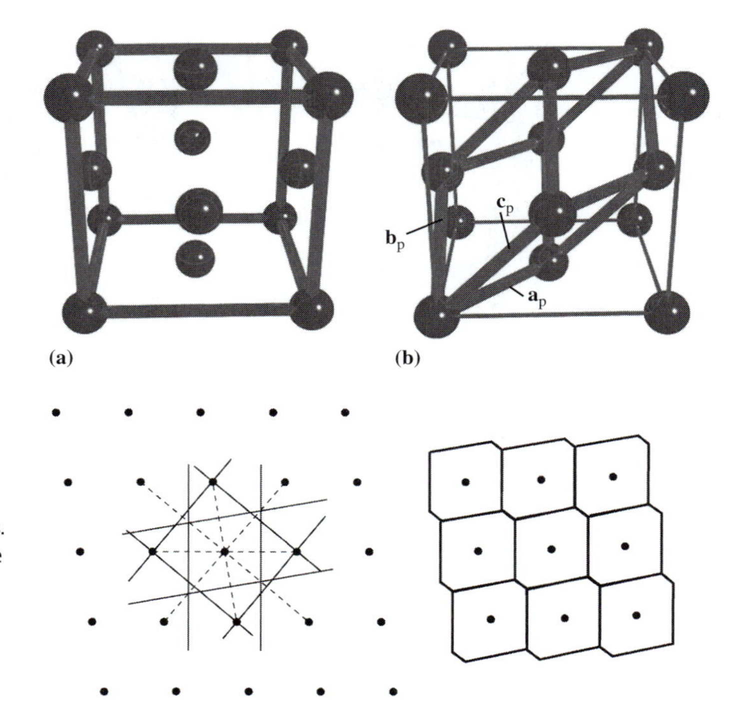

Fig. 3.12. (a) Unit cell of the *cF* lattice, (b) primitive rhombohedral unit cell, with edge length $a = a_{cF}2^{-1/2}$ and angle $\alpha = 60°$.

Fig. 3.13. Example of the Wigner–Seitz cell construction for an oblique net. Solid lines are perpendicular to the dashed lines connecting nodes. The gray region represents the WS cell. One can build the complete lattice by stacking WS cells in a regular way, as shown on the right.

There is yet another useful way to define a unit cell: the *Wigner–Seitz cell*. The Wigner–Seitz (WS) cell corresponding to a particular lattice point is *the region of space which is closer to that particular lattice point than to any other lattice point*. It is straightforward to construct the WS cell (see Fig. 3.13): construct the vector between the origin and one of the neighboring lattice points. Draw the perpendicular plane through the midpoint of this vector. This plane separates space into two regions, each of which contains all of the points closer to one of the endpoints of the vector than to the other endpoint. Repeat this construction for all other lattice points. The smallest volume around the selected point enclosed by all these planes is the Wigner–Seitz cell. Note that the WS cell can have more than six sides in 3-D, or more than 4 in 2-D. All WS cells are primitive by construction and they do display the true symmetry of the underlying lattice. In 3-D, it can be shown that there are 24 topologically different Wigner–Seitz cells for the 14 Bravais lattices (Burns and Glazer, 1990). These cells have different shapes, depending on the actual values of the lattice parameters. An example of the WS cell for the *cI* Bravais lattice is shown in Fig. 3.14. The WS cell is also known as the *Voronoi domain*, the *Dirichlet domain*, or the *domain of influence* of a given lattice point. It can be shown on theoretical grounds that the number of faces of a 3-D WS cell is always between 6 and 14 (inclusive). In 2-D, the number of edges of the WS cell lies between 4 and 6. Inspection of Fig. 3.14 reveals that the WS cell has the

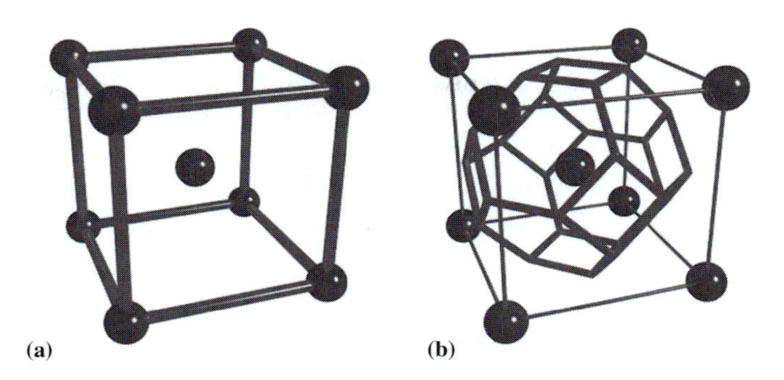

Fig. 3.14. The Wigner–Seitz cell (b) for the body centered cubic Bravais lattice (a). This shape has fourteen faces, six of them are squares, the other eight are hexagons.

same symmetry as the Bravais lattice. This is true for the general WS cell as well.

In spite of the possibly complicated shape of the Wigner–Seitz cell, it is often very easy to compute its volume. The difficult way would be to actually use geometry to determine the volume. However, there is a much easier method. We know that the cI Bravais lattice is a cubic lattice, so that the volume of the unit cell is a^3, where a is the edge length. We also know that there are 2 nodes in the unit cell (the one in the center counts as a whole, whereas the 8 at the corners count for 1/8 each). So, the volume per node is equal to $a^3/2$. If we take the WS cell for this Bravais lattice, then two of those WS cells must still be equal to the volume a^3; the shape is different, but the available volume must be the same. Therefore, in spite of its complicated shape, the volume of the WS cell for the cI Bravais lattice is simply $a^3/2$.

In summary, there are fourteen Bravais lattices and we can define three types of unit cells to describe them: the conventional unit cell, the primitive unit cell, and the Wigner–Seitz cell. Of these three, only the conventional cell and the Wigner–Seitz cell display the true symmetry of the underlying lattice.

3.7 Historical notes

Moritz Frankenheim (1801–69) was a German crystallographer who was the first to enumerate the 32 crystal classes. He was also the first to enumerate the 14 three-dimensional lattices, but his list contained an error. In 1850, **August Bravais** (1811–63), a French naval officer and scientist, showed that two of Frankenheim's lattices were identical, and he subsequently correctly derived the 14 lattices that now carry his name (Bravais, 1850). After the classification of crystals into seven axial systems, the question of which symmetry operations were compatible with these crystal systems was addressed and first solved correctly by Frankenheim. In 1830, **J. F. C. Hessel** (1796–1872, Fig. 9.15(b) on page 228) independently solved the problem of the symmetries compatible with the seven axial systems, i.e., he found that only 2-, 3-, 4-, and

Fig. 3.15. (a) A. Bravais (1811–1863) (picture courtesy of J. Lima-de-Faria), and (b) E.P. Wigner (1902–1995) (picture courtesy of the Nobel Museum).

6-fold rotation axes were compatible with the translational lattice symmetry. Neither his work nor the work by Frankenheim were noticed by scientists at the time.

Eugene Paul Wigner (1902–95) was a Hungarian scientist. While at the Technische Hochschule in Berlin, he learned about the role of symmetry in crystallography. At about the same time, the new quantum mechanics was being developed, and Wigner immediately realized the importance of symmetry principles in quantum mechanics. His work in this area earned him the 1963 Nobel prize in physics. After a short stay at the University of Göttingen, he moved to Princeton, where he worked on solid state physics, along with his first graduate student, **Frederick Seitz**. The Wigner–Seitz cell, as introduced in this chapter, results from their joint research. Wigner applied the mathematics of irreducible representations of groups to a variety of physics problems; he became especially well known for his ground-breaking paper on the relativistic Lorentz transformation and for his work on the algebra of angular momentum coupling in quantum mechanics. Wigner's interest in nuclear physics and his knowledge of chemistry were instrumental in his design of a full scale nuclear reactor, which was to become the basis for the commercial Dupont reactors in the post World War II years. In his later years, Wigner founded the quantum theory of chaos.

3.8 Problems

(i) *Bravais lattices I:* Show that a face centered tetragonal lattice (tF) can be reduced to one of the 14 Bravais lattices. Write the basis vectors of this Bravais lattice in terms of those of the tF lattice.

(ii) *Bravais lattices II:* Consider the cubic Bravais lattices *cP*, *cI*, and *cF*, each with lattice parameter a. Make a table showing for each lattice the number of first nearest neighbour lattice sites N_1, the distance to those neighbours d_1, the number of second nearest neighbour lattice sites N_2, and the distance to those neighbours d_2.

(iii) *Bravais lattices III:* Describe the consecutive deformations that need to be applied to a cubic unit cell to turn it into a monoclinic unit cell; repeat the question for the deformation of a tetragonal cell into a triclinic cell.

(iv) *Bravais lattices IV:* Show, using a graphical example, that is not possible to create a new Bravais lattice which has two centered faces (e.g., both A and B centering).

(v) *Other unit cells:* Determine graphically the 3-D primitive unit cell corresponding to the *cI* Bravais lattice and express its lattice parameters in terms of the cubic ones.

(vi) *Wigner–Seitz cells I:* Make a drawing of the Wigner–Seitz unit cell for the *hP* lattice and compute the volume of this cell. (Hint: this does not require any actual computations. The volume can be derived simply by thinking about the definition of the WS cell.)

(vii) *Wigner–Seitz cells II:* Compute the volume of the largest sphere that can be inscribed in the Wigner–Seitz cell of the *cI* lattice. (Hint: As in the previous question, this does not really require any significant computations.)

(viii) *fcc Wigner–Seitz cell:* Construct the Wigner–Seitz cell for the *fcc* lattice.

(ix) *bcc Wigner–Seitz cell:* Show that the fractional coordinates of the vertices of the Wigner–Seitz cell of the *bcc* lattice in the $x = 0$ plane are $(0, 1/2, 1/4)$, $(0, 1/2, 3/4)$, $(0, 1/4, 1/2)$, and $(0, 3/4, 1/2)$.

(x) *fcc molecular solid:* Fullerites (discussed in more detail in Chapter 25) have *Buckminsterfullerene* C_{60} molecules decorating the sites of an *fcc* Bravais lattice. The reported low temperature lattice constant, $a_0 = 1.404(1)$ nm for *fcc* C_{60}.

(a) Calculate the number of C atoms contained in the cubic cell.
(b) Calculate the touching molecular sphere radius of C_{60} in the structure.
(c) What is the coordination number, CN of C_{60} molecules about another in this structure.

(xi) *Cubic lattices packing fractions:* Determine directions in which hard spheres touch, and the volume fractions occupied by them in three cubic structures:

$$sc \quad \frac{\pi}{6}; \qquad bcc \quad \frac{\pi\sqrt{3}}{8}; \qquad fcc \quad \frac{\pi\sqrt{2}}{6}.$$

Fig. 3.16. Monoclinic unit cell.

$\alpha = \gamma = 90°$
$\{a, b, c, 90, \beta, 90\}$

MONOCLINIC

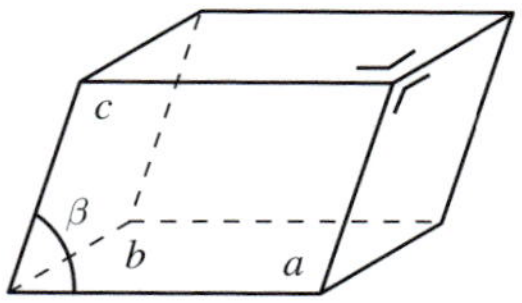

(xii) *Monoclinic crystal system:* Consider the monoclinic unit cell illustrated in Fig. 3.16. Give an example of a 2-fold rotational symmetry and mirror plane that leaves this lattice invariant. (i.e., show an axis about which you can rotate the cell and a plane through which you can reflect the cell and not tell it apart.)

CHAPTER

4 Crystallographic computations

> *"We are told such a number as the square root of two worried Pythagoras and his school almost to exhaustion. Being used to such queer numbers from early childhood, we must be careful not to form a low idea of the mathematical intuition of these ancient sages; their worry was highly credible."*
>
> Erwin Schrödinger

In this chapter, we introduce the *metric tensor*, a computational tool that simplifies calculations related to distances, directions, and angles between directions. First, we illustrate the importance of the metric tensor with a 2-D example. Then, we introduce the 3-D metric tensor and discuss how it can be used for simple lattice calculations in all crystal systems. We end this chapter with a few worked examples.

4.1 Directions in the crystal lattice

We know that a vector has two attributes: a *length* and a *direction*. By selecting a translation vector **t** in the space lattice, we are effectively selecting a *direction* in the crystal lattice, namely the direction of the line segment connecting the origin to the endpoint of the vector **t**. Directions in crystal lattices are used so frequently that a special symbol has been developed to describe them. The direction parallel to the vector **t** is described by the symbol $[uvw]$, where (u, v, w) are the smallest integers proportional to the components of the vector **t**. Note the *square brackets* and the *absence of commas* between the components. If a component is negative, then the minus sign is always written *above* the corresponding integer, e.g., $[1\bar{1}2] = [1 -1\,2]$. If one or more of

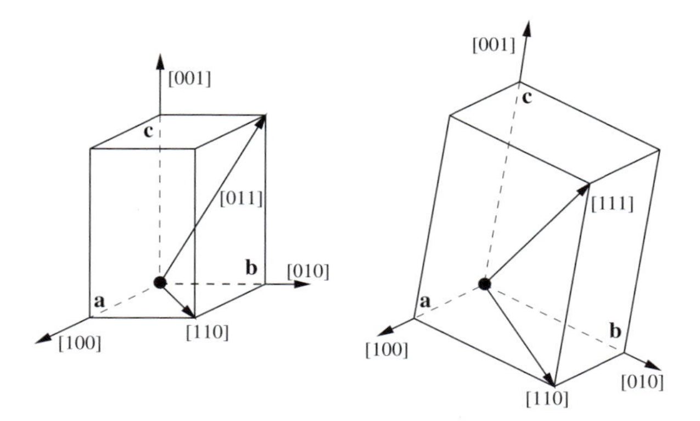

Fig. 4.1. Illustration of directional indices in two different crystal lattices.

the integer components is larger than 9, then one usually introduces a narrow space between the numbers, e.g., [1 12 4], to avoid ambiguities. The symbol $\bar{u}$ may be pronounced in two different ways: *bar-u* (as in negative-u or minus-u) or *u-bar* (which would correspond to the order in which most people would write down this symbol, first the u, then the minus sign on top). Both are commonly used and one must be careful not to confuse them, in particular when there are multiple negative numbers. For instance, the direction $[\bar{1}1\bar{2}]$ can be pronounced as bar-one, one, bar-two, or as one-bar, one, two-bar. It is good practice to leave a short *audible* space between the numbers to avoid incorrect interpretations.

Figure 4.1 shows a few examples of directions in lattices. Directions are always defined with respect to the crystallographic basis vectors $\mathbf{a}_i$. Note that the [100] direction is *always* parallel to the **a** axis, *regardless of the crystal system*. Similarly, the [111] direction is always parallel to the body diagonal of the unit cell. Since the indices $[uvw]$ are always integers, we find that every direction vector is also a lattice translation vector. However, while the directions [123] and [246] are identical, the corresponding lattice translation vectors are not the same.

In the following sections, we will introduce a computational tool that will allow us to compute the following quantities for an arbitrary Bravais lattice:

- the distance between two arbitrary points (e.g., a bond length);
- the angle between two directions (e.g., a bond angle);
- the volume of the unit cell.

4.2 Distances and angles in a 3-D lattice

4.2.1 Distance between two points

Consider the following problem: What is the distance D between the origin and the lattice point with coordinates (1, 1, 1) in an arbitrary Bravais lattice (see Fig. 4.2)? In a Cartesian (orthonormal) reference frame this would be

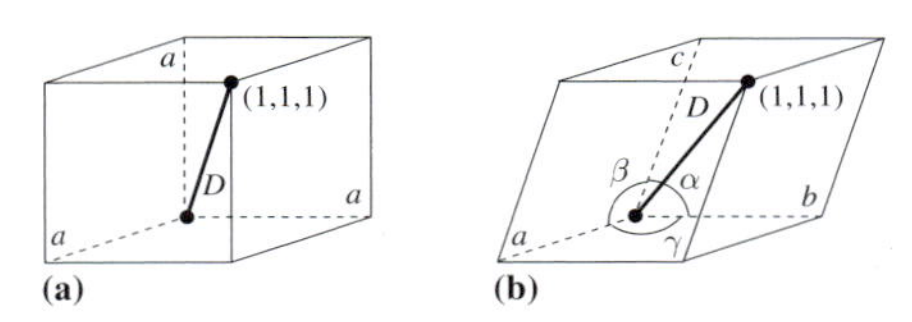

Fig. 4.2. The distance between (0, 0, 0) and (1, 1, 1) is readily computed in a Cartesian reference frame (a), but is a bit more complicated in an arbitrary reference frame (b).

an easy question to answer: simply use *Pythagoras's theorem* to find that $D = \sqrt{1^2 + 1^2 + 1^2} = \sqrt{3}$. For a cubic lattice with lattice parameter a, shown in Fig. 4.2(a), the distance is simply multiplied by a, i.e., $D = a\sqrt{3}$. For a triclinic lattice, however, the computation becomes a bit more complicated. One way to solve the problem would be to transform all triclinic coordinates into Cartesian coordinates, and then use the standard formula. This is possible but it can be quite cumbersome to find the actual coordinate transformation, and one would have to do this over and over again for different sets of lattice parameters. In this section, we will derive an alternative method which uses only the lattice parameters of the Bravais lattice, and does not rely on any other reference frames. For simplicity, we will first work out the answer for a two dimensional net, and then generalize the solution to three dimensional lattices.

The unit vectors $\{\mathbf{e}_x, \mathbf{e}_y\}$ form a Cartesian reference frame[1] (see Fig. 4.3) and the components of two arbitrary vectors $\mathbf{p}$ and $\mathbf{q}$ are given by (p_x, p_y) and (q_x, q_y), respectively. The distance between the points P and Q is given by the length of the vector $\mathbf{PQ}$, or $D = |\mathbf{q} - \mathbf{p}|$. In a Cartesian reference frame, we know that we can use the Pythagorean equation:

$$D = \sqrt{(q_x - p_x)^2 + (q_y - p_y)^2} \tag{4.1}$$

which for the points $P = (0, 1/2)$ and $Q = (1/2, 0)$ reduces to $D = 1/\sqrt{2}$.

If the reference frame is not orthogonal, then equation (4.1) for D is no longer correct. Let us consider the second reference frame (indicated by primes) $\{\mathbf{e}'_x, \mathbf{e}'_y\}$. Note that these vectors do not have unit length, and they

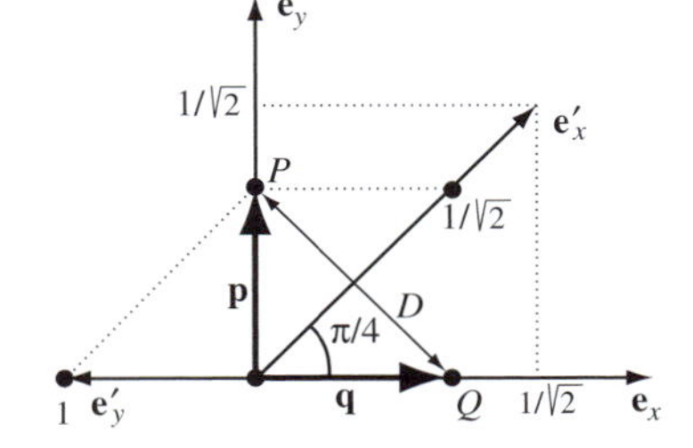

Fig. 4.3. Schematic of the Cartesian and non-Cartesian reference frames used for the example in the text.

[1] In this book, we will always denote the Cartesian basis vectors by symbols of the type $\mathbf{e}_x$, $\mathbf{e}_y, \ldots$ This notation is fully equivalent to the frequently used notation $\hat{\mathbf{i}}, \hat{\mathbf{j}}, \hat{\mathbf{k}}$. We will always use the following correspondence: $\mathbf{e}_x \equiv \hat{\mathbf{i}}$, $\mathbf{e}_y \equiv \hat{\mathbf{j}}$, $\mathbf{e}_z \equiv \hat{\mathbf{k}}$.

are not orthogonal to each other. The coordinates in the primed reference frame are given by $P = (1/\sqrt{2}, 1)$ and $Q = (0, -1)$, which would lead to the incorrect distance $D' = 3\sqrt{2}/2$ if the Cartesian equation were used. To get the correct answer, we must express the old basis vectors in terms of the new ones; we can see from the drawing that

$$\begin{aligned} \mathbf{e}_x &= -2\mathbf{e}'_y; \\ \mathbf{e}_y &= \sqrt{2}\mathbf{e}'_x + 2\mathbf{e}'_y. \end{aligned}$$

It is also easy to show, and we leave this as an exercise for the reader, that the vector components in the primed reference frame are given by:

$$\begin{aligned} p'_x &= \sqrt{2}\, p_y; \\ p'_y &= -2p_x + 2p_y. \end{aligned}$$

From elementary vector calculus we know that the length of a vector is also equal to the square root of the dot-product of that vector with itself (see Box 4.1):

$$|\mathbf{a}| = \sqrt{\mathbf{a} \cdot \mathbf{a}},$$

so we find that the distance D can be expressed quite generally as:

$$D = |\mathbf{q} - \mathbf{p}| = \sqrt{(\mathbf{q} - \mathbf{p}) \cdot (\mathbf{q} - \mathbf{p})}.$$

Box 4.1 The vector dot-product

The vector dot-product, $\mathbf{a} \cdot \mathbf{b}$, also known as the *scalar product*, is defined geometrically as the projection of the vector $\mathbf{a}$ onto the direction of $\mathbf{b}$, multiplied by the length of $\mathbf{b}$. Mathematically, this means that

$$\mathbf{a} \cdot \mathbf{b} = |\mathbf{a}|\,|\mathbf{b}| \cos\gamma.$$

The length of a vector $\mathbf{a}$ is then equal to the square root of the dot-product $\mathbf{a} \cdot \mathbf{a}$:

$$\mathbf{a} \cdot \mathbf{a} = |\mathbf{a}|\,|\mathbf{a}| \cos 0 \rightarrow |\mathbf{a}| = \sqrt{\mathbf{a} \cdot \mathbf{a}}.$$

We can write this expression explicitly in terms of the vector components and the basis vectors for both the Cartesian and the primed reference frames:

Cartesian

$$D = \sqrt{(q_x - p_x)^2 \mathbf{e}_x \cdot \mathbf{e}_x + (q_y - p_y)^2 \mathbf{e}_y \cdot \mathbf{e}_y + 2(q_x - p_x)(q_y - p_y)\mathbf{e}_x \cdot \mathbf{e}_y};$$

Primed reference frame

$$D' = \sqrt{(q'_x - p'_x)^2 \mathbf{e}'_x \cdot \mathbf{e}'_x + (q'_y - p'_y)^2 \mathbf{e}'_y \cdot \mathbf{e}'_y + 2(q'_x - p'_x)(q'_y - p'_y)\mathbf{e}'_x \cdot \mathbf{e}'_y}.$$

We thus find that the expressions for the distance between two points *require calculation of the dot-products between the basis vectors*. For the Cartesian reference frame, the dot-products are $\mathbf{e}_x \cdot \mathbf{e}_x = 1$, $\mathbf{e}_x \cdot \mathbf{e}_y = 0$ and $\mathbf{e}_y \cdot \mathbf{e}_y = 1$, so that the equation for the distance reduces to the standard Pythagorean expression:

$$D = \sqrt{(q_x - p_x)^2 + (q_y - p_y)^2}.$$

For the primed reference frame, we find from inspection of Fig. 4.3 that $\mathbf{e}'_x \cdot \mathbf{e}'_x = 1$, $\mathbf{e}'_x \cdot \mathbf{e}'_y = \cos \frac{3\pi}{4}/2 = -1/2\sqrt{2}$ and $\mathbf{e}'_y \cdot \mathbf{e}'_y = 1/4$ and, hence, the distance is expressed by

$$D' = \sqrt{(q'_x - p'_x)^2 + \frac{1}{4}(q'_y - p'_y)^2 - \frac{\sqrt{2}}{2}(q'_x - p'_x)(q'_y - p'_y)}.$$

Inserting the coordinates of the primed reference frame we find that $D = D' = 1/\sqrt{2}$, as it should be.

We conclude that, to measure or compute distances in a non-orthonormal reference frame, we need to know the dot-products between the basis vectors. In 3-D, this is still the case. For the three basis vectors **a**, **b**, and **c**, there are six terms, and in the following section we introduce a convenient shorthand notation for these products.

4.2.2 The metric tensor

For the 2-D example of the previous section, we can rewrite the equation for the distance between two points, which is now valid in *every* non-orthonormal reference frame, in terms of 2×2 matrices and column and row vectors:

$$D^2 = \begin{bmatrix} q_x - p_x & q_y - p_y \end{bmatrix} \begin{bmatrix} \mathbf{e}_x \cdot \mathbf{e}_x & \mathbf{e}_x \cdot \mathbf{e}_y \\ \mathbf{e}_y \cdot \mathbf{e}_x & \mathbf{e}_y \cdot \mathbf{e}_y \end{bmatrix} \begin{bmatrix} q_x - p_x \\ q_y - p_y \end{bmatrix}.$$

The 2×2 matrix in this expression contains the dot products and is symmetric with respect to the main diagonal. It is straightforward to work out all matrix multiplications. The explicit expressions for the two reference frames considered above are given by:

$$D^2 = \begin{bmatrix} q_x - p_x & q_y - p_y \end{bmatrix} \begin{bmatrix} 1 & 0 \\ 0 & 1 \end{bmatrix} \begin{bmatrix} q_x - p_x \\ q_y - p_y \end{bmatrix},$$

and

$$D'^2 = \begin{bmatrix} q'_x - p'_x & q'_y - p'_y \end{bmatrix} \begin{bmatrix} 1 & \frac{-1}{2\sqrt{2}} \\ \frac{-1}{2\sqrt{2}} & \frac{1}{4} \end{bmatrix} \begin{bmatrix} q'_x - p'_x \\ q'_y - p'_y \end{bmatrix}.$$

Note that we have separated the vector components from the quantities involving the basis vectors. The 2×2 matrix defines the geometrical characteristics of the reference frame and is important for all computations that involve the distance between two points. Since this matrix *defines* how distances are measured, it is known as the *metric matrix*, or, more commonly, the *metric tensor*. At this point in the book it is not necessary to understand exactly what a tensor is. For now, simply think of a tensor as a mathematical object, the components of which can be represented by a matrix, just as vector components can be represented by a row or column matrix. We will use square brackets to surround the components of the matrix whenever this matrix represents a tensor or a vector. Regular matrices will be surrounded by round brackets.

Next, we extend our discussion to the three-dimensional case. For a crystallographic reference frame with basis vectors $\mathbf{a}_i$, the metric tensor is represented by a 3×3 matrix. It is defined by:

$$g \equiv \begin{bmatrix} \mathbf{a}\cdot\mathbf{a} & \mathbf{a}\cdot\mathbf{b} & \mathbf{a}\cdot\mathbf{c} \\ \mathbf{b}\cdot\mathbf{a} & \mathbf{b}\cdot\mathbf{b} & \mathbf{b}\cdot\mathbf{c} \\ \mathbf{c}\cdot\mathbf{a} & \mathbf{c}\cdot\mathbf{b} & \mathbf{c}\cdot\mathbf{c} \end{bmatrix} = \begin{bmatrix} a^2 & ab\cos\gamma & ac\cos\beta \\ ba\cos\gamma & b^2 & bc\cos\alpha \\ ca\cos\beta & cb\cos\alpha & c^2 \end{bmatrix}.$$

The metric tensor contains the same information about the lattice as the lattice parameters, but in a form that is directly suited for geometric computations.

We can rewrite the expression for the distance squared between two points in a somewhat more compact notation:

$$D^2 = \sum_{i,j=1}^{N} (\mathbf{q}-\mathbf{p})_i g_{ij} (\mathbf{q}-\mathbf{p})_j,$$

where N is the dimensionality of the vector space (the cases $N = 2$ and $N = 3$ are sufficient for most of this book). In many situations it is not necessary to write the summation sign explicitly (since it is clear that a summation must be carried out) and we will drop the summation sign. In other words, we will use the *Einstein summation convention*, introduced in the previous chapter. The shorthand notation for the distance squared then becomes:

$$D^2 = (\mathbf{q}-\mathbf{p})_i g_{ij} (\mathbf{q}-\mathbf{p})_j.$$

Note that i is the row index of the matrix representation of g, therefore the vector components $(\mathbf{q}-\mathbf{p})_i$ must be written as a row when carrying out the matrix multiplication. Similarly, since j is the column index, the vector components $(\mathbf{q}-\mathbf{p})_j$ must be written as a column. In the 3-D case, we have for the dimensions of the matrices in the expression above: 1×3, 3×3, and 3×1. The result of the product is a 1×1 matrix, i.e., a scalar.

At this point, we can answer the question posed at the beginning of the previous section (page 80): given three basis vectors, what is the distance between the origin and the point with coordinates (1, 1, 1)? We can take the origin to be point $\mathbf{p}$ and $\mathbf{q} = (1, 1, 1)$, so that $\mathbf{q}-\mathbf{p} = (1, 1, 1)$; this leads to

$$\begin{aligned} D^2 &= [1 \quad 1 \quad 1] \begin{bmatrix} a^2 & ab\cos\gamma & ac\cos\beta \\ ba\cos\gamma & b^2 & bc\cos\alpha \\ ca\cos\beta & cb\cos\alpha & c^2 \end{bmatrix} \begin{bmatrix} 1 \\ 1 \\ 1 \end{bmatrix}; \\ &= a^2 + b^2 + c^2 + 2ab\cos\gamma + 2ac\cos\beta + 2bc\cos\alpha. \end{aligned} \tag{4.2}$$

The metric tensor thus provides us with a general way to measure the distance between two points in an arbitrary crystal lattice.

4.2.3 The dot-product in a crystallographic reference frame

We can also use the metric tensor to describe the dot-product between two arbitrary vectors. For a crystal, it is not always advisable to work in a Cartesian reference frame, and we need to use the full definition of the dot-product

Box 4.2 About the metric tensor

The metric tensor is an important concept: it allows for the description of the metric properties (i.e., how distances are measured) of any kind of space with any kind of coordinate system (orthogonal, curvilinear) in N dimensions and it simplifies vector and tensor operations in the most general coordinate frames. An important part of Einstein's *General Theory of Relativity* is concerned with the derivation of the metric tensor for a space containing a distribution of masses; the presence of mass is incorporated in the definition of the basis vectors, thereby introducing the concept of *curved* space. The crystallographic use of the metric tensor is restricted to use as a computational tool.

with the appropriate metric tensor for the crystallographic reference frame. The dot-product of two vectors is then defined as:

$$\mathbf{p}\cdot\mathbf{q}=\sum_{i,j=1}^{N}(p_i\mathbf{a}_i)\cdot(q_j\mathbf{a}_j)=\sum_{i,j=1}^{N}p_i g_{ij} q_j = p_i g_{ij} q_j, \tag{4.3}$$

where N is the dimension of the space in which the vectors are defined; in the last equality, the *Einstein summation convention* is used. Let us now discuss two important uses of the metric tensor: the Cartesian reference frame, and the crystallographic reference frames.

In a Cartesian reference frame the dot-product reduces to a very simple expression, because the metric tensor is equal to the identity matrix, i.e., a matrix with 1s along the diagonal and all other entries equal to 0. We introduce a new shorthand symbol for the identity matrix: δ_{ij}. This symbol is known as the *Kronecker delta*, and it is defined:

$$\delta_{ij}=\begin{cases}1 & i=j\\ 0 & i\neq j\end{cases} \tag{4.4}$$

The dot-product in a Cartesian reference frame is therefore:

$$\mathbf{p}\cdot\mathbf{q}=p_i\mathbf{e}_i\cdot\mathbf{e}_j q_j = p_i\delta_{ij}q_j = p_i q_i = p_1q_1+p_2q_2+\dots \tag{4.5}$$

Note that the summation $\delta_{ij}q_j = q_i$, is easily verified by explicitly writing down the individual terms. Only the term with $j = i$ contributes to the sum.

Since we know the general expressions for the lattice parameters of the seven crystal systems, we can readily write down the explicit expressions for the metric tensors of all crystal systems:

$$g_{\text{triclinic}}=\begin{bmatrix}a^2 & ab\cos\gamma & ac\cos\beta\\ ba\cos\gamma & b^2 & bc\cos\alpha\\ ca\cos\beta & cb\cos\alpha & c^2\end{bmatrix},\quad g_{\text{hexagonal}}=\begin{bmatrix}a^2 & -\frac{a^2}{2} & 0\\ -\frac{a^2}{2} & a^2 & 0\\ 0 & 0 & c^2\end{bmatrix},$$

$$g_{\text{monoclinic}}=\begin{bmatrix}a^2 & 0 & ac\cos\beta\\ 0 & b^2 & 0\\ ac\cos\beta & 0 & c^2\end{bmatrix},\quad g_{\text{tetragonal}}=\begin{bmatrix}a^2 & 0 & 0\\ 0 & a^2 & 0\\ 0 & 0 & c^2\end{bmatrix},$$

$$g_{\text{orthorhombic}}=\begin{bmatrix}a^2 & 0 & 0\\ 0 & b^2 & 0\\ 0 & 0 & c^2\end{bmatrix},\quad g_{\text{cubic}}=\begin{bmatrix}a^2 & 0 & 0\\ 0 & a^2 & 0\\ 0 & 0 & a^2\end{bmatrix},$$

$$g_{\text{rhombohedral}}=\begin{bmatrix}a^2 & a^2\cos\alpha & a^2\cos\alpha\\ a^2\cos\alpha & a^2 & a^2\cos\alpha\\ a^2\cos\alpha & a^2\cos\alpha & a^2\end{bmatrix}.$$

If the lattice parameters are known, then a simple substitution in one of these expressions results in the explicit metric tensor for that particular lattice. In the following section, we will use the metric tensor to compute the distance between two atoms and the angle between two directions in an arbitrary reference frame. All these quantities can be computed directly from Equation (4.3).

4.3 Worked examples

4.3.1 Computation of the length of a vector

Compute, for a crystal with lattice parameters $\{3, 4, 6, 90, 120, 90\}$ (i.e., a monoclinic crystal), the length of the main body diagonal.

Answer The main body diagonal is the line connecting the origin to the point $(1, 1, 1)$. The length of the vector $\mathbf{t} = \mathbf{a} + \mathbf{b} + \mathbf{c}$ is computed via the metric tensor:

$$g_{ij} = \begin{bmatrix} a^2 & ab\cos\gamma & ac\cos\beta \\ ab\cos\gamma & b^2 & bc\cos\alpha \\ ac\cos\beta & bc\cos\alpha & c^2 \end{bmatrix} = \begin{bmatrix} 9 & 0 & -9 \\ 0 & 16 & 0 \\ -9 & 0 & 36 \end{bmatrix}.$$

Using Equation (4.3) we find:

$$|\mathbf{t}| = \sqrt{[1\ \ 1\ \ 1]\begin{bmatrix} 9 & 0 & -9 \\ 0 & 16 & 0 \\ -9 & 0 & 36 \end{bmatrix}\begin{bmatrix} 1 \\ 1 \\ 1 \end{bmatrix}};$$

$$= \sqrt{[1\ \ 1\ \ 1]\begin{bmatrix} 0 \\ 16 \\ 27 \end{bmatrix}} = \sqrt{43} = 6.557.$$

4.3.2 Computation of the distance between two atoms

A crystal with lattice parameters $\{2, 2, 3, 90, 90, 90\}$ contains, among others, atoms at the positions $(1/2, 1/3, 1/4)$ and $(1/3, 1/2, 3/4)$. Compute the distance between these two atoms.

Answer First we compute the metric tensor for this tetragonal crystal system:

$$g_{ij}^{\text{tetragonal}} = \begin{bmatrix} a^2 & 0 & 0 \\ 0 & b^2 & 0 \\ 0 & 0 & c^2 \end{bmatrix} = \begin{bmatrix} 4 & 0 & 0 \\ 0 & 4 & 0 \\ 0 & 0 & 9 \end{bmatrix}.$$

The distance between any two points in a crystal equals the length of the vector connecting those two points; in this case $\mathbf{r} = (1/3 - 1/2, 1/2 - 1/3,$

$3/4 - 1/4) = (-1/6, 1/6, 1/2)$. The length of a vector is equal to the square root of the dot-product of that vector with itself, so that:

$$
\begin{aligned}
|\mathbf{r}| &= \sqrt{\mathbf{r}\cdot\mathbf{r}} = \sqrt{r_i g_{ij} r_j};\\
&= \sqrt{\left[-\frac{1}{6}\ \ \frac{1}{6}\ \ \frac{1}{2}\right]\begin{bmatrix}4 & 0 & 0\\ 0 & 4 & 0\\ 0 & 0 & 9\end{bmatrix}\begin{bmatrix}-\frac{1}{6}\\ \frac{1}{6}\\ \frac{1}{2}\end{bmatrix}};\\
&= \sqrt{\left[-\frac{1}{6}\ \ \frac{1}{6}\ \ \frac{1}{2}\right]\begin{bmatrix}-\frac{2}{3}\\ \frac{2}{3}\\ \frac{9}{2}\end{bmatrix}} = \frac{\sqrt{89}}{6} = 1.572.
\end{aligned}
$$

The lattice parameters have units (usually nanometers or Ångströms; 1 nm = 1 Å), so if a, b, and c are given in nanometers, then the distance is also expressed in nanometers. We will use the nanometer as the basic unit for distances in crystals.

4.3.3 Computation of the angle between atomic bonds

In a cubic crystal with lattice parameter a, an oxygen atom is present at the position $(0, 0, 0)$; this atom is bonded to two titanium atoms, located at the positions $(1/2, 1/2, 0)$ and $(1/2, 0, 1/2)$. Compute the angle between these two bonds.

Answer The angle between two bonds corresponds to the angle between the two direction vectors parallel to those bonds; in this case, the direction vectors are $\mathbf{s} = (1/2, 1/2, 0)$ and $\mathbf{t} = (1/2, 0, 1/2)$. Since the crystal is cubic, the metric tensor is given by:

$$
g_{ij}^{\text{cubic}} = \begin{bmatrix}a^2 & 0 & 0\\ 0 & a^2 & 0\\ 0 & 0 & a^2\end{bmatrix} = a^2 \times \begin{bmatrix}1 & 0 & 0\\ 0 & 1 & 0\\ 0 & 0 & 1\end{bmatrix} = a^2\delta_{ij}.
$$

The cosine of the angle between two directions is computed via the normalized dot-product:

$$
\mathbf{s}\cdot\mathbf{t} = |\mathbf{s}|\,|\mathbf{t}|\cos\theta,
$$

or

$$
\cos\theta = \frac{s_i g_{ij} t_j}{\sqrt{s_i g_{ij} s_j}\sqrt{t_i g_{ij} t_j}} = \frac{a^2 s_i \delta_{ij} t_j}{\sqrt{a^2 s_i \delta_{ij} s_j}\sqrt{a^2 t_i \delta_{ij} t_j}} = \frac{s_i t_i}{\sqrt{s_i s_i}\sqrt{t_i t_i}}.
$$

This leads to:

$$\cos\theta = \frac{[\tfrac{1}{2}\ \tfrac{1}{2}\ 0]\begin{bmatrix}\tfrac{1}{2}\\0\\\tfrac{1}{2}\end{bmatrix}}{\sqrt{[\tfrac{1}{2}\ \tfrac{1}{2}\ 0]\begin{bmatrix}\tfrac{1}{2}\\\tfrac{1}{2}\\0\end{bmatrix}}\sqrt{[\tfrac{1}{2}\ 0\ \tfrac{1}{2}]\begin{bmatrix}\tfrac{1}{2}\\0\\\tfrac{1}{2}\end{bmatrix}}} = \frac{1}{2},$$

and hence $\theta = 60°$.

4.3.4 Computation of the angle between lattice directions

Consider a monoclinic crystal with lattice parameters $a = 4$ nm, $b = 6$ nm, $c = 5$ nm and $\beta = 120°$. What is the angle between the [101] and $[\bar{2}01]$ directions ?

Answer First we derive the metric tensor for this crystal system:

$$g_{ij}^{\text{monoclinic}} = \begin{bmatrix} a^2 & 0 & ac\cos\beta \\ 0 & b^2 & 0 \\ ac\cos\beta & 0 & c^2 \end{bmatrix} = \begin{bmatrix} 16 & 0 & -10 \\ 0 & 36 & 0 \\ -10 & 0 & 25 \end{bmatrix}.$$

The dot product is then obtained from the product $p_i g_{ij} q_j$ or, explicitly:

$$\mathbf{t}_{[101]} \cdot \mathbf{t}_{[\bar{2}01]} = [101]\begin{bmatrix} 16 & 0 & -10 \\ 0 & 36 & 0 \\ -10 & 0 & 25 \end{bmatrix}\begin{bmatrix}-2\\0\\1\end{bmatrix} = [101]\begin{bmatrix}-42\\0\\45\end{bmatrix} = 3.$$

The dot-product of two vectors is also equal to the product of the lengths of the vectors times the cosine of the angle between them:

$$\mathbf{t}_{[101]} \cdot \mathbf{t}_{[\bar{2}01]} = t_{[101]} t_{[\bar{2}01]} \cos\theta.$$

The length of a vector is the square root of the dot-product of a vector with itself, hence we can use the metric tensor again to compute the lengths of $\mathbf{t}_{[101]}$ and $\mathbf{t}_{[\bar{2}01]}$:

$$t_{[101]}^2 = [101]\begin{bmatrix} 16 & 0 & -10 \\ 0 & 36 & 0 \\ -10 & 0 & 25 \end{bmatrix}\begin{bmatrix}1\\0\\1\end{bmatrix} = [101]\begin{bmatrix}6\\0\\15\end{bmatrix} = 21;$$

$$t_{[\bar{2}01]}^2 = [-201]\begin{bmatrix} 16 & 0 & -10 \\ 0 & 36 & 0 \\ -10 & 0 & 25 \end{bmatrix}\begin{bmatrix}-2\\0\\1\end{bmatrix} = [-201]\begin{bmatrix}-42\\0\\45\end{bmatrix} = 129.$$

The angle between the two vectors is therefore:

$$\theta = \cos^{-1}\left(\frac{\mathbf{t}_{[101]}\cdot\mathbf{t}_{[\bar{2}01]}}{t_{[101]}t_{[\bar{2}01]}}\right) = \cos^{-1}\left(\frac{3}{\sqrt{129\times 21}}\right) = 86.69^\circ.$$

4.3.5 An alternative method for the computation of angles

The angle between two direct space vectors can be computed in a single operation, instead of using the three individual dot products described in the previous example. In this section, we present an alternative procedure for the computation of the angle θ based on a 2×3 matrix containing the two vectors $\mathbf{p}$ and $\mathbf{q}$.

Consider the following formal relation:

$$\begin{pmatrix}\mathbf{p}\\ \mathbf{q}\end{pmatrix}\cdot(\mathbf{p}\ \mathbf{q}) = \begin{pmatrix}\mathbf{p}\cdot\mathbf{p} & \mathbf{p}\cdot\mathbf{q}\\ \mathbf{q}\cdot\mathbf{p} & \mathbf{q}\cdot\mathbf{q}\end{pmatrix}.$$

The resulting 2×2 matrix contains all three dot-products needed for the computation of the angle θ, and only one set of matrix multiplications is needed. We can apply this shortcut to the previous example:

$$\begin{pmatrix}1 & 0 & 1\\ -2 & 0 & 1\end{pmatrix}\begin{bmatrix}16 & 0 & -10\\ 0 & 36 & 0\\ -10 & 0 & 25\end{bmatrix}\begin{pmatrix}1 & -2\\ 0 & 1\\ 1 & 1\end{pmatrix} = \begin{pmatrix}21 & 3\\ 3 & 129\end{pmatrix},$$

from which we find the same angle of $\theta = 86.69^\circ$.

4.3.6 Further comments

We have seen that we can use the metric tensor to compute distances between lattice points and angles between lattice vectors. It is important that you familiarize yourself with this kind of computation; many computer programs use the metric tensor concept for crystallographic computations and, with a minor effort and some knowledge of computer programming, you should be able to implement these equations yourself.

Many textbooks on crystallography do not mention the metric tensor at all. It is possible to do all crystallographic computations without the metric tensor, but then one needs to have the explicit equations for the length of a vector and the angles between vectors for all crystal systems. It is sometimes useful to do this in order to appreciate the geometry, but for purely computational efficiency, the metric tensor is the preferred tool. For completeness, we list all the relevant equations for the length of a vector (Table 4.1) and the angles between two vectors (Table 4.2) for the seven crystal systems. The values for the length of a vector are denoted by the symbol ${}^{s}l$ where s stands for the crystal system. Use the appropriate values from Table 4.1 for the

Table 4.1. Expressions for the length l of a vector $[uvw]$ in the seven crystal systems.

System	l	Expression
Cubic	${}^{c}l$	$a\left(u^2+v^2+w^2\right)^{1/2}$
Tetragonal	${}^{t}l$	$\left(a^2(u^2+v^2)+c^2w^2\right)^{1/2}$
Orthorhombic	${}^{o}l$	$\left(a^2u^2+b^2v^2+c^2w^2\right)^{1/2}$
Hexagonal	${}^{h}l$	$\left(a^2(u^2+v^2-uv)+c^2w^2\right)^{1/2}$
Rhombohedral	${}^{r}l$	$a\left(u^2+v^2+w^2+2\cos\alpha\,[uv+uw+vw]\right)^{1/2}$
Monoclinic	${}^{m}l$	$\left(a^2u^2+b^2v^2+c^2w^2+2acuw\cos\beta\right)^{1/2}$
Triclinic	${}^{a}l$	$\left(a^2u^2+b^2v^2+c^2w^2+2bcvw\cos\alpha\right.$ $\left.+2acuw\cos\beta+2abuv\cos\gamma\right)^{1/2}$

Table 4.2. Expressions for the cosine of the angle θ between two vectors $[u_1v_1w_1]$ and $[u_2v_2w_2]$ in the seven crystal systems. The quantities l_1 and l_2 should be taken from Table 4.1 for the appropriate crystal system, with $[uvw]$ substituted by $[u_1v_1w_1]$ or $[u_2v_2w_2]$.

System	$l_1 \times l_2 \times \cos\theta$
Cubic	$a^2(u_1u_2+v_1v_2+w_1w_2)$
Tetragonal	$a^2(u_1u_2+v_1v_2)+c^2w_1w_2$
Orthorhombic	$a^2u_1u_2+b^2v_1v_2+c^2w_1w_2$
Hexagonal	$a^2(u_1u_2+v_1v_2-\frac{1}{2}(u_1v_2+v_1u_2))+c^2w_1w_2$
Rhombohedral	$a^2(u_1u_2+v_1v_2+w_1w_2$ $+\cos\alpha\,[u_1(v_2+w_2)+v_1(u_2+w_2)+w_1(u_2+v_2)])$
Monoclinic	$a^2u_1u_2+b^2v_1v_2+c^2w_1w_2+ac(w_1u_2+u_1w_2)\cos\beta$
Triclinic	$a^2u_1u_2+b^2v_1v_2+c^2w_1w_2+bc(v_1w_2+v_2w_1)\cos\alpha$ $+ac(u_1w_2+u_2w_1)\cos\beta+ab(u_1v_2+u_2v_1)\cos\gamma$

denominators of the expressions in Table 4.2. It should be noted, however, that it is much more efficient to implement the metric tensor equations in a computer program, because in that case there is only one equation instead of two sets of seven rather complicated relations.

4.4 Historical notes

In the seventeenth century, the work of **Nicolaus Steno** (1638–86, Fig. 4.4(a)) represented an important early contribution to the field of crystallography (Steno, 1669). Nicolaus Steno (whose name was Latinized from Niels Stensen) was a Danish scientist and physician (and later a priest) born in Copenhagen, Denmark, in 1638. He argued that crystals were formed by the accretion of congruent units. Steno studied quartz crystals and noted that, despite

Fig. 4.4. (a) Nicolaus Steno (1638–1686), and (b) Jean Baptiste Louis Romé de l'Isle (1736–1790) (pictures courtesy of J. Lima-de-Faria).

differences in size, origin, or habit, the angles between corresponding faces were constant. **Domenico Guglielmini** (1655–1710), an Italian physician and mathematician, restated Steno's law of the constancy of interfacial angles and applied it to other crystals, such as potassium nitrate (nitre), sodium chloride (common salt), alum, and blue vitriol (Guglielmini, 1688, 1705). He proposed four basic forms for salt particles (cube, hexagonal prism, rhombohedron, and octahedron) and his work can be considered to be the first geometrical theory of crystal structure.

The law of constancy of interfacial angles was later restated by **Jean Baptiste Louis Romé de l'Isle** (1736–90, Fig. 4.4(b)) after studying a variety of crystals and became the important central tenet in geometrical crystallography. Steno proposed that crystals were built by aggregation of very small particles and that crystals grew from solution by successive addition of particle layers. He opposed the early view of vegetative growth of crystals. Romé de l'Isle was a French scientist who made important contributions to the field of crystallography (Romé de L'Isle, 1772, 1783). In 1783, he used a contact goniometer developed by his student **Arnould Carangeot** (1742–1806), to make angular measurements on crystals (Carangeot, 1783), confirming Steno's earlier work on quartz. He formulated the *law of constancy of interfacial angles*, now known as the first law of crystal habit. He taught the first course in crystallography in Paris in about 1783, which formally defined crystallography as a new science. Romé de l'Isle was the first to describe the geometric importance of twins. He also determined six primitive crystal forms and showed that they could be modified to produce secondary forms. His work was very influential in the shaping of the field of crystallography.

Ruggero Giuseppe Boscovich (1711–87), an Italian scientist, proposed the extraordinarily perceptive notion that crystals were formed of points linked by attractive and repulsive forces (Boscovich, 1758). This new concept of *point charges* was well ahead of its time. He viewed atoms as being replaced by a discrete set of points (poles) whose positions were determined by a balance of attractive and repulsive forces. Space could be partitioned into regions that surrounded the discrete points. **Tobers Olof Bergman** (1735–84), a Swedish chemist and mineralogist, deduced the shape of crystal faces using geometrical constructions which were later stated mathematically by Haüy in his law of simple rational intercepts (Bergman, 1784, 1773). He published his construction of a *scalenohedron* built from a rhombohedral nucleus by superposition of lamellae (rhombuses).

Johannes Carl Gehler (1732–96) was a German crystallographer and a proponent of the importance of the external characteristics of crystals as tenets of geometrical crystallography (Gehler, 1757). **Abraham Gottlob Werner** (1750–1817) wrote *Von den Ausserlichen Kennzeichen der Fossilien* ("On the External Character of Minerals" Werner, 1774) which was translated into several languages. This was essentially a translation of the work of Gehler, but with his own new ideas added in. He based his classification on practical considerations such as color, cohesion, external shape, luster, fracture, transparency, hardness, specific weight, etc. Many of these classifications are used in geological field books today. His initial work in mineralogy evolved into his seminal efforts in founding the modern field of geology. Werner stated that the morphology of crystals was based on seven (rather than the pervasive four) primary forms: regular dodecahedron, icosahedron, hexahedron, prism, pyramid, plate, and lens. He organized one of the first courses on mineralogy at Freiberg, in about 1775 (Werner, 1775). Several of his students, including **Christian Samuel Weiss** and **Friedrich Mohs**, also made significant contributions (Mohs, 1822).

4.5 Problems

(i) *Directions*: Draw the following direction vectors in a cubic unit cell : [110], [112] and $[\bar{3}2\bar{1}]$. Then repeat the question for an orthorhombic lattice with lattice parameters $\{2, 3, 4, 90, 90, 90\}$.

(ii) *Lattice geometry I*: Consider vectors $\mathbf{q} = (q_1, q_2, q_3)$ and $\mathbf{p} = (p_1, p_2, p_3)$ each pointing to an atom. Show that in the triclinic crystal system the square distance between the atoms is given by:

$$\begin{aligned} D^2 =& (q_1 - p_1)^2 a^2 + (q_2 - p_2)^2 b^2 + (q_3 - p_3)^2 c^2 \\ &+ 2(q_1 - p_1)(q_2 - p_2)ab\cos\gamma + 2(q_1 - p_1)(q_3 - p_3)ac\cos\beta \\ &+ 2(q_2 - p_2)(q_3 - p_3)bc\cos\alpha. \end{aligned}$$

Show what this reduces to in the monoclinic system.

(iii) *Lattice geometry II*: Compute, for a crystal with lattice parameters $\{2, 2, 6, 90, 90, 120\}$ (i.e., a hexagonal crystal):

(a) The length of the main body diagonal;
(b) The length of the basal plane diagonal.

(iv) *Lattice geometry III*: Show that for a hexagonal lattice, the length of a vector with components (u, v, w) is given by:

$$\left(a^2(u^2+v^2-uv)+c^2w^2\right)^{1/2}.$$

(v) *Lattice geometry IV*: Show, using the metric tensor, that for a rhombohedral lattice, the angle between two vectors $[u_1v_1w_1]$ and $[u_2v_2w_2]$ is given by the inverse cosine of the following expression:

$$\frac{(u_1u_2+v_1v_2+w_1w_2+\cos\alpha\,[u_1(v_2+w_2)+v_1(u_2+w_2)+w_1(u_2+v_2)])}{l_1\times l_2},$$

where

$$l_i=\left(u_i^2+v_i^2+w_i^2+2\cos\alpha\,[u_iv_i+u_iw_i+v_iw_i]\right)^{1/2}.$$

(vi) *Lattice geometry V*: Write down the metric tensor for a tetragonal lattice with lattice parameters $a=0.2$ nm and $c/a=1.5$.

(a) Compute, using the metric tensor, the distance between the origin and the body center, $(1/2, 1/2, 1/2)$.
(b) Compute, again using the metric tensor, the angle between the directions [100] and [111].

(vii) *Lattice geometry VI*: What is the angle between the **a** axis and the direction [221] in a monoclinic lattice with lattice parameters $\{1, 3, 2, 90, 45, 90\}$?

(viii) *Lattice geometry VII*: A triclinic lattice has lattice parameters $\{1, 2, 3, 45, 60, 90\}$.

(a) What is the distance between the center of the cell and the point with coordinates $(1, 1, 1)$?
(b) What is the angle between the **b** axis and the [112] direction?

(ix) *Lattice geometry VIII*: A monoclinic lattice has lattice parameters $\{1, 3, 2, 90, 45, 90\}$.

(a) What is the distance between the origin and the point with coordinates $(1, 1, 1)$?
(b) What is the angle between the **a** axis and the direction [221]?
(c) There are three atoms in this unit cell, with fractional coordinates $\mathbf{r}_1=(0, 1/2, 0)$, $\mathbf{r}_2=(1/2, 0, 0)$, and $\mathbf{r}_3=(1/2, 1/2, 1/2)$. What is the angle between the bonds $\mathbf{r}_2\leftrightarrow\mathbf{r}_1$ and $\mathbf{r}_2\leftrightarrow\mathbf{r}_3$?

(x) *Lattice geometry IX*: A primitive orthorhombic crystal has $\{2, b=3, 4, 90, 90, 90\}$ as lattice parameters. Compute the following quantities, using the metric tensor formalism:

(a) The distance between the origin and the point with fractional coordinates $(1, 1, 0)$;
(b) The angle between the [110] and [101] directions;

(c) For which value of the lattice parameter b above, is the angle between [100] and [111] equal to 60°?

(xi) *Lattice geometry X:* For this problem consider a rhombohedral crystal for which the angle between the **a** and **b** axes is $2\pi/6$.

(a) Write an expression for the lattice constants of a rhombohedral crystal.
(b) Derive the general form of the metric tensor for a rhombohedral crystal.
(c) Determine the general length of the $[uvw]$ vector in this crystal system.
(d) Determine the projection of a [110] vector onto a [111] vector (as a function of the lattice parameters).
(e) Determine the bond angle between atoms in the positions $(0, 0, 0)$, $(1, 1, 0)$ and $(1, 1, 1)$.

(xii) *Diamond cubic structure*: The diamond cubic structure is a crystal structure adopted by C and many semiconducting materials such as Si, Ge, etc. It is illustrated in Chapter 17. The diamond structure has the cF Bravais lattice with C atoms at the origin (0,0,0) and at $(1/4, 1/4, 1/4)$, the diamond site. All the C atoms in the structure are 4-fold tetrahedrally coordinated by other C atoms.

(a) Express the direction of a bond between a C atom at the origin and the diamond site.
(b) Express the direction of a bond between a C atom at the origin and the **C** face center.
(c) Use the metric tensor to calculate the bond angle between the previous two bonds.

(xiii) β-Sn : β-Sn is the high-temperature polymorph of tin, stable at temperatures above 286.4 K. It assumes the tI (body centered tetragonal) Bravais lattice with lattice constants $a = 0.58315$ nm and $c = 0.31814$ nm. Because of its tetragonal structure, as opposed to the diamond cubic structure of other group IV elements, β-Sn is a semi-metal (near metal) rather than a semiconductor.

(a) Compute, using the metric tensor, the distance between the origin and the body center, $(1/2, 1/2, 1/2)$.
(b) Determine the angle between the directions [100] and [111].

(xiv) *Crystallographic computations*: Write Mathematica or MATLAB scripts (functions) for the following operations (all in an arbitrary Bravais lattice):

- compute the metric tensor, with as input the six lattice parameters;

- write a function for the vector dot-product (the function should take two vectors and the metric tensor as input, and return the scalar product);
- write a similar function to compute the length of a vector (i.e., the distance between two points);
- write a function to compute the angle between two vectors.

CHAPTER

5 Lattice planes

"The description of right lines and circles, upon which geometry is founded, belongs to mechanics. Geometry does not teach us to draw these lines, but requires them to be drawn."

Sir Isaac Newton

5.1 Miller indices

In the previous chapters, we have seen how *directions* in a crystal lattice can be labeled, and how we can compute the distance between points, and the angle between lattice directions. What about planes? Figure 5.1 shows $2 \times 2 \times 2$ unit cells of the cF Bravais lattice. In (a), the central horizontal plane of lattice sites is highlighted in gray. In (b), a different plane is highlighted. We can take any three non-collinear lattice points, and create a plane through those points. Such a plane is known as a *lattice plane*.

One way to identify a lattice plane would be to write down its algebraic equation. This is relatively straightforward in a Cartesian reference frame but it becomes tedious in other systems, when the coordinate axes are not at right angles. We must, therefore, look for a method which is valid for all Bravais lattices, regardless of the lattice parameters.

Consider the drawing in Fig. 5.2. This is a portion of a crystal described by the basis vectors **a**, **b**, and **c**. Several faces of the crystal are outlined by solid lines in (a). The reference frame is a general Bravais reference frame. In (b), the largest face is extended in its plane, until the plane intersects the three reference axes. The intersection points are indicated by solid circles.

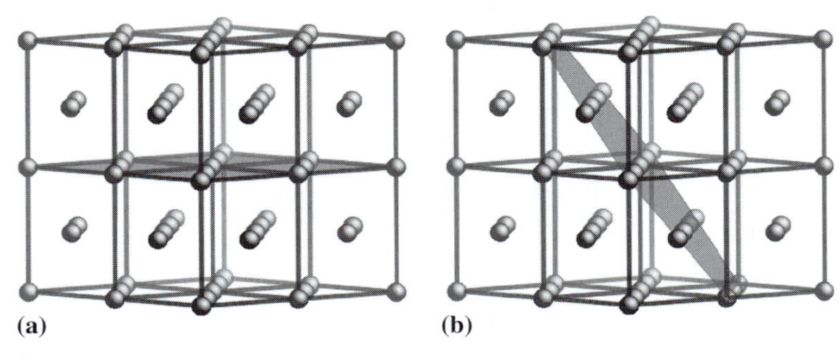

Fig. 5.1. 2 × 2 × 2 unit cells of the *cF* Bravais lattice: (a) and (b) show two different lattice planes (in gray).

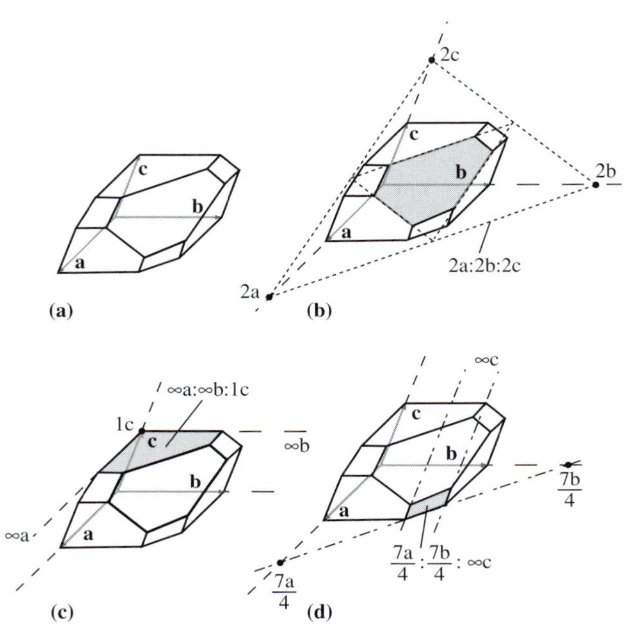

Fig. 5.2. Illustration of the labeling of individual crystal faces: (a) shows the original crystal, while (b), (c), and (d) show the determination of the intersection of a plane with the three basis directions.

The distance between each intersection point and the origin is also indicated. We can take these three numbers, $(2a, 2b, 2c)$, and construct a symbolic representation of the plane: $2a : 2b : 2c$. Similarly, when the plane is parallel to one (d) or more (c) of the axes, we determine the intersections to be $\infty a : \infty b : 1c$ for (c) and $7a/4 : 7b/4 : \infty c$ for (d).

This is still a rather cumbersome way to label the planes, so let us try to simplify the notation a little. First of all, we will measure the intercepts in units of the corresponding basis vector length. This means that we divide the first intercept by a, the second by b, and the third by c. The resulting symbols are then: $2 : 2 : 2$, $\infty : \infty : 1$, and $7/4 : 7/4 : \infty$, respectively. Next, we attempt to get rid of the factors of infinity. We know that $1/\infty = 0$, so we can take the inverse of all the numbers, resulting in: $1/2 : 1/2 : 1/2$, $0 : 0 : 1$, and $4/7 : 4/7 : 0$, respectively. Finally, it would be nice if we could have integers to describe planes, just like we have the integers $[uvw]$ for directions.

So, we reduce all the fractions above to the smallest integers. This leads to $1:1:1$, $0:0:1$, and $1:1:0$, respectively. These triplets of integers are known as the *Miller indices* of the corresponding planes. They are written between parentheses without commas, as in (111), (001), and (110), respectively. If an intercept is negative, then we write the minus sign above the corresponding index, just as we did for direction indices. The general notation is typically written as (hkl).

Let us now summarize the procedure for obtaining the Miller indices of a lattice plane:

(i) If the plane goes through the origin, then displace it parallel to itself so that it no longer contains the origin.
(ii) Determine the intercepts of the plane with the three basis vectors. Call those intercepts s_1, s_2, and s_3. The intercepts must be measured in units of the basis vector length. If a plane is parallel to one or more of the basis vectors, then the corresponding intercept value(s) must be taken as ∞.
(iii) Invert all three intercepts. If one of the intercepts is ∞, then the corresponding number is zero.
(iv) Reduce the three numbers to the smallest possible integers (relative primes).
(v) Write the three numbers surrounded by round parentheses, i.e., (123).

Figure 5.3 shows three different examples of planes in three different Bravais lattices. Note that the Miller indices of the plane parallel to both the **a** and **b** axes are always given by (001), regardless of the Bravais lattice type. This independence of the reference frame turns Miller indices into very useful numbers.

The indices of the planes in Fig. 5.3 are derived as follows:

(a) The intercepts are given by (1, 1, 1); the reciprocal values are obviously 1, 1, and 1 so that the Miller indices of this plane are (111).
(b) This plane intercepts the **a** axis at 1, the **b** axis at $-1/2$ and the **c** axis at $-1/2$. The reciprocal values are 1, -2, and -2 which leads to the Miller indices $(1\bar{2}\bar{2})$.
(c) This plane is parallel to the **a** and **b** axes and intercepts the **c** axis at 1. The reciprocals of the intercept values are $(1/\infty, 1/\infty, 1) = (0, 0, 1)$ and hence the Miller indices are (001).

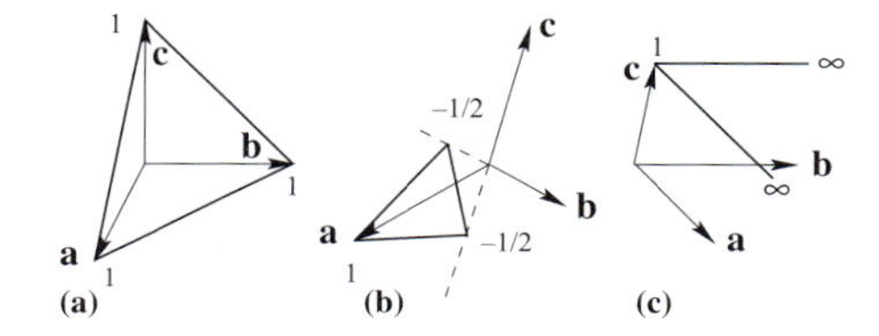

Fig. 5.3. Three examples of planes in different lattices, discussed in the text.

Note that if a plane is translated parallel to itself, the three intercept values are increased or decreased proportionally. Hence, a plane with Miller indices $(nh\,nk\,nl)$ with n integer is parallel to the plane (hkl). This will become important when we talk about X-ray diffraction in Chapter 11.

5.2 Families of planes and directions

The Miller indices can be used to describe planes in all seven crystal systems. Let us take a closer look at the (110) plane in a cubic unit cell. Figure 5.4(a) is a drawing of this plane. From (b) it is clear that the $(1\bar{1}0)$ plane is a similar plane, i.e., it also cuts diagonally through opposite edges of the cube. As a matter of fact, by permuting the indices 1, 1, and 0 and their negatives in the Miller symbol (110) we can generate five other planes $(1\bar{1}0)$, (101), $(10\bar{1})$, (011), and $(01\bar{1})$, as shown in Figure 5.4(c). Each of those planes cuts diagonally through the cube and hence they are *equivalent*. We will see later on, in Chapter 8, that they are related to each other by a symmetry operator. Planes that are related to each other by symmetry form a *family* of planes. A family of planes is denoted by curly braces, i.e., the family of planes equivalent to (110) is denoted by {110}. In general, the family of the plane (hkl) is $\{hkl\}$. The external shape of a crystal often consists of planes belonging to one or more families; families that make up the external shape of a crystal are known as *forms*. In Section 5.4, we will introduce the possible crystal forms.

It is important to realize that the number of planes belonging to a certain family is determined by the crystal system (or, more precisely, by the symmetry). In the cubic unit cell shown in Figure 5.4(a)–(c), all planes of the type (110) (including permutations and negative values of the indices) are

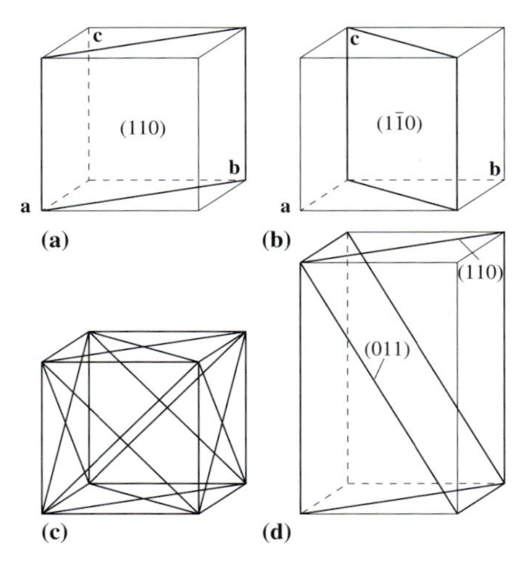

Fig. 5.4. Equivalence of the {110} planes in a cubic crystal; in (d) the lattice is tetragonally distorted, and the (110) and (101) planes are no longer equivalent.

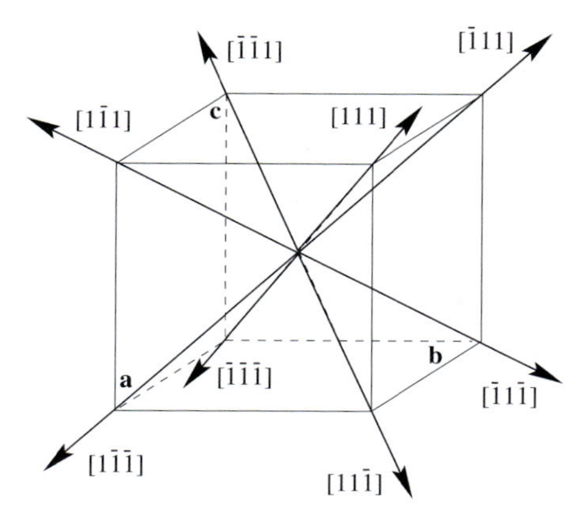

Fig. 5.5. The eight $\langle 111 \rangle$ directions in the cubic crystal system.

equivalent and hence belong to the family $\{110\}$. However, if the unit cell is distorted along the **c** axis (in other words, the unit cell becomes tetragonal) then the (110) and (011) planes are no longer equivalent, since the angles of the plane with respect to the three basis vectors are different. In this case, there are two families, $\{110\}$, consisting of (110), and ($1\bar{1}0$), and $\{011\}$, consisting of (011), ($0\bar{1}1$), (101), and ($\bar{1}01$). When talking about a family of planes one must therefore mention the crystal system. The number of planes in a family $\{hkl\}$ is called the *multiplicity* of the plane (hkl) and is usually denoted by p_{hkl}. Multiplicities are integer numbers and can vary from 1 to 48 (in 3-D). We will return to the concept of multiplicity in Chapter 8.

Directions also belong to families. The example in Figure 5.5 shows the directions of the type [111] in a cubic unit cell. All eight directions are equivalent (i.e., the cube looks exactly the same when viewed from any one of those directions) and, hence, they belong to the *family of directions* $\langle 111 \rangle$. In general, a family of directions is denoted by $\langle uvw \rangle$; the number of directions in a family, its *multiplicity*, again depends on the lattice type and symmetry and can vary from 1 to 48.

5.3 Special case: the hexagonal system

We have seen in the previous section that, for a cubic system, we can list all the members of a family $\{hkl\}$ by writing down all the permutations of the three numbers h, k, and l and their negatives. If the symmetry of the system is lower than cubic, then the members of a family are still given by permutations, but not all permutations belong to the same family. For instance, in the rhombohedral system we have $\{100\} = \{(100), (\bar{1}00), (010), (0\bar{1}0), (001), (00\bar{1})\}$ as a family. In the orthorhombic system $\{100\} = \{(100), (\bar{1}00)\}$ contains just two elements. The only exception to this rule of index permutations is the *hexagonal crystal system*.

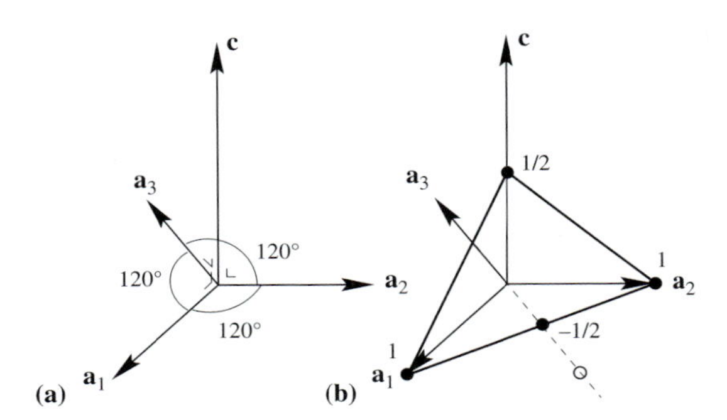

Fig. 5.6. (a) Schematic representation of the four basis vectors used to describe the hexagonal crystal system; (b) outline of the (112) or $(11\bar{2}2)$ plane.

The hexagonal system is conveniently described by *four* basis vectors, three of which are coplanar. In other words, those three vectors are not linearly independent. The vectors are chosen as $\mathbf{a}_1$, $\mathbf{a}_2$, and $\mathbf{a}_3$, shown in Figure 5.6(a). The angle between any two vectors is 120°. One defines the *Miller–Bravais indices* of a plane as the relative primes corresponding to the reciprocals of the intercepts of that plane with *all four axes*. For the plane shown in Figure 5.6(b), the intercepts on the four axes are $(1, 1, -1/2, 1/2)$. Inverting these numbers we find the indices $(11\bar{2}2)$. Since three indices are sufficient to identify any plane in 3-D space, there must be a relation between the four indices. If we denote the general Miller–Bravais index as $(hkil)$, with i the extra index corresponding to the vector $\mathbf{a}_3$, then it is easy to show (reader exercise) that the following relation must hold:

$$i = -(h+k). \tag{5.1}$$

Since the third index is not really necessary to unambiguously determine the plane, it is often not written explicitly, but represented by a period (dot), i.e., $(hkil) = (hk.l)$. The third index *is* useful to determine the members of a family in the hexagonal system. The members of the family $\{11\bar{2}0\}$ can be derived by permuting the first three indices including negative values and results in $\{11\bar{2}0\} = \{(11\bar{2}0), (\bar{1}2\bar{1}0), (\bar{2}110), (\bar{1}\bar{1}20), (1\bar{2}10), (2\bar{1}\bar{1}0)\}$. If we had used the standard three-index notation, then the family members would be: $\{110\} = \{(110), (\bar{1}20), (\bar{2}10), (\bar{1}\bar{1}0), (1\bar{2}0), (2\bar{1}0)\}$. It is not a-priori clear how one could write down these members without the four-index system, since they are clearly not permutations of $\{110\}$.

For directions, the situation becomes even more complicated. Direction indices in the Miller–Bravais notation are described by the symbol $[uvtw]$, where t is an additional index corresponding to the vector $\mathbf{a}_3$. Since a direction is described by a linear combination of the basis vectors of the crystal lattice, we find that, in the hexagonal system, we need *four* successive displacements parallel to each of the four basis vectors to describe a direction vector. According to the international conventions the translation along $\mathbf{a}_3$ must be equal to $-(u+v)$, similar to the expression for i above. Additionally, we have $\mathbf{a}_3 = -(\mathbf{a}_1 + \mathbf{a}_2)$, as is obvious from Figure 5.6(a). We can now derive the relation between the three-index and four-index systems for directions.

If a direction is described by the indices $[uvtw]$, then the corresponding three-index symbol $[u'v'w']$ is derived as follows:

$$u\mathbf{a}_1 + v\mathbf{a}_2 + t\mathbf{a}_3 + w\mathbf{c} = u'\mathbf{a}_1 + v'\mathbf{a}_2 + w'\mathbf{c};$$
$$u\mathbf{a}_1 + v\mathbf{a}_2 + (u+v)(\mathbf{a}_1 + \mathbf{a}_2) + w\mathbf{c} = u'\mathbf{a}_1 + v'\mathbf{a}_2 + w'\mathbf{c}.$$

from which follows:

$$u' = 2u + v;$$
$$v' = 2v + u;$$
$$w' = w.$$

The inverse relations are given by:

$$u = \frac{1}{3}(2u' - v');$$
$$v = \frac{1}{3}(2v' - u');$$
$$t = -\frac{1}{3}(u' + v') = -(u+v);$$
$$w = w'.$$

Figure 5.7 illustrates how this conversion works. The three-index [120] direction is equivalent to $1\mathbf{a}_1 + 2\mathbf{a}_2$. If we include the extra index, then we must take $1\mathbf{a}_2 - 1\mathbf{a}_3$ or $[01\bar{1}0] = [01.0]$. Table 5.1 lists some examples of equivalent directions in the Miller and Miller–Bravais indexing systems. Remember that all indices must be reduced to relative primes! Note that, for directions, one cannot just leave out the third index to convert from four-index

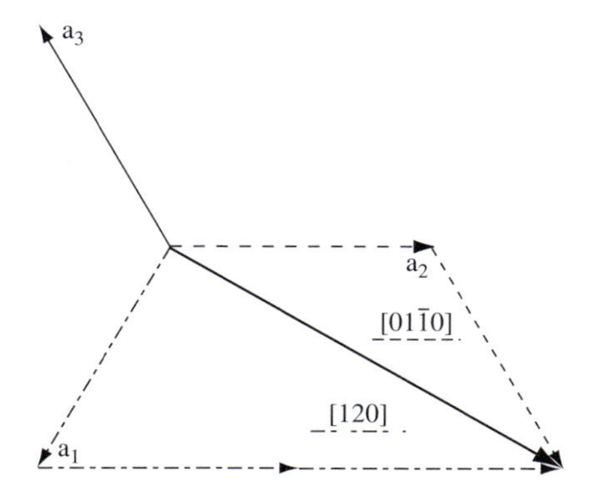

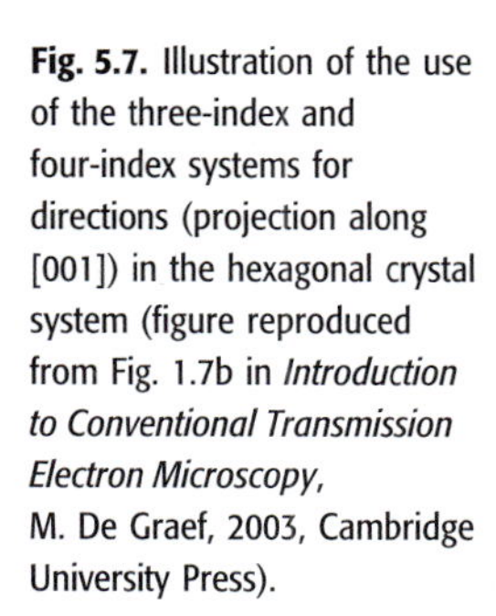
Fig. 5.7. Illustration of the use of the three-index and four-index systems for directions (projection along [001]) in the hexagonal crystal system (figure reproduced from Fig. 1.7b in *Introduction to Conventional Transmission Electron Microscopy*, M. De Graef, 2003, Cambridge University Press).

Table 5.1. Equivalent indices in the Miller and Miller–Bravais indexing systems for hexagonal directions.

Miller	Miller–Bravais	Miller	Miller–Bravais
[100]	$[2\bar{1}\bar{1}0]$	[010]	$[\bar{1}2\bar{1}0]$
[110]	$[11\bar{2}0]$	$[\bar{1}10]$	$[\bar{1}100]$
[001]	[0001]	[101]	$[2\bar{1}\bar{1}3]$
[011]	$[\bar{1}2\bar{1}3]$	[111]	$[11\bar{2}3]$
[210]	$[10\bar{1}0]$	[120]	$[01\bar{1}0]$
[211]	$[10\bar{1}1]$	[112]	$[11\bar{2}6]$

to three-index notation; for planes the third index can always be left out without causing any ambiguities.

5.4 Crystal forms

The concept of a family of planes is important when it comes to describing the external shape of a crystal. In this context, we usually refer to the *form* instead of the family. A form is a group of crystal faces that belong to the same family; in other words, a form is the collection of crystal faces that are equivalent to each other. In particular, this means that all the faces in a form have precisely the same shape. We have already seen that the {100} family in a cubic crystal system consists of six planes. Similarly, the {111} family consists of eight planes. Consider next an object with as external faces the planes from the {100} family. It is obvious that such an object is a cube. In that case, we say that an object bounded by the planes of the {100} family has the *cube crystal form*. An object bounded by the {111} family of planes has the *octahedron crystal form*. Both of these objects have an underlying crystal lattice that belongs to the cubic crystal system.

The *International Tables for Crystallography* identify 47 different possible forms for crystals belonging to one of the 7 crystal systems. Each form has a specific name, listed in Table 5.2. The table lists the official name, an alternative name (following the Groth–Rogers naming convention (Hurlbut and Klein, 1977)), and the number of faces in the form (i.e., the multiplicity of the family). Note that, in most cases, the names of the two naming schemes are identical, but there are quite a few differences for the forms belonging to the cubic crystal class (forms 33 through 47). From this table, we learn, for instance, that a cubic crystal with {100} planes as faces is known as a *hexahedron*, whereas a cubic crystal with {111} faces is an *octahedron*.

It is often easier to visualize the crystal forms; Fig. 5.8 shows all 47 crystal forms as wireframe drawings. The numbers correspond to the entries in Table 5.2. For the monohedron (form 1), there is only one plane, with no

Table 5.2. The names, multiplicities, and point group symmetries of the 47 different crystal forms. The second number in the first column is the sequential number according to the International Tables for Crystallography; the first number corresponds to the Roth–Rogers nomenclature.

#	International name	Roth–Rogers name	Multiplicity	Point group
1 (1)	Monohedron	Pedion	1	$\infty\mathbf{m}$
2 (2)	Parallelohedron	Pinacoid	2	$\frac{\infty}{\mathrm{m}}\mathbf{m}$
3 (3)	Dihedron	Dome/Sphenoid	2	$\mathbf{mm2}$ (C_{2v})
4 (6)	Rhombic prism	Rhombic prism	4	$\mathbf{mmm}$ (D_{2h})
5 (18)	Trigonal prism	Trigonal prism	3	$\bar{\mathbf{6}}\mathbf{m2}$ (D_{3h})
6 (22)	Ditrigonal prism	Ditrigonal prism	6	$\bar{\mathbf{6}}\mathbf{m2}$ (D_{3h})
7 (10)	Tetragonal prism	Tetragonal prism	4	$\mathbf{4/mmm}$ (D_{4h})
8 (15)	Ditetragonal prism	Ditetragonal prism	8	$\mathbf{4/mmm}$ (D_{4h})
9 (25)	Hexagonal prism	Hexagonal prism	6	$\mathbf{6/mmm}$ (D_{6h})
10 (30)	Dihexagonal prism	Dihexagonal prism	12	$\mathbf{6/mmm}$ (D_{6h})
11 (5)	Rhombic pyramid	Rhombic pyramid	4	$\mathbf{mm2}$ (C_{2v})
12 (17)	Trigonal pyramid	Trigonal pyramid	3	$\mathbf{3m}$ (C_{3v})
13 (20)	Ditrigonal pyramid	Ditrigonal pyramid	6	$\mathbf{3m}$ (C_{3v})
14 (8)	Tetragonal pyramid	Tetragonal pyramid	4	$\mathbf{4mm}$ (C_{4v})
15 (12)	Ditetragonal pyramid	Ditetragonal pyramid	8	$\mathbf{4mm}$ (C_{4v})
16 (23)	Hexagonal pyramid	Hexagonal pyramid	6	$\mathbf{6mm}$ (C_{6v})
17 (28)	Dihexagonal pyramid	Dihexagonal pyramid	12	$\mathbf{6mm}$ (C_{6v})
18 (7)	Rhombic dipyramid	Rhombic dipyramid	8	$\mathbf{mmm}$ (D_{2h})
19 (24)	Trigonal dipyramid	Trigonal dipyramid	6	$\bar{\mathbf{6}}\mathbf{m2}$ (D_{3h})
20 (29)	Ditrigonal dipyramid	Ditrigonal dipyramid	12	$\bar{\mathbf{6}}\mathbf{m2}$ (D_{3h})
21 (14)	Tetragonal dipyramid	Tetragonal dipyramid	8	$\mathbf{4/mmm}$ (D_{4h})
22 (16)	Ditetragonal dipyramid	Ditetragonal dipyramid	16	$\mathbf{4/mmm}$ (D_{4h})
23 (31)	Hexagonal dipyramid	Hexagonal dipyramid	12	$\mathbf{6/mmm}$ (D_{6h})
24 (32)	Dihexagonal dipyramid	Dihexagonal dipyramid	24	$\mathbf{6/mmm}$ (D_{6h})
25 (19)	Trigonal trapezohedron	Trigonal trapezohedron	6	$\mathbf{32}$ (D_3)
26 (11)	Tetragonal trapezohedron	Tetragonal trapezohedron	8	$\mathbf{422}$ (D_4)
27 (27)	Hexagonal trapezohedron	Hexagonal trapezohedron	12	$\mathbf{622}$ (D_6)
28 (13)	Tetragonal scalenohedron	Tetragonal scalenohedron	8	$\bar{\mathbf{4}}\mathbf{2m}$ (D_{2d})
29 (26)	Ditrigonal scalenohedron	Hexagonal scalenohedron	12	$\bar{\mathbf{3}}\mathbf{m}$ (D_{3d})
30 (21)	Rhombohedron	Rhombohedron	6	$\bar{\mathbf{3}}\mathbf{m}$ (D_{3d})
31 (4)	Rhombic tetrahedron	Rhombic disphenoid	4	$\mathbf{222}$ (D_2)
32 (9)	Tetragonal tetrahedron	Tetragonal disphenoid	4	$\bar{\mathbf{4}}\mathbf{2m}$ (D_{2d})
33 (34)	Hexahedron	Cube	6	$\mathbf{m}\bar{\mathbf{3}}\mathbf{m}$ (O_h)
34 (35)	Octahedron	Octahedron	8	$\mathbf{m}\bar{\mathbf{3}}\mathbf{m}$ (O_h)
35 (40)	Rhomb-dodecahedron	Dodecahedron (rhombic)	12	$\mathbf{m}\bar{\mathbf{3}}\mathbf{m}$ (O_h)
36 (46)	Tetrahexahedron	Tetrahexahedron	24	$\mathbf{m}\bar{\mathbf{3}}\mathbf{m}$ (O_h)
37 (43)	Tetragon-trioctahedron	Trapezohedron	24	$\mathbf{m}\bar{\mathbf{3}}\mathbf{m}$ (O_h)
38 (42)	Trigon-trioctahedron	Trisoctahedron	24	$\mathbf{m}\bar{\mathbf{3}}\mathbf{m}$ (O_h)
39 (47)	Hexaoctahedron	Hexoctahedron	48	$\mathbf{m}\bar{\mathbf{3}}\mathbf{m}$ (O_h)
40 (33)	Tetrahedron	Tetrahedron	4	$\bar{\mathbf{4}}\mathbf{3m}$ (T_d)
41 (39)	Trigon-tritetrahedron	Tristetrahedron	12	$\bar{\mathbf{4}}\mathbf{3m}$ (T_d)
42 (38)	Tetragon-tritetrahedron	Deltoid dodecahedron	12	$\bar{\mathbf{4}}\mathbf{3m}$ (T_d)
43 (45)	Hexatetrahedron	Hextetrahedron	24	$\bar{\mathbf{4}}\mathbf{3m}$ (T_d)
44 (44)	Pentagon-trioctahedron	Gyroid	24	$\mathbf{432}$ (O)
45 (37)	Dihexahedron	Pyritohedron	12	$\mathbf{m}\bar{\mathbf{3}}$ (T_h)
46 (41)	Didodecahedron	Diploid	24	$\mathbf{m}\bar{\mathbf{3}}$ (T_h)
47 (36)	Pentagon-tritetrahedron	Tetartoid	12	$\mathbf{23}$ (T)

equivalent planes. In other words, the family consists of only one plane. This can only happen in the triclinic crystal system. The parallelohedron has two parallel faces, whereas the dihedron consists of two intersecting faces. Forms 4 through 10 consist of planes that have one direction in common. The top and

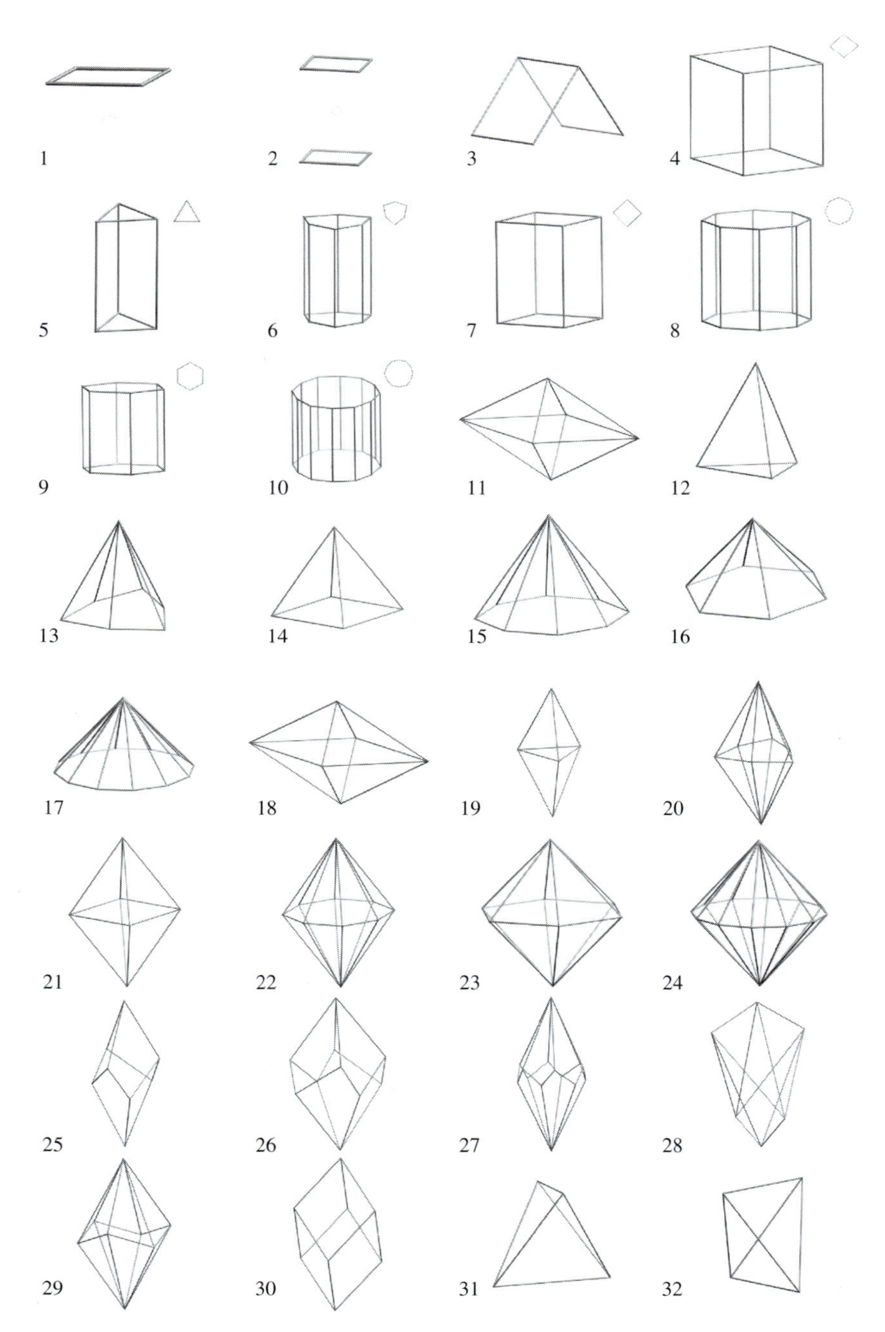

Fig. 5.8. Graphical representation of the 47 crystal forms.

Fig. 5.8. (cont.).

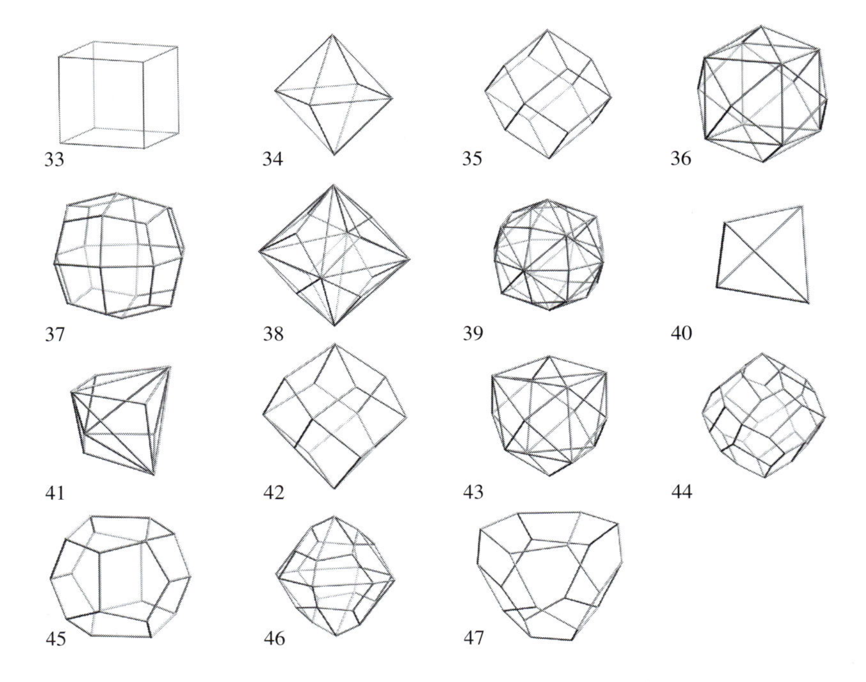

bottom planes in each of these figures do not correspond to planes of the form. The shape of the cross section of the form is shown in the upper right-hand corner of each of the forms. Forms 11 through 17 consist of planes meeting in a single point (the top of the pyramid). The horizontal bottom plane does not belong to the form. This means that forms 1 through 17 do not enclose a finite volume. Crystals that display one of these forms, must have additional forms in their description, in order for the description to be complete.

It is not uncommon for crystals to occur in shapes that correspond to multiple forms. For instance, consider the hexahedron form and the octahedron form. Fig 5.9(a) shows a regular octahedron. Imagine now that this octahedron sits completely inside a hexahedron (cube), and that the size of the hexahedron decreases. At some point, the hexahedron will begin to cut through the octahedron at all six vertices. When the hexahedron shrinks even more, the intersections will be small square planes, as shown in Fig. 5.9(b). This is hence known as a *truncated octahedron*. When the hexahedron shrinks even more, the

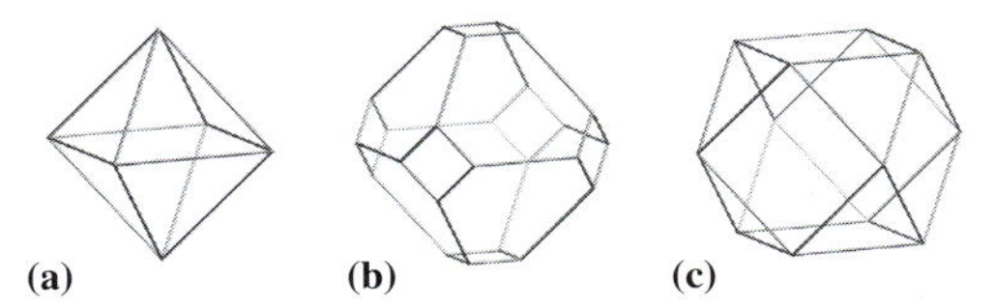

Fig. 5.9. Example of the truncation of a form. An octahedral form (a) is gradually truncated by a shrinking hexahedron (b), until the square sections touch (c) in what is known as a cuboctahedron.

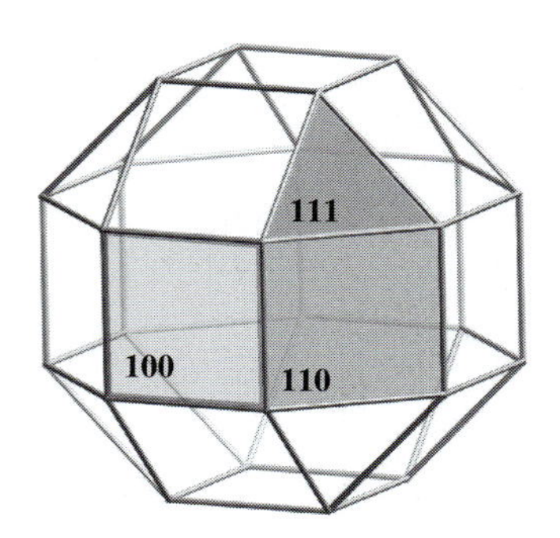

Fig. 5.10. Example of an octahedron form, truncated by both a hexahedron, labeled by the {100}-type faces, and a rhomb-dodecahedron, labeled by the {110}-faces.

truncated shape will eventually look like the one in Fig. 5.9(c), where the squares touch each other, leaving small triangles where the original full octahedron faces were. The resulting shape is known as a *cuboctahedron*; it consists of the hexahedron and octahedron forms. There is a complex nomenclature for such compound shapes, but this would lead us too far from the main topic. The interested reader is referred to Rogers (1935) for further information.

When more than two forms are present, the shape of the crystal can become even more complex. Figure 5.10 shows the resulting shape, when an octahedron form (indicated by the triangular {111}-type faces) is truncated by both a hexahedron ({100}-faces) and a rhomb-dodecahedron ({110}-faces). It is clear that a very large variety of crystal shapes can be obtained by combining two or more of the 47 crystal forms. On the other hand, the identification of a crystal shape is helped by the fact that there are *only* 47 different forms. Shape identification requires a careful measurement of the interfacial angles. In addition, not all crystals have perfect shapes; there may be distorted faces, for instance, when a face that should be square is not perfectly square. We say that the crystal is *malformed*. However, the angle between the malformed face and its neighbors is the same as for the perfect crystal shape. This is known as the law of *constancy of interfacial angle*, which is one of the basic rules of the description and classification of crystal shapes. Since the angles between planes are so important, we will need a tool to graphically represent them in a two-dimensional drawing (i.e., on a sheet of paper). Most people, the authors included, are not very good at drawing 3-D shapes by hand on a piece of paper, in such a way that the perspective is correct, and those parts of the shape that should be hidden from the observer are indeed hidden. To alleviate this problem, William H. Miller (yes, the same Miller as the one who proposed the *Miller indices*) proposed a technique which is now known as the *stereographic projection*. We will discuss this technique in Section 7.1.

5.5 Historical notes

Carolus Linnaeus (1707–78, Fig. 5.11(a)) was a Swedish naturalist best known for his classification of plants. His classification of crystals was less

Fig. 5.11. (a) Carolus Linnaeus (1707–78), and (b) William H. Miller (1801–80) (pictures courtesy of J. Lima-de-Faria).

successful, in that he was a proponent of a vegetative growth model in which he classified crystals (like plants) into classes, orders, and genera. However, he was the first to emphasize the importance of the shape of crystals. In his third volume of *Systema Naturae* (Linnaeus, 1768), he proposed a classification of minerals based on their thermal behavior. He also categorized crystals by their morphological features using four categories: (1) nitre-type hexagonal prisms, (2) alum-type octahedra, (3) blue-vitriol rhombic dodecahedra, and (4) common salt cubes, in agreement with previous ideas of **Guglielmini**. He drew many accurate pictures of the geometrical forms of crystals. This insight into geometrical form was later used by **René-Just Haüy**, and Romé de l'Isle, who both credited Linnaeus as being the father of the science of crystallography.

William Hallowes Miller (1801–80, Fig. 5.11(b)) was a British mineralogist and crystallographer. Although the so-called *Miller indices* were proposed earlier by other crystallographers, they were attributed to him because of his use of them in his book and educational efforts. The Miller indices are the inverse of the so-called *Weiss indices*. Miller developed the familiar hkl notation for referring to these indices. Miller also developed the first two-circle goniometer (Miller, 1839).

5.6 Problems

(i) *Miller indices I:* Make a 3D sketch of an orthorhombic reference frame with lattice parameters $\{2, 3, 4, 90, 90, 90\}$. On this sketch, draw the outline of the following planes: (110), $(0\bar{3}2)$, $(\bar{1}1\bar{1})$, and $(06\bar{4})$.

(ii) *Miller indices II:* Show that the third index in the Miller–Bravais system of indexing for hexagonal crystals equals $i = -(h+k)$.

(iii) *Family of planes I:* Show graphically that the $\{110\}$ and $\{011\}$ families of planes are not equivalent in a tetragonal unit cell (see Fig. 5.4(d)). If the unit cell were to become orthorhombic, what would be the families corresponding to the cubic family $\{110\}$? List the members of each family.

(iv) *Family of planes II:* An attribute of a family of planes in a given crystal system is that the spacing between planes in a family must be the same. Consider a tetragonal crystal system having the lattice constants (normalized by the a-lattice constant) $\{1, 1, c/a, 90, 90, 90\}$.

   (a) Determine the distance between (110) planes as a function of c/a.
   (b) Determine the distance between (101) planes as a function of c/a.
   (c) Are (110) and (101) planes in the same family?

(v) *Miller indices and forms I*: Make a 2-D drawing of a possible cross section in the (001) plane of a crystal belonging to the cubic crystal system. The external planes all belong to the $\{120\}$ family of planes. Note that there are many different solutions, depending on the relative sizes of all the planes. (Hint: first, draw the intersections with the (001) plane of all the planes of the family. Then, rearrange these intersection lines to form a closed polyhedral figure.)

(vi) *Miller indices and forms II*: Repeat the previous exercise for the case where the crystal system is the hexagonal system, and the cross section is made in the (00.1) plane. Use the $\{120\} = \{12.0\}$ family of planes.

(vii) *Miller–Bravais direction indices*: Assuming that equivalent directions in the Miller–Bravais four-index notation can be obtained by permutations of the first three indices, make a 2-D drawing in the (00.1) plane of the direction families $\langle 120 \rangle$ (three-index) and $\langle 1010 \rangle$ (four-index). (Hint: to determine the family members of the first family, first convert the indices to four-index notation. Then, superimpose on that drawing the $\{120\}$ planes and the $\{100\}$ planes.)

(viii) Make a sketch of a tetragonal dipyramid and write down the general Miller indices of its faces.

(ix) Make a sketch of a rhomb-dodecahedron and write down the general Miller indices of its faces. Which family/families do these faces belong to?

(x) Find an expression for the ratio of the surface area of the $\{100\}$ planes to the surface area of the $\{111\}$ planes for the truncated octahedron of Fig. 5.9.

CHAPTER

6 Reciprocal space

> *"The scientist describes what is; the engineer creates what never was."*
>
> Theodore von Karman, quoted in A. L. Mackay, *Dictionary of Scientific Quotations* (London 1994)

6.1 Introduction

In the previous chapter, we introduced a compact notation for an arbitrary plane in an arbitrary crystal system. The Miller indices (hkl) form a triplet of integer numbers and fully characterize the plane. It is tempting to interpret the Miller indices as the components of a vector, similar to the components $[uvw]$ of a lattice vector $\mathbf{t}$. This raises a few questions: if h, k, and l are indeed the components of a vector, then how does this vector relate to the plane (hkl)? Furthermore, since vector components are always taken with respect to a set of basis vectors, we must ask which are the relevant basis vectors for the components (h, k, l)? In this chapter, we will introduce the concept of *reciprocal space*. We will show that reciprocal space allows us to interpret the Miller indices h, k, and l as the components of a vector; not just any vector, but the *normal* to the plane (hkl). We will also show that the length of this vector is related to the spacing between consecutive (hkl) planes.

At first, you will probably find this whole reciprocal space business a bit abstract and difficult to understand. This is normal. It will take a while for you to really understand what is meant by reciprocal space. So, be patient; reciprocal space is probably one of the most complicated topics in this book, which means that an understanding will not come immediately. It is important, however, that you persist in trying to understand this topic, because it is of fundamental importance for everything that has to do with diffraction experiments.

It is interesting to note that these abstract physical concepts find widespread use in many areas of physics and engineering. The originators of some of the early reciprocal space ideas were able to move around successfully in seemingly disparate areas of science, applied mathematics, and engineering.

6.2 The reciprocal basis vectors

A unit cell is defined by the crystallographic basis vectors $\mathbf{a}_i$. We know from our discussions in previous chapters, that the choice of basis vectors is really an arbitrary one. There are an infinite number of possible choices for the basis vectors. Usually, we select those that reflect the symmetry of the underlying crystal system, and that is a very convenient choice. However, let us now see what happens when we select a different set of basis vectors.

Let us define three new vectors, denoted by the symbol $\mathbf{a}_j^*$ (the asterisk is used to indicate that we are talking about a different set of basis vectors), and defined such that

$$\boxed{\mathbf{a}_i \cdot \mathbf{a}_j^* = \delta_{ij},} \tag{6.1}$$

where δ_{ij} is the *Kronecker delta* (i.e., equal to 1 for $i = j$ and 0 for $i \neq j$). This is basically a fancy name for the unit matrix. Let us rewrite this equation explicitly:

$$\begin{pmatrix} \mathbf{a}_1 \cdot \mathbf{a}_1^* & \mathbf{a}_1 \cdot \mathbf{a}_2^* & \mathbf{a}_1 \cdot \mathbf{a}_3^* \\ \mathbf{a}_2 \cdot \mathbf{a}_1^* & \mathbf{a}_2 \cdot \mathbf{a}_2^* & \mathbf{a}_2 \cdot \mathbf{a}_3^* \\ \mathbf{a}_3 \cdot \mathbf{a}_1^* & \mathbf{a}_3 \cdot \mathbf{a}_2^* & \mathbf{a}_3 \cdot \mathbf{a}_3^* \end{pmatrix} = \begin{pmatrix} 1 & 0 & 0 \\ 0 & 1 & 0 \\ 0 & 0 & 1 \end{pmatrix}. \tag{6.2}$$

From the definition 6.1 (or the first column in the matrix above), we find that the vector $\mathbf{a}_1^*$ must be perpendicular to the vectors $\mathbf{a}_2$ and $\mathbf{a}_3$, since both dot-products vanish. If a vector is normal to two other vectors, then it must be parallel to the cross-product of those two vectors. Hence, we can write:

$$\mathbf{a}_1^* = K\,\mathbf{a}_2 \times \mathbf{a}_3,$$

where K is a proportionality factor. Similarly, we find that

$$\mathbf{a}_2^* = L\,\mathbf{a}_3 \times \mathbf{a}_1,$$

and

$$\mathbf{a}_3^* = M\,\mathbf{a}_1 \times \mathbf{a}_2.$$

Equation 6.1 also tells us that the dot-products $\mathbf{a}_i \cdot \mathbf{a}_i^*$ must be equal to unity, or:

$$\begin{aligned}
\mathbf{a}_1 \cdot \mathbf{a}_1^* &= K\mathbf{a}_1 \cdot (\mathbf{a}_2 \times \mathbf{a}_3) = 1; \\
\mathbf{a}_2 \cdot \mathbf{a}_2^* &= L\mathbf{a}_2 \cdot (\mathbf{a}_3 \times \mathbf{a}_1) = 1; \\
\mathbf{a}_3 \cdot \mathbf{a}_3^* &= M\mathbf{a}_3 \cdot (\mathbf{a}_1 \times \mathbf{a}_2) = 1.
\end{aligned} \tag{6.3}$$

Since the mixed product of three vectors does not depend on the cyclic order of the three vectors (i.e., $\mathbf{a} \cdot (\mathbf{b} \times \mathbf{c}) = \mathbf{b} \cdot (\mathbf{c} \times \mathbf{a}) = \mathbf{c} \cdot (\mathbf{a} \times \mathbf{b})$), we find that:

$$K = L = M = \frac{1}{\mathbf{a}_1 \cdot (\mathbf{a}_2 \times \mathbf{a}_3)}.$$

The scalar $\mathbf{a}_1 \cdot (\mathbf{a}_2 \times \mathbf{a}_3)$ is the volume of the cell created by the three vectors, i.e., the volume V of the unit cell.

We conclude that, if we define the new basis vectors as:

$$\boxed{\begin{aligned}
\mathbf{a}_1^* &= \frac{\mathbf{a}_2 \times \mathbf{a}_3}{V}; \\
\mathbf{a}_2^* &= \frac{\mathbf{a}_3 \times \mathbf{a}_1}{V}; \\
\mathbf{a}_3^* &= \frac{\mathbf{a}_1 \times \mathbf{a}_2}{V},
\end{aligned}} \tag{6.4}$$

then Equation 6.1 is satisfied. The basis vectors $\mathbf{a}_i$ are the *direct basis vectors*, the vectors $\mathbf{a}_i^*$ are known as the *reciprocal basis vectors*. In many textbooks on crystallography, the reciprocal basis vectors are denoted by special symbols: $\mathbf{a}_1^* = \mathbf{a}^*$, $\mathbf{a}_2^* = \mathbf{b}^*$, and $\mathbf{a}_3^* = \mathbf{c}^*$.

The new vectors form a new basis for the crystal lattice. An arbitrary vector $\mathbf{p}$ can be expressed in terms of these new basis vectors as follows:

$$\mathbf{p} = p_i^* \mathbf{a}_i^*. \tag{6.5}$$

A vector is a quantity that exists independent of the reference frame; therefore, the vector $\mathbf{p}$ must also have components with respect to the regular basis vectors $\mathbf{a}_i$:

$$\mathbf{p} = p_j \mathbf{a}_j. \tag{6.6}$$

Since this is the same vector, we find:

$$\mathbf{p} = p_j \mathbf{a}_j = p_i^* \mathbf{a}_i^*. \tag{6.7}$$

We could, in principle, select any three linearly independent vectors to form a basis set. The definition above (Equation 6.1) imposes special conditions on the reciprocal basis vectors. In other words, out of the infinite set of possible basis vector selections, we have picked the one that satisfies equation 6.1. We will see in the following sections that this choice has important (and useful) consequences.

Before we do so, let us first examine the reciprocal basis vectors by means of an example. Consider a monoclinic lattice with lattice parameters $\{1, 1, 1, 90, 45, 90\}$, shown in Fig. 6.1(a). This is not a realistic set of lattice parameters, but it is perfectly suited to illustrate the reciprocal basis vectors. First of all, we compute the volume, V, of the unit cell: if $\mathbf{e}_y$ is the unit vector normal to the plane of the drawing going into the drawing, then we have $\mathbf{b} = b\mathbf{e}_y = \mathbf{e}_y$, since $b = 1$. In order to have a right-handed reference frame, $\mathbf{b}$ must be normal to the plane of the drawing, going into the drawing. Therefore,

$$
\begin{aligned}
V &= \mathbf{a} \cdot (\mathbf{b} \times \mathbf{c}); \\
&= \mathbf{b} \cdot (\mathbf{c} \times \mathbf{a}); \\
&= \mathbf{b} \cdot (|\mathbf{c}|\,|\mathbf{a}| \sin \frac{\pi}{4} \mathbf{e}_y); \\
&= \frac{\sqrt{2}}{2} \mathbf{e}_y \cdot \mathbf{e}_y; \\
&= \frac{\sqrt{2}}{2}.
\end{aligned}
$$

First, we determine the reciprocal basis vector $\mathbf{a}^*$. From the definition, we have:

$$
\begin{aligned}
\mathbf{a}^* &= \frac{\mathbf{b} \times \mathbf{c}}{V}; \\
&= \sqrt{2}|\mathbf{b}|\,|\mathbf{c}| \sin \frac{\pi}{2} \mathbf{e}_{bc}; \\
&= \sqrt{2}\mathbf{e}_{bc},
\end{aligned}
$$

where $\mathbf{e}_{bc}$ is a unit vector normal to both $\mathbf{b}$ and $\mathbf{c}$, and we use the fact that $V^{-1} = \sqrt{2}$. This vector is indicated in Fig. 6.1(b), as well as the reciprocal vector $\mathbf{a}^*$.

Next, we determine $\mathbf{c}^*$ in a similar fashion:

$$
\begin{aligned}
\mathbf{c}^* &= \frac{\mathbf{a} \times \mathbf{b}}{V}; \\
&= \sqrt{2}|\mathbf{a}|\,|\mathbf{b}| \sin \frac{\pi}{2} \mathbf{e}_{ab}; \\
&= \sqrt{2}\mathbf{e}_{ab}.
\end{aligned}
$$

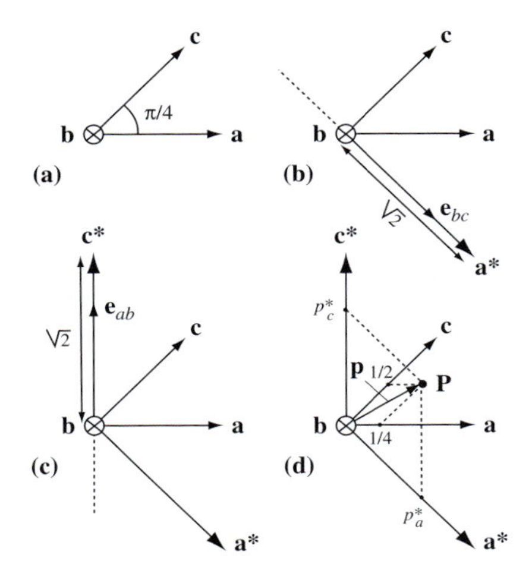

Fig. 6.1. Schematic illustration of the reciprocal basis vectors for a monoclinic unit cell with lattice parameters {1, 1, 1, 90, 45, 90}. In (d), the vector **p** has components with respect to both sets of basis vectors.

Fig. 6.1(c) shows the location of these vectors. This completes the construction of the reciprocal basis vectors for this particular lattice.[1]

If we take a vector $\mathbf{p} = \mathbf{a}/4 + \mathbf{c}/2$, as shown in Fig. 6.1(d), then it is clear that **p** also has components with respect to the reciprocal basis vectors:

$$\mathbf{p} = \frac{1}{4}\mathbf{a} + \frac{1}{2}\mathbf{c} = p_a^*\mathbf{a}^* + p_c^*\mathbf{c}^*. \tag{6.8}$$

We could measure both parameters p_a^* and p_c^* from the drawing (doing so results in $p_a^* \approx 0.6$ and $p_c^* \approx 0.7$). This is not the most accurate way, so we anticipate that there should be a way to compute them directly. We will consider a general method later on in this book; for now, we can proceed from the drawing itself.

It is easy to see that the direct basis vectors can be expressed as linear combinations of the reciprocal basis vectors. We find:

$$\mathbf{a} = \mathbf{a}^* + \frac{1}{\sqrt{2}}\mathbf{c}^*;$$

$$\mathbf{c} = \frac{1}{\sqrt{2}}\mathbf{a}^* + \mathbf{c}^*.$$

Then, we substitute these expressions in Equation 6.8:

$$\begin{aligned}\mathbf{p} &= \frac{1}{4}\mathbf{a} + \frac{1}{2}\mathbf{c};\\ &= \frac{1}{4}(1+\sqrt{2})\mathbf{a}^* + \frac{1}{8}(4+\sqrt{2})\mathbf{c}^*;\end{aligned}$$

[1] We leave it to the reader to determine the location of $\mathbf{b}^*$.

$$= 0.604\mathbf{a}^* + 0.677\mathbf{c}^*;$$

$$= p_a^*\mathbf{a}^* + p_c^*\mathbf{c}^*.$$

Before we look at more general (and easier) methods to perform these computations, we must first consider a few more properties and uses of the reciprocal basis vectors.

6.3 Reciprocal space and lattice planes

Now that we have defined the three reciprocal basis vectors, let us take a look at the properties of an arbitrary vector, expressed in this reference frame. Consider the vector $\mathbf{g}$, with components g_i^*, i.e.,

$$\mathbf{g} = g_i^*\mathbf{a}_i^*.$$

If we restrict the values of g_i^* to be integers, as we did when we introduced lattice translation vectors, then we find that the set of all vectors $\mathbf{g}$ expressed in this way constructs a new lattice, a lattice based on the reciprocal basis vectors. This lattice is called the *reciprocal lattice*. It is customary to write the individual components of the reciprocal lattice vector $\mathbf{g}$ as (h, k, l) (i.e., $g_1^* = h$, $g_2^* = k$, and $g_3^* = l$), or:

$$\mathbf{g} = h\mathbf{a}_1^* + k\mathbf{a}_2^* + l\mathbf{a}_3^* = h\mathbf{a}^* + k\mathbf{b}^* + l\mathbf{c}^*.$$

Next, we must ask the question: what do these vectors $\mathbf{g}$ represent? What can we do with them? To answer these questions, we proceed as follows: we will look for all the vectors $\mathbf{r}$ (with components $r_i = (x, y, z)$) which are perpendicular to the vector $\mathbf{g}$. We already know that two vectors are perpendicular to each other if their dot-product vanishes; in this case, we find:

$$\mathbf{r}\cdot\mathbf{g} = (r_i\mathbf{a}_i)\cdot(g_j^*\mathbf{a}_j^*) = r_i g_j^*(\mathbf{a}_i\cdot\mathbf{a}_j^*) = 0.$$

We know from Equation 6.1, that the last dot-product is equal to δ_{ij}, so:

$$\mathbf{r}\cdot\mathbf{g} = r_i g_j^*\delta_{ij} = r_i g_i^* = r_1 g_1^* + r_2 g_2^* + r_3 g_3^* = hx + ky + lz = 0. \qquad (6.9)$$

The last equality represents the equation of a plane through the origin in the direct crystal lattice. If a plane intersects the basis vectors at intercepts s_i along the vectors $\mathbf{a}_i$, then the equation of the plane is given by:

$$\frac{x}{s_1} + \frac{y}{s_2} + \frac{z}{s_3} = 1, \qquad (6.10)$$

where (x, y, z) is an arbitrary point in the plane. The right-hand side of this equation changes to different values when we translate the plane perpendicular to itself, and in particular is equal to zero when the plane goes through the origin. Comparing:

$$hx + ky + lz = 0,$$

with:

$$\frac{x}{s_1} + \frac{y}{s_2} + \frac{z}{s_3} = 0,$$

we find that the integers h, k, and l are reciprocals of the intercepts of a plane with the direct lattice basis vectors. This is exactly the definition of the *Miller indices* of a plane! We thus find the important result:

> The reciprocal lattice vector $\mathbf{g}$, with components (h, k, l), is perpendicular to the plane with Miller indices (hkl).

For this reason, a reciprocal lattice vector is often denoted with the Miller indices as subscripts, e.g., $\mathbf{g}_{hkl}$.

To illustrate this fact, we return to the monoclinic lattice of Figure 6.1. Figure 6.2(a) shows how the reciprocal lattice vector $\mathbf{g}_{102}$ is related to the (102) plane. This plane intersects the direct basis vectors at $1\mathbf{a}$ and $\mathbf{c}/2$. The normal to the plane is given by the vector:

$$\mathbf{g}_{102} = \mathbf{a}^* + 2\mathbf{c}^*.$$

This vector is also shown on the figure. The larger black filled circles indicate the locations of the reciprocal lattice points or nodes. The $\mathbf{g}$-vectors in reciprocal space are reciprocal lattice vectors, in the same way that the translation

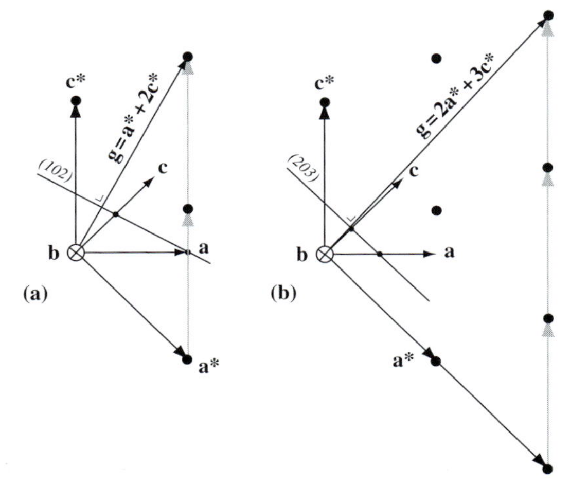

Fig. 6.2. Schematic illustration of the relation between the reciprocal lattice and the normals to various lattice planes, based on the unit cell of Fig. 6.1.

vectors **t** are direct lattice vectors. A similar illustration is shown in Fig. 6.2(b) for the (203) plane. From these illustrations we learn that each point or node of the reciprocal lattice is the end-point of a vector that is normal to the plane with the corresponding Miller indices. Therefore, the reciprocal lattice is a useful tool, since it allows us to describe the plane normals with simple integers, even in a triclinic crystal system. We note also that the reciprocal lattice vectors $\mathbf{g}_{102}$ and $\mathbf{g}_{203}$ have different lengths. To interpret what this means, we must introduce the *reciprocal metric tensor*.

6.4 The reciprocal metric tensor

A vector is characterized by a direction and a magnitude; the direction of the reciprocal lattice vectors is given by the normals to the lattice planes. What about the magnitude $|\mathbf{g}|$? We know that the length of a vector is given by the square root of the dot-product of the vector with itself. Thus, the length of $\mathbf{g}$ is given by:

$$|\mathbf{g}| = \sqrt{\mathbf{g}\cdot\mathbf{g}} = \sqrt{(g_i^*\mathbf{a}_i^*).(g_j^*\mathbf{a}_j^*)} = \sqrt{g_i^* g_j^*(\mathbf{a}_i^*\cdot\mathbf{a}_j^*)}.$$

We find, once again, that the dot-product involves knowledge of the dot-products of the basis vectors, in this case the reciprocal basis vectors. Note that the equation above is identical in form to the equations for the distance between two lattice points that we have seen in Chapter 4. The only difference is that all the quantities now have an asterisk on them, indicating that we are working with respect to different basis vectors. At this point, we introduce the *reciprocal metric tensor*:

$$\boxed{g_{ij}^* \equiv \mathbf{a}_i^*\cdot\mathbf{a}_j^*.} \tag{6.11}$$

We will often denote this tensor by the symbol $g^* = g_{ij}^*$. Explicitly, the reciprocal metric tensor is given by:

$$g^* = \begin{bmatrix} \mathbf{a}^*\cdot\mathbf{a}^* & \mathbf{a}^*\cdot\mathbf{b}^* & \mathbf{a}^*\cdot\mathbf{c}^* \\ \mathbf{b}^*\cdot\mathbf{a}^* & \mathbf{b}^*\cdot\mathbf{b}^* & \mathbf{b}^*\cdot\mathbf{c}^* \\ \mathbf{c}^*\cdot\mathbf{a}^* & \mathbf{c}^*\cdot\mathbf{b}^* & \mathbf{c}^*\cdot\mathbf{c}^* \end{bmatrix};$$

$$= \begin{bmatrix} a^{*2} & a^*b^*\cos\gamma^* & a^*c^*\cos\beta^* \\ b^*a^*\cos\gamma^* & b^{*2} & b^*c^*\cos\alpha^* \\ c^*a^*\cos\beta^* & c^*b^*\cos\alpha^* & c^{*2} \end{bmatrix}, \tag{6.12}$$

where $\{a^*, b^*, c^*, \alpha^*, \beta^*, \gamma^*\}$ are the *reciprocal lattice parameters*; in other words, a^* is the length of the reciprocal basis vector $\mathbf{a}^*$, α^* is the angle between $\mathbf{b}^*$ and $\mathbf{c}^*$, and so on. In the following chapter, we will introduce a simple procedure to determine these parameters.

Using the reciprocal metric tensor, we can rewrite the length of the reciprocal lattice vector **g** as:

$$|\mathbf{g}| = \sqrt{\mathbf{g}\cdot\mathbf{g}} = \sqrt{g_i^* g_{ij}^* g_j^*}.$$

Note that there should not be any confusion between the vector components (one subscript) and the tensor components (two subscripts). The reciprocal metric tensors for the seven crystal systems are given by:

$$g^*_{\text{cubic}} = \begin{bmatrix} \frac{1}{a^2} & 0 & 0 \\ 0 & \frac{1}{a^2} & 0 \\ 0 & 0 & \frac{1}{a^2} \end{bmatrix}, \quad g^*_{\text{tetragonal}} = \begin{bmatrix} \frac{1}{a^2} & 0 & 0 \\ 0 & \frac{1}{a^2} & 0 \\ 0 & 0 & \frac{1}{c^2} \end{bmatrix},$$

$$g^*_{\text{orthorhombic}} = \begin{bmatrix} \frac{1}{a^2} & 0 & 0 \\ 0 & \frac{1}{b^2} & 0 \\ 0 & 0 & \frac{1}{c^2} \end{bmatrix}, \quad g^*_{\text{hexagonal}} = \begin{bmatrix} \frac{4}{3a^2} & \frac{2}{3a^2} & 0 \\ \frac{2}{3a^2} & \frac{4}{3a^2} & 0 \\ 0 & 0 & \frac{1}{c^2} \end{bmatrix},$$

$$g^*_{\text{rhombohedral}} = \frac{1}{W^2}\begin{bmatrix} 1+\cos\alpha & -\cos\alpha & -\cos\alpha \\ -\cos\alpha & 1+\cos\alpha & -\cos\alpha \\ -\cos\alpha & -\cos\alpha & 1+\cos\alpha \end{bmatrix},$$

with

$$W^2 = a^2(1+\cos\alpha - 2\cos^2\alpha),$$

$$g^*_{\text{monoclinic}} = \begin{bmatrix} \frac{1}{a^2\sin^2\beta} & 0 & -\frac{\cos\beta}{ac\sin^2\beta} \\ 0 & \frac{1}{b^2} & 0 \\ -\frac{\cos\beta}{ac\sin^2\beta} & 0 & \frac{1}{c^2\sin^2\beta} \end{bmatrix}$$

$$g^*_{\text{triclinic}} = \frac{1}{V^2}\begin{bmatrix} b^2c^2\sin^2\alpha & abc^2\mathcal{F}(\alpha,\beta,\gamma) & ab^2c\mathcal{F}(\gamma,\alpha,\beta) \\ abc^2\mathcal{F}(\alpha,\beta,\gamma) & a^2c^2\sin^2\beta & a^2bc\mathcal{F}(\beta,\gamma,\alpha) \\ ab^2c\mathcal{F}(\gamma,\alpha,\beta) & a^2bc\mathcal{F}(\beta,\gamma,\alpha) & a^2b^2\sin^2\gamma \end{bmatrix}$$

with:

$$\mathcal{F}(\alpha,\beta,\gamma) = \cos\alpha\cos\beta - \cos\gamma$$

and:

$$V^2 = a^2b^2c^2(1-\cos^2\alpha - \cos^2\beta - \cos^2\gamma + 2\cos\alpha\cos\beta\cos\gamma).$$

One can show (reader exercise; see also Chapter 7) that the matrices representing the direct and reciprocal metric tensors are each other's inverse.

6.4.1 Computation of the angle between planes

The angle between two planes can now also be computed in the standard way: this angle must be equal to the angle α between the two plane normals $\mathbf{g}$ and $\mathbf{h}$, or:

$$\alpha = \cos^{-1}\left(\frac{\mathbf{g}\cdot\mathbf{h}}{|\mathbf{g}|\,|\mathbf{h}|}\right) = \cos^{-1}\left(\frac{g_i^* h_j^* g_{ij}^*}{\sqrt{g_i^* g_j^* g_{ij}^*}\sqrt{h_i^* h_j^* g_{ij}^*}}\right).$$

Note that this equation is identical to the one for the angle between two directions, except that now we are using the reciprocal basis vectors as the reference frame.

6.4.2 Computation of the length of the reciprocal lattice vectors

We know that the vector $\mathbf{g} = g_i^* \mathbf{a}_i^*$ is perpendicular to the plane with Miller indices $g_i^* = (hkl)$. Therefore, the unit normal, $\mathbf{n}$, to the plane is given by:

$$\mathbf{n} = \frac{\mathbf{g}_{hkl}}{|\mathbf{g}_{hkl}|}.$$

The perpendicular distance from the plane intersecting the direct basis vectors at the points $1/h$, $1/k$ and $1/l$ to the origin is given by the projection of any vector $\mathbf{t}$ from the origin to a point in the plane onto the plane normal $\mathbf{n}$ (see Figure 6.3). This distance is also, by definition, the *interplanar spacing* d_{hkl}. Thus,

$$\mathbf{t}\cdot\mathbf{n} = \mathbf{t}\cdot\frac{\mathbf{g}_{hkl}}{|\mathbf{g}_{hkl}|} = d_{hkl}.$$

We can arbitrarily select $\mathbf{t} = \mathbf{a}/h$, which leads to:

$$\begin{aligned}\frac{\mathbf{a}}{h}\cdot(h\mathbf{a}^* + k\mathbf{b}^* + l\mathbf{c}^*) &= d_{hkl}|\mathbf{g}_{hkl}|;\\ \frac{\mathbf{a}}{h}\cdot h\mathbf{a}^* &= d_{hkl}|\mathbf{g}_{hkl}|;\\ 1 &= d_{hkl}|\mathbf{g}_{hkl}|,\end{aligned}$$

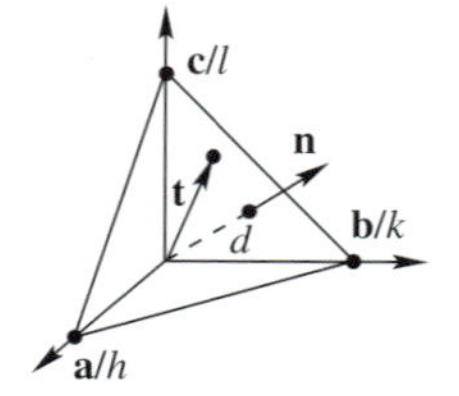

Fig. 6.3. The distance, d, of a plane to the origin equals the projection onto the unit plane normal, $\mathbf{n}$, of any vector $\mathbf{t}$ from the lattice point in the origin to a point in this plane.

from which we find that:

$$\boxed{|\mathbf{g}_{hkl}| = \frac{1}{d_{hkl}}.} \qquad (6.13)$$

> The length of a reciprocal lattice vector is equal to the inverse of the interplanar spacing of the corresponding lattice planes.

This result can be rewritten in subscript notation:

$$g_i^* g_{ij}^* g_j^* = \frac{1}{d_{hkl}^2}. \tag{6.14}$$

As an example we work out the interplanar spacing for the monoclinic crystal system. The expression on the left-hand side can be written in matrix notation as:

$$g_i^* g_{ij}^* g_j^* = [h\,k\,l] \begin{bmatrix} \frac{1}{a^2 \sin^2\beta} & 0 & -\frac{\cos\beta}{ac\sin^2\beta} \\ 0 & \frac{1}{b^2} & 0 \\ -\frac{\cos\beta}{ac\sin^2\beta} & 0 & \frac{1}{c^2\sin^2\beta} \end{bmatrix} \begin{bmatrix} h \\ k \\ l \end{bmatrix} \tag{6.15}$$

$$= [h\,k\,l] \begin{bmatrix} \frac{h}{a^2\sin^2\beta} - \frac{l\cos\beta}{ac\sin^2\beta} \\ \frac{k}{b^2} \\ -\frac{h\cos\beta}{ac\sin^2\beta} + \frac{l}{c^2\sin^2\beta} \end{bmatrix}, \tag{6.16}$$

or:

$$\frac{1}{d_{hkl}^2} = \frac{h^2}{a^2\sin^2\beta} + \frac{k^2}{b^2} + \frac{l^2}{c^2\sin^2\beta} - \frac{2hl\cos\beta}{ac\sin^2\beta}.$$

Let us now apply these equations to the example that we started in Fig. 6.1 and continued in Fig. 6.2. We have already determined that the reciprocal lattice parameters of the monoclinic lattice with direct lattice parameters $\{1, 1, 1, 90, 45, 90\}$ are given by $\{\sqrt{2}, 1, \sqrt{2}, 90, 135, 90\}$. The distance between the reciprocal lattice point (102) and the origin is given by the length of the vector $\mathbf{g}_{102}$. To compute this length, we can use the previous equation:

$$\begin{aligned} |\mathbf{g}_{102}|^2 &= \frac{h^2}{a^2\sin^2\beta} + \frac{k^2}{b^2} + \frac{l^2}{c^2\sin^2\beta} - \frac{2hl\cos\beta}{ac\sin^2\beta}; \\ &= \frac{1}{1\times\frac{1}{2}} + 0 + \frac{4}{1\times\frac{1}{2}} - \frac{2\times 1\times 2\times\frac{1}{\sqrt{2}}}{1\times 1\times\frac{1}{2}}; \\ &= 10 - 4\sqrt{2}. \end{aligned}$$

Alternatively (and more efficiently), we could have used the reciprocal metric tensor directly:

$$\begin{aligned} |\mathbf{g}_{102}|^2 &= [1\,0\,2] \begin{bmatrix} 2 & 0 & -\sqrt{2} \\ 0 & 1 & 0 \\ -\sqrt{2} & 0 & 2 \end{bmatrix} \begin{bmatrix} 1 \\ 0 \\ 2 \end{bmatrix}; \\ &= [1\,0\,2] \begin{bmatrix} 2-2\sqrt{2} \\ 0 \\ 4-\sqrt{2} \end{bmatrix}; \\ &= 10 - 4\sqrt{2}, \end{aligned}$$

Table 6.1. Expressions for the length $|\mathbf{g}| = 1/d_{hkl}$ of a reciprocal lattice vector $\mathbf{g}_{hkl}$ in the seven crystal systems.

System	g	Expression
Cubic	${}^{c}g$	$\frac{1}{a}\left\{h^2+k^2+l^2\right\}^{1/2}$
Tetragonal	${}^{t}g$	$\left\{\frac{1}{a^2}(h^2+k^2)+\frac{1}{c^2}l^2\right\}^{1/2}$
Orthorhombic	${}^{o}g$	$\left\{\frac{1}{a^2}h^2+\frac{1}{b^2}k^2+\frac{1}{c^2}l^2\right\}^{1/2}$
Hexagonal	${}^{h}g$	$\left\{\frac{4}{3a^2}(h^2+k^2+hk)+\frac{1}{c^2}l^2\right\}^{1/2}$
Rhombohedral	${}^{r}g$	$\frac{1}{a}\left\{\frac{(1+\cos^2\alpha)(h^2+k^2+l^2)-(1-\tan^2\frac{\alpha}{2})(hk+kl+lh)}{1+\cos\alpha-2\cos^2\alpha}\right\}^{1/2}$
Monoclinic	${}^{m}g$	$\left\{\frac{1}{a^2}\frac{h^2}{\sin^2\beta}+\frac{1}{b^2}k^2+\frac{1}{c^2}\frac{l^2}{\sin^2\beta}-\frac{2hl\cos\beta}{ac\sin^2\beta}\right\}^{1/2}$
Triclinic	${}^{a}g$	$\frac{1}{V}\left\{h^2b^2c^2\sin^2\alpha+k^2a^2c^2\sin^2\beta+l^2a^2b^2\sin^2\gamma\right.$ $+2hkabc^2\mathcal{F}(\alpha,\beta,\gamma)+2kla^2bc\mathcal{F}(\beta,\gamma,\alpha)$ $\left.+2lhab^2c\mathcal{F}(\gamma,\alpha,\beta)\right\}^{1/2}$ with $V=\left\{a^2b^2c^2(1-\cos^2\alpha-\cos^2\beta-\cos^2\gamma\right.$ $\left.+2\cos\alpha\cos\beta\cos\gamma)\right\}^{1/2}$

so that $|\mathbf{g}_{102}| = \sqrt{10-4\sqrt{2}} = 2.084$. If we measure the distance between (102) and the origin in Fig. 6.2(a) (in units of the length of **a**), we find good agreement. Furthermore, since $d_{hkl} = |\mathbf{g}_{hkl}|^{-1}$ we also have $d_{102} = 1/2.084 = 0.488$, which is the shortest distance between the plane (102) and the origin. We leave it to the reader to repeat this computation for the (203) plane. What we learn from this exercise is that the reciprocal lattice description allows us to compute the closest distance between a plane and the origin by computing the length of a vector that is normal to that plane. The mathematics involved in the computation is not all that difficult, since the plane is described by integers, and the reciprocal metric tensor components can be derived from the direct space lattice parameters. In the next chapter, we will see an easy method to compute the components of g^*.

To conclude this section on the reciprocal lattice we list all relevant equations for the seven crystal systems in Table 6.1 (reciprocal of the interplanar spacings) and Table 6.2 (cosine of the angle between plane normals). The same remarks about the ease of implementation of the metric tensor formalism (see page 90) hold for the reciprocal metric tensor.

Table 6.2. Expressions for the cosine of the angle α between two vectors $(h_1k_1l_1)$ and $(h_2k_2l_2)$ in the seven crystal systems.

System	$\cos\alpha$
Cubic	$\frac{1}{a^2}\frac{h_1h_2+k_1k_2+l_1l_2}{{}^c g_1\times{}^c g_2}$
Tetragonal	$\frac{\frac{1}{a^2}(h_1h_2+k_1k_2)+\frac{1}{c^2}l_1l_2}{{}^t g_1\times{}^t g_2}$
Orthorhombic	$\frac{\frac{1}{a^2}h_1h_2+\frac{1}{b^2}k_1k_2+\frac{1}{c^2}l_1l_2}{{}^o g_1\times{}^o g_2}$
Hexagonal	$\frac{\frac{4}{3a^2}(h_1h_2+k_1k_2+\frac{1}{2}(h_1k_2+k_1h_2))+\frac{1}{c^2}l_1l_2}{{}^h g_1\times{}^h g_2}$
Rhombohedral	$\frac{(1+\cos\alpha)(h_1h_2+k_1k_2+l_1l_2)-\frac{1}{2}(1-\tan^2\frac{\alpha}{2})(h_1(k_2+l_2)+k_1(h_2+l_2)+l_1(h_2+k_2))}{a^2(1+\cos\alpha-2\cos^2\alpha)\times{}^r g_1\times{}^r g_2}$
Monoclinic	$\frac{\frac{1}{a^2\sin^2\beta}h_1h_2+\frac{1}{b^2}k_1k_2+\frac{1}{c^2\sin^2\beta}l_1l_2+(l_1h_2+h_1l_2)\frac{\cos\beta}{ac\sin^2\beta}}{{}^m g_1\times{}^m g_2}$
Triclinic	$\frac{h_1h_2b^2c^2\sin^2\alpha+k_1k_2a^2c^2\sin^2\beta+l_1l_2a^2b^2\sin^2\gamma+abc^2(k_1h_2+k_2h_1)\mathcal{F}(\alpha,\beta,\gamma)+ab^2c(h_1l_2+h_2l_1)\mathcal{F}(\beta,\gamma,\alpha)+a^2bc(k_1l_2+k_2l_1)\mathcal{F}(\gamma,\alpha,\beta)}{V^2\times{}^a g_1\times{}^a g_2}$

6.5 Worked examples

(i) Compute, for a crystal with lattice parameters $\{2, 2, 2, 90, 90, 90\}$ (i.e., cubic), the distance between the (110) planes.

Answer The distance between subsequent (110) planes is the inverse of the length of the reciprocal lattice vector $\mathbf{g}_{110}$. The length of the vector $\mathbf{g}_{110}$ is computed via the reciprocal metric tensor for the cubic crystal system (which is the inverse of the direct metric tensor):

$$g^*{}_{\text{cubic}} = (g_{\text{cubic}})^{-1} = (4\delta_{ij})^{-1} = \frac{1}{4}\delta_{ij}.$$

The length of the vector $\mathbf{g}_{110}$ is given by:

$$|\mathbf{g}| = \sqrt{[1 \quad 1 \quad 0]\frac{1}{4}\begin{bmatrix} 1 & 0 & 0 \\ 0 & 1 & 0 \\ 0 & 0 & 1 \end{bmatrix}\begin{bmatrix} 1 \\ 1 \\ 0 \end{bmatrix}};$$

$$= \sqrt{[1 \quad 1 \quad 0]\begin{bmatrix} \frac{1}{4} \\ \frac{1}{4} \\ 0 \end{bmatrix}} = \sqrt{\frac{1}{2}} = 0.7071,$$

from which the distance $d_{110} = 1/|\mathbf{g}_{110}| = 1.414$.

(ii) Compute, for a crystal with lattice parameters $\{3, 4, 6, 90, 90, 120\}$, the distance between the (111) planes.

Answer The distance between subsequent (111) planes is the inverse of the length of the reciprocal lattice vector $\mathbf{g}_{111}$. The length of the vector $\mathbf{g}_{111}$ can be computed via the general (triclinic) reciprocal metric tensor:

$$g^*_{\text{triclinic}} = \frac{1}{V^2}\begin{bmatrix} b^2c^2\sin^2\alpha & abc^2\mathcal{F}(\alpha,\beta,\gamma) & ab^2c\mathcal{F}(\gamma,\alpha,\beta) \\ abc^2\mathcal{F}(\alpha,\beta,\gamma) & a^2c^2\sin^2\beta & a^2bc\mathcal{F}(\beta,\gamma,\alpha) \\ ab^2c\mathcal{F}(\gamma,\alpha,\beta) & a^2bc\mathcal{F}(\beta,\gamma,\alpha) & a^2b^2\sin^2\gamma \end{bmatrix},$$

with

$$\mathcal{F}(\alpha,\beta,\gamma) = \cos\alpha\cos\beta - \cos\gamma,$$

and

$$V^2 = a^2b^2c^2(1 - \cos^2\alpha - \cos^2\beta - \cos^2\gamma + 2\cos\alpha\cos\beta\cos\gamma).$$

Filling in the values of the lattice parameters we find that:

$$V^2 = 9 \times 16 \times 36 \times (1 - 0 - 0 - \frac{1}{4} + 0) = 27 \times 36 \times 4,$$

and:

$$\mathcal{F}(\alpha,\beta,\gamma)=\frac{1}{2},\ \mathcal{F}(\gamma,\alpha,\beta)=\mathcal{F}(\beta,\gamma,\alpha)=0,$$

which leads to:

$$g^* = \frac{1}{27\times 36\times 4}\begin{bmatrix} 16\times 36\times 1 & \frac{12\times 36}{2} & 0 \\ \frac{12\times 36}{2} & \frac{9\times 36\times 1}{4} & 0 \\ 0 & 0 & \frac{9\times 16\times 3}{4} \end{bmatrix};$$

$$= \begin{bmatrix} \frac{4}{27} & \frac{1}{18} & 0 \\ \frac{1}{18} & \frac{1}{12} & 0 \\ 0 & 0 & \frac{1}{36} \end{bmatrix}.$$

The length of the vector is computed from:

$$|\mathbf{g}| = \sqrt{[1\quad 1\quad 1]\begin{bmatrix} \frac{4}{27} & \frac{1}{18} & 0 \\ \frac{1}{18} & \frac{1}{12} & 0 \\ 0 & 0 & \frac{1}{36} \end{bmatrix}\begin{bmatrix} 1 \\ 1 \\ 1 \end{bmatrix}};$$

$$= \sqrt{[1\quad 1\quad 1]\begin{bmatrix} \frac{11}{54} \\ \frac{25}{180} \\ \frac{1}{36} \end{bmatrix}} = \sqrt{\frac{10}{27}} = 0.6085.$$

The distance between the (111) planes is thus given by $d_{111} = 1/|\mathbf{g}_{111}| = 1.643$.

(iii) Let us consider a monoclinic crystal with lattice parameters $a = 4\,\text{nm}$, $b = 6\,\text{nm}$, $c = 5\,\text{nm}$ and $\beta = 120°$. What is the angle between the normals to the (101) and ($\bar{2}01$) planes?

Answer: First we derive the reciprocal metric tensor for this crystal system:

$$g^*_{\text{monoclinic}} = \begin{bmatrix} \frac{1}{a^2\sin^2\beta} & 0 & -\frac{\cos\beta}{ac\sin^2\beta} \\ 0 & \frac{1}{b^2} & 0 \\ -\frac{\cos\beta}{ac\sin^2\beta} & 0 & \frac{1}{c^2\sin^2\beta} \end{bmatrix} = \begin{bmatrix} \frac{1}{12} & 0 & \frac{1}{30} \\ 0 & \frac{1}{36} & 0 \\ \frac{1}{30} & 0 & \frac{4}{75} \end{bmatrix}.$$

The dot-product is then obtained from the product $g_i^* g_{ij}^* h_j^*$ or, explicitly:

$$\mathbf{g}_{(101)}\cdot\mathbf{g}_{(\bar{2}01)} = [101]\begin{bmatrix} \frac{1}{12} & 0 & \frac{1}{30} \\ 0 & \frac{1}{36} & 0 \\ \frac{1}{30} & 0 & \frac{4}{75} \end{bmatrix}\begin{bmatrix} -2 \\ 0 \\ 1 \end{bmatrix} = [101]\begin{bmatrix} -\frac{2}{15} \\ 0 \\ -\frac{1}{75} \end{bmatrix} = -\frac{11}{75}.$$

The dot-product of two vectors is also equal to the product of the lengths of the vectors multiplied by the cosine of the angle between them:

$$\mathbf{g}_{101} \cdot \mathbf{g}_{\bar{2}01} = g_{101} g_{\bar{2}01} \cos\theta.$$

The length of a vector is the square root of the dot-product of a vector with itself, hence we can use the reciprocal metric tensor again to compute the lengths of $\mathbf{g}_{101}$ and $\mathbf{g}_{\bar{2}01}$:

$$g_{101}^2 = [101]\begin{bmatrix} \frac{1}{12} & 0 & \frac{1}{30} \\ 0 & \frac{1}{36} & 0 \\ \frac{1}{30} & 0 & \frac{4}{75} \end{bmatrix}\begin{bmatrix} 1 \\ 0 \\ 1 \end{bmatrix} = [101]\begin{bmatrix} \frac{7}{60} \\ 0 \\ \frac{26}{300} \end{bmatrix} = \frac{61}{300};$$

$$g_{\bar{2}01}^2 = [-201]\begin{bmatrix} \frac{1}{12} & 0 & \frac{1}{30} \\ 0 & \frac{1}{36} & 0 \\ \frac{1}{30} & 0 & \frac{4}{75} \end{bmatrix}\begin{bmatrix} -2 \\ 0 \\ 1 \end{bmatrix} = [-201]\begin{bmatrix} -\frac{2}{15} \\ 0 \\ -\frac{1}{75} \end{bmatrix} = \frac{19}{75}.$$

The angle between the two vectors is therefore given by:

$$\theta = \cos^{-1}\left(\frac{\mathbf{g}_{101} \cdot \mathbf{g}_{\bar{2}01}}{g_{101} g_{\bar{2}01}}\right) = \cos^{-1}\left(\frac{-22}{\sqrt{61 \times 19}}\right) = 130.25°.$$

6.6 Historical notes

Paul Peter Ewald (1888–1985) was a German crystallographer. He studied under Arnold Sommerfeld at the University of Munich. He received his doctorate in 1912. Ewald contributed importantly in the development and the application of the concept of the reciprocal lattice to the field of X-ray crystallography. He developed this concept in 1913 and applied it to the geometrical interpretation of Bragg's law in 1921. Its widespread use came about after the 1926 development of Bernal for the interpretation of X-ray diffraction pictures. The *Ewald construction* and the *Ewald sphere* are both named after him.

Paul Ewald served until 1960 as the editor of *Acta Crystallographica* and he founded the IUCr, the International Union of Crystallography. He was awarded the Max Planck medal in 1978. He was the father-in-law of the famous physicist Hans Bethe. Ewald wrote several books on the mechanics of solids and liquids, including one with Ludwig Prandtl.

Theodore Von Karman (1881–1963) was a mathematical prodigy who contributed to the diverse fields of crystal dynamics, fluid mechanics, and aerodynamics. Von Karman was born in Budapest in 1881. He graduated

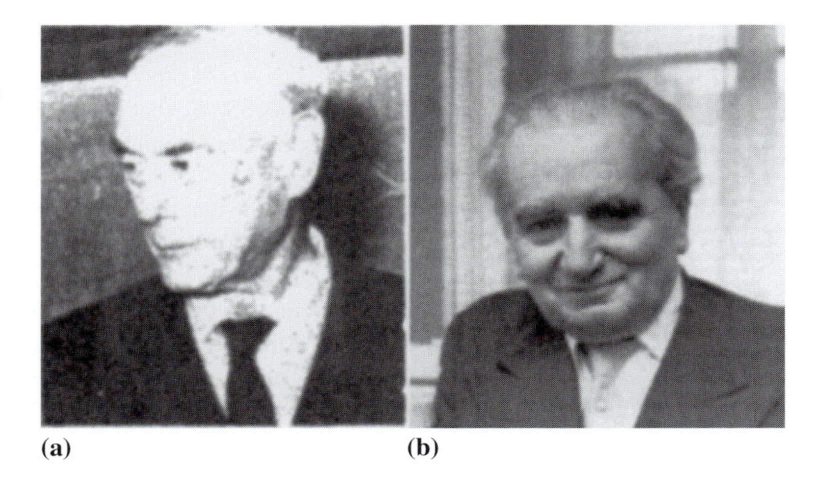

Fig. 6.4. (a) P. P. Ewald (1888–1985), and (b) T. Von Karman (1881–1963) (pictures courtesy of J. Lima-de-Faria).

in 1902, with a degree in mechanical engineering, from the Palatine Joseph Polytechnic. In 1906 he received a two year fellowship from the Hungarian Academy of Sciences. This was used to study at Göttingen, where he was influenced by scientists and mathematicians such as Klein, Hilbert and Prandtl. Von Karman's interest in fluid dynamics and aerodynamics was evident in early work modeling effects in a wind tunnel for the Zeppelin airship company. *Karman's vortex street* is a fluid flow concept derived from that work. It describes the alternating double row of vortices behind a flat body in fluid flow.

In 1913, Von Karman was named director of the Aeronautical Institute at Aachen, Germany. He also served as the chair of aeronautics and mechanics at the technical university in Aachen. In 1930, after serving in the Austro-Hungarian army and subsequent years at Aachen, Von Karman took the position of full-time director of the Aeronautical Laboratory at California Institute of Technology. In 1933 he founded the US Institute of Aeronautical Sciences. During World War II, Von Karman made important contributions to the United States' efforts to develop rockets. Von Karman was awarded the United States Medal for Merit in 1946. He was the first recipient of the National Medal for Science in 1963.

Von Karman also made significant contributions to the field of crystal physics. With **Max Born** (1882–1970), a German physicist, Von Karman expressed the famous periodic boundary conditions for the solution of Schrödinger's equation in a periodic crystalline lattice. In this work, they developed three-dimensional Fourier analysis and periodic boundary conditions to treat the problem of lattice dynamics and the normal modes of vibration of a crystal.

6.7 Problems

(i) *Reciprocal metric tensor*: Show that the product of the direct and reciprocal metric tensors for the triclinic crystal system results in the unit matrix.

(ii) *Reciprocal unit cell*: Show that the volume of the reciprocal unit cell is the inverse of the volume of the direct space unit cell (for an arbitrary crystal system).

(iii) *Reciprocal basis vectors*: For a monoclinic unit cell with lattice parameters $\{2, 3, 4, 90, 75, 90\}$, work out explicit expressions for the reciprocal lattice vectors $\mathbf{a}_1^*$, $\mathbf{a}_2^*$, and $\mathbf{a}_3^*$, using Equations (6.4).

(iv) *Reciprocal lattice*: Repeat the drawings of Figs. 6.1 and 6.2 for a hexagonal lattice with lattice parameters $\{2, 2, 1, 90, 90, 120\}$. Make all your drawings in the plane formed by **a** and **b**.

(a) Determine the reciprocal lattice vectors and place them in your drawing.
(b) Draw the reciprocal lattice points for the range $-2 \leq h \leq 2$, $-2 \leq k \leq 2$.
(c) Draw the (11.0) and (12.0) lattice planes and show graphically that these planes are perpendicular to the corresponding $\mathbf{g}_{hkl}$ vectors.
(d) Compute the direct and reciprocal metric tensors.
(e) Compute the interplanar spacing for the (11.0) and (12.0) lattice planes, using the reciprocal metric tensor. Confirm graphically the inverse relation between interplanar spacing and length of the reciprocal lattice vectors.
(f) Compute the angle between the (11.0) and (12.0) lattice planes, and verify graphically that this angle is correct.

(v) *Angles and interplanar spacings*: Consider a monoclinic lattice with lattice parameters $\{1, 3, 2, 90, 45, 90\}$.

(a) What is the distance between the (111) planes?
(b) What is the angle between the $\mathbf{a}^*$ axis and the normal to the (221) plane?
(c) What is the angle between the plane normals (101) and $(\bar{1}01)$?
(d) What is the volume of the reciprocal unit cell?

(vi) C-*graphite: reciprocal metric tensor*: Graphite is an important solid lubricant and catalyst. Its structure, discussed in Chapter 17, consists of hexagonal networks of C atoms stacked along the c-axis; atoms in one layer are located above the center of the hexagons in the surrounding layers. Graphite has the lattice constants: $\{0.246, 0.246, 0.67, 90, 90, 120\}$ (with a and c in nm).

(a) Determine the reciprocal metric tensor for graphite.
(b) Determine the angle between (111) and (100) planes for graphite using the reciprocal metric tensor.

(vii) *Density of solid benzene*: Benzene, C_6H_6, is a molecular hydrocarbon discussed in Chapter 25. Benzene is a liquid at standard atmospheric pressure and temperature. The lowest pressure allotrope of solid benzene crystallizes at 0.7 kbar at room temperature and is orthorhombic with four formula units per unit cell. Given X-ray diffraction determined literature values for the spacing between the (020), (200) and (111) planes of 0.478, 0.372, and 0.448 nm, respectively, determine the following:

(a) The values of the a, b, and c lattice constants.
(b) The volume of the unit cell.
(c) The density of solid benzene.

CHAPTER

7 Additional crystallographic computations

> *"Nature is an infinite sphere of which the center is everywhere and the circumference nowhere."*
>
> Blaise Pascal

In this chapter, we introduce a few important tools for crystallography. We begin with the stereographic projection, an important graphical tool for the description of 3-D crystals. Then, we discuss briefly the vector cross product, which we used in Chapter 6 to define the reciprocal lattice. We introduce general relations between different lattices (coordinate transformations), a method to convert crystal coordinates to Cartesian coordinates, and we conclude the chapter with examples of stereographic projections for cubic and monoclinic crystals.

7.1 The stereographic projection

In Chapter 5, we defined the *Miller indices* as a convenient tool to describe lattice planes. We also defined the concept of a *family*. Since real crystals are 3-D objects, we should, in principle, make 3-D drawings to represent planes and plane normals. This is tedious, in particular for the lower symmetry crystal systems, such as the triclinic and monoclinic systems. W. H. Miller devised a graphical tool to simplify the representation of 3-D objects such as crystals. The tool is known as the *stereographic projection*.

A stereographic projection is a 2-D representation of a 3-D object located at the center of a sphere. Figure 7.1 shows a sphere of radius R; to obtain the stereographic projection (SP) of a point on the sphere, one connects the point with the south pole of the sphere and then determines the intersection of this connection line with the equatorial plane. The resulting point is the

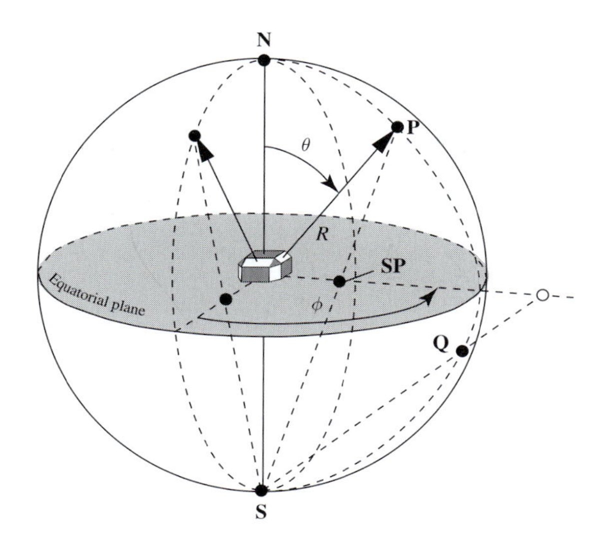

Fig. 7.1. Stereographic projection of the normals on crystal faces (figure reproduced from Fig. 1.9 in *Introduction to Conventional Transmission Electron Microscopy*, M. De Graef, 2003, Cambridge University Press).

SP of the original point. The point on the sphere could represent the normal to a crystal plane, as shown in the figure. The stereographic projection itself is then only the equatorial plane of Fig. 7.1. The projection is represented by a circle, representing the equatorial circle. Inside the circle, the projections from points in the Northern hemisphere are represented by small solid circles.

If a point lies *above* the equator plane (e.g., the point **P** in Fig. 7.1), then its SP will be a point *inside* the great circle in the equator plane. If a point lies *below* the equator plane (e.g., point **Q**), then the projection will fall outside the circle in the equatorial plane. In Fig. 7.1, the SP of **Q** is represented by an open circle. To avoid projection points being too far from the circle, it is customary to project points that lie in the southern hemisphere from the north pole instead of from the south pole. To distinguish between those points, a point projected from the north pole is represented by an open circle, whereas a point projected from the south pole is represented by a closed circle. This is represented in the 2-D drawing of Fig. 7.2.

The location of the SP of a point is most easily expressed using spherical coordinates. If the original point has coordinates (R, ϕ, θ), with ϕ measured from a fixed axis in the equatorial plane and θ measured from the north pole (see Fig. 7.1), then the stereographic coordinates are easily shown to be $(\phi, \rho) = (\phi, R\tan(\theta/2))$. From this, we see that a point on the equatorial circle, i.e. with coordinates $(R, \phi, \pi/2)$, will have an SP on the equatorial circle with coordinates (ϕ, R).

One can show mathematically that stereographic projections conserve angles, i.e., measurement of an angle on the projection will always correspond to the real 3-D angle. It is this property that renders SP an extremely useful technique for crystallography. In the following paragraphs, we will discuss several basic SP operations, using the so-called *Wulff net*, shown in Fig. 7.3.

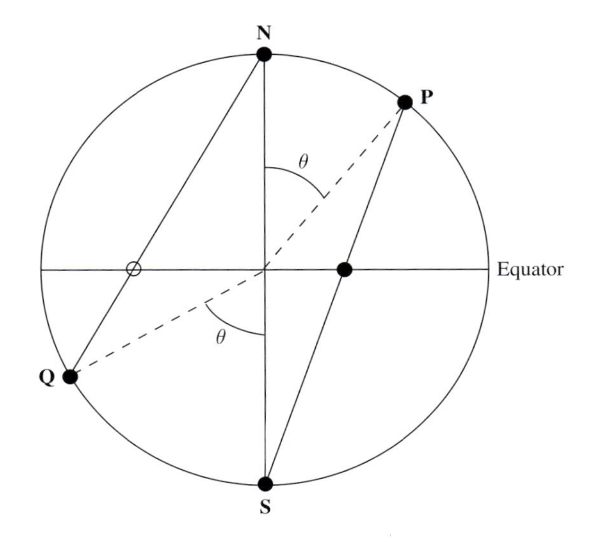

Fig. 7.2. Projection of a point below the equatorial plane is usually done from the north pole; the projection is then represented by an open circle.

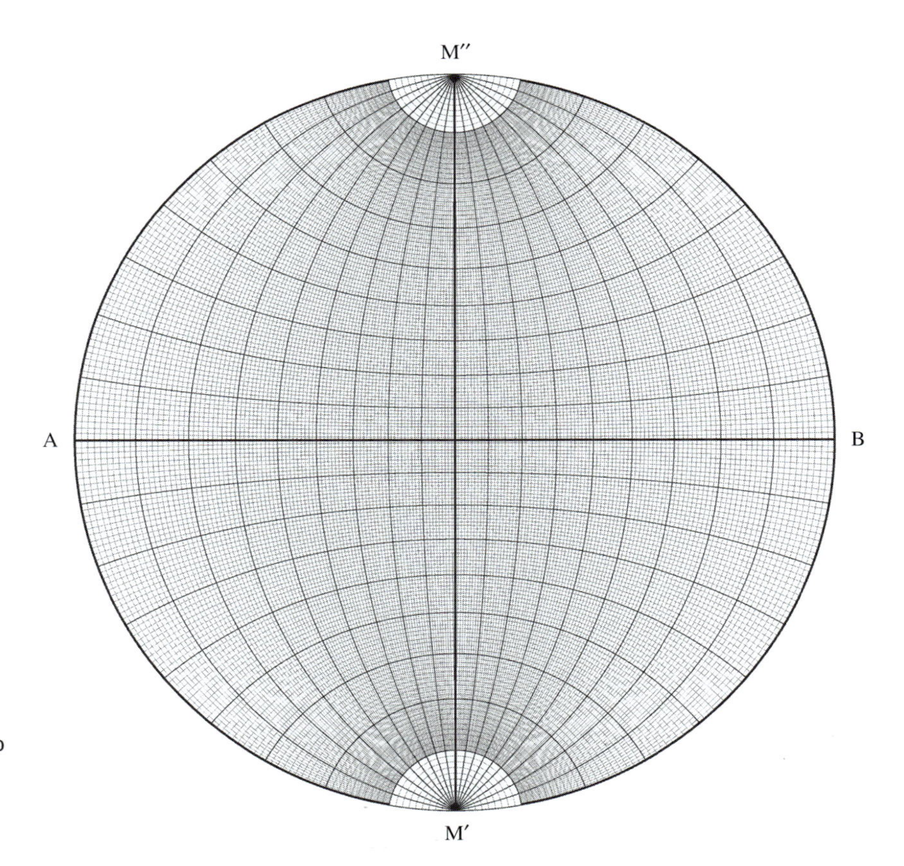

Fig. 7.3. Wulff net used for stereographic projections. The size of the net is sized down to fit the page. Normally one uses a net with a standard diameter of 20 cm.

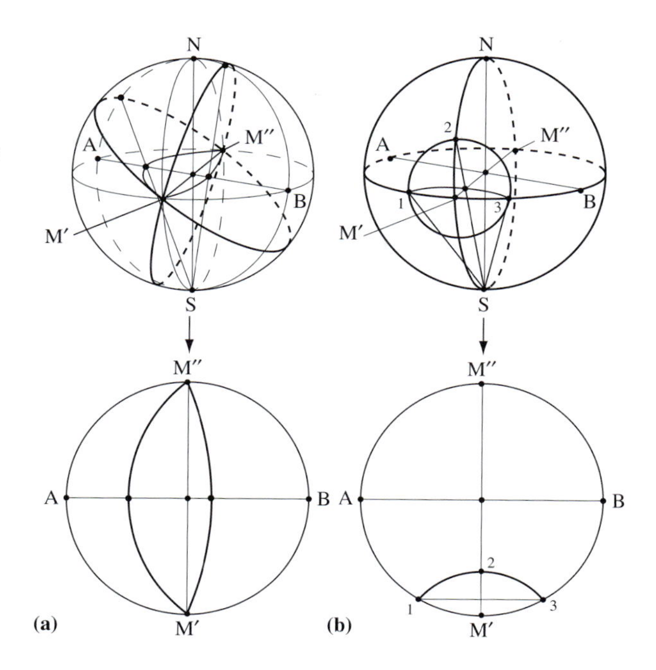

Fig. 7.4. Illustration of the origin of the arcs on the Wulff net. (a) shows how great circles through the points M' and M'' give rise to arcs in the projection plane. In (b), the projection sphere is cut by a plane normal to the M'–M'' axis. The projections of the points 1, 2, and 3 are indicated in the drawing, as well as in the stereographic projection in the lower portion of the drawing.

A standard Wulff net has a diameter of 20 cm. The net shows two sets of arcs : the first set intersects the points M′ and M″ and represents the projections of *great circles*, i.e., circles with the same diameter as the projection sphere. Figure 7.4(a) shows how these arcs are related to the great circles. If the line M′–M″ is taken as the origin for measurement of ϕ, then one can read the value of θ from the line A–B for each of these great circles. There is a great circle per degree, and every tenth circle is drawn with a slightly thicker line.

The second set of arcs on the net corresponds to the projection of a set of parallel planes, intersecting the projection sphere in circles. These planes are perpendicular to the equatorial plane and to the M′–M″ axis, as shown in Fig. 7.4(b). If the projection sphere is rotated around this axis, then a point on the surface will trace a circular path; the projection of this path is given by the second set of arcs, which, on the standard Wulff net, are again spaced by one degree.

Before we illustrate the uses of the stereographic projection, we must first introduce a number of additional concepts. We will return to the stereographic projection with a set of examples in Section 7.5.

7.2 About zones and zone axes

In this section, we will define how to compute the vector cross product in an arbitrary crystal system. This will then lead to the concepts of *zone* and *zone axis*.

7.2.1 The vector cross product

The *vector product*, also known as the *cross product*, of two vectors $\mathbf{p}$ and $\mathbf{q}$, is a third vector $\mathbf{z}$. This vector is perpendicular to both $\mathbf{p}$ and $\mathbf{q}$, with magnitude:

$$|\mathbf{z}| = |\mathbf{p} \times \mathbf{q}| = |\mathbf{p}|\,|\mathbf{q}|\,\sin\alpha,$$

with α the angle between the two vectors. This definition is independent of the particular reference frame in which these vectors are described (see below).

The cross product between two vectors $\mathbf{p}$ and $\mathbf{q}$ expressed in a Cartesian reference frame can be determined by means of a simple determinant equation:

$$\begin{aligned}\mathbf{p}\times\mathbf{q} &= (p_1\mathbf{e}_1 + p_2\mathbf{e}_2 + p_3\mathbf{e}_3) \times (q_1\mathbf{e}_1 + q_2\mathbf{e}_2 + q_3\mathbf{e}_3);\\ &= (p_2q_3 - p_3q_2)\mathbf{e}_1 + (p_3q_1 - p_1q_3)\mathbf{e}_2 + (p_1q_2 - p_2q_1)\mathbf{e}_3;\\ &= \begin{vmatrix} \mathbf{e}_1 & \mathbf{e}_2 & \mathbf{e}_3 \\ p_1 & p_2 & p_3 \\ q_1 & q_2 & q_3 \end{vmatrix}.\end{aligned} \tag{7.1}$$

In a non-Cartesian reference frame this equation becomes somewhat more complicated. We need to define the so-called *permutation symbol* to generalize the definition of the cross product to an arbitrary reference frame.

We define the *permutation symbol* ϵ_{ijk} as follows:

$$\epsilon_{ijk} = +V \quad \text{even permutations of 123;} \tag{7.2}$$

$$\epsilon_{ijk} = -V \quad \text{odd permutations of 123,} \tag{7.3}$$

where V is the volume of the unit cell. We define the *reciprocal permutation symbol* ϵ^*_{ijk} as follows:

$$\epsilon^*_{ijk} = +\frac{1}{V} = V^* \quad \text{even permutations of 123;} \tag{7.4}$$

$$\epsilon^*_{ijk} = -\frac{1}{V} = -V^* \quad \text{odd permutations of 123,} \tag{7.5}$$

where V^* is the volume of the reciprocal unit cell. Note that one sometimes uses the *normalized permutation symbol* e_{ijk} which is defined by:

$$e_{ijk} = \frac{\epsilon_{ijk}}{V} \quad \text{and} \quad e^*_{ijk} = \frac{\epsilon^*_{ijk}}{V^*}.$$

An even permutation is one of the following combinations of indices: 123, 231, or 312. An odd permutation is one of 132, 213, or 321. For all other combinations (such as the ones where two or more indices are equal to each other) the permutation symbol is equal to zero. The meaning of the permutation symbol can best be illustrated with an example; consider the following expression:

$$F = \epsilon_{ijk} p_i q_j r_k.$$

We know that the summation convention is used, so this expression is equal to:

$$F = \sum_{i,j,k=1}^{3} \epsilon_{ijk} p_i q_j r_k,$$

i.e., there are a total of 27 terms in the summation. The permutation symbol is only different from zero for 6 combinations of the indices, so we find:

$$F = V(p_1q_2r_3 + p_2q_3r_1 + p_3q_1r_2 - p_2q_1r_3 - p_1q_3r_2 - p_3q_2r_1).$$

We can regroup these terms by separating the components of $\mathbf{r}$:

$$F = V\,[(p_2q_3 - p_3q_2)r_1 + (p_3q_1 - p_1q_3)r_2 + (p_1q_2 - p_2q_1)r_3].$$

Comparing this expression with the one in Equation 7.1, we find that the terms in parentheses are identical. This means that the permutation symbol can be used to rewrite the expression for the cross product of two vectors. For the Cartesian system above, we have $V = 1$. Replacing the components r_k by the Cartesian basis vectors $\mathbf{e}_k$, we find for the cross product in a Cartesian reference frame:

$$\mathbf{p} \times \mathbf{q} = \epsilon_{ijk} p_i q_j \mathbf{e}_k. \tag{7.6}$$

It is easy to show that Equation 7.6 is identical to Equation 7.1.

We can now generalize this equation to non-Cartesian reference frames. We begin by noting that the vector cross product results in a vector that is normal to the plane formed by the two original vectors, $\mathbf{p}$ and $\mathbf{q}$. We know that the normal to a plane can be described by a vector in reciprocal space. So, we conclude that the vector cross product of two vectors in real space is a vector in reciprocal space! The general definition of the vector cross product is then:

$$\boxed{\mathbf{p} \times \mathbf{q} = \epsilon_{ijk} p_i q_j \mathbf{a}^*_k.} \tag{7.7}$$

For the Cartesian reference frame, the direct and reciprocal basis vectors are identical, so the distinction between reciprocal and direct basis vectors is not necessary.

The cross product is now written in determinant notation as:

$$\mathbf{p}\times\mathbf{q} = V\begin{vmatrix} \mathbf{a}_1^* & \mathbf{a}_2^* & \mathbf{a}_3^* \\ p_1 & p_2 & p_3 \\ q_1 & q_2 & q_3 \end{vmatrix}. \tag{7.8}$$

Let us now look at a few examples. First of all, consider the cross product between $\mathbf{a}_1$ and $\mathbf{a}_2$. Since $\mathbf{a}_1 = [100]$ and $\mathbf{a}_2 = [010]$, we have:

$$\mathbf{a}_1\times\mathbf{a}_2 = V\begin{vmatrix} \mathbf{a}_1^* & \mathbf{a}_2^* & \mathbf{a}_3^* \\ 1 & 0 & 0 \\ 0 & 1 & 0 \end{vmatrix} = V\mathbf{a}_3^* = V\mathbf{g}_{001}. \tag{7.9}$$

In fact, we already knew this from the definition of the reciprocal basis vectors, but it is nice to see that our definition of the general cross product is consistent with the definition of the reciprocal space basis vectors.

Another example: consider the vectors $\mathbf{p} = [120]$ and $\mathbf{q} = [011]$. Their cross product is given by:

$$\mathbf{p}\times\mathbf{q} = V\begin{vmatrix} \mathbf{a}_1^* & \mathbf{a}_2^* & \mathbf{a}_3^* \\ 1 & 2 & 0 \\ 0 & 1 & 1 \end{vmatrix} = V(2\mathbf{a}_1^* - \mathbf{a}_2^* + \mathbf{a}_3^*) = V\mathbf{g}_{2\bar{1}1}. \tag{7.10}$$

In other words, the cross product between the lattice translation vectors [120] and [011] is normal to the $(2\bar{1}1)$ plane. In most cases, we are only interested in the direction of the cross product vector, not its length, so it is common practice to drop the volume factor. The reader should realize that the true length of the cross product vector *always* involves the volume of the unit cell!

Next, let us try to answer the question: what is the cross product between two plane normals? This is equivalent to asking: which direction is common to two planes? We can start from the general definition of the vector cross product, and simply replace all quantities by their starred (i.e., reciprocal) counterparts:

$$\mathbf{g}\times\mathbf{h} = \epsilon_{ijk}^* g_i^* h_j^* \mathbf{a}_k^{**}. \tag{7.11}$$

Since the reciprocal of the reciprocal basis vector is the direct space basis vector, $\mathbf{a}_k^{**} = \mathbf{a}_k$ (reader exercise), this relation reduces to:

$$\mathbf{g}\times\mathbf{h} = \epsilon_{ijk}^* g_i^* h_j^* \mathbf{a}_k. \tag{7.12}$$

We find that the cross product of two plane normals is parallel to the direction normal to the plane formed by the two plane normals, i.e., the direction common to the two planes. In determinant form we have:

$$\mathbf{g} \times \mathbf{h} = V^* \begin{vmatrix} \mathbf{a}_1 & \mathbf{a}_2 & \mathbf{a}_3 \\ g_1 & g_2 & g_3 \\ h_1 & h_2 & h_3 \end{vmatrix}. \tag{7.13}$$

As an example: what is the direction common to the (111) and (120) planes?

$$\mathbf{g}_{111} \times \mathbf{g}_{120} = V^* \begin{vmatrix} \mathbf{a}_1 & \mathbf{a}_2 & \mathbf{a}_3 \\ 1 & 1 & 1 \\ 1 & 2 & 0 \end{vmatrix} = V^*(-2\mathbf{a}_1 + \mathbf{a}_2 + \mathbf{a}_3) = V^* \mathbf{t}_{\bar{2}11}. \tag{7.14}$$

At this point, we introduce a shorthand notation for the computation of cross products. Write down the indices of the two vectors twice in horizontal rows, as follows (for two plane normals $(h_1k_1l_1)$ and $h_2k_2l_2)$):

$$\begin{matrix} h_1 & k_1 & l_1 & h_1 & k_1 & l_1 \\ h_2 & k_2 & l_2 & h_2 & k_2 & l_2 \end{matrix}.$$

Then, remove the first and the last column, i.e.,

$$\begin{matrix} \not{h}_1 & k_1 & l_1 & h_1 & k_1 & \not{l}_1 \\ \not{h}_2 & k_2 & l_2 & h_2 & k_2 & \not{l}_2 \end{matrix}.$$

Next, compute the three 2×2 determinants formed by the 8 numbers above, as in:

$$\begin{matrix} \not{h}_1\ k_1 & & l_1 & & h_1 & & k_1\ \not{l}_1 \\ & \times & & \times & & \times & \\ \not{h}_2\ k_2 & & l_2 & & h_2 & & k_2\ \not{l}_2 \end{matrix}.$$

This leads to the following components for the direction vector $[uvw]$, consistent with the definition of the general cross product:

$$\begin{aligned} u &= k_1 l_2 - k_2 l_1; \\ v &= l_1 h_2 - l_2 h_1; \\ w &= h_1 k_2 - h_2 k_1. \end{aligned} \tag{7.15}$$

Note that, in these equations, we have consistently dropped the volume of the unit cell (or the reciprocal volume), since we are only interested in directions, not actual vector lengths.

What is the direction common to the (231) and (111) planes, using the shorthand notation? Write down the rows of components and remove the first and last columns:

$$\begin{array}{cccccc} \not{2} & 3 & 1 & 2 & 3 & \not{1} \\ \not{1} & 1 & 1 & 1 & 1 & \not{1} \end{array}.$$

Then compute the determinants:

$$u = 3 - 1 = 2;$$
$$v = 1 - 2 = -1;$$
$$w = 2 - 3 = -1.$$

Therefore, the common direction is the $[2\bar{1}\bar{1}]$ direction.

Before we end this section with another example of the use of the vector cross product, we should point out that the general equation of the vector cross product is independent of the crystal system that is being used. In other words, the [001] direction is normal to the plane formed by the (100) and (010) plane normals *in every crystal system*! Similarly, the (100) plane normal is perpendicular to the [010] and [001] directions *in every crystal system*. This is a direct consequence of the way we have defined the reciprocal basis vectors in the previous chapter. Any other definition would have led to a much more complicated relation between direct and reciprocal space quantities.

As a final example of the use of the vector cross product, let us show that the volume of the unit cell is given by:

$$\boxed{V = \mathbf{a} \cdot (\mathbf{b} \times \mathbf{c}).} \tag{7.16}$$

This equation can be derived as follows:

$$\begin{aligned} \mathbf{a} \cdot (\mathbf{b} \times \mathbf{c}) &= \mathbf{a}_1 \cdot (\mathbf{a}_2 \times \mathbf{a}_3); \\ &= \mathbf{a}_1 \cdot \left[\epsilon_{ijk} \mathbf{a}_{2,i} \mathbf{a}_{3,j} \mathbf{a}_k^*\right]; \\ &= \epsilon_{ijk} \delta_{2i} \delta_{3j} \mathbf{a}_1 . \mathbf{a}_k^*; \\ &= \epsilon_{23k} \delta_{1k}; \\ &= \epsilon_{231}; \\ &= V. \end{aligned}$$

In this derivation we have used the fact that the i-th component of the basis vector $\mathbf{a}_j$ with respect to the direct basis vectors is $\mathbf{a}_{j,i} = \delta_{ij}$. We have also used the definition of the reciprocal basis vectors. One can show (reader exercise) that the volume of the unit cell is also given by the square root of the determinant of the metric tensor:

$$V = \sqrt{\det |g|} \quad (\text{and } V^* = \sqrt{\det |g^*|}). \tag{7.17}$$

7.2.2 About zones and the zone equation

It is frequently useful to talk about crystallographic directions that are common to two or more planes. As we have seen in the previous section, the direction common to two planes can be written as the cross product of the two normals to the planes. Sometimes we will need to determine *all* the planes that contain a given direction $[uvw]$. Such a set of planes is known as a *zone* (see Fig. 7.5). The direction $[uvw]$ is then the *zone axis*. Zones play an important role in the discussion of symmetry (see Chapter 8).

It is easy to see that a plane belongs to a zone only when the plane normal is perpendicular to the zone axis. This means that the dot-product $\mathbf{g}\cdot\mathbf{t}$, with $\mathbf{t} = [uvw]$ must vanish.

$$\mathbf{g}\cdot\mathbf{t} = (g_i\mathbf{a}_i^*)\cdot(u_j\mathbf{a}_j) = g_i u_j(\mathbf{a}_i^*\cdot\mathbf{a}_j) = g_i u_j \delta_{ij} = g_i u_i = 0,$$

where we have used the definition of the reciprocal basis vectors. We find that the condition for a plane (hkl) to belong to a zone with zone axis $[uvw]$ is:

$$\boxed{hu + kv + lw = 0.} \tag{7.18}$$

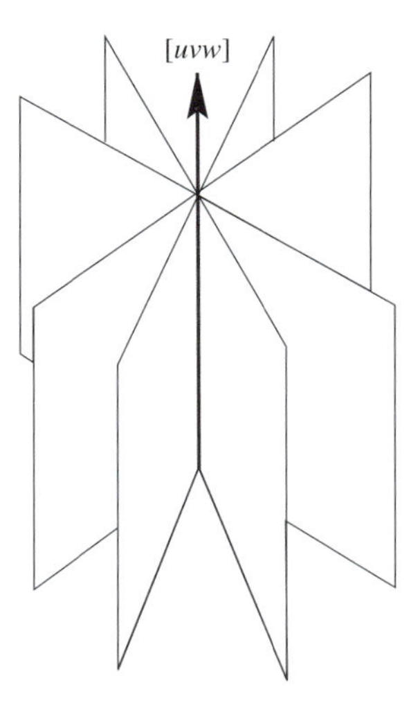

Fig. 7.5. Schematic illustration of a zone.

This equation is again valid for all crystal systems. The equation is known as the *zone equation*.

Consider the direction [101] in an arbitrary crystal. Inserting these values into the zone equation we find

$$h + l = 0.$$

All planes (hkl) satisfying this equation contain the [101] direction. Sometimes one inserts the zone equation explicitly into the Miller indices, which in this case would lead to planes of the form $(hk\bar{h})$, since $l = -h$.

Alternatively, we can ask for all the directions $[uvw]$ that are contained within a plane (hkl). The zone equation can again be used. The directions in the (111) plane all satisfy the equation:

$$u + v + w = 0.$$

To conclude this section, we need to establish the relation between a zone and its stereographic projection. If we select a zone axis, say [001] in a cubic crystal, then all the planes that belong to this zone will have their normals in the plane normal to [001], in this case the (001) plane. The stereographic projection of the [001] direction is a point in the equatorial plane. The (001) plane intersects the stereographic projection sphere in a great circle. We have seen in Section 7.1 that the projection of a great circle is again the arc of a circle, similar to the arcs between the points M' and M''. This arc corresponds to the projection of all the plane normals that are normal to the [001] zone axis. In other words, the SP of a zone axis is a point, the SP of a zone is a great circle in the equatorial plane (see Fig. 7.6). The angle between the projection of the zone axis and the projection of the great circle is 90°.

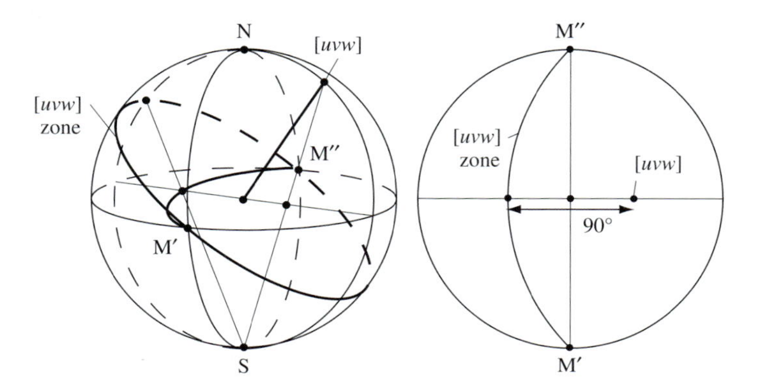

Fig. 7.6. Schematic illustration of the relation between a zone and zone axis and their stereographic projections.

7.2.3 The reciprocal lattice and zone equation in the hexagonal system

For the hexagonal crystal system, one can use the four-index notation to derive the following zone equation (Okamoto and Thomas, 1968):

$$hu + kv + it + lw = 0. \tag{7.19}$$

It is important to point out that this relation can only be valid if the reciprocal lattice vectors for the hexagonal crystal system are defined in a way *which differs from the standard definition for the other crystal systems.* If a plane in a hexagonal crystal is represented by the three indices (hkl), then the normal to that plane would be given in the usual way by:

$$h\mathbf{a}_1^* + k\mathbf{a}_2^* + l\mathbf{c}^*.$$

When we use the four-index notation, then we must define new reciprocal lattice vectors, $\mathbf{A}_i^*$ with $i = 1, 2, 3$ and $\mathbf{C}^*$, such that the following relation is valid:

$$h\mathbf{a}_1^* + k\mathbf{a}_2^* + l\mathbf{c}^* = h\mathbf{A}_1^* + k\mathbf{A}_2^* + i\mathbf{A}_3^* + l\mathbf{C}^*. \tag{7.20}$$

From the usual definition of the reciprocal lattice vectors we find:

$$\mathbf{a}_1^* = \frac{2(2\mathbf{a}_1 + \mathbf{a}_2)}{3a^2};$$
$$\mathbf{a}_2^* = \frac{2(\mathbf{a}_1 + 2\mathbf{a}_2)}{3a^2};$$
$$\mathbf{c}^* = \frac{\mathbf{c}}{c^2}.$$

Substituting these relations into Equation 7.20, we find that the new reciprocal lattice vectors are given by:

$$\mathbf{A}_1^* = \frac{2}{3a^2}\mathbf{a}_1;$$
$$\mathbf{A}_2^* = \frac{2}{3a^2}\mathbf{a}_2;$$
$$\mathbf{A}_3^* = \frac{2}{3a^2}\mathbf{a}_3;$$
$$\mathbf{C}^* = \frac{1}{c^2}\mathbf{c}.$$

The zone equation for the hexagonal system can now be derived by taking the dot-product between a direction $[uvtw]$ and a plane normal $(hkil)$ expressed in the new reciprocal lattice:

$$\begin{aligned}
\mathbf{t}\cdot\mathbf{g} &= (u\mathbf{a}_1+v\mathbf{a}_2+t\mathbf{a}_3+w\mathbf{c})\cdot(h\mathbf{A}_1^*+k\mathbf{A}_2^*+i\mathbf{A}_3^*+l\mathbf{C}^*);\\
&= (u\mathbf{a}_1+v\mathbf{a}_2+t\mathbf{a}_3+w\mathbf{c})\cdot(\frac{2h}{3a^2}\mathbf{a}_1+\frac{2k}{3a^2}\mathbf{a}_2+\frac{2i}{3a^2}\mathbf{a}_3+\frac{l}{c^2}\mathbf{c});\\
&= \frac{1}{3}(u\ v\ t\ w)\begin{pmatrix} 2 & \bar{1} & \bar{1} & 0\\ \bar{1} & 2 & \bar{1} & 0\\ \bar{1} & \bar{1} & 2 & 0\\ 0 & 0 & 0 & 3\end{pmatrix}\begin{pmatrix} h\\ k\\ i\\ l\end{pmatrix};\\
&= hu+kv+it+lw,
\end{aligned}$$

which is the zone equation in the hexagonal crystal system.

7.3 Relations between direct space and reciprocal space

We know that a vector is a mathematical object that exists independent of the reference frame. This means that every vector defined in the direct lattice must also have components with respect to the reciprocal basis vectors and vice versa. In this section, we will devise a tool that will permit us to transform vector quantities back and forth between direct and reciprocal space.

Consider the vector $\mathbf{p}$:

$$\mathbf{p} = p_i\mathbf{a}_i = p_j^*\mathbf{a}_j^*,$$

where p_j^* are the reciprocal space components of $\mathbf{p}$. Multiplying both sides by the direct basis vector $\mathbf{a}_m$, we have:

$$\begin{aligned}
p_i\mathbf{a}_i\cdot\mathbf{a}_m &= p_j^*\mathbf{a}_j^*\cdot\mathbf{a}_m,\\
p_i g_{im} &= p_j^*\delta_{jm} = p_m^*,
\end{aligned} \qquad (7.21)$$

or:

$$p_m^* = p_i g_{im}. \qquad (7.22)$$

It is easily shown that the inverse relation is given by:

$$p_i = p_m^* g_{mi}^*. \qquad (7.23)$$

We thus find that *post-multiplication by the metric tensor transforms vector components from direct space to reciprocal space, and post-multiplication by the reciprocal metric tensor transforms vector components from reciprocal to direct space.* These relations are useful because they permit us to determine the components of a direction vector $\mathbf{t}_{[uvw]}$ with respect to the reciprocal basis vectors, or the components of a plane normal $\mathbf{g}_{hkl}$ with respect to the direct basis vectors.

Now we have all the tools we need to express the reciprocal basis vectors in terms of the direct basis vectors. Consider again the vector $\mathbf{p}$:

$$\mathbf{p} = p_i \mathbf{a}_i.$$

If we replace p_i by $p^*_m g^*_{mi}$, then we have:

$$\mathbf{p} = p^*_m g^*_{mi} \mathbf{a}_i = p^*_m \mathbf{a}^*_m,$$

from which we find:

$$\mathbf{a}^*_m = g^*_{mi} \mathbf{a}_i, \tag{7.24}$$

and the inverse relation:

$$\mathbf{a}_m = g_{mi} \mathbf{a}^*_i. \tag{7.25}$$

In other words, *the rows of the metric tensor contain the components of the direct basis vectors in terms of the reciprocal basis vectors, whereas the rows of the reciprocal metric tensor contain the components of the reciprocal basis vectors with respect to the direct basis vectors.*

Finally, from Equation 7.25 we find after multiplication by the vector $\mathbf{a}^*_k$:

$$\begin{aligned} \mathbf{a}_m \cdot \mathbf{a}^*_k &= g_{mi} \mathbf{a}^*_i \cdot \mathbf{a}^*_k, \\ \delta_{mk} &= g_{mi} g^*_{ik}. \end{aligned} \tag{7.26}$$

In other words, the matrices representing the direct and reciprocal metric tensors are each other's inverse. This leads to a simple procedure to determine the reciprocal basis vectors of a crystal:

(i) compute the direct metric tensor;
(ii) invert it to find the reciprocal metric tensor;
(iii) apply Equation 7.24 to find the reciprocal basis vectors.

Let us illustrate this procedure using an example based on the monoclinic unit cell $\{1, 1, 1, 90, 45, 90\}$ that we used before in Chapter 6. The direct metric tensor of this cell is given by:

$$g_{ij} = \begin{bmatrix} 1 & 0 & \frac{\sqrt{2}}{2} \\ 0 & 1 & 0 \\ \frac{\sqrt{2}}{2} & 0 & 1 \end{bmatrix}.$$

The inverse of this matrix is equal to the reciprocal metric tensor:

$$g^*_{ij} = (g_{ij})^{-1} = \begin{bmatrix} 2 & 0 & -\sqrt{2} \\ 0 & 1 & 0 \\ -\sqrt{2} & 0 & 2 \end{bmatrix}.$$

Therefore, the reciprocal basis vectors can be written as:

$$\begin{pmatrix} \mathbf{a}^* \\ \mathbf{b}^* \\ \mathbf{c}^* \end{pmatrix} = \begin{bmatrix} 2 & 0 & -\sqrt{2} \\ 0 & 1 & 0 \\ -\sqrt{2} & 0 & 2 \end{bmatrix} \begin{pmatrix} \mathbf{a} \\ \mathbf{b} \\ \mathbf{c} \end{pmatrix} = \begin{pmatrix} 2\mathbf{a} - \sqrt{2}\mathbf{c} \\ \mathbf{b} \\ -\sqrt{2}\mathbf{a} + 2\mathbf{c} \end{pmatrix},$$

in agreement with our findings in Chapter 6.

7.4 Coordinate transformations

The mathematical relations derived in the preceding chapters allow us to compute any geometrical quantity in any of the seven crystal systems. One may now ask the question: how do these relations change when we change the reference frame? The need to change from one reference frame to another frequently arises in the study of solid state phase transformations, when the crystal structure changes with temperature or applied field (electric or magnetic). In this section, we will describe in detail how one can convert vectors and the metric tensors from one reference frame to another.

7.4.1 Transformation rules

Let us consider two crystallographic reference frames, $\{\mathbf{a}_1, \mathbf{a}_2, \mathbf{a}_3\}$ and $\{\mathbf{a}'_1, \mathbf{a}'_2, \mathbf{a}'_3\}$. In general, the relation between the two sets of basis vectors can be written as:

$$\begin{aligned} \mathbf{a}'_1 &= \alpha_{11}\mathbf{a}_1 + \alpha_{12}\mathbf{a}_2 + \alpha_{13}\mathbf{a}_3; \\ \mathbf{a}'_2 &= \alpha_{21}\mathbf{a}_1 + \alpha_{22}\mathbf{a}_2 + \alpha_{23}\mathbf{a}_3; \\ \mathbf{a}'_3 &= \alpha_{31}\mathbf{a}_1 + \alpha_{32}\mathbf{a}_2 + \alpha_{33}\mathbf{a}_3. \end{aligned} \tag{7.27}$$

This is a *linear* relation, known as a *coordinate transformation*. The nine numbers α_{ij} can be grouped as a square matrix:

$$\alpha_{ij} = \begin{pmatrix} \alpha_{11} & \alpha_{12} & \alpha_{13} \\ \alpha_{21} & \alpha_{22} & \alpha_{23} \\ \alpha_{31} & \alpha_{32} & \alpha_{33} \end{pmatrix}, \tag{7.28}$$

and the transformation equations can be rewritten in short form as:

$$\mathbf{a}'_i = \alpha_{ij}\mathbf{a}_j. \tag{7.29}$$

The *inverse* transformation must also exist and is described by the inverse of the matrix α_{ij}:

$$\mathbf{a}_i = \alpha^{-1}_{ij}\mathbf{a}'_j. \tag{7.30}$$

Consider the position vector $\mathbf{p}$. This vector is independent of the reference frame, and has components in both the unprimed and primed reference frames. We must have the following relation:

$$\mathbf{p} = p_i\mathbf{a}_i = p'_j\mathbf{a}'_j. \tag{7.31}$$

Using the inverse coordinate transformation we can rewrite the first equality as:

$$p_i\mathbf{a}_i = p_i\alpha^{-1}_{ij}\mathbf{a}'_j,$$

and after comparison with the last equality of Equation 7.31 we find:

$$p'_j = p_i\alpha^{-1}_{ij}. \tag{7.32}$$

Note the order of the indices of the matrix α; the summation index is the index i, which means that we must *pre-multiply* the matrix by the *row vector* p_i. Similarly, one can readily show that:

$$p_i = p'_j\alpha_{ji}. \tag{7.33}$$

We interpret Equations 7.32 and 7.33 as follows: the vector $\mathbf{p}$ is independent of the chosen reference frame if its components with respect to two different reference frames are related to each other by Equations 7.32 and 7.33. This relation obviously also holds for direction vectors $[uvw]$, since they are a special case of position vectors $\mathbf{p}$ (integer components instead of rational).

It is now straightforward to derive the transformation relation for the direct metric tensor:

$$\begin{aligned} g'_{ij} &= \mathbf{a}'_i \cdot \mathbf{a}'_j; \\ &= \alpha_{ik}\mathbf{a}_k \cdot \alpha_{jl}\mathbf{a}_l; \\ &= \alpha_{ik}\alpha_{jl}\mathbf{a}_k \cdot \mathbf{a}_l, \end{aligned}$$

and hence:

$$g'_{ij} = \alpha_{ik}\alpha_{jl}g_{kl}. \tag{7.34}$$

The inverse relation is given by:

$$g_{ij} = \alpha^{-1}_{ik} \alpha^{-1}_{jl} g'_{kl}. \tag{7.35}$$

One can use these relations to *define* a second-rank tensor: any mathematical quantity h_{ij} that satisfies the above transformation rules is a second-rank tensor. Similarly, any mathematical quantity p_i, satisfying the transformation rules 7.32 and 7.33, is a vector.

Next, we will derive the transformation relations for quantities in reciprocal space. We have seen in the preceding section that, if the components of a vector **p** are known in direct space, then its components in the reciprocal reference frame are given by:

$$p^*_i = g_{ij} p_j.$$

Using Equation 7.33 we have:

$$p^*_i = g_{ij} \alpha_{kj} p'_k. \tag{7.36}$$

From Equation 7.35 we find, after multiplying both sides of the equation by α_{kj}:

$$g_{ij} \alpha_{kj} = \alpha^{-1}_{il} g'_{lk},$$

and substitution in Equation 7.36 leads to:

$$\begin{aligned} p^*_i &= \alpha^{-1}_{il} g'_{lk} p'_k, \\ &= \alpha^{-1}_{il} p'^*_l, \end{aligned}$$

where we have once again used the properties of the direct metric tensor. The components of a vector in reciprocal space thus transform as follows:

$$p^*_i = \alpha^{-1}_{il} p'^*_l, \tag{7.37}$$

and the inverse relation is given by:

$$p'^*_l = \alpha_{li} p^*_i. \tag{7.38}$$

In particular these equations are valid for the reciprocal lattice vectors **g**. The reciprocal basis vectors satisfy similar transformation relations which are derived as follows:

$$\begin{aligned} \mathbf{g} &= g^*_i \mathbf{a}^*_i = g'^*_l \mathbf{a}'^*_l, \\ &= \alpha_{li} g^*_i \mathbf{a}'^*_l, \end{aligned}$$

Table 7.1. Overview of all transformation relations for vectors and the metric tensor in direct and reciprocal space. Pay close attention to the order of the indices!

Quantity	Old to new	New to old
direct basis vectors	$\mathbf{a}'_i = \alpha_{ij}\mathbf{a}_j$	$\mathbf{a}_i = \alpha^{-1}_{ij}\mathbf{a}'_j$
direct metric tensor	$g'_{ij} = \alpha_{ik}\alpha_{jl}g_{kl}$	$g_{ij} = \alpha^{-1}_{ik}\alpha^{-1}_{jl}g'_{kl}$
direct space vectors	$p'_i = p_j\alpha^{-1}_{ji}$	$p_i = p'_j\alpha_{ji}$
reciprocal basis vectors	$\mathbf{a}'^{*}_i = \mathbf{a}^{*}_j\alpha^{-1}_{ji}$	$\mathbf{a}^{*}_i = \mathbf{a}'^{*}_j\alpha_{ji}$
reciprocal metric tensor	$g'^{*}_{ij} = g^{*}_{kl}\alpha^{-1}_{ki}\alpha^{-1}_{lj}$	$g^{*}_{ij} = g'^{*}_{kl}\alpha_{ki}\alpha_{lj}$
reciprocal space vectors	$k'^{*}_i = \alpha_{ij}k^{*}_j$	$k^{*}_i = \alpha^{-1}_{ij}k'^{*}_j$

from which we find:

$$\mathbf{a}^{*}_i = \mathbf{a}'^{*}_l\alpha_{li}. \tag{7.39}$$

The corresponding inverse relation is given by:

$$\mathbf{a}'^{*}_i = \mathbf{a}^{*}_j\alpha^{-1}_{ji}. \tag{7.40}$$

Finally, it is again easy to show that the reciprocal metric tensor transforms according to the rules:

$$g^{*}_{ij} = \alpha_{ki}\alpha_{lj}g'^{*}_{kl}, \tag{7.41}$$

and:

$$g'^{*}_{ij} = \alpha^{-1}_{ki}\alpha^{-1}_{lj}g^{*}_{kl}. \tag{7.42}$$

The transformation rules derived in this section are summarized in Table 7.1. All that is required to carry out *any* coordinate transformation is the matrix α_{ij}, expressing the new basis vectors in terms of the old ones. The relations in Table 7.1 require, in addition to α_{ij}, the inverse α^{-1}_{ij} and transpose α^{T}_{ij} matrices, and the transpose of the inverse matrix $(\alpha^{-1}_{ij})^{T}$. These transformation rules seem easy enough, but one must actually pay close attention to the indices in order to avoid mistakes. In the following subsection we will illustrate coordinate transformations by means of a few examples.

7.4.2 Example of a coordinate transformation

Consider the face-centered cubic lattice shown in Fig. 7.7; we can define a primitive rhombohedral unit cell for this structure, as indicated by the primed basis vectors. Determine the transformation matrix α_{ij}, and express the reciprocal basis vectors of the new reference frame in terms of those of the

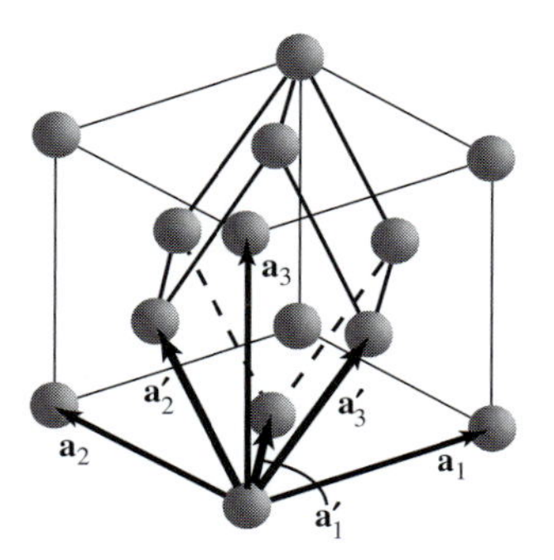

Fig. 7.7. Unit cell drawing of the face-centered cubic lattice, along with its primitive unit cell (Figure reproduced from Fig. 1.21 in *Introduction to Conventional Transmission Electron Microscopy*, M. De Graef, 2003, Cambridge University Press).

old reference frame. Then compute the direct metric tensor for the primitive cell using the transformation equations.

The transformation matrix α_{ij} is easily derived from a visual inspection of Fig. 7.7:

$$\alpha_{ij} = \frac{1}{2}\begin{pmatrix} 1 & 1 & 0 \\ 0 & 1 & 1 \\ 1 & 0 & 1 \end{pmatrix}.$$

The reciprocal basis vectors transform according to the inverse of this matrix, or:

$$\begin{aligned}
(\mathbf{a}_1^{\prime *}\, \mathbf{a}_2^{\prime *}\, \mathbf{a}_3^{\prime *}) &= (\mathbf{a}_1^*\, \mathbf{a}_2^*\, \mathbf{a}_3^*)\, \alpha_{ji}^{-1}; \\
&= (\mathbf{a}_1^*\, \mathbf{a}_2^*\, \mathbf{a}_3^*) \begin{pmatrix} 1 & -1 & 1 \\ 1 & 1 & -1 \\ -1 & 1 & 1 \end{pmatrix}; \\
&= (\mathbf{a}_1^* + \mathbf{a}_2^* - \mathbf{a}_3^* \,|\, -\mathbf{a}_1^* + \mathbf{a}_2^* + \mathbf{a}_3^* \,|\, \mathbf{a}_1^* - \mathbf{a}_2^* + \mathbf{a}_3^*)\,.
\end{aligned}$$

The direct metric tensor transforms according to $g'_{ij} = \alpha_{ik} g_{kl} \alpha_{jl}$, or (note that the matrix α_{jl} must be transposed before multiplication since the summation index l must be the row index!)

$$\begin{aligned}
g'_{ij} &= \frac{1}{4}\begin{pmatrix} 1 & 1 & 0 \\ 0 & 1 & 1 \\ 1 & 0 & 1 \end{pmatrix} \begin{bmatrix} a^2 & 0 & 0 \\ 0 & a^2 & 0 \\ 0 & 0 & a^2 \end{bmatrix} \begin{pmatrix} 1 & 0 & 1 \\ 1 & 1 & 0 \\ 0 & 1 & 1 \end{pmatrix}; \\
&= \frac{a^2}{4}\begin{pmatrix} 1 & 1 & 0 \\ 0 & 1 & 1 \\ 1 & 0 & 1 \end{pmatrix} \begin{bmatrix} 1 & 0 & 1 \\ 1 & 1 & 0 \\ 0 & 1 & 1 \end{bmatrix}; \\
&= \frac{a^2}{4}\begin{bmatrix} 2 & 1 & 1 \\ 1 & 2 & 1 \\ 1 & 1 & 2 \end{bmatrix}.
\end{aligned}$$

The rhombohedral metric tensor is given by (see page 86):

$$g_{ij} = b^2 \begin{bmatrix} 1 & \cos\alpha & \cos\alpha \\ \cos\alpha & 1 & \cos\alpha \\ \cos\alpha & \cos\alpha & 1 \end{bmatrix},$$

where b and α are the lattice parameters of the primitive unit cell. From the drawing one can easily show that $b = a\sqrt{2}$ and $\cos\alpha = 1/2$ which leads to the same expression for g_{ij}.

The [001] direction in the cubic reference frame can be transformed into the rhombohedral frame as follows:

$$[u\ v\ w]_\mathrm{r} = [0\ 0\ 1]_\mathrm{c} \begin{pmatrix} 1 & -1 & 1 \\ 1 & 1 & -1 \\ -1 & 1 & 1 \end{pmatrix} = [-1\ 1\ 1]_\mathrm{r},$$

as is easily verified in Fig. 7.7. The (110) cubic plane transforms as follows:

$$\begin{pmatrix} h \\ k \\ l \end{pmatrix}_\mathrm{r} = \frac{1}{2} \begin{pmatrix} 1 & 1 & 0 \\ 0 & 1 & 1 \\ 1 & 0 & 1 \end{pmatrix} \begin{pmatrix} 1 \\ 1 \\ 0 \end{pmatrix}_\mathrm{c} = \frac{1}{2} \begin{pmatrix} 2 \\ 1 \\ 1 \end{pmatrix},$$

so that the $(110)_\mathrm{c}$ plane has Miller indices (211) in the rhombohedral reference frame.

7.4.3 Converting vector components into Cartesian coordinates

We have seen in previous sections, that there is a distinct advantage to working in crystal coordinates (i.e., in a non-Cartesian reference frame) in both direct and reciprocal space. However, at the end of a simulation or calculation, the results are almost invariably represented on a computer screen or on a piece of paper, both of which are 2-D media with essentially Cartesian reference frames. We must, therefore, provide a way to transform direct and reciprocal crystal coordinates into Cartesian coordinates. It is not difficult to carry out such a conversion for the crystal systems of high symmetry (cubic, tetragonal, and orthorhombic) since their coordinate axes are already at right angles to each other. However, for a monoclinic or triclinic system the conversion to Cartesian coordinates is a bit more difficult and it becomes important to have an algorithm that will do the conversion, independent of the crystal system. Such a conversion exists and is derived below. The derivation is somewhat tedious, but the resulting transformation is quite general and can be used for both direct and reciprocal space quantities.

The transformation can be carried out by means of the so-called *direct and reciprocal structure matrices*. Let us assume a crystal reference frame $\mathbf{a}_i$, and the corresponding reciprocal reference frame $\mathbf{a}^*_j$. From these two reference

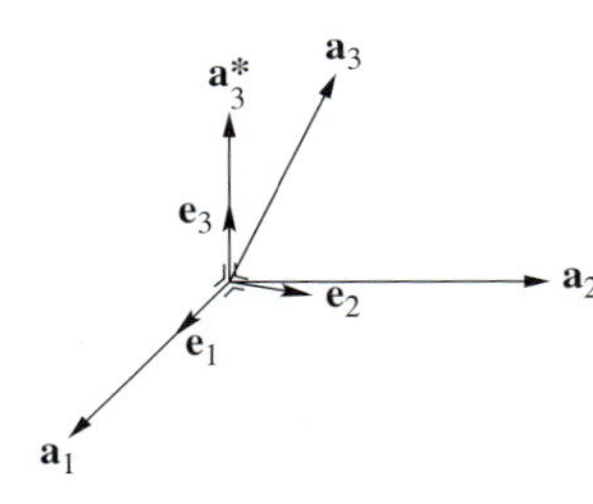

Fig. 7.8. Definition of the Cartesian reference frame from the direct and reciprocal reference frames (Figure reproduced from Fig. 1.23 in *Introduction to Conventional Transmission Electron Microscopy*, M. De Graef, 2003, Cambridge University Press).

frames, we can construct a Cartesian reference frame $\mathbf{e}_i$ as follows: $\mathbf{e}_1$ is the unit vector along $\mathbf{a}_1$, $\mathbf{e}_3$ is the unit vector along the *reciprocal* basis vector $\mathbf{a}_3^*$ (and is, therefore, by construction normal to $\mathbf{e}_1$), and $\mathbf{e}_2$ completes the right-handed Cartesian reference frame (see Fig. 7.8).

$$\boxed{\begin{aligned} \mathbf{e}_1 &= \frac{\mathbf{a}_1}{|\mathbf{a}_1|}; \\ \mathbf{e}_2 &= \mathbf{e}_3 \times \mathbf{e}_1; \\ \mathbf{e}_3 &= \frac{\mathbf{a}_3^*}{|\mathbf{a}_3^*|} = \frac{\mathbf{a}_1 \times \mathbf{a}_2}{V|\mathbf{a}_3^*|}. \end{aligned}} \tag{7.43}$$

We will refer to this reference frame as the *standard Cartesian frame*. Now consider a vector $\mathbf{r}$ with components r_i with respect to the basis vectors $\mathbf{a}_i$. The components of $\mathbf{r}$ in the Cartesian reference frame are given by x_j, or:

$$\mathbf{r} = r_i\mathbf{a}_i = x_j\mathbf{e}_j.$$

The components r_i and x_j are related to one another by a linear coordinate transformation represented by the matrix a_{ij}:

$$x_i = a_{ij}r_j.$$

The elements of the transformation matrix can be determined as follows: Equations 7.43 are rewritten in terms of the direct and reciprocal metric tensors as:

$$\begin{aligned} \mathbf{e}_1 &= \frac{\mathbf{a}_1}{\sqrt{g_{11}}}; \\ \mathbf{e}_2 &= \frac{\mathbf{a}_3^* \times \mathbf{a}_1}{\sqrt{g_{11}g_{33}^*}}; \\ \mathbf{e}_3 &= \frac{\mathbf{a}_3^*}{\sqrt{g_{33}^*}} = \frac{g_{3m}^*\mathbf{a}_m}{\sqrt{g_{33}^*}}. \end{aligned}$$

From the definition of $\mathbf{a}_3^*$ we derive:

$$V(\mathbf{a}_3^* \times \mathbf{a}_1) = (\mathbf{a}_1 \times \mathbf{a}_2) \times \mathbf{a}_1 = -\mathbf{a}_1 \times (\mathbf{a}_1 \times \mathbf{a}_2).$$

The triple vector product can be simplified using the vector identity:

$$\mathbf{u} \times (\mathbf{v} \times \mathbf{w}) = (\mathbf{u} \cdot \mathbf{w})\mathbf{v} - (\mathbf{u} \cdot \mathbf{v})\mathbf{w},$$

which leads to:

$$V(\mathbf{a}_3^* \times \mathbf{a}_1) = g_{11}\mathbf{a}_2 - g_{12}\mathbf{a}_1,$$

and finally:

$$\mathbf{e}_2 = \frac{g_{11}\mathbf{a}_2 - g_{12}\mathbf{a}_1}{V\sqrt{g_{11}g_{33}^*}}.$$

The vector $\mathbf{r}$ can now be written as follows:

$$\mathbf{r} = x_j\mathbf{e}_j = \left[\frac{x_1}{\sqrt{g_{11}}} - \frac{g_{21}x_2}{V\sqrt{g_{11}g_{33}^*}} + \frac{g_{31}^*x_3}{\sqrt{g_{33}^*}}\right]\mathbf{a}_1 + \left[\frac{g_{11}x_2}{V\sqrt{g_{11}g_{33}^*}} + \frac{g_{32}^*x_3}{\sqrt{g_{33}^*}}\right]\mathbf{a}_2 + \frac{g_{33}^*x_3}{\sqrt{g_{33}^*}}\mathbf{a}_3.$$

Using the fact that the direct and reciprocal metric tensors are each other's inverse, we can explicitly write the matrix a_{ij} as:

$$a_{ij} = \begin{pmatrix} \sqrt{g_{11}} & \frac{g_{21}}{\sqrt{g_{11}}} & \frac{g_{31}}{\sqrt{g_{11}}} \\ 0 & V\sqrt{\frac{g_{33}^*}{g_{11}}} & -\frac{Vg_{32}^*}{\sqrt{g_{33}^*g_{11}}} \\ 0 & 0 & \frac{1}{\sqrt{g_{33}^*}} \end{pmatrix};$$

$$= \begin{pmatrix} a & b\cos\gamma & c\cos\beta \\ 0 & b\sin\gamma & -\frac{c\mathcal{F}(\beta,\gamma,\alpha)}{\sin\gamma} \\ 0 & 0 & \frac{V}{ab\sin\gamma} \end{pmatrix}, \tag{7.44}$$

where:

$$\mathcal{F}(\alpha,\beta,\gamma) = \cos\alpha\cos\beta - \cos\gamma.$$

The matrix a_{ij} is known as the *direct structure matrix* and it transforms crystal coordinates to Cartesian coordinates. Note that its elements depend both on the direct and reciprocal metric tensors and, thus, on the lattice parameters $\{a, b, c, \alpha, \beta, \gamma\}$, as shown by the second equality in 7.44. The inverse transformation is given by the inverse matrix:

$$r_i = a_{ij}^{-1}x_j \quad \text{with} \quad a_{ij}^{-1} = \begin{pmatrix} \frac{1}{a} & \frac{-1}{a\tan\gamma} & \frac{bc\mathcal{F}(\gamma,\alpha,\beta)}{V\sin\gamma} \\ 0 & \frac{1}{b\sin\gamma} & \frac{ac\mathcal{F}(\beta,\gamma,\alpha)}{V\sin\gamma} \\ 0 & 0 & \frac{ab\sin\gamma}{V} \end{pmatrix}. \tag{7.45}$$

The direct structure matrix is particularly useful if one wants to create a drawing of a crystal structure, and for the computation of stereographic projections, as demonstrated in the next section.

As an example, compute the Cartesian coordinates of the lattice point $(2, 3, 1)$ in a tetragonal lattice with lattice parameters $\{\frac{1}{2}, \frac{1}{2}, 1, 90, 90, 90\}$. From the lattice parameters $a = 1/2$ and $c = 1$ we find for the direct structure matrix:

$$a_{ij} = \begin{pmatrix} \frac{1}{2} & 0 & 0 \\ 0 & \frac{1}{2} & 0 \\ 0 & 0 & 1 \end{pmatrix}.$$

Hence the Cartesian components of the vector $(2, 3, 1)$ are $(1, 3/2, 1)$.

One can use the same formalism to determine the Cartesian coordinates of a reciprocal lattice point; such coordinates would be used to draw a representation of reciprocal space. To preserve the relative orientation of crystal and reciprocal space, we look for a second structure matrix b_{ij} which represents the transformation from the reciprocal reference frame to *the same* Cartesian reference frame. Consider the reciprocal space vector $\mathbf{k}$, with components k_j with respect to the reciprocal basis vectors $\mathbf{a}_j^*$ and Cartesian components q_i.

$$\mathbf{k} = q_i\mathbf{e}_i = k_j\mathbf{a}_j^*.$$

This can be rewritten in terms of the direct basis vectors $\mathbf{a}_l$ as:

$$\mathbf{k} = q_i\mathbf{e}_i = k_j g_{jl}^*\mathbf{a}_l = r_l\mathbf{a}_l \qquad \text{with} \qquad r_l = k_j g_{jl}^*.$$

We can now use the direct structure matrix a_{il} to relate q_i to r_l:

$$q_i = a_{il}r_l = a_{il}g_{jl}^*k_j = a_{il}g_{lj}^*k_j = b_{ij}k_j.$$

The *reciprocal structure matrix* b_{ij} is thus defined by:

$$b_{ij} = a_{il}g_{lj}^*. \tag{7.46}$$

This matrix converts reciprocal space coordinates into Cartesian coordinates. The inverse relation is given by:

$$k_i = b_{ij}^{-1}q_j.$$

One can use the fact that the length of a vector must be independent of the reference frame to show that the transpose of the reciprocal structure matrix b is equal to the inverse of the direct structure matrix a, or:

$$b^{\mathrm{T}} = a^{-1}.$$

The reciprocal structure matrix is thus given by the transpose of the matrix in Equation 7.45. Note that only the lattice parameters are used to compute the structure matrices. In addition, the lattice parameters are used to compute the direct and reciprocal metric tensors. In a computer implementation it is convenient to have a single routine which computes all four matrices.

For the tetragonal crystal of the previous example, what are the Cartesian components of the reciprocal lattice point (221)? The Cartesian components q_i require the reciprocal structure matrix b_{ij}, which is the transpose of the inverse of a_{ij}:

$$b_{ij} = \begin{pmatrix} 2 & 0 & 0 \\ 0 & 2 & 0 \\ 0 & 0 & 1 \end{pmatrix},$$

from which the Cartesian components of (221) follow as $(4, 4, 1)$.

7.5 Examples of stereographic projections

In Chapter 5, we introduced the concept of a *family* of planes or directions, $\{hkl\}$ and $\langle uvw \rangle$. At this point, we can make use of nearly everything that we have learned in this and the preceding chapters to create stereographic projections of arbitrary crystals. We will begin with the simplest case, the cubic crystal system.

7.5.1 Stereographic projection of a cubic crystal

Consider a cubic crystal system. We are asked to draw the stereographic projection containing the $\langle 100 \rangle$, $\langle 110 \rangle$, and $\langle 111 \rangle$ families of directions. Let us start with the $\langle 100 \rangle$ family, which consists of the directions: [100], $[\bar{1}00]$, [010], $[0\bar{1}0]$, [001], and $[00\bar{1}]$. We will orient the crystal so that its [001] direction points from the south pole to the north pole of the projection sphere. We also know that the angle between the [001] and [100] directions is 90°, so that the SP of the [100] direction must lie on the equatorial circle. We place the projection of [100] to lie along the line A–B (Fig. 7.9(a)). The positions of all other projections are now fixed. [010] lies at 90° from both [100] and [001]. There are two possibilities: [010] pointing towards M' or M''. To create a right-handed reference frame, we must have $[100] \cdot ([010] \times [001]) > 0$, which means that [010] must point towards M''.

The negative directions must lie on the opposite side of the projection sphere, which means that $[\bar{1}00]$ points towards A and $[0\bar{1}0]$ towards M'. Since $[00\bar{1}]$ points towards the south pole (which lies in the southern hemisphere), we represent its SP by an open circle at the center of the projection. The full projection of the $\langle 100 \rangle$ family of directions is shown in Fig. 7.9(b).

To draw the other two families, we will keep the crystal in the same orientation. There are twelve members in the $\langle 110 \rangle$ family. Of these twelve, we

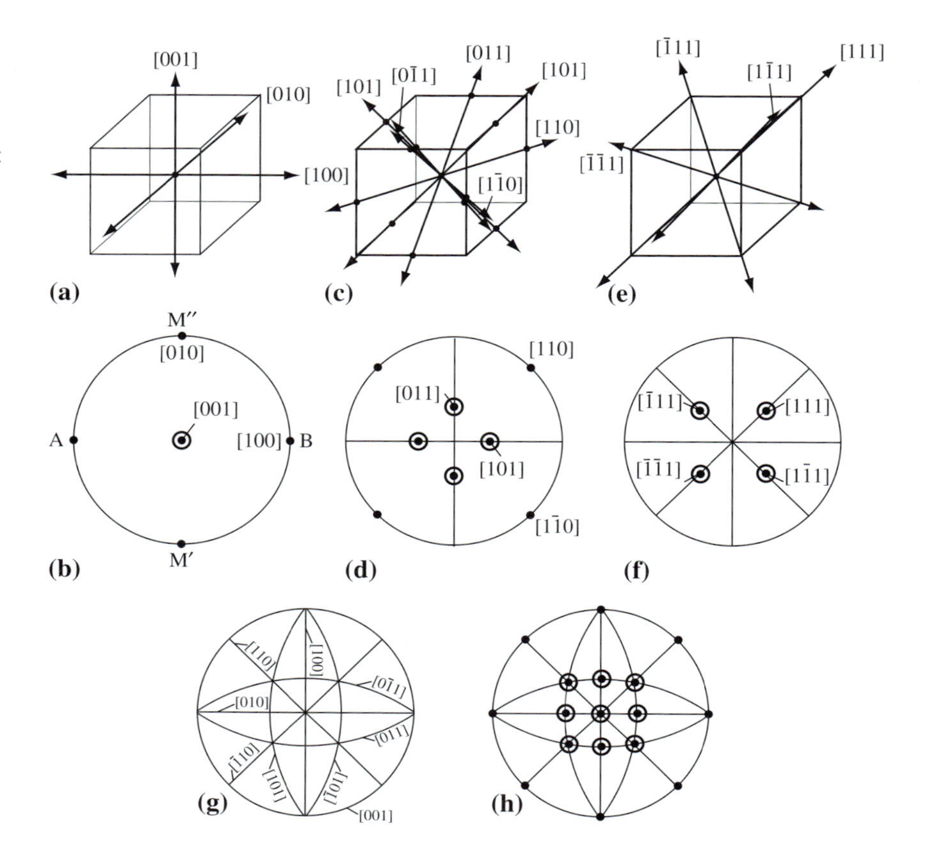

Fig. 7.9. Stereographic projection of the ⟨100⟩ (a,b), ⟨110⟩ (c,d), and ⟨111⟩ (e,f) families of directions. (g) shows the most important zones in the [001] projection. The three projections are superimposed into a single one in (h).

know that four belong to the [001] zone, since the zone equation for the [001] zone states that l must be zero. These are the [110], [1$\bar{1}$0], [$\bar{1}$10], and [$\bar{1}\bar{1}$0] directions. We can use the metric tensor formalism to compute the angle between [110] and [100] (or simply read the angle from a drawing, which is easy for cubic symmetry), and we find 45°. The stereographic coordinates are then (45°, R), since $\theta = 90°$. So, the SP of [110] lies on the projection circle, halfway between [100] and [010], as shown in Fig. 7.9(d). The other directions belonging to the [001] zone result in projections in the other three quadrants of the projection circle.

We can repeat this procedure for the other members of the ⟨110⟩ family. For instance, we know that there are four family members that belong to the [100] zone, namely those directions with zero as the first index. The zone [100] corresponds to all directions normal to [100], and on the stereographic projection this is the great circle between M' and M''. The angle between [011] and [001] is again equal to 45°, so that the stereographic coordinates are given by (90°, $R\tan(45°/2)$), which results in the point labeled [011] in Fig. 7.9(d). The other members of the family follow in the same way.

Finally, the $\langle 111 \rangle$ family consists of eight members, shown in Fig. 7.9(e). None of these directions belongs to a zone from the $\langle 001 \rangle$ family. However, if we take the [110] direction and consider its zone, then we know that all the directions with $u+v=0$ belong to this zone. There are four members of the $\langle 111 \rangle$ family in this zone: $[1\bar{1}1]$, $[1\bar{1}\bar{1}]$, $[\bar{1}11]$, and $[\bar{1}1\bar{1}]$. So, we know that the projections of these four directions must lie on the [110] zone, indicated in Fig. 7.9(f). The angle between $[1\bar{1}1]$ and [001], which also belongs to the [110] zone, can be computed using the metric tensor formalism, which results in $\cos\theta = 1/\sqrt{3}$, or $\theta = 54.74°$. The angle between $[1\bar{1}1]$ and $[1\bar{1}0]$ is, in similar fashion, $\theta' = 35.26°$. The stereographic coordinates of $[1\bar{1}1]$ are then given by $(-45°, R\tan 27.37°)$, which results in the location shown in Fig. 7.9(f). The other members of the $\langle 111 \rangle$ family can be treated in the same way.

Note that the direction [111] belongs to a number of $\langle 110 \rangle$-type zones: it belongs to the zones $[1\bar{1}0]$, $[\bar{1}01]$, and $[0\bar{1}1]$. This means that the projection of [111] must lie at the intersection of three great circles, one corresponding to each of these zones. The $[1\bar{1}0]$ zone is a straight line in the projection, because it also contains the $[\bar{1}\bar{1}0]$, [001] and [110] directions. The $[0\bar{1}1]$ zone is normal to the $[0\bar{1}1]$ direction and contains the $[\bar{1}00]$, $[\bar{1}11]$, [011], [111], and [100] directions. Therefore, its projection must be an arc of a circle going from A to B through all these points. Repeating this for all the members of the [110] family, we arrive at the zone drawing of Fig. 7.9(g), which shows all the zones labeled. At the intersection of zones, we have zone axes, which are directions in the crystal lattice, and the resulting stereographic projection of the cubic crystal is shown in Fig. 7.9(h). Since the center of the projection corresponds to [001], this is known as the [001] stereographic projection of the cubic crystal. It is left as a reader exercise to obtain the [110] and [111] stereographic projections for the cubic crystal.

What about the plane normals in this crystal system? What would be the stereographic projection of the {100}, {110}, and {111} families of plane normals? We have seen in the previous chapter, that the reciprocal basis vectors are parallel to the direct space basis vectors:

$$\mathbf{a}_i^* = \frac{\mathbf{a}_i}{a^2}. \tag{7.47}$$

This means that

$$\mathbf{g}_{hkl} = h\mathbf{a}^* + k\mathbf{b}^* + l\mathbf{c}^* = \frac{1}{a^2}(h\mathbf{a} + k\mathbf{b} + l\mathbf{c}) \parallel h\mathbf{a} + k\mathbf{b} + l\mathbf{c} = \mathbf{t}_{hkl}; \tag{7.48}$$

In other words, in the cubic crystal system, the plane normal for the plane (hkl) is parallel to the direction with indices $[hkl]$, so that the [001] stereographic projection of the directions is identical to the [001] stereographic projection of the plane normals. All we need to do is replace the direction symbols

$[uvw]$ by plane normals (uvw). Note that this is only the case for the cubic crystal system; for all other systems, the direct space and reciprocal space projections will be different from each other, as we will illustrate next for the case of a monoclinic crystal.

7.5.2 Stereographic projection of a monoclinic crystal

In Chapter 6, we used as an example a monoclinic crystal with lattice parameters $\{1, 1, 1, 90, 45, 90\}$. In this section, we will obtain stereographic projections for both the directions and the plane normals in this crystal. Before we do so, we must first determine how to compute the stereographic projection coordinates in a more standardized way.

We already know from Section 7.4.3 how to convert crystal or reciprocal coordinates into Cartesian coordinates. Since the SP represents directions, the next step is to *normalize* the Cartesian components so that the direction corresponds to a point on the unit sphere. Let us assign the coordinates (x, y, z) to this point P, as illustrated in Fig. 7.10. If we connect the point P with the south pole S, then the triangle PQS is congruent with $P'OS$, from which we derive:

$$x' = \frac{x}{1+z};$$
$$y' = \frac{y}{1+z}.$$

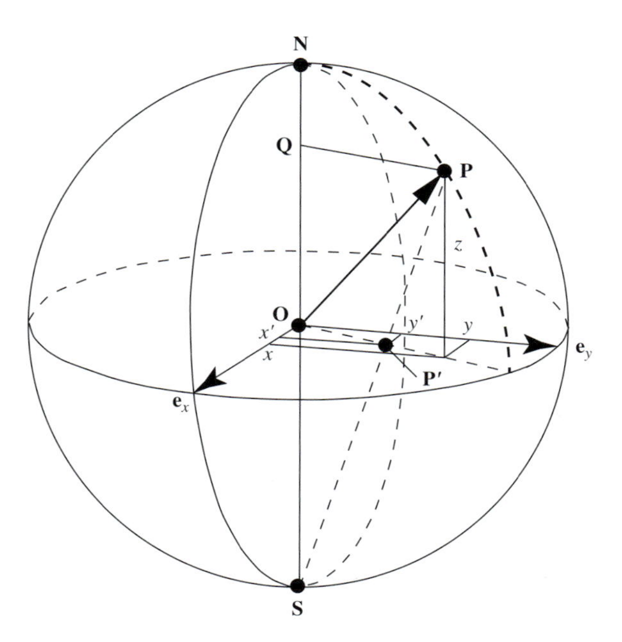

Fig. 7.10. Schematic representation of the relation between 3-D Cartesian coordinates and stereographic projection coordinates (Figure reproduced from Fig. 1.25 in *Introduction to Conventional Transmission Electron Microscopy*, M. De Graef, 2003, Cambridge University Press).

If the projection circle has radius R, then the above equations are simply multiplied by R to arrive at the proper 2-D coordinates. If z is negative, then we replace z by $|z|$ in the equations above, and we represent the corresponding point by an open circle, to indicate that the projection was done from the north pole instead of the south pole. We will now illustrate this procedure for both the real space and reciprocal space stereographic projections of a monoclinic crystal with lattice parameters $\{1, 1, 1, 90, 45, 90\}$.

7.5.2.1 Direct space stereographic projection

First of all, we determine the direct structure matrix a_{ij} from the lattice parameters and the definition in Equation 7.44:

$$a_{ij} = \begin{pmatrix} 1 & 0 & \frac{\sqrt{2}}{2} \\ 0 & 1 & 0 \\ 0 & 0 & \frac{\sqrt{2}}{2} \end{pmatrix}.$$

This means that the direction [001] is represented by:

$$(x\, y\, z) = \begin{pmatrix} 1 & 0 & \frac{\sqrt{2}}{2} \\ 0 & 1 & 0 \\ 0 & 0 & \frac{\sqrt{2}}{2} \end{pmatrix} \begin{pmatrix} 0 \\ 0 \\ 1 \end{pmatrix} = (\frac{\sqrt{2}}{2}\, 0\, \frac{\sqrt{2}}{2}).$$

It is easy to verify that this vector is already normalized, so that the point with Cartesian coordinates $(1/\sqrt{2}, 0, 1/\sqrt{2})$ lies on the unit projection sphere. Then we take the ratios introduced above to compute the stereographic projection coordinates:

$$x' = R\frac{\frac{\sqrt{2}}{2}}{1+\frac{\sqrt{2}}{2}} = R(\sqrt{2}-1);$$

$$y' = 0.$$

This point is shown on the stereographic projection in Fig. 7.11(a).

The same procedure can be followed for all other crystal directions. It is clear that such a repetitive procedure is perfectly suited for implementation in a spreadsheet program. The results of such a computation for selected directions are shown in Table 7.2, and the corresponding projections are indicated in Fig. 7.11. We leave it to the reader to verify that this projection is indeed correct.

Table 7.2. Normalized Cartesian coordinates and stereographic coordinates for selected directions in a monoclinic crystal with lattice parameters $\{1, 1, 1, 90, 45, 90\}$.

$[uvw]$	Normalized Cartesian	Stereographic
[100]	(1.000, 0.000, 0.000)	R(1.000, 0.000)
[010]	(0.000, 1.000, 0.000)	R(0.000, 1.000)
[001]	(0.707, 0.000, 0.707)	R(0.414, 0.000)
[110]	(0.707, 0.707, 0.000)	R(0.707, 0.707)
[101]	(0.924, 0.000, 0.383)	R(0.668, 0.000)
[011]	(0.500, 0.707, 0.500)	R(0.333, 0.471)
[111]	(0.813, 0.476, 0.337)	R(0.608, 0.356)

7.5.2.2 Reciprocal space stereographic projection

The reciprocal space stereographic projection of the monoclinic crystal can be derived in exactly the same way, by using the reciprocal structure matrix b_{ij} instead of a_{ij}. We find that:

$$b_{ij} = \begin{pmatrix} 1 & 0 & 0 \\ 0 & 1 & 0 \\ -1 & 0 & \sqrt{2} \end{pmatrix}.$$

The (001) plane normal then has the Cartesian coordinates:

$$(x\,y\,z) = \begin{pmatrix} 1 & 0 & 0 \\ 0 & 1 & 0 \\ -1 & 0 & \sqrt{2} \end{pmatrix} \begin{pmatrix} 0 \\ 0 \\ 1 \end{pmatrix} = (0, 0, \sqrt{2}).$$

Normalized, this becomes the point with coordinates $(0, 0, 1)$, and the stereographic projection coordinates are $(0, 0)$, i.e., the center of the projection (Fig. 7.11(b)). This is to be expected, since the (001) plane is formed by the [100] and [010] directions, which lie in the plane of the stereographic

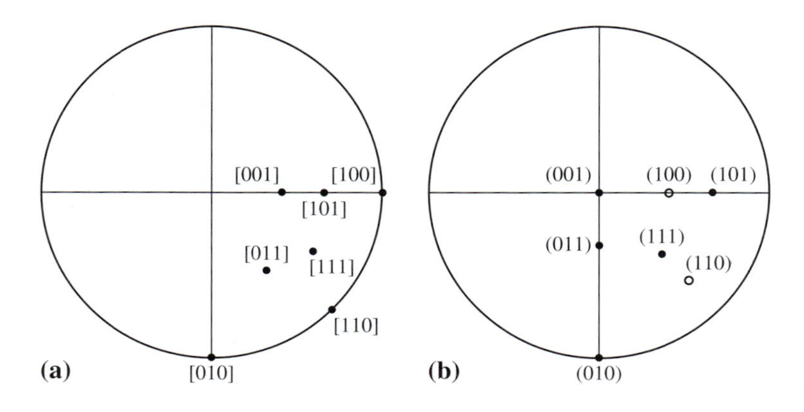

Fig. 7.11. Stereographic projections of selected directions (a) and plane normals (b) of a monoclinic crystal with lattice parameters $\{1, 1, 1, 90, 45, 90\}$.

Table 7.3. Normalized Cartesian coordinates and stereographic coordinates for selected plane normals in a monoclinic crystal with lattice parameters $\{1, 1, 1, 90, 45, 90\}$. Points labeled "north pole projection" are represented by open circles in the stereographic projection of Fig. 7.11.

$[uvw]$	Normalized Cartesian	Stereographic
(100)	(0.707, 0.000, −0.707)	R(0.414, 0.000) (north pole projection)
(010)	(0.000, 1.000, 0.000)	R(0.000, 1.000)
(001)	(0.000, 0.000, 1.000)	R(0.000, 0.000)
(110)	(0.577, 0.577, −0.577)	R(0.366, 0.366) (north pole projection)
(101)	(0.924, 0.000, 0.383)	R(0.668, 0.000)
(011)	(0.000, 0.577, 0.816)	R(0.000, 0.318)
(111)	(0.679, 0.679, 0.281)	R(0.530, 0.530)

projection (see Fig. 7.11(a)). The stereographic coordinates for selected plane normals are shown in Table 7.3, and the corresponding projections are indicated in Fig. 7.11(b).

Note that the plane normal (101) and the direction [101] have the same stereographic coordinates, which indicates that they are parallel, i.e., the [101] direction is normal to the (101) plane *in this particular crystal system*. If this were a cubic crystal system, then the two tables would show identical entries for each given direction $[uvw]$ and plane normal (uvw).

7.6 Historical notes

Yuri Victorovich (George) Wulff (1863–1925, Fig. 7.12a) was a Russian crystallographer known in crystal growth theory for his construction of the ideal *equilibrium form* (Wulff, 1902). So-called Wulff plots are constructions

Fig. 7.12. (a) Yuri Victorovich Wulff (1863–1925), and (b) Anton Van Leewenhoek (1623–1723) (pictures courtesy of J. Lima-de-Faria).

used to predict the equilibrium form of crystals. He developed the construction of the stereographic net as an equivalent to the stereographic projection (Wulff, 1908). **Paul Heinrich Ritter Von Groth** (1843–1927) was a German scientist and a contemporary of Wulff, who devoted his career to the study of crystallography. In particular, he was responsible for microscopic observations of growing crystals and for demonstrating the constancy of interfacial angles in gypsum. In 1895, he published his *Arcana Natura Detecta* (Von Groth, 1895) describing his observations of crystal growth.

In 1783, **Arnould Carangeot** (1742–1806) developed the *contact goniometer*, a device to measure the angle between crystal faces (Carangeot, 1783). This tool was later used by his teacher, Romé de l'Isle (see Historical notes in Chapter 4) to make angular measurements on crystals, confirming Steno's earlier work on quartz.

There are several additional important tools in crystallography. **Robert Hooke** (1635–1703) was one of the fathers of optical microscopy. This British scientist and member of the Royal Society wrote the monograph *Micrographia* (Hooke, 1665) which contains detailed sketches of his observations with the microscope, including some of crystalline solids. Like Christian Huyghens, **Anton Van Leewenhoek** (1623–1723, Fig. 7.12(b)) was also a Dutch scientist who devoted his career to the pursuit of microscopic observations of materials (Van Leewenhoek, 1685a,b). As a draper's apprentice, he used a simple magnifying glass to count threads in cloth. He later learned the art of lens grinding which he used to make a succession of hand-held microscopes (he used these to study biological micro-organisms leading to his identification as the father of microbiology). In his career he built 247 microscopes and designed more than 419 lenses, most of which were double convex lenses.

David Brewster (1781–1868) invented the quartz compensator in 1830; the compensator is an important tool in physical crystallography and crystal optics (Brewster, 1830a,b,c). **Giovanni Battista Amici** (1786–1863) was an Italian scientist who developed the first polarizing microscope and the tilting stage (Amici, 1844). His polarizing microscope used lenses in conjunction with a polarizer and analyzer. The British crystallographer and Cambridge professor, **William Hyde Wollaston** (1766–1828), in 1809 invented the reflecting goniometer which permitted accurate and precise measurements of the positions of crystal faces (Wollaston, 1809, 1813). The reflecting goniometer allowed extensive measurements on both naturally occurring and artificial crystals. **Dominique Francois Jean Arago** (1786–1853) was a French scientist specializing in crystal optics. In 1811, Arago discovered the rotation of the plane of polarization of light traveling through quartz. This phenomenon was referred to as the *optical activity* of quartz crystals (Arago, 1811). **Moritz Frankenheim** (1801–69) developed a polarizing microscope with a Nicol prism. **Evgraf Stepanovich Federov** (1853–1919) was responsible for the universal stage for the polarizing microscope.

7.7 Problems

(i) *Vector cross product I*: Two pairs of directions are given in a cubic crystal system: [100]—[121] and [011]—[111].

  (a) Compute the Miller indices of the planes formed by each pair of directions.
  (b) What is the direction common to those two planes?
  (c) Repeat the exercise for a triclinic crystal system with lattice parameters $\{1, 2, 3, 40, 60, 80\}$.

(ii) *Vector cross product II*: Consider an orthorhombic crystal with lattice parameters $\{3, 4, 5, 90, 90, 90\}$.

  (a) Use Equation 7.8 to compute the cross product between the vectors $\mathbf{p} = (1/2, 1/3, 1/4)$ and $\mathbf{q} = (1, 1, 0)$.
  (b) Use Equation 7.12 to compute the cross product between the vectors $\mathbf{g} = (1/2, 1/3, 1/4)$ and $\mathbf{h} = (1, 1, 0)$.

(iii) *Vector cross product III*: In a hexagonal crystal system, how would you compute the normal to the plane formed by two direction vectors, when these vectors are expressed in four-index notation?

(iv) *Unit cell volume*: Consider a unit cell with $\{2, 3, 4, 90, 60, 90\}$ as lattice parameters. Show, through explicit computation, that the volume of the cell computed using Equation 7.16 is identical to that found using Equation 7.17.

(v) *Zone equation I*: Given are two zones, described by the sets of planes of the types $(hk\bar{h})$ and $(hh0)$. What is the normal to the plane formed by the two corresponding zone axes?

(vi) *Zone equation II*: List at least four planes that belong to the hexagonal zone $[11\bar{2}1]$.

(vii) *Coordinate transformations I*: If a vector has direct space components $(1, 0, 3)$ with respect to the basis vectors of a lattice with lattice parameters $\{2, 2, 3, 90, 90, 90\}$, then what are the components of that vector with respect to the reciprocal basis vectors?

(viii) *Coordinate transformations II*: Use the metric tensor formalism to determine the reciprocal basis vectors of a cell with lattice parameters $\{1, 4, 2, 90, 60, 90\}$.

(ix) *Coordinate transformations III*: Determine the coordinate transformation matrix α_{ij} that expresses the basis vectors of the primitive unit cell of the body centered cubic lattice in terms of those of the conventional unit cell of this lattice. If we denote quantities in the cubic reference frame by a subscript c, and quantities in the primitive reference frame by a subscript p, then answer the following questions:

  (a) What are the p components of the direction vectors $[110]_c$ and $[111]_c$?

(b) What are the c components of the direction vectors $[101]_p$ and $[012]_p$?
(c) What are the p components of the plane normals $(112)_c$ and $(\bar{2}01)_c$?
(d) What are the c components of the plane normals $(112)_p$ and $(\bar{2}01)_p$?
(e) What are the p components of the position vector $(1/3, 1/3, 1/3)$?
(f) Express the p reciprocal basis vectors in terms of the c direct basis vectors.
(g) Express the c reciprocal basis vectors in terms of the p direct basis vectors.
(h) What is the volume of the p unit cell?
(i) Write down the p zone equation for the $[121]_c$ zone axis.
(j) What are the standard Cartesian coordinates for the point $(1, 2, -1)_c$. Show that they are identical to the standard Cartesian coordinates of the corresponding point in the p reference frame.

(x) *Structure matrices*: Show by direct computation that the product $a^T a$ (with T indicating the transpose of the matrix and a the direct structure matrix) is equal to the direct space metric tensor g.
(xi) *Stereographic projections I*: Repeat the construction of Fig. 7.9, but this time, place the [110] direction at the center of the projection, and $[1\bar{1}0]$ at the point B.
(xii) *Stereographic projections II*: Consider a unit cell with the following lattice parameters: $\{2, 2, 3, 90, 90, 120\}$.

(a) Create a table similar to Table 7.2, for all the directions of the types [10.0], [11.0], [00.1], and [10.2] (include all appropriate permutations and negative signs). Then, create a stereographic projection with radius 10 cm, and plot all the points.
(b) On this projection, identify as many zones as you can find.
(c) Repeat the same for the plane normals of the types (10.0), (11.0), (00.1), and (10.2) (and all permutations and negatives), and plot these points on a similar stereographic projection.
(d) Again, identify as many zones as you can find.

CHAPTER

8 Symmetry in crystallography

> *"Mathematics possesses not only cold truth but supreme beauty, a beauty cold and austere, like that of a sculpture, sublimely pure and capable of stern perfection, such that only the greatest art can show."*
>
> Bertrand Russell

8.1 Symmetry of an arbitrary object

Many objects encountered in nature show some form of symmetry, in many cases only an approximate symmetry; e.g. the human body shows an approximate mirror symmetry between the left and right halves, many flowers have five- or seven-fold rotational symmetry, etc . . . In the following paragraphs, we will discuss the classical theory of symmetry, which is the theory of symmetry transformations of space into itself.

> If an object can be (1) rotated, (2) reflected, or (3) displaced, without changing the distances between its material points and so that it comes into self-coincidence, then that object is symmetric.

A transformation of the type (1), (2), or (3) or combinations thereof that preserve distances and bring the object into coincidence is called a *symmetry operation*. It should be clear that *translations* can only be symmetry operations for infinite objects. The word "symmetric" stems from the Greek word for "commensurate." Note that the identity operator (i.e., not doing anything) is also considered to be a symmetry property; therefore, each object has at least one symmetry property.

Throughout this chapter and many of the following chapters, we will need to write down shorthand notations for all the symmetry operations that we will introduce. We will follow the *International Tables for Crystallography* (Volume A, Hahn, 1996) for all notational conventions. In particular, there are two major schools of notation: the *international notation* (also known as the *Hermann–Mauguin notation*) is the standard, which we will follow throughout this text. An alternative notational convention, used primarily by physicists and chemists, is the *Schönflies notation*. Since this latter notation is widely used in the scientific literature, we will, whenever possible, always give both notations. The Hermann–Mauguin notation will always be used first, followed by the Schönflies notation in parenthesis; for instance, the identity operator mentioned in the previous paragraph will be denoted by the symbol 1 (E), where 1 is the Hermann–Mauguin symbol and E the Schönflies symbol. Note that we use a sans serif font for all operator symbols. While this dual notation is longer than either one used individually, the advantage is that the reader will become familiar with both notational schemes. Most symmetry operations can also be represented by a graphical symbol, and we will introduce those symbols at the appropriate locations throughout the text.

Before we begin a detailed overview of symmetry operations, let us first approach the problem with some simple two-dimensional considerations. Consider a hexagon with an edge length a, centered in the origin of a Cartesian reference frame, as shown in Fig. 8.1(a). The hexagon has six vertices (corners), labeled 1 through 6. It is obvious that a rotation of 60° around the origin, or any multiple of 60°, will generate a new hexagon that coincides with the first one (i.e., it cannot be distinguished from the first one). According to our definition of symmetry, this means that the hexagon has rotational symmetry around an axis normal to the hexagon, passing through

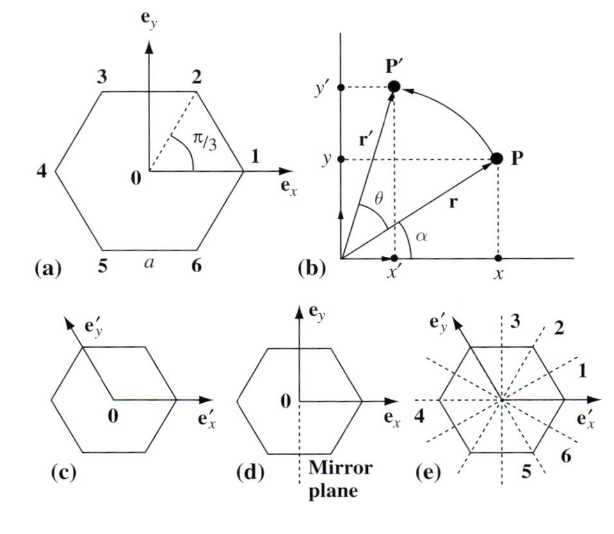

Fig. 8.1. Illustration of the six-fold symmetry of the regular hexagon with respect to (a) a Cartesian reference frame and (c) a hexagonal reference frame. In (b), the point **P** is rotated by an angle θ to the point **P′**. (d) and (e) show the presence of mirror planes, again with respect to Cartesian and hexagonal reference frames.

the origin. If we consider one of the vertices, say point 1, then we can ask the question: what will be the location of this point after we carry out a (counterclockwise) rotation of 60°?[1] From the drawing, it is clear that point 1 will move to point 2, 2 will move to 3 and so on; the final point 6 will move to 1. The coordinates of each point can be read from the drawing, but what we really need is a mathematical method to *compute* the coordinates, starting from the initial coordinates. In the mathematical literature this can be written as a permutation matrix of the vertices: (123456) → (612345).

Consider the two-dimensional Cartesian reference frame shown in Fig. 8.1(b). A general point **P** has coordinates (x, y), and is represented by the position vector **r**. We apply a counterclockwise rotation around an axis perpendicular to the plane of the drawing and going through the origin O. The rotation angle is θ. The point **P** is now rotated to the position $\mathbf{P}'$, with position vector $\mathbf{r}'$ and components (x', y'). The angle between the original position vector **r** and the $\mathbf{e}_x$ axis is α. From the drawing, we find that the components of the new position vector are given by:

$$
\begin{aligned}
x' &= r' \cos(\theta + \alpha);\\
y' &= r' \sin(\theta + \alpha),
\end{aligned}
$$

where $r' = |\mathbf{r}'| = |\mathbf{r}| = r$. Applying the addition theorems for trigonometric functions we find:

$$
\begin{aligned}
x' &= r(\cos\theta\cos\alpha - \sin\theta\sin\alpha);\\
y' &= r(\sin\theta\cos\alpha + \cos\theta\sin\alpha).
\end{aligned}
$$

From the drawing, we also find that $x = r\cos\alpha$ and $y = r\sin\alpha$, which leads to:

$$
\begin{aligned}
x' &= x\cos\theta - y\sin\theta;\\
y' &= x\sin\theta + y\cos\theta,
\end{aligned}
$$

which can be rewritten in matrix form as:

$$
\begin{pmatrix} x' \\ y' \end{pmatrix} = \begin{pmatrix} \cos\theta & -\sin\theta \\ \sin\theta & \cos\theta \end{pmatrix} \begin{pmatrix} x \\ y \end{pmatrix}. \tag{8.1}
$$

[1] It is important to note that a counterclockwise rotation is the common and International crystallographic convention for defining a rotation operation. In literature from other fields, the opposite clockwise rotation has been chosen as the convention. In reading the literature it is, therefore, important to identify the convention being used.

We find that a rotation around an axis can be represented by a *rotation matrix* $\mathsf{D}_2(\theta)$. In two dimensions, we have:

$$\mathsf{D}_2(\theta) = \begin{pmatrix} \cos\theta & -\sin\theta \\ \sin\theta & \cos\theta \end{pmatrix}.$$

In 3-D, the rotation matrix will become a 3×3 matrix. The easiest case is the one that we already worked out, since the third dimension $\mathbf{e}_z$ can be chosen along the rotation axis. The rotation axis itself does not change during the rotation, and we say that the rotation axis is *invariant*. The 3-D rotation matrix is therefore given by:

$$\mathsf{D}_3(\theta) = \begin{pmatrix} \cos\theta & -\sin\theta & 0 \\ \sin\theta & \cos\theta & 0 \\ 0 & 0 & 1 \end{pmatrix}. \qquad (8.2)$$

For simplicity, we will, throughout this book, drop the subscript 3 on D_3, i.e., $\mathsf{D} \equiv \mathsf{D}_3$. The coordinate transformation in 3-D is explicitly given by:

$$\begin{pmatrix} x' \\ y' \\ z' \end{pmatrix} = \begin{pmatrix} \cos\theta & -\sin\theta & 0 \\ \sin\theta & \cos\theta & 0 \\ 0 & 0 & 1 \end{pmatrix} \begin{pmatrix} x \\ y \\ z \end{pmatrix}, \qquad (8.3)$$

or, alternatively:

$$x' = x\cos\theta - y\sin\theta;$$
$$y' = x\sin\theta + y\cos\theta;$$
$$z' = z.$$

As expected, we find that the third component z does not change during the rotation.

Now that we have an expression for the coordinate transformation matrix $\mathsf{D}_2(\theta)$, we can determine the coordinates of the rotated points in Fig. 8.1(a) for $\theta = \pi/3$. The matrix is given by:

$$\mathsf{D}_2\left(\frac{\pi}{3}\right) = \begin{pmatrix} \frac{1}{2} & -\frac{\sqrt{3}}{2} \\ \frac{\sqrt{3}}{2} & \frac{1}{2} \end{pmatrix},$$

so that the point 1 with coordinates $(a, 0)$ is rotated into the point with coordinates $a(1/2, \sqrt{3}/2)$, which corresponds to point 2 (Fig. 8.1(a)).

Rotation matrices have a number of special properties:

(i) *the inverse of a rotation matrix is equal to the transpose of that matrix.* This property makes it extremely easy to compute the inverse of any rotation matrix. Any matrix whose inverse is equal to its transpose is generally known as an *orthonormal* matrix. As an example, consider the 2×2 rotation matrix $\mathsf{D}_2(\pi/4)$:

$$\mathsf{D}_2\left(\frac{\pi}{4}\right) = \begin{pmatrix} \frac{1}{\sqrt{2}} & \frac{-1}{\sqrt{2}} \\ \frac{1}{\sqrt{2}} & \frac{1}{\sqrt{2}} \end{pmatrix} = \frac{1}{\sqrt{2}}\begin{pmatrix} 1 & -1 \\ 1 & 1 \end{pmatrix}.$$

The transpose of this matrix is:

$$\mathsf{D}_2^{\mathrm{T}}\left(\frac{\pi}{4}\right) = \frac{1}{\sqrt{2}}\begin{pmatrix} 1 & 1 \\ -1 & 1 \end{pmatrix}.$$

Upon multiplication of $\mathsf{D}_2(\pi/4)$ with its transpose we find:

$$\frac{1}{2}\begin{pmatrix} 1 & -1 \\ 1 & 1 \end{pmatrix}\begin{pmatrix} 1 & 1 \\ -1 & 1 \end{pmatrix} = \frac{1}{2}\begin{pmatrix} 1 & 1 \\ -1 & 1 \end{pmatrix}\begin{pmatrix} 1 & -1 \\ 1 & 1 \end{pmatrix} = \begin{pmatrix} 1 & 0 \\ 0 & 1 \end{pmatrix}$$

which proves that $\mathsf{D}_2(\pi/4)$ is an orthonormal matrix.

(ii) *For an orthonormal matrix, the sum of the squares of all elements on any row or column is equal to* 1. This can easily be seen from Equation 8.1, since $\cos^2\theta + \sin^2\theta = 1$.

(iii) *The determinant of a rotation matrix is always equal to* $+1$. It is easy to show that $\det(\mathsf{D}_2(\theta)) = \det(\mathsf{D}(\theta)) = \cos^2\theta + \sin^2\theta = 1$.

At this point it is useful to connect what we have just derived with the theory of coordinate transformations presented in the previous chapter (Section 7.4 on page 144). In that section, we defined a transformation matrix, α_{ij}, that transforms the old basis vectors, $\mathbf{a}_j$, into the new ones, $\mathbf{a}'_i$, as follows:

$$\mathbf{a}'_i = \alpha_{ij}\mathbf{a}_j.$$

For the components of vectors, we found, in Table 7.1, that the transformation relation is given by:

$$p'_i = p_j\alpha^{-1}_{ji};$$

in other words, to transform the coordinates, p_j, of a point, we must post-multiply by the transpose of the inverse of α_{ij}. In the present section, Equation (8.3) can be rewritten as:

$$p'_i = \mathsf{D}_{ij}p_j.$$

These two relations describe the same coordinate transformation, so that we must have:

$$\mathsf{D} = \alpha^{-1}. \tag{8.4}$$

This means that we have two equivalent descriptions for the coordinate transformation (in this case a counterclockwise rotation by $\pi/3$): the description with the matrix $\mathsf{D}(\theta)$ operates directly on the coordinates p_i, whereas the description with the matrix α operates on the basis vectors $\mathbf{a}_j$. Both descriptions describe the same transformation, and we will return to them in Section 8.2 on page 170.

In all examples above, we have worked with the Cartesian reference frame $(\mathbf{e}_x, \mathbf{e}_y)$ defined in Fig. 8.1(b). As a consequence of this choice, the entries of the rotation matrices are typically non-integer numbers, such as $1/2$ or $\sqrt{3}/2$. If we select a different reference frame, for instance $(\mathbf{e}'_x, \mathbf{e}'_y)$ as shown in Fig. 8.1(c), then the rotation matrix changes to a much simpler form, which we will now derive for the 3-D case.

To determine the entries of the rotation matrix $\mathsf{D}(\pi/3)$ with respect to the primed basis vectors, all we need to do is figure out what happens to the basis vectors themselves. For instance, the vector $\mathbf{e}'_x$ is rotated to $\mathbf{e}'_x + \mathbf{e}'_y$, and $\mathbf{e}'_y$ is rotated to $-\mathbf{e}'_x$. The vector $\mathbf{e}'_z$ remains unchanged. These three relations are sufficient to determine the entire matrix, since we can write:

$$\left(\mathbf{e}'_x + \mathbf{e}'_y \;\; -\mathbf{e}'_x \;\; \mathbf{e}'_z\right) = \left(\mathbf{e}'_x \;\; \mathbf{e}'_y \;\; \mathbf{e}'_z\right) \begin{pmatrix} 1 & -1 & 0 \\ 1 & 0 & 0 \\ 0 & 0 & 1 \end{pmatrix}.$$

The matrix on the right-hand side is the rotation matrix, D, with respect to the primed reference frame. We see that it contains only the integers 0, -1 and $+1$. The standard procedure for the determination of a transformation matrix is then described in Box 8.1. It is straightforward to show that the points 2 through 6 of Fig. 8.1(a) can be obtained by repeated application of the matrix D on the initial point 1.

Let us consider another example: a mirror plane containing the Cartesian $\mathbf{e}_y$ and $\mathbf{e}_z$ basis vectors (dashed line in Fig. 8.1(d)). This mirror plane leaves

Box 8.1 Determination of a coordinate transformation matrix

Here is how to determine a coordinate transformation matrix based on a drawing:

(i) determine from a drawing how each of the basis vectors transforms under the symmetry operation, and write the transformed basis vector as a linear combination of the untransformed basis vectors;
(ii) take the coefficients of this linear combination and write them as the columns of the matrix D. The coefficients of the first transformed basis vector go in the first column, and so on.

the vectors in the plane unchanged, and replaces $\mathbf{e}_x$ by $-\mathbf{e}_x$. Therefore, the transformation matrix is given by:

$$\mathsf{D}(\mathrm{m}) = \begin{pmatrix} -1 & 0 & 0 \\ 0 & 1 & 0 \\ 0 & 0 & 1 \end{pmatrix}$$

where the argument m of the matrix represents the mirror plane. We will introduce a more complete notation in a later section. Note that this matrix has a negative determinant, equal to -1. This indicates that this symmetry operation results in a reversal of handedness, which we will discuss in detail in Section 8.2.5. The entries in the above matrix are simple integers, because the mirror plane contains two of the basis directions. It is left as an exercise for the reader to determine what the matrix would look like if the mirror plane were rotated clockwise by 30°.[2]

If we describe the mirror planes of the hexagon (all six of them are shown in Fig. 8.1(e)) *with respect to the primed basis vectors*, then once again all mirror planes can be represented by matrices which contain only the entries 0, -1 and $+1$. The matrices for the mirrors 1, 3, and 5 are given by:

$$\mathsf{D}(\mathrm{m}_1) = \begin{pmatrix} 1 & 0 & 0 \\ 1 & -1 & 0 \\ 0 & 0 & 1 \end{pmatrix};$$

$$\mathsf{D}(\mathrm{m}_3) = \begin{pmatrix} -1 & 1 & 0 \\ 0 & 1 & 0 \\ 0 & 0 & 1 \end{pmatrix};$$

$$\mathsf{D}(\mathrm{m}_5) = \begin{pmatrix} -1 & 0 & 0 \\ -1 & 1 & 0 \\ 0 & 0 & 1 \end{pmatrix}.$$

We leave it to the reader to determine the remaining three matrices $\mathsf{D}(m_{2,4,6})$. In the remainder of this book, we will nearly always use symmetry transformation matrices that are defined with respect to the basis vectors of the Bravais unit cell, so that the entries of those matrices will be the integers 0, -1, and $+1$. This will allow us to simplify the description of the symmetry of all crystals. In the following section, we will discuss in more detail the nature of symmetry operations, in particular those that are relevant to crystallography.

[2] The entries in this new matrix will no longer be simple integers.

8.2 Symmetry operations

There are two ways to view the action of a symmetry operator on an object: one can either interpret the symmetry operator as *acting on the object*, leaving the reference frame unchanged in space or, equivalently, one can let the operator *act on the reference frame*, leaving the object unchanged in space. If we represent the object by the symbol $\mathcal{F}$, and the vector $\mathbf{r}$ represents a point of the object, then this relation can be expressed mathematically as follows:

$$\boxed{\mathcal{F}(\mathcal{O}[\mathbf{r}]) = \mathcal{O}^{-1}[\mathcal{F}(\mathbf{r})],} \tag{8.5}$$

where $\mathcal{O}$ represents the symmetry operator and $\mathcal{O}^{-1}$ its inverse. This can be understood easily when the operator is a simple rotation: rotation of an object through $+60°$ is the same as rotating the reference frame through an angle of $-60°$ (see Fig. 8.2). If the object is moved in space, leaving the reference frame constant, then the operator is said to be an *active operator*. If the reference frame is moved in space, leaving the object unchanged, then the operator is said to be a *passive operator*. In this text, we will always consider a symmetry operator as acting on the coordinates of the material points, rather than on the reference frame. In other words, we will always take the *active* interpretation of a symmetry operation.

We saw in the previous section, that there are two equivalent matrix descriptions for the coordinate transformation corresponding to the symmetry operation: the matrix $\mathsf{D}(\theta)$ represents the transformation of the coordinates, p_i, and is, hence, a representation of the *active* interpretation. The other matrix, α, transforms the basis vectors, and is, hence, a representation of the *passive* interpretation.

Any physical property of the object/crystal will be invariant under the symmetry operator:

$$\mathcal{F}(\mathbf{r}) = \mathcal{F}(\mathcal{O}[\mathbf{r}]) = \mathcal{F}(\mathbf{r}') \tag{8.6}$$

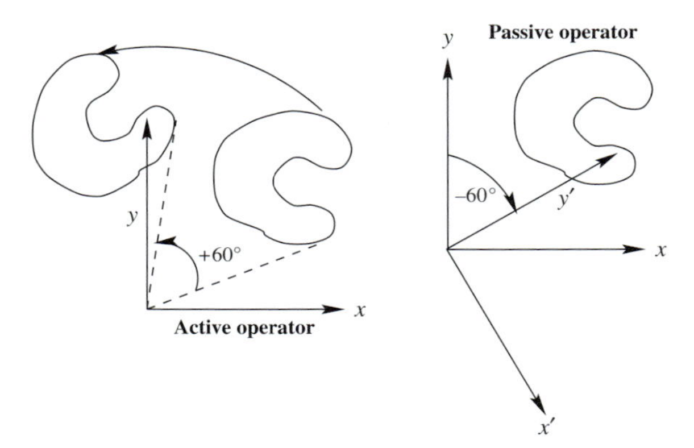

Fig. 8.2. Illustration of an *active* and a *passive* operator.

This implies that the inverse transformation, from $\mathbf{r}'$ to $\mathbf{r}$ exists and is also a symmetry operation. If several symmetry properties exist, then it can be shown that any combination of them will also be a symmetry property. We will study such combinations of symmetry operators in the next chapter, when we introduce the concept of *point group*.

Symmetry is an inherent attribute of an object; without a certain external measure, symmetric orientations of an object are indistinguishable. Note that such an external measure would destroy the symmetry because it would require the attachment of a reference point to the object. Hence, symmetry can be a very subtle aspect of an object; the more intricate symmetry properties, such as time reversal symmetry in magnetic materials, can be difficult to detect and understand. For now, we will restrict ourselves to study basic symmetry operations, namely the isometric transformations of an object.

8.2.1 Basic isometric transformations

> An isometric transformation is a transformation that leaves the metric properties of space unaltered, i.e., no stretching, twisting, or bending is involved.

Obviously, empty space is invariant under any kind of symmetry operator, even the non-isometric ones. In order to distinguish between isometric and non-isometric operators, we have to provide a gauge by which distances and coordinates can be measured. In 3-D-space, this can be done by providing three basis vectors or four non-coplanar points, i.e., an *asymmetric tetrahedron*. The four points are the origin and the three endpoints of the basis vectors. Any fifth point can then be located with respect to this tetrahedron and, hence, any object can be fully and unambiguously defined, except for its *symmetric variants*.

It can be shown that *any isometric transformation can be reduced to either a translation, a rotation, or a reflection, or a combination of these.* The mathematical proof uses the concept of the *Chasles-center* but this is beyond the scope of this text. The proof involves consideration of two identical asymmetric tetrahedra in different orientations on different locations and determination of the ways to bring them into self-coincidence.

There are two different correspondences between an object and its symmetric variant(s):

- *coincidental or congruent equality* = a transformation of the first kind
- *mirror equality* = a transformation of the second kind

Transformations of the first kind are also known as *proper motions*, since they can be physically realized.[3] The mirror operation is not a proper motion, because it is physically impossible to change the handedness of an object without actually deforming it. Therefore, this type of operation is distinguished from the proper motions. Any transformation of the first kind is either a *translation* or a *rotation* or a combination of both. A transformation of the second kind is either a *reflection* or an *inversion*. Before discussing the various fundamental symmetry operations in more detail, we will first derive some important constraints on the crystallographically permitted rotational symmetries.

8.2.2 Compatibility of rotational symmetries with crystalline translational periodicity

In this section, we address the question of compatibility (or incompatibility) of rotational symmetry axes with the translational symmetry of crystalline solids. As an example, consider the hexagonal tiling of the 2-D plane in Fig. 8.3(a). There are no gaps in between the tiles, so that no area is left uncovered. If we attempt to repeat this with a five-fold tile, then we quickly find (Fig. 8.3(b)) that we cannot tile the 2-D plane with tiles of five-fold symmetry. The same is true of higher order polygons, such as the heptagon, the octagon, and so on. Figure 8.3(c) shows an example of what may be a *molecular solid* in which pentagonal molecules decorate the sites of a square lattice. Although the local units are pentagons, the five-fold symmetry axis is lost since it is inconsistent with the square lattice. Furthermore, the four-fold symmetry, which would exist when spherical atoms decorate the square lattice, is lost by virtue of the lack of four-fold symmetry in the pentagonal

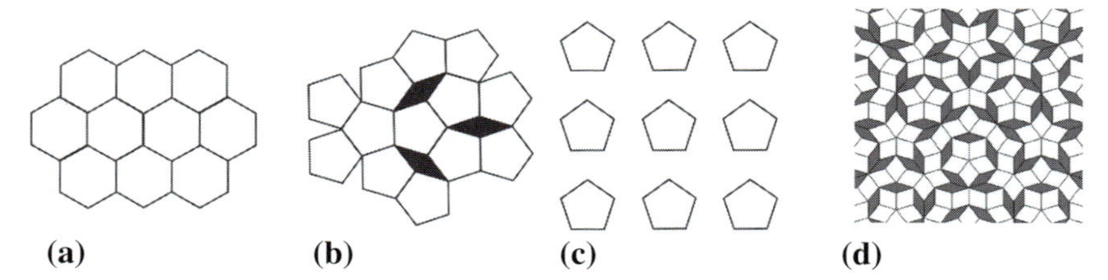

Fig. 8.3. (a) Tiling of the 2-D plane with hexagons illustrating the compatibility of a six-fold symmetry axis and crystalline periodicity; (b) illustration of the fact that the 2-D plane cannot be tiled with pentagons; (c) square lattice decorated with pentagons; and (d) 2-D Penrose tiling which preserves five-fold rotational symmetry.

[3] We all know from experience that we can translate and rotate arbitrary objects, but we cannot change their handedness without temporarily changing their shape. For instance, to turn a left handed glove into a right handed glove, we must turn it inside-out, which is not an isometric operation.

basis. Figure 8.3(d) shows an example of what may be called a *2-D quasi-crystal*, in which two different rhombuses are used to fill the plane, in a such a way that five-fold rotational symmetry is preserved. In Chapter 16, we will discuss in more detail quasi-periodic motifs which tile the plane and preserve five-fold rotational symmetries but do not involve tilings with pentagons. This notion can be generalized to include quasi-periodic tilings preserving other symmetries, either non-crystallographic or crystallographic.

The fact that only 1-, 2-, 3-, 4-, and 6-fold axes are compatible with crystalline axes was elegantly demonstrated using the *rule of rational indices* (Haüy, 1784, 1801, 1822). The rule of rational indices is illustrated geometrically in Fig. 8.4. Consider the two lattice points A and A′ which are separated by a unit translation vector, **t**. Consider also a rotation operator, D, with an axis normal to the plane containing the vector AA′ and passing through the point A. If D represents a counterclockwise rotation by the angle α, it takes lattice point A′ into a symmetrically equivalent lattice point B. Since every rotational symmetry operation has an inverse operation, we may also consider a rotation by $-\alpha$ about the point A′ which yields the lattice point B′ starting from A. Finally, if the rotations are consistent with the translational symmetry of the lattice, then the distance t' (the length of $\mathbf{t}'$) between B and B′ must be an integral number of unit translations t (the length of **t**), i.e.:

$$t' = mt, \tag{8.7}$$

where m is an integer. From examination of the geometry shown in Fig. 8.4, it is also clear that:

$$t' = t + 2t\cos\alpha. \tag{8.8}$$

Combining the previous equations, we conclude that:

$$\cos\alpha = \frac{1-m}{2}. \tag{8.9}$$

If m is an integer, then so is $(1-m)$. Furthermore, we know that $|\cos\alpha| \leq 1$, and, therefore, $|1-m| \leq 2$. This, in turn, means that $1-m = -2, -1, 0, 1, 2$

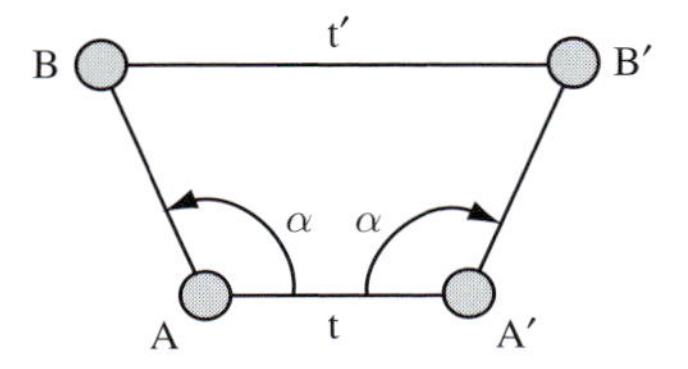

Fig. 8.4. Lattice translation vector, **t**, connecting points A and A′. The lattice points B and B′ are generated by a rotational symmetry axis (rotation by α) normal to the plane and passing through sites A or A′, respectively.

and, hence, $\alpha = \pi, 2\pi/3, \pi/2, \pi/3$, or 0, respectively. Thus, the only rotations that are consistent with the translational symmetry found in periodic crystals are the 1-fold ($\alpha = 2\pi$), 2-fold ($\alpha = \pi$), 3-fold ($\alpha = 2\pi/3$), 4-fold ($\alpha = \pi/2$), and 6-fold ($\alpha = \pi/3$) rotations. Note in particular the absence of five-fold rotations; in Chapter 16, we will describe in detail how crystals with five-fold rotational symmetry *but no translational symmetry* can be constructed.

The derivation of the compatible rotational symmetries can be repeated more elegantly by considering the *trace* of the rotation matrices, as explored in a problem at the end of this chapter.

8.2.3 Operations of the first kind: pure rotations

A pure rotation is characterized by a *rotation axis* and a *rotation angle* that is chosen to be positive for a counterclockwise rotation. The rotation axis can be described by a direction vector $[uvw]$, or by the equation of the line that represents the rotation axis. It is customary to write the rotation angle as a fraction of 2π, i.e., $\alpha = 2\pi/n$. The number n is the *order* of the rotation and we say that a rotation is n-fold if its angle is given by $2\pi/n$. The order n can take any integer value from $n = 1$ (the identity operator, 1 (E)), to $n = \infty$ for a circle. Examples of rotation axes of orders 3, 4, and 6 are shown in Fig. 8.5. Note that the object that is used to illustrate the rotational symmetry,

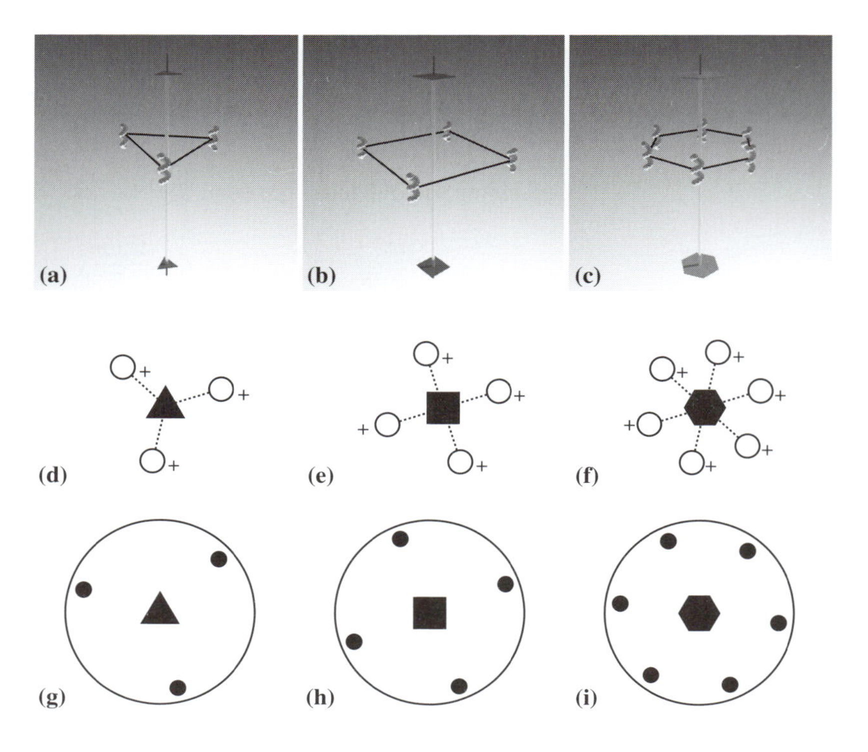

Fig. 8.5. Rendered 3-D drawings of (a) three-fold, (b) four-fold, and (c) six-fold rotational symmetry, along with 2-D drawings of the symmetry operator (second row), and the corresponding stereographic projections.

in this case a distorted helical string of spheres, should always be taken to be an object which itself does not have any symmetry other than the identity operator. If we had used a sphere as object, we would not have discovered that the orientation of the object changes while it is rotated.

A pure rotation is denoted by the symbol n (C_n), with n the order of the rotation.[4] In many cases, the rotation axis will be parallel to the third basis direction, i.e., parallel to the crystallographic **c** axis. In those cases, we do not need to state explicitly the orientation of the rotation axis. For all other cases, however, the rotation symbol must be augmented by information about the orientation of the rotation axis. This can be done in two ways: we can use the direction indices $[uvw]$ of the rotation axis, or we can use the equation of the line coinciding with the axis. Hence, a three-fold rotation axis aligned along the [111] direction of a cubic reference frame can be represented as **3** (C_3) [111]. The equation of the line along this direction is given by $x = y = z$, so that an alternative notation would be **3** (C_3) x, x, x. This latter notation is the one used in the *International Tables for Crystallography* (Hahn, 1996); the former notation is used in many textbooks, e.g., Burns and Glazer (1990). Another example can be found in Box 8.2.

In drawings, we use a standard symbol to indicate an n-fold rotation axis. The symbol is a filled regular polygon with n sides. A six-fold rotation axis perpendicular to the drawing plane is then indicated by the ⬢ symbol, a three-fold axis by ▲, and a two-fold axis by ⬮. The use of these symbols is illustrated in Fig. 8.5(d) through (i); the center row shows the graphical symbol surrounded by open circles with a plus sign next to them. The rotation operator takes any one of these points and rotates it onto the next one (in a counterclockwise sense). During the rotation, the elevation of the point

Box 8.2 Determination of a rotation symbol

A two-fold rotation axis is oriented so that it is normal to the **c** axis of a cubic reference frame, and it bisects the angle between the **a** and **b** directions. What are the two full symbols for this rotation operator?

Answer: The two-fold rotation axis is represented by the symbol **2** (C_2), but since it is not oriented along the **c** axis, we must add information about the orientation. It is easy to verify that the axis lies along the [110] direction, so that the symbol is written as **2** (C_2) [110]. The equation for the line along this direction is given by $x = y$ and $z = 0$, so that the international symbol is given by **2** (C_2) $x, x, 0$.

[4] Recall that we state the Hermann–Mauguin symbol first, followed by the Schœnflies symbol between parentheses.

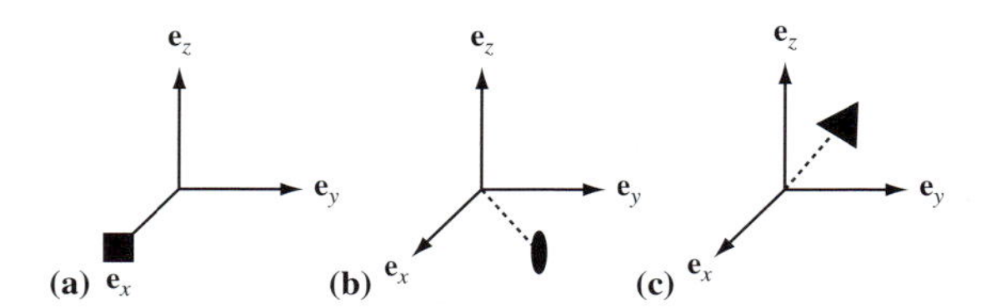

Fig. 8.6. (a) four-fold, (b) two-fold, and (c) three-fold rotations used as examples in the text.

does not change, so that if one point is at some positive distance above the plane of the drawing, indicated by the + sign, then all of them are. This is consistent with the fact that rotations occur in a plane normal to the rotation axis. The bottom row of Fig. 8.5 shows a commonly used representation of symmetry operations by means of stereographic projections. The rotation axis is aligned along the north–south center line, which is normal to the plane of the drawing, and a point in the northern hemisphere, represented by a filled circle, is rotated to its equivalent positions, all of them in the northern hemisphere. In the remainder of this book, we will make frequent use of both graphical representation schemes.

In Section 8.1, we have seen that rotations can be represented by orthogonal matrices $\mathsf{D}(\theta)$, and we considered examples of rotations around the z axis of the reference frame. Additional examples are shown in Fig. 8.6. Fig. 8.6(a) shows a four-fold rotation around the x axis of a cubic reference frame. The $\mathbf{e}_x$ basis vector is obviously invariant, whereas $\mathbf{e}_y$ is rotated onto $\mathbf{e}_z$, and $\mathbf{e}_z$ is rotated to $-\mathbf{e}_y$. Using the method described in Box 8.1, we find for the transformation matrix:

$$\mathsf{D} = \begin{pmatrix} 1 & 0 & 0 \\ 0 & 0 & -1 \\ 0 & 1 & 0 \end{pmatrix}. \tag{8.10}$$

For a two-fold rotation $\mathbf{2}$ (C_2) $x, x, 0$ (or $\mathbf{2}$ (C_2) [110]), shown in Fig. 8.6(b), we find:

$$\mathsf{D} = \begin{pmatrix} 0 & 1 & 0 \\ 1 & 0 & 0 \\ 0 & 0 & -1 \end{pmatrix}, \tag{8.11}$$

and for a $\mathbf{3}$ (C_3) x, x, x rotation ($\mathbf{3}$ (C_3) [111], Fig. 8.6(c)) we have:

$$\mathsf{D} = \begin{pmatrix} 0 & 0 & 1 \\ 1 & 0 & 0 \\ 0 & 1 & 0 \end{pmatrix}. \tag{8.12}$$

8.2.4 Operations of the first kind: pure translations

A pure translation is characterized by a translation vector $\mathbf{t}$. We have already discussed translations in Chapter 4. In crystals, a translation vector is limited to the lattice vectors and, if present, the lattice centering vectors. We will see later that certain fractions of lattice translation vectors may also be allowed. In drawings, translation vectors are indicated by arrowed lines, i.e., $\longrightarrow$. The international symbol for a translation is $t(u, v, w)$, as in $t(1/2, 0, 1/2)$; alternatively, one could also use $t[uvw]$, since a translation implies a direction, which is described by a direction symbol. We remind the reader that a translation is only a true symmetry operation in an infinite solid.

In the preceding sections, we have seen that rotations can be represented mathematically by matrices. In the remainder of this section, we will explore the possibility of representing translations also by means of matrices. This may seem a bit odd, since a translation is usually represented as a vector addition, e.g., translation by a vector **t** is described by the equation:

$$\mathbf{r}' = \mathbf{r} + \mathbf{t}. \tag{8.13}$$

It is clear from this relation that we cannot simply multiply **r** by a matrix to get the same result. It would be useful to have a matrix representation for a translation, since then we could represent *all* symmetry operations by matrices. It turns out that we can do this by working with four-dimensional (4-D) vectors instead of the regular three-dimensional vectors. This sounds complicated but it is really very simple: to each triplet of vector components (x_1, x_2, x_3) we simply add the number 1 as fourth component, i.e., $(x_1, x_2, x_3, 1)$. These types of coordinates are called *normal coordinates* or *homogeneous coordinates* (see Box 8.3). The fourth component *must always be equal to* 1, and is denoted by x_4.

Box 8.3 About normal or homogeneous coordinates

Four-dimensional coordinates are often used in the world of computer graphics (e.g., Salmon and Slater (chapter 13, 1987)) and make it relatively easy to carry out a large number of different coordinate transformations, such as rotations and translations (which we discuss in more detail in the main text), but also scaling, orthogonal, and perspective projections from 3-D to 2-D. In particular, viewing transformations, which convert 3-D objects into 2-D representations on a computer screen, are conveniently expressed in normal coordinates. A general homogeneous coordinate corresponding to a Cartesian point (x, y, z) is described by (wx, wy, wz, w), with $w \neq 0$. Equivalently, the 4-D coordinate (x, y, z, w) corresponds to the point with Cartesian coordinates $(x/w, y/w, z/w)$. In all cases of importance for crystallography, we may take $w = 1$, so that conversion between conventional and homogeneous coordinates is simply a matter of dropping (or adding) the fourth component $x_4 = 1$.

Consider the component notation for Equation 8.13 above:

$$x'_i = x_i + u_i,$$

where the components of $\mathbf{t}$ are denoted by u_i. This relation is valid when the index i takes on the values 1, 2, and 3. If we assume that it is also valid when the index, i, goes from 1 to 4, then we can write out the following explicit relations (recall that $x_4 = u_4 = 1$):

$$\begin{aligned} x'_1 &= 1 \times x_1 + 0 \times x_2 + 0 \times x_3 + u_1 \times x_4; \\ x'_2 &= 0 \times x_1 + 1 \times x_2 + 0 \times x_3 + u_2 \times x_4; \\ x'_3 &= 0 \times x_1 + 0 \times x_2 + 1 \times x_3 + u_3 \times x_4; \\ x'_4 &= 0 \times x_1 + 0 \times x_2 + 0 \times x_3 + u_4 \times x_4. \end{aligned}$$

The last equation is trivial, since it simply states that $1 = 1$. We can rewrite this set of equations in matrix form, as follows:

$$\begin{pmatrix} x'_1 \\ x'_2 \\ x'_3 \\ 1 \end{pmatrix} = \begin{pmatrix} 1 & 0 & 0 & u_1 \\ 0 & 1 & 0 & u_2 \\ 0 & 0 & 1 & u_3 \\ 0 & 0 & 0 & 1 \end{pmatrix} \begin{pmatrix} x_1 \\ x_2 \\ x_3 \\ 1 \end{pmatrix}. \tag{8.14}$$

It is clear that this equation is completely equivalent to $\mathbf{r}' = \mathbf{r} + \mathbf{t}$, because when we write it out in components, we recover the correct relation. By going to 4-D vectors instead of 3-D, we have been able to include the translation components u_i as part of a 4×4 matrix. We don't have to pay any attention to the last equation of this set of four, since it is a trivial one $(1 = 1)$.[5]

We will introduce the symbol $\mathcal{W}$ for the 4×4 matrix defined above. In other words:

$$\tilde{x}'_i = \mathcal{W}_{ij} \tilde{x}_j,$$

where the tilde $\tilde{}$ indicates that normal coordinates are being used. The indices i and j both take on values from 1 to 4. Note that $\mathcal{W}$ consists of three

[5] It is good practice to always carry out all computations including this fourth row; at the end of a computation this row should always consist of three zeros and a one. When this is not the case, then there must have been a computational error somewhere!

parts: a 3×3 matrix in the upper left corner, a 3×1 column in the top rightmost column, and a 1×4 row at the bottom:

$$\mathcal{W} = \left(\begin{array}{ccc|c} 1 & 0 & 0 & u_1 \\ 0 & 1 & 0 & u_2 \\ 0 & 0 & 1 & u_3 \\ \hline 0 & 0 & 0 & 1 \end{array} \right).$$

Since the bottom row does not contain any information, we can define a new notation, which is equivalent to the matrix $\mathcal{W}$. This notation is known as the *Seitz symbol*, written as $(\mathsf{D}|\mathbf{t})$. The symbol D is the 3×3 sub-matrix of $\mathcal{W}$, and $\mathbf{t}$ represents the translation. For a translation, D is the identity matrix, which we can represent by the symbol E. Therefore, the Seitz symbol for a translation is given by $(\mathsf{E}|\mathbf{t})$. It is always straightforward to construct the matrix $\mathcal{W}$ when the Seitz symbol is known.

We can also use this notation for rotations. In that case, there is no translation, so that $\mathbf{t} = \mathbf{0}$, and the Seitz symbol for a pure rotation becomes $(\mathsf{D}|\mathbf{0})$. The $\mathcal{W}$ matrix for a rotation is then given by:

$$\mathcal{W} = \begin{pmatrix} \mathsf{D}_{11} & \mathsf{D}_{12} & \mathsf{D}_{13} & 0 \\ \mathsf{D}_{21} & \mathsf{D}_{22} & \mathsf{D}_{23} & 0 \\ \mathsf{D}_{31} & \mathsf{D}_{32} & \mathsf{D}_{33} & 0 \\ 0 & 0 & 0 & 1 \end{pmatrix}. \qquad (8.15)$$

The Seitz symbol for the identity operator is simply $(\mathsf{E}|\mathbf{0})$.

8.2.5 Operations of the second kind: pure reflections

In Fig. 8.7 the two different types of operation of the second kind are depicted: a pure reflection, and an inversion. A pure *reflection*, also known as a *mirror*, is characterized by a plane, indicated in a drawing by a thick solid line ——— if the plane is at an angle with respect to the plane of the drawing. If the mirror plane lies in the plane of the drawing, then the symbol is ⌉. Examples are shown in Fig. 8.7(c) and (d). The open circle with a plus sign is copied into the second circle, which now has opposite handedness. This is indicated by means of a comma inside the circle. When the two circles coincide in the drawing, as in (d), then a vertical line divides the circle into two halves, each with its own + or − sign, and the appropriate comma. The stereographic projections for these two cases are shown in Fig. 8.7(f) and (g). Note that in stereographic projections, one does not usually distinguish between points of opposite handedness.

The Hermann–Mauguin symbol for a mirror plane is m, while the Schönflies symbol is the Greek letter σ, resulting in the notation m (σ). Sometimes, the Schönflies symbol has a subscript which indicates the orientation of the mirror plane. The symbol m (σ_h) refers to a mirror plane that is normal to a given direction (typically a rotation axis), whereas m (σ_v) represents a mirror plane which contains that direction. An additional symbol, m (σ_d), is also used as a variant of m (σ_v).

The international notation for a mirror plane is sometimes extended by providing information on the orientation of the plane. This can again be done in two possible ways, either by specifying the direction of the normal to the mirror plane, or by writing down the equation of the plane. A mirror plane parallel to the (110) plane in a cubic system would be written as either m (σ) [110] or m (σ) $x, -x, 0$.

We have already seen how we can represent a mirror operation by a matrix of the type $\mathsf{D}(m)$ (see page 169). It is then straightforward to write down the Seitz symbol for a mirror operation: $(\mathsf{D}(m)|\mathbf{0})$.

8.2.6 Operations of the second kind: inversions

An *inversion* is a point symmetry operation that takes all the points $\mathbf{r}$ of an object and projects them into the new points given by $-\mathbf{r}$. The operation is denoted by the symbols $\bar{1}$ (i). The inversion operation is illustrated in Fig. 8.7(b); the

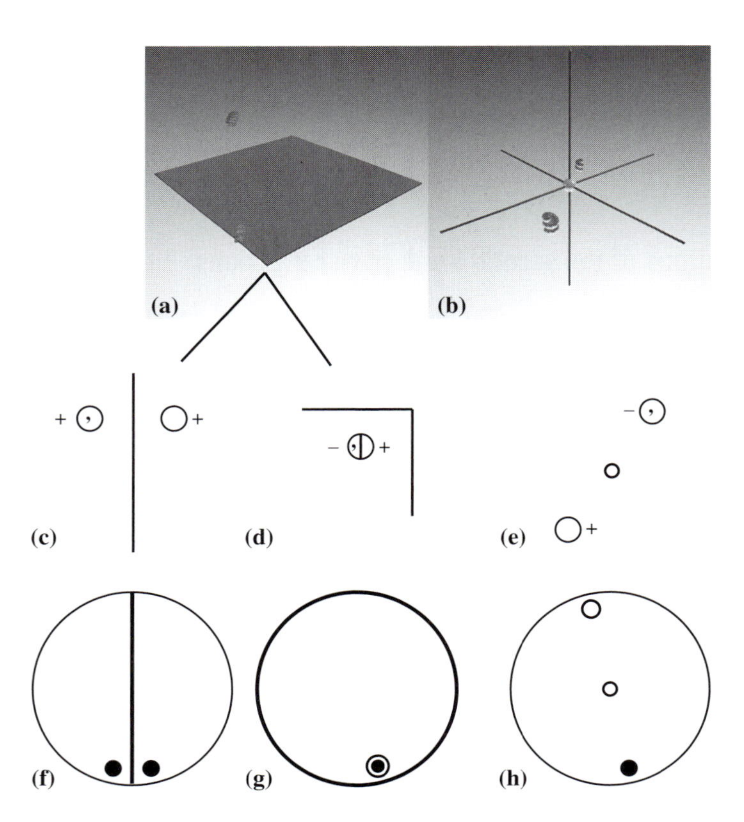

Fig. 8.7. The two different types of second kind operation: pure reflection, and inversion center.

helix in the front of the image is inverted to the back. Note that this changes the handedness of the helix. In drawings, such as Fig. 8.7(e), the inversion center is denoted by the symbol ∘ (i.e., a small open circle). It is impossible to bring an inverted object into coincidence with the original object by rotations and translations alone. Inversion symmetry only exists in three dimensions, and is equivalent to a 180° rotation in two dimensions (proof left for the reader). In stereographic projections, the inversion is represented as shown in Fig. 8.7(h).

It is easy to see that the matrix representing the inversion operator is simply $-\mathsf{E}$, i.e., a matrix with -1 along the diagonal and zero elsewhere. The Seitz symbol for the inversion operation is then $(-\mathsf{E}|\mathbf{0})$.

8.2.7 Symmetry operations that do not pass through the origin

All of the operators introduced so far always passed through the origin of the reference frame. Obviously, it is possible for a symmetry operator to be located at some other point (this will become important when we talk about space groups in Chapter 10). We must then ask the question: how do we compute the matrix $\mathcal{W}$ for a symmetry element that is not located at the origin? As an example, consider the six-fold axis ⬢ located at the point $(1/2, 1/2, 0)$ and parallel to the $\mathbf{c}$ direction of the reference frame shown in Fig. 8.8. Rotation of point 1 over an angle of 60° can be decomposed into three elementary steps (elementary in the sense that the decomposition consists only of pure operations, as introduced in the preceding sections). If we call the position vector of the rotation axis $\boldsymbol{\tau}$, then the following three steps are equivalent to the original rotation:

(i) translate the point 1 by a vector $-\boldsymbol{\tau}$ to the point 2;
(ii) rotate counterclockwise by an angle of 60° around an axis through the origin to the point 3;
(iii) translate by a vector $\boldsymbol{\tau}$ to the point 4.

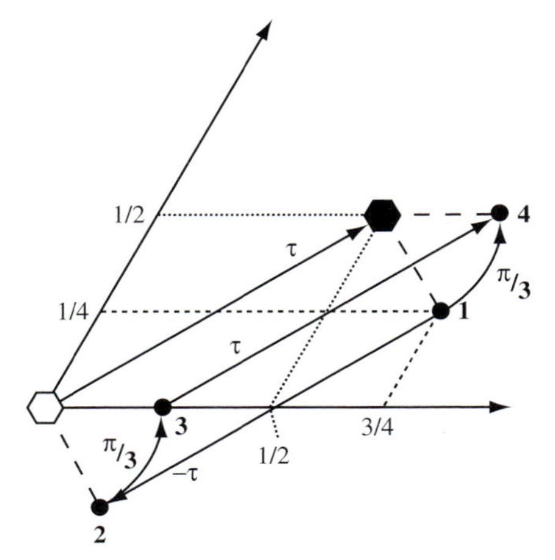

Fig. 8.8. Decomposition of a six-fold rotation at location $(1/2, 1/2, 0)$ into two translations and a rotation about a six-fold axis through the origin.

This is the same as going directly from 1 to 4 by application of the rotation axis at $\boldsymbol{\tau}$. We already know that a translation is described by a matrix $\mathcal{W}_{\boldsymbol{\tau}}$ of the form:

$$\mathcal{W}_{\boldsymbol{\tau}} = \begin{pmatrix} 1 & 0 & 0 & \tau_1 \\ 0 & 1 & 0 & \tau_2 \\ 0 & 0 & 1 & \tau_3 \\ 0 & 0 & 0 & 1 \end{pmatrix},$$

so that the complete transformation is given by the ordered matrix product:

$$\begin{aligned} \mathcal{W} &= \mathcal{W}_{\boldsymbol{\tau}} \mathcal{W}_{\blacklozenge} \mathcal{W}_{-\boldsymbol{\tau}} \\ &= \begin{pmatrix} 1 & 0 & 0 & \frac{1}{2} \\ 0 & 1 & 0 & \frac{1}{2} \\ 0 & 0 & 1 & 0 \\ 0 & 0 & 0 & 1 \end{pmatrix} \begin{pmatrix} 0 & -1 & 0 & 0 \\ 1 & 1 & 0 & 0 \\ 0 & 0 & 1 & 0 \\ 0 & 0 & 0 & 1 \end{pmatrix} \begin{pmatrix} 1 & 0 & 0 & -\frac{1}{2} \\ 0 & 1 & 0 & -\frac{1}{2} \\ 0 & 0 & 1 & 0 \\ 0 & 0 & 0 & 1 \end{pmatrix}; \\ &= \begin{pmatrix} 0 & -1 & 0 & 1 \\ 1 & 1 & 0 & -\frac{1}{2} \\ 0 & 0 & 1 & 0 \\ 0 & 0 & 0 & 1 \end{pmatrix}. \end{aligned}$$

If we multiply this matrix by a general position vector $(x, y, z, 1)$, then the result is $(1 - y, x + y + 1/2, z, 1)$. For the point 1 in Fig. 8.8, with normal coordinates $(3/4, 1/4, 0, 1)$, we find the new coordinates $(3/4, 1/2, 0, 1)$, which is in agreement with the graphical result. This decomposition technique is generally valid for any symmetry operation which does not pass through the origin.

8.3 Combinations of symmetry operations

8.3.1 Combination of rotations with the inversion center

The combination of a rotation axis with an inversion center located somewhere on that axis is called a *roto-inversion* operation. As an example, we combine a three-fold rotation 3 (C_3) with an inversion center $\bar{1}$ (i), as shown in the stereographic projections of Fig. 8.9. On the left-hand side, we show the

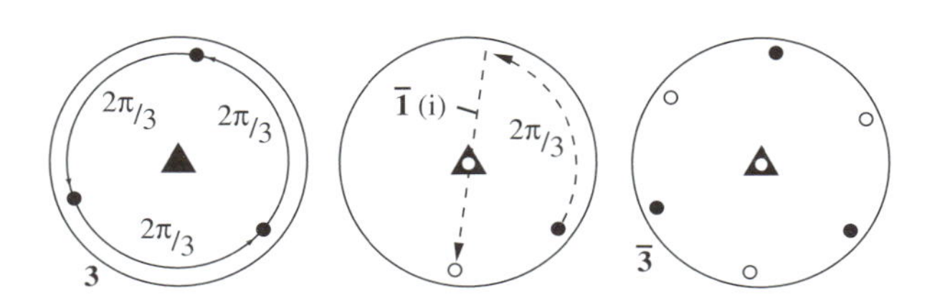

Fig. 8.9. Regular three-fold rotation axis, operation of a three-fold roto-inversion and result of the repeated operation of the roto-inversion $\bar{3}$.

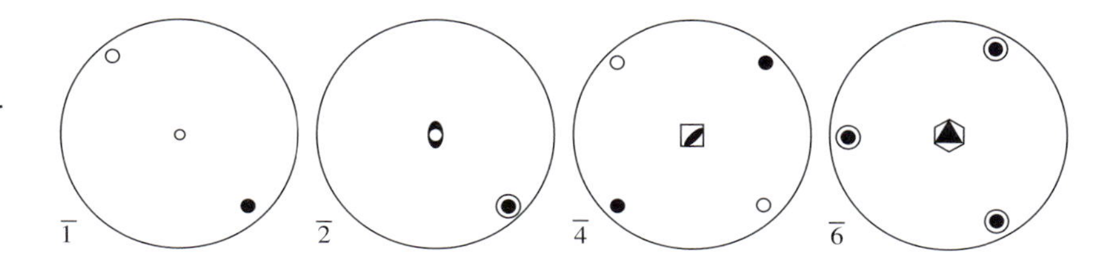

Fig. 8.10. Stereographic projections for the roto-inversions $\bar{1}$, $\bar{2}$, $\bar{4}$, and $\bar{6}$.

projection of the standard three-fold rotation. On the right-hand side, we show the roto-inversion. The roto-inversion rotates a point over an angle $2\pi/3$ and inverts the resulting point through the inversion center, as shown in the middle projection.

The standard notation for a roto-inversion axis of order n is the symbol $\bar{\mathsf{n}}$, e.g., $\bar{3}$ (the Schönflies notation will be introduced in the following subsection). The roto-inversions relevant to crystallography are $\bar{1}$, $\bar{2}$, $\bar{3}$, $\bar{4}$, and $\bar{6}$. The corresponding stereographic projections are shown in Fig. 8.10. Note that for all values of n which can be written as $n = 4k + 2$, the inversion actually generates a mirror plane perpendicular to the rotation axis and the effective order of the rotation axis is only $n/2$. In other words, the roto-inversion $\bar{6}$ is equivalent to the rotation 3 (C_3) with a perpendicular mirror plane m (σ_h). Note also that the two-fold roto-inversion $\bar{2}$ is equivalent to a simple mirror plane m (σ). On drawings, the roto-inversions are represented by special symbols: ∘ for $\bar{1}$, ⬮ for $\bar{2}$, ▲ for $\bar{3}$, ◆ for $\bar{4}$, and ▲ for $\bar{6}$.

The transformation matrix associated with a roto-inversion is obtained by simply multiplying the rotation matrix with the matrix representing the inversion operator. As an example, consider the four-fold roto-inversion $\bar{\mathbf{4}}$ oriented along the z-axis of a Cartesian reference frame:

$$
\begin{aligned}
W_{\lozenge} &= W_{\circ} W_{\blacklozenge}; \\
&= \begin{pmatrix} -1 & 0 & 0 & 0 \\ 0 & -1 & 0 & 0 \\ 0 & 0 & -1 & 0 \\ 0 & 0 & 0 & 1 \end{pmatrix} \begin{pmatrix} 0 & -1 & 0 & 0 \\ 1 & 0 & 0 & 0 \\ 0 & 0 & 1 & 0 \\ 0 & 0 & 0 & 1 \end{pmatrix}; \\
&= \begin{pmatrix} 0 & 1 & 0 & 0 \\ -1 & 0 & 0 & 0 \\ 0 & 0 & -1 & 0 \\ 0 & 0 & 0 & 1 \end{pmatrix}.
\end{aligned}
\tag{8.16}
$$

A point with coordinates (x, y, z) is hence moved to the position $(y, -x, -z)$.

8.3.2 Combination of rotations and mirrors

A *mirror-rotation* is the combination of a rotation axis with a perpendicular mirror plane. Each point is first rotated over an angle $2\pi/n$, and the resulting

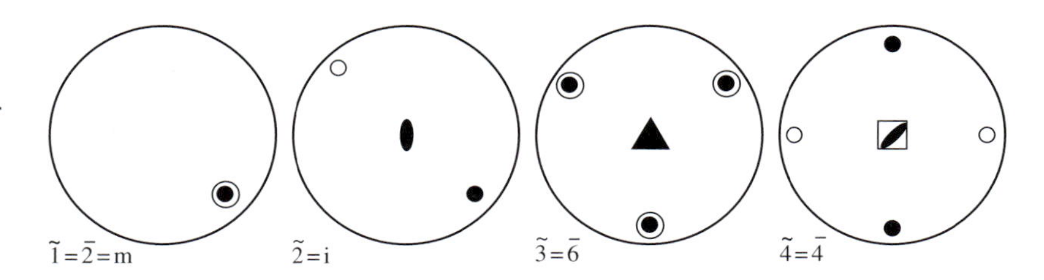

Fig. 8.11. Stereographic projections for the mirror-rotations $\tilde{1}$, $\tilde{2}$, $\tilde{3}$, and $\tilde{4}$.

point is then mirrored through the plane. We have already encountered an example in the previous section. Mirror-rotations of order n are usually indicated with the symbol $\tilde{\mathrm{n}}$. Note that for odd values of n, the symmetry element $\tilde{\mathrm{n}}$ is actually equivalent to $\overline{2\mathrm{n}}$; for values of $n = 4N$, with N an integer, there is no difference between the symbol $\tilde{\mathrm{n}}$ and $\bar{\mathrm{n}}$, as is easily verified with a stereographic projection (reader exercise). Several mirror-rotations relevant to crystallography are shown in Fig. 8.11. For historical reasons, the Schönflies notation makes use of mirror-rotations, while the Hermann–Mauguin notation uses the roto-inversions, so that the combined symbol is a little more difficult to remember. We introduce the following new Schönflies symbol for the mirror-rotation:

$$\mathsf{S}_\mathsf{n} \equiv \sigma_\mathsf{h}\mathsf{C}_\mathsf{n},$$

where the h subscript on the mirror plane indicates that the plane is *horizontal*, i.e., perpendicular to the rotation axis. From the general rules above and the illustrations in Figs. 8.10 and 8.11, it is not difficult to derive the following list of relations:

Hermann–Mauguin		S_n	Schönflies
roto-inversion	mirror-rotation		
$\bar{1}$	$\tilde{2}$	S_2	C_i
$\bar{2}\,(=\mathsf{m})$	$\tilde{1}$	S_1	C_h
$\bar{3}$	$\tilde{6}$	S_6	$\mathsf{C}_{3\mathsf{i}}$
$\bar{4}$	$\tilde{4}$	S_4	S_4
$\bar{5}$	$\tilde{10}$	S_{10}	$\mathsf{C}_{5\mathsf{i}}$
$\bar{6}\,(=3/\mathsf{m})$	$\tilde{3}$	S_3	$\mathsf{C}_{3\mathsf{h}}$
$\bar{7}$	$\tilde{14}$	S_{14}	$\mathsf{C}_{7\mathsf{i}}$
$\bar{8}$	$\tilde{8}$	S_8	S_8
$\bar{9}$	$\tilde{18}$	S_{18}	$\mathsf{C}_{9\mathsf{i}}$
$\bar{10}\,(=5/\mathsf{m})$	$\tilde{5}$	S_5	$\mathsf{C}_{5\mathsf{h}}$
⋮	⋮	⋮	⋮

While the Schönflies notation for these symmetry elements (third column) is easy enough to remember, it is, unfortunately, *not* the standard notation listed in the *International Tables for Crystallography* (Volume A, Hahn, 1996).

The standard Schönflies notation is listed in the last column of the table above. The general rule used to construct this table takes the order of the roto-inversion axis, n, and considers three distinct cases:

- $n = 4N$, with N an integer: in these cases, there is no difference between the roto-inversion and the mirror-rotation. The official Schönflies symbol is then S_n, and the complete symbol is denoted by $\bar{\mathsf{n}}\,(\mathsf{S}_n)$.
- $n = 2N + 1$ (i.e., n is odd): in all these cases, the equivalent mirror-rotation has *twice* the order of the roto-inversion, i.e., $\bar{\mathsf{n}} \equiv \widetilde{2\mathsf{n}}$. The official Schönflies symbol is C_{ni}, so that the complete symbol becomes $\bar{\mathsf{n}}\,(\mathsf{C}_{ni})$ for n odd.
- $n = 4N + 2$ (i.e., even numbers that are not a multiple of 4): in such cases, the equivalent mirror-rotation has *half* the order of the roto-inversion axis, i.e., $\bar{\mathsf{n}} \equiv \widetilde{\tfrac{1}{2}\mathsf{n}}$. It is easy to show, using stereographic projections, that we have $\bar{\mathsf{n}} = \tfrac{1}{2}\mathsf{n}/\mathsf{m}$ (where the symbol n/m means that there is a mirror plane normal to the rotation axis). The official Schönflies symbol is $\mathsf{C}_{\frac{1}{2}nh}$, so that the complete symbol becomes $\tfrac{1}{2}\mathsf{n}/\mathsf{m}\,(\mathsf{C}_{\frac{1}{2}nh})$.

The 4-D transformation matrices associated with the mirror-rotations can be obtained easily by multiplying the matrix for a mirror plane by the rotation matrix (reader exercise). In the remainder of this book, we will always work with roto-inversions instead of mirror-rotations. However, the Schönflies symbol for these operations will always be based on the mirror-rotation.

8.3.3 Combination of rotations and translations

It is a general property of the symmetry operations we have discussed so far that, with the exception of the *translation*, they all return to the same point when applied repeatedly. For instance, when we apply a six-fold rotation six consecutive times, we recover the initial point; when we apply a mirror operation twice, we recover the same point. This means that, for each symmetry element discussed so far, there exists an integer k, such that $\mathcal{O}^k = 1$ (E). If we include translations, then this statement is no longer true, since repeated operation of the translation operator takes us ever further away from the initial point.

Combining translations with rotations yields a new type of symmetry element: the *screw axis*. Consider a rotation axis of order n, and apply it several times in a row. After every rotation step, we also *translate* the resulting point by a certain vector $\boldsymbol{\tau}$, parallel to the rotation axis. If we apply this operation n times, then the resulting point will be translated with respect to the original point by a vector $n\boldsymbol{\tau}$. Since we are working in a 3-D lattice, this new point must again be a lattice point, which means that $n\boldsymbol{\tau} = m\mathbf{t}$, where $\mathbf{t}$ is the shortest lattice vector parallel to the rotation axis. We have, hence, created a new symmetry operation, the *screw axis*, which consists of an n-fold rotation combined with a translation parallel to the rotation axis $\mathbf{t}$ by a vector $\boldsymbol{\tau} = m\mathbf{t}/n$. The distance $\tau = |\boldsymbol{\tau}|$ is called the *pitch* of the screw axis.

The resulting symmetry element is generally denoted by the symbol $\mathsf{n_m}$. If $m = 0$, then there is no translation associated with the rotation and we recover the proper rotations as a subset of the screw rotations. The value of m must lie between 0 and n.

> A screw axis $\mathsf{n_m}$ consists of a counterclockwise rotation through $2\pi/n$, followed by a translation $m\mathbf{t}/n$ in the positive direction along the rotation axis.

An example of the screw axis 6_1 is shown in Fig. 8.12(a): the vertical line indicates the screw axis. The helix labeled 0 is the starting point. After a rotation of $2\pi/6$, the resulting helix is translated by a vector $\boldsymbol{\tau} = \mathbf{t}/6$ to the point 1. Another rotation followed by a translation leads to point 2 and so on, until point 6, which is located at a lattice vector $\mathbf{t} = 6\boldsymbol{\tau}$ from point 0. The 6_2 operation, shown in Fig. 8.12(b), consists of a 60° rotation and a translation by $\boldsymbol{\tau} = 2\mathbf{t}/6 = \mathbf{t}/3$. Both vectors $\mathbf{t}$ and $\boldsymbol{\tau}$ are indicated on the drawing. The helix at point 0 is first rotated over 60° counterclockwise (indicated by the curved arrow), then translated over $\boldsymbol{\tau}$, rotated again, translated, etc. After three rotations, the resulting point has been translated by a distance $3\boldsymbol{\tau} = \mathbf{t}$. The following three points are located in the next unit cell along the rotation axis. Since $\mathbf{t}$ is a lattice vector, there must also be corresponding points in the lower unit cell (points 4, 5, and 6). We can construct the same drawing for the 6_4 axis (reader exercise); the resulting screw is identical to 6_2, but has opposite rotation direction. We thus find that 6_4 produces the mirror image of 6_2. This is true for all screw axes: the screw axes of the type $\mathsf{n_n}$ and $\mathsf{n_{n-m}}$ are each other's mirror image. A screw axis is called *right-handed* if $m < n/2$,

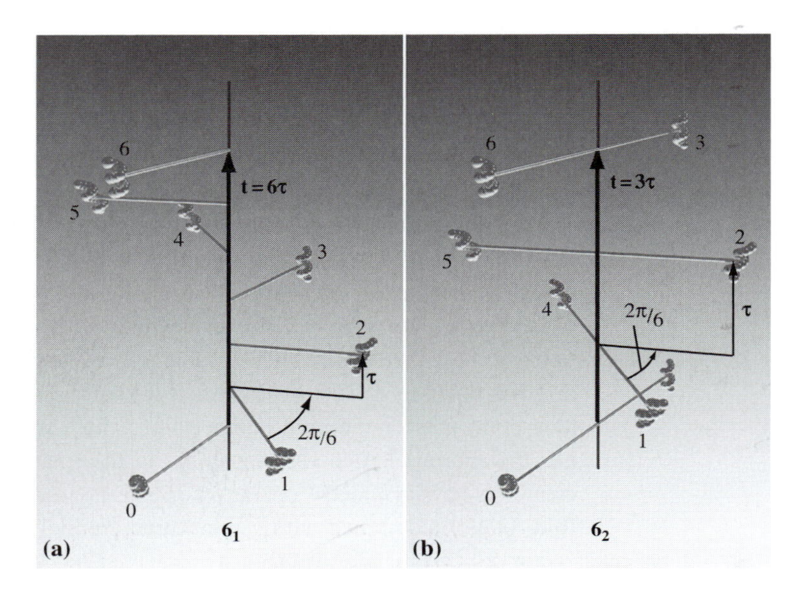

Fig. 8.12. Rendered perspective drawings of the operations involved in the construction of the screw axes 6_1 (a) and 6_2 (b).

left-handed if $m > n/2$ and *without hand* if $m = 0$ or $m = n/2$. Screw axes related to each other by a mirror operation are called *enantiomorphous*.

We have seen in Section 8.2.1 that the only rotations of importance to classical crystallography are $2, 3, 4$, and 6-fold rotations; hence, the only screw axes that we will consider in this text are (with their official graphical symbols): 2_1, 3_1, 3_2, 4_1, 4_2, 4_3, 6_1, 6_2, 6_3, 6_4, and 6_5. All screw axes are shown in Fig. 8.13 as both rendered drawings and standard graphical representations: the number next to each circle refers to the height of the circle above the plane of the drawing. The axes are all perpendicular to the drawing.

To conclude this section on screw axes, we determine how we can represent this symmetry operator by means of a matrix. We have already seen all the ingredients that we need to come up with a compact notation. The screw axis has both a rotation and a translation, so that the Seitz symbol is given by $(\mathrm{D}(\theta)|\boldsymbol{\tau})$. The 4×4 matrix $\mathcal{W}$ is then easily constructed by taking the 3×3 rotation matrix and combining it with the column vector of the translation components. As an example, a screw axis of the type 6_1 along the **c** axis of a hexagonal reference frame is described by the matrix:

$$\mathcal{W} = \begin{pmatrix} 1 & -1 & 0 & 0 \\ 1 & 0 & 0 & 0 \\ 0 & 0 & 1 & \frac{1}{6} \\ 0 & 0 & 0 & 1 \end{pmatrix}. \tag{8.17}$$

We leave it as an exercise for the reader to determine the matrix representation of a 4_2 operator parallel to the **c** axis of a tetragonal reference frame, when the screw axis goes through the point with coordinates $(1/4, 3/4, 0)$.

8.3.4 Combination of mirrors and translations

The last class of symmetry operators involves combinations of mirror planes and translations. As indicated in Fig. 8.14(a), one could combine a mirror plane with a translation over a *lattice vector* parallel to the plane. This does not give rise to a new symmetry operation. One then defines the *glide plane* as the combined operation of a mirror with a translation over *half* a lattice vector parallel to the mirror plane. An example is shown in Fig. 8.14(b).

The allowed glide vectors must be equal to one half of the lattice vectors. In the case of centered Bravais lattices, there may be additional glide vectors, equal to one half of the centering vectors **A**, **B**, **C**, and **I**. Table 8.1 lists the various possibilities and names for all types of glide planes.[6]

[6] In the *International Tables for Crystallography* (Volume A, Hahn, 1996), the concept of *double glides* is used. This refers to the fact that sometimes both *a and b* glides are simultaneously present. A special graphical symbol and notation are used to describe these cases (see page 6), but, in this book, we will not make any use of double glide planes.

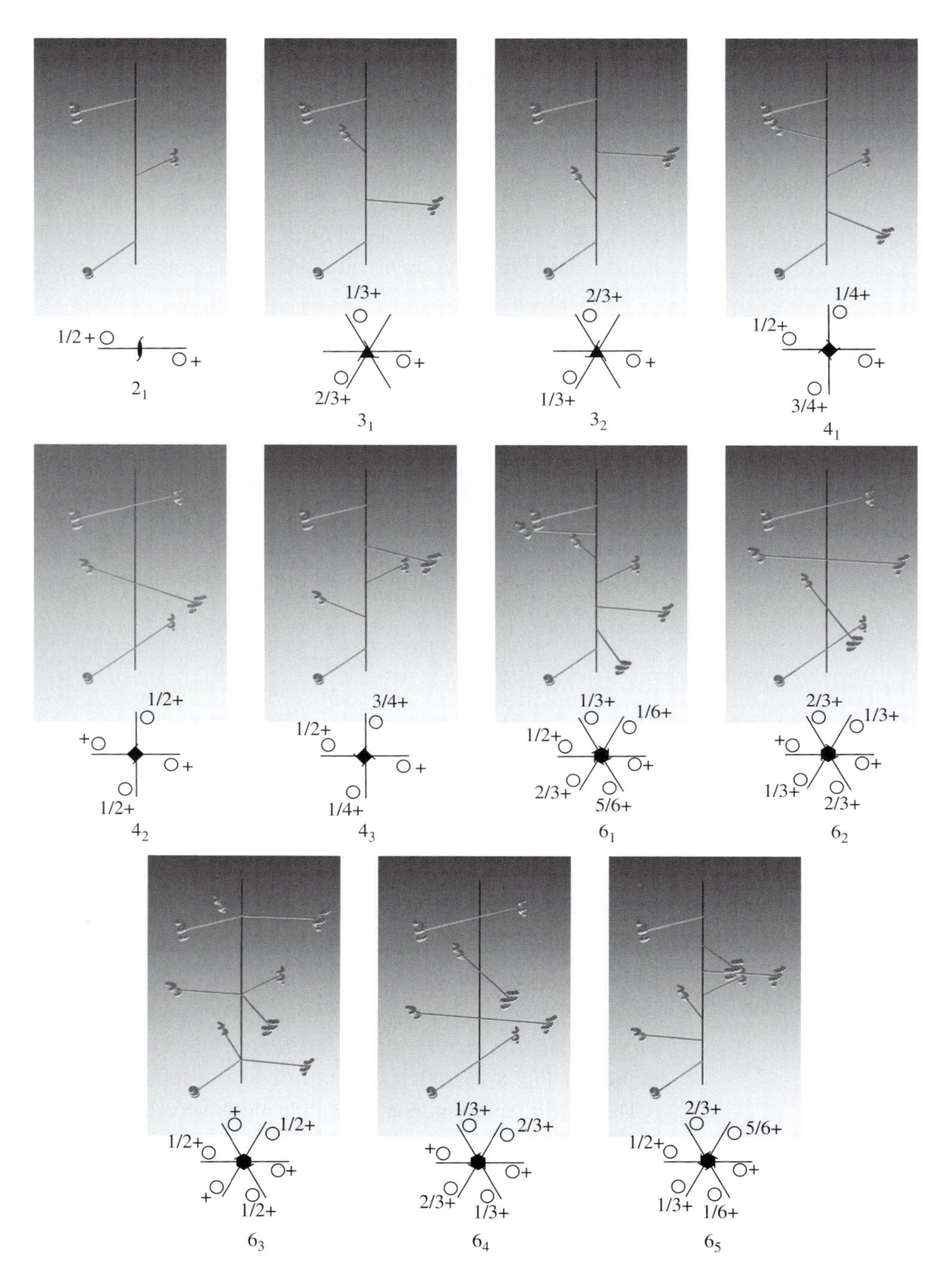

Fig. 8.13. Rendered perspective drawings of the crystallographic screw axes.

The official drawing symbols for glide planes depend on the orientation of the plane with respect to the drawing. If the plane is perpendicular or at an angle to the drawing, then the plane is indicated by a *solid bold* line for m (σ), a *dashed bold* line for a glide with translation *in* the plane of the drawing, a

Table 8.1. Different types of mirror and glide planes, their symbols and glide vectors. Glide vectors with a (†) symbol are only possible in cubic and tetragonal systems.

Name	Symbol	Glide vector(s)
mirror	m	none
axial glide	a	$\mathbf{a}/2$
	b	$\mathbf{b}/2$
	c	$\mathbf{c}/2$
diagonal glide	n	**A**, **B**, **C**
		I (†)
diamond glide	d	$(\mathbf{a}\pm\mathbf{b})/4$, $(\mathbf{b}\pm\mathbf{c})/4$, $(\mathbf{c}\pm\mathbf{a})/4$
		$(\mathbf{a}\pm\mathbf{b}\pm\mathbf{c})/4$ (†)

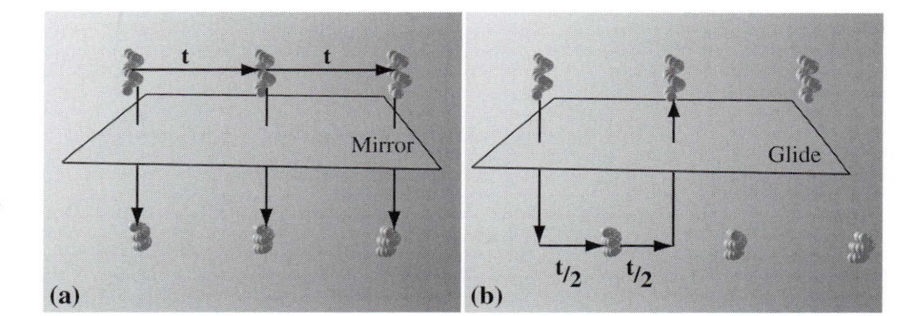

Fig. 8.14. (a) the combination of a mirror plane with an integer-valued translation does not give rise to a new symmetry operation; (b) a glide plane, consisting of a mirror plane and a half-integer translation.

dotted bold line for a glide with translation *perpendicular to* the plane of the drawing, and a *dash-dotted bold* line for a *diagonal* glide (see Fig. 8.15). Note that, once again, a circle reflected in a mirror plane is represented by a circle with a comma in the center, to indicate that an odd number of reflections relate that point to the original point. For glide planes parallel to the plane of the drawing, one uses a symbol based on ⌉ which represents a pure mirror. For an *axial glide plane*, one adds an arrow to the symbol in the direction of the glide vector, i.e., ←⌉. For a *diagonal glide plane*, the arrow points at an angle away from the corner, as in ↙⌉. Examples are shown in the right side column of Fig. 8.15.

The matrix representation for a glide plane is readily obtained by combining the 3×3 mirror operation with the appropriate translation components. For instance, an a-glide reflection in the plane $z = 0$ converts the z-coordinate of any point into $-z$ and translates that point by $(1/2, 0, 0)$. The resulting matrix is hence given by:

$$\mathcal{W} = \begin{pmatrix} 1 & 0 & 0 & \frac{1}{2} \\ 0 & 1 & 0 & 0 \\ 0 & 0 & -1 & 0 \\ 0 & 0 & 0 & 1 \end{pmatrix}. \tag{8.18}$$

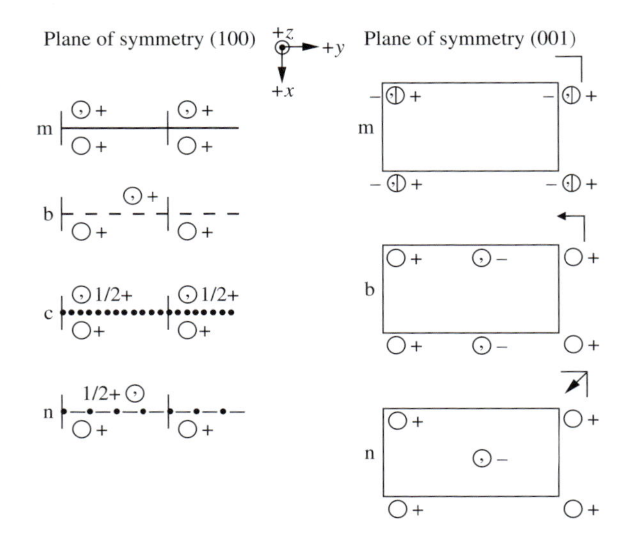

Fig. 8.15. Graphical symbols for glide planes perpendicular (left) and parallel (right) to the plane of the drawing (figure reproduced from Fig. 1.17 in *Introduction to Conventional Transmission Electron Microscopy*, M. De Graef, 2003, Cambridge University Press).

The Seitz symbol is also shown readily to be $(\mathsf{D}(m)|\boldsymbol{\tau})$, where $\boldsymbol{\tau}$ is the glide vector. We leave it to the reader to determine the transformation matrix $\mathcal{W}$ for a diagonal **B** glide reflection parallel to a (010) mirror plane going through the point $(1/2, 1/2, 1/2)$ (cubic reference frame).

8.3.5 Relationships and differences between operations of first and second type

In this section, we will take a closer look at the differences between operators of the first and second kind. From the examples in the previous sections, we can infer that the proper motions (operations of the first kind) are described by orthogonal matrices with positive unit determinant, i.e.,

$$\det |{}^{\prime}\mathsf{D}| = +1, \tag{8.19}$$

whereas all transformations which change the handedness of an object (operations of the second kind) have:

$$\det |{}^{\prime\prime}\mathsf{D}| = -1. \tag{8.20}$$

Successive products of any number q of operators of the first kind leave the determinant unchanged:

$$\det |{}^{\prime}\mathsf{D}_q {}^{\prime}\mathsf{D}_{q-1} \dots {}^{\prime}\mathsf{D}_1| = +1. \tag{8.21}$$

However, for operators of the second kind, the handedness of the product is determined by the number of operations: if q is even, then the resulting

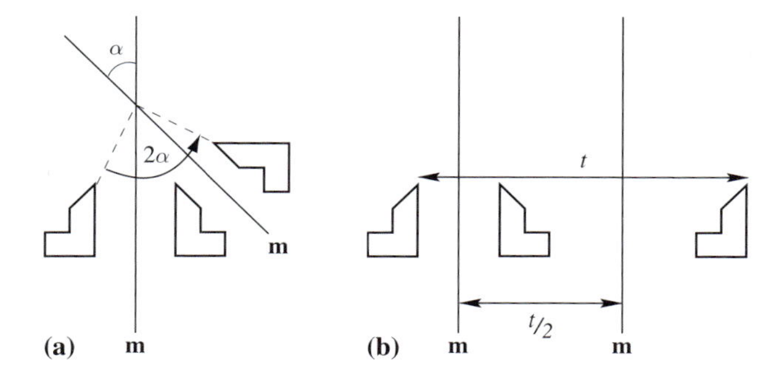

Fig. 8.16. Combination of intersecting symmetry elements creates new elements, according to Euler's theorem. Some limiting cases for intersecting and parallel mirror planes are shown in (a) and (b), respectively.

operation is of the first kind, otherwise it remains of the second kind. The determinant of the composite operation $\mathsf{C} = {}^{II}\mathsf{D}_q {}^{II}\mathsf{D}_{q-1} \ldots {}^{II}\mathsf{D}_1$ equals:

$$\det|\mathsf{C}| = (-1)^q = \begin{cases} +1 & q = 2n & \rightarrow \mathsf{C} \text{ is first kind operator;} \\ -1 & q = 2n+1 & \rightarrow \mathsf{C} \text{ is second kind operator.} \end{cases}$$

Since the product of two orthogonal matrices is again an orthogonal matrix, it follows that the combination of two symmetry operations must again be a symmetry operation. Figure 8.16 shows two special cases: on the left, two intersecting mirror planes are shown to be equivalent to a rotation over twice the angle between the mirror planes around their intersection line. On the right, it is shown that the combination of two parallel mirror planes with a spacing $t/2$ is equivalent to a translation over a distance t.

We thus find that intersecting mirror-planes always imply the existence of rotational symmetry. Parallel mirror planes always imply the existence of translational symmetry. In general, it can be shown that the combination of two rotations about intersecting axes can always be replaced by a single rotation; this is known as *Euler's theorem* and the complete proof of this theorem can be found in McKie and McKie (1986), page 43–47.[7] This theorem is important for the derivation of all the point groups in Chapter 9.

8.4 Point symmetry

It is instructive to take a closer look at the 3×3 sub-matrix of the 4×4 symmetry matrices. If we ignore all translations, then all symmetry matrices of the type:

[7] Euler has many theorems that carry his name, and this is just one of them.

$$\mathcal{W} = \begin{pmatrix} D_{11} & D_{12} & D_{13} & 0 \\ D_{21} & D_{22} & D_{23} & 0 \\ D_{31} & D_{32} & D_{33} & 0 \\ 0 & 0 & 0 & 1 \end{pmatrix}$$

represent symmetry elements which *all intersect in one common point, namely the origin.* This is easy to see: when the matrix D is multiplied by the position vector $(0, 0, 0)$, the result is always $(0, 0, 0)$, regardless of the particular details of D. Hence, the origin is the only point that is *invariant* under all symmetry operations D. For this reason, these operations are called *point symmetries*.

The question we must answer next is the following: is it possible for an object to have multiple rotational symmetries, and, if so, which combinations of rotation axes are allowed? The first part of this question is easy to answer: take a cylinder for example. The cylinder has a rotation axis of infinite order parallel to its long axis, and an infinite number of two-fold rotation axes perpendicular to the long axis, going through the center plane. Thus, it *is* possible for an object to have multiple rotational symmetries.

In the case of crystals, we must refine the second part of the question to: which combinations of rotation axes are compatible with translational periodicity? In other words, under which conditions can two (or more) rotation axes simultaneously be compatible with the Bravais lattice? Consider the stereographic projection in Fig. 8.17: there are three rotation axes, represented by the poles A, B, and C. Then Euler's theorem states the following (without proof):

> If the angle between the great circles AB and AC is α, the angle between the great circles BA and BC is β, and the angle between the great circles CA and CB is γ, then a clockwise rotation about A through the angle 2α, followed by a clockwise rotation about B through the angle 2β is equivalent to a counterclockwise rotation about C through the angle 2γ.

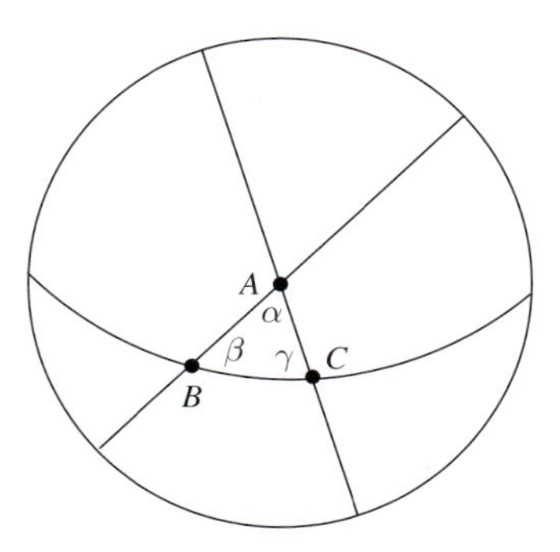

Fig. 8.17. Three rotation axes A, B, and C on a stereographic projection, used to illustrate Euler's theorem.

If the rotation axes at A, B, and C are n-fold rotation axes, then the angles 2α, 2β, and 2γ must be of the type $2\pi/n$ with, in the case of crystals, $n = 1, 2, 3, 4, 6$. We can turn this statement around so that the angles between the great circles connecting different rotation axes must be of the type $2\pi/2n$, or π/n. Note that these are angles between great circles, i.e., they are measured on the surface of the projection sphere. The angles between the rotation axes themselves can then be derived by means of spherical trigonometry. It can be shown that the angles between the rotation axes are given by:

$$\begin{aligned} \cos\widehat{BC} &= \frac{\cos\alpha + \cos\beta\cos\gamma}{\sin\beta\sin\gamma}; \\ \cos\widehat{CA} &= \frac{\cos\beta + \cos\alpha\cos\gamma}{\sin\alpha\sin\gamma}; \\ \cos\widehat{AB} &= \frac{\cos\gamma + \cos\beta\cos\alpha}{\sin\beta\sin\alpha}. \end{aligned} \tag{8.22}$$

If we substitute all possible combinations for α, β, and γ, and compute the angles $\widehat{AB}$, $\widehat{BC}$, and $\widehat{CA}$, then we arrive at the values shown in Table 8.2.

We find that only six different combinations of multiple rotation axes are compatible with the Bravais lattices. Those combinations are: 222, 223, 224, 226, 233, and 234. They are represented as rendered perspective drawings and stereographic projections in Fig. 8.18. Other combinations, such as $22n$ with $n = 5$ (also shown in the table) or $n > 6$, are possible for isolated objects, but are not compatible with the Bravais lattice translational symmetry. The combination 235 (last line in Table 8.2), known as the *icosahedral symmetry*,

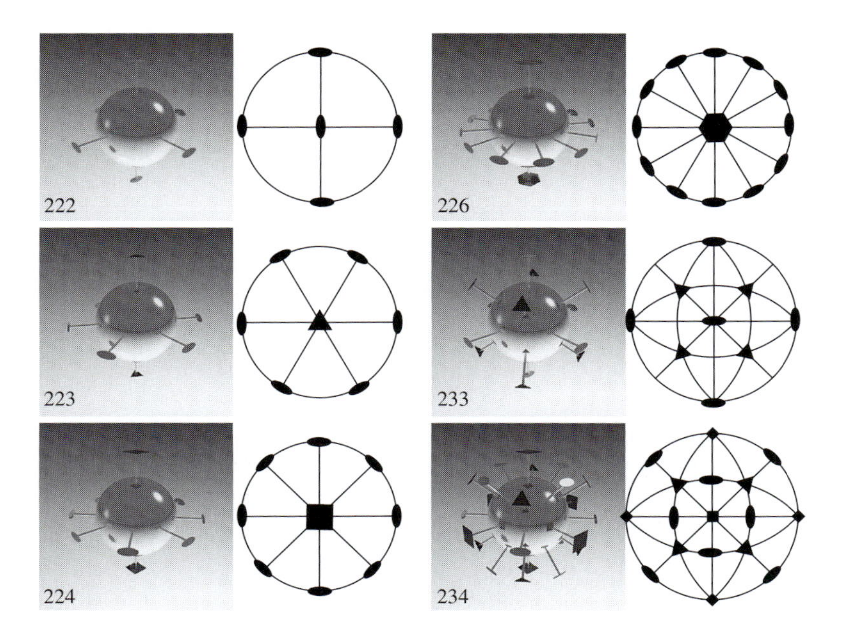

Fig. 8.18. Rendered perspective drawings and stereographic projections showing all six crystallographic combinations of multiple rotation axes.

Table 8.2. Combinations of three rotations axes for crystallographic symmetries. A $*$ indicates that the cosine of one of the angles falls outside the range $[-1, 1]$. The bottom two lines show combinations involving the non-crystallographic five-fold rotation axis; these will become important in Chapter 15.

n_A	n_B	n_C	$\widehat{BC}$	$\widehat{CA}$	$\widehat{AB}$	possible?
2	2	2	90°	90°	90°	yes
2	2	3	90°	90°	60°	yes
2	2	4	90°	90°	45°	yes
2	2	6	90°	90°	30°	yes
2	3	3	70°32′	54°44′	54°44′	yes
2	3	4	54°44′	45°	35°16′	yes
2	3	6	0	0	0	trivial
2	4	4	0	0	0	trivial
2	4	6	*	*	*	no
2	6	6	*	*	*	no
3	3	3	0	0	0	trivial
3	3	4	*	*	*	no
3	3	6	*	*	*	no
3	4	4	*	*	*	no
3	4	6	*	*	*	no
3	6	6	*	*	*	no
4	4	4	*	*	*	no
4	4	6	*	*	*	no
4	6	6	*	*	*	no
6	6	6	*	*	*	no
2	2	5	90°	90°	36°	not in crystals
2	3	5	37°23′	31°43′	20°54′	not in crystals

is shown as a rendered drawing in Fig. 8.19. This and other examples of non-crystallographic symmetries will be discussed in Chapter 15.

8.5 Historical notes

Arthur Moritz Schönflies (1853–1928) was a German mathematician. He studied at the University of Berlin from 1870 until 1875 and subsequently received a doctorate from Berlin in 1877. Schönflies was appointed as chair of applied mathematics at Göttingen, set up by Felix Klein in 1892. In 1899,

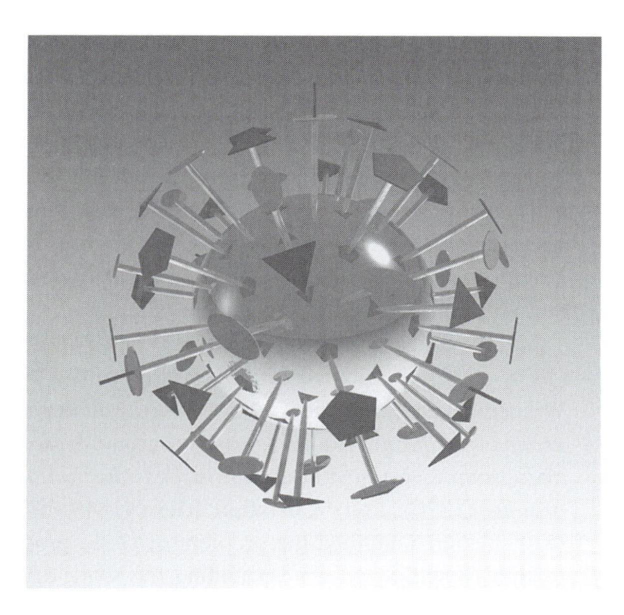

Fig. 8.19. Rendered drawing of the icosahedral combination of rotation axes 235.

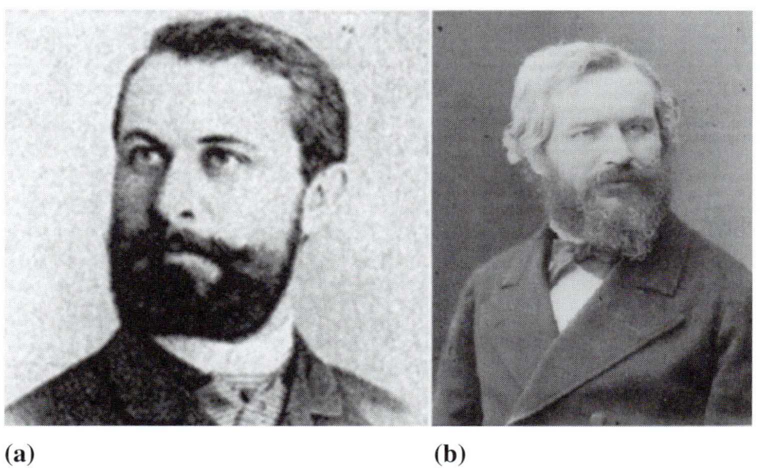

Fig. 8.20. (a) Arthur Moritz Schönflies (1853–1928) (picture courtesy of J. Lima de Faria), and (b) Camille Jordan (1838–1922) (picture courtesy of J. Lima de Faria).

he was named chair at Königsberg. In 1911, he became Professor at the Academy for Social and Commercial Sciences in Frankfurt. Schönflies served as professor from 1914 until 1922 at the University of Frankfurt and was rector during the period 1920–21.

Schönflies was one of three men who, in the 1890s, almost simultaneously derived the existence of the 230 space groups (which will be discussed in Chapter 10). Schönflies also developed the famous Schönflies notation for the crystal forms index. He applied his space group theory to crystal structures and to the problem of the division of space into congruent regions related by the symmetry of the group. His interest in symmetry guided many of his scientific endeavors.

Many of the initial efforts in describing the way that atoms and molecules are arranged in crystals were developed in the eighteenth, nineteenth and twentieth centuries prior to the development of techniques to actually probe the atomic structure. The simple packing models were precursors of the monumental developments that occurred as a result of considerations of symmetry and the mathematical properties of groups (in this case groups of symmetry operations). In 1844, Cauchy studied the group properties of permutations, a mathematical construct that was a precursor to symmetry group operations that will be discussed in subsequent chapters.

Mathematicians such as **Christan Wiener** (1826–96) and **Camille Jordan** (1838–1922) contributed to the development of the theory of space groups. Wiener developed a theory of *symmetrical repetitions*, and Jordan studied *groups of motions*. Jordan's work on group theory was published in the first ever group theory text book: *Traité des substitutions and des équations algebraique* (1870). **Felix Christian Klein** (1849–1925), an influential German mathematician, studied the problem of space groups from the point of view of transformation groups, extending the work of Jordan to consider adding improper rotations to the discrete group of proper rotations (1892). He also published the important work, *Lectures on the Icosahedron* in 1876, in which he laid out the group theory for icosahedral groups. The German mathematician **Georg Frobenius** (1849–1917) was also known for his work in group theory. His famous congruence is the basis for an algorithm for the construction of the space groups.

8.6 Problems

(i) *Inversion operation*: Show that the inversion operator in 2-D is equivalent to a two-fold rotation.

(ii) *Roto-inversions and mirror-rotations*: Use stereographic projections to show that, for values of $n = 4N$, with N integer, there is no difference between the symbol $\tilde{\mathsf{n}}$ and $\bar{\mathsf{n}}$.

(iii) *Mirror-rotations*: Determine the 4-D transformation matrix for the mirror-rotation $\tilde{\mathbf{6}}$ aligned along the **c** axis of a hexagonal reference frame, and show that it is equal to the matrix for $(\bar{3})^2\mathrm{C}_{3\mathrm{i}}^2$. What are the coordinates of all points generated from the point $(x, x, 0)$ (with respect to the hexagonal reference frame)?

(iv) *Screw axes*: Show by means of a drawing that the 6_4 screw axis is the mirror image of 6_2.

(v) *Glide planes*: Determine the transformation matrix $\mathcal{W}$ for a diagonal **B** glide reflection parallel to a (010) mirror plane going through the point (1/2, 1/2, 1/2) (cubic reference frame).

(vi) *Seitz symbol*: Show that the inverse of the Seitz symbol $(\mathsf{D}|\mathbf{t})$ is given by the symbol $(\mathsf{D}^{-1}|-\mathsf{D}^{1}\mathbf{t})$.

(vii) *4-D transformation matrices I*: Determine the coordinates of a point (x, y, z), after it has undergone transformation by a screw axis 6_3 along the hexagonal **c**-axis, followed by a mirror operation through the (11.0) plane going through the origin.

(viii) *4-D transformation matrices II*: Determine the 4-D transformation matrix and the action of the matrix on the coordinates (x, y, z) for the following operations:

(a) A 4_2 screw operator about the [001] in a cubic system.
(b) A b-glide operator with a reflection across a plane perpendicular to the a-axis at the $x = 0$ position in a cubic system.
(c) A $\bar{3}$ roto-inversion operation about the [111] in a cubic system.
(d) A 4_3 screw operator about the [100] in a cubic system. What would happen if you operated with this screw operator twice?

(ix) *4-D transformation matrices III*: Consider a 4_2 screw operation not passing through the origin:

(a) Determine the matrix representation $\mathcal{W}$ of a pure translation from the origin to the point with coordinates $(1/4, 3/4, 0)$.
(b) Determine the matrix representation $\mathcal{W}$ of a 4_2 operator parallel to the **c**-axis of a tetragonal reference frame.
(c) Determine the matrix representation $\mathcal{W}$ of a 4_2 operator parallel to the **c**-axis of a tetragonal reference frame, when the screw axis goes through the point with coordinates $(1/4, 3/4, 0)$.

(x) *Permutations*: Label the vertices of an equilateral triangle 1, 2, and 3, respectively.

(a) Determine all of the permutations of the numbers, 1, 2, and 3.
(b) Describe a symmetry operation associated with each permutation.
(c) Write down a 2-D rotation matrix representing each symmetry operation.

CHAPTER

9 Point groups

> *"Group theory is a branch of mathematics in which one does something to something and then compares the results with the result of doing the same thing to something else, or something else to the same thing."*
>
> James Newman, mathematician

9.1 What is a group?

In this section, we will give a practical example of a *group* and illustrate some of the important group properties; then we will define a group in precise mathematical terms.

9.1.1 A simple example of a group

Consider the crystal depicted in Fig. 9.1. This is a typical shape for a quartz crystal. Looking at the drawing, we find that the only symmetry elements present in this crystal are the ones corresponding to the configuration 223, discussed in the previous chapter. If we denote a general symmetry operator by the symbol $\mathcal{O}$, then we can define the following six symmetry operators for the quartz crystal:

$$\begin{aligned}
\mathcal{O}_0 &= e &&= \text{the identity operator;}\\
\mathcal{O}_1 &= 3 &&= \text{rotation through an angle } \alpha = 2\pi/3;\\
\mathcal{O}_2 &= 3^2 &&= \text{rotation through an angle } \alpha = 4\pi/3;\\
\mathcal{O}_3 &= 2_x &&= \text{rotation through } \pi \text{ about the } x\text{-axis;}\\
\mathcal{O}_4 &= 2_y &&= \text{rotation through } \pi \text{ about the } y\text{-axis;}\\
\mathcal{O}_5 &= 2_u &&= \text{rotation through } \pi \text{ about the } u\text{-axis.}
\end{aligned} \tag{9.1}$$

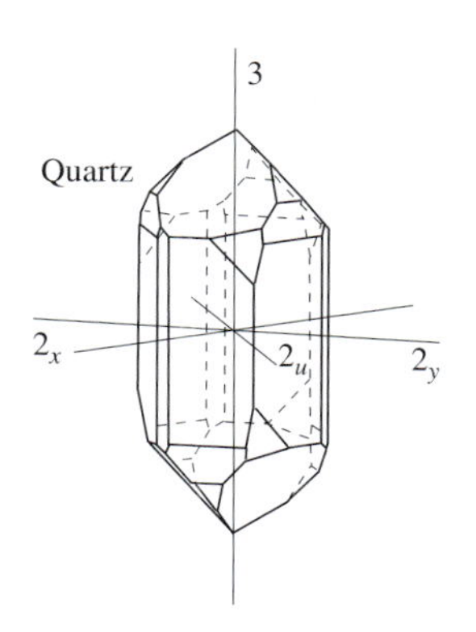

Fig. 9.1. A typical quartz crystal and its symmetry elements.

The rotation axes $2_{x,y,u}$ are perpendicular to a central axis 3. If 3 and e.g., 2_x are given, then, by Euler's theorem, the other two 2-fold axes are automatically determined. We define the successive operation of two operators by writing the operators in the reverse order, i.e., the operator that is applied first is written to the right:

$$\mathcal{O}_i.\mathcal{O}_j : \quad \mathcal{O}_j \text{ is applied first, followed by } \mathcal{O}_i \tag{9.2}$$

The symbol . stands for the multiplication or successive application of the two operators.

The successive operation of two symmetry operators is again a symmetry operator. Some examples are:

- $\mathcal{O}_1.\mathcal{O}_1 = \mathcal{O}_1^2 = \mathcal{O}_2$
- $\mathcal{O}_1.\mathcal{O}_4 = \mathcal{O}_3$

The inverse of every operator is also an operator:

- $\mathcal{O}_1$ and $\mathcal{O}_2$ are mutually reverse operators.
- $\mathcal{O}_3$, $\mathcal{O}_4$ and $\mathcal{O}_5$ are self-inverse operators.

And, finally, the identity operator e can also be obtained by combining the different operators, e.g., $e = \mathcal{O}_i.\mathcal{O}_i^{-1} = \mathcal{O}_1.\mathcal{O}_2 = \mathcal{O}_1^3 = \ldots$

9.1.2 Group axioms

From the quartz example in the preceding section, we learn that the combination of two symmetry operators is again a symmetry operator, that there exists a *neutral* operator or *identity* operator, and that each operator has an inverse. In this section, we will formalize these statements and properly define what is a group.

The following four rules must be satisfied, in order for a set $\mathcal{G}$ to become a group under a certain multiplicative operation:

(i) in the set $\mathcal{G}$, the multiplicative operation is defined such that the product of any pair of elements $\mathcal{O}_i, \mathcal{O}_j$ is also an element of $\mathcal{G}$, i.e., the set is *closed*:

$$\boxed{\forall \mathcal{O}_i, \mathcal{O}_j \in \mathcal{G}, \exists \mathcal{O}_k \in \mathcal{G} : \mathcal{O}_i.\mathcal{O}_j = \mathcal{O}_k} \tag{9.3}$$

(ii) the multiplicative operation is associative:

$$\boxed{\mathcal{O}_i.(\mathcal{O}_j.\mathcal{O}_k) = (\mathcal{O}_i.\mathcal{O}_j).\mathcal{O}_k} \tag{9.4}$$

(iii) there exists a unit element e:

$$\boxed{\exists e \in \mathcal{G} \rightarrow \forall \mathcal{O}_i \in \mathcal{G} : e\mathcal{O}_i = \mathcal{O}_i = \mathcal{O}_i e} \tag{9.5}$$

(iv) each element has its inverse element:

$$\boxed{\forall \mathcal{O}_i \in \mathcal{G}, \exists \mathcal{O}_k \in \mathcal{G} : \mathcal{O}_i.\mathcal{O}_k = \mathcal{O}_k.\mathcal{O}_i = e \quad ; \quad \mathcal{O}_k = \mathcal{O}_i^{-1}} \tag{9.6}$$

If a set of elements $\{\mathcal{O}_1, \mathcal{O}_2, \ldots, \mathcal{O}_n\}$ satisfies rules 1 through 4 for a given multiplication rule, then that set is called a group.

(v) the order of the operations is unimportant:

$$\boxed{\forall \mathcal{O}_i, \mathcal{O}_j \in \mathcal{G} : \mathcal{O}_i.\mathcal{O}_j = \mathcal{O}_j.\mathcal{O}_i} \tag{9.7}$$

If, in addition, the set satisfies rule 5, then the group $\mathcal{G}$ is an Abelian or commutative group.

It is easy to show that the symmetry operators of the quartz example in the previous section satisfy all four principal rules; therefore, the set of symmetry operators of quartz forms a group. In general, the symmetry operators of any crystal form a group; as we will discuss in a later section, the crystallographic groups have definite designations (although there are different systems in use). The quartz group is named **32** (D_3); the first name is the internationally agreed upon name for this point group, i.e., the so-called *Hermann–Mauguin* symbol. The second name is the *Schönflies* symbol, which is used primarily by solid state physicists and chemists.

The properties of any abstract group are fully determined if all possible multiplications between its elements are defined; for this purpose the concept of a *multiplication table* (or *Cayley's square*) is introduced. The products of any two group-elements are written in a 2-dimensional table; if every element appears only once in each row and column, then the set forms a group. If furthermore the square is symmetric with respect to the main diagonal, the group is *Abelian*. The multiplication order goes from the left-most column to the top row (i.e., elements of the top row are executed first).

As an example, we give the full multiplication table of the quartz group:

32 (D_3)	e	3	3^2	2_x	2_y	2_u
e	e	3	3^2	2_x	2_y	2_u
3	3	3^2	e	2_u	2_x	2_y
3^2	3^2	e	3	2_y	2_u	2_x
2_x	2_x	2_y	2_u	e	3^2	3
2_y	2_y	2_u	2_x	3	e	3^2
2_u	2_u	2_x	2_y	3^2	3	e

9.1.3 Principal properties of groups

The following properties refer to all abstract groups.

- **Order of a group**: A group may contain one, several, or an infinite number of elements; this is called the *order* n of the group. If n is finite then the group is called a finite group.
  Example: the order of the quartz point group is $n = 6$.
- **Isomorphism**:

> If a one-to-one correspondence exists between the elements of two groups, then those groups are isomorphous.

Obviously, the order of isomorphous groups is the same. This relation between two groups is denoted by a double-headed arrow $\leftrightarrow$, i.e., if $\mathcal{G} = \{g_1, g_2, \ldots, g_n\}$ and $\mathcal{H} = \{h_1, h_2, \ldots, h_n\}$ then

$$\mathcal{G} \leftrightarrow \mathcal{H} \text{ if } \quad \forall g_i, g_j G, \forall h_i, h_j \epsilon H : g_i . g_j \leftrightarrow h_i . h_j \ . \tag{9.8}$$

Example : The group $\mathcal{G}$ denoting the space rotations through an angle $2\pi/N$ is isomorphous with the group of complex numbers $\exp(2\pi i n/N)$ where $0 \leq n < N$.

Isomorphous groups are, therefore, different representations of one and the same abstract group and consequently have identical multiplication tables.

- **Homomorphous groups**:

> Two groups are homomorphous if there exists a unidirectional correspondence between them, i.e., with every element of one group there corresponds one or more elements of the other group.

This relation is denoted by a single-headed arrow $\rightarrow$, pointing from the group with the highest order to the one with the lowest order, i.e. :

$$\mathcal{G} \rightarrow \mathcal{H} \text{ if } \{g_{i_1}, \ldots, q_{i_s}\} \rightarrow h_i, \{g_{j_1}, \ldots, q_{j_t}\} \rightarrow h_j, \quad (9.9)$$
$$g_{i_p} \cdot g_{j_q} = h_i . h_j \qquad (1 \leq p \leq s, 1 \leq q \leq t).$$

Mapping several elements of one group onto one element of another group can be useful in the classification of elements: one obvious example is the mapping of symmetry operators according to their handedness, i.e., a symmetry group $\mathcal{G}$ can be mapped onto the group $\{1, -1\}$. Another interpretation might be the following: if a symmetry operator changes the sign of e.g., the z-component of an atom coordinate vector, then this operator is mapped on -1, else on $+1$. In the case of the quartz group **32** (D_3), this would produce the following mapping:

$$\mathbf{32}\ (D_3) \rightarrow \{1, -1\} : \text{with } \{\mathcal{O}_0, \mathcal{O}_1, \mathcal{O}_2\} \rightarrow 1 \text{ and } \{\mathcal{O}_3, \mathcal{O}_4, \mathcal{O}_5\} \rightarrow -1 \quad (9.10)$$

It is easy to verify that the set $\{1, -1\}$ is a group under ordinary multiplication.

- **Cyclic groups** :

> If a group contains an element $\mathcal{O}$, such that the powers of $\mathcal{O}$ exhaust all group elements, then this group is called a cyclic group, and $\mathcal{O}$ is called the generating element.

The group can thus be written as:

$$\mathcal{G} = \{\mathcal{O}, \mathcal{O}^2, \ldots, \mathcal{O}^l, \ldots, \mathcal{O}^n = e\} \quad (9.11)$$

This type of group represents rotations through an angle $2\pi/n$ and is denoted by C_n (Schönflies notation) or **n** (Hermann–Mauguin symbol). For this type of group, it is sufficient to give the generating element instead of the full multiplication table.

- **Group generators**: Since all elements in a group can be written as products of the other elements (this is essentially shown in the group multiplication table) there will be a minimum number of elements from which all the others can be reconstructed. This minimal set of group elements is called the set of *group generators*. One can then reconstruct the complete group by (1) taking all the powers of each generator until either the identity or the generator itself is found, (2) take all the products between the resulting elements and the generators until no new elements are found.

- **Subgroups and supergroups**: Assume a group $\mathcal{G}$ of order n; if a subset of n_k elements $\mathcal{O}_k$ also forms a group, then this set $\mathcal{G}_k$ is called a *subgroup* of $\mathcal{G}$ and is denoted as $\mathcal{G}_k \subset \mathcal{G}$. If, in addition, this subgroup is different from either $\{e\}$ or $\mathcal{G}$ itself, then the subgroup is called a *proper* subgroup.

  The quartz group **32** (D_3) contains two proper subgroups:

$$\text{subgroup } \mathbf{3}\ (C_3) = \{\mathcal{O}_0, \mathcal{O}_1, \mathcal{O}_2\} \tag{9.12}$$

$$\text{subgroup } \mathbf{2}\ (C_2) = \{\mathcal{O}_0, \mathcal{O}_3\}. \tag{9.13}$$

  These groups describe pure three- and two-fold rotations, respectively, and are both cyclic.

  Some groups do not have any proper subgroups; in that case only the *trivial* subgroups $\{e\}$ and $\mathcal{G}$ exist. The order of a subgroup is always an integer divider of the order of the main group (this is called the *Theorem of Lagrange*); the ratio

$$p = \frac{n}{n_k}$$

  is called the subgroup *index*.

  In a similar way, $\mathcal{G}$ is a *supergroup* of $\mathcal{G}_k$; a large part of group theory is devoted to the generation of complicated supergroups from simple, low order groups.

9.2 Three-dimensional crystallographic point symmetries

In the preceding chapter, we derived all rotational symmetry operations compatible with the 14 Bravais lattices, and we have seen that the combination of symmetry elements generates new symmetry elements. These sets of symmetry elements which are mutually consistent are called *groups*. In this section, we will derive all 32 crystallographic *point groups*, i.e., all the symmetry groups that are compatible with translational periodicity in three dimensions. They are known as point groups because all symmetry elements intersect each other in a single point.

The 32 crystallographic point groups can be derived in a very simple way by counting the ways in which the symmetry elements can be combined to form closed sets. In order to derive all groups in an unambiguous way, we will proceed in seven steps, from the simplest groups to the highest order cubic groups. In the following, all point groups will be denoted with both the international (or Hermann–Mauguin) symbol and the Schönflies symbol (between brackets). Note that all point groups will be discussed, and the crystallographic ones will be highlighted.

To clarify the meaning of the various combinations of symmetry elements, rendered 3-D illustrations are included in the following sections; these images

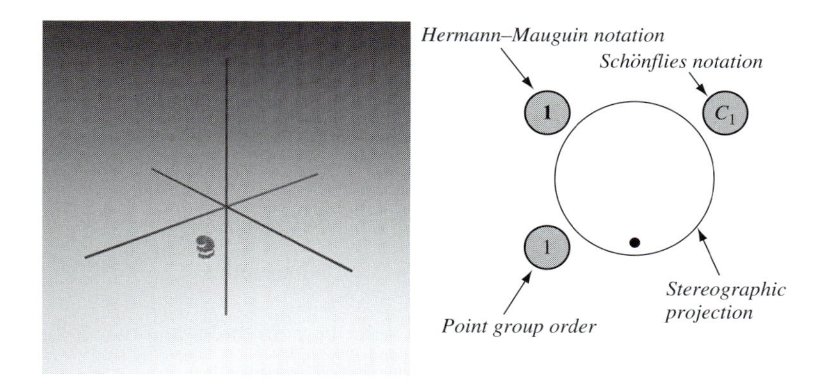

Fig. 9.2. Graphic representation of the identity point group **1** (C_1).

were created with a ray-tracing program (*Rayshade*, De Graef, 1998), and show the symmetry elements in the proper orientation in space, along with sets of equivalent points. These points are organized as a short helix, to highlight the fact that some of the symmetry elements change the handedness of the object, whereas others leave it unchanged. For each point group, the corresponding stereographic projection is also shown. Note that the handedness of the equivalent points is not usually shown in these stereographic projections; they only distinguish between points in the northern and southern projection hemispheres.

The simplest point group contains only the identity operator 1 (E); its schematic representation is shown in Fig. 9.2. The corresponding stereographic projection is shown to the right of the figure: three labels are shown, one for the Hermann–Mauguin symbol (top left), one for the Schönflies notation (top right), and one for the order of the point group (lower left).

Before we proceed with the derivation, it is useful to make a few comments on the point group notation, in particular for the international or Hermann–Mauguin notation. In general, the Hermann–Mauguin notation consists of (at most) three symbols; each symbol corresponds to a particular direction in the Bravais lattice. The relevant directions are listed in Table 9.1. When we name the crystallographic point groups in the following subsections, we will refer back to this table.

9.2.1 Step I: the proper rotations

The simplest groups are the proper rotation groups, formed by the identity and all powers of an n-fold rotation axis, which is the generating element.

Notation: n [C_n] [C for Cyclic]

All rotational groups are obviously cyclic groups of order n; they contain a single invariant line (the rotation axis itself). All odd groups are polar in

Table 9.1. Primary, secondary, and tertiary symmetry directions in each of the seven crystal systems. For the tetragonal system, the symbol ⟨*uvw*] refers to the fact that equivalent directions are obtained from permutations of the first two indices only.

Crystal system	Primary [uvw]	Secondary [uvw]	Tertiary [uvw]
Cubic	⟨100⟩	⟨111⟩	⟨110⟩
Hexagonal	[00.1]	[10.0]	[12.0]
Tetragonal	[001]	⟨100]	⟨110]
Orthorhombic	[100]	[010]	[001]
Trigonal	[111]	[010]	[$1\bar{1}0$]
Monoclinic	[010]	—	—
Triclinic	—	—	—

all directions whereas the even groups are only polar along the rotation axis.[1] If $N = nm$ then both **n** and **m** are subgroups of **N**; as an example, consider the group **6** (C_6) of order 6, which has both **2** (C_2) and **3** (C_3) as subgroups. For $n \rightarrow \infty$, the rotation group describes the symmetry of a rotating cone. The crystallographic point (rotation) groups are **1** (C_1), **2** (C_2), **3** (C_3), **4** (C_4) and **6** (C_6) and they are shown graphically in Figs. 9.2 and 9.3. Using the symmetry directions of Table 9.1, we find for the monoclinic system, for which the two-fold rotation is the highest order rotational symmetry, that the rotation axis is parallel to the primary direction, which is the [010] direction. For the trigonal system, on the other hand, the primary direction is the [111] direction, so that the three-fold axis in this system is parallel to [111]. In the hexagonal system, the six-fold axis lies along the primary [00.1] direction, and in the tetragonal system, the four-fold rotation axis lies along [001].

9.2.2 Step II: combining proper rotations with two-fold rotations

According to Euler's theorem, the presence of one two-fold axis perpendicular to a rotation axis, will result in the generation of a set of equivalent two-fold axes. This generates the so-called *dihedral* point groups.

Notation: n2 [D_n] [D for Dihedral]

Once again, there is a difference between the odd and the even groups: for the odd groups the operator n generates all two-fold axes, as shown in Fig. 9.4 for the group **32** (D_3). For all these groups the n-fold axis is non-polar, whereas

[1] A group is polar if the directions **t** and $-$**t** are not related to each other by a symmetry operation.

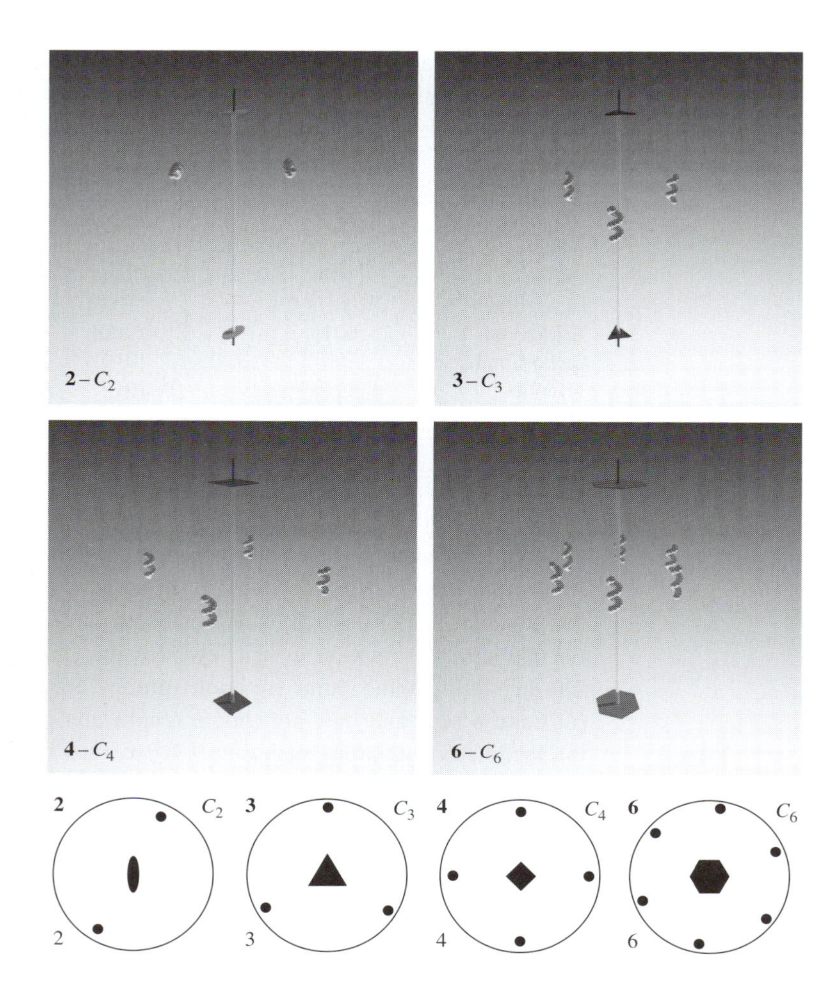

Fig. 9.3. Graphic representation of the crystallographic cyclic point groups.

the two-fold axes are polar. The group **12** =**2** (C_2) is already contained in the cyclic rotation groups.

For the even groups on the other hand, not all two-fold axes are generated from just one initial axis; a second two-fold axis needs to be provided in order to generate the complete set. Therefore, the Hermann–Mauguin symbol contains two **2** (C_2)s. The even dihedral groups are all non-polar.

The order of all dihedral groups is $2n$. The crystallographic dihedral point groups are **222** (D_2), **32** (D_3), **422** (D_4) and **622** (D_6) (see Fig. 9.4).

In the trigonal system, the three-fold axis lies along the [111] direction, as before, and the two-fold axis lies along the tertiary $\langle 1\bar{1}0 \rangle$ directions, since these directions are at right angles to [111]. In the tetragonal and hexagonal systems, the two-fold axes lie along both secondary and tertiary directions (see Table 9.1), whereas the orthorhombic system has a two-fold axis along each of the three symmetry directions.

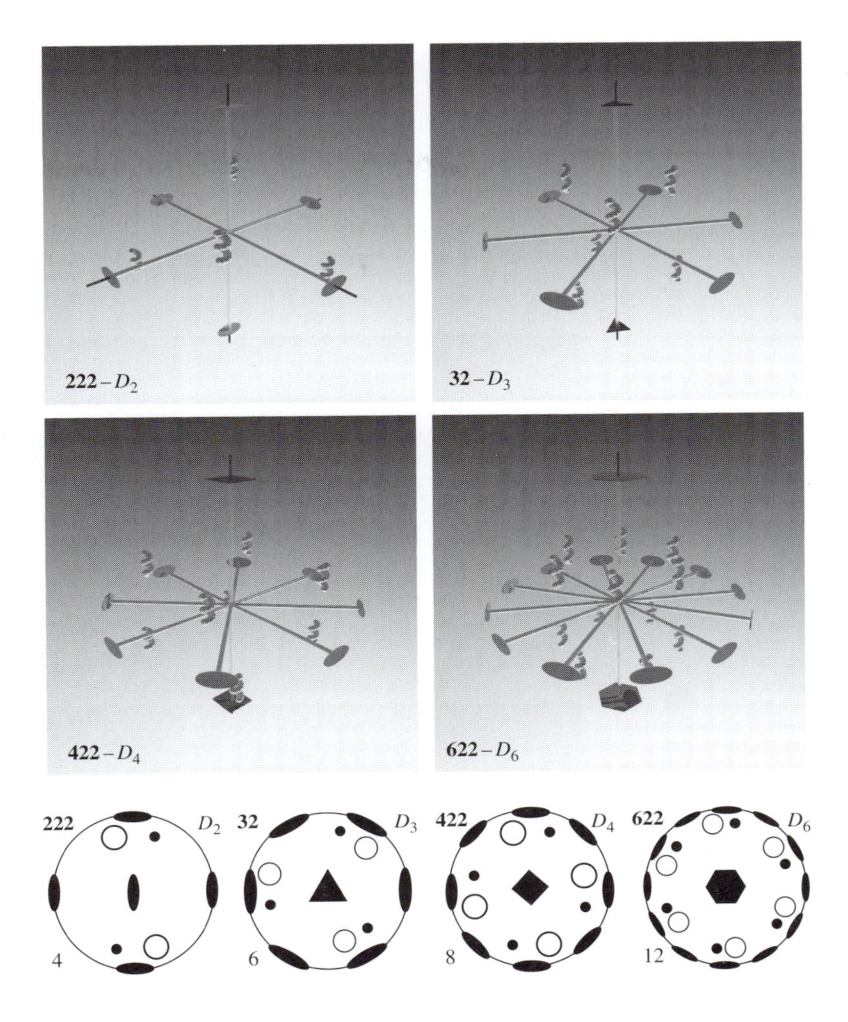

Fig. 9.4. Graphic representation of the crystallographic dihedral point groups.

9.2.3 Step IIIa: combining proper rotations with inversion symmetry

These groups are obtained by combination of an n-fold rotation axis with a center of symmetry in the origin. Each inversion rotation group has equivalence with a mirror rotation group $\tilde{n}$: $\bar{n}_{\mathrm{odd}} \leftrightarrow \widetilde{2}n$ and $\tilde{n}_{\mathrm{odd}} \leftrightarrow \bar{2}n$.

Notation: $\bar{\mathbf{n}}$ $[S_n]$

Note that these groups have one single axis but contain operators of the second kind. This is particularly clear in the 3-D renderings, which show the presence of both left-handed and right-handed helices.

For $n = 4k$, the inversion groups and mirror rotation groups are identical, e.g., $\bar{4} = \tilde{4}$, $\bar{8} = \tilde{8}$, . . . For $n = 4k + 2$, the inversion element actually generates a simple mirror plane perpendicular to the rotation axis; these groups are the odd groups of step IIIb (see next section). Normally, $\bar{\mathbf{6}}$ (C_{3h}) is written as **3/m**.

The crystallographic inversion rotation groups are $\bar{\mathbf{1}}$ (C_i), $\bar{\mathbf{2}}$ (=**m** (C_s)), $\bar{\mathbf{3}}$ (C_{3i}), $\bar{\mathbf{4}}$ (S_4), and $\bar{\mathbf{6}}$ (C_{3h}) (=**3/m**) and they are shown schematically in Fig. 9.5. The order of these point groups is $2n$ when n is odd, and n when n is even. The direction of the roto-inversion axes in the various crystal systems is the same as that of the proper rotation axes discussed in Section 9.2.1.

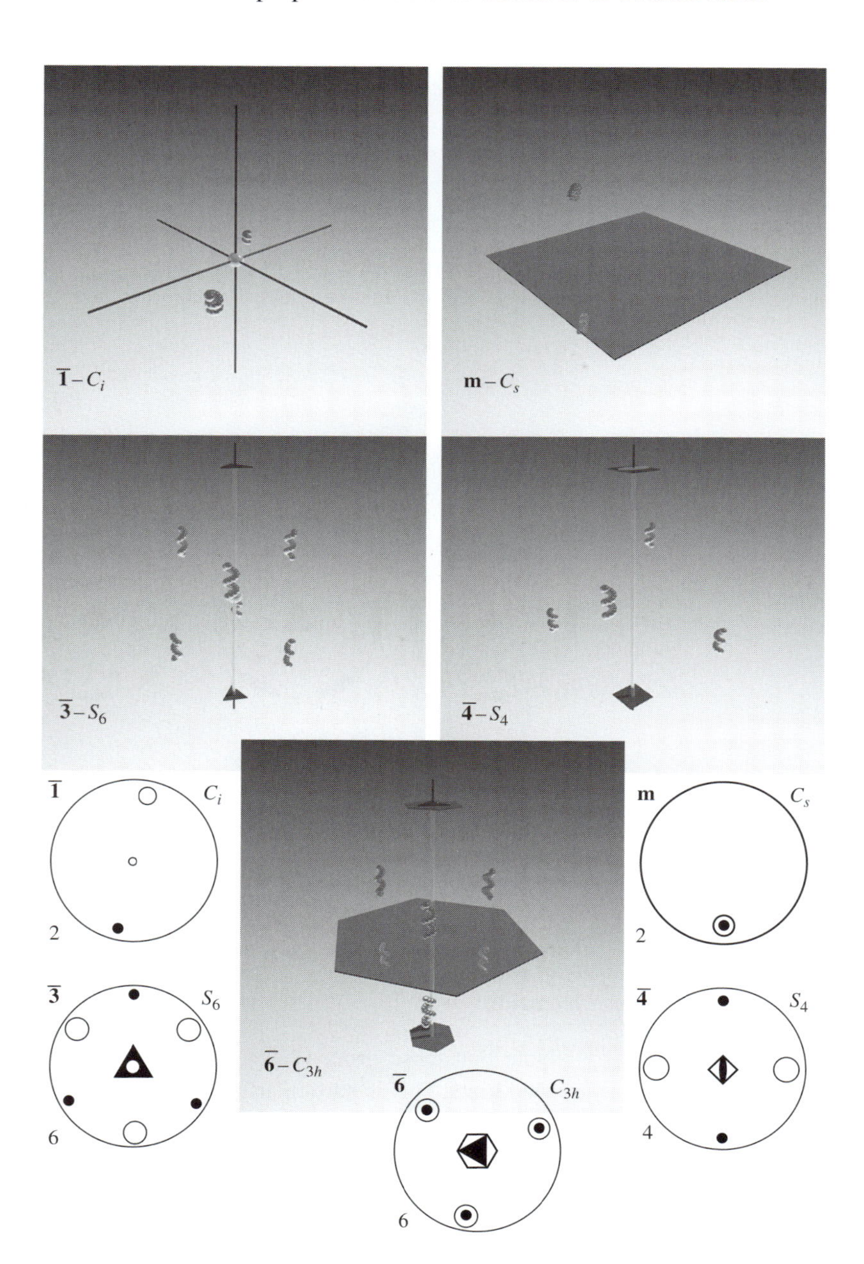

Fig. 9.5. Graphic representation of the crystallographic $\bar{n}$ point groups.

9.2.4 Step IIIb: combining proper rotations with perpendicular reflection elements

These groups contain a principal axis and a mirror plane perpendicular to the axis; hence, they contain operations of the second kind and both left-handed and right-handed helices in the 3-D representations.

Notation: n/m $[C_{nh}]$ [h for horizontal]

The odd members of this type of point group were already discussed in the previous section; the point groups $\mathbf{n_{odd}/m}$ are equivalent to $\bar{\mathbf{2n}}$. The order of all these groups is $2n$. The limiting group of this type describes a rotating cylinder $\infty/\mathbf{m}$.

The crystallographic point groups of type **n/m** are **2/m** (C_{2h}), **4/m** (C_{4h}) and **6/m** (C_{6h}) (see Fig. 9.6). The direction of the rotation axis is given by the primary direction for each crystal system in Table 9.1, and the Miller indices of the mirror plane correspond to the plane perpendicular to the rotation axis in each crystal system. For instance, for the monoclinic point group **2/m** (C_{2h}), the two-fold axis lies along the [010] direction and the (010) plane is the mirror plane.

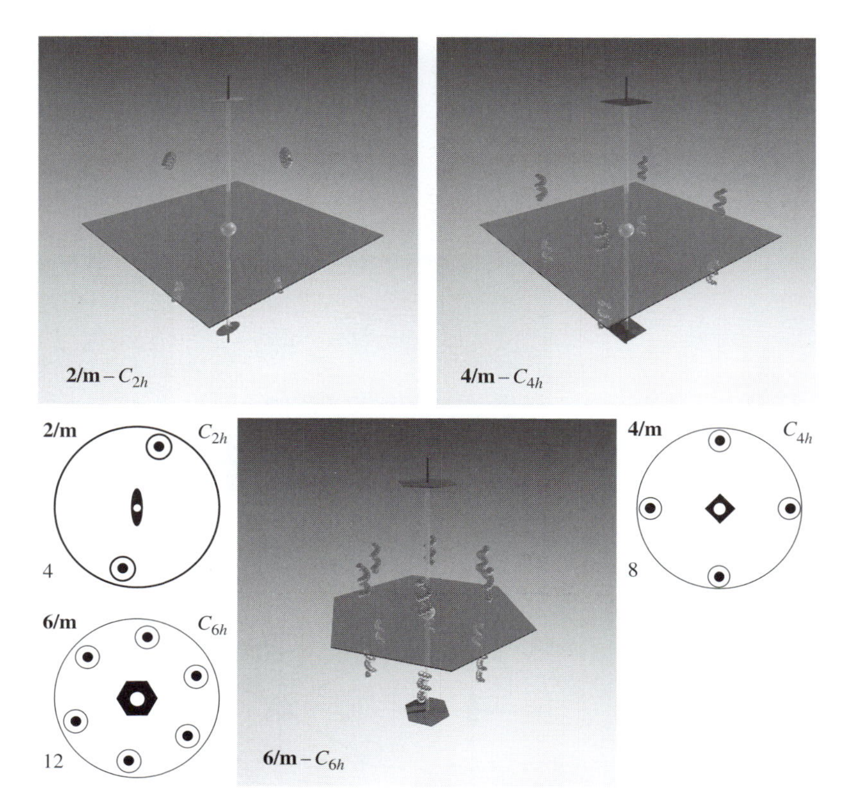

Fig. 9.6. Graphic representation of the crystallographic **n/m** point groups.

9.2.5 Step IV: combining proper rotations with coinciding reflection elements

Mirror planes containing a rotation axis of order n generate a new type of point group; the generating elements are **n** and **m**.

Notation: nm $[C_{nv}]$ [v for vertical]

The situation here is very similar to the one in Step II: odd rotation axes generate all the mirror planes from a single starting plane, whereas for even rotation axes a second generating mirror plane must be provided, as is illustrated in Fig. 9.7. The order of all these groups is $2n$. The group **1m** is identical to **m** $(C_s) = \bar{\mathbf{2}}$. The limiting symmetry ∞**mm** describes a regular cone.

The crystallographic point groups of type **nm** are **mm2** (C_{2v}), **3m** (C_{3v}), **4mm** (C_{4v}) and **6mm** (C_{6v}). Note that the two-fold axis in the point group **mm2** (C_{2v}) is written as the third symbol, not the first one. This is in agreement

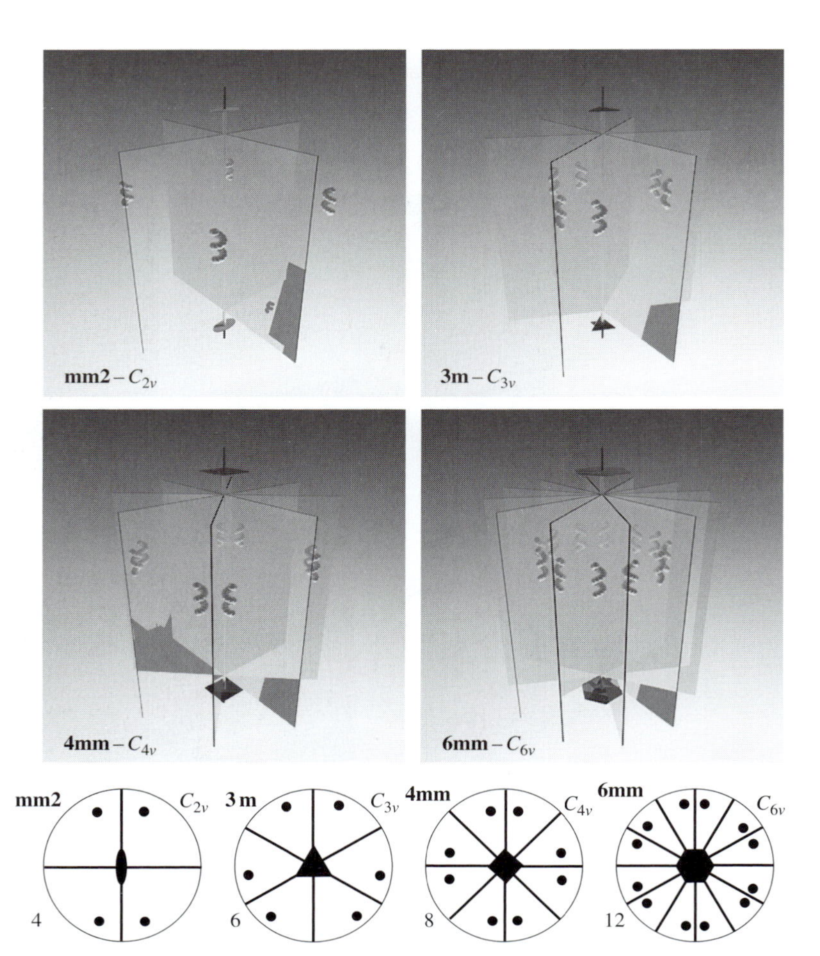

Fig. 9.7. Graphic representation of the crystallographic **nm** point groups.

with the entries for the orthorhombic crystal system in Table 9.1, which shows that the [001] axis (parallel to the two-fold axis) is the tertiary direction. For all other crystal systems, the rotation axis falls along the primary direction for this class of point groups.

9.2.6 Step Va: combining inversion rotations with coinciding reflection elements

The combination of an inversion point and a reflection plane through this point is equivalent to a two-fold axis. The generators are the inversion rotation and the mirror plane.

Notation: $\bar{\mathbf{n}}\mathbf{m}$ $[D_{nd}]$ [d for diagonal]

In odd and "twice-even" ($n = 4k$) groups, the two-fold axes bisect the angles between the mirror planes. For odd groups only one mirror plane is needed in the international symbol, for even groups a two-fold axis and a mirror plane are considered to be the generators. The simplest groups $\bar{\mathbf{1}}\mathbf{m}$ and $\bar{\mathbf{2}}\mathbf{m}$ are equivalent to $\mathbf{2/m}$ (C_{2h}) and **mm2** (C_{2v}) respectively.

The crystallographic point groups of type $\bar{\mathbf{n}}\mathbf{m}$ are $\bar{\mathbf{3}}\mathbf{m}$ (D_{3d}), $\bar{\mathbf{4}}\mathbf{2m}$ (D_{2d}), and $\bar{\mathbf{6}}\mathbf{m2}$ (D_{3h}) and are shown in Fig. 9.8. In $\bar{\mathbf{4}}\mathbf{2m}$ (D_{2d}), the two-fold axes are

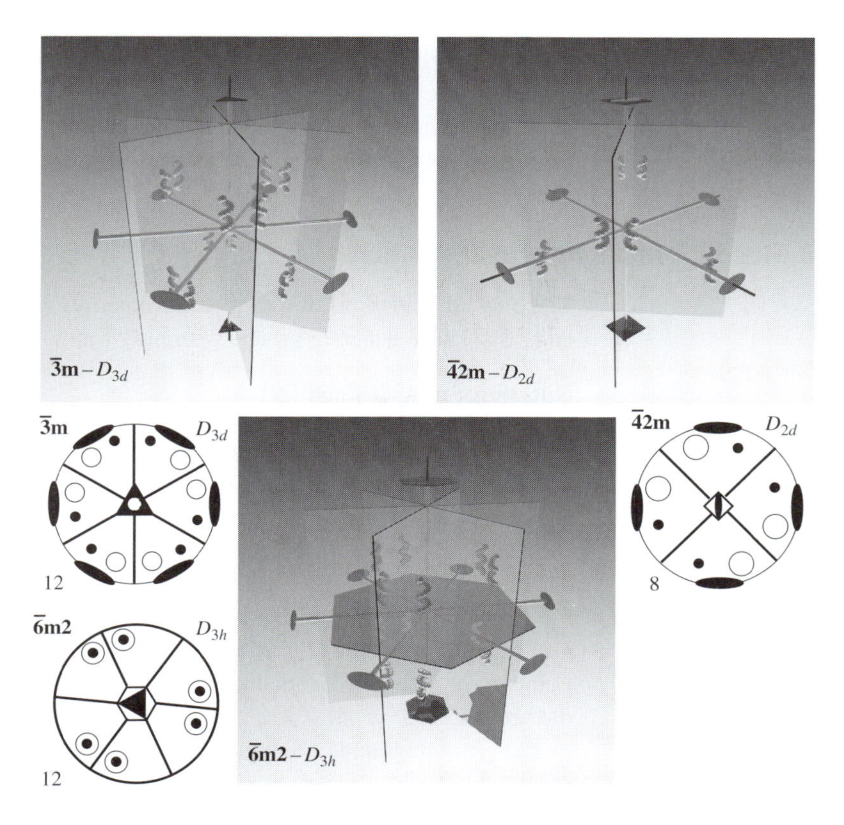

Fig. 9.8. Graphic representation of the crystallographic $\bar{\mathbf{n}}\mathbf{m}$ point groups.

considered to lie along the secondary directions of the type ⟨100], whereas the mirror planes lie perpendicular to the tertiary directions (see Table 9.1). For $\mathbf{\bar{6}m2}$ (D_{3h}) on the other hand, the mirror planes are taken to be normal to the secondary directions ⟨10.0] and the two-fold axes lie along the tertiary directions, hence the order of the symbols **m** and **2** is reversed in the two point group symbols. Note that for $\mathbf{\bar{6}m2}$ (D_{3h}), the two-fold axes lie in the mirror planes, whereas in $\mathbf{\bar{4}2m}$ (D_{2d}), the mirror planes bisect the two-fold axes.

9.2.7 Step Vb: combining proper rotations with coinciding and perpendicular reflection elements

If both perpendicular and coinciding mirror planes are present, then only the even rotations create new groups; the rotation axis and one mirror plane of each type are needed as generators.

Notation: $\mathbf{\frac{n}{m}m}$ $[D_{nh}]$

The intersection of two mirror planes at right angles generates a two-fold axis along the intersection line. Again, two independent coinciding mirror planes need to be defined for the even rotation groups. The full symbols of the groups are $\mathbf{\frac{2}{m}\frac{2}{m}\frac{2}{m}}$, $\mathbf{\frac{4}{m}\frac{2}{m}\frac{2}{m}}$, and $\mathbf{\frac{6}{m}\frac{2}{m}\frac{2}{m}}$. The order of all these groups is $4n$.

The crystallographic point groups of type $\mathbf{\frac{2}{m}m}$ are **mmm** (D_{2h}), **4/mmm** (D_{4h}) and **6/mmm** (D_{6h}) (in shorthand notation) and the graphical representations are shown in Fig. 9.9. The rotation axes are, as before, oriented along the primary directions of the crystal systems, and the mirror planes are perpendicular to both the secondary and tertiary directions.

9.2.8 Step VI: combining proper rotations

The only combinations of rotational symmetries we have not used yet are 233 and 432. These generate groups with only rotational elements present.

Notation: $\mathbf{n_1n_2}$ $[T]$ and $[O]$

There are only three point groups of this type, namely **23** (T), **432** (O) and the (non-crystallographic) icosahedral point group **532** (I). The Schönflies symbols for the crystallographic point groups stand for tetrahedral and octahedral symmetry respectively. The orders of these point groups are 12, 24, and 60 respectively. Note that they are groups with only operators of the first kind. Because of the presence of multiple three-fold axes, the point groups **23** (T) and **432** (O) belong to the cubic symmetry type. All equivalent points have the same handedness, since there are no symmetry operators of the second kind.

The crystallographic point groups of type $\mathbf{n_1n_2}$ are **23** (T) and **432** (O) and schematic drawings are shown in Fig. 9.10. In point group **23** (T), the two-fold

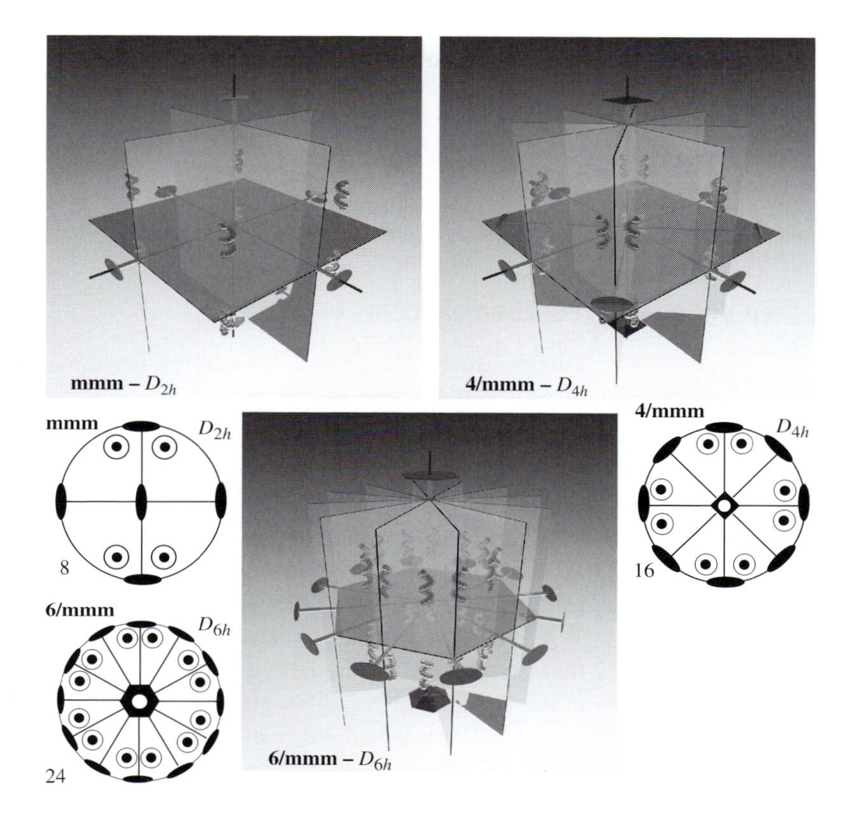

Fig. 9.9. Graphic representation of the crystallographic $\frac{\mathbf{n}}{\mathbf{m}}\mathbf{m}$ point groups.

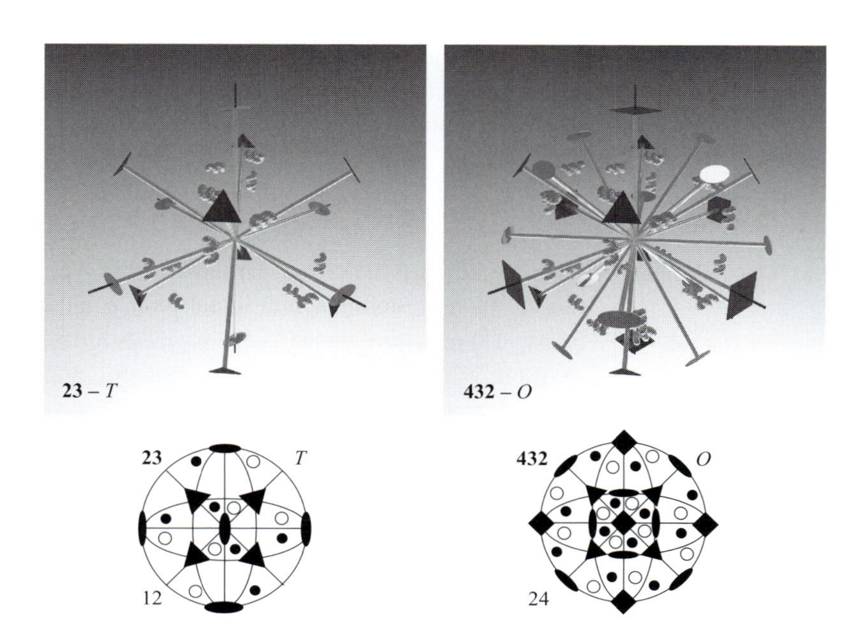

Fig. 9.10. Graphic representation of the crystallographic $\mathbf{n_1 n_2}$ point groups.

axis is oriented along the primary direction of the cubic crystal system (see Table 9.1), with the three–fold axes along the secondary directions. Note that there is also a point group **32** (D_3), so the order of the symbols is important! In point group **432** (O), the four-fold axes lie along the primary directions, the three-fold axes, as before, along the secondary directions, and the two-fold axes lie along the tertiary directions of the cubic crystal system.

9.2.9 Step VII: adding reflection elements to Step VI

Finally, we can add mirror planes (or inversion symmetry) to the groups of Step VI; because of the mutual orientation of the rotation axes, these symmetry planes will in general be arranged in an "oblique" way.

Notation: $\bar{\mathbf{n}}_1\mathbf{n}_2$ $[T_h]$, $[T_d]$, and $[O_h]$.

If we add mirror planes through the two-fold axes of **23** (T), the three-fold axes will become inversion axes and the resulting point group is of order 24 and is denoted by the symbol $\mathbf{m\bar{3}}$ (T_h).

Adding mirror planes through the three-fold axes of **23** (T) results in four-fold inversion axes (but note that the inversion element itself is not present in this group!); the resulting group is of order 24 and denoted by $\mathbf{\bar{4}3m}$ (T_d).

Finally, we can add either mirror planes or the inversion symmetry to the group **432** (O) to obtain the crystallographic group of the highest order (48) which is denoted by $\mathbf{m\bar{3}m}$ (O_h). In full notation this group is known as $\frac{\mathbf{4}}{\mathbf{m}}\mathbf{\bar{3}}\frac{2}{m}$. Combination of mirror planes with **532** (I) results in the icosahedral group $\mathbf{m\overline{35}}$ (I_h) of order 120.[2]

The crystallographic point groups of type $\bar{\mathbf{n}}_1\mathbf{n}_2$ are $\mathbf{m\bar{3}}$ (T_h), $\mathbf{\bar{4}3m}$ (T_d) and $\mathbf{m\bar{3}m}$ (O_h) and they are shown in Fig. 9.11. The orientation of the rotation axes is identical to that of the point groups **23** (T) and **432** (O). This concludes the enumeration of all 32 crystallographic point groups.

9.2.10 General remarks

9.2.10.1 Classes of point groups

Out of the 32 crystallographic point groups, only 11 have a center of symmetry. These *centrosymmetric* point groups are shown in the second column of Table 9.2. The other non-centrosymmetric point groups are subgroups of these 11 groups. It is not difficult to verify that each of the groups in the last column of Table 9.2 becomes equivalent to the centrosymmetric one when combined with the inversion operator. All the groups in a single row of the

[2] The icosahedral symmetry groups will be discussed in much more detail in the chapter on non-crystallographic symmetry (Chapter 15).

Table 9.2. The seven crystal systems with the centrosymmetric point groups (Laue classes), and the remaining point groups that are subgroups of the 11 centrosymmetric groups.

Crystal system	Laue class	Lower symmetry class members
Triclinic	$\bar{\mathbf{1}}$ (C_i)	**1** (C_1)
Monoclinic	**2/m** (C_{2h})	**2** (C_2), **m** (C_s)
Orthorhombic	**mmm** (D_{2h})	**222** (D_2), **mm2** (C_{2v})
Tetragonal	**4/m** (C_{4h})	**4** (C_4), $\bar{\mathbf{4}}$ (S_4)
	4/mmm (D_{4h})	**422** (D_4), **4mm** (C_{4v}), $\bar{\mathbf{4}}$**2m** (D_{2d})
Trigonal	$\bar{\mathbf{3}}$ (C_{3i})	**3** (C_3)
	$\bar{\mathbf{3}}$**m** (D_{3d})	**32** (D_3), **3m** (C_{3v})
Hexagonal	**6/m** (C_{6h})	**6** (C_6), $\bar{\mathbf{6}}$ (C_{3h})
	6/mmm (D_{6h})	**622** (D_6), **6mm** (C_{6v}), $\bar{\mathbf{6}}$**m2** (D_{3h})
Cubic	**m**$\bar{\mathbf{3}}$ (T_h)	**23** (T)
	m$\bar{\mathbf{3}}$**m** (O_h)	**432** (O), $\bar{\mathbf{4}}$**3m** (T_d)

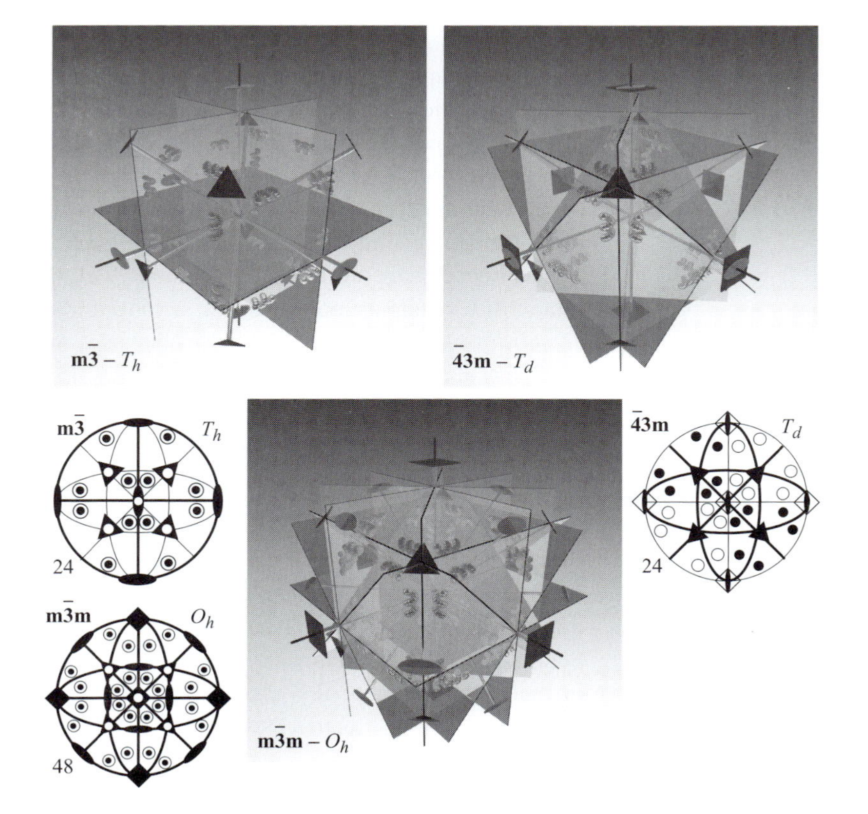

Fig. 9.11. Graphic representation of the crystallographic $\bar{\mathbf{n}}_1\mathbf{n}_2$ point groups.

table belong to a so-called *Laue class*.[3] The centrosymmetric point group in a Laue class is used as the class symbol.

Polar point groups are groups in which there exists at least one direction that has no symmetrically equivalent directions. It is easy to see that this can only happen in non-centrosymmetric point groups in which there is at the most a single rotation axis. There are 10 such polar groups: they are **4mm** (C_{4v}) and **6mm** (C_{6v}) and all their subgroups **6** (C_6), **4** (C_4), **3m** (C_{3v}), **3** (C_3), **mm2** (C_{2v}), **m** (C_s), **2** (C_2), and **1** (C_1).

The presence of symmetry at the atomic length scale has far-reaching consequences for the properties of a given material, whether they be mechanical, optical, electrical, magnetic, or thermal. It can be shown in very general terms that the symmetry group of *any* property of a crystal must include the point group symmetry operations of that crystal. This is known as *Neumann's principle* (Newnham, 2004). Certain material properties, such as piezoelectricity and optical activity, can occur only in non-centrosymmetric point groups. The application of symmetry to material properties is rich and extensive but outside the scope of this text; we refer the interested reader to the literature (Newnham, 2004, Nye, 1957).

9.2.10.2 Chirality and enantiomorphism

The stereographic projection is the standard method to represent the point groups graphically. The rendered images shown in Figs. 9.2 through 9.11 have the added advantage that the handedness of each equivalent point is easily observed. If all equivalent points have the same handedness, then the object with the corresponding point group can exist in two different versions: a left-handed version and a right-handed version. The point groups for which this can occur have no improper rotations, i.e., no roto-inversions or mirror rotations. Crystals or molecules that belong to these point groups are known as *chiral objects*, i.e., they have a handedness. Such objects are also known as *dissymmetric objects* (Hahn, 1996, p. 787). In chemistry and biology, the terms *enantiomerism* and *chirality* are used, whereas in crystallography the term *enantiomorphism* is more common.

9.2.10.3 Matrix representation of point groups

In the previous chapter, we have seen that every symmetry operation can be represented as a matrix. Since the point group operations do not contain any translations, we can restrict ourselves to the 3×3 sub-matrix D of the full Seitz symbol $(\mathsf{D}|\mathbf{t})$. Returning to the quartz point group **32** (D_3) introduced in the first section of this chapter, we can associate with each symmetry operation $\mathcal{O}_i$ a 3×3 matrix D_i. If we take the axes 2_x, 2_y and 3 of Fig. 9.1 as

[3] The reason for this terminology will become clear when we discuss diffraction in Chapter 11.

the reference axes, then it is easy to show that the six transformation matrices are given by:

$$\begin{aligned} \mathsf{D}_0 &= \begin{pmatrix} 1 & 0 & 0 \\ 0 & 1 & 0 \\ 0 & 0 & 1 \end{pmatrix} & \mathsf{D}_1 &= \begin{pmatrix} 0 & -1 & 0 \\ 1 & -1 & 0 \\ 0 & 0 & 1 \end{pmatrix} \\ \mathsf{D}_2 &= \begin{pmatrix} -1 & 1 & 0 \\ -1 & 0 & 0 \\ 0 & 0 & 1 \end{pmatrix} & \mathsf{D}_3 &= \begin{pmatrix} 1 & -1 & 0 \\ 0 & -1 & 0 \\ 0 & 0 & -1 \end{pmatrix} \\ \mathsf{D}_4 &= \begin{pmatrix} -1 & 0 & 0 \\ -1 & 1 & 0 \\ 0 & 0 & -1 \end{pmatrix} & \mathsf{D}_5 &= \begin{pmatrix} 0 & 1 & 0 \\ 1 & 0 & 0 \\ 0 & 0 & -1 \end{pmatrix}. \end{aligned} \quad (9.14)$$

Using these matrices, we can construct a multiplication table, similar to the one on page 201, by taking all possible products between two matrices:

32 (D_3)	D_0	D_1	D_2	D_3	D_4	D_5
D_0	D_0	D_1	D_2	D_3	D_4	D_5
D_1	D_1	D_2	D_0	D_5	D_3	D_4
D_2	D_2	D_0	D_1	D_4	D_5	D_3
D_3	D_3	D_4	D_5	D_0	D_2	D_1
D_4	D_4	D_5	D_3	D_1	D_0	D_2
D_5	D_5	D_3	D_4	D_2	D_1	D_0 .

Note that these two tables have exactly the same structure! This is an example of an *isomorphism* between the group of the operators $\mathcal{O}_i$ and the group of the six matrices shown above. Both of these are *representations* of an underlying abstract group, $\mathcal{G}$, with elements $\{a, b, c, d, e, f\}$, which can be defined by its own multiplication table; the same can be done for all other point groups as well. It is possible for multiple point groups to be isomorphous with a single abstract group.

$\mathcal{G}$	a	b	c	d	e	f
a	a	b	c	d	e	f
b	b	c	a	f	d	e
c	c	a	b	e	f	d
d	d	e	f	a	c	b
e	e	f	d	b	a	c
f	f	d	e	c	b	a .

9.2.10.4 Group–subgroup relations; descent in symmetry

Figure 9.12 shows all the group–subgroup relationships between the crystallographic point groups. The two highest order point groups are the hexagonal group **6/mmm** (D_{6h}) and the cubic group **m$\bar{3}$m** (O_h); all other groups can be considered as subgroups of these two. In some cases, there are several possible orientations of a subgroup with respect to the larger group. For instance,

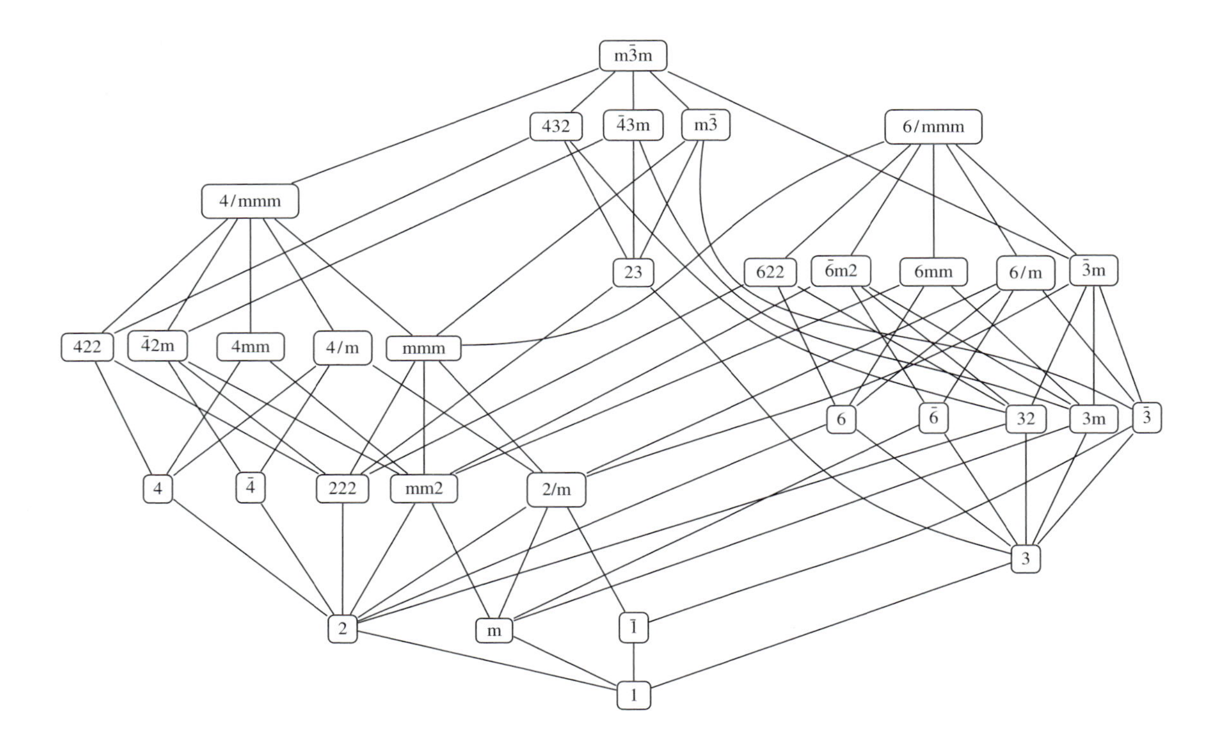

Fig. 9.12. Descent in symmetry for the crystallographic point groups, illustrating group–subgroup relationships and the point group orders (left column). Only the international point group symbols are shown.

point group **3** (C_3), a simple three-fold rotation axis, has four possible orientations with respect to point group **23** (T). On the other hand, **222** (D_2) has the same orientation as **23** (T), as is readily seen from a comparison of the two stereographic projections. The table can also be used to determine all the centrosymmetric point groups; simply start from $\bar{\mathbf{1}}$ (C_i) and list all the point groups that $\bar{\mathbf{1}}$ (C_i) is a subgroup of (i.e., follow all the lines originating in $\bar{\mathbf{1}}$ (C_i)). The centrosymmetric point groups are: $\bar{\mathbf{1}}$ (C_i), **2/m** (C_{2h}), **4/m** (C_{4h}), **mmm** (D_{2h}), **4/mmm** (D_{4h}), $\mathbf{m\bar{3}m}$ (O_h), $\bar{\mathbf{3}}$ (C_{3i}), $\mathbf{m\bar{3}}$ (T_h), **6/m** (C_{6h}), $\bar{\mathbf{3}}\mathbf{m}$ (D_{3d}), and **6/mmm** (D_{6h}), in agreement with the 11 Laue classes listed in Table 9.2. Figure 9.12 is known as a *descent in symmetry*, and we will use this representation again when we discuss the non-crystallographic point groups in Chapter 15.

9.2.10.5 Special positions and orbits

All point group illustrations in the preceding sections use as general object a right-handed helix. Care has been taken to make sure that this helix does not intersect any of the symmetry elements of the group. When the object *is* allowed to intersect one or more of the symmetry elements, then something interesting happens. Fig. 9.13(a) shows a rendered representation of point group **4/mmm** (D_{4h}). The helix has been replaced by a small sphere. When the sphere does not intersect any of the symmetry elements, there are a total of 16 equivalent spheres, which is the order of the point group. These positions

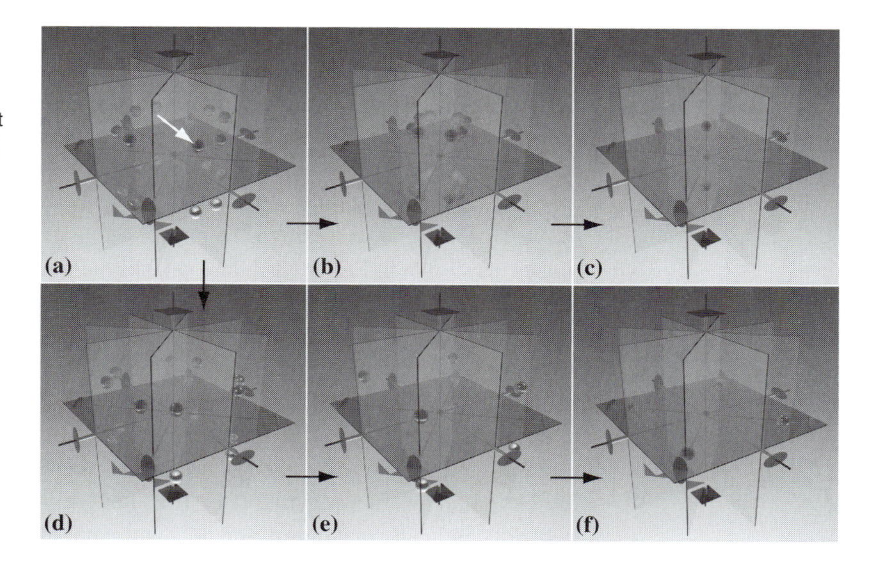

Fig. 9.13. Rendered representation of point group **4/mmm** (D_{4h}) and a general point orbit (a). In (b), the point moves towards the z-axis, and in (c) the special position $(0, 0, z)$ is shown. In (d), the general point moves towards one of the mirror planes and results in special point (x,x,z) in (e). In (f), the special point of the type (x,x,0) is shown.

are known as *general positions*. We define the *site symmetry* of a particular location as the point group of the object with respect to that location. For the sphere arrowed in Fig. 9.13(a), there are no symmetry elements going through the center of the sphere, so the site symmetry for that sphere is equal to **1** (C_1). This is always the case for a general position, and, in fact, we can define the general position for any point group symmetry as a position with site symmetry **1** (C_1).

The set of all points that are symmetrically equivalent to a point with coordinates (x, y, z) with respect to a particular point group $\mathcal{G}$ is called the *crystallographic orbit* (or, simply, the *orbit*) of (x, y, z) with respect to $\mathcal{G}$. The number of points in the orbit of a general point with respect to a point group is equal to the order of that point group. For the example in Fig. 9.13(a), the orbit consists of the following points:

$$\begin{array}{cccc} (x, y, z) & (x, -y, z) & (-x, y, z) & (-x, -y, z) \\ (x, y, -z) & (z, -y, -z) & (-z, y, -z) & (-z, -y, -z) \\ (y, x, z) & (-y, x, z) & (y, -x, z) & (-y, -x, z) \\ (y, x, -z) & (-y, x, -z) & (y, -x, -z) & (-y, -x, -z)\,. \end{array} \tag{9.15}$$

These coordinates can be obtained by operating with all the symmetry matrices of the point group **4/mmm** (D_{4h}) on the point with general coordinates (x, y, z). The general position used for Fig. 9.13(a) is equal to $(0.65, 0.15, 0.40)$.

When we move the general point towards the z-axis, as shown in Fig. 9.13(b) and (c), then the 16 equivalent points will "merge" into only two points located on the z-axis. This means that for the point $(0, 0, z)$, the orbit consists of only two points, $(0, 0, z)$ and $(0, 0, -z)$. This is also obvious from

Equation 9.15: when we put $x = 0$ and $y = 0$, then there are only two distinct points left over. Note that the point $(0, 0, z)$ is located on the four-fold rotation axis, and also on the four mirror planes that contain the four-fold axis, but not on the horizontal mirror plane. Hence, the site symmetry of $(0, 0, z)$ is **4mm** (C_{4v}). Points like this are known as *special points*, and their site symmetry is higher than that of point group **1** (C_1).

Alternatively, we can move the point (x, y, z) to one of the vertical mirror planes, as shown in Fig. 9.13(d) and (e). On the mirror plane, the coordinate of the point is (x, x, z), and it is clear that there are only eight equivalent points. The site symmetry is simply a single mirror plane, represented by the point group **m** (C_s). However, we can say something more about this mirror plane, namely that it is oriented at 45° with respect to the coordinate axes. Referring to Table 9.1, we see that the $\langle 110]$ directions in the tetragonal crystal system represent the tertiary symmetry directions. We include this information in the Hermann–Mauguin symbol for the site symmetry as follows: $\cdot\cdot\mathbf{m}$. The two dots indicate that there are no symmetry elements along the primary and secondary directions. If we had moved the general point to the $(x, 0, z)$ position, then there would again be eight equivalent positions, but this time the site symmetry would be $\cdot\mathbf{m}\cdot$, since the mirror plane is normal to the secondary symmetry direction for the tetragonal system.

Finally, we move the special point (x, x, z) down to the horizontal mirror plane in Fig. 9.13(f). The result is an orbit with four equivalent points. The site symmetry consists of two intersecting mirror planes (at right angles). We know that this corresponds to point group **mm2** (C_{2v}), but once again we can include a little additional information in the site symmetry symbol. The horizontal mirror plane is normal to the primary tetragonal symmetry direction. Both the vertical mirror plane and the two-fold axis along the intersection of the two mirror planes are located along the tertiary direction, so that the site symmetry is written as $\mathbf{m}\cdot\mathbf{m2}$. If we had moved the special point to the $(x, 0, 0)$ position, then the orbit would again contain four points, this time with site symmetry $\mathbf{m2m}\cdot$.

Let us now summarize the general and special positions that we have found for the point group **4/mmm** (D_{4h}). We start with the orbit with the largest number of members, which is the general position (x, y, z), and then list the special positions in descending order:

$$\begin{array}{llll} 16\ g & \mathbf{1} & (x, y, z) & (hkl) \\ 8\ f & \cdot\mathbf{m}\cdot & (x, 0, z) & (h0l) \\ 8\ e & \cdot\cdot\mathbf{m} & (x, x, z) & (hhl) \\ 8\ d & \mathbf{m}\cdot\cdot & (x, y, 0) & (hk0) \\ 4\ c & \mathbf{m2m}\cdot & (x, 0, 0) & (100) \\ 4\ b & \mathbf{m}\cdot\mathbf{m2} & (x, x, 0) & (110) \\ 2\ a & \mathbf{4mm} & (0, 0, z) & (001)\ . \end{array} \tag{9.16}$$

The last column contains Miller indices; for the (hhl)-type planes, there are eight equivalent planes, just as there are 8 equivalent positions for the $(x, 0, z)$ point. The first column is then known as the *multiplicity* of the plane in the point group **4/mmm** (D_{4h}).

The letter symbols in the second column are added once all the special positions are listed. By convention, the lowest order site symmetry is labeled by the letter a, the next one by b, and so on, until all site symmetries have been labeled. If we combine the first two columns, then we obtain what is known as a *Wyckoff position*. For instance, we can talk about the $8e$ position, or the $4b$ position. Chapter 10 in the *International Tables for Crystallography*, volume A, standardizes all possible site symmetries for all crystallographic point groups. This means that the $8e$ position for point group **4/mmm** (D_{4h}) is, by international convention, a position of the type (x, x, z). We refer the reader to the tables for further information.

As a final remark, note that all the special site symmetries are subgroups of the original group **4/mmm** (D_{4h}). This can be verified easily, using Fig. 9.12.

9.2.10.6 Crystallographic and non-crystallographic point groups

Table 9.3 illustrates the relations between the crystallographic point groups and the larger set of general 3-D point groups, in particular the icosahedral point groups **532** (I) and $\mathbf{m\bar{3}\bar{5}}$ (I_h). The limiting point groups of infinite order are shown schematically on the bottom row of the table. They are based on the highly symmetric shapes of the sphere, the cone, and the cylinder. Arrows on these shapes indicate that the object is rotating; for instance, point group ∞ is represented by a rotating cone, which has a rotation axis of infinite order along the cone axis. If the cone were at rest, then there would also be mirror planes containing the rotation axis (which is the case for point group ∞**mm**), but the rotation motion eliminates these mirror planes. Similarly, the top and bottom planes of the cylinder rotate in opposite directions in point group ∞**2**, so that there are an infinite number of two-fold axes normal to the cylinder axis, but no mirror planes. A single mirror plane normal to the cylinder axis results in point group ∞/**m** (the rotation once again prevents the presence of mirror planes containing the rotation axis). The full cylinder symmetry of ∞/**mm** is obtained by eliminating all rotation motions. Finally, the sphere symmetries $\infty\infty$ and $\infty\infty$**m** are distinguished from each other by the fact that, in the former, every point on the sphere surface rotates around the axis connecting it to the center of the sphere.

9.2.10.7 Examples of shapes, molecules, and crystals

Table 9.4 lists, for each crystallographic point group, the names of a geometric shape, a molecule, and a mineral with that symmetry. In addition, in the second column, it lists two numbers, which represent the percentage of a population of 127 000 inorganic and 156 000 organic compounds that have that particular point group symmetry. For instance, for **2/m** (C_{2h}) the

Table 9.3. Crystallographic and non-crystallographic point groups.

I	II	IIIa	IIIb	IV	Va	Vb	VI	VII
n	n2	$\bar{n}$	n/m	nm	$\bar{n}$m	$\frac{n}{m}$m	n_1n_2	$\bar{n}_1n_2$
1			m					
2	222	$\bar{1}$	2/m	mm2		mmm		
3	32		$\bar{6}$	3m	$\bar{6}$m2		23	m$\bar{3}$
4	422	$\bar{4}$	4/m	4mm	$\bar{4}$2m	4/mmm	432	$\bar{4}$3m
6	622	$\bar{3}$	6/m	6mm	$\bar{3}$m	6/mmm		m$\bar{3}$m
							532	m$\overline{35}$
↓	↓		↓	↓		↓	↓	↓
∞	∞2		∞/m	∞ mm		∞/mm	∞∞	∞∞ m

population numbers are 34.63% for inorganic compounds and 44.81% for organic compounds. This means that out of 127 000 inorganic compounds, approximately 43 980 compounds have the monoclinic **2/m** (C_{2h}) point group symmetry. Note that the 11 point groups corresponding to the Laue classes have the highest population numbers, indicating that the majority of inorganic compounds, around 80%, are centrosymmetric. Amongst the organic compounds, about 74% are centrosymmetric. Despite the fact that the highest cubic symmetry, **m$\bar{3}$m** (O_h), only accounts for about 6.7% of all the inorganic compounds, the cubic materials play a very important role in our technological world.

9.2.10.8 Definition of generator matrices

We know from the multiplication table of a group that each element of the group can be written as the product of other elements. In practice, we need to know only a few elements, and the complete group can be constructed from these by matrix multiplication. The minimum symmetry operators needed to generate the complete point group are known as the *generators*. There are only 14 matrices, D, from which all possible crystal symmetries can be derived. We will represent these matrices by symbols of the form $D^{(x)}$, where (x) is a letter ranging from (a) through (n).[4] The 14 matrices are defined in Table 9.5.

[4] Additional generator matrices can be added to this list to describe non-crystallographic point groups.

Table 9.4. Examples of molecules and crystals with symmetries belonging to the various point groups. The % population (taken from Table 3.2 in (Newnham, 2004)) consists of the percentage of 127 000 inorganic and 156 000 organic compounds (inorganic/organic) that are found to belong to each point group.

Point group	% population	Shape	Molecule	Crystal
1 (C_1)	0.67/1.24	Pedion	CHFClBr §FClO	—
$\bar{\mathbf{1}}$ (C_i)	13.87/19.18	Pinacoid	$C_2H_2Cl_2Br_2$	Anorthite Turquoise Wollastinite
2 (C_2)	2.21/6.70	Sphenoid	H_2O_2 $C_2H_2Cl_2$	—
m (C_s)	1.30/1.46	Dome	NOCl	—
2/m (C_{2h})	34.63/44.81	Rhombic prism	H_2O_2 planar $C_2H_2Cl_2$ planar	Chlorite Datolite Epidote Gypsym Orpiment Realgar Talc Titanite
222 (D_2)	3.56/10.13	Rhombic disphenoid	C_2H_4	Edingtonite
mm2 (C_{2v})	3.32/3.31	Rhombic pyramid	CH_2Cl_2 H_2O H_2O_2 NO_2	Hemimorphite
mmm (D_{2h})	12.07/7.84	Rhombic dipyramid	$C_{10}H_8$ C_2H_4	Andalusite Aragonite Barite Chrysobalite Enstatite Goethite Marcasite Sillimanite Stibnite Sulfur
3 (C_3)	0.36/0.32	Trigonal pyramid	$C_2H_3Cl_3$	—
$\bar{\mathbf{3}}$ (C_{3i})	1.21/0.58	Rhombohedron Hexagonal prism	—	Dolomite Ilmenite Phenakite
32 (D_3)	0.54/0.22	Trigonal trapezohedron	C_2H_6	Cinnabar Low Quartz
3m (C_{3v})	0.74/0.22	Ditrigonal pyramid	NH_3 SF_3Cl_3	Tourmaline
$\bar{\mathbf{3}}$**m** (D_{3d})	3.18/0.25	Ditrigonal scalenohedron	C_2H_6	Arsenic Brucite Calcite Corundum Hematite
4 (C_4)	0.19/0.25	Tetragonal pyramid	—	—
$\bar{\mathbf{4}}$ (S_4)	0.25/0.18	Isosceles tetrahedron Tetragonal disphenoid	$C_{12}H_16$ $C_8F_4Cl_{14}$	—

Table 9.4. (cont.).

Point group	% population	Shape	Molecule	Crystal
4/m (C_{4h})	1.17/0.67	Tetragonal dipyramid	—	Scapolite Scheelite
422 (D_4)	0.40/0.48	Tetragonal trapezohedron	—	—
4mm (C_{4v})	0.30/0.09	Ditetragonal pyramid Tetragonal prism	$Co(NH_3)_4ClBr$ SF_4ClBr	—
$\bar{4}$2m (D_{2d})	0.82/0.34	Ditetragonal scalenohedron	C_2H_4 C_3H_4 C_8H_8	Chalcopyrite
4/mmm (D_{4h})	4.53/0.69	Ditetragonal dipyramid Ditetragonal prism	SF_4Cl_2 XeF_4 $[AuCl_4]^-$ $[Ni(CN)_4]^{-2}$	Rutile Zircon
6 (C_6)	0.41/0.22	Hexagonal pyramid	—	—
$\bar{6}$ (C_{3h})	0.07/0.01	Trigonal pyramid	$Fe(OH)_3$	Nepheline
6/m (C_{6h})	0.82/0.17	Hexagonal dipyramid	—	Apetite
622 (D_6)	0.24/0.05	Hexagonal trapezohedron	—	High quartz
6mm (C_{6v})	0.45/0.03	Dihexagonal pyramid	$(C_6H_6)_2Cr$	Wurzite
$\bar{6}$m2 (D_{3h})	0.41/0.02	Ditrigonal dipyramid	BF_3 C_2H_6 C_3H_3 PF_5 SO_3 $[NO_3]^-$	Benitoite
6/mmm (D_{6h})	2.82/0.05	Dihexagonal dipyramid Hexagonal prism	C_6H_6	Beryl Niccolite
23 (T)	0.44/0.09	Tetartoid	—	—
m$\bar{3}$ (T_h)	0.84/0.15	Diploid	—	Pyrite
432 (O)	0.13/0.01	Gyroid	—	—
$\bar{4}$3m (T_d)	1.42/0.11	Hexakistetrahedron	CH_4	Tetrahedrite Spalerite
m$\bar{3}$m (O_h)	6.66/0.12	Cuboctahedron Cube Tetrahexahedron Truncated octahedron Rhombic dodecahedron Octahedron	SF_6 $[Fe(CN)_6]^{-3}$ $[PF_6]^-$	Halite Copper Cuprite Diamond Flourite Galena Garnet Gold Silver Spinel

Table 9.6 lists for all 32 crystallographic point groups which generator matrices are needed to create the entire group. Note that the identity matrix, $\mathsf{D}^{(a)}$, is always an element of the group but does not appear explicitly in Table 9.6. Note also that the selection of generators is not unique. For instance, for point group **6** (C_6), we could select just one generator (a 60° rotation around the z axis). Instead, the table only lists generators selected from the 14 matrices in Table 9.5; the matrix representing the 6-fold rotation is not one

Table 9.5. Definition of the 14 fundamental crystallographic point symmetry matrices.

$$\mathsf{D}^{(a)} \equiv \begin{pmatrix} 1 & 0 & 0 \\ 0 & 1 & 0 \\ 0 & 0 & 1 \end{pmatrix} \quad \mathsf{D}^{(b)} \equiv \begin{pmatrix} -1 & 0 & 0 \\ 0 & -1 & 0 \\ 0 & 0 & 1 \end{pmatrix} \quad \mathsf{D}^{(c)} \equiv \begin{pmatrix} -1 & 0 & 0 \\ 0 & 1 & 0 \\ 0 & 0 & -1 \end{pmatrix}$$

$$\mathsf{D}^{(d)} \equiv \begin{pmatrix} 0 & 0 & 1 \\ 1 & 0 & 0 \\ 0 & 1 & 0 \end{pmatrix} \quad \mathsf{D}^{(e)} \equiv \begin{pmatrix} 0 & 1 & 0 \\ 1 & 0 & 0 \\ 0 & 0 & -1 \end{pmatrix} \quad \mathsf{D}^{(f)} \equiv \begin{pmatrix} 0 & -1 & 0 \\ -1 & 0 & 0 \\ 0 & 0 & -1 \end{pmatrix}$$

$$\mathsf{D}^{(g)} \equiv \begin{pmatrix} 0 & -1 & 0 \\ 1 & 0 & 0 \\ 0 & 0 & 1 \end{pmatrix} \quad \mathsf{D}^{(h)} \equiv \begin{pmatrix} -1 & 0 & 0 \\ 0 & -1 & 0 \\ 0 & 0 & -1 \end{pmatrix} \quad \mathsf{D}^{(i)} \equiv \begin{pmatrix} 1 & 0 & 0 \\ 0 & 1 & 0 \\ 0 & 0 & -1 \end{pmatrix}$$

$$\mathsf{D}^{(j)} \equiv \begin{pmatrix} 1 & 0 & 0 \\ 0 & -1 & 0 \\ 0 & 0 & 1 \end{pmatrix} \quad \mathsf{D}^{(k)} \equiv \begin{pmatrix} 0 & -1 & 0 \\ -1 & 0 & 0 \\ 0 & 0 & 1 \end{pmatrix} \quad \mathsf{D}^{(l)} \equiv \begin{pmatrix} 0 & 1 & 0 \\ 1 & 0 & 0 \\ 0 & 0 & 1 \end{pmatrix}$$

$$\mathsf{D}^{(m)} \equiv \begin{pmatrix} 0 & 1 & 0 \\ -1 & 0 & 0 \\ 0 & 0 & -1 \end{pmatrix} \quad \mathsf{D}^{(n)} \equiv \begin{pmatrix} 0 & -1 & 0 \\ 1 & -1 & 0 \\ 0 & 0 & 1 \end{pmatrix}$$

Table 9.6. The generator matrices for the crystallographic point groups. The letters x in the generator columns refer to the transformation matrices $\mathsf{D}^{(x)}$ of Table 9.5.

Point group	Generators	Point group	Generators	Point group	Generators	Point group	Generators
1 (C_1)	—	**4** (C_4)	g	$\bar{\mathbf{3}}$ (C_{3i})	h, n	**6mm** (C_{6v})	b, k, n
$\bar{\mathbf{1}}$ (C_i)	h	$\bar{\mathbf{4}}$ (S_4)	m	**32** (D_3)	e, n	$\bar{\mathbf{6}}$**m2** (D_{3h})	i, k, n
2 (C_2)	c	**4/m** (C_{4h})	g, h	**3m** (C_{3v})	k, n	**6/mmm** (D_{6h})	b, e, n, h
m (C_s)	j	**422** (D_4)	c, g	$\bar{\mathbf{3}}$**m** (D_{3d})	f, h, n	**23** (T)	c, d
2/m (C_{2h})	c, h	**4mm** (C_{4v})	g, j	**6** (C_6)	b, n	**m**$\bar{\mathbf{3}}$ (T_h)	c, d, h
222 (D_2)	b, c	$\bar{\mathbf{4}}$**2m** (D_{2d})	c, m	$\bar{\mathbf{6}}$ (C_{3h})	i, n	**432** (O)	d, g
mm2 (C_{2v})	b, j	**4/mmm** (D_{4h})	c, g, h	**6/m** (C_{6h})	b, h, n	$\bar{\mathbf{4}}$**3m** (T_d)	g, m
mmm (D_{2h})	b, c, h	**3** (C_3)	n	**622** (D_6)	b, e, n	**m**$\bar{\mathbf{3}}$**m** (O_h)	d, g, h

of the 14 listed, and we need 2 generators from the list of 14 to generate the entire group **6** (C_6). The main reason for defining the 14 matrices is that they are also used as generators for the 230 space groups, to be introduced in the next chapter.

As an example, consider point group $\bar{\mathbf{3}}$**m** (D_{3d}), which has generators $\mathsf{D}^{(f)}$, $\mathsf{D}^{(h)}$, and $\mathsf{D}^{(n)}$. We begin by multiplying the generators with themselves until we obtain the identity matrix. For $\mathsf{D}^{(f)}$ and $\mathsf{D}^{(h)}$ this does not lead to a new symmetry operation, since $\mathsf{D}^{(ff)} = \mathsf{D}^{(hh)} = \mathsf{D}^{(a)}$; for $\mathsf{D}^{(n)}$ we obtain a new matrix:[5]

$$\mathsf{D}^{(nn)} = \begin{pmatrix} -1 & 1 & 0 \\ -1 & 0 & 0 \\ 0 & 0 & 1 \end{pmatrix},$$

[5] We use the obvious notation that $\mathsf{D}^{(x)}\mathsf{D}^{(y)} \equiv \mathsf{D}^{(xy)}$.

and also $\mathsf{D}^{(nnn)} = \mathsf{D}^{(a)}$. This is to be expected, since this matrix represents a three-fold rotation.

Next, we compute the product of the two generator matrices $\mathsf{D}^{(f)}$ and $\mathsf{D}^{(n)}$: we find two new matrices (recall that matrix multiplication is, in general, not commutative, so we must multiply the matrices together in both possible orders):

$$\mathsf{D}^{(fn)} = \begin{pmatrix} -1 & 1 & 0 \\ 0 & 1 & 0 \\ 0 & 0 & -1 \end{pmatrix}, \quad \text{and} \quad \mathsf{D}^{(nf)} = \begin{pmatrix} 1 & 0 & 0 \\ 1 & -1 & 0 \\ 0 & 0 & -1 \end{pmatrix}.$$

This brings our total to six matrices, including the identity matrix: $\mathsf{D}^{(a)}$, $\mathsf{D}^{(f)}$, $\mathsf{D}^{(n)}$, $\mathsf{D}^{(nn)}$, $\mathsf{D}^{(fn)}$, and $\mathsf{D}^{(nf)}$. Finally, multiply all six matrices with the generator $\mathsf{D}^{(h)}$, which leads to the following six new matrices:

$$\mathsf{D}^{(ha)} = \begin{pmatrix} -1 & 0 & 0 \\ 0 & -1 & 0 \\ 0 & 0 & -1 \end{pmatrix}, \qquad \mathsf{D}^{(hf)} = \begin{pmatrix} 0 & 1 & 0 \\ 1 & 0 & 0 \\ 0 & 0 & 1 \end{pmatrix},$$

$$\mathsf{D}^{(hn)} = \begin{pmatrix} 0 & 1 & 0 \\ -1 & 1 & 0 \\ 0 & 0 & -1 \end{pmatrix}, \qquad \mathsf{D}^{(hnn)} = \begin{pmatrix} 1 & -1 & 0 \\ 1 & 0 & 0 \\ 0 & 0 & -1 \end{pmatrix},$$

$$\mathsf{D}^{(hfn)} = \begin{pmatrix} 1 & -1 & 0 \\ 0 & -1 & 0 \\ 0 & 0 & 1 \end{pmatrix}, \qquad \mathsf{D}^{(hnf)} = \begin{pmatrix} -1 & 0 & 0 \\ -1 & 1 & 0 \\ 0 & 0 & 1 \end{pmatrix}.$$

It is straightforward to verify that any other product of these 12 matrices will generate one of the 12, in other words, we have generated the entire group of order 12. For instance, consider the product $\mathsf{D}^{(hnhnn)}$: since $\mathsf{D}^{(h)}$ is a diagonal matrix, we can change the order of the matrix multiplications to $\mathsf{D}^{(hhnnn)} = \mathsf{D}^{(aa)} = \mathsf{D}^{(a)}$ which is one the 12 group elements. Similarly, for $\mathsf{D}^{(hnfhfn)}$ we have: $\mathsf{D}^{(hnfhfn)} = \mathsf{D}^{(hhnffn)} = \mathsf{D}^{(anan)} = \mathsf{D}^{(nn)}$ which belongs to the 12 that we derived. The same procedure can be applied to the other 31 point groups, and is not too difficult to implement in a computer algorithm.

9.3 Two-dimensional crystallographic point symmetries

In 2-D, there are fewer possible point groups. We can derive the 2-D crystallographic point groups starting from the 3-D groups by eliminating those groups that contain operators that are inconsistent with two dimensions. For instance, a horizontal mirror plane has no meaning in 2-D; neither

Table 9.7. 2-D crystal systems and corresponding 2-D crystallographic point groups.

Crystal system	Compatible point groups
Oblique	**1** (C_1), **2** (C_2)
Rectangular	**m** (C_s), **mm2** (C_{2v})
Square	**4** (C_4), **4mm** (C_{4v})
Hexagonal	**3** (C_3), **3m** (C_{3v}), **6** (C_6), **6mm** (C_{6v})

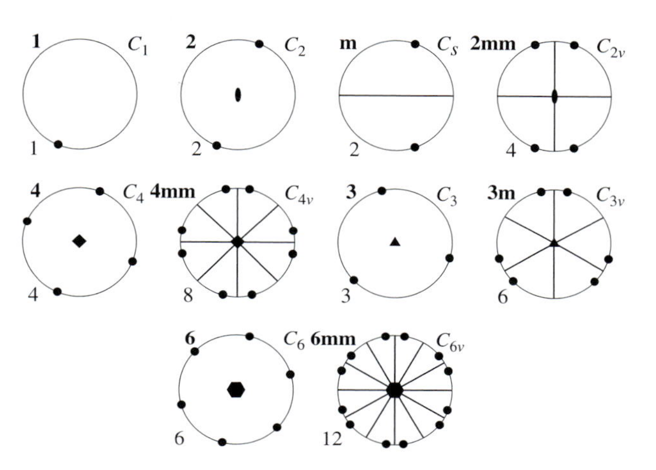

Fig. 9.14. Graphical representations of the ten 2-D crystallographic point groups. Note that the general points all lie in the equatorial plane, hence they correspond to points on the projection circle.

does an inversion operation. Mirror planes inclined to the horizontal plane must also be excluded. Instead of mirror planes, we only have *mirror lines*. This leaves only ten 2-D crystallographic point groups: **1** (C_1), **m** (C_s), **2** (C_2), **mm2** (C_{2v}), **3** (C_3), **3m** (C_{3v}), **4** (C_4), **4mm** (C_{4v}), **6** (C_6), and **6mm** (C_{6v}). It is customary to use the same point group symbols in 2-D and 3-D; it will be clear from the context which of the groups one refers to. Table 9.7 shows the four 2-D crystal systems, along with the point groups that belong to each system. Graphical representations of the groups are shown in Fig. 9.14.

9.4 Historical notes

Group theory currently occupies an important position amongst the mathematical theories. Group theory was invented by the gifted French mathematician **Evariste Galois** (1811–32). Galois lived during a tumultuous period in French history. He was born at the peak of Napoleon's power in 1811,

Fig. 9.15. (a) Evariste Galois (1811–32) and (b) J.F.C. Hessel (1796–1872) (picture courtesy of J. Lima-de-Faria).

just a few years before his historic loss at Waterloo. As a teenager, Galois became interested in mathematics, in particular in the theory of equations, and he published his first paper on continued fractions in 1829, at age 17! Mathematicians at that time were very interested in the solubility of polynomial equations. While the solutions to the quadratic equation had been known for a very long time (going back to the Babylonians and the Greeks), the general cubic and quartic equations had been solved only recently (see Livio (2005) for an entertaining account), and the next equation, the quintic, appeared to resist all attempts at finding a mathematical expression for its solutions. Galois developed what is now known as the *Galois theory*, a precursor of modern group theory, which uses certain symmetry properties of the polynomial equations to decide whether or not the solutions can be written down using rational functions and n-th order roots. He showed from the symmetry of the quintic equation that its solution cannot be written down using only additions, subtractions, multiplications, divisions, and roots.

Despite his young age, Galois was one of the most influential mathematicians of the nineteenth century. His scientific work includes results on elliptic functions and Abelian integrals. He died at age 20 from wounds received during a pistol duel (Livio, 2005). After his death, his brother, Alfred, and a friend collected all of Galois' writings and, eventually, another influential French mathematician, Joseph Liouville, published them in 1846. From then on, group theory became an important mathematical field, with numerous applications in symmetry, physics, biology, language, music, and so on.

The contributions of **J. F. C. Hessel** (Fig. 9.15(b)) were discussed in the historical section of Chapter 3, on page 75.

9.5 Problems

(i) *Generating relationship*: Express a generating relationship for **622** (D_6).

(ii) *Subgroups*: Determine the operations that are lost in reducing the symmetry from **4/mmm** (D_{4h}) to **422** (D_4).

(iii) *Multiplication table*: Express the multiplication table for **6** (C_6). Is it cyclic?

(iv) *Laue classes*: Show that the point groups **2** (C_2) and **m** (C_s) both become equal to **2/m** (C_{2h}) when they are combined with the inversion operator.

(v) *Polar point groups*: Determine the polar directions for all 10 polar groups described in Section 9.2.10.1 on page 214. (Note: for the lowest symmetry point groups there is no unique polar direction.)

(vi) *Group – subgroup relations*: Show graphically, using stereographic projections, that **mm2** (C_{2v})$\subset$**4mm** (C_{4v}) and **3m** (C_{3v})$\subset$**6mm** (C_{6v}).

(vii) *Point group operations*: Consider the **23** (T) tetrahedral point group:

(a) Derive 3-D matrix representations for the two generators of this group (use Table 9.6 to determine the generators).
(b) Determine the matrices representing the inverses of the generators.
(c) Show that you can represent 2_y and 2_z (i.e., the two-fold rotations around the y and z axes, respectively) by cyclic permutations of the diagonal elements of 2_x.
(d) By repeated operation of the generators, determine the rest of the matrix representations for the elements of the **23** (T) point group.

(viii) *Point group multiplication table I*: Consider the **422** (D_4) dihedral point group:

(a) Express 3-D matrix representations for the two generators of the group (use Table 9.6 to determine the generators).
(b) By repeated operation with these matrices, determine the matrix representation for all the elements of the **422** (D_4) dihedral point group.
(c) Determine the entries in a group a *multiplication table* (*Cayley's square*) for this group by depicting repeated operations or by explicit matrix multiplication.

(ix) *Point group multiplication table II*: Consider the **4/m** (C_{4h}) point group.

(a) Express 3-D matrix representations for the two generators of this group.
(b) By repeated operation with these two matrices, determine matrix representation for all elements of the group.
(c) Determine entries in a group *multiplication table* for the group using a stereographic projection and explicit matrix multiplication.

CHAPTER

10 Plane groups and space groups

> *"The presentation of mathematics in schools should be psychological and not systematic. The teacher should be a diplomat. He must take account of the psychic processes in the boy in order to grip his interest, and he will succeed only if he presents things in a form intuitively comprehensible. A more abstract presentation is only possible in the upper classes."*
>
> Felix Klein, quoted in D. MacHale, *Comic Sections* (Dublin, 1993)

10.1 Introduction

In the previous chapter, we derived the 32 point group symmetries that are compatible with the translational symmetry of the 14 Bravais lattices. Now we can ask the following question: *what happens when we place a molecule (or a motif) with a certain point group symmetry $\mathcal{G}$ on each lattice point of a certain Bravais lattice $\mathcal{T}$?* To fully answer this question, we would need to take every point group that belongs to a given crystal system and combine it with the translational symmetries of each of the Bravais lattices belonging to the same crystal system. For instance, for the cubic point group $\mathbf{m\bar{3}m}$ (O_h), we would need to combine it with the three cubic Bravais lattices cP, cI, and cF. For each of these combinations, we would need to ask the question: is this a new symmetry group? Furthermore, for each combination we would need to replace each mirror plane by all possible compatible glide planes, and each rotation axis by all possible screw axes and again ask the question: is this a new symmetry group?

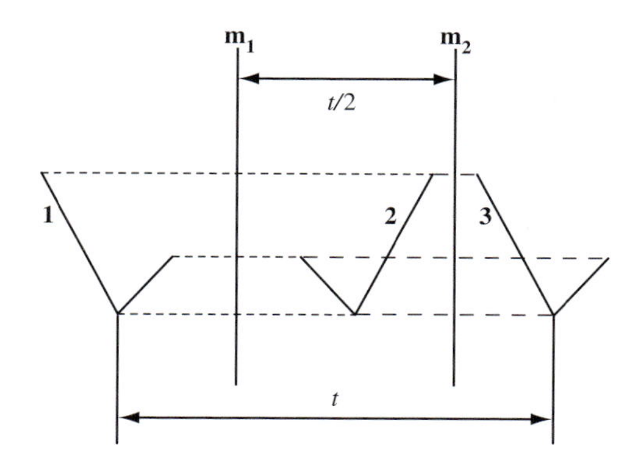

Fig. 10.1. Illustration of the equivalence of two parallel mirror planes at a distance $t/2$ from each other to a lattice translation by a distance t.

The reader might wonder where all these new symmetry elements come from? Consider a simple example. If we have two parallel mirror planes, $\mathbf{m}_1$ and $\mathbf{m}_2$, separated by a distance $t/2$, as shown in Fig. 10.1, then a consecutive mirror operation in both mirrors is seen easily to be equivalent to a translation over a distance t. Object 1 is mirrored into object 2 by the first mirror, and then again mirrored into object 3 by the second mirror. The distance between objects 1 and 3 is equal to twice the distance between the mirror planes. Conversely, since the combination of any two symmetry operations must again be a symmetry operation, we also find that a mirror plane combined with a lattice translation perpendicular to that mirror plane *must* give rise to a new mirror plane at a distance half that of the translation length. This simple example illustrates that, as we combine symmetry operations from the point groups with lattice translations, we will create many additional symmetry operations. Glide planes and screw axes are amongst these newly generated symmetry elements.

Figuring out which of these combinations gives rise to a new symmetry group is clearly a formidable task. This task was first completed by the end of the nineteenth century by Federov in Russia (Federov, 1891), Barlow in England (Barlow, 1894), and Schönflies in Germany (Schönflies, 1891). They independently concluded that there are "only" 230 distinct symmetries. This is a very important result! It essentially tells us that *every* crystal in nature, and *every* crystalline material we can fabricate, *must* have one of those 230 symmetries. Note that this result was obtained by pure group theoretical reasoning, and that very few, if any, experimental observations were used in its derivation. The 230 resulting symmetry groups are known as the 3-D *space groups*.

Instead of systematically enumerating all 230 space groups, a somewhat lengthy task, we will, in this chapter, present a few examples of space group symmetry in 2-D and 3-D, and explain the international notation for the space groups. Then we will discuss the two types of space groups (symmorphic and non-symmorphic), and we will conclude with a description of the space group entries in the *International Tables for Crystallography*, volume A.

10.2 Plane groups

Since there are only ten 2-D point groups, and five different 2-D Bravais lattices, it should not be too difficult to consider all possible combinations. A *plane group* is the infinite group obtained by combining point group symmetries (and glides) with the translational symmetries of a 2-D lattice. The plane groups are obtained as the union of the translational symmetries of the 2-D lattice with the point group symmetry about the lattice point when decorated with an atomic or molecular basis.

The assignment of the ten point groups to their respective crystal system results in ten plane groups determined by their combination with the primitive Bravais lattice in each system. Two additional plane groups are obtained by combining **m** (C_s) and **mm2** (C_{2v}) with the centered rectangular lattice. The last five plane groups are determined by the addition of glide operations where compatible with the preceding 12 plane groups. This yields a total of **17 plane groups** with their distribution among the 5 2-D Bravais lattices summarized in Table 10.1.

As an example of a procedure to determine the symmetry operations of a plane group, we consider Fig. 10.2. This figure illustrates the decoration of an oblique lattice with objects having the **2** (C_2) 2-D point group symmetry. First we draw a primitive oblique unit cell. Then we consider the **2** (C_2) point group, represented graphically by the ⬮ symbol. This point group has two operations, the identity and a two-fold proper rotation. At every lattice point of the oblique cell, we place a copy of the point group symbol. Then we look for all implied symmetry elements. If we were to take an arbitrary object at a point (x, y) near a lattice point $(0, 0)$, the action of the two-fold rotation operator is to replicate it at the position $(-x, -y)$. A translation along the a-axis (horizontal direction) would then replicate it at the position $(1-x, -y)$. An equivalent operation that takes an object from the position (x, y) to the position $(1-x, -y)$ is a two-fold rotation about the point $(1/2, 0)$. Therefore, we conclude that this plane group has two-fold rotation operators at all cell edge centers. Hence, we draw the ⬮ symbol at each edge center. A similar analysis allows us to conclude that the cell center also has two-fold symmetry, so we draw the ⬮ symbol there as well. This exhausts all of the symmetries of this plane group.

The result of this derivation is the plane group **p2**. Note the nomenclature used for the plane groups: first we use the Bravais lattice centering symbol (**p** for "primitive" in this case), followed by the point group symbol (**2**). Note that the oblique lattice is the only primitive lattice that we can combine with the **2** (C_2) point group. Therefore, it is not necessary to explicitly state the crystal system in the plane group symbol **p2**.

The prior procedure can also be depicted in terms of how a general point is "copied" into equivalent points in the lattice. This is shown in the lower left side of Fig. 10.2. Here an object, represented by an open circle, is placed

Table 10.1. 2-D Bravais lattices, point groups, and the number of plane groups for each Bravais lattice type.

2-D Bravais lattice	Point groups	# Plane groups
primitive oblique	**1** (C_1), **2** (C_2)	2
primitive rectangular	**m** (C_s), **mm2** (C_{2v})	5
centered rectangular	**m** (C_s), **mm2** (C_{2v})	2
primitive square	**4** (C_4), **4mm** (C_{4v})	3
primitive hexagonal	**3** (C_3), **3m** (C_{3v}), **6** (C_6), **6mm** (C_{6v})	5

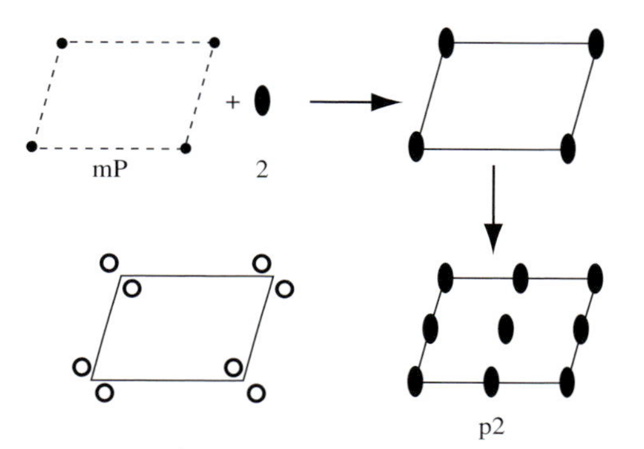

Fig. 10.2. Construction of the **p2** plane group, one of the 17 2-D plane groups from the primitive oblique Bravais lattice and the point group 2 (C_2).

near a lattice point. The action of the two-fold operator replicates this object at a position rotated by an angle π about this point. The pair of objects is then replicated by adding translations to repeat it within the cell and adjacent cells. After this is complete, it is apparent that the resulting collection of objects possesses two-fold symmetries about the cell center and edge centers, in addition to the two-fold symmetry about the vertices of the cell. Note that the orbit of a general point in a plane group is an infinite set of points, due to the fact that the lattice translations continue from $-\infty$ to $+\infty$ in both directions. Note also that there are only two equivalent points per unit cell.

The *site symmetry* of a general point is by definition equal to **1** (C_1). In the plane group **p2**, there are four special positions, namely the locations of the two-fold axes, (0, 0), (0, 1/2), (1/2, 0), and (1/2, 1/2). The site symmetry at each of these locations is, obviously, **2** (C_2). The Wyckoff positions are then $1a$, $1b$, $1c$, and $1d$, respectively for the four locations of the two-fold axes, and $2e$ for the general position (x, y). The *International Tables of Crystallography* (Hahn, 1996) lists all of the general and special points for all 17 plane groups (Chapter 6).

Let us consider another example of the construction of a plane group. Figure 10.3 illustrates the decoration of a square lattice with objects having the **4mm** (C_{4v}) point group symmetry. This results in the **p4mm** plane group, as follows. First, we draw a primitive square unit cell. Then we consider the **4mm** (C_{4v}) point group. To the ◆ symbol for a four-fold rotation axis we

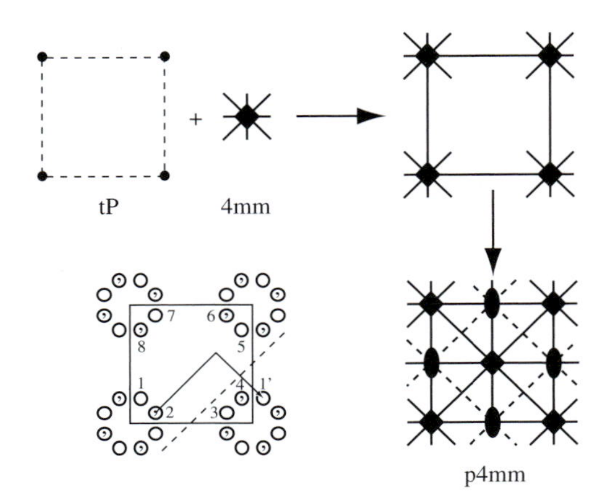

Fig. 10.3. Construction of **p4mm** plane group, one of the 17 2-D plane groups from the *tP* Bravais lattice and the point group **4mm** (C_{4v}).

add the four mirror planes to represent the 2-D point group. This point group has eight operations: the identity, the three four-fold proper rotation operations, and four mirror lines. If we take an arbitrary object at a point (x, y) near the lattice point $(0, 0)$, the action of the four-fold rotation operator is to replicate it at the positions $(-y, x)$, $(-x, -y)$, and $(y, -x)$, respectively. The mirror lines give rise to the positions $(-x, y)$, $(x, -y)$, (y, x), and $(-y, -x)$.

We replicate the point group symbol at every lattice point of the square cell. We align the mirror lines along the cell edges and the cell diagonals. Then we look for all implied symmetry elements. It is clear that the square lattice must have a four-fold axis at the center of each cell, so we draw the ◆ symbol at the cell center. The mirrors along the cell diagonal pass through the center as well. The four-fold axis at the center rotates these mirrors to produce new mirrors which pass through the center of the cell edges. This indicates that the point at the center of the cell also has **4mm** (C_{4v}) point group symmetry. Since each edge has a pair of intersecting mirrors, there must be a two-fold axis at each edge center. The presence of parallel diagonal mirror lines through the origin and the cell center implies the presence of another mirror line centered halfway between the original two mirror lines. This new diagonal mirror line turns out to be a glide line; this is easy to show when we consider the drawing in the lower left-hand corner of Fig. 10.3. Here an object, represented by an open circle, is placed near a lattice point. The action of the four-fold operator replicates this object at positions rotated by $\pi/2$, π, and $3\pi/2$ about this point. The mirror lines replicate each of the four right-handed objects into left-handed objects, represented by an open circle with a comma in the center. The eight objects are then replicated by adding translations to repeat them within the cell and adjacent cells. We see that there are eight equivalent points inside the unit cell. There must be precisely one symmetry element connecting each pair of two equivalent points. For instance, point 1 is converted into point 3 by the four-fold rotation at the center of the cell, and points 5 and 7 through the repeated application of the four-fold rotation. Point 1 can be mapped onto point 1′ by a regular lattice translation along the horizontal direction. We can also reach point 1′ starting from point 2, by means of a diagonal glide line going through the points with coordinates (0, 1/2) and (1/2, 1) (dashed line). The hooked arrow indicates the translate–mirror motion of point 2. It is clear from this simple illustration that in an infinite space group, there is precisely one symmetry operation for each pair of points, no matter how many unit cells apart these points are. This completes the derivation of the **p4mm** plane group. Note that once again the crystal system is not mentioned in the plane group symbol, because the point group **4mm** (C_{4v}) can only be combined with the square lattice.

The same procedure can be applied to derive the other 15 plane groups. Without going into all the details, here is how the general derivation works. We start by considering the most general (lowest symmetry) of the 2-D Bravais lattices, the oblique lattice, which is simply a parallelogram with angles between the bases that are not $\pi/2$ or $2\pi/3$. This lattice is compatible only with the **1** (C_1) and **2** (C_2) point groups. Combination of **1** (C_1) with the oblique lattice gives rise to the primitive plane group labeled **p1**. Combination of **2** (C_2) with the oblique lattice gives rise to the plane group **p2** (also primitive) which was derived above. Figure 10.4 shows the crystallographic representation of all 17 plane groups. As in the illustrations above, this consists of (a) decoration of the lattice points with the appropriate point group operations; and (b) inclusion of new symmetry elements implied by the first. For additional illustrations and lists of general and special point positions the reader is referred to Chapter 6 in the *International Tables for Crystallography* (Hahn, 1996).

Combination of a mirror plane with the rectangular lattice gives rise to the primitive plane group, **pm**. Replacing the mirror line by an axial glide line we find a second plane group in the rectangular Bravais lattice: **pg**. Combination

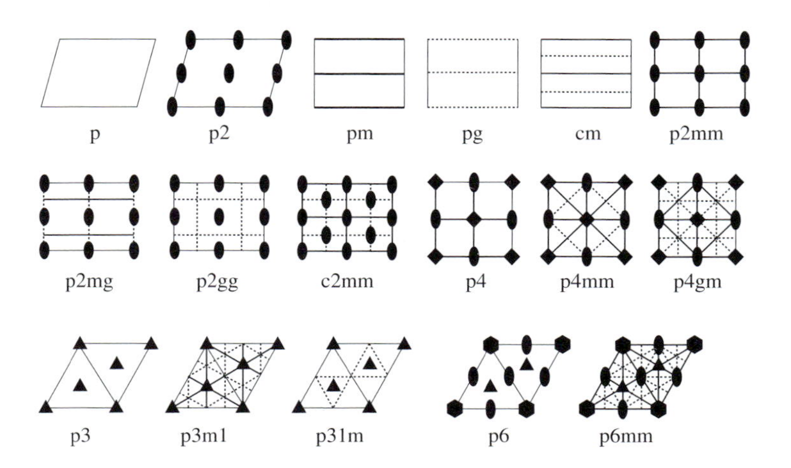

Fig. 10.4. The 17 2-D plane groups: **p1**, **p2**,**pm**,**pg**,**cm**, **p2mm**,**p2mg**, **p2gg**, **c2mm**,**p4**,**p4mm**,**p4gm** **p3**,**p3m1**,**p31m**,**p6**, and **p6mm**.

of a mirror with the centered rectangular lattice also gives rise to a primitive plane group, however, it is again convention to represent this as an equivalent centered rectangular plane group with symbol **cm**. It is also possible to combine two orthogonal mirror planes with the rectangular Bravais lattice giving rise to the **p2mm** plane group. Of course, additional 2-fold axes are required at the intersections between the orthogonal mirror planes. Axial and diagonal glides give rise to the **p2mg** and **p2gg** plane groups, respectively. Diagonal glide lines in combination with the centered rectangular lattice results in the **c2mm** plane group.

The plane group **p4** is obtained by adding the **4** (C_4) point group to the square Bravais lattice. The **p4mm** plane group, derived above, has two orthogonal sets of diagonal glide lines that are implied and therefore not included in the plane group symbol. Addition of new axial and diagonal glide lines to the group **p4** gives rise to the plane group **p4gm**.

Three- and six-fold rotational axes are only compatible with the hexagonal Bravais lattice. The primitive unit cell is represented as a parallelogram (rhombus). In plane groups with three-fold rotational symmetry, the three-fold operation is replicated at all lattice sites. In plane groups with six-fold rotational symmetries, the six-fold operation is replicated at all lattice sites. In both cases, the three-fold axes are replicated also at positions $(1/3, 2/3)$ and $(2/3, 1/3)$ along the cell diagonal.

Combination of the hexagonal lattice and a three-fold axis gives rise to the primitive plane group **p3**. There are two ways of adding mirror planes to the plane group **p3**. The three-fold axis replicates the mirror planes along the edges of an underlying triangular lattice. The plane group **p3m1** has an additional mirror line along the cell diagonal, mirror lines passing through the cell vertices and the midpoint of opposite edges as well as implied glide lines. The **p31m** plane group does not have the additional mirror lines but does have a triangular network of glide lines around the three-fold axes located on the cell diagonal. Combination of the triangular lattice and a six-fold axis gives rise to the primitive plane group **p6**. Addition of a mirror plane to **p6mm** gives rise to the plane group **p6m**. In both cases, two-fold axes on the cell edges are implied.

As a final exercise, we consider the 2-D structure illustrated in Fig. 10.5(a). This is an example of a *Kockel diagram* (Borchardt-Ott, 1995) which can be used to illustrate the symmetry of a particular plane group using a structural motif. Such diagrams can provide interesting exercises to test the comprehension of the subject. In such an exercise, one is typically asked to (1) determine the unit cell (unit mesh); (2) show all symmetry elements in the cell; and (3) determine the 2-D plane group for the structure. The example in Figure 10.5 is a square mesh that has four-fold and two-fold rotation axes, vertical and diagonal mirror lines, and diagonal glide lines. It is, therefore, an example of the **p4mm** plane group illustrated above.

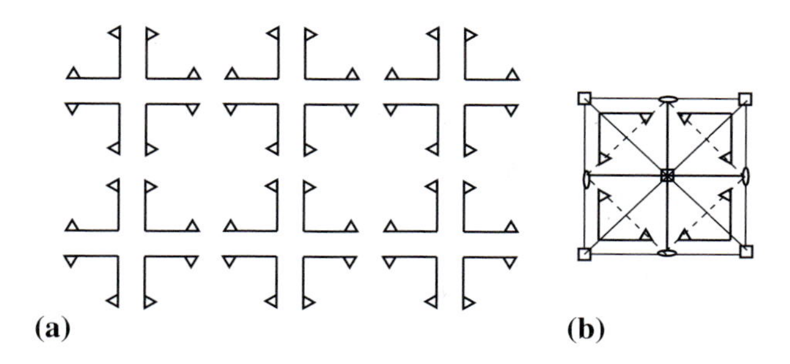

Fig. 10.5. (a) Kockel diagram of a symmetric 2-D structure and (b) its unit mesh and symmetry operations.

10.3 Space groups

Consider the 3-D orthorhombic point group **mm2** (C_{2v}). It has four elements: the identity, a two-fold axis parallel to the z-direction, and two mirror planes m_1 and m_2 at right angles to each other. Since orthorhombic point groups can only be compatible with the orthorhombic Bravais lattices (of which there are four), we need to investigate the combination of these symmetry elements with the various orthorhombic centering operators. Figure 10.6 shows how we can construct a space group based on the point group **mm2** (C_{2v}) and the Bravais lattice oP. At every lattice point of oP, indicated by filled circles on the drawing, we put a copy of the point group drawing. We notice that the mirror planes line up with the unit cell boundaries, so we draw solid lines to indicate that they are indeed mirror planes. We also know that two parallel mirror planes at a distance t give rise to an additional mirror plane at the distance $t/2$. Hence we draw two new mirror planes half-way through the unit cell. Finally, we know that the intersection of two mirror planes is equivalent to a two-fold axis along the intersection line, so we draw the ⬮ symbol at each

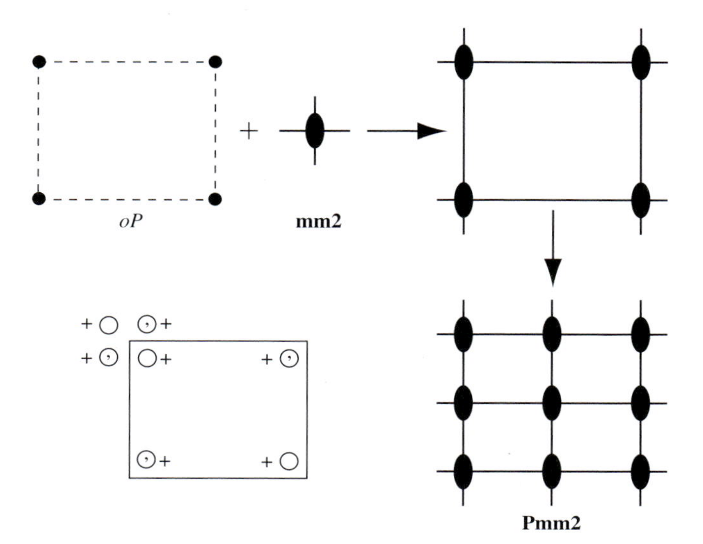

Fig. 10.6. Construction of the space group **Pmm2** (C^1_{2v}) from the Bravais lattice oP and the point group **mm2**(C_{2v}).

intersection. This completes the drawing. This configuration of symmetry elements is of course repeated in every unit cell. The infinite set of symmetry elements thus obtained forms the *space group* **Pmm2** (C^1_{2v}). The space group symbol is formed by combining the centering information of the Bravais lattice with the point group symbol (in International or Hermann–Mauguin notation). It is not necessary to include the o for orthorhombic, since the point group symbol already indicates that the group must have orthorhombic symmetry. Note that the centering symbol is written as an uppercase letter, as opposed to the centering symbols for the 2-D plane groups, which are written as lowercase letters. The Schönflies notation for the space groups will also be shown between parentheses; this symbol is based on the point group Schönflies symbol.

Finally, we can indicate how a general point is "copied" into the equivalent points. This is shown on the lower right corner of the drawing. The open circle with the + sign next to it indicates a general point *above* the plane of the drawing, a minus sign would be below. The symbol $\frac{1}{2}+$ indicates that the point is half a unit cell above the original point, etc. The circles with a comma inside indicate that the point is related to the original point by an improper symmetry operation. In other words, those points are left-handed versions of the original right-handed point. This type of drawing is for space groups what the stereographic representations are for point groups.

The space group **Pmm2** (C^1_{2v}) has four equivalent points inside the unit cell; these points have coordinates:

$$(x, y, z) \quad (\bar{x}, \bar{y}, z) \quad (x, \bar{y}, z) \quad (\bar{x}, y, z).$$

The positions with the highest site symmetry are found along the two-fold rotation axes, located at $(0, 0, z)$, $(0, 1/2, z)$, $(1/2, 0, z)$, and $(1/2, 1/2, z)$. These four locations have site symmetry **mm2**, and are denoted by the Wyckoff positions $1a$, $1b$, $1c$, and $1d$, respectively. In addition, there are four mirror planes in this space group, leading to the following special positions: $(x, 0, z)$, $(x, 1/2, z)$, $(0, y, z)$, and $(1/2, y, z)$. The site symmetry for the first two of these is ·**m**·, and for the other two it is **m**··. The Wyckoff positions are labeled $2e$, $2f$, $2g$, and $2h$, respectively. The general position, (x, y, z), then gets the notation $4i$.

Let us repeat this exercise for the orthorhombic oC lattice, using the same point group symmetry. Figure 10.7 shows the steps in the construction of space group **Cmm2** (C^{11}_{2v}). The centering vector of the oC Bravais lattice copies every point with coordinates (x, y, z) onto an equivalent point $(x + 1/2, y + 1/2, z)$. Looking at the equivalent sites inside the unit cell (lower left of Fig. 10.7), we find that there are eight equivalent positions. If we take any pair of points, there must be a symmetry operator that converts one into the other. For instance, point 1 is converted into point 2 by means of the two-fold rotation axis located at position $(1/4, 1/4, 0)$; 2 goes to 7 by means

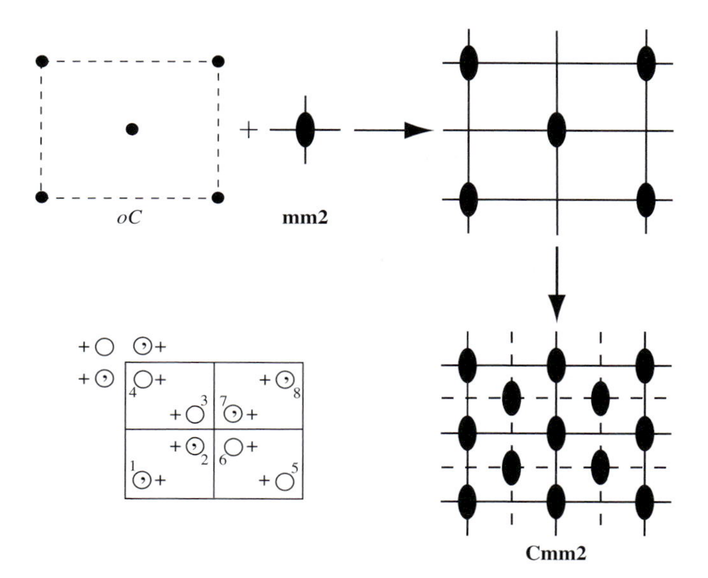

Fig. 10.7. Construction of the space group **Cmm2** (C_{2v}^{11}) from the Bravais lattice oC and the point group **mm2** (C_{2v}).

of the two-fold rotation axis at the center of the cell, as do 4 to 5, 1 to 8, and 3 to 6. Which symmetry operator brings 1 onto 3? To answer this question we start from the presence of two parallel horizontal mirror planes (top right of the figure) going through the points $(0, 0, 0)$ and $(1/2, 0, 0)$. Due to the translational symmetry, there must be a third parallel mirror plane half-way in between them, going through the point $(1/4, 0, 0)$. From the positions of the equivalent points, we see that this mirror plane is not just a simple mirror plane, but instead a glide plane with a translation vector $(0, 1/2, 0)$. This glide plane takes point 1, translates it to the position $(x, y + 1/2, z)$, and then mirrors it into position 3, with coordinates $(1/2 - x, y + 1/2, z)$.

The full set of eight equivalent points has the following coordinates:

$$
\begin{array}{cccc}
 & (0,0,0)+ & (\frac{1}{2},\frac{1}{2},0)+ & \\
(x,y,z) & (\bar{x},\bar{y},z) & (x,\bar{y},z) & (\bar{x},y,z).
\end{array}
$$

The first line contains the zero-vector and the centering vector **C**. To find all the equivalent point coordinates we must add these vectors (hence the + sign) to all four coordinates on the second row. Those coordinates are found by matrix multiplication of the symmetry matrices of the point group **mm2** (C_{2v}) with the general position vector (x, y, z). The *multiplicity* of a general point is equal to the number of equivalent positions for a general point inside a single unit cell. For this space group, **Cmm2** (C_{2v}^{11}), the multiplicity is 8, whereas it is 4 for **Pmm2** (C_{2v}^{1}). Note that the multiplicity for a space group is similar to the *order* for a point group; since all space groups are of infinite order, there is no point in using the order of the space group to say something about the number of equivalent points inside a single unit cell.

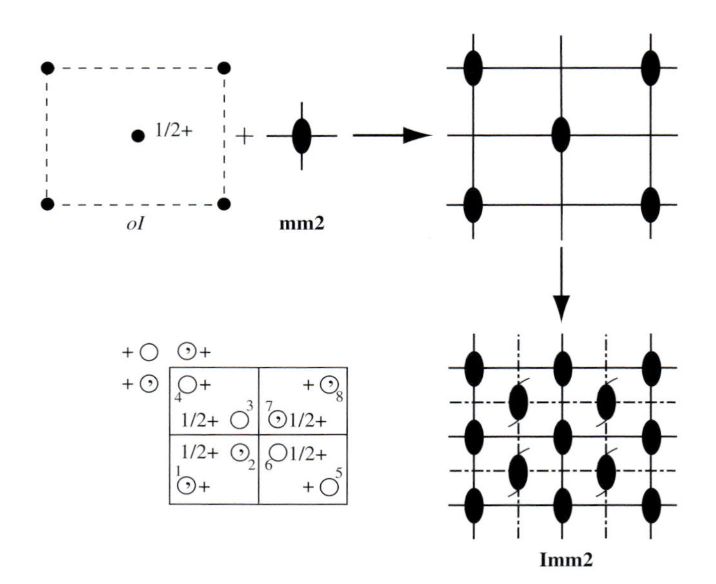

Fig. 10.8. Construction of the space group **Imm2** (C_{2v}^{20}) from the Bravais lattice *oI* and the point group **mm2** (C_{2v}).

As a third example we will combine the Bravais lattice *oI* with the point group **mm2** (C_{2v}). The construction is shown in Fig. 10.8 and results in the space group **Imm2** (C_{2v}^{20}). We must remember that the point group symbol is copied at every lattice site, but now the central lattice site is at a position 1/2 *above* the plane of the drawing. The two-fold axis which for the space group **Cmm2** (C_{2v}^{11}) is generated at the position (1/4, 1/4, 0) now becomes a two-fold screw axis, indicated by the symbol ⌠. This is easy to understand when we consider how to go from point 1 to point 2: we rotate 180° around the point (1/4, 1/4, 0), and then we must translate by (0, 0, 1/2) along the axis to obtain point 2. This combined operation is a 2_1 screw axis. A similar analysis for the points 1 and 3 reveals that the glide plane must be a diagonal glide rather than the axial glide that we found for space group **Cmm2** (C_{2v}^{11}). The dash-dotted lines represent the four *diagonal* glide planes in the unit cell. The multiplicity of a general point is again 8, but the coordinates are now given by:

$$(0, 0, 0)+ \quad (\tfrac{1}{2}, \tfrac{1}{2}, \tfrac{1}{2})+$$
$$(x, y, z) \quad (\bar{x}, \bar{y}, z) \quad (x, \bar{y}, z) \quad (\bar{x}, y, z).$$

We leave it to the reader to determine the special positions for both **Cmm2** (C_{2v}^{11}) and **Imm2** (C_{2v}^{20}) space groups.

We have found that the space group **Cmm2** (C_{2v}^{11}) contains axial glide planes amongst its elements, and that **Imm2** (C_{2v}^{20}) has two-fold screw axes and diagonal glide planes. Since those symmetry elements are the result of the combinations of other, "regular" symmetry elements, we do not need to include the symbols for the axial glides or the screw-rotation symbol in the space group symbol. If a space group symbol does not contain any screw-rotation symbols or glide plane symbols, then that space group is called *symmorphic*. If the symbol does contain screw-rotation symbols, glide plane symbols, or any combination of both, then that space group is called *non-symmorphic*. In the following sections, we will discuss both classes of space groups in more detail.

10.4 The symmorphic space groups

As mentioned above, a symmorphic space group is a space group with a symbol that does not contain any screw-rotations or glide planes. One can enumerate all the symmorphic space groups by taking the combinations of all Bravais lattices with all point groups of the same crystal class, just as we did in the previous section. In other words, for the lattice aP we have two triclinic point groups, **1** (C_1) and $\bar{\mathbf{1}}$ (C_i); for mP there are three monoclinic point groups, for mC there are also three monoclinic point groups, and so on. This results in the formation of 61 space groups, listed in Table 10.2.[1] Note that each space group has been assigned a sequential number, from 1 to 230. The International notation is shown along with the Schönflies notation; it is clear that the Schœnflies notation is simply a sequential numbering added to the original point group Schönflies symbol. These symbols are not as instructive as the Hermann–Mauguin symbols.

This does not conclude the enumeration of the symmorphic space groups, however. There are 12 additional symmorphic space groups, indicated in Table 10.2 by an underlined space group number. The additional space groups are formed in a number of different ways:

- We have assumed that there is only one possible *relative orientation* between the point group symmetry elements and the Bravais lattice directions. For several point groups there is more than one way to copy the point group onto the lattice. A typical example would be the space groups $\mathbf{P\bar{4}2m}$ (D^1_{2d}) and $\mathbf{P\bar{4}m2}$ (D^5_{2d}). For space group $\mathbf{P\bar{4}2m}$ (D^1_{2d}), the mirror planes are positioned along the diagonals of the tetragonal unit cell, whereas for group $\mathbf{P\bar{4}m2}$ (D^5_{2d}) they are parallel to the faces of the unit cell. This is indicated by reversing the order of the two-fold axis and the mirror plane in the Hermann–Mauguin symbol. A similar situation occurs for space groups $\mathbf{I\bar{4}m2}$ (D^9_{2d}) and $\mathbf{I\bar{4}2m}$ (D^{11}_{2d}), and for $\mathbf{P\bar{6}m2}$ (D^1_{3h}) and $\mathbf{P\bar{6}2m}$ (D^3_{3h}).
- For the orthorhombic point group **mm2** (C_{2v}) we can position the two-fold axis perpendicular to the C-centered plane, which results in **Cmm2** (C^{11}_{2v}), or we can position the two-fold axis parallel to the centered plane, in which case one uses the A-centered unit cell. The latter case gives rise to the symmorphic space group **Amm2** (C^{14}_{2v}).

[1] It is easy to verify that there are 61 groups resulting from the combination of Bravais lattices and point groups: 1×2 (triclinic) $+ 2 \times 3$ (monoclinic) $+ 4 \times 3$ (orthorhombic) $+ 2 \times 7$ (tetragonal) $+ 1 \times 5$ (trigonal) $+ 1 \times 7$ (hexagonal) $+ 3 \times 5$ (cubic) $= 61$.

Table 10.2. The 73 symmorphic space groups, with sequential number, International and Schönflies symbols, and corresponding point group.

S.G.#	PG	Symbol	S.G.#	PG	Symbol
1	**1** (C_1)	**P1** (C_1^1)	143	**3** (C_3)	**P3** (C_3^1)
2	$\bar{\mathbf{1}}$ (C_i)	$\mathbf{P\bar{1}}$ (C_i^1)	146		**R3** (C_3^4)
3	**2** (C_2)	**P2** (C_2^1)	147	$\bar{\mathbf{3}}$ (C_{3i})	$\mathbf{P\bar{3}}$ (C_{3i}^1)
5		**C2** (C_2^3)	148		$\mathbf{R\bar{3}}$ (C_{3i}^2)
6	**m** (C_s)	**Pm** (C_s^1)	149	**32** (D_3)	**P312** (D_3^1)
8		**Cm** (C_s^3)	150		**P321** (D_3^2)
10	**2/m** (C_{2h})	**P2/m** (C_{2h}^1)	155		**R32** (D_3^7)
12		**C2/m** (C_{2h}^3)	156	**3m** (C_{3v})	**P3m1** (C_{3v}^1)
16	**222** (D_2)	**P222** (D_2^1)	157		**P31m** (C_{3v}^2)
21		**C222** (D_2^6)	160		**R3m** (C_{3v}^5)
22		**F222** (D_2^7)	162	$\bar{\mathbf{3}}$**m** (D_{3d})	$\mathbf{P\bar{3}1m}$ (D_{3d}^1)
23		**I222** (D_2^8)	164		$\mathbf{P\bar{3}m1}$ (D_{3d}^3)
25	**mm2** (C_{2v})	**Pmm2** (C_{2v}^1)	166		$\mathbf{R\bar{3}m}$ (D_{3d}^5)
35		**Cmm2** (C_{2v}^{11})	168	**6** (C_6)	**P6** (C_6^1)
38		**Amm2** (C_{2v}^{14})	174	$\bar{\mathbf{6}}$ (C_{3h})	$\mathbf{P\bar{6}}$ (C_{3h}^1)
42		**Fmm2** (C_{2v}^{18})	175	**6/m** (C_{6h})	**P6/m** (C_{6h}^1)
44		**Imm2** (C_{2v}^{20})	177	**622** (D_6)	**P622** (D_6^1)
47	**mmm** (D_{2h})	**Pmmm** (D_{2h}^1)	183	**6mm** (C_{6v})	**P6mm** (C_{6v}^1)
65		**Cmmm** (D_{2h}^{19})	187	$\bar{\mathbf{6}}$**m2** (D_{3h})	$\mathbf{P\bar{6}m2}$ (D_{3h}^1)
69		**Fmmm** (D_{2h}^{23})	189		$\mathbf{P\bar{6}2m}$ (D_{3h}^3)
71		**Immm** (D_{2h}^{25})	191	**6/mmm** (D_{6h})	**P6/mmm** (D_{6h}^1)
75	**4** (C_4)	**P4** (C_4^1)	195	**23** (T)	**P23** (T^1)
79		**I4** (C_4^5)	196		**F23** (T^2)
81	$\bar{\mathbf{4}}$ (S_4)	$\mathbf{P\bar{4}}$ (S_4^1)	197		**I23** (T^3)
82		$\mathbf{I\bar{4}}$ (S_4^2)	200	$\mathbf{m\bar{3}}$(T_h)	$\mathbf{Pm\bar{3}}$ (T_h^1)
83	**4/m** (C_{4h})	**P4/m** (C_{4h}^1)	202		$\mathbf{Fm\bar{3}}$ (T_h^3)
87		**I4/m** (C_{4h}^5)	204		$\mathbf{Im\bar{3}}$ (T_h^5)
89	**422** (D_4)	**P422** (D_4^1)	207	**432** (O)	**P432** (O^1)
97		**I422** (D_4^9)	209		**F432** (O^3)
99	**4mm** (C_{4v})	**P4mm** (C_{4v}^1)	211		**I432** (O^5)
107		**I4mm** (C_{4v}^9)	215	$\bar{\mathbf{4}}$**3m** (T_d)	$\mathbf{P\bar{4}3m}$ (T_d^1)
111	$\bar{\mathbf{4}}$**2m** (D_{2d})	$\mathbf{P\bar{4}2m}$ (D_{2d}^1)	216		$\mathbf{F\bar{4}3m}$ (T_d^2)
115	$\bar{\mathbf{4}}$**m2** (D_{2d})	$\mathbf{P\bar{4}m2}$ (D_{2d}^5)	217		$\mathbf{I\bar{4}3m}$ (T_d^3)
119	$\bar{\mathbf{4}}$**m2** (D_{2d})	$\mathbf{I\bar{4}m2}$ (D_{2d}^9)	221	$\mathbf{m\bar{3}m}$ (O_h)	$\mathbf{Pm\bar{3}m}$ (O_h^1)
121	$\bar{\mathbf{4}}$**2m** (D_{2d})	$\mathbf{I\bar{4}2m}$ (D_{2d}^{11})	225		$\mathbf{Fm\bar{3}m}$ (O_h^5)
123	**4/mmm** (D_{4h})	**P4/mmm** (D_{4h}^1)	229		$\mathbf{Im\bar{3}m}$ (O_h^9)
139		**I4/mmm** (D_{4h}^{17})			

- The trigonal point groups can be combined with the R Bravais lattice which gives rise to the space groups **R3** (C_3^4), $\mathbf{R\bar{3}}$ (C_{3i}^2), **R32** (D_3^7), **R3m** (C_{3v}^5), and $\mathbf{R\bar{3}m}$ (D_{3d}^5). The trigonal point groups can also be combined with a *hexagonal* primitive Bravais lattice, which gives rise to space groups **P312** (D_3^1), **P3m1** (C_{3v}^1), and $\mathbf{P\bar{3}1m}$ (D_{3d}^1) (these are part of the regular 61 space groups obtained by combining the hexagonal Bravais lattice with the various point groups). In addition, one can change the orientation of the point group elements with respect to the hexagonal basis vectors which creates the space groups **P321** (D_3^2), **P31m** (C_{3v}^2), and **P$\bar{3}$m1** (D_{3d}^3).

This concludes the enumeration of the 73 symmorphic space groups.

10.5 The non-symmorphic space groups

The remaining 157 (= 230 − 73) space groups can be derived by systematically replacing one or more of the symmetry elements in the Hermann–Mauguin symbols of the point groups by screw-rotations and/or glide planes. Let us consider an example. We have seen in Fig. 10.6 that the combination of **mm2** (C_{2v}) with *oP* produces the space group **Pmm2** (C_{2v}^1). If we replace the two-fold axis by a two-fold screw axis 2_1, then we can see from the drawing in Fig. 10.9 that a general point is transformed into 7 other points, 4 of them at height +, the other four at height $\frac{1}{2}$+. Comparing this arrangement of points with the points in Fig. 10.6 we find that they are identical. In other words, if we combine the symmetry **mm2$_1$** with the Bravais lattice *oP*, then we find the symmorphic space group **Pmm2** (C_{2v}^1), with a unit cell doubled along the two-fold axis. This is not a new symmetry, so the space group **Pmm2$_1$** is not a new space group.

The next combination we can try is **mc2$_1$**, i.e., we convert the mirror plane perpendicular to the **b**-axis into a *c*-glide plane. This combination results in the construction shown in Fig. 10.10. The presence of the *c*-glide plane removes half of the points that were introduced by the mirror plane in Fig. 10.9. The resulting space group is called **Pmc2$_1$** (C_{2v}^2).

This procedure must be repeated for all possible combinations of screw axes and glide planes in all point groups and Bravais lattices. This is a tedious task and we will only list the resulting non-symmorphic space groups in Table 10.3. Interested readers may wish to consult Buerger (1956) for a detailed and complete derivation of all 230 space groups.

Note that it is straightforward to determine which point group corresponds to a given non-symmorphic space group: simply replace all screw axes by a regular rotation of the same order, and all glide planes by a mirror plane. For instance, consider space group **I4$_1$/amd** (D_{4h}^{19}). If we replace the 4_1 screw axis by a four-fold rotation, and both glides *a* and *d* by mirrors *m*, then we obtain point group **4/mmm** (D_{4h}). If we use the Schönflies notation, we can simply drop the superscript from the space group symbol.

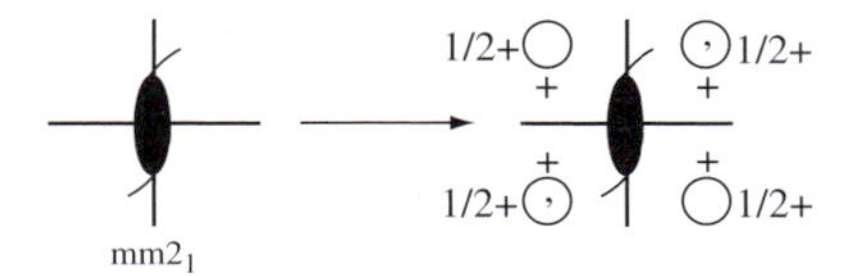

Fig. 10.9. The symmetry combination $mm2_1$ gives rise to 8 equivalent points, which correspond to twice the arrangement of points for the space group **Pmm2** (C_{2v}^1).

Table 10.3. The non-symmorphic space groups, with sequential number and corresponding point group.

S.G.#	PG	Symbol	S.G.#	PG	Symbol
4	**2** (C_2)	**P2$_1$** (C_2^2)	66		**Cccm** (D_{2h}^{20})
7	**m** (C_s)	**Pc** (C_s^2)	67		**Cmma** (D_{2h}^{21})
9		**Cc** (C_s^4)	68		**Ccca** (D_{2h}^{22})
11	**2/m** (C_{2h})	**P2$_1$/m** (C_{2h}^2)	70		**Fddd** (D_{2h}^{24})
13		**P2/c** (C_{2h}^4)	72		**Ibam** (D_{2h}^{26})
14		**P2$_1$/c** (C_{2h}^5)	73		**Ibca** (D_{2h}^{27})
15		**C2/c** (C_{2h}^6)	74		**Imma** (D_{2h}^{28})
17	**222** (D_2)	**P222$_1$** (D_2^2)	76	**4** (C_4)	**P4$_1$** (C_4^2)
18		**P2$_1$2$_1$2** (D_2^3)	77		**P4$_2$** (C_4^3)
19		**P2$_1$2$_1$2$_1$** (D_2^4)	78		**P4$_3$** (C_4^4)
20		**C222$_1$** (D_2^5)	80		**I4$_1$** (C_4^6)
24		**I2$_1$2$_1$2$_1$** (D_2^9)	84	**4/m** (C_{4h})	**P4$_2$/m** (C_{4h}^2)
26	**mm2** (C_{2v})	**Pmc2$_1$** (C_{2v}^2)	85		**P4/n** (C_{4h}^3)
27		**Pcc2** (C_{2v}^3)	86		**P4$_2$/n** (C_{4h}^4)
28		**Pma2** (C_{2v}^4)	88		**I4$_1$/a** (C_{4h}^6)
29		**Pca2$_1$** (C_{2v}^5)	90	**422** (D_4)	**P42$_1$2** (D_4^2)
30		**Pnc2** (C_{2v}^6)	91		**P4$_1$22** (D_4^3)
31		**Pmn2$_1$** (C_{2v}^7)	92		**P4$_1$2$_1$2** (D_4^4)
32		**Pba2** (C_{2v}^8)	93		**P4$_2$22** (D_4^5)
33		**Pna2$_1$** (C_{2v}^9)	94		**P4$_2$2$_1$2** (D_4^6)
34		**Pnn2** (C_{2v}^{10})	95		**P4$_3$22** (D_4^7)
36		**Cmc2$_1$** (C_{2v}^{12})	96		**P4$_3$2$_1$2** (D_4^8)
37		**Ccc2** (C_{2v}^{13})	98		**I4$_1$22** (D_4^{10})
39		**Abm2** (C_{2v}^{15})	100	**4mm** (C_{4v})	**P4bm** (C_{4v}^2)
40		**Ama2** (C_{2v}^{16})	101		**P4$_2$cm** (C_{4v}^3)
41		**Aba2** (C_{2v}^{17})	102		**P4$_2$nm** (C_{4v}^4)
43		**Fdd2** (C_{2v}^{19})	103		**P4cc** (C_{4v}^5)
45		**Iba2** (C_{2v}^{21})	104		**P4nc** (C_{4v}^6)
46		**Ima2** (C_{2v}^{22})	105		**P4$_2$mc** (C_{4v}^7)
48	**mmm** (D_{2h})	**Pnnn** (D_{2h}^2)	106		**P4$_2$bc** (C_{4v}^8)
49		**Pccm** (D_{2h}^3)	108	**4mm** (C_{4v})	**I4cm** (C_{4v}^{10})
50		**Pban** (D_{2h}^4)	109		**I4$_1$md** (C_{4v}^{11})
51		**Pmma** (D_{2h}^5)	110		**I4$_1$cd** (C_{4v}^{12})
52		**Pnna** (D_{2h}^6)	112	**$\bar{4}$2m** (D_{2d})	**P$\bar{4}$2c** (D_{2d}^2)
53		**Pmna** (D_{2h}^7)	113		**P$\bar{4}$2$_1$m** (D_{2d}^3)
54		**Pcca** (D_{2h}^8)	114		**P$\bar{4}$2$_1$c** (D_{2d}^4)
55		**Pbam** (D_{2h}^9)	116		**P$\bar{4}$c2** (D_{2d}^6)
56		**Pccn** (D_{2h}^{10})	117		**P$\bar{4}$b2** (D_{2d}^7)
57		**C2** (C_2^3)7	118		**P$\bar{4}$n2** (D_{2d}^8)
58		**Pnnm** (D_{2h}^{12})	120		**I$\bar{4}$c2** (D_{2d}^{10})
59		**Pmmn** (D_{2h}^{13})	122		**I$\bar{4}$2d** (D_{2d}^{12})
60		**Pbcn** (D_{2h}^{14})	124	**4/mmm** (D_{4h})	**P4/mcc** (D_{4h}^2)
61		**Pbca** (D_{2h}^{15})	125		**P4/nbm** (D_{4h}^3)
62		**Pnma** (D_{2h}^{16})	126		**P4/nnc** (D_{4h}^4)
63		**Cmcm** (D_{2h}^{17})	127		**P4/mbm** (D_{4h}^5)
64		**Cmca** (D_{2h}^{18})	128		**P4/mnc** (D_{4h}^6)

Table 10.3. (cont.).

S.G.#	PG	Symbol	S.G.#	PG	Symbol
129		**P4/nmm** (D_{4h}^{7})	180		**P6$_2$22** (D_6^4)
130		**P4/ncc** (D_{4h}^{8})	181		**P6$_4$22** (D_6^5)
131		**P4$_2$/mmc** (D_{4h}^{9})	182		**P6$_3$22** (D_6^6)
132		**P4$_2$/mcm** (D_{4h}^{10})	184	**6mm** (C_{6v})	**P6cc** (C_{6v}^2)
133		**P4$_2$/nbc** (D_{4h}^{11})	185		**P6$_3$cm** (C_{6v}^3)
134		**P4$_2$/nnm** (D_{4h}^{12})	186		**P6$_3$mc** (C_{6v}^4)
135		**P4$_2$/mbc** (D_{4h}^{13})	188	**$\bar{6}$m2** (D_{3h})	**P$\bar{6}$c2** (D_{3h}^2)
136		**P4$_2$/mnm** (D_{4h}^{14})	190		**P$\bar{6}$2c** (D_{3h}^4)
137		**P4$_2$/nmc** (D_{4h}^{15})	192	**6/mmm** (D_{6h})	**P6/mcc** (D_{6h}^2)
138		**P4$_2$/ncm** (D_{4h}^{16})	193		**P6$_3$/mcm** (D_{6h}^3)
140		**I4/mcm** (D_{4h}^{18})	194		**P6$_3$/mmc** (D_{6h}^4)
141		**I4$_1$/amd** (D_{4h}^{19})	198	**23** (T)	**P2$_1$3** (T^4)
142		**I4$_1$/acd** (D_{4h}^{20})	199		**I2$_1$3** (T^5)
144	**3** (C_3)	**P3$_1$** (C_3^2)	201	**m$\bar{3}$** (T_h)	**Pn$\bar{3}$** (T_h^2)
145		**P3$_2$** (C_3^3)	203		**Fd$\bar{3}$** (T_h^4)
151	**32** (D_3)	**P3$_1$12** (D_3^3)	205		**Pa$\bar{3}$** (T_h^6)
152		**P3$_1$21** (D_3^4)	206		**Ia$\bar{3}$** (T_h^7)
153		**P3$_2$12** (D_3^5)	208	**432** (O)	**P4$_2$32** (O^2)
154		**P3$_2$21** (D_3^6)	210		**F4$_1$32** (O^4)
158	**3m** (C_{3v})	**P3c1** (C_{3v}^3)	212		**P4$_3$32** (O^6)
159		**P31c** (C_{3v}^4)	213		**P4$_1$32** (O^7)
161		**R3c** (C_{3v}^6)	214		**I4$_1$32** (O^8)
163	**$\bar{3}$m** (D_{3d})	**P$\bar{3}$1c** (D_{3d}^2)	218	**$\bar{4}$3m** (T_d)	**P$\bar{4}$3n** (T_d^4)
165		**P$\bar{3}$c1** (D_{3d}^4)	219		**F$\bar{4}$3c** (T_d^5)
167		**R$\bar{3}$c** (D_{3d}^6)	220		**I$\bar{4}$3d** (T_d^6)
169	**6** (C_6)	**P6$_1$** (C_6^2)	222	**m$\bar{3}$m** (O_h)	**Pn$\bar{3}$n** (O_h^2)
170		**P6$_5$** (C_6^3)	223		**Pm$\bar{3}$n** (O_h^3)
171		**P6$_2$** (C_6^4)	224		**Pn$\bar{3}$m** (O_h^4)
172		**P6$_4$** (C_6^5)	226		**Fm$\bar{3}$c** (O_h^6)
173		**P6$_3$** (C_6^6)	227		**Fd$\bar{3}$m** (O_h^7)
176	**6/m** (C_{6h})	**P6$_3$/m** (C_{6h}^2)	228		**Fd$\bar{3}$c** (O_h^8)
178	**622** (D_6)	**P6$_1$22** (D_6^2)	230		**Ia$\bar{3}$d** (O_h^{10})
179		**P6$_5$22** (D_6^3)			

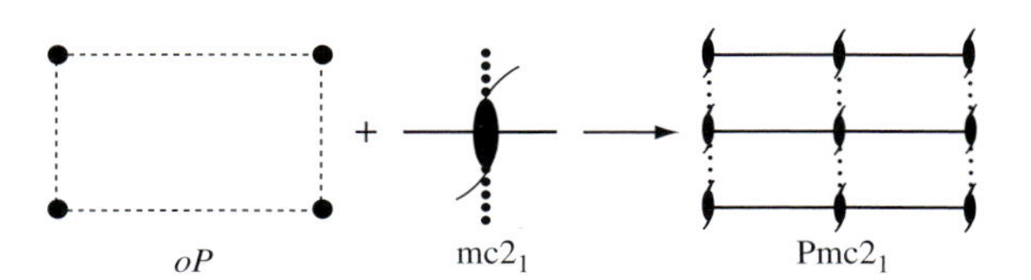

Fig. 10.10. The symmetry combination **mc2$_1$** gives rise to a new non-symmorphic space group **Pmc2$_1$** (C_{2v}^1)

10.6 General remarks

We conclude this section with some remarks about space groups. Thus far we have considered space groups in 3-D and plane groups in 2-D. A more general approach to symmetry theory shows that there are four different types of symmetry groups in 3D:

- $\mathcal{G}_3^3$: the **space** groups, describing periodic 3-D structures (infinite crystals).
- $\mathcal{G}_2^3$: the **layer** groups, describing the symmetries of objects, infinitely extending in two directions but finite in the third direction (one plane is invariant).
- $\mathcal{G}_1^3$: the **rod** groups, describing the symmetries of line-type objects, infinite in one direction (one line remains invariant).
- $\mathcal{G}_0^3$: the **point** groups, describing the symmetries of finite figures in 3-D space (only one point remains invariant).

In this text, only $\mathcal{G}_0^3$ and $\mathcal{G}_3^3$ have been discussed extensively because they are of central importance for crystallography; the other groups are used in specific research fields. In two dimensions, one distinguishes between three different types of symmetries:

- $\mathcal{G}_2^2$: the 2-D **plane** groups, describing the symmetries of two dimensional planar figures (NOT planes in 3-D space!).
- $\mathcal{G}_1^2$: the 2-D **rod** groups, describing the symmetries of line figures in a plane.
- $\mathcal{G}_0^2$: the 2-D **point** groups, describing the symmetries of finite figures in a plane.

It is interesting to look at the number of possible groups as a function of the dimension of the space; in the following table, the number of possible groups is given for one, two, and three dimensions. In four dimensions, there are 4250 space groups $\mathcal{G}_4^4$; all of these are known and tabulated!

	3	2	1	0
3	230	80	75	32
2	—	17	7	10
1	—	—	2	2 .

In the remainder of this book, we will use space groups to describe important crystal structures. Before doing so, it is instructive to discuss the information that is tabulated in the *International Tables for Crystallography, Volume A* (Hahn, 1996). This book presents, among many other things, a complete listing of all 230 space groups. Two typical space group descriptions are shown on the following pages for the symmorphic space group **Cmm2** (C_{2v}^{11}) (Fig. 10.11), and for the non-symmorphic space group **Pmna** (D_{2h}^{7}) (Fig. 10.12).

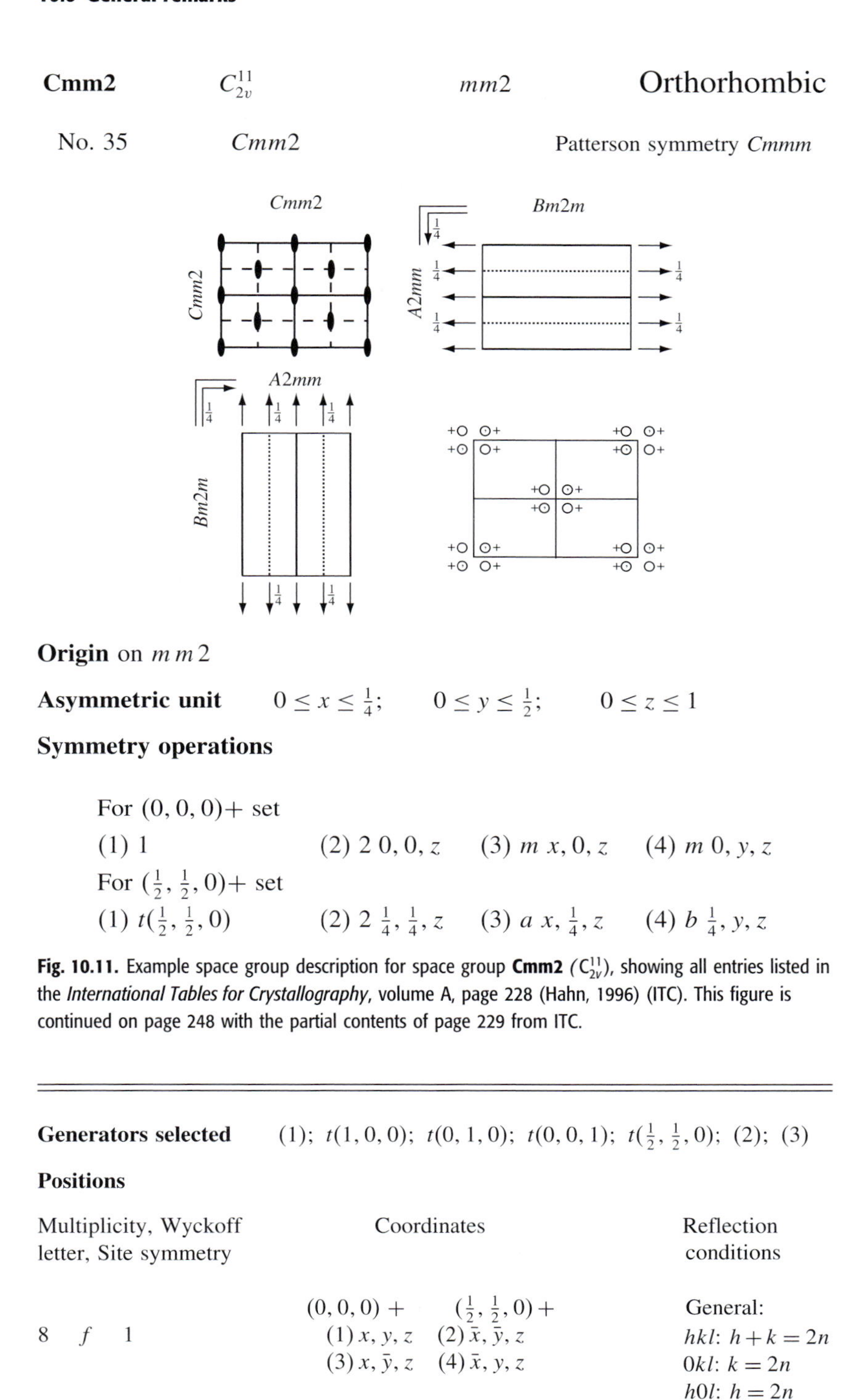

Cmm2 C_{2v}^{11} $mm2$ Orthorhombic

No. 35 $Cmm2$ Patterson symmetry $Cmmm$

Origin on $m\,m\,2$

Asymmetric unit $0 \le x \le \frac{1}{4}$; $0 \le y \le \frac{1}{2}$; $0 \le z \le 1$

Symmetry operations

For $(0, 0, 0)+$ set

(1) 1 (2) $2\ 0, 0, z$ (3) $m\ x, 0, z$ (4) $m\ 0, y, z$

For $(\frac{1}{2}, \frac{1}{2}, 0)+$ set

(1) $t(\frac{1}{2}, \frac{1}{2}, 0)$ (2) $2\ \frac{1}{4}, \frac{1}{4}, z$ (3) $a\ x, \frac{1}{4}, z$ (4) $b\ \frac{1}{4}, y, z$

Fig. 10.11. Example space group description for space group **Cmm2** (C_{2v}^{11}), showing all entries listed in the *International Tables for Crystallography*, volume A, page 228 (Hahn, 1996) (ITC). This figure is continued on page 248 with the partial contents of page 229 from ITC.

Generators selected (1); $t(1,0,0)$; $t(0,1,0)$; $t(0,0,1)$; $t(\frac{1}{2}, \frac{1}{2}, 0)$; (2); (3)

Positions

Multiplicity, Wyckoff letter, Site symmetry	Coordinates	Reflection conditions
	$(0,0,0)+$ $(\frac{1}{2}, \frac{1}{2}, 0)+$	General:
8 f 1	(1) x, y, z (2) $\bar{x}, \bar{y}, z$	hkl: $h+k=2n$
	(3) $x, \bar{y}, z$ (4) $\bar{x}, y, z$	$0kl$: $k=2n$
		$h0l$: $h=2n$
		$hk0$: $h+k=2n$
		$h00$: $h=2n$
		$0k0$: $k=2n$

					Special: as above, plus
4	*e*	$m..$	$0, y, z$	$0, \bar{y}, z$	no extra conditions
4	*d*	$.m.$	$x, 0, z$	$\bar{x}, 0, z$	no extra conditions
4	*c*	$..2$	$\frac{1}{4}, \frac{1}{4}, z$	$\frac{1}{4}, \frac{3}{4}, z$	hkl: $h = 2n$
2	*b*	$mm2$	$0, \frac{1}{2} z$		no extra conditions
2	*a*	$mm2$	$0, 0, z$		no extra conditions

Symmetry of special projections

Along [001] $p2mm$	Along [100] $p2gm$	Along [010] $c2mm$
$\mathbf{a}' = \frac{1}{2}\mathbf{a}$ $\mathbf{b}' = \mathbf{b}$	$\mathbf{a}' = \mathbf{b}$ $\mathbf{b}' = \mathbf{c}$	$\mathbf{a}' = \mathbf{c}$ $\mathbf{b}' = \mathbf{a}$
Origin at $0, 0, z$	Origin at $x, 0, 0$	Origin at $0, y, 0$

Pmna D_{2h}^{7} mmm Orthorhombic

No. 53 $P\,2/m\,2/n\,2_1/a$ Patterson symmetry $Pmmm$

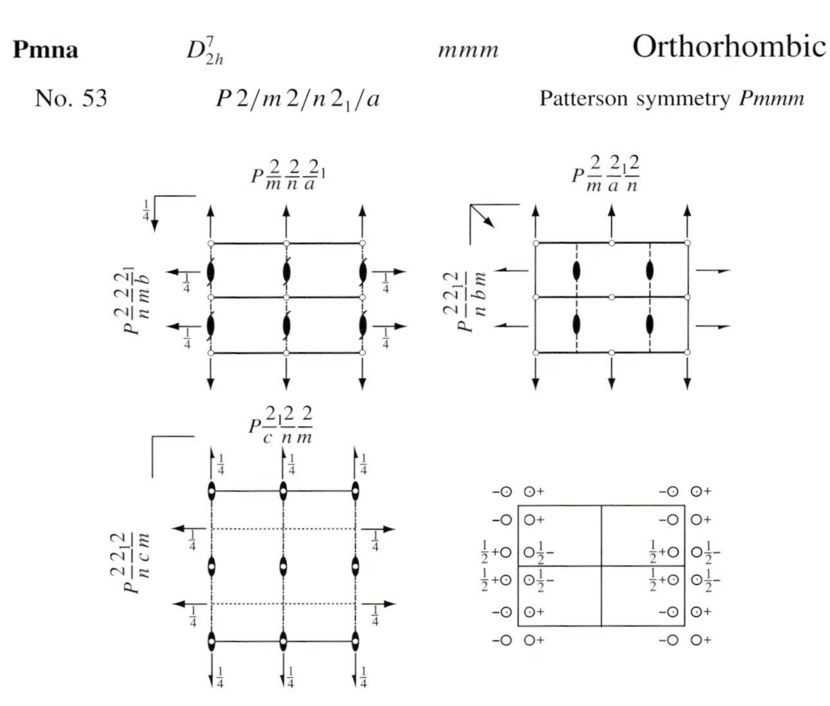

Origin at center $(2/m)$ at $2/m\ n\ 1$

Asymmetric unit $0 \le x \le \frac{1}{2}$; $0 \le y \le 1$; $0 \le z \le \frac{1}{4}$

Symmetry operations

(1) 1	(2) $2(0, 0, \frac{1}{2})\ \frac{1}{4}, 0, z$	(3) $2\ \frac{1}{4}, y, \frac{1}{4}$	(4) $2\ x, 0, 0$
(5) $\bar{1}\ 0, 0, 0$	(6) $a\ x, y, \frac{1}{4}$	(7) $n(\frac{1}{2}, 0, \frac{1}{2})\ x, 0, z$	(8) $m\ 0, y, z$

Fig. 10.12. Example space group description for space group **Pmna** (D_{2h}^{7}), showing all entries listed in the *International Tables for Crystallography*, volume A, page 268 (Hahn, 1996). This figure is continued on page 249 with the partial contents of page 269 from ITC.

Generators selected (1); $t(1,0,0)$; $t(0,1,0)$; $t(0,0,1)$; (2); (3); (5)

Positions

Multiplicity, Wyckoff letter, Site symmetry			Coordinates				Reflection conditions
							General:
8	i	1	(1) x, y, z	(2) $\bar{x}+\frac{1}{2}, \bar{y}, z+\frac{1}{2}$	(3) $\bar{x}+\frac{1}{2}, y, \bar{z}+\frac{1}{2}$		$h0l$: $h+l=2n$
			(4) $x, \bar{y}, \bar{z}$	(5) $\bar{x}, \bar{y}, \bar{z}$	(6) $x+\frac{1}{2}, y, \bar{z}+\frac{1}{2}$		$hk0$: $h=2n$
			(7) $x+\frac{1}{2}, \bar{y}, z+\frac{1}{2}$	(8) $\bar{x}, y, z$			$h00$: $h=2n$
							$00l$: $l=2n$
							Special: as above, plus no extra conditions
4	h	$m\cdot\cdot$	$0, y, z$	$\frac{1}{2}, \bar{y}, z+\frac{1}{2}$	$\frac{1}{2}, y, \bar{z}+\frac{1}{2}$	$0, \bar{y}, \bar{z}$	
4	g	$\cdot 2 \cdot$	$\frac{1}{4}, y, \frac{1}{4}$	$\frac{1}{4}, \bar{y}, \frac{3}{4}$	$\frac{3}{4}, \bar{y}, \frac{3}{4}$	$\frac{3}{4}, y, \frac{1}{4}$	hkl: $h=2n$
4	f	$2\cdot\cdot$	$x, \frac{1}{2}, 0$	$\bar{x}+\frac{1}{2}, \frac{1}{2}, \frac{1}{2}$	$\bar{x}, \frac{1}{2}, 0$	$x+\frac{1}{2}, \frac{1}{2}, \frac{1}{2}$	hkl: $h+l=2n$
4	e	$2\cdot\cdot$	$x, 0, 0$	$\bar{x}+\frac{1}{2}, 0, \frac{1}{2}$	$\bar{x}, 0, 0$	$x+\frac{1}{2}, 0, \frac{1}{2}$	hkl: $h+l=2n$
2	d	$2/m\cdot\cdot$	$0, \frac{1}{2}, 0$	$\frac{1}{2}. \frac{1}{2}, \frac{1}{2}$			hkl: $h+l=2n$
2	c	$2/m\cdot\cdot$	$\frac{1}{2}, \frac{1}{1}, 0$	$0, \frac{1}{2}, \frac{1}{2}$			hkl: $h+l=2n$
2	b	$2/m\cdot\cdot$	$\frac{1}{2}, 0, 0$	$0, 0, \frac{1}{2}$			hkl: $h+l=2n$
2	a	$2/m\cdot\cdot$	$0, 0, 0$	$\frac{1}{2}, 0, \frac{1}{2}$			hkl: $h+l=2n$

Symmetry of special projections

Along [001] $p2mm$	Along [100] $p2gm$	Along [010] $c2mm$
$\mathbf{a}'=\frac{1}{2}\mathbf{a}$ $\mathbf{b}'=\mathbf{b}$	$\mathbf{a}'=\mathbf{b}$ $\mathbf{b}'=\mathbf{c}$	$\mathbf{a}'=\mathbf{c}$ $\mathbf{b}'=\mathbf{a}$
Origin at $0, 0, z$	Origin at $x, 0, 0$	Origin at $0, y, 0$

The space group description consists of the following items (referring to Figs. 10.11 and 10.12):

- At the top of the page, we find the shorthand space group symbol **Cmm2**, the space group number (35), the Schönflies symbol C_{2v}^{11}, the complete space group symbol **Cmm2**, the point group symmetry **mm2** (C_{2v}), the crystal system (*orthorhombic*), and the Patterson symmetry which will be defined in Chapter 11. For space group **Pmna** (D_{2h}^{7}), the full space group symbol is given by $P\frac{2}{m}\frac{2}{n}\frac{2_1}{a}$.
- Next, we have a set of drawings showing the relative positions of all symmetry elements, projected along the three main directions of the orthorhombic reference frame. The drawing in the lower right corner indicates the equivalent positions with a notation similar to the one introduced earlier in this chapter. Each symmetry operator has a specific graphical symbol. We have introduced many of them in Chapter 8; for a complete listing we refer the reader to Chapter 1 in ITC-A (Hahn, 1996).

  Note that the drawings for **Pmna** (D_{2h}^{7}) are significantly more complicated than those for **Cmm2** (C_{2v}^{11}). The main reason for this is the presence of multiple glide planes. The arrows with only a "half head" represent two-fold screw axes that lie in the plane of the drawing; the number next to the symbol indicates the height of the screw axes above the plane of the drawing.

- For space group **Cmm2** (C_{2v}^{11}), the origin of the reference frame has point symmetry **mm2** (C_{2v}), i.e., the origin is taken at the intersection of two mirror planes and a two-fold axis. For **Pmna** (D_{2h}^{7}), the origin is taken at a point with symmetry **2/m** (C_{2h}), where a two-fold axis (lying in the plane of the figure at upper right) intersects a mirror plane. Note that this point also coincides with an inversion operator. For many space groups, the International Tables list two possible origin choices; in those cases, the International Tables show all the information in Fig. 10.11 for both origin choices.
- The *asymmetric unit* is the smallest part of space from which, by application of all symmetry operations, the whole of space is filled exactly. In other words, it is that portion of the unit cell that will completely fill all of space when it is copied by the various symmetry operators. Recall that space groups are infinite groups, by virtue of the lattice translations. The volume of the asymmetric unit is equal to the volume of the unit cell divided by the product of the order, n, of the point group corresponding to the space group, and one plus the number of centering operations. For **Cmm2** (C_{2v}^{11}), we find $n = 4$ and there is a C-centering operation, so that the the volume, V_a, of the asymmetric unit equals $V_a = V/(4 \times (1+1)) = V/8$. For **Pmna** ($D_{2h}^{7}$), we have $V_a = V/(8 \times 1) = V/8$.
- A complete list of all *symmetry operations*, grouped by centering vector (if any). In the case of **Cmm2** (C_{2v}^{11}), there is a C-centering operation, whereas **Pmna** (D_{2h}^{7}) is a primitive space group. The notation for the symmetry elements includes the location of each of the elements, and a more complete explanation can be found in Chapter 11 of ITC-A (Hahn, 1996). There are eight symmetry operators in space group **Pmna** (D_{2h}^{7}); in addition to the identity, we have a two-fold screw axis, two two-fold rotation axes along the two directions normal to the screw axis, an inversion center at the origin, an a glide plane, a diagonal glide plane, and a regular mirror plane.

For each space group, a similar display can be found in ITC-A. For tetragonal, hexagonal, and rhombohedral symmetry, the Tables show only one graphical representation of the symmetry operators, projected along the [001], [00.1] and [111] directions, respectively. For many of the cubic space groups, only one quadrant of the unit cell is shown, because of the large number of symmetry operators. The highest possible number of symmetry operations in a space group, not counting the Bravais lattice translation vectors, but including the centering vectors, is 192; this corresponds to the combination of the highest order point group, **m$\bar{3}$m** (O_h), with 48 elements, and the face-centered cubic Bravais lattice, which has three centering operations. The resulting space group is **Fm$\bar{3}$m** (O_h^5).

On the second page of each space group description, we find additional information. The most important portion is reproduced in the second part of Figs. 10.11 and 10.12.

- The generators are the symmetry elements from which all others can be generated by matrix multiplication. Note that the basis vectors are always among the generators, as are the centering vectors. For **Cmm2** (C_{2v}^{11}), the entire space group can be created if the three basis vectors, the centering vector, and the two-fold axis and one of the mirror planes are provided. Every other symmetry operator of the infinite space group can be created by an appropriate matrix multiplication (using the 4×4 matrices introduced in Chapter 8).
- The *positions* table indicates the most general position and all its equivalent positions. The most general position is always indicated by x, y, z. Application of a symmetry element with number (j) from the first page of the space group description then results in the position (j) x', y', z', where the primes indicate some combination of the coordinates x, y, and z. The first set of coordinates refer to the most general point. The number at the beginning of the top line of the table gives the multiplicity of the general position. The third symbol on the first line indicates the site symmetry of the general point, which is always equal to **1** (C_1). The remaining entries in the table indicate *special positions*, for which the point lies on one or more of the symmetry elements. For instance, the second entry for space group **Cmm2** (C_{2v}^{11}) indicates a point of the type $0, y, z$, which lies in the mirror plane formed by the **a** and **b** vectors. Because of this, there are only half as many equivalent points, which results in a multiplicity of $8/2 = 4$, as shown in the first column. The *site symmetry* for this position is then equal to **m**··. One can progressively move the special point to the intersection of two or more symmetry elements, which further reduces the multiplicity. The highest symmetry is obtained for points along the **c** axis, for which the site symmetry is **mm2** (C_{2v}). The third column of the table indicates the site symmetry of the general (always **1** (C_1)) and special positions. Note that these site symmetries will always be subgroups of the point group corresponding to the space group. It is customary to refer to the special positions in the following way: one provides a letter as a label for each entry in the table, starting with the last entry (highest point symmetry). One refers to a special site by the combination of the multiplicity and the letter, as in $8f$, or $4d$. The letter is known as the *Wyckoff* letter, and the symbol $16j$ is the *Wyckoff symbol*. A crystal structure may then be described by listing the Wyckoff symbols for all the atoms in the asymmetric unit. The last column in the *Positions* table contains diffraction information, and this will be discussed in more detail in Chapter 11.
- The next section shows the *symmetry of special projections*. If the unit cell is projected along the [001] direction, then the resulting 2-D figure will have the plane group symmetry **p 2m m** for both space groups **Cmm2**

(C_{2v}^{11}) and **Pmna** (D_{2h}^{7}). The 2-D unit cell will have lattice parameters $\mathbf{a}'$ and $\mathbf{b}'$ that are, in general, fractions of linear combinations of the original basis vectors. The location of the origin of the 2-D unit cell is also specified.

- There are several additional entries in the space group descriptions in ITC-A; these are related to subgroup–supergroup relations, but we will not need them for this text. The interested reader is referred to Section 2.15 in ITC-A (Hahn, 1996).

With the exception of the last column under the heading *Positions*, the reader should now be able to understand the basic space group information listed in the *International Tables for Crystallography*.

10.7 *Space group generators

The section on space groups in ITC-A is more than 600 pages long, and contains a lot of information. The true power of group theory can be shown by considering what it would take to encode all space groups in a computer program. The naïve way would be to key in all special positions for each space group. This would be a tremendous task, and there would be a lot of opportunities for mistakes. Group theory provides us with a much easier way of accomplishing this task. We know from the point group examples in Section 9.2.10.8 that we can generate all the symmetry matrices of a point group starting from a list of generators. This list can be rather short for each point group, assuming that we make use of a predefined set of 14 symmetry matrices, listed in Table 9.5. Since we must also allow for translational symmetry elements in space groups, we introduce another set of symbols representing the fractions that need to be used to complete the generators elements. These symbols are shown in Table 10.4.

Since a symmetry operator is represented in matrix form by a 4×4 matrix, we must combine the generator matrices $\mathsf{D}^{(x)}$ with the components of the translations to fully identify the space group generators. We should note that, to create the entire (infinite) space group, we must include the Bravais translation vectors as generators. We will ignore these generators in this section, since it is clear that we can go from one unit cell to another one by means of translations.

It is surprising that it is possible to compile all the information on the generators of all 230 space groups in a short ASCII file that is only 4104 bytes long! This is how it works: for each space group, we create a string of characters. The first character is either a 1 or a 0, depending on whether or not there is an inversion operator present in the group. The second character indicates how many generators there are, in addition to the inversion operator

Table 10.4. Encoding scheme for the components of translation vectors.

$A = \frac{1}{6}$	$B = \frac{1}{4}$	$C = \frac{1}{3}$	$D = \frac{1}{2}$	$E = \frac{2}{3}$	$F = \frac{3}{4}$
$G = \frac{5}{6}$	$O = 0$	$X = -\frac{3}{8}$	$Y = -\frac{1}{4}$	$Z = -\frac{1}{8}$	

(if present). As an example, consider the space group **Cmm2** (C_{2v}^{11}) that we encountered in the previous section. Its generator string is given by:

$$03aDDObOOOjOOO0.$$

The first character is a 0, indicating that the inversion operator is not a generator. Then, the 3 indicates that there are three generator matrices. Each generator is described by four characters: the first one determines which of the 14 standard matrices is to be used (from Table 9.5). The subsequent three characters indicate the three translation components (from Table 10.4). So, for **Cmm2** (C_{2v}^{11}) the generators are $aDDO$, $bOOO$ and $jOOO$. The first generator is represented by matrix $\mathsf{D}^{(a)}$, and has translation components $(DDO)= (1/2, 1/2, 0)$. This corresponds to the 4×4 matrix:

$$\mathcal{W} = \begin{pmatrix} 1 & 0 & 0 & \frac{1}{2} \\ 0 & 1 & 0 & \frac{1}{2} \\ 0 & 0 & 1 & 0 \\ 0 & 0 & 0 & 1 \end{pmatrix}.$$

We recognize this matrix as the C-centering translation. The next generator is $bOOO$, which does not have a translational component. The corresponding 4×4 matrix is:

$$\mathcal{W} = \begin{pmatrix} -1 & 0 & 0 & 0 \\ 0 & -1 & 0 & 0 \\ 0 & 0 & 1 & 0 \\ 0 & 0 & 0 & 1 \end{pmatrix},$$

corresponding to a two-fold rotation around the z-axis. And, finally, the third generator is given by $jOOO$, which corresponds to:

$$\mathcal{W} = \begin{pmatrix} 1 & 0 & 0 & 0 \\ 0 & -1 & 0 & 0 \\ 0 & 0 & 1 & 0 \\ 0 & 0 & 0 & 1 \end{pmatrix},$$

which represents a mirror plane normal to the y-direction.

The last symbol in the generator string is a 0, which indicates that for this space group there is no alternative choice for the origin. If there were a second origin choice listed in ITC-A, then this character would be a 1, and it would be followed by three more characters indicating the location of the second origin choice with respect to the first (i.e., a translation). An example of this can be found for space group **P4/nbm** (D^3_{4h}) with generator string:

$$04bOOOgOOOcOOOhDDO1YYO.$$

There is no inversion operator, there are four generators (meaning 4 four-character substrings *bOOO*, *gOOO*, *cOOO*, and *hDDO*), followed by a 1 and *YYO*, indicating that the second origin choice is located at $(-1/4, -1/4, 0)$ with respect to the first origin. When the complete space group is generated by a procedure similar to that described in Section 9.2.10.8, then one can change between the two origin locations by multiplying each symmetry matrix by the 4×4 matrix representing the translation $\pm(-1/4, -1/4, 0)$; the $\pm$ sign is taken to be $+$, when going from the first choice to the second and $-$ when going the opposite way.

For the highest order space group, **Fm$\bar{3}$m** (O^5_h), with 192 elements, we only need six generator matrices (plus the inversion operator), and the generator string is given by:

$$16aODDaDODbOOOcOOOdOOOeOOO0.$$

There are two space groups that require seven generators, **Fd$\bar{3}$m** (O^7_h) and **Fd$\bar{3}$c** (O^8_h). The complete listing of all 230 space group generator strings can be found in an appendix on the book's web site.

Once all symmetry operators of a space group have been determined, it is a simple matter to determine the equivalent positions, starting from a general position (x, y, z). For special positions, one would have to eliminate positions that occur twice or more in the list generated by multiplying all matrices with the general position. We leave it as an exercise for the reader to determine the complete set of symmetry matrices for the space groups **P2/c** (C^4_{2h}) and **P6$_3$/m** (C^2_{6h}).

10.8 Historical notes

One of the striking developments in the field of crystallography was the fact that **William Barlow** (1845–1934), **Evgraf Stepanovich Federov** (1853–1919) and **Arthur Moritz Schönflies** (1853–1928), men of different nationalities and technical interests, nearly simultaneously derived the existence of the 230 space groups in the 1890s.

Federov was a Russian crystallographer. He introduced the concept of *regularity* to describe configurations of objects, and in 1885 he published *The Elements of the Study of Configurations*, an influential work on fundamental

Fig. 10.13. (a) William Barlow (1845–1934), and (b) Evgraf Stepanovich Federov (1853–1919) (pictures courtesy of J. Lima-de-Faria).

geometry (Federov, 1885). He then went on to apply the concept of regularity to the structure of the atom, which eventually led to the creation, by D. I. Mendeleev, of the Periodic Table of the Elements. Subsequently, he applied the concept of regularity to crystal structures which led, in 1890, to his derivation of the 230 space groups (known as *Federov groups* in the Russian literature). Around 1870, Camille Jordan, in France, had two students working on the study of continuous (Marius Sophus Lie) and discrete (Felix Klein) groups. About two decades later, Arthur Schönflies picked up on this work, applied it to crystal structures, and published intermediate results, which were read with great interest by Federov. Federov then sent his results to Schönflies, which started a "lively correspondence" (Galiulin, 2003) between the two researchers. Federov completed his derivation and published it in 1891 (Federov, 1891), while Schönflies published his results a few months later (Schönflies, 1891). The reader will find an interesting account of Federov's life and work in the article by Galiulin (2003).

Barlow was an English amateur geologist, specializing in the field of crystallography. He derived all 230 space groups using an approach that was quite different from that used by Schönflies and Federov (Barlow, 1894). Barlow correctly predicted the crystal structure of many compounds, including NaCl and CsCl, long before they were confirmed by means of X-ray diffraction.

By allowing each point in a crystallographic space to have a color (black or white), the 230 space groups can be extended to the 1651 Heesch–Shubnikov black–white space groups (Shubnikov and Belov, 1964). These groups can be used to describe all possible 3-D magnetic symmetries. It is also possible to extend the concept of space group to higher dimensional spaces; it has been established that there are 4250 4-D space groups. In addition, De Wolff *et al.* (1981) published the theory of *superspace groups*, which allows for the description of commensurately and in-commensurately modulated 3-D crystals in terms of 4-D groups. The study of groups as applied to crystallography is still a very active field of research that was set in motion around 1830 by the young mathematician Evariste Galois.

10.9 Problems

(i) *Plane groups I*: For each of the 2-D patterns shown in the figure below, identify a periodic repeat unit cell; the equipoints in projection with the symmetry elements; and the plane groups.

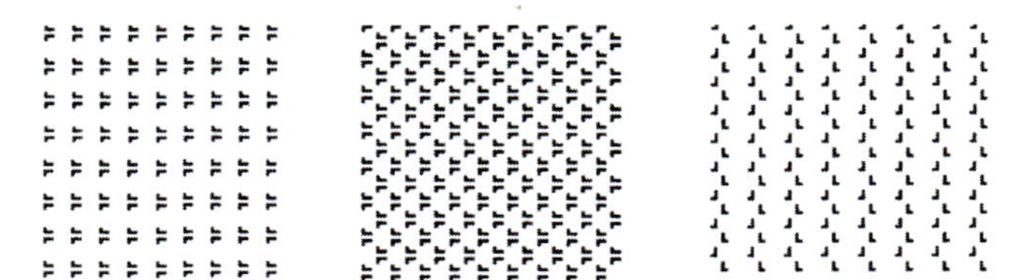

Fig. 10.14. 2-D patterns tiling the plane.

(ii) *Plane groups II*: Consider plane groups with a square lattice:

(a) Derive the **p4** plane group by adding the **4** (C_4) point group to a lattice point. List the equipoints, site symmetries, and multiplicities.
(b) Derive the **p4mm** plane group by adding the **4mm** (C_{4v}) point group to a lattice point. List the equipoints, site symmetries, and multiplicities.

(iii) *Plane groups III* : Consider the symmetric 2-D structures illustrated in Fig. 10.15(a).

(a) Determine the unit cell (unit mesh).
(b) Show all symmetry elements in the cell.
(c) Determine the 2-D plane group for this structure.

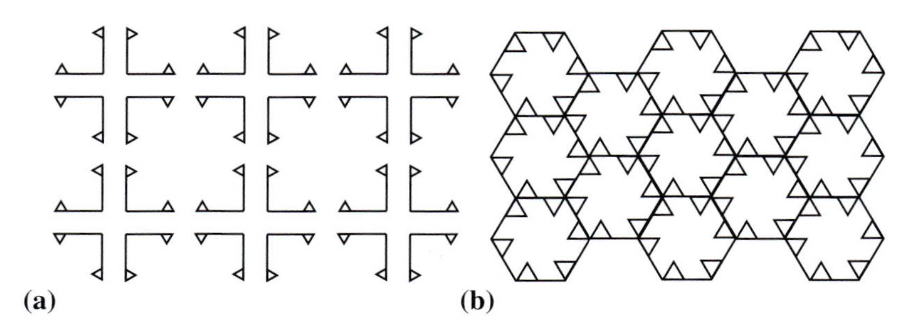

Fig. 10.15. Symmetric 2-D structures.

(iv) *Plane groups IV*: Consider the 2-D tiling in Fig. 10.15(b).

(a) Determine the unit cell (unit mesh).
(b) Show all symmetry elements in the cell.
(c) What is the 2-D plane group for this structure?

(v) *Space groups I*: The *spinel-type magnetic oxides* include the compound *magnetite*, Fe_3O_4, which is one of the important oxides of iron. The space group for spinel is **Fd$\bar{3}$m** (O_h^7). In magnetite O^{2-} anions occupy the $32e$ special position (x, x, x) with $x = 0.3799$, Fe^{3+} cations occupy the $16c$ special position $(1/8, 1/8, 1/8)$ and Fe^{2+} cations occupy the $8b$ special position $(1/2, 1/2, 1/2)$. The magnetite lattice constant is 0.83941 nm.

(a) Determine the spinel Bravais lattice and list the site symmetry and equivalent positions for each ion in the Fe_3O_4 structure. How many point group operations leave each site invariant?
(b) Show that the site multiplicity is consistent with the Fe_3O_4 composition. How many formula units are there per unit cell?
(c) Use the atomic weights for Fe and O to calculate the density of Fe_3O_4.

(vi) *Space groups II*: The *Heusler alloys* include Ni_2MnGa, which is an important new ferromagnetic shape memory alloy. Heusler alloys are discussed in more detail in Chapter 17 of the text. The famous isostructural *Heusler alloy*, Cu_2MnAl, is the prototype for the $L2_1$ structure. The space group for Ni_2MnGa is **Fm$\bar{3}$m** (O_h^5). In Ni_2MnGa, Ni atoms occupy the $8c$ $(1/4, 1/4, 1/4)$ positions, Mn atoms occupy the $4b$ $(1/2, 1/2, 1/2)$ sites, and Ga atoms are found on the $4a$ $(0, 0, 0)$ special positions. The cubic lattice constant is 0.5825 nm.

(a) Use the *International Tables for Crystallography* to determine the coordinates of all atoms in a unit cell of Ni_2MnGa.
(b) Determine the Bravais lattice and list the site symmetry and equivalent positions for each atom in the Ni_2MnGa structure. How many point group operations leave each site invariant?
(c) Show that the site multiplicity is consistent with the Ni_2MnGa composition. How many formula units are there per unit cell?
(d) Explain how to construct this structure as a $2 \times 2 \times 2$ superlattice structure (hint: consult a picture of the $L2_1$ structure in Chapter 17).

CHAPTER

11 X-ray diffraction: geometry

> *"In trying to think of some way in which diffraction effects with X-rays might be found, and the question of their true nature answered, he [von Laue] came to the realization that Nature had provided, in a crystal, a diffraction grating exactly suited for that purpose."*
>
> William L. Bragg, Nobel Lecture, 1922

11.1 Introduction

The first ten chapters of this book provide an in-depth description of the crystallographic concepts used to describe crystals and to perform crystallographic computations. Armed with these skills, we are now ready to begin a discussion of commonly used experimental X-ray diffraction methods. First, we will discuss what X-rays are and how we can generate them. Then, we will talk about the interaction of X-rays with crystal lattices and introduce the concept of *diffraction*. This will lead to *Bragg's law*, a central theorem for diffraction. We will convert Bragg's law from its usual direct space formulation to a reciprocal space form, and introduce a graphical tool, known as the *Ewald sphere*, to describe diffraction events. We conclude the chapter with a brief overview of a few commonly used experimental methods. In Chapter 12, we will continue our discussion of X-rays, and consider in detail how an X-ray photon interacts with a single atom, then with a unit cell, and finally with an entire crystal. This will lead to a few important concepts, such as *atomic scattering factors*, *structure factors*, *systematic absences*, and so on. We will then apply these concepts to the technique of *powder diffractometry*, and show that there is a precise relation between the experimental powder diffraction pattern and the crystal structure of the sample. In Chapter 13, we expand the description of diffraction to include *neutron diffraction* and *electron diffraction*, both important materials characterization techniques. Finally, in Chapter 14, we will apply the concepts we learned in Chapters 11 and 12 to some of the materials introduced in Section 1.6 of Chapter 1.

11.2 Properties and generation of X-rays

In this section, we will discuss some of the fundamental properties of X-rays, and show how we can generate X-rays experimentally. We will introduce the concept of a *wave vector*, and describe how one can experimentally select a particular wave length.

X-rays are electromagnetic waves with a wave length in the range 0.01–1.0 nm. They travel in a straight line at the velocity of light, $c = 299\,792\,458$ m/s, and they have enough energy to travel through sufficiently thin solids. Before the exact nature of X-rays was fully understood, they were put to use in the medical field in the early part of the twentieth century. Their discovery is attributed to the German physicist Röntgen (1896).

Electrodynamics has taught us that X-rays are located between ultra-violet rays and gamma rays in the electromagnetic spectrum. Quantum mechanics has taught us that electromagnetic radiation can be regarded as either a wave or a particle. In the case of a wave, we typically talk about an X-ray wave, whereas the particle description employs the term *X-ray photon*. Since the two are equivalent, we will use them interchangeably.

Graphically, one can represent an X-ray wave as a sinusoidally changing electric field, with a perpendicular magnetic field, as shown in Fig. 11.1. The distance between two consecutive peaks in the magnitude of the electric field is known as the *wave length*, λ. Both the electric and magnetic field vectors are perpendicular to the propagation direction of the X-ray wave, i.e., parallel to $\mathbf{E} \times \mathbf{B}$. We will indicate the direction of propagation by means of a unit vector $\mathbf{e}_k$. If we take the x-axis to be parallel to this direction, then we can write down the following expression for the amplitude of the electric field vector (using x as the position and t as time):

$$E(x, t) = A \cos\left(2\pi(kx - \nu t)\right),$$

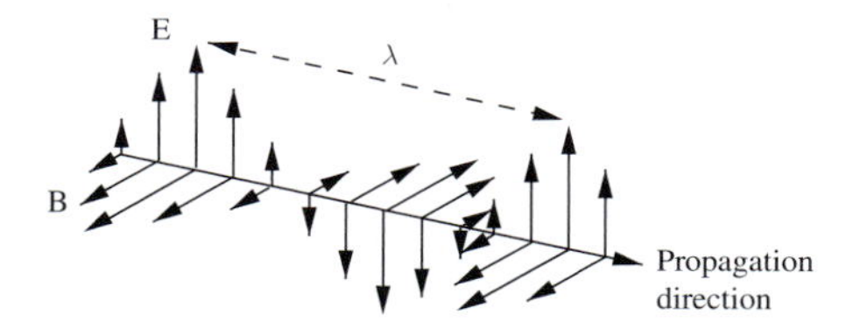

Fig. 11.1. Schematic drawing of an electromagnetic ray, consisting of an alternating set of orthogonal electric and magnetic fields.

where k is the inverse of the wave length λ, (i.e., $k = 1/\lambda$) and ν is the frequency of the oscillating field. For electromagnetic radiation, one can show that the relation between wave length and frequency, the so-called *dispersion relation*, is given by:

$$\lambda\nu = c.$$

Substitution in the electric field expression above leads to:

$$E(x, t) = A\cos(2\pi k(x - ct)).$$

The magnetic field associated with this changing electric field has the same spatial and temporal behavior. It is customary to use complex number notation for periodic phenomena, and we replace the previous notation by a complex exponential notation, using the Euler formula:[1]

$$E(x, t) = A\mathrm{e}^{2\pi \mathrm{i}k(x-ct)},$$

with the understanding that only the real part of this expression is physically relevant. Using the properties of exponentials, we can separate the spatial and temporal parts of this expression and we find:

$$E(x, t) = A\mathrm{e}^{2\pi \mathrm{i}kx}\mathrm{e}^{\mathrm{i}2\pi\nu t}.$$

In the remainder of this book, we will not be interested in the temporal behavior of X-rays, so we will usually omit this term from all equations.

We must make two important observations at this point:

(i) electromagnetic radiation can, for all practical purposes, be regarded as an oscillating electric field, and
(ii) since the argument of an exponential function must be dimensionless, the dimensions of the wavenumber k are the *inverse of a length*.

This means that the wave vector $\mathbf{k} = k\mathbf{e}_k$ must be a vector *in reciprocal space*! If we generalize the equation above to an arbitrary orientation for the electromagnetic propagation direction, then the product kx becomes $\mathbf{k}\cdot\mathbf{r}$. The components of the wavevector $\mathbf{k}$ are measured with respect to the reciprocal basis vectors, whereas the position vector $\mathbf{r}$ is measured with respect to the direct basis vectors. From the definition of direct and reciprocal basis vectors we know that the dot product consists of only three terms:

$$\mathbf{k}\cdot\mathbf{r} = k_x x + k_y y + k_z z.$$

[1] Euler's formula is $e^{\mathrm{i}x} = \cos(x) + \mathrm{i}\sin(x)$, which can easily be shown by writing down the Taylor expansions for all three functions.

When we introduced the reciprocal lattice, we showed that each reciprocal lattice vector corresponds to a vector normal to a certain crystal plane, and the length of the vector equals the inverse of the distance between subsequent planes. In the case of electromagnetic radiation represented by the complex exponential expression, we can find an analogous interpretation: the complex exponential $Ae^{2\pi i\mathbf{k}\cdot\mathbf{r}}$ has a real value, A, whenever $\mathbf{k}\cdot\mathbf{r}$ is equal to zero. This equation is satisfied for all the vectors $\mathbf{r}$ that are perpendicular to $\mathbf{k}$. In other words, the electric field has a constant value in every point in a plane perpendicular to the wave vector $\mathbf{k}$. The distance between subsequent maxima of the electric field is equal to the wave length λ, and we can represent the wave by the wave vector $\mathbf{k}$, which is oriented perpendicularly to a plane with constant field amplitude and which has length equal to the inverse between subsequent planes of constant field amplitude.[2] Thus, $\mathbf{k}$ is similar to a reciprocal lattice vector, and we will always express $\mathbf{k}$ in terms of the reciprocal basis vectors. We are now prepared to express an arbitrary electromagnetic wave with respect to the basis vectors of an arbitrary crystal lattice. This will become important later on in this and the following chapter.

11.2.1 How do we generate X-rays?

The theory of electromagnetism tells us that electromagnetic radiation will be produced whenever a charged particle is accelerated or decelerated. One way to accelerate and decelerate an electron, for instance, is to make it oscillate in an electric field. This is easy to understand if one realizes that a field is, by definition, a force per unit of something, in this case electric charge.[3] So, an alternating electric field will make an electron go up and down, which will cause it to emit electromagnetic waves *with the same frequency as the driving field*. This is the principle of broadcasting, where one forces the electrons in a piece of metal, called an antenna, to oscillate at a fixed frequency. To produce X-rays, we must make an electron oscillate at a very high frequency, about 10^{18} Hz. It would be very difficult to produce a driving field with this frequency, so we need to resort to simpler means.

If we can create a beam of high energy electrons, and then abruptly bring them to a halt, then those electrons will emit part or all of their energy in the form of X-rays. This is called *braking-radiation* or *Bremsstrahlung*. The easiest way to do this is shown schematically in Fig. 11.2: a cathode (on the left) is heated to a high temperature, and a voltage V is applied between the cathode and the anode (on the right). If the temperature is sufficiently high, but not so high that the cathode will melt, then the electric field $E = V/d$, with d the spacing between the electrodes, will pull the electrons with the

[2] It is for this reason that the expression $e^{2\pi i\mathbf{k}\cdot\mathbf{r}}$ is often referred to as a *plane wave*.
[3] Other example: the gravitational field is the gravitational force per unit of mass.

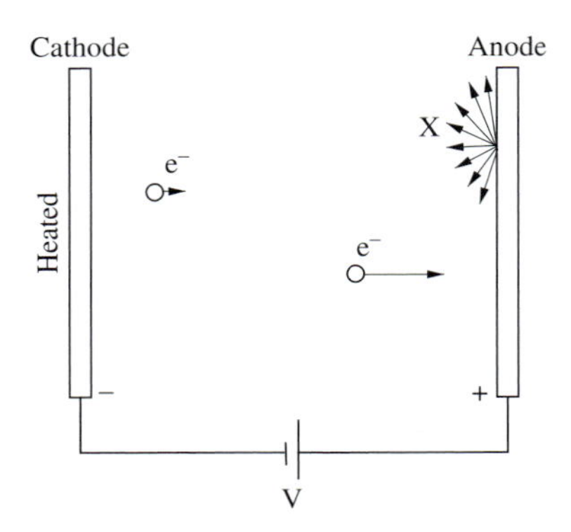

Fig. 11.2. Schematic drawing of an X-ray generator.

highest energies out of the cathode and accelerate them towards the anode (this process is known as *thermal emission*). The potential energy of such an electron is equal to the electron charge multiplied by the potential drop:

$$E_\mathrm{p} = eV.$$

When the electron reaches the anode, it has been accelerated to its maximum velocity, and the potential energy has been converted completely into kinetic energy:

$$eV = \frac{1}{2}mv^2.$$

For a potential drop of $V = 10\,000$ volts, the final velocity will be 59.3×10^6 m/s, or close to 20% of the velocity of light. At that point, the electron collides with the anode, and loses all of its energy in a very short time. About 99% of this kinetic energy is converted into thermal energy in the form of lattice vibrations, and about 1% is transferred to atomic and conduction electrons. These electrons will be excited into higher energy levels, and when they revert back to their original energy levels, they will emit electromagnetic radiation with a frequency in the X-ray range. Needless to say, such a setup requires a water-cooling system, because most of the energy is lost in the form of heat. The anode is thus either kept cool by means of a coolant, or it is continuously rotated so that the exposed region is only briefly heated.

Each electron hitting the anode (or *target*) may lose either a fraction of its energy, or it may lose all energy at once. Since the energy of an X-ray photon is given by Planck's constant multiplied by the frequency, we can determine the maximum X-ray frequency, or, equivalently, the shortest wave length that will be generated, as a function of the applied potential V:

$$eV = \mathrm{h}\nu = \frac{\mathrm{h}c}{\lambda},$$

from which we find:

$$\boxed{\lambda_{\text{min}} = \frac{1239.8}{V} \quad [\text{nm}]} \tag{11.1}$$

if V is measured in volts. This minimum value is known as the *short wave length limit*, and it represents the smallest wave length that can be generated by an X-ray tube, for a given potential drop. A minimum wave length of 0.1 nm thus requires a potential of 12 398 volts.

If the electron does not pass on all of its energy at once, but only a fraction, then a longer wave length X-ray photon will be created. The curves in Fig. 11.3 represent schematically the X-ray intensity as a function of wave length and tube voltage, for a molybdenum target. For sufficiently high voltages, the curves will show, in addition to a broad maximum, sharp peaks, which are known as *characteristic radiation*.

The total X-ray intensity generated by the target is equal to the integral over one of these curves, and one can show that it is proportional to the target current, the target atomic number, and roughly the square of the applied voltage. To generate a high intensity of X-rays one thus needs a heavy atom target, say tungsten, and as high a voltage as one can generate.

The characteristic peaks can be understood by considering the electronic structure of the target atoms. Suppose that the incident electrons have enough energy to knock a K electron out of its shell. The atom is left in a highly excited state and one of the electrons from the higher levels will fall into the low energy state, emitting a quantum of energy in the form of an X-ray photon. The energy of this photon is equal to the energy difference between

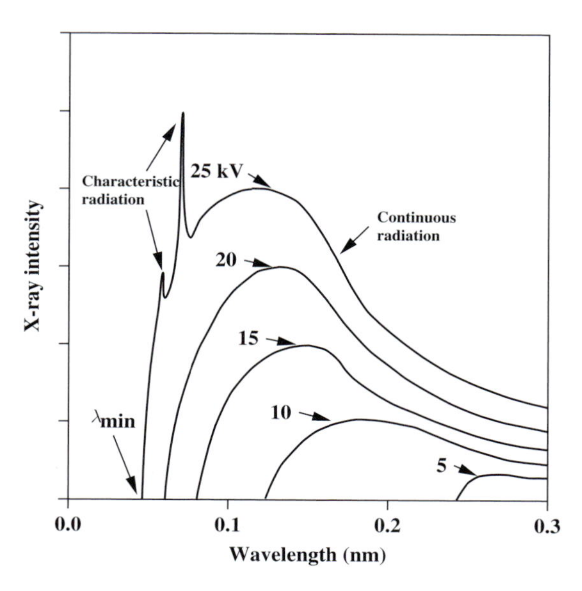

Fig. 11.3. Schematic X-ray spectrum for a molybdenum target as a function of voltage and wave length. Based on Fig. 1.4 in Cullity (1978).

Table 11.1. List of the more commonly used *K* wave lengths in nm (Cullity and Stock, 2001).

Target	Kα	Kα_1	Kα_2	Kβ_1
Cr	0.229 100 0	0.229 360 6	0.228 970	0.208 487
Fe	0.193 735 5	0.193 998 0	0.193 604 2	0.175 661
Co	0.179 026 0	0.179 285 0	0.178 896 5	0.163 079
Cu	0.154 183 8	0.154 439 0	0.154 056 2	0.139 221 8
Mo	0.071 073 0	0.071 359 0	0.070 930 0	0.063 228 8

the initial and final energy state of the atom, and thus depends on the type of atom. If the beam of electrons does not have enough energy to knock one of the inner electrons out of its shell, then there will be no X-ray emission caused by de-excitation of the atom, simply because it can never be excited in that way to begin with. As the tube voltage is increased, the short wave length limit decreases and when the electron energy becomes larger than a threshold value, a characteristic peak will appear on top of the Bremsstrahlung background. The intensity of this characteristic peak can be many times higher than that of the continuous radiation, and depends on the difference between the tube voltage and the minimum voltage necessary to excite the target atoms. The higher the tube voltage, the more intense are the characteristic peaks with wave length larger than λ_{min}.

The characteristic peaks have labels which consist of three parts: a capital letter indicates *the level into which the atom will de-excite*, a Greek letter indicates *the level from which the atom de-excites*, and a subscript number distinguishes between various energy levels of the excited state. An example would be Kα_1, which represents radiation caused by de-excitation from the first level of the L-shell into the K-shell, Kβ_2 which corresponds to de-excitation from the second energy level in the M-shell into the K-shell, and so on. The energies (and X-ray wave lengths) associated with such de-excitation processes have been measured for all elements. A few of the more important ones (wave lengths) are listed in Table 11.1. Since the Kα_1 and Kα_2 wave lengths are closely spaced, and they are the ones used most frequently for X-ray diffraction experiments, one can compute a weighted average wave length, with the Kα_1 wave length counting for 2/3 and Kα_2 for 1/3. If the subscript is not used, as in Kα, then it is understood that the weighted average is being used. Unless mentioned otherwise, all examples in this chapter will use Cu Kα radiation.

In summary, an X-ray source employs a stream of high energy electrons, generated by a heated filament wire, which is directed towards a metal target. The kinetic energy of the electrons is converted mostly into heat, but a small percentage leaves the target as X-ray photons. These photons have a continuous range of wave lengths, with a few superimposed characteristic

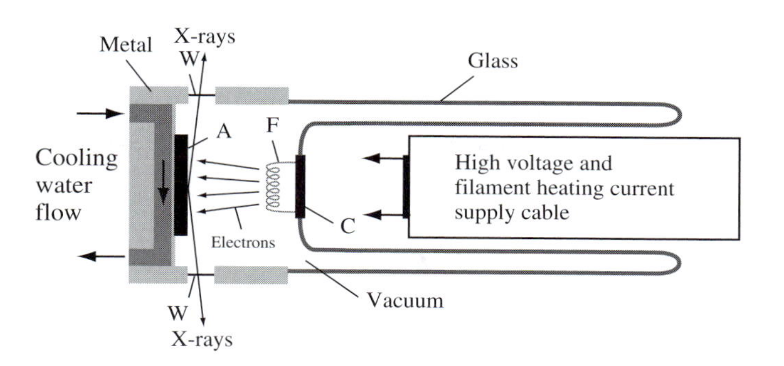

Fig. 11.4. Schematic cross section of an X-ray tube, showing the filament (F), the target (T), and the windows (W) through which the X-rays exit the tube.

high intensity peaks. The higher the accelerating voltage, the larger the wave length range of the X-ray source.

X-ray sources have many different uses, apart from diffraction experiments, so their design has become more or less standardized. A cross section through a typical X-ray tube is shown in Fig. 11.4. The tube consists of a cylindrical metal part (light gray) with an inlet and outlet for cooling water. The anode (A) is located right next to the cooling water channel. The cathode (C) is located at the bottom of a hollow cylindrical glass tube that has been evacuated (vacuum). The filament (F) is typically a tungsten coil. When the filament is heated and a voltage is applied between the anode and cathode, a beam of electrons is emitted from the filament and accelerated towards the anode. The cooling water removes the thermal energy generated by the sudden deceleration of the electrons. The X-rays escape from the tube through a number of thin windows, typically made of beryllium or a polymeric material. Depending on the design of the tube, there can be one to four such windows. The glass cylinder provides electrical insulation for the high voltage supply cable which is inserted into the tube. This cable also carries wiring for the filament heating current.

11.2.2 Wave length selection

If we want to study crystals with X-rays, then we must first find a way to select one particular wave length with which we will illuminate the crystal. From the continuous spectrum shown in Fig. 11.3 we should select only the high intensity characteristic peak, and remove all other wave lengths. This appears to be a non-trivial task, but fortunately nature provides us with a straightforward solution. To understand how we can remove a large fraction of the wave length spectrum it is instructive to determine what happens to a beam of X-ray photons when they travel through a medium with a certain density ρ. Part of the X-ray beam will be *absorbed*, and the amount of absorption increases with increasing thickness of the material, with its density, and with the wave length of the X-rays. The latter is easy to understand if one realizes that a longer wave length means a lower photon energy and hence a reduced penetration depth.

Mathematically, the absorption process is expressed by an exponential function, known as *Beer's law*:

$$I_x = I_0 e^{-\mu x},$$

where μ is the *linear absorption coefficient*, with units of cm^{-1}; x is the distance travelled through the solid, and I_0 is the incident X-ray intensity. This equation states that, if a beam of X-rays travels a distance $1/\mu$ through a solid, then its intensity will have decreased to 36.78% ($= e^{-1}$) of its original value. The absorption coefficient depends on the material and the X-ray wave length. It is customary to normalize μ with respect to the density (since μ is proportional to the density) and one defines the *mass absorption coefficient* μ/ρ. The equation for absorption then reads:

$$\boxed{I_x = I_0 e^{-\left(\frac{\mu}{\rho}\right)\rho x}}. \tag{11.2}$$

Values of the mass absorption coefficients for the elements are tabulated as a function of X-ray wave length. Coefficients for a compound can be computed from the values for the N individual elements by taking a weighted average, with the weight fractions w_i as weight factors, i.e.:

$$\left(\frac{\mu}{\rho}\right) = \sum_{i=1}^{N} w_i \left(\frac{\mu}{\rho}\right)_i.$$

Table 11.2 lists the mass absorption factors for about a dozen elements, for the most frequently used X-ray wave lengths. If we draw the mass absorption coefficient for a particular element as a function of increasing wave length (or decreasing photon energy), then we find a curve similar to that shown in Fig. 11.5. Starting at the large wave length end of the figure, we find that the mass absorption coefficient decreases as the photon energy increases. If the photon energy becomes high enough to knock out an electron from the K-shell of the absorbing atoms, then μ/ρ will suddenly increase by several orders of magnitude. This is similar to the absorption of energy from the incident electrons in an X-ray tube, except that now the energy is provided by the X-ray photon instead of by the incident electron. With increasing energy, we find again a decrease of the mass absorption coefficient for wave lengths shorter than the critical wave length λ_K. The sharp feature in the absorption versus wave length curve is called an *absorption edge*, in this case the K absorption edge. From the values of the edge wave lengths, one can deduce information about the energy levels of the atoms in the absorbing material.

Table 11.2. Mass absorption coefficients for selected elements (in cm^2/g) (taken from Appendix 8 in Cullity (1978)).

Element	Density (g/cm^3)	Mo		Cu		Co		Cr	
		Kα 0.0711 nm	Kβ 0.0632 nm	Kα 0.1542 nm	Kβ 0.1392 nm	Kα 0.1790 nm	Kβ 0.1621 nm	Kα 0.2291 nm	Kβ 0.2085 nm
C	2.27	0.5348	0.4285	4.219	3.093	6.683	4.916	14.46	10.76
Al	2.70	5.043	3.585	50.23	37.14	77.54	58.08	158.0	120.7
Si	2.33	6.533	4.624	65.32	48.37	100.4	75.44	202.7	155.6
Ti	4.51	23.25	16.65	202.4	153.2	300.5	231.0	571.4	449.0
Fe	7.87	37.74	27.21	304.4	233.6	56.25	345.5	113.1	86.77
Co	8.8	41.02	29.51	338.6	258.7	62.86	47.71	124.6	96.06
Ni	8.91	47.24	34.18	48.83	282.8	73.75	56.05	145.7	112.5
Cu	8.93	49.34	35.77	51.54	38.74	78.11	59.22	155.2	119.5
Mo	10.22	18.44	13.29	158.3	119.7	236.6	181.0	457.4	356.5
Ag	10.50	26.38	19.10	218.1	165.8	323.5	248.9	617.4	483.5
Pt	21.44	108.6	80.23	198.2	151.2	295.2	226.4	571.6	443.9
Au	19.28	111.3	82.33	207.8	160.6	303.3	235.7	568.0	446.7
Pb	11.34	122.8	90.55	232.1	178.6	340.8	263.8	644.5	504.9

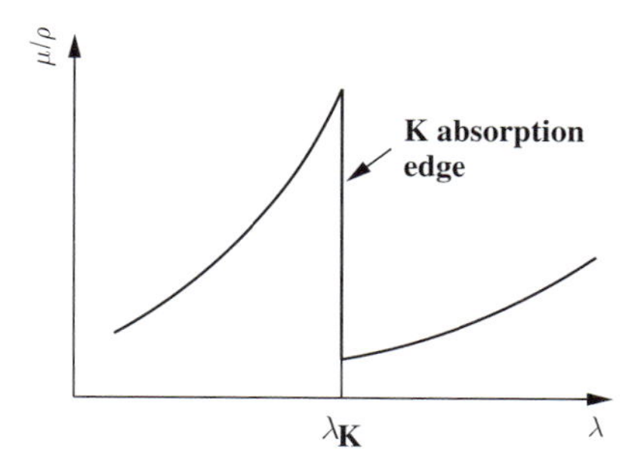

Fig. 11.5. Schematic representation of the mass absorption coefficient versus wave length, showing a K absorption edge.

The general variation of the mass absorption coefficient with wave length (away from the absorption edges) is of the form:

$$\left(\frac{\mu}{\rho}\right) = k\lambda^3 Z^3,$$

with k a constant and Z the atomic number of the absorbing atom. Absorption is thus sensitive to both the wave length and the atomic number (and hence density) of the absorbing material.

The location of the absorption edges for various elements is particularly useful in that it allows us to construct *X-ray filters*. Consider the continuous spectrum of a copper target, as shown in Fig 11.6(a). It consists of two peaks, Kα (split in two separate peaks) and Kβ at a slightly shorter wave length. If we superimpose the mass absorption coefficient of nickel onto this drawing (dashed line) then we find that the edge lies in between the copper Kα and Kβ peaks. If we place a sufficiently thick layer of nickel in an X-ray beam generated with a copper target, then, because of absorption, the beam emerging from the nickel layer will have a wave length spectrum similar to

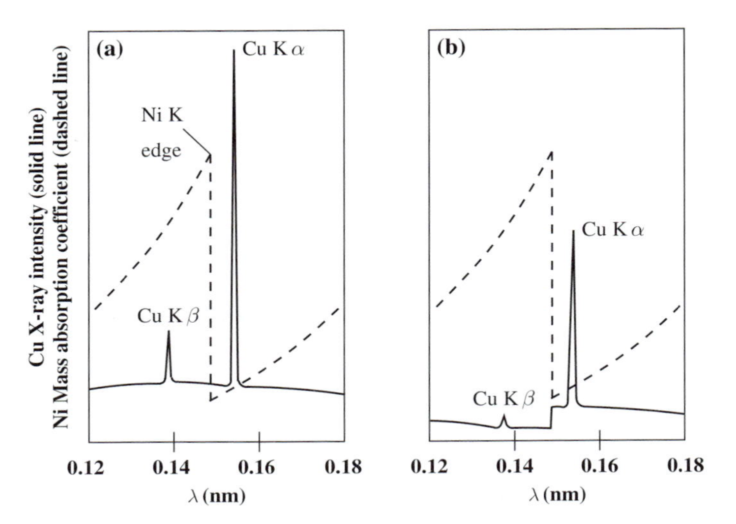

Fig. 11.6. Cu X-ray intensity and Ni mass absorption coefficient versus wave length, before (a) and after (b) application of the filter. Based on Fig. 1.13 in Cullity (1978).

the one shown in Fig. 11.6(b). In other words, if we place a thin foil with atomic number $Z-1$ in a beam of X-rays generated with a target of atomic number Z, then the Kβ peak of the target radiation will be almost completely absorbed, and the background intensity around the Kα peak will also be attenuated. The ratio of Kα to Kβ copper radiation intensity will change from 7.5 : 1 before filtering to 500 : 1 after filtering, if the thickness of the nickel filter is about 20 microns (see problems at the end of the chapter).

11.3 X-rays and crystal lattices

In this section, we will analyze how X-ray waves interact with crystalline matter. Before we begin, it is useful to consider a simple example, using visible light. Fig. 11.7 shows a pattern obtained by shining a bright green laser beam through a thin piece of finely woven fabric onto a white wall. The 2-D fabric, with its regular pattern of fibers and open spaces between the fibers, is analogous to a 3-D crystal, which has atoms and open spaces in between the atoms. The monochromatic laser beam provides a reasonable approximation to a plane wave and the wave interacts with the regular fiber pattern to produce a *diffraction pattern*.[4] Such diffraction patterns can sometimes be observed when looking through a thin curtain at a distant street light, in particular when the street light is a bright yellow Na light, which has only two strong wave lengths in the yellow range of the visible spectrum. The patterns can also be observed with a regular light point source, but in that case different wave lengths will be scattered by different angles, so that the diffraction pattern contains rainbow-like streaks instead of well-defined points.

[4] Recall that a plane wave is a wave described by $e^{2\pi i \mathbf{k}\cdot\mathbf{r}}$.

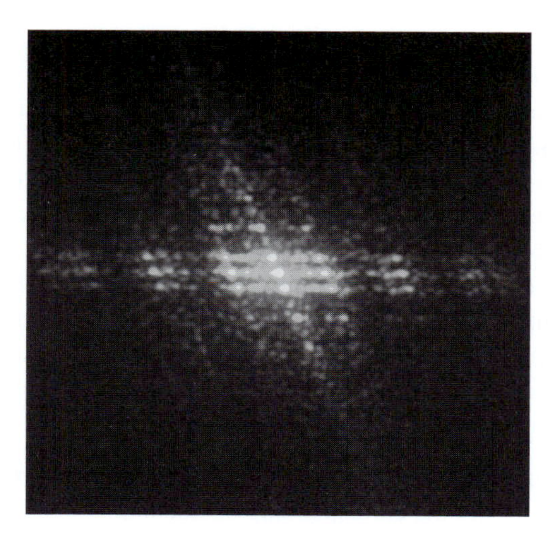

Fig. 11.7. Optical diffraction pattern obtained by shining a green laser through a piece of finely woven fabric onto a wall.

Instead of using fabrics and street lights to do our experiments, let us now focus on regular patterns of dots on a transparent slide. We can use a simple drawing program to create 2-D regular arrays of dark dots on a white background, and then photograph the array using 35 mm slide film. If we take a laser (say, a red He-Ne laser, with a wave length of 670 nm) and shine it through the slide onto a white wall, then we will observe a diffraction pattern on the wall. This diffraction pattern is very similar to that observed with fabrics, but the advantage is that we can now measure the spacings between the clearly defined intensity maxima on the wall.

We can change the pattern of the original array to other 2-D lattices, and instead of dots we can use squares, triangles, or any other shape. Fig. 11.8 shows a set of eight different patterns, taken from the *Optical Transform Kit*, published by the Institute for Chemical Education.[5]

The observed diffraction patterns are shown below each individual pattern.[6] There are several important things we should point out about these patterns:

- Patterns (**a**) and (**b**) are centered and primitive square unit cells of the same dimensions. Note that the diffraction pattern of the primitive cell has more diffracted spots than that of the centered cell.
- Patterns (**c**) and (**d**) have the same lattice parameters, but the unit cell is filled with a different arrangement of squares. Note that the positions of the diffracted spots are identical, but their intensity distribution is somewhat different.
- Patterns (**b**) and (**d**) are both primitive square lattices, and the lattice parameter of pattern (**b**) is smaller than that of (**d**); comparing the diffraction patterns, we find that the spacing between the diffraction spots is larger for

[5] URL: *http://ice.chem.wisc.edu*.

[6] We did not actually take a picture of the patterns on the wall, but, instead, we computed the mathematical/numerical equivalent, which is known as a *Fourier transform*.

(**b**) than for (**d**), which reminds us of the properties of objects in reciprocal space.

- Pattern (**e**) is a rectangular pattern with the same a-parameter as pattern (**b**); note that the *horizontal* axis of the diffraction pattern is longer than the vertical one, while the opposite is true for the original pattern.
- Pattern (**f**) has an oblique unit cell; note that the diffraction pattern is also oblique, but with a different angle.
- Pattern (**g**) contains a unit cell with a vertical glide plane. Note that in the diffraction pattern some of the reflections along the central vertical line are missing (arrows).
- Pattern (**h**) is a hexagonal pattern, which produces a hexagonal diffraction pattern.

Since there appear to be definite relations between the original lattice patterns and the corresponding diffraction patterns, we can hope that, given a diffraction pattern, we might be able to *reconstruct* the original pattern. It is, in fact, possible to derive rules that relate the diffraction pattern to the original pattern and vice versa. We have already seen the mathematical framework needed to express these rules: it is the framework of the *reciprocal lattice*.

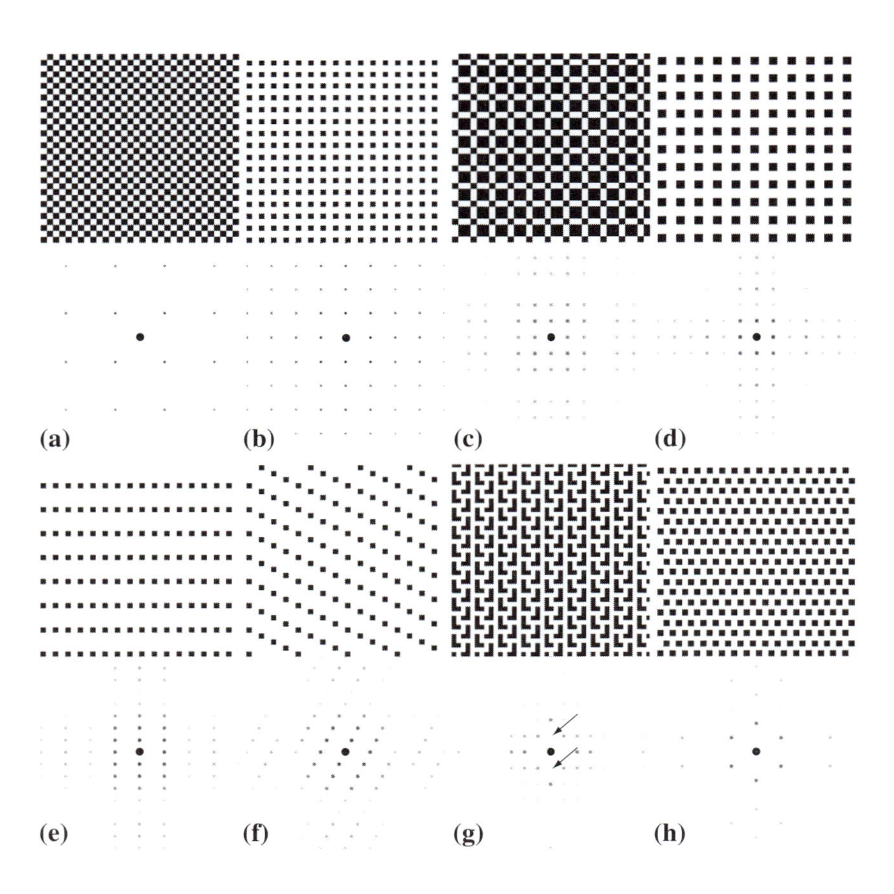

Fig. 11.8. Eight periodic patterns with different lattice parameters and/or different unit cell contents, and the corresponding diffraction patterns. The patterns are part of the *Optical Transform Kit* available from the Institute for Chemical Education, and are reproduced here with permission.

It turns out that the diffraction pattern is, in fact, an "image" of the reciprocal lattice of the original pattern.

When we shine a bright beam of X-rays onto a crystal, then that crystal will act very much like the demagnified lattice on the filmstrip, and a diffraction pattern will be formed. From the positions of the individual diffracted *beams*, we will be able to deduce information about the *size* and *shape* of the unit cell. This is the subject of the remainder of this chapter. From the relative intensities of the diffracted beams we will derive information about the *positions* of the atoms within the unit cell, which forms the subject of the next chapter.

Before we clarify the relation between diffraction phenomena and crystal structures, we must first discuss one of the fundamental properties of any form of wave-motion: the *phase* of the wave. If we have two periodic waves with wave length λ but shifted with respect to each other, and we take a point at which the first wave has zero wave amplitude as the origin, then we can measure the *phase difference* between the two waves as follows: determine the distance Δx to the closest zero crossing of the second wave (see Fig. 11.9). The value of Δx will be between $-\lambda/2$ and $+\lambda/2$. Then scale that value by dividing it by the wave length λ. Finally, convert it into radians by multiplication by 2π. The phase of the second wave relative to the first wave is then given by:

$$\phi = 2\pi\frac{\Delta x}{\lambda}.$$

If a wave is represented by a cos-function, then the phase shift can be represented by the addition of ϕ to the argument of the cosine, i.e.:

$$A(x) = \cos\left(2\pi\frac{x}{\lambda} + \phi\right).$$

If two or more waves are added to each other, then the relative phase will determine if the waves reinforce each other (*constructive interference*) or if they cancel each other (*destructive interference*) (see Fig. 11.10). Consider two waves, with a relative phase difference ϕ. The sum of the two waves

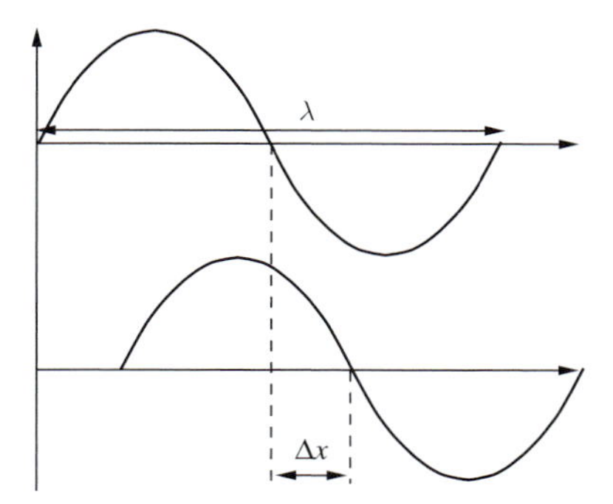

Fig. 11.9. Measurement of the relative phase of two sinusoidal waves.

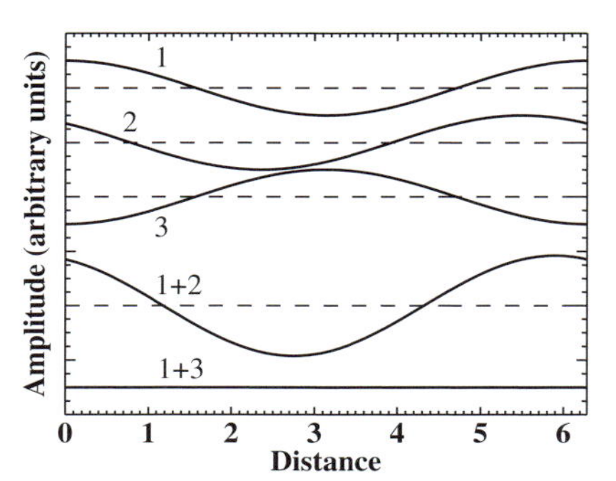

Fig. 11.10. Three sinusoidal waves, (1 = zero phase, 2 = $\pi/4$, and 3 = π relative phases), and at the bottom the sum of waves 1 + 2 and 1 + 3. If the phase difference is π, then the waves completely cancel each other out.

can be computed using the sum and difference equations for trigonometric functions:

$$\mathcal{S}(x) = \cos(x) + \cos(x + \phi) = 2\cos\left(x + \frac{\phi}{2}\right)\cos\frac{\phi}{2},$$

from which we find that, if the phase difference is 180° (or π), then the sum of the two waves will vanish (since $\cos(\pi/2) = 0$). This is illustrated by waves 1 and 3 in Fig. 11.10. If the phase difference is different from an odd integer multiple of π, then the result of the wave addition will be similar to that shown for waves 1 and 2.

11.3.1 Scattering of X-rays by lattice planes

The notion of relative phase or phase difference between two waves is of central importance to the diffraction phenomenon. We will see in the next chapter that every electron in every atom in a solid will oscillate when a beam of X-rays is sent through that solid. Remember that an X-ray photon can be regarded as an oscillating electric field, which drives all electrons into a periodic motion; the result of that motion is that each electron will emit radiation, at the same frequency as the incident radiation. This radiation is emitted in all directions and an external observer will detect the *sum* of many periodic waves.[7] Since all waves have the same frequency, they will interfere with each other, and only in directions for which all interferences are constructive can one detect radiation. The phase differences between waves depend sensitively on the exact location of the *scattering centers* which give rise to those waves. This means that a careful study of the spatial distribution of diffracted radiation can provide information on the location of all scattering centers, in our case the atoms.

To use the phenomenon of diffraction, we must first understand the relation between the crystal structure and the directions in which radiation can

[7] We will quantify this statement in the next chapter.

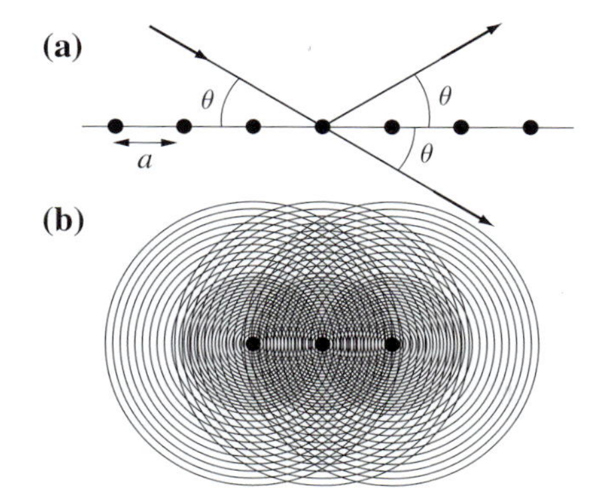

Fig. 11.11. (a) Reflection and refraction of an electromagnetic ray from a single plane of atoms; (b) interference pattern created by spherical waves emanating from three scattering centers.

be diffracted by that crystal. Let us first consider a single plane of atoms, represented in Fig. 11.11(a) by a line with dots at a regular spacing a. If a beam of electromagnetic radiation is directed at this surface under an angle θ, then part of the radiation will go through the surface (i.e., will be *refracted*) without a change in direction and part will be *reflected* by the plane, at an angle equal to the incidence angle θ.[8] This is essentially Snell's law for the reflection of light from a mirror surface. Reflection of light is possible for any incidence angle.

If we use a coherent light source, then all electrons in all atoms on the line in Fig. 11.11(a) will oscillate with the same frequency. As a consequence, each atom will emit concentric spherical waves, and those waves will interact with each other and set up a complex interference pattern, as shown in Fig. 11.11(b) for just three scattering centers. Constructive interference will occur in only a few directions, with a sensitive dependence on the ratio between the atom spacing and the wave length of the radiation. X-ray diffraction methods are all based on the direct observation of such interference patterns and the subsequent extraction of information on the atom spacings and positions.

In a 3-D crystal, we have many parallel planes, and we have to account for reflection not only from the top plane, but also from *all* underlying planes as well. Consider the situation in Fig. 11.12: wave 1 is incident on the first plane at the point O with an incidence angle θ. Part of the wave is reflected at an angle θ, part is transmitted without a change in direction.[9] Wave 2 undergoes the same process at point O′ where part of the intensity is reflected at an angle θ. The waves 1′ and 2′ then leave the crystal and travel in the same

[8] If the refractive index above the plane is different from that below the plane, then there will be a change in direction; this occurs for instance at the surface of water.

[9] The reader might wonder why there is no refraction of X-rays when they enter a crystal. Refraction *does* occur, but the difference in refractive index for X-rays inside and outside the crystal is so small, of the order of 10^{-4} to 10^{-5}, that the change in direction is virtually zero. As a consequence, it is not possible to build lenses for X-rays (recall that it is the refractive property of glass that allows a lens to bend the light to a focal point; if there is no refraction, then one cannot build a lens).

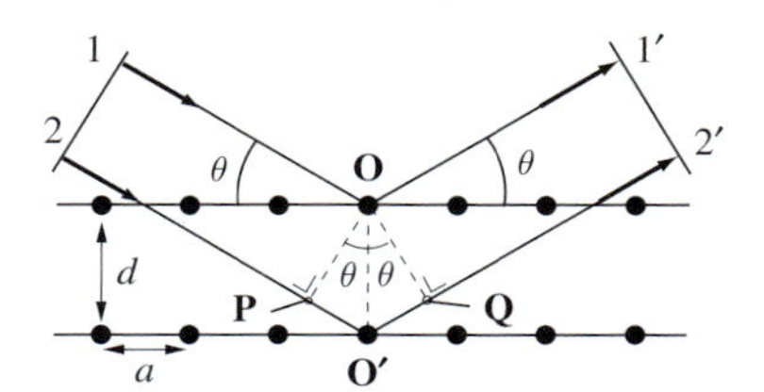

Fig. 11.12. Reflection of an electromagnetic ray from a set of parallel planes.

direction towards an observer, located far away from the crystal (i.e., at a distance many times larger than the interplanar spacing d).

Since the waves 1′ and 2′ have the same wave length, they will interfere with each other, and the interference will depend on the phase difference between the two waves. The phase difference in turn is derived from the *path difference*, and it is clear from the drawing that wave 2 has traveled further through the crystal than wave 1. The path difference is equal to the sum of the distances PO′ and O′Q. Constructive interference will only occur if this path difference equals *a multiple of the wave length* λ. If we denote the interplanar spacing by d, then we find:

$$\begin{aligned} \mathrm{PO'} + \mathrm{O'Q} &= n\lambda; \\ \mathrm{OO'}(\sin\theta + \sin\theta) &= n\lambda, \end{aligned}$$

from which we derive:

$$\boxed{2d\sin\theta = n\lambda.} \tag{11.3}$$

This is the fundamental equation of diffraction and it is known as the *Bragg equation*, after W. L. Bragg, who first derived this relation in 1912. This equation states that constructive interference from a set of consecutive parallel planes can only occur for certain angles θ, and that θ is determined by both the X-ray wave length and the interplanar spacing. The angle θ is known as the *Bragg angle*. Knowing the wave length, a measurement of the Bragg angle can thus provide a direct measurement of the interplanar spacing, and if we repeat this measurement for many different sets of planes, we can ultimately determine the dimensions of the unit cell.

The integer number n defines the *order* of the diffraction process. If $2d\sin\theta = 2\lambda$, then the diffracted beam is known as a *second-order beam*, or, in general, as the *n-th order beam*. One can, however, interpret the diffraction order in a different way. Consider the following way to rewrite the Bragg equation:

$$2\frac{d}{n}\sin\theta = \lambda. \tag{11.4}$$

If d represents the spacing for the planes (hkl), i.e., $d_{hkl} = 1/|g_{hkl}|$, then d/n represents the spacing for the planes $(nh\,nk\,nl)$. This follows because

$n\mathbf{g}_{hkl} = nh\mathbf{a}^* + nk\mathbf{b}^* + nl\mathbf{c}^* = \mathbf{g}_{nh\,nk\,nl}$. An n-th order diffracted beam from the plane (hkl) can thus be regarded as a first-order diffracted beam from the plane $(nh\,nk\,nl)$.

From this point on we will only consider first-order diffraction (i.e., we will drop the factor n in the Bragg equation). This means that from this point on we must distinguish, for instance, between the planes (100) and (200). Even though the planes are parallel, for diffraction purposes *they are no longer equivalent*. The interplanar spacing for the planes (100) is twice that for the (200) planes, i.e., $d_{100} = 2d_{200}$. The Bragg equation will thus be written as:

$$2d_{hkl} \sin\theta = \lambda. \tag{11.5}$$

We have discussed explicit techniques to compute d_{hkl} for an arbitrary crystal system. Using the Bragg equation and the metric tensor formalism, we can thus compute the allowed diffraction angles for any known crystal, provided the X-ray wave length is known. Let us consider an example: pure copper. Copper is the prototype for the face-centered cubic structure, with a lattice parameter $a = 0.36148$ nm. From Chapter 6 we know that the explicit expression for the interplanar spacing in a cubic crystal is given by:

$$d_{hkl} = \frac{a}{\sqrt{h^2+k^2+l^2}},$$

and the Bragg equation is then rewritten as:

$$\theta_{hkl} = \sin^{-1}\left(\frac{\lambda}{2a}\sqrt{h^2+k^2+l^2}\right). \tag{11.6}$$

For Cu Kα radiation, with a wave length $\lambda = 0.1541838$ nm, this equation reduces to:

$$\theta_{hkl} = \sin^{-1}\left(0.213267 \times \sqrt{h^2+k^2+l^2}\right).$$

Note that the angle between the incident beam and the diffracted beam is equal to twice the Bragg angle, 2θ; Table 11.3 lists the *diffraction angle* $2\theta_{hkl}$ for the lowest-index planes. Note also that the Bragg equation does not guarantee that a diffracted beam will be present; it merely states the geometrical constraints which must be satisfied before an X-ray beam can be diffracted by a certain set of crystal planes. In the next chapter, we will discuss why certain diffracted beams (the ones indicated by an asterisk in Table 11.3) cannot be observed for copper, even when the geometrical conditions are satisfied.

11.3.2 Bragg's Law in reciprocal space

Bragg's law predicts the geometrical conditions that need to be satisfied in order to observe diffraction from sets of lattice planes in a crystal. For a

Table 11.3. Interplanar spacings and diffraction angles for the lowest order planes in copper. The planes marked with an asterisk do not give rise to an experimentally observed diffracted beam, even when they satisfy the Bragg equation.

(hkl)	$h^2+k^2+l^2$	d_{hkl} [nm]	2θ [°]
(100)*	1	0.036148	24.6278
(110)*	2	0.025560	35.1081
(111)	3	0.020870	43.3559
(200)	4	0.016166	50.4957
(210)*	5	0.014757	56.9636
(211)*	6	0.013663	62.9861
(220)	8	0.012049	79.5540
(221)*	9	0.011431	84.8166
(300)*	9	0.011431	84.8166
(301)*	10	0.010899	90.0356

given wave length, the equation defines the Bragg angle, θ, for all sets of planes in the crystal. In the first section of this chapter, we have seen that the incident X-ray photon can be represented by a wave vector $\mathbf{k}$, parallel to the propagation direction of the wave and with length equal to the inverse of the wave length. This vector has the dimension of a reciprocal length and is thus a vector in reciprocal space. We have also seen that each set of parallel planes (hkl) can be represented by a reciprocal lattice vector $\mathbf{g}_{hkl}$. Let us now consider the diffraction process from a reciprocal space point of view.

Consider the X-ray beam characterized by the wave vector $\mathbf{k}$, as shown in Fig. 11.13(a). This vector makes an angle θ with the plane (hkl), which is represented by the normal vector $\mathbf{g}_{hkl}$. In the figure, we have chosen the origin of reciprocal space to be at *the endpoint of the wave vector* $\mathbf{k}$. The diffracted wave $\mathbf{k}'$ also encloses an angle θ with the diffracting plane, and is drawn with its initial point in the origin. Since vectors can be translated parallel to themselves, we can translate the vector $\mathbf{k}'$ so that its initial point coincides with the initial point of $\mathbf{k}$. The angle between $\mathbf{k}$ and $\mathbf{k}'$ is thus equal to 2θ. Since the wave length does not change during the diffraction process, the length of the vector $\mathbf{k}'$ must also be equal to $1/\lambda$. Therefore, the endpoints of both wave vectors are located on a circle, with center at the initial point of $\mathbf{k}$ and radius $1/\lambda$, as shown in Fig. 11.13(b). This circle is known as the

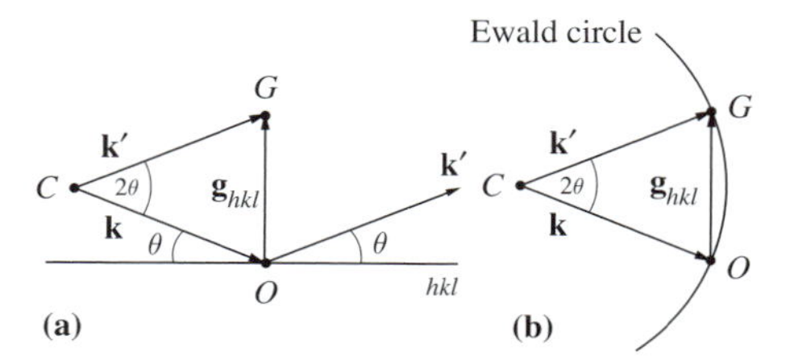

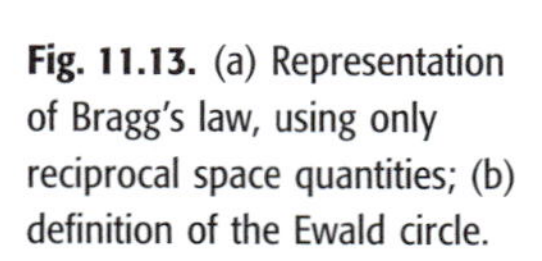
Fig. 11.13. (a) Representation of Bragg's law, using only reciprocal space quantities; (b) definition of the Ewald circle.

Ewald circle (or the *Ewald sphere* in 3-D), after the Austrian crystallographer Paul Peter Ewald (1888–1985), who first introduced this circle (sphere) into crystallography (Ewald, 1913, 1962). From the drawing, we also find that the endpoint of $\mathbf{k}'$ lies on the same normal to the plane (hkl) as the endpoint of $\mathbf{k}$. The distance between the two endpoints can be written as:

$$OG = |\mathbf{k}|\sin\theta + |\mathbf{k}'|\sin\theta = \frac{2\sin\theta}{\lambda}.$$

According to the Bragg equation, this ratio must be equal to $1/d$. This means that the distance between the points O and G is equal to the length of the reciprocal lattice vector $\mathbf{g}_{hkl}$. Therefore, we can simplify the drawing to the one shown in Fig. 11.13(b). The Bragg equation can be expressed in reciprocal space as:

$$\boxed{\mathbf{k}' = \mathbf{k} + \mathbf{g}.} \qquad (11.7)$$

The reciprocal space version of the Bragg equation states that *a diffracted beam with wave vector* $\mathbf{k}'$ *will be present if and only if the endpoint of the vector* $\mathbf{k}+\mathbf{g}$ *lies on the Ewald sphere*. The direction of the diffracted beam is then given by the direction of $\mathbf{k}+\mathbf{g}$.

A particular plane (hkl) in a crystal is said to be *in Bragg orientation*, if the corresponding reciprocal lattice point $\mathbf{g}_{hkl}$ lies on the Ewald sphere. The Ewald sphere thus provides a particularly simple interpretation for the geometry of the whole diffraction process: all we need to do to diffract X-rays from a given set of planes is rotate the crystal (or the incident beam) until the corresponding reciprocal lattice point falls on the Ewald sphere. The various experimental techniques presented in the following sections are all based on this simple observation.

Since the Ewald sphere is a central concept for diffraction experiments, it will be useful to give a few additional examples, both in 2-D and 3-D. First, consider the following problem: a reciprocal lattice from an orthorhombic crystal is shown in Fig. 11.14(a). The reciprocal lattice spacings are 3 nm^{-1} for the (100) planes, and 2 nm^{-1} for the (010) planes. The reciprocal lattice points shown in the figure all belong to the [001] zone. The question is then the following: if an X-ray beam with wave number 4 nm^{-1} is incident on this crystal, then what should be its direction to result in diffraction from the (120) planes? We will assume for now that the X-ray beam will lie in the plane of the drawing.

To answer this question, we look back at Fig. 11.13(b): the center of the Ewald sphere, C, lies on the perpendicular bisector line of the reciprocal lattice vector. If we change the wave length, λ, then, in order to maintain the diffraction condition, the point C must move along this line, so that the distance between C and O is kept at $1/\lambda$. This is shown schematically in Fig. 11.14(b). To determine the location of the point C for which the (120) planes will

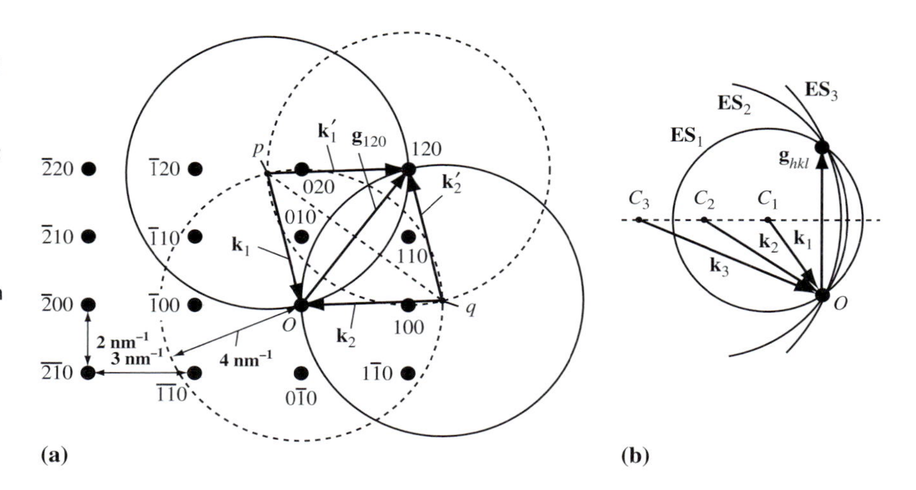

Fig. 11.14. (a) Orthorhombic [001] zone with superimposed incident beam directions $\mathbf{k}_1$ and $\mathbf{k}_2$ for which the planes (120) are in Bragg orientation; (b) illustration of the fact that the center, C, of the Ewald sphere always lies on the perpendicular bisector of the reciprocal lattice vector $\mathbf{g}$ when $\mathbf{g}$ lies on the Ewald sphere.

satisfy the Bragg condition, we proceed as follows: draw two circles (dashed in Fig. 11.14(a)) with radius $4\,\mathrm{nm}^{-1}$, one centered on the origin O, and one centered on the reciprocal lattice point $\mathbf{g}_{120}$. The circles intersect each other at two points, p and q, and these points lie on the perpendicular bisector line for $\mathbf{g}_{120}$ (dashed line). Either one of these two intersection points can be chosen as the center of the Ewald circle, which leads to two possible Ewald circles (full lines) and two corresponding incident beam directions $\mathbf{k}_1$ and $\mathbf{k}_2$. The diffracted beams, $\mathbf{k}'_i$, are then easily found by adding the reciprocal lattice vector $\mathbf{g}_{120}$.

As a second example, consider again the orthorhombic lattice of Fig. 11.14(a): is it possible to have both the (120) and $(\bar{1}00)$ planes *simultaneously* in Bragg orientation? In order for this to happen, we must place the center of the Ewald sphere on both of the perpendicular bisectors, i.e., on the intersection C of the perpendicular bisectors, as shown in Fig. 11.15. Note that there is only one wave length for which both of the reciprocal lattice points (along with the origin of reciprocal space) can lie on the Ewald sphere. We leave it to the reader to determine the wave length for which this is the case.

The previous two examples were drawn in 2-D to keep things simple, but, in reality, diffraction is a 3-D process. This is illustrated in Fig. 11.16, which shows how the Bragg condition for a particular plane is satisfied on a conical surface. This means that there are an infinite number of incident beam directions (those on the surface of the cone, pointing towards the apex) for which the geometrical condition for diffraction is satisfied. This is quite important, because it means that there are potentially many sample orientations for which a diffracted beam from a given set of planes can be observed. For the true 3-D situation, the Ewald sphere construction becomes a bit more difficult to visualize, but the main ideas are identical to those depicted in Figs. 11.14 and 11.15. In the case of Fig. 11.15, the perpendicular bisector

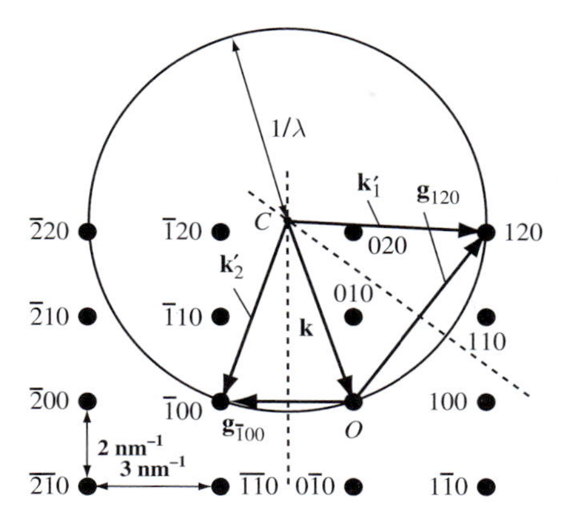

Fig. 11.15. Illustration of the geometrical condition to have both (120) and ($\bar{1}00$) simultaneously in Bragg orientation.

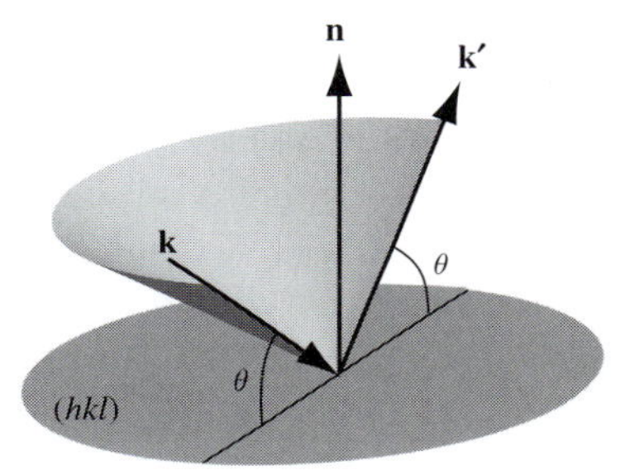

Fig. 11.16. The Bragg condition is satisfied for every incident beam direction $\mathbf{k}$ that lies on the cone surface (figure reproduced from Fig. 2.2b in *Introduction to Conventional Transmission Electron Microscopy*, M. De Graef, 2003, Cambridge University Press).

line becomes a perpendicular bisector *plane*, so that the Bragg orientation for (120) and ($\bar{1}00$) planes is simultaneously satisfied along the intersection line of the two bisector planes; this line is normal to the plane of the drawing and goes through the point C.

Before we conclude this section, it is useful to consider the concept of the *limiting sphere*. Consider the Bragg equation, rewritten as follows:

$$\sin\theta = \frac{\lambda}{2d} \leq 1. \tag{11.8}$$

The inequality is valid because the sin-function must always give a result between -1 and $+1$. From this condition, we can derive the range of interplanar spacings for which a diffracted beam can be generated. We rewrite the equation as follows

$$\lambda \leq 2d \quad \leftrightarrow \quad \frac{1}{d} \leq \frac{2}{\lambda} \quad \leftrightarrow \quad |\mathbf{g}| \leq 2|\mathbf{k}|. \tag{11.9}$$

Since $2|\mathbf{k}|$ is the diameter of the Ewald sphere, we find that diffraction can only occur from those reciprocal lattice points that lie inside a sphere with radius $2|\mathbf{k}|$, as shown in Fig. 11.17. All reciprocal lattice points outside of this sphere (represented by open circles) can never give rise to a diffracted beam (for the particular wave length selected for this drawing). Obviously, in 3-D, the limiting circle becomes a limiting sphere.

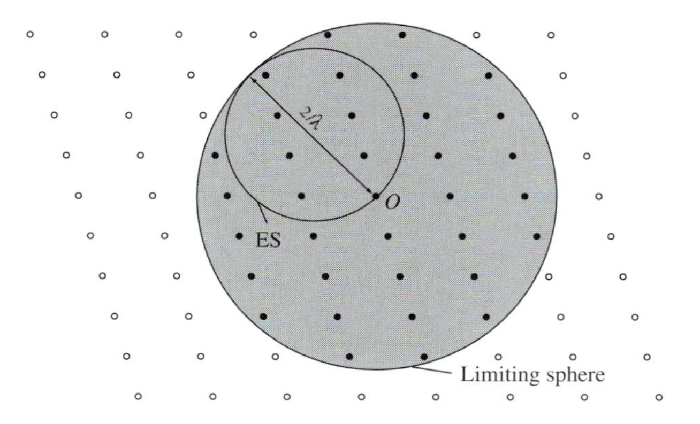

Fig. 11.17. Illustration of the limiting sphere; reciprocal lattice points outside the limiting sphere can never give rise to a diffracted beam for the selected wavelength.

This illustration also suggests an experimental approach for guaranteeing that a maximum number of diffracted beams can be observed: if the orientation of the incident beam is rotated around, so that the Ewald sphere sweeps across the entire volume of the limiting sphere, then each reciprocal lattice point inside the limiting sphere will, at some point, be in Bragg orientation. Alternatively, we could rotate the crystal, and hence the reciprocal lattice, so that each reciprocal lattice point would coincide with the Ewald sphere twice.[10]

11.4 Basic experimental X-ray diffraction techniques

In the previous section, we have seen that we can bring a particular plane in Bragg orientation in two different ways:

(i) rotate the crystal until the reciprocal lattice point $\mathbf{g}_{hkl}$ falls on the Ewald sphere;
(ii) rotate the incident beam direction (and hence the Ewald sphere) until the sphere intersects the reciprocal lattice point.

The direction of the wave vector is determined by the position and orientation of the X-ray tube. The X-ray beam that leaves one of the windows of the tube is rather wide and diverging and must first be collimated into the proper shape and size. This is usually done by means of collimating slits (or a collimating cylinder) which eliminate a portion of the beam and produce a more parallel incident beam. For more information on collimators, we refer the interested reader to Arndt and Wonacott (1977). Once the beam is collimated, it can be represented by a single incident wave vector, $\mathbf{k}$. The Ewald sphere, with its origin at the starting point of $\mathbf{k}$ and radius $1/\lambda$, is "attached" to the wave vector; in other words, the Ewald sphere moves along with the wave vector

[10] Why twice?

Table 11.4. Commonly used X-ray diffraction techniques and the types of materials they can be applied to.

technique	wavelength	crystal	application
Laue	polychromatic	single/large grained poly	orientation determination
Diffractometer	monochromatic	poly (powder)	phase identification
Weissenberg	monochromatic	single	lattice parameters
Precession	monochromatic	single	lattice parameters
Three-/four-circle	monochromatic	single	structure determination

when the incident beam direction is changed. Often, it is experimentally more practical to keep the X-ray beam direction constant; changing the incident beam direction requires that the complete X-ray tube, with watercooling and all attachments, be moved. Many standard X-ray diffraction techniques use a stationary beam. In experiments where the X-ray tube is stationary, the Ewald sphere does not change its position during the experiment.

For a given orientation of **k**, it is rather improbable that one or more than one reciprocal lattice points will fall on the Ewald sphere for an arbitrary crystal orientation. In fact, almost all reciprocal lattice points will *not* fall on the Ewald sphere, *unless we make them fall on the Ewald sphere*. If the sphere is fixed in space, then the only thing one can do to improve the chances of observing a diffracted beam is to rotate the crystal such that each reciprocal lattice point will, at some time, cross the Ewald sphere. The techniques presented in the next paragraphs accomplish this in different ways.[11]

Table 11.4 lists the most commonly used X-ray diffraction geometries, along with the types of crystals for which they are used. In the remainder of this chapter, we will discuss the standard powder diffractometer, the Debye–Sherrer camera, and the Laue camera. For the other methods we refer the interested reader to Volume C of the *International Tables for Crystallography*.

11.4.1 The X-ray powder diffractometer

In this section, we will consider only *polycrystalline* materials, meaning that the sample consists of many thousands or millions of *grains*, each with a random orientation. Consider, for example, pure copper: we have computed

[11] The discussion in this chapter is not intended to describe exhaustively all the details of X-ray diffraction methods, but, instead, deals only with the most important features of a number of important experimental techniques. For a more detailed description, we refer the interested reader to one of the many excellent textbooks available, e.g., Cullity and Stock (2001) (this is an updated version of Cullity's original 1978 text (Cullity, 1978)), Fultz and Howe (2002), Giacovazzo (2002b).

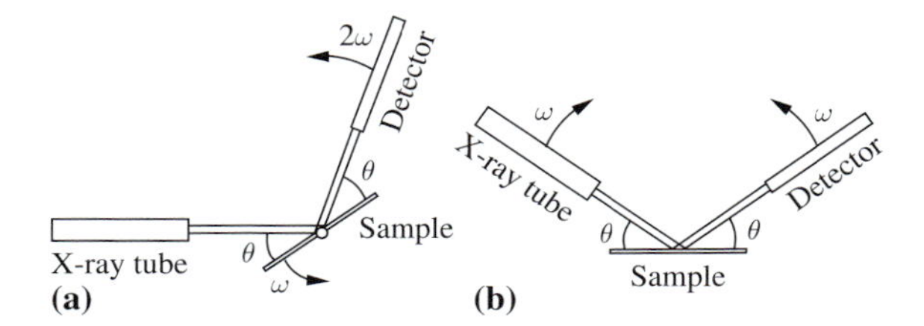

Fig. 11.18. Schematic experimental setup for the X-ray powder diffractometer: in (a), the X-ray tube is stationary, and both sample and detector move at angular rates ω and 2ω, respectively. In (b), the sample is stationary, and both X-ray tube and detector move towards each other at angular rate ω.

the diffraction angles 2θ for copper in Table 11.3. We can measure those angles experimentally in the following way: we know from the diffraction theory discussed in the previous sections that the angle between the incident and diffracted wave vectors must always be equal to 2θ. The angle between the diffracting plane and both of the wave vectors is equal to half of that, or θ. This suggests the experimental geometries shown in Fig. 11.18. In (a), the X-ray tube is mounted such that the incident wave vector lies in the horizontal plane. The sample is mounted on a platform that can rotate around an axis perpendicular to the plane of the drawing; the rotation angle is equal to θ and the angular rate is ω. The X-ray detector is mounted on a movable arm, which rotates around the same axis, but at twice the angular rate, 2ω. This means that the angle between the detector axis and the incident beam direction is always equal to twice the angle between the plane of the specimen and the incident beam direction. This geometry is known as the Bragg–Brentano geometry, or, more commonly, the θ–2θ geometry. A variant of the Bragg–Brentano diffractometer is shown in Fig. 11.18(b); in this setup, known as the θ–θ geometry, the sample is stationary, and both X-ray tube and detector move at angular rate ω towards each other.

The angular relations between the X-ray tube, the specimen and the detector guarantee that the conditions for diffraction are at all times satisfied. Since the crystal is polycrystalline, every possible orientation of every set of planes can be found somewhere in the sample. This means that, when the geometry is correct for diffraction from the (hkl) plane, there will be some grains in the sample for which the Bragg equation is satisfied, and they will give rise to diffracted intensity in the direction of the detector. The diffraction pattern is then formed by plotting the detector signal as a function of the diffraction angle 2θ. Note that only those planes for which the plane normal lies in the plane formed by the X-ray tube and the detector will give rise to a diffracted beam; therefore, only a small percentage of all grains in the sample will contribute to the final pattern. This technique only gives good statistical results if the average grain size is relatively small, say tens of microns. Typical diffraction patterns obtained with the Bragg–Brentano geometry are shown in Fig. 1.7.

If we consider the $\theta-2\theta$ diffraction experiment from a reciprocal space point of view, we come to the following interesting observation: we know that a diffracted beam will be present whenever a reciprocal lattice point intersects the (in this case stationary) Ewald sphere. Since the detector is mounted in a single plane (it can rotate around the sample axis but does not move in any other way) this means that it cannot detect diffraction from planes that are in Bragg orientation, but whose plane normal does not lie in the plane of the detector. Since it is more likely for a plane normal to lie on the Ewald sphere but not in the detector plane, than it is for a plane normal to lie in the intersection of both, this means that there are a large number of diffracted beams going off in directions other than that of the detector. This means that (1) we must make sure that the detector only "looks" at the diffracting region of the sample and (2) we must shield the region around the diffractometer so that no radiation will leak into the room. The first goal is accomplished by adding "collimators" in front of the detector; a collimator is a narrow slit that limits the field of view of the detector to only the area of the sample that is illuminated by the incident X-ray beam. One also employs collimators to define the shape of the illuminated area on the sample. Shielding is accomplished by surrounding the complete diffractometer with a lead-glass chamber; to prevent a distracted operator from opening the chamber while the beam is on, the sliding glass doors are interlocked with the beam shutter.

Powder diffractometry has become one of the standard tools of materials science. Often, one of the first characterization steps undertaken after a new material has been fabricated is to record a powder pattern, to identify which phases or crystal structures are present. To facilitate the task of identifying a structure, databases have been created, that contain tens of thousands of crystal structures. These databases used to be printed on small index cards but are now available on CD-ROM. The *International Center for Diffraction Data* (ICDD) publishes a number of databases known as the *Powder Diffraction Files*. The largest of these contains 271 813 material data sets;[12] several smaller versions, limited to organic entries or minerals, are also available. One can use these databases to search for a particular structure; often the input to the search program consists of the 2θ values for the three most intense peaks of the spectrum. That information, combined with some knowledge of the chemistry of the sample, is often sufficient to identify the crystal structure. The *Powder Diffraction Files* have become essential tools in the study of the structure of materials.

11.4.1.1 The Debye–Scherrer camera

Before the advent of automated diffractometers, powder diffraction patterns were obtained routinely using the *Debye–Scherrer camera*. The camera

[12] This number refers to the PDF4+ database in March of 2006.

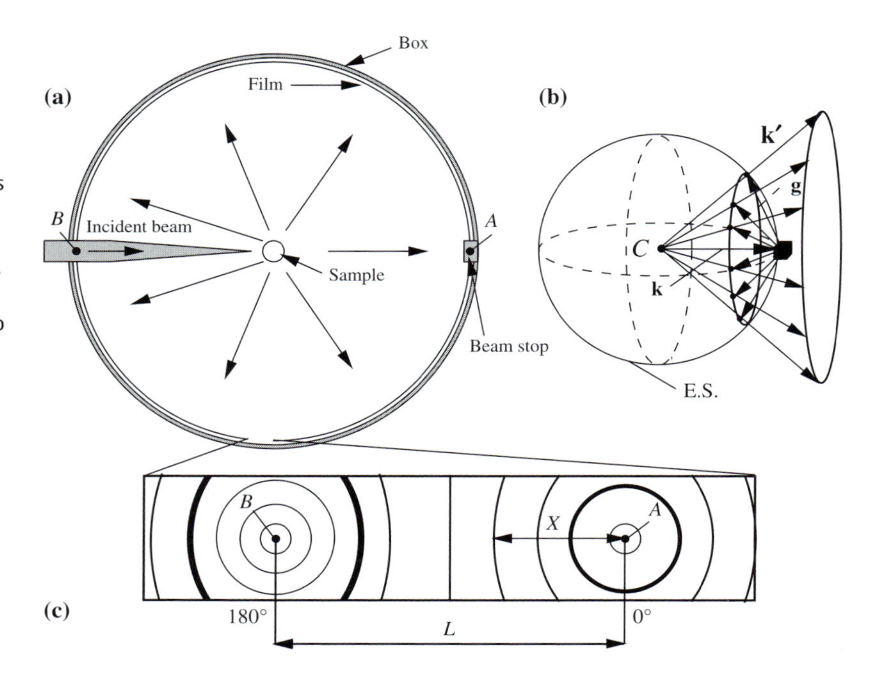

Fig. 11.19. (a) Experimental setup for the Debye–Scherrer camera. The drawing in (b) represents the Ewald sphere construction for this camera. When the photographic strip is straightened out, a diffraction pattern consisting of ring segments can be observed (c). From the measurement of the ring positions x with respect to the distance A–B, the diffraction angle 2θ can be derived.

consists of a flat cylindrical box, with radius R of about 8 cm, and thickness of about 3 cm. Inside this box, against the outside wall, a strip of photographic film is mounted. The sample is mounted in the center of the box (see Fig. 11.19). The X-ray beam enters the box from the left, through a collimator tube, and hits the sample. A large fraction of the intensity goes straight through the sample and exits the box on the opposite side, or, alternatively, the beam hits a *beam stop*, an X-ray absorbing material.

The diffraction geometry for this setup is very similar to that for the modern diffractometer. The shape of the sample is different: where the diffractometer uses a flat sample, the Debye–Scherrer camera uses a cylindrical sample. Since every set of crystallographic planes in the sample is present with a random orientation, there will be a large number of grains for which the (hkl) plane is in Bragg orientation. The incident wave vector $\mathbf{k}$ is fixed (which also fixes the Ewald sphere), all the planes of the type (hkl) that are in Bragg orientation will have their plane normals $\mathbf{g}_{hkl}$ intersect the Ewald sphere along a circle (Fig. 11.19(b)). The *diffracted* wave vectors $\mathbf{k}'$ will, therefore, lie on the surface of a *cone*, with top in the center C of the Ewald sphere and surface going through the endpoints of all vectors $\mathbf{g}_{hkl}$.

When this conical surface of diffracted radiation intersects the film, mounted against the outside wall of the camera, ring-segments will appear on the film. An example of such a ring pattern is shown in Fig. 11.19(c). The hole on the right corresponds to the position A of the beam stop (and hence $2\theta = 0$), and the hole on the left corresponds to the position of the incident beam (B, or $2\theta = 180°$).

The value of 2θ for a particular diffraction ring can be measured as follows: determine the distance L (in mm) between the centers of the holes in the film. Label each ring with a sequential number, and measure the distance between the rightmost hole and the intersection of each ring with the line connecting the centers of the holes (call this distance x_i, where i numbers the rings). The ratio x_i/L then determines the value of 2θ as:

$$2\theta_i = 180^\circ \frac{x_i}{L}. \tag{11.10}$$

One can again use the Bragg equation to convert these angles into interplanar spacings. Note that the filmstrip only intersects with part of the conical surface. The fraction of the conical surface that intersects with the film depends on the angle 2θ: for values of 2θ close to 0° or 180°, a larger fraction of the conical surface is intercepted by the film than for angles around 90°. In the next chapter, we will define a correction factor to account for this angular dependence.

11.4.1.2 The Laue methods

The first X-ray diffraction experiments were conducted by von Laue around the beginning of the twentieth century (Friedrich *et al.*, 1912). His experiments were of fundamental importance because he showed simultaneously that (1) X-rays are electromagnetic *waves* and (2) crystals are made up of regular arrangements of atoms. In a sense, this was the first direct experimental evidence that atoms exist. The experimental setup to obtain a so-called *Laue pattern* is rather simple. A crystal (preferably a single crystal but a polycrystal with large grains will also work) is mounted in the path of an X-ray beam. Instead of using a monochromatic X-ray beam, one removes the filter from the beam path, so that the complete X-ray spectrum of wave lengths is present in the beam. This means that, instead of having only one wave vector **k** in the beam, one now has a range of wave vectors with lengths from very small up to $1/\lambda_{\text{swl}}$ (swl = short wavelength limit).

Fig. 11.20(a) shows the experimental setup for the *transmission Laue* method. The polychromatic beam of X-rays goes through the crystal and the diffracted beams are intercepted by a planar detector (originally a photographic negative, but nowadays typically a CCD camera or image plate), mounted at right angles to the beam path at some fixed distance L. The film has a hole in the center to allow the direct beam to pass through (a direct beam hitting a photographic negative would completely overexpose the negative, particularly for this setup which often requires one or more hours of exposure time). In Fig. 11.20(b) an alternative setup, known as *reflection Laue*, is shown. In this case, the detector records the diffracted beams for which 2θ is close to 180°.

The geometry of a Laue pattern is somewhat more complicated than that of a powder or ring-pattern. Let us first consider the transmission Laue method.

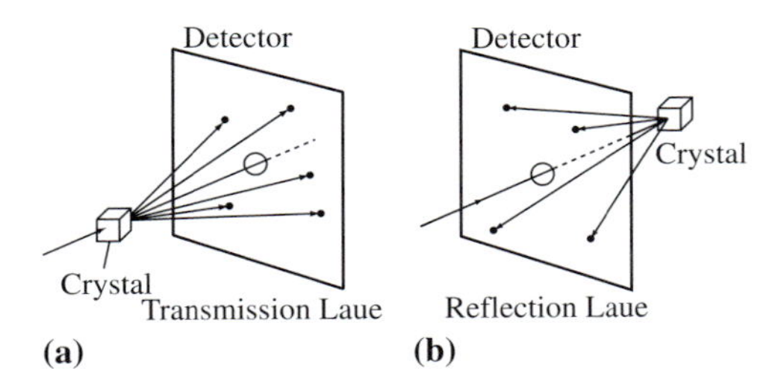

Fig. 11.20. Transmission and reflection Laue methods.

The single crystal has a fixed orientation with respect to the fixed incident beam. This means that, in addition to **k**, all the vectors $\mathbf{g}_{hkl}$ are fixed in space. If there were only one value for the length of **k** (i.e., monochromatic radiation), then the probability that any of the reciprocal lattice points would fall onto the Ewald sphere would be extremely small. However, for the Laue method one uses *polychromatic* radiation, which means that instead of a single Ewald sphere, we now have a "Ewald volume" available. This can be understood by considering two different wave lengths in the continuous spectrum, $\lambda_1 < \lambda_2$. The corresponding wave vectors $\mathbf{k}_1$ and $\mathbf{k}_2$ are parallel and with each vector we can associate a Ewald sphere, shown in Fig. 11.21. Since all wave lengths between λ_1 and λ_2 are also present in the beam, the complete area (volume) between the two Ewald spheres (shaded in the figure) can give rise to a diffracted beam. Every reciprocal lattice point that falls inside this volume will give rise to a diffracted beam. This means that the number of beams simultaneously excited can be rather large. Since a photographic emulsion reacts to all wave lengths in an almost identical way, one cannot determine which wave length gives rise to which reflection. It is thus impossible to determine interplanar spacings using the Laue methods. It is possible, however, to determine the *orientation* of the crystal with respect to the incident beam.

The positions of the diffracted beams on a Laue photograph can be predicted in the following way: in Chapter 3, we derived a relation between the interplanar spacing and the length of the reciprocal lattice vectors, $|\mathbf{g}_{hkl}| = 1/d_{hkl}$.

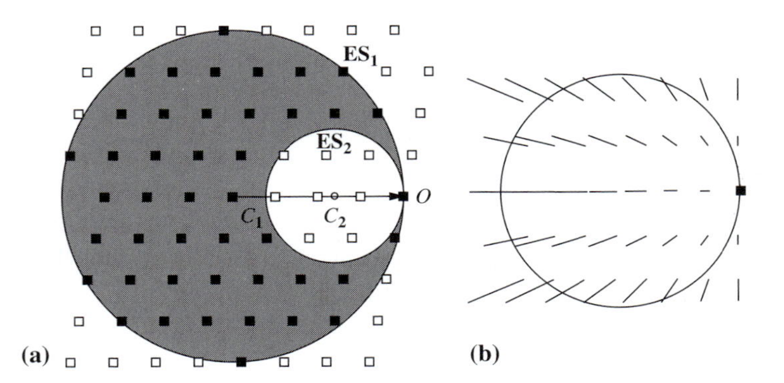

Fig. 11.21. (a) Ewald sphere and diffracting volume for continuous radiation (drawing for K = 1). (b) Alternative interpretation, where the Ewald sphere has unit radius, and the reciprocal lattice points are streaks (case for K = λ).

We can generalize this expression by writing $|\mathbf{g}_{hkl}| = \mathrm{K}/d_{hkl}$, where K is a constant. We recover the traditional equation by setting $\mathrm{K} = 1$. We must introduce the same factor K into the definition of the wave vector:

$$|\mathbf{k}| = \frac{\mathrm{K}}{\lambda}$$

and again $\mathrm{K} = 1$ results in the standard definition. The geometry of a Laue diffraction pattern can readily be understood by setting $\mathrm{K} = \lambda$, in other words, we rescale all dimensions in units of the wave length, rather than in Ångstroms or nanometers. This means that the length of the wave vector becomes equal to 1 ($= \lambda/\lambda$), and the reciprocal lattice point $\mathbf{g}_{hkl}$ is now located at a distance λ/d_{hkl} from the origin of reciprocal space. If we use poly-chromatic radiation, then each lattice point becomes a *line segment*, as shown in Fig. 11.21(b), since there is a range of λ values. The Ewald volume defined on the left side of the figure collapses into a single Ewald sphere, since each wave length is measured in units of itself. Whenever a line segment intersects the Ewald sphere, a diffracted beam will occur. Since this intersection point can occur anywhere along the line segment, we have no information about the actual wave length of the diffracted beam, as already mentioned above. We will call the line segments *extended reciprocal lattice points*.

From the zone equation we know that all planes belonging to a zone have their plane normals in the plane perpendicular to the zone axis. This plane is a plane in reciprocal space, going through the origin. All the extended reciprocal lattice points in this plane intersecting the normalized Ewald sphere will give rise to a diffracted beam; those beams lie on the surface of a cone, with top in the center, C, of the Ewald sphere. The intersection of a cone with the detector plane is an *ellipse*, as shown in Fig. 11.22(a). The incident beam direction is along the line segment CO. Planes belonging to one zone axis will thus give rise to reflections which are arranged in ellipses on a transmission Laue photograph. On a reflection Laue photograph the same interpretation is valid and the intersection of the diffraction cone with a plane on the other side of the sample gives rise to *hyperbolic* curves (Fig. 11.22(b)). Planes belonging to a zone axis thus give rise to hyperbolic sets of spots on a reflection Laue photograph.

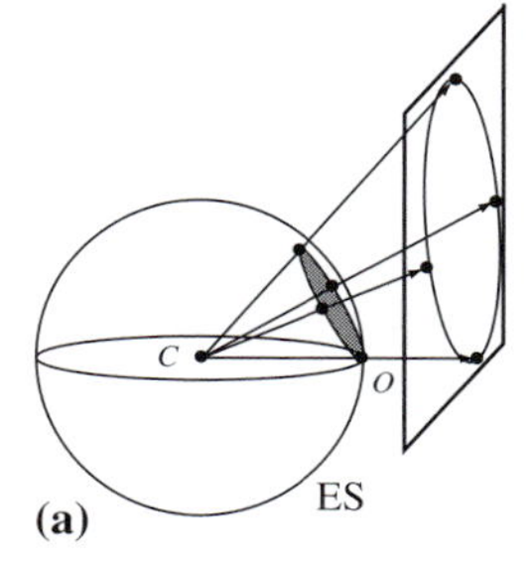

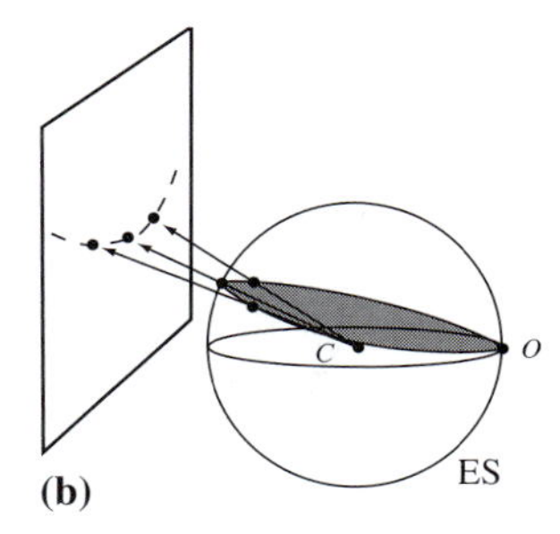

Fig. 11.22. $\mathrm{K} = \lambda$ interpretation of zone axis diffraction in (a) transmission and (b) reflection Laue photography. The shaded planes are normal to zone axis directions.

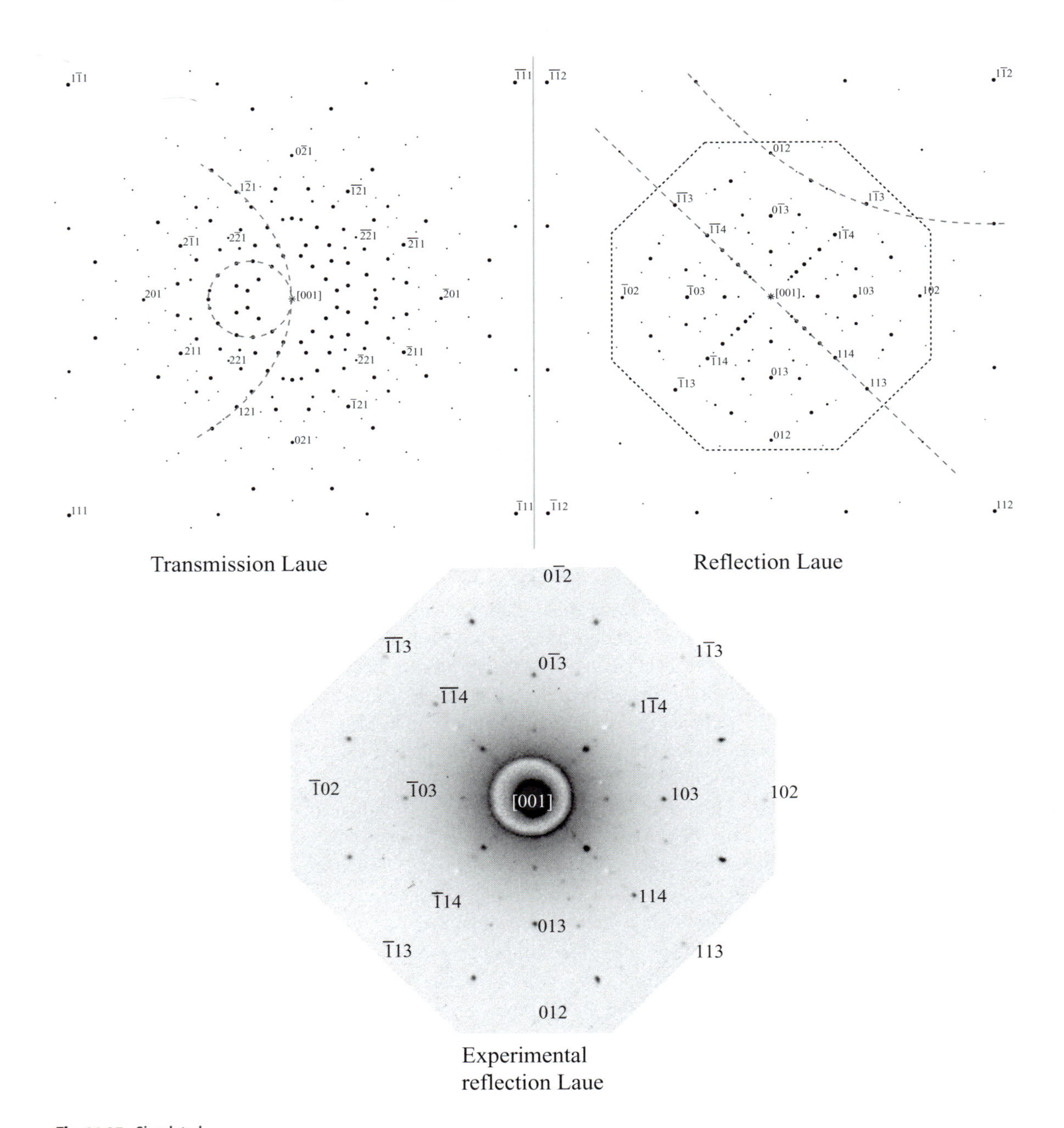

Fig. 11.23. Simulated transmission and reflection Laue patterns for the [001] orientation of a Si single crystal. At the bottom, an experimental reflection Laue pattern is shown with the beam along the [001] orientation.

Fig. 11.23 shows examples of computed Laue patterns for the [001] zone axis orientation of a Si crystal (i.e., the incident beam was taken to be parallel to the [001] direction). On the transmission Laue pattern one can clearly see the ellipses corresponding to planes that belong to the same zone (two ellipses are indicated in dashed lines). The reflection Laue pattern, on the other hand, has reflections that lie along hyperbolic curves, as indicated by the dashed line, and straight lines through the origin; the straight lines are degenerate hyperbolae. Multiple such straight lines occur only when the incident beam is aligned with a (low index) crystallographic direction. Note that both Laue patterns reveal the symmetry of the projection of the cubic crystal; both patterns have four-fold rotational symmetry and two pairs of mirror planes normal to the plane of the figure (i.e., a 4 mm planar point group). An experimental Laue pattern corresponding to the octagonal area outlined in the simulated reflection Laue pattern is shown at the bottom of Fig. 11.23. Only reflections with low Miller indices are indexed. There is good agreement between the experimental and simulated patterns.

To facilitate the interpretation of reflection Laue photographs a tool is used to convert the positions of the spots on the photograph to a stereographic projection. This tool is the *Greninger Chart*, shown in Fig. 11.24; by superimposing this chart onto a Laue photograph one can directly read the stereographic coordinates from the set of curved lines and transpose the spots onto a Wulff net. The Greninger chart is computed for a given source–sample distance (typically 2 or 3 cm). The angles γ and δ are read from the chart, and then transferred onto a stereographic projection. From the angles between reflections on the Wulff net one can then deduce the Miller indices of the reflections, provided the crystal structure is known. Laue recordings are predominantly used to determine the orientation of a single crystal (or a grain in a large-grained sample) with respect to some external reference frame. This is particularly useful if the single crystal does not have any well defined facets. We leave it to the reader to derive the mathematical relations between the stereographic projection, the Greninger chart, and the reflection Laue pattern.

This concludes our brief discussion of some of the more important X-ray diffraction methods. There are many other techniques, in particular for dealing with single crystals, and we refer the interested reader to volume C of the *International Tables for Crystallography* for detailed descriptions. In the next chapter, we will introduce methods to compute the

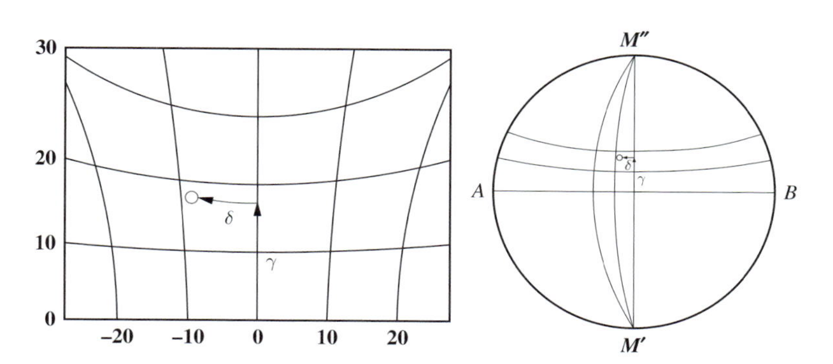

Fig. 11.24. Greninger chart (left) for conversion of a reflection Laue photograph to a stereographic projection (right).

intensity of diffracted beams for the powder diffraction (Bragg–Brentano) geometry.

11.5 Historical notes

William Henry Bragg (1862–1942) was the pioneer in the development of X-ray diffraction as a tool in crystallography. He was a British physicist and father of **William Lawrence Bragg** (1890–1971). The elder Bragg studied the ionizing properties of α-, β-, γ-, and X-rays, and developed the idea of X-rays and γ-rays as consisting of neutral-pair particles. In 1915, W. H. Bragg showed that the diffraction of X-rays by crystals could be interpreted in terms of reflection from atomic lattice planes of the crystal. When a beam of parallel, monochromatic X-rays is scattered off a crystal, the reflected waves emerge in phase if the so-called Bragg condition, relating the fixed wavelength, λ, the interplanar spacing, d, and the angle of reflection, θ, is satisfied. For waves satisfying the Bragg condition, maxima in the scattered intensity are observed. W. H. Bragg was also the first to use Fourier series to represent electron densities in crystalline solids. W. L. Bragg was a pioneer in the development of X-ray diffraction as a tool in crystallography. He was responsible for the first structural determinations of crystals using X-rays, and an example of his writings will be shown in the historical section of Chapter 14. The Braggs used scattering results to infer electron density maps around atoms in crystalline solids.

(a) (b)

Fig. 11.25. (a) William H. Bragg (1862–1942), and (b) William L. Bragg (1890–1971) (pictures courtesy of the Nobel Museum).

11.6 Problems

(i) *Absorption of X-rays I*: Compute the mass absorption coefficient for the alloy Cu_3Au, for copper $K\alpha$ radiation.

(ii) *Absorption of X-rays II*: If the ratio $I(K\alpha)/I(K\beta)$ before filtering is 7.5 : 1 for a copper target, then compute the thickness of a nickel filter, that would increase this ratio to 500 : 1.

(iii) *Absorption of X-rays III*: You are an engineer working on the design of lead shielding for medical applications of X-rays. How thick should a lead shield be if it has to attenuate the shortest wavelength of a 30 000 volt copper tube by a factor of 10 000? (Hint: use the equation $\mu/\rho = k\lambda^2 Z^3$ with $k = 7.80 \times 10^{20}$ per gram per centimeter.)

(iv) *Diffraction angles I*: Compute the diffraction angles 2θ for the (100), (010), and (001) reflections of a crystal with lattice parameters $\{0.2, 0.3, 0.4, 90, 60, 45\}$ (in nanometers), assuming Mo $K\alpha$ radiation.

(v) *Diffraction angles II*: One measures the powder diffraction pattern of an unknown crystal. Suppose that one knows the Miller indices corresponding to the various reflections; in particular, the diffraction angles associated with three reflections are given by

$$2\theta_{100} = 29.78^\circ$$

$$2\theta_{220} = 93.24^\circ$$

$$2\theta_{310} = 108.70^\circ$$

Could this be a cubic crystal? If so, what is the lattice parameter (assuming Cu $K\alpha$ radiation)?

(vi) *Diffraction angles III*: For a cubic crystal with lattice parameter $a = 0.408$ nm, compute the Miller indices of the plane for which the diffraction angle 2θ will be largest (i.e., closest to 180°), assuming Cu $K\alpha$ radiation. What will this value be for Mo $K\alpha$ radiation?

(vii) *Diffraction angles IV*: Consider a cubic crystal structure with lattice parameter $a = 0.5$ nm. X-rays are generated using a Fe target ($K\alpha$ radiation).

(a) Draw all reciprocal lattice points with indices up to 3 for the zone [001]. Make sure the points are drawn in the correct relative position.

(b) Draw to scale, on the same drawing, the radiation wave vector for an incident beam directed along the [110] direction. Also draw the corresponding Ewald sphere.

(c) What would be a possible direction for the incident beam so that the (320) reciprocal lattice point would fall on the Ewald sphere? (You may draw the answer or compute it.)

(d) What is the diffraction angle for the reflection (320)?

(viii) *X-ray powder diffraction analysis for a cubic crystal*: KI is a salt. Suppose one measures a powder diffraction pattern for cubic crystals of KI using Cu Kα radiation and determines the positions of the first nine reflections to occur at:

$2\theta = 21.80°; 25.20°; 36.00°; 42.50°; 44.45°; 51.75°; 56.80°; 58.45°; 64.65°$

(a) Determine the d-spacing for each of these peaks using Bragg's law.
(b) It is suspected that KI crystallizes in the rocksalt (NaCl) structure described in Chapter 22. The ionic radii for I^{-1} and K^{+1} are 0.22 and 0.138 nm, respectively. Predict the cubic lattice constant, a, for KI.
(c) Determine the Miller indices for each of the reflections and refine the lattice constant, a, for KI based on the experimental data. Comment as to whether there are any indices that are a mixture of even and odd numbers.
(d) The density of KI is determined to be 3.13 g/cm^3. Determine the number of formula units per unit cell.
(e) Show that the number of formula units per cell agrees with the suspected rocksalt (NaCl) structure.

(ix) *X-ray powder diffraction limits on reflections*: For a cubic crystal with lattice parameter $a = 0.408$ nm, compute the Miller indices of the plane for which the diffraction angle 2θ will be largest (i.e., closest to 180°), assuming Cu Kα radiation. What will this value be for Mo Kα radiation?

(x) *Debye–Scherrer X-ray pattern*: The figure below shows a Debye–Scherrer X-ray pattern for W metal using Mo Kα radiation.

(a) Describe the differences between an X-ray powder pattern and a Debye–Scherrer pattern.
(b) Use a spreadsheet program to tabulate the following quantities in columns: x, θ, $\sin^2\theta = \lambda^2(h^2+k^2+l^2)/4a^2$ and d_{hkl};
(c) Index the pattern and determine $\lambda^2/4a^2$ and the lattice constant, a, for W.

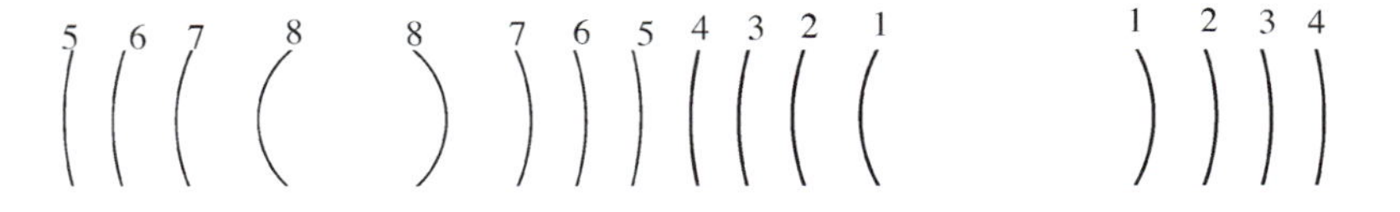

Fig. 11.26. Debye–Scherrer pattern for W using Mo Kα radiation.

(xi) *X-ray diffraction*: Pt is a noble metal catalyst. Metallic Pt has an *fcc* crystal structure with a lattice constant $a = 0.3924$ nm. Calculate d-spacings and predict values of 2θ for the four lowest angle peaks in a powder XRD pattern, using Cu Kα radiation.

(xii) *Laue diffraction pattern I*: Derive the mathematical relations that relate the stereographic coordinates to the coordinates in a reflection Laue pattern (i.e., the mathematical relations that describe the curves on a Greninger chart).

(xiii) *Laue diffraction pattern II*: A transmission Laue pattern is made of a cubic crystal with a lattice parameter of 0.36 nm. The X-ray beam is horizontal. The $[0\bar{1}0]$ axis of the crystal points along the beam towards the X-ray tube, the [100] axis points vertically upward. The film is 4 cm from the crystal.

(a) What is the wave length of the radiation diffracted from the $(\bar{3}10)$ planes?

(b) How high above the horizontal plane (the plane containing the incident X-ray beam) will the $\bar{3}10$ reflection strike the film (in cm)? You may assume that the diffracted beam is created at the origin of the reference frame (or, equivalently, that the crystal dimensions are very small compared to the scale of the experiment).

CHAPTER

12 X-ray diffraction: intensities

> *"The flickering greenish light, crackling and smell of ozone were sufficiently terrifying to impress the incident deeply in a child's mind. When I think, however, of the early experiments, the interest which they aroused in medical men is not their chief significance to me! I see them as fore-runners of my father's interest in the ionization of gases leading to his experiments with X-rays from radium and finally the experiments on the diffraction of X-rays by matter which we carried out together."*
>
> W. L. Bragg, foreword to "Salute to the X-ray Pioneers of Australia"

12.1 Scattering by electrons, atoms, and unit cells

We have seen in the previous chapter how X-rays are generated when an electron is accelerated or decelerated. If a beam of X-rays is incident upon a collection of electrons, either bonded to atoms or in a conduction band, then the electric field associated with the X-rays will force those electrons into oscillation. Because of the forced oscillation, they will emit their own X-rays, and this phenomenon is known as *X-ray scattering*. In the following subsections, we will describe quantitatively how first a single electron, then an atom, and finally a complete unit cell scatters an incident beam of X-rays.

12.1.1 Scattering by a single electron

Consider a single electron located in the origin of a reference frame (Fig. 12.1). Assume that an X-ray beam goes from the negative x-direction towards the electron. An observer is located at the point P, in the $x-z$ plane, at a distance r from the origin, and at an angle 2θ above the $x-y$ plane (one can always

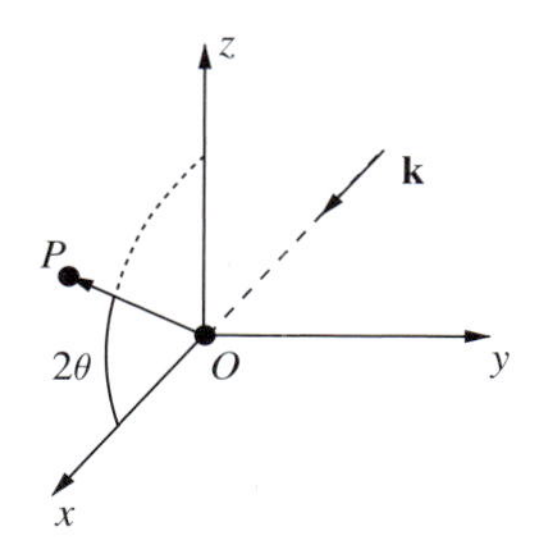

Fig. 12.1. Reference frame used for the computation of single electron scattering.

rotate the reference frame so that this setup is realized). If the incident beam has intensity I_0, then the scattered radiation at the point P can be computed from the *Thomson* equation:

$$\boxed{I = I_0 \frac{\mathrm{K}}{r^2} \sin^2 \alpha,} \tag{12.1}$$

where α is the angle between the scattering direction and the direction of acceleration of the electron. The constant K is given by:

$$\mathrm{K} = \left(\frac{\mu_0}{4\pi}\right)^2 \times \left(\frac{e^4}{m^2}\right) = 7.94 \times 10^{-30}\mathrm{m}^2.$$

This is a very small number, which indicates that the scattering due to a single electron is rather weak. It is only when one brings together large numbers of electrons (of the order of 10^{23} or more) that the scattering becomes easily measurable.

The electric field of an incident X-ray beam will cause the electron to oscillate along a direction parallel to the electric field vector **E**. Since this vector is perpendicular to the propagation direction (which we have taken along the positive x-axis), the vector must be located in the $y-z$ plane. The direction of **E** is known as the *polarization direction*. For a normal X-ray beam, the polarization is random. This means that, on average, the y component of **E** must be equal to the z component. The intensity of an X-ray beam depends quadratically on the magnitude of the electric field, and hence the components of the intensity along the y and z directions must be, on average, equal to each other and equal to half the total intensity. Mathematically this is stated as follows:

$$I_{0y} = I_{0z} = \frac{1}{2} I_0.$$

The average incident X-ray photon can thus be decomposed in a component along y, and a component along z. Let us determine how each of these components contributes to the scattering at the point P. First, we consider the y component. The angle between the y direction and the scattering direction is $\alpha = \pi/2$, from which we find:

$$I_y(P) = I_{0y} \frac{\mathrm{K}}{r^2}.$$

For the z component, the angle α becomes equal to $\pi/2 - 2\theta$ and we find

$$I_z(P) = I_{0z} \frac{\mathrm{K}}{r^2} \cos^2 2\theta.$$

The total scattered intensity at the point P is equal to the sum of the two components:

$$I_P = I_0 \frac{\mathrm{K}}{r^2}\left(\frac{1+\cos^2 2\theta}{2}\right). \tag{12.2}$$

Note that the scattered intensity is strongest for $\theta = 0$ and $\theta = \pi$; it is weakest for $\theta = \pi/2$. The angular factor in Equation 12.2 is known as the *polarization factor*. We will return to this factor later on in this chapter.

From here on, we will assume that the observer is located far from the sample (far compared to the atomic scale) and the factor K/r^2 will not always be explicitly written. Most diffraction experiments do not work with absolute diffracted intensities, but with relative values; one takes the strongest diffraction peak to be the value 100, and re-scales all peaks with respect to this peak. In the re-scaling process, pre-factors such as K/r^2 cancel out against each other.

There is another way in which X-rays can be scattered by an electron. From quantum mechanics we know that a particle can have both particle-like and wave-like properties. The reverse is also true: a wave can in certain situations behave as if it were a particle. This is the reason why we regard electromagnetic rays as particles, or *photons*. As a particle, a photon has a definite momentum and it can transfer part of this momentum in a billiard-like collision with the electron. In doing so, the photon loses part of its energy and therefore it changes its wavelength. Radiation scattered in this manner is known as *Compton modified radiation* and the "collision process" is known as *Compton scattering*. Since the X-ray photon loses energy during this process, the process is referred to as an *inelastic scattering event*. During such an event, the *phase* of the X-ray photon is changed in a random way, so that the photon no longer carries phase sensitive (i.e., diffraction) information. Compton modified radiation is thus useless from a diffraction point of view, but it does contribute to the signal in an X-ray detector. Photons which are scattered through the normal "Thomson" process do not undergo a random phase change, but have their phases modified by half a wavelength (phase shift of π). Thomson scattering is also known as *coherent* scattering.

12.1.2 Scattering by a single atom

When an X-ray beam is incident upon an atom with atomic number Z, each of its electrons will scatter the X-ray photons according to the Thomson equation. In addition, some of the X-ray intensity will be scattered *incoherently*, or via the Compton process. In this section, we will only regard the coherent scattering of X-rays by a single atom.

In the forward direction, $\theta = 0$, each of the Z electrons will scatter the beam with an identical phase change π. Since there is no difference between the path lengths of the X-rays scattered in this direction (see Fig. 12.2(a)), there is also no destructive interference, and the total scattering in the forward direction is equal to Z times that of a single electron.

In all other directions, $\theta \neq 0$, there will be some path length difference between X-rays scattered by the different electrons and the total scattered intensity will decrease from the level of the forward scattered beam (see Fig. 12.2(b)). The exact mathematical theory for scattering from a single atom is described in some detail in Chapter 21. For now, it is sufficient to say that it involves an integral of the electron wave-function (multiplied by an appropriate quantum mechanical operator) over the entire volume of the atom. These calculations have been done for all atoms and the results are tabulated in the *International Tables for Crystallography*. From the mathematical treatment, one finds that the important variable governing the diffraction process is the ratio of the sine of half the diffraction angle to the wavelength, i.e., $\sin\theta/\lambda$. One defines the *atomic scattering factor*, f for a given angle θ and wavelength λ as the ratio of the amplitude scattered by the entire atom to the amplitude scattered in the same direction by a single electron. The atomic scattering factor is thus a function of the variable, $\sin\theta/\lambda$. The atomic scattering factors for Cu and Au are shown in Fig. 12.3. Note that the value for $\theta = 0$ is

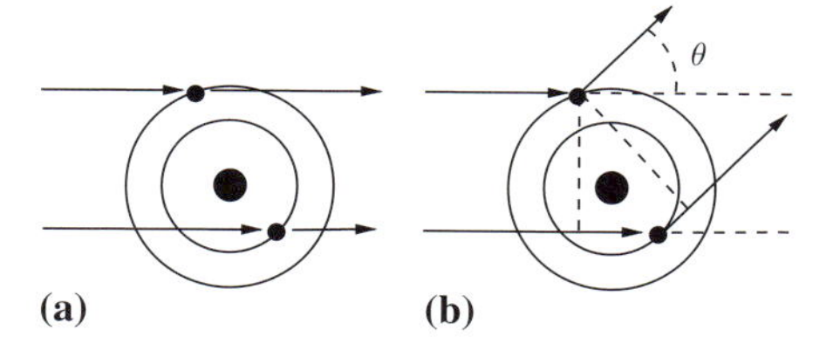

Fig. 12.2. (a) Forward scattering and (b) scattering at an angle θ. Note that there is no path length difference for forward scattering.

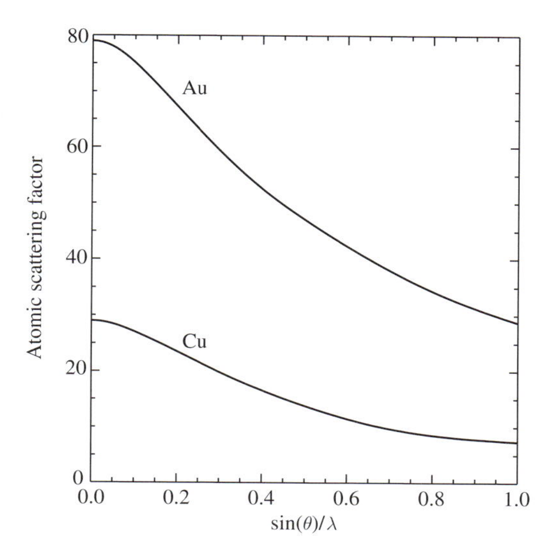

Fig. 12.3. Atomic scattering factors for copper and gold as a function of $\sin\theta/\lambda$.

indeed equal to the atomic number, and that the curve decreases rapidly with increasing scattering angle or decreasing wavelength.

Most tables list the atomic scattering factors in terms of curve fitting parameters. The curves represented in Fig. 12.3 are fitted with a sum of either three or four exponential functions. The scattering factor for a particular value of $\sin\theta/\lambda$ can then be computed using the following equation:

$$f(s) = Z - 41.782\,14 \times s^2 \times \sum_{i=1}^{N} a_i e^{-b_i s^2} \tag{12.3}$$

where $s = \sin\theta/\lambda$. The number of terms in the summation, N, is either 3 or 4. Table 12.1 lists the values of the coefficients a_i and b_i for all the elements in the periodic table. The elements that have a + in front of their atomic number use four terms in the expansion, the others use three terms. The numbers are the result of computations by Doyle & Turner (Doyle and Turner, 1968) and Smith & Burge (Smith and Burge, 1962). Note that the equation assumes that the wavelength is expressed in angstroms.

This table can be used in the following way: suppose one wants to compute the contribution of a single tungsten atom to diffraction of copper Kα radiation from the (222) plane of a body-centered cubic crystal with lattice parameter $a = 3.1653$Å.[1] We need to evaluate the scattering parameter s for this particular situation. From the Bragg equation we know that:

$$s = \frac{\sin\theta}{\lambda} = \frac{1}{2d_{hkl}}.$$

The value for d_{222} in a cubic crystal is easily found using:

$$d_{hkl} = \frac{a}{\sqrt{h^2+k^2+l^2}}$$

from which we find $d_{222} = 3.1653/\sqrt{12} = 0.9137$Å. The scattering parameter s is then equal to 0.5472 Å^{-1}. Substitution of this value, and the parameters for a_j and b_j for tungsten, into the atomic scattering factor equation results in:

$$\begin{aligned} f_W(0.5472) &= 74 - 41.78214 \times (0.5472)^2 \times \Big[5.709e^{-28.782(0.5472)^2} + \\ &\quad 4.677e^{-5.084(0.5472)^2} + 2.019e^{-0.572(0.5472)^2}\Big] \\ &= 74 - 34.0642 \\ &= 39.9358. \end{aligned}$$

[1] Note that we work in angstroms instead of nanometers, since the values in Table 12.1 are listed in angstrom units.

Table 12.1. Atomic scattering parameters for all elements $Z = 1$–98: Doyle & Turner parameters have a + in front of the atomic number, the others are Smith & Burge parameters. This table assumes that s is expressed in Å^{-1}; if s is expressed in nm^{-1}, then all entries must be multiplied by 0.01.

Name	Z	a_1	b_1	a_2	b_2	a_3	b_3	a_4	b_4
Ac	89	6.278	28.323	5.195	4.949	2.321	0.557	—	—
Ag	+47	2.036	61.497	3.272	11.824	2.511	2.846	0.837	0.327
Al	+13	2.276	72.322	2.428	19.773	0.858	3.080	0.317	0.408
Am	95	6.378	29.156	5.495	5.102	2.495	0.565	—	—
Ar	+18	1.274	26.682	2.190	8.813	0.793	2.219	0.326	0.307
As	+33	2.399	45.718	2.790	12.817	1.529	2.280	0.594	0.328
At	85	6.133	28.047	5.031	4.957	2.239	0.558	—	—
Au	+79	2.388	42.866	4.226	9.743	2.689	2.264	1.255	0.307
B	+05	0.945	46.444	1.312	14.178	0.419	3.223	0.116	0.377
Ba	+56	7.821	117.657	6.004	18.778	3.280	3.263	1.103	0.376
Be	+04	1.250	60.804	1.334	18.591	0.360	3.653	0.106	0.416
Bi	+83	3.841	50.261	4.679	11.999	3.192	2.560	1.363	0.318
Bk	97	6.502	28.375	5.478	4.975	2.510	0.561	—	—
Br	+35	2.166	33.899	2.904	10.497	1.395	2.041	0.589	0.307
C	+06	0.731	36.995	1.195	11.297	0.456	2.814	0.125	0.346
Ca	+20	4.470	99.523	2.971	22.696	1.970	4.195	0.482	0.417
Cd	+48	2.574	55.675	3.259	11.838	2.547	2.784	0.838	0.322
Ce	58	5.007	28.283	3.980	5.183	1.678	0.589	—	—
Cf	98	6.548	28.461	5.526	4.965	2.520	0.557	—	—
Cl	+17	1.452	30.935	2.292	9.980	0.787	2.234	0.322	0.323
Cm	96	6.460	28.396	5.469	4.970	2.471	0.554	—	—
Co	+27	2.367	61.431	2.236	14.180	1.724	2.725	0.515	0.344
Cr	+24	2.307	78.405	2.334	15.785	1.823	3.157	0.490	0.364
Cs	+55	6.062	155.837	5.986	19.695	3.303	3.335	1.096	0.379
Cu	+29	1.579	62.940	1.820	12.453	1.658	2.504	0.532	0.333
Dy	66	5.332	28.888	4.370	5.198	1.863	0.581	—	—
Er	68	5.436	28.655	4.437	5.117	1.891	0.577	—	—
Eu	+63	6.267	100.298	4.844	16.066	3.202	2.980	1.200	0.367
F	+09	0.387	20.239	0.811	6.609	0.475	1.931	0.146	0.279
Fe	+26	2.544	64.424	2.343	14.880	1.759	2.854	0.506	0.350
Fr	87	6.201	28.200	5.121	4.954	2.275	0.556	—	—
Ga	+31	2.321	65.602	2.486	15.458	1.688	2.581	0.599	0.351
Gd	64	5.225	29.158	4.314	5.259	1.827	0.586	—	—
Ge	+32	2.447	55.893	2.702	14.393	1.616	2.446	0.601	0.342
H	01	0.202	30.868	0.244	8.544	0.082	1.273	—	—
He	+02	0.091	18.183	0.181	6.212	0.110	1.803	0.036	0.284
Hf	72	5.588	29.001	4.619	5.164	1.997	0.579	—	—
Hg	+80	2.682	42.822	4.241	9.856	2.755	2.295	1.270	0.307
Ho	67	5.376	28.773	4.403	5.174	1.884	0.582	—	—
I	+53	3.473	39.441	4.060	11.816	2.522	2.415	0.840	0.298
In	+49	3.153	66.649	3.557	14.449	2.818	2.976	0.884	0.335
Ir	77	5.754	29.159	4.851	5.152	2.096	0.570	—	—
K	+19	3.951	137.075	2.545	22.402	1.980	4.532	0.482	0.434
Kr	+36	2.034	29.999	2.927	9.598	1.342	1.952	0.589	0.299
La	57	4.940	28.716	3.968	5.245	1.663	0.594	—	—
Li	+03	1.611	107.638	1.246	30.480	0.326	4.533	0.099	0.495
Lu	71	5.553	28.907	4.580	5.160	1.969	0.577	—	—

Table 12.1. (cont.).

Name	Z	a_1	b_1	a_2	b_2	a_3	b_3	a_4	b_4
Mg	+12	2.268	73.670	1.803	20.175	0.839	3.013	0.289	0.405
Mn	+25	2.747	67.786	2.456	15.674	1.792	3.000	0.498	0.357
Mo	+42	3.120	72.464	3.906	14.642	2.361	3.237	0.850	0.366
N	+07	0.572	28.847	1.043	9.054	0.465	2.421	0.131	0.317
Na	+11	2.241	108.004	1.333	24.505	0.907	3.391	0.286	0.435
Nb	41	4.237	27.415	3.105	5.074	1.234	0.593	—	—
Nd	60	5.151	28.304	4.075	5.073	1.683	0.571	—	—
Ne	+10	0.303	17.640	0.720	5.860	0.475	1.762	0.153	0.266
Ni	+28	2.210	58.727	2.134	13.553	1.689	2.609	0.524	0.339
Np	93	6.323	29.142	5.414	5.096	2.453	0.568	—	—
O	+08	0.455	23.780	0.917	7.622	0.472	2.144	0.138	0.296
Os	76	5.750	28.933	4.773	5.139	2.079	0.573	—	—
P	+15	1.888	44.876	2.469	13.538	0.805	2.642	0.320	0.361
Pa	91	6.306	28.688	5.303	5.026	2.386	0.561	—	—
Pb	+82	3.510	52.914	4.552	11.884	3.154	2.571	1.359	0.321
Pd	46	4.436	28.670	3.454	5.269	1.383	0.595	—	—
Pm	61	5.201	28.079	4.094	5.081	1.719	0.576	—	—
Po	84	6.070	28.075	4.997	4.999	2.232	0.563	—	—
Pr	59	5.085	28.588	4.043	5.143	1.684	0.581	—	—
Pt	78	5.803	29.016	4.870	5.150	2.127	0.572	—	—
Pu	94	6.415	28.836	5.419	5.022	2.449	0.561	—	—
Ra	88	6.215	28.382	5.170	5.002	2.316	0.562	—	—
Rb	+37	4.776	140.782	3.859	18.991	2.234	3.701	0.868	0.419
Re	75	5.695	28.968	4.740	5.156	2.064	0.575	—	—
Rh	45	4.431	27.911	3.343	5.153	1.345	0.592	—	—
Rn	+86	4.078	38.406	4.978	11.020	3.096	2.355	1.326	0.299
Ru	44	4.358	27.881	3.298	5.179	1.323	0.594	—	—
S	+16	1.659	36.650	2.386	11.488	0.790	2.469	0.321	0.340
Sb	+51	3.564	50.487	3.844	13.316	2.687	2.691	0.864	0.316
Sc	+21	3.966	88.960	2.917	20.606	1.925	3.856	0.480	0.399
Se	+34	2.298	38.830	2.854	11.536	1.456	2.146	0.590	0.316
Si	+14	2.129	57.775	2.533	16.476	0.835	2.880	0.322	0.386
Sm	62	5.255	28.016	4.113	5.037	1.743	0.577	—	—
Sn	+50	3.450	59.104	3.735	14.179	2.118	2.855	0.877	0.327
Sr	+38	5.848	104.972	4.003	19.367	2.342	3.737	0.880	0.414
Ta	73	5.659	28.807	4.630	5.114	2.014	0.578	—	—
Tb	65	5.272	29.046	4.347	5.226	1.844	0.585	—	—
Tc	43	4.318	28.246	3.270	5.148	1.287	0.590	—	—
Te	52	4.785	27.999	3.688	5.083	1.500	0.581	—	—
Th	90	6.264	28.651	5.263	5.030	2.367	0.563	—	—
Ti	+22	3.565	81.982	2.818	19.049	1.893	3.590	0.483	0.386
Tl	81	5.932	29.086	4.972	5.126	2.195	0.572	—	—
Tm	69	5.441	29.149	4.510	5.264	1.956	0.590	—	—
U	+92	6.767	85.951	6.729	15.642	4.014	2.936	1.561	0.335
V	+23	3.245	76.379	2.698	17.726	1.860	3.363	0.486	0.374
W	74	5.709	28.782	4.677	5.084	2.019	0.572	—	—
Xe	+54	3.366	35.509	4.147	11.117	2.443	2.294	0.829	0.289
Y	39	4.129	27.548	3.012	5.088	1.179	0.591	—	—
Yb	70	5.529	28.927	4.533	5.144	1.945	0.578	—	—
Zn	+30	1.942	54.162	1.950	12.518	1.619	2.416	0.543	0.330
Zr	40	4.105	28.492	3.144	5.277	1.229	0.601	—	—

Note that this number is independent of the wave length of the X-rays being used since the number $1/2d_{hkl}$ is independent of the wave length.

12.1.3 Scattering by a single unit cell

The scattering of X-rays due to a complete unit cell can be computed by taking into account the *relative* positions of all the atoms in the unit cell. We know that scattering from electrons belonging to the same atom can give rise to destructive interference because of the relative positions of the electrons inside the electron cloud. A similar thing happens for scattering from a unit cell. Consider the simple example shown in Fig. 12.4. Rays 1 and 2 are diffracted from the planes (hkl) if they satisfy the Bragg equation, which means that the path length difference between the two waves must be equal to the wave length λ. Suppose that the interplanar spacing d is equal to one of the lattice parameters, say a. In that case we would be talking about the (100) planes. If we add an atom to the unit cell, say at position $(1/2, 0, 0)$, exactly in between the atoms at O and O', then X-rays will also be scattered by this atom. The diffracted waves $1'$ and $2'$ are in phase (i.e., path length difference equal to λ), and from the drawing it is easy to see that the pathlength difference between $1'$ and $3'$ (and also between $2'$ and $3'$) is equal to *half of the wave length*, $\lambda/2$. This means that the waves are out-of-phase, which means that they will cancel each other out, despite the fact that geometrically the Bragg equation for the planes (100) is satisfied! The additional atom is located on the (200) plane. If we were to construct the diffraction condition for this plane, then we would find that diffracted beams from the (200) planes, with diffraction angle θ different from that for the (100) planes, are in-phase, which means that the (200) planes *will* give rise to a diffracted beam.

This example illustrates that diffraction from a certain set of planes (hkl) not only depends on the particular orientation of the incoming beam with respect to the plane, but also on the particular position of atoms within the unit cell with respect to the plane. One can show graphically, that the phase difference between waves $1'$ and $3'$ does not depend on the position of the extra atom within the (200) plane; i.e., moving the atom within the plane does not change the destructive interference between the two beams. We thus

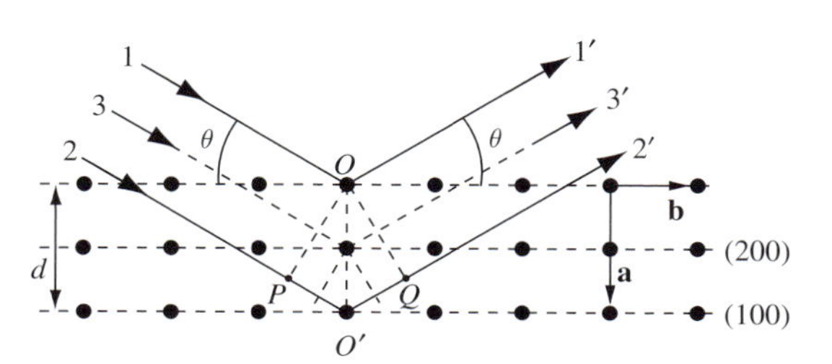

Fig. 12.4. The waves diffracted from the (100) planes interfere destructively when an atom is added in the (200) plane.

conclude that the only thing that matters is the distance between the extra atom and the (100) plane. This distance can be expressed by the projection of the position vector $\mathbf{r}$ of the atom onto the normal to the plane (100). We have seen in Chapter 6 that the plane normal is given by $\mathbf{g}_{100}$, and the projection of $\mathbf{r}$ onto $\mathbf{g}_{100}$ is equal to the dot-product $\mathbf{g}_{100} \cdot \mathbf{r}$. If we take the extra atom to be at the position $(1/2, 0, 0)$, then this dot-product becomes equal to:

$$\mathbf{g}_{100} \cdot \mathbf{r} = 1\mathbf{a}^*. \frac{1}{2}\mathbf{a} = \frac{1}{2}.$$

Here we have used the definition of the reciprocal lattice vectors. We can translate the projection of $\mathbf{r}$ onto the plane normal into a phase difference by multiplying the dot-product with 2π. This leads to a phase difference of π, since:

$$2\pi\mathbf{g}_{100} \cdot \mathbf{r} = \pi,$$

and, therefore, waves scattered by the extra atom and the (100) planes are out-of-phase. If the extra atom is at position $(1/2, y, z)$, i.e., still in the (200) plane, then the phase difference becomes:

$$2\pi\mathbf{g}_{100} \cdot \mathbf{r} = 2\pi\mathbf{a}^* \cdot \frac{1}{2}\mathbf{a} + 2\pi\mathbf{a}^* \cdot y\mathbf{b} + 2\pi\mathbf{a}^* \cdot z\mathbf{c} = \pi,$$

where again we have used the properties of the reciprocal basis vectors. We find that adding an atom to a unit cell affects the diffraction from all lattice planes (hkl) in a way determined by the phase difference $2\pi\mathbf{g}_{hkl} \cdot \mathbf{r}$, where $\mathbf{r}$ is the position vector of the atom with respect to the direct basis vectors. In general, the phase difference is expressed by:

$$\boxed{\phi = 2\pi\mathbf{g}_{hkl} \cdot \mathbf{r} = 2\pi(hx + ky + lz).} \tag{12.4}$$

In the previous section, we have seen how strongly a single atom scatters X-rays in a particular direction. In the present section, we have determined the relative phase for scattering from two atoms. We can combine these two numbers, amplitude and phase, into a single complex number:

$$f(s)e^{i\phi} = f(\frac{\sin\theta}{\lambda})e^{2\pi i\mathbf{g}_{hkl} \cdot \mathbf{r}}. \tag{12.5}$$

This expression states how an atom at position $\mathbf{r}$ contributes to diffraction of X-rays of wavelength λ from the plane (hkl). Scattering from a complete unit cell is then described by adding together these factors for all atoms in the unit cell.

12.2 The structure factor

The quantity describing scattering from a complete unit cell is known as the *structure factor*, and is represented by the symbol F_{hkl}. The formal definition of the structure factor is:

$$F_{hkl} = \sum_{j=1}^{N} f_j\left(\frac{\sin\theta_{hkl}}{\lambda}\right) e^{2\pi i \mathbf{g}_{hkl}\cdot\mathbf{r}_j} = \sum_{j=1}^{N} f_j\left(\frac{\sin\theta_{hkl}}{\lambda}\right) e^{2\pi i(hx_j+ky_j+lz_j)}, \quad (12.6)$$

with N the number of atoms in the unit cell. The intensity in the diffracted beam from the planes (hkl) is proportional to the *modulus squared of the structure factor*:

$$I_{hkl} = |F_{hkl}|^2 = F_{hkl}F^*_{hkl}, \quad (12.7)$$

where the asterisk indicates complex conjugation.

This structure factor also has a geometrical interpretation: scattering from each atom is represented by a complex number $fe^{i\phi}$. We know that a complex number can be represented by a vector in the complex plane, as shown in Fig. 12.5(a), which shows the complex number $2e^{i\pi/3}$. Scattering from an individual atom is represented by such a number or vector, and, therefore, the addition of all complex numbers in the structure factor is equivalent to vector addition of all the corresponding vectors in the complex plane (Fig. 12.5(b)). This is known as an *Argand diagram*. If the positions of atoms in the unit cell are such that for a particular set of planes (hkl) the total sum of complex vectors ends up in the origin, then there will be no diffracted beam for that particular plane, *even when the Bragg condition is satisfied.* This is known as an *extinction*. There are several possible reasons for extinctions to occur, and we will discuss the most important ones in the following sections.

12.2.1 Lattice centering and the structure factor

We have seen in the previous chapter that the geometry of the diffraction process is completely determined by the shape and dimensions of the unit

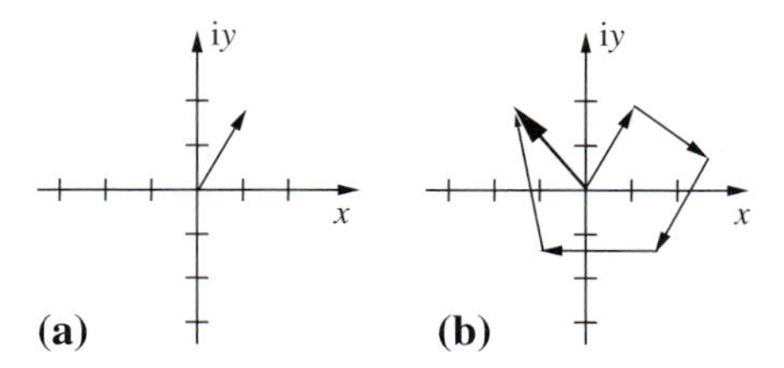

Fig. 12.5. (a) Graphical representation of the complex number $2e^{i\pi/3}$; (b) The scattering from all atoms in a unit cell is computed by adding all corresponding vectors; the thicker vector is equal to the structure factor. The square of the length of this vector is proportional to the total diffracted intensity.

cell. The structure factor is independent of the lattice parameters. Note that this is the case because of the particular definition we use for the reciprocal lattice vectors. In the following subsections, we will look at the four possible types of centering (P, C, I, and F) and determine the extinctions for each of them.

12.2.1.1 Primitive lattice

A primitive lattice is characterized by the absence of any centering vectors. This means that for the most general atom position $\mathbf{r} = (x, y, z)$ (general in the sense that the atom does not lie on a symmetry element of the structure) *there is no equivalent atom located at any of the positions* $\mathbf{r}+\mathbf{A}$, $\mathbf{r}+\mathbf{B}$, $\mathbf{r}+\mathbf{C}$, or $\mathbf{r}+\mathbf{I}$ (using the notation from Chapter 3). For a primitive structure with only one atom in the unit cell, say at $\mathbf{r} = (0, 0, 0)$, we find:

$$F_{hkl} = \sum_{j=1}^{1} f_j \mathrm{e}^{2\pi\mathrm{i}(h0+k0+l0)} = f,$$

and, therefore, the diffracted intensity is proportional to $I_{hkl} = f^2$. Remember that the value of f does depend on the particular lattice plane (hkl). In other words, for a primitive lattice there are no extinctions; all lattice planes give rise to a diffracted beam.

12.2.1.2 C-centered lattice

A C-centered lattice is characterized by the fact that, for every atom at position $\mathbf{r}$, there is an *identical* atom at position $\mathbf{r}+\mathbf{C}$. The structure factor for this situation (for $\mathbf{r} = (0, 0, 0)$) is given by:

$$F_{hkl} = \sum_{j=1}^{2} f_j \mathrm{e}^{2\pi\mathrm{i}(hx_j+ky_j+lz_j)} = f(1+\mathrm{e}^{\pi\mathrm{i}(h+k)}).$$

Using the properties of exponentials and Euler's formula we can rewrite this as:

$$\begin{aligned} F_{hkl} &= f\mathrm{e}^{\frac{\pi}{2}\mathrm{i}(h+k)}\left(\mathrm{e}^{-\frac{\pi}{2}\mathrm{i}(h+k)}+\mathrm{e}^{\frac{\pi}{2}\mathrm{i}(h+k)}\right); \\ &= 2f\mathrm{e}^{\frac{\pi}{2}\mathrm{i}(h+k)}\cos\frac{\pi}{2}(h+k). \end{aligned}$$

The intensity in the diffracted beams is then proportional to:

$$\begin{aligned} |F_{hkl}|^2 &= \left(2f\mathrm{e}^{\frac{\pi}{2}\mathrm{i}(h+k)}\cos\frac{\pi}{2}(h+k)\right)\left(2f\mathrm{e}^{-\frac{\pi}{2}\mathrm{i}(h+k)}\cos\frac{\pi}{2}(h+k)\right); \\ &= 4f^2\cos^2\frac{\pi}{2}(h+k). \end{aligned} \tag{12.8}$$

This intensity vanishes whenever the cosine becomes zero, and this happens whenever $h+k = 2n+1$. We conclude that for a C-centered lattice all diffracted beams with $h+k=$ odd, will vanish. These extinctions are called *systematic absences* or *systematic extinctions*. Note again that this is true independent of the shape and dimensions of the unit cell; in particular, it is true for the mC and oC Bravais lattices.

12.2.1.3 Body-centered lattice

The body-centered lattice is characterized by the presence of an *identical* atom at position $\mathbf{r}+\mathbf{I}$ for every atom at position $\mathbf{r}$. The structure factor is thus written as:

$$F_{hkl} = f\left(1+\mathrm{e}^{\pi\mathrm{i}(h+k+l)}\right).$$

and, using the same mathematical steps as before, we find for the intensity:

$$I_{hkl} = 4f^2\cos^2\frac{\pi}{2}(h+k+l), \tag{12.9}$$

from which we derive that for the body-centered lattices, all reflections with $h+k+l=$ odd will vanish. This is in particular true for the oI, tI, and cI Bravais lattices.

12.2.1.4 Face-centered lattice

The face-centered lattice is characterized by the simultaneous presence of three centering vectors, **A**, **B**, and **C**. The structure factor thus contains four terms:

$$F_{hkl} = f\left(1+\mathrm{e}^{\pi\mathrm{i}(h+k)}+\mathrm{e}^{\pi\mathrm{i}(h+l)}+\mathrm{e}^{\pi\mathrm{i}(k+l)}\right).$$

To compute the intensity we can now proceed in three different ways: (1) we multiply this expression with its complex conjugate and work out all the terms (a tedious task); (2) we attempt to rewrite the expression as the product of trigonometric functions (product, because it is easy to determine when a product will be zero); or (3) we use the fact that $\mathrm{e}^{\pi\mathrm{i}}=-1$. First we follow the second method. Let us take a closer look at the following product:

$$\left(1+\mathrm{e}^{\pi\mathrm{i}(h+k)}\right)\left(1+\mathrm{e}^{\pi\mathrm{i}(h+l)}\right) = 1+\mathrm{e}^{\pi\mathrm{i}(h+k)}+\mathrm{e}^{\pi\mathrm{i}(h+l)}+\mathrm{e}^{\pi\mathrm{i}(2h+k+l)}.$$

This expression is equal to the structure factor above, except for the factor $2h$ in the third exponential. However, from Euler's formula we know that:

$$\mathrm{e}^{2\pi\mathrm{i}h} = \cos 2\pi h + \mathrm{i}\sin 2\pi h = 1,$$

for all integers h. We can thus replace the structure factor by:

$$\begin{aligned} F_{hkl} &= f\left(1+\mathrm{e}^{\pi\mathrm{i}(h+k)}\right)\left(1+\mathrm{e}^{\pi\mathrm{i}(h+l)}\right); \\ &= 4f\mathrm{e}^{\frac{\pi}{2}\mathrm{i}(h+k)}\cos\frac{\pi}{2}(h+k)\mathrm{e}^{\frac{\pi}{2}\mathrm{i}(h+l)}\cos\frac{\pi}{2}(h+l), \end{aligned}$$

Table 12.2. Comparison of the extinctions in the three cubic Bravais lattices.

(hkl)	$h^2+k^2+l^2$	cP	cI	cF
(100)	1	f^2	0	0
(110)	2	f^2	$4f^2$	0
(111)	3	f^2	0	$16f^2$
(200)	4	f^2	$4f^2$	$16f^2$
(210)	5	f^2	0	0
(211)	6	f^2	$4f^2$	0
(220)	8	f^2	$4f^2$	$16f^2$
(221)	9	f^2	0	0
(300)	9	f^2	0	0
(310)	10	f^2	$4f^2$	0
(311)	11	f^2	0	$16f^2$
(222)	12	f^2	$4f^2$	$16f^2$
(320)	13	f^2	0	0
(321)	14	f^2	$4f^2$	0
(400)	16	f^2	$4f^2$	$16f^2$
(322)	17	f^2	0	0
(410)	17	f^2	0	0

from which we find for the intensity

$$I_{hkl} = 16f^2 \cos^2 \frac{\pi}{2}(h+k) \cos^2 \frac{\pi}{2}(h+l). \tag{12.10}$$

This expression is equal to zero whenever the Miller indices h, k, and l have different parity; in other words, for a face-centered lattice only reflections with all Miller indices even or all odd will be present. Mixed reflections will be absent. For all allowed reflections the cosine functions are equal to 1, and we find $I_{hkl} = 16f^2$.

We can also follow the third method, which goes as follows: we use the following property of the exponential function:

$$e^{\pi i(h+k)} = \left(e^{\pi i}\right)^{h+k} = (-1)^{h+k},$$

which allows us to rewrite the structure factor as

$$F_{hkl} = f\left(1+(-1)^{h+k}+(-1)^{h+l}+(-1)^{k+l}\right). \tag{12.11}$$

For mixed indices, two of these factors are equal to -1, and two are equal to $+1$, so that their sum vanishes and $F_{hkl} = 0$. For indices of equal parity, all terms are equal to $+1$ and hence $F_{hkl} = 4f$, or $I_{hkl} = 16f^2$. The results for all different centering variants of the cubic lattice are summarized in Table 12.2. This table can be compared with Table 11.3 on page 276, which lists the Bragg angles for a number of planes in face-centered cubic copper; only a few of the planes will actually give rise to a diffracted beam due to centering extinctions.

12.2.2 Symmetry and the structure factor

In the previous section we have seen how lattice centering causes systematic absences. Symmetry elements can also give rise to extinctions. Let us consider three examples:

(i) **Inversion symmetry**: An inversion center is characterized by the fact that for every atom at position $\mathbf{r}$ there is an equivalent atom at position $-\mathbf{r}$. If we have a unit cell with N atoms, then we can split the structure factor for that cell in two terms:

$$\begin{aligned} F_{hkl} &= \sum_{j=1}^{N} f_j e^{2\pi i \mathbf{g}\cdot\mathbf{r}_j}; \\ &= \sum_{j=1}^{N/2} f_j \left(e^{2\pi i \mathbf{g}\cdot\mathbf{r}_j} + e^{-2\pi i \mathbf{g}\cdot\mathbf{r}_j}\right); \\ &= 2\sum_{j=1}^{N/2} f_j \cos\left(2\pi \mathbf{g}\cdot\mathbf{r}_j\right). \end{aligned}$$

The last summation is a sum of *real* numbers, and therefore we conclude that the structure factor for a unit cell with inversion symmetry is always a real number.

(ii) **Screw axis symmetry**: A screw axis can be represented by a 4×4 transformation matrix which indicates how atom coordinates of equivalent atoms are related to one another. As an example, we consider the presence of a 4_1 screw axis, parallel to the c-axis and going through the origin of the unit cell. For every atom at position $\mathbf{r} = (x, y, z)$, there are three additional equivalent atoms, at positions $(-y, x, z+1/4)$, $(-x, -y, z+1/2)$, and $(y, -x, z+3/4)$. The structure factor can thus be rewritten as:

$$\begin{aligned} F_{hkl} &= \sum_{j=1}^{N/4} f_j \Big[e^{2\pi i(hx_j+ky_j+lz_j)} + e^{2\pi i(-hy_j+kx_j+l(z_j+\frac{1}{4})} \\ &\quad + e^{2\pi i(-hx_j-ky_j+l(z_j+\frac{1}{2})} + e^{2\pi i(hy_j-kx_j+l(z_j+\frac{3}{4})} \Big]. \end{aligned}$$

This equation can be simplified substantially if we only consider reflections of the type $(00l)$. In that case we can write:

$$F_{00l} = \left(\sum_{s=0}^{3} e^{\pi i \frac{sl}{2}}\right) \sum_{j=1}^{N/4} f_j e^{2\pi i l z_j}.$$

Table 12.3. Systematic absences for screw axes parallel to the **c** direction.

Screw axis	$\lvert\boldsymbol{\tau}\rvert$	extinction for
2_1	$c/2$	$l = 2n+1$
$3_1, 3_2$	$\pm c/3$	$l \neq 3n$
$4_1, 4_3$	$\pm c/4$	$l \neq 4n$
4_2	$c/2$	$l = 2n+1$
$6_1, 6_5$	$\pm c/6$	$l \neq 6n$
$6_2, 6_4$	$\pm c/3$	$l \neq 3n$
6_3	$c/2$	$l = 2n+1$

The first factor is a simple finite geometric series and it is easy to show that it can be rewritten as:

$$\sum_{s=0}^{3} \mathrm{e}^{\pi\mathrm{i}\frac{sl}{2}} = \frac{1-\mathrm{e}^{2\pi\mathrm{i}l}}{1-\mathrm{e}^{\pi\mathrm{i}\frac{l}{2}}}.$$

The numerator of this expression is always equal to zero; we have to be careful with the denominator, however, because the ratio 0/0 is not defined. The denominator becomes equal to zero whenever $l = 4n$, with n an integer. The value of the ratio is then determined by the de l'Hopital rule, which states that the value for l approaching $4n$ is equal to the ratio of the derivatives of nominator and denominator, evaluated at $l = 4n$. In mathematical terms this means:

$$\begin{aligned}\lim_{l\to 4n} \frac{1-\mathrm{e}^{2\pi\mathrm{i}l}}{1-\mathrm{e}^{\pi\mathrm{i}\frac{l}{2}}} &= \lim_{l\to 4n} \frac{-2\pi\mathrm{i}\mathrm{e}^{2\pi\mathrm{i}l}}{-\frac{\pi\mathrm{i}}{2}\mathrm{e}^{\pi\mathrm{i}\frac{l}{2}}};\\ &= \lim_{l\to 4n} 4\mathrm{e}^{\pi\mathrm{i}\frac{3l}{2}};\\ &= 4\mathrm{e}^{6\pi\mathrm{i}n} = 4.\end{aligned}$$

Summarizing, we find that the reflections of the type $(00l)$, with $l \neq 4n$ are absent in the presence of a screw axis of the type 4_1 parallel to the c-axis. One can derive similar extinction conditions for all other screw axes and the results are summarized in Table 12.3.

(iii) **Glide plane symmetry**: For a glide plane we can apply a similar method to determine which reflections will be absent. Let us consider an n glide plane, parallel to the (001) plane, going through the origin, with translation vector $\boldsymbol{\tau} = (1/2, 1/2, 0)$. For each atom at position (x, y, z) there is an equivalent atom at position $(x+1/2, y+1/2, \bar{z})$. The structure factor can thus be written as:

$$F_{hkl} = \sum_{j=1}^{N/2} f_j \left(\mathrm{e}^{2\pi\mathrm{i}(hx_j+ky_j+lz_j)} + \mathrm{e}^{2\pi\mathrm{i}(hx_j+\frac{h}{2}+ky_j+\frac{k}{2}-lz_j)}\right);$$

Table 12.4. Systematic absences in the $(hk0)$ reflections for glide planes parallel to the (001) plane.

glide type	$\vert\boldsymbol{\tau}\vert$	extinction for
a	$\frac{a}{2}$	$h = 2n+1$
b	$\frac{b}{2}$	$k = 2n+1$
n	$\frac{a+b}{2}$	$h+k = 2n+1$
d	$\frac{a\pm b}{4}$	$h+k = 4n+2$ with $h = 2n$ and $k = 2n$

$$= \sum_{j=1}^{N/2} f_j \mathrm{e}^{2\pi\mathrm{i}(hx_j+ky_j)} \left(\mathrm{e}^{2\pi\mathrm{i}lz_j} + \mathrm{e}^{2\pi\mathrm{i}(\frac{h+k}{2}-lz_j)}\right)$$

For reflections of the type $(hk0)$ we find:

$$F_{hk0} = \sum_{j=1}^{N/2} f_j \mathrm{e}^{2\pi\mathrm{i}(hx_j+ky_j)} \left(1 + \mathrm{e}^{2\pi\mathrm{i}(\frac{h+k}{2})}\right);$$

$$= 2\mathrm{e}^{\pi\mathrm{i}\frac{h+k}{2}} \cos\pi\left(\frac{h+k}{2}\right) \sum_{j=1}^{N/2} f_j \mathrm{e}^{2\pi\mathrm{i}(hx_j+ky_j)},$$

and therefore the structure factor becomes zero whenever $h+k = 2n+1$. Summarizing, we find that the reflections of the type $(hk0)$, with $h+k = 2n+1$, are absent in the presence of a glide plane parallel to (001), going through the origin, with glide vector $\boldsymbol{\tau} = (1/2, 1/2, 0)$. One can derive similar extinction conditions for all other glide planes and the results for glide planes parallel to (001) are summarized in Table 12.4.

This concludes the discussion of the effect of symmetry elements on the structure factor. As a final remark we should mention that diffraction from a given crystal structure will always have an intrinsic symmetry. This is caused by the following observation: since we can only measure intensities, and not phases, we cannot distinguish between the (hkl) plane and the $(\bar{h}\bar{k}\bar{l})$ plane. For structures without inversion symmetry, the structure factor for the (hkl) plane is equal to the complex conjugate of that of the $(\bar{h}\bar{k}\bar{l})$ plane, or:

$$F_{hkl} = F^*_{\bar{h}\bar{k}\bar{l}}.$$

Therefore, we also have:

$$F^*_{hkl} = F_{\bar{h}\bar{k}\bar{l}}.$$

Combining these relations we have:

$$I_{hkl} = F_{hkl}F^*_{hkl} = F_{\bar{h}\bar{k}\bar{l}}F^*_{\bar{h}\bar{k}\bar{l}} = I_{\bar{h}\bar{k}\bar{l}}. \tag{12.12}$$

Therefore, an X-ray diffraction data set will *always* display a center of symmetry, even when the crystal structure does not. This is known as *Friedel's law*. As a consequence, the point group symmetry of a diffraction data set must belong to one of the 11 *Laue classes* described in section 9.2.10.1 on page 214.

12.2.3 Systematic absences and the International Tables for Crystallography

In Chapter 10 we described how the *International Tables for Crystallography* list all 230 space groups. In particular, we showed a portion of the actual entries for space groups **Cmm2** (C_{2v}^{11}) and **Pmna** (D_{2h}^{7}) in Figs. 10.11 and 10.12. Under the entry "**Positions**", which lists the Wyckoff positions and the coordinates of the equivalent positions for general and special sites, we find information about the "reflection conditions," i.e., the conditions that need to be satisfied by the Miller indices h, k, and l in order to have a diffracted beam (non-zero structure factor). Note that these conditions are the opposite of the extinction conditions.

Consider space group **Cmm2** (C_{2v}^{11}) as an example. The space group is C-centered, so that the extinction condition derived previously reads $h+k=2n+1$, i.e., the structure factor vanishes for all odd $h+k$. The International Tables then state the reflection condition $h+k=2n$, i.e., $h+k$ must be even to have a non-zero structure factor. The general condition is typically simplified for special combinations of the indices, such as $h00$ for which the reflection condition simplifies to $h=2n$. For the special positions all general reflection conditions apply, and sometimes there are additional conditions. For instance, for **Cmm2** (C_{2v}^{11}) we have for the $4c$ Wyckoff position the additional reflection condition $hkl : h=2n$. In other words, if we have a structure with space group **Cmm2** (C_{2v}^{11}), and with atoms only on the $4c$ site, then only planes for which h is even can give rise to a diffracted beam. Since we must also satisfy the general reflection condition $h+k=2n$, and h is even, we find that k must also be even.

For space group **Pmna** (D_{2h}^{7}), with a diagonal glide plane n normal to the **b** direction, we find the general reflection condition $h0l : h+l=2n$, similar to the derivation in the present chapter. For nearly all special positions, there is an additional condition: $hkl : h+l=2n$, i.e., the general condition is not only valid for $k=0$, but must be valid for all values of k. The exception is the $4h$ Wyckoff position, for which no extra conditions apply.

From these examples we see how the space group symmetry, which is a combination of lattice centering and point group symmetry, dictates which planes can give rise to a diffracted beam. Conversely, by studying a diffraction data set, we can determine which planes do not give rise to a diffracted beam, and from this information we can, in principle, determine to which space group the structure belongs. We will return to the topic of space group determination in the next chapter, when we talk about convergent beam electron diffraction.

12.2.4 Examples of structure factor calculations

In this section, we will carry out a few simple structure factor calculations for the CsCl structure, the NaCl structure, and the diamond structure. In later chapters, the reader will find additional examples of structure factor computations.

Example 1: CsCl. The unit cell of CsCl contains only two atoms, Cs in the origin and Cl at the center of the cell, so that the structure factor is given by:

$$F_{hkl} = f_{\text{Cs}} + f_{\text{Cl}}(-1)^{h+k+l}.$$

For the reflections with $h+k+l = 2n$ we find that $F_{hkl} = f_{\text{Cs}} + f_{\text{Cl}}$; for all other reflections we have $F_{hkl} = f_{\text{Cs}} - f_{\text{Cl}}$. The observed intensities will hence be equal to:

$$\begin{aligned} I_{hkl} &= (f_{\text{Cs}} + f_{\text{Cl}})^2 \quad \text{for } h+k+l = 2n; \\ I_{hkl} &= (f_{\text{Cs}} - f_{\text{Cl}})^2 \quad \text{for } h+k+l = 2n+1. \end{aligned}$$

This means that we now have two sets of reflections: reflections with intensities proportional to the square of the *sum* of the atomic scattering factors, and reflections proportional to the square of the *difference* of the atomic scattering factors. The former reflections are known as *fundamental reflections*, the weaker ones as *superlattice reflections*.

If Cs and Cl were *randomly* distributed over the two sites of a body-centered cubic unit cell, then the atomic scattering factor for each site would be the average of those of the atoms, i.e., $f = (f_{\text{Cs}} + f_{\text{Cl}})/2$, and then the reflections with $h+k+l = 2n+1$ would have zero intensity, as required for a body-centered cell. Any deviation from the random arrangement of the atoms results in a non-zero intensity for these reflections, and, therefore, the superlattice reflections give information on the *degree of order* in the material.

Example 2: NaCl. The sodium chloride structure can be regarded as two interpenetrating *fcc* lattices, one filled with Na and the other with Cl. The structure factor for each individual lattice is equal to that for a regular *fcc* lattice. The Cl lattice is shifted with respect to the Na lattice by a vector $\boldsymbol{\tau} = (1/2, 1/2, 1/2)$. This means that the total structure factor can be written as:

$$\begin{aligned} F_{hkl} &= f_{\text{Na}}(1 + (-1)^{h+k} + (-1)^{h+l} + (-1)^{k+l}) \\ &\quad + f_{\text{Cl}}\mathrm{e}^{\pi i(h+k+l)}(1 + (-1)^{h+k} + (-1)^{h+l} + (-1)^{k+l}). \end{aligned}$$

Since $\mathrm{e}^{\pi i n} = (-1)^n$ we find:

$$F_{hkl} = (f_{\text{Na}} + f_{\text{Cl}}(-1)^{h+k+l})(1 + (-1)^{h+k} + (-1)^{h+l} + (-1)^{k+l}).$$

The corresponding intensity is thus given by:

$$I_{hkl} = (f_{\mathrm{Na}} + f_{\mathrm{Cl}}(-1)^{h+k+l})^2 \times (1 + (-1)^{h+k} + (-1)^{h+l} + (-1)^{k+l})^2.$$

We already know from the *fcc* example that only reflections for which all indices have the same parity are allowed. In addition, the presence of the second *fcc* lattice introduces a new condition: if $h+k+l = 2n$, then the two atomic scattering factors must be added, if $h+k+l = 2n+1$ then they are subtracted. The intensities for NaCl are thus as follows:

$$I_{hkl} = 0 \quad \text{for } h, k, l \text{ different parity;}$$

$$I_{hkl} = 16(f_{\mathrm{Na}} + f_{\mathrm{Cl}})^2 \quad \text{for } h, k, l \text{ same parity } \underline{\text{and}} \ h+k+l = 2n;$$

$$I_{hkl} = 16(f_{\mathrm{Na}} - f_{\mathrm{Cl}})^2 \quad \text{for } h, k, l \text{ same parity } \underline{\text{and}} \ h+k+l = 2n+1.$$

Once again we find two different sets of reflections: fundamental reflections and superlattice reflections.

Example 3: Diamond. The diamond structure can also be regarded as two interpenetrating *fcc* lattices, but this time with translation vector $\boldsymbol{\tau} = (1/4, 1/4, 1/4)$. The structure factor thus becomes:

$$\begin{aligned} F_{hkl} &= f_{\mathrm{C}}(1 + (-1)^{h+k} + (-1)^{h+l} + (-1)^{k+l}) \\ &\quad + f_{\mathrm{C}}\,\mathrm{e}^{\frac{\pi}{2}\mathrm{i}(h+k+l)}(1 + (-1)^{h+k} + (-1)^{h+l} + (-1)^{k+l}). \end{aligned}$$

This can be rewritten as:

$$F_{hkl} = 2f_{\mathrm{C}}\,\mathrm{e}^{\frac{\pi}{4}\mathrm{i}(h+k+l)} \cos\left(\frac{\pi}{4}(h+k+l)\right) \times (1 + (-1)^{h+k} + (-1)^{h+l} + (-1)^{k+l}),$$

from which we find for the intensity:

$$I_{hkl} = 4f_{\mathrm{C}}^2 \cos^2\left(\frac{\pi}{4}(h+k+l)\right) \times (1 + (-1)^{h+k} + (-1)^{h+l} + (-1)^{k+l})^2.$$

In addition to being zero for all reflections for which h, k, l are of different parity, this factor is also zero whenever $h+k+l = 4n+2$ with n an integer.

From these examples we conclude that absent reflections in a diffraction pattern give valuable information about the location of atoms in the unit cell, and about the presence of certain symmetry elements.

12.3 Intensity calculations for diffracted and measured intensities

When we perform an X-ray diffraction experiment, we typically measure the diffracted intensity for a certain period of time and with a detector with a certain aperture. The measured intensity thus represents a time average (or an integration) of the scattered intensity and we only measure a small fraction of the total scattered intensity, because of the finite dimensions of the detector. In the following sections, we will describe a number of correction factors that must be included to compute the *measured* intensity, rather than the *diffracted* intensity. We will limit ourselves to the standard powder diffraction geometry described in the previous chapter, since that is the most commonly used phase identification method. Intensity computations for Scherrer patterns proceed along similar lines, whereas the intensities of reflections in Laue patterns require a more complicated approach (e.g., (Marín and Diéguez, 1999)).

12.3.1 Description of the correction factors

12.3.1.1 Temperature factor

Atoms in a crystal are not rigidly attached to their lattice sites. They move around their lattice sites in a (mostly) random fashion. The amplitude of this motion is determined by the available energy, which in turn is determined by the temperature. If the temperature of a solid increases, then the atoms will vibrate with a larger amplitude. At very low temperatures, the available energy is much smaller and therefore the atoms will be, on average, closer to their *equilibrium* positions.

If an atom vibrates with a certain amplitude, then its electron cloud will, on average, appear to be much larger and more diffuse than if the atom were stationary at one point. A larger electron cloud with the same number of electrons means that the electron density becomes slightly smaller, and this affects the value of the atomic scattering factor, f. This can be understood by considering the definition of f, as the ratio of the scattered amplitude of the total atom to that of one single electron. The theory of lattice vibrations is very complex and requires sophisticated mathematical techniques far beyond the level of this book. For our purposes, it will be sufficient to state the result: the atomic scattering factor must be multiplied by an exponential attenuation or damping factor, generally known as the *Debye–Waller factor*. Mathematically stated, we find:

$$f_T\left(\frac{\sin\theta}{\lambda}\right) = f_0\left(\frac{\sin\theta}{\lambda}\right)\mathrm{e}^{-B(T)\left(\frac{\sin\theta}{\lambda}\right)^2} = f_0(s)\mathrm{e}^{-B(T)s^2},$$

where the subscript 0 on the scattering factor indicates that the value at temperature $T = 0\,\mathrm{K}$ must be taken. The factor B is a function of temperature T and is proportional to the mean square displacement of the atom in a

Table 12.5. Debye–Waller factors $B(T)$ in Å^2 for a few elemental crystals (Peng *et al.*, 1996). These numbers must be multiplied by 0.01 to convert them to nm^2.

T (K)	Al *fcc*	Ti *hcp*	Fe *fcc*	Fe *bcc*	Cu *fcc*	Ag *fcc*	W *bcc*	Au *fcc*
90.0	0.3374	0.1579	0.1493	0.1715	0.1692	0.2259	0.0491	0.1908
130.0	0.3465	0.2281	0.1443	0.2476	0.2444	0.3262	0.0709	0.2755
170.0	0.4531	0.2982	0.1886	0.3238	0.3196	0.4265	0.0927	0.3602
210.0	0.5596	0.3684	0.2330	0.3999	0.3947	0.5267	0.1145	0.4448
260.0	0.6928	0.4560	0.2884	0.4950	0.4886	0.6517	0.1417	0.5503
270.0	0.7194	0.4735	0.2995	0.5140	0.5073	0.6767	0.1471	0.5714
280.0	0.7460	0.4911	0.3106	0.5330	0.5261	0.7017	0.1526	0.5925

direction normal to the reflecting plane.[2] The net effect of temperature is that every atom scatters less strongly than it would at absolute zero. The exponential attenuation factor is often written as e^{-M}, with $M = B(T)s^2$. The intensity of a diffracted beam is thus reduced by a factor e^{-2M} with respect to the intensity of that beam at absolute zero.

The theory behind the Debye–Waller factor is quite involved and requires knowledge of the *phonon density of states*, i.e., the number of lattice vibrations (or phonons) with a given frequency or wave length. Such computations are far beyond the level of this textbook, so, instead, we simply list $B(T)$ for a few pure elements with different crystal structures in Table 12.5. A more complete listing can be found in Peng *et al.* (1996).

On page 298, we computed the atomic scattering factor for tungsten, when Cu Kα radiation diffracts from the (222) planes of a *bcc* crystal with lattice parameter 3.1653Å; the result was $f_0 = 39.9358$. If we include the Debye–Waller correction factor, for $T = 290$ K (room temperature), then we find that

$$f = f_0 e^{-B(270)s^2} = 39.9358 \times 0.9538 = 38.0908.$$

The Debye–Waller factors are unknown for most crystal structures. If experimental values for the Debye–Waller factors are unavailable, then one could use the elemental values as rough estimates. From the data listed in Peng *et al.* (1996), it can be seen that, in general, the Debye–Waller factor is larger for elements in the left-most columns of the periodic table. At room temperature, values around 0.1 nm^2 would be quite reasonable for first column

[2] We are assuming that the atomic vibration amplitude is isotropic, i.e., it is the same in all directions. While this is a good approximation for close-packed (metallic) structures, in many other cases we must allow for the vibration amplitude to be different in different directions. The Debye–Waller factor is then described by a *vibration ellipsoid*, a 3-D shape that indicates in which direction(s) the maximal vibration amplitude occurs. For our purposes, we will always assume an isotropic atomic vibration pattern.

elements (Li, Na, K, Rb, Cs), whereas for second column elements (Be, Mg, Ca, Sr, Ba) values around 0.01–0.03 nm^2 are reasonable. For most other elements, values in the range 0.003–0.007 nm^2 are acceptable. At liquid nitrogen temperature, the Debye–Waller factors are typically about one-third of their room temperature values. The Debye–Waller factors for the elemental solids decrease down a column of the periodic table. If experimental values for $B(T)$ are available, then these should be used instead of the estimated values. Moreover, it is probably not a good idea to use an isotropic Debye–Waller factor in all situations.

12.3.1.2 Absorption factor

As X-rays travel through a sample, they are partially absorbed. Mathematically, this means that the total diffracted intensity must be multiplied by an *absorption factor* A. The value of A depends on the thickness of the material through which the beam has travelled and, in general, also on the shape of the sample. In addition, the absorption factor can depend on the diffraction angle θ, and one usually writes $A = A(\theta)$. Note that it is the *intensity*, not the amplitude, that must be multiplied by A.

One can show (see, for instance, (Cullity, 1978), page 134 for a detailed proof) that for the standard powder diffractometer, the absorption factor is independent of the diffraction angle and equal to:

$$A = \frac{1}{2\mu},$$

with μ the linear absorption coefficient of the specimen.

For a Debye–Scherrer camera, the absorption constant is more complicated to compute; for a cylindrical sample, however, one can show that the absorption is large for small diffraction angles and small for large angles. The thermal effect discussed in the previous section gives rise to an opposite behavior of the correction factors, so that the thermal correction and the absorption correction nearly cancel each other out. For other methods, in particular single crystal methods, absorption corrections can become rather involved since the absorption factor depends strongly on the sample shape, which is not always a simple shape, such as a sphere or a cylinder. For more details on absorption corrections, we refer the interested reader to (Cullity, 1978, Giacovazzo, 2002a).

12.3.1.3 Multiplicity factor

The modulus squared of the structure factor is proportional to the total diffracted intensity, scattered in a certain direction. However, for scattering from, say, the (200) plane in a Cu powder sample, there will also be scattering *in the same direction* from the (020) and (002) planes and their negatives. For planes with larger Miller indices, there are in general more possibilities. The total number of equivalent planes is known as the *multiplicity* of that

Table 12.6. Multiplicities for general and special planes in all crystal systems. The notation lists the Miller indices above and the multiplicity below the line: hkl/p_{hkl}.

Cubic	$\frac{hkl}{48}$	$\frac{hhl}{24}$	$\frac{0kl}{24}$	$\frac{0kk}{12}$	$\frac{hhh}{8}$	$\frac{00l}{6}$	
Hex./Rhom.	$\frac{hk.l}{24}$	$\frac{hh.l}{12}$	$\frac{0k.l}{12}$	$\frac{hk.0}{12}$	$\frac{hh.0}{6}$	$\frac{0k.0}{6}$	$\frac{00.l}{2}$
Tetragonal	$\frac{hkl}{16}$	$\frac{hhl}{8}$	$\frac{0kl}{8}$	$\frac{hk0}{8}$	$\frac{hh0}{4}$	$\frac{0k0}{4}$	$\frac{00l}{2}$
Orthorhombic	$\frac{hkl}{8}$	$\frac{0kl}{4}$	$\frac{h0l}{4}$	$\frac{hk0}{4}$	$\frac{h00}{2}$	$\frac{0k0}{2}$	$\frac{00l}{2}$
Monoclinic	$\frac{hkl}{4}$	$\frac{h0l}{2}$	$\frac{00l}{2}$				
Triclinic	$\frac{hkl}{2}$						

plane. Multiplicity is represented by the integer number p_{hkl}, and depends on the crystal symmetry. Table 12.6 lists the multiplicities for all planes in all crystal systems. As an example, consider the (220) planes in a cubic crystal. The table states that for reflections of the type $(0kk)$, the multiplicity is equal to 12; for the (224) planes we have $p_{224} = 24$. The total intensity scattered from a plane (hkl) must be multiplied by p_{hkl} to obtain the total intensity scattered in the direction corresponding to the angle $2\theta_{hkl}$.

12.3.1.4 Lorentz polarization factor

We have seen in the first section of this chapter that an unpolarized beam of X-rays is scattered differently in different directions, even by a single electron. The trigonometric factor describing this effect is given by:

$$P(\theta) = \frac{1 + \cos^2 2\theta}{2}.$$

There are three additional geometric factors that influence the total intensity in a diffracted beam. A powder crystal diffracts X-rays onto conical surfaces with tops in the center of the Ewald sphere and opening angles equal to the diffraction angles 2θ (see Fig. 11.19(b) on page 284). The total diffracted intensity scattered by the (hkl) planes is thus distributed over a conical surface. However, when we use a detector, either for powder diffractometry or in the form of the Debye–Scherrer camera, then we only intercept a fraction of this total diffracted intensity. For instance, for a Debye–Scherrer camera of radius R, the radius of the cone at the point where it intersects the photographic film is $R \sin 2\theta$. The total length of the diffraction line (circumference of the circle) is then $2\pi R \sin 2\theta$. The intensity per unit length of diffraction line is, therefore, proportional to $1/\sin 2\theta$; close to $2\theta = 0°$ or $180°$, the diffraction circles are small (see Fig. 11.19(c)) and the intensity per unit line length is high, whereas for angles close to $90°$ the intensity per unit line length becomes much smaller. This provides a first trigonometric correction factor of $1/\sin 2\theta$.

Consider next a powder sample with randomly oriented grains. The number of grains that are oriented close to a particular Bragg angle, θ, depends on the

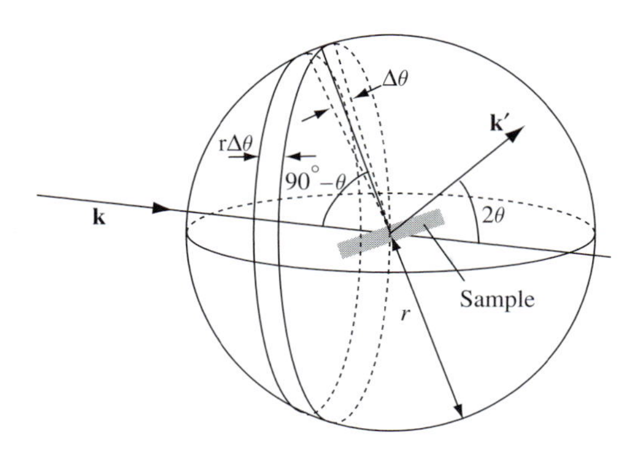

Fig. 12.6. Schematic illustration of the dependence of the number of grains close to a particular Bragg angle on that angle (see text for explanation; this figure is based on Fig. 4.16 in (Cullity and Stock, 2001)).

value of that angle. Let us assume that we are measuring the total intensity of a reflection with diffraction angle 2θ (see Fig. 12.6). The corresponding planes have plane normals that make an angle $90^\circ - \theta$ with respect to the incoming beam. If the X-ray detector measures all the intensity over an angular interval $2\theta \pm \Delta\theta/2$, then the normal to the planes may vary from $90^\circ - \theta - \Delta\theta/2$ to $90^\circ - \theta + \Delta\theta/2$. If we consider the intersection of the plane normals with a sphere of radius r, centered in the origin of reciprocal space, then the plane normals of the planes giving rise to measurable diffracted intensity lie within a band of width $r\Delta\theta$ on the surface of this sphere. For randomly oriented grains, the end points of the plane normals will be uniformly distributed over the entire surface of this sphere. Therefore, the fraction of grains diffracting into the detector is equal to the ratio of the number of grains with normals inside the band, ΔN, to the total number of grains, N. This ratio, in turn, is equal to the ratio of the surface area of the band to the total area of the sphere:

$$\frac{\Delta N}{N} = \frac{r\Delta\theta.2\pi r \sin(90^\circ - \theta)}{4\pi r^2} = \frac{\Delta\theta \cos\theta}{2}.$$

The total number of grains oriented favorably for diffraction with an angle 2θ is thus proportional to $\cos\theta$, which is the second trigonometric correction factor.

The third correction factor is due to the fact that a set of planes does not diffract X-rays only at the exact Bragg orientation, but also when the orientation deviates slightly from the correct angle. This is easy to understand as follows: assume that a set of planes has Bragg angle θ. When the incident beam is in Bragg orientation, X-rays reflected from consecutive planes in the crystal are completely in-phase. If the incident beam is then tilted by a small angle $\Delta\theta$, the phase difference between consecutive planes in the crystal will change by a small amount as well. When $\Delta\theta$ is small, this phase difference will not be large enough to cause complete destructive interference, but the interference is not completely constructive either. Therefore, the intensity of the diffracted beam will be non-zero, but less than the intensity at exact Bragg orientation. The more planes that are present in the crystal, i.e., the larger the grain size, the smaller the range of $\Delta\theta$ values for which some intensity can be observed. As we move the X-ray detector through the Bragg angle, we will begin to measure some intensity at $2(\theta-\theta')$; this intensity will reach a maximum value at 2θ, and then decrease again until it vanishes at $2(\theta+\theta'')$ (see Fig. 12.7). We define the *integrated intensity* as the total intensity over the entire angular range.

There are two commonly used mathematical functions that describe the peak shape: the *Gaussian function* and the *Lorentzian function*. They are defined as:

$$I^{\text{Gaussian}}(\alpha) = I_0 \exp\left[-4\ln 2\left(\frac{\alpha-2\theta}{w}\right)^2\right]; \tag{12.13}$$

$$I^{\text{Lorentzian}}(\alpha) = \frac{I_0}{1+4\left(\frac{\alpha-2\theta}{w}\right)^2}, \tag{12.14}$$

where I_0 is the maximum intensity, w is the full-width-at-half-maximum (FWHM), and α is the diffraction angle. The curves in Fig. 12.7 are shown for the following parameter values: $[2\theta = 40°, w = 2°, I_0 = 100]$ for the Gaussian function and $[2\theta = 50°, w = 1°, I_0 = 100]$ for the Lorentzian function. The Gaussian function drops off rapidly away from the maximum, whereas the Lorentzian peak has longer tails.

It can be shown (e.g., Cullity and Stock, 2001) that the FWHM value of a diffraction peak is related to the size of the grains that give rise to that peak and also to the value of the Bragg angle, θ. The relation is known as *Scherrer's formula* and reads as follows for grains with average diameter D:

$$w = \frac{0.9\lambda}{D\cos\theta}. \tag{12.15}$$

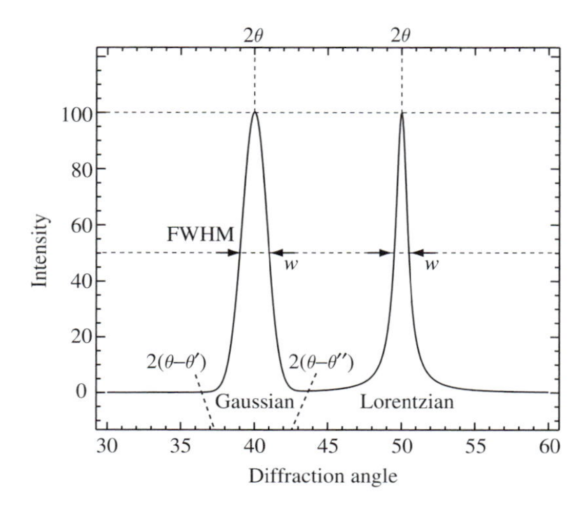

Fig. 12.7. Schematic illustration of the shape of a diffraction peak for two peak functions: Gaussian (left) and Lorentzian (right).

The smaller the grain size, the wider the diffraction peak. This relation can be used to determine approximate grain sizes for nano-crystalline materials.

If we approximate the integrated intensity by a rectangle, wI_0, we find from Eq. 12.15 that the width of the rectangle varies as $1/\cos\theta$. The height, I_0, of the rectangle also depends on θ, and it can be shown (Cullity and Stock, 2001, Fultz and Howe, 2002) that the dependence takes on the form $1/\sin\theta$. The integrated intensity of a diffraction peak is, therefore, proportional to the product of $1/\cos\theta$ and $1/\sin\theta$, so that the third correction factor is $1/\sin 2\theta$ (ignoring constant factors). Combining all three correction factors with the polarization factor we find that the intensity diffracted by an angle 2θ is proportional to:

$$\boxed{L_p(\theta) = \frac{1+\cos^2 2\theta}{\sin^2\theta\cos\theta}.} \qquad (12.16)$$

This expression is known as the *Lorentz polarization factor*, and it is shown graphically in Fig. 12.8. Note that this factor is significantly different from 1, and must be taken into account in any intensity computation.

12.3.2 Expressions for the total measured intensity

The total intensity can now be computed by putting together all contributing factors and correction factors. It is standard practice in most diffraction experiments to measure all intensities, and then re-scale them so that the most intense peak has intensity 100. In doing so, one effectively cancels the incident intensity I_0, the constant K, the distance r from the observer to the sample, and a number of other factors that are common to all reflections. The relevant part of the intensity of a diffracted beam is, therefore, given by:

$$\boxed{I_{hkl} = |F_{hkl}|^2 p_{hkl} L_p(\theta) A(\theta) \mathrm{e}^{-2M}.} \qquad (12.17)$$

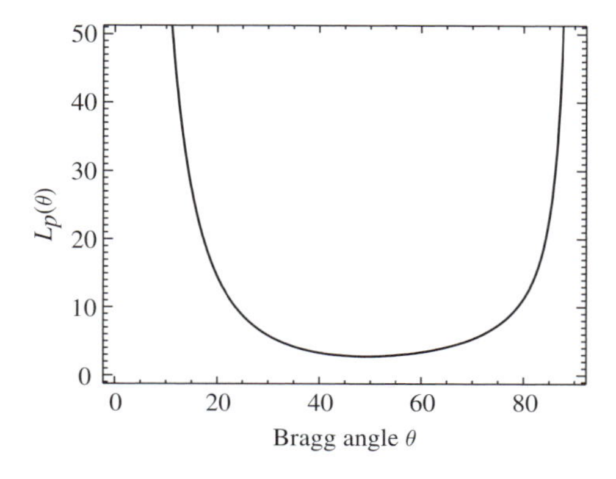

Fig. 12.8. Lorentz polarization factor $L_p(\theta)$.

Table 12.7. Computation of integrated intensities for polycrystalline tungsten.

Line	(hkl)	d_{hkl} (Å)	$\sin\theta$	θ (rad)	s (Å^{-1})	$f_{hkl}(s)$	e^{-M}	$L_p(\theta)$	p
1	110	2.2381	0.3445	0.3517	0.2234	59.5101	0.9921	14.1997	12
2	200	1.5826	0.4871	0.5088	0.3159	52.9576	0.9843	6.1572	6
3	211	1.2921	0.5966	0.6393	0.3870	48.2587	0.9766	3.7912	24
4	220	1.1190	0.6889	0.7600	0.4468	44.6851	0.9689	2.9144	12
5	310	1.0009	0.7702	0.8792	0.4996	41.9907	0.9613	2.7349	24
6	222	0.9137	0.8437	1.0042	0.5472	39.9344	0.9538	3.0871	8
7	321	0.8459	0.9114	1.1466	0.5911	38.3054	0.9463	4.2034	48
8	400	0.7913	0.9743	1.3435	0.6319	36.9457	0.9388	8.4479	6

Table 12.8. Computation of integrated intensities for polycrystalline tungsten (continued).

Line	$\lvert F_{hkl}\rvert^2 e^{-2M}$	Intensity	2θ (°)	Relative intensity	Experiment (PDF# 040806)
1	13943.997	2376009	40.30	100.0	100
2	10869.570	401555	58.31	16.9	21
3	8884.827	808416	73.26	34.0	40
4	7498.409	262237	87.09	11.0	16
5	6517.732	427815	100.75	18.0	25
6	5802.711	143309	115.08	6.0	10
7	5255.362	1060339	131.39	44.6	48
8	4812.331	243924	153.95	10.3	6

For a powder diffractometer one can remove the absorption factor $A(\theta)$ from the expression since it is constant and will cancel out when the integrated intensities are re-scaled to the most intense reflection.

Let us consider an explicit example: tungsten, body-centered cubic with lattice parameter $a = 0.31653$ nm, for Cu-Kα radiation at $T = 290$ K; the Debye–Waller factor B is equal to 0.1581 Å^2 (Peng *et al.*, 1996). To compute the diffracted intensities for a powder diffraction pattern, it is convenient to work in table format, as shown in Tables 12.7 and 12.8. The computation is relatively easy if one works through the tables one column at a time. Spreadsheet programs are useful for these types of calculations.

The last column in Table 12.8 is taken from the Powder Diffraction File, card # 040806, which lists the experimental relative intensities for the eight reflections observed in the 2θ range $[0° - 180°]$.[3] The agreement between

[3] Since the calculated values assume that the illuminated area on the sample remains constant for the entire angular range, the values from the Powder Diffraction File were corrected so that they represent a variable slit diffractometer (i.e., we used *Int-v* values).

the relative intensities is reasonably good, considering that the experimental values represent peak intensities, not integrated intensities. In Chapter 14, we will consider a more explicit example where we compare the calculated integrated intensities with experimental integrated intensities for an NaCl powder sample.

The computation of diffracted intensities for the Laue geometry is more complicated than for the powder pattern, due to the continuous wave length range. For a detailed description of Laue intensity computations, we refer the interested reader to Marín and Diéguez (1999).

12.4 Historical notes

Wilhelm Konrad Röntgen (1845–1923) was a German scientist who, in 1895, discovered X-rays. He was born in Lennep in the German Rhineland and was the son of a cloth merchant and manufacturer. He moved to Holland at an early age and in 1862 attended the Ultrecht Technical School and later the Polytechnical School in Zürich. He was a professor of physics at the University of Würzburg and director of its physical institute when, in 1895, he discovered X-rays using a Crookes vacuum cathode tube (Röntgen, 1896). Röntgen also pursued the study of the electrical conductivity and heat expansion of crystals. It was his discovery of X-rays, however, that paved the way for future scientists to study the atomic structure in the crystalline solid state.

Max Theodor Felix von Laue (1879–1960) was a German physicist and crystallographer who made many contributions to the theory and practice of X-ray diffraction. He was a professor at the University of Munich. He was the first to observe X-ray diffraction (1912) from copper sulfate. This discovery

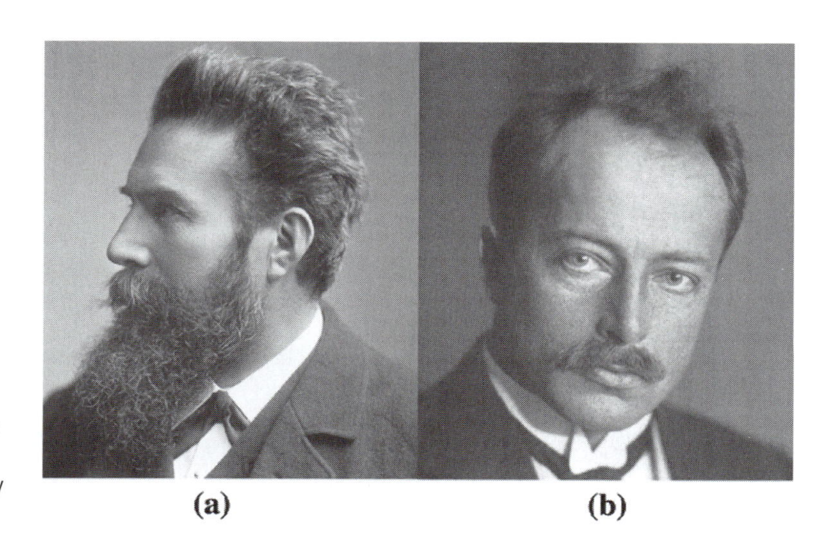

Fig. 12.9. (a) William Konrad Röntgen (1845–1923), and (b) Max Theodor Felix von Laue (1879–1960) (pictures courtesy of the Nobel Museum).

opened the door to many future studies of the structure of the solid state. Laue's discussions with Ewald on scattering of X-rays from 3-D gratings with a periodicity close to that of the X-ray wave length, stimulated his experiments to demonstrate the diffraction of X-rays by crystals. He developed what is now known as the Laue method for X-ray diffraction.

12.5 Problems

(i) *Symmetry related extinctions*: Derive the extinctions that are implied for the following symmetry operations:

(a) A 2_1 screw axis parallel to the a-direction.
(b) Absences in $(hk0)$ reflections for a b-glide parallel to the (001) plane.

(ii) *Structure factor I*: Consider an *hcp* cell with identical atoms in the $2c$ position of space group **P6$_3$/mmc** (D^4_{6h}) at $(1/3, 2/3, 1/4)$ and $(2/3, 1/3, 3/4)$.

(a) Show that the atomic positions can be expressed equivalently as: (0,0,0) and $(1/3, 2/3, 1/2)$.
(b) Show that the structure factor can be expressed as:

$$F_{hkl} = f_{atom}\left(1 + e^{2\pi i\left[\frac{h+2k}{3}+\frac{l}{2}\right]}\right).$$

(c) Calculate the square modulus of this structure factor $F^2_{hkl} = F_{hkl}F^*_{hkl}$.
(d) Express the square modulus of the structure factor for each of the following four cases:

- $h + 2k = 3n$, $l =$ even;
- $h + 2k = 3n \pm 1$, $l =$ odd;
- $h + 2k = 3n \pm 1$, $l =$ even;
- $h + 2k = 3n$, $l =$ odd.

(e) Co has two polymorphic forms, *hcp* and *fcc*. Describe extinction conditions that could be used to distinguish the reflections from the *hcp* and *fcc* phases, respectively.

(iii) *Structure factor II*: GaAs adopts a *fcc* structure with Ga atoms at the $(0, 0, 0)$ and As at the $(1/4, 1/4, 1/4)$ special positions.

(a) Express the positions of all atoms in the unit cell.
(b) Express the structure factor for GaAs.
(c) Express the square modulus of the structure factor.

(d) Simplify the square modulus under the following conditions:

1. $h+k+l=2n+1$
2. $h+k+l=2(2n+1)$
3. $h+k+l=2n$.

(iv) *Structure factor III*: Consider the face-centered cubic BiF_3 structure, with Bi atoms on the $(0,0,0)$ special position, and F on $(1/2,1/2,1/2)$ and $(1/4,1/4,1/4)$ special positions of the space group **Fm$\bar{3}$m** (O_h^5). Derive a simple expression for the structure factor; are there any systematic absences other than those caused by the face centering?

(v) *Structure factor IV*: Consider the face-centered cubic CaF_2 structure, with Ca atoms on the $(0,0,0)$ special position, and F on $(1/4,1/4,1/4)$ special positions of the space group **Fm$\bar{3}$m** (O_h^5). Derive a simple expression for the structure factor; are there any systematic absences other than those caused by the face centering? How does this structure factor differ from that of the BiF_3 structure in the previous problem?

(vi) *Integrated intensities I*: Repeat the computation of the integrated intensities of Tables 12.7 and 12.8 for the structures described in the preceding two problems. You may ignore the Debye–Waller factors (i.e., put $e^{-M}=1$).

(vii) *Integrated intensities II*: Consider the NaCl structure, with lattice parameter 0.5639 nm, space group **Fm$\bar{3}$m** (O_h^5), Na at $(0,0,0)$ and Cl at $(1/2,1/2,1/2)$. Compute the ratio of the integrated intensities for the 111 and 200 reflections as a function of the position of the Cl atom, when this atom is translated linearly from the position $(1/2,1/2,1/2)$ to the position $(1/4,1/4,1/4)$. Hint: define a parameter q so that $q=1$ corresponds to $(1/2,1/2,1/2)$ and $q=0$ to $(1/4,1/4,1/4)$; then express the structure factor as a function of q and compute the integrated intensities.

CHAPTER

13 Other diffraction techniques

> *". . . carriers of negative electricity are bodies, which I have called corpuscles, having mass very much smaller than that of the atom of any known element . . . "*
>
> Joseph J. Thomson, Nobel Lecture, 1906

13.1 Introduction

Experimental techniques used to study the structure of materials nearly always involve the scattering of electromagnetic radiation or particle waves from atomic configurations. The Bragg equation along with the concept of the structure factor form the basis of a well developed theory that enables us to understand these scattering processes and the information that can be derived from them as to the positions of atoms in a material. The most common waves used for diffraction experiments are X-rays. Other important and widely used techniques involve the wave properties of electrons and neutrons, charged and uncharged particles respectively, in scattering experiments.

Diffraction experiments can be compared and contrasted on several levels. X-ray diffraction experiments are typically the most economical means of determining crystal structures. X-ray diffractometers are commonly found in university, national and industrial laboratories. Electron diffraction is typically performed using transmission electron microscopes, which are considerably more expensive than typical X-ray diffractometers, but still common in competitive laboratory facilities. Neutron diffraction is typically performed at national or international reactor facilities. High energy, high flux X-ray scattering experiments are also used to study materials, but they too require advanced reactor facilities. In this chapter, we will describe briefly diffraction experiments involving neutron reactors, electron microscopes and high energy synchrotron X-ray facilities, as examples of other common diffraction techniques.

It is important to distinguish the different radiation sources for diffraction experiments in terms of charge, magnetic moment, and wavelength. X-ray photons are high energy electromagnetic radiation particles with a wavelength of about 0.1 nm. X-rays are uncharged and do not have a magnetic dipole moment, but they carry orthogonal electric and magnetic induction components, so that they are scattered by electronic charges and (very weakly) by magnetic dipoles. The most important atomic scattering object for X-rays is the atomic charge density, which is also spatially distributed over a 0.1 nm length scale. Electrons are charged particles and have a magnetic dipole moment. They have wavelengths around 0.002 nm and can be scattered by atomic charges or spins. The scattering objects for electrons are both the electronic and nuclear charge densities, again spatially distributed over a 0.1 nm length scale. For neutrons, wavelengths of about 0.1 nm are common. As uncharged particles, the scattering object is the atomic nucleus, which is spatially distributed over a 0.0001 nm length. Neutrons also have magnetic dipole moments that give rise to significant magnetic scattering.

13.2 *Neutron diffraction

Neutron diffraction refers to interference effects when neutrons are scattered by a crystalline solid.[1] The neutron, discovered by Sir James Chadwick in 1932, is an electrically neutral particle with mass 1.67×10^{-27} kg (0.14% heavier than the proton mass). The neutron is a constituent particle of the atomic nucleus. For an atom with atomic number Z and atomic weight A, there are $A - Z$ neutrons in its nucleus. The neutron is not a stable fundamental particle; it can be regarded as a proton (positively charged) to which a negatively charged π meson is bound. This π meson is exchanged with other protons in the nucleus, so that protons and neutrons continuously transform into one another. The opposite charges of the proton and the π meson leave the neutron electrically neutral, but the orbit of the π meson about the proton gives rise to a magnetic dipole moment that is quantized in units of the Bohr magneton, similar to the moment of the electron.

[1] One of the authors (MEM) gratefully acknowledges the course notes from a 1979 course by Professor Linn Hobbs at the Case Western Reserve University (now at MIT) as influencing the discussion of neutron diffraction in this chapter.

A "free" neutron decays with a half-life of about 886 seconds (roughly 15 minutes) to produce a proton, an electron, and an anti-neutrino (Hodgson *et al.*, 1999).

Neutron scattering is a powerful tool for the study of the structure of materials. Since neutrons are uncharged particles, they will penetrate deeply into most materials. As a consequence, samples with a large volume can be analyzed (a volume of several cubic centimeters is not unusual), and the sample preparation requirements are often not too demanding. Neutrons interact with atomic nuclei and also with the magnetic dipole moments of the nuclei. Since they are neutral particles, they do not interact with the electron cloud. The interactions between the neutron magnetic moment and the magnetic moments in magnetic materials can be used to determine how the moments in that material are oriented. Neutrons can have a *de Broglie wavelength* that is comparable to atomic spacings. The kinetic energy of neutrons is comparable to the energy of vibrational waves (phonons) in solids. Neutron scattering has revolutionized several areas of physics. The 1994 Nobel prize in physics was awarded to Clifford Glenwood Shull for development of the field of neutron scattering.

Neutron scattering can be used in a number of different modes to probe the solid state (Richter and Rowe, 2003). These are:

(i) *Elastic nuclear scattering of neutrons*: Bragg scattering can be used to determine the structure of crystalline solids, in much the same way that X-rays or electrons are used. The short range of the neutron-atomic scatterer interaction (the neutron interacts only with the small atomic nucleus) and the seemingly arbitrary variations of the atomic neutron scattering factors across the periodic table (as compared to the quite regular atomic scattering factors for X-rays, for instance) make neutrons quite attractive and unique probes of the crystalline structure. Since the scattering does not have a simple dependence on the atomic number, it is often possible to locate accurately light atomic species in the presence of heavy atoms; X-ray and electron diffraction techniques are typically not sensitive to low atomic number elements in the presence of much heavier elements.

(ii) *Elastic magnetic scattering of neutrons*: Neutrons are excellent probes of the magnetic structure of crystalline solids, due to the interactions between the magnetic dipole moment and the magnetic moments in the solid. In addition, neutrons can be polarized to facilitate the study of magnetic materials.

(iii) *Inelastic neutron scattering*: Low energy neutrons can interact with the vibrating crystal lattice (i.e., through neutron scattering by phonons) as well as with spin waves (magnons) associated with spatially varying magnetic dipole moments. This type of inelastic scattering is a particularly useful way to probe magnetic phase transitions.

(iv) *Isotopic substitution*: Most elements exist in various isotopic forms, i.e., they have nuclei with different numbers of neutrons, and, hence, atomic weights. Isotopes can have dramatically different neutron scattering powers. Since isotopes have identical chemical properties, isotopic substitution may be used in some cases to change the atomic scattering factors in the material, i.e., to tailor the scattering factors so that the diffraction experiment becomes more (or less) sensitive to a particular element. This is especially useful for the structure determination of organic molecules; for example, substitution of deuterium for hydrogen may make it easier to locate the hydrogen sites in the molecule. This makes neutron scattering a very useful technique in the study of polymers and biomolecules.

The first type of elastic scattering can be important in probing chemical order in both non-magnetic and magnetic crystalline solids. As we have seen, the scattering power of X-rays is proportional to the atomic number, Z, of the scatterer. This can cause problems when the sample being studied is composed of atoms with nearly the same atomic number. Under these circumstances, the scattering powers are nearly identical and quantitative measurements of atomic ordering are difficult. For neutrons, however, the main scattering source is the nucleus rather than the electrons orbiting the nucleus. The neutron scattering lengths do not vary regularly with Z, so that neutron diffraction becomes an attractive method for studying chemical ordering in certain transition metal alloys, such as the Fe–Co described in this chapter.

The interaction of the neutron spin with nuclear spins in a material depends on their relative orientation. The above mentioned inelastic scattering events are of particular interest for probing dynamic magnetic spin excitations (magnons). Magnetic scattering of neutrons can probe both static magnetic order and, in temperature dependent scattering experiments, magnetic phase transitions. The inelastic scattering of neutrons also provides a sensitive probe of the magnetic ordering transition. Although it will not be discussed in detail here, *small angle neutron scattering, SANS*, observed at angles between approximately 0.15° and 15°, is a powerful technique for the study of macromolecules, small defects in crystalline materials, as well as magnetic domains.

13.2.1 Neutrons: generation and properties

The properties of neutrons, protons, and electrons are compared in Table 13.1. Neutrons can be produced from a variety of sources, including the decay of radioactive elements, and nuclear reactions (fission). A typical weak source of neutrons comes from the decay of radioactive elements such as $^{210}_{84}Po$ or $^{226}_{88}Ra$ to produce α particles ($^{4}_{2}He$). The energetic α particles can then be

Table 13.1. Properties of electrons, neutrons, and protons.

Particle	Electron	Proton	Neutron
Symbol	e^-	p^+	n^0
Relative charge	-1	$+1$	0
Actual charge	-1.6×10^{-19} C	1.6×10^{-19} C	0
Relative mass	1/1837	1	1
Actual mass	9.1055×10^{-24} kg	1.678×10^{-27} kg	1.675×10^{-27} kg

used to bombard Be atoms ($^{9}_{4}$Be or $^{11}_{4}$Be) to produce neutrons through the reactions (Hodgson *et al.*, 1999):

$$^{9}_{4}\text{Be} + ^{4}_{2}\text{He} \rightarrow ^{12}_{6}\text{C} + ^{1}_{0}\text{n} + 5.7\,\text{MeV};$$
$$^{11}_{4}\text{Be} + ^{4}_{2}\text{He} \rightarrow ^{14}_{7}\text{N} + ^{1}_{0}\text{n}.$$

More intense and energetic sources of neutrons come from *fission reactions* involving isotopes of U or Pu. A typical self-sustaining nuclear fission reaction is:

$$^{235}_{92}\text{U} + ^{1}_{0}\text{n} \rightarrow ^{236}_{92}\text{U} \rightarrow ^{144}_{56}\text{Ba} + ^{89}_{36}\text{Kr} + 3\,^{1}_{0}\text{n} + 177\,\text{MeV}.$$

Neutrons are classified by their kinetic energy, E_{kin}, as:

(i) *Cold neutrons*: $E_{\text{kin}} \ll 0.025$ eV;
(ii) *Thermal neutrons*: in thermal equilibrium with the atmosphere at 293 K, with $E_{\text{kin}} \sim 0.025$ eV;
(iii) *Slow neutrons*: $0.025 < E_{\text{kin}} < 100$ eV;
(iv) *Intermediate neutrons*: 100 eV $< E_{\text{kin}} <$ 10 keV;
(v) *Fast neutrons*: 10 keV $< E_{\text{kin}} <$ 10 MeV, and
(vi) *Ultra-fast neutrons*: $E_{\text{kin}} >$ 10 MeV.

Because the interaction of neutrons with matter is weak, only high flux neutron sources, i.e., those available from nuclear reactors, are useful for neutron scattering experiments. A typical research reactor might use the fission products associated with the absorption of neutrons by atoms and the subsequent chain reaction. Such a reactor typically generates 20–60 MW of energy and a neutron-flux between 10^{16} and 10^{19} n/m^2/s, with an energy distribution peaked near 1 MeV. More intense neutron sources can be generated by bombarding heavy nuclei with high energy light particles (e.g., protons or alpha particles). These sources, known as *spallation sources*, can achieve fluxes in the range 10^{18}–10^{22} n/m^2/s. Spallation is the process in which a heavy atom is bombarded by an intense beam of high-energy protons. These protons reach energies of several GeV and velocities of about 90% of the speed of light.

The fast neutrons produced by reactors or pulsed sources have fluxes in a useful range, but are too penetrating for use in diffraction experiments. For this reason, these neutrons are often first *thermalized* through interaction with a *moderator*. This involves a loss of kinetic energy through collisions with moderator atoms, which eventually results in neutrons with a kinetic energy about equal to the thermal energy, i.e.:

$$\frac{1}{2}m_n v^2 = \frac{3}{2}k_B T, \tag{13.1}$$

where m_n is the neutron mass, v the velocity, T the absolute temperature (in Kelvin) and k_B the Boltzmann constant ($k_B = 1.38 \times 10^{-23}$ J/K). Solving for v yields:

$$v = \sqrt{\frac{3k_B T}{m}}. \tag{13.2}$$

Using the de Broglie relation, the neutron wavelength for thermal neutrons is found to be:

$$\boxed{\lambda = \frac{h}{mv} = \frac{h}{\sqrt{3mk_B T}}} \tag{13.3}$$

for non-relativistic particles. At 300 K, the neutron kinetic energy, E_{kin}, is about $1/40$ eV, and the wavelength, $\lambda = 0.18$ nm; the neutron velocity is about 2200 m/s. Longer wavelengths (slower neutrons) can be accessed at lower temperatures. Typical moderator materials are water, *heavy water* (containing hydrogen and *deuterium*) and pure C graphite.

Thermal neutrons are more correctly described as an ideal gas of particles possessing a *Maxwell–Boltzmann distribution* of kinetic energies:

$$N(E_{kin}) = N_{tot}(8mk_B T)^{\frac{1}{2}}\left(\frac{E_{kin}}{k_B T}\right)^{\frac{5}{2}} \exp\left(\frac{-E_{kin}}{k_B T}\right), \tag{13.4}$$

where $N(E_{kin})$ is the number of neutrons with kinetic energy, E_{kin}, and N_{tot} is the total number of neutrons.

13.2.2 Neutrons: wave length selection

For diffraction experiments, it is desirable to have *monochromatic* neutrons (i.e., a single wave length λ). Monochromatization of neutrons can be accomplished through:

- *Time of flight monochromatization*: Time of flight monochromatization relies on the fact that the velocity distribution of the neutrons is described by the Maxwell–Boltzmann distribution; therefore, neutrons with a fixed energy will travel a predictable distance over a given amount of time.

- *Use of neutron velocity selectors*: Velocity selectors consist of neutron absorbing materials cut with helical channels. When the helical channel velocity selector is rotated at a constant velocity only those neutrons with a specific range of velocities will traverse the channels without colliding with the absorber. These velocity selectors are also known as *choppers* because they "chop" the neutron beam into specific velocity (wave length) ranges.
- *Use of single crystal monochromators*: Single crystal monochromatization relies on collecting neutrons of a fixed wave vector after Bragg scattering from a single crystal grating. This type of monochromatization is the most accurate method to obtain a monochromatic neutron beam.

A variety of sample geometries are possible in neutron diffraction. These can range from powders to single crystals to thin films.

13.2.3 Neutrons: atomic scattering factors

Neutrons, as uncharged particles, interact with atoms only at very short distances. Unlike X-rays, which scatter off the electron cloud, neutrons interact only with the nuclei. Neutron scattering involves nearly head-on collisions with the nuclei, so that the probability of neutron scattering, the *neutron scattering cross section*, is related to the size of the nucleus. The number of nucleons (protons and neutrons) in a nucleus can be determined from the atomic weight, since protons and neutrons have nearly equal weights and the electron mass is negligible. As the neutron and proton sizes are comparable, the volume of a nucleus of radius r_{N} is:

$$\frac{4}{3}\pi r_{\mathrm{N}}^3 = \frac{4}{3}\pi r_{\mathrm{n}}^3 A, \tag{13.5}$$

where A is the atomic weight and r_{n} is about 1.5×10^{-15} m. For all atomic nuclei, A is less than 250, so that $r_{\mathrm{N}} < 10^{-14}$ m. Compared with the room temperature wave length of thermal neutrons this radius is four to five orders of magnitude smaller. Since the size of the nucleus is so small, it can, therefore, be treated as a δ-function scattering source (i.e., a point source). This is extremely important, since it implies that the scattered intensity is uniform, i.e., it is not a function of the scattering angle, θ. The nucleus acts as a source of a spherical scattering wave with an angle-independent amplitude; for electron and X-ray scattering, the scattering amplitude depends strongly on the scattering angle.

The scattered wave functions for neutrons, electrons, and X-rays take on the following functional forms:

$$\psi_n = (\tfrac{-b}{r_p})\exp[-i\mathbf{k}' \cdot \mathbf{r}_p];$$

$$\psi_x(\theta) = (\tfrac{f^x(\theta)}{r_p})[\tfrac{-e^2}{4\pi m_e c^2}][\tfrac{1+cos^2(\theta)}{2}]\exp[-i\mathbf{k}' \cdot \mathbf{r}_p]; \qquad (13.6)$$

$$\psi_e(\theta) = (\tfrac{f^e(\theta)}{r_p})\exp[-i\mathbf{k}' \cdot \mathbf{r}_p],$$

where $\mathbf{k}'$ is the scattered wave vector, $\mathbf{r}_p$ is the position of the wavefront, b is the neutron scattering length and θ the scattering angle.

It is customary to relate the neutron scattering length b to the apparent size of the atomic nucleus. When a neutron approaches a nucleus, this nucleus will have an apparent size, similar to (but much smaller than) the size of a bull's eye target in an archery competition. Each nucleus – in fact, each isotope – has a different size denoted by the symbol σ; σ is known as the *scattering cross section*, and one can show that:

$$\boxed{\sigma = 4\pi b^2.} \qquad (13.7)$$

Since b does not depend on the scattering angle θ, we would also expect that $\sigma \sim 4\pi r_N^2$ (where the factor of 4 comes from a quantum mechanical treatment). Therefore, we conclude that $b \approx r_N$ and:

$$b = r_n A^{\frac{1}{3}} = (1.5 \times 10^{-15}) A^{\frac{1}{3}} \text{m} = 15 A^{\frac{1}{3}} (\text{barns})^{\frac{1}{2}} \qquad (13.8)$$

where 1 barn $= 10^{-28}$ m^2.[2] Equation 13.8 also implies only a small variation in scattering cross section for heavy and light nuclei. While the general $A^{1/3}$ dependence of the scattering cross section is indeed observed, considerable variation in b is found from element to element or even isotope to isotope, with some scattering lengths even being negative. These variations in b are principally due to resonance absorption in compound nucleus formation which reduces the cross section, and, therefore, the scattering length.

A list of neutron scattering lengths for scattering of thermal neutrons is shown in Table 13.2 and graphically in Fig. 13.1. The neutron scattering lengths for Po, At, Rn, Fr, and Ac are not available. Some of the numbers in this table have an imaginary component: B, Cd, In, Sm, Eu, Gd, and Dy. This indicates that the nuclei of these elements have a strong tendency to capture neutrons, i.e., to absorb neutrons. Vanadium (V) has the smallest (absolute) value for b; for this reason, it is mostly transparent to neutrons and is hence used as a sample holder material (e.g., a thin-walled cylinder to hold a powder sample).

[2] The unit barn is named after the expression "hitting the broad side of a barn."

Table 13.2. Neutron scattering lengths (in femtometer, fm) for all naturally occurring elements; thermal neutrons are assumed. This list is taken from a longer list, including all isotopes, at the web site of the National Institute of Standards and Technology (NIST, URL: http://www.ncnr.nist.gov/resources/n-lengths/list.html). The values in the table are averages over all isotopes, weighted by the natural isotope abundances. Dashes indicate that the scattering length has not been determined.

Atom	*b*	Atom	*b*	Atom	*b*	Atom	*b*
H	−3.7390	He	3.26(3)	Li	−1.90	Be	7.79
B	5.30−0.213i	C	6.6460	N	9.36	O	5.803
F	5.654	Ne	4.566	Na	3.63	Mg	5.375
Al	3.449	Si	4.1491	P	5.13	S	2.847
Cl	9.5770	Ar	1.909	K	3.67	Ca	4.70
Sc	12.29	Ti	−3.438	V	−0.3824	Cr	3.635
Mn	−3.73	Fe	9.45	Co	2.49	Ni	10.3
Cu	7.718	Zn	5.680	Ga	7.288	Ge	8.185
As	6.58	Se	7.970	Br	6.795	Kr	7.81
Rb	7.09	Sr	7.02	Y	7.75	Zr	7.16
Nb	7.054	Mo	6.715	Tc	6.8	Ru	7.03
Rh	5.88	Pd	5.91	Ag	5.922	Cd	4.87−0.7i
In	4.065−0.0539i	Sn	6.225	Sb	5.57	Te	5.80
I	5.28	Xe	4.92	Cs	5.42	Ba	5.07
La	8.24	Ce	4.84	Pr	4.58	Nd	7.69
Pm	12.6	Sm	0.80−1.65i	Eu	7.22−1.26i	Gd	6.5−13.82i
Tb	7.38	Dy	16.9−0.276i	Ho	8.01	Er	7.79
Tm	7.07	Yb	12.43	Lu	7.21	Hf	7.7
Ta	6.91	W	4.86	Re	9.2	Os	10.7
Ir	10.6	Pt	9.60	Au	7.63	Hg	12.692
Tl	8.776	Pb	9.405	Bi	8.532	Po	—
At	—	Rn	—	Fr	—	Ra	10.0
Ac	—	Th	10.31	Pa	9.1	U	8.417

The incoherent scattering of neutrons is different from that of X-rays (*Compton scattering*) due to the fact that neutrons can have their spins oriented to be in one of two directions, denoted spin-up and spin-down. As a result, the scattering length will be different for each and can be denoted as b_+ and b_-, respectively. Scattered neutrons can have differences in phase which arise from the different total nuclear angular momentum of the compound nucleus with the neutron. These phase differences can give rise to incoherent scattering analogous to that of isotope disorder in X-ray scattering.

The neutron's spin angular momentum gives rise to a net magnetic moment of 1.04×10^{-3} Bohr magnetons or 1.91 nuclear magnetons. As a result, in addition to nuclear scattering, there will also be a magnetic contribution to the

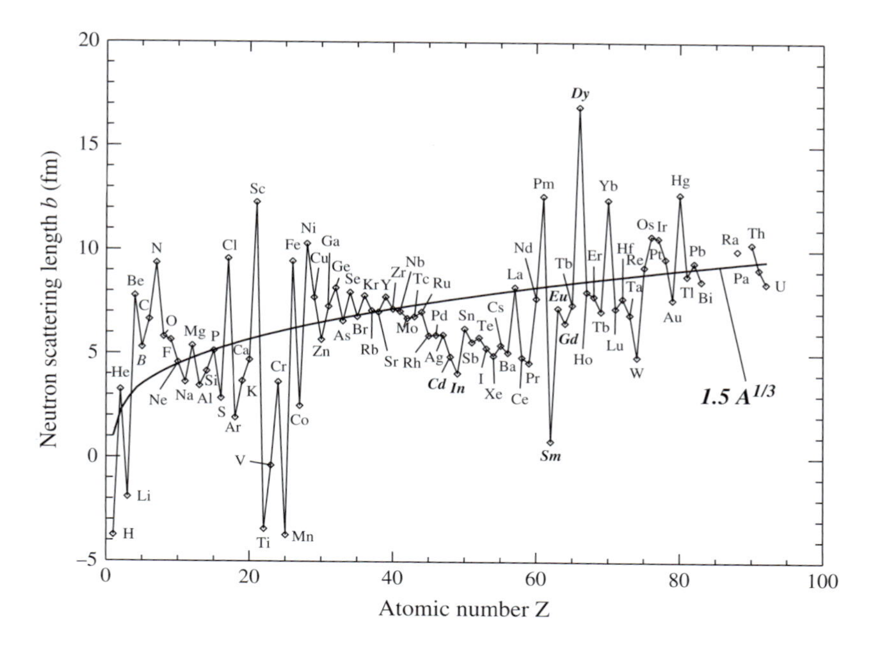

Fig. 13.1. Neutron scattering lengths of Table 13.2 as a function of the atomic number. The solid curve indicates the the approximate relation 13.8. Elements in bold italic font indicate that the scattering length has a significant imaginary component.

scattered intensity. Magnetic scattering arises from the interactions between neutron spins and the net atomic spin due to electrons occupying incomplete shells. This is especially important for systems containing transition metal or rare earth elements. For ionic systems, the net atomic magnetic dipole moment can be calculated by quantum mechanical rules. Cations have magnetic dipole moments determined by *Hund's rules* (see Box 13.1).

The magnetic scattering of neutrons is treated similarly to X-ray scattering in terms of the sine of the angle, β, between the incident neutron wave vector and the direction of the net atomic dipole moment arising from the unfilled electronic shells (as given by Hund's rules). A *magnetic scattering length*, b_{M} is defined as:

$$b_{\mathrm{M}} = \sin\beta\left(\frac{e^2\gamma}{m_e c^2}\right) S f_{\mathrm{S}}, \tag{13.9}$$

where S is the total spin angular momentum (Table 13.3), f_{S} is a *magnetic form factor*, and γ is the magnetic dipole moment of the neutron. It is, therefore, possible to determine both the magnitude and direction of the atomic magnetic dipole moment using neutron scattering.

One of the most important implications of magnetic scattering is diffraction from ordered, often collinear systems of atomic dipole moments. This includes *ferromagnetic*, *antiferromagnetic*, and *ferrimagnetic* materials. Ferromagnetism is an example of correlated or collective magnetism. To define ferromagnetism, we begin with permanent atomic dipole moments. In a ferromagnetic material, the local atomic moments remain aligned, even

Box 13.1 Magnetic dipole moments and Hund's rules

Magnetic dipole moments result from orbital and spin angular momentum of electrons in an unfilled atomic shell. The relationship between magnetic dipole moment vector, $\boldsymbol{\mu}$; half the charge to mass ratio of an electron, $\frac{e}{2m}$; and the angular momentum vector, $\boldsymbol{\Pi}$, is given by:

$$\boldsymbol{\mu} = g\frac{e}{2m}\boldsymbol{\Pi},$$

where $\boldsymbol{\Pi}$ can refer to the orbital angular momentum ($\mathbf{L}$) or the spin angular momentum ($\mathbf{S}$) and g is the gyromagnetic factor. In ferrites, the d-shells of the transition metal cations are of interest, and the orbital angular momentum is quenched (i.e., $\mathbf{L} = 0$) in the crystal. The spin angular momentum for a single electron is quantized by the spin quantum number, $m_s = \pm 1/2$, to be $m_s\hbar = \pm\hbar/2$. For spin only, the gyromagnetic factor is $g = 2$, and the single electron dipole moment is:

$$\mu = \pm g\frac{e}{2m}\frac{\hbar}{2} = \pm\frac{\hbar e}{2m} \equiv \pm\mu_B \qquad (= 9.27 \times 10^{-24}\ \text{A m}^2),$$

where μ_B is the Bohr magnetron, the fundamental unit of magnetic dipole moment.

For a multi-electron atom, the total spin angular momentum is:

$$S = \sum_{i=1}^{n}(m_s)_i$$

with the sum over all electrons in the outer shell. Hund's first rule states that, for an open shell multi-electron atom, we fill the $(2l + 1)$-fold degenerate (for d-electrons we have $(2l + 1) = 5$) orbital angular momentum states so as to maximize total spin. To do so, we must fill each of the five d-states with a positive (spin-up) spin before returning to fill the negative (spin-down) spin. The total spin angular momentum for the $3d$ transition metal ions is summarized in Table 13.3.

in the absence of an applied field, below a temperature, T_c, known as the *Curie temperature*. As a result, a ferromagnetic material possesses a non-zero net magnetic dipole moment over a macroscopic volume, called a *magnetic domain*, containing many atomic sites. Ferromagnetic materials give rise to coherent magnetic scattering of neutrons.

For a simple antiferromagnet, like *bcc* Cr, for example, equal spin dipole moments on adjacent nearest neighbor atomic sites are arranged in an antiparallel fashion below an ordering temperature, T_N, called the *Néel*

Table 13.3. Transition metal ion spins and dipole moments ($\mathbf{L}=0$).

# d electrons	Cations	S	μ/μ_B
1	Ti^{3+}, V^{4+}	1/2	1
2	V^{3+}	1	2
3	V^{2+}, Cr^{3+}	3/2	3
4	Cr^{2+}, Mn^{3+}	2	4
5	Mn^{2+}, Fe^{3+}	5/2	5
6	Fe^{2+}	2	4
7	Co^{2+}	3/2	3
8	Ni^{2+}	1	2
9	Cu^{2+}	1/2	1
10	Cu^{+}, Zn^{2+}	0	0

temperature. For a simple ferrimagnet, unequal spin dipole moments on adjacent nearest neighbor atomic sites are arranged in an antiparallel fashion below the Néel temperature. These give rise to the possibility of the space group for magnetic scattering being different for neutron scattering as opposed to scattering by X-rays or by electrons. In fact, if the assumption of spherical atoms is replaced with one for which a vector representing the atomic magnetic dipole moment is attached to each atom, then the number of possibilities for different space groups (the *magnetic space groups*) greatly exceeds the 230 space groups previously enumerated. We forgo a complete discussion of the 1651 magnetic space groups in this text. In 1949, antiferromagnetic MnO was the first material for which neutron diffraction was used to study the magnetic order.

13.2.4 Neutrons: scattering geometry

The basic geometry of a neutron scattering experiment consists of a neutron source, a sample, and one or more detectors (Fig. 13.2). Neutrons from the source are scattered by the sample and are collected by the detector(s). As neutrons are scattered from the sample, they will undergo a change in either momentum or energy (or both). Measurement of these changes, together with an application of the physical theory of scattering, leads to an understanding of the structural and dynamic properties of the sample. Neutrons with a wide wave length spectrum are produced in a typical nuclear reactor source. A crystal monochromator is then used to deflect neutrons of only one wavelength, resulting in a monochromatic source of incident neutrons. Elastic neutron scattering (neutrons are scattered without energy change) results in the standard equality of the angles of incidence and reflection, $\theta = \theta'$, as shown in Fig. 13.2.

Neutrons can be described as particle waves with a wave vector $\mathbf{k} = k\mathbf{e}_k$; as before, $\mathbf{k}$ is a vector in reciprocal space. The particle nature of the neutron also

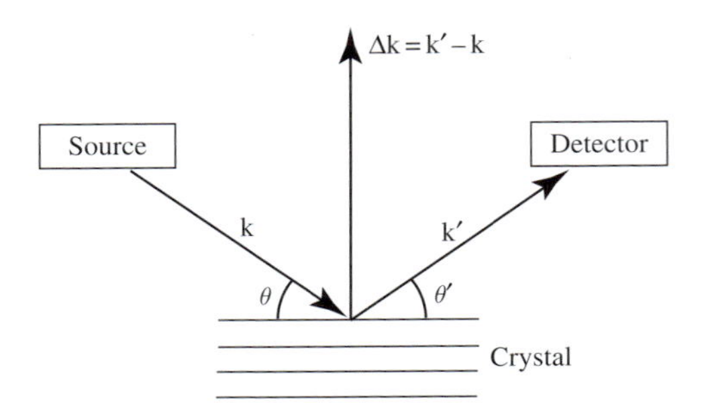

Fig. 13.2. Scattering geometry for a neutron diffraction experiment.

means that $\mathbf{k}$ multiplied by Planck's constant represents the vector momentum of the neutron, i.e., $\mathbf{p} = \mathrm{h}\mathbf{k}$. The *scattering vector*, $\Delta\mathbf{k}$, is defined as the difference between the momentum $\mathbf{k}'$ of the scattered neutron and the momentum $\mathbf{k}$ of the incident neutron, i.e., $\Delta\mathbf{k} = \mathbf{k}' - \mathbf{k}$. The absolute value of the scattering vector $|\Delta\mathbf{k}|$ equals $2\sin\theta/\lambda$. Bragg diffraction occurs when the scattering vector is equal to a reciprocal lattice vector, $\mathbf{g}_{hkl}$, of the crystal being studied; in that case we also have $\theta = \theta'$. Inelastic scattering events can be probed by changing the position of the detector while keeping the sample orientation relative to the source constant; in other words, the exit angle θ' is varied while θ is kept constant.

The fact that a nuclear reactor is needed to provide the neutron beam means that neutron scattering experiments are typically carried out only at dedicated, often national, user facilities; most universities cannot afford to run and maintain such a facility. Fig. 13.3 shows block schematics for instruments used for neutron diffraction at the NIST National Center for Neutron Research (NCNR) in Gaithersburg, MD. Figure 13.3 (a) shows the main reactor and instruments used to study materials with thermal neutrons. Figure 13.3 (b) shows the Guide Hall at NCNR, housing various beam lines for using cold

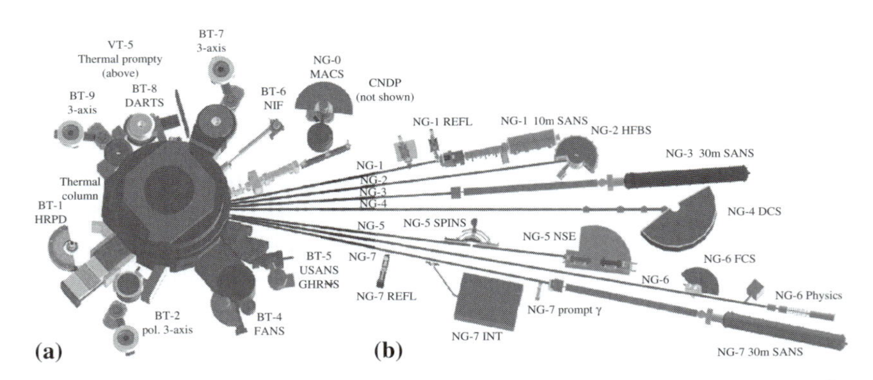

Fig. 13.3. Instruments used for neutron diffraction at the NIST National Center for Neutron Research (NCNR) in Gaithersburg, MD. (a) Detectors close to the neutron reactor source for use of thermal neutrons, (b) plan view of the Guide Hall at NCNR, housing various beam lines for using cold neutrons including small angle neutron scattering. (Figure reproduced with permission.)

neutrons including *small angle neutron scattering*. The use of cold neutrons requires cryogenic cooling. Neutrons with lower temperatures (energies) have longer wave lengths appropriate for the study of polymer or biomolecule crystals with large lattice constants.

13.2.5 Neutrons: example powder pattern

One advantage of the use of *neutron diffraction* over X-ray diffraction is illustrated in Fig. 13.4: the determination of the superlattice reflections for B2 α'-FeCo. Since *neutron scattering cross sections* show considerable variation from element to element or even isotope to isotope, neutron diffraction is attractive for the study of chemical ordering in certain transition metal alloys. Fig. 13.4(b) shows a neutron diffraction pattern for an ordered FeCo alloy using neutrons of wavelength 0.154 nm, the same wavelength as Cu K_α X-rays used in Fig. 13.4(a). Because of the large difference in the neutron cross sections for Fe and Co (see Table 13.2), the superlattice reflections are easily observed, whereas for X-rays they are not, unless anomalous scattering is considered, as discussed in Section 13.4.2.

Using the same approach as in Chapter 12, Table 13.4 lists the calculated relative integrated intensities for X-ray diffraction and the calculated intensities for neutron diffraction. The structure factor squared for this compound is simply:

$$|F^x_{hkl}|^2 = (f_{\text{Fe}} + (-1)^{h+k+l} f_{\text{Co}})^2$$

for X-rays and

$$|F^n_{hkl}|^2 = (b_{\text{Fe}} + (-1)^{h+k+l} b_{\text{Co}})^2$$

for neutrons. Using a lattice parameter of $a = 0.28571$ nm, and Debye-Waller factors for both Fe and Co of 0.0055 nm^2, we can compute the relative integrated intensity I^x_{hkl} for X-rays and I^n_{hkl} for neutrons, as shown in Table 13.4;

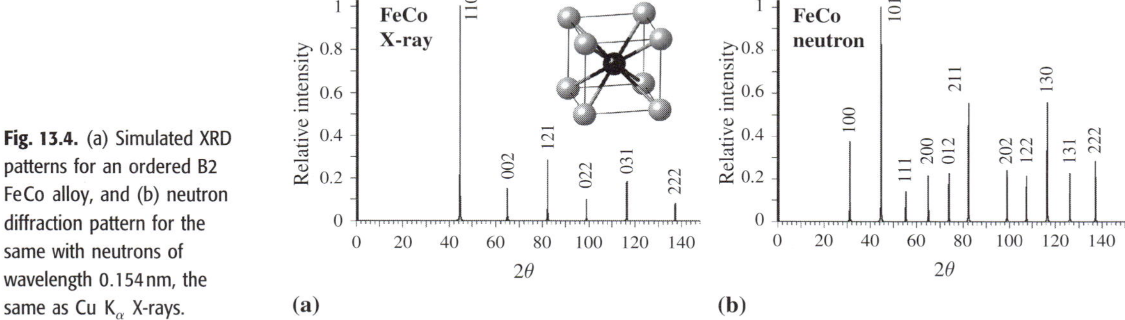

Fig. 13.4. (a) Simulated XRD patterns for an ordered B2 FeCo alloy, and (b) neutron diffraction pattern for the same with neutrons of wavelength 0.154 nm, the same as Cu K_α X-rays.

Table 13.4. Comparison of the integrated relative intensities for Fe Co, using Cu Kα radiation and neutrons of the same wave length.

# line	hkl	s	F^x_{hkl}	L_p	p_{hkl}	I^x_{rel}	F^n_{hkl}	I^n_{hkl}
1	100	0.1750	1.0424	24.722	6	0.09	6.96	37.6
2	110	0.2475	36.597	11.184	12	100.00	11.94	100.0
3	111	0.3031	0.9142	6.8353	8	0.03	6.96	13.8
4	200	0.3500	29.318	4.7995	6	13.77	11.94	21.5
5	210	0.3913	0.7804	3.7061	24	0.03	6.96	22.5
6	211	0.4287	24.579	3.1034	24	25.05	11.94	55.5
7	220	0.4950	21.030	2.7289	12	8.07	11.94	24.4
8	221	0.5250	0.5869	2.8465	24	0.01	6.96	17.3
9	310	0.5534	18.250	3.1743	24	14.18	11.94	56.8
10	311	0.5804	0.5027	3.7959	24	0.01	6.96	23.1
11	222	0.6062	16.073	4.9846	8	5.77	11.94	29.7

Lorentz polarization and multiplicities are taken into account. The X-ray intensities for reflections with $h+k+l=2n+1$ are nearly zero, so that X-ray diffraction is not an appropriate technique to detect ordering in this system; the powder pattern looks exactly the same as the pattern for the disordered Fe Co compound. In the neutron powder pattern, however, all reflections have significant intensities, so that the presence of ordering can be detected easily.

13.3 ∗Electron diffraction

13.3.1 The electron as a particle and a wave

Electrons are traditionally considered to be point-like charged particles that can be accelerated by an electric field, e.g., in the X-ray tube, or in cathode ray tubes (i.e., TV tubes). A single electron has a mass of $m_0 = 9.109\,389\,7 \times 10^{-31}$ kg, and a charge of $-e = -1.602\,177\,33 \times 10^{-19}$ C. The electron mass is about 2000 times lighter than that of the proton or neutron, and it is correspondingly easier to accelerate the electron using an electric field. Recall that the interaction of X-ray photons (which are, essentially, oscillating electric fields) with atoms makes the electrons oscillate at a much larger amplitude than the protons, precisely because of this mass difference. When a single electron interacts with matter, we expect it to interact with both the negatively charged electron clouds of the atoms *and* with the positively charged atomic nuclei. The interaction of electrons with matter is, therefore, expected to be significantly stronger than that of X-rays with matter. This has important consequences for electron diffraction, as we will see in the following sections.

According to quantum theory, every moving particle has a wavelength associated with it (see Equation 1.3 on page 8). Because an electron can be

Table 13.5. Electron wavelengths, mass ratio, and fractional velocity (fraction of c) for various accelerating voltages.

V (volts)	λ (pm)	m/m_0	$\beta = \frac{v}{c}$
100	122.6	1.00019	0.0198
1 000	38.76	1.00196	0.0625
10 000	12.20	1.01957	0.1950
100 000	3.701	1.19570	0.5482
200 000	2.508	1.3914	0.6953
400 000	1.644	1.7828	0.8279
1 000 000	0.872	2.9569	0.9411

easily accelerated to a velocity that is a significant fraction of the velocity of light, we must describe the electron wave length using *relativistic* physics. The equation relating the electron's wavelength (in pm) to the accelerating potential V (in volts) is then given by:

$$\lambda = \frac{\mathrm{h}}{mv} = \frac{\mathrm{h}}{\sqrt{2m_0eV(1+\frac{e}{2m_0c^2}V)}} = \frac{1226.39}{\sqrt{V+0.97845\times 10^{-6}V^2}}.$$

For an accelerating voltage of 400 000 volts, electrons have a wave length of 1.644 pm, they travel at 83% of the speed of light and they appear to be 78% heavier than an electron at rest! Values for other accelerating voltages can be found in Table 13.5.

Since the electron can be regarded as a wave, this means that electrons can be diffracted by crystal lattices. The fact that the electron is a charged particle complicates the diffraction process substantially: an electron traveling at high speed through a crystal interacts *strongly* with all other charges in the material, including the nuclear charges. The probability of an electron being scattered by a single atom is about four orders of magnitude larger than that for X-rays and, therefore, the *atomic scattering factor for electrons*, f^{el}, (defined in a manner analogous to that for X-rays and neutrons) is also much larger than f^{X}. It is, in fact, possible to express the electron scattering factor in terms of the X-ray atomic scattering factors; the relation, which is known as the *Moth–Bethe formula*, and which we will state without an explicit derivation, reads:

$$f^{\mathrm{el}}(s) = \frac{|e|}{16\pi^2\epsilon_0|s|^2}\left[Z - f^{\mathrm{X}}(s)\right], \qquad (13.10)$$

where $s = \sin\theta/\lambda$ has the same meaning as in Section 12.1.2. The same expansion coefficients listed in Table 12.1 can be used to compute the electron scattering factors as well.

The most important consequence of this increased scattering probability is that an electron, after it has been scattered once, can diffract again and again, from different lattice planes. For X-rays, the probability of *multiple* scattering events is very small, so that we can usually ignore it. For electrons, this would lead to major errors. If the intensity of a diffracted beam can be computed by simply taking the modulus squared of the structure factor (as we have done in Chapter 12), then one refers to the scattering process as *kinematical scattering*; if the intensity is no longer proportional to the structure factor squared, as is the case for multiple scattering, then we talk about *dynamical scattering*. Dynamical scattering theory *must* be used to describe electron diffraction, although, under certain conditions, the kinematical approach is a reasonable approximation. While this dynamical theory is well beyond the scope of this book, we will briefly discuss its consequences in the following sections.

13.3.2 The geometry of electron diffraction

The small value for the electron wave length has important consequences for the geometry of the scattering process. Bragg's law tells us that the typical diffraction angle θ for a 200 000 eV electron with wave length 0.002 508 nm, in a crystal with lattice spacing $d = 0.2$ nm, is about 6 milliradians or 0.36°! This should be compared to the 22.7° angle for diffraction of Cu Kα X-rays from the same planes. Since the diffraction angle θ is small, we can expand the trigonometric function in Bragg's law as $\sin\theta \approx \theta$, so that we have the approximate expression for electron diffraction:

$$2d_{hkl}\theta = \lambda.$$

This means that almost all electron diffraction will occur close to the forward direction, or, in other words, we only need to look close to the incident beam direction to find the diffracted beams.

Alternatively, we can describe the electron scattering process in reciprocal space: the electron wavelength is about 100 to 1000 times shorter than the typical X-ray wave length. This means that the radius of the Ewald sphere, $1/\lambda$, will be 100 to 1000 times larger. The Ewald sphere is thus huge compared to the reciprocal lattice spacings. Figure 13.5 shows a to-scale drawing of the Ewald sphere and a reciprocal lattice for a cubic crystal with a lattice parameter of 0.4 nm (i.e., $a^* = 2.5\,\text{nm}^{-1}$). The small circle near the center is the Ewald sphere for Cu Kα X-rays; the two large circle segments represent the Ewald spheres for 200 keV and 1 MeV electrons. The corresponding Ewald sphere radii are $6.486\,\text{nm}^{-1}$ for the X-rays, and 398.73 and $1146.89\,\text{nm}^{-1}$, respectively, for the electrons. Since the ES by definition goes through the origin of reciprocal space, we find that we can orient our crystal such that a *whole plane of reciprocal lattice points* is tangent to the sphere (which can

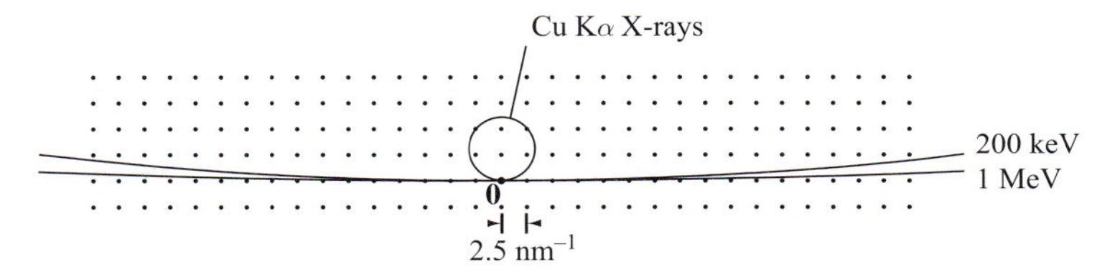

Fig. 13.5. Schematic comparison of the Ewald spheres for Cu Kα X-rays (small circle) and 200 keV and 1 MeV electrons (large circle segments). The underlying reciprocal lattice corresponds to a cubic crystal with lattice parameter $a = 0.4$ nm (figure reproduced from Fig. 2.7 in *Introduction to Conventional Transmission Electron Microscopy*, M. De Graef, 2003, Cambridge University Press).

be approximated by a plane close to the origin). Therefore, in general, there will be many diffracted beams *simultaneously*, all close to the transmitted beam.

Another fundamental difference between X-ray or neutron diffraction and electron diffraction is the fact that the strong interaction of electrons with matter necessitates the use of very thin or small samples. Whereas X-rays and neutrons can easily traverse samples with dimensions in the range of several millimeters, electrons are quickly absorbed by matter, so that sample thicknesses are limited to a few hundred nanometers. This has two important consequences:

- Sample preparation becomes a time consuming and difficult task, since it is not straightforward to prepare thin foils with a thickness of less than about 100 nm; furthermore, handling these thin foils can also become quite difficult;
- On a more fundamental level, the small thickness of the foil has an effect on the reciprocal lattice points. So far, in our discussion of X-ray diffraction, we have tacitly assumed that our crystal is effectively infinitely large. It can be shown mathematically that the shape of a reciprocal lattice point is "the reciprocal" of the shape of the crystal; for an infinite crystal, we obtain reciprocal lattice points with zero volume, i.e., mathematical points. For the clearly finite thickness thin foils used for electron diffraction, we must allow for the reciprocal lattice points to have a shape and associated volume. Each reciprocal lattice point becomes a cylindrical rod, extended in the direction normal to the thin foil.[3] Since each reciprocal lattice point now has a finite extent, it is possible for the reciprocal lattice point to intersect the Ewald sphere, even when the actual point (i.e., the center of the volume) is not actually on the Ewald sphere. In other words, for electron diffraction, we can have a diffracted beam even when the reciprocal lattice

[3] In general, one can show that the shape of a reciprocal lattice point is equal to the *Fourier transform* of the shape of the crystal. The mathematical formulation of this relation is beyond the scope of this text.

point is not exactly on the Ewald sphere. This increases the probability of the diffraction process.

There is one more basic difference between electrons and X-rays/neutrons: electron trajectories; electrons being charged particles, can be modified by the influence of magnetic fields. Hence, we can build *lenses* for electrons and use them to "look" at the internal structure of materials. A *transmission electron microscope* (TEM) combines the phenomenon of electron diffraction with the ability to form images. The best microscopes today have a resolving capacity of about 0.05 nm, i.e., they can distinguish between two objects (atoms in this case) separated by about 0.05 nm. In the following section, we will take a closer look at the structure of a TEM, and at the ways in which it can be used to obtain electron diffraction patterns.

13.3.3 The transmission electron microscope

Now that we know the basic differences between electron diffraction and other types of diffraction, we are ready to study the basic structure of a transmission electron microscope (TEM). Figure 13.6 shows a diagram of a typical TEM. It consists of four sections:

(i) The top of the TEM contains the electron gun, which is essentially a heated tungsten filament that emits electrons, which are then accelerated down the column;
(ii) The illumination stage consists of a set of *condenser lenses* which allow the user to focus and direct the electron beam onto the sample;
(iii) The *objective lens* is the main image-forming lens of the microscope; the sample is placed inside the lens on a special specimen stage. This is where the electrons interact with the specimen.
(iv) The bottom section of the microscope consists of the magnifying lenses and a viewing chamber and/or camera.

The column is always mounted vertically and can be as much as three stories high for the highest accelerating voltages. The main reason for the tall cylindrical shape of the microscope is the fact that electron diffraction angles are very small, so that all the diffracted beams travel close to the incident beam direction. In addition to the main components described above, the TEM also has a vacuum system, since electrons cannot travel very far through air; the magnetic lenses are water-cooled, so that the heat generated by the electrical resistance effect is removed; the column is surrounded by radiation shields, to prevent dangerous X-rays from escaping; and a high-voltage tank is usually present in the room (a separate tank is used for voltages up to about 300 kV; for higher voltages, the accelerator is actually placed on top of the column). For more details on the structure and components of the microscope we refer the interested reader to one of many textbooks (e.g., Williams and Carter (1996), Fultz and Howe (2002), De Graef (2003)).

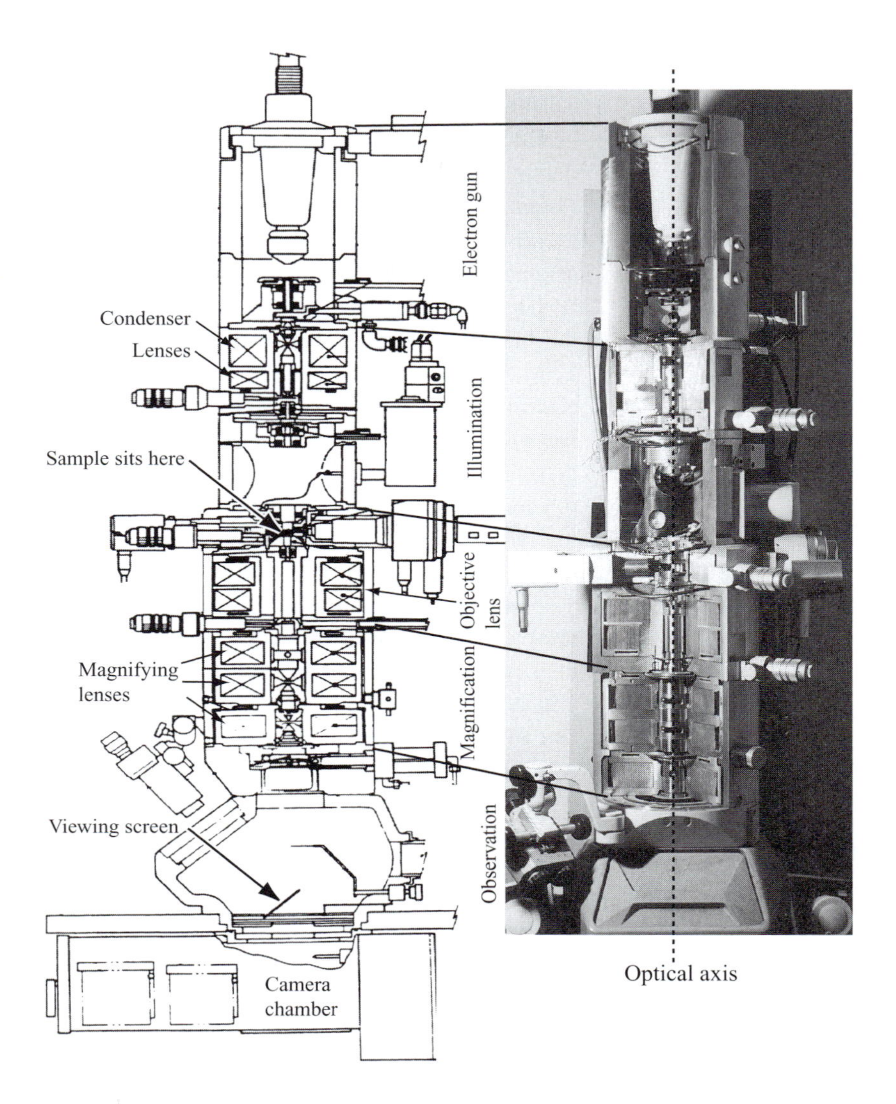

Fig. 13.6. Schematic diagram of a 120 keV transmission electron microscope, with corresponding ray diagram (figure reproduced from Fig. 3.2 in *Introduction to Conventional Transmission Electron Microscopy*, M. De Graef, 2003, Cambridge University Press).

The TEM uses round magnetic lenses to affect the trajectory of the electrons. There are typically six or more such lenses in the column: two (or more) condenser lenses, which form the beam that is incident on the sample; an objective lens, with the sample immersed in the lens magnetic field; and three (or more) imaging lenses, which take the image or diffraction pattern produced by the objective lens and further magnify it. The final lens, the projector lens, then projects the image onto a fluorescent screen or onto a digital detector, such as a CCD camera. Inside the TEM column there are several additional magnetic or electrostatic elements that can be used to change the direction of the electron beam (deflection coils) and to change the shape of the beam (stigmator coils).

13.3.4 Basic observation modes in the TEM

To understand how the TEM functions as an elaborate diffractometer, it is useful to consider how a single lens works. We know from optical physics that a standard glass lens can be characterized by a number of special planes. Figure 13.7(a) shows the essential lens elements: the object, represented by an arrow, is located in the *object plane*, and the lens has a *focal plane* and an *image plane*. A typical ray diagram is superimposed onto this drawing. A magnified image of the object is projected into the image plane; note that the image is inverted with respect to the object. For the objective lens in a TEM we can create a similar drawing, shown in Fig. 13.7(b). A beam of electrons parallel to the optical axis is incident on the sample. For simplicity, we will assume that there is a small area on the sample (indicated in grey) that gives rise to a diffracted beam; the electrons that are diffracted into this direction move through the lens and are focused into the back focal plane of the lens at a point that is removed from the optical axis. The electrons that leave the sample in the same direction as the incident beam are focused into a point at the intersection of the optical axis and the back focal plane. From the back focal plane, all electrons then continue to the image plane where they recombine to form an image.

It is important to understand the connection between the back focal plane and the Ewald sphere. The incident electron beam can be represented by a wave vector, **k**, with length equal to $1/\lambda$. The direction of this vector is parallel to the optical axis. If we draw this vector so that its end point coincides with the intersection of the optical axis and the back focal plane, then we can employ the standard Ewald sphere construction introduced in

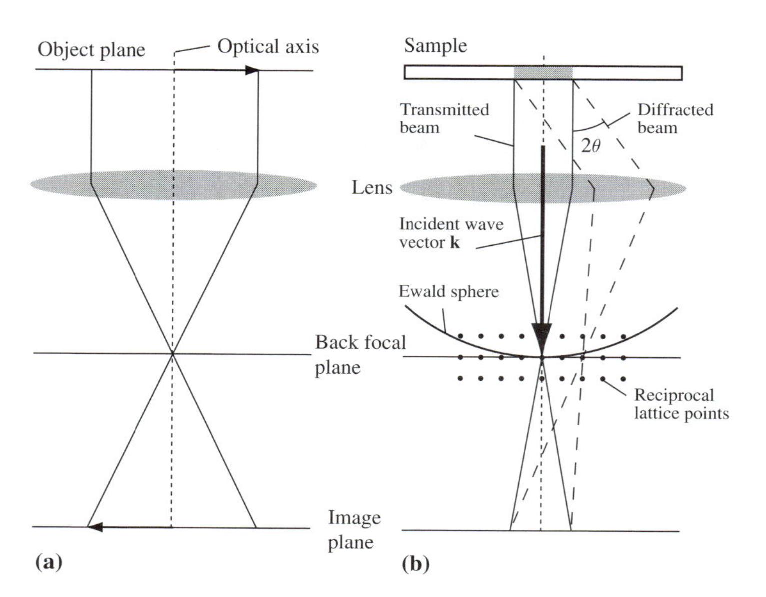

Fig. 13.7. (a) Schematic illustration of the important elements of a lens; (b) illustration of the geometry of electron diffraction superimposed on the lens drawing of (a).

Chapter 11. For every reciprocal lattice point close to the Ewald sphere there will be a diffracted beam (only one is shown in the figure). We conclude from this drawing that, in a TEM, the Ewald sphere is tangent to the back focal plane of the objective lens. If we use the magnifying lenses that follow the objective lens to magnify the image plane of the objective lens, then we will observe an image on the viewing screen. If, on the other hand, we use the objective lens back focal plane as the object plane for the magnifying lenses, then the viewing screen will display a magnified version of a planar section of the reciprocal lattice. Such a pattern is known as an *electron diffraction pattern*.

Electron diffraction patterns are representations of 2-D sections through the reciprocal lattice of the sample. An example of an electron diffraction pattern for a Ti thin foil, oriented such that the (200 keV) incident electron beam is parallel to the [11.0] direction, is shown in Fig. 13.8(a). This figure clearly shows the 2-D nature of electron diffraction patterns. Furthermore, this pattern also shows that electron diffraction is different from X-ray diffraction when it comes to systematic absences. For the Ti structure, we know from a structure factor analysis, that the reflections of the type $(00l)$, with $l = 2n+1$ must vanish (this is due to the presence of a 6_3 screw axis in the crystal structure). In a standard X-ray powder diffraction pattern, the reflections (001), (003), . . . are always absent. However, in the electron diffraction pattern shown in Fig. 13.8, these forbidden reflections are clearly present! This is a prime example of the fact that electron diffraction is governed by dynamical diffraction theory instead of the kinematical theory; in other words, the intensity of a diffracted electron beam is not necessarily proportional to the modulus squared of the corresponding structure factor. In the Ti structure, the reflections $(\bar{1}10)$ and $(1\bar{1}1)$ are both allowed (i.e., they have a non-zero structure factor). Because of the strong interaction of electrons with the atoms

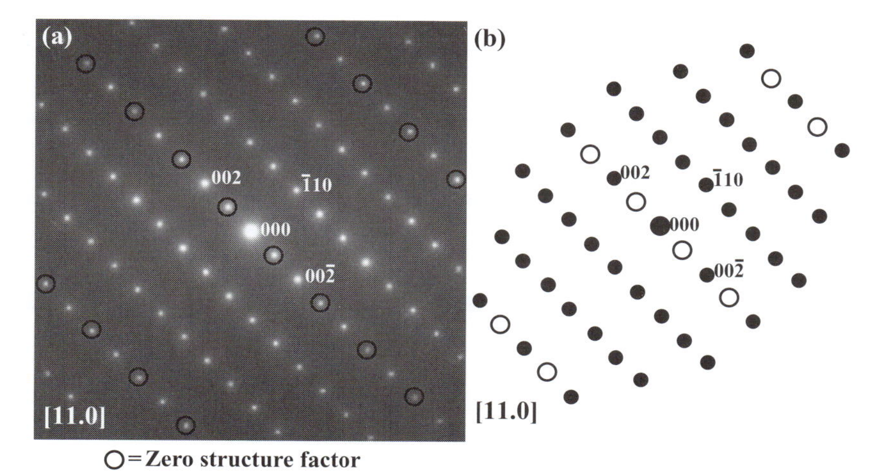

Fig. 13.8. (a) Electron diffraction pattern of *hcp* Ti, taken with the electron beam parallel to the [11.0] direction (figure reproduced from Fig. 4.18b in *Introduction to Conventional Transmission Electron Microscopy*, M. De Graef, 2003, Cambridge University Press). (b) A schematic representation of this diffraction pattern. Open circles correspond to reciprocal lattice points for which the structure factor vanishes.

in the crystal structure, an electron that is first diffracted by the $(\bar{1}10)$ planes, can be diffracted again, this time by, for instance, the $(1\bar{1}1)$ planes. If we write down the Bragg equation in reciprocal space we find:

$$\mathbf{k}' = \mathbf{k} + \mathbf{g}_{\bar{1}10} + \mathbf{g}_{1\bar{1}1} = \mathbf{k} + \mathbf{g}_{001}$$

so that it appears as though the electron was diffracted by the (001) planes, despite their vanishing structure factor! This process is commonly known as *double diffraction*. The intensity in each of the diffracted beams can be calculated using a quantum mechanical approach. In essence, the computation requires solving the Schrödinger equation for the interaction of the beam electron with the electrostatic lattice potential. We refer the interested reader to the literature for more information on these types of computation (De Graef, 2003).

The TEM can also be used in imaging mode. In this mode, the magnifying lenses take the image plane of the objective lens as their object plane, and project a magnified image onto the viewing screen. Typical magnifications range from a few hundred times to more than 1 000 000 times. In the back focal plane of the objective lens, one can physically introduce an aperture (a metal foil with a tiny hole in the center). This aperture blocks all reflections, except one (see Fig. 13.9(a) and (b)). If the transmitted beam is allowed to continue through the aperture, then the image thus obtained is called a *bright field* image. If one of the diffracted beams is allowed to pass through the aperture, then the resulting image is called a *dark field* image. By selecting one particular diffracted beam to create an image, we can select one particular set of lattice planes in the crystal; this is an extremely powerful imaging technique for the study of crystal defects, such as dislocations and stacking faults. For examples of these types of images, we refer the interested reader to Williams and Carter (1996) and De Graef (2003).

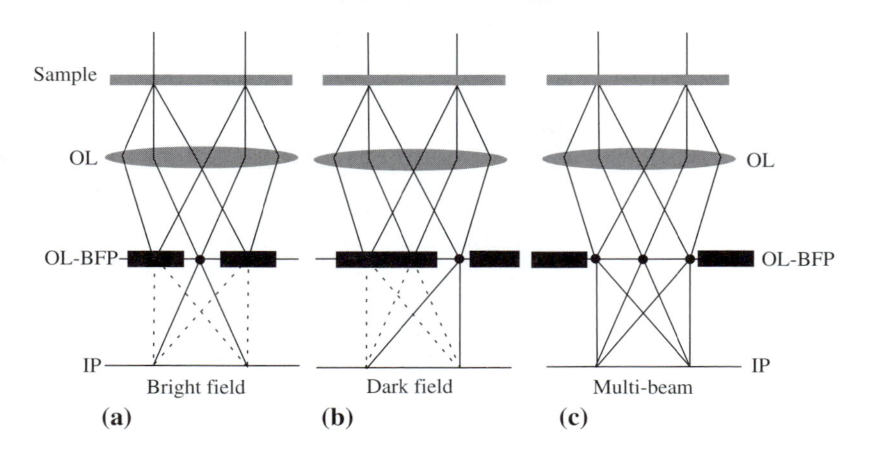

Fig. 13.9. Schematic representation of the bright field, dark field, and multi-beam imaging modes in a TEM.

A third imaging mode makes use of a larger aperture hole, so that multiple diffracted beams can contribute simultaneously to the image. This mode allows for interference between the diffracted beams and is known as *multi-beam imaging* or *high resolution imaging*. An example of a high resolution image is shown in Fig. 13.10 for the [100] orientation of tetragonal $BaTiO_3$. Figure 13.10(a) shows the schematic electron diffraction pattern for this incident beam orientation. The crystal structure is shown in (b). The high resolution multi-beam image in (c) shows a pattern of white dots; the brightest dots correspond to the locations of the Ba atoms, whereas the weaker dots in between correspond to the Ti locations. The O atoms are not visible in this particular image. For more information about high resolution imaging we refer the interested reader to Spence (1988) and De Graef (2003).

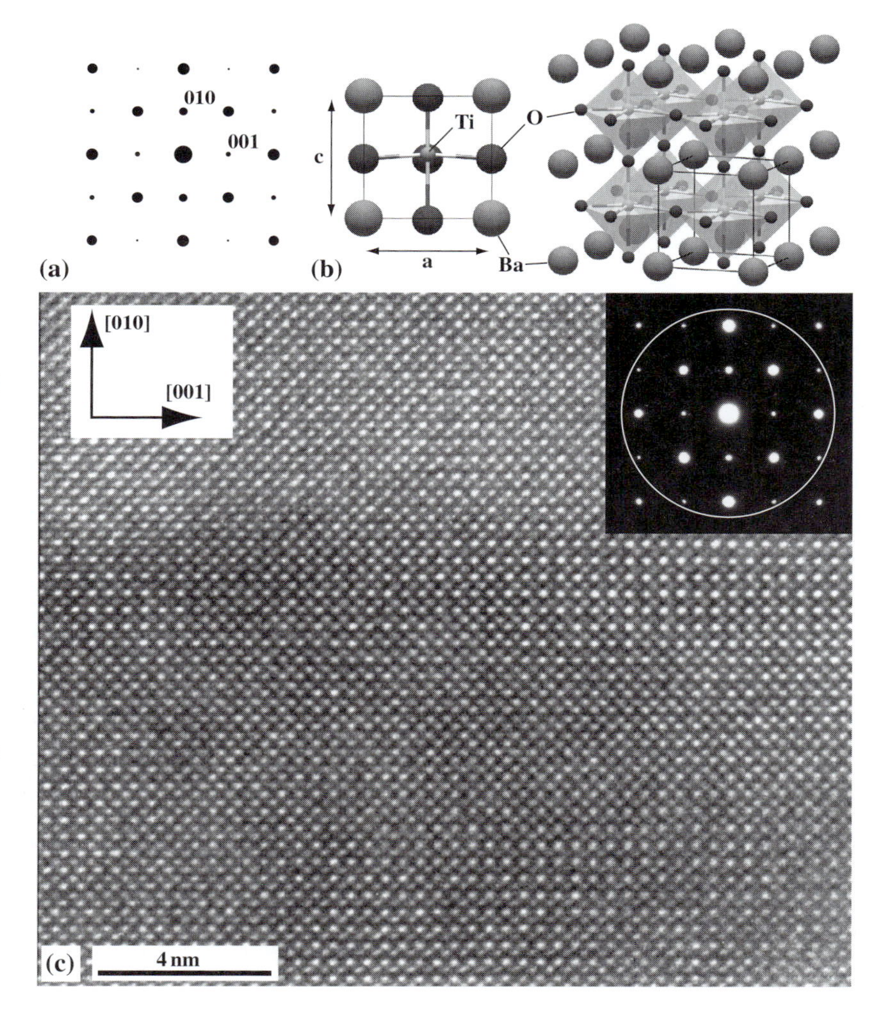

Fig. 13.10. (a) Schematic [100] electron diffraction pattern for the structure of tetragonal $BaTiO_3$, shown in (b). (c) shows a high resolution multi-beam image using the reflections inside the white circle in the inset diffraction pattern. The brightest white dots in this image correspond to the positions of Ba, the weaker spots in between four bright spots correspond to Ti positions. The O atoms are not visible in this image. (Figure reproduced from Fig. 4.7, 10.1a and 10.1d in *Introduction to Conventional Transmission Electron Microscopy*, M. De Graef, 2003, Cambridge University Press).

13.3.5 Convergent beam electron diffraction

In addition to enabling the acquisition of standard electron diffraction patterns, the TEM provides another diffraction mode that deserves to be mentioned in a textbook on crystallography: *convergent beam electron diffraction (CBED)*. In a standard electron diffraction experiment, the incident beam electrons all travel in exactly the same direction, i.e., the incident beam is a parallel beam. In CBED, the incident beam is focused onto a small point on the sample surface, and a wide range of incident directions is present, as illustrated in Fig. 13.11(a). The incident beam directions lie inside a cone with opening angle θ_c and apex on the sample surface. The angle θ_c is rather large in the figure, but is, in reality, of the order of a few milli-radians.[4] If we assume that the sample is in Bragg orientation for a particular set of planes represented by the reciprocal lattice vector **g**, then the beam direction that lies at the center of the cone will give rise to a diffracted beam in the direction $\mathbf{k}+\mathbf{g}$. The transmitted electrons end up in the point O (Fig. 13.11(a)), whereas the diffracted electrons end up at the location **g**. We have seen that the thin foil nature of the TEM sample gives rise to reciprocal lattice points with a finite volume and shape. This means that the diffraction condition can be satisfied approximately even when the reciprocal lattice point does not lie exactly on the Ewald sphere. As a consequence, the other incident beam directions inside the cone will also give rise to diffracted beams, albeit with a different intensity than that for perfect Bragg orientation. The result is that the diffraction pattern consists of circular disks (Fig. 13.11(b)), one for each reciprocal lattice point; the intensity distribution inside each

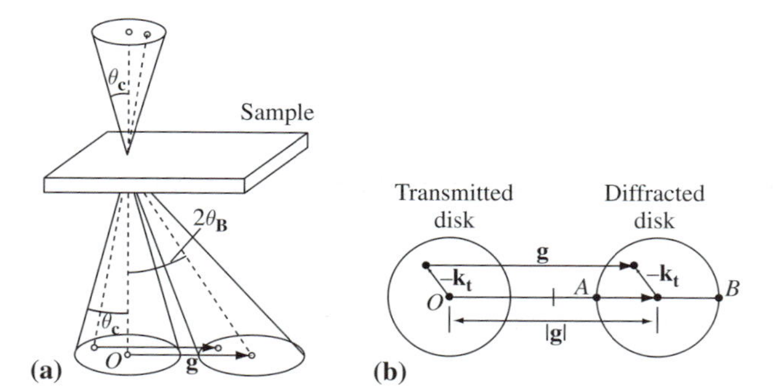

Fig. 13.11. (a) Schematic representation of the incident beam geometry for convergent beam electron diffraction. (b) each reciprocal lattice point becomes a circular disk. The vector $\mathbf{k}_t$ is the tangential component (tangential to the objective lens back focal plane) of the tilted incident beam wave vector **k** (figure reproduced from Fig. 6.15 in *Introduction to Conventional Transmission Electron Microscopy*, M. De Graef, 2003, Cambridge University Press).

[4] Recall that one milli-radian equals $0.0573° = 3'26''$.

disk need not be uniform, and, in most cases, is highly non-uniform due to complex dynamical scattering events inside the crystal. Each point inside the disks corresponds to a different incident beam direction. As shown in Fig. 13.11(b), points in different disks connected by the vector **g** correspond to the same incident beam direction; the vector $\mathbf{k}_t$ is the tangential component of the incident wave vector that contributes to that particular point in the disk.

Figure 13.12 shows two examples of experimental CBED patterns. Figure 13.12(a) is a CBED pattern obtained at 120 kV of the [110] orientation

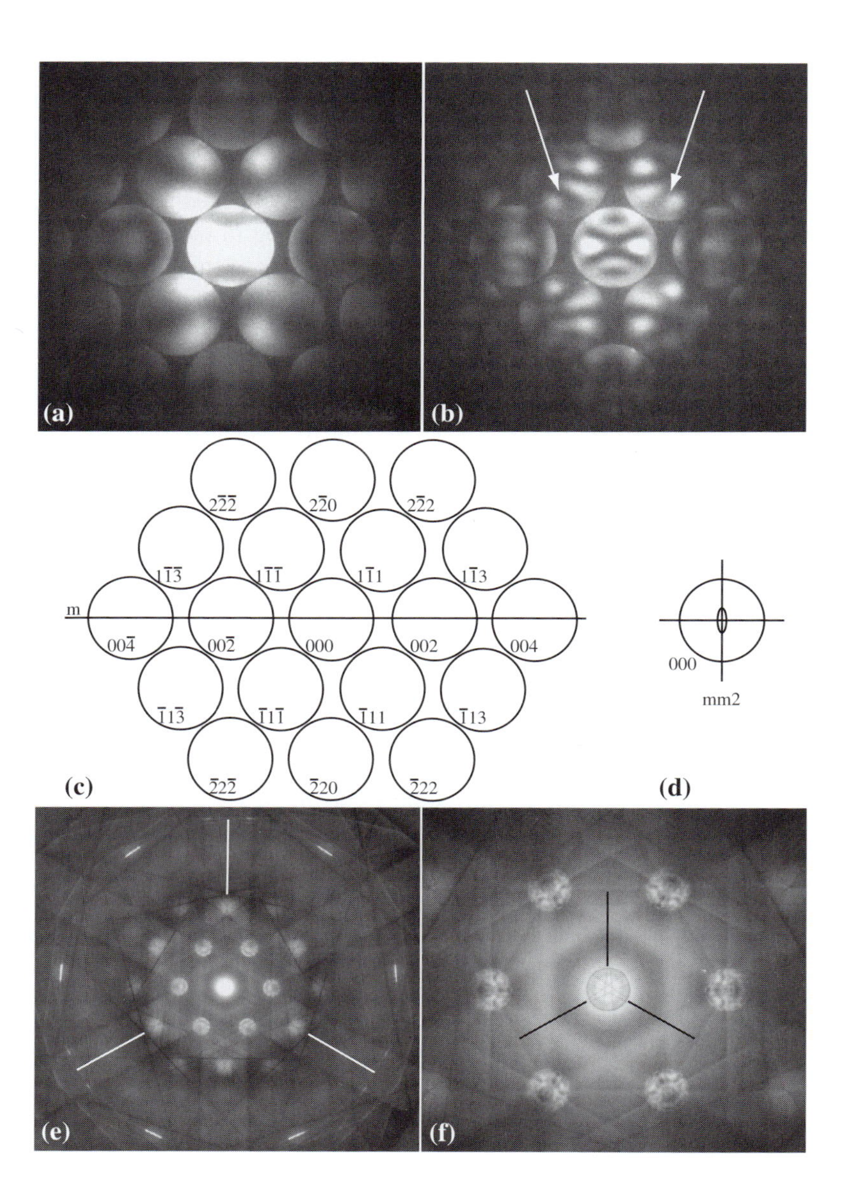

Fig. 13.12. (a) and (b) are [110] CBED patterns, obtained at 120kV, of GaAs, for two different foil thicknesses. (c) shows the entire pattern of disks schematically, along with the Miller indices of each disk. The overall pattern symmetry is **m** (C_s), whereas the symmetry of the central disk (d) is mm2 (C_{2v}). (e) and (f) show an 80kV CBED pattern for the [111] orientation of Cu-15 at% Al at two different magnifications (figure reproduced from Figs. 4.25 and 9.26 in *Introduction to Conventional Transmission Electron Microscopy*, M. De Graef, 2003, Cambridge University Press).

of Ga As. This pattern consists of nearly touching diffraction disks; the beam convergence angle θ_c is equal to 5.1 milli-radians in this case. Pattern (b) is identical to (a) except for the crystal foil thickness, which is larger in (b) than in (a). If we analyze the intensity distributions in the individual disks of (a) and (b), we find that the overall pattern symmetry, known as the "whole pattern symmetry," is given by the point group **m** (C_s). At first glance, one might be inclined to see a two-fold axis in the central disk, which along with a horizontal mirror plane would result in point group **mm2** (C_{2v}), but a careful comparison of the intensities at the arrowed locations in (b) shows that there is no vertical mirror plane, and hence no two-fold axis. The whole pattern symmetry is thus **m** (C_s). If we restrict our attention to the central disk only, then we see that both horizontal and vertical mirror planes are present, so that the central disk symmetry is **mm2** (C_{2v}). The difference between these two symmetries (whole pattern versus central disk) indicates that this crystal structure does not have inversion symmetry. When we combine this information with similar CBED patterns obtained for different crystal orientations, we can show that Ga As must belong to the non-centrosymmetric point group $\bar{\mathbf{4}}\mathbf{3m}$ (T_d). There is an extensive set of rules that must be applied to CBED patterns in order to extract the point group of the crystal; for a more detailed discussion of these rules we refer the interested reader to De Graef (2003) and references therein.

The CBED patterns in Fig. 13.12(e) and (f) were obtained at 80 kV on a Cu-15 at% Al thin foil oriented along the [111] zone axis. Pattern (e) was obtained at a smaller magnification than pattern (f). We have seen in the previous chapter that Friedel's law states that a diffraction pattern must always have inversion symmetry. This means that for every reflection **g** there must be an equivalent reflection at −**g**. The resulting [111] diffraction pattern then appears to have six-fold symmetry, since there are six {110}-type reflections symmetrically positioned around the central disk. In reality, however, the symmetry of the [111] CBED pattern is not six-fold, as can be seen by carefully analyzing all the contrast features in both CBED patterns. For both patterns, we find that the symmetry is **3m** (C_{3v}). Application of the rules mentioned above then leads to the overall crystal point group $\mathbf{m}\bar{\mathbf{3}}\mathbf{m}$ (O_h). The extraction of the 3-D point group based on 2-D CBED patterns involves the concept of *diffraction groups*, a set of 31 groups that are used to connect the 2-D CBED symmetry to the 3-D crystal symmetry. A full discussion of diffraction groups in the context of CBED can be found in Buxton *et al.* (1976), Williams and Carter (1996) and De Graef (2003). It is also possible to use CBED patterns to determine the complete space group of a material. These methods are explained in detail in Tanaka *et al.* (1983); in 185 out of 230 space groups, these methods will determine unambiguously the correct space group.

13.4 *Synchrotron X-ray sources for scattering experiments

Diffraction studies of crystalline and amorphous materials have been aided by the availability of high intensity X-ray sources, produced by the acceleration of charged particles to relativistic velocities. When the acceleration results from the circular motion of charged particles, the resulting radiation is known as *synchrotron radiation*. In Fig. 13.13, we illustrate the characteristic dipolar radiation pattern emitted by an accelerating electron moving at a non-relativistic velocity (a) and a relativistic velocity (b). The radiation pattern reflects contours of the flux of radiated power density as given by the *Poynting vector*, $\mathbf{S} = \mathbf{E} \times \mathbf{B}$.

Consider a point charge, q, subjected to an acceleration, $\mathbf{a}$. The angle, θ, is the angle between the instantaneous acceleration and the direction of wave propagation, $\mathbf{k}$, at a time, t. The instantaneous emitted power per unit area in the direction $\mathbf{k}$, the *Poynting flux*, is given by:[5]

$$\mathbf{S}(\mathbf{r}, t) = \frac{q^2|\mathbf{a}|^2 \sin^2\theta}{16\pi\epsilon_0 r^2 c^3}\mathbf{k}. \tag{13.11}$$

The radiation pattern is a polar diagram of $\mathbf{S}$ as a function of θ. For non-relativistic particles, Fig. 13.13(a), this radiation is anisotropic with the emission strongly weighted at right angles, $\theta = \pi/2$, to the direction of acceleration. Fig. 13.13(b) shows that for relativistic charged particles the radiation pattern becomes strongly peaked in the forward direction as observed in the laboratory frame of reference, i.e., the radiation is strongly peaked in the direction of the instantaneous velocity.[6] This radiation pattern can be thought of as a

Fig. 13.13. (a) Radiation pattern for an accelerating electron with a circular trajectory for (a) non-relativistic and (b) relativistic velocities and (c) floor plan of the Advanced Photon Source (APS) synchrotron reactor at the Argonne National Lab (ANL) (courtesy of ANL).

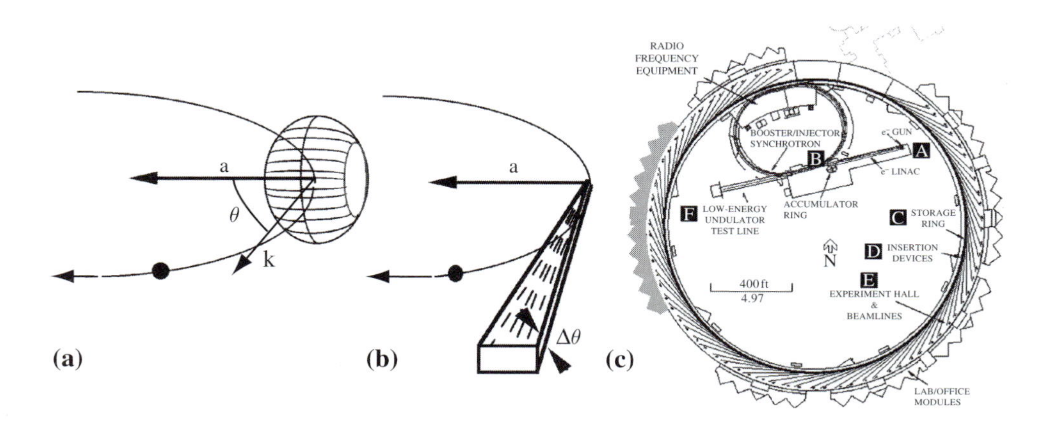

[5] The total power radiated can be obtained by integrating the Poynting flux over the area of a sphere centered on the instantaneous position of the charged particle.

[6] Relativistic velocities, v, are velocities that approach the speed of light, c; in other words, the ratio v/c approaches unity.

searchlight beam tangent to the trajectory of the charged particle. The flux of the radiation, $|\mathbf{S}|$, is confined to be inside a cone of angular width:

$$\Delta\theta = \sqrt{1 - \frac{v^2}{c^2}}. \tag{13.12}$$

Using the fact that the total relativistic energy of the charged particle is $E = m_0c^2/\sqrt{(1 - v^2/c^2)}$, this angle can be rewritten as:

$$\Delta\theta = \frac{m_0c^2}{E}, \tag{13.13}$$

where, for an electron, we have $m_0c^2 \approx 5 \times 10^5$ eV. In high energy particle accelerators, $E \approx 1$ GeV (10^9 eV) and, therefore, small angles around 5×10^{-4} radians are attainable for forward emitted radiation. This radiation can be concentrated on a very small area. High flux densities on small targets result in X-ray intensities that are orders of magnitude larger than those obtainable in powder diffractometers or rotating anode machines. The *brilliance* of the source is a combined measure of the radiation flux, the angular divergence of the light, and the source size. A high brilliance implies a high x-radiation flux, a small angular divergence, and a small source size.

13.4.1 Synchrotron accelerators

Figure 13.13(c) shows a schematic of the *Advanced Photon Source (APS)* synchroton reactor at Argonne National Lab (ANL). The APS is used to generate synchrotron radiation for high flux X-ray experiments. Figure 13.14 (a) shows an aerial view of the APS facility; the outer diameter of the main experiment hall (the doughnut shaped building) is 390 meters. There are 35 individual *beam lines* in the experiment hall; a beam line is a point along the circumference of the accelerator ring where radiation is allowed to exit the ring; each such beam is then directed towards an experimental setup for a

Fig. 13.14. (a) Aerial view of the Argonne National Laboratory (ANL) Advanced Photon Source (APS) facility (photo courtesy of ANL); (b) a schematic of the Brookhaven National Laboratory (BNL) National Synchrotron Light Source (NSLS, picture courtesy of BNL).

variety of scattering experiments. Figure 13.14 (b) shows a schematic of the Brookhaven National Lab (BNL) National Synchrotron Light Source (NSLS).

In the storage ring, charged particles are accelerated, in an evacuated pipe, to velocities approaching the speed of light.[7] Quadrupole, sextupole, and octupole magnets are used to control the beam shape as the charged particles circulate around the ring. Bending magnets guide the beams and the centrifugal acceleration provides the useful radiation. In most electron accelerators, an electron gun provides a continuous stream of electrons, which are first accelerated to a medium energy in a linear accelerator, then boosted up to higher energy in a booster ring, and finally merged into the main accelerator ring.

Synchrotron radiation has many properties that are attractive for scattering experiments, including:

(i) *a high beam intensity*: Synchrotron radiation is a continuous source with five or more orders of magnitude higher intensity than rotating anode X-ray tubes. The total power (in kilowatts) is given by:

$$P = \frac{88E^4 I}{R}, \tag{13.14}$$

where E is the electron energy (GeV), I is the electron current (A), and R is the radius of curvature of the storage ring (m). Figure 13.15 shows the spectral distribution of radiation as a function of electron energy E for the 15 GeV, 12.7 m SPEAR facility. Several megawatts of power is radiated in a smooth featureless continuum. Highly monochromatic radiation can

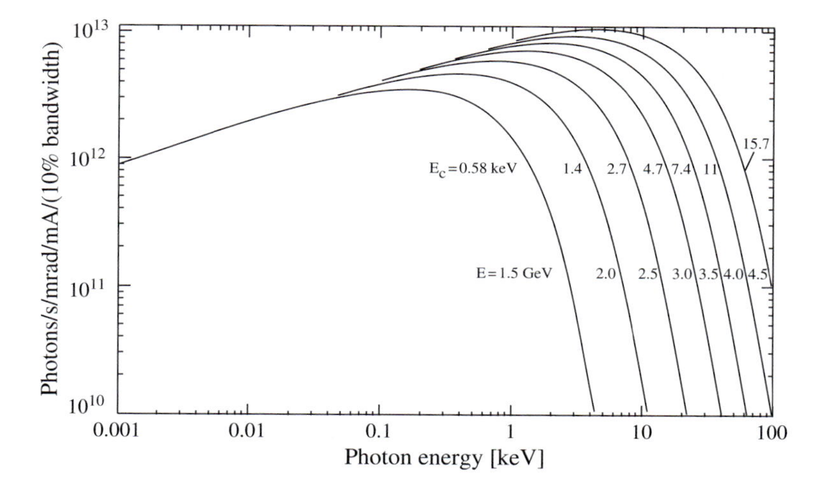

Fig. 13.15. Photon brilliance as a function of photon energy for different values of the stored electron beam energy, E, of the Stanford Positron Accelerating Ring (SPEAR); E_c is the critical energy of Eq. 13.15.

[7] Both APS and NSLS use electrons, whereas the *Stanford Positron Accelerating Ring (SPEAR)* uses positrons.

be produced using a grating or crystal monochromator. Band widths after monochromatization are typically $\Delta E/E = \Delta\lambda/\lambda \approx 10^{-3}$.

(ii) *a broad radiation spectrum*: Figure 13.15 shows the photon intensity as a function of photon energy for different values of the stored electron beam energy. This spectrum can also be parameterized in terms of a critical energy, E_c:

$$E_c = \frac{2.2E^3}{R}\,\text{KeV}. \tag{13.15}$$

(iii) *a strong polarization*: In the plane of the electron orbit and for energies near the critical energy, the synchrotron radiation is nearly 100% linearly polarized. Linear, circular, and elliptical polarizations are all possible.

The X-rays produced in a synchrotron can be provided to a number of beam lines which continue as tangents to the storage ring itself. Figure 13.16(a) shows an example of such a beam line. Additional magnets are used in *insertion devices* which tailor the radiation spectrum for specific experiments; these include *wigglers* and *undulators*. Figure 13.16(b) shows them to consist of arrays of permanent magnets (typically the NdFeB magnets discussed in Chapter 19) with alternating N/S polarity. The alternating magnetic field causes oscillations in the electron trajectory which, through relativistic effects, shift the radiation spectrum to higher energies. Wigglers do not change the spectrum substantially but aid in tailoring beam shape. Undulators cause the radiated power to be further focused, increasing the brilliance of the source.

13.4.2 Synchrotron radiation: experimental examples

In this section, we illustrate examples of the use of synchrotron radiation in the study of *nanocrystallization*. Figure 13.17 (a) shows constant heating rate, 3-D synchrotron XRD patterns at different temperatures during *in-situ* crystallization of a NANOPERM amorphous ribbon (Hsiao, 2001, Hsiao *et al.*, 2002).

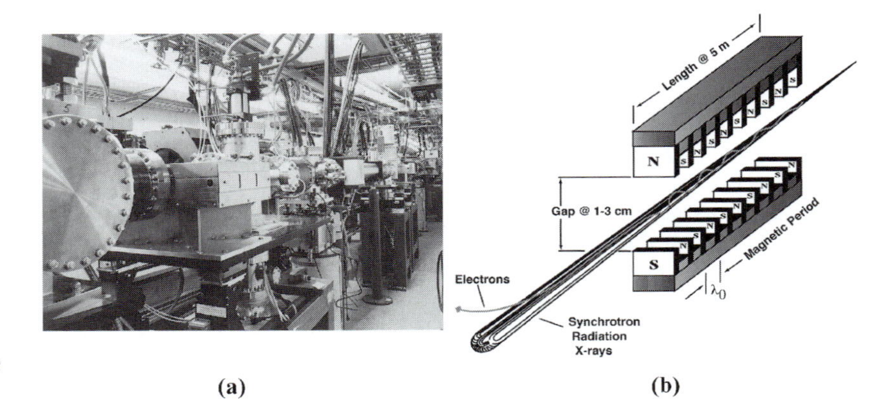

Fig. 13.16. (a) Example of a beam line tangent to the storage ring at the APS; (b) arrays of permanent magnets in an insertion device (courtesy of ANL).

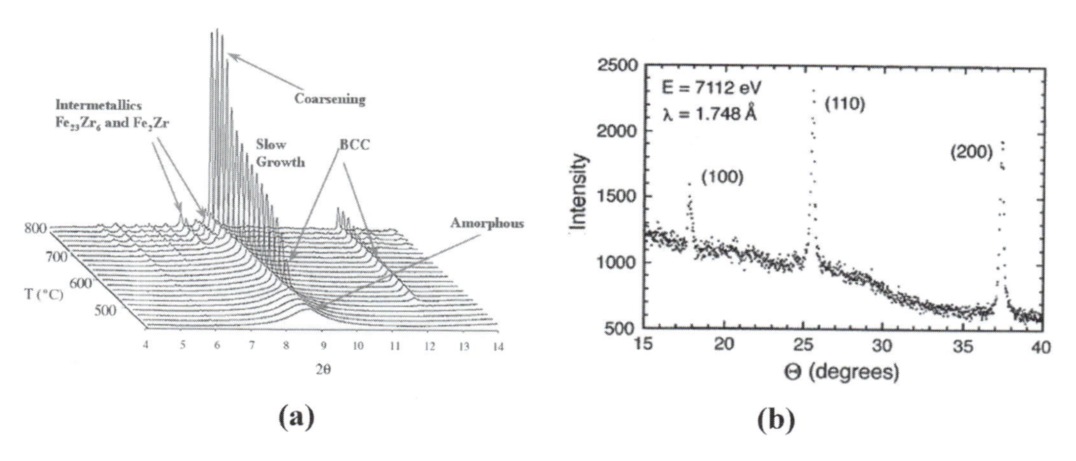

Fig. 13.17. (a) Constant heating rate 3-D synchrotron XRD patterns for *in-situ* crystallization of an Fe-Zr based (NANOPERM) alloy (courtesy of A. Hsiao) and (b) synchrotron XRD of an FeCo-based HITPERM magnet, showing superlattice reflections of the ordered α-(FeCo) phase (courtesy of M. Willard).

The figure shows the primary nanocrystallization product, the secondary crystallization products, and coarsening phenomena. Features include: (a) the appearance of Fe (110) and Fe (200) peaks during the primary crystallization of α-Fe at a temperature T_{x1} near 510 °C, (b) secondary crystallization occurs at a temperature T_{x2} near 710 °C with Fe_2Zr and $Fe_{23}Zr_6$ appearing, and (c) the narrowing of the Fe (110) and Fe (200) peaks after secondary crystallization. The high intensity synchrotron radiation allows for diffraction patterns to be taken in short times (Kramer *et al.*, 1978) during a constant heating rate experiment so that crystallization kinetics and products can be studied. More information on this alloy system can be found in Chapter 21.

Synchrotron XRD experiments on HITPERM (Willard *et al.*, 1998, Willard, 2000) (Fig. 13.17(b)) were used to identify superlattice reflections identifying the ordered α'-(FeCo) phase. X-rays with a wavelength of 0.1748 nm, corresponding to an energy between the Co and Fe K_α edges, were chosen to take advantage of *anomalous scattering*.

Figure 13.4(a) shows a simulated XRD pattern for the ordered B2 FeCo alloy. As discussed earlier, the difficulty in distinguishing between α-FeCo and α'-FeCo using conventional XRD is due to the similarity of the atomic scattering factors of Fe and Co. Since the structure factor for a superlattice reflection is related to differences between the two atomic scattering factors, observation of superlattice reflections in conventional XRD is difficult. The *superlattice reflections* are several orders of magnitude less intense than the *fundamental reflections*. In order to detect the superlattice reflections, we must use long counting times or very intense X-ray sources. Using the high intensity of the synchrotron source and choosing the radiation wavelength to take advantage of anomalous scattering allows for the direct observation of superlattice reflections. Figure 13.17(b) shows synchrotron anomalous X-ray scattering superlattice reflections, demonstrating that FeCo nanoparticles in HITPERM magnets possess the ordered (B2) CsCl structure.

Superlattice reflections are determined by tuning the wavelength (energy) of the X-rays to an energy for which the atomic scattering factors of Fe and

Co have a more appreciable difference (Willard *et al.*, 1998). Because of *dispersion corrections* (previously attributed to *anomalous scattering*) near an *X-ray absorption edge*, we can significantly increase the difference between f_{Co} and f_{Fe}. On approaching an absorption edge, X-ray absorption dispersion corrections to the atomic scattering factors acquire a complex contribution:

$$f = f^0 + f' + \mathrm{i}f'', \tag{13.16}$$

where f' is the real part of the dispersion correction, and f'' is the loss part. Both f' and f'' are functions of the X-ray wavelength. With corrections for dispersion, the structure factor for the α'-FeCo B2 superlattice reflections is:

$$F_{hkl} = (f^0_{\mathrm{Co}} - f^0_{\mathrm{Fe}}) + (f'_{\mathrm{Co}} - f'_{\mathrm{Fe}}) + \mathrm{i}(f''_{\mathrm{Co}} - f''_{\mathrm{Fe}}) = \Delta f^0 + \Delta f' + \mathrm{i}\Delta f''. \tag{13.17}$$

If we select the X-ray wavelength so that $\Delta f'$ is maximized, then the structure factor for the superlattice reflections is also maximized, so that these reflections can be observed in the powder pattern.

13.5 Historical notes

Clinton J. Davisson (1881–1958) was an American physicist. In 1927, while at Bell Laboratories, Davisson and **Lester H. Germer** demonstrated electron

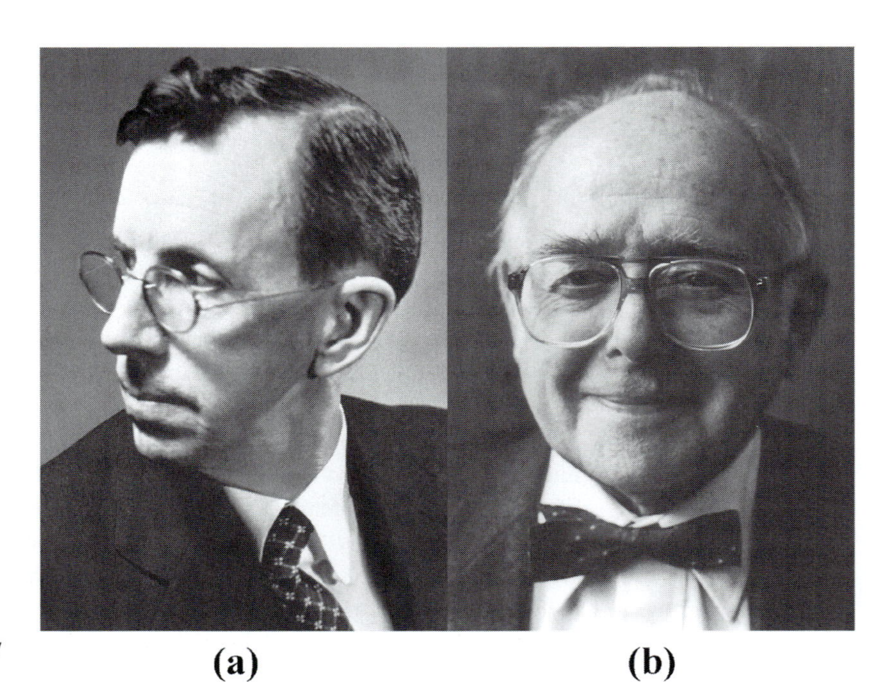

Fig. 13.18. (a) Clinton J. Davisson (1881–1958), and (b) Clifford G. Shull (1915–2001) (pictures courtesy of the Nobel Museum).

Table 13.6. Partial chronology of the history of the electron and the electron microscope, and other events which have had a significant impact on the microscopy field, with references to fundamental papers.

Year	Event
1871	Cromwell Fleetwood Varley suggests that the carriers of electricity are corpuscular, with a negative charge (Varley, 1871).
1876	Eugene Goldstein studies discharges in gases, and coins the name *cathode rays*, starting a long debate about their nature (Goldstein, 1876).
1891	George Johnstone Stoney coins the word *electron* for the unit of charge (Stoney, 1891).
1897	Emil Wiechert is the first to obtain reasonable bounds on the magnitude of e/m (January) (Wiechert, 1897).
1897	Walter Kaufmann and J. J. Thomson independently measure e/m (April) (Kaufmann, 1897, Thomson, 1897).
1899	J. J. Thomson determines the value of e which makes him the discoverer of the electron (Thomson, 1899).
1905	Albert Einstein publishes the *Special Theory of Relativity* and establishes the equivalence of mass and energy (Einstein, 1905).
1913	Niels Bohr introduces a model for the structure of the hydrogen atom (Bohr, 1913a,b,c).
1923	Louis de Broglie establishes the wave–particle duality (de Broglie, 1923).
1925	Wolfgang Pauli discovers the exclusion principle; Werner Heisenberg develops matrix quantum mechanics (Pauli, 1925, Heisenberg, 1925).
1926	Erwin Schrödinger develops quantum mechanics based on differential equations; Hans Busch develops the theory of magnetic lenses (Schrödinger, 1926, Busch, 1926).
1927	Clinton Davisson and Lester Germer discover electron diffraction (Davisson and Germer, 1927b).
1928	Paul Dirac formulates the relativistic theory of the electron; Hans Bethe develops the first dynamical theory of electron diffraction (Dirac, 1928, Bethe, 1928).
1931–4	Ernst Ruska and Max Knoll build the first electron microscope (Knoll and Ruska, 1932a,b, Ruska, 1934a,b).

diffraction by scattering electrons (cathode rays) off a single crystal of nickel. In the same year, **George Paget Thomson** (1892–1975) and A. Reid demonstrated electron diffraction effects in scattering from thin gold films. These first observations were quickly followed by electron diffraction experiments by **Seishi Kikuchi** (1902–74). Davisson and Thomson won the 1937 Nobel prize in physics for their experimental discovery of the diffraction of electrons by crystals.

Table 13.6 (taken from De Graef (2003)) lists a partial chronology of the history of the discovery of the electron and how these events led up to the design and construction of the first transmission electron microscope by Ernst Ruska and Max Knoll in the early 1930's. Transmission electron microscopes can now be found in many laboratories around the world, and are considered to be fundamental research instruments.

Clifford Glenwood Shull (1915–2001) was born in the Glenwood section of Pittsburgh, Pennsylvania, which was the reason for the selection of his middle name.[8] He developed his first interest in the study of physics while attending Schenley high school in Pittsburgh, PA. He received his undergraduate degree in physics at the Carnegie Institute of Technology (now Carnegie Mellon University) in 1937. Shull continued with graduate school at New York University, where he worked with the nuclear physics group headed by Frank Myers and Robert Huntoon, and he had his thesis supervised by Richard Cox.

Shull used thermalized neutrons from a Ra-Be neutron source to search for paramagnetic scattering from materials to confirm predictions by O. Halpern and M. Johnson and their students at NYU. From 1941 to 1946 he worked for The Texas Company in Beacon, NY. In 1946 he moved with his family to the Oak Ridge National Laboratory to work with Ernest Wollan in the study of neutron diffraction from crystalline materials. In 1955, he moved to take up an academic career at the Massachusetts Institute of Technology. His career at MIT included studies of magnetization in crystals, development of polarized beam technology, dynamical scattering, interferometry, and the fundamental properties of the neutron.

Shull won the 1994 Nobel Prize in Physics, with **Bertram N. Brockhouse** (1918–2003), for the development of the neutron diffraction technique. His son, Robert, is a successful materials scientist at the National Institute of Standards and Technology (NIST) in Gaithersburg, MD, and one of the authors of the US Nanotechnology policy.

13.6 Problems

(i) *Neutron wave length*: Calculate the wave length of thermal neutrons at (a) room temperature and (b) cooled to 4 K (liquid He temperature). How do these compare with 400 keV electrons and to typical *Brehmmstrahlung radiation* from an X-ray tube?

(ii) *Thermal velocities, de Broglie wave length*: A thermal neutron has a kinetic energy of 3/2 kT, where T is room temperature, 300 K. i.e., these neutrons are in thermal equilibrium with their surroundings. (a) Calculate the thermal velocity of such a neutron. Compare this with the thermal

[8] A complete biography can be found at /http://nobelprize.org/nobel (Nobel Lectures, Physics 1981–1990, Tore Frängsmyr, and Gösta Ekspång, editors, World Scientific Publishing Co., Singapore, 1993).

velocity of an electron at 300 K. (b) Calculate the de Broglie wave length of the neutron and the electron of part (a). (c) Would either of these be useful in resolving atomic positions?

(iii) *Neutron flux and fluence*: (a) Consider a neutron source capable of providing a flux of 10^{19} n/m^2-s. Calculate the **neutron fluence** over a 24 hour period. (b) How many neutrons will impinge upon a 1 cm^3 cubic Au sample during this period? (c) If every neutron is scattered once what fraction of the Au atoms provide a scattering site for a neutron during this time period?

(iv) *Time of flight monochromatization*: Consider a source of thermal neutrons. Calculate the following properties:

(a) The mean velocity of neutrons at 300 K.
(b) The fraction of neutrons having a velocity 1% greater than and 1% less than the mean velocity at 300 K.
(c) The time of flight for a neutron traveling at the mean velocity to traverse 10 m (a distance typical for a beam line).
(d) The wavelength of a 300 K thermal neutron.
(e) The spread in wave length for neutrons having velocities $\pm$ 1% of the mean velocity.

(v) *Single crystal monochromator*: Consider the use of an NaCl single crystal as a neutron monochromator. What is the energy of neutrons scattered at 12° for a first order ($n = 1$) reflection? What is the energy of X-rays?

(vi) *Cold neutrons*: Calculate the mean kinetic energy, de Broglie wave length and mean velocity of neutrons in equilibrium with liquid N_2 at 77 K.

(vii) *Neutron wave length I*: Calculate the typical range of wavelengths for the following neutrons:

(a) Cold;
(b) Thermal;
(c) Slow;
(d) Intermediate;
(e) Fast;
(f) Ultra-fast.

(viii) *Neutron wave length II*: Calculate the energy and temperature of neutrons required to produce a wave length of 0.154 nm, identical to that of Cu K_α radiation.

(ix) *Neutron wave length III*: Polymer crystals are unique in that the basis consists of macromolecules that may in fact be very large and contain many light atoms. The lattice constant of such crystals may also be very large. Neutron diffraction may offer several advantages for the study of such crystals. This problem explores the relative advantages of neutron diffraction for the study of polymers.

(a) Describe the relative merits of neutrons for scattering from light atoms as compared with heavy atoms.
(b) Consider *small angle scattering* using thermal neutrons. Using the wave length for neutrons thermalized to 300 K, determine the scattering angle for the (111) reflection for a cubic crystal with a 1 and 10 nm lattice constant, respectively.
(c) What energy neutrons would be chosen to have a wave length equivalent to the spacing between (111) planes for a cubic crystal with a 1 and 10 nm lattice constant, respectively.

(x) *Neutron scattering factors:*

(a) Calculate the ratio of the X-ray scattering factors for Pu and H.
(b) Calculate the ratio of the neutron scattering factors for Pu and H.
(c) Explain why neutron diffraction would be a more appropriate tool for studying the structure of plutonium hydride.

(xi) *Synchrotron radiation*: An electron moving in a circular orbit emits synchrotron radiation. The energy that is radiated in turn can be expressed in terms of $\beta = v/c$ as:

$$E = \frac{4\pi}{3} \frac{e^2\beta^2}{(1-\beta^2)^2 R} \tag{13.18}$$

where R is the radius of curvature of the storage ring. Determine the energy radiated for electrons in a 1 km storage ring for the following accelerating voltages:

(a) 1 GeV;
(b) 5 GeV;
(c) 10 GeV;

(xii) *Synchrotron power spectrum*: Using a simple dimensional analysis verify that:

$$P = \frac{88E^4 I}{R} \tag{13.19}$$

where E is the electron energy in GeV, I is the electron current in A and R is the radius of curvature of the storage ring in meters.

(xiii) *Electron diffraction I*: Consider the *fcc* structure of Cu. Electron diffraction patterns can be approximated by planar sections through the reciprocal lattice of this structure. Create a drawing of the planar sections normal to the [001], [110], and [111] zone axes.

(xiv) *Electron diffraction II*: An electron beam is incident upon a crystal; the incident beam direction is specified by the wave vector **k**. The crystal is oriented such that an entire plane of reciprocal lattice points is tangent to the Ewald sphere. Derive an expression for the distance, measured parallel to the incident beam direction, between each reciprocal lattice point and the Ewald sphere. This distance is commonly known as the *deviation parameter* or *excitation error*, and plays an important role in the theory of image formation in the TEM.

(xv) *Electron diffraction III*: Create a drawing of the [110] reciprocal lattice plane for diamond. Indicate which reflections have zero structure factor. Is it possible that some of these reflections will have a non-zero intensity due to double diffraction?

CHAPTER

14 About crystal structures and diffraction patterns

> *"That which we must learn to do, we learn by doing."*
>
> Aristotle, *Nicomachean Ethics*

The preceding chapters have provided us with the tools necessary to describe crystal structures in an unambiguous way, and to perform any kind of crystallographic computation. In this chapter, we will first review several common graphical representation techniques for crystal structures. Then we take a closer look at the relation between crystal structure and diffraction pattern, and we will consider the structures of NaCl and *fcc* Ni as examples. We conclude this chapter with a reproduction of one of the seminal papers in the field of X-ray diffraction: the 1913 paper by W. H. and W. L. Bragg on the structure determination of diamond.

14.1 Crystal structure descriptions

14.1.1 Space group description

It has become standard practice in the international literature to describe a crystal structure by stating the space group, the lattice parameters, and the Wyckoff positions for all atoms in the asymmetric unit. From this information, one can deduce all atom positions in the unit cell by application of the space group elements listed in the *International Tables for Crystallography* (Hahn, 1989). These parameters provide the minimum information needed to unambiguously describe the structure.

A typical example would be the structure of the element Cu:

Space Group: Fm$\bar{3}$m (O_h^5) (# 225);
Lattice parameter: $a = 0.36147$ nm;
Atoms: Cu in $4a$.

This is indeed sufficient to describe fully the crystal structure since the space group symbol indicates that the unit cell is *face-centered*, with the *cubic* point group **m$\bar{3}$m** (O_h). For a cubic unit cell we need only state one *lattice parameter*, a. From the space group tables we find that the Wyckoff position $4a$ corresponds to the position $(0, 0, 0)$. The face-centering vectors then copy this point into the points $(1/2, 1/2, 0)$, $(1/2, 0, 1/2)$, and $(0, 1/2, 1/2)$. Note that the Wyckoff symbol also tells us how many equivalent atoms of that particular type there are in a unit cell.

A more complex example is *rutile*, a form of TiO_2:

Space Group P4$_2$/mnm (D_{4h}^{14}) (# 136);
Lattice parameters: $a = 0.4594$ nm, $c = 0.2958$ nm;
Atoms: Ti in $2a$, O in $4f$ with $x = 0.3$.

This structure is *primitive*, with a *tetragonal* point group **4/mmm** (D_{4h}) and lattice parameters a and c. Referring to the Tables, we find that the $2a$ position has coordinates $(0, 0, 0)$ and $(1/2, 1/2, 1/2)$. The $4f$ position corresponds to the positions $(x, x, 0)$, $(\bar{x}, \bar{x}, 0)$, $(\bar{x}+1/2, x+1/2, 1/2)$, and $(x+1/2, \bar{x}+1/2, 1/2)$. The value of x must also be given and for rutile it is equal to 0.3. The total number of atoms per unit cell is equal to the sum of the numbers in the Wyckoff symbols, i.e., 6. Since one chemical formula TiO_2 contains three atoms, there are two *formula units per unit cell*. One often states this explicitly as part of the structure description. Since most researchers do not usually have a copy of the Tables on their desk, one often specifies a little more than the minimum information when describing a structure. A more complete description for *rutile* would be :

Formula Unit: TiO_2, titanium dioxide
Space Group: P4$_2$/mnm (D_{4h}^{14}) (# 136)
Lattice parameters: $a = 0.4594$ nm, $c = 0.2958$ nm
Cell Content: two formula units
Atoms: Ti in $2a\,[(0, 0, 0)$ and $(1/2, 1/2, 1/2)]$;
O in $4f\,[(x, x, 0), (\bar{x}+1/2, x+1/2, 1/2), (\bar{x}, \bar{x}, 0)$
and $(x+1/2, \bar{x}+1/2, 1/2)]$,
with $x = 0.3$.

14.1.2 Graphical representation methods

In addition to the mathematical descriptions in terms of space groups and coordinates it is often instructive to provide the reader with a drawing of the crystal structure. Since drawings must, of necessity, be two-dimensional, there is really no unique drawing method that always provides an unambiguous

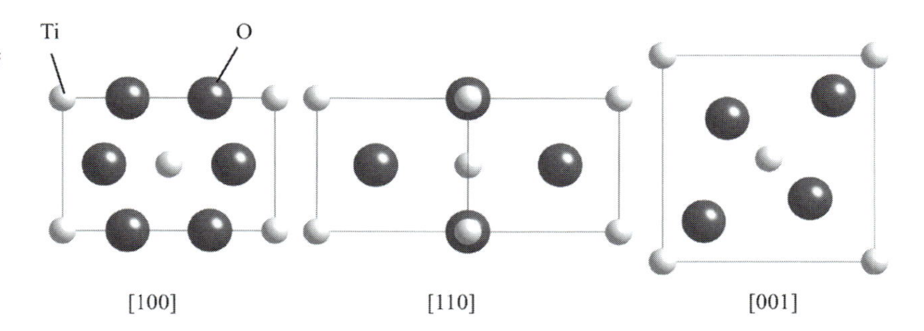

Fig. 14.1. [100], [110], and [001] orthogonal projections of the unit cell of rutile.

and clear representation. Instead, there are several different techniques that are frequently used and we will discuss the more important ones. Almost all of these techniques use computer programs to generate the drawings. Unfortunately, there is no single input file format for all of these programs, so depending on the type of drawing one wishes to create, one usually has to create a new input file. For large structures this can be a tedious and time consuming task. All of the examples below will use the tetragonal *rutile* unit cell described above.[1] Structure input files for all crystal structures visualized in this book are available from the book web site.

14.1.2.1 Orthogonal projection

The *orthogonal projection* is obtained by specifying direction indices $[uvw]$, and projecting the complete unit cell parallel to that direction onto a plane perpendicular to $[uvw]$. The advantage of this projection is that the linear dimensions perpendicular to the projection axes are conserved; one of the disadvantages is that atoms can obscure other atoms. One usually needs at least two orthogonal projections to create an understanding of the 3-D structure. An example of orthogonal projections of rutile along the [100], [110], and [001] directions is shown in Fig. 14.1.

14.1.2.2 Perspective projection

The *perspective projection* introduces a 3-D aspect into the structure drawing and this is often sufficient to display the structure clearly. Usually, the best views are obtained for directions close to low index directions. The examples in Fig. 14.2 indicate perspective views close to the directions of Fig. 14.1. Usually, the amount of perspective effect can be adjusted; some programs will also allow for *depth shading*, where the contrast between an atom and the background is decreased the further away the atom is located from the observation point.

[1] All crystal structure drawings in this book were created with *CrystalMaker®*, a crystal/molecular structures visualization program for Mac and Windows. CrystalMaker is a registered trademark of CrystalMaker Software Limited (http://www.crystalmaker.com).

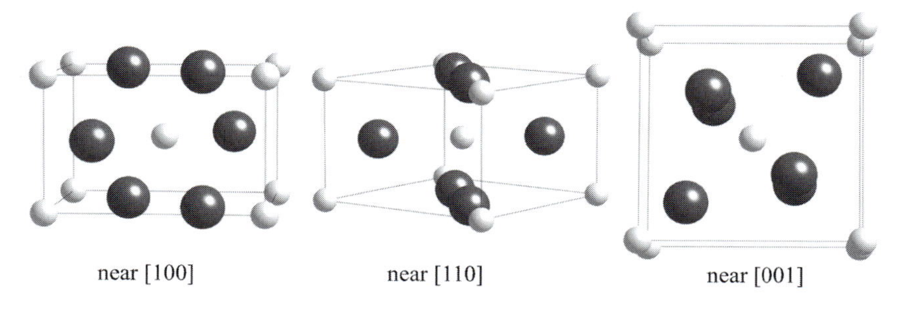

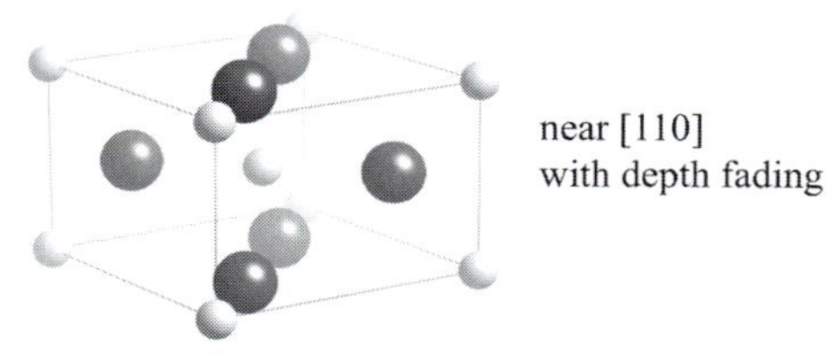

Fig. 14.2. Perspective projections near the [100], [110], and [001] directions for the unit cell of rutile. The bottom image illustrates the use of depth shading.

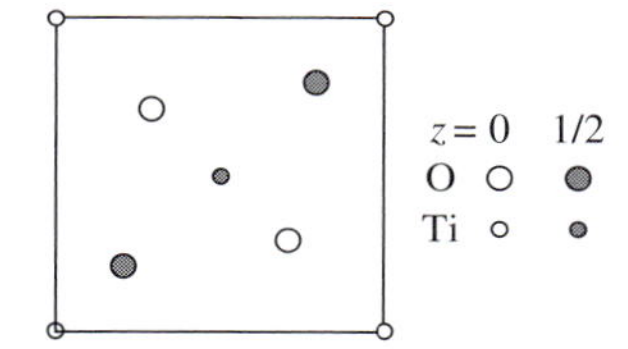

Fig. 14.3. Orthogonal [001] projection for the unit cell of rutile. Atoms are indicated with the symbols described in the legend.

14.1.2.3 Height labels

One can add information to an orthogonal projection by drawing atoms at different levels in the unit cell with different symbols. An example is shown in Fig. 14.3: there are symbols with two different shades and two different sizes. The open circles are at the level $z = 0$, where z is measured parallel to the projection axis. The filled symbols are at height $z = 1/2$. The size of the symbols indicates whether the corresponding atom is Ti or O. This type of representation is often used to indicate the relative 3-D locations of atoms without having to resort to 3-D representations.

14.1.2.4 Ball-and-stick representations

One of the most common representations for crystal models is the so-called *ball-and-stick representation*. Atoms are represented by spheres of different diameters (often one uses radii proportional to the appropriate ionic, metallic, or covalent radii) and they are connected to their immediate neighbors by rods or sticks. The advantage of this representation is that one can easily visualize the coordination of each atom. Fig. 14.4 shows an orthogonal projection and a perspective drawing of rutile, using the ball-and-stick display mode.

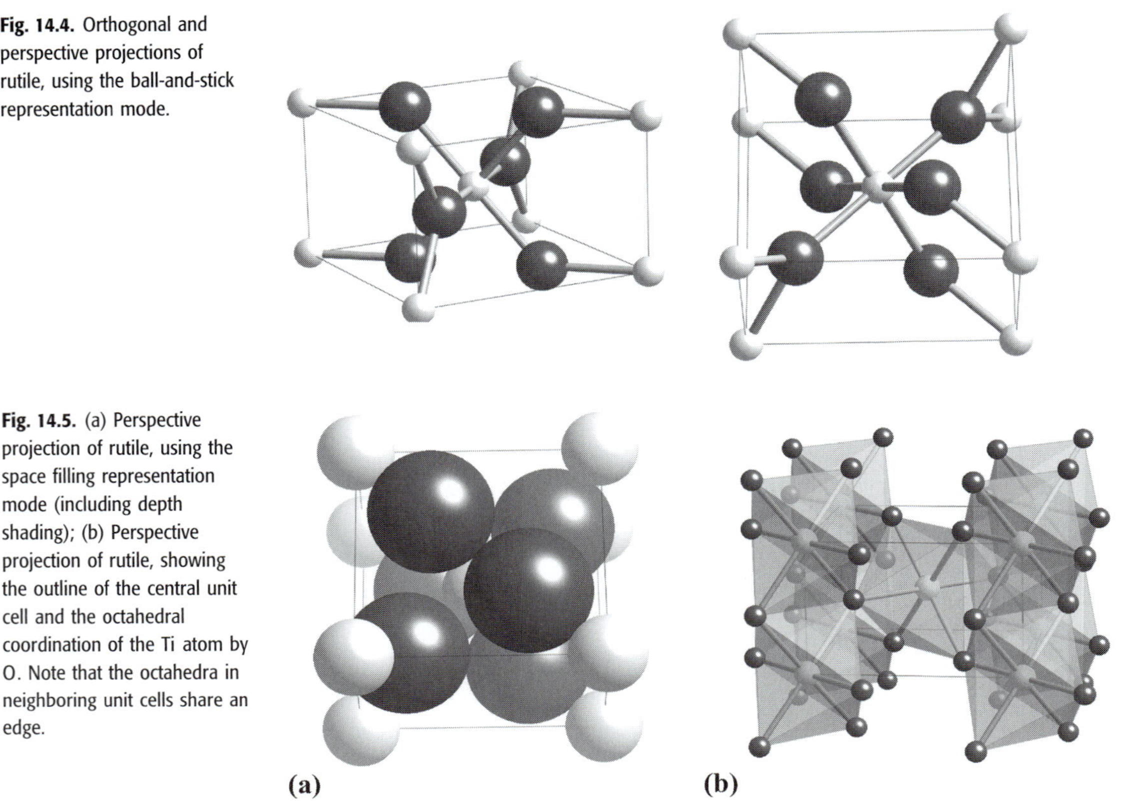

Fig. 14.4. Orthogonal and perspective projections of rutile, using the ball-and-stick representation mode.

Fig. 14.5. (a) Perspective projection of rutile, using the space filling representation mode (including depth shading); (b) Perspective projection of rutile, showing the outline of the central unit cell and the octahedral coordination of the Ti atom by O. Note that the octahedra in neighboring unit cells share an edge.

14.1.2.5 Space filling models

Space filling models employ the correct atomic, ionic, metallic, or covalent radii and draw the atoms as touching spheres. This representation mode can provide some insight into the *density* of packing, and it can, as usual, be represented in orthogonal and perspective projection modes. The main disadvantage of this representation is that only atoms close to the faces of the unit cell can be distinguished, and the space filling model is hence more useful for molecular biologists and chemists than for material science. An example of a space filling representation for rutile is shown in Fig. 14.5(a).

14.1.2.6 Polyhedral models

It is often useful, especially in structures with ionic or covalent bonds, to emphasize the *coordination* of the cations by the anions. This results in *polyhedral* drawings, such as the one shown in Fig. 14.5(b), which shows the octahedral coordination of Ti by O, and the edge sharing of neighboring octahedra.

14.2 Crystal structures ↔ powder diffraction patterns

In this section, we will take some of the concepts and techniques that we learned about in the first half of this book and apply them to the Ni and NaCl powder diffraction patterns shown in Fig. 1.7 on page 20. We will attempt to predict the diffraction patterns, starting from the crystal structures, and, then, we will attempt to determine the crystal structures of these two materials, starting from the diffraction patterns. While the former is a relatively simple application of the theoretical considerations in Chapters 11 and 12, the latter will prove to be quite a bit more challenging. The examples are both for cubic crystal structures; in section 14.2.5, we will comment briefly on procedures that deal with arbitrary crystal structures.

14.2.1 The *Ni* powder pattern, starting from the known structure

We have seen that the unit cell dimensions (the lattice parameters) determine the location of the diffraction peaks, whereas the atom positions inside the cell determine the diffracted intensities. We begin by considering the locations of the peaks in the nickel coin powder pattern, shown in Fig. 1.7 and, for a larger 2θ range, 14.6; there are eight peaks at the following diffraction angles (estimated from the peak maxima):

$$43.67^\circ, 50.84^\circ, 74.76^\circ, 90.74^\circ, 96.08^\circ, 118.24^\circ, 138.66^\circ, \text{and } 147.26^\circ.$$

These angles correspond to the Cu Kα_1 wave length. This pattern was obtained on a Rigaku $\theta - \theta$ diffractometer with a Cu target, operated at 35 kV with a current of 25 mA. The angular step size was 0.05°, and the intensity was integrated for 2 seconds at each angle. The nickel coin surface was cleaned

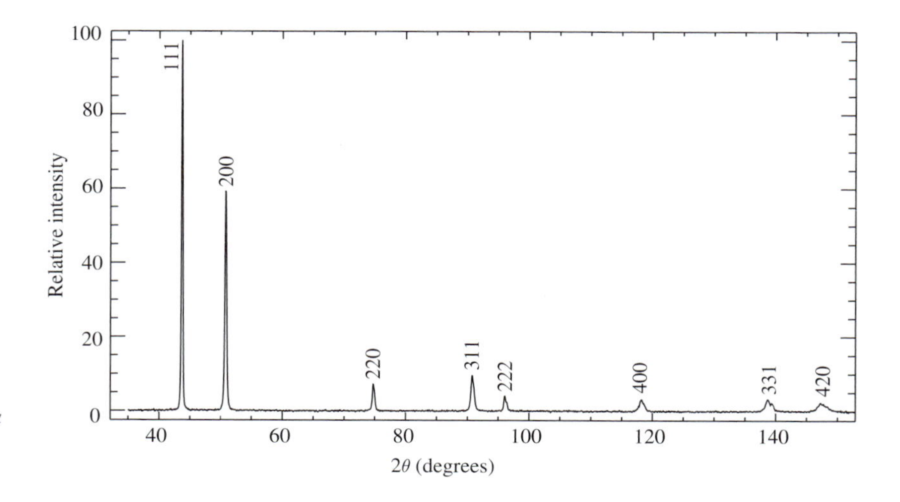

Fig. 14.6. Experimental Cu Kα powder diffraction pattern for a US 5 cent "nickel" coin.

Table 14.1. Lattice parameter computation for the nickel coin, assuming a *cF* Bravais lattice.

2θ (deg) $K\alpha_1$	*cF*	
	(hkl)	a (nm)
43.67	111	0.35871
50.84	200	0.35890
74.76	220	0.35887
90.74	311	0.35898
96.08	222	0.35883
118.24	400	0.35900
138.66	331	0.35886
147.26	420	0.35903

before data acquisition, but no other sample preparation was carried out. From the US mint web site we find that the nickel coin is made from an alloy with 75 at % Cu, and 25 at % Ni.[2] Both Ni and Cu have the *fcc* structure, and they are fully soluble in one another, so that we expect the alloy to have the same structure, with a lattice parameter intermediate between that of the two end members. In fact, *Vegard's law*, an empirical rule, states that there exists an approximate linear relationship between the lattice parameters of a solid solution alloy and the composition of this alloy. So, if we consider the lattice parameters of pure Ni and pure Cu:

$$a_{\mathrm{Ni}} = 0.352\,36\,\mathrm{nm} \qquad \text{and} \qquad a_{\mathrm{Cu}} = 0.360\,78\,\mathrm{nm},$$

then we expect to find $a \approx 0.35868$ nm for the alloy. We also know that the relation between the lattice parameter and the Bragg angle is given by:

$$\boxed{a = \frac{\lambda\sqrt{h^2+k^2+l^2}}{2\sin\theta}.} \tag{14.1}$$

Using this relation, we can convert the measured diffraction angles into an estimate of the lattice parameter. This is shown in Table 14.1, resulting in an average observed lattice parameter of $a = 0.35890 \pm 0.00275$ nm.

This procedure assigns equal weight to each of the reflections, and is only a first-order estimate. A better estimate would take into account the fact that the relation between interplanar spacing and Bragg angle is a non-linear relation. If we consider the Bragg relation:

$$2d\sin\theta = \lambda,$$

[2] URL: http://www.usmint.gov/about_the_mint/index.cfm?action=coin_specifications

and we increase d by a small amount Δd, then θ will change by an amount $\Delta\theta$, such that:

$$2(d+\Delta d)\sin(\theta+\Delta\theta)=\lambda.$$

If we use the trigonometric relation $\sin(a+b)=\sin a\cos b+\sin b\cos a$, $\cos\Delta\theta\approx 1$, $\sin\Delta\theta\approx\Delta\theta$, and we drop terms of order $\Delta d\Delta\theta$, then we find:

$$\frac{\Delta d}{d}=-\cot\theta\,\Delta\theta.$$

In cubic crystals, we also have $\Delta d/d=\Delta a/a$, so that:

$$\frac{\Delta a}{a}=-\cot\theta\,\Delta\theta.$$

In other words, the fractional error in the lattice parameter is smallest when θ approaches 90°, i.e., for diffracted beams with 2θ close to 180°. This means that the reflections with the largest 2θ values should receive a larger weight in the averaging procedure. Using the FINAX program (Hovestreydt, 1983), we find for the lattice parameter of the nickel coin: $a=0.35896\pm 0.0003$ nm.[3] This is in reasonable agreement with the expected lattice parameter (0.35868 nm) based on Vegard's law.

Now that we have determined the lattice parameter, we can turn our attention to the peak intensities. The structure factor for the *fcc* structure was derived in Chapter 12, and, for a solid solution of Ni and Cu, it is given by:

$$F_{hkl}=4\left(\frac{1}{4}f_{\text{Ni}}\,\mathrm{e}^{-B_{\text{Ni}}s^2}+\frac{3}{4}f_{\text{Cu}}\,\mathrm{e}^{-B_{\text{Cu}}s^2}\right),$$

for reflections with Miller indices of equal parity. The Debye–Waller factors for Cu and Ni (at $T=290$ K) are: $B_{\text{Ni}}=0.0035\,\text{nm}^2$ and $B_{\text{Cu}}=0.0054\,\text{nm}^2$ (Peng *et al.*, 1996). Table 14.2 lists the computation of the integrated intensities for the eight observed reflections. The experimental pattern shows a clear peak splitting due to the presence of the $\text{K}\alpha_1$ and $\text{K}\alpha_2$ wave lengths; the splitting becomes larger with increasing diffraction angle. Therefore, we will compute the integrated intensities for all reflections, using both the $\text{Cu}\,\text{K}\alpha_1$ and $\text{Cu}\,\text{K}\alpha_2$ wave lengths, and then combine those intensities using

$$I_{hkl}=p_{hkl}|F_{hkl}|^2\left(\frac{2}{3}L_{\text{p}}(\theta_{hkl,\text{K}\alpha_1})+\frac{1}{3}L_{\text{p}}(\theta_{hkl,\text{K}\alpha_2})\right),$$

[3] The FINAX program is a computer program than can be used to refine lattice parameters based on experimental measurements of diffraction angles. A web interface to this program can be found on this book's web site.

Table 14.2. Tabular computation of the integrated intensities for the nickel coin.

#	hkl	d_{hkl}(Å)	s(Å^{-1})	F_{hkl}	p	L_p (Kα_1)	L_p (Kα_2)	Relative I (comp.)	Relative I (exp.)
1	111	2.0725	0.2413	84.2163	8	11.8816	11.8120	100.0	100.0
2	200	1.7948	0.2786	78.1220	6	8.4089	8.3581	45.7	80.1
3	220	1.2691	0.3940	61.0456	12	3.6524	3.6330	24.2	12.6
4	311	1.0823	0.4620	52.4271	24	2.8109	2.8046	27.5	22.3
5	222	1.0362	0.4825	50.0049	8	2.7354	2.7335	8.1	8.8
6	400	0.8974	0.5572	41.9161	6	3.2386	3.2638	5.1	11.1
7	331	0.8235	0.6072	37.1922	24	5.0484	5.1562	25.1	15.6
8	420	0.8027	0.6229	35.8288	24	6.5993	6.8272	30.6	14.4

in accordance with the relative intensities of the two Kα lines. The only difference between the two contributions is the Lorentz polarization factor, which is listed for both Kα_1 and Kα_2 wave lengths in Table 14.2. The relative integrated intensities are listed in column 9 of this table. To compare these values with the experimental pattern we must first convert the pattern to integrated intensities. There are several ways to do this; in this section, we will use the method of the cumulant function, and in the next section we will use a curve-fitting procedure.

The *cumulant function* is essentially the integral of the powder diffraction pattern. This can be computed quite easily by first subtracting the background from the pattern, and then replacing the value for each 2θ by the integral of the spectrum (i.e., the sum of all the intensities) up to that value of 2θ. If the background subtraction is carried out properly, then the resulting cumulant function should show steps corresponding to each diffracted beam, separated by horizontal segments, as shown in Fig. 14.7. The difference between two consecutive horizontal segments is equal to the integrated intensity of the corresponding peak. The last column of Table 14.2 lists the relative integrated intensities obtained in this way.

The agreement between the experimental and calculated relative integrated intensities shown in Table 14.2 is not particularly good. One would typically expect to find agreement to within a few percent, not a factor of two, as is the case for the (200), (220), and (400) reflections. In this case, the explanation is that one of the assumptions made in the derivation of the theoretical expression for the integrated intensity is not valid. We have assumed that all crystallite orientations are present with equal probability. That is not the case for the nickel coin, since it is made from a rolled sheet that is subsequently heat treated to recrystallize the grain structure. As a result of this operation, the orientation of the individual grains is no longer random but displays a *texture*. Texture is defined as the presence of a deviation from the random distribution of grain orientations. In the case of the CuNi alloy, it turns out that the individual grains are more likely to have their cube planes, i.e., the {100}-type planes, lie parallel to the plane of the sheet (which is also

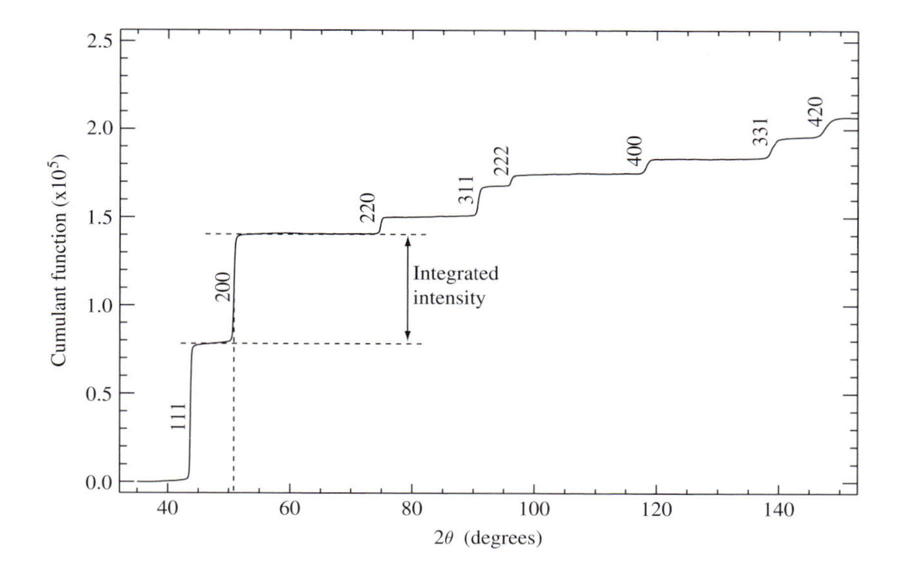

Fig. 14.7. Cumulant function for the nickel coin powder diffraction pattern of Fig. 14.6.

parallel to the top and bottom surfaces of the coin). As a consequence, the probability for diffraction from the cube planes is larger than in the randomly oriented microstructure, so that the 200 diffraction peak has a significantly higher intensity than predicted. We conclude that, in the presence of texture, the observed integrated intensities will deviate from the predicted values. Conversely, by analyzing the deviations between the observed and predicted intensities, it is possible to derive the preferential orientation of the grains. This approach is known as *texture analysis*, and the interested reader is referred to the literature for more information; basic information can be found in Bunge (1983), Randle (2000), and Kocks *et al.* (2001).

14.2.2 The *NaCl* powder pattern, starting from the known structure

The NaCl powder pattern shown in Fig. 1.7 is actually a section of a pattern obtained over the angular range [10°–120°]. The complete experimental pattern is shown in Fig. 14.8. There are 14 peaks in this range, corresponding to values of $h^2+k^2+l^2$ between 3 (111) and 40 (620). The pattern was recorded in a Rigaku $\theta-\theta$ diffractometer with an angular step-size of 0.05° and 2 second integration at each step. The X-ray tube was operated at 35 kV with a current of 25 mA. The NaCl powder sample was ground with a mortar and pestle, and subsequently passed through a 325 mesh sieve (this sieve removes all particles larger than about 45 μm diameter). The powder was then back-loaded in a standard powder holder. The reason for this somewhat complicated powder preparation procedure is the fact that NaCl usually occurs in a highly facetted form. When these crystallites are placed on the diffractometer sample holder, most of them will orient themselves with one of their facets parallel to the plane of the holder, so that the orientation is no longer completely random. This would give rise to integrated relative intensities that do not agree with the computed values. Larger crystals are more likely to orient themselves parallel to the holder, hence the sieving procedure.

The computation of integrated intensities for the NaCl structure follows the same procedure as described in Table 12.7 in Chapter 12. We start with the crystal structure information:

Space Group: Fm$\bar{3}$m (O_h^5) (# 225);
Lattice parameter: $a = 0.56407$ nm;
Atoms: Na in $4a$ [(0,0,0)]; Cl in $4b$ [(1/2,1/2,1/2)].

There are eight atoms per unit cell, and the structure factor was derived in Chapter 12 as:

$$F_{hkl} = (f_{\mathrm{Na}} + f_{\mathrm{Cl}}(-1)^{h+k+l})(1 + (-1)^{h+k} + (-1)^{h+l} + (-1)^{k+l}),$$

so that

$$|F_{hkl}|^2 = 16(f_{\mathrm{Na}} + f_{\mathrm{Cl}}(-1)^{h+k+l})^2$$

for reflections with Miller indices of the same parity. There are two types of reflections: those with $h+k+l = 2n$, for which we have:

$$|F_{hkl}|^2 = 16(f_{\mathrm{Na}} + f_{\mathrm{Cl}})^2,$$

and those with $h+k+l = 2n+1$ for which:

$$|F_{hkl}|^2 = 16(f_{\mathrm{Na}} - f_{\mathrm{Cl}})^2.$$

The presence of two sets of reflections is also clearly visible in the powder pattern of Fig. 14.8.

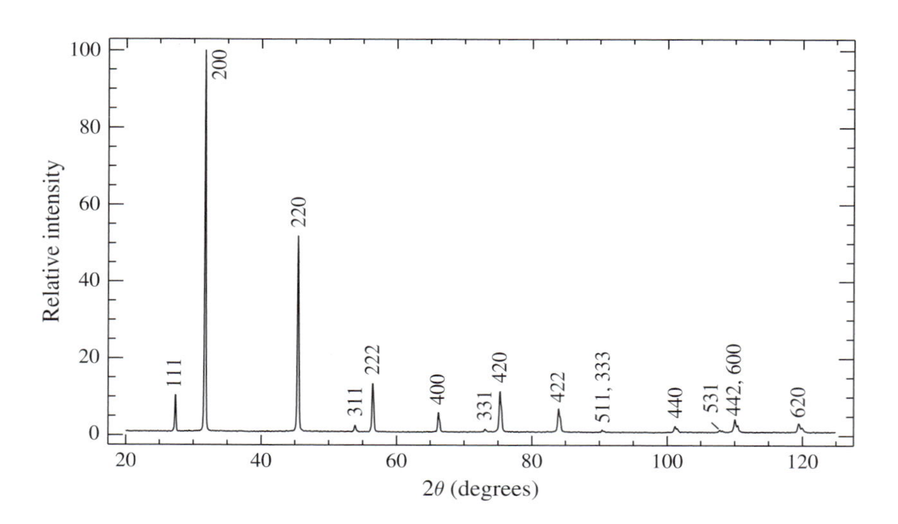

Fig. 14.8. Experimental powder diffraction pattern for randomly oriented NaCl powder, using Cu Kα radiation.

Table 14.3. Tabular computation of the integrated intensities for NaCl.

#	hkl	d_{hkl} (Å)	s (Å^{-1})	f_{Na}	f_{Cl}	e^{-M}-Na	e^{-M}-Cl	F_{hkl}	p
1	111	3.2567	0.1535	8.9867	13.4761	0.9603	0.9673	17.6237	8
2	200	2.8204	0.1773	8.6498	12.6974	0.9474	0.9567	81.3663	6
3	220	1.9943	0.2507	7.5874	10.6150	0.8975	0.9152	66.0983	12
4	311	1.7007	0.2940	6.9823	9.6709	0.8619	0.8853	10.1742	24
5	222	1.6283	0.3071	6.8019	9.4248	0.8503	0.8755	56.1406	8
6	400	1.4102	0.3546	6.1454	8.6649	0.8055	0.8376	48.8313	6
7	331	1.2941	0.3864	5.7038	8.2574	0.7735	0.8102	9.1117	24
8	420	1.2613	0.3964	5.5649	8.1430	0.7632	0.8013	43.0860	24
9	422	1.1514	0.4343	5.0504	7.7602	0.7230	0.7665	38.3992	24
10a	511	1.0856	0.4606	4.7074	7.5285	0.6943	0.7415	9.2554	24
10b	333	1.0856	0.4606	4.7074	7.5285	0.6943	0.7415	9.2554	8
11	440	0.9971	0.5014	4.2152	7.2020	0.6489	0.7015	31.1501	12
12	531	0.9535	0.5244	3.9647	7.0264	0.6231	0.6786	9.1898	48
13a	442	0.9401	0.5318	3.8881	6.9699	0.6148	0.6711	28.2708	24
13b	600	0.9401	0.5318	3.8881	6.9699	0.6148	0.6711	28.2708	6
14	602	0.8919	0.5606	3.6135	6.7503	0.5824	0.6420	25.7533	24

The atomic scattering factors of Na and Cl can be derived using the table on page 299. The Debye–Waller factors were determined from phonon-dispersion computations by Gao *et al.* (1999): at room temperature (290 K), we have $B_{Na} = 0.0172\,\text{nm}^2$ and $B_{Cl} = 0.0141\,\text{nm}^2$. The relevant factors are tabulated in Tables 14.3 and 14.4. Note that lines 10 and 13 have contributions from two sets of planes.

The final column in Table 14.4 is derived from the experimental powder pattern by means of a simple peak-fitting routine. A cursory comparison of the experimental peak shape with the Gaussian and Lorentzian functions defined in Chapter 12 shows that the actual peak shape is in between those two; in other words, the experimental peaks fall off more slowly than the Gaussian profile, but faster than the Lorentzian. The combination of these two shapes is known as a *pseudo-Voigt peak shape* (Giacovazzo, 2002a), and is described mathematically as:

$$I_k(2\theta) = \tilde{I}_k\left(\eta\frac{\sqrt{C_0}}{\pi w_k}\frac{1}{1+C_0X_k^2} + (1-\eta)\frac{\sqrt{C_1}}{\sqrt{\pi}w_k}e^{-C_sX_k^2}\right), \tag{14.2}$$

where the subscript k labels the individual diffraction peaks; $\tilde{I}_k$ is the maximum peak intensity, $X_k \equiv 2(\theta-\theta_k)/w_k$, w_k is the full width at half maximum (FWHM) of peak k, $2\theta_k$ is the diffraction angle for the reflection k, and η is a mixing parameter ($0 \leq \eta \leq 1$). The constants C_i are given by $C_0 = 4$ and $C_1 = 4\ln 2$. When η equals 1, the peak shape is a pure Lorentzian, for $\eta = 0$

Table 14.4. Tabular computation of the integrated intensities for NaCl (continued).

#	2θ, $K\alpha_1$	2θ, $K\alpha_2$	$I_{K\alpha_1}$	$I_{K\alpha_2}$	Relative I (Comp.)	Relative I (Exp.)
1	27.3630	27.4324	81767	81329	8.57	9.50
2	31.6995	31.7803	954316	949083	100.00	100.00
3	45.4419	45.5612	568559	565200	59.57	59.57
4	53.8611	54.0059	18309	18198	1.92	2.06
5	56.4649	56.6179	166922	165902	17.49	17.27
6	66.2176	66.4035	66555	66159	6.97	7.05
7	73.0596	73.2708	7593	7551	0.80	0.9
8	75.2809	75.5008	160599	159759	16.83	17.68
9	83.9788	84.2355	107552	107139	11.28	11.18
10a	90.4006	90.6878	5795	5782	0.81	0.59
10b			1932	1927		
11	101.1554	101.5026	31877	31909	3.35	3.09
12	107.7797	108.1712	11521	11561	1.21	0.93
13a	110.0387	110.4469	55691	55929	7.32	7.36
13b			13923	13982		
14	119.4615	119.9514	52576	53014	5.53	6.06

we have a Gaussian peak, and for all other values the shape is intermediate. It is relatively straightforward to fit this peak shape to each of the experimentally observed diffraction peaks. Since the experimental peaks have two sub-peaks, we use two pseudo-Voigt functions for each reflection. The fitting routine considers the parameters $2\theta_k$, w_k, and η along with the peak height to be variable parameters, and returns the best fit for each. From these values one can immediately compute the integrated intensity of each peak, either by direct numerical integration, or by analytical integration of Eq. 14.2 over the width of each peak. The fitted value for the mixing parameter η is 0.25.

The experimental relative integrated intensities are in excellent agreement with the theoretical predicted values, as can be seen by comparing columns 6 and 7 in Table 14.4. The fitted pattern is shown in Fig. 14.9, along with the difference pattern (bottom of figure). It is clear that the fit is quite reasonable over the full range of diffraction angles. To make this a more quantitative statement, one usually defines a few *agreement indices* or *residuals*. For our purposes, we define the *profile* R_p agreement index:

$$R_\mathrm{p} = \frac{\sum |I_{i\mathrm{o}} - I_{i\mathrm{c}}|}{\sum I_{i\mathrm{o}}},$$

where $I_{i\mathrm{o}}$ and $I_{i\mathrm{c}}$ are the observed and calculated intensities for diffraction angle $2\theta_i$, and the sums run over all values of i. The *weighted profile* R_wp agreement index is defined as:

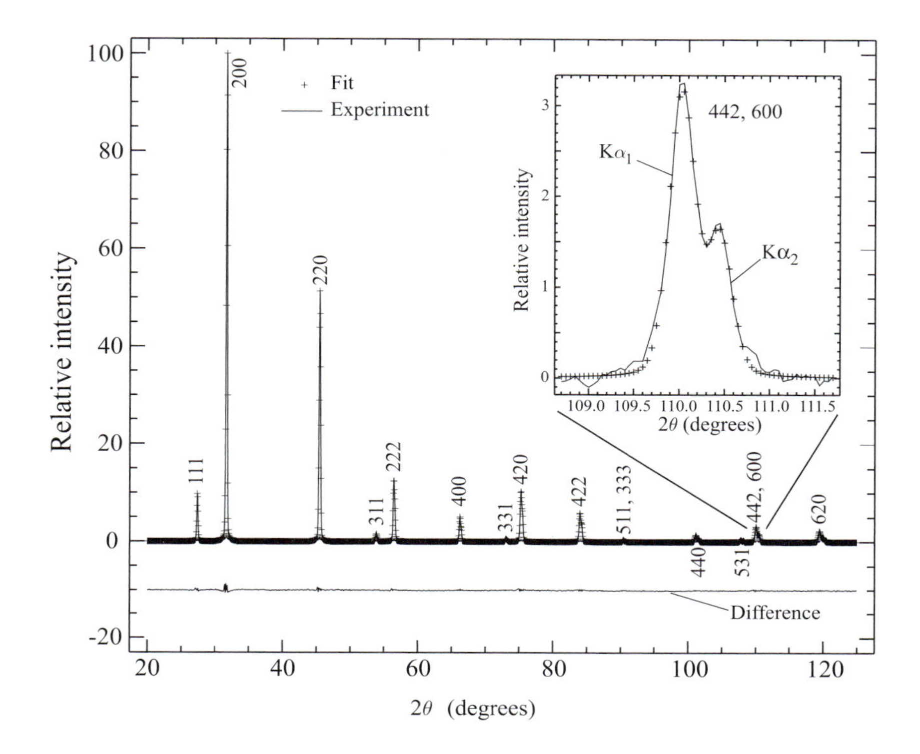

Fig. 14.9. Fitted intensity profile for the NaCl powder pattern of Fig. 14.8. The inset shows the detailed fit for the (442), (600) diffraction peak.

$$R_{wp} = \left[\frac{\sum w_i (I_{io} - I_{ic})^2 |}{\sum w_i I_{io}^2} \right]^{\frac{1}{2}},$$

where the weight factors w_i are usually computed from:

$$\frac{1}{w_i} = \sigma_i^2,$$

and σ_i is the standard deviation associated with the intensity measured at each value $2\theta_i$. The agreement indices for the curve fit in Fig. 14.9 are $R_p = 12.9\%$ and $R_{wp} = 4.5\%$. While there is no absolute standard to compare the agreement indices against, these values are generally considered to be satisfactory.

The analysis in this section shows that it is not too difficult to compute the integrated intensities for a given crystal structure, and compare them with an experimental data set. The present analysis is little more than a curve-fitting exercise, and there are far more sophisticated approaches to the analysis of powder diffraction patterns. A commonly used method for both X-ray and neutron powder spectra is known as the *Rietveld method*. In this method, the entire spectrum, including the background intensity, is considered to be a single discrete function against which a multi-parameter model must be fitted. This model accounts for the background intensity, the peak shape (location, intensity, FWHM, and tails), the atom positions and occupations in the unit cell, the cell parameters, thermal Debye–Waller factors, and so on. All of these parameters can be refined together or separately, and a series of agreement indices are computed which provide information on the "goodness of fit." The Rietveld method is a powerful tool for the quantitative interpretation of powder diffraction patterns. For more information on the Rietveld method, we refer the interested reader to the following references: Rietveld (1967, 1969), Hill and Howard (1986), and Young (2000).

14.2.3 The *Ni* structure, starting from the experimental powder diffraction pattern

The diffraction angles for the Ni powder pattern were listed in Section 14.2.1. In this section, we will try to answer the question: can we derive the structure of Ni directly from the diffraction pattern and a few additional pieces of information, such as the density? First of all, the density of pure Ni is $8.912\,\text{g/cm}^3$; for pure Cu we have $8.933\,\text{g/cm}^3$; these are very similar, so that we can take the average to represent the alloy, i.e., $8.923\,\text{g/cm}^3$. The molar mass of $Cu_{0.75}Ni_{0.25}$ is $62.3328\,\text{g/mol}$, so that there are $8.923/62.3328\,\text{mol} = 8.6205 \times 10^{22}$ formula units per cubic centimeter, or, equivalently, $1\ \text{FU} = 0.0116\,\text{nm}^3$.

Next, we need to find out how many formula units there are per unit cell. It is a general rule-of-thumb, that the number of peaks in a powder pattern is inversely proportional to the complexity of the structure, which includes the symmetry. This is obviously not a rule that can be proven with mathematical equations, but it does hold approximately for a wide range of structures. Since the nickel coin pattern has only a few widely spaced peaks, we anticipate that the structure will be simple, with a high symmetry unit cell. We will begin with the assumption of a cubic lattice. There are three cubic Bravais lattices, cP, cI, and cF, so we must check all three of them to see which centering corresponds to the experimental pattern. Table 14.5 lists, in the first column, the Bragg angle θ (in radians), followed by the three sets of Miller indices and calculated cubic lattice parameters for the three centering operations. To compute a, we make use of Equation 14.1.

From the table we see that the cP and cI lattices have a rather large standard deviation, whereas the cF lattice has a small standard deviation, as well as a nearly integer number of formula units per unit cell. Therefore, we conclude that the Bravais lattice is the cF lattice. One formula unit equals $Cu_{0.75}Ni_{0.25}$, so that there are three Cu atoms and one Ni atom per unit cell, for a total of four atoms per unit cell. Since the lattice is cF, these four atoms must be equivalent to each other, so that there is only one atom position that must be determined; the others are fixed by the lattice centering operations. The easiest choice for that atom position is the origin, so that the entire structure is determined.

The derivation above is deceptively simple, and might give the wrong impression to the reader. Structure determination is almost never straightforward, for the simple reason that, in most cases, the unit cell is not known.

Table 14.5. Lattice parameter computation for the nickel coin, assuming that the Bravais lattice is *cP*, *cI*, and *cF*. The angles in the first column are derived from the experimental powder diffraction pattern.

θ (rad) $K\alpha_1$	cP		cI		cF	
	(hkl)	a (nm)	(hkl)	a (nm)	(hkl)	a (nm)
0.3811	100	0.2710	110	0.2929	111	0.35871
0.4437	110	0.2538	200	0.3589	200	0.35890
0.6524	111	0.2198	211	0.3108	220	0.35887
0.7919	200	0.2165	220	0.3061	311	0.35898
0.8385	210	0.2316	310	0.3276	222	0.35883
1.0318	211	0.2198	222	0.3109	400	0.35900
1.2100	220	0.2329	321	0.3080	331	0.35886
1.2851	221	0.2408	400	0.3211	420	0.35903
av. $\pm\sigma$	0.2278 ± 0.0150 nm		0.3170 ± 0.0198 nm		0.35890 ± 0.00001 nm	
# FU/cell	1.019		2.7472		3.9852	

The example in this section was intentionally chosen to be cubic, so that we would have to determine only one lattice parameter. In general, however, there are six lattice parameters, three lengths and three angles. Furthermore, we know from the discussions in Chapter 3, in particular Fig. 3.7, that there is an infinite number of possible choices for the unit cell! So, given a diffraction pattern, which unit cell do we try out? Obviously, it is desirable to select the most symmetric unit cell, which is what we did above, but if we were to pick another cell, for instance, the rhombohedral cell introduced in Section 7.4.2 on page 147, then we should still be able to consistently assign Miller indices to all experimental peaks, and find the correct rhombohedral lattice parameters. Let us analyze briefly how this can be done.

First of all, we know that the rhombohedral cell is a primitive cell, with only one atom per unit cell. We will need to determine two lattice parameters, a and α. For a primitive cell, we know that there are no systematic absences due to centering, so for arbitrary a and α, we expect to see a large number of peaks. Yet, the experimental pattern has only eight peaks in the observed angular range. In other words, we are looking for a particular combination of a and α for which there are relatively few peaks.

The relation between the diffraction angle and the Miller indices and lattice parameters for the rhombohedral system is given by:

$$\sin^2\theta = \frac{\lambda^2}{4a^2}\left((h^2+k^2+l^2)(1+\cos\alpha) - 2(hk+hl+kl)\cos\alpha\right).$$

This relation can be rewritten as:

$$\sin^2\theta = A\sigma^2 + B(\sigma - 2\tau), \tag{14.3}$$

where $\sigma = h^2 + k^2 + l^2$, $\tau = hk + hl + kl$, and A and B are constants related to the lattice parameters:

$$A = \frac{\lambda^2}{4a^2} \quad \text{and} \quad B = A \cos \alpha.$$

The lattice parameters can then be determined by comparing the experimental values for $\sin^2 \theta$ with calculated values for a range of A and B parameters and all possible sets of Miller indices. In practice, the number of lattice planes that can give rise to diffraction is rather limited, since the X-ray wave length is of the same order of magnitude as the lattice parameter. For a given choice of a (and hence A), we can plot all the possible values of $\sin^2 \theta$ for all possible values of α and allowed combinations of the Miller indices. An example is shown in Fig. 14.10; each curve represents the values of $\sin^2 \theta$ according to Equation 14.3. Along the horizontal axis, the value of α varies. Each curve is labeled with the corresponding Miller indices. The value of a determines the vertical position of all the curves; when a is decreased, all curves move upwards, whereas they move downwards with increasing a. We know from the experimental results that the first peak does not occur until $\theta \approx 20°$, and

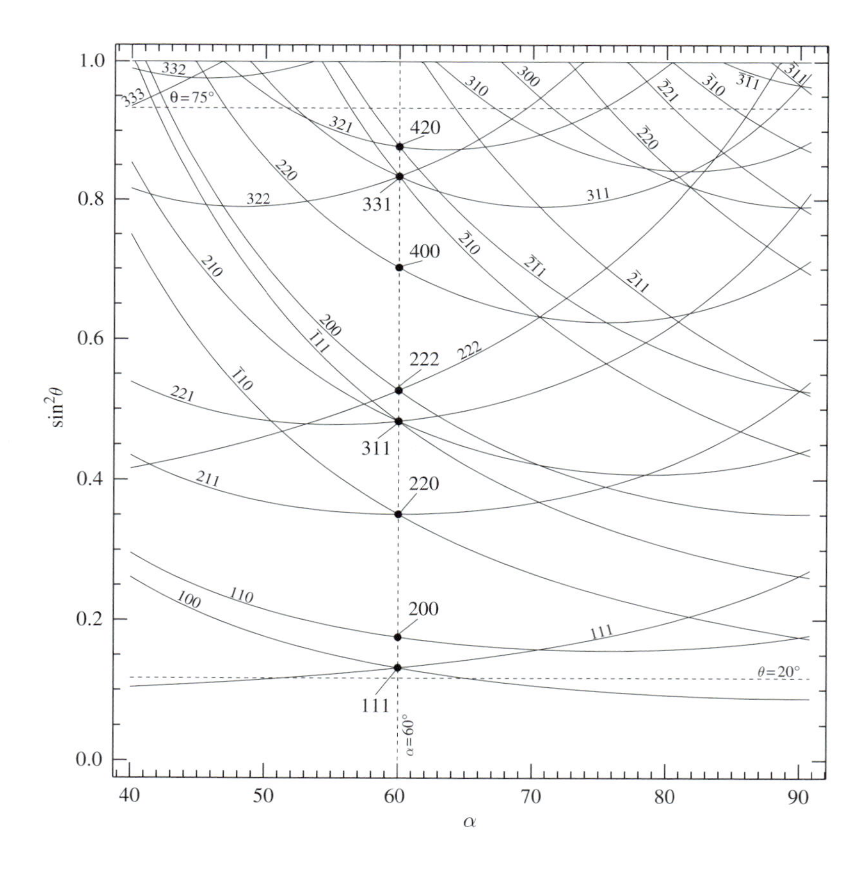

Fig. 14.10. Curves of $\sin^2 \theta$ versus α (from Equation 14.3) for a range of Miller indices; smaller Miller indices labeling the curves are with respect to the rhombohedral reference frame. Larger Miller indices along the line $\alpha = 60°$ are with respect to the cubic system.

this places an upper limit on the value of a; similarly, if a becomes too small, then the lines will move upward and disappear off the top of the drawing. We know that we need eight lines between $\theta = 20°$ and $\theta = 75°$, so that places a lower limit on a. By trial and error, we find that $a = 0.26$ nm is a reasonable value; Fig. 14.10 is drawn for that particular value of a and Cu Kα radiation, so that $A \approx 0.08776$.

To determine the value of B, we find that, for most values of α, there are about a dozen lines present in the interval between the two dashed lines. There are two values of α for which there are fewer lines: $\alpha = 60°$ and $\alpha = 90°$. The latter one would result in a primitive cubic unit cell, which can be excluded based on the foregoing discussion. The former produces eight diffraction peaks, as is needed to explain the experimental data. From a non-linear least-squares fit of the experimental data to Equation 14.3 we find $A = 0.092118$ and $B = 0.046059$, or $a = 0.25378$ and $\alpha = 60°$. These lattice parameters are in good agreement with the cubic ones derived earlier. The rhombohedral Miller indices can be converted to the cubic ones by means of the relation (derived in Chapter 7):

$$\begin{pmatrix} h \\ k \\ l \end{pmatrix}_{\mathrm{r}} = \frac{1}{2}\begin{pmatrix} 1 & 1 & 0 \\ 0 & 1 & 1 \\ 1 & 0 & 1 \end{pmatrix}\begin{pmatrix} h \\ k \\ l \end{pmatrix}_{\mathrm{c}}.$$

Similar procedures are available for unit cells belonging to the tetragonal, hexagonal, and orthorhombic crystal systems. While the graphical procedures based on figures similar to Fig. 14.10 were quite popular before the advent of desktop computing, nowadays unit cell determination is essentially a non-linear least-squares problem that is readily solved by means of standard numerical algorithms.

We conclude from this example that the determination of the unit cell parameters is not always straightforward, but it is always the first step in structure determination. Once the lattice parameters are known, then the positions of all diffracted beams are known, and one can focus on the intensities of the individual peaks, i.e., on the determination of the atom positions. There are several algorithms available for the computation of the so-called *reduced basis*, the basis with the shortest possible non-coplanar lattice vectors. Such algorithms compute the shortest possible lattice parameters based on an initial guess of the lattice parameters. We refer the interested reader to Chapter 9 of the *International Tables of Crystallography* for more information on unit cell determination (Hahn, 1996).

14.2.4 The *NaCl* structure, starting from the experimental powder diffraction pattern

In this section, we will attempt to *derive* the NaCl crystal structure from the powder diffraction pattern; i.e., we will try to show that there is no other configuration of Na and Cl atoms that gives rise to the same diffraction pattern. First of all, we will assume that we don't know the crystal structure at all. Let us list the facts that we *do* know:[4]

(i) We know that NaCl crystals occur in nearly perfect cube shapes, which indicates that the crystal system is most likely the cubic system.
(ii) We also know the density (which can be measured in a variety of ways): $2.16\,\mathrm{g/cm^3}$.
(iii) From chemical analysis, we might be able to figure out that there are two different atomic species, Na and Cl, and that there are identical amounts of each element in the compound (i.e., the formula unit is NaCl). The Periodic Table of the Elements, first constructed in the last decade of the nineteenth century, would show us that Na is in the first column, whereas Cl is in the column next to the inert elements. The molar masses of the elements are 22.989 g/mol for Na, and 35.452 g/mol for Cl.

Combining the molar masses with the density, we can find out how many formula units (FU) there are in a unit volume, or, conversely, the volume of one formula unit. We have $1\,\mathrm{cm^3} = 2.16\,\mathrm{g} = 2.16/58.441\,\mathrm{mol} = 2.2257 \times 10^{22}$ FU, so that $1\ \mathrm{FU} = 0.04493\,\mathrm{nm^3}$.

Next, we need to find out how many formula units there are in a unit cell. To do so, we turn to the powder diffraction pattern of Fig. 14.8. If we assume that the structure is cubic, then we have three possible Bravais lattices, *cP*, *cI*, and *cF*. Combining Bragg's equation with the equation for the interplanar spacing in a cubic crystal we find, as before:

$$a = \frac{\lambda\sqrt{h^2+k^2+l^2}}{2\sin\theta}.$$

We can obtain the diffraction angles of the $K\alpha_1$ peaks directly from the curve fits to the experimental powder pattern presented in the previous section. The results (in radians) are shown in the first column of Table 14.6. Then we use Table 12.2 and the equation above to determine, for each of the three possible Bravais lattices, what the lattice parameter a would be; to do so, we simply assume that the first reflection is a (100) reflection in the *cP* case, a (110) reflection in the *cI* case, and a (111) reflection in the *cF* case. When we work our way through the table, we find that only in the case of the *cF* Bravais lattice are all entries close to a single number. The average of these values yields $a = 0.5639 \pm 0.0002$ nm, which is the cubic lattice parameter.[5] Along the way, we also figured out that the structure must be cubic face-centered.

The unit cell volume is then equal to $0.17931\,\mathrm{nm^3}$, so that the number of formula units per unit cell equals $0.17931/0.04493 = 3.9909 \approx 4$. There

[4] These are facts that a late nineteenth century scientist would be able to find out.

[5] The accepted room temperature lattice parameter for NaCl is 0.56402 nm (Rohrer, 2001), which is well within the standard deviation of the current measurement.

Table 14.6. Lattice parameter computation for NaCl, assuming that the Bravais lattice is *cP*, *cI*, and *cF*. The angles in the first column are derived from the curve-fitting approach in the previous section.

θ (rad)	*cP*		*cI*		*cF*	
Kα_1	(hkl)	a (nm)	(hkl)	a (nm)	(hkl)	a (nm)
0.239	100	0.3254	110	0.4602	111	0.5636
0.277	110	0.3983	200	0.5633	200	0.5633
0.397	111	0.3451	211	0.4880	220	0.5635
0.470	200	0.3402	220	0.4811	311	0.5641
0.493	210	0.3639	310	0.5147	222	0.5638
0.578	211	0.3453	222	0.4884	400	0.5639
0.638	220	0.3658	321	0.4839	331	0.5637
0.657	221	0.3784	400	0.5045	420	0.5640
0.733	300	0.3454	411	0.4884	422	0.5640
0.789	310	0.3432	420	0.4854	511	0.5640
0.883	311	0.3306	332	0.4676	440	0.5640
0.941	222	0.3302	422	0.4669	531	0.5639
0.960	320	0.3390	431	0.4795	442	0.5642
1.043	321	0.3336	440	0.5044	620	0.5639
av. $\pm\sigma$	0.3489 ± 0.0206 nm		0.4912 ± 0.0257 nm		0.5639 ± 0.0002 nm	
# FU/cell	0.9452		2.6371		3.9909	

are hence four formula units of NaCl per unit cell. Since the Bravais lattice is face-centered cubic, this means that for each Na atom, there are three equivalent ones at positions given by the centering vectors, and the same for Cl. To determine the atom positions, we proceed as follows. We assume that there is an Na atom located in the origin; this is a reasonable assumption, since we can place the origin anywhere we want, so why not on an atom position? This means that *all four* Na atoms are now fixed. That leaves the position of Cl to be determined. We must only determine the position of one of the four Cl atoms, since the others are related to the first by the face centering vectors. For now, let us assume that the Cl atom is located at the position (x, y, z).[6]

The structure factor for this unit cell is then given by:

$$F_{hkl} = (f_{\text{Na}} + f_{\text{Cl}}\,\mathrm{e}^{-2\pi\mathrm{i}(hx+ky+lz)})(1 + (-1)^{h+k} + (-1)^{h+l} + (-1)^{k+l}).$$

[6] We know that Cl sits in the center of the unit cell, but we would like to be able to derive that directly from the data!

Note that the structure factor can be rewritten (for reflections with Miller indices of equal parity) as:

$$F_{hkl} = 4(f_{\mathrm{Na}} + f_{\mathrm{Cl}}\,\mathrm{e}^{-\mathrm{i}\phi_{hkl}}),$$

where $\phi_{hkl} = 2\pi(hx + ky + lz)$ is a phase factor. Writing the structure factor in this way reveals clearly the nature of the problem that we have to solve. In order to determine the location of the Cl atom (or, more generally, the entire crystal structure) we must determine the phases ϕ_{hkl}. Once we know the phases, we can simply solve a system of linear equations to retrieve the coordinates (x, y, z). The modulus squared of the structure factor is given by:

$$|F_{hkl}|^2 = 16(f_{\mathrm{Na}}^2 + f_{\mathrm{Cl}}^2 + 2f_{\mathrm{Na}}f_{\mathrm{Cl}}\cos\phi_{hkl}).$$

One way to solve this problem would be to vary the positions (x, y, z) all over the unit cell, and to compute the agreement indices R_{p} and R_{wp}, as we have done in Section 14.2.2; where these indices reach their minimum values is where the Cl atom should be located. Alternatively, let us simplify the problem a little by assuming that the Cl atom must be located in one of the interstitial sites in the Na *fcc* lattice. There are two such sites, the tetrahedral site at $(1/4, 1/4, 1/4)$ and the octahedral site at $(1/2, 1/2, 1/2)$. The corresponding values of the phase factor are:

$$\text{octahedral} \quad \phi^{\mathrm{o}}_{hkl} = \pi(h + k + l);$$
$$\text{tetrahedral} \quad \phi^{\mathrm{t}}_{hkl} = \frac{\pi}{2}(h + k + l).$$

If we consider the first two reflections of the powder pattern, 111 and 200, then we have:

$$\phi^{\mathrm{o}}_{111} = 3\pi; \qquad \phi^{\mathrm{t}}_{111} = \frac{3\pi}{2};$$
$$\phi^{\mathrm{o}}_{200} = 2\pi; \qquad \phi^{\mathrm{t}}_{200} = \pi.$$

Substitution in the structure factor expression results in:

$$|F^{\mathrm{o}}_{111}|^2 = 16(f_{\mathrm{Na}} - f_{\mathrm{Cl}})^2;$$
$$|F^{\mathrm{o}}_{200}|^2 = 16(f_{\mathrm{Na}} + f_{\mathrm{Cl}})^2,$$

for the octahedral site, and:

$$|F^{\mathrm{t}}_{111}|^2 = 16(f_{\mathrm{Na}}^2 + f_{\mathrm{Cl}}^2);$$
$$|F^{\mathrm{t}}_{200}|^2 = 16(f_{\mathrm{Na}} - f_{\mathrm{Cl}})^2,$$

In the tetrahedral case, we find that the 111 reflection should have a larger intensity than the 200 peak (even after correcting for the multiplicity and the Lorentz polarization factor), which does not agree with the experiment; the octahedral position does give good agreement with the experimental powder pattern, so that we conclude that the Cl atom is located in the octahedral interstitial positions of the Na *fcc* lattice. This concludes the structure determination for NaCl.

14.2.5 *General comments about crystal structure determination

In the previous sections, we have discussed the structure determination of two very simple cubic materials, a Ni-Cu alloy and NaCl. In Fig. 1.7, we also show the powder pattern for sucrose, with chemical formula $C_{12}H_{22}O_{11}$. This is a much more complex pattern! First of all, we note that there are many reflections, even for small values of 2θ. This most likely means that the structure has a large unit cell with a low symmetry. In fact, the structure of sucrose can be considered as an example of a *molecular structure*, i.e., a structure in which the individual sucrose molecules remain identifiable. We will take a closer look at a large variety of molecular solids in Chapter 25, the final chapter of this book. For now, we start from the structure of the sucrose molecule.

Sucrose is a *disaccharide*, i.e., a sugar molecule made from two monosaccharides, *fructose* and *glucose*, through a condensation reaction. Glucose and fructose are simple sugars with chemical formula $C_6H_{12}O_6$, but with a different structure.[7] In glucose, a six-membered ring contains five carbon atoms and one oxygen atom, with a single CH_2OH group; fructose has a five-membered ring with four carbon atoms and one oxygen atom, and two CH_2OH groups. When both monosaccharides react, they form sucrose and release one water molecule. The structure of the sucrose molecule is shown in Fig. 14.11(a); the two component molecules are clearly identifiable.

The sucrose crystal structure is monoclinic, with lattice parameters (in nm) $\{1.08633, 0.8705, 0.77585, 90.0°, 102.945°, 90.0°\}$. The structure is primitive, and for each atom at position (x, y, z), there is an equivalent atom at position $(-x, y+1/2, -z)$, i.e., all atoms occupy the general $2a$ positions of the space group $\mathbf{P2_1}$ (C_2^2). This space group has as its only symmetry element (other than the identity), a two-fold screw axis 2_1 parallel to the [010] direction. There are two formula units per cell, for a total of 90 atoms per unit cell. The fractional atom coordinates are listed in Table 14.7, and the structure is shown projected along two directions in Fig. 14.11(b) and (c).[8]

[7] Sugars belong to the larger category of *carbohydrate* molecules with general composition $(CH_2O)_n$.

[8] The atom coordinates for the sucrose structure were extracted from a data file located at the URL: http://www.ccp14.ac.uk/ccp/ccp14/ftp-mirror/platon-spek/pub/special/sucrose.cif

Table 14.7. Fractional atom coordinates for the sucrose crystal structure.

atom	x	y	z	atom	x	y	z
C1	0.70039	0.85792	0.51513	H1	0.66530	0.74510	0.46120
C2	0.68747	0.97474	0.36400	H2	0.58840	0.96930	0.28830
C3	0.71455	0.13673	0.43553	H3	0.81290	0.14480	0.51030
C4	0.62596	0.17095	0.55802	H4	0.52830	0.16810	0.47820
C5	0.64075	0.05107	0.70471	H5	0.73620	0.06130	0.79070
C6	0.54246	0.07083	0.81545	H6	0.54690	0.18730	0.86560
C7	0.89699	0.63110	0.45620	H7	0.94800	0.52240	0.47840
C8	0.87554	0.69262	0.63105	H8	0.80530	0.60910	0.36700
C9	0.99282	0.69075	0.78515	H9	0.72810	0.87230	0.15340
C10	0.93522	0.66653	0.94524	H10	0.04950	0.58840	0.77100
C11	0.82365	0.56133	0.87136	H11	0.76820	0.26600	0.25750
C12	0.71073	0.58194	0.95332	H12	0.90160	0.77720	0.98450
O1	0.82857	0.84630	0.60835	H13	0.65930	0.39130	0.56750
O2	0.77046	0.93550	0.25234	H14	0.85350	0.44010	0.88680
O3	0.69199	0.24770	0.29720	H15	0.56280	0.99260	0.92810
O4	0.65120	0.31410	0.64370	H16	0.73490	0.53990	0.08870
O5	0.62281	0.89878	0.63136	H17	0.39850	0.93830	0.71340
O6	0.41856	0.04530	0.71380	H18	0.91550	0.82470	0.34580
O7	0.96983	0.73550	0.37881	H19	0.03100	0.90890	0.82390
O8	0.78795	0.59445	0.68428	H20	0.01670	0.65380	0.19610
O9	0.07367	0.81776	0.79548	H21	0.65230	0.77800	0.83970
O10	0.02123	0.59734	0.08904	H22	0.63280	0.51190	0.88050
O11	0.67356	0.73800	0.95965				

Note that the structure determination for this compound requires the locations of 45 atoms ($3 \times 45 = 135$ coordinates) to be determined. That means that there are 135 phase factors ϕ_{hkl} to be determined, one for each atom!

Since we already know the structure parameters, it is relatively straightforward to compute the powder pattern, using the same approach as in earlier sections. The resulting pattern is shown in Fig. 14.12. The top profile is the computed one, the bottom shows the experimental pattern. The most intense reflections are labeled with the corresponding Miller indices. Note that the overall agreement is reasonably good, although the relative intensities are not always the same, indicating a possible preferential orientation in the powder sample. It is interesting to note that this complex structure was first solved in the 1950s (e.g., Beevers *et al.*, 1952), well before the advent of modern computers!

Structure determination is, in the most general sense, a phase problem; i.e., if the phases of all the terms in the structure factor are known, then all atom positions are known and the structure is considered to be solved (note that the atom types follow from the amplitudes in the structure factor). From the experimental observations, one can only derive the modulus squared of the structure factor, $|F_{hkl}|^2$ for each reflection. While the phase, ϕ_{hkl}, is unknown,

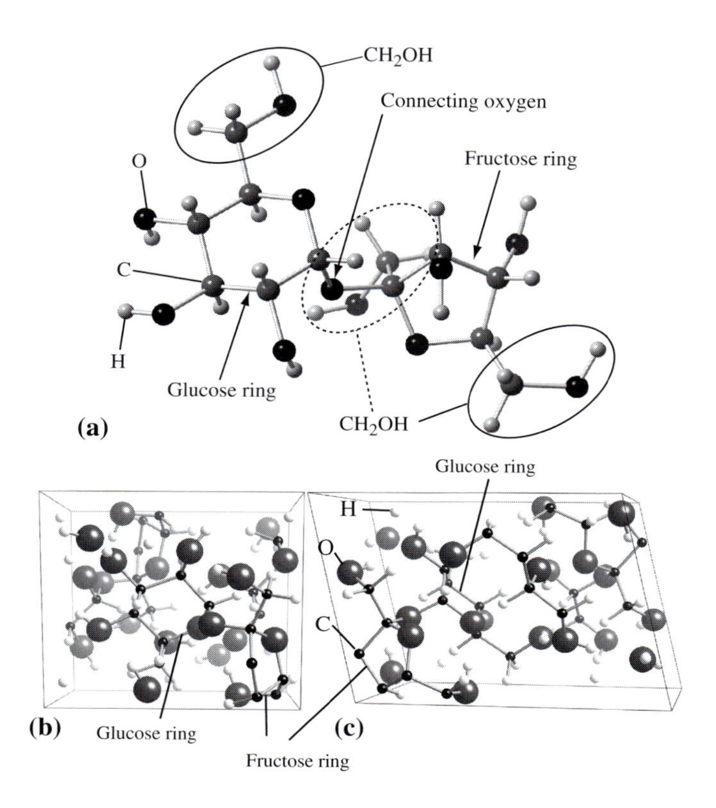

Fig. 14.11. (a) Structure of the sucrose molecule; the individual monosaccharides are indicated as well as the three CH_2OH groups (the dashed ellipse indicates a group *behind* the fructose ring). The crystal structure of sucrose is viewed near the $[\bar{1}00]$ (b) and $[1\bar{8}0]$ (c) directions. The glucose and fructose rings are clearly identifiable.

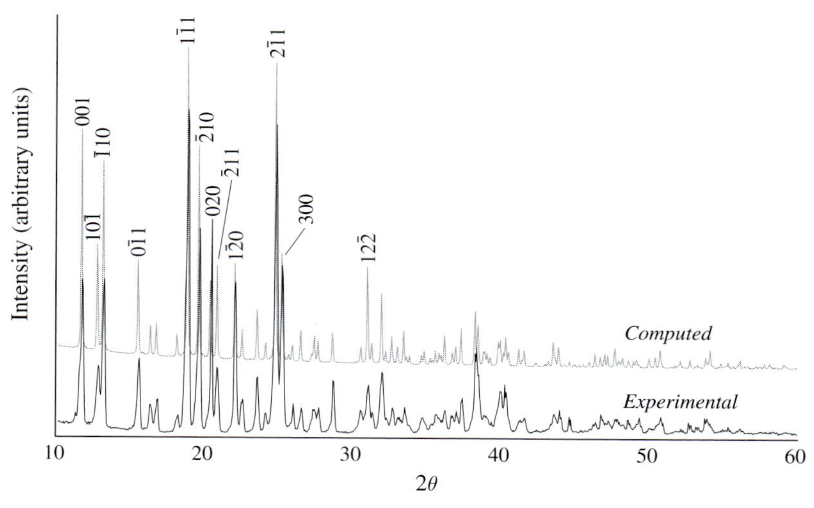

Fig. 14.12. Simulated and experimental powder diffraction patterns for sucrose. A few of the most intense reflections are labeled.

it is still possible to derive useful information from the intensities alone. To understand how this works, we must first realize that X-ray diffraction analysis of a crystal structure is essentially a *Fourier analysis* of the electronic charge density. In Fourier analysis, a 3-D function, $f(\mathbf{r})$, is *decomposed* into a superposition of sinusoidal waves by means of the *Fourier transform* operation:

$$f(\mathbf{k}) = \mathcal{F}[f(\mathbf{r})] = \int \mathrm{d}^3\mathbf{r}\, f(\mathbf{r})\, \mathrm{e}^{2\pi \mathrm{i}\mathbf{k}\cdot\mathbf{r}}; \tag{14.4}$$

the function $f(\mathbf{k})$ is a function in Fourier space (note that Fourier space and reciprocal space are identical), and is said to be the Fourier space representation of $f(\mathbf{r})$. The *inverse Fourier transform* is defined by:

$$f(\mathbf{r}) = \mathcal{F}^{-1}[f(\mathbf{k})] = \int \mathrm{d}^3\mathbf{k}\, f(\mathbf{k})\, \mathrm{e}^{-2\pi \mathrm{i}\mathbf{r}\cdot\mathbf{k}}; \tag{14.5}$$

the functions $f(\mathbf{r})$ and $f(\mathbf{k})$ form a Fourier transform pair. The value of the function $f(\mathbf{k})$ for a particular $\mathbf{k}$ represents how much the plane wave $\exp(-2\pi \mathrm{i}\mathbf{k}\cdot\mathbf{r})$ contributes to the function $f(\mathbf{r})$.

We can use the concept of the Fourier transform to describe the diffraction process. Consider the charge density $\rho(\mathbf{r})$; the Fourier transform of the charge density is given by:

$$\rho(\mathbf{k}) = \mathcal{F}[\rho(\mathbf{r})] \quad \text{or} \quad \rho(\mathbf{r}) = \mathcal{F}^{-1}[\rho(\mathbf{k})].$$

The charge density can then be written as:

$$\rho(\mathbf{r}) = \frac{1}{V}\sum_{\mathbf{g}} F_{\mathbf{g}}\mathrm{e}^{-2\pi \mathrm{i}\mathbf{g}\cdot\mathbf{r}}; \tag{14.6}$$

V is the volume of the unit cell. In other words, the charge density consists of a sum of plane waves, one for each set of planes in the crystal, and each plane wave contribution is equal to the structure factor for that set of planes. In explicit coordinate notation we have:

$$\rho(x, y, z) = \frac{1}{V}\sum_{h=-\infty}^{+\infty}\sum_{k=-\infty}^{+\infty}\sum_{l=-\infty}^{+\infty} F_{hkl}\mathrm{e}^{-2\pi \mathrm{i}(hx+ky+lz)}.$$

The structure factor is defined as before:

$$F_{\mathbf{g}} = \sum_{j=1}^{N} f_j \mathrm{e}^{2\pi \mathrm{i}\mathbf{g}\cdot\mathbf{r}_j} = \mathcal{F}[\rho(\mathbf{r})].$$

If we know both the amplitude and phase of all structure factors $F_{\mathbf{g}}$, then Eq. 14.6 allows us to compute the complete electron density $\rho(\mathbf{r})$. Knowledge of the electron density at each point in the unit cell then reveals where each atom is located, since the density shows maxima at the atom locations. This means that structure determination is equivalent to computing the inverse Fourier transform of the structure factors.

From an experimental point of view, we know that we are limited by the fact that we can only determine the modulus, $|F_{\mathbf{g}}|$, of the structure factor and not its phase. This means that we cannot, in general, perform the inverse Fourier transform. Based on these moduli, we can define a new function, $P(\mathbf{r})$, as follows:

$$P(\mathbf{r}) = \frac{1}{V} \sum_{\mathbf{g}} |F_{\mathbf{g}}|^2 \mathrm{e}^{-2\pi i \mathbf{g}\cdot\mathbf{r}}. \tag{14.7}$$

This function is known as the *Patterson function*, and it can be computed directly from the experimental data, without the need for any of the phases (Patterson, 1934). It can be shown that the Patterson function has maxima at locations **r** corresponding to all interatomic vectors. The height of these maxima is related to the atomic scattering factors. So, while the Patterson function does not directly provide the structure solution, it is a very useful tool to narrow down the search for the correct structure, since it allows for the determination of all the interatomic vectors in the structure.

A significant research effort over most of the past hundred years has resulted in several methods to determine the phases of the structure factors; none of these methods provides the solution to the general phase problem, and all these methods need some kind of "first guess" for the phases, after which they will, often iteratively, solve for the correct phases. The techniques used for solving the phase problem for a particular structure can be quite involved, and we refer the reader to the following texts for more detailed information: Giacovazzo (2002a), Drenth (2002), Rhodes (2000), Warren (1990), Glusker and Trueblood (1985) and many others.

Solving crystal structures, in particular those of complicated proteins which crystallize in huge unit cells with thousands of atoms, is quite an involved task, and is usually carried out in specialized, dedicated laboratories equipped with three-or four-circle diffractometers (i.e., the sample can be oriented with three or four angular degrees of freedom, so that every possible orientation can be obtained). The reader may consult the journal *Acta Crystallographica* to find many examples of modern structure determinations.

To discuss these specialized structure solution methods in detail would lead us too far away from the main purpose of this book: to describe the structure of materials. In the following chapters (15 through 25), we will provide many examples of important and interesting crystal structures. On occasion, we will refer back to this chapter when we show diffraction patterns and such. The reader should not forget that every single crystal structure discussed in this book was, at one point in time, the subject of an experimental study by means of X-ray, neutron, or electron diffraction (sometimes even a combination of two or more of these techniques). The number of known (solved) crystal structures is very large; more than a quarter of a million different structures have been solved, and that number is increasing at a steady rate. The small selection of structures in the second half of this book reflects to some extent the interests of the authors, but is kept sufficiently general so that most readers will find something of interest.

14.3 Historical notes

In this section, we reproduce in its entirety one of the seminal papers of X-ray crystallography, a 1913 paper by W. H. and W. L. Bragg on the structure determination of diamond. While there are several early papers by father and son Bragg that we could have selected, we choose this one because it is the first example of the use of systematic absences due to symmetry elements (what we now call the *diamond glide planes*). Other important papers include: Bragg (1912), Bragg and Bragg (1913), Bragg (1914, 1915a,b, 1920, 1929, 1930).

The original citation is: W. H. Bragg and W. L. Bragg (The Structure of the Diamond) *Proc. R. Soc. A*, **89**, pp. 277–291 (1913), and the article is reproduced with permission from The Royal Society.

The Structure of the Diamond.

By W. H. Bragg, M.A., F.R.S., Cavendish Professor of Physics in the University of Leeds, and W. L. Bragg, B.A., Trinity College, Cambridge.

(Received July 30, 1913.)

There are two distinct methods by which the X-rays may be made to help to a determination of crystal structure. The first is based on the Laue photograph and implies the reference of each spot on the photograph to its proper reflecting plane within the crystal. It then yields information as to the positions of these planes and the relative numbers of atoms which they contain. The X-rays used are the heterogeneous rays which issue from certain bulbs, for example, from the commonly used bulb which contains a platinum anticathode.

The second method is based on the fact that homogeneous X-rays of wave-length λ are reflected from a set of parallel and similar crystal planes at an angle θ (and no other angle) when the relation $n\lambda = 2d\sin\theta$ is fulfilled. Here d is the distance between the successive planes, θ is the glancing angle which the incident and reflected rays make with the planes, and n is a whole number which in practice so far ranges from one to five. In this method the X-rays used are those homogeneous beams which issue in considerable intensity from some X-ray bulbs, and are characteristic radiations of the metal of the anticathode. Platinum, for example, emits several such beams in addition to the heterogeneous radiation already mentioned. A bulb having a rhodium anticathode, which was constructed in order to obtain a radiation having about half the wave-length of the platinum characteristic rays, has been found to give a very strong homogeneous radiation consisting of one main beam of wave-length $0{\cdot}607 \times 10^{-8}$ cm.,*, and a much less intense beam of wave-length $0{\cdot}533 \times 10^{-8}$ cm. It gives relatively little heterogeneous radiation. Its spectrum, as given by the (100) planes of rock-salt, is shown in fig. 1. It is very convenient for the application of the second method. Bulbs having nickel, tungsten, or iridium anticathodes have not so far been found convenient; the former two because their homogeneous radiations are relatively weak, the last because it is of much the same

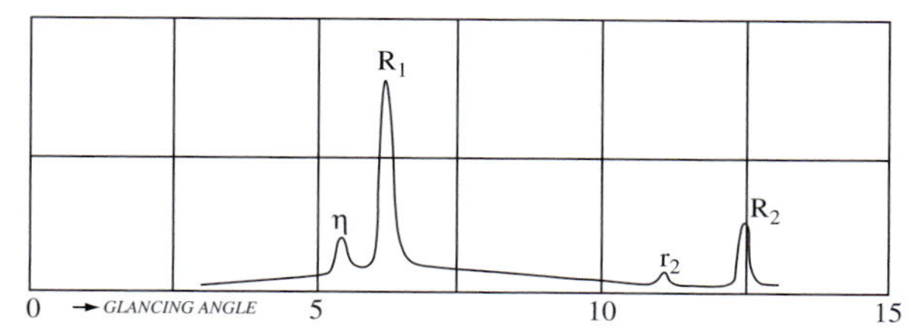

FIG. 1.—Spectra of rhodium rays : 100 planes of rock-salt.

wave-length as the heterogeneous rays which the bulb emits, while it is well to have the two sets of rays quite distinct. The platinum homogeneous rays are of lengths somewhat greater than the average wave-length of the general heterogeneous radiation; the series of homogeneous iridium rays are very like the series of platinum rays raised one octave higher. For convenience, the two methods may be called the method of the Laue photograph, or, briefly, the photographic method, and the reflection method. The former requires heterogeneous rays, the latter homogeneous. The two methods throw light upon the subject from very different points and are mutually helpful.

The present paper is confined almost entirely to an account of the application of the two methods to an analysis of the structure of the diamond.

The diamond is a crystal which attracts investigation by the two new methods, because in the first place it contains only one kind of atom, and in the second its crystallographic properties indicate a fairly simple structure. We will consider, in the first place, the evidence given by the reflection method.

The diagram of fig. 2 shows the spectrum of the rhodium rays thrown by the (111) face, the natural cleavage face of the diamond. The method of obtaining such diagrams, and their interpretation, are given in a preceding

* This value is deduced from the positions of the spectra of the rhodium rays in the (100) planes of rock-salt on the assumption that the structure of rock-salt is as recently described (see preceding paper).

paper.* The two peaks marked R_1, r_1 constitute the first order spectrum of the rhodium rays, and the angles at which they occur are of importance in what follows. It is also a material point that there is no second order spectrum. The third is shown at R_3, r_3; the strong line of the fourth order is at R_4, and of the fifth at R_5.

The first deduction to be made is to be derived from the quantitative measurements of the angle of reflection. The sines of the glancing angles

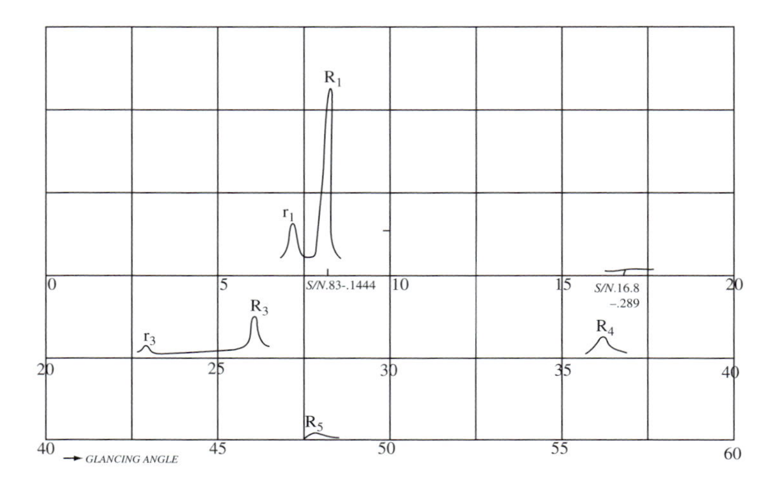

Fig. 2.—Spectra of rhodium rays: 111 planes of diamond.

for R_1, R_3, R_4, R_5 are (after very slight correction for errors of setting) 0·1456, 0·4425, 0·5941, 0·7449. Dividing these by 1, 3, 4, 5 respectively, we obtain 0·1456, 0·1475, 0·1485, 0·1490. These are not exactly equal, as they might be expected to be, but increase for the larger angles and tend to a maximum. The effect is due to reasons of geometry arising from the relatively high transparency of the diamond for X-rays, and the consequent indefiniteness of the point at which reflection takes place. The true value is the maximum to which the series tends, and may with sufficient accuracy be taken as 0·1495. In order to keep the main argument clear, the consideration of this point is omitted.

We can now find the distance between successive (111) planes.

We have

$$\lambda = 2d \sin\theta, \qquad 0{\cdot}607\times10^{-8} = 2d\times0{\cdot}1495, \qquad d = 2{\cdot}03\times10^{-8}.$$

The structure of the cubic crystals which have so far been investigated by

* 'Roy. Soc. Proc.,' vol. 88, p. 428.

these methods may be considered as derived from the face-centred lattice (fig. 3): that is to say, the centres which are effective in causing the reflection of the X-rays are placed one at each corner and one in the middle of each face of the cubical element of volume. This amounts to assigning

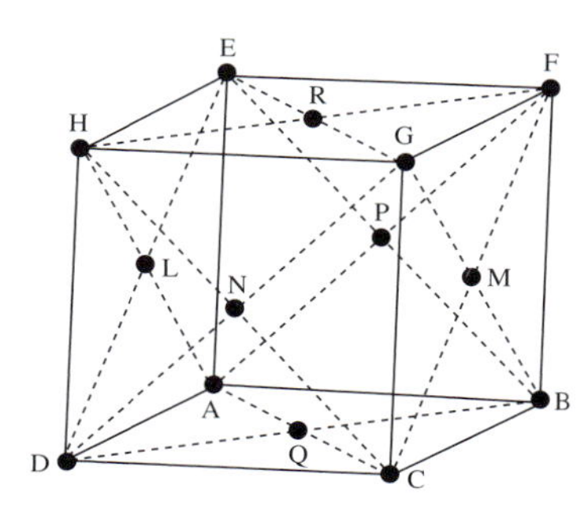

Fig. 3.

four molecules to each such cube, for in general one atom in each molecule is so much more effective than the rest that its placing determines the structure from our point of view. There are four, because the eight atoms at the corners of the cube only count as one, each of them belonging equally to eight cubes, and the six atoms in the centres of the faces only count as three, each of them belonging equally to two cubes. The characteristics of the reflection are then as follows:—

Let ABCDEFGH be the cubical element. There are effective centres at all the corners and at L, M, N, P, Q, R, the middle points of the faces. The edge of the cube being denoted by $2a$, the reflecting planes which are parallel to a cube face, called generally the (100) planes, are spaced regularly, the distance from plane to plane being a. All the planes contain equal numbers of centres.

The (110) planes, of which the plane through ACGE is a type, are regularly spaced at a distance $a/\sqrt{2}$, and also are all equally strewn with effective centres.

The (111) planes, of which the planes through EDB, HCF are types, are regularly spaced at a distance $2a/\sqrt{3}$, and again are all similar to each other.

In what may for the present be called the normal case, any one of these sets of planes gives a series of spectra which diminish rapidly in intensity as we proceed from lower to higher orders, as, for example, the spectra of the rhodium rays given by the (100) planes of rock-salt. (Fig. 1 shows the spectra of the first two orders.)

The relative spacings of the spectra given by these three sets of planes are shown in fig. 4. Spectra of the (100) planes being supposed to occur at values of $\sin\theta$ proportional to $1, 2, 3, \ldots$, it follows from the above argument that the (110) planes will give spectra at 1·414, 2·828, 4·242, ..., and the (111) planes at 0·866, 1·732, 2·598 ....

The position of the first spectrum of the (111) planes (fig. 4) is a peculiarity of the face-centred lattice. If the effective centres were at the corners only

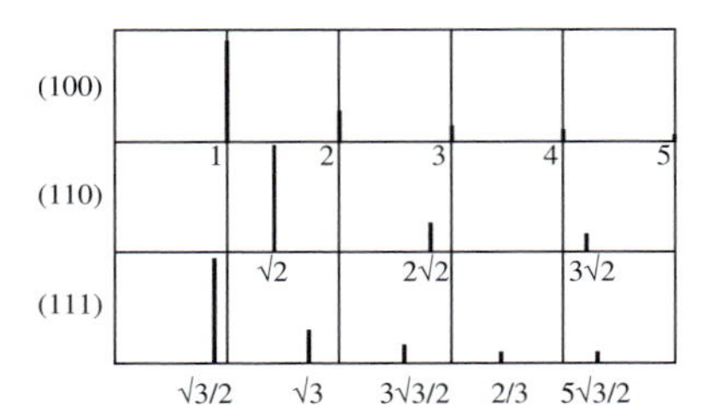

Fig. 4.—Spectra of face-centred lattice.

of a cube whose length of side was a, the spacings of the three sets of planes would be a, $a/\sqrt{2}$, and $a/\sqrt{3}$, and the three sets of spectra would occur at 1, 2, 3; $\sqrt{2}$, $2\sqrt{2}$, $3\sqrt{2}$; $\sqrt{3}$, $2\sqrt{3}$, $3\sqrt{3}$.

The cubical crystals which we have so far examined give results which resemble the diagram of fig. 4 more or less closely. Individual cases depart so little from the type of the diagram that the face-centred lattice may be taken as the basis of their structure and the departures considered to reveal their separate divergencies from the standard. For convenience of description we will speak of the first, second, third spectra of the (100) or (111) planes and so on, with reference to fig. 4. We may then, for example, describe the peculiarity of the rock-salt (111) spectrum* by saying that the first order spectrum is weak and the second strong. The interpretation (*loc. cit.*) is that the sodium atoms are to be put at the centres of the edges of the cubic element of volume, and the chlorine atoms at the corners and in the middle of each face or *vice versâ*: for then the face-centred lattice (cube edge $2a$) is brought half way to being the simple cubic lattice (edge a) having an effective centre at every corner. The first (111) spectrum tends to disappear, the second to increase in importance. In the case of potassium chloride, the atoms are all of equal weight and the change is complete: the first order spectrum of the (111) planes disappears entirely. In zincblende or iron pyrites one atom is so much more effective than the other that the diagram of spectra is much more nearly characteristic of the face-centred

* See preceding paper.

lattice: at least so far as regards the spectra of the lower orders. We hope to deal with these cases later.

Let us now consider the case of the diamond. The spectrum given by the (111) planes is shown in some detail in fig. 2. It should be stated that the ordinates represent the gross currents observed; nothing has been subtracted for natural leak, scattered radiation, and so forth.

We first use the angular measurements to enable us to determine the number of carbon atoms in the elementary cube of side $2a$. Let us assume provisionally that there are four carbon atoms to each cube, making the face-centred lattice. The density of the diamond is 3·51, and the weight of each atom is 12 times the weight of each hydrogen atom or $12 \times 1{\cdot}64 \times 10^{-24}$.

The volume of the cube is therefore

$$\frac{4 \times 12 \times 1{\cdot}64 \times 10^{-24}}{3{\cdot}51} = 22{\cdot}4 \times 10^{-24}.$$

The length of each edge (*i.e.* $2a$) will then be

$$\sqrt[3]{(22{\cdot}4 \times 10^{-24})} = 2{\cdot}82 \times 10^{-8}.$$

The distance between consecutive (111) planes

$$= 2a/\sqrt{3} = 1{\cdot}63 \times 10^{-8}.$$

Now we have found experimentally that the right value is $2{\cdot}03 \times 10^{-8}$. These two numbers are very nearly in the ratio of $1 : \sqrt[3]{2}$. It is clear that we must put eight, not four, carbon atoms in the elementary cube; we then obtain $2a/\sqrt{3} = 2{\cdot}05 \times 10^{-8}$, and this close agreement with the experimental value suggests that we are proceeding in the right way. The value of $2a$ is $3{\cdot}55 \times 10^{-8}$.

We have therefore four carbon atoms which we are to assign to the elementary cube in such a way that we do not interfere with the characteristics of the face-centred lattice.

It is here that the absence of the second order spectrum gives us help. The interpretation of this phenomenon is that in addition to the planes spaced at a distance apart $2{\cdot}03 \times 10^{-8}$ there are other like planes dividing the distances between the first set in the ratio 1 : 3. In fact there must be parallel and similar planes as in fig. 5, so spaced that $AA' = A'B/3$, and so on. For if waves fall at a glancing angle θ on the system ABC, and are reflected in a second order spectrum we have $2\lambda = 2\,AB \sin\theta$. The planes A′B′C′ reflect an exactly similar radiation which is just out of step with the first, for the difference of phase of waves reflected from A and B is 2λ, and therefore the difference of phase of waves reflected from A and A′ is $\lambda/2$. Consequently the four atoms which we have

A A′ B B′ C C′

Fig. 5.

at our disposal are to make new (111) planes parallel to the old and related to them as A′B′C′ are to ABC. When we consider where they are to go we are helped by the fact that being four in number they should go to places which are to be found in the cubes in multiples of four. The simplest plan is to put them in the centres of four of the eight smaller cubes into which the main cube can be divided. We then find that this gives the right spacing because the perpendicular from each such centre on the two (111) planes which lie on either side of it are respectively $a/2\sqrt{3}$ and $\frac{1}{2}(a\sqrt{3})$, where a is the length of the side of one of the eight smaller cubes. For symmetry it is necessary to place them at four centres of smaller cubes which touch each other along edges only: *e.g.* of cubes which lie in the A, C, H and F corners of the large cube. If this is done in the same way for all cubes like the one taken as unit it may be seen on examination that we arrive at a disposition of atoms which has the following characteristics:—

(1) They are arranged similarly in parallel planes spaced alternately at distances $a/2\sqrt{3}$ and $a\sqrt{3}/2$, or in the case of the diamond $0{\cdot}508 \times 10^{-8}$ and $1{\cdot}522 \times 10^{-8}$ cm.: the sum of these being the distance $2{\cdot}03 \times 10^{-8}$ which we have already arrived at.

(2) The density has the right value.

(3) There is no second order spectrum in the reflection from (111) planes.

It is not very easy to picture these dispositions in space. But we have come to a point where we may readjust our methods of defining the positions of the atoms as we have now placed them, and arrive at a very simple result indeed. Every carbon atom, as may be seen from fig. 5, has four neighbours at distances from it equal to $a\sqrt{3}/2 = 1{\cdot}522 \times 10^{-8}$ cm., oriented with respect to it in directions which are parallel to the four diagonals of the cube. For instance, the atom at the centre of the small cube *Abcdefgh*, fig. 6, is related in this way to the four atoms which lie at corners of that cube (A, *c*, *f*, *h*), the atom at the centre of the face ABFE is related in the same way to the atoms at the centres (P, Q, R, S) of four small cubes, and so on for every other atom. We may take away all the structure of cubes and rectangular axes, and leave only a design into which no elements enter but one length and four directions equally inclined to each other. The characteristics of the design may be realised from a consideration of the accompanying photographs (figs. 7 and 8) of a model, taken from different points of view. The very simplicity of the result suggests that we have come to a right conclusion.

The appearance of the model when viewed at right angles to a cube diagonal is shown in fig. 7. The (111) planes are seen on edge, and the

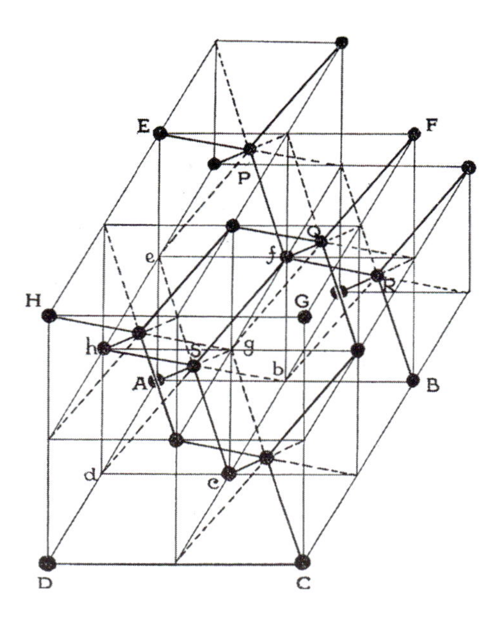

Fig. 6.

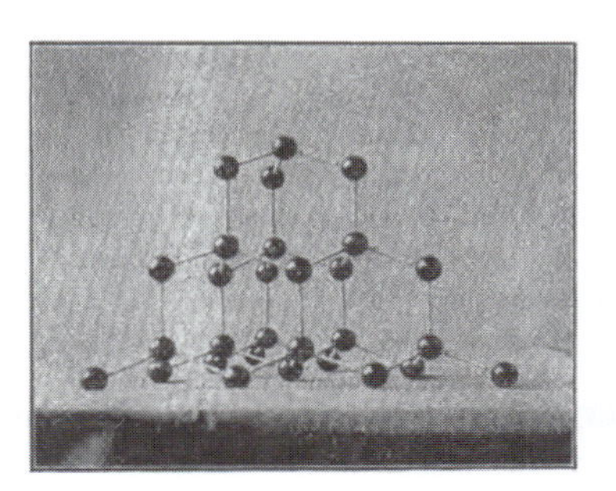

Fig. 7.—View perpendicular to a (111) axis.

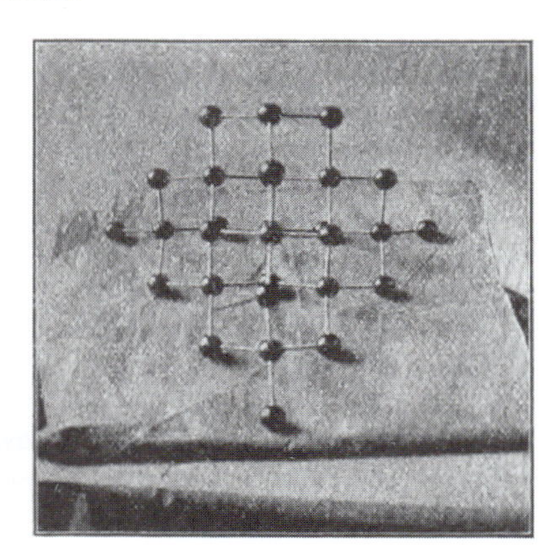

Fig. 8.—The (110) planes are vertical and horizontal.

1 : 3 spacing is obvious. The union of every carbon atom to four neighbours in a perfectly symmetrical way might be expected in view of the persistent tetravalency of carbon. The linking of six carbon atoms into a ring is also an obvious feature of the structure. But it would not be right to lay much stress on these facts at present, since other crystals which do not contain carbon atoms possess, apparently, a similar structure.

We may now proceed to test the result which we have reached by examining the spectra reflected by the other sets of planes. One of the diamonds which we used consisted of a slip which had cleavage planes as surfaces; its surface was about 5 mm. each way and its thickness 0·8 mm. By means of a Laue photograph, to be described later, it was possible to determine the orientation of its axes and so to mount it in the X-ray spectrometer as to give reflection from the (110) or the (100) planes as desired.

As regards the former there should be no special features, for the four carbon atoms which we placed at the centres of four of the eight smaller cubes all now lie in (110) planes. The latter are equally spaced and all alike, the space distance being $a/\sqrt{2}$ or $1{\cdot}25 \times 10^{-8}$. The first glancing angle at which reflection occurs is, therefore, $\sin^{-1}\frac{0{\cdot}607 \times 10^{-8}}{2{\cdot}5 \times 10^{-8}} = 14{\cdot}15°$. The experimental value was 14·35°. The spectra of higher orders occurred at 29·3° and 47·2°. The sines of these three angles are 0·2478, 0·4894, and 0·7325, or nearly as 1 : 2 : 3. Great precision was not attempted; to attain it would have been needlessly troublesome. The intensity of the different orders fell off in the usual way.

On the other hand, the (100) spectrum might be expected to show certain peculiarities. By placing four atoms at the centres of the four small cubes we have, in fact, interleaved the 100 planes, as it were: and these now consist of similar planes regularly spaced at a distance $a/2$ or $0{\cdot}885 \times 10^{-8}$. The first spectrum should therefore occur at an angle $\sin^{-1}\frac{0{\cdot}607 \times 10^{-8}}{1{\cdot}77 \times 10^{-8}}$ $= \sin^{-1} 0{\cdot}343 = 20{\cdot}0$. Using the language already explained, we may say that the first (100) spectrum has disappeared, and, indeed, all the spectra of odd order. Spectra were actually found at 20·3° and 43·8°: the sines of these angles being 0·3469 and 0·6921, the latter being naturally much less intense than the former. A careful search in the neighbourhood of 10° showed that there was no reflection at all at that angle.

The results for all three spectra are shown diagrammatically in fig. 9, which should be compared with fig. 4.

It is instructive to compare the reflection effects of the diamond with those of zincblende. Our results seem to show that it is built up in exactly the same way, except that the (111) planes contain alternately zinc atoms only and sulphur atoms only. If the zinc atoms are placed at each corner of the cube and at the centre of each face, the sulphur atoms lie at four of the eight centres of the smaller cubes. The (100) planes, like the (111) planes, contain

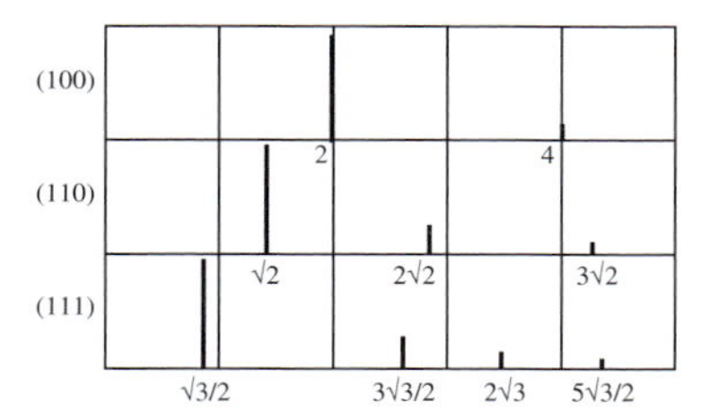

FIG. 9.—Spectra of diamond.

alternately zinc and sulphur atoms. These alternations of constitution modify the forms of the various spectra, so that they lie between the forms of the space-centred lattice (fig. 4) and the forms of the diamond (fig. 9). The first (100) spectrum is not entirely absent but is much smaller than the second, and in the same way the second (111) spectrum, though it is to be seen, is smaller even than the third. The scheme of the zincblende spectra is shown in fig. 10. Their actual positions agree perfectly with those which

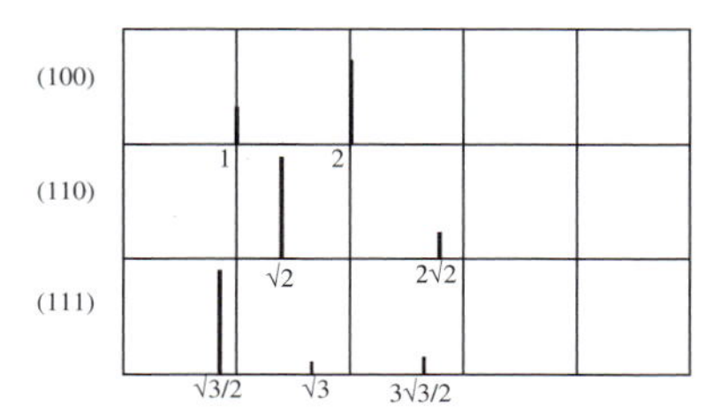

FIG. 10.—Spectra of zincblende.

can be calculated from a knowledge of the density of the crystal, the weight of the ZnS molecule, and the wave-lengths employed. In consequence of the alternation of zinc and sulphur planes at unequal spacings along the (111) axis, the crystal ceases to be symmetrical about a plane perpendicular to that axis. It becomes hemihedral, and acquires polarity.

We now go on to consider the Laue photograph of the diamond. A photograph taken with a section of diamond cut parallel to the cleavage plane (111) is shown in fig. 11. The experimental arrangement was similar to the original arrangement of Laue, the distance from diamond to photographic plate being 1·80 cm., and the time of exposure four hours. A test photograph was taken first, which made it possible to calculate the exact orientation to be given to the diamond in order that the incident X-rays might be truly parallel to a trigonal axis. The symmetry of fig. 11 shows

Fig. 11.

that a close approximation to this orientation has been obtained. The X-ray bulb had a platinum anticathode.

In fig. 12 is given the stereographic projection of this pattern.* The spots of the photograph are represented in the diagram by dots of corresponding magnitude, and several circles, each passing through the spots reflected by the planes of one zone, are drawn. The indices placed next the spots are the Millerian indices of the planes which reflect these spots, the planes being referred to three equal axes making 60° with each other as in the case of the examples zincblende and fluorspar given in the above paper. Imagining a

* See preceding paper.

cube with one corner at the diamond and the long diagonal of the cube parallel to the incident X-rays, the three cube edges would meet the photographic plate at the points marked X, Y, Z. The spot (110) is thus reflected in the cube face, meeting the plate along XY, (110) being the indices of a cube face referred to the axes employed.

It will now be shown that on analysis the photograph appears to be in accordance with the structure which we have assigned to the diamond on the

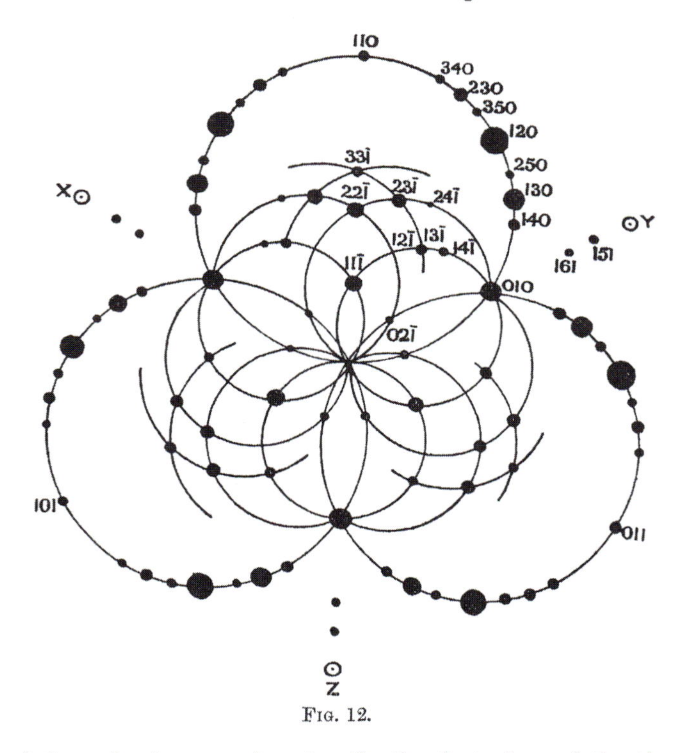

Fig. 12.

result of the reflection experiments. In the first place, of the three cubic space lattices it is evidently that which has points at cube corners and at the centres of the cube faces which is most characteristic of the diffracting system. For our purpose this space lattice is most conveniently referred to three axes which are diagonals of the cube faces meeting in a corner. The co-ordinates of any point of the system may then be written

$$pc, \qquad qc, \qquad rc,$$

where p, q, r are any integers, positive or negative, and c is half the diagonal of the square of edge $2a$.

The indices of the reflecting plane are given for each spot of the photograph, and it will be seen that they could not possibly have a more simple form. If referred to the cubic axes they become much more complex. Along the axes chosen, the interval between successive points of the lattice is the smallest possible, and these axes are very important point-rows of the system. The remarkable series of spots lying on the three circles in the diagram which culminate at the points (110), (101), (011), are due to planes which pass through these point-rows, and this alone is good evidence of the paramount importance of the cube face diagonals as axes.

It is thus clear that a simple analysis of the pattern can be made if the planes are referred to axes of the face-centred cubic lattice. It is also evident, however, that the pattern is more complex than it should be if due to a set of identical points arranged in this lattice, of which examples have been given in a former paper. For instance, there are spots reflected by the planes $(11\bar{1})$, (131), (141), and $(22\bar{1})$, $(02\bar{1})$, and yet none by the plane $(12\bar{1})$ (see diagram, fig. 12). In the case of zincblende and fluorspar no complications of this kind occur, although in these cases the presence of the lighter atoms of sulphur and fluorine must affect somewhat the diffraction pattern given by the lattice arrangement of heavy atoms of zinc and calcium. Yet here, where carbon atoms alone are present, the pattern is not as straightforward as those given by zincblende and fluorspar. We thus come to the conclusion that the carbon atoms are not arranged on a single space lattice.

If the structure assigned to diamond in the former part of this paper is correct, a simple explanation of the diffraction pattern can be arrived at. According to this structure the carbon atoms are not arranged on a space lattice, but they may be regarded as situated at the points of two interpenetrating face-centred space lattices. These lattices are so situated in relation to each other that, calling them A and B, each point of lattice B is surrounded symmetrically by four points of lattice A, arranged tetrahedron-wise and *vice versâ*. This can be seen by reference to the diagram of fig. 6.

It is now clear why the pattern must be referred to the axes of the face-centred lattice, for if the structure is to be regarded as built up of points arranged on the simple cubic lattice, with three equal axes at right angles, no fewer than eight interpenetrating lattices must be used to give all the points.

Consider lattice A referred to the cube face diagonals as axes. Then all the points of that lattice have indices

$$pc, \quad qc, \quad rc,$$

p, q, r being any integers. The relative position of lattice B is arrived at if we imagine lattice A to suffer a translation along the trigonal axis which is the long diagonal both of the elementary parallelepiped and of the cube, the amount of this translation being one-fourth of the long diagonal. Reference to one of the diagrams will make this more clear than any explanation which could be given here. The points of lattice B then have co-ordinates

$$(p+\tfrac{1}{4})\,c, \qquad (q+\tfrac{1}{4})\,c, \qquad (r+\tfrac{1}{4})\,c.$$

The planes of lattice A which have Millerian indices (lmn) are given by

$$lx+my+nz = \mathrm{P}c,$$

where P is any integer. The corresponding planes of lattice B are given by

$$l\,(x-\tfrac{1}{4}c)+m\,(y-\tfrac{1}{4}c)+z\,(n-\tfrac{1}{4}c) = \mathrm{Q}c,$$

or

$$lx+my+nz = \left(\mathrm{Q}+\frac{l+m+n}{4}\right)c.$$

When the (lmn) planes of both lattices are considered together, three cases present themselves :—

(1) When $l+m+n$ is a multiple of four, the planes of lattice B are coincident with those of lattice A, both being given by

$$lx+my+nz = (\text{integer} \times c).$$

An example of this is found in the plane $(1\bar{1}0)$ or (130).

(2) When $l+m+n$ is a multiple of two but not of four, the planes of lattice A are given by

$$lx+my+nz = \mathrm{P}c.$$

Those of lattice B are given by

$$lx+my+nz = (\mathrm{P}+\tfrac{1}{2})\,c,$$

and are thus half-way between the planes of lattice A.

Examples.—Planes such as (110) and $(12\bar{1})$.

(3) When $l+m+n$ is odd, the equations of the two sets of planes are

$$lx+my+nz = \mathrm{P}c,$$

and

$$lx+my+nz = (\mathrm{P}+\tfrac{1}{4}c),$$

or

$$lx+my+nz = (\mathrm{P}-\tfrac{1}{4})\,c,$$

and the planes occur in pairs, in such a way that the two planes of a pair are separated by one-fourth of the distance between the successive pairs.

Examples.—Octahedron faces (100), (010), (001), and (111).

It is now clear wherein lies the difference between planes $(11\bar{1})$ and (131), on the one hand, and $(12\bar{1})$ on the other. The $(12\bar{1})$ planes of the one lattice alone would probably give a strong reflection of a part of the X-ray spectrum in which there was a large amount of energy, but the presence half-way between them of the planes of the other lattice $(1+2-1=2)$ annuls their effect. On the other hand, though the $(13\bar{1})$ and $(11\bar{1})$ planes now occur in pairs, the wave-length reflected from them is the same as that for a single lattice. On looking over the indices of the reflecting planes, it will be seen how large a proportion of them have $l+m+n$ either odd or a multiple of four; in fact, the departure of the pattern from simplicity is just that which would be expected from the nature of the point system, which differentiates the planes into these three sets.

A more complete analysis of the pattern would be of little interest here because the positions of the reflection peaks afford a much simpler method of analysing the structure. In comparison with the examples given in the former paper, this is a case where the diffraction is caused by a point system as against a space lattice, both a translation and a rotation being necessary to bring the system into self-coincidence. This gives special interest to the photograph.

We have to thank both Prof. S. P. Thompson, F.R.S., and Dr. Hutchinson, of the Mineralogical Laboratory, Cambridge, for their kindness in lending us diamonds which were used in these experiments.

CHAPTER

15 Non-crystallographic point groups

> *"When I am working on a problem I never think about beauty. I only think about how to solve the problem. But when I have finished, if the solution is not beautiful, I know it is wrong."*
>
> Buckminster Fuller (1895–1983)

15.1 Introduction

In Chapter 9, we considered the group of symmetry operations for the quartz crystal. We used this example to define what a group is (in terms of the group axioms). Then we derived the 32 crystallographic point group symmetries. These are the only point groups compatible with translational periodicity. This restriction on crystallographic point groups was important for the development of space groups in Chapter 10. A less restrictive view of point group symmetry may be necessary, however, to understand the structure of many non-traditional materials. In this chapter, we consider examples of *non-crystallographic point groups*.[1]

Non-crystallographic point groups are useful to understand more complicated structures. In *molecular solids*, the Bravais lattice is decorated by molecules (i.e., the unit cell has a *molecular basis*). These solids may possess symmetries not belonging to a crystallographic point group. In decorating the

[1] We cannot enumerate them all because there are an infinite number of them.

Table 15.1. 5 (C_5) group multiplication table. The notation D^n stands for either the symmetry operation 5^n (C_5^n) or for the corresponding transformation matrix $D(5^n\ (C_5^n))$; D^0 is the identity operator/matrix.

5 (C_5)	D^0	D^1	D^2	D^3	D^4
D^0	D^0	D^1	D^2	D^3	D^4
D^1	D^1	D^2	D^3	D^4	D^0
D^2	D^2	D^3	D^4	D^0	D^1
D^3	D^3	D^4	D^0	D^1	D^2
D^4	D^4	D^0	D^1	D^2	D^3

crystal lattice, molecules may lose some of their symmetry elements. Knowledge of the non-crystallographic symmetries can be helpful to understand the properties of molecular solids. In the *Frank–Kasper phases* introduced in Chapter 18, structural motifs include distorted polyhedral units (such as the *icosahedron*) that, in their undistorted forms, have non-crystallographic point group symmetries. A recently discovered solid, known as a *quasicrystal*, has a *quasi-periodic* arrangement of atoms, giving rise to diffraction patterns with non-crystallographic point group symmetries! Quasicrystals are discussed in detail in Chapter 20.

15.2 Example of a non-crystallographic point group symmetry

Consider all the symmetry operations that are written as powers of the 5-fold rotation: 5 (C_5), 5^2 (C_5^2), 5^3 (C_5^3), 5^4 (C_5^4), 5^5 (C_5^5) = 1 (E). These five operations form a point group, labeled **5** (C_5).[2] Since each operator is written as a power of a single operator, 5 (C_5), the resulting group is a *cyclic group* of order 5. The operator 5 (C_5) is the *generator* of the group. Group elements are represented by transformation matrices:

$$D(5^n\ (C_5^n)) = \begin{pmatrix} \cos\frac{2\pi n}{5} & -\sin\frac{2\pi n}{5} & 0 \\ \sin\frac{2\pi n}{5} & \cos\frac{2\pi n}{5} & 0 \\ 0 & 0 & 1 \end{pmatrix} \quad (n = 0\ldots4), \qquad (15.1)$$

for a counterclockwise rotation by $2\pi n/5$ around the z-axis.

If we use the shorthand notation D^n to indicate the symmetry operation 5^n (C_5^n), then we can establish the *group multiplication table* for the **5** (C_5) group, as shown in Table 15.1. Each of the operations has an inverse. For example, the inverse of the 5 (C_5) operation is 5^4 (C_5^4) and vice versa. One can

[2] Once again, we denote point groups by their *Hermann–Mauguin symbols*, with the corresponding *Schönflies notation* in parenthesis.

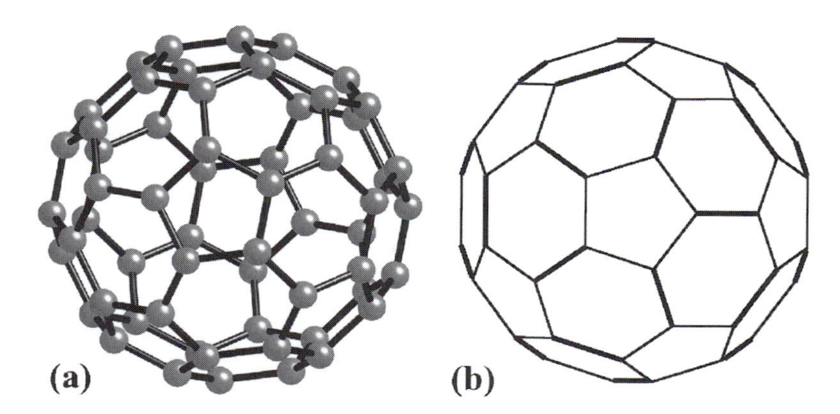

Fig. 15.1. (a) Ball-and-stick model of the C_{60} fullerene showing C atoms at vertices of a truncated icosahedron. (b) Projected structure along a five-fold axis showing single and double bonds.

see that the product 5 (C_5)· 5^4 (C_5^4) is equal to the identity operator, 1 (E), as is the product 5^4 (C_5^4)·5 (C_5). Similarly, the inverse of 5^2 (C_5^2) is 5^3 (C_5^3). The *order* of the **5** (C_5) group is 5.

The trace of each transformation matrix is an *invariant*; i.e., it is independent of the choice of coordinate system. For the matrix in Equation 15.1, the trace is equal to $1+2\cos(2\pi n/5)$. All operations of a point group that have the same trace belong to the same *equivalence class* (*class*). For the **5** (C_5) point group, the operations 5 (C_5) and 5^4 (C_5^4) have the same trace, $1+2\cos(2\pi/5)$, and belong to the same equivalence class. Operators 5^2 (C_5^2) and 5^3 (C_5^3) have a different trace, $1+2\cos(4\pi/5)$, and, hence, belong to a second class. The identity operation, with trace equal to 3, is in its own class. The traces of symmetry operators are important for the description of physical properties.

When the rotation operators of the point group **5** (C_5) are combined with other symmetry operators, such as a two-fold axis normal to 5 (C_5), or a mirror plane, we can derive a number of new point groups. In the following sections, we will take a closer look at symmetry groups with rotations that have non-crystallographic orders, such as 5 (C_5), 8 (C_8), 10 (C_{10}), and 12 (C_{12}).

15.3 Molecules with non-crystallographic point group symmetry

Molecular point groups are not restricted to the 32 crystallographic point groups of Chapter 9. The *icosahedral groups* $\mathbf{m\overline{3}\overline{5}}$ (I_h) and **532** (I), which have 5-, 3-, and 2-fold rotational symmetry axes, describe the symmetry of the C_{60} molecule (Fig. 15.1). Sub-groups of the icosahedral groups include the *pentagonal groups* that have five-fold rotation axes but no three-fold rotation axes. A decagonal group, $\mathbf{\overline{10}m2}$ (D_{5h}), describes the symmetry of the C_{70} molecule. In this section, we will use C_{60} to illustrate icosahedral group theory and discuss related molecules.

The C_{60} molecule is a third allotrope of C, in addition to the diamond cubic and hexagonal graphitic forms. Box 15.1 summarizes the symmetries

Box 15.1 Symmetry operations for the icosahedral group of C_{60}.

Rotational symmetries of the $\mathbf{m\bar{3}\bar{5}}$ (I_h) and **532** (I) groups, with reference to C_{60} are:

(i) The identity operator, 1 (E).

(ii) Each 5-fold rotation axis (a) has five operations: 1 (E), 5 (C_5), 5^2 (C_5^2), 5^3 (C_5^3), and 5^4 (C_5^4). The 5 (C_5) axes pass through pairs of the 12 pentagonal faces. Six pairs, each with four 5-fold rotations (excluding the identity operator), yield a total of 24 operations.

(iii) Each 3-fold rotation axis (b) has two operations (in addition to 1 (E)): 3 (C_3) and 3^2 (C_3^2). The 3 (C_3) axes pass through pairs of the 20 hexagonal faces. Ten pairs of hexagonal faces, each with two 3-fold rotations, yield a total of 20 operations.

(iv) Each 2-fold rotation axis (c) has a single 2 (C_2) operation (in addition to 1 (E)). The 2 (C_2) axes pass through pairs of the 30 edges shared between hexagonal faces. Fifteen pairs of edges, each with a single 2-fold rotation, yield 15 operations.

The icosahedral rotational group, **532** (I), thus has $1+24+20+15=60$ symmetry operations, resulting in a group order of 60. Sixty additional improper rotations can be added to the proper rotations of the **532** (I) group to yield the 120 operations of $\mathbf{m\bar{3}\bar{5}}$ (I_h). Mathematically, this is accomplished by the *direct product* operation:

$$\mathbf{m\bar{3}\bar{5}} = \mathbf{532} \oplus \bar{\mathbf{1}} \qquad (I_h = I \oplus C_i). \tag{15.2}$$

The *direct product* of two groups G_1 and G_2 is denoted by $G_1 \oplus G_2$. The direct product group contains all operations formed by taking the pairwise products of the elements of one group and those of the other. The resulting group order, h, is equal to the product of the orders of the two groups, i.e., $h = h_1 h_2$. The $\mathbf{m\bar{3}\bar{5}}$ (I_h) group is the direct product of the icosahedral group, **532** (I), and the $\bar{\mathbf{1}}$ (C_i) group containing the identity and inversion operators. The 60 new operations include 15 mirror planes, 24 $\bar{5}$ (C_{5i}) rotoinversions, 20 $\bar{3}$ (C_{3i}) rotoinversions, and the inversion operation, $\bar{1}$ (i).

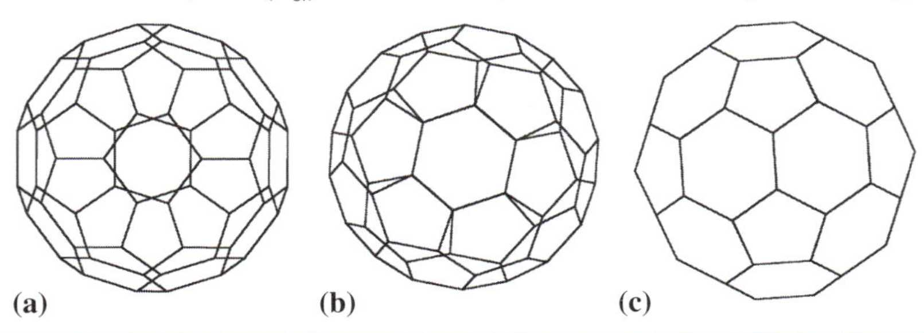

of the icosahedral group. The five-fold, three-fold, and two-fold axes of the icosahedral **532** (I) group are visible in the wire frame drawings of C_{60} in Box 15.1. Each C atom has an identical environment. The $\mathbf{m\bar{3}\bar{5}}$ (I_h) and **532** (I) groups have important effects on the electronic structure (Johnson *et al.*, 1991) and magnetic properties of C_{60} and C_{60}-based solids (McHenry and Subramoney, 2000).

15.3.1 Fullerene molecular structures

Sir Harry Kroto and co-workers (Kroto *et al.*, 1985) determined the structure of C_{60} which received much attention due to its aesthetically pleasing, highly symmetric arrangement of C atoms, a configuration similar to a soccer ball. This molecule was named *Buckminsterfullerene*, after the American architect, R. Buckminster Fuller (1895–1983).[3] Scientists asserted that many C clusters, previously (Rohlfing *et al.*, 1984) and subsequently observed, might have similar geodesic structures. Other caged structures with even numbers of C atoms, empty central cavities, closed shells, and exclusively hexagonal and pentagonal faces were named *fullerenes*. Fullerenes describe the class of C_{2n} structures with $n \geq 16$.[4] Examples of these C nanostructures are shown in Box 15.2.

Box 15.2 C nanostructures: the higher fullerenes

The figure below shows examples of higher fullerenes with 60–80 carbon atoms and a variety of crystallographic and non-crystallographic point group symmetries.

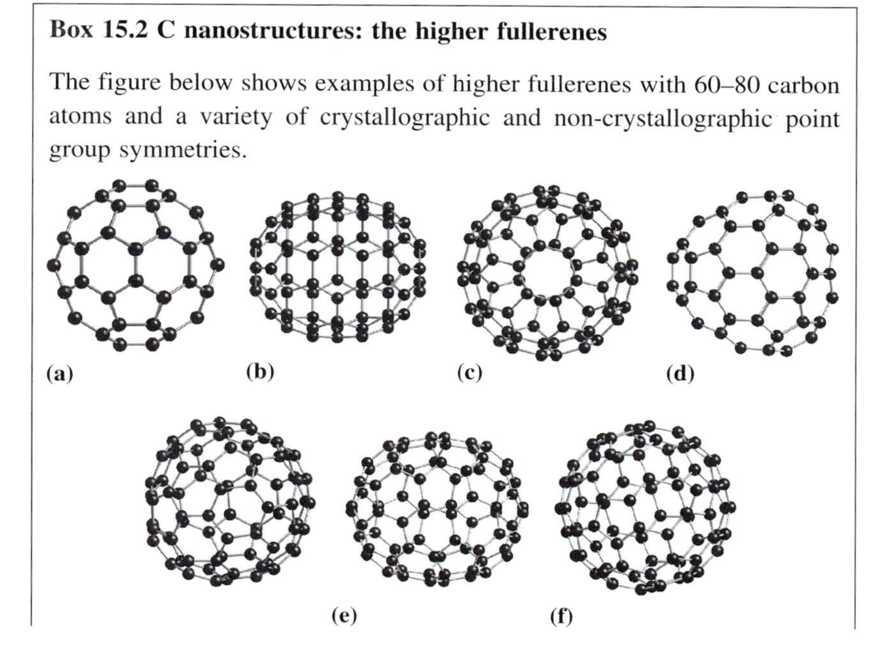

[3] Fuller was a proponent of *geodesic* structures as prominent building design components.
[4] The affectionate term *Buckyballs* is also widely used.

(a) and (b) show the C_{60} and C_{70} clusters with $\mathbf{m\overline{35}}$ (I_h) and $\mathbf{\overline{10}m2}$ (D_{5h}) point group symmetries; (c) and (d) show the C_{72} and C_{74} fullerenes, with $\mathbf{\overline{12}2m}$ (D_{6d}) and $\mathbf{\bar{6}m2}$ (D_{3h}) point group symmetries; (e) shows the two isomers of the C_{76} fullerene with $\mathbf{\bar{4}3m}$ (T_d) (upper) and **222** (D_2) (lower) point group symmetries. Many of the larger fullerenes have multiple isomers. The C_{78} fullerene has five isomers, one with **32** (D_3), two with $\mathbf{\bar{6}m2}$ (D_{3h}) and two with **mm2** (C_{2v}) point group symmetries. The C_{80} fullerene has seven isomers, one of which has $\mathbf{m\overline{35}}$ (I_h) symmetry and is depicted in (f).[a]

[a] The coordinates of known higher fullerenes have been graciously made available at the website of Dr. M. Yoshida; http://www.cochem2.tutkie.tut.ac.jp/Fuller/Fuller.html.

Synthesis of macroscopic amounts of C_{60} molecules was made possible by a graphitic arc technique (Kratschmer *et al.*, 1990). Fullerenes were identified for even-numbered compositions from C_{32} to C_{600} by Robert Curl *et al.* (Curl and Smalley, 1991). Researchers progressed by (a) identifying molecules by the number of constituent C atoms (by mass spectroscopy) and (b) determining their atomic coordinates and symmetries. New discoveries included:

- trapping of atoms inside the fullerenes to produce *endohedral fullerenes*;
- growth of macroscopic single crystals consisting of a particular fullerene molecule;
- chemical substitution of C atoms by other elements; and
- attachment of atoms to the outside of fullerenes, to produce *exohedral* structures.

To stick with the "bucky" naming convention, cylindrical structures constructed from hexagonal C units (essentially folded-up sheets of graphite) were called *Bucky tubes* (Iijima, 1991). They are discussed further in Chapter 25.

Researchers in Richard Errett Smalley's group postulated that the structural stability of geodesics, that completely tile a 2-D space (e.g., the surface of a sphere), could explain the notable stability of the C_{2n} ($n \geq 16$) clusters. Hexagonal networks of C were not surprising, because planar hexagonal nets are stabilized by the sp^2 bonding between C atoms in graphite. Pentagons are required to provide the curvature needed for closure. The minimal closed structure constructed exclusively from regular pentagons is the 20 vertex Platonic solid, the *pentagonal dodecahedron*.

The eighteenth century mathematician Leonhard Euler showed that any closed tiling involving regular pentagons must contain the 12 pentagons of the pentagonal dodecahedron.

Thus, a fullerene must have 12 pentagonal faces and an even number of hexagonal faces as described by the chemical formula C_{20+2H}, where $2H$ is the number of hexagonal faces. Fullerenes can only contain even numbers of C atoms.

The smallest fullerene is C_{60} for which none of the 12 pentagons share edges. Larger fullerenes are called *higher fullerenes*. For the higher fullerenes, many possible isomers are possible, even with the constraint of 12 pentagonal and 2H hexagonal faces.

> The *isolated pentagon rule (IPR)* states that pentagons in fullerenes prefer not to share edges.

This rule is almost universally obeyed for the higher fullerenes. A second rule prescribes the avoidance of diametrically positioned pentagons on any given hexagon. This constraint allows alternating bonds to be maintained on the hexagons. There are few instances in C_{2n} structures where the second rule is broken (Guo *et al.*, 1992).

Bond distances on C_{60} molecules were inferred by Liu *et al.* (1991) (and later by Burgi *et al.* (1994)) from crystal structure refinements of C_{60} solids (discussed in Chapter 25). Distances of 0.1355(9) nm for the $C = C$ double bonds and 0.1467 nm for $C - C$ single bonds were found. Many *lower fullerenes* are illustrated in an entertaining article by Curl and Smalley (1991). They arrived at the concept of *deflated structures*, by selectively removing C atom pairs from C_{60}.

The *isolated pentagon rule (IPR)* provides a topological criterion to predict which clusters will have icosahedral symmetry (if not distorted by an electronic *Jahn–Teller effect*). Those clusters that maintain icosahedral symmetry are special cases of polyhedra known as *Goldberg polyhedra*.

> Goldberg polyhedra are built from $20(b^2 + bc + c^2)$ vertices where b and c are non-negative integers.

If $b = c$ or $bc = 0$, then the undistorted Goldberg polyhedron will have icosahedral symmetry. The first few icosahedral fullerenes are predicted to be C_{20} ($b = 1$, $c = 0$), C_{60} ($b = c = 1$), C_{80} ($b = 2$, $c = 0$), C_{180} ($b = 3$, $c = 0$), C_{240} ($b = c = 2$), C_{320} ($b = 4$, $c = 0$), C_{500} ($b = 5$, $c = 0$), and C_{540} ($b = c = 3$) (Fowler and Steert, 1987, Goldberg, 1934, 1937).

15.4 Icosahedral group representations

The point group **532** (I) describes the symmetry of two *Platonic solids*: the regular *icosahedron*, and its dual, the *pentagonal dodecahedron*. It is also

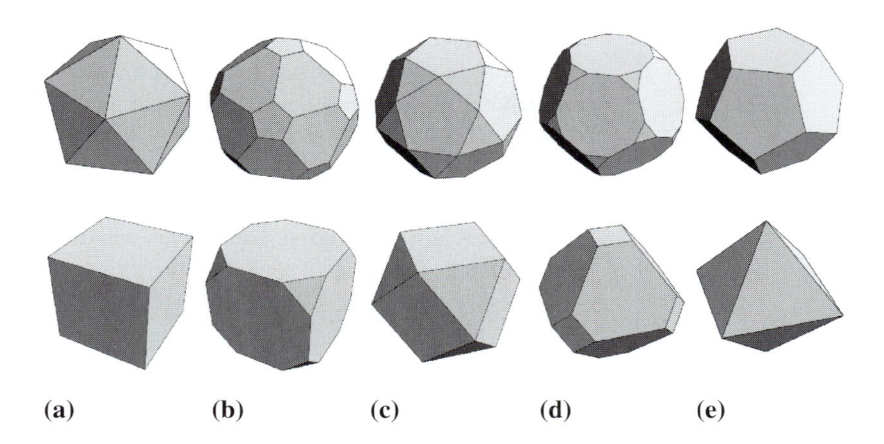

Fig. 15.2. Icosahedral (top) and cuboctahedral (bottom) solids: (a) icosahedron, cube, (b) truncated icosahedron, truncated cube, (c) icosidodecahedron, cuboctahedron, (d) truncated pentagonal dodecahedron, truncated octahedron, and (e) pentagonal dodecahedron, octahedron.

the point group for three *Archimedean solids*: the *truncated icosahedron*, the *truncated dodecahedron*, and the *icosidodecahedron*.[5] Figure 15.2 (top) illustrates that truncating (cutting or shaving) the icosahedron normal to the five-fold axes creates new pentagonal faces and turns the triangular faces into irregular 6-sided polyhedra.

The term *truncated icosahedron* refers to the Archimedean solid in which the 6-sided faces become regular hexagons.[6] Further truncation leads to the icosidodecahedron. A similar truncation of the pentagonal dodecahedron along three-fold axes passing through its vertices leads to a structure with triangular faces along with irregular 10-sided polyhedra. The term *truncated dodecahedron* refers to the shape where the 10-sided face is a regular decagon. Further truncation again leads to the icosidodecahedron. While the Platonic solids have faces with *either* five-fold *or* three-fold symmetry, the Archimedean solids have *both* five-fold *and* three-fold faces. We refer the interested reader to Wenninger (1971) for structures with more than two types of faces (i.e., beyond the Archimedean solids).

It is instructive to compare the symmetry of icosahedral and cubic solids. $\mathbf{m\bar{3}m}$ (O_h) is the point group symmetry of highest order, 48 (with inversion symmetry), and **432** (O) is a point group symmetry of order 24 (without inversion). These point groups describe the symmetry of the regular *octahedron*, and its dual, the *cube*, both *Platonic solids*. $\mathbf{m\bar{3}m}$ (O_h) is also the symmetry group for three *Archimedean solids*: the *truncated octahedron*, the *truncated cube*, and the *cuboctahedron* (Fig. 15.2 bottom).

Truncating the cube in directions normal to the three-fold axes creates new triangular faces and turns the square faces into irregular 8-sided polyhedra. The *truncated cube* is the Archimedean solid for which the

[5] The Platonic and Archimedean solids will be described in more detail in the next chapter.

[6] This is nearly the shape taken by the C_{60} molecule. Even though there are 6-sided faces in this shape, there is no six-fold symmetry. Can you explain why?

8-sided face becomes a regular octagon (however, the 3-D shape does not possess eight-fold symmetry!). Further truncation leads to the cuboctahedron. Similarly, truncation of the octahedron along four-fold axes passing through its vertices leads to a structure with square faces along with irregular 6-sided polyhedra. The *truncated octahedron* is the shape for which the 6-sided faces become regular hexagons. Note that the truncated shapes always involve structures with {100} and {111}-type faces.

The 60 proper and 60 improper rotation axes of the icosahedral group, $\mathbf{m\bar{3}\bar{5}}$ (I_h), describe the symmetry of the *truncated icosahedron*, the *icosahedron*, the *pentagonal dodecahedron*, and many other polyhedra. The symmetry elements of the icosahedral groups are depicted in the icosahedral stereographic projections of Fig. 15.3 (Hahn, 1989). These stereographic projections use the convention that the Cartesian x, y, and z directions are aligned along the three orthogonal two-fold axes in the icosahedral symmetry group. The icosahedral groups are the only point groups for which both five-fold and three-fold axes are simultaneously present as group operations, just as some of the cubic point groups are the only ones having both four-fold and three-fold axes. Figure 15.3(c) and (d) show rendered 3-D representations of the two icosahedral point groups.

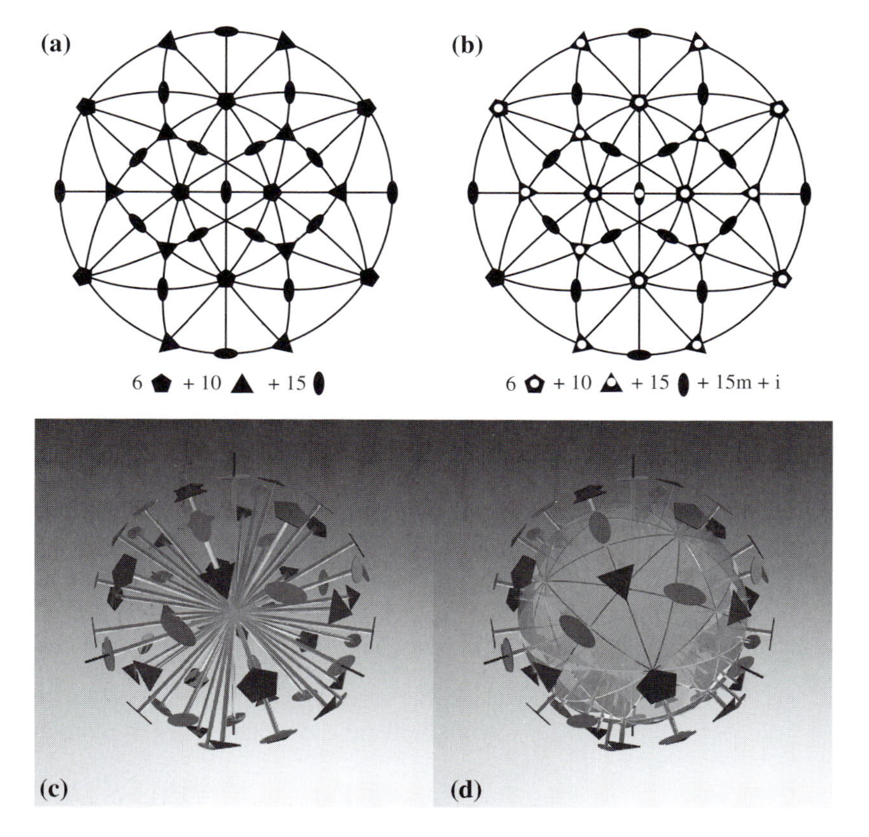

Fig. 15.3. Icosahedral group stereographic projections (a) **532** (I), and (b) $\mathbf{m\bar{3}\bar{5}}$ (I_h), along with 3-D rendered representations.

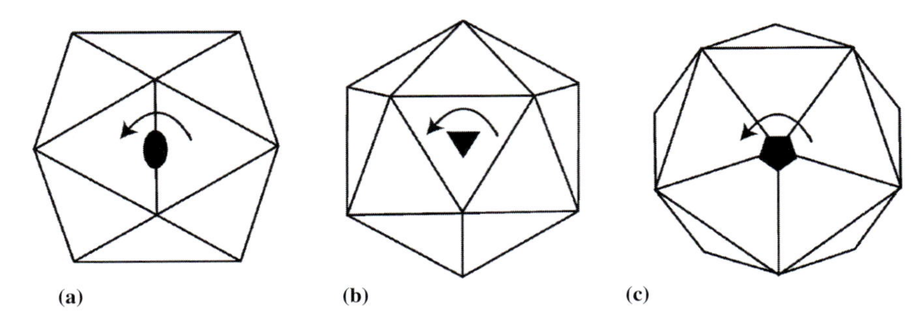

Fig. 15.4. Two-, three-, and fivefold axes of the icosahedron in useful coordinate systems.

Convenient representations of the icosahedral group operations include the geometry where the five-fold axis corresponds to the z-axis and another where three orthogonal sets of two-fold axes are aligned with respect to the x, y, and z coordinate axes, Fig. 15.4. The representation where the three-fold axis corresponds to the z-axis is rarely used. A Cartesian coordinate system where three orthogonal two-fold axes are chosen as axes offers an orthogonal basis for which the three axes have the same symmetry. It is, therefore, compatible with standard orientations for the cubic groups. This is also the choice made by the *International Tables for Crystallography* (Hahn, 1989) in depicting the stereographic projections shown in Fig. 15.3(a) and (b).

A *generating relationship* for the icosahedral group was first proposed by Speiser (1937), subsequently discussed by McLellan (1961) and is summarized in Box 15.3. This generating relationship starts by building a subgroup [H] of all of the rotation operations about a single five-fold axis. New operations result from multiplication with two other generators which are orthogonal **2** (C_2) operations. We can conveniently use the geometry where the five-fold axis corresponds to the z-axis to express the generating matrices. However, we can easily rotate these matrices into the other convenient coordinate system with orthogonal sets of two-fold axes as the basis.

Box 15.3 Generating relationship for the icosahedral group

The icosahedral rotational group, $\mathbf{m\bar{3}\bar{5}}$ (I_h), can be generated as follows:

(i) Identify a five-fold rotation operation (around the z-axis) as operation A and two orthogonal two-fold rotations as operations B and C.

(ii) Construct the cyclic subgroup, [H], as five successive five-fold rotations:

$$[H] = (A, A^2, A^3, A^4, E). \tag{15.3}$$

This subgroup is the same as the **5** (C_5) point group.

(iii) Construct a set, [K], by adding to [H] five new elements, obtained by multiplying each rotation of [H] by the first **2** (C_2) operation, B:

$$[K] = [H] + B \cdot [H]. \tag{15.4}$$

(iv) Use the generating relationship to construct the set of operations [I] for the icosahedral rotational group, **m$\bar{3}\bar{5}$** (I_h):

$$[I] = [K] + (C)\cdot[K] + \sum_{i=1}^{4}(A^iC)\cdot[K]. \tag{15.5}$$

This operation results in the generation of all 120 symmetry operators of the **m$\bar{3}\bar{5}$** (I_h) point group.

Box 15.4 Generating matrices for the icosahedral group

The generating matrices for the icosahedral group are listed below. On the left, with the five-fold axis along z, we have three generating matrices: $D(C_5)$, $D(C_2)$, and $D(C_2')$; on the right, with three orthogonal two-fold axes acting as the basis, we have as generators: $D(C_2)$, $D(C_5)$, and $D(C_2')$.

$$D(C_5) = \frac{1}{2}\begin{pmatrix} \tau-1 & -\sqrt{\tau+2} & 0 \\ \sqrt{\tau+2} & \tau-1 & 0 \\ 0 & 0 & 1 \end{pmatrix};\quad D(C_2) = \begin{pmatrix} -1 & 0 & 0 \\ 0 & 1 & 0 \\ 0 & 0 & -1 \end{pmatrix};$$

$$D(C_2) = \frac{1}{2\tau}\begin{pmatrix} \tau & 1 & \tau^2 \\ 1 & \tau^2 & -\tau \\ \tau^2 & \tau & 1 \end{pmatrix};\quad D(C_5) = \frac{1}{\sqrt{5}}\begin{pmatrix} 1 & 0 & -2 \\ 0 & 1 & 0 \\ -2 & 0 & -1 \end{pmatrix};$$

$$D(C_2) = \begin{pmatrix} -1 & 0 & 0 \\ 0 & 1 & 0 \\ 0 & 0 & -1 \end{pmatrix};\quad D(C_2') = \begin{pmatrix} 1 & 0 & 0 \\ 0 & -1 & 0 \\ 0 & 0 & -1 \end{pmatrix},$$

where $\tau = (1+\sqrt{5})/2$ is the *golden mean* of pentagonal symmetry.

Box 15.4 gives matrices for the generators A, B, and C in each of two geometries. The first of these has the five-fold axis oriented along z (counter-clockwise rotation) and the two-fold axis B aligned along y. The generating matrices in this geometry are designated $D(C_5)$, $D(C_2)$, and $D(C_2')$, respectively. For the second choice of coordinate axes, the generating matrices are designated $D(C_2)$, $D(C_5)$, and $D(C_2')$. One may also generate the icosahedral group using only two generators, a **5** (C_5) and a **3** (C_3) rotation axis.

It is interesting to note that the icosahedral rotation group, **532** (I) is *isomorphic* with, i.e., has the same multiplication table as, the alternating group of five elements (A_5, all odd permutations of five numbers). The icosahedral group shares this type of isomorphism with two other high symmetry groups (associated with the *Platonic solids*). The tetrahedral group **23** (T) is isomorphic with the alternating group A_4 (all odd permutations of four numbers), and the octahedral group maps one to one onto the symmetric group S_4 (the group of all even and odd permutations of four numbers). The *isomorphism* of the **532** (I) and A_5 groups is useful in deriving the generating relationship, described in Box 15.3, for the icosahedral group.

15.5 Other non-crystallographic point groups with five-fold symmetries

The icosahedral groups **532** (I) and **m$\bar{3}\bar{5}$** (I_h) have subgroups that are crystallographic point groups, and others that are examples of non-crystallographic point groups. For example, the **m$\bar{3}$** (T_h) group is a subgroup of **m$\bar{3}\bar{5}$** (I_h), and **23** (T) is a subgroup of **532** (I). These crystallographic point groups represent the intersections of the icosahedral group with the cubic groups **m$\bar{3}$m** (O_h) and **432** (O), respectively. Other subgroups of the icosahedral group, as shown in the *descent in symmetry* in Fig. 15.5, include **$\bar{3}$m** (D_{3d}), **32** (D_3), **222** (D_2), **3** (C_3), **2/m** (C_{2h}), **2** (C_2), **$\bar{1}$** (C_i), **m** (C_s), and **1** (C_1) among the crystallographic point groups and **$\bar{5}$m** (D_{5d}), **52** (D_5), **5m** (C_{5v}), **$\bar{5}$** (C_{5i}), and **5** (C_5) among non-crystallographic point groups.

Figure 15.6 illustrates fullerenes with pentagonal or decagonal point group symmetries, **52** (D_5) and **$\bar{5}$m** (D_{5d}), respectively. These are (a) an isomer of a C_{100} fullerene with **52** (D_5) point group symmetry, (b) the C_{70} fullerene,

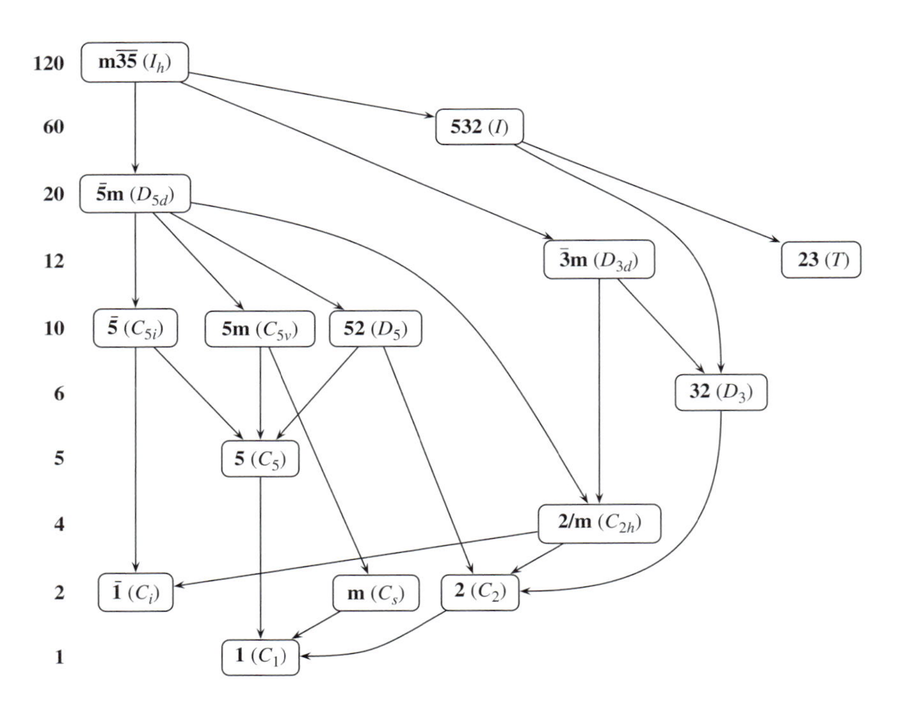

Fig. 15.5. Descent in symmetry with subgroups and their orders for the icosahedral point groups.

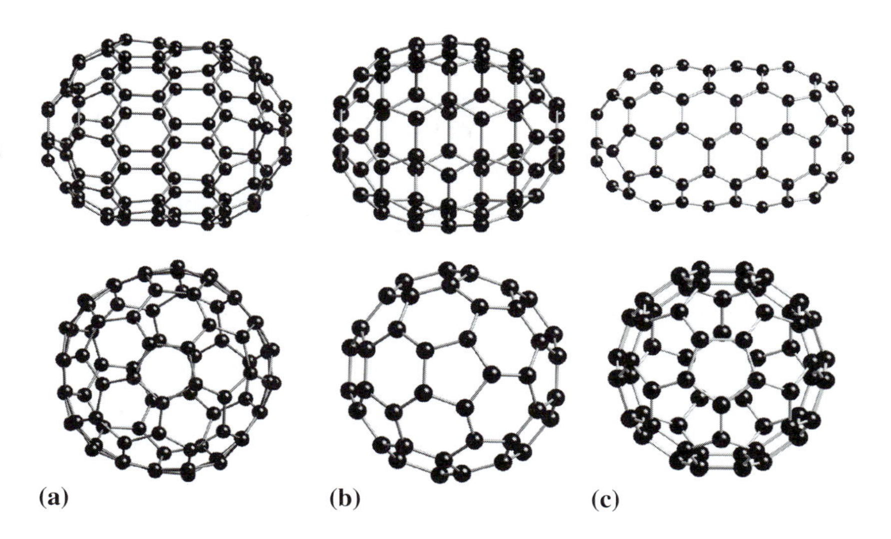

Fig. 15.6. Fullerenes possessing **52** (D_5), $\overline{\mathbf{10}}$**m2** (D_{5h}), and $\bar{\mathbf{5}}$**m** (D_{5d}) symmetries (a) an isomer of C_{100} (**52** (D_5)), (b) C_{70} ($\overline{\mathbf{10}}$**m2** (D_{5h})), and (c) a second isomer of C_{100} ($\bar{\mathbf{5}}$**m** (D_{5d})).

and (c) a second isomer of a C_{100} fullerene with $\bar{\mathbf{5}}$**m** (D_{5d}) point group symmetry. Each of these is shown projected with the five-fold axis lying in the plane (top) and pointing out of the plane (bottom) of the drawing. Each of these molecules has the cyclic group operations: 1 (E), 5 (C_5), 5^2 (C_5^2), 5^3 (C_5^3), and 5^4 (C_5^4), and 5 2 (C_2) axes perpendicular to the main 5 (C_5) axis. These are the only symmetry elements for the C_{100} fullerene with the **52** (D_5) point group, Fig. 15.6 (a). The other isomer of C_{100}, with $\bar{\mathbf{5}}$**m** (D_{5d}) the point group (Fig. 15.6(c)) has inversion symmetry. The pseudo-decagonal symmetry shown in the projection with the five-fold axis out of the plane results from a five-fold roto-inversion axis, $\bar{5}$ (C_{5i}), and 5 m (σ_d) dihedral mirror planes.

The $\overline{\mathbf{10}}$**m2** (D_{5h}) group is the symmetry group of the C_{70} fullerene (Fig. 15.6 (b)) (Heath *et al.*, 1985) and the *truncated pentagonal prism*. This molecule is constructed by separating the two hemispheres of the C_{60} molecule and adding a belt of 10 carbon atoms between them. This construction destroys all but one of the five-fold axes of the original icosahedron, as well as all of the three-fold axes. The oval structure of C_{70} contains 25 hexagonal faces. Its structure was first verified by ^{13}C NMR (nuclear magnetic resonance). C_{70} is the second most abundant fullerene formed in graphitic arcs and one of the few fullerenes produced in sufficient abundance to grow single crystals. The structure of the C_{70} molecular solid is discussed further in Chapter 25.

In addition to the **52** (D_5) point group operations, the C_{70} molecule has a horizontal mirror plane and associated roto-inversion operations along with 5 m (σ_v) vertical mirror planes. Figure 15.7 illustrates the symmetry operations for the $\overline{\mathbf{10}}$**m2** (D_{5h}) group with reference to a pentagonal prism. This group has inversion symmetry, horizontal and vertical mirror planes, and a five-fold symmetry axis.

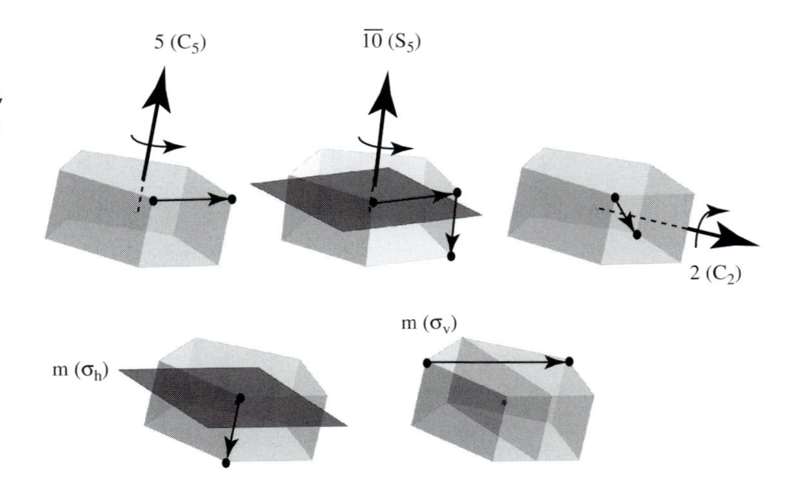

Fig. 15.7. Symmetry operations for a pentagonal prism: (a) 5 (C_5), (b) $\bar{1}\bar{0}$ (C_{5h}), (c) 2 (C_2), (d) m (σ_h), and (e) m (σ_v).

Figure 15.7 (a) shows the action of a five-fold rotation, 5 (C_5). Its repeated action yields the 1 (E), 5 (C_5), 5^2 (C_5^2), 5^3 (C_5^3), and 5^4 (C_5^4) operations. Figure 15.7(b) shows the action of a $\bar{1}\bar{0}$ (C_{5h}) roto-inversion operation. Figure 15.7(c) shows the action of a 2 (C_2) rotation operation perpendicular to the main 5 (C_5) axis. This axis can be rotated using each of the 5 (C_5) operations to yield 5 2 (C_2) operations. Figures 15.7(d) and (e) show m (σ_h) and m (σ_v) mirror planes, respectively. The m (σ_v) mirror can be rotated using each of the 5 (C_5) operations to yield 5 m (σ_v) operations. It is also instructive to label the vertices of the pentagonal prism and write the symmetry operations in terms of permutations of the vertices.

15.6 Descents in symmetry: decagonal and pentagonal groups

Figure 15.8 (a) illustrates the descent in symmetry for the general $(4N+2)$-gonal groups. General descents in symmetry allow us to see similarities between point groups. For $N = 1$ (i.e., $4N + 2 = 6$), we can determine the descent in symmetry for the hexagonal groups. For the decagonal groups we have $N = 2$ or $4N + 2 = 10$. The pattern of the descent is the same as for the hexagonal groups. The non-crystallographic point groups with five-fold rotational axes are summarized in Fig. 15.8 (b). These are all the groups with five- or ten-fold symmetry, with the exception of the icosahedral groups. Some are subgroups of the icosahedral groups, while others have symmetry elements not present in the icosahedral groups. The decagonal point group symmetry occurs in several quasicrystalline phases to be discussed in Chapter 20.

Stereographic projections for the decagonal and pentagonal point groups are illustrated in Fig. 15.9. These follow the descent in symmetry for a $(4N+2)$-gonal symmetry group. They are separated into sets having decagonal symmetry (a), and those having pentagonal symmetry (b). Note the similarity

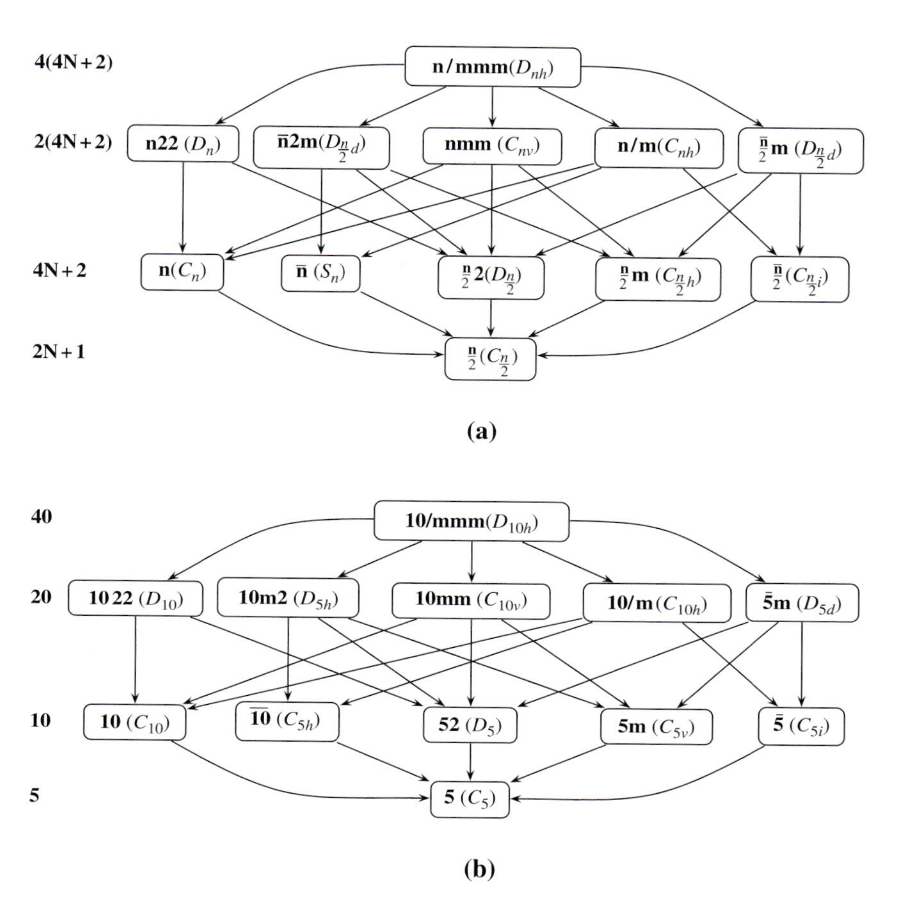

Fig. 15.8. Descent in symmetry for the (a) $(4N+2)$-gonal and (b) decagonal and pentagonal ($N = 2$) groups.

between the stereographic projections for the decagonal (pentagonal) and the hexagonal (trigonal) groups. These similarities reflect that groups of order $4(2N+1)$ and $4(4N+2)$, with N an integer, have a common descent in symmetry with similar numbers of sub- and supergroups. Groups of order $4(4N)$, where N is an integer, have a common descent in symmetry different from the $(4N+2)$-gonal symmetry groups. These common descents imply that octagonal and tetragonal groups will have similarities.

Figure 15.10 illustrates shapes following the descent in symmetry of Fig. 15.8(b). The gray shapes belong to the pentagonal groups $2N+1$, $N=2$. We illustrate the symmetry of the point groups beginning with the most symmetric decagonal prism, and successively destroying some of its symmetry by eliminating, rotating, shearing, or translating vertices.

A decagonal prism, Fig. 15.10 (top), possesses **10/mmm** (D_{10h}) symmetry. It is the only decagonal group of order 40. If the two decagons are rotated relative to one another (but not by an integer multiple of $\frac{\pi}{10}$), a twisted decagonal prism with **10 22** (D_{10}) symmetry results. This rotation destroys

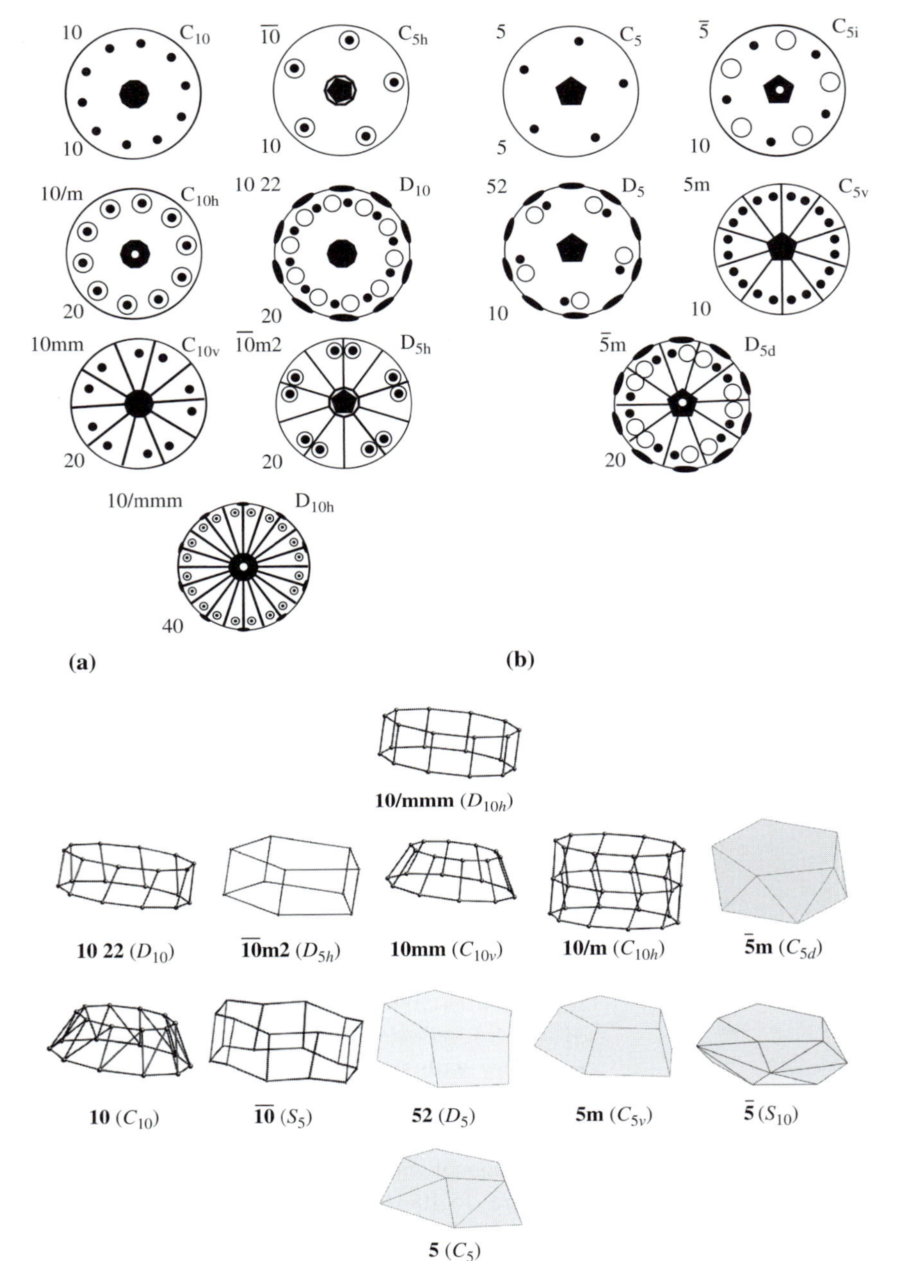

Fig. 15.9. Stereographic projections of the (a) decagonal and (b) pentagonal point groups. The top two symbols indicate the Hermann–Mauguin and Schönflies notation, the number at the left bottom of each projection is the order of the point group.

Fig. 15.10. Shapes (described in the text) having the symmetries of the pentagonal or decagonal groups in the same descent in symmetry illustrated in Fig. 15.8(b).

the horizontal and vertical mirror planes and all roto-inversion operations of the decagonal prism, yielding a group of order 20.

The pentagonal prism has **$\overline{10}$m2** (D_{5h}) symmetry. A truncated decagonal pyramid possesses 10 vertical mirror planes but no perpendicular two-fold rotations or roto-inversions; hence, it has the **10mm** (C_{10v}) symmetry. **10/m**

Table 15.2. Symmetry elements of the pentagonal and decagonal point groups.

Hermann–Mauguin	Schönflies	Symmetry elements (International notation only)
5	C_5	$1, 5^{1-4}$
$\bar{\mathbf{5}}$	C_{5i}	$1, 5^{1-4}, \bar{5}^{1,3,7,9}$, m
5/m	C_{5h}	same as $\bar{\mathbf{5}}$ (C_{5i})
5m	C_{5v}	$1, 5^{1-4}, 5\times$m
52	D_5	$1, 5^{1-4}, 5\times2$ ($\perp$ to 5)
10	C_{10}	$1, 10^{1-9}$
$\overline{\mathbf{10}}$	C_{5h}	$1, 5^{1-4}, \overline{10}^{1,3,7,9}, \bar{1}$
$\bar{\mathbf{5}}$**m**	D_{5d}	$1, 5^{1-4}, \overline{10}^{1,3,7,9}, \bar{1}, 5\times$m, 5×2
10mm	C_{10v}	$1, 10^{1-9}, 10\times$m
10/m	C_{10h}	$1, 10^{1-9}, \bar{1}, \overline{10}^{1-9}$
$\overline{\mathbf{10}}$**m2**	D_{5h}	$1, 5^{1-4}, \bar{5}^{1,3,7,9}$, m ($\perp$ to 5), $5\times$m, 5×2 ($\perp$ to 5)
10 22	D_{10}	$1, 10^{1-9}, 10\times2$ ($\perp$ to 10)
10/mmm	D_{10h}	$1, 10^{1-9}, 10\times2$ ($\perp$ to 10), $\bar{1}, \overline{10}^{1-9}, 10\times$m

(C_{10h}) symmetry is obtained by stacking twisted decagonal prisms of different chirality. This structure has inversion symmetry but no orthogonal two-fold axes, nor vertical or horizontal mirrors. The pentagonal antiprism has $\bar{\mathbf{5}}$**m** (D_{5d}) symmetry, which is the only pentagonal group of order 20 ($4(2N+1)$ with $N=2$). It has ten-fold roto-inversion operations, dihedral mirror planes, and perpendicular two-fold rotation axes.

A twisted truncated decagonal pyramid has the symmetry group **10** (C_{10}). A puckered decagonal prism has point group $\overline{\mathbf{10}}$ (C_{5h}). The twisted pentagonal prism possesses **52** (D_5) point group symmetry. The truncated pentagonal pyramid possesses **5m** (C_{5v}) symmetry. A pentagonal prism bisected by a rotated pentagon possesses $\bar{\mathbf{5}}$ (C_{5i}) symmetry, completing the subgroups of order 10. Finally, a twisted pentagonal pyramid has point group symmetry **5** (C_5).

One can determine the matrix representations of the symmetries of the other decagonal and pentagonal point groups by multiplying the matrices of the **5** (C_5) group or the **10** (C_{10}) group by that of another generator for the point group in question. A second generator will be the inversion operator and/or a two-fold rotation operator. We end our discussion of the decagonal and pentagonal point groups by summarizing the symmetry operations for each of the groups. Table 15.2 lists all of the symmetry elements for each of the 12 decagonal and pentagonal point groups. In this table, the notation $5^{1,\ldots,N}$ is shorthand for the operators $5, 5^2, \ldots 5^N$, and there is similar notation for other repeated operators. Note that the $\bar{\mathbf{5}}$ (C_{5i}) and **5/m** (C_{5h}) groups are identical.

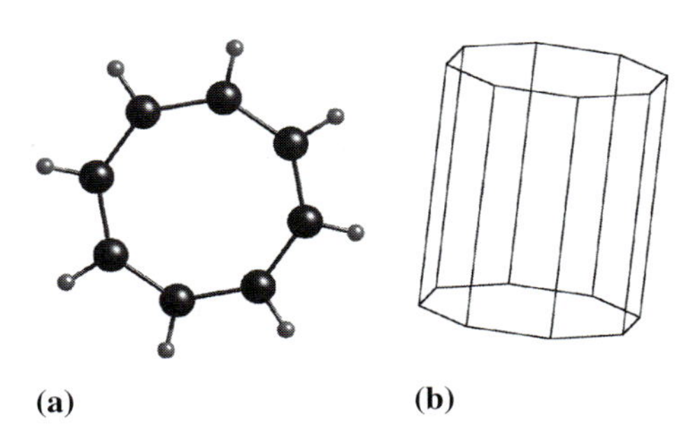

Fig. 15.11. The (a) $(C_8H_8)^{2-}$ molecule and (b) octagonal prism.

15.7 Non-crystallographic point groups with octagonal symmetry

In this section, we consider point groups with an eight-fold rotational axis, **8** (C_8) and **8/mmm** (D_{8h}). The octagonal groups are also among the point groups for which 2-D quasi-periodic structures have been observed in alloy systems (see Chapter 20). Figure 15.11 illustrates two structures with **8/mmm** (D_{8h}) symmetry: (a) the *cyclooctatetraene dianion* $(C_8H_8)^{2-}$, and (b) the *octagonal prism*. While the $(C_8H_8)^{2-}$ dianion is planar with **8/mmm** (D_{8h}) symmetry, the neutral molecule, cyclooctatetraene (C_8H_8), is not planar or aromatic.

Many of the octagonal point groups can be expressed as direct products of lower order groups. For instance, the **8/m** (C_{8h}) point group is the direct product of the **8** (C_8) and $\bar{\mathbf{1}}$ (C_i) point groups:

$$\mathbf{8/m} = \mathbf{8} \oplus \bar{\mathbf{1}}. \tag{15.6}$$

It is a point group of order 16. As before, octagonal groups can be represented by stereographic projections. These are illustrated in Fig. 15.12. This figure shows that the stereographic projection for the **8/mmm** (D_{8h}) point group, for example, consists of the eight-fold rotational axis, a set of eight two-fold axes orthogonal to this axis, eight roto-inversion operations and eight mirror planes for a total of 32 operations. The **822** (D_8), $\bar{\mathbf{8}}$**2m** (D_{4d}), **8mm** (C_{8v}), and **8/m** (C_{8h}) point groups each have 16 operations and can be derived as subgroups of **8/mmm** (D_{8h}) by destroying half of the symmetry operations. The derivations are left as exercises. The **8** (C_8) and $\bar{\mathbf{8}}$ (S_8) subgroups are each of order 8. It is left as a reader exercise to construct the multiplication tables for the groups **8** (C_8) and $\bar{\mathbf{8}}$ (S_8).

15.8 Descents in symmetry: octagonal and dodecagonal groups

Descents in symmetry are generalized based on the order of the main symmetry axis. If the main symmetry axis is of order $n = 4N$ (where N is an integer), the generalization described above for the octagonal group

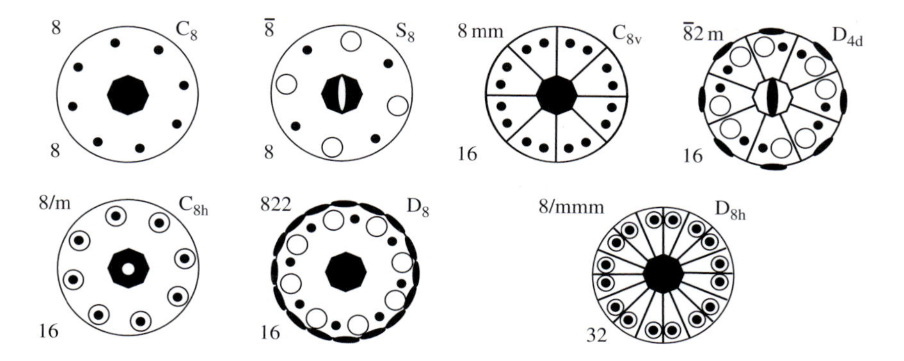

Fig. 15.12. Stereographic projections for the octagonal point groups.

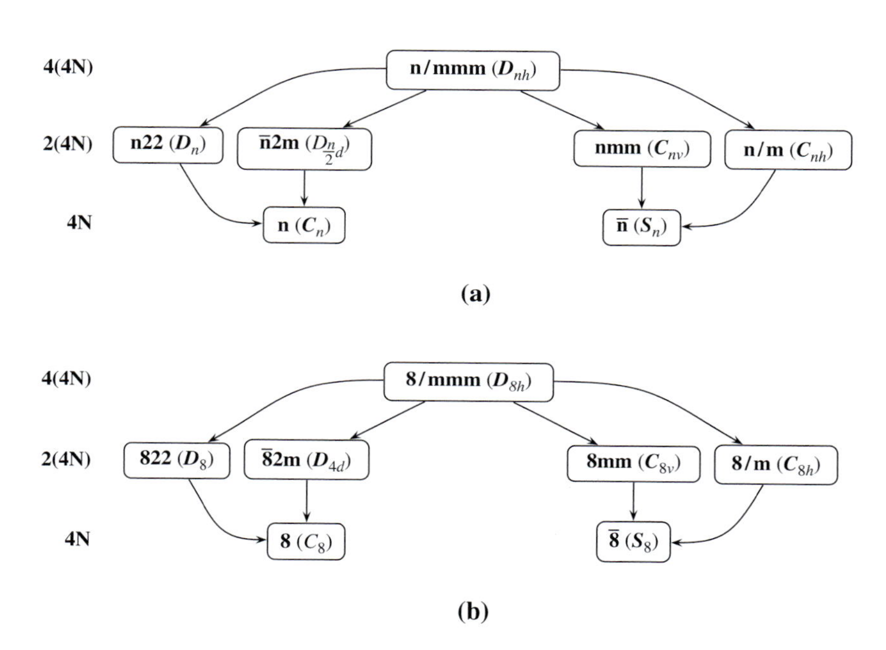

Fig. 15.13. Descent in symmetry for the (a) ($4N$)-gonal and (b) octagonal ($N = 2$) groups.

($N = 2$) produces the descent in symmetry illustrated in Fig. 15.13(a). If the main symmetry axis is of order $n = 4N + 2$ or $n = 2N + 1$ (where N is an integer), a generalization produces descents identical to those of the decagonal and pentagonal groups, illustrated in Fig. 15.8(a). Note that $N = 2$ and thus $n = 2N + 1 = 5$ represents the pentagonal groups and $n = 4N + 2 = 10$ the decagonal groups. Here one can see the super- and subgroup relationships between the ($2N + 1$)-gonal and ($4N + 2$)-gonal groups, respectively.

Fig. 15.13(b) shows the octagonal groups of orders 32, 16, and 8. It can be shown quite generally that, for a main symmetry axis of order $n = 4N$ (where

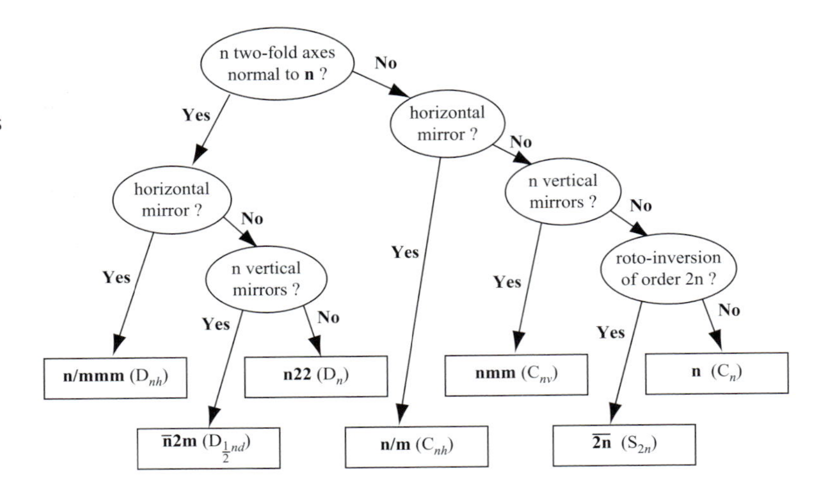

Fig. 15.14. Flowchart for determining the non-crystallographic point group of a molecule or objects having a C_n rotational axis.

N is an integer), there will be one group of order $4n$, four subgroups of order $2n$[7] and two subgroups of order n.

Other important non-crystallographic point groups that will be studied in the context of quasi-periodic tilings include the *dodecagonal* point groups. The dodecagonal groups have twelve-fold symmetry axes. The main symmetry axis is of order $n = 4N$, with $N = 3$. Thus, one can determine the descent in symmetry by referring to Fig. 15.13(a). This is left as a reader exercise at the end of the chapter.

Fig. 15.14 represents a systematic method for determining the point group symmetry (Bishop, 1973). With the exception of the icosahedral groups, non-crystallographic point groups cannot have two or more n (C_n) axes when $n \geq 3$, so that the highest symmetry is a single n (C_n) axis. The flowchart is used as follows: if the order, n, of the rotation is known, then start with the question: *are there any two-fold rotation axes normal to the n-fold axis?* If the answer is *yes*, then follow the left branch and determine if there is a mirror plane normal to the rotation axis. If so, then the point group is of the type **n/mmm** (D_{nh}); otherwise, find out if there are multiple mirror planes that contain the n-fold axis. If *yes*, the point group symbol is **$\bar{n}$2m** ($D_{\frac{1}{2}nd}$), otherwise it is **n22** (D_n). The top right branch of the flowchart asks about mirror planes and roto-inversions, and leads to an additional four different point group symbols.

In summary, we have tabulated examples of the icosahedral, decagonal and octagonal group symmetries that one may use to describe solid shapes and molecules with non-crystallographic symmetries. The point groups treated in this chapter are summarized in Table 15.3, along with the names of shapes

[7] One of these subgroups actually represents two subgroups with the same symbol, indicating that they have the same multiplication table.

Table 15.3. Example of shapes and molecules with symmetries belonging to selected non-crystallographic point groups.

Point group [order]	Face form	Point form	Molecule
m$\bar{3}\bar{5}$ (I_h) [120]	pentagon-hexecontahedron	snub pentagon-dodecahedron	C_{60}, C_{80}
532 (I) [60]	hecatonicosahedron	{ pentagon-dodecahedron truncated by icosahedron and by rhomb-triacontahedron	polyoma virus
5 (C_5) [5]	pentagonal pyramid	regular pentagon	
$\bar{5}$ (C_{5i}) [10]	pentagonal streptohedron	pentagonal antiprism	
5/m (C_{5h}) [10]	(same as **$\bar{5}$** (C_{5i}))		
5m (C_{5v}) [10]	dipentagonal pyramid	truncated pentagon	
52 (D_5) [10]	pentagonal trapezohedron	twisted pentagonal antiprism	C_{100}
$\bar{5}$m (D_{5d}) [20]	dipentagonal scalenohedron	pentagonal antiprism sliced by pinacoid	{ C_{80}, C_{100}, $[Fe(C_5H_5)_2]$
8 (C_8) [8]	octagonal pyramid	regular octagon	
$\bar{8}$ (S_8) [8]	square streptohedron	square antiprism	
8/m (C_{8h}) [16]	octagonal dipyramid	octagonal prism	
822 (D_8) [16]	octagonal trapezohedron	twisted octagonal antiprism	
8mm (C_{8v}) [16]	di-octagonal pyramid	truncated octagon	
$\bar{8}$2m (D_{4d}) [16]	octagonal scalenohedron	square antiprism sliced by pinacoid	S_8
8/mmm (D_{8h}) [32]	di-octagonal dipyramid	edge-truncated octagonal prism	$(C_8H_8)^{2-}$
10 (C_{10}) [10]	decagonal pyramid	regular decagon	
$\overline{10}$ (C_{5h}) [10]	pentagonal dipyramid	pentagonal prism	
10/m (C_{10h}) [20]	decagonal dipyramid	decagonal prism	
1022 (D_{10}) [20]	decagonal trapezohedron	twisted decagonal antiprism	
10mm (C_{10v}) [20]	didecagonal pyramid	truncated decagon	
$\overline{10}$m2 (D_{5h}) [20]	dipentagonal dipyramid	truncated pentagonal prism	{ C_{70}, C_{80}, C_{90}, $Ru(C_5H_5)_2$
10/mmm (D_{10h}) [40]	didecagonal dipyramid	edge-truncated decagonal prism	
12 (C_{12}) [12]	dodecagonal pyramid	regular dodecagon	

Table 15.3. (cont.).

Point group [order]	Face form	Point form	Molecule
$\overline{\mathbf{12}}$ (S_{12}) [12]	hexagonal streptohedron	hexagonal antiprism	
12/m (C_{12h}) [24]	dodecagonal dipyramid	dodecagonal prism	
12 22 (D_{12}) [24]	dodecagonal trapezohedron	twisted dodecagonal antiprism	
12mm (C_{12v}) [24]	didodecagonal pyramid	truncated dodecagon	
$\overline{\mathbf{12}}$**2m** (D_{6d}) [24]	dodecagonal scalenohedron	hexagonal antiprism sliced by pinacoid	C_{72}, C_{96}
12/mmm (D_{12h}) [48]	didodecagonal dipyramid	edge-truncated dodecagonal prism	

with these symmetries, and a few representative molecules. The *face form* is the name of an object bounded by flat faces, whereas the *point form* is the dual shape of the face form. For more details we refer the interested reader to Chapter 10 in the *International Tables for Crystallography* (Hahn, 1996).

15.9 Historical notes

Felix Christian Klein (1849–1925) was an influential German mathematician born in Düsseldorf, Prussia (now Germany) on April 25, 1849. As a mathematician he liked to point out that his birthday was 2^2, 5^2, 43^2 (all numbers in the birthdate are the squares of prime numbers). Klein studied mathematics and physics at the University of Bonn from 1865–66, where he continued, receiving a doctorate in 1868. Klein's doctoral research was supervised by Julius Plücker, Chair of Mathematics and Experimental Physics at Bonn. Klein moved to Göttingen in 1868, where he was made lecturer in 1871. He was appointed professor at Erlangen, in Bavaria, in 1872. In 1875 he accepted a chair at the Technische Hochschule at Munich. From 1880–86, he served as Chair of Geometry at Leipzig. In 1886 Klein accepted a chair at the University of Göttingen, where he remained until retirement in 1913.

Klein established a mathematics research center in Göttingen and served as editor (after **Alfred Clebsch**) of the journal *Mathematische Annalen*. Klein's area of expertise was in analytical geometry. In particular, he contributed to the study of the properties of figures that are invariant under a transformation group. This work, which explored connections between geometry and group theory, was influential in the development of crystallography. Klein explicitly influenced the field of crystallography when he suggested that Schönflies

Fig. 15.15. (a) Felix Klein (1849–1925) (picture courtesy of J. Lima-de-Faria), and (b) Richard Smalley (1943–2005) (picture courtesy of the Nobel e-Museum)

study the problem of space groups by considering transformation groups. This suggestion led to an extension of the work of Jordan adding improper rotations to the discrete group of proper rotations (1892) (Lima-de-Faria, 1990).

Klein also published the important work *Lectures on the Icosahedron and the Solution of Equations of the Fifth Degree* in 1876. In this book, he showed how rotation groups could be applied to solve algebraic problems. He also laid out the group theory for icosahedral groups (Klein, 1876), which are prominent among the non-crystallographic point groups discussed in this chapter. In 1884, this work was published as a book which was reprinted in 1956 (Klein, 1956). One may find an interesting extended biography of Klein at the history website of the School of Mathematics and Statistics at the University of St. Andrews, Scotland.[8]

Richard Errett Smalley (1943–2005) was born in Akron, Ohio, on June 6, 1943.[9] In 1946, his family moved to Kansas City, Missouri, where he stayed until his university days. Smalley received his B.S. degree from the University of Michigan in 1965. He worked for four years as a research chemist with Shell and received his Ph.D. from Princeton in 1973. He pioneered supersonic beam laser spectroscopy as a post-doctoral associate at the University of Chicago working for Lennard Wharton and Donald Levy. In 1976, he joined the faculty at Rice University where he discovered the C_{60} molecule. He remained at Rice as University Professor, Gene and Norman Hackerman Professor of Chemistry, and Professor of Physics and Astronomy. He continued research on continuous carbon nanofibers until his death in 2005.

[8] See http://www-history.mcs.st-andrews.ac.uk/history/Mathematicians/Klein.html.

[9] Smalley's autobiography, written for Le Prix Nobel (1997) is reproduced at his Rice University website: http://smalley.rice.edu.

At Rice University, in 1976, Smalley began collaborations with **Robert Curl**. He set up both a supersonic beam apparatus and a second generation apparatus with pulsed supersonic nozzles to study large molecules, radicals, and clusters. During the late 1970s, he collaborated with Andrew Kaldor and his group at Exxon on laser-based uranium isotope separation processes. Kaldor's group also observed the clusters with even numbers of C atoms in a laser vaporization cluster beam that are now known as fullerenes. These experiments were repeated on the apparatus of Richard Smalley in 1985. The discovery of fullerenes and the subsequent explanation for their structures opened a new field of C chemistry which is still growing today.[10] Smalley was named the 1996 Nobel Laureate in Chemistry for the discovery of fullerenes along with Robert Curl and Sir **Harry Kroto**. Kroto was active in microwave spectroscopy, and his measurements determined the structure of the C_{60} molecule.[11]

15.10 Problems

(i) *Subgroups of the icosahedral point groups*: List all of the subgroups of the icosahedral groups and their orders.

(ii) *Icosahedral fullerenes*: Determine the number of hexagons on the first few icosahedral fullerenes: C_{20}, C_{60}, C_{80}, C_{180}, C_{240}, C_{320}, C_{500}, and C_{540}.

(iii) *Icosahedron*: Show that an icosahedron, with vertices a unit distance from the origin and 2-fold axes on the Cartesian coordinate axes, has vertex coordinates: $(\pm 1, 0, \pm\tau)/\sqrt{1+\tau^2} + cp$, ($cp$ denotes *cyclic permutations*).

(iv) *Pentagonal dodecahedron I*: Show that a pentagonal dodecahedron, with vertices a unit distance from the origin and two-fold axes along the Cartesian coordinate axes, has vertices at: $(0, \pm\tau, \pm\frac{1}{\tau})/\sqrt{3}$; $(\pm 1, \pm 1, \pm 1)/\sqrt{3} + cp$.

(v) *Pentagonal dodecahedron II*: Show that the pentagonal dodecahedron is the dual of the icosahedron, i.e., show a mapping of the faces and vertices of the icosahedron into the vertices and faces of the pentagonal dodecahedron. Show that the pentagonal dodecahedron is a $b = 1, c = 0$ Goldberg polyhedron.

[10] An interesting account of the discovery of the fullerenes and the period after their discovery is given in the book *Perfect Symmetry: The Accidental Discovery of the Fullerenes* (Baggott, 1994).

[11] Another account of the discovery of the fullerenes is given by Kroto at the URL: http://invention.smithsonian.org/centerpieces/ilives/kroto/kroto.html.

(vi) *Cube*: Show that the cube is the dual of the octahedron, i.e., that there is a mapping of the faces and vertices of the cube into the vertices and faces of the octahedron.

(vii) *Tetrahedron*: Show that the coordinates for a tetrahedron with unit edge length and with two-fold axes oriented on the Cartesian coordinate axes are given by: $[\pm 1, 0, \surd 3]/2$ and $[0, \pm 1, -\surd 3]/2$.

(a) Express the coordinates of a dual with unit edges obtained by decorating the faces of the original tetrahedron.

(b) Construct the shape resulting from connecting the closest vertices of a tetrahedron and its interpenetrating dual (the *tetraheder-stern*).

(viii) *Characters*: Show that the trace (*character*) of a matrix for an n-fold rotation axis is $1 + 2\cos(2\pi/n)$. For which of the one-through twelve-fold axes is the trace an irrational number? Discuss the result in light of the *law of rational indices*.

(ix) *Golden mean, τ*: Show that $\tau^2 = \tau + 1$. Show that the trace of a 5 (C_5) rotation matrix about the z-axis is τ. Show that the trace of a 5^2 (C_5^2) rotation matrix about the z-axis is $-1/\tau$.

(x) *Cyclic groups*: Show that the groups of simple rotations n (C_n) are cyclic. Show that any group whose order is a prime number must be cyclic.

(xi) *Alternating group A_3*: Consider three elements: (1 2 3) mapped into the vertices of an equilateral triangle. Construct the symmetry operations of the *alternating group A_3*. Construct a group multiplication table for this group. To what crystallographic point groups is this group isomorphous?

(xii) *Symmetric group S_3*: Consider three elements: (1 2 3) mapped into the vertices of an equilateral triangle. Construct the symmetry operations of the *symmetric group S_3*. Construct a group multiplication table for this group. To what crystallographic point groups is this group isomorphous?

(xiii) *Alternating group A_4*: Consider four elements: (1 2 3 4) mapped into the vertices of a tetrahedron. Construct the symmetry operations of the *alternating group A_4*. Construct the group multiplication table for this group. To what crystallographic point groups is this group isomorphous?

(xiv) *Symmetric group S_4*: Consider four elements: (1 2 3 4) mapped into the vertices of a tetrahedron. Construct the symmetry operations of the *symmetric group S_4*. Construct the group multiplication table for this group. To what crystallographic point groups is this group isomorphous?

(xv) *Symmetric groups*: Show that the order of the symmetric group S_n is $h = n!$ Show that S_n $(n \geq 2)$ has two subgroups, one of all even

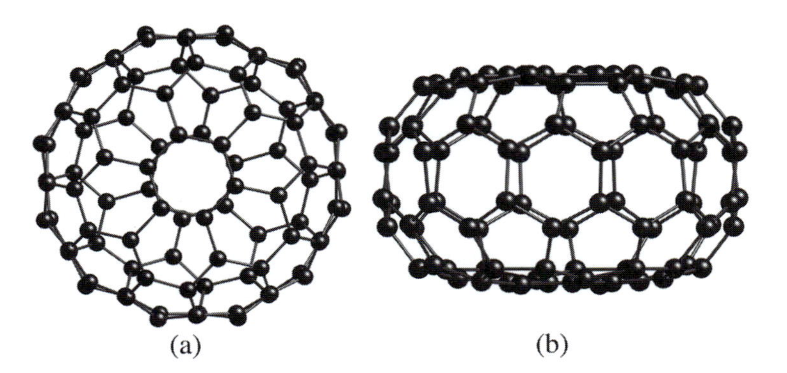

Fig. 15.16. An isomer of C_{96} (a) along a six-fold axis and (b) orthogonal to the six-fold axis.

and one of all odd permutations of n numbers, and both have order $h = n!/2$. Show this decomposition for the icosahedral and cubic groups.

(xvi) *Pentagonal point group*: Determine symmetry operations for the *ferrocene* $[(C_5H_5)_2Fe]$ molecule. What point group does this represent?

(xvii) *Decagonal point group*: List all operations of the decagonal group. What is the group's order? Construct a shape with decagonal symmetry.

(xviii) *Direct product* **5** (C_5) *and* $\bar{\mathbf{1}}$ (C_i): Consider a five-fold axis for the pentagonal point group **5** (C_5) oriented along the z-axis in a Cartesian coordinate system. List rotation matrices for the five operations of this group. Express the rotation matrix for the inversion operation. Identify new operations obtained in the direct product of the five-fold rotation matrices of the **5** (C_5) and the inversion group $\bar{\mathbf{1}}$ (C_i). What is the new group obtained?

(xix) *Direct product* **5** (C_5) *and* **2/m** (C_{2h}): Consider a five-fold axis, oriented along the z-axis in a Cartesian coordinate system, for the **5** (C_5) group. Express the rotation matrix for a horizontal mirror plane. Take the direct product of the five-fold rotation matrices of the $\bar{\mathbf{5}}\mathbf{m}$ (D_{5d}) and the **2/m** (C_{2h}) group. Identify the new operations obtained. What is the new group obtained?

(xx) C_{96} *isomer*: Fig. 15.16 illustrates one of the isomers of a C_{96} fullerene along a six-fold axis (a) and orthogonal to the six-fold axis (b). List the symmetry elements for this molecule and identify its point group.

(xxi) **7** (C_7) *point group*: Express the multiplication table for the heptagonal group. Is it cyclic? What is the character (trace) of its generating matrix?

(xxii) **7** (C_7) *point group*: What are the intersections of heptagonal and cubic groups?

(xxiii) *Truncated polyhedra*: Explain why the octagonal faces on a truncated cube do not possess eight-fold symmetry. Do the same for the hexagonal faces on the truncated icosahedron to show that they do not possess six-fold symmetry.

(xxiv) *Octagonal groups I*: Express a generating relationship for **8/mmm** (D_{8h}).

(xxv) *Octagonal groups II*: List the intersections of the octagonal and cubic groups.

(xxvi) *Octagonal groups III*: Determine the operations that are lost in reducing the symmetry from **8/mmm** (D_{8h}) to **8mm** (C_{8v}).

(xxvii) *Octagonal groups IV*: Determine the classes of operations for the **8** (C_8) group by determining the trace of the 3-D rotation matrices for each element.

(xxviii) *Octagonal groups V*: Determine the generator for the point group **8** (C_8). Construct the multiplication table for this group. Identify the powers of the generator element that can be reduced to rotation axes with smaller n.

(xxix) *Decagonal groups*: Express a generating relationship for **10/mmm** (D_{10h}).

(xxx) *Dodecagonal groups*: Construct a descent in symmetry for groups with twelve-fold rotational axes. What are their intersections with the cubic groups? And with the hexagonal groups?

(xxxi) *Dodecagonal point group classes*: Determine the classes of operations for the **12** (C_{12}) group by determining the trace of the 3-D Cartesian coordinate rotation matrices for each element.

(xxxii) *I_h point group*: Develop a recursive relationship to generate rotation matrices for the 120 operations of the I_h group in a coordinate system in which the x, y, and z axes correspond to the two-fold rotational axes. Write a program using matrix generators and the generating relationship to list the operations of the I_h group. Organize the matrices into *classes* for which the *trace* of the matrices is the same. Develop a group multiplication table for the I_h group.

CHAPTER

16 Periodic and aperiodic tilings

> *"The diversity of the phenomena of nature is so great, and the treasures hidden in the heavens so rich, precisely in order that the human mind shall never be lacking in fresh nourishment."*
>
> Johannes Kepler 1571–1630

16.1 Introduction

Crystalline solids have been described in terms of a Bravais lattice and a basis. For more complex crystal structures, it is instructive to describe a crystal in terms of the stacking of crystalline planes. While there are 230 *space groups*, there are only 17 *plane groups*, which simplifies classification. However, the number of possible plane stacking sequences is infinite.

In Chapters 3 and 9, we have introduced the concepts of 2-D Bravais lattices and 2-D plane groups, respectively. In the present chapter, we build upon these concepts to introduce the mathematics, nomenclature, and classification schemes often encountered in the materials or crystallographic literature of *2-D periodic tilings*. Since *quasi-periodic* and *aperiodic tilings* such as the *Penrose tile* have become important in crystallography (e.g., *quasicrystallography*), we describe these important tilings in this chapter. A detailed discussion of quasicrystallography is left for Chapter 20. Finally, we end with a discussion of the construction of 3-D structures from the stacking of 2-D tiles, and the tiling of an n-dimensional space with polyhedra (in 3-D) or *polytopes* (in higher dimensional spaces, i.e., $n > 3$).

16.2 2-D plane tilings

In the mathematical literature, a tiling is synonymous with a *tessellation*. The theory of tilings is rich, and we will introduce concepts that are useful for the classification of crystal structures. The text *Mathematical Models* (Cundy and Rollet, 1952), was cited in the original definitions of *Frank–Kasper phases* to be discussed in Chapter 18. The more recent book, *Tilings and Patterns* (Grünbaum and Shepard, 1987), is an authoritative treatment of this subject. *Quasicrystals and Geometry* (Senechal, 1995), is an excellent review of aperiodic tilings and quasicrystals. Box 16.1 defines a plane tiling, and its *prototiles*.[1]

16.2.1 2-D regular tilings

We begin by discussing the three *regular tilings* of the 2-D plane illustrated in Fig. 16.1. These are examples of *monohedral tilings*, i.e., all the tiles are the same size and shape. These tilings are also *edge-to-edge*, meaning that all tiles share edges. For regular tilings, the prototiles are *regular polygons*. A regular polygon has identical sides and interior angles. We prove that there are only three regular tilings in Box 16.2.

The three possible regular tilings (edge-to-edge, monohedral regular polygons as prototiles) are shown in Fig. 16.1. Their tiles are an equilateral triangle, a square, and a regular hexagon, respectively. The tilings are labeled by the *Schläfli symbols*, 3^6, 4^4, and 6^3.

Box 16.1 Definitions of plane tilings and their tiles

A *plane tiling*, $\mathcal{T}$, is a countable family of closed sets :

$$\mathcal{T} = \{T_1, T_2, \ldots\}$$

which covers the plane without gaps or overlaps (Grünbaum and Shepard, 1987). The elements:

$$T_1, T_2, \ldots$$

are the *tiles* of $\mathcal{T}$. The interiors of the tiles, T_i, are taken as being pairwise disjoint, i.e., they have no area in common. The union of the sets is, therefore, the entire plane.

[1] Prototiles are the tiles used as the basis for the tilings.

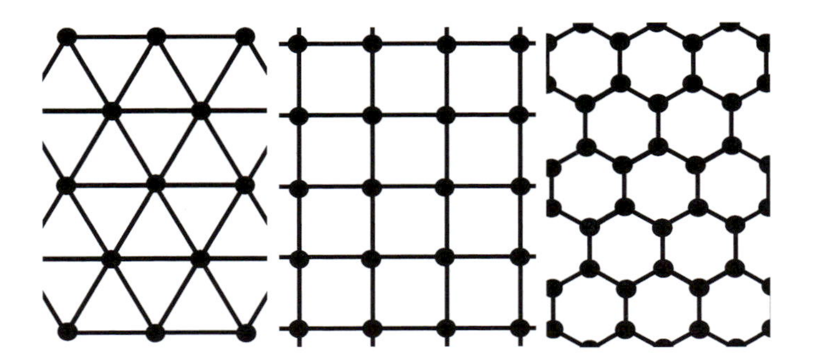

Fig. 16.1. The three regular tilings 3^6, 4^4, and 6^3.

Box 16.2 Derivation of the Regular Tilings

Consider a regular polygon with r sides. The angle between the two sides meeting at a single vertex is $(1-2/r)\pi$. The *valence*, v, of a vertex in the tiling is the number of r-gons meeting at that vertex. As we travel the 2π radians around any vertex, we encounter v identical polygons. Therefore, the angle between any two consecutive sides meeting at this vertex is $2\pi/v$. Equating the two expressions for these angles:

$$\left(1-\frac{2}{r}\right)\pi=\frac{2\pi}{v},$$

we can solve for integers r and v, that satisfy this criterion. Rearranging yields:

$$vr-2r-2v=0 \quad \text{or} \quad (r-2)(v-2)=4.$$

By inspection, one can determine that the only integer solutions to this equation are $(r=3, v=6)$, $(r=4, v=4)$, and $(r=6, v=3)$.

A Schläfli symbol describes the *number* and *type* of polygons (n-gons) that meet at a vertex in the tiling.

Note that in tiling drawings, we will often highlight the points at which neighboring tiles meet with a filled circle; these circles do not belong to the tiles, and are only used to clarify the drawings. In the regular triangular tiling, 6 equilateral triangles (3-gons) meet at any vertex. This tiling is then described as 3^6 in the Schläfli notation. The square tiling has 4 squares (4-gons) meeting at a point and is, hence, given the symbol 4^4. Finally, the hexagonal tiling has 3 hexagons (6-gons) meeting at a vertex and receives the symbol 6^3. In tilings with regular polygons that are not monohedral, we may have more than one type of regular polygon meeting at a vertex and the tiling may have more than one type of vertex.

16.2.2 2-D Archimedean tilings

If we relax the restriction of a monohedral tiling, but require: (a) that the tiling be edge-to-edge; (b) that the tiles be regular polygons; and (c) that all vertices are of the same type, then we can show that 11 distinct tilings result (including the three regular tilings). These tilings are known as *uniform tilings* or *Archimedean tilings*.

All vertices in a uniform tiling are symmetrically equivalent.

They can be labeled by a Schläfli symbol that describes (1) each polygon type and (2) the degeneracy of each type which meets at an equivalent vertex.

As illustrated in Fig. 16.2, the eight uniform tilings are: $(3^4 \cdot 6)$, $(3^3 \cdot 4^2)$, $(3^2 \cdot 4 \cdot 3 \cdot 4)$, $(3 \cdot 4 \cdot 6 \cdot 4)$, $(3 \cdot 6 \cdot 3 \cdot 6)$ (known as the *Kagome tiling*), $(3 \cdot 12^2)$, $(4 \cdot 6 \cdot 12)$, and $(4 \cdot 8^2)$. The $3^4 \cdot 6$ tiling occurs in two enantiomorphic forms (i.e., right-and left-handed). Consider the $4 \cdot 8^2$ tiling, as an example. At each vertex of this tiling, a single square (4) and two regular octagons (8^2) meet, resulting in the Schläfli symbol $(4 \cdot 8^2)$. Note that for the Archimedean tiling $(3 \cdot 4 \cdot 6 \cdot 4)$, as we travel 2π radians around any vertex we come upon a regular triangle, a square, a regular hexagon, and finally another square. Thus the designation is $(3 \cdot 4 \cdot 6 \cdot 4)$, and not $(3 \cdot 6 \cdot 4^2)$.

The Archimedean tilings are also called *Kepler tilings*. Kepler proved that these were the only tilings of the plane by regular polygons with all vertices surrounded identically. The procedure for the proof is sketched out in Box 16.3. It is instructive to consider the uniform tiles within the context of the 2-D plane groups that have been introduced previously. An example is shown in Box 16.4.

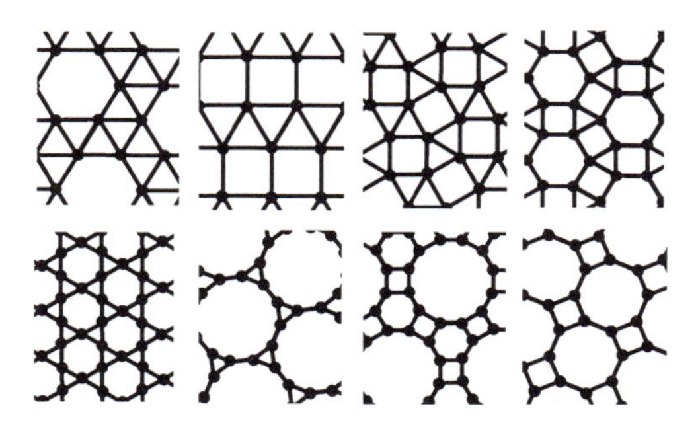

Fig. 16.2. The eight additional uniform (Archimedean) tilings (top left to bottom right): $(3^4 \cdot 6)$, $(3^3 \cdot 4^2)$, $(3^2 \cdot 4 \cdot 3 \cdot 4)$, $(3 \cdot 4 \cdot 6 \cdot 4)$, $(3 \cdot 6 \cdot 3 \cdot 6)$, $(3 \cdot 12^2)$, $(4 \cdot 6 \cdot 12)$ and $(4 \cdot 8^2)$.

Box 16.3 How to determine the 11 Kepler tiles

Consider a regular polygon with r sides and the angle $(1-2/r)\pi$ between the sides. As we travel the 2π radians around any vertex, we encounter n_3 equilateral triangles, n_4 squares, n_5 pentagons, n_6 hexagons, ... n_r r-gons. The total angle 2π equals the sum of the angles of all the r-gons meeting at the vertex:

$$\sum_{r=3}^{N_r}\left(1-\frac{2}{r}\right)\pi n_r = 2\pi \quad \text{or} \quad \sum_{r=3}^{N_r}\left(1-\frac{2}{r}\right)n_r = 2,$$

where N_r denotes the highest order of polygon present. There are 17 solutions to this equation, and four have two ways of arranging the r-gons around the vertex yielding 21 vertex types. Because proving this is a long process we will regard this as a fact. To show that only 11 of these allow repeated tiling of the entire plane, we must show that 10 out of 21 vertex types do not allow for the tiling of the plane without gaps. The reader can prove that the 11 Kepler tiles are solutions to this equation as an exercise.

Box 16.4 Symmetry, 2-D point group, Bravais lattice and dual of a regular tiling

Question: Describe the translational and rotational symmetries of the $3^2 \cdot 4 \cdot 3 \cdot 4$ Archimedean tiling and determine its plane group. Construct cells that are closer to a vertex than to any other.

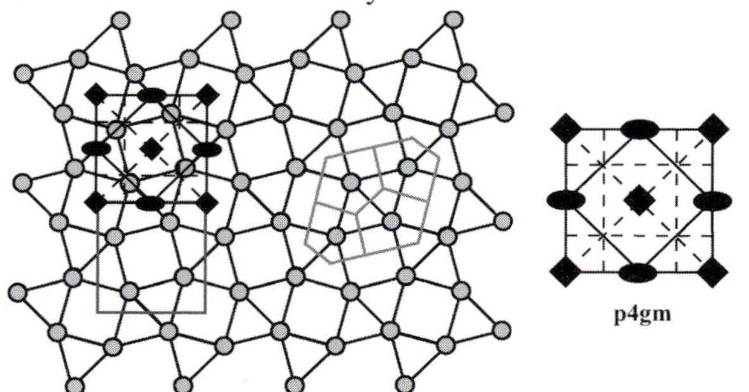

Solution: The solution is illustrated in the figure above. The unit cell is a square containing two square tiles and four triangular tiles (lower left). The plane group is **p4gm** (far right) with four-fold and two-fold symmetry axes, mirror and glide planes (upper left). The construction of four Wigner–Seitz cells closer to one vertex than to any other is shown (middle). These give rise to a (dual) tiling of the plane, which is an example of a Laves tiling.

16.2.3 *k*-uniform regular tilings

If we relax the restriction of a monohedral tiling by still requiring edge-to-edge tiling of regular polygons, but now allowing for two types of vertices, then we have *2-uniform tilings*. There are symmetry operations relating all the vertices of one type or the other, but no symmetry operations that take a vertex of the first kind into a vertex of the second. We can see that there are 20 distinct types of 2-uniform, edge-to-edge tilings by regular polygons. These are shown in Fig. 16.3, and are described by the following Schläfli symbols: $(3^6; 3^4\cdot 6)_1$, $(3^6; 3^2\cdot 4\cdot 12)$, $(3^3\cdot 4^2; 3\cdot 4\cdot 6\cdot 4)$, $(3\cdot 4\cdot 3\cdot 12; 3\cdot 12^2)$, $(3^6; 3^4\cdot 6)_2$, $(3^6; 3^2\cdot 6^2)$, $(3^3\cdot 4^2; 4^4)_1$, $(3\cdot 4^2\cdot 6; 3\cdot 4\cdot 6\cdot 4)$, $(3^6; 3^3\cdot 4^2)_1$, $(3^4\cdot 6; 3^2\cdot 6^2)$, $(3^3\cdot 4^2; 4^4)_2$, $(3\cdot 4^2\cdot 6; 3\cdot 6\cdot 3\cdot 6)_1$, $(3^6\cdot 3^3\cdot 4^2)_2$, $(3^3\cdot 4^2; 3^2\cdot 4\cdot 3\cdot 4)_1$, $(3^2\cdot 4\cdot 3\cdot 4; 3\cdot 4\cdot 6\cdot 4)$, $(3\cdot 4^2\cdot 6; 3\cdot 6\cdot 3\cdot 6)_2$, $(3^6; 3^2\cdot 4\cdot 3\cdot 4)$, $(3^3\cdot 4^2; 3^2\cdot 4\cdot 3\cdot 4)_2$, $(3^2\cdot 6^2; 3\cdot 6\cdot 3\cdot 6)$, and $(3\cdot 4\cdot 6\cdot 4; 4\cdot 6\cdot 12)$.

As a natural extension of the 2-uniform tilings, we can describe *k-uniform tiles*, in which there are k symmetrically distinct vertex types in the tiling. An example of a 3-uniform tiling, $(3^3\cdot 4^3; 3^2\cdot 4\cdot 3\cdot 4; 4^4)$ is shown in Fig. 16.4 (a). For the remainder of this text, we will not study k-uniform tilings where $k > 2$. Figure 16.4 (b) illustrates both a tiling and its superimposed dual tiling as discussed below.

16.2.4 Dual tilings – the Laves tilings

There is another class of tilings in which we do not require regular edges, yet we require that the vertices be regular. If we have v edges which meet at a vertex, we define the *valence* of the vertex as v.

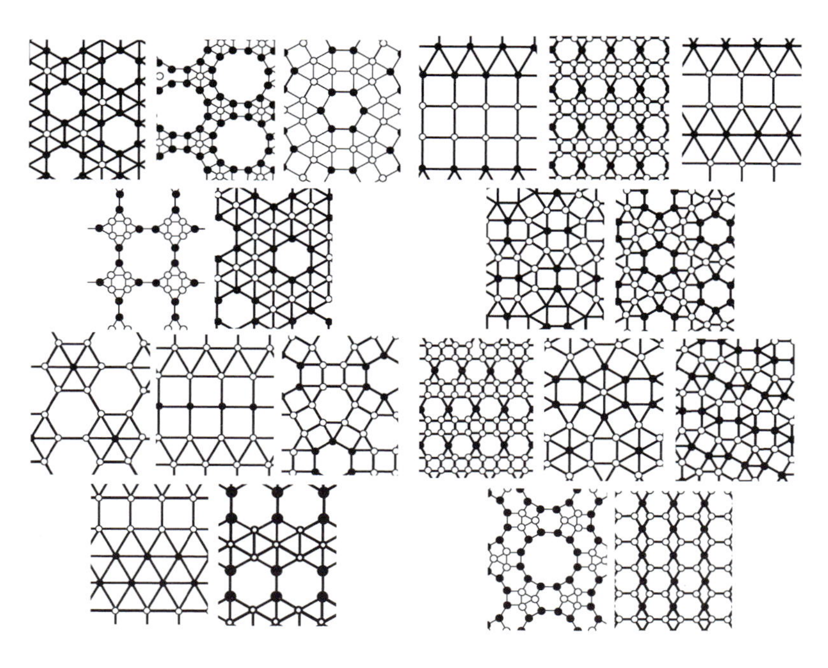

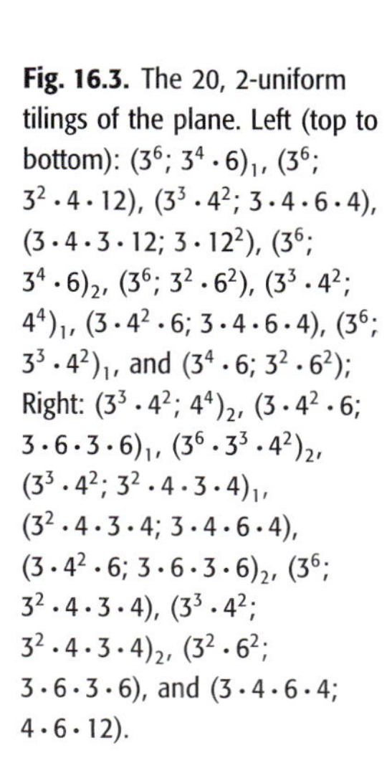

Fig. 16.3. The 20, 2-uniform tilings of the plane. Left (top to bottom): $(3^6; 3^4\cdot 6)_1$, $(3^6; 3^2\cdot 4\cdot 12)$, $(3^3\cdot 4^2; 3\cdot 4\cdot 6\cdot 4)$, $(3\cdot 4\cdot 3\cdot 12; 3\cdot 12^2)$, $(3^6; 3^4\cdot 6)_2$, $(3^6; 3^2\cdot 6^2)$, $(3^3\cdot 4^2; 4^4)_1$, $(3\cdot 4^2\cdot 6; 3\cdot 4\cdot 6\cdot 4)$, $(3^6; 3^3\cdot 4^2)_1$, and $(3^4\cdot 6; 3^2\cdot 6^2)$; Right: $(3^3\cdot 4^2; 4^4)_2$, $(3\cdot 4^2\cdot 6; 3\cdot 6\cdot 3\cdot 6)_1$, $(3^6\cdot 3^3\cdot 4^2)_2$, $(3^3\cdot 4^2; 3^2\cdot 4\cdot 3\cdot 4)_1$, $(3^2\cdot 4\cdot 3\cdot 4; 3\cdot 4\cdot 6\cdot 4)$, $(3\cdot 4^2\cdot 6; 3\cdot 6\cdot 3\cdot 6)_2$, $(3^6; 3^2\cdot 4\cdot 3\cdot 4)$, $(3^3\cdot 4^2; 3^2\cdot 4\cdot 3\cdot 4)_2$, $(3^2\cdot 6^2; 3\cdot 6\cdot 3\cdot 6)$, and $(3\cdot 4\cdot 6\cdot 4; 4\cdot 6\cdot 12)$.

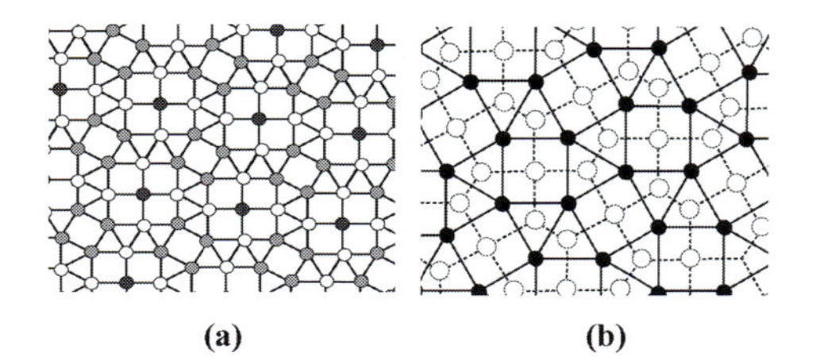

Fig. 16.4. (a) Example of a 3-uniform tiling, $(3^3 \cdot 4^2; 3^2 \cdot 4 \cdot 3 \cdot 4; 4^4)$; (b) Illustration of the Archimedean tile $(3^2, 4, 3, 4)$ and, superimposed, its dual Laves tile $[3^2, 4, 3, 4]$.

A vertex is a *regular vertex* if the angle between all consecutive pairs of edges is $2\pi/v$, i.e., the angular distribution of edges about the vertex is regular.

There is a one-to-one correspondence between these new tiles and the Archimedean tiles: each Archimedean tile has a dual in the set of tiles with regular vertices. This *duality* is a significant concept. For a given tile $(v_1, v_2, \ldots, v_r)$, we construct its dual, $[v_1, v_2, \ldots, v_r]$, by mapping the tile centers of the first tiling into the vertices of its dual and the vertices of the first into the tile centers in its dual.[2] Thus, the dual preserves the symmetry of the original tile. A tiling and its dual are completely analogous to a lattice and its reciprocal lattice.

Consider tiles that are polygons with r sides and vertex valences $v_1, \ldots, v_r$. The computation in Box 16.5 explains how to derive an equation for the possible tilings with regular vertices. There are 17 solutions to this equation

Box 16.5 Derivation of an equation to determine tilings with regular vertices

The sum of the angles at the corners of the regular polygons is $(r-2)\pi$, so that:

$$\frac{2\pi}{v_1} + \frac{2\pi}{v_2} + \cdots + \frac{2\pi}{v_r} = (r-2)\pi;$$

after rearranging the terms, we obtain:

$$\frac{(v_1-2)}{v_1} + \frac{(v_2-2)}{v_2} + \cdots + \frac{(v_r-2)}{v_r} = 2.$$

It is not difficult to find solutions for this equation by trial and error. For instance, the tiling $[3 \cdot 6 \cdot 3 \cdot 6]$ obviously satisfies this equation.

[2] A tile is designated by (...) and its dual by [...].

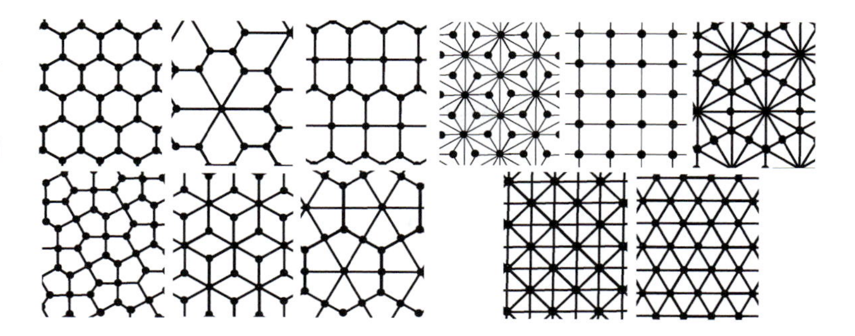

Fig. 16.5. The 11 Laves tilings, Top Row: $[3^6]$, $[3^4 \cdot 6]$, $[3^3 \cdot 4^2]$, $[3 \cdot 12^2]$, $[4^4]$, $[4 \cdot 6 \cdot 12]$; Bottom Row: $[3^2 \cdot 4 \cdot 3 \cdot 4]$, $[3 \cdot 6 \cdot 3 \cdot 6]$, $[3 \cdot 4 \cdot 6 \cdot 4]$, $[4 \cdot 8^2]$ and $[6^3]$. The tiling $[3^4 \cdot 6]$ occurs in 2 enantiomorphic forms.

(Grünbaum and Shepard, 1987), giving rise to 21 possibilities for the valences $v_1, v_2, \ldots, v_r$ taken around a tile.[3] Of the 21 possibilities, only 11 are monohedral.

> The *Laves tilings* are the eleven monohedral tilings with regular vertices that are duals to the Archimedean tilings.

The tiles and their duals have identical rotational symmetries. The Laves tiles, which are important in the study of complex metallic alloys, are illustrated in Fig. 16.5.

The issue of duality has further significance. For lattices with unit lattice parameters, the reciprocal lattice and the dual tiling are identical. Fig. 16.4(b) illustrates the Archimedean tile $(3^2, 4, 3, 4)$ and, superimposed, its dual Laves tile $[3^2, 4, 3, 4]$. We can see that all edges of the tile and its dual are orthogonal to each other. The cells of the dual tile are equivalent to the *Wigner–Seitz cell* of the original tiling; i.e., the dual tiles represent the locus of points closer to one vertex than to any other in the original tiling. It is apparent that the regular tiles (3^6) and (6^3) are duals of one another, i.e., $[3^6] = (6^3)$ and $[6^3] = (3^6)$. The regular tile (4^4) is self-dual, i.e., $[4^4] = (4^4)$.

16.2.5 Tilings without regular vertices

In addition to the regular tilings introduced in the previous sections, we can imagine an infinite number of tilings that involve tiles *without* regular vertices. Among these are interesting tiles that encompass simple symmetry reductions of regular tiles, accomplished by anisotropic deformation. For example, a distortion of the 4^4 tiling by stretching along the x- or y-axis will give rise to a rectangular tiling.

[3] Some solutions lead to more than one possibility because of *enantiomorphism*.

16.3 *Color tilings

In this section, we illustrate one aspect of *color tilings*, the *uniform coloring of regular tilings*. We can assign colors to each of the vertices or we can assign one color to the entire tile. If we color the vertices with two colors, we can represent important magnetic symmetries, where the two colors denote spin up and spin down. In what follows, we will consider concepts that are also useful in quilting, wallpapering, and floor tiling with tiles of more than one color.

> A *colored tiling* is a plane tiling, $\mathcal{T} = \{T_1, T_2, \ldots\}$ that covers the plane without gaps or overlaps and each of the tiles, $T_1, T_2, \ldots$ is assigned one of a finite number of *colors*.

A complete description of color groups is beyond the scope of this text. The interested reader is referred to Grünbaum and Shepard for a detailed discussion (Grünbaum and Shepard, 1987). The question "how can one color a tiling so that it remains uniform or just Archimedean?" is a more manageable problem which we will discuss here.

Figure 16.6 distinguishes between a uniform and an Archimedean colored tiling of the regular tiling (3^6). These tilings are designated the symbols (3^6) 111112 and (3^6) 111112-A, where the sequence 111112 assigns the coloring of the six tiles around a vertex. The numeral 1 represents the color *white* and 2 represents *black*. Every vertex has five white and one black tile connected to it. Both are Archimedean because each vertex is surrounded by tiles of the same colors arranged in the same way. In Fig. 16.6 (a), all of the black triangles are pointing down, therefore all are equivalent. Thus, this colored tiling is also uniform. Because in Fig. 16.6 (b) half of the triangles point down and half point up, this not a uniform color tiling.

There are six possible uniform colored tilings of the original (3^6) tiling (Grünbaum and Shepard, 1987). If we consider all the ways of coloring the tiles around a vertex and determine whether the coloring is uniform or Archimedean, we can list all such tiles. With two colors of tiles (i.e., black

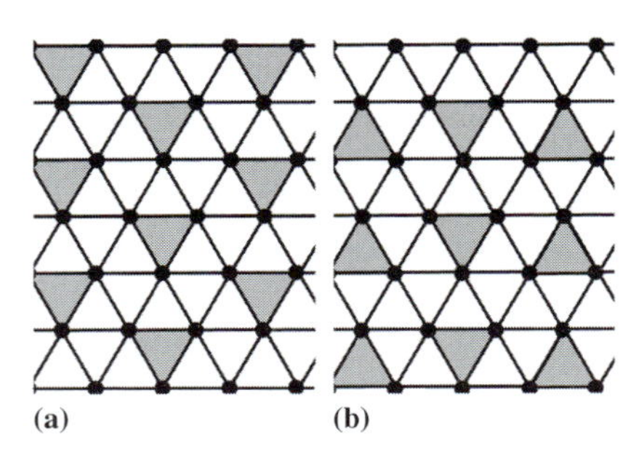

Fig. 16.6. (a) (3^6) 111112 regular and (3^6) 111112-A Archimedean two-colored tilings.

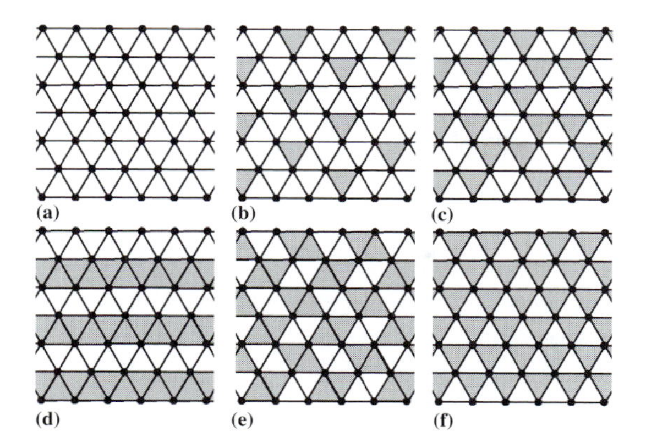

Fig. 16.7. All regular single and two-colored tilings of the plane tiling (3^6): (a) (3^6) 111111; (b) (3^6) 111112; (c) (3^6) 111212; (d) (3^6) 111222; (e) (3^6) 112122, and (f) (3^6) 121212.

and white), we can have six regular two-colored tilings of the plane tiling (3^6), as illustrated in Fig. 16.6. These regular single or two-colored tilings are (a) (3^6) 111111; (b) (3^6) 111112; (c) (3^6) 111212; (d) (3^6) 111222; (e) (3^6) 112122, and (f) (3^6) 121212. Note that the trivial examples where the color 1 is mapped into 2 or 2 into 1 (e. g., (3^6) 222222, (3^6) 222221, etc.) are degenerate tilings. If we allow for more colors, we can introduce additional regular n-colored tilings. It is left as an exercise for the reader to show that the only regular 3-colored tilings of (3^6) are (3^6) 111213 and (3^6) 121213 and the only regular 4-colored tiling of (3^6) is (3^6) 121314. There are no uniform 5- or 6-colored regular tilings of (3^6).

We can similarly determine the regular n-colored tilings of (4^4) and (6^3). However, there is an added complication in the case of (4^4), where two distinct tilings have identical labels. The additional regular single and 2-colored tilings are (4^4) 1111, (4^4) 1112 (i), (4^4) 1112 (ii), (4^4) 1122, (4^4) 1212, (6^3) 111 and (6^3) 112. The additional regular 3-colored tilings are (4^4) 1123 (i), (4^4) 1123 (ii), (4^4) 1213 and (6^3) 123, respectively. Only one additional regular 4-colored tiling is possible: the (4^4) 1234. The reader is encouraged to make color drawings of these tilings.

We can determine the regular n-colored tilings of the other Archimedean plane tilings. These include the 2-colored tilings $(3^3 \cdot 4^2)$ 11122, $(3^2 \cdot 4 \cdot 3 \cdot 4)$ 11212, $(3 \cdot 6 \cdot 3 \cdot 6)$ 2121, $(3 \cdot 12^2)$ 211, and $(4 \cdot 8^2)$ 211; and the 3-colored tilings $(3^4 \cdot 6)$ 11213, $(3^2 \cdot 4 \cdot 3 \cdot 4)$ 11213, $(3 \cdot 4 \cdot 6 \cdot 4)$ 2131, $(3 \cdot 6 \cdot 3 \cdot 6)$ 2131, $(4 \cdot 6 \cdot 12)$ 123, and $(4 \cdot 8^2)$ 312. An additional Archimedean, but not regular tiling, exists as $(3^3 \cdot 4^2)$ 11123-A. Once again, the reader is encouraged to make color drawings of these tilings as an exercise.[4]

[4] The reader is referred to Grünbaum and Shepard (Grünbaum and Shepard, 1987) for illustrations of these additional regular color tilings.

16.4 *Quasi-periodic tilings

A recent development in crystallography is the discovery of *quasicrystalline alloys*, i.e., alloys that exhibit *quasi-periodic* structures. The motifs in quasicrystals are typically built out of units with non-crystallographic symmetries, and consist of more than one tile (in 2-D) or *brick* (in 3-D). The dimensions (lengths, areas, or volumes) of these pairs of tiles are related to each other by an *irrational* number. While the motifs are not periodic in the traditional (translational) sense, there is a set of rules for constructing a space-filling tiling. If *quasi-lattices* are decorated with atoms, they diffract under Bragg conditions. The resulting diffraction pattern has discrete peaks in the scattered X-ray intensity and symmetrically oriented spots in electron diffraction.

Figure 16.8 illustrates the famous Penrose tiling (Penrose, 1974, 1978, Gardner, 1977), an example of a plane tiling which preserves global five-fold symmetry. We can identify *two Penrose rhombs* with areas in the ratio of the *golden mean* $= (1+\sqrt{5})/2 = 1.618034$. τ is an irrational number often encountered in pentagonal geometry. We can construct *pseudo-translations* in the Penrose tile by adding vectors pointing to the five vertices of the pentagon. This set of five basis vectors allow us to think of the Penrose tile as a crystal projected from a higher-dimensional (5-D in this case) space, where the irrational relationship between the two tiles is maintained.

Examples of 3-D quasicrystals are *decagonal alloys* and *icosahedral alloys*, first observed in Al-T and Al-T-Si alloys (T = transition metal atom). Icosahedral alloys exhibit global icosahedral symmetry. Thus, electron diffraction patterns conform to the icosahedral stereographic projections illustrated in Chapter 15. Icosahedral quasicrystals are likened to 3-D Penrose tiles, with Penrose "brick" volumes in the ratio of the golden mean. We can construct pseudo-translations in icosahedral quasicrystals by adding vectors pointing to

Fig. 16.8. Two-dimensional Penrose tiling.

6 of the 12 vertices of an icosahedron. This set of six basis vectors indicates that we can think of icosahedral quasicrystals as crystals projected from a 6-D space, in such a way that the irrational relationship between the two tiles results.[5]

16.5 *Regular polyhedra and *n*-dimensional regular polytopes

While there are three regular tiles in 2-D, the *equilateral triangle*, the *hexagon*, and the *square*, there are five regular or *Platonic solids* in 3-D. These are illustrated in Fig. 16.9 in a ball-and-stick format with "atoms" at the vertices of the polyhedron surrounding a central atom.

> The Platonic solids are: the *tetrahedron*, which is its own dual; the *cube* and *octahedron*, which are duals; and the *icosahedron* and *pentagonal dodecahedron*, which are duals.

Of the Platonic solids, only the cube can be used to tile 3-D Euclidian space. Although the other four solids often occur as coordination polyhedra in 3-D solids, they occur mostly as distorted units or clusters. Because these other polyhedra tile higher dimensional Euclidian spaces, we can consider projections from higher-dimensional spaces which preserve the symmetries of these solids. We will explore this further in the discussion of quasicrystals and amorphous metals.

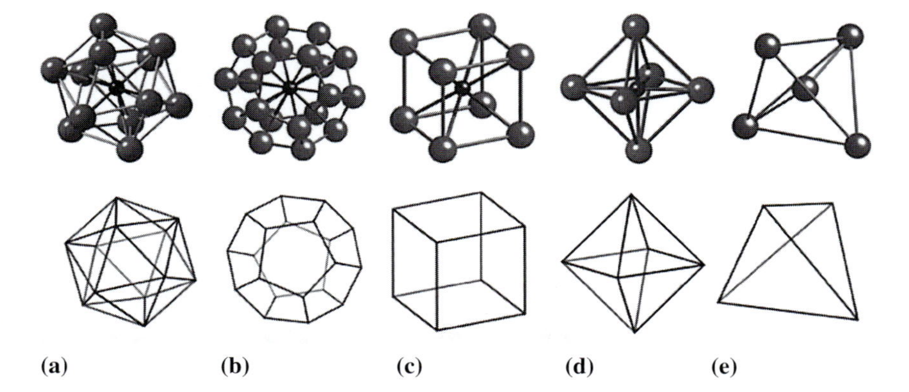

Fig. 16.9. Ball-and-stick and wire frame representations of the Platonic solids: (a) icosahedron, (b) pentagonal dodecahedron, (c) cube, (d) octahedron, and (e) tetrahedron.

[5] We discuss quasi-periodic tilings and quasicrystals in great detail in Chapter 20. Quasi-periodic tilings can be generated using the *QuasiTiler* from the Geometry Center at the University of Minnesota. *QuasiTiler* generates quasi-periodic tilings by projecting sections of higher-dimensional integer lattices onto a plane, an approach introduced by de Bruijn (1981). Originally, *QuasiTiler* was written for NeXTSTEP, for Marjorie Senechal, a visitor at the Geometry Center. Senechal is the author of an authoritative book on the geometry of quasicrystals (Senechal, 1995).

In the mathematics of tilings, we can consider tilings or tessellations of n-D spaces, such as the n-dimensional *Euclidean space*, or curved spaces, such as *spherical surfaces* or *hyperbolic surfaces*. We denote these spaces as E, S, and H, respectively, and the dimension of the space is indicated as a superscript. For the most part, we are interested in Euclidean and, sometimes, spherical spaces. Hyperbolic spaces are used in the description of lattice *defects*, in particular *disclinations*.

If we have curved arcs instead of lines as edges, we can show that the Platonic solids are polyhedra that tile the *spherical space,* S^2, i.e., the surface of a sphere. These curved edge structures are called *geodesic* structures. Figure 16.10 shows examples of geodesic tilings: the (partly truncated) icosahedral (a) and pentagonal dodecahedral (b) tilings.[6] The partial truncation emphasizes the five-fold symmetry axes in (a) and the three-fold symmetry axes in (b). If we replace the curved edges of regular geodesics by straight lines, then we obtain the Platonic solids. The cube is the only regular solid that will tile the 3-D Euclidian space, E^3. These progressive geometric ideas allow us to extend the concept of regular tilings to higher-dimensional spaces such as the 3-D spherical space, S^3, and 4-D Euclidean space, E^4. Such higher-dimensional tiles are known as *regular polytopes* (Coxeter, 1973). This nomenclature extends the sequence described in Table 16.2.

We can also describe n-D tiles by Schläfli symbols. Anticipating the need for three or more symbols to describe a node, we make a slight modification to the nomenclature to describe these tiles; in particular, we will no longer use superscripts. For the regular tiles of S^2 , i.e., the Platonic solids, we use the Schläfli symbol $\{p, q\}$, where p denotes the regular polygon that forms the faces and q the number of regular polygons meeting at a node. The tetrahedron is then designated as $\{3, 3\}$, the octahedron as $\{3, 4\}$, the cube as $\{4, 3\}$, the icosahedron as $\{3, 5\}$, and the pentagonal dodecahedron as $\{5, 3\}$.

Truncated and semi-regular solids that have two types of regular polyhedra as faces are called *Archimedean* solids.

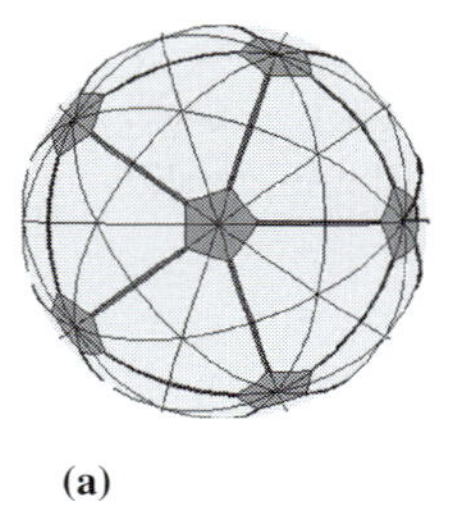

(a)

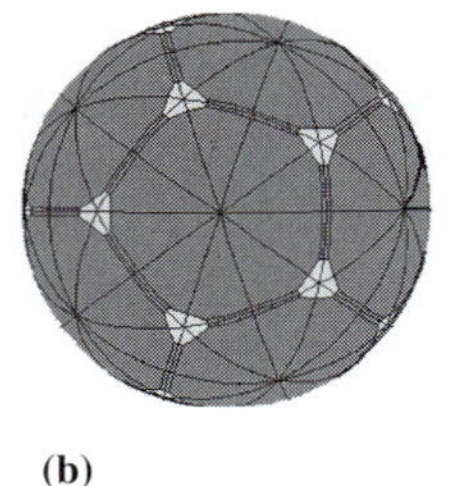

(b)

Fig. 16.10. Geodesics (partly truncated): (a) icosahedron and (b) pentagonal dodecahedron.

[6] These were generated using Kaleidotile (Version 1.5) written by J. Weeks for The Geometry Center http://geometrygames.org.

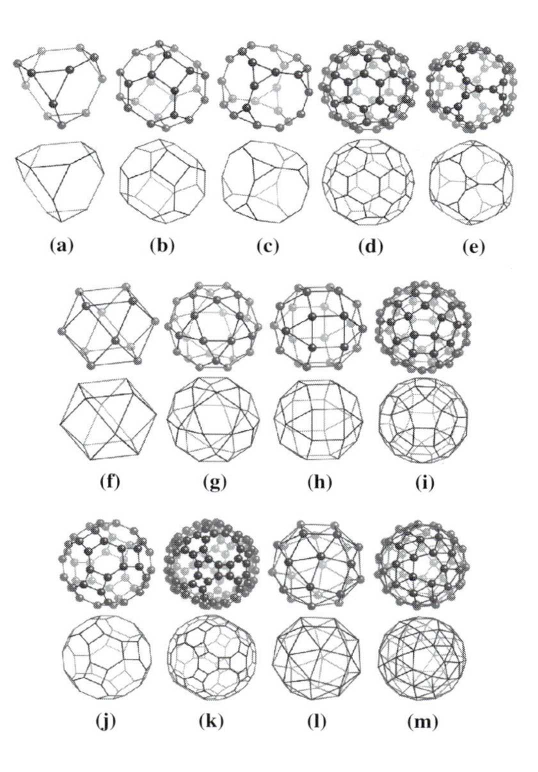

Fig. 16.11. Ball-and-stick and wire frame drawings of the Archimedean solids: (a) truncated tetrahedron; (b) truncated octahedron; (c) truncated cube; (d) truncated icosahedron; (e) truncated dodecahedron; (f) cubeoctahedron; (g) icosidodecahedron; (h) rhombicuboctahedron; (i) rhombicosidodecahedron; (j) rhombitruncated cuboctahedron; (k) rhombitruncated icosidodecahedron; (l) snub cube; and (m) snub dodecahedron.

The 13 Archimedean solids are illustrated in Fig. 16.11. Table 16.1 summarizes the geometrical features of each of the Platonic and Archimedean solids. This table gives the number of vertices, V, edges, E, and faces, F for the 5 Platonic and 13 Archimedean solids. The table also lists a volume factor, V_s, and the Schläfli symbol. The total volume of each solid is given by $V_s L^3$, where L is the edge length.

The five solids that are obtained by simple *truncation* of the Platonic solids are given the Schläfli symbol $t\{p, q\}$, where p and q denote the regular polygons that form the two types of faces. Truncation results in polygons with the symmetry of the original vertex; e.g., pentagons replace vertices for the truncated icosahedron, and the original triangular faces become hexagons after sufficient truncation to form a regular polygon. The five Archimedean solids resulting from truncation of the Platonic solids are the *truncated tetrahedron*, designated, $t\{3, 3\}$; *truncated octahedron*, $t\{3, 4\}$; *truncated cube*, $t\{4, 3\}$; *truncated icosahedron*, $t\{3, 5\}$; and *truncated dodecahedron*, $t\{5, 3\}$.

Two additional Archimedean solids are the *quasi-regular solids*. These have two types of regular polyhedral faces, where each face of one kind is surrounded completely by faces of the other kind. They are given Schläfli symbols of the type $\left\{ \begin{smallmatrix} p \\ q \end{smallmatrix} \right\}$, where again p and q denote the regular polygons that form the two types of faces. The quasi-regular solids are the *cuboctahedron* $\left\{ \begin{smallmatrix} 3 \\ 4 \end{smallmatrix} \right\}$ and *icosidodecahedron* $\left\{ \begin{smallmatrix} 3 \\ 5 \end{smallmatrix} \right\}$.

Table 16.1. Number of vertices V, edges E, and faces F for the 5 Platonic and 13 Archimedean solids, along with the volume factor V_s and Schläfli symbol.

Polyhedron	V	E	F	V_s	Schläfli symbol
tetrahedron	4	6	4	0.118	$\{3,3\}$
octahedron	6	12	8	0.471	$\{3,4\}$
cube	8	12	6	1.000	$\{4,3\}$
icosahedron	12	30	12	2.182	$\{3,5\}$
pentagonal dodecahedron	20	30	12	7.663	$\{5,3\}$
truncated tetrahedron	12	18	8	2.711	$t\{3,3\}$
truncated octahedron	24	36	14	11.31	$t\{3,4\}$
truncated cube	24	36	14	13.60	$t\{4,3\}$
truncated icosahedron	60	90	32	55.29	$t\{3,5\}$
truncated dodecahedron	60	90	32	85.04	$t\{5,3\}$
cuboctahedron	12	24	14	2.357	$\left\{ {3 \atop 4} \right\}$
icosidodecahedron	30	60	32	13.84	$\left\{ {3 \atop 5} \right\}$
rhombicuboctahedron	24	48	26	8.714	$r\left\{ {3 \atop 4} \right\}$
rhombicosidodecahedron	60	120	62	41.61	$r\left\{ {3 \atop 5} \right\}$
rhombitruncated cuboctahedron	48	72	26	41.80	$t\left\{ {3 \atop 4} \right\}$
rhombitruncated icosidodecahedron	120	180	62	206.8	$t\left\{ {3 \atop 5} \right\}$
snub cube	24	60	38	7.890	$s\left\{ {3 \atop 4} \right\}$
snub dodecahedron	60	150	92	37.62	$s\left\{ {3 \atop 5} \right\}$

The Schläfli symbol for the remaining solids employ the characters r, s, and t for rhombic, snub, and (rhombi)truncated, respectively. The rhombic solids are given the Schläfli symbols $r\left\{ {p \atop q} \right\}$, where rhombic implies additional square faces. There are two of these: the *rhombicuboctahedron* $r\left\{ {3 \atop 4} \right\}$ and *rhombicosidodecahedron* $r\left\{ {3 \atop 5} \right\}$, respectively. The rhombitruncated solids, given the Schläfli symbols $t\left\{ {p \atop q} \right\}$, are the *rhombitruncated cuboctahedron* $t\left\{ {3 \atop 4} \right\}$ and *rhombitruncated icosidodecahedron* $t\left\{ {3 \atop 5} \right\}$. Snub implies additional triangular faces; the snub solids are given the Schläfli symbols $s\left\{ {p \atop q} \right\}$. They are the *snub cube* $s\left\{ {3 \atop 4} \right\}$ and *snub dodecahedron* $s\left\{ {3 \atop 5} \right\}$.

A *simple closed surface* separates space into interior, surface, and exterior points. A polyhedron is a simple closed surface made of polygonal regions. *Euler's formula*, which relates the number of edges, vertices, and faces of a simply connected polyhedron is:

$$F - E + V = 2. \tag{16.1}$$

It is easy to verify that this relation is satisfied by all the Platonic and Archimedean solids in Table 16.1. This formula was discovered around 1750 by the Swiss mathematician Leonhard Euler (1707–1783), and first proven by Legendre in 1794.

Table 16.2. Summary of regular polytopes in E^4. V is the number of vertices, E is the number of edges, F is the number of faces and P is the number of polyhedra.

Name	Tile	Tessellation	V	E	F	P
Regular Simplex	$\{3, 3\}$	$\{3, 3, 3\}$	5	10	10	5
16-cell	$\{3, 3\}$	$\{3, 3, 4\}$	8	24	32	16
600-cell	$\{3, 3\}$	$\{3, 3, 5\}$	120	720	1200	600
24-cell	$\{3, 4\}$	$\{3, 4, 3\}$	24	96	96	24
Hypercube	$\{4, 3\}$	$\{4, 3, 3\}$	16	32	24	8
120-cell	$\{5, 3\}$	$\{5, 3, 3\}$	600	1200	720	120

We describe tessellations in higher-dimensional spaces by the arrangement of r regular blocks $\{p, q\}$ in an edge-to-edge arrangement with the Schläfli symbol $\{p, q, r\}$. The Schläfli symbol for the cube is $\{4, 3\}$. The only tilings of the spherical space S^3 are $\{3, 3, 3\}$, $\{3, 3, 4\}$, $\{3, 3, 5\}$, $\{4, 3, 3\}$, $\{3, 4, 3\}$, and $\{5, 3, 3\}$. There are six tessellations of E^4, which are summarized in Table 16.2 (Coxeter, 1973). We find several of these tessellations to be important in higher-dimensional spaces from which we project quasicrystalline and amorphous structures.

We can tile another curved 2-D space by taking any of the plane tilings discussed in this chapter and wrapping it around the circumference of a cylinder so that the vertices at the beginning and the end of the cylinder coincide. Such a construction is important in the discussion of the structure of *carbon nanotubes (CNT)* in Chapter 25. We describe the results of such tilings of the cylinder in terms of a *chiral angle*, ϕ, that describes the direction of the plane tile's wrapping around the cylinder.

16.6 Crystals with stacking of 3^6 tilings

We can decompose 3-D crystal structures into stackings of 2-D tilings. In this section, we give examples of the ordered stacking of atomic *close-packed planes*. Crystals that are *fcc* or *hcp* have low index planes with simple 3^6 tilings. We use stacking sequences of these tilings to describe the *polytypes* of the *wide bandgap semiconductor* SiC. Polytypes are structures of the same compound that differ only in their stacking sequence.

16.6.1 Simple close-packed structures: *ABC* stacking

The *fcc* and *hcp* structures can be described in terms of regular *triangular* 3^6 tilings, decorated with atoms at all vertices. The distance between vertices of the tiling coincides with twice the atomic radius, $2r$. In the *hcp* structure,

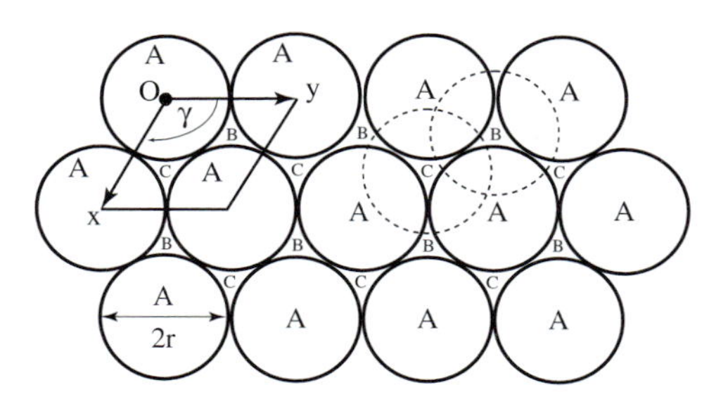

Fig. 16.12. (a) The close-packing of spheres on a 3^6 tile showing *A*-, *B*-, and *C*-sites.

these tilings make up the (00.1) planes, whereas in the *fcc* structure they correspond to the (111) planes. The *fcc* structure is also known as the *cubic close-packed structure, ccp*. The hexagonal coordination of atoms in close-packed planes about a central atom of equal size is illustrated in Fig. 16.12.

The 3^6 plane tiling has a **p6mm** plane group with lattice parameters $a = b = 2r$ and $\gamma = 2\pi/3$. Touching sphere atoms sit on each vertex of the 3^6 tiling; we refer to these sites in the first close-packed layer as A-sites. Close packing leaves two types of voids or *interstices* that manifest themselves as triangles in the plane. Triangles with their apex pointing up sit at B-sites, and those with apex pointing downward sit at C-sites. The A-, B-, and C-sites have basal plane coordinates $(0, 0)$, $(1/3, 2/3)$, and $(2/3, 1/3)$, respectively. The B- and C-sites are displaced, in the basal plane, with respect to A-sites by vectors $+\mathbf{S}$ and $-\mathbf{S}$, respectively, where $\mathbf{S} = a[\bar{1}010]/3$ in the Miller–Bravais notation (Krishna and Pandey, 2001).

We can construct 3-D *close-packed structures* with particular *stacking sequences* of the close-packed planes, as illustrated in Fig. 16.13. For touching spheres, each subsequent plane is at an elevation $z = a\sqrt{2/3}$ with respect to the previous one. Figure 16.13 shows stacking sequences for the *hcp* and *fcc* structures. The tilings are the same in stacked layers, but translated so that the $ABAB\ldots$ stacking constitutes the *hcp* structure, and the $ABCABC\ldots$ stacking sequence constitutes the *fcc* structure.

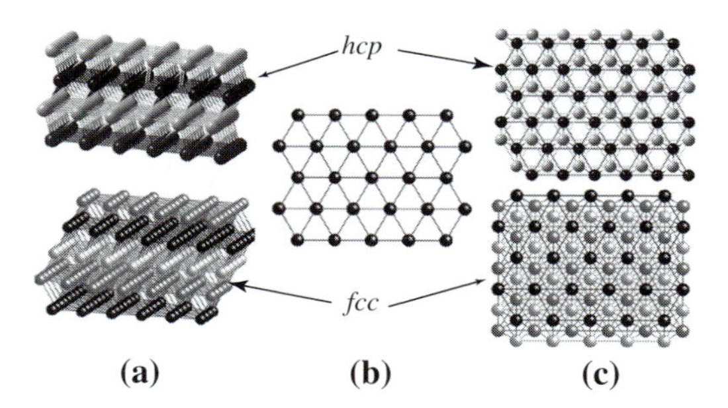

Fig. 16.13. (a) Close-packed planes in the *hcp* structure (above), *A*-site atoms are black and *B*-site atoms are grey. In the *fcc* structure (below) an additional layer of *C*-site atoms are light gray; (b) close-packed (00.1) *hcp* plane ((111) *fcc*); (c) projection of the *hcp* structure into a (00.1) plane; and the *fcc* structure projected into the (111) plane.

Figure 16.13 (a) (top) shows *ABAB*... stacking of close-packed planes in the *hcp* structure. Atoms on the *A*-sites are colored black, and those on the *B*-sites are colored gray. Figure 16.13 (a) (bottom) illustrates the *ABCABC*... stacking of close-packed planes in the *fcc* structure. Atoms on the *C*-sites are colored light gray. Figure 16.13(b) shows a close-packed (00.1) *hcp* plane (or (111) in *fcc*), decorated with *A*-site atoms, for reference. This is equivalent to the (111) plane for the *fcc* lattice. We distinguish the two structures by projecting all of the atoms into the close-packed planes. Figure 16.13(c) (top) shows the projection of all atoms in the *hcp* structure into an (00.1) plane, illustrating the occupancy of the *A*- and *B*-sites and empty *C*-sites. Figure 16.13 (c) (bottom) shows the projection of the *fcc* structure into a (111) plane with all *A*-, *B*-, and *C*-sites occupied.

16.6.2 Interstitial sites in close-packed structures

The *interstices* in a close-packed structure can be occupied by other atoms to form new compound structures. Figure 16.14 shows the two types of interstices that exist in close-packed structures. If the triangular void in a close-packed structure has an atom directly above it, then the four atoms surrounding the interstice form a regular *tetrahedron* (Fig. 16.14(a)). This is a *tetrahedral interstice*, or *tetrahedral interstitial site*.

Figure 16.14(b) shows an *octahedral interstice*, or *octahedral interstitial site*. Here the triangular void does not have an atom directly above it. The six atoms surrounding the interstitial site form an *octahedron*. If the two atomic layers are *A* and *B*, then this is a *C*-site interstice and can be labeled γ (α and β interstitial sites are similarly defined). Because the *octahedral interstitial site* is larger than the *tetrahedral interstitial site* it can be occupied by a larger atom.

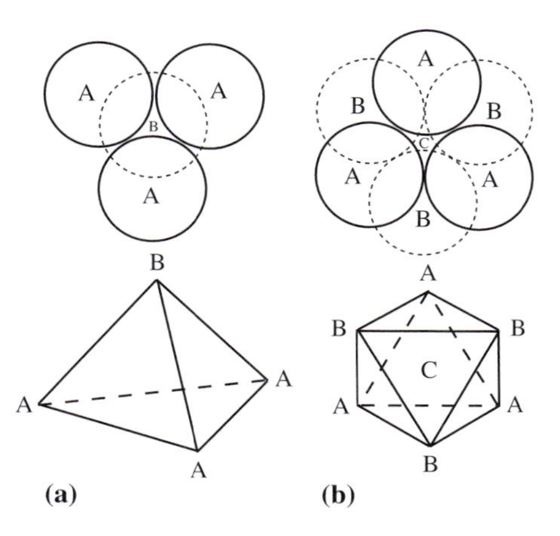

Fig. 16.14. Atoms and polyhedra about (a) tetrahedral and (b) octahedral interstices.

16.6.3 Representation of close-packed structures

In this section, we summarize a few of the more commonly used notations for the representation of close-packed structures. These are the *ABC notation*, the *Ramsdell notation* (Ramsdell, 1947), the *Zhdanov notation* (Iglesias, 2006), and the *h–c notation* (Jagodzinski, 1949).

16.6.3.1 *ABC* notation

We used the *ABC notation* in preceding paragraphs to describe the stacking sequences in simple *hcp* and *fcc* structures. This notation can be modified for a binary or multicomponent interstitial compound to describe the stacking sequence of each atom type. For example, CdI_2 has a structure in which the larger I ions are close-packed. The ions follow the stacking sequence $ABC\ldots$ The smaller Cd ions occupy the octahedral interstices between alternate close-packed I planes. The structure can be denoted as $\alpha BC\ldots$ where the Greek letters identify the interstitial cations.

Silicon carbide, (SiC) is a material that is known to have many *polytypes* which are distinguished by their stacking sequences (Frank, 1951). In all cases, the carbon atoms occupy tetrahedral interstices between all consecutive Si planes. One polytype has the stacking sequence $ABCACB\ldots$ for the Si atoms. This structure can in principle be written as $A\alpha B\beta C\gamma A\alpha C\gamma B\beta\ldots$ However, this is redundant because all of the interstitial sites are fixed by the positions of the Si atoms; therefore, we customarily omit the Greek symbols. We can also modify the notation so the layer-to-layer translations, A, B, and C always describe positions in a unit cell. Negative translations are then designated as $\underline{A}$, $\underline{B}$, and $\underline{C}$ (Pearson, 1972).

In Chapters 17 and 18, we will consider the stacking of 3^6, 6^3, and 3636 nets within the same structures. We will reserve the ABC notation to describe the stacking of close-packed 3^6 layers; while the same interlayer translation vectors apply to the 6^3 and 3636 nets, we will denote their stacking sequences using the symbols a, b, and c (for 6^3), and α, β, and γ (for 3636).

16.6.3.2 Ramsdell notation

While the *ABC notation* gives a complete description of the stacking sequence in close-packed structures, it does not specify the lattice type. It also requires long character strings for large repeat units. The *Ramsdell notation* is a shorthand notation that specifies the total number of close-packed layers followed by a letter that indicates whether the lattice type is cubic (C), hexagonal (H), or rhombohedral (R). If two or more structures have the same lattice type and the same repeat period, we use a subscript a, b, c, or 1, 2, 3 to distinguish between the structures. For example, SiC has two hexagonal polytypes with ABC stacking sequences $ABCACB\ldots$ and $ABCBAB\ldots$ for the Si atoms. Both of these structures have a hexagonal lattice with a six-layer repeat along the c-axis. They are then distinguished by their subscripts as $6H_1$

and $6H_2$, respectively. Although this notation is simply compact, it does not specify the actual stacking sequence.

16.6.3.3 Zhdanov notation

The positions of *A*-, *B*-, and *C*-sites have coordinates $(0, 0)$, $(1/3, 2/3)$, and $(2/3, 1/3)$ in the basal plane. Atomic translations from *A* to *B* (*B* to *C*) and *A* to *C* (*C* to *B*) are $(1/3, 2/3)$ and $(-1/3, -2/3)$, respectively. A classification scheme for the stacking sequence replaces all *AB* and *BC* pairs with a + symbol and all *AC* and *CB* pairs with a − symbol. We can also view the steps in terms of clockwise or counterclockwise rotations with respect to the (00.1) plane normal to the layers. This vision led Frank (1951) to label the transitions $\triangle$ and $\triangledown$, respectively. The 9H polytype of SiC has an *ABACACBCB*... stacking sequence that can be written as $+--+--+--$ or $\triangle\triangledown\triangledown\triangle\triangledown\triangledown\triangle\triangledown\triangledown$. This has the same number of characters as does the *ABC notation*. Zhdanov suggested a more compact notation that just records the sum of consecutive + ($\triangle$) and − ($\triangledown$) signs in the sequence. Thus the 9H polytype becomes (121212) or just (12) as this is the repeat sequence. The *Zhdanov notation* completely describes stacking sequences of close-packed structures. We can create an even more compact notation in multiple repeating sequences by using superscripts or subscripts to designate the number of repeating symbols (Pearson, 1972); in other words, (33333332) can be written as $(3_7 2)$ and (33323332333212133323332332) as $((3_3 2)_3 121(3_3 2)_3)$.

16.6.3.4 *h*–*c* notation

To use the *h–c notation* we must look at an *ABC* stacking sequence and label individual layers according to whether the layers above and below it are the same or different. In the *hcp ABAB*... stacking, the layers above and below are the same, so each layer is hexagonally surrounded, which is labeled by an *h*. Because the repeat is a single *h*, this structure is denoted by *h* using the *h*−*c* notation. In the *fcc ABCABC*... stacking, the layers above and below each plane are different and labeled as *c* (for cubic). Because the repeat is a single *c*, this structure is denoted as *c*. The 9H polytype of SiC is *hhchhchhc*... When we recognize the repeat unit, we obtain the symbol *hhc*.

16.6.3.5 Defects: stacking faults

While the *polytypes* of SiC are distinguished by their stacking sequences (Frank, 1951), a crystal of a single polytype may have local regions where the stacking is not perfect. A *stacking fault* is an example of a planar defect in a crystal where locally the stacking sequence deviates from that in a perfect crystal. An interesting example of a stacking fault in 4H-SiC is discussed in Box 16.6 (Liu *et al.*, 2002); 4H is the most common polytype of SiC. Potential stacking faults in SiC include: an *intrinsic Frank stacking fault*, an *extrinsic Frank stacking fault*, and a *Shockley stacking fault*. Frank faults are created by removal (or insertion) of a single bi-layer into a perfect crystal. A Shockley fault is created by displacing the crystal above the shear plane.

Box 16.6 Semiconductor nanostructure – planar fault in SiC

Scientists are currently actively studying SiC wide bandgap semiconductors for applications in high power and high temperature electronics. In all semiconducting materials, structural defects are a concern as they can degrade the electrical and optical properties of devices. Defects must be understood so as to minimize their occurrence in the crystal growth process. Common defects in SiC include a variety of stacking faults. Their identification by high resolution TEM allows scientists to understand microstructure properties relationships in these important semiconductors.

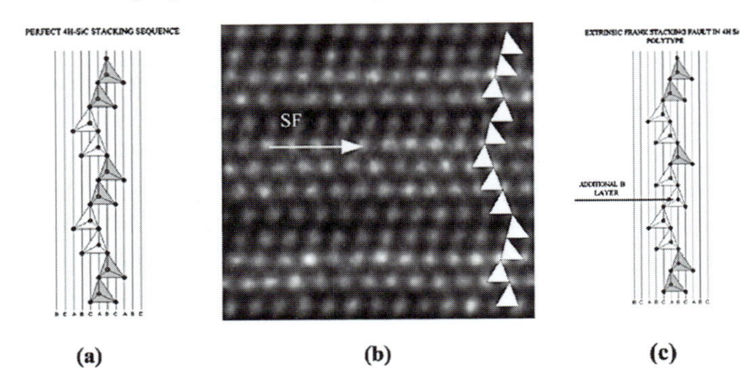

The figure above, courtesy of M. Skowronski, shows the stacking sequence of a perfect 4H-SiC crystal (a). A cross-sectional high resolution TEM (HRTEM) image of a stacking fault in 4H-SiC (b) shows white triangles on the right-hand side of the image to illustrate the stacking sequence. A white arrow on the left side indicates the position of a shear plane. This Shockley fault is created by displacing the crystal above the shear plane. All experimentally observed faults in SiC have a structure corresponding to a single layer Shockley fault. An example of a hypothetical extrinsic Frank stacking fault (not observed in SiC crystals) is shown in (c).

16.6.4 Polytypism and properties of SiC semiconductors

In *semiconducting materials*, electron states exist in *energy bands* separated by gaps with no electronic states. The origin of these energy gaps is explained readily by making use of the concepts of band theory and *hybridization gaps*. As shown schematically in Fig. 16.15(a), two atoms can lower

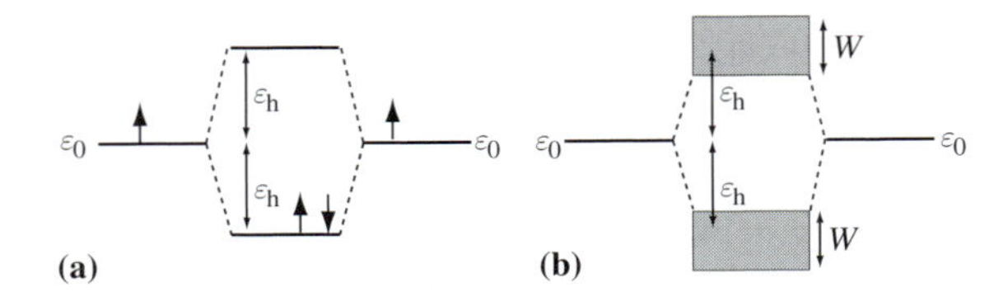

Fig. 16.15. (a) Hybridization between two atomic orbitals. (b) Hybridization in a solid separating bonding and antibonding bands with bandwidths W.

Table 16.3. Energy gaps (direct and indirect) for polytypes of SiC.

Polytype	3C	6H	4H	2H
Direct energy gap (eV)	5.14	4.4	4.6	4.46
Indirect energy gap (eV)	2.39	2.4	2.8	3.35

their energy through hybridization to form *bonding orbitals* and *antibonding orbitals* separated by twice the *hybridization energy*, ϵ_h. If we bring many atoms together to form a crystalline solid, the bonding and antibonding states broaden into *energy bands* of width W. The reason for the formation of energy bands is the Pauli exclusion principle. By comparing the size of the hybridization gap and the band widths of bonding and anti-bonding states, we can determine whether a gap persists or the two bands overlap. We can make a distinction between metallic, semiconducting, or insulating behavior by measuring the size of the *bandgap*, ϵ_g (the bandgap is measured from the top of the lower band to the bottom of the upper band). Energy bands for metals, semiconductors, and insulators, have $\epsilon_g \sim 0$, ~ 2 and ~ 10 eV, respectively.

SiC has a *wide bandgap* that ranges from $\epsilon_g = 2.4\text{–}5.1$ eV, (as compared to 1.1 eV for Si). The ϵ_g of SiC varies because this compound crystallizes into a large number of polytypic structures with different stacking sequences of hexagonal SiC double layers. Bandgaps are summarized in Table 16.3 for four of the more important SiC polytypes. We use the Ramsdell notation to label the polytypes. Energy gaps determine, among other things, the optical absorption spectrum of the semiconductor.

16.7 3^6 close-packed tilings of polyhedral faces

Given the efficiency of stacking close-packed layers of atoms, we can consider structures that arise from the tiling of the triangular faces of the Platonic solids with sections of a triangular tile. From this, we define new tilings which are important, for example, in understanding the structure of viruses in Chapter 25 (Caspar and Klug, 1963). We consider Platonic solids with exclusively equilateral triangular faces, i.e., the tetrahedron, the octahedron, and the icosahedron. The faces can be tiled recursively, whereby along any edge of the triangular face we place f equilateral triangles, where f is an integer. We see that this tiling can be continued to where there are f^2 new self-similar tiles in the original face; i.e., if we subdivide a linear dimension by f, we subdivide the areal dimension by f^2. This leads to the tilings illustrated in Fig. 16.16 (top). The resulting polyhedron, with faces that are all equilateral triangles, is known as a *deltohedron*.

This is not the only triangular subtiling of a triangular face. We can replace each equilateral triangle with a tetrahedron, whose base replaces the original triangular face. Three equilateral triangles are then inclined with respect to the plane of the base triangle. This new structure has three exposed faces for each

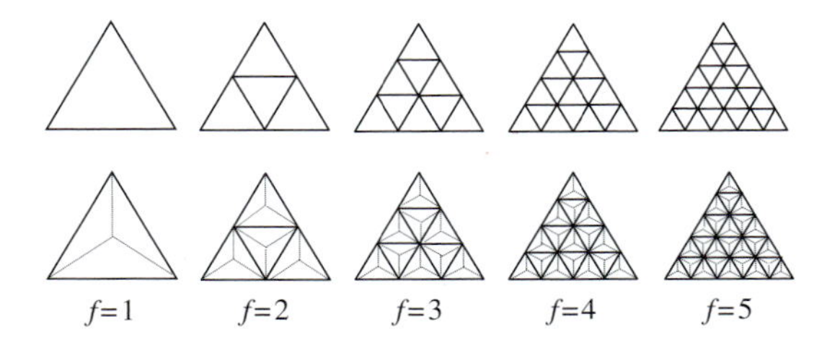

Fig. 16.16. Succesive tiling of an equilateral triangle with f^2 self-similar equilateral triangles (top) and decoration of the triangles with tetrahedra (bottom).

tetrahedron, replacing the triangular face of the original tile. This replacement has been done for all of the tiles in Fig. 16.16 (top) and is shown in Fig. 16.16 (bottom). We can see that there are now $3f^2$ equilateral triangles replacing the original triangular polyhedral face.

We define the *triangulation number* that describes the new *deltohedron* formed by these tilings. In general, the triangulation number is given by:

$$T = Pf^2 \quad \text{where} \quad P = 1, 3, 7, 13, 19, 21, 34, 37\ldots \tag{16.2}$$

where the numbers for P are given by $P = b^2 + bc + c^2$, with b and c non-negative integers having no common factors.[7] The number of facets on the deltohedron is $4T$, $8T$, and $20T$, for the tetrahedron, octahedron, and icosahedron, respectively. In Fig. 16.16, the illustrations are for $P = 1$ (top) and $P = 3$ (bottom). Larger values of P give rise to deltohedra that do not have planes of symmetry.

As an example, let us determine the number of individual molecular units (mers) that can decorate icosahedral deltohedra. Figure 16.17 (top) shows $P = 1$ (left) and $P = 3$ (right) tilings of the triangular faces of the icosahedron, schematically, with each of five faces showing the respective, $f = 1, 2, 3, 4$, and 5 subtriangulations. By decorating the vertices of the $P = 1$ (middle) and $P = 3$ (bottom) subtriangulations, we can count the number of molecular entities that decorate the deltohedra. In each case, there will be 12 entities corresponding to the decoration of the vertices of the original icosahedron. Because each entity that lies on an edge is shared between two faces we can count it as half, whereas an entity in the interior of a face is fully counted. In total, there are $12, 42, 92, 162$, and 252 entities in the $f = 1, 2, 3, 4$, and 5 variants of the $P = 1$ deltohedron, and $32, 122, 272, 482$, and 752 entities in the variants of the $P = 3$ deltohedron.

16.8 Historical notes

Many of today's crystallographic concepts can be traced to the 1611 writings of **Johannnes Kepler** (1571–1630) on the snowflake (Kepler, 1611). Kepler

[7] Notice the similarity with the Goldberg polyhedra used to describe fullerenes.

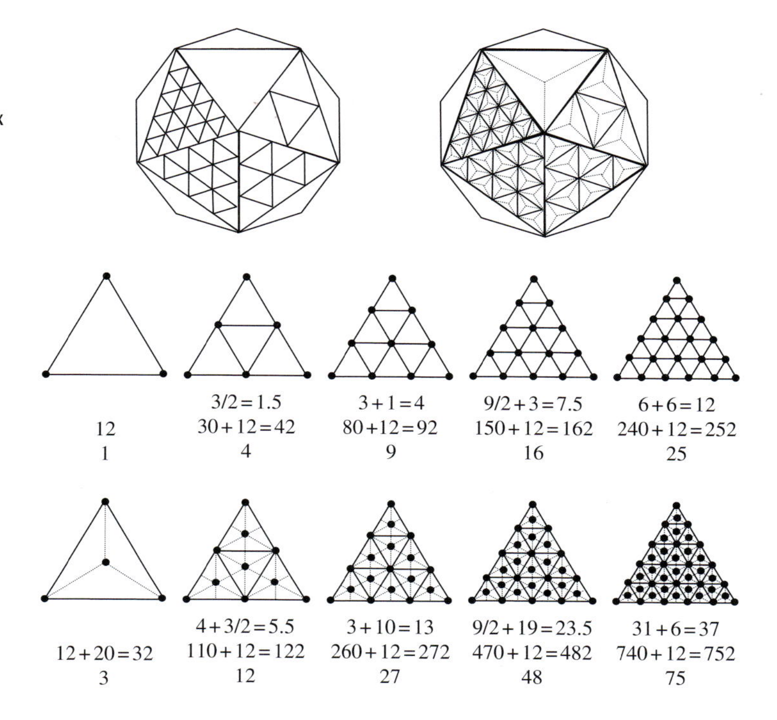

Fig. 16.17. $P = 1$ (left) and $P = 3$ (right) icosahedral tilings showing $f = 1, 2, 3, 4,$ and 5 subtriangulations. Vertex decorations in the $P = 1$ (middle) and $P = 3$ (bottom) subtriangulations.

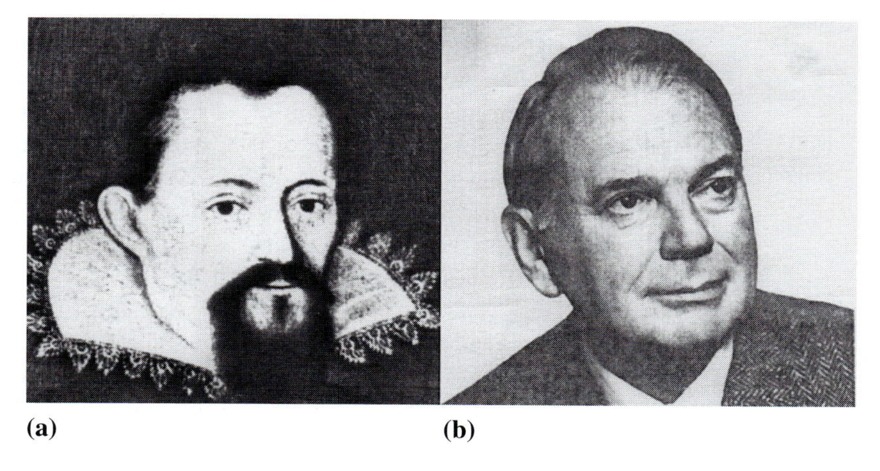

Fig. 16.18. (a) Johannnes Kepler (1571–1630) (picture courtesy of J. Lima-de-Faria and originally from Taton (1969)) and (b) Frederich Laves (1906–78) (picture courtesy of J. Lima-de-Faria and originally from Jagodzinski (1971)).

was a German scientist, astronomer, and mathematician; he was born in Weil der Stadt in Swabia, Germany. He studied theology at the Protestant University of Tübingen and passed the M. A. examination in 1591. Kepler was instructed by Michael Maestlin (1580–1635) in mathematical subjects, including Copernicus' heliocentric theory of astronomy.

Kepler was Professor of Mathematics at the Protestant seminary in Graz from 1594–1600. He left Graz during the Counter Reformation and moved to Prague in 1600 to serve as assistant to Tycho Brahe. When Brahe died in 1601, Kepler was appointed his successor as Imperial Mathematician, a position he held until 1612. A notable scientific accomplishment was Kepler's famous law of planetary motion. In 1612 Kepler's wife, Barbara, died and in the same year Emperor Rudolph II was deposed. From 1612–25, Kepler was the district mathematician in the city of Linz. Kepler and his family left Linz in 1626 during a peasant rebellion. He died in Regensburg in 1630.[8]

Kepler's 1611 treatise, *Strena Seu de Nive Sexangula* (Kepler, 1611), is considered the first serious work on geometrical crystallography (Lima-de-Faria, 1990).[9] In this work, he explained the observation of six-cornered snowflakes (as opposed to ones with five or seven corners) in terms of close-packing of minute ice spheres. He discussed simple cubic, face- and body-centered cubic, and hexagonal close-packing of spherical units, all of which are illustrated as motifs in the simple crystal structures of Chapter 17.

Kepler made important contributions to the theory of tiling. By considering combinations of regular polygons and how they tiled the plane, he determined the 11 plane networks in which the arrangements of polygons at each vertex are congruent. Kepler's fascination with the Platonic solids led him to assert, in his book *Mysterium Cosmographicum* (1596), that these "spherical harmonics" were integral to an explanation of the structure of the Solar System, an assertion that was later found to be mistaken (Mackay, 1981). Nonetheless, the notion of spherical harmonics proved to be of great mathematical importance as eigenfunction solutions to important partial differential equations, including the Schrödinger equation discussed in Chapter 2.

George Ludwig Friedrich (Fritz) Laves (1906–78) was a Swiss crystallographer who gave a mathematical derivation for the 11 plane networks first shown by Kepler. Though he was born in Hannover, Laves grew up in Göttingen. He was a descendant of Georg Ludwig Friedrich Laves (1788–1864), the court architect of the King of Hannover and England. In 1924, Fritz Laves began university studies in Innsbruck, Göttingen, and finally Zürich. Laves was influenced by the work of another crystallographer, **Paul Niggli** (1888–1953), whose ideas convinced him to study crystallography in Zürich. Laves used concepts such as sphere packings, geometrical space, and plane partitioning to describe silicates, AB_2 compounds, metals, and intermetallic compounds. He considered the influence of chemical factors, such as valence electrons and ionic bonding, on structures.

Laves classified crystal structures on the basis of topological concepts. He described structural units in a series of topologically closed-packed phases, now known as the *Friauf–Laves phases*.[10] For much of his career, Laves worked in two main fields of research: metals and intermetallic compounds

[8] For a complete biography consult Caspar (1993)

[9] (translated as "The Six-Cornered Snowflake" Kepler, 1966).

[10] *Friauf–Laves phases* will be discussed in detail in Chapter 18.

and order/disorder phenomena in alloys. He worked as an assistant to **Victor Moritz Goldschmidt** (1853–1933).[11] Goldschmidt and Laves developed a topological approach to structure derivation. Laves remained in Zürich in several positions until 1948 when he moved to the University of Chicago. In Chicago, he teamed with **Julian Royce Goldsmith** (1918–99) to investigate order/disorder in silicates. They performed important experiments on the alkali feldspars, studying Al-Si order and disorder. In this work, they developed an X-ray diffraction method to measure the degree of atomic order in alkali feldspars, a method that is still used today. In 1954, he returned to Zürich where he was made the Chair of Mineralogy at the ETH (as successor to P. Niggli).

16.9 Problems

(i) *Regular tilings I*: Determine the plane group for each of the regular uniform tilings.

(ii) *Regular tilings II*: Determine the unit cell for each of the regular uniform tilings.

(iii) *Regular tilings III*: Decorate the edges of the 3^6, 6^3, and 4^4 tilings with new vertices. Connect the vertices and identify the duals to the regular tilings.

(iv) *Regular tilings IV*: Consider decorating the vertices of the 3^6, 6^3, and 4^4 with touching circles. Calculate the fractional area covered by circles for each tiling.

(v) *Archimedean tiling I*: Show that the Archimedean $(4 \cdot 8^2)$ and Laves $[4 \cdot 8^2]$ tilings are duals. Compare the areas of the two prototiles.

(vi) *Archimedean tiling II*: Draw the dual to the Archimedean tiling $(3 \cdot 12^2)$. Identify the new tiling (i.e., assign a Schläfli symbol to the tiling).

(vii) *Archimedean tiling III*: Show that the Archimedean tiling $(3^4 \cdot 6)$ has two enantiomorphic forms; i.e., construct a right-and left-handed tile.

(viii) *Archimedean tiling IV*: Determine the 2-D Bravais lattice for the $(3 \cdot 4 \cdot 6 \cdot 4)$ Archimedean tiling and its dual. Show that the corresponding lattices are reciprocal to each other.

(ix) *Archimedean tiling V*: Consider the $(3 \cdot 12^2)$ Archimedean tiling:

(a) Calculate the fractional area covered by touching circles at the vertices.

(b) Determine the size of the largest touching circle that can be placed in the center of the 12-gon.

[11] Some of Goldschmidt's contributions will be discussed in subsequent chapters.

(c) Calculate the fractional area covered by the two circles. How does it compare with the one circle fractional coverage of the 3^6 tiling?

(x) *Kepler's tiles I*: For each of the Kepler tiles, identify the r-gons and the number, n_r present at a vertex. Show that the r-gons meeting at a vertex satisfy:

$$\sum_{r=3}^{r} \frac{(r-2)}{r} n_r = 2$$

(xi) *Kepler's tiles II*: Consider the following vertex types: $3 \cdot 7 \cdot 42$, $3 \cdot 8 \cdot 24$, $3 \cdot 9 \cdot 18$, $3 \cdot 10 \cdot 15$, $4 \cdot 5 \cdot 20$, $5^2 \cdot 10$, $3^2 \cdot 4 \cdot 12$, and $3 \cdot 4 \cdot 3 \cdot 12$.

(a) Show that they all satisfy the relation stated in the previous question.

(b) Explain why these vertex types do not yield Kepler tiles.

(xii) *Laves tiles*: Pick two of the Laves tiles and show that they satisfy:

$$\frac{(v_1-2)}{v_1} + \frac{(v_2-2)}{v_2} + \cdots + \frac{(v_r-2)}{v_r} = 2$$

(xiii) *Aperiodic Penrose tiling*: Consider the tiling of a decagon with 5 each of the oblate and prolate *Penrose rhombs* with unit edge lengths.

(a) Determine the pairs of interior angles for each of the Penrose rhombs.

(b) Determine the area and the ratio of the areas of the two rhombs.

(xiv) *k-uniform tilings*: Fig. 16.19 shows an example of a k-uniform tiling.

(a) What is the the Schläfli symbol for this tiling? Identify the value of k and show examples of distinct nodes.

(b) Draw a "unit cell" for this tile. Identify the plane group of the tile.

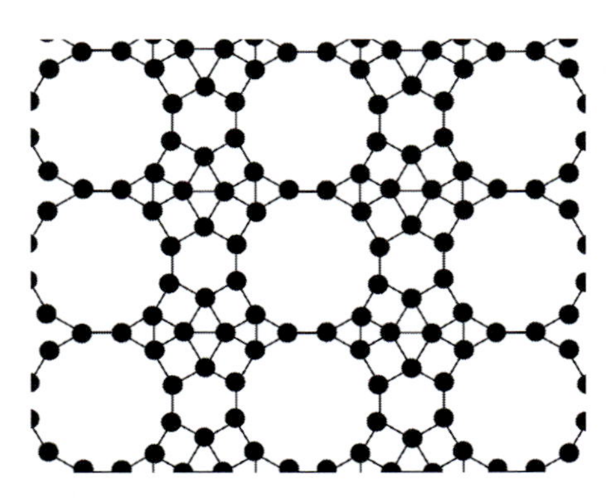

Fig. 16.19. Example of a k-uniform tiling.

(xv) *Colored tilings I*: Show the following for colored tilings of (3^6):

(a) The only regular 3-colored tilings are (3^6) 111213 and (3^6) 121213.
(b) The only regular 4-colored tiling is (3^6) 121314.
(c) There are no uniform 5- or 6-colored regular tilings of (3^6).

(xvi) *Colored tilings II*: Show that (3^6) 111232 is not a regular color tile of (3^6).

(xvii) *Platonic solids I*: Determine the point group for each of the five Platonic solids. Which of the Archimedean solids have the same point groups?

(xviii) *Platonic solids II*: Consider the cube:

(a) Construct the cube's dual by decorating the face centers with new vertices and connecting them.
(b) What structure results when you decorate the edge centers of the cube?

(xix) *Platonic solids III*: Determine the ratio of the center to vertex distance and edge length for an icosahedron.

(xx) *Platonic solids IV*: Determine the ratio of the center to vertex distance and edge length for a pentagonal dodecahedron.

(xxi) *Archimedean solid I*: Determine the coordinates of the 12 vertices of a cuboctahedron on the basis that the coordinate axes correspond to the 4-fold rotation axes and the edges are of unit length.

(xxii) *Archimedean solid II*: Determine the number of each type of face for the rhombitruncated icosidodecahedron.

(xxiii) *Tilings of curved space*: Consider tiling of curved spaces:

(a) Explain how to use 6^3 to cover a finite cylinder and ensure that the vertices coincide at the start and end of the circumference?
(b) Rationalize why it is not possible to cover a sphere with the 6^3 tiling but it is possible to tile the sphere with pentagons.

(xxiv) *Close packing I*: Show that for touching spheres in an *fcc* structure each subsequent (111) plane is at an elevation $z = a\sqrt{2/3}$ with respect to the previous.

(xxv) *Close packing II*: Determine the number of octahedral and tetrahedral interstices per atom in the *fcc* and *hcp* structures.

(xxvi) *Tetrahedral interstitial sites*: Calculate the ratio of radii for small and large spheres when the small spheres just fit into the tetrahedral sites in an *hcp* arrangement of the large spheres.

(xxvii) *Octahedral interstitial sites*: Calculate the ratio of the radii for small and large spheres when the small spheres just fit into the octahedral sites in an *fcc* arrangement of the large spheres.

(xxviii) *ABC notation*: Show that the *hcp* structure can equivalently be described by the sequence $A\underline{B}C$ if it is required that all translations are to positions in the same basal plane cell. The underlined symbol indicates that the translation is in the opposite direction.

(xxix) SiC *polytypes*: Determine the Ramsdell notation, the Zhdanov number and the h–c symbol for the following SiC polytypes: (a) 2H *ABAB*, (b) 3C *ABC*, (c) 4H *ABAC*, (d) 6H *ABCACB*, (e) *ABCB*, (f) *ABACACBCB*, and (g) *ABCBACABACBCACB*.

(xxx) Zn*S polytypes*: The hexagonal wurtzite and cubic sphalerite forms of ZnS, have ABABAB... and ABCABC... stackings, respectively. Determine the Ramsdell symbol for each.

(xxxi) *Deltohedra I*: Determine values of b and c that give rise to $P = 1, 3, 7, 13, 19, 21, 34, 37$ in the formula for triangulation numbers.

(xxxii) *Deltohedra II*: Determine how many entities exist in the f= 1, 2, 3, 4, and 5 variants of the $P = 1$ and $P = 3$ deltohedra for decorations of the faces of

(a) the octahedron;
(b) the tetrahedron.

CHAPTER

17 Metallic structures I: simple, derivative, and superlattice structures

> *"The important thing in science is not so much to obtain new facts as to discover new ways of thinking about them."*
>
> Sir William H. Bragg 1862–1942

17.1 Introduction

It is often useful to go beyond the standard description of a crystal in terms of the Bravais lattice and the unit cell decoration. In this chapter, we will look at various ways to *disassemble* and understand a crystal structure in terms of other concepts, including:

- *derivative structures*: new structures can often be derived from simpler structures by substitutions of one atom for another. Examples include the ordered occupation of body center and face center sites by different atoms in the *fcc*, *hcp*, or *bcc* derived structures.
- *interstitial structures*: new structures can be derived by the ordered occupation of subsets of the *interstitial sites* in a simpler structure. In particular, we will illustrate the occupation of octahedral and/or tetrahedral interstices in close-packed structures.
- *stacking variations*: new structures derived by the 1-D, 2-D, or 3-D stacking of substructures, e.g., ordered substitutions into an $m \times n \times o$ (m, n, o integers) superlattice of a parent structure.
- *decomposition into 2-D tilings*: we will give further examples of the ordered stacking of atomic planes. The *fcc*, *hcp*, and *bcc* structures are typically low index planes with simple tilings. Chapter 18 will extend

this to more complicated tiles with the introduction of the Frank–Kasper phases.

- *incommensurate and long-period stackings*: Sometimes, structures cannot be defined in terms of standard unit cells, but require the introduction of commensurate or incommensurate long-period modulations and superlattices.
- *polyhedral connectivity*: the types of atomic coordination polyhedra in a structure and the manner in which these coordination polyhedra connect (i.e., by sharing vertices, edges, or faces) can reveal important aspects of the crystalline structure. This will become important when we discuss Pauling's rules for ionic structures (Pauling, 1946).

We begin this chapter with a brief description of the most important parent structures (for metallic materials) and introduce the Hume–Rothery rules and some basic phase diagrams. Then we cover a more systematic approach to the description of derivative and superlattice structures in *fcc*, *bcc*, diamond, and *hcp*-derived structures. We conclude the chapter with a discussion of structures with interstitial alloys, alternative stacking sequences, and natural and artificial (commensurate and incommensurate) long-period superlattices, including how they can be identified by means of X-ray diffraction methods.

17.2 Classification of structures

In classifying crystal structures, we can make use of several classification schemes. One is the *StrukturBericht symbol* (Table 17.1), defined in the next section; another is the *Pearson symbol*. We also often use a common name for the structure type (e.g., diamond cubic, zinc blende), or the name of a mineral or compound with that structure type and the name of a prototype material. As a consequence, many structures have multiple symbols, depending on the classification scheme in use.

17.2.1 StrukturBericht symbols

Common StrukturBericht symbols (Hahn, 1989) are listed in Table 17.1. The StrukturBericht symbols begin with a letter followed by a number. A designates the structure of pure elements and B designates equiatomic *AB* compounds, and so on. The number following the letter gives the sequential order of the discovery of the particular structure-type. In some cases, we choose the number to parallel the structure first found in elemental form. One such instance is when the new structure is a compound or an alloy with a derivative or a superlattice of the elemental structure. For example, the A2 structure refers to body-centered cubic (*bcc*) and B2 refers to an ordered *AB* alloy with *A* atoms on the vertex sites and *B* atoms on body-centered sites.

Table 17.1. StrukturBericht symbols for various crystal structure types.

Structure	Types
A types	Elements
B types	*AB* compounds
C types	AB_2 compounds
D types	A_mB_n compounds
E...K types	More complex compounds
L types	Alloys
O types	Organic compounds
S types	Silicates

In some cases, there is more than one derivative of an elemental structure within a crystal structure type. These are designated by an additional subscript number that follows the number of the elemental structure-type. For example, two derivatives of the *fcc* (A1) structure are the $L1_0$ and $L1_2$ structures. The L designates that these are alloy structure-types while the 1 signifies that they are derivatives of the A1 elemental *fcc* structure-type. The 0 and 2 subscripts distinguish the two types of derivative structures. StrukturBericht symbols are used extensively in the materials literature, where, often, they are the only symbol listed to identify a particular structure-type.

17.2.2 Pearson symbols

A Pearson symbol uses the Bravais lattice symbol (cubic, c; tetragonal, t; hexagonal and rhombohedral, h; orthorhombic, o; monoclinic, m; or triclinic/anorthic, a, followed by a symbol designating the lattice centering (primitive, P; single face-centered, A, B, or C; face-centered, F; body-centered, I; or rhombohedral, R). The final character(s) indicate the number of atoms in the unit cell.

A few examples will clarify the use of the Pearson symbol: With 8 atoms in a face-centered (F) cubic (c) cell, NaCl is designated cF8. With a single atom in a hexagonal (h) primitive (P) cell, *hcp*-Co is designated hP1. With two atoms in a tetragonal (t) primitive cell, CuAu is designated tP2. The compound FeB has eight atoms in an orthorhombic (o) primitive cell and is designated oP8. α-Plutonium (Pu) has 16 atoms in a monoclinic (m) primitive cell and is designated mP16. Californium (Cf) has four atoms in an asymmetric (a) primitive cell so is designated aP4.

The Pearson symbol does not define a single structure uniquely; there may be several different structures that have the same Pearson symbol. Nevertheless, the Pearson symbol is useful when used in conjunction with the *Pearson Handbook of Crystallographic Data for Intermetallic Phases* (Villars and Calvert, 1991). This handbook contains about 50 000 entries of intermetallic

Table 17.2. Representative elements for **Structure 1**. Pearson's tables list 485 intermetallic compounds (mostly solid solutions) with this structure-type.

Element	a	Element	a	Element	a	Element	a
Cu	0.3615	Ag	0.4086	Au	0.4078	Al	0.4049
Ni	0.3524	Pd	0.3891	Pt	0.3924	Pb	0.4950

structures. A smaller, desktop edition (Villars, 1997) lists 27 686 structures, covering the literature from 1913 until 1995. The handbook is accompanied by a four-volume *Atlas of Crystal Structure Types for Intermetallic Phases* (Daams *et al.*, 1991), detailing the atom coordinates, coordination polyhedra, structure drawings, and so on. These books are valuable reference works for researchers in the area of intermetallics, although many other closely related crystal structures are also listed.

17.2.3 Structure descriptions in this book

Throughout the following chapters, we will introduce more than a hundred important structure types. The tables containing the description of all the structure types introduced in this book can be found as an on-line appendix on the book's web site.[1] The structures appendix is formatted as follows: all structure types receive a sequential number, in the order in which they are introduced in the text. For each structure, the following information is listed: prototype, StrukturBericht symbol (SBS), Pearson symbol (PS), space group number and symbol (SG), and the lattice complex (the atom positions in the asymmetric unit). In addition, a table is shown listing other compounds with the same structure and the corresponding lattice parameters. An example of such an entry is shown in Table 17.2 for the first structure, the *fcc* structure with prototype Cu:

Structure 1 *Prototype:* Cu
SBS/PS: A1/cF4 *SG # 225:* $\mathbf{Fm\bar{3}m}$ (O_h^5)
Lattice complex: Cu @ $4a(0, 0, 0)$

Assuming that we can use the space group symmetry to generate the atom positions outside the asymmetric unit, the information listed in the structures appendix should be sufficient to generate the entire unit cell. For each of the structure types, the reader can also find a *CrystalMaker* input file on the web site, as well as a color illustration of the structure.

[1] The structure descriptions are available as an on-line appendix rather than a real appendix because they would take up a large number of additional pages in an already voluminous text. The appendix can be downloaded as the file *StructuresAppendix.pdf*.

17.3 Parent structures

For metals, many important derivative and superlattice structures are based on the fundamental *fcc*, *bcc*, and *hcp* parent structures illustrated in Fig. 17.1. The *simple cubic structure, sc*, is rarely observed in nature; one particular example is the element α-polonium (Beamer and Maxwell, 1949). The *bcc* and *fcc* structures are derivatives of the *sc* structure. These three structures can alternatively be described in terms of stacking of simple atomic layers with triangular, 3^6, or square, 4^4, *regular tilings*. Here, the *bcc* structure has an AB stacking of regular *square* tilings, which constitute the (001) planes of the structure. The vertices of the B layer tiling sit above the centers of the squares in the A layer tiling.[2]

It is useful to consider the XRD patterns for these parent structures, and to compare them with superlattice diffraction patterns that we will discuss later on in this chapter. Figure 17.2 (a), (b), and (c) show simulated patterns for *fcc* Fe, *bcc* Fe, and *hcp* Co, respectively. These patterns were generated assuming Cu-Kα radiation and equilibrium lattice constants (for large, untextured polycrystalline grains). These simple structures also have relatively simple XRD patterns. The *fcc* structure has reflections corresponding to the planes, (111), (020), (022), (113), (222), (004), (313), and (240), respectively, in the range $0 \leq 2\theta \leq 150°$. These satisfy the *extinction rules* that h, k, and l have the same parity, as we derived in Section 12.2.1.4. Elemental Cu is the prototype for the StrukturBericht symbol A1 (**Structure 1**).[3]

The *bcc* structure has reflections corresponding to the (110), (002), (121), (022), (031), and (222) planes respectively, in the range $0 \leq 2\theta \leq 150°$. These satisfy the extinction rule that $h + k + l$ must be even, as derived in Section 12.2.1.3. While the *fcc* structure (with Cu as its prototype) was given the StrukturBericht symbol A1, the *bcc* structure (with W as its prototype) is given the symbol A2 (**Structure 2**).

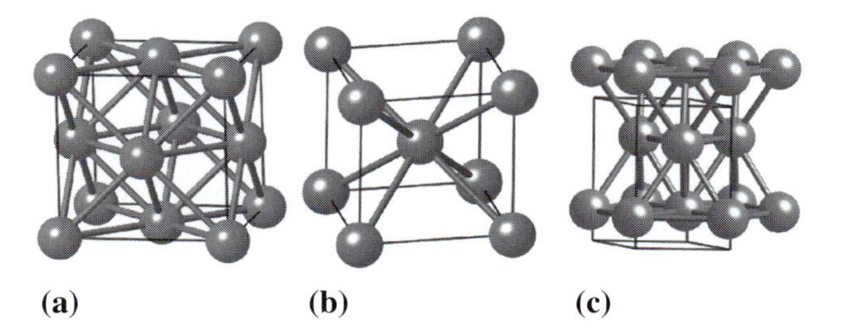

Fig. 17.1. The basic *fcc*, *bcc*, and *hcp* parent structures.

[2] We discussed the stacking of close-packed layers in Chapter 16.

[3] Recall that the structure numbers refer to the entries in the *StructuresAppendix.pdf* file on the web site.

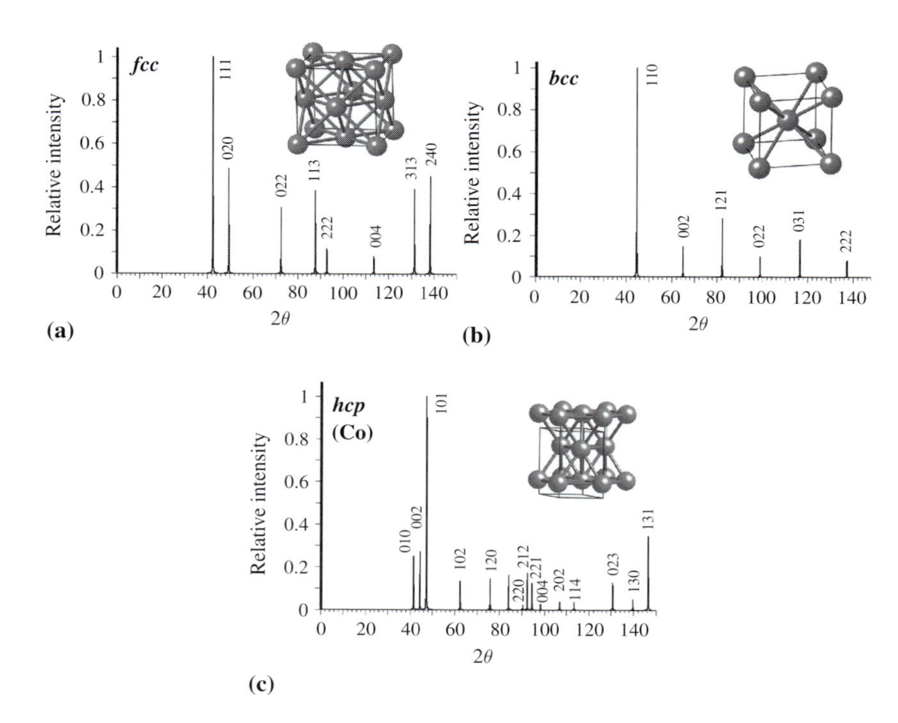

Fig. 17.2. Simulated XRD patterns assuming Cu-Kα radiation and equilibrium lattice constants for the (a) *fcc* Fe, (b) *bcc* Fe and (c) *hcp* Co structures.

The *hcp* structure has reflections corresponding to the planes, (00.1), (00.2), $(10.\bar{1})$, $(10.\bar{2})$, $(\bar{1}2.0)$, $(0\bar{1}.3)$, $(\bar{2}2.0)$, $(\bar{2}1.2)$, $(\bar{2}2.1)$, (00.4), $(20.\bar{2})$, $(1\bar{1}.4)$, $(0\bar{2}.3)$, $(\bar{1}3.0)$ and $(1\bar{3}.1)$, in the angular range $0 \leq 2\theta \leq 150°$. Note that this *hcp* diffraction pattern was computed for Co, which does not have an ideal c/a ratio of $\sqrt{8/3}$. The *hcp* structure (with Mg as its prototype) is given the symbol A3 (**Structure 3**).

17.3.1 Geometrical calculations for cubic structures

We can calculate the *number of lattice points* in a unit cell in any arbitrary crystal system. If we know the lattice constants and types of atoms decorating each of the lattice points, we can calculate properties such as the *atomic volume*, the *atomic packing fraction* and the *theoretical density* of the crystalline solid. Next, we illustrate such calculations for the *sc*, *bcc*, and *fcc* structures. Similar computations for other crystal systems are provided as exercises at the end of the chapter. We already know from Section 3.5 on page 69 how to compute the number of atoms per unit cell. Using that procedure, we find that there is one lattice site per unit cell for the *sc* structure, two sites per cell for the *bcc* structure, and four sites per cell for the *fcc* structure.

17.3.1.1 Atomic sizes

In materials with predominantly metallic bonds, we can learn about atomic sizes by analyzing the close-packing of hard atomic spheres. We can determine the metallic radius in a pure elemental metallic solid from a simple touching sphere consideration, such as taking half the bond distance along a close-packed direction. This concept was proposed by Bragg (1920) and extended to metals by Goldschmidt (1928).

For elements having two (or more) *allotropic* forms (meaning that there is more than one crystal structure for the same chemical composition), we can compare the touching sphere atomic radii to assess atomic size changes between structures. An example is the *bcc* and *fcc* allotropic forms of Fe; the *bcc* lattice constant is 0.28664 nm, and the *fcc* lattice constant is 0.36468 nm. We compute the *bcc* and *fcc* metallic radii by considering the geometry of touching spheres and the close-packed directions in the *fcc* and *bcc* structures, as shown in Box 17.1. Next, we will consider how these radii differ if the volume per atom is conserved.

The *packing fraction* is defined as the volume of space that is occupied by atoms divided by the total volume of the unit cell. In an exercise at the end of the chapter, we explore the packing fraction of hard spheres in various lattices. With atoms of the same radius, there is an approximately 8% difference in the packing density between the *bcc* and *fcc* structures. If we assume that the *atomic volume* is conserved, we can compute the expected value for the difference in metallic radii for the *fcc* and *bcc* structures (see Box 17.2).

Box 17.2 shows that the metallic radius in the higher coordinated structure, i.e., *fcc* with coordination number 12, is larger than that of the lower coordinated structure, i.e., *bcc* CN = 8, at constant atomic volume. The difference between the atomic radii in *fcc* and *bcc* structures is nearly 3%. The observed difference for Fe is 4%, indicating that the atomic volume is not the same in the two allotropic forms. We can account for atomic volume differences in different allotropes of the same element through considerations such as thermal expansion, directional bonding differences, differences in the magnetic state, and so on.

Box 17.1 Computation of the metallic radius of Fe in *fcc* and *bcc* structures

For an *fcc* metal, atoms touch along [110] directions and the atomic radius is given by $a\sqrt{2}/4$ (a is the cubic lattice constant). For a *bcc* metal, atoms touch along [111] directions: the atomic radius is hence given by $a\sqrt{3}/4$. The atomic radii of Fe in the *fcc* and *bcc* structures are then computed by substituting the equilibrium lattice constants:

$$r_{fcc} = \frac{\sqrt{2}}{4}a = 0.1289\,\text{nm} \quad \text{and} \quad r_{bcc} = \frac{\sqrt{3}}{4}a = 0.1242\,\text{nm},$$

showing that the *fcc* metallic radius is roughly 4% larger than in the *bcc* structure.

Box 17.2 The volume conserving *fcc* and *bcc* atomic radii of Fe

The volume per atom for each of these structures is given by:

$$\Omega_{fcc} = \frac{(a)^3_{fcc}}{4}; \qquad \Omega_{bcc} = \frac{(a)^3_{bcc}}{2}.$$

To conserve the atomic volume (i.e., to have $\Omega_{fcc} = \Omega_{bcc}$) we must have:

$$(a)_{fcc} = 2^{\frac{1}{3}}(a)_{bcc}.$$

and we find:

$$r_{fcc} = \frac{\sqrt{2}}{4}(a)_{fcc} = \frac{2^{\frac{1}{3}}\sqrt{2}}{4}(a)_{bcc}; \quad r_{bcc} = \frac{\sqrt{3}}{4}(a)_{bcc} \quad \text{and} \quad \frac{r_{fcc}}{r_{bcc}} = 1.029.$$

The *theoretical density*, ρ, of a material can also be calculated by considerations of the unit cell volume, V_{cell}, of the crystal structure, the atomic mass, M, and Avogadro's number, N_{A}. The general formula can be expressed as:

$$\rho = \frac{\frac{\text{atoms}}{\text{cell}} \times M}{V_{\text{cell}} \times N_{\text{A}}} = \frac{\Omega \times M}{N_{\text{A}}}. \tag{17.1}$$

17.4 Atomic sizes, bonding, and alloy structure

When we analyze a crystal structure of a multi-component alloy, we must understand the relative sizes of the atoms in the structure (Barrett and Massalski, 1980). This notion is not so simple, because the same atom can have a different size in a different crystal structure. While atomic size generally scales with atomic number, it also depends on the degree of filling of the outer electronic shells, the valence number, and the coordination number.[4] The influence of bonding on the atomic size of atoms in crystals is important because it leads to the concept of metallic bond radii, covalent bond radii, ionic bond radii, and van der Waals bond radii, first described by Pauling (1946). These distinctions between radii follow the distinctions made for *chemical bonding* in Chapter 2.

It is useful to develop an intuition about atomic sizes and how they vary as a function of the type of bonding, coordination number, valence, and valence differences in alloys and compounds. In this chapter, we will consider mainly

[4] The coordination number is the number of bonds formed with neighboring atoms.

metallic bonding and covalently bonded solids. In Chapter 22, we will consider ionic solids for which Pauling's rules predict a preference for structures based on the relative anion and cation sizes. Because some illustrated structures are common to metallic, covalent, and ionic structures, we will see some overlap in the structures described in these chapters.

In alloys, we can use the lattice constants of the elemental form of the components to estimate the atomic radii, and then use these radii to predict the lattice constants of the alloy. The simplest case is a *substitutional solid solution* of two components, as it has the same crystal structure as the elemental components, with relatively small differences in atomic number and valence. The lattice constant for an AB alloy is predicted by *Vegard's law*:

$$a_{\text{alloy}} = X_A a_A + X_B a_B. \tag{17.2}$$

where a_A and a_B are the lattice constants of the pure components, and X_A and X_B their *atomic fractions* in the alloy. If the differences in atomic number and valence electrons are not small, corrections need to be made on the basis of the electronegativity difference and/or lattice strain on substitution, yielding modifications to *Vegard's law*. Such corrections are beyond the scope of this introductory discussion.

In AB alloys where one of the atoms is significantly smaller than the other (e.g., Fe-C), we can have *interstitial solid solutions* where the smaller interstitial solute atom fits comfortably in the interstices of the solvent atom lattice. Lattice constant changes are then determined by strain effects.

17.4.1 Hume-Rothery rules

The extent of solid solubility is of considerable interest to a materials scientist studying an alloy system. There are certain composition ranges in alloys where the range of solid solubility is limited and new alloy phases or compounds occur. Empirical rules for the extent of solid solubility were first enumerated by Hume-Rothery (1926). The *Hume-Rothery rules* state the factors limiting the extent of solid solution:

(i) *atomic size factor*: The range of solid solubility will be restricted if the atomic radii differ by more that about 15%. Defining atomic diameters is simple in *fcc* and *bcc* structures, as illustrated above, but can become complicated in others.

(ii) *electronegativity valency effect*: Large *electronegativity* differences between components of a binary alloy can promote charge transfer and differences in the covalency, ionicity, or metallicity of the bonds. This leads to bond energy differences between A–A, A–B, and B–B bond energies in the alloy. Electronegativity is an empirical parameter introduced by Linus Pauling and extended by Mulliken (as discussed in

Chapter 2). A strong proclivity for A–B bond formation can lead to the formation of stable compounds.

(iii) *relative valency effect*: A metal of lower valency is more likely to dissolve in a metal of higher valency than vice versa. This rule is not universally obeyed.

Hume-Rothery rules identify the need to consider compound formation and intermediate phases in discussing the crystal structures observed in an alloy system.

An ideal (line) compound has a single stoichiometry prescribed by the ratio of A to B in the compound formula.

A compound at higher temperatures will usually be able to dissolve additional A or B to form a non-stoichiometric *intermediate solid solution phase*. Intermediate phases are also known as *intermetallic compounds*. Examples of intermediate phases include normal valency and electron compounds, Laves phases, and Frank–Kasper phases.

Normal valency compounds form when there are large electronegativity differences between the elements. Examples include: Mg_2Sn, Mg_3Sb_2, $MgTe$, etc. where the valences are 2, 3, 3(5), and 6 for Mg, Sn, Sb, and Te, respectively. *Band theory* explains *electron compounds* based on specific *electron-to-atom ratios*, (e/a). At certain e/a values, many alloys will have identical crystal structures. Prototype electron compounds include: CuZn $(e/a = 3/2)$, Cu_5Zn_8 $(e/a = 21/13)$, $CuZn_3$ $(e/a = 7/4)$, and α-Mn $(e/a \sim 1)$. Box 17.3 explains the calculation of e/a values for some CuZn alloys. These are unambiguous in that the number of free electrons in Cu can be assigned as 1 and for Zn as 2 (corresponding to a 4s and $4s^2$ valence shell configuration, respectively). In alloys with transition metal atoms for which the d-shell is not full, there is ambiguity in the varying number of free s-electrons across the transition series.

Box 17.3 Calculation of electron-to-atom ratios in Cu-Zn alloys

Here is how to compute the electron-to-atom ratios in several CuZn alloys:

$$\mathrm{CuZn}\ : \frac{e}{a} = \frac{1(1)+1(2)}{1+1} = \frac{3}{2};$$

$$\mathrm{Cu_5Zn_8} : \frac{e}{a} = \frac{5(1)+8(2)}{5+8} = \frac{21}{13};$$

$$\mathrm{CuZn_3} : \frac{e}{a} = \frac{1(1)+3(2)}{1+3} = \frac{7}{4}.$$

Laves phases, or *Friauf–Laves phases* (Laves and Witte, 1935), have the chemical formula AB_2. Their atomic diameters are predicted to be in the ratio of 1.2:1. Friauf first discovered the prototype $MgZn_2$ material (Friauf, 1927a,b) while Laves did extensive work in describing the structure of this and the related $MgNi_2$ and $MgCu_2$ phases. These are examples of a larger class of *topologically close-packed phases* (TCPs) discussed in Chapter 18, known as the *Frank–Kasper phases*. The TCP structures can be understood qualitatively using the *free electron theory* and quantitatively using the *band theory of solids*. The free electron theory assumes an isotropic, uniformly dense *electron gas*.

The *cohesive energy* is the minimum total potential energy occurring at the equilibrium spacing R_0, i.e., the energy per atom required to break all of the bonds in the solid.

A large portion of the cohesive energy in metals exists in the electron gas and depends on the electron density and its spatial variation. A general rule for metallic structures is that atoms in metals should fill space to maximize the electron density. Metals usually have high symmetry to maintain high coordination and uniform electron densities. Thus, they choose structures with tetrahedral interstices, which leads to a close packing of atoms. The TCPs have *only* tetrahedral interstices and are derived from the packing of *Kasper triangulated coordination polyhedra*. This packing allows distorted tetrahedra to pack without problems of steric constraints.[5]

17.4.2 Bonding in close-packed rare gas and metallic structures

Calculating binding energies in solids is a complicated quantum mechanical problem that is generally solved through numerical solutions to Schrödinger's equation. In special cases, such as the rare gas solids, crystalline binding can be treated empirically by considering a simple pair potential such as the *Lennard-Jones potential* or the *Morse potential*. These potentials are isotropic (have radial but no angular dependences) and are instructive because of their simplicity. A Lennard-Jones potential (introduced in Chapter 2) is applicable to inert gas solids and describes the potential energy, V, of an atom pair as a function of separation, r, as:

$$V(r) = 4\epsilon\left[\left(\frac{\sigma}{r}\right)^{12} - \left(\frac{\sigma}{r}\right)^{6}\right], \qquad (17.3)$$

[5] It is interesting to note that perfect tetrahedra cannot be used to fill 3-D space (just as pentagons cannot be used to fill a 2-D plane) but they will tile 4-D space (Sadoc and Mosseri, 1984).

where the parameters ϵ and σ set the scale of the potential energy and interatomic spacing, respectively. The force between the two atoms is given by the negative gradient, $-\mathrm{d}V/\mathrm{d}r$. As an atom pair is in equilibrium only when the force between them is zero, we differentiate the potential energy with respect to r and set this equal to zero (see Box 17.4).

Pair potential analysis often aims to express the interactions in a reduced (or universal) form. We define a *reduced Lennard-Jones potential*, $v(r/r_0)$, by normalizing $V(r)$ by -4ϵ and expressing it as function of $\bar{r} \equiv r/r_0$. This potential then has the reduced form:

$$v(\bar{r}) = \left(\frac{1}{\bar{r}}\right)^6 - 2\left(\frac{1}{\bar{r}}\right)^{12}. \tag{17.6}$$

The Morse potential assumes an exponential dependence on the interatomic spacing, r. A *reduced Morse potential* is written as:

$$v(\bar{r}) = [1 - \mathrm{e}^{-\alpha(1-\frac{1}{\bar{r}})}]^2 - 1 \tag{17.7}$$

where the parameter α is typically of the order of 1.5 (Hoare, 1978), and determines the *compressibility* of the spherical atoms as they bond in the pair. In what follows, we will restrict our discussion to the Lennard-Jones pair potential.

We can calculate the total energy in a crystalline solid by summing the Lennard-Jones potential over all atomic pairs to yield (Kittel, 1990) :

$$V_{\mathrm{TOT}} = \frac{N}{2}(4\epsilon)\left[\sum_{ij}{}' \left(\frac{\sigma}{p_{ij}R}\right)^{12} - \sum_{ij}{}' \left(\frac{\sigma}{p_{ij}R}\right)^{6}\right], \tag{17.8}$$

where $\sum'_{ij}$ refers to a standard summation over all pairs except those that pair with themselves (i.e., excluding $i = j$) and, $p_{ij} = r_{ij}/R$ is the distance between

Box 17.4 Derivation of the Lennard-Jones equilibrium pair separation

We determine the equilibrium separation by differentiating the potential and setting the derivative equal to zero:

$$\frac{\mathrm{d}V}{\mathrm{d}r} = 4\epsilon\left[\left(\frac{-12}{r}\right)\left(\frac{\sigma}{r}\right)^{12} - \left(\frac{-6}{r}\right)\left(\frac{\sigma}{r}\right)^{6}\right] = 0. \tag{17.4}$$

From this equation, we find the equilibrium distance, r_0, between the two atoms to be:

$$r_0 = 2^{\frac{1}{6}}\sigma = 1.12\sigma. \tag{17.5}$$

atoms i and j, expressed in units of the nearest neighbor distance, R. Note that we do not sum the reduced potential, $v(r)$, because we are interested in determining the new (different) equilibrium interatomic spacing for atoms in the lattice, not just an isolated pair. The factor $N/2$ accounts for N atoms in the system without double counting the pair interactions. The summations of the p_{ij} in the previous expression are known as *lattice sums*. For the *fcc* structure, they are:

$$A_{12}^{fcc} = \sum_{ij}{}' \left(\frac{1}{p_{ij}}\right)^{12} = 12.13188; \quad A_{6}^{fcc} = \sum_{ij}{}' \left(\frac{1}{p_{ij}}\right)^{6} = 14.45392. \quad (17.9)$$

For the *hcp* structure, they are:

$$A_{12}^{hcp} = \sum_{ij}{}' \left(\frac{1}{p_{ij}}\right)^{12} = 12.13229; \quad A_{6}^{hcp} = \sum_{ij}{}' \left(\frac{1}{p_{ij}}\right)^{6} = 14.45489. \quad (17.10)$$

We determine the equilibrium value of the interatomic spacing $R = R_0$ by differentiating the total potential energy with respect to R and setting it equal to 0. By setting the derivative equal to 0, we satisfy the requirement of mechanical equilibrium where the sum of the forces is equal to zero. Box 17.5 derives the equilibrium atomic separation for the *fcc* lattice.

This expression agrees with experimental results for the rare gas solids, using independently determined values of σ, as shown in Table 17.3. The

Box 17.5 Derivation of the *fcc* Lennard-Jones equilibrium atomic separation

We determine the equilibrium atomic separation by first differentiating the pair potential and setting the derivative equal to zero. For the *fcc* structure, this yields:

$$\frac{\mathrm{d}V_{\mathrm{TOT}}}{\mathrm{d}R} = 0 = -\frac{2N\epsilon}{R} \times \left(\frac{\sigma}{R}\right)^6 \times \left[12 \times 12.132 \times \left(\frac{\sigma}{R}\right)^6 - 6 \times 14.454\right], \quad (17.11)$$

which can be solved to yield the equilibrium spacing, R_0, between the two atoms:

$$R_0 = 1.09\sigma. \quad (17.12)$$

Table 17.3. R_0 and the cohesive energy per atom for inert gases. Values for σ, ϵ, R_0, and the cohesive energy per atom V_{TOT}/N are experimental (Rohrer, 2001).

Element	σ nm	ϵ meV	R_0 nm	1.12σ nm	1.09σ nm	(V_{TOT}/N) meV/atom	-8.6ϵ meV/atom
Ne	0.274	3.1	0.313	0.308	0.299	−20	−27
Ar	0.340	10.4	0.376	0.382	0.371	−80	−89
Kr	0.365	14.0	0.401	0.410	0.398	−116	−120
Xe	0.398	20.0	0.435	0.447	0.434	−170	−172

Lennard-Jones potential predicts that the cohesive energy will be of the same form for all *fcc* rare gas solids. The explicit expression for the cohesive energy obtained by evaluating the total potential energy at the equilibrium spacing, R_0, is:

$$V_{TOT}(R_0) = -2.15 \times 4N\epsilon. \tag{17.13}$$

The cohesive energy per atom is then equal to -8.6ϵ. As this cohesive energy is in significant error for lighter rare gas species (see Table 17.3), it requires quantum mechanical corrections for greater accuracy. For larger rare gas species, the cohesive energy is predicted to within a few per cent accuracy.

The Lennard-Jones potential predicts physical properties, such as the cohesive energy and equilibrium lattice spacing, from a few simple parameters. A drawback to using these simple pair potentials is that they have no angular dependence: therefore, they do not capture the angularly dependent bonding existing in many solids. An important example is the sp^3 hybrid bonding, prevalent in semiconducting solids, which causes a preference for diamond cubic and related structures. In simple metals isotropic pair potentials have some predictive value.

Rose *et al.* (1984) (later extended by Smith *et al.* (1991)) showed that suitably scaled equations of state for metals, derived from first principles quantum mechanical calculations, follow a *universal behavior*. This universal behavior (Fig. 17.3) is common for bulk metals, metal–metal adhesion, and chemisorption of selected materials. Binding energy curves derived from first principles calculations are often conveniently expressed as a function of the *Wigner–Seitz radius*, r_{ws}. For N atoms per unit volume:

$$N\frac{4}{3}\pi r_{ws}^3 = 1. \tag{17.14}$$

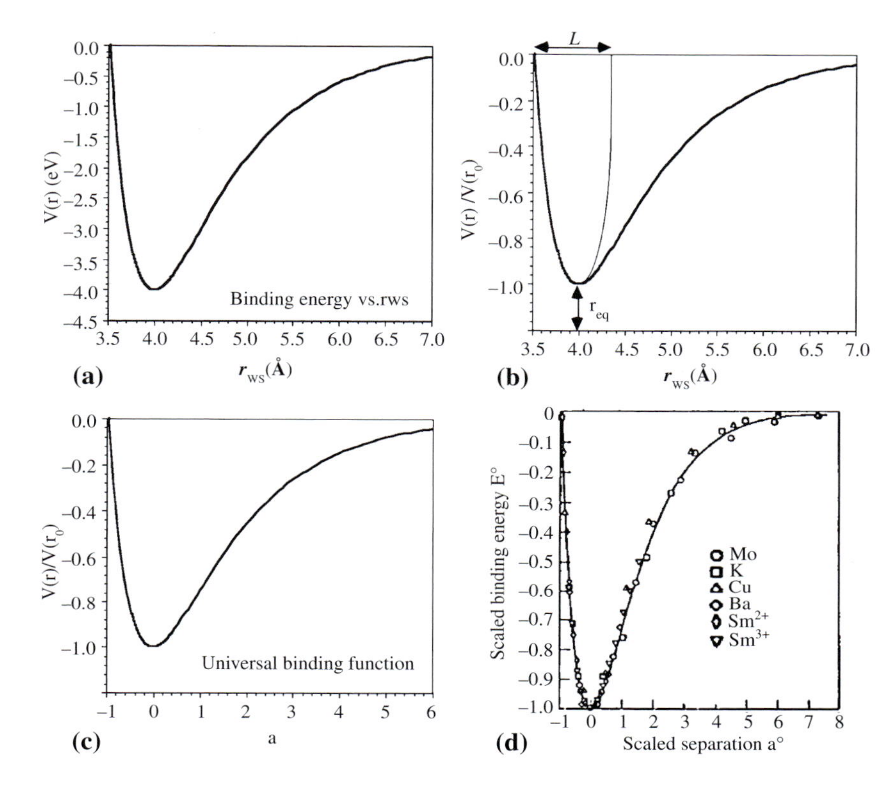

Fig. 17.3. (a) Bonding energy versus Wigner–Seitz radius, (b) energy normalized by the equilibrium cohesive energy showing the range of the potential L, (c) scaled energy versus separation. (d) Universal bonding curve for metals determined from first principles calculations. (J. H. Rose, *et al.*, Phys. Rev. B, **29**(6), 2963–2969, 1984; Copyright (1984) by the American Physical Soc.).

The Wigner–Seitz sphere is a sphere of volume $\frac{4}{3}\pi r_{ws}^3$, equal to the atomic volume of the material at the equilibrium spacing.

Figure 17.3 (a) illustrates a typical bonding energy curve showing the potential energy as a function of the Wigner–Seitz radius, r_{ws}. When we construct a *universal binding curve*, we first normalize the energy scale by dividing by the equilibrium cohesive energy, $V(r_0)$, as illustrated in Fig. 17.3 (b). The harmonic portion of the $V(r_{ws})$ curve near the equilibrium spacing, $r_{ws,0}$, is now fitted to a parabola that extends until it intersects the $V = 0$ axis. The width of the parabola, at $V = 0$, is called the *range* of the potential and designated L.[6]

Figure 17.3 (c) shows the scaled energy versus separation curve, which illustrates the universal binding relationship. The scaled length is given by $(r - r_{ws,0})/L$. Figure 17.3 (d) shows the universal bonding curve for a variety of metals, as determined from first principles calculations. This curve illustrates

[6] It can be shown that L is related to the *bulk modulus* of the material at the equilibrium spacing (in general, elastic moduli are related to curvatures of potential functions).

that scaled bonding energy versus scaled separation data (Rose *et al.*, 1984) for a variety of metals fall on the same universal curve! Ferrante *et al.* (1991) have shown that similar binding curves fit the bonding of diatomic molecules. Through first principles calculations (using *local density functional theory*) one can determine lattice constants to an accuracy of about 1%, and also which crystal structure has the lowest total energy.

17.4.3 Phase diagrams

A *phase diagram* is a graphical representation of alloy crystal structure stability ranges as a function of temperature and composition. Figure 17.4 shows examples of binary phase diagrams in two-component AB alloy systems. These are (a) a *eutectic phase diagram* (b) a *peritectic phase diagram*, and (c) a phase diagram with compound formation. The axes of these diagrams are temperature, T, along the vertical axis and the atomic fraction of B, X_B, along the horizontal axis. Note that we have $X_A = 1 - X_B$. The melting temperature of pure A is T_A, and of pure B is T_B. We assume that the crystal structure of solid A differs from that of B in these phase diagrams. The α-phase refers to solid solutions of B dissolved in A (dilute in B) and the β-phase refers to a solid solution that is dilute in A. Both phases show a limited range of

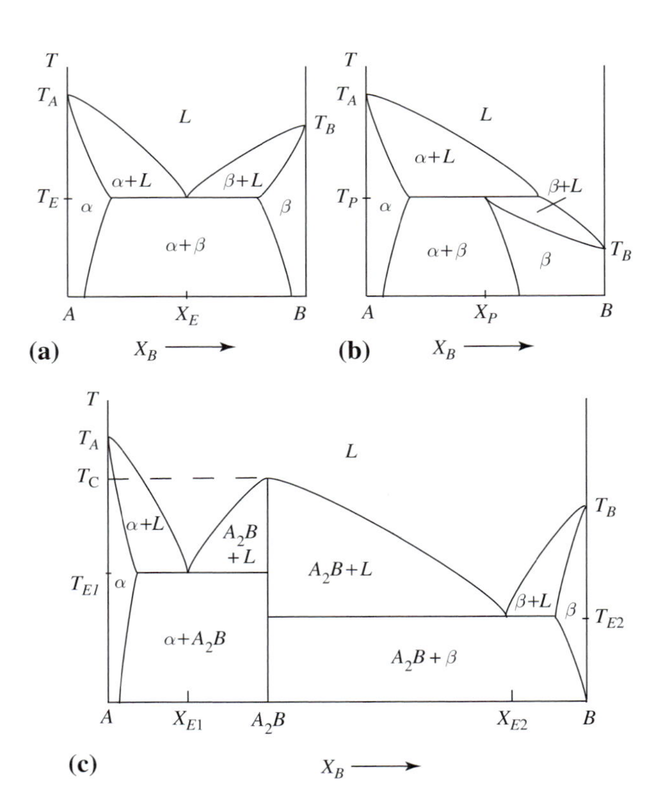

Fig. 17.4. Binary AB phase diagrams: (a) eutectic, (b) peritectic, and (c) with an A_2B line compound.

solid solubility at all temperatures. For a range of values of X_B and low temperatures, the two solid α and β phases coexist.

In all three phase diagrams, the *liquidus curves* have negative slopes. On cooling, a *eutectic transformation* (Fig. 17.4(a)) refers to the intersection of two liquidus curves is at a point called the *eutectic point*. The eutectic point denotes the coexistence of the (high temperature) liquid and a (low temperature) 2-phase solid region. The coexistence point (T_E, X_E) is the point at which the liquid-solid phase transformation, $L \rightarrow \alpha + \beta$, occurs.

A *peritectic transformation* (Fig. 17.4(b)) refers to the intersection of two liquidus curves at a point called the *peritectic point*. The peritectic point denotes the coexistence of the (high temperature) liquid and a (low temperature) 2-phase liquid + solid region. The coexistence point represents the $L \rightarrow L + \beta$ phase transformation.

A phase diagram with an A_2B phase *line compound* is shown in Fig. 17.4(c). If the compound is ideal, there will be no solubility of either A or B in the compound, and the compound has a single composition at all temperatures. If a range of compositions is possible, the line will be replaced by a phase field. Figure 17.4(c) shows an example of a phase diagram with a line compound and a double eutectic structure. If a compound has strong bonding, the melting point of the compound may be larger than either or both of the pure phases. The melting temperature of the A_2B compound is denoted by T_C in Fig. 17.4(c). Because the A_2B phase melts without a change in composition, T_C is called a *congruent melting point*; A_2B is said to undergo *congruent melting*.

17.5 Superlattices and sublattices: mathematical definition

In Chapters 3 and 6, we defined the basis vectors $\mathbf{a}_i$ and $\mathbf{a}_j^*$ for the real and reciprocal lattices of a structure. This was followed by a discussion of how to describe general vectors in real and reciprocal space using these bases. In the present section, we discuss the special circumstance that occurs when the basis vectors for a given lattice can be written as a linear combination of the basis vectors of another lattice. If this transformation *increases* the volume of the unit cell by an integer factor, then the new lattice is called a *superlattice* of the original lattice. If the transformation *decreases* the volume of the unit cell (by dividing by an integer), then the new lattice is called a *sublattice* of the original lattice.

Let us now consider the mathematical description of such a transformation (Giacovazzo, 2002a). A lattice, $\mathcal{T}$, is described by its basis vectors $(\mathbf{a}_1, \mathbf{a}_2, \mathbf{a}_3)$. For a transformed lattice, $\mathcal{T}'$, the basis vectors are written as $(\mathbf{a}_1', \mathbf{a}_2', \mathbf{a}_3')$.

$$A = \begin{pmatrix} \mathbf{a}_1 \\ \mathbf{a}_2 \\ \mathbf{a}_3 \end{pmatrix}; \qquad A' = \begin{pmatrix} \mathbf{a}_1' \\ \mathbf{a}_2' \\ \mathbf{a}_3' \end{pmatrix}. \tag{17.15}$$

The linear transformation that takes the lattice, $\mathcal{T}$, with primitive unit cell basis vectors A into the lattice $\mathcal{T}'$, with primitive unit cell basis vectors A' is given by:

$$A' = \begin{pmatrix} \mathbf{a}_1' \\ \mathbf{a}_2' \\ \mathbf{a}_3' \end{pmatrix} = \begin{pmatrix} M_{11} & M_{12} & M_{13} \\ M_{21} & M_{22} & M_{23} \\ M_{31} & M_{32} & M_{33} \end{pmatrix} \begin{pmatrix} \mathbf{a}_1 \\ \mathbf{a}_2 \\ \mathbf{a}_3 \end{pmatrix} = MA. \tag{17.16}$$

Thus, the transformation can be written as $A' = MA$, and the reverse transformation as $A = M^{-1}A'$. These equations relate the primitive unit cell in $\mathcal{T}'$ to that in $\mathcal{T}$ and vice versa. Now, let $\|M\|$ be the determinant of the transformation matrix. There are several special cases:

(i) If the matrix elements M_{ij} are integers and $\|M\| = 1$, then the lattices $\mathcal{T}$ and $\mathcal{T}'$ coincide.
(ii) If the matrix elements M_{ij} are integers and $\|M\| > 1$, then the lattice $\mathcal{T}'$ is a superlattice of the lattice $\mathcal{T}$, and the volume of the primitive cell in $\mathcal{T}'$ is $\|M\|$ times greater than the volume of the primitive cell in $\mathcal{T}$.
(iii) If the matrix $M = Q^{-1}$, where the matrix elements q_{ij} are integers and $\|Q\| > 1$, then the lattice $\mathcal{T}'$ is a sublattice of the lattice $\mathcal{T}$ with the volume of the primitive cell in $\mathcal{T}'$ being $\|Q\|$ times smaller than the volume of the primitive cell in $\mathcal{T}$.

An additional case of interest occurs when the matrix M is rational, in which case the transformation describes so-called *coincident site lattices*.

17.6 Derivative structures and superlattice examples

Many alloy solid solutions are disordered at high temperature, i.e., there is an equal probability for any atom to occupy any lattice site. When the temperature is lowered, some of these alloys will undergo a *disorder–order* transition, resulting in an *ordered solid solution* or, synonymously, a *superlattice* or *superstructure*. In this section, we consider derivative and superlattice structures based on the *fcc*, *bcc*, diamond, and *hcp* parent structures.

17.6.1 *fcc*-derived structures and superlattices

The *fcc* structure derivatives and superlattices include structures in which atoms order on the original *fcc* atomic sites, making sites that were originally symmetrically equivalent now inequivalent. Other structures include occupation of the interstices (some or all) by the same or different atomic species. A third type involves $m \times n \times o$ *fcc* cells and site occupation patterns that reduce the structure's symmetry.

17.6.1.1 *fcc* ordered structures

Examples of *fcc*-based ordered structures include the $\mathsf{L1}_0$ (**Structure 4**) and $\mathsf{L1}_2$ (**Structure 5**) structures. Because these are alloy structures, they are given Strukturbericht symbols beginning with L. In fact, they are both designated as $\mathsf{L1}_x$, where L1 indicates that these alloy structures are derived from the first of the elemental structures, A1. Figure 17.5 illustrates the *fcc* structure (A1) and its ordered counterparts, the tetragonal $\mathsf{L1}_0$ and the cubic $\mathsf{L1}_2$ structures. We find the $\mathsf{L1}_0$ structure in the CuAu prototype and the $\mathsf{L1}_2$ structure in ordered Cu_3Au.

The tetragonal $\mathsf{L1}_0$ structure is found in ordered *AB* alloys (*binary* alloys), where *A* atoms order on alternate (001) planes and *B* atoms on the others. The ordering gives rise to a two-layered structure, modulated along the *c*-axis normal to the (001) planes. The *A* atoms occupy the $1a(0, 0, 0)$ and $1c(1/2, 1/2, 0)$ special positions of the space group **P4/mmm** (D^1_{4h}), while *B* atoms occupy the $2e(0, 1/2, 1/2)$ positions. Some examples of alloys with this structure include CuAu, FePt, FePd, CuTi.

Figure 17.6(a) shows a simulated XRD pattern for an equiatomic *fcc* FePt alloy in which the Fe and Pt atoms are distributed randomly on the *fcc* sites. Figure 17.6(b) shows a simulated XRD pattern for an equiatomic ordered $\mathsf{L1}_0$ FePt alloy with an equilibrium *a*-lattice constant, and an idealized c/a

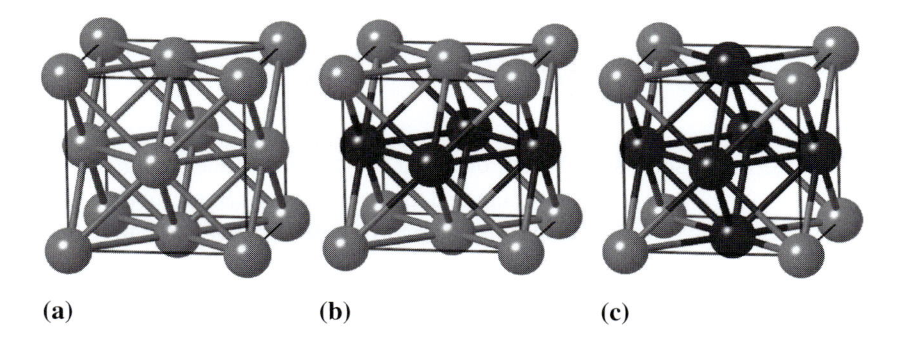

Fig. 17.5. (a) *fcc* (A1), (b) ordered $\mathsf{L1}_0$, and (c) ordered $\mathsf{L1}_2$ structures.

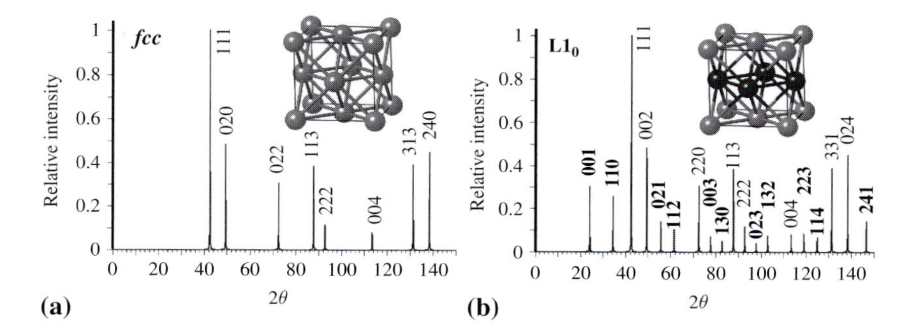

Fig. 17.6. Simulated XRD patterns for (a) random *fcc* and (b) ordered $\mathsf{L1}_0$ FePt structures.

ratio of 1.[7] The *fcc* structure has reflections that obey the standard extinction condition that h, k, and l must have the same parity. For the *fcc* structure, any permutation of the three hkl indices yields an identical reflection. This defines the *multiplicity* of the reflection. The **L1$_0$** alloy structure has additional reflections corresponding to the planes (001), (110), (021), (112), (003), (130), (023), (132), (223), (114), and (241). As the Bravais lattice of the **L1$_0$** structure is tetragonal, the h and k indices are no longer equivalent to the l index.

The magnetic recording industry is interested in **L1$_0$** alloys, such as the (near) equiatomic CoPt and FePt compounds, for data storage applications. Layered A and B atoms of different sizes give rise to a *tetragonal distortion* of the parent cubic structure, in addition to the lowering of symmetry.[8] The web structures appendix lists the c/a ratios for several representative compounds. An interesting application of **L1$_0$** alloys is in self-assembled nanoparticle arrays, as discussed in Box 17.6.

We observe the cubic **L1$_2$** structure in ordered A_3B alloys where A orders on the face centers and B on the cube vertices. With reference to the center of the original *fcc* unit cell, one set of atoms occupies the octahedron formed by the face center atoms and the other set occupies the cube formed by the vertex atoms, as illustrated in Fig. 17.5(c). As a consequence, the Bravais lattice is no longer face centered, and the space group of this structure is $\mathbf{Pm\bar{3}m}$ (O_h^1). The B atoms occupy the $1a(0,0,0)$ special position while A atoms occupy the $3c(0,1/2,1/2)$ special positions. Examples of compounds with this structure include Au_3Cd, $AlCo_3$, Pt_3Sn, $FeNi_3$, $FePd_3$, etc.

17.6.1.2 *fcc* interstitial substitution: diamond cubic and rocksalt structures

The diamond cubic structure (**Structure 6**) is an *fcc* derivative structure, consisting of two *interpenetrating fcc* lattices, where half of the tetrahedral sites in either lattice are occupied (Fig. 17.7). The origin of the second *fcc* lattice is at a tetrahedral interstitial site of the first. This is the structure for important semiconductors like Si and Ge. This structure is given the Strukturbericht symbol **A4**. The prototype for this structure is the diamond allotrope of C. If the two lattices have different atoms, we have an ordered zinc-blende structure, an AB compound, which will be discussed further in section 17.6.3.

[7] Unless stated otherwise, all simulated XRD patterns are generated assuming Cu-Kα radiation and the equilibrium lattice constants found in the *StructuresAppendix.pdf* file on the book's web site.

[8] Note that four-fold rotation axes of the cubic cell, parallel to the a and b axes, disappear when the structure becomes ordered. Can you identify all other vanishing symmetry operations?

Box 17.6 Magnetic nanostructures – self-assembled nanoparticle arrays

Scientists are currently actively studying arrays of magnetic nanoparticles for data storage applications. $L1_0$ materials figure prominently in this research. The goal is to store a single bit of information on a single magnetic nanoparticle; the implications are enormous potential increases in the areal storage density (potentially approaching terabits per square inch). In order to store information on a single particle, the particle must be stable with respect to thermally activated switching of the magnetization, which would cause the stored information to be lost. The thermal stability is determined by the magnetocrystalline anisotropy of the material, which is directly related to the anisotropy of the crystal structure.

Magnetic materials with the $L1_0$ structure have a large magnetocrystalline anisotropy and a preference for the magnetization to lie along the c-axis of the structure. There are several important technical considerations for the use of $L1_0$ nanoparticles. Monodisperse magnetic nanoparticles having the $L1_0$ structure must be synthesized. This typically involves the synthesis of *fcc* particles in a system such as FePt, followed by annealing to achieve atomic ordering into the $L1_0$ structure. In order to address the information that is eventually to be stored, the nanoparticles must be arranged periodically. This is currently being achieved over short length scales using a process called *self-assembly*. Periodic arrangements of nanoparticles over longer length scales are being actively pursued.

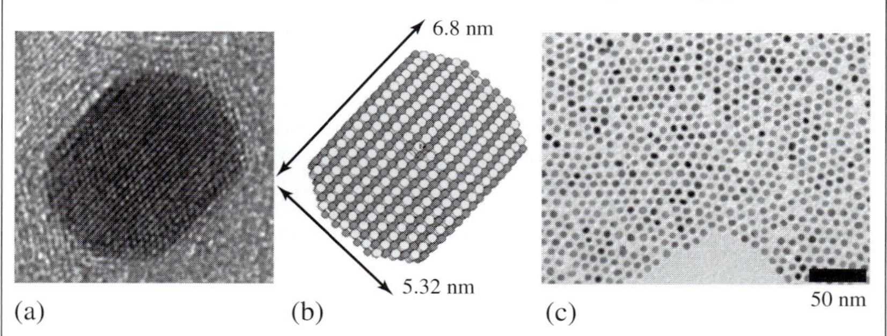

The figure above, courtesy of M. Tanase, D. E. Laughlin and J.-G. Zhu, shows a high resolution transmission electron microscopy (HRTEM) image (a) of a truncated cubo-octahedral FePt nanoparticle (produced at Seagate Research) used in self-assembled arrays. Frame (b) shows a cartoon of the Fe and Pt atoms in the nanoparticle of (a), illustrating shape and orientation of the particle. Frame (c) shows a TEM image of a self-assembled array of these nanoparticles.

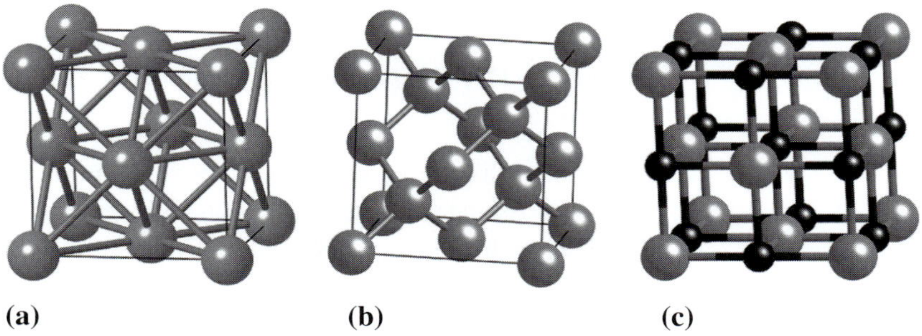

Fig. 17.7. Ball-and-stick models of (a) the *fcc*; (b) the diamond cubic (DC), showing a tetrahedral network of C atoms, and (c) the rocksalt NaCl structures.

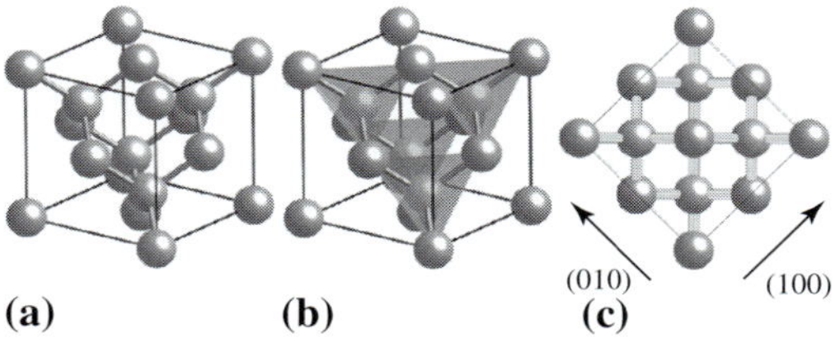

Fig. 17.8. Diamond cubic structure displayed as (a) a ball-and-stick figure, (b) a ball-and-stick figure with tetrahedral coordination polyhedra, and (c) a projection of bonds onto (100) planes.

The diamond cubic structure is common in many semiconductors.

> Prototype semiconductors are *Group IV* elements like Si or Ge, having four electrons in their outer shells.

In semiconductors, the atoms are typically located on sites with four-fold coordination; this way they can participate in four covalent bonds with a total of eight shared electrons. The four-fold coordination is typically tetrahedral, because of the hybrid sp^3 bonding described in Chapter 2. The diamond cubic structure can be viewed in terms of *vertex sharing tetrahedral units*. Figure 17.8(b) shows the diamond cubic structure of Si and its four-fold tetrahedral coordination. To illustrate this bonding in a 2-D picture, we project the sp^3 bonds into a (100) plane, as depicted in Fig. 17.8(c). Here, the even number of electrons and strong hybridization in covalent bonds creates a large hybridization gap between *bonding energy states* and *antibonding energy states*.

The rocksalt structure (**Structure 7**, Fig. 17.7(c)) is an example of an *fcc* derivative structure consisting of two interpenetrating *fcc* lattices, where each of the *octahedral* sites in either lattice is occupied by the other. The origin of the second *fcc* lattice is at an octahedral interstitial site of the first. Note that, if both atoms were the same, then we would just arrive at a simple cubic (*sc*) lattice with a lattice constant of half that of the *fcc* cell from which it was derived. This structure is found for many ionic compounds, including rocksalt, NaCl, which is its prototype. This *AB* compound, based on the first elemental structure (A1), has the StrukturBericht symbol B1. The connectivity of octahedral coordination polyhedra is discussed further in Chapter 22.

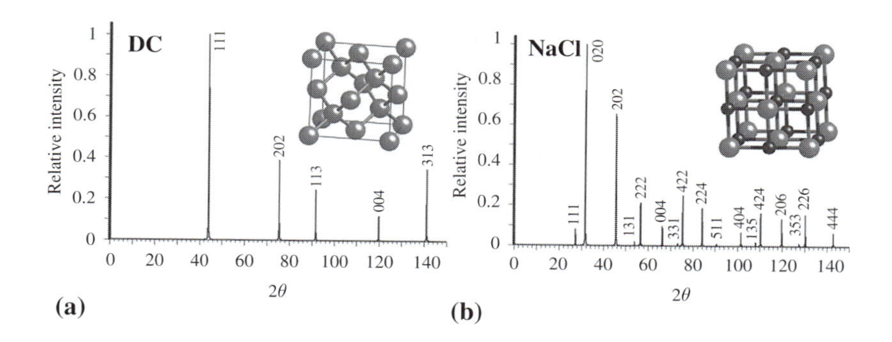

Fig. 17.9. Simulated XRD patterns for the (a) diamond cubic C and (b) NaCl structures.

Figure 17.9 shows simulated XRD patterns for the diamond cubic C structure (a), and the NaCl structure (b). These should be compared with the parent *fcc* structure of Fig. 17.2. Further *fcc* derivative structures can be considered in terms of ordered and/or combined occupancy of octahedral and tetrahedral interstices.

17.6.1.3 *fcc*-derived superlattices

The DO_3 (**Structure 8**) and $L2_1$-type superlattices (**Structure 9**) are ordered *fcc* structures. The DO_3 structure has a conventional *fcc* cell of *A*-type atoms with *B* atoms occupying all tetrahedral and octahedral sites. There are two tetrahedral and one octahedral interstices per *A* atom, resulting in an AB_3 stoichiometry. Stoichiometric phases with the DO_3 structure include Fe_3Si and $AlFe_3$; BiF_3 is the prototype. Non-stoichiometric alloys with the DO_3 structure are possible by having imperfect order between the sites. The $L2_1$ superlattice structure is based on a $2 \times 2 \times 2$ supercell of the B2 structure, described in the next section. Both DO_3 and $L2_1$ structures have *fcc* Bravais lattices and typical lattice constants range from 0.55 to 0.75 nm. A lattice constant of 0.5670(5) nm has been reported for Fe_3Si collected from cosmic dust (Zuxiang, 1986).

These superlattice derivatives of the *fcc* structure can also be described as *bcc* derivatives. We should view these structures in settings with origins at the (0, 0, 0) and at the (1/4, 1/4, 1/4) sites, respectively, as shown in Fig. 17.10 (in (b) the outline of the cell with origin at (0, 0, 0) is shown for reference).

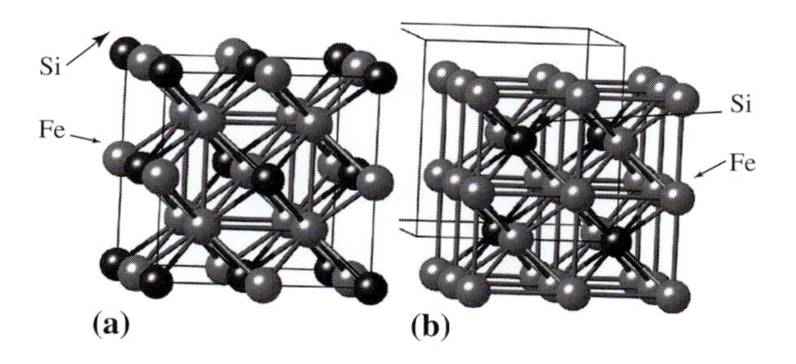

Fig. 17.10. The DO_3 structure of ordered Fe_3Si in a ball and stick representation of (a) the (0,0,0) setting and (b) in the alternative unit cell with origin at the $(\frac{1}{4}, \frac{1}{4}, \frac{1}{4})$ position.

We recognize that this structure has features characteristic of both the *fcc* and *bcc* structures.

17.6.2 *bcc*-derived superlattices

The *bcc* structure is the structure of elemental iron. Iron (Fe) is of widespread engineering importance because of its abundance and its mechanical and magnetic properties. In the following subsections, we will review a number of *bcc*-derived ordered structures and superlattices.

17.6.2.1 *bcc*-derived ordered structures

The *β-brass* superlattice structure, named for the ordered CuZn alloy, has Cu atoms occupying either of the body centered or vertex sites in the *bcc* lattice and Zn occupying the other. The ordered *AB* compound has the StrukturBericht symbol B2, and is represented by the prototype CsCl (**Structure 10**). We observe this structure in many *AB* alloys and ionic compounds. We discuss this compound in terms of coordination polyhedra and their connectivity in Chapter 22.

Figure 17.11 illustrates (a) the simple cubic (*sc*) structure, (b) the *bcc* structure (A2 type), and (c) the B2-CsCl structure. The B2 structure can be viewed in terms of interpenetrating simple cubic lattices. The B2 structure has *A* atoms on the $(0, 0, 0)$ special position and *B* atoms on the $(1/2, 1/2, 1/2)$ sites. Examples of phases with the B2 structure include β-brass (CuZn), β-AuCd, β-AlNi, β-NiZn, AlFe, LiTl, α′-CoFe, etc.; CoFe is an example of an important soft magnetic material with the largest known magnetic induction in any system and a high Curie temperature.

Superlattice reflections distinguish between the A2 (*bcc*) and B2 (CsCl) structures.

> Reflections shared between ordered and disordered structures are known as *fundamental reflections*. Extra reflections found in the ordered phase are called superlattice reflections.

Superlattice reflections arise when the primitive unit cell of the ordered structure is larger than that of the disordered structure. Consequently, the

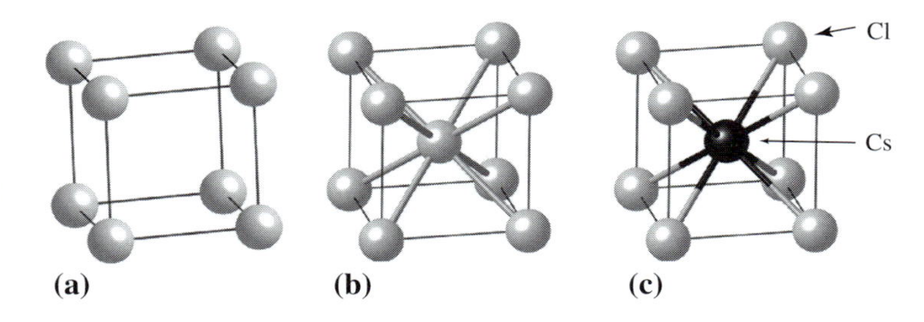

Fig. 17.11. The simple cubic structure (a), the *bcc* structure (A2 type) (b), and the ordered counterpart of the *bcc* structure, the B2-CsCl structure (c).

density of reciprocal lattice points is higher for the ordered structure. The *structure factor* of the *fundamental reflections* of an AB alloy with an **A2** structure is:

$$F_{hkl} = f_A + f_B \quad \text{where} \quad h+k+l = \text{even},$$

and for the superlattice reflections of the **B2** structure we have:

$$F_{hkl} = f_A - f_B \quad \text{where} \quad h+k+l = \text{odd}.$$

For the ionic compound CsCl, the two atoms in the structure have different X-ray *atomic scattering factors*; therefore, the superlattice reflections in the CsCl structure are easily resolved by XRD (because $f_{Cs} - f_{Cl}$ is large). However, this is often not the case in transition metal alloy systems that occur in the **B2** structure; in such systems, the superlattice reflections may be very weak. An example is the **B2** structure of ordered CoFe. Iron and Cobalt are directly next to one another in the periodic table and have nearly identical atomic scattering factors, so that $f_{Fe} - f_{Co} \approx 0$. Therefore, the superlattice reflections are very difficult to observe by conventional XRD. The FeCo diffraction pattern was introduced earlier, in Chapter 13.

17.6.2.2 *bcc* derivative 2 × 2 × 2 superlattices

In this section, we will reconsider the $\mathsf{D0_3}$ and $\mathsf{L2_1}$-type superlattices as *bcc* derivative structures, shown in Fig. 17.12. If we consider a $2 \times 2 \times 2$ cubic cell in an AB binary alloy, where the cube edge sites and half of the body centered sites are decorated with A atoms and the remaining body centered sites with B atoms, we arrive at the $\mathsf{D0_3}$ structure. We can partition the body centered sites into two interpenetrating tetrahedra (the X and Y sites, respectively).

In the set of 8 body-centered cells, there are 16 positions (a body center and vertex site for each cell). The B atoms occupy the positions: (1/4, 1/4, 3/4), (1/4, 3/4, 1/4), (3/4, 1/4, 1/4), and (3/4, 34, 3/4), with A atoms occupying all of the other positions for a composition A_3B. For compositions between AB and A_3B, A atoms occupy vertex sites as well as (preferentially) the X sites. In the composition A_3B, A atoms occupy cube vertex sites as well as *all*

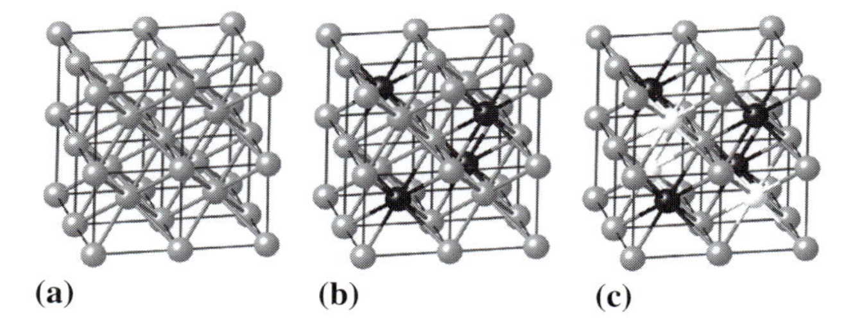

Fig. 17.12. *bcc* derivative superlattices : 2 × 2 × 2 *bcc* cells (a), and the $\mathsf{D0_3}$ (b) and $\mathsf{L2_1}$-type (c) superlattices.

the X sites and B atoms occupy all the Y sites. The **D** in the **D0$_3$** Strukturbericht symbol indicates an A_mB_n compound. When B atoms are on X sites, but not on any Y sites, the stoichiometry corresponds to $m = 3$ and $n = 1$.

The **L2$_1$** structure is a variant of the **D0$_3$** structure in ternary alloys of composition A_2BC, where A atoms occupy cube vertex sites, B atoms occupy all the X sites, and C atoms, all the Y sites. The famous *Heusler alloy*, Cu_2MnAl, is the prototype **L2$_1$** structure.[9] Even though Cu and Al are non-magnetic atoms, this alloy is ferromagnetic. This alloy was studied extensively because it provided an important example of a structure for which Mn atoms, when kept at large distances from each other (in this case $a/\sqrt{2}$), couple to one another ferromagnetically.[10] The Heusler alloy $GaNi_2Mn$ is another material of current interest because of its large magnetoelastic response. It is an example of a *ferromagnetic shape memory alloy* (FSMA). Other Heusler alloys containing Mn have large room temperature *magneto-optic Kerr rotations*. This can be related to a large orbital moment on Mn in these materials, which makes them interesting for magneto-optic recording media. Heusler alloys with interesting magneto-optic effects include Ni_2MnSb, $MnPt_2Sb$, $MnPt_2Sn$, etc. There are hundreds of compounds with the Heusler structure.

17.6.3 Diamond cubic derived superlattices

17.6.3.1 Diamond cubic derived ordered structure: zinc blende

Compounds between group *III* and group *V* elements (or group *II* and group *VI* elements), which have an average of four electrons in their outer shells, are also typical semiconducting materials. These include AsGa, an example of a *III–V* material, and InSb, a *II–VI* material. Similar bonding and crystallography is displayed in *III–V* semiconductors that have the ordered *zinc-blende* or *sphalerite* structure, for which cubic ZnS is the prototype (**Structure 11**). This structure represents an ordered AB compound and, therefore, it is designated by the StrukturBericht symbol **B3**. The zinc-blende structure, its coordination polyhedra connectivity and [001] projection are illustrated in Fig. 17.13.

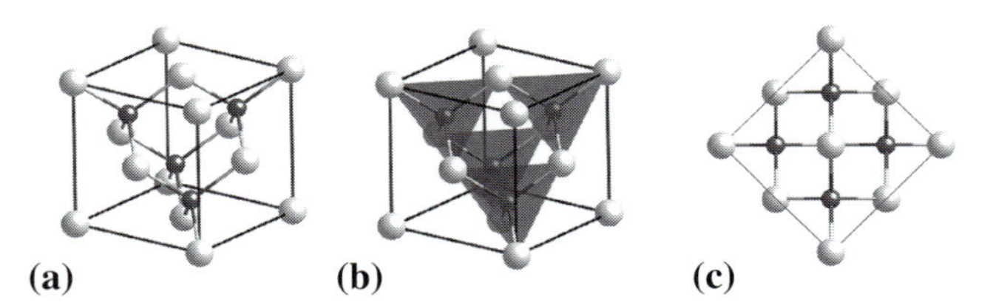

Fig. 17.13. Zinc-blende structure in (a) space filling, (b) ball-and-stick representation, showing the tetrahedral network, and (c) projection on (001) planes.

[9] Recall that for prototype structures, we use the historical name of the alloy, not the alphabetized version, i.e., we use Cu_2MnAl, not $AlCu_2Mn$.

[10] The magnetic coupling in pure Mn is antiferromagnetic.

Zinc-blende has an ordered structure in which the Zn atoms are tetrahedrally coordinated by the S atoms and vice versa, as is readily apparent from the projection of the structure onto (001) planes. We can view Zn atoms as occupying half of the tetrahedral sites in the *fcc* sulfur anion sublattice. As a *II–VI* material, its bonding is more ionic than in the *III–V* materials. The zinc-blende structure will be considered further in Chapter 22, in a discussion of *Pauling's rules* for ionic structures.

17.6.3.2 Interstitial substitution in the diamond cubic structure: fluorite

When B atoms occupy all otherwise empty tetrahedral sites in the zinc-blende structure, the resulting compound has AB_2 stoichiometry. Compounds with this structure are most often ionic; however, important intermetallics with this structure have also been discovered. The first intermetallic with this structure was Mg_2Sn, which was solved by Pauling (1923). Others include Mg_2Si and Mg_2Pb. These particular compounds are textbook examples of *line compounds*. The binary Mg–Sn, Mg–Si, and Mg–Pb phase diagrams resemble the hypothetical diagram of Fig. 17.4 (c).

An ionic structure of this type is the *fluorite* (**C1**) structure, of which CaF_2 is the prototype (**Structure 12**). In this structure, the A atom occupies the $(0, 0, 0)$ special position in the *fcc* lattice and the B atoms decorate the eight interstitial positions of the type $(\frac{1}{4}, \frac{1}{4}, \frac{1}{4})$. Again, the StrukturBericht symbol **C** is used for compounds with AB_2 stoichiometry. Figure 17.14 (a)–(c) illustrates the zinc-blende structure, the occupation of the other tetrahedral sites to yield the fluorite structure, and the ordering between the tetrahedral sites to yield the AlAsMg structure.

The fluorite structure is discussed further in Chapter 22 in terms of Pauling's rules for ionic structures. A structure of stoichiometry A_2B, of which K_2O

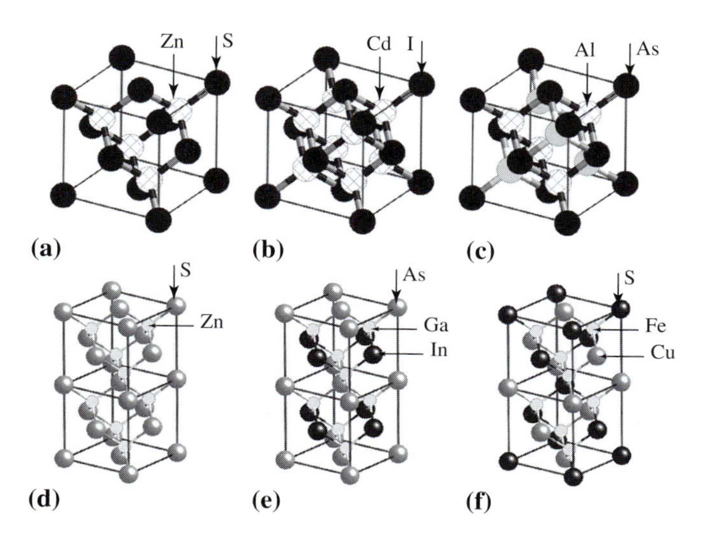

Fig. 17.14. (a) Zinc-blende, (b) fluorite, and (c) AlAsMg structures; 1 × 1 × 2 supercells (d) zinc-blende, (e) $InGaAs_2$ and (f) chalcopyrite $CuFeS_2$ structures.

is an example, where A atoms occupy all of the tetrahedral sites in an *fcc* B atom lattice, is called the *antifluorite structure*. We can expand this structure type to the stoichiometry ABC, in which, for example, B atoms occupy half of the tetrahedral sites and C atoms occupy the other half in an *fcc* A lattice. AlAsMg is an example of this structure type, shown in Fig. 17.14(c).

The occupation of the empty octahedral sites in the fluorite structure (i.e., at the special position (1/4, 1/4, 1/4) and those related by symmetry) by B atoms results in a compound with stoichiometry AB_3, of which $BiLi_3$ and $AlFe_3$ are examples and the ionic BiF_3 is the prototype. We already described this structure as **Structure 8** in an earlier section. In this structure, not all of the B atoms are symmetrically equivalent, consistent with the supercell derivation in the discussion of *bcc* derivatives, where the B atoms occupied both vertices and half of the body-center sites. As an exercise, the reader may wish to show that, if the two sets of tetrahedral sites in fluorite are occupied by B and C atoms, and octahedral sites by B, the resulting compound has a stoichiometry of AB_2C, and its structure corresponds to the previously discussed $L2_1$ structure type.

17.6.3.3 Diamond cubic derivative 1 × 1 × 2 superlattices

We may construct additional superlattices by modulating the diamond cubic (or derivative) structures along one direction. Figure 17.14 (d) illustrates two zinc-blende cubic cells stacked along the c-axis. While this structure can be ordered into a layered $L1_0$-type structure of which $InGaAs_2$ is the prototype (shown in Fig. 17.14(e)), a more complicated ordering gives rise to the *chalcopyrite* structure (Fig. 17.14(f)). The chalcopyrite structure is given the StrukturBericht symbol $E1_1$ and has the chemical formula $CuFeS_2$ (**Structure 13**).

17.6.4 Hexagonal close-packed derived superlattices

In this section we describe ordered structures based on *hcp* structure. These include interstitial substitutions into the *hcp* structure and other *hcp* derivatives.

17.6.4.1 Interstitial occupation: *hcp* structure

New structures can be derived by the occupation of the octahedral and/or interstitial sites in the *hcp* structure. Depending on the occupation, the resulting cell can have rhombohedral or lower symmetry. Although we can view these structures in their primitive cells, we can visualize these structures more clearly by viewing them in a hexagonal prismatic setting. This setting depicts three cells bounded by a hexagonal prism. We can conveniently represent the structures in terms of the stacking sequence of close-packed planes using the familiar ABC notation.

Occupation of one of the tetrahedral sites in the *hcp* structure yields an AB compound with the StrukturBericht notation B4 (**Structure 14**). This is

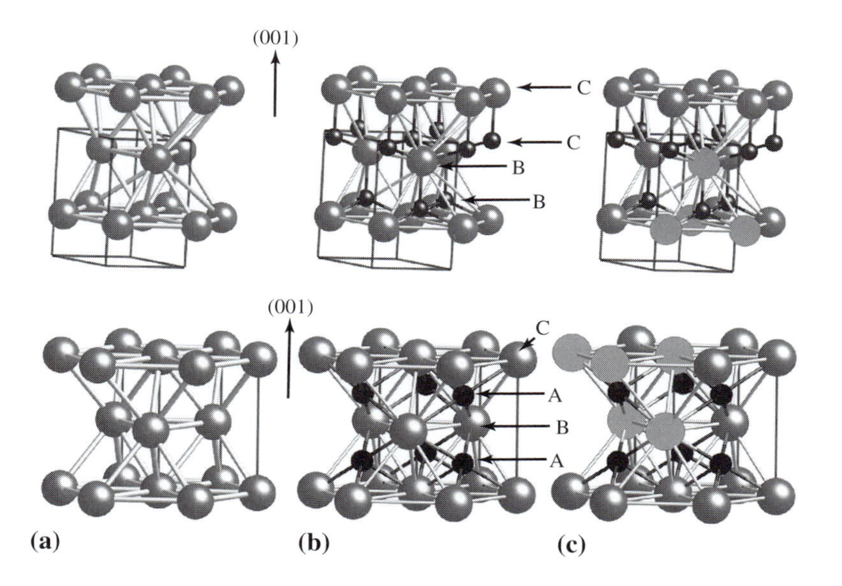

Fig. 17.15. (a) Close-packed oxygen sublattice in a hexagonal prismatic representation; (b) B4 ZnO-type wurtzite structure (top) and NiAs structure (bottom), and (c) same as (b) with an O anion tetrahedron (top) or octahedron (bottom) highlighted.

the *wurtzite* structure, named for a polymorph of ZnS. Its prototype is ZnO (*zincite*), with *hcp* O anions and the Zn cations occupying half the tetrahedral interstices. Figure 17.15 (top) illustrates *hcp* O anions in a hexagonal prismatic setting. The wurtzite structure, with tetrahedral sites occupied by Zn cations, is shown in (b). Figure 17.15 (c) highlights an O anion tetrahedron. This structure has the stacking sequence *BBCC* with Zn at $z = 0$ and $1/2$ (*BC*) and O at $z = 3/8$ and $7/8$ (*BC*).

Occupation of the octahedral sites in the *hcp* structure yields an *AB* compound with the StrukturBericht notation $B8_1$. A related elemental A8 structure (that of Te) is discussed below. A prototype for the $B8_1$ structure is the compound NiAs (**Structure 15**) in which Ni atoms occupy octahedral sites in an *hcp* As sublattice.

Figure 17.15 (bottom) illustrates an *hcp* cell in hexagonal prismatic representation. This figure also illustrates the filling of the octahedral interstices in the *hcp* As lattice to construct the $B8_1$ NiAs-type structure. Atoms in the top cap of an As octahedron are highlighted in frame (c). Note that the origin is displaced by $(\frac{1}{3}, \frac{2}{3}, \frac{1}{4})$. This structure has the stacking sequence *ABAC* with Ni at $z = 1/4$ and $3/4$ (*AA*) and As at $z = 1/2$ and 0 (*BC*). The NiAs structure is the analog of the NaCl structure because both structures are derived from the filling of octahedral interstices in the *fcc* and *hcp* derivatives. The connectivity of the coordination polyhedra in the NiAs structure will be discussed in greater detail in Chapter 22.

17.6.4.2 Other *hcp* derivative structures

A new structure related to the $L1_1$-type *fcc* derivative structure is the $D0_{19}$ Ni_3Sn-type (also Mg_3Cd) structure (**Structure 16**). We can describe this

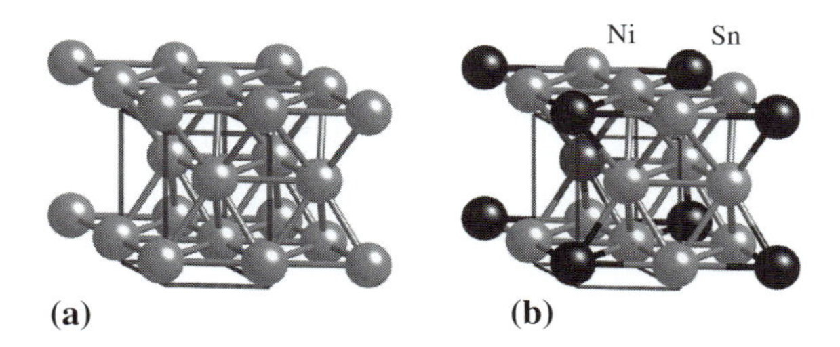

Fig. 17.16. Hexagonal close-packed derivative structure based on a 2 × 2 × 1 supercell of the *hcp* cell: (a) 2 × 2 × 1 *hcp* supercell and decoration to yield the (b) $D0_{19}$ (Ni_3Sn)-type structure.

structure with 4 interpenetrating *hcp* lattices, of which one is typically occupied by B atoms and three by A atoms – favoring the stoichiometry A_3B. In the StrukturBericht notation, the **D** types are reserved for A_mB_n compounds. We can see that for the $\mathsf{D0}_{19}$ structure, $m = 3$ and $n = 1$.

Figure 17.16(a) illustrates a $2 \times 2 \times 1$ supercell of the *hcp* cell, and its decoration (b) to yield the $\mathsf{D0}_{19}$ structure. The new $2 \times 2 \times 1$ supercell contains two A_3B formula units. In the prototype Ni_3Sn structure, the Ni and Sn atoms share the close-packed planes with the atomic ratios of 3 : 1 and the typical BC stacking of close-packed layers. However, the two layers can be decomposed into a larger 3^6 tiling of the Sn atoms and a 3636 Kagome tiling of the Ni atoms.

> The ABC notation for the stacking of close-packed planes (3^6 tilings) can be generalized to stacking of 6^3 tiles using the notation abc and to 3636 tiles using the notation $\alpha\beta\gamma$.

In the NiAs structure, the stacking sequence of the 3^6 tiles can be denoted as BC and that of the Kagome tiles as $\beta\gamma$. In total, this structure has the stacking sequence $[B\beta][C\gamma]$ with Sn at $z = 1/4$ and $3/4$ (BC) and Ni at $z = 1/4$ and $3/4$. The square brackets [ ] denote atoms in the same plane. We will illustrate this nomenclature further in Chapter 18.

Materials scientists have been interested in Co–Pt alloys because the $\mathsf{L1}_0$ phase of this material possesses a high *magnetocrystalline anisotropy*. They also studied ordered *hcp* derivative structures in the Co–Pt system (Willoughby *et al.*, 2003) as materials for use in high density magnetic recording. $Co_{1-x}Pt_x$ alloys are examples of binary systems that exhibit a complete range of *fcc* solid solutions at elevated temperatures. Such isotropic cubic phases are not of interest for magnetic recording media, because they lack magnetic anisotropy.

The equilibrium $Co_{1-x}Pt_x$ phase diagram has cubic (α) and *hcp* (ϵ) phases along with the ordered *fcc* derivative structures CoPt ($\mathsf{L1}_0$) and $CoPt_3$ ($\mathsf{L2}_1$) at room temperature. $Co_{1-x}Pt_x$ alloys can be quenched from high temperature and retain a disordered *fcc* α-phase crystal structure. Upon annealing alloys with $x \leq 0.23$ at low temperatures, a transformation to the stable *hcp* ϵ-phase structure occurs. Of these structures, only the *hcp* ϵ-phase (Fig. 17.17(a)) and

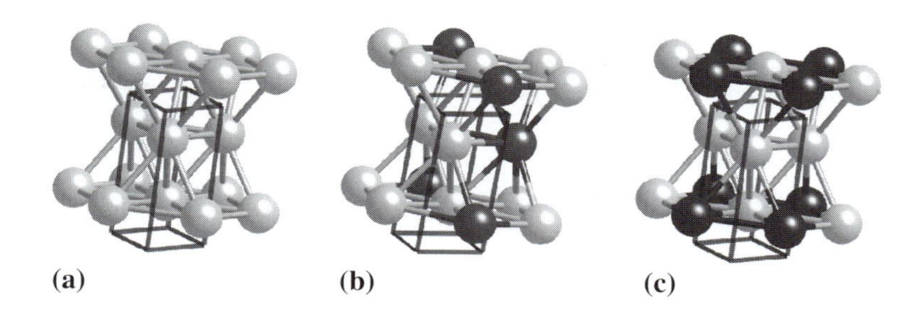

Fig. 17.17. *hcp* derivative structure: (a) 2 × 2 × 1 *hcp* supercell for disordered ϵ-Co_3Pt and decoration to yield the DO_{19} Co_3Pt (b) and orthorhombic (c) structures.

the tetragonal $L1_0$ phase have uniaxial magnetocrystalline anisotropy. Thus, only these two equilibrium phase crystal structures are of interest for either bulk permanent magnets or magnetic recording applications.

Researchers have observed new *hcp* derivative structures of chemically ordered Co_3Pt in thin films produced by *molecular beam epitaxy* (MBE) (Harp *et al.*, 1993, Maret *et al.*, 1996). Co_3Pt is appealing as a material for data storage because of a decreased cost compared to the equiatomic CoPt (due to the lower Pt content in Co_3Pt). The *hcp* derivative has been suggested to be an orthorhombic derivative of the DO_{19} structure with space group **Pmm2** (C^1_{2v}). The DO_{19} structure is shown in Fig. 17.17 (b), and the orthorhombic derivative structure is shown in Fig. 17.17 (c). The DO_{19} phase consists of mixed layers of 75 at % Co and 25 at % Pt. The *fcc* derivative $L1_2$ phase can also be decomposed into (111) planes stacked as all mixed layers. In contrast, the orthorhombic phase possesses a pure Co layer, alternating with a mixed layer containing half Co and half Pt. Willoughby and co-workers have calculated magnetic dipole moments and magnetocrystalline anisotropy energy densities for both the DO_{19} hexagonal crystal structure and the orthorhombic **Pmm2** (C^1_{2v}) derivative (Willoughby *et al.*, 2003).

17.7 Elements with alternative stacking sequences or lower symmetry

17.7.1 Elements with alternative stacking sequences

There can be many variations of superlattices and stacking sequences of close-packed layers. Some of these are easily derived from symmetry-lowering distortions of one of the previously described structures.

The structure of α-La has a $\mathbf{P6_3/mmc}$ (D^4_{6h}) space group with an a-lattice constant of 0.377 nm and a c-lattice constant of 1.2159 nm (Spedding *et al.*, 1956). The La atoms occupy two special positions, (0, 0, 0) and (1/3, 2/3, 1/4). As before, we will depict hexagonal and rhombohedral structures in a *hexagonal prismatic representation*. Figure 17.18(a) shows the the hexagonal prismatic representation of the α-La structure. This consists of three unit cells in a hexagon in the basal plane. Figure 17.18(a) shows the atoms, in the basal plane of the three cells, in sequence, highlighted as light

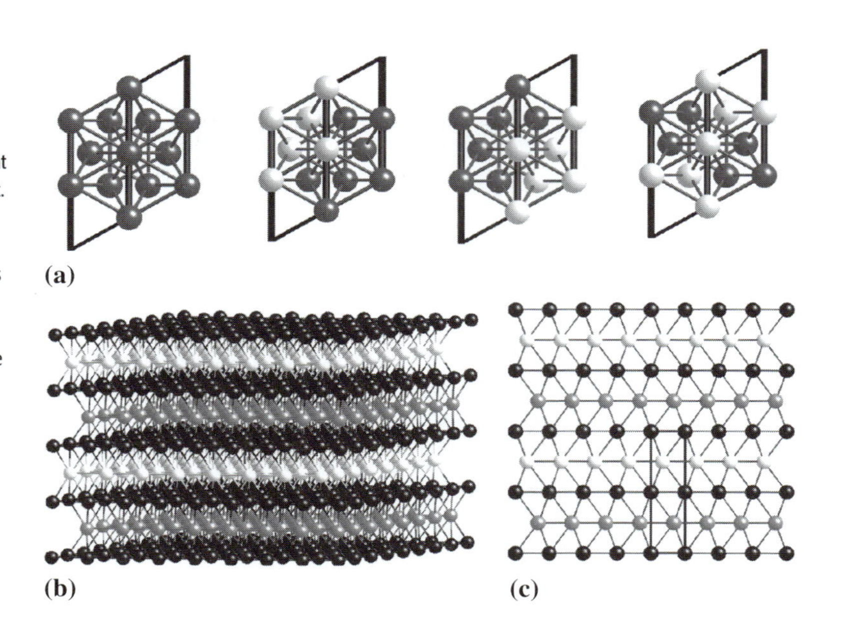

Fig. 17.18. (a) α-La crystal structure in a hexagonal prismatic representation. The atoms in the three cells are highlighted sequentially as light gray in the frames to the right. (b) Stacking of close-packed planes in the α-La crystal structure, atoms on the *A*-sites black, on the *C*-sites are gray, and on the *B*-sites light gray; (c) structure projected into the (01.0) plane.

gray. The third cell shown at the right of Fig. 17.18(a) consists of two half cells that complete the hexagon.

Figure 17.18(b) illustrates the stacking of close-packed planes in the α-La crystal structure: atoms on the A-sites are black, those on the C-sites are gray, and on the B-site light gray. Examining Fig. 17.18(c), we can see that the stacking of the close-packed layers is $ACAB$ (the same as NiAs), and the structure contains regions of *hcp*-like, *h*, stacking and *fcc*-like, *c*, stacking. Using the h–c notation of Chapter 16, we have the stacking sequence *hc*. If the structure is projected into a (00.1) plane, as in the *hcp* structure, all three A, B, and C sites are occupied; but, unlike the *hcp* structure, this structure requires the projection of *four* layers to sample all three types of site. In the Strukturbericht notation, this is known as the A3$'$ structure (**Structure 17**).

17.7.2 Elements with lower symmetry structures

The next elemental structure defined in the StrukturBericht notation, is the A5 structure, which is the structure of β-Sn. β-Sn is a high-temperature semi-metallic polymorph of metallic tin, and is stable above 286.4 K. This structure has space group $\mathbf{I4_1/amd}$ (D_{4h}^{19}), with lattice constants $a = 0.58315$ nm and $c = 0.31814$ nm (Swanson and Tatge, 1953) (**Structure 18**).

Figure 17.19(a) shows a unit cell for the β-Sn crystal structure in a ball-and-stick representation, with the origin at the tetrahedral site, offset $(0, 1/4, -1/8)$ from the center of symmetry. This structure is a diamond cubic derivative structure that is highly tetragonally distorted. There are four shorter Sn–Sn bonds at approximately 0.302 nm and two more bonds at 0.318 nm.

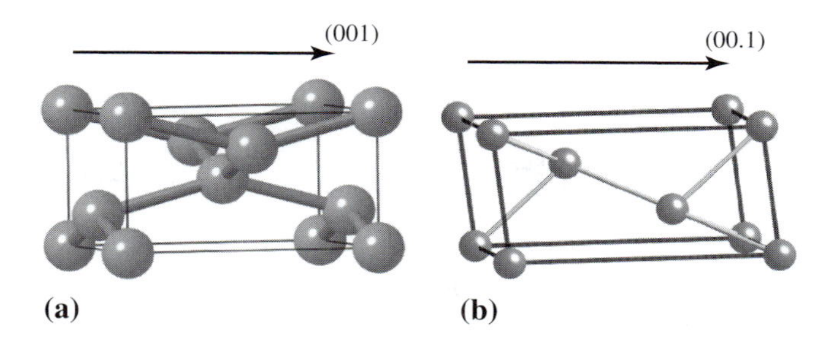

Fig. 17.19. (a) Unit cell for β-Sn crystal structure with the origin at the tetrahedral site (0, 1/4, −1/8) from the center of symmetry and (b) rhombohedral cell of the Hg A10 structure.

We can consider Sn to be octahedrally coordinated (in a distorted octahedron) by setting it at origin choice (1). However, we typically plot the structure of Sn using a tetrahedral origin (origin choice 2 for the space group $\mathbf{I4_1/amd}$ (D_{4h}^{19})) to show the relationship to the low-temperature, α-Sn polymorph, which has a diamond cubic structure.

Elemental In is an example of an element with the A6 tetragonal structure, which we can derive by distorting the *fcc* or, alternatively, *bcc* structures. We leave it to the reader to derive this structure as an exercise. The full structural information can be found as **Structure 19** in the web structures appendix.

The A7 structure is rhombohedral with space group $\mathbf{P\bar{3}c1}$ (D_{3d}^{4}), for which allotropes of Bi are examples and As is the prototype (**Structure 20**). For an allotrope of Bi, with lattice constants $a = 0.4546$ nm and $c = 1.1862$ nm, Bi occupies the position (0, 0, 0.2339) at room temperature. The z component of this position is further observed to be strongly dependent upon temperature.

Figure 17.20(a) shows four-unit cells of the Bi structure in a ball-and-stick representation. The Bi structure consists of sheets of puckered six-fold rings that are cross-connected to the sheets above and below it. This structure was first determined by Cucka and Barrett (1962). Bismuth is metallic and less toxic than As and Sb; it is an important component in high temperature superconductors. A rhombohedral polymorph of C (graphite) is also reported to have the A7 structure, although hexagonal graphite has the A9 structure discussed below.

The A8 structure has γ-Se as its prototype. It is also the structure of Te which has space group $\mathbf{P3_121}$ (D_3^4) with $a = 0.4527$ nm and $c = 0.5921$ nm

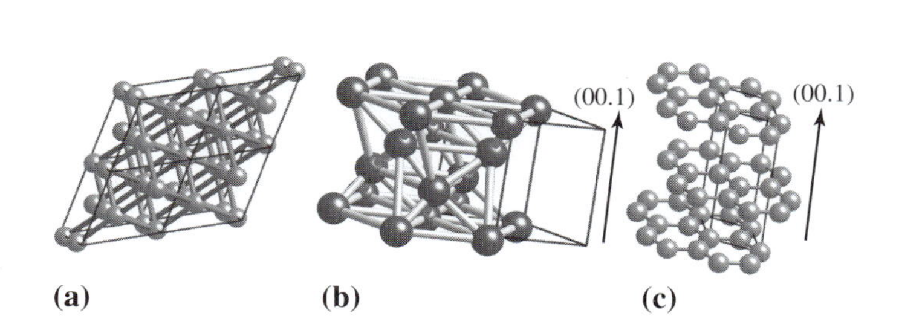

Fig. 17.20. (a) four-unit cells (hexagonal representation) of the rhombohedral Bi (A7) structure in a ball-and-stick representation; (b) Te (A8) structure in a hexagonal prismatic representation (three cells); (c) sheets of hexagonal C nets displaced from layer to layer in the graphite (A9) structure (Wyckoff, 1963).

and Te at the position (0.7364, 0, 1/3) at room temperature (**Structure 21**). The structure consists of Te atoms, along the z-axis, that snake around a 3_1 screw axis (Adenis *et al.*, 1989). Figure 17.20(b) shows this structure in a hexagonal prismatic representation.

The A9 structure is that of graphite (C). It has displaced hexagonal nets, 6^3 tiles, which are arranged in alternating layers. Atoms in alternate layers sit above the center of the hexagon in the previous layer as depicted in Fig. 17.20(c). Graphite has space group $\mathbf{P6_3/mmc}$ (D^4_{6h}), with $a = 0.2456$ nm and $c = 0.6696$ nm. (**Structure 22**). The bonding between the alternating layers is of the van der Waals type, which accounts for the relative ease with which individual layers can be peeled off (used in pencil leads and graphitic lubricants).

The A10 structure (**Structure 23**) is also a rhombohedral structure with space group $\mathbf{P\bar{3}c1}$ (D^4_{3d}), for which solid Hg is the prototype. The idealized structure with $a = 0.3464$ nm, $c = 0.6677$ nm and with Hg at position (0, 0, 0) is depicted in Fig. 17.19(b). This structure is similar to that of Bi (i.e., it belongs to the same space group), but the atoms are located at $z = 0$. Figure 17.19(b) shows a single rhombohedral cell of the Hg structure.

The A11 structure is that of Ga (**Structure 24**). Gallium has an orthorhombic space group **Cmca** (D^{18}_{2h}) with lattice constants $a = 0.4517$ nm, $b = 0.7645$ nm, and $c = 0.4511$ nm and Ga in the position (0, 0.1525, 0.079) (Villars and Calvert, 1991), as originally reported by Bradley (1935). The a and c lattice constants are very close in size.

Figure 17.21(a) shows a space-filling depiction of the atoms in a single orthorhombic unit cell for Ga. Figure 17.21(b) shows a ball-and-stick representation of the short 0.24 nm bonds between "dimerized" Ga atoms. This tendency to dimerize (i.e., form pairs with short bonds) is perhaps an indicator as to why Ga melts at such a low temperature (roughly room temperature) and what type of clustering persists in the liquid state. Figure 17.21(c) shows a ball-and-stick depiction including the longer 0.27 nm bonds between other Ga atom pairs. This depiction shows a staircase arrangement of the Ga

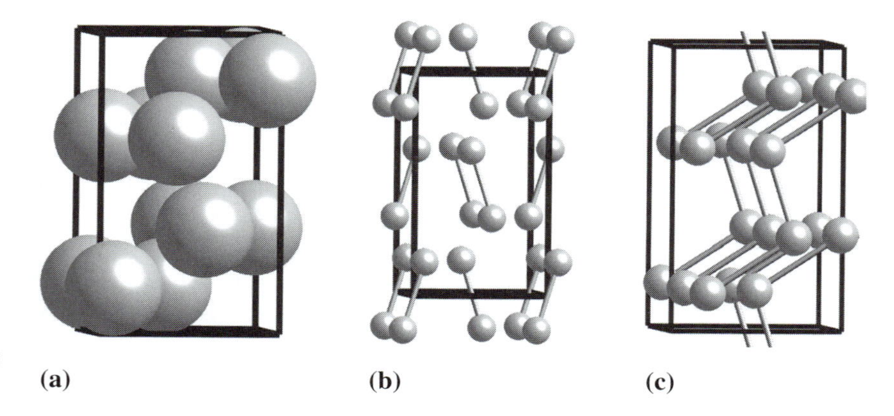

Fig. 17.21. (a) Space filling A11 orthorhombic crystal structure of Ga; ball-and-stick depictions (b) of the short 0.24 nm bonds between "dimerized" Ga atoms; (c) the longer 0.27 nm bonds between other Ga atom pairs.

bonds in this structure. This structure is similar to the **A14** structure of I_2 and the **A17** structure of P, which differ only in their degree of dimerization. These structures are not illustrated here.

The **A12** (**Structure 25**) and **A13** (**Structure 26**) structures are those of α- and β-Mn, respectively. These complicated structures result because Mn has a half-filled d-shell and can be stabilized in many different symmetries. As such, Mn will occupy different sites in these structures. The coordination polyhedra in these phases are very complicated and are similar to the Frank–Kasper phases discussed in the next chapter.

A cubic unit cell of α-Mn has 58 atoms (Wyckoff, 1963). This structure consists of four crystallographically distinct Mn atoms. In the α-Mn structure, one of the distinct Mn atoms sits in 12-fold coordination, one Mn sits in 13-fold coordination, and the last in 16-fold coordination. The β-Mn cubic unit cell has 20 atoms (Shoemaker, 1978). This structure has two crystallographically distinct Mn atoms, and both are 12-fold coordinated. Because these complicated structures are not simplified easily to a level appropriate for an introductory text, we do not illustrate them here.

The **A15** structure will be discussed in more detail in Chapter 18. First identified as a structure of an allotropic form of W, **A15** was incorrectly identified; its prototype is not an element, but the structure of W_3O.

The **A20** structure is that of α-U, the stable phase of uranium at room temperature and another orthorhombic structure with space group **Cmcm** (D_{2h}^{17}) (**Structure 27**). It has lattice constants $a = 0.2854$ nm, $b = 0.5869$ nm, and $c = 0.4955$ nm, with a U atom in the position (0.0, 0.1025, 1/4) (Wyckoff, 1963). A space-filling depiction of the atoms in a single orthorhombic cell is shown in Fig. 17.22(a). Figure 17.22(b) shows a projection of two planes of atoms on an (025) plane, where we can see that the U atoms form a (puckered) hexagonal network. Figure 17.22(c) shows the stacking of these networks normal to the (025) plane. The crystal structure of β-U is not illustrated here, but is similar to that observed for the σ-phase found in CrFe intermetallics (Chapter 18).

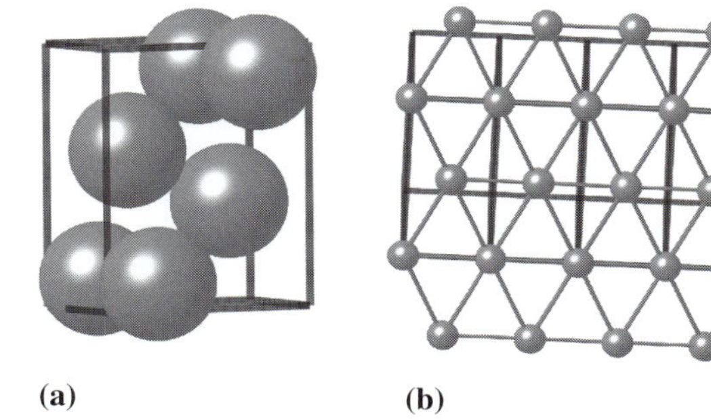

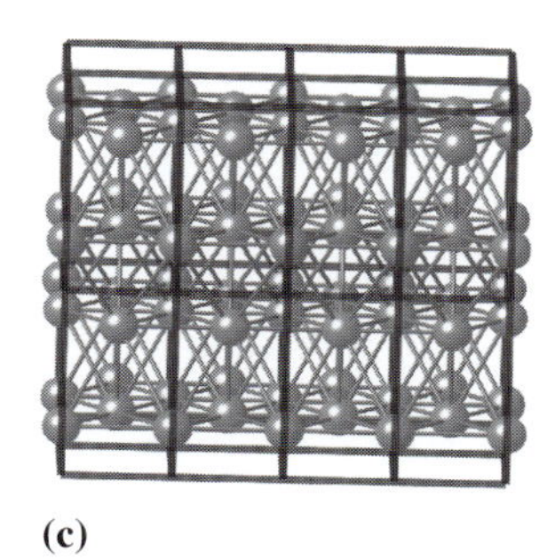

Fig. 17.22. Crystal structure of α-U; (a) 6 atoms in the orthorhombic unit cell. Projection of 2 layers of atoms into (b) and the stacking of hexagonal networks normal to (c) the (025) planes.

17.8 *Natural and artificial superlattices (after Venkataraman *et al.*, 1989)

17.8.1 Superlattice structures based on the $L1_2$ cell

Every structure type can serve as the starting point for the construction of a new structure type. We have seen this repeatedly in the previous sections, where we derived new structure types starting from the *hcp*, *fcc*, and *bcc* structures. In this section, we will take the $L1_2$ structure type, and use it to create a new type known as $D0_{22}$. Then we will illustrate a series of structures known as long period superlattices.

We begin with the $L1_2$ structure. Consider an A_3B alloy in the disordered state at high temperature. The atoms randomly occupy the sites of an *fcc* solid solution. When the temperature decreases, the structure orders into the $L1_2$ structure type, meaning that the B atoms preferentially occupy the cube corners. However, because every lattice site in the *fcc* lattice is equivalent, there are four possible choices for the B atoms to occupy! In other words, any of the four sites in the unit cell can be selected as the corner of the new ordered, unit cell. Likewise, there are four possible *sublattices* for the B atom to choose from, as shown in Fig. 17.23. The structures in (b), (c), and (d) are shifted with respect to the one in (a) by one of the three face centering vectors, **A**, **B**, or **C**.

In different regions of a macroscopic crystal, the B atom may select different sublattices to occupy; and when those ordered regions grow, they will eventually meet each other and form an interface. This interface is known as an *anti-phase boundary (APB)*, because the structures on either side of the boundary are "out-of-phase." It costs energy to form such an interface because the bonds across the interface are not entirely ordered as those in the $L1_2$ structure. The energy per unit area of interface is known as the *APB energy*, γ_{APB}. This energy may be an isotropic quantity, meaning that the energy does not depend on the orientation of the interface with respect to the crystal lattice, or this energy may be highly anisotropic. In some material systems, such as Al_3Ti and $AlCu_3$, the APB energy is strongly anisotropic, and only APBs along {001}-type planes are found. (see Box 17.7 for more detailed information on the number of different sublattices).

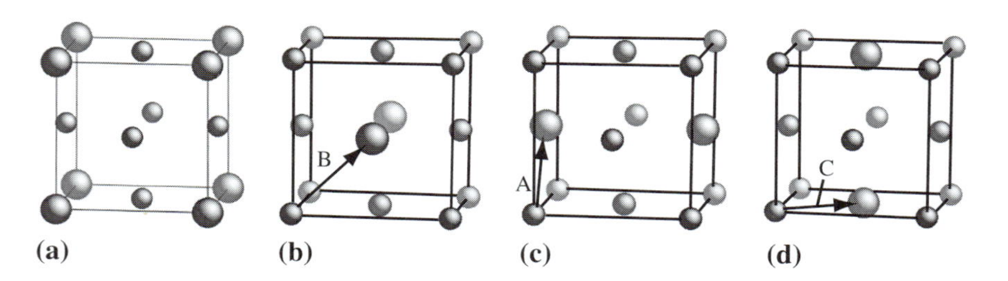

Fig. 17.23. Illustration of the four sublattices for the $L1_2$ structure.

Box 17.7 Orientation variants, translation variants, and group theory

Consider a high temperature phase described by a point group of order h. Generally, the ordered phase has low symmetry and is a subgroup of the high temperature group. The order of the low temperature point group is represented by l. It can then be shown (e.g., (Van Tendeloo and Amelinckx, 1974)) that the total number of possible *orientations* of the ordered unit cell with respect to the disordered one is given by the ratio $\frac{h}{l}$. For instance, for a disorder–order transition from *fcc* to $\mathsf{L1}_0$, we have $h = 48$ for the point group $\mathbf{m\bar{3}m}$ (O_h), and $l = 16$ for **4/mmm** (D_{4h}), so that there are three possible orientation relations. Intuitively, this is rather straightforward, because the c-axis of the tetragonal $\mathsf{L1}_0$ structure can be oriented along any one of the three $\langle 001 \rangle$-type directions of the *fcc* cell.

For the $\mathsf{L1}_2$ ordered structure, the point group is $\mathbf{m\bar{3}m}$ (O_h), so that the ratio $\frac{h}{l} = 1$. There are, however, several possible sublattices on which the ordering may begin. We may determine the total number of possible sublattices in a given crystal structure by considering the volume of the *primitive* unit cells of each phase. The parent *fcc* phase has a rhombohedral primitive unit cell with volume $a^3/4$, where a is the cubic lattice parameter. The volume of the primitive ordered $\mathsf{L1}_2$ unit cell is equal to a^3, as the space group $\mathbf{Pm\bar{3}m}$ (O_h^1) implies a primitive lattice. The ratio of the primitive volume of the ordered superlattice to that of the primitive parent lattice is the number of sublattices for the ordering transition, the number of *translation variants*. For the *fcc* $\rightarrow$ $\mathsf{L1}_2$ transition we find four sublattices. For the *fcc* $\rightarrow$ $\mathsf{L1}_0$ transition, we find only two sublattices, as the smallest primitive unit cell for $\mathsf{L1}_0$ has volume $a^3/2$ (reader exercise).

Figure 17.24(a) shows a planar APB in an $\mathsf{L1}_2$ structure; the location of the APB is indicated by a dashed line in this [100] projection. In most compounds, the APB is considered to be a *defect* in the otherwise perfect order of the crystal structure. There are compounds for which γ_{APB} is vanishingly small, so it does not take much energy to introduce APBs. When this is the case, an interesting phenomenon may occur: the APBs may form a periodic array! This is illustrated in Fig. 17.24(b), which shows two APBs separated by a distance $2a$, where a is the $\mathsf{L1}_2$ lattice parameter. When this defect periodically repeats we can easily recognize a new unit cell, in this case with a repeat distance of $4a$ along the former c-axis. The lattice parameters of this new unit cell are hence a, a, and $4a$. When the distance between neighboring APBs shortens, we will reach the point where each $\mathsf{L1}_0$ cell has an APB, as shown in Fig. 17.24(c). The resulting structure type is known as the $\mathsf{D0}_{22}$ structure type, with Al_3Ti as prototype structure (**Structure 28**). This is a body-centered tetragonal structure with eight atoms per unit cell.

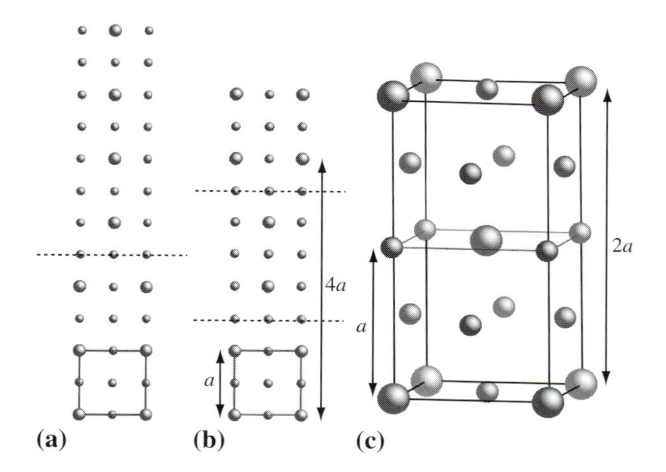

Fig. 17.24. Illustration of the formation of an anti-phase boundary in the $L1_2$ structure. In (a), a single boundary is present, indicated by the dashed line ([100] projection). In (b), multiple APBs are present in a periodic arrangement. When the APBs are present in every $L1_2$ unit cell, as shown in (c), then the $D0_{22}$ structure is obtained.

From the preceding discussion, we can derive a parameter, M, which is related to the "density" of APBs. If p is the number of APBs in the repeat unit of the superlattice, and q is the total number of $L1_2$ unit cells in that same repeat unit, then the ratio $M \equiv \frac{q}{p}$ is the inverse of the APB density, or, equivalently, the average distance between the APBs. For the structure in Fig. 17.24(b), we find $p = 2$ and $q = 4$, so $M = 2$. For the $D0_{22}$ structure, we have $p = 2$ and $q = 2$, so $M = 1$. The APB-free $L1_2$ structure has $M = \infty$. As shown by the superlattice structures of Fig. 17.25, M does not need to be an integer. While these structures typically do not have a special StrukturBericht symbol, they are found in many alloy systems, such as Al_3Ti, $AlCu_3$, Cu_3Pd, Au_3Zn, . . . The occurrence of these long period superlattice structures requires the use of elaborate statistical mechanics models.

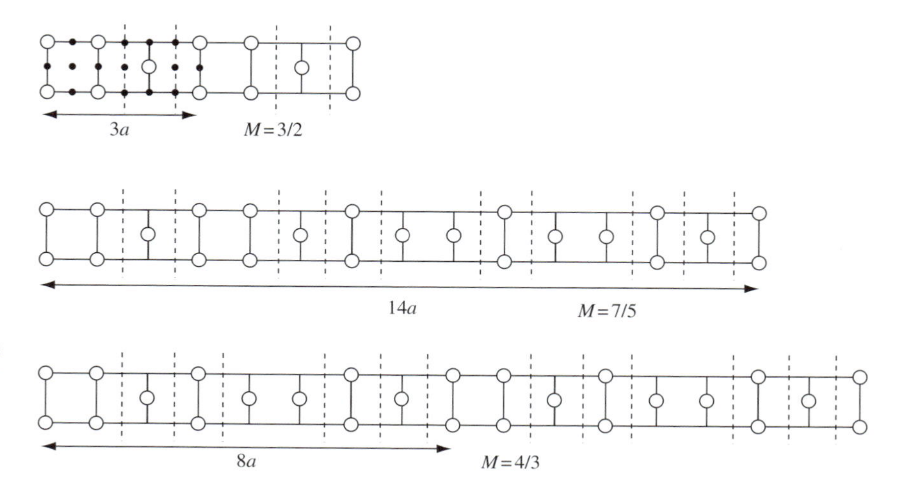

Fig. 17.25. Examples of long period superlattice structures with non-integer M values, based on the $L1_2$ unit cell. Each open circle represents the location of a B atom in a [100] projection; solid circles are A atoms, and are removed from most of the figure for clarity.

17.8.2 Artificial superlattices

Modern advanced deposition techniques, such as *molecular beam epitaxy* (MBE) and *pulsed laser deposition* (PLD), allow for the precise synthesis of artificially layered structures. A large number of new artificially structured materials have been synthesized in recent years, and the possibilities seem to be limited only by the researcher's imagination.

The term *synthetic modulated structure* (synonymous with *artificially modulated structure*) describes any periodically configured material with a repetition wavelength greater than the unit cell dimensions for the equilibrium material. These structures have made an impact in a variety of fields, including the magnetic materials discussed in Box 17.8 below. Scientists are interested in these materials for their multilayer periods which strongly impact properties, especially when the size approaches quantum mechanical length scales. For example, synthetic modulated semiconductor structures referred to as semiconductor superlattices or *quantum well structures*, have revolutionized a variety of semiconductor devices.[11] Seminal work on the development of satellite peaks in Cu_mNb_n multilayers as a function of the modulation wavelength was published by Schüller (1980).

A *sandwich structure*, in which a monolayer of a material of interest is deposited on a substrate and capped with one or many layers of the substrate material, allows us to study two monolayer–substrate interfaces. Among the most widely studied sandwich structures are the magnetic transition metal–noble metal systems (TM–NM). In these, similar atomic spacings can be chosen to reduce the influence of interfacial strain on properties, and the chemical interactions between the monolayer and the supporting layer are small. We can study more than two interfaces by synthesizing multilayer structures as illustrated in Box 17.8.

17.8.3 X-ray scattering from long period multilayered systems

X-ray scattering is a powerful tool that can be used to determine structural parameters of multilayers. The 1D periodicity of planar multilayers gives rise to *satellite reflections*, which can be used to calculate lattice constants and modulation wave lengths. Fluctuations in the Bragg and satellite peak positions and widths give information about the coherency and interdiffusion between the layers. Figure 17.26(b) shows satellite peaks in $FePt_n$ multilayers. The $FePt_n$ multilayers mimic systems studied by first principles calculations (McHenry *et al.*, 1991) in which Fe monolayers were embedded in Pt (or Pd) hosts with an odd number of layers between them, maintaining the stacking of (001) planes in the $n = 1$ $L1_0$ structure.

[11] First synthesized by Esaki and Tsu (1970).

Box 17.8 Magnetic nanostructures – multilayer structures

The study of magnetism at interfaces in sandwiches and multilayers is a technologically important area of research. Novel materials properties in non-equilibrium configurations of dissimilar materials has also been a driving force behind the study of artificially derived (man-made) structures. Scientists have been interested in the behavior of single magnetic monolayers and 2-D magnetism. It is possible to produce magnetic monolayers supported by a substrate or sandwiched between two substrate layers. The modulated superlattice structures illustrated below are examples of systems with interesting 2-D magnetism (McHenry *et al.*, 1990). These are the artificially constructed superlattices Au_1Fe (a), Au_3Fe (b), Au_5Fe (c), . . . , $Au_\infty Fe$ (d).

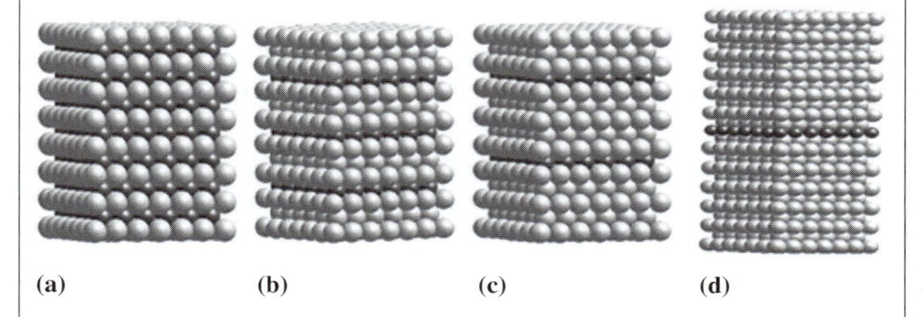

Scientists have studied implications of 2-D magnetism on magnetic dipole moments (MacLaren *et al.*, 1990), exchange coupling between the dipole moments, and magnetocrystalline anisotropy (McHenry *et al.*, 1991). Equiatomic FePt, FePd, and ternary $FePt_xPd_{1-x}$ alloys crystallize in the $L1_0$ phase structure. Their *c*-axis orientation (texture) with respect to the substrate is important for recording magnetic information permanently in *perpendicular recording media*. As materials advance, high density magnetic recording is done at ever decreasing bit sizes (Jeong, 1994). This is important in the miniaturization of hard disk drives and to increase storage capacity at fixed size (Weller *et al.*, 2002).

Large uniaxial magnetic anisotropies reflect a strong preference for magnetization vectors to lie along the *c*-axis in $L1_0$ phase magnets, resulting in unprecedented *magnetic anisotropy* (Klemmer *et al.*, 1995). This, along with their notable corrosion resistance, makes the $L1_0$ materials among the most attractive permanent magnet materials for thin film magnetic recording applications. The *c*-axis texture in $L1_0$ phase magnets aligns the natural 1×1 superlattice with alternating Fe and Pd and/or Pt layers, repeating in a direction normal to the substrate plane. Ternary alloys are investigated using first principles calculations for their potential, even larger, anisotropies and to understand the influence of alloying on the atomic ordering transition that takes a disordered *fcc solid solution* to the ordered $L1_0$ phase (Willoughby *et al.*, 2002, Willoughby, 2002).

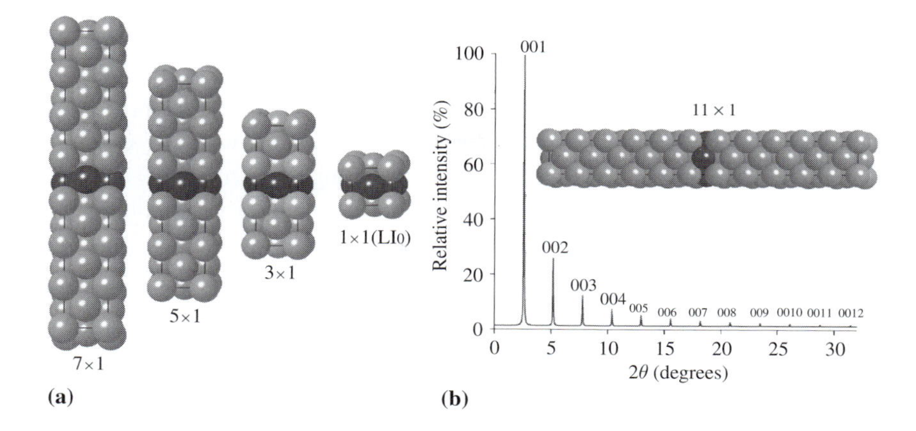

Fig. 17.26. (a) $FePt_7$, $FePt_5$, $FePt_3$, and FePt superlattice structures. (b) Simulated intensity versus 2θ X-ray diffraction pattern for an $FePt_{11}$ superlattice structure.

Figure 17.26(a) illustrates a sequence of structures constructed by adding an additional two Pt layers to the previous structure. This yields the $FePt_7$, $FePt_5$, $FePt_3$, and FePt superlattice structures. These are 7×1, 5×1, 3×1, and 1×1 superlattice structures, respectively. Each of these has a **P4/mmm** (D^1_{4h}) space group with a c-lattice constant of approximately 7, 5, 3, and 1 times the a lattice constant.[12]

Figure 17.26(b) shows a simulated intensity versus 2θ X-ray diffraction pattern for an $FePt_{11}$ superlattice structure, scattering Cu-Kα radiation near the (001) reflection. The $FePt_{11}$ superlattice structure (rotated by $\pi/2$) is shown in the inset. This XRD pattern shows several interesting features. First, because of the large (≈ 4.2 nm) c lattice constant for this system, the (001) reflection occurs at a small value of $2\theta \approx 2.5°$. The most dramatic feature, however, is the appearance of the $(00l)$ ($l = 2, \ldots, 12$) superlattice reflections. From these two observations we can infer that *low angle X-ray scattering* is an important tool for the study of superlattices.

AlNb multilayers have been sputter deposited (Barmak *et al.*, 1998) and used as diffusion couples to monitor the formation of equilibrium phases upon subsequent annealing. The as-deposited structures offer good examples of systematic changes in the X-ray scattering as a function of the multilayer composition, period, and individual layer thicknesses. An example of X-ray scattering from such multilayer structures is described in Box 17.9.

17.8.4 Incommensurate superlattices

Thus far, we have always described the translational periodicity of crystal structures in terms of a lattice and its basis vectors. In a one-dimensional (1-D) lattice, we can describe the positions of the lattice points as follows:

[12] Approximately, because of a slight tetragonal distortion in the parent 1×1 structure, which has $a = 0.3806$ nm and $c = 0.3684$ nm.

Box 17.9 Interface engineering in Nb Al multilayer structures

AlNb multilayers preserving a 3:1 Nb-to-Al stoichiometry have been sputter deposited as thin films with varying periodicities between 10 and 500 nm. X-ray scattering experiments (a) and microstructural observations (b) of the modulated superlattice structures are illustrated below (figure courtesy of K. Barmak).

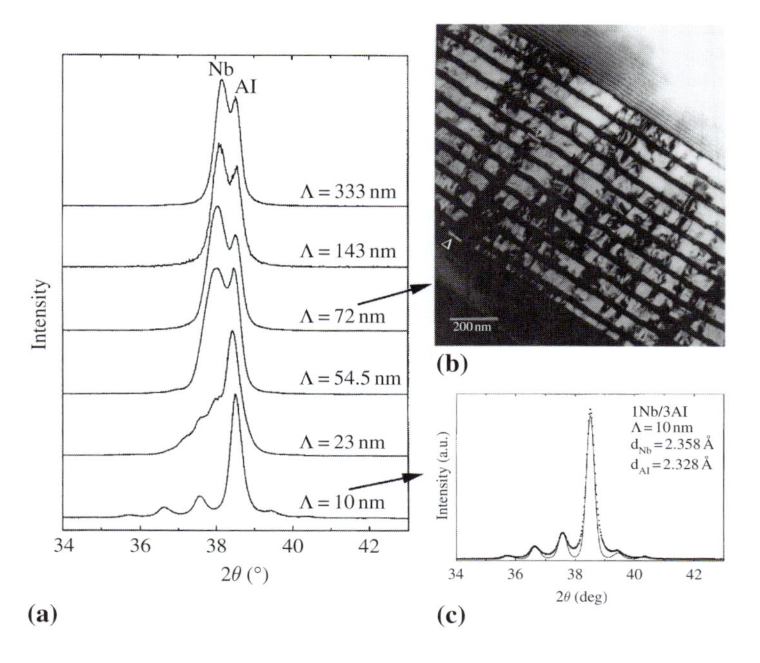

The XRD patterns for AlNb multilayer films (a) show several systematic changes with periodicity, $\Lambda = t_{\mathrm{Nb}} + t_{\mathrm{Al}}$:

(i) At large Λ, the most intense peaks observed are for *bcc* Nb and *fcc* Al.
(ii) At small Λ, the most intense peaks merge indicating that the layers are strained and exhibit a single common lattice constant.
(iii) Superlattice reflections are observed as satellite peaks around the main peak at the smallest multilayer periodicities.

A cross-sectional TEM micrograph (b) of an as-deposited multilayer film with $\Lambda = 72$ nm shows sharp interfaces and uniform thicknesses of the constituent layers. The XRD pattern for the 10 nm multilayer film fits well to a simulated pattern assuming a single Bragg peak with an average lattice spacing and superlattice reflections predicted by a square wave compositional profile for the constituent layers.

$$x_n = na, \tag{17.17}$$

where a is the 1-D lattice constant and n is an integer. The concept of a *modulated structure* can be illustrated by the introduction of *atomic displacements* in this 1-D lattice. For simplicity, we will consider first a monatomic basis in which the atoms sit on the lattice points defined above. Then, we consider displacements of the original atoms to new sites, X_n, given by:

$$X_n = x_n + f \sin\left(\frac{2\pi}{a} q x_n\right) = x_n + f \sin(2\pi n q), \tag{17.18}$$

where f is the modulation amplitude, and q describes the modulation wavelength. If q is a non-zero rational number, then the new structure is also periodic, but with a larger unit cell (i.e., a/q). This structure is referred to as a *commensurate superlattice*, or a *commensurate long period structure*. If q is an irrational number, such as $1/\sqrt{2}$, then the structure does not possess traditional periodicity but, instead, possesses *quasi-periodicity*. Such structures are known as *incommensurate superlattices*. Incommensurate superlattices can be constructed artificially, but they are also found in nature.

Figure 17.27(a) and (b) illustrate the transformation of a periodic 1-D lattice, with lattice constant a', to a commensurate, long period lattice with $q = 1/4$ and 1/6, respectively ($f = \frac{1}{2}$). Note that this transformation leads to the long period lattice constant of $a = 4a'$ and $a = 6a'$, respectively. Fig. 17.27(c) and (d) shows a similar transformation with $f = 1/4$, but with $q = 1/\sqrt{2}$ or $q = 1/\sqrt{3}$. This transformation leads to an incommensurate structure without a traditional lattice constant. Because the sin function is clearly periodic, but with an irrational period, the modulated lattice is incommensurate with the first, i.e., the lattice parameter of the modulated lattice is not a rational multiple of that of the original lattice. This irrational periodicity is, however, recognized and given the name *quasi-periodicity*.

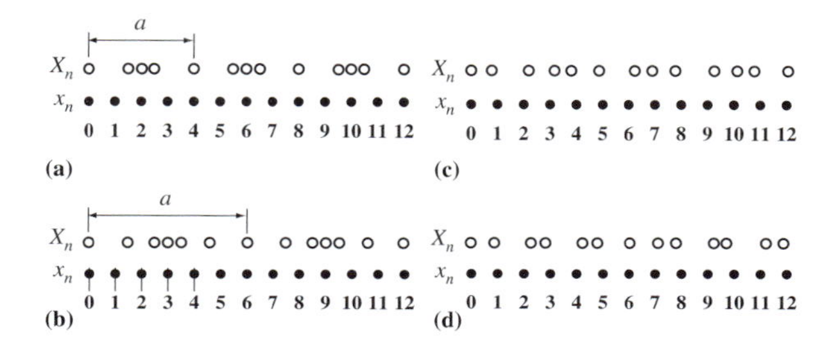

Fig. 17.27. Transformation of a periodic 1-D lattice to long period superlattices with lattice constants (a) $a = 4a'$ and (b) $a = 6a'$, respectively ($f = 0.5$). Transformation of a periodic 1-D lattice to incommensurate lattices with period (c) $1/\sqrt{2}$ and (d) $1/\sqrt{3}$, respectively, ($f = 0.25$).

A quasi-periodic function is a superposition of periodic functions whose periods are incommensurate with one another.

These ideas are easily generalized to describe a 3-D monatomic lattice, where we can introduce one or more commensurate or incommensurate modulations in different directions. Incommensurate structures can also be constructed in a non-displacive manner. For example, we could have a traditional periodic lattice for which the charge density, or spin density, or even the chemical composition is modulated in an incommensurate manner. Waves that interact with the charge or spin densities (i.e., neutrons) will be scattered as if they originate from an incommensurate lattice.

17.9 Interstitial alloys

In substitutional alloying in an AB binary system, it is possible that the A and B atoms will substitute randomly for one another on the same crystalline lattice. Alternatively, a new phase can be formed in which the two atoms are ordered on the original lattice. The *Hume-Rothery rules* predict that when there is a large difference in atomic sizes, the lattice of the larger species may remain intact and the smaller atoms may occupy interstitial sites in that lattice; such systems are known as *interstitial alloys*. Two important examples of classes of interstitial alloys are transition metals which dissolve small amounts of smaller atoms such as C, N, H, etc. and ionic solids where large anions occupy sites in a close-packed lattice and the smaller cations occupy the interstitial sites. We discuss these latter solids in Chapter 22.

Interstitial alloys are among the most technologically important alloys. Notable among these are Fe–C steel alloys, for which the dissolved C is crucial to the properties of the steel. For pure Fe, the low temperature and room temperature phase has a *bcc* (A2) crystal structure, known as *ferrite*. At high temperatures, it has the *fcc* (A1) crystal structure, called *austenite*. While C is much smaller than Fe and can be dissolved interstitially, it does strain the *bcc* lattice; only about 1% C can be dissolved in ferrite at room temperature. For higher concentrations of C, a mixture of ferrite and a carbide, Fe_3C, called *cementite* exists in equilibrium at room temperature. The structure of cementite is discussed in Chapter 21 in the context of metal–metalloid alloys. C, B, P, Si, etc. are examples of *metalloid* elements.

At higher temperatures, much more C (up to ≈ 2.1 %) can be dissolved in austenite. The quenching of austenitic alloys containing larger amounts of dissolved C to room temperature gives rise to the formation of metastable phases of *bcc* Fe with larger C content existing at room temperature.

Martensite is a tetragonally distorted body-centered cubic variant of iron with carbon dissolved at a non-equilibrium concentration level. Martensite is named after the German metallurgist Adolf Martens (1850–1914), who first studied the phase, which forms during quenching without the precipitation of cementite. While martensite can be considered as ferrite supersaturated with carbon, the additional dissolved carbon and its strain on the *bcc* ferrite structure cause a tetragonal distortion of the structure. On cooling, the *fcc* austenite smoothly deforms into the tetragonally distorted *bcc* martensite by a shear deformation, expanding in one direction and contracting in the other two. This strain and its resulting distortion lead to a significant hardening of martensite as compared to the equilibrium ferrite.

Other interstitial alloys of technological importance include hydrides (simple metal, transition metal, or rare earth metal). These include light metal hydrides like Li hydride because of their large H capacities per unit weight. Renewed interest in the AlH_4Na phase (Bogdanovic and Schwickhardi, 1997) followed reports on how doping with early transition metals, Ti and Zr, leads to decomposition and release of H at low temperatures, an important reaction for *hydrogen storage* technologies. Similar results with small rare earth element additions have also been reported. Again, low temperature H decomposition kinetics are important for viable H storage applications.

The crystal structure of the AlH_4Na phase is the tetragonal structure illustrated in Fig. 17.28. This model is based on crystal structure data for the phase as taken from Pearson's tables (attributed to Bel'skii *et al.* (1983)). The structure can be understood in terms of $(AlH_4)^-$ tetrahedral units, forming chains along the [010] directions that are connected by bridging Na^+ ions. The Na^+ ions are also tetrahedrally coordinated by H, but with noticeably distorted tetrahedra. Hydrogen atoms also sit in tetrahedral coordination by the metal atoms. These infinite planar arrays are stacked along the [001] direction. As transition metal or rare earth atom substitutions expand the *c*-axis lattice parameter, more room for hydrogen motion

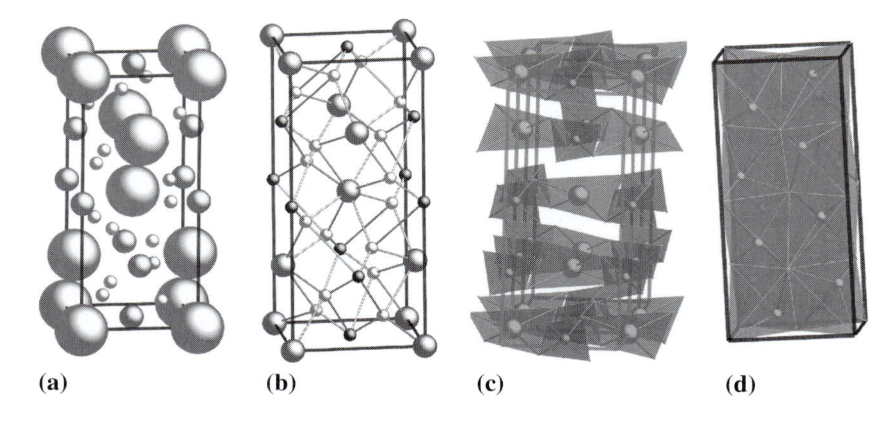

Fig. 17.28. a) Atomic and (b) ball-and-stick depictions of the AlH_4Na phase structure; (c) connectivity of tetrahedral AlH_4 tetrahedra and NaH_4 tetrahedra; (d) connectivity of the H tetrahedral coordination polyhedra (Bel'skii *et al.*, 1983).

in the lattice is provided. The structure of the intermediate AlH_6Na_3 phase is still the subject of scientific debate; using XRD studies, a monoclinic phase has been identified (Gross *et al.*, 2000) whereas first principles calculations have been employed to identify two possible structural variants (Opalka and Anton, 2003).

17.10 Historical notes

William Hume-Rothery (1899–1968) was born in 1899 in Worcester Park, Surrey. Hume-Rothery studied at Cheltenham College from 1912–16 before entering the Royal Military Academy at Woolwich, where he originally planned on a military career. When cerebrospinal meningitis left him deaf, he transferred to Magdalen College, Oxford in 1918. Hume-Rothery completed his education at Oxford in Chemistry in the Honour School of Natural Science. From 1922–5 he worked under H. C. H. Carpenter, the Professor of Metallurgy, at the Royal School of Mines of Imperial College, London. There he studied structure/property relationships of intermetallic compounds.

In 1925 Hume-Rothery returned to Oxford, where from 1925–52 he held various research fellowships that supported his work at Oxford. Hume-Rothery was instrumental in establishing metallurgy as a discipline at Oxford. He founded the Department of Metallurgy at Oxford in 1956. From 1955–8, he was the first holder of the George Kelley Readership in Metallurgy, 1955–8, from 1958–66 he was the first Isaac Wolfson Professor of Metallurgy. Hume-Rothery is internationally known for his work on the formation of alloys and intermetallic compounds.

(a)

(b)

Fig. 17.29. (a) William Hume-Rothery (1899–1968) (picture courtesy of Department of Materials, University of Oxford) and (b) Thaddeus B. Massalski (1929–) courtesy of Ted Massalski.

During World War II, he supervised a group that performed important work on aluminium and magnesium alloys. Hume-Rothery and co-workers established stability rules for alloys and developed the equilibrium diagrams for many alloy systems. An entertaining short biography of Hume-Rothery written by **Jack Christian** (Christian, 1997) is reprinted at the University of Oxford web site.[13] Christian and **W. B. Pearson**, working with Hume-Rothery, did some of the early studies of the σ phase, an embrittling agent in steels and one of the Frank–Kasper phases discussed in Chapter 18.

Hume-Rothery's investigations of alloy equilibria led to the famous *Hume-Rothery rules for alloy stability*. These rules are discussed in this chapter. They offer an empirical guide to deciding when two metals will be completely miscible (i.e., form a single crystalline phase at all concentrations of one dissolved in the other). Hume-Rothery was a proponent and early pioneer in the *electron theory of metals* (Hume-Rothery, 1952). He was made an honorary member of the American Society of Metals (now ASM International) in 1957. He published the influential book "The Structure of Metals and Alloys" in 1954 (Hume-Rothery and Raynor, 1954).

Thaddeus B. Massalski (1929–), was born in Warsaw, Poland and lived there until the middle of World War II, when he and his family left for Switzerland to seek haven from the wartime ravages in Poland (Hutton, 2004). Massalski began his college studies after the war at the Reggio Politecnico di Torino in Italy followed by work at Imperial College, London. Finally, he studied physical and theoretical metallurgy at the University of Birmingham, and received a B.Sc. He received a Ph.D. from the University of Birmingham in 1954 under the direction of G. V. Raynor, who was a former student of Hume-Rothery. His thesis work considered interactions between the Fermi surface of metals and Brillouin zones.

After a post-doctoral appointment at the University of Chicago and serving as a lecturer at the University of Birmingham (Department of Physical Metallurgy), he joined the faculty at Carnegie Mellon in 1959. Professor Massalski served in several visiting professor appointments with Pol Duwez at the California Institute of Technology, at Stanford with Professor Flory, at Oxford University with Professor Jack Christian and Professor Hume-Rothery, and at Harvard University with Professor Chalmers. Ted Massalski is currently Professor Emeritus of Materials Science and Engineering at Carnegie Mellon University and a valued colleague of the authors of this book.

Massalski has contributed in the areas of stability of alloy phases, imperfections in crystals, phase transformations, and amorphous structures. He proposed thermodynamic criteria for metallic glass formation, as discussed in Chapter 18. He contributed to the development of the theory of solid state phase transformations including displacive and diffusional with many

[13] See http://www.materials.ox.ac.uk/infoandnews/history/goldenyears.html.

contributions in the area of massive transformations (Laughlin, 2004). He was also involved in the early studies of phase transformations in actinide metals. Professor Massalski has served as Editor-in-Chief of the ASM Binary Phase Diagram program. With Charles Barrett, Massalski authored the book "Structure of Metals, Crystallographic Methods, Principles and Data" (Barrett and Massalski, 1980), which has influenced much of the presentation in this chapter. Professor Massalski is a Fellow of TMS and ASM International and also a Fellow of APS. He has served as Guggenheim Fellow at Oxford and obtained the Alexander von Humboldt Prize in Germany. He is a foreign member of the Göttingen Academy and the Polish Academy of Science. He won the ASM International Gold Medal in 1993 and the Acta Metallurgica Gold Medal in 1995 as well as the British Hume-Rothery Prize and the American Hume-Rothery Award.

17.11 Problems

(i) *hcp structure, $\frac{c}{a}$ ratio*: Express the primitive translation vectors for the *hcp* lattice. Show that for an ideal *hcp* structure the c/a ratio is equal to $\sqrt{8/3}$.

(ii) *Packing fractions*: Show that the volume fractions occupied by hard spheres in the following structures are:

$$sc\ \frac{\pi}{6};\quad bcc\ \frac{\pi\sqrt{3}}{8};\quad fcc\ \frac{\pi\sqrt{2}}{6};\quad hcp\ \frac{\pi\sqrt{2}}{6};\quad \text{diamond}\ \frac{\pi\sqrt{3}}{16}.$$

(iii) *Interstitial sites I*: Determine fractional coordinates of octahedral and tetrahedral sites in *hcp* and *fcc* structures.

(iv) *Interstitial sites II*: Determine coordinates of octahedral and tetrahedral interstices of a *bcc* lattice. How many of each of these are there per atom? Are the coordination polyhedra for each of these sites regular?

(v) *Hume-Rothery rules*: *fcc* Cu has a lattice constant $a_0 = 0.361$ nm. *hcp* Mg has a lattice constant of 0.321 nm. *fcc* Ni has a lattice constant $a_0 = 0.352$ nm. Consider binary alloy pairs of these elements. Which are likely to exhibit complete solubility? Which are predicted to form compound phases?

(vi) *Structure factor*: Consider an *hcp* cell with identical atoms in the $2c$ position of space group #194 at $(1/3, 2/3, 1/4)$, and $(2/3, 1/3, 3/4)$.

(a) Show that atom positions can also be expressed as: (0,0,0) and $(1/3, 2/3, 1/2)$.

(b) Show that the structure factor can be expressed as:

$$F_{hkl} = f_{\text{atom}}\left(1 + \mathrm{e}^{2\pi\mathrm{i}\left[\frac{h+2k}{3} + \frac{l}{2}\right]}\right) \tag{17.19}$$

(c) Calculate the square modulus of this structure factor $F_{hkl}^2 = F_{hkl}F_{hkl}^*$.

(d) Express the structure factor for each of the following four cases:

- $h+2k=3n$, $l=$ even $\qquad h+2k=3n\pm1$, $l=$ odd;
- $h+2k=3n\pm1$, $l=$ even $\qquad h+2k=3n$, $l=$ even

(vii) *Lennard-Jones potential I*: Using the Lennard-Jones potential, derive the equilibrium interatomic spacing for the *hcp* lattice. Why is this spacing slightly different from the equilibrium spacing for the *fcc* lattice, derived in Box 17.5?

(viii) *Lennard-Jones potential II*: The *bcc* lattice sums are:

$$A_{12}^{bcc} = \sum_{ij}{}' \left(\frac{1}{p_{ij}}\right)^{12} = 9.11418;\ A_6^{bcc} = \sum_{ij}{}' \left(\frac{1}{p_{ij}}\right)^{6} = 12.2533. \quad (17.20)$$

Calculate the *bcc* to *fcc* cohesive energy ratio for a rare gas solid.

(ix) Fe *allotrope*: At 1700 K, δ-Fe the *bcc* high temperature allotrope of Fe has a lattice constant, $a_0 = 0.293$ nm. Calculate the metallic radius and atomic volume of Fe in δ-Fe and compare it with the room temperature allotrope.

(x) CuAu $L1_0$ *structure*: This structure is derived from the *fcc* structure through a disorder–order transition. Determine all the point group symmetry operations of the disordered phase that vanish during this transformation.

(xi) CsCl *superlattice structure*: Determine the matrix M for the linear transformation that takes the primitive basis vectors for the *bcc* structure into those for the primitive basis for the CsCl structure.

(xii) CuAu $L1_0$ *superlattice structure*: Determine the matrix M for the linear transformation that takes the primitive basis vectors for the *fcc* structure into those for the primitive basis for the $L1_0$ structure.

(xiii) *bcc structure factor*: List atomic positions for atoms in the *bcc* unit cell. Derive a simple expression for the structure factor and use it to predict the extinction rules for the *bcc* structure.

(xiv) CsCl *structure factor*: List atomic positions in the CsCl unit cell. Express the structure factor and predict the extinction rules. How would it differ if atoms on the Cs and Cl sites had equal atomic scattering factors?

(xv) NaCl *structure*: How many formula units are there in a unit cell of NaCl?

(xvi) *Diamond cubic* C: Consider the diamond cubic C allotrope with $a_0 = 0.356$ nm. Calculate the C–C bond length and C–C–C tetrahedral bond angle.

(xvii) *Graphite*: Consider the hexagonal graphite allotrope of C with $a = 0.2456$ and $c = 0.6696$ nm. Calculate the C–C bond lengths and C–C–C in and out of (basal) plane bond angles.

(xviii) *Diamond-graphite*: Calculate the density of C in graphite and diamond.

(xix) *Diamond cubic structure factor*: The diamond structure has an *fcc* Bravais lattice and a 2-atom basis with coordinates $(0, 0, 0)$ and $(1/4, 1/4, 1/4)$. List the atomic positions for the remaining atoms in the unit cell, derive a simple expression for the structure factor, and predict the extinction rules. Do these agree with the reflections in Fig. 17.9(a)? Predict the relative intensity of the (222) and (202) reflections if the (111) reflection is the most intense.

(xx) *Fluorite I*: CaF_2 in the *fluorite* (C1) structure has a density of 3.18 g/cm^3 at room temperature. Calculate the lattice constant for CaF_2.

(xxi) *Fluorite II*: Show that if the two sets of tetrahedral sites in fluorite are occupied by B and C atoms, and the empty octahedral sites by B atoms, the resulting compound has a stoichiometry of AB_2C and the structure is $L2_1$.

(xxii) $D0_{19}$ *structure I*: Plot the atom positions in the (001) for Ni_3Sn in the $D0_{19}$ structure. Show that the Sn atoms tile the plane in a larger 3^6 net and the Ni atoms tile the plane in a 3636 Kagome net. Show that there are three Ni atoms and a single Sn atom in the hexagonal unit cell. Describe how with a BC stacking sequence, there are two formula units per unit cell.

(xxiii) *Kagome tile stacking*: Draw atom positions, with respect to the primitive *hcp* cell, for a Kagome tiling centered on the α, β, and γ, positions.

(xxiv) 6^3 *tile stacking*: Draw the atoms in the primitive unit cell for the *hcp* lattice for a 6^3 tiling centered on the a, b, and c, positions, respectively.

(xxv) $D0_{19}$ *structure-II*: Show that the stacking sequence for Ni_3Sn with the $D0_{19}$ structure is $[B\beta][C\gamma]$.

(xxvi) *Long period superlattices based on* $L1_2$: Make a drawing similar to Fig. 17.25 for the following average APB spacings: $M = 9/7$, $M = 7/3$, and $M = 11/3$. Draw only the B-type atom locations. [Hint: Note that the largest distance between APBs is given by Na, where N is the nearest integer larger than M.]

(xxvii) α-La *structure*: Determine the Ramsdell notation, the Zhdanov number, and the h–c symbol for the α-La structure.

(xxviii) α-La: Calculate the density of α-La using its lattice constant. How does this compare with the density of a transition metal.

(xxix) In *structure*: Elemental In has the A6 tetragonal structure. The full structural information can be found as Structure 19 in the on-line structures appendix. Show how this can be derived from a distortion of the *fcc* or, alternatively, *bcc* structure.

(xxx) *Structure factor: graphite*: Determine the structure factor for C graphite.

(xxxi) α-U: Calculate the density of α-U using its lattice constants.

(xxxii) *Interstital compound*: Determine the atomic radius for C in diamond. Calculate the strain required to place C in a tetrahedral interstice in *bcc* Fe.

(xxxiii) *Commensurate and incommensurate lattices*: Consider a 1-D lattice with lattice constant a, and atomic positions: $x_n = na$ with $n = 0, 1, 2, \ldots$ Construct a new crystal by displacing the atoms to the positions:

$$X_n = x_n + \frac{1}{2}\sin(2\pi nq).$$

(a) Plot the atomic positions for $q = 1/3$. Is this a commensurate structure? What is the new lattice constant? How many atoms are in a cell?

(b) Plot atomic positions for $q = 1/\sqrt{3}$. Is this a commensurate structure?

(xxxiv) *Incommensurate spin density waves*: Consider a 1-D crystal, $x_n = na$, with a monatomic basis, but with each atomic site having magnetic moment:

$$\mu(x_n) = \mu_B \sin\left(\frac{\sqrt{3}\pi x_n}{a}\right).$$

Given that the neutron scattering power is proportional to $\mu(x_n)$, describe the magnetic diffraction of neutrons from such a structure.

(xxxv) *Superlattice reflections*: Estimate the special positions of the Fe and Pt atoms and the c lattice constant in a $FePt_9$ superlattice structure. Use Bragg's law to predict the angle of the $(00l)$ superlattice reflections. What is the angle for the (110) reflection? Do any superlattice reflections exceed this angle?

CHAPTER

18 Metallic structures II: topologically close-packed phases

> *"Let no one destitute of geometry enter my doors."*
>
> Plato 427–347 BC.

18.1 Introduction: electronic states in metals

The *free electron theory* of metals assumes the presence of an isotropic, uniformly dense, electron gas in a metallic material. Of course, this is an idealization because the charge density of crystalline solids is restricted by the periodicity of the lattice. Despite this assumption, the free electron theory provides a few guiding principles to obtain an understanding of metallic structures. A large portion of the cohesive energy in metals is derived from the energy of the *electron gas*. This energy depends sensitively on the electron density and its spatial variation.

We use the concept of *density of states* to describe the distribution of electron energies in a solid. In Box 18.1, the density of states is defined for a free electron system. The electronic filling level, the *Fermi energy*, ϵ_F, depends only on the number of conduction electrons per atom and the atomic volume, Ω. We determine the number of electrons per atom by the metal species and its atomic volume in the crystal structure. ϵ_F is the most important electronic energy level: electrons with energies near ϵ_F are the ones that respond to an applied electric field and, hence, they provide *electrical conductivity* in metals.

If we modify the free electron theory to account for a non-zero periodic crystal potential, then we can calculate the electronic states in the context of

Box 18.1 The free electron theory and the density of states

The free electron theory in metals treats a "gas" of independent electrons with discrete quantum energy states. Electron pairs occupy states as dictated by the Pauli exclusion principle (pairs with opposite spin state). The lowest energy states are occupied, and sequentially higher energy states are filled until all electrons have been assigned to a state. "Free" refers to those conduction electrons that experience a zero potential in the metal. This approximation allows us to quantum mechanically calculate the density of states. The number of free electrons, N_e, in a mole of metal is:

$$N_e = \frac{\#\text{ electrons}}{\text{atom}} \times \frac{N_A\text{atoms}}{\text{mole}} \quad \text{and} \quad \frac{N_e}{V} = \frac{\#\text{ electrons}}{\text{atom}} \times \frac{1}{\Omega}.$$

If we define by $n(\epsilon)$ the total number of states with energy below ϵ, then a quantum mechanical calculation yields the following form for a free electron system:

$$n(\epsilon) = \frac{V}{3\pi^2} \times \left(\frac{2m\epsilon}{\hbar^2}\right)^{\frac{3}{2}},$$

where m is the electron mass, and $\hbar$ is Planck's constant divided by 2π. We can calculate the *Fermi energy*, ϵ_F, by equating the number of states with energy below the Fermi level to the total number of electrons:

$$N_e = \int_0^{\epsilon_F} n(\epsilon)\mathrm{d}\epsilon \quad \text{so that} \quad \epsilon_F = \frac{\hbar^2}{2m} \times \left(3\pi^2 \frac{N_e}{V}\right)^{\frac{2}{3}}.$$

band theory. For a crystalline solid, the potential will have the same periodicity as the lattice. In band theory, the density of states function is different from that of the free electron theory and it depends strongly on the symmetry of the crystalline lattice. We can visualize conduction electrons as scattering off the periodically arranged nuclei in the crystalline solid; these scattering events determine the intrinsic properties of the metal, such as electron conduction.

The free electron theory implies that metal atoms will fill space so as to maximize the electron density. Metallic crystals usually have a high symmetry, maintain high coordination numbers, and have relatively uniform electron densities. Metals prefer structures with tetrahedral interstices, which provide the most efficient packing of atoms. The local environments in 12-fold coordinated structures allow for three possibilities: that observed in *fcc* structures (the *cuboctahedron*, Fig. 18.1(a)); that observed in *hcp* structures (the *twinned*

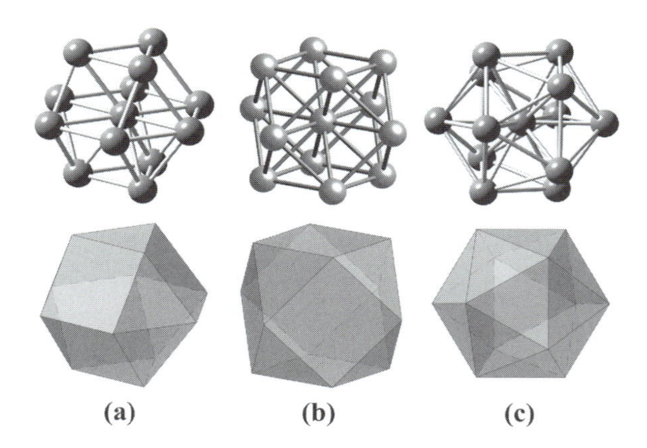

Fig. 18.1. 12-atom coordination clusters (top) and polyhedra (bottom) illustrating first nearest neighbor environments for (a) *hcp*, (b) *fcc*, and (c) icosahedral configurations.

cuboctahedron, Fig. 18.1(b)); and that observed in topologically close-packed structures (the *icosahedral coordination*, Fig. 18.1(c)).[1] The corresponding coordination polyhedra are shown in the bottom row. The *fcc* and *hcp* coordination polyhedra commonly occur in elemental and alloy metal structures. Icosahedra appear in a large number of intermetallic phases and as local structural units in amorphous alloys.

Sir (Frederick) Charles Frank (Frank, 1952) demonstrated that 12-fold icosahedral coordination about a central atom, interacting through a *Lennard-Jones potential*, yielded a *lower* total energy than the *fcc* and *hcp* arrangements of the same 13 atoms. This observation is rationalized geometrically because the distance from the central atom to the vertex atoms in the icosahedral cluster is about 5% smaller than that of neighboring vertices, and all surface atoms have five (versus four) neighbors. Frank asserted that icosahedral coordination might prevail in densely packed liquids, and that the stability of the icosahedral arrangements may account for the large supercooling possible in liquids. We will further detail Frank's pair potential calculation in Chapter 21 where we discuss amorphous metals.

Frank and Kasper (Frank and Kasper, 1958, 1959) used the notable structural stability of icosahedral units to understand the structures of many complex alloy structures. Of the three possible 12-fold coordination motifs, the icosahedral arrangement is found in these complex alloy phases. The icosahedral arrangement occurs because of its lower energy and its exclusively tetrahedral interstices. These interstices maximize the interstitial electron density.

[1] 12 is the largest number of equal spheres that can contact one central sphere.

18.2 Topological close packing

A common feature in many technologically important intermetallics (including Laves phases, A15 alloys, σ-phase, μ-phase, etc.) is *topological close packing (TCP)*.

Structures with TCP have exclusively tetrahedral interstices.

In TCP phases, a limited number of coordination polyhedra are possible. Topological close packing often results in an alloy electronic structure with sharp peaks in the density-of-states near the Fermi level. These peaks explain physical phenomena such as *itinerant ferromagnetism* (free electron ferromagnetism) in the Laves phase, $ZrZn_2$, and *superconductivity* in the Frank–Kasper phase, Nb_3Sn.

Neither the *hcp* nor the *fcc* structure is topologically close packed because for every two *tetrahedral interstices* in the structure there exists one *octahedral interstice*. Figure 18.2(a) and (b) show these interstices in the *fcc* and *hcp* structures, respectively. The region around an octahedral interstice is not as densely packed as the region around a tetrahedral interstice. In structures with two or more atom types of different size, close-packed structures can be constructed with exclusively tetrahedral interstices. These have a large, uniform packing density. In the *fcc* and *hcp* structures, tetrahedral interstices share triangular faces with octahedral interstices; in TCP materials, tetrahedral interstices share triangular faces *only* with other tetrahedral interstices.

The *tetrahederstern*, shown in Fig. 18.2(c), is an important structural element in Frank–Kasper phases. The tetrahederstern is constructed by starting with a tetrahedron of smaller A atoms. The four faces of the A-atom tetrahedron are then decorated with larger B atoms. The B-atom tetrahedron, i.e., the tetrahedron formed by connecting the four B atoms, is a dual of the A-atom tetrahedron. The tetrahederstern of Fig. 18.2(c) is projected along a 2-fold axis of both A and B tetrahedra. By observing the connectivity of tetrahedersterns we may understand the crystallographic structures of the Frank–Kasper phases.

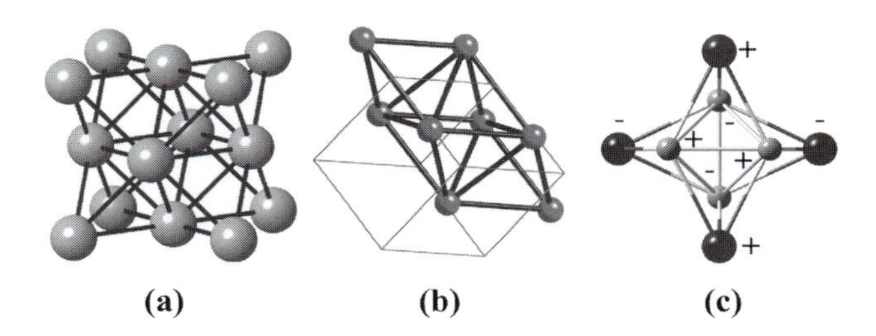

Fig. 18.2. Attachment of tetrahedral and octahedral interstices (a) in the *fcc* structure; in the *hcp* structure (b), and (c) a tetrahederstern (+ and − denote atoms above and below the plane).

Topological close packing can also be motivated by the 3-D atomic coordination polyhedra. In *fcc* and *hcp* structures, atoms of equal size are 12-fold coordinated. The *fcc* coordination polyhedron is the *cuboctahedron*, and in the *hcp* structure it is the *twinned cuboctahedron*. Both polyhedra have square and triangular faces, and octahedral and tetrahedral interstices. The third 12-fold coordination polyhedron, the *icosahedron*, has all triangular faces and only tetrahedral interstices. The icosahedron is a *triangulated coordination polyhedron*, one of the Kasper polyhedra (Frank and Kasper, 1958) shown in Fig. 18.3 and described in the next section.

18.2.1 The Kasper polyhedra

Frank and Kasper (1958) derived ways in which distorted icosahedra can be accommodated in crystals by packing with other polyhedra that have larger coordination numbers and atoms. These coordination polyhedra were constructed to maintain *topological close packing (TCP)*. This requires that the coordination polyhedra have all triangular faces; hence, they are called *triangulated coordination polyhedra*. In total there are four possible triangulated coordination polyhedra, the *Kasper polyhedra*, with coordination numbers of 12, 14, 15, and 16, as illustrated in Fig. 18.3.

The *domain* of an atom (Frank and Kasper, 1958) is defined as the region of space containing all points nearer to that atom than to any other. The "domain" is hence equivalent to a *Wigner–Seitz cell* in a monatomic crystal and a *Voronoi polyhedron* in an amorphous structure. We can construct the domain by considering all planes that bisect the lines joining the central atom with its neighbors, and selecting the innermost polyhedron bounded by these

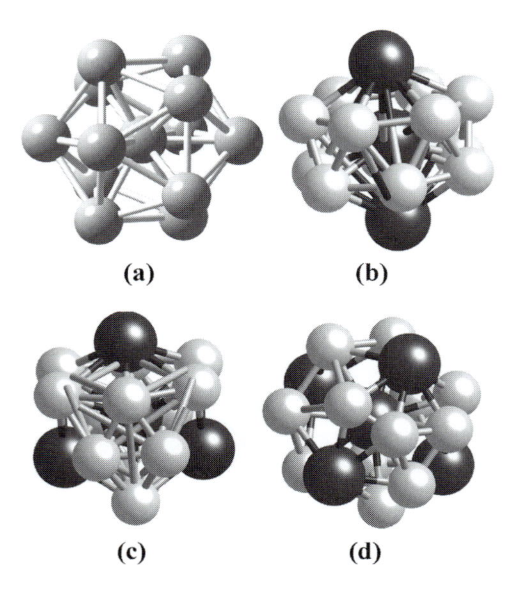

Fig. 18.3. The Kasper polyhedra (a) CN12 (icosahedron); (b) CN14; (c) CN15; and (d) CN16.

planes. The number of neighbors or, equivalently, the number of faces of this polyhedron, is called the *coordination number*; the set of neighbors is known as the *coordination shell*; and the polyhedron whose edges connect nearest neighbors in the coordination shell is the *coordination polyhedron*.[2]

The *surface coordination number*, S_q, is the number of surface neighbors of an atom. Atoms on the surface of the cuboctahedron (*fcc*) or twinned cuboctahedron (*hcp*) have exclusively S_4 coordination (Fig. 18.1(a) and (b)). The icosahedron has S_5. Among the $Z = 12$ structures, only the icosahedron is a triangulated coordination polyhedron. To construct triangulated coordination polyhedra, Frank and Kasper identified polyhedra with only S_5 and S_6 surface coordination numbers. They derived an equation, shown in Box 18.2, that

Box 18.2 Regular polyhedra with triangular faces satisfying Euler's theorem

Euler's formula relates the number of vertices, V, edges, E, and faces, F, of a polyhedron:

$$V - E + F = 2.$$

Define v_q as the number of vertices connected by q edges to neighboring vertices, then:

$$V = \Sigma_q v_q \quad \text{and} \quad E = \frac{1}{2}\Sigma_q q v_q.$$

Additionally, if we consider only polyhedra with triangular faces:

$$F = \frac{1}{3}\Sigma_q q v_q.$$

Finally, substituting these relations into Euler's equation results in:

$$\Sigma_q (6 - q) v_q = 12.$$

This can be satisfied for $q = 3$ ($V = 4$) for the tetrahedron with four triangular faces, for the octahedron with $q = 4$ ($V = 6$), and for the icosahedron with $q = 5$ ($V = 12$).

[2] The *hcp* structure has $Z = 12$ if c/a is precisely $\sqrt{8/3}$; otherwise, $Z = 6$. For small deviations from the ideal c/a, the coordination number remains $Z = 12$. The coordination number $Z = 14$ can be chosen for *bcc* metals, relaxing the definition to include both first and second nearest neighbors.

describes regular polyhedra with exclusively triangular faces and satisfying *Euler's formula*.

In considering triangulated coordination polyhedra, consistent with TCP, we find that only three, four, five, or six equilateral triangles intersect at a point on a non-reentrant polyhedron. If there are three, we have a single closed tetrahedron that we connect to form triangulated coordination polyhedra. If there are four, we have the octahedron with octahedral interstices incompatible with TCP. If there are five triangles, S_5, meeting in a point we have the icosahedron. If there are six, the equilateral triangles must be coplanar.

Next, we will consider polyhedra that are not regular, i.e., polyhedra that have combinations of vertices with different surface coordinations. For polyhedra with S_4 and S_5-type vertices, we have: $2v_4 + v_5 = 12$. In general, we can exclude polyhedra with S_4 vertices because they do not give rise to TCP (they give rise to octahedral interstices). We are, therefore, left with polyhedra with S_5 and S_6 vertices. Recognizing that V is equivalent to the coordination number, Z, and considering only S_5 and S_6 vertices, we have $v_5 = 12$ and $v_6 = Z - 12$.

The *Kasper polyhedra* are triangulated coordination polyhedra with only S_6 and S_5 vertices. Only the icosahedron satisfies $v_5 = 12$, $v_6 = 0$. For $Z = 13$ there is no possible triangulated coordination polyhedron. There is only one $Z = 14$ coordination polyhedron, with $v_5 = 12$ and $v_6 = 3$, and only one $Z = 16$ polyhedron, with $v_5 = 12$ and $v_6 = 4$. For $Z > 16$ and only S_6 and S_5-type surface sites, there must be at least one pair of adjacent (planar) six-fold vertices (Frank and Kasper, 1958). The CN14 polyhedron consists of a hexagonal antiprism with two additional atoms at the poles of the six-fold axis. The CN15 polyhedron has three larger atoms that are coplanar with the central atom. The CN16 Kasper polyhedron has four larger atoms in a tetrahedral arrangement.

The geometry of the Kasper polyhedra, including the point group of the ideal polyhedron, the number of vertices with surface coordination 5 and 6, respectively, and the number of faces and edges (Sinha, 1972) are summarized in Table 18.1. Note the non-crystallographic point groups for the CN12 and CN14 polyhedra. In TCP alloy systems, the *Frank–Kasper phases* contain icosahedral coordination polyhedra with one or more of the other Kasper polyhedra. The commonality of the icosahedral clusters in Frank–Kasper phases emphasizes their significance as efficiently packed local units.

18.2.2 Connectivity of Kasper polyhedra

The TCP polyhedra, with coordination ≥ 12, are the icosahedron (CN12) and the CN14, CN15, and CN16 Kasper polyhedra. For the icosahedron (CN12), the central atom has no atom neighbor with which it shares six common neighbors. The latter three polyhedra have two, three, or four non-adjacent surface atoms,

Table 18.1. Geometrical features of the Kasper polyhedra.

Item	CN12	CN14	CN15	CN16
Ideal point group symmetry	$\mathbf{m\overline{35}}$ (I_h)	$\mathbf{\overline{12}2m}$ (D_{6d})	$\mathbf{\bar{6}m2}$ (D_{3h})	$\mathbf{\bar{4}3m}$ (T_d)
# S_q = five vertices	12	12	12	12
# S_q = six vertices	0	2	3	4
# Edges	30	36	39	42
# Faces	20	24	26	28

with six-fold surface coordination. The six-fold coordinated surface atoms are of a larger size. The smaller atoms are connected by *minor ligand lines*.

Lines connecting the large, six-fold coordinated, peripheral atoms are called *major ligand lines*. The networks of major ligand lines are known as the *major skeleton*.

The main atomic layers are *tessellated* (see Chapter 16). They contain arrays (*primary layers*) of hexagons, pentagons, and triangles where triangular meshes correspond to nearest neighbor atoms. These 2-D layers normally consist of double layers made up of a primary layer and a simple secondary layer in which the coordination number does not correspond to the nearest neighbor coordination number. Connectivity of the Kasper polyhedra results in many pentagonal and hexagonal anti-prisms. Because the large atoms destroy the high symmetry of the icosahedra, major ligand lines can represent directions of strongly anisotropic properties. Likewise, we are not surprised to see strong magnetocrystalline anisotropy and directionality of superconducting properties along these lines.

18.2.3 Metallic radii

Table 18.2 summarizes metallic separation radii calculated from elemental crystal structures (Barrett and Massalski, 1980) for some of the metallic species. These are the components in the prototypical alloy structures discussed here and in Chapter 19. They are used as a guide to compare the relative atomic sizes in the solid state.

In Chapter 17, we determined the atomic sizes in metals from considerations of close-packed directions and lattice constants. In elemental metallic solids, the metallic radius is half the bond distance along a close-packed direction, using touching sphere arguments (Bragg, 1920, Goldschmidt, 1928). While packing arrangements, coordination, and anisotropy in the bonding can cause changes in the metallic radii from the ones reported in Table 18.2, the differences are often small, and we will not analyze them any further.

Table 18.2. Metallic radii (from Barrett and Massalski (1980)).

Atom	R (nm)	Atom	R (nm)	Atom	R (nm)	Atom	R (nm)
Al	0.143	Co	0.125	Cr	0.125	Cu	0.128
Dy	0.175	Fe	0.124	Mg	0.160	Mn	0.112
Mo	0.136	Nb	0.143	Nd	0.183	Ni	0.125
Pr	0.182	Sm	0.179	Sn	0.140	Tb	0.176
U	0.138	W	0.137	Zn	0.133	Zr	0.159

For ionic solids (Chapter 22), however, we do need to calculate different *ionic radii* for ions in different valence states and/or in different polyhedral environments.

18.3 *Frank–Kasper alloy phases

In the following subsections, we illustrate structural prototypes for examples of Frank–Kasper phases and we discuss their symmetries, coordination polyhedra, and structural connectivity. Table 18.3 gives examples of Frank–Kasper alloy phases (adapted from Shoemaker and Shoemaker (1969)). In some more complicated structures, such as α-Mn, Kasper polyhedra are observed along with coordination polyhedra (such as CN13) that are inconsistent with topological close packing. We can decompose Frank–Kasper phases into a stacking of the 2-D tilings described in Chapter 16. In the Frank–Kasper phases, we will encounter: the regular tilings, 3^6, 4^4, and 6^3; the 2-D Archimedean tilings, $(3^3 \cdot 4^2)$, $(3^2 \cdot 4 \cdot 3 \cdot 4)$, and $(3 \cdot 6 \cdot 3 \cdot 6)$ (Kagome tiling); and the 2-uniform tilings $(3^3 \cdot 4^2; 3^2 \cdot 4 \cdot 3 \cdot 4)_2$, and $(3^2 \cdot 6^2; 3 \cdot 6 \cdot 3 \cdot 6)$.

18.3.1 A15 phases and related structures

18.3.1.1 The A15 structure

The A15 phases are intermetallic alloys that have an A_3B stoichiometry with B atoms in 12-fold icosahedral coordination and the A atoms in a 14-fold Kasper coordination polyhedron. As shown in Fig. 18.4, the major skeleton is a *bcc* lattice. The Strukturbericht symbol A15 is ambiguous because the common A15 compounds have an A_3B stoichiometry while the A designation is typically used for elements.[3] In this TCP phase, tetrahedra are distorted (both to fill space and to accommodate atoms of two different sizes). The A15

[3] The original prototype was thought to be *β-tungsten*. This identification was mistaken and it was later learned that the A15 structure corresponded to an oxide of W with the W_3O stoichiometry (Sinha, 1972).

Table 18.3. Frank–Kasper alloys (Shoemaker and Shoemaker, 1969), structure type, alloy (compound) examples, Pearson symbol, space group number, and approximate frequency of the Kasper polyhedra.

Structure symbol	Example(s)	Pearson symbol	Space group #	% CN12/14/15/16
A15	Nb_3Sn	cP8	223	25/75/0/0
Zr_4Al_3	$Zr_{57}Mo_{43}$	hP7	174	43/28/28/0
Fe_2B	$Fe_{67}B_{33}$	tI12	140	
C15	$MgCu_2$	cF24	227	67/0/0/33
C14	$MgZn_2$	hP12	194	67/0/0/33
C36	$MgNi_2$	hP24	194	67/0/0/33
W_2CoB_2	W_2CoB_2	oI10	71	
σ	CoCr, $Fe_{54}Cr_{46}$	tP30	136	33/54/13/0
μ	Fe_7W_6, $Co_{54}Mo_{46}$	R13	166	55/15/15/15

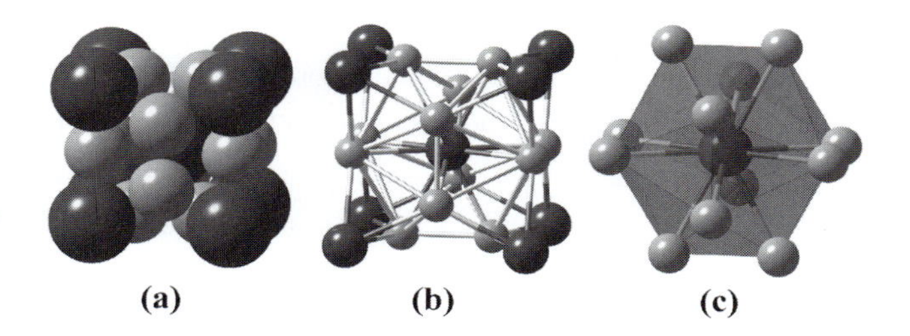

Fig. 18.4. (a) Unit cell for Nb_3Sn in close-packed rendering, (b) as a ball-and-stick model, and (c) CN12 (icosahedral) coordination polyhedron of Nb atoms about Sn.

structure has only CN12 (1) and CN14 (3) Kasper polyhedra, leading to an *average coordination number (ACN)* of 13.5.[4]

Figure 18.4 shows a single unit cell for the **A15** structure of Nb_3Sn in both close packing (a) and ball-and-stick models (b). Figure 18.4 (c) shows the icosahedral CN12 polyhedron of the smaller Nb atoms about the larger Sn atoms. The icosahedra are distorted in that the faces are not equilateral triangles. In the **A15** compounds, *A* atoms occupy two positions in each cube face, linking *A* atoms in infinite chains along the [100], [010], and [001] directions. The central *B* atom belongs exclusively to its cell. The vertex *B* atoms are shared among eight cells. The *A* atoms are shared between two cells.

A15 compounds belong to space group $\mathbf{Pm\bar{3}n}$ (O_h^3). The **A15** prototype structure is Cr_3Si (Boren, 1933), which has a lattice constant, $a = 0.4555$ nm, *A* (Si) atoms in the $2a$ (0, 0, 0) special position and *B* (Cr) atoms in $6c$ positions at (1/4, 0, 1/2). Full structure information is available as **Structure 29** in the on-line structures appendix. We calculate the structure factor for Cr_3Si

[4] The ACN is determined by summing the fraction of CN12, CN14, CN15, and CN16 sites, each multiplied by the respective coordination numbers (12, 14, 15, and 16).

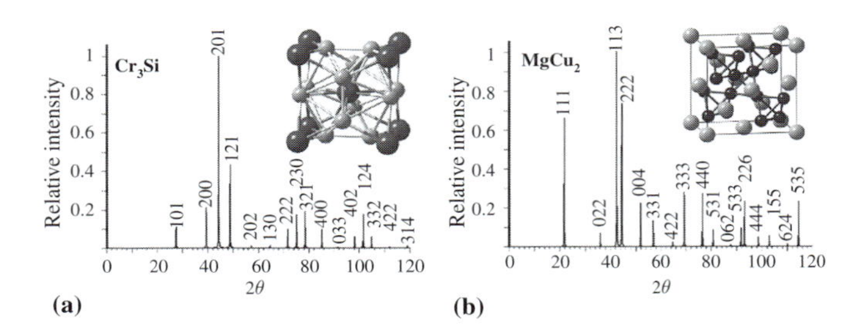

Fig. 18.5. Simulated XRD patterns for (a) the Cr_3Si A15 and (b) Cu_2Mg Laves phases.

in Box 18.3. Figure 18.5(a) shows a simulated XRD pattern for the Cr_3Si A15 phase with evident extinctions for odd h, k, and l; Fig. 18.5(b) is discussed in a later section.

Box 18.3 Calculation of the structure factor and extinctions for the Cr_3Si structure.

Consider Cr_3Si with Si atoms positions $(0, 0, 0)$ and $(1/2, 1/2, 1/2)$ and Cr atoms with fractional coordinates $(1/4, 0, 1/2)$. The contribution of the Si atoms to the structure factor F_{hkl}^{Si} is just like that of a *bcc* lattice:

$$F_{hkl}^{\text{Si}} = f_{\text{Si}}\left(1 + e^{\pi i(h+k+l)}\right) = f_{\text{Si}}\left(1 + (-1)^{(h+k+l)}\right).$$

The contribution of Cr to the structure factor F_{hkl}^{Cr} consist of six terms:

$$F_{hkl}^{\text{Cr}} = f_{\text{Cr}}\left[e^{\pi i(\frac{h}{2}+l)} + e^{\pi i(\frac{3h}{2}+l)} + e^{\pi i(\frac{k}{2}+h)} + e^{\pi i(\frac{3k}{2}+h)} + e^{\pi i(\frac{l}{2}+k)} + e^{\pi i(\frac{3l}{2}+k)}\right].$$

This can be rewritten using Euler's formula as:

$$F_{hkl}^{\text{Cr}} = 2f_{\text{Cr}}\left[(-1)^h \cos\frac{\pi k}{2} + (-1)^k \cos\frac{\pi l}{2} + (-1)^l \cos\frac{\pi h}{2}\right].$$

Conditions that simultaneously make F_{hkl}^{Si} and F_{hkl}^{Cr} vanish occur when h, k, and l are all odd, because all three cos functions vanish in the expression for F_{hkl}^{Cr}. As an exercise the reader can verify that (hhl) reflections are also extinct if l is odd. In addition, the only non-zero values possible for the structure factor F_{hkl}^{Cr} are $\pm 2f_{\text{Cr}}$, $\pm 4f_{\text{Cr}}$ and $\pm 6f_{\text{Cr}}$. The structure factor is real, reflecting the fact that this is a centro-symmetric structure. There are five different types of reflections in this structure; the possible intensities are: $4(3f_{\text{Cr}} + f_{\text{Si}})^2$, $4(3f_{\text{Cr}} - f_{\text{Si}})^2$, $16f_{\text{Cr}}^2$, $4(f_{\text{Cr}} + f_{\text{Si}})^2$, and $4(f_{\text{Cr}} - f_{\text{Si}})^2$. The reader is also encouraged to compute the full diffracted intensities, including multiplicities and the Lorentz polarization factor, and compare the resulting intensities with the pattern displayed in Fig. 18.5(a).

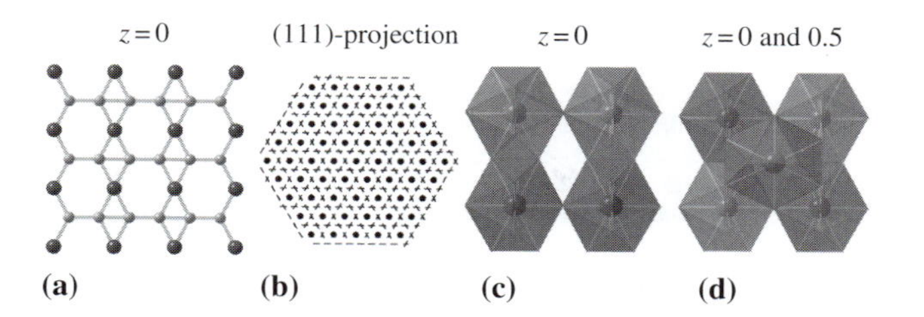

Fig. 18.6. (001) planes in the A15 structure (a) 2-uniform $3^2 6^2$ tiling at $z = 0$, and structure projected into a (111) plane (b). Icosahedra surrounding B atoms (c), and their stacking (d).

If we "disassemble" the A15 structure into lattice planes, we can understand this structure in terms of a stacking of the tilings discussed in Chapter 16. The atomic positions of the A15 structure in a (001) plane at $z = 0$ are shown in Fig. 18.6(a). Here we see that 2-uniform ($3^2 6^2$) tilings are stacked along the [001] with 4^4 secondary layer at $z = 0.25$ (not shown) to build the structure. Each subsequent $3^2 6^2$ net is rotated by $\pi/2$ in each (001) plane a half lattice constant up along the z-axis. In the 3-D structure, edge sharing tetrahedra (compare with the *tetrahederstern* in Fig. 18.2) result in edge sharing triangles in the projection.

The A15 structure can also be decomposed by considering projections into a (111) plane as illustrated in Fig. 18.6(b). Note that such a projection is of the entire structure and not a single (111) plane. This shows the A atoms (open diamonds) form $(3 \cdot 6 \cdot 3 \cdot 6)$ Kagome tilings with the larger B atoms (filled circles) filling the hexagonal sites in this tiling. We can also describe the networks in Fig. 18.6(a) and (b) in terms of the connectivity of the icosahedra surrounding B atoms in the A_3B structure. Figure 18.6(c) shows edge-sharing icosahedra, connected in chains along [010] directions. Figure 18.6(d) shows the icosahedron surrounding the B atom at (1/2, 1/2, 1/2) projected onto the structure depicted in Fig. 18.6(c). This icosahedron is rotated by $\pi/2$.

The A15 compounds include the important intermetallic superconductors Nb_3Sn, Nb_3Zr, and Nb_3Ti. Thin-film $GeNb_3$ (developed in the Westinghouse research lab in Pittsburgh, PA) has a transition temperature of 23 K, which was a record prior to the discovery of high temperature oxide superconductors (HTSCs) in the late 1980s (Gavaler, 1973). Nb_3Sn is a commonly used A15 superconductor. A majority of superconducting magnets are constructed out of a Nb_3Ti alloy (Pandey, 2000). While A15 compounds remain important materials in wires for superconducting solenoids, the vigorous efforts to develop the HTSCs have resulted in significant competition. We will discuss the HTSC materials in detail in Chapter 23.

18.3.1.2 Pearson's notation for stacking of 3^6, 6^3, and Kagome tiles

We can subdivide the close-packed 3^6 tiling into a 6^3 and a larger 3^6 tiling, with sites in the ratio of 2:1. We can also subdivide this tiling into a $3 \cdot 6 \cdot 3 \cdot 6$ (Kagome) and a larger 3^6 tiling, with sites in the ratio of 3:1 (Pearson, 1972). The ordered occupation of these subdivided tiles allows us to construct ordered phases with specific stoichiometries. It also highlights the compatibility of

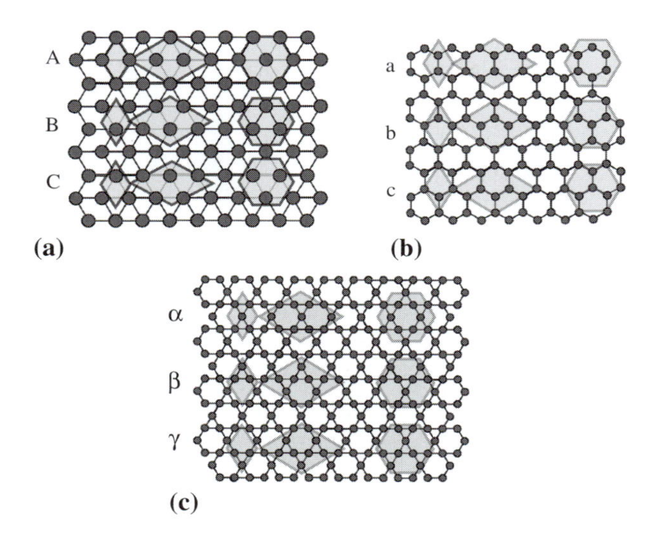

Fig. 18.7. Stacking positions in (a) 3^6, (b) 6^3, and (c) $3 \cdot 6 \cdot 3 \cdot 6$ (Kagome) tiles.

these three tiles in stacking sequences associated with close-packed planes. Figure 18.7 illustrates geometrical relationships between three tiles and rhombuses associated with a single or three primitive hexagonal unit cells.

A method for describing the stacking sequence in TCPs and other related phases has been described by Pearson (1972) as an extension of the *ABC* notation of Chapter 16. In the Laves phases, the stacking of 3^6 tiles is denoted by *ABC* as before; the stacking of 6^3 tiles is denoted by the lower-case symbols *abc*; and the stacking of $(3 \cdot 6 \cdot 3 \cdot 6)$ Kagome tiles is denoted by the Greek symbols $\alpha\beta\gamma$. Figure 18.7 shows the relationships between the positions in these networks and the stacking nomenclature. Figure 18.7 (a) shows a 3^6 tile, (b) a 6^3 tile, and (c) a $(3 \cdot 6 \cdot 3 \cdot 6)$ Kagome tile. Each frame shows a rhombus representing the basal plane of a primitive hexagonal unit cell; a rhombus rotated by $\frac{\pi}{2}$, representing the basal plane containing three hexagonal unit cells (rhombi) and the basal plane containing three hexagonal unit cells in a hexagonal prismatic representation. Each of these is in turn reproduced at the *A*, *B*, and *C* positions of the close-packed structures.

We now describe a systematic method for labeling a structure by the stacking of these three tiles. Although this is a complicated nomenclature, we find it to be useful for the description of most TCPs and permanent magnets (described in Chapter 19). The stacking sequence is reported for a structure by providing a label (*A*, *B*, *C*, *a*, *b*, *c*, α, β, and γ) for each layer in succession. We will use the square brackets [] to denote (different components or different tiles with the same component) atoms in the same plane (Pearson, 1972). We also use the round brackets () to denote groups of planes (puckered) that, while not coplanar, are closer together than the average interplanar distances in the structure. Finally, a prime $'$ will denote layers in a structure whose stacking sequence is defined with respect to the larger basal plane containing three hexagonal unit cells (rhombi). We will see several examples of this notation in the following subsections.

18.3.1.3 Shear-related structures: Al_3Zr_4, Al_2Zr_3 and BFe_2

We can obtain structures related to the **A15** structure by stacking 3^6, 6^3, and Kagome tiles in a different order or by applying conservative and *non-conservative shears* (Fig. 18.8). Shears are translations parallel to the close-packed planes. We distinguish between conservative and non-conservative shears; conservative shears do not involve local changes of the chemistry, whereas non-conservative shears have a shear component normal to the plane, so that the composition of the alloy changes as a result of the shear operation. The normal component either adds or removes planes from the stacking sequence. Al_3Zr_4, Al_2Zr_3, and BFe_2 are structures related to **A15** by shear.

The Al–Zr alloy system has a complex binary transition-metal aluminide phase diagram with ten intermediate phases; of these, we will illustrate only the Al_3Zr_4 and Al_2Zr_3 Frank–Kasper phases. We can relate the structure of Al_3Zr_4 to the **A15** structure by a non-conservative shear or change in stacking sequence. This structure has space group $\mathbf{P\bar{6}}$ (C^1_{3h}) (**Structure 30**, (Wilson *et al.*, 1960)). The related Al_2Zr_3 structure has the space group $\mathbf{P4_2/mnm}$ (D^{14}_{4h}) and lattice constants $a = 0.763$ nm and $c = 0.700$ nm (**Structure 31**, (Wilson and Spooner, 1960)). These structures are illustrated in Fig. 18.8 along with the shear-related structure of BFe_2. We can easily understand these structures by considering the stacking tiles and/or projections of atoms into single planes shown in Fig. 18.9.

The structure of Al_3Zr_4 can be understood in terms of a stacking of tilings, similar to those of the **A15** structure. The (001) planes at $z = 0$ and $z = 1/2$ are shown in Fig. 18.9(a) and (b). These show that a $(3\cdot 6\cdot 3\cdot 6)$ Kagome tiling describes the Al atom network in the $z = 0$ plane, while the $z = 1/2$ plane consists of a 6^3 tiling of Zr atoms. The network of Zr atoms in the $z = 1/4$ and $z = 2/4$ planes (not shown) is a larger simple triangular 3^6 tiling, which one can infer by viewing the structure projected into a (001) plane. Figure 18.9(c) shows the [00.1] projection of the three planes. Here, the solid

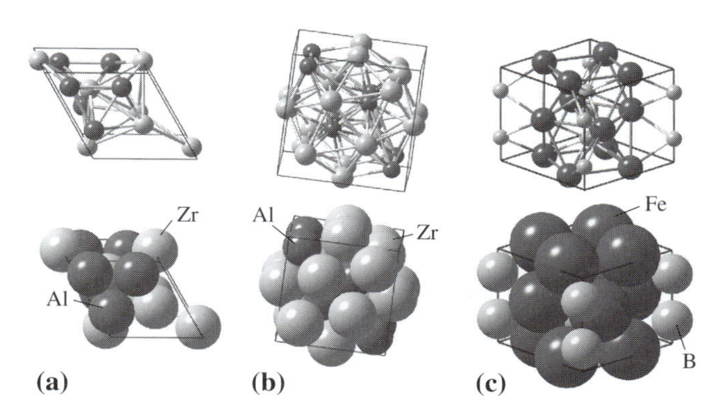

Fig. 18.8. Ball-and-stick (top) and space-filling (bottom) models for (a) the rhombohedral unit cell of Al_3Zr_4, (b) the tetragonal unit cell of Al_2Zr_3, and (c) the tetragonal unit cell of BFe_2.

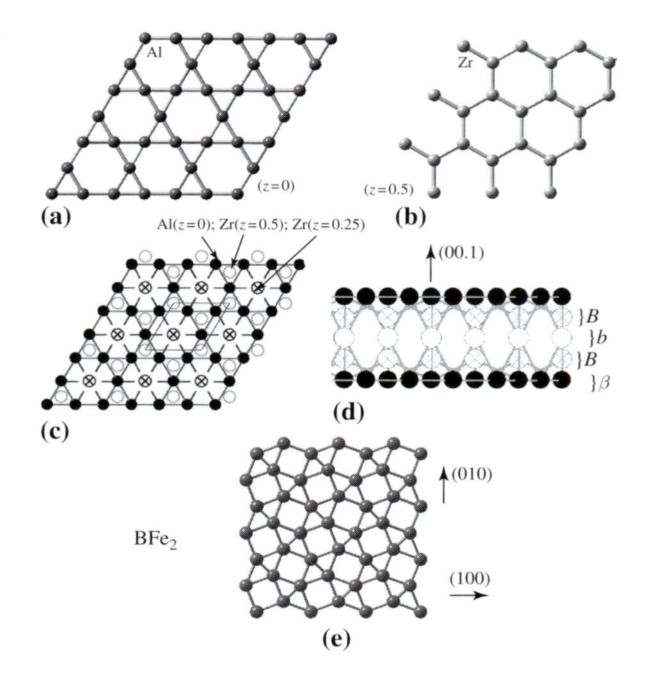

Fig. 18.9. Al_3Zr_4 (001) planes: (a) Al atom net at $z = 0$; (b) Zr atom net at $z = 1/2$; (c) projection of the structure into a (00.1) plane; (d) stacking sequence along the [00.1] direction; and (e) Fe atoms in a (001)-type plane at $z = 0$ in BFe_2.

circles represent Al atoms at $z = 0$, the crossed circles Zr atoms at the $z = 1/4$ elevation, and the open circles Zr atoms at the $z = 1/2$ elevation. We can describe the stacking sequence in Al_2Zr_3 in the Pearson notation as βBbB (Fig. 18.9 (d)).

The BFe_2 structure can be obtained by a non-conservative shear operating on the Al_3Zr_4 structure. Figure 18.8 (c) shows the BFe_2 structure in ball-and-stick (top) and space-filling (bottom) representations. Figure 18.9(e) shows the network of the Fe atoms in the $z = 0$ plane. This network is a 2-uniform $(3^2 \cdot 4 \cdot 3 \cdot 4)$ Schläfli tiling. This $z = 0$ plane tiling is repeated at $z = 1/2$, but rotated by $\pi/2$. The B planes are not illustrated, but they form a simple 4^4 square lattice. Although this is a TCP, there is a preponderance of squares in the tilings. The structure contains the relatively uncommon square anti-prismatic coordination polyhedra.

Anti-prismatic coordination polyhedra are common motifs in amorphous metals and their crystallization products. These are illustrated further in Chapter 21. BFe_2 is a notable crystallization product in Fe–B-based metallic glasses. We often consider the structure of crystallization products to infer the local coordination in amorphous metals. We consider relationships between the Al_3Zr_4 and BFe_2 structures in Fig. 18.10.

If we view the Al_3Zr_4 structure as projected into the (110) plane, we see chains of mixed Zr and Al tetrahedersterns in Fig. 18.10 (a). Crossed and open circles represent the same crystallographically distinct Zr atoms as in Fig. 18.9(c). If we now shear this structure, replacing the planes with Zr atoms

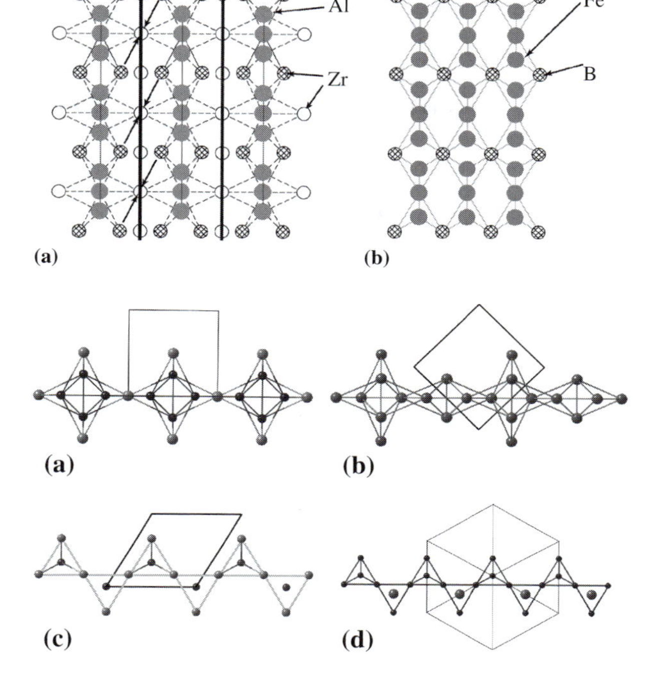

Fig. 18.10. Projection of Al_3Zr_4 structure into a (110) plane (a) (solid lines are non-conservative shear planes) and the resulting BFe_2 structure projected into a (110) plane.

Fig. 18.11. Shear-related structural units: (a) A15 structure [100] projection; (b) projection of Al_3Zr_4 along *c*-axis; (c) BFe_2 structure; (d) Cu_2Mg Laves phase.

(open circles) with half planes of the Zr atoms (crossed circles) we change the stoichiometry from 4:3 to 1:2. By this shear mechanism vertex-sharing tetrahedersterns are formed. The resulting BFe_2 structure is illustrated as a projection into the (110) plane, in Fig. 18.10(b). The reader may wish to draw the stacking sequence of the BFe_2 structure as an exercise.

It is instructive to summarize shear transformations based on the projected A15 and related structures. Previously, we stated that certain shears give rise to the structures of the Al_3Zr_4 and BFe_2 phases, shown in a simpler depiction in Fig. 18.11. The last shear gives rise to a structural unit of the cubic Laves phase structure, which we will discuss below. If we consider structures formed as intergrowths of these sheared structures, then we can describe the σ and μ phases as discussed in section 18.3.3.

18.3.2 The Laves phases and related structures

Laves phases (Laves *et al.*, 1934, Laves and Witte, 1935)), or *Friauf–Laves phases* (Friauf, 1927a,b) are intermetallic compounds observed in many binary metal systems. Friauf discovered the prototype $MgZn_2$ materials and identified the *Friauf polyhedron*. Laves phases are predicted to have an AB_2 stoichiometry, based on considerations of relative metallic radii. Because the range of solid solubility about AB_2 is usually small, Laves phases are nearly *line compounds*. They form when:

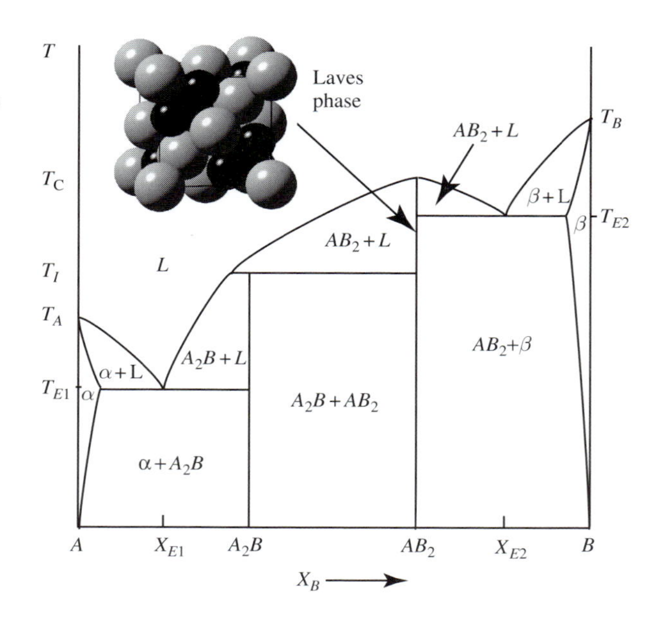

Fig. 18.12. *AB* phase diagram for a system with an A_2B intermetallic and an AB_2 Laves phase.

(i) A has a strong tendency for metallic bonding;
(ii) the next to outermost electron shell of B is incomplete (i.e., B is a transition metal);
(iii) B is smaller than A with size differences larger than 20%.

A typical AB phase diagram with a Laves phase compound is shown in Fig. 18.12. The AB_2 Laves phase has a limited range of solubility and is typically accompanied by another A_2B intermetallic phase (usually not a Laves phase). The A species has a lower melting temperature than the B (transition metal) species.

Figure 18.12 is an example of a double eutectic phase diagram. The first eutectic, at (x_{E1}, T_{E1}), is between A and the intermetallic compound A_2B. The second, at (x_{E2}, T_{E2}), is between B and the AB_2 Laves phase. While the AB_2 Laves phase has a *congruent melting point*, T_C, the A_2B intermetallic melts incongruently. Incongruent melting requires a composition change. We should also note the two-phase field, where the A_2B phase and the AB_2 Laves phase coexist.

We begin the description of Laves phase crystal structures by drawing a close packed array of the smaller B atoms and replacing two of every four B atoms with a large A atom. Now we have an AB_2 stoichiometry. The Laves phases have thus been called *hemisubstitutional* alloys (Rudman, 1965b). The *Hume-Rothery rules* in Chapter 17 state that ideal binary substitutional solid solutions occur when the ratio of the atomic volumes, $v_B/v_A = 1$. We predict hemisubstitutional alloys to occur if $2v_B/v_A = 1$. With this prediction, we have an optimum size ratio in the Laves phases (Laves *et al.*, 1934), given by: $r_A/r_B = 2^{1/3} \approx 1.26$. This ratio agrees well with experimental observations.

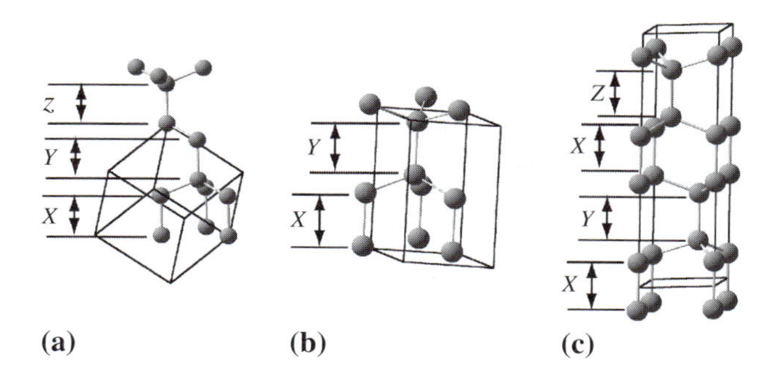

Fig. 18.13. Comparison of the Cu_2Mg, $MgZn_2$, and $MgNi_2$ Laves phase structures showing connectivity of *A* atom major skeletons and tetrahedral double layers (Barrett and Massalski, 1980).

The atomic radii in Laves phases are typically different from their elemental values (Rudman, 1965a), and the differences provide information about bonding (Rudman, 1965b).

Prototype structures include the cubic Cu_2Mg (**C15**), hexagonal $MgZn_2$ (**C14**), and $MgNi_2$ (**C36**) phases. Laves phases are Frank–Kasper (TCP) phases that have only icosahedral CN12 and CN16 coordination polyhedra. As shown in Table 18.3, $\frac{2}{3}$ of the atoms have CN12 coordination and $\frac{1}{3}$ have CN16 coordination for an ACN of 13.33. The CN16 coordination polyhedron was first associated with the Laves phases and given the name *Friauf polyhedron*. The Friauf polyhedron, Fig. 18.17(a), consists of four larger *A* atoms forming a tetrahedron and hexagonal rings of *B* atoms.

18.3.2.1 Stacking sequences in Laves phases

The connectivity of the *A* atoms describes the *major skeleton* of the structure. For each of the structures the major skeletons are compared in Fig. 18.13.[5] Figure 18.13 shows major skeleton linkages between major ligands for Laves phase prototypes. The *A* atoms occur only in the CN16 Friauf polyhedra. We can describe the connectivity of major ligands in the major skeleton by the stacking of the *A* atom interpenetrating tetrahedra. For Cu_2Mg, the major skeleton is a diamond cubic lattice. For $MgZn_2$, the arrangement of Mg atoms is in a wurtzite structure; and for $MgNi_2$, the major skeleton consists of a mixture (intergrowth) of diamond cubic and wurtzite structures.

We can see the major skeletons by viewing the stacking of major ligand tetrahedra. In Fig. 18.13, we use X, Y, and Z to describe stacking of tetrahedral double layers in analogy to A, B, and C sites in close-packed atomic systems. Figure 18.13 shows that the stacking for Cu_2Mg is $XYZXYZ\ldots$ for $MgZn_2$ it is $XY\,XY\ldots$ and for $MgNi_2$ we have $XYXZXYXZ\ldots$ The double layer stacking sequence of Laves phases is similar to single layer stacking in the ordered compounds $AuCu_3$, Cd_3Mg, and Ni_3Ti. Below, we will explore the Pearson notation to describe stacking of single atomic layers.

[5] Major skeletons in CN16-containing alloy phases consist of tetrahedral networks.

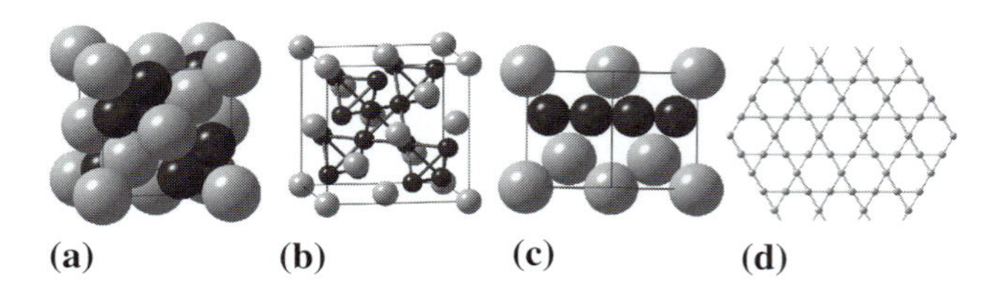

Fig. 18.14. Single unit cell of Cu_2Mg in a space filling (a) and a ball-and-stick (b) representation; (c) the positions of atoms in a (110) plane and (d) Cu atoms in a (111) plane.

18.3.2.2 The Cu_2Mg-type C15 Laves phase structure

The Cu_2Mg-type C15 Laves phase (**Structure 32**) has the cubic **Fd$\bar{3}$m** (O_h^7) space group. The Cu_2Mg prototype has a lattice constant $a = 0.7048$ nm, with Mg in $8a$ and Cu in $16d$ special positions. Figure 18.14(a) and (b) illustrate a single unit cell of the Cu_2Mg Laves phase in space filling and ball-and-stick representations.

While this Cu_2Mg structure has a diamond cubic arrangement of the larger B (Mg) atoms, the smaller A (Cu) atoms form tetrahedra that collectively occupy the empty tetrahedral sites in the B atom diamond sublattice. The distance between Cu atoms in the Cu_2Mg Laves phase structure is $a\sqrt{2}/4 = 0.249$ nm and between Mg atoms is $a\sqrt{3}/4 = 0.305$ nm, while the Cu–Mg distance is $a\sqrt{11}/8 = 0.292$ nm. We can easily show these distances by considering the positions of atoms in a (110) plane, as illustrated in Fig. 18.14(c). Figure 18.5(b) shows the simulated XRD pattern for the cubic Cu_2Mg Laves phase structure. We can see that this pattern satisfies the *extinction rules* for *fcc* lattices – namely h, k, and l are all even or all odd.

Figure 18.14(c) shows chains of B atoms in a (110) plane of the unit cell. If the A atoms touch one another, and the B atoms touch one another, then the radius ratio is given by: $r_A/r_B = \sqrt{3/2} = 1.225$, which differs only slightly (2.5%) from the Laves atomic volume criterion. This difference complicates the assignment of metallic radii for the A and B atoms. In fact, if we calculate the $A-B$ bond distance from the $A-A$ and $B-B$ distances, the resulting metallic radii would be too small. However, if we add the metallic radii of Mg and Cu based on their elemental crystal structures, we have better agreement with the $A-B$ bond distance.

Figure 18.14 (d) shows a single (111) plane of Cu atoms. This is the Archimedean tiling $(3\cdot6\cdot3\cdot6)$, the now familiar *Kagome net*. We can further analyze this structure by considering a three-layer stacking ($\alpha\beta\gamma$) of Kagome nets and three additional mixed (and puckered) Cu and Mg planes sandwiched between them, for a total of six layers (Fig. 18.15). The Mg atoms (below) sit above the large hexagonal holes in the Kagome net capped by Cu triangles below (above).

Figure 18.15 (a) illustrates the six-layer stacking of (111) planes of the Cu_2Mg Laves phase structure. Combining the fact that the Kagome nets follow an ABC stacking and the intervening mixed planes all have the same orientation, Frank assigned this structure's stacking sequence as $A^+B^+C^+$.

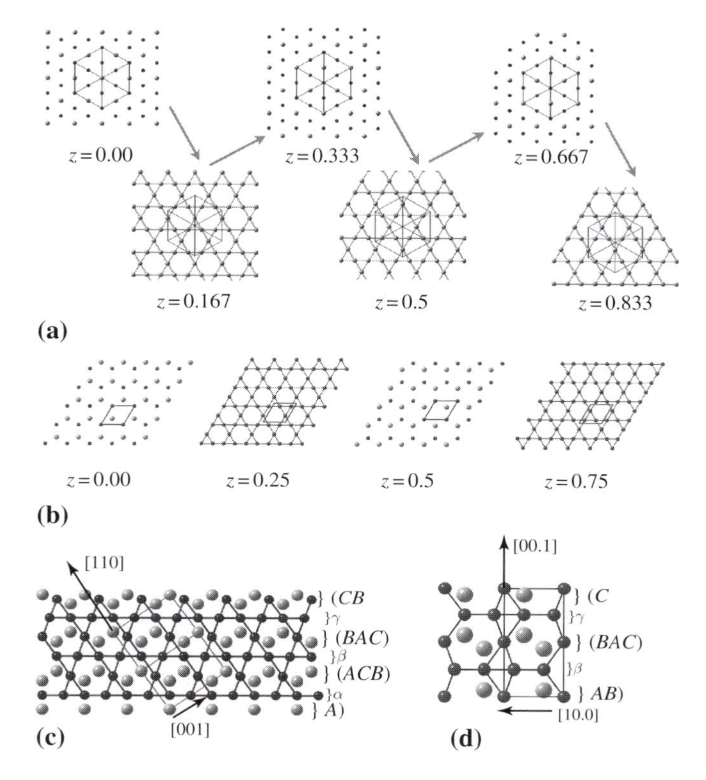

Fig. 18.15. (a) Six-layer stacking of (111) planes in the Cu_2Mg Laves phase structure. Top, the *ABC* stacking of Kagome nets is shown; bottom, the mixed (puckered) (111) planes. (b) Four-layer stacking of (00.1) planes of the $MgZn_2$ structure, and projected structures showing the derivation of the Pearson notation stacking sequences for (c) Cu_2Mg and (d) $MgZn_2$.

However, distinguishing between 3^6, 6^3, and $(3 \cdot 6 \cdot 3 \cdot 6)$ tiles for individual atomic species in the Pearson notation, we see that the stacking sequence for the Cu_2Mg-type (C15) Laves phase is actually $A)\alpha(ACB)\beta(BAC)\gamma(CB$, as illustrated in Fig. 18.15(c).

Following the Kagome networks is equivalent to following the major ligand lines in the structure. Each six member ring is a portion of the CN16 Friauf polyhedron, which is the other coordination polyhedron that coexists with the icosahedron in this topologically close-packed structure. The Cu_2Mg-type (C15) Laves phase is the most commonly occurring Laves phase structure and has a variety of magnetic and superconducting phases among its examples.

18.3.2.3 The $MgZn_2$-type (C14) Laves phase structure

The $MgZn_2$-type (C14) Laves phase is the next most commonly occurring Laves phase after the Cu_2Mg-type. $MgZn_2$ (C14) has the space group $\mathbf{P6_3/mmc}$ (D^4_{6h}) with lattice constants $a = 0.518\,\text{nm}$ and $c = 0.852\,\text{nm}$. The atom positions are described as **Structure 33** in the on-line structures appendix. Figure 18.16(a) and (b) illustrate a single unit cell of $MgZn_2$ in both space filling and ball-and-stick representations. This structure has a wurtzite arrangement of the larger *B* (Mg) atoms. Figure 18.16(c) illustrates the positions of atoms in three unit cells projected into a (00.1) plane.

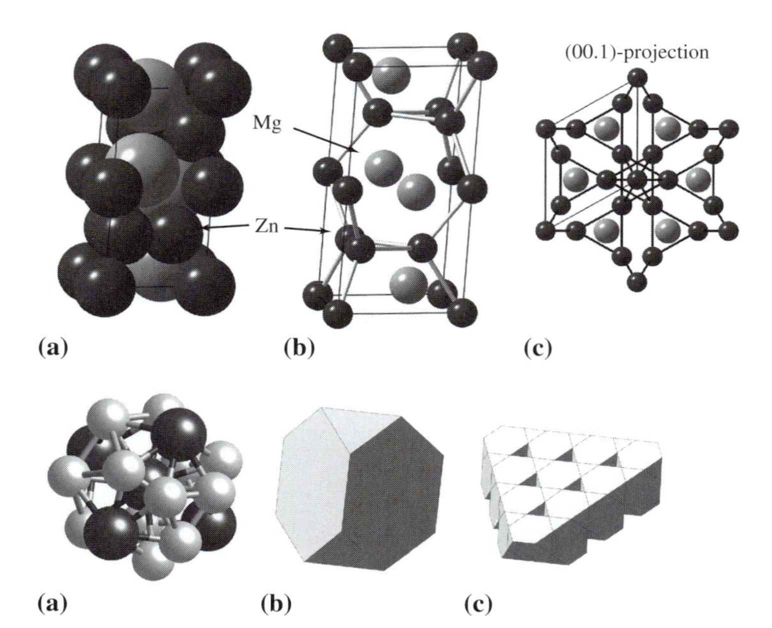

Fig. 18.16. $MgZn_2$ unit cell in the space filling (a) and ball-and-stick (b) representations, projection into a (00.1) plane (c).

Fig. 18.17. (a) illustration of the CN16 Friauf polyhedra, (b) truncated tetrahedron formed by the 12 Zn atoms, (c) stacking of the truncated tetrahedra in a layer of the $MgZn_2$ structure.

The $MgZn_2$ Laves phase structure can also be understood in terms of (00.1) plane stacking. We can see (Fig. 18.15(b)) that the (111) planes at the $z = \frac{1}{4}$ and $z = \frac{3}{4}$ elevations are $(3 \cdot 6 \cdot 3 \cdot 6)$ Archimedean tilings of Zn atoms. We can understand the structure by considering this two layer (AB) stacking of Kagome nets along with two additional mixed (and puckered) Zn and Mg planes sandwiched between them at the $z = 0$ and $z = \frac{1}{2}$ elevations, for a total of four layers. Figure 18.15 illustrates the four layer stacking of (00.1) planes that describes the $MgZn_2$ structure. This structure's stacking sequence is described in Frank's notation as A^+B^-. The **C14** structure is the hexagonal variant of the **C15** structure. In the Pearson notation, the stacking sequence for the $MgZn_2$ (**C14**) Laves phase is $AB)\beta(BAC)\gamma(C$, as illustrated in Fig. 18.15 (d). This figure illustrates atoms projected into a (01.0) plane.

The stacking of (00.1) planes in the $MgZn_2$ structure lends itself to an additional way of understanding the structure in terms of the connectivity of the CN16 Friauf polyhedra or in terms of the stacking of truncated tetrahedra, as shown in Fig. 18.17. Fig. 18.17(a) shows the familiar CN16 Friauf polyhedron which has 16 vertices and 28 triangular faces. If we ignore the four larger Mg atoms we see (Fig. 18.17(b)) that the polyhedron formed by the 12 smaller Zn atoms is a *truncated tetrahedron*. As the CN16 Friauf polyhedra are interpenetrating in the $MgZn_2$ structure, their connectivity can be understood by the stacking of these truncated tetrahedra shown in Fig. 18.17(c), for a single layer of the $MgZn_2$ structure. A final example of the $MgZn_2$ Laves phase structure occurs in interesting self-assembled nanostructures as illustrated in Box 18.4.

Box 18.4 Self-assembled nanostructures with a Laves phase lattice

Self-assembly of nanoparticles has been an active area of recent research and is discussed in Chapter 25. In self-assembly, nanoparticles, as opposed to atoms, can be organized into a crystalline lattice. A beautiful example of a self-assembled structure is illustrated below for a system of 6.2 nm PbSe and 3.0 nm Pd nanoparticles.

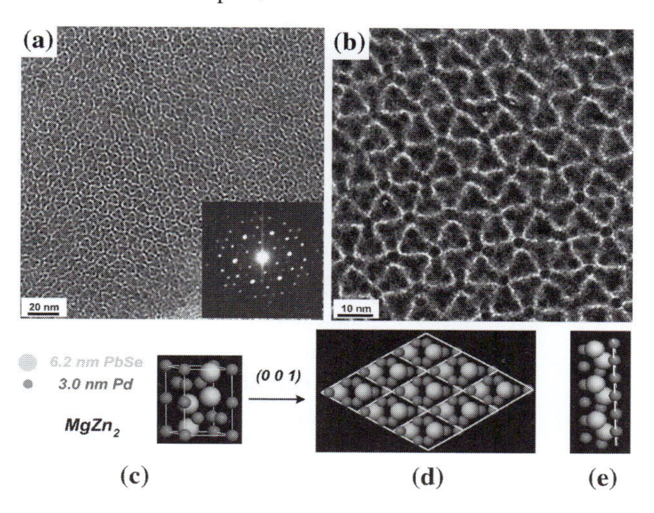

The figure shows TEM micrographs (courtesy of E. Shevshenko and Chris Murray, IBM) of nanoparticle superlattices isostructural with $MgZn_2$: (a) TEM overview of (00.1) plane; (a, insert) – small angle electron diffraction pattern of an (00.1) plane. (b) As (a) but at high magnification; (c) the $MgZn_2$ crystal structure. (d) Depiction of the (00.1) plane. (e) The minimum number of layers in (00.1) planes, leading to the formation of the patterns identical to that observed.

18.3.2.4 The $MgNi_2$-type (C36) Laves phase structure

The $MgNi_2$-type (C36) Laves phase has space group **P6$_3$/mmc** (D^4_{6h}) and lattice constants $a = 0.4815$ and $c = 1.580$ nm. The atom positions are described as **Structure 34**. Figure 18.18(a) and (b) show a single unit cell of $MgNi_2$ in both space filling and ball-and-stick representations. This structure has a combination of stacking in the diamond and wurtzite structures of the larger *B* (Mg) atoms. Figure 18.18(c) illustrates the positions of atoms projected into an (00.1) plane with three cells, again showing attractive hexagonal symmetries for *A* and *B* atoms in projection. Figure 18.18(d) illustrates the positions of atoms projected into a (01.0) plane (two cells are shown). The $MgNi_2$-type (C36) Laves phase is the least common Laves phase structure.

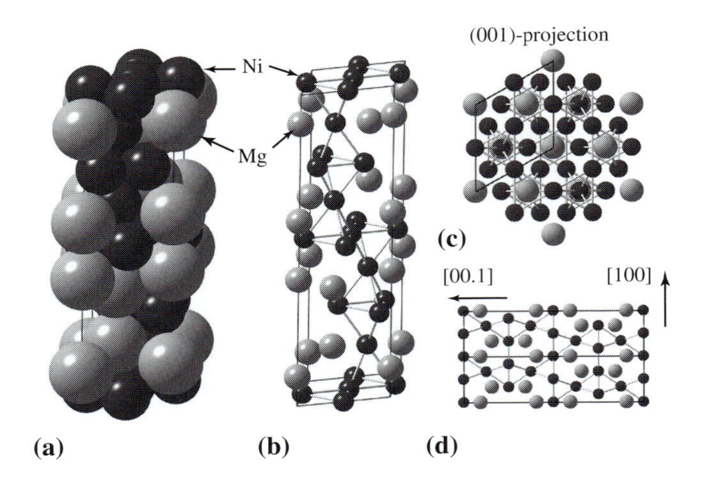

Fig. 18.18. $MgNi_2$ unit cell in the space filling (a) and ball-and-stick (b) representations, (c) projection of all atoms into a (001) plane, and (d) into a (010) plane.

The $MgNi_2$ structure can also be understood in terms of the stacking of (00.1) planes. It is left as an exercise to show that this is an eight-layer stacking sequence with an *ABAC* stacking of the Kagome nets. Combining the Kagome nets stacking with the intermediate mixed planes, which follow a $+--+$ sequence in successive planes, this structure's stacking sequence is $A^+B^-A^-C^+$. As an exercise, the reader may show that this stacking has the Pearson notation $\alpha(ABC)\gamma(CBA)\alpha(ACB)\beta(BCA)$.

18.3.2.5 Shear derivatives of the Laves phases: B_2CoW_2 structure

We can compare different Laves phase structures by considering the stacking of their Kagome nets with mixed planes sandwiched between them. We can also construct interesting structures related to the Laves phases through a non-conservative shear mechanism. Figure 18.19 shows an idealized structure of the Cu_2Mg Laves phase, projected into a (111) plane, including Cu atoms and Mg atoms from adjacent (111) planes. Through a simple shear, this structure can be transformed into the idealized B_2CoW_2 structure (Rieger *et al.*, 1966) shown in Fig. 18.19(b). The shear planes shown are $(11\bar{2})$ planes.

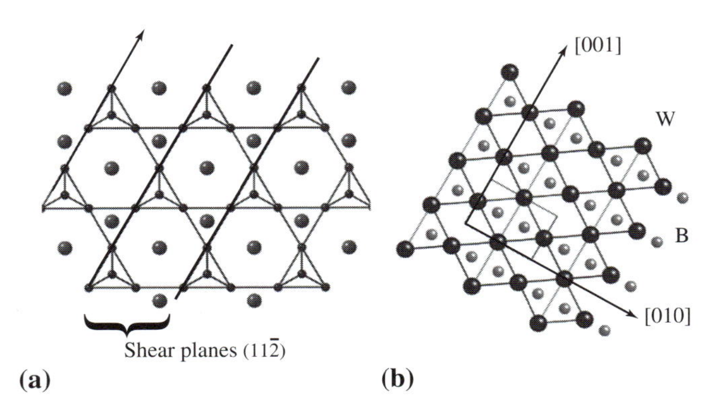

Fig. 18.19. Shear of Cu_2Mg structure, projected (111) planes (a) can be transformed into the idealized B_2CoW_2 projected structure (100) planes (b).

In Fig. 18.19(b) only W and B atoms are shown from adjacent (100) planes (W at $x = 0.205$ and B at $x = 0.0$). A detailed analysis of the latter structure would include the positions of the Co atoms in the framework of Fig. 18.19(b). These details are omitted for clarity in illustrating the shear mechanism. The B_2CoW_2 structure is described as **Structure 35** in the on-line structure appendix.

18.3.3 The sigma phase

The sigma (σ) phase, Fig. 18.20, is an example of an *AB* intermetallic compound that is important to many transition metal alloys. It is very hard and, as a second phase, was known to embrittle steels even before its crystal structure was determined in early studies at the California Institute of Technology. Sigma phases in binary transition metal alloys were studied by Hume-Rothery's group at Oxford as early as 1952 (Pearson *et al.*, 1951). The σ-phase appears to form when the electron/atom ratio is in the range from 6.2 to 7 (see Box 17.3). Typically, one end member has a *bcc* structure and the other an *fcc* or *hcp* structure; yet, their metallic radii are similar. A σ-phase has also been observed in the Al–Nb system, where it is a second phase in the **A15** $AlNb_3$ superconductor. Recent experimental observations suggest that this σ-phase is the product of *eutectoid decomposition* of a high temperature *bcc* Al–Nb phase (Buta *et al.*, 2003).[6]

The σ-phase structure is related to that of β-uranium (Tucker, 1950). In the early 1950s, researchers at the California Institute of Technology, under

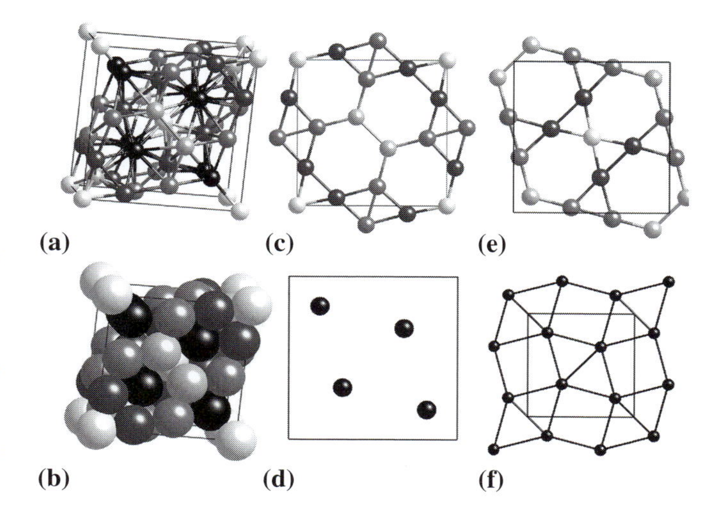

Fig. 18.20. Structure of CrFe (a) ball-and-stick and (b) space filling models; (c) primary layer 2-uniform tiling at the $z = 0$ (001) plane; (d) a secondary layer of $M(5)$ atoms near $z = 1/4$; (e) primary layer at $z = \frac{1}{2}$; and (f) a secondary layer of $M(5)$ atoms near $z = \frac{3}{4}$.

[6] A eutectoid reaction is equivalent to a eutectic reaction, but the parent phase is a solid rather than a liquid. The general reaction is $\gamma \rightarrow \alpha + \beta$.

the direction of Linus Pauling (Bergman and Shoemaker, 1954); (Shoemaker, 1950) attempted to determine the structure of the σ-phase. Their efforts followed those made by P. Pietrokowsky and P. Duwez, who synthesized CrFe and FeMo powders containing about 0.1 mm σ-phase single crystals, large enough for X-ray studies. The σ-phase structure in CoCr was solved independently by Dickins *et al.* (1951, 1956) and Kasper *et al.* (1951).

Figure 18.20 illustrates a single unit cell in the CrFe σ-phase in both ball-and-stick (a) and space filling representations (b). Note that this structure is complicated by the similar sizes of the A and B atoms; likewise, we may find it difficult to determine their preference for various atomic positions. In Fig. 18.20, the five atomic sites (special positions) $M(1)$, $M(2)$, $M(3)$, $M(4)$, and $M(5)$ are illustrated as increasingly darker gray scales. The decomposition of the structure into stacked tiling is shown in Fig. 18.20(c) through (f), where we can see that the primary layers at $z = 0$ and $z = \frac{1}{2}$ exhibit the 2-uniform tiling $(3^2 \cdot 6^2; 3 \cdot 6 \cdot 3 \cdot 6)$. Figure 18.20(f) shows that the secondary tiling is the 2-uniform tiling $(3^2 \cdot 4 \cdot 3 \cdot 4)_2$.

CrFe is a prototypical alloy that crystallizes in the σ-phase structure at the equiatomic composition. The two metals (with atomic numbers 24 and 26, respectively), have similar X-ray scattering factors and therefore cannot be distinguished easily by XRD. CrFe has been determined to have space group $\mathbf{P4_2/mnm}$ (D_{4h}^{14}). This tetragonal structure has lattice constants $a = 0.8800$ nm and $c = 0.44544$ nm, and its atom positions can be found as **Structure 36** in the on-line structure appendix. The Strukturbericht symbol for this structure is $D8_b$. The σ-phase can precipitate at grain boundaries as a secondary phase in stainless steels, where it degrades the corrosion resistance. Cr (and Mo) and Fe (and Ni) are major components in stainless steels, in which the composition of σ-phase is nearer to $(Fe,Ni)_3(Cr,Mo)_2$.

The coordination polyhedra are icosahedra around atoms $M(1)$ and $M(4)$, CN14 around $M(3)$ and $M(5)$, and CN15 around $M(2)$. The σ-phase can also be viewed as an intergrowth of structural units of two of the previously discussed TCP phases. This is illustrated in Fig. 18.21, where the σ-phase can be projected as an intergrowth of tetrahedersterns. Here, we illustrate with reference to a portion of an [001] projection of the cubic A15 structure and a portion of a [111] projection of the cubic Cu_2Mg structure. In both, we show the projected unit cells for reference.

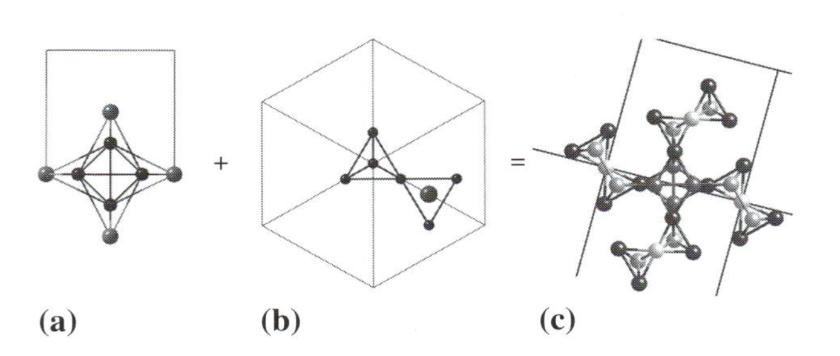

Fig. 18.21. Structural units found in (a) A15 [001] projection and (b) Cu_2Mg [111] projection, and (c) intergrowth of the two units to form an [001] projection of the σ-phase structure.

18.3.4 The μ-phase and the M, P, and R phases

The μ-phase has an ideal stoichiometry A_6B_7. The prototype W_6Fe_7 has space group **R$\bar{3}$m** (D^5_{3d}), and the rhombohedral cell contains 13 atoms. The W_6Fe_7 structure was determined to have lattice constants $a = 0.4757$ nm and $c = 2.584$ nm (Arnfelt, 1935). The atom positions are listed as **Structure 37** in the on-line structure appendix. Figure 18.22(a) and (b) show the μ-phase structure of W_6Fe_7 in ball-and-stick and space filling representations. Figure 18.22(c) shows the derivation of the Pearson notation for the stacking sequence. We can see that the stacking sequence is rather complicated and given by the string $AB)\beta BbB\beta(BCA)\alpha AaA\alpha(ABC)\gamma CcC\gamma(C$. This is a mixture of regions with Laves phase and Al_3Zr_4 stacking.

Figure 18.22(d), (e), and (f) show the primary and secondary tilings from the decomposition of the μ-phase structure. The top frame shows the secondary tiling of W atoms in the $z = 0.167$ plane, forming a hexagonal 6^3 net. The primary tiling of Fe atoms in the $z = 0.257$ plane forms the familiar Kagome net. Finally, the secondary tiling of W atoms in the $z = 0.346$ plane forms a triangular 3^6 net. In Fig. 18.23, we see how idealized A15 tetrahedersterns and Cu_2Mg structural units can be connected to yield a projection of the μ-phase.

While many other Frank–Kasper alloy types have been identified, more continue to be found. Sinha (1972) published an excellent review from which many of the examples in this chapter are taken. There are several other important structures that we will list without illustration, namely the M, P, and R phases. The alloy $Nb_{10}Ni_9Al_3$ is the prototype for the M-phase

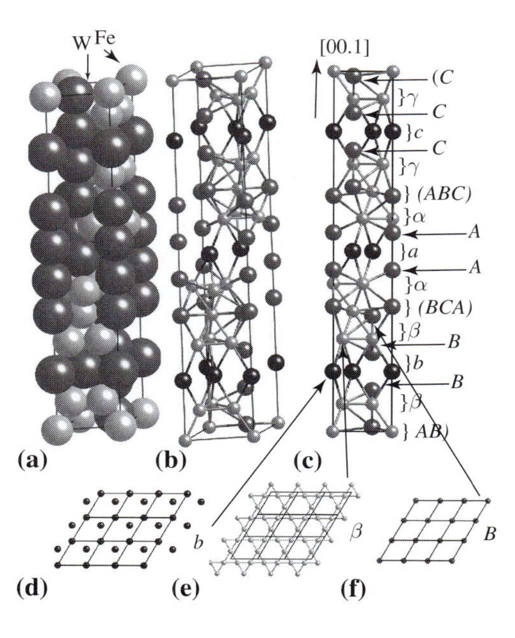

Fig. 18.22. Structure of W_6Fe_7 in (a) ball-and-stick and (b) space filling representations; (c) projection labeling layers using the Pearson notation; (d) secondary (6^3) W atom tile at $z = 0.167$, (e) primary Kagome tile at $z = 0.257$ and (f) secondary (3^6) W atom tile at $z = 0.346$.

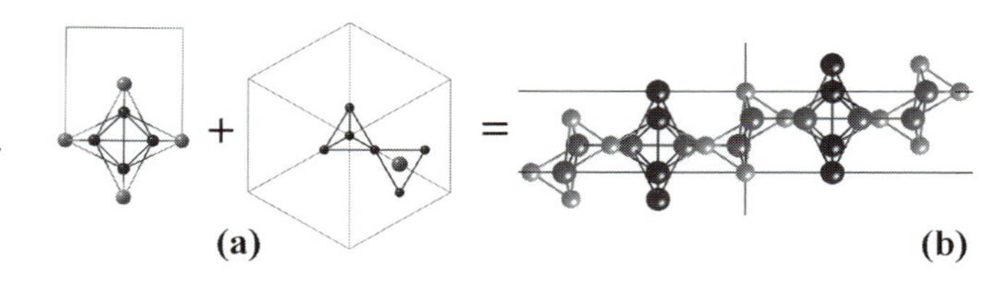

Fig. 18.23. (a) Structural units found in an A15 [001] projection and Cu_2Mg Laves phases (unit cell projections for reference) and (b) connectivity of the two structural units in a (110) plane projection of a portion of the W_6Fe_7 μ-phase structure.

(see **Structure 38**). It has the orthorhombic space group **Pnma** (D_{2h}^{16}) with 52 atoms per unit cell. Of the 11 crystallographically distinct sites, 9 belong to the $4c$ special position and 2 to the $8d$ special position. The alloy $Cr_9Mo_{21}Ni_{20}$ is the prototype for the P-phase. It has a primitive orthorhombic cell with 56 atoms and the space group is also **Pnma** (D_{2h}^{16}) (see **Structure 39**). There are 12 crystallographically distinct sites; ten belong to the $4c$ special positions and two to the $8d$ special position. The alloy $Co_5Cr_2Mo_3$ is the prototype for the R-phase. It belongs to the rhombohedral space group $\mathbf{R\bar{3}}$ (C_{3i}^2) with 53 atoms per rhombohedral cell (see **Structure 40**). There are 11 crystallographically distinct sites, one belongs to the $3b$, two to the $6c$, and eight to the $18f$ special positions, respectively.

Also interesting, but beyond the scope of this text, are the *giant cell structures*, described by Samson (1969). The structures of Cd_2Na, β-Al_3Mg_2, and Cd_3Cu_4 are reported to have large structural units with cells containing more than 1100 atoms. The structure of these phases, though considerably more complicated, can also be analyzed in terms of the principles discussed in this chapter. Pauling studied the Cd_2Na phase as one of the initial intermetallic phases studied by X-ray diffraction (Pauling, 1923). It was later determined that this phase has an $\mathbf{Fd\bar{3}m}$ (O_h^7) space group with a lattice constant of 3.056 nm and 1192 atoms in the unit cell. β-Al_3Mg_2 has the same space group with 1832 atoms per unit cell and a lattice constant of 2.8239 nm with Al atoms in 14 different special positions and Mg atoms in 9 different special positions (Samson, 1965). Cd_3Cu_4 has the $\mathbf{F\bar{4}3m}$ (T_d^2) space group with Cu atoms in 13 different special positions, Cd atoms in 11 different special positions and mixtures of atoms in 5 different special positions (Samson, 1967). There are 1124 atoms in this cubic unit cell with a 2.5871 nm lattice constant. The structural information for each of these giant cell structures can be found in Samson's original work.

18.4 *Quasicrystal approximants

Interesting intermetallic phases, such as $Mg_{32}(Al,Zn)_{49}$ (Fig. 18.24), are those in which icosahedral coordination persists to several coordination shells. Strictly speaking, these are not Frank–Kasper phases because a subset of the

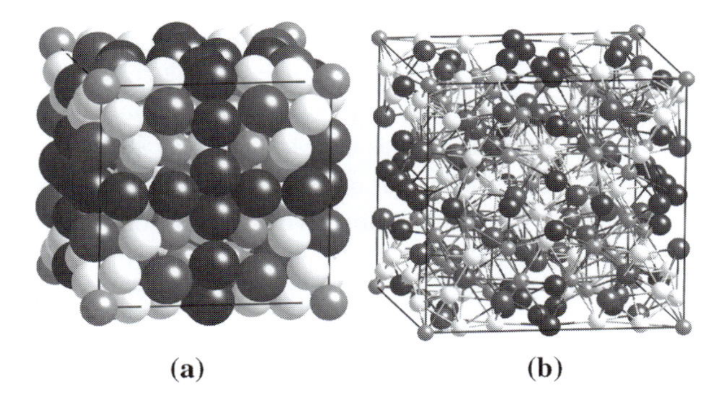

Fig. 18.24. $Mg_{32}(Al,Zn)_{49}$ structure, showing a single unit cell in the space filling (a) and ball-and-stick (b) representations.

coordination polyhedra are not Kasper polyhedra. In these phases, icosahedral clusters grow to several coordination shells and then organize to form a crystal. These phases can pack icosahedral units of a Bravais lattice with or without sharing of atoms in the outer shell of the icosahedral cluster. The propagation of icosahedral order is an important consideration in the discussion of the atomic positions in quasicrystals, discussed in Chapter 20.

In some ways intermetallic phases, with clusters decorating sites of a Bravais lattice, resemble molecular crystals, except that the impetus for maintaining icosahedral environments seems to maximize the interstitial electron density rather than promote a favorable type of molecular bonding. Recently, scientists have discovered that phases with atomic configurations consisting of several icosahedral coordination shells are often crystalline *approximants* (not surprisingly) of icosahedral quasicrystalline phases. To introduce further illustrations of quasicrystals in a later chapter, we discuss here the structure of two such phases, $Mg_{32}(Al,Zn)_{49}$ and α-Al–Mn–Si (Fig. 18.25).

18.4.1 $Mg_{32}(Al,Zn)_{49}$ and alpha-Al–Mn–Si crystal structures

The structure of $Mg_{32}(Al,Zn)_{49}$ was first determined in 1952 by Linus Pauling's group (Bergman *et al.*, 1952). It is a cubic phase with 162 atoms in the unit cell. The structure is based on the *bcc* lattice with a lattice parameter

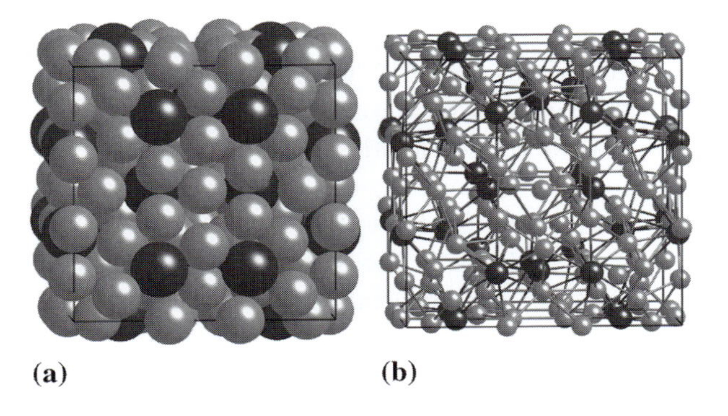

Fig. 18.25. α-Al–Mn–Si structure, showing a single unit cell in the space filling (a) and ball-and-stick (b) representations.

of 1.416 nm and space group **Im$\bar{3}$** (T_h^5). Mg atoms are located in four special positions, and (with some disorder) Al and Zn atoms are located in four additional special positions (Bergman *et al.*, 1957) (see **Structure 41**). Figure 18.24 illustrates the $Mg_{32}(Al,Zn)_{49}$ structure, showing a single unit cell in the space filling (a) and ball-and-stick (b) representations; this is clearly a complex structure! Below, we will carry out a decomposition into icosahedral clusters, which will shed light on the origin of this complexity.

α-Al–Mn–Si (Cooper and Robinson, 1966) is also a cubic phase with 138 atoms in the unit cell. The structure is based on the primitive cubic (*sc*) lattice with a lattice parameter of 1.268 nm and space group **Pm$\bar{3}$** (T_h^1). Mn atoms are located in two special positions and Al(Si) atoms in nine other special positions (Cooper and Robinson, 1966) (see **Structure 42**). Figure 18.25 illustrates the α-Al–Mn–Si structure, showing a single unit cell in the space filling (a) and ball-and-stick (b) representations. Again, this structure is best understood in terms of icosahedral cluster growth.

18.4.2 $Mg_{32}(Al,Zn)_{49}$ and alpha-Al–Mn–Si shell models

We can understand the structure of $Mg_{32}(Al,Zn)_{49}$ by considering coordination shells, beginning with an Al(Zn) atom at the vertex or body-centered site of the *bcc* lattice. Figure 18.26 illustrates the atomic clusters about this site (left column), the ball-and-stick depiction of the n-th shell (center column), and a wire frame drawing of the n-th shell polyhedron (right column). Figure 18.26(a)–(c) illustrates the Al(Zn) first coordination shell, an icosahedral cluster, about the Al(Zn) atom in the $(1/2, 1/2, 1/2)$ site of the *bcc* lattice. Figure 18.26(d)–(f) illustrate the pentagonal dodecahedral Mg second coordination shell. Each Mg atom sits above an Al(Zn) atom triangular face, forming a tetrahedral interstice preserving topological close packing. Figure 18.26(g)–(i) show the addition of another 12-atom Al(Zn) icosahedron to the previous cluster. These atoms, combined with the 20 atom Mg *pentagonal dodecahedron* unit, form a nearly spherical shell. Together, these 32 atoms comprise the solid form known as a *rhombic triacontahedron*, each face of which is a rhombus comprising orthogonal pairs of Al(Zn) and Mg atoms. This solid will be significant in the later discussion of quasicrystals. Figure 18.26(j)–(l) illustrate the last coordination shell, consisting of an ordered mixture of 12 Mg atoms and 48 Al(Zn) atoms sitting at the vertices of a *truncated icosahedron*.

As we have seen earlier, 60 C atoms also decorate the vertices of a truncated icosahedron in the C_{60} molecule. Extending out to this fourth shell, the entire cluster consists of $1+12+20+12+48+12=105$ atoms. The 48 Al(Zn) atoms in the outer shells are, however, shared with the clusters centered at the $(0, 0, 0)$ and equivalent vertex sites of the *bcc* lattice. Each cluster shares an Al(Zn) hexagon with a neighboring cluster. Therefore, the total number of atoms belonging to any one cluster is $1+12+20+12+24+12=81$ atoms.

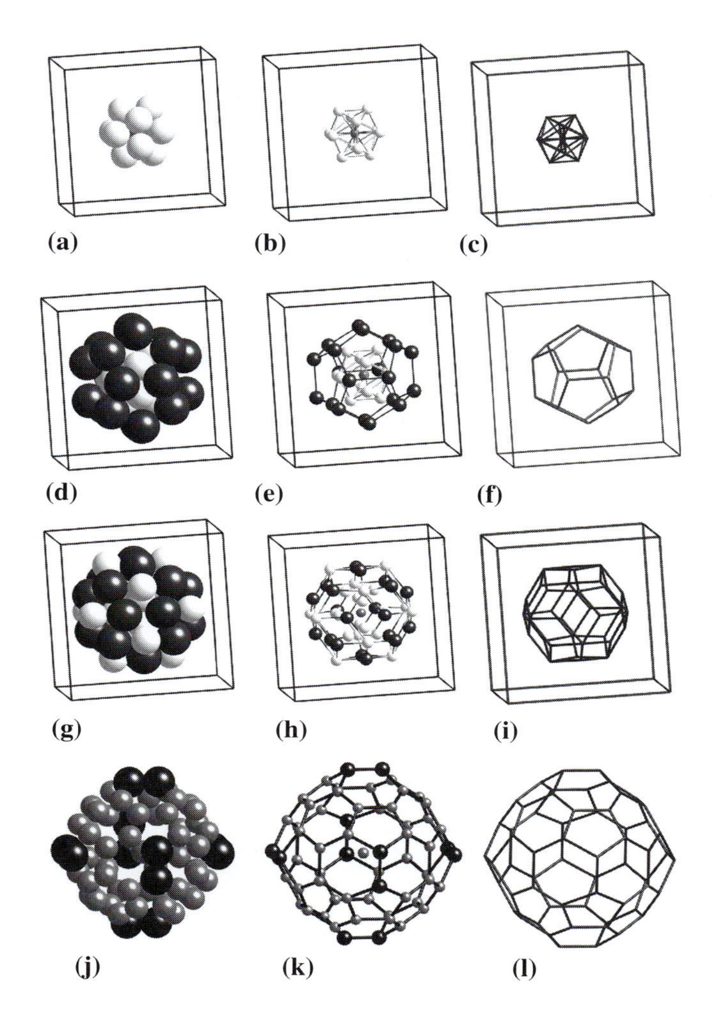

Fig. 18.26. $Mg_{32}(Al,Zn)_{49}$ structure. Each column shows a space filling depiction of the cluster, a ball-and-stick depiction, and a wire frame drawing of the n-th shell polyhedron. (a)–(c) icosahedral Al(Zn) first coordination shell; (d)–(f) pentagonal dodecahedral second shell of Mg atoms; (g)–(i) a mixed Mg–Al rhombic triacontahedron; (j)–(l) a mixed Mg–Al truncated icosahedron.

In the *bcc* lattice, there are two such clusters per cell, resulting in 162 atoms per unit cell!

The structure of α-Al–Mn–Si (Cooper and Robinson, 1966) can be understood by considering coordination shells, beginning with a vacant site at a vertex site in a primitive simple cubic lattice. The top frames of Fig. 18.27 illustrate an empty 12-atom icosahedron of Al(Si) atoms at this vacant site. The next three frames illustrate the decoration of the 30 edges of the first icosahedron with 30 additional Al(Si) atoms. This structure forms a coordination shell known as an *icosidodecahedron*, which is an Archimedean solid possessing both pentagonal and triangular faces. The gaps in the pentagonal faces of the icosidodecahedron provide space for decoration with the larger Mn atoms, forming a large Mn icosahedron as the third shell. At this point, the middle row shows each cubic site that will be decorated with a *Mackay icosahedron* (Mackay, 1962, 1982).

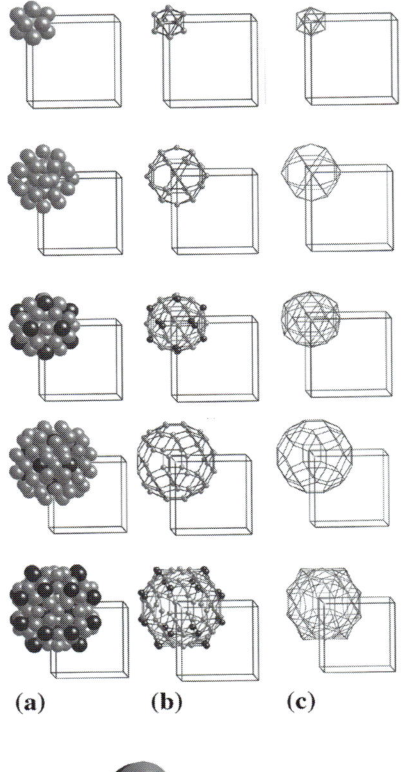

Fig. 18.27. α-Al–Mn–Si structure decomposed in terms of icosahedral shells. Each column illustrates a (a) space filling depiction of the cluster of the *n*-th shell, (b) a ball-and-stick depiction of the *n*-th shell, and (c) a wire frame drawing of the *n*-th shell polyhedron. Top to bottom shows clusters for the 1st through 5th coordination shells.

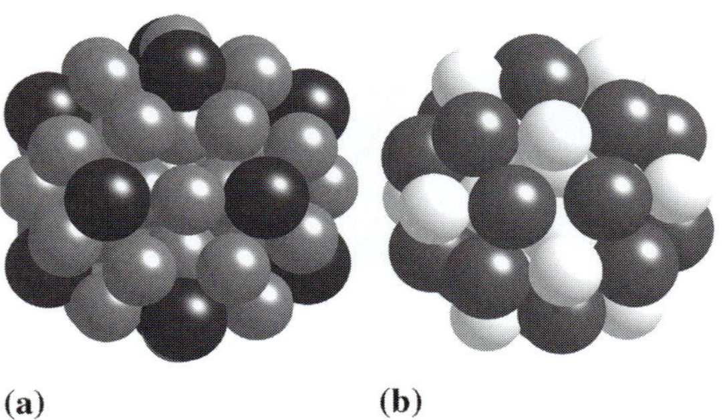

Fig. 18.28. (a) The 54-atom Mackay icosahedron of the α-Al–Mn–Si structure and (b) the 45-atom Bergman polyhedron of the $Mg_{32}(Al,Zn)_{49}$ structure.

The Mackay icosahedron, illustrated in Fig. 18.28, is a 54-atom unit, consisting of the following coordination shells: an inner icosahedron (12), an icosidodecahedron (30), and an outer icosahedron (12). Typically, Al and Si decorate the sites of the inner two coordination shells, and the transition metal atoms (Mn) decorate the outer icosahedron. The decoration of the Mackay icosahedron is illustrated in the next two rows of frames of Fig. 18.27. First, an additional 60 Al atoms are added, forming a coordination polyhedron with triangular, square, and pentagonal faces. This polyhedron is known as a *rhombicosidodecahedron* (Wenninger, 1971), another one of the Archimedean solids. This solid has 20 triangular, 30 square, and 12 pentagonal faces. A subgroup of square faces is then aligned to conform with the cubic symmetry of the lattice. This polyhedron is, in turn, decorated by 24 additional Mn atoms in the final shell (an icosahedron with Mn dumbbells at its vertices). The total number of atoms in the final cluster is, therefore, $12+30+12+60+24=138$ atoms. All the 138 atoms in the cell are accounted for as there is one cluster per primitive cubic cell.

In anticipation of future discussions of quasicrystals, we will make a few comparisons of the $Mg_{32}(Al,Zn)_{49}$ and α-Al–Mn–Si structures in terms of icosahedral clusters. Considering both structures up to the third coordination shell, we find that icosahedral symmetry is preserved. In the $Mg_{32}(Al,Zn)_{49}$ phase, it is preserved around an occupied Al(Zn) central site with the coordination first by a 12 atom Al(Zn) icosahedron, followed by a 20 Mg-atom pentagonal dodecahedron, and then a 12-atom Al(Zn) icosahedron. This $1+12+20+12=45$-atom cluster is called the *Bergman polyhedron*. In the case of the α-Al–Mn–Si structure, icosahedral symmetry is preserved around a vacant site with the coordination first by a 12-atom Al(Si) icosahedron, followed by a 30 Al(Si)-atom icosiododecahedron, and then a 12-atom Mn icosahedron to yield the 54-atom Mackay icosahedron.

Figure 18.28 (a) illustrates the 54-atom Mackay icosahedron of the α-Al–Mn–Si phase structure and (b) the 45-atom Bergman polyhedron of the $Mg_{32}(Al,Zn)_{49}$ structure, respectively. In Chapter 20, we will use these structures to understand the origin of the 3-D Penrose bricks used to tile Al–Mn–Si and Al–Mg–Zn quasicrystals (Henley and Elser, 1986), respectively. Interestingly, the proximity of the α-Al–Mn–Si phase composition to that of the Al–Mn–Si quasicrystals has been used to argue the similarities between the crystalline and quasi crystalline phases. Additionally, scientists have pointed out that other Frank–Kasper phases with large fractions of sites of CN12 coordination could be related to the observed quasicrystalline phases. In particular, the structure of the μ-Al_4Mn phase, also with a composition close to that of quasicrystalline Al–Mn phases, was studied by Shoemaker, *et al.* (Shoemaker *et al.*, 1989).

With this, we end our discussion of Frank–Kasper phases in transition metal systems. In Chapter 19, we turn our attention to rare earth/transition metal systems, for which the A atom to B atom radius ratio can be very large. In many alloy systems of interest, AB_2 Laves phases are observed. Many other interesting structures are also possible because of this large radius ratio. Some of these will be discussed in the context of important permanent magnet materials.

18.5 Historical notes

Sir (Frederick) Charles Frank (1911–98): Sir Charles Frank contributed to the advancement of knowledge in the growth and structure of crystals, defects

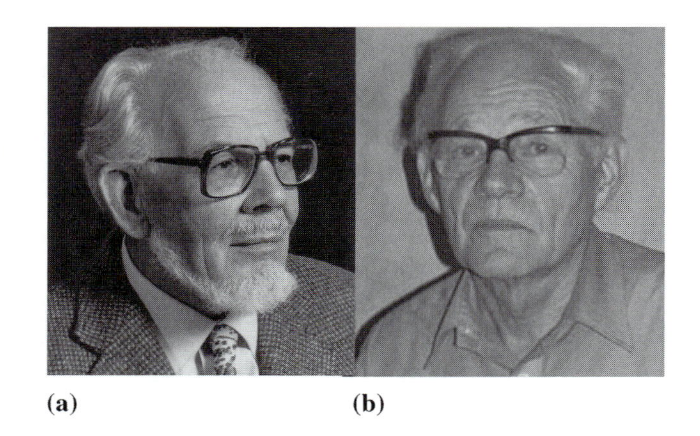

Fig. 18.29. (a) Sir Frederick Charles Frank (1911–1998) (picture courtesy of the Materials Research Society, Warrendale, PA) and (b) Alan L. Mackay (1926–) (picture courtesy of Alan L. Mackay).

in crystals, and the molecular alignments within liquid crystals, among other areas. He obtained his B.A. and B.Sc. in 1933 and D.Phil. in 1937 at Oxford University. In 1946, Frank joined the H. H. Wills Physics Laboratory at the University of Bristol and worked under Professor Nevill Mott. At Bristol, Frank studied crystal growth and the plastic deformation of metals. He led a group that studied growth steps on beryl, cadmium iodide, and other crystals.

Frank's contributions to the theory of dislocations were among his most important accomplishments (Frank, 1949, 1951). In the field of plastic deformation, Frank, in collaboration with W. T. Read, proposed a model of the origin of slip bands. Frank developed the theory for dislocation multiplication, the so-called Frank–Read source, while he was visiting Carnegie Mellon University in Pittsburgh. Local lore has it that the Frank–Read source was first drawn on a napkin at a local bar near the university.

The Frank–Kasper phases discussed in this chapter were also among Frank's notable contributions to the field of crystallography. An enormous number of Frank–Kasper alloys have been discovered and continue to be discovered to this day. The icosahedral arrangements of atoms in Frank–Kasper phases and the notable stability of icosahedral arrangements of atoms in pair potentials also led Frank to consider their importance in the liquid phase. This notion is still pursued vigorously in the study of liquid metals and amorphous metallic solids.

In 1985, Frank won the prestigious Von Hippel award of the Materials Research Society for which it was stated: "(Frank) has had wide-ranging impact on modern materials science through seminal contributions in areas of inorganic crystals, metals, polymers, and liquid crystals. His outstanding research in crystallography, chemistry, physics, and materials science exemplifies the interdisciplinary approach." Four years before his death, Frank won the Copley Medal, the premier award of the British Royal Society: "In recognition of his fundamental contribution to the theory of crystal morphology, in particular to the source of dislocations and their consequences in interfaces and crystal growth; to fundamental understanding of liquid crystals and the concept of disclination; and to the extension of crystallinity concepts to aperiodic crystals." Frank has also contributed to a number of physical problems through a variety of remarkable insights.

Alan L. Mackay (1926–) is a British crystallographer and mathematician. Alan Mackay completed his Ph.D. at Birkbeck College, University of London and joined the staff as an assistant lecturer in 1951. He became a professor of crystallography in 1988, the same year he was made a fellow of the British Royal Society. In 2002, he was made a fellow of Birbeck College. He is currently Professor emeritus at Birkbeck College, University of London.

Alan Mackay is a proponent of a broader scope for crystallography (Mackay, 1975). His contributions to the field of icosahedral structures and quasicrystals are of note; the 54-atom Mackay icosahedron bears his name. He generalized the Penrose tiling to three dimensions, using two types of rhombohedra in 1981 (Mackay, 1981). In 1982, he dubbed 2-D and 3-D Penrose tiles *quasilattices*. He suggested decorations of the quasilattice points with atoms. He was one of the first to demonstrate that these non-periodic point sets could produce crystal-like diffraction patterns with "forbidden" symmetries (Mackay, 1982), by showing that the structures give rise to optical diffraction patterns with ten-fold symmetry. Fifty five atom clusters of Au (an atom-centered Mackay icosahedron) have been observed in nature; a theoretical analysis shows them to have a lower total energy than similar cubic (*fcc*) clusters.

18.6 Problems

(i) *Fermi level and effective number of free electrons*: ϵ_F for Li is 4.72 eV at T = 0 K. Li crystallizes in a *bcc* lattice with lattice constant $a = 0.349$ nm. Calculate the number of conduction electrons per unit volume in Li. How many electrons per Li atom does this correspond to?

(ii) *Electron density and* ϵ_F *for* Al: Al has an *fcc* crystal structure with lattice parameter $a = 0.405$ nm and three conduction electrons per atom. Calculate the following:

(a) The conduction electron density, N_e/V, and Fermi energy, ϵ_F, for Al.

(b) The dimensionless radius parameter, r_{WS}, is defined as r_0/a_H where r_0 is the radius of a sphere containing one electron and a_H is the radius of the first Bohr orbit (0.0529 nm) of a H atom. Calculate r_{WS} for Al.

(iii) *Variation in electron density by alloying*: Crystalline silver, Ag, has an *fcc* crystal structure with lattice parameter $a = 0.409$ nm. For all practical purposes, Ag and Al can be considered as being the same size. Assuming Ag to have one conduction electron per atom, calculate the variation in free electron density, N_e/V, with composition x in an $Al_{1-x}Ag_x$ solid solution. Calculate the variation in the Fermi energy, ϵ_F, with composition in the same alloy.

(iv) *Kasper polyhedra (icosahedron)*: The vertex coordinates of a regular icosahedron with three orthogonal two-fold axes along the Cartesian coordinate axes are cyclic permutations of $[\pm 1, \pm\tau, 0]$. Use this to calculate:

(a) The center-to-vertex, center-to-edge, edge-to-edge, and center-to-face distances for the icosahedron.
(b) Comment on the regularity of the tetrahedra in the icosahedron.

(v) *Kasper polyhedra (CN14)*: The CN14 Kasper polyhedra is a hexagonal antiprism with two large atoms sitting in "nests" provided by the hexagons.

(a) Determine the center to small atom vertex distance for this polyhedron in terms of the hexagonal edge length.
(b) Show that the triangular faces formed between a hexagonal edge and the two edges connecting the vertices of the first edge with the large atom vertex are isoceles.

(vi) *Kasper polyhedra (CN14, CN15, CN16)*: With the exception of the icosahedron, all the other Kasper polyhedra have one or more larger atoms sitting in a planar hexagonal nest. Determine an expression for the position of the large atom along the z-axis normal to the nest plane and the angle of the isoceles triangle as a function of the size of the large atom.

(vii) *A15 phase: atom positions*: Assume that the origin of the A_3B A15 crystal structure is at the **m3** center of symmetry. List the position of all B and A atoms in the unit cell. How many atoms in total are in the unit cell?

(viii) *A15 phase: structure factor*: Show that (hhl) reflections will be extinct if l is odd (and cyclic permutations) for the A_3B A15 crystal structure.

(ix) *A15 phase: neighbor distances*: Show that the B atoms in the A15 structure are coordinated by 12 A atoms at a distance $\sqrt{5}a/4$ ($= r_A + r_B$), where a is the cubic lattice constant. Show that the CN14 Kasper polyhedron has $2A$ atoms at a distance $a/2$; $4B$ atoms at a distance $a\sqrt{5}/4$; and $8A$ atoms at a distance $a\sqrt{6}/4$.

(x) *A15 phase: packing*: Considerations of touching spheres in the two coordination polyhedra in the A15 structure show that in the ideal structure, $a\sqrt{5}/4 = r_A + r_B$ and $a\sqrt{6}/4 = 2r_A$. Calculate an ideal touching sphere, r_A/r_B, radius ratio for this structure assuming B

atoms to be larger than A atoms. What is the ideal atomic packing fraction for this structure?

(xi) Nb_3Sn *density*: The lattice constant for Nb_3Sn is 0.529 nm (Müller, 1977), calculate its density. Calculate the density of pure Nb and pure Sn, based on their crystal structures and lattice constants. How does this compare with the atom weighted densities of Nb and Sn?

(xii) *Projection of the* Al_2Zr_3 *structure I*: Refer to the discussion in Wilson and Spooner (1960) of the Al_2Zr_3 structure. Construct a model of the atoms in the tetragonal Al_2Zr_3 unit cell. Identify the tiling of Zr atoms in the $z = \frac{1}{2}$ plane.

(xiii) Al_2Zr_3 *structure II*: Describe the connectivity of the tetrahedersterns in the Al_2Zr_3 structure. Determine the composition of each of the (00.1) planes stacked in the Al_2Zr_3 structure.

(xiv) BFe_2 *structure*: Show that the square antiprisms surrounding B in the BFe_2 structure share faces. Describe (001) plane stacking in this structure.

(xv) Cu_2Mg *Laves phase I*: Cu_2Mg has lattice constant $a = 0.7034$ nm. Consider atoms in the (110) plane of this structure to show that the distance between B atoms in the structure is $a\sqrt{2}/4 = 0.249$ nm and between A atoms is $a\sqrt{3}/4 = 0.305$ nm, while the $A-B$ distance is given by $a\sqrt{11}/8 = 0.292$ nm. What do you conclude about the A and B radii in this structure? How do these compare with the metallic radii of Mg and Cu in their α-phase crystal structure for the pure single component materials?

(xvi) Cu_2Mg *Laves phase II*: Show that starting with a close-packed B lattice (e.g., *fcc* for Cu_2Mg) and substituting A atoms for pairs of B atoms for an ideal AB_2 Laves phase, the radius ratios are such that $V_A/V_B = 2$. Calculate the packing fraction in the AB_2 Laves phase.

(xvii) Cu_2Mg *Laves phase III*: Using the lattice constant for Cu_2Mg, calculate the density of this phase. Compare the density with that of Cu and Mg.

(xviii) $MgNi_2$ *Laves phase IV*: Show that the $MgNi_2$ Laves phase structure's stacking sequence can be described as $A^+B^-A^-C^+$ in Frank's notation.

(xix) $MgNi_2$ *Laves phase V*: Determine the Pearson notation for the stacking sequence of the $MgNi_2$ Laves phase.

(xx) *Laves phase-stacking sequence I*: Given that puckered planes correspond to permutations of (ABC) stacking, show that + and − Frank symbols imply cyclic and anti-cyclic permutations of the (ABC) symbol.

(xxi) *Laves phase-stacking sequence II*: Determine the stoichiometry of the Kagome and puckered planes in the Laves phase structures.

(xxii) *Laves phase-stacking sequence III*: Show that for Kagome nets stacked on puckered layers, α must follow an A layer; β, a B layer; and γ, a C layer.

(xxiii) μ*-phase-stacking*: Show that the stacking sequence for the μ-phase, as described in the Pearson notation, can be decomposed into regions that are like that of the Laves and Al_3Zr_4 phases, respectively.

(xxiv) *Rhombic triacontahedron I*: Consider a rhombic triacontahedron with vertices a unit length from the origin and two-fold axes on the Cartesian coordinate axes. Show that the vertex coordinates are $(\pm 1, \pm 1, \pm 1)/\sqrt{3}$, $(0, \pm 1/\tau, \pm \tau)/\sqrt{3}$ and all cyclic permutations. What is the edge length of the polyhedron? Show that eight of the vertices represent an inscribed cube.

(xxv) *Rhombic triacontahedron II*: Consider the rhombic triacontahedron.

(a) Show that the short diagonals of the faces of the rhombic triacontahedron define the edges of a pentagonal dodecahedron.
(b) Show that the long diagonals define the edges of an icosahedron.
(c) Show that the ratio of the dimension of the long diagonal to that of the short diagonal is the golden mean, τ.

(xxvi) *Icosidodecahedron I*: Consider an icosidodecahedron with all vertices a unit length from the origin and with two-fold axes on the Cartesian coordinate axes.

(a) Show that vertex coordinates are $(\pm 1,0,0)$ and $(\pm \tau, \pm 1/\tau, \pm 1)/2$ and all cyclic permutations. What is the length of an edge of this polyhedron?
(b) Show that six of the vertices represent an inscribed octahedron.

(xxvii) *Icosidodecahedron II*: Show that decorating the 30 faces of the rhombic triacontahedron yields a icosidodecahedron, establishing the two as duals. Show that the icosidodecahedron has 20 equilateral triangles and 12 regular pentagonal faces. What is the Schläfli symbol for the icosidodecahedron?

(xxviii) *Quasicrystal approximants*: Explain why you might expect quaiscrystal approximants to be line compounds.

(xxix) *Mackay icosahedron*: Compare the Mackay icosahedron outer shell with the $P = 1$, $f = 2$ deltohedron. How is it different?

(xxx) *Bergman polyhedron*: Compare the Bergman polyhedron outer shell with the $P = 3$, $f = 1$ deltohedron. How is it different?

CHAPTER

19 Metallic structures III: rare earth–transition metal systems

> *"MAGNET, n. Something acted upon by magnetism.*
> *MAGNETISM, n. Something acting upon a magnet.*
> *The two definitions immediately foregoing are condensed from the works of one thousand eminent scientists, who have illuminated the subject with a great white light, to the inexpressible advancement of human knowledge."*
>
> Ambrose Bierce 1842–1914, *The Devil's Dictionary*

19.1 Introduction

Rare earth–transition metal (RT) alloy systems are important for many technological applications. Among these are permanent magnets and hydrogen storage materials. RT alloy systems have magnetic properties that depend on the T to R ratio. For example, when the ratio is 2, alloys with the cubic Laves phase structure show interesting *magnetostrictive properties* and a strong dependence of the magnetic properties on *hydrogenation*. In the case of alloys with ratios above 2, a large uniaxial magnetocrystalline anisotropy can be developed that is important for permanent magnet materials. A major development in the evolution of high energy permanent magnets has been the discovery of rare earth permanent magnet (REPM) materials. These materials are notable for their large *magnetocrystalline anisotropy*, which, in turn, lies at the origin of their large magnetic coercivities. The *coercivity* is the field required to reduce the macroscopic permanent magnetic dipole moment, or *remnant magnetization*, to zero upon application of a field in the opposite direction and is, thus, a figure of merit for a permanent magnet.

Table 18.2 on page 518 shows that R metals such as Dy, Nd, Pr, Sm, and Tb, have metallic radii in the range of 0.175–0.185 nm. This is about 30% larger than early transition metals and 50 to 60% larger than late transition metals. Rare earth–transition metal systems display a wide variety of interesting structures, often occurring at very precise stoichiometries. Examples include, RT_2, RT_3, R_6T_{23}, RT_5, R_2T_{17}, R_3T_{29}, RT_{12}, and several others. Their phase diagrams are typically very rich, with many line compounds or compounds with limited solubility ranges. Figure 19.1 illustrates a schematic RT phase diagram (one approximating that of the Sm–Co permanent magnet system). Note the line compounds (or near line compounds) at the stoichiometries: R_2T_{17}, RT_5, R_2T_7, RT_3, RT_2, R_9T_4, and R_3T. While the R-rich end of RT phase diagrams has line compounds with interesting structures and magnetic properties, in this chapter we will focus our attention on the technologically important T-rich compounds.

Structural relationships with the *Frank–Kasper phases* will also be pointed out. In many RT systems of interest, AB_2 *Laves phases* are observed, with the large rare earth atom as the A element and the small transition metal atom as the B element. This is despite the fact that the radius ratio is, in fact, much larger than the 1.26 originally calculated by Laves. In metallic alloys, these radius rules should be viewed cautiously. It may be more productive to study the *major ligand lines* and the packing of the B atoms to understand these structures.

Mixtures of rare earth and transition metal species also result in large *saturation* and *remnant* magnetizations, which reflect the size of the magnetization in a saturating field (one that aligns all of the atomic magnetic

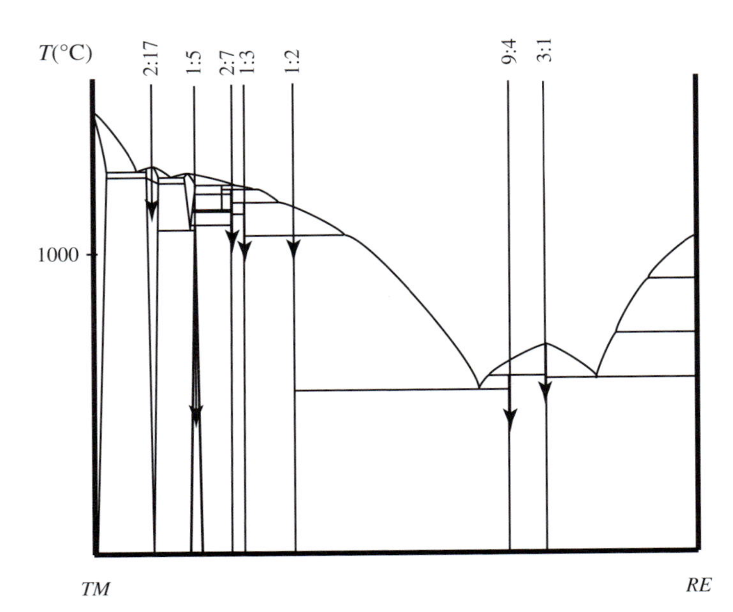

Fig. 19.1. Schematic RT phase diagram (approximating the Sm–Co system).

dipole moments) and after returning to zero field after saturation, respectively. These are figures of merit for a permanent magnet material. The remnant magnetization defines the strength of the magnet and the coercivity, how hard it is to remove the permanent magnetization. The potential magnitude of the coercivity depends intimately on the crystal structure, since it reflects how hard it is to rotate magnetic dipole moments out of crystallographically preferred "easy" directions. Coercivity is an extrinsic property and, therefore, its observed magnitude depends on the microstructure of the magnet.

The *magnetocrystalline anisotropy* is the most important material parameter influencing the properties of a magnet and it cannot be understood without understanding the underlying crystal structure. In RT systems, the rare earth atoms typically contribute greatly to the magnetic anisotropy, while the transition metal atoms contribute to the dipole moment. The major skeleton in these magnetic structures determines the topology of the rare earth connectivity and is, therefore, important to understand in any analysis of the magnetic anisotropy. In this chapter, we summarize some of the important crystal structures in RT systems, emphasizing permanent magnet materials.

19.2 *RT* Laves phases

Many RT systems have an AB_2 Laves phase structure that is often stable only over a very narrow range of compositions. We will consider some compounds for which the B atoms are Fe or Co, which are transition metals with large magnetic dipole moments. For example, the Fe compounds $CeFe_2$, $PrFe_2$, $PmFe_2$, $SmFe_2$, $GdFe_2$, $DyFe_2$, $HoFe_2$, $ErFe_2$, $TmFe_2$, and $LuFe_2$, and the cobalt compounds $CeCo_2$, $PrCo_2$, $NdCo_2$, $GdCo_2$, $TbCo_2$, $DyCo_2$, $HoCo_2$, $ErCo_2$, $TmCo_2$, $YbCo_2$, and $LuCo_2$ have the cubic $MgCu_2$-type Laves phase structure. Others, like $PmMn_2$ and $SmCo_2$, adopt the hexagonal $MgZn_2$-type Laves phase structure.

An important example of a magnetic Laves phase is found in the ternary $Tb_{1-x}Dy_xFe_2$ system, studied for its large *magnetostriction coefficient*, λ. *Magnetostriction* refers to the change in the shape of a ferromagnetic material in response to an applied field; it is the result of magnetoelastic interactions in that it involves the coupling of the magnetic anisotropy and the elastic properties of the material. The shape change can be expressed as a strain.[1] The average strain, $\Delta l/l$, for a typical transition metal ferromagnet is in the order of 10^{-5}–10^{-6}. However, for these Laves phase compounds, λ can be quite large. For example, $\Delta l/l$ is about 10^{-3} for $Tb_{1-x}Dy_xFe_2$. This compound is commonly known as *Terfenol-D*, which refers to its components *ter*bium, iron

[1] Strain is defined as the change in length divided by the original length and is, hence, a dimensionless quantity.

(*fe*), and *Dy*sprosium; the suffix *nol* is short for Naval Ordinance Laboratory, where the alloy was first developed (Clark *et al.*, 1982).

Magnetostriction can be a deleterious property for certain applications (e.g., it is the source of energy loss in electrical transformers which is accompanied by the typical "transformer hum," vibrations caused by cyclic straining); as a consequence, efforts to reduce magnetostriction coefficients are often the focus of alloy studies in soft magnetic materials. On the other hand, large magnetostriction coefficients, like those in Terfenol-D, are critical to the operation of other devices; the generation of magnetostatic waves and consequent magnetoacoustic emission is of paramount importance in sonar applications, for example.

19.3 Cubic UNi_5, Th_6Mn_{23}, and $LaCo_{13}$ phases

In this section, we describe three cubic *RT* compounds. UNi_5 and $LaCo_{13}$ have interesting structural relationships with the $MgCu_2$ Laves phase; Th_6Mn_{23} has an interesting cubic structure with 116 atoms in each unit cell. As cubic phases, these structures do not have the anisotropy required of good permanent magnets, but they are illustrated here for completeness as examples of interesting structure types. Table 19.1 summarizes the structural information for the cubic *RT* phases illustrated below.

19.3.1 The UNi_5 phase

As we have seen, there are many RT_2 Laves phase structures and it is, therefore, desirable to understand the relationship between AB_2 and AB_5 crystal structures, of which there are fewer. An AB_5 stoichiometry can be obtained starting from an AB_2 compound by substitution of *B* atoms for *half* of the *A* atoms in the parent phase. The prototype cubic AB_5 phase is $AuBe_5$ (**Structure 43**). As illustrated in Fig. 19.2, the *A* atoms residing at the diamond tetrahedral sites in the Laves phase structure are replaced by *B* atoms to yield the UNi_5-type structure. Figure 19.2(a) and (b) compare unit cells of the $MgCu_2$ and UNi_5-type structures. Figure 19.2(c) and (d) compare projections of all atoms from a cell into the (100) planes of both structures. The projected structures clearly illustrate the structural differences between the two phases.

It is an instructive exercise to examine the stacking of (111) planes in the UNi_5 structure. The stacking sequence for the Cu_2Mg-type (**C15**) Laves phase was shown to be $A)\alpha(ACB)\beta(BAC)\gamma(CB$ in the Pearson notation. For UNi_5 it is the same but for the substitutions of *B* for *A* atoms, which can be represented by the "reaction":

$$2RT_2 - R + T = RT_5. \quad (19.1)$$

Table 19.1. Structural information for the cubic *RT* phases discussed in this chapter.

Compound	Space group	Z	a (nm)	Structures appendix
UNi_5	**F$\bar{4}$3m** (T_d^2)	4	0.6783	**Structure 43**
Sm_6Mn_{23}	**Fm$\bar{3}$m** (O_h^5)	4	1.2558	**Structure 44**
$LaCo_{13}$	**Fm$\bar{3}$c** (O_h^6)	4	1.14	**Structure 45**

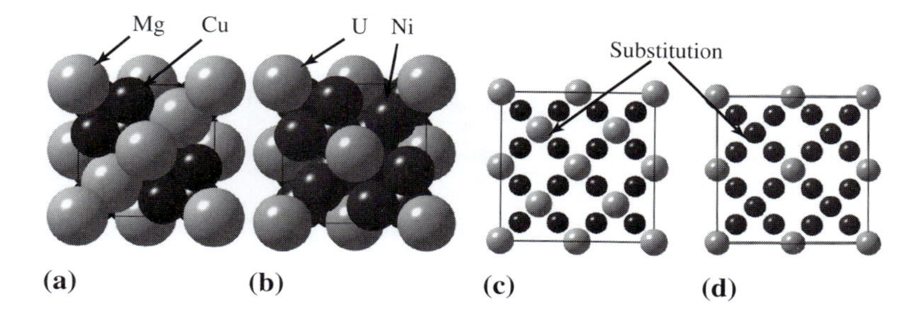

Fig. 19.2. Space filling depictions of the atoms in the cubic unit cell for the (a) $MgCu_2$ Laves phase and (b) the UNi_5-type compound. Comparison of the projections of all atoms into the (100) planes for the $MgCu_2$ (c) and UNi_5-type structures (d) are also shown.

Figure 19.2 shows that this substitution occurs on diamond sites of the cubic R sublattice.

19.3.2 The Th_6Mn_{23} phase

The Th_6Mn_{23} phase is another cubic prototype and as such does not hold promise as a permanent magnet system. It does have an interesting crystal structure (**Structure 44**) and is found in X_6Fe_{23} with X = Yb, Tm, Er, Ho, Gd, Tb, and Dy; and X_6Mn_{23} with X = Pr, Nd, Pm, Sm, Gd, Tb, Dy, Ho, Er, and Tm. The phase has a cubic **Fm$\bar{3}$m** (O_h^5) space group with four formula units in the unit cell. The site occupations are listed in the on-line structures appendix.

The structural decomposition of Th_6Mn_{23} is illustrated in Fig. 19.3. From Table 19.1 we know that there are four different special positions for Mn; we will represent them as Mn-1, Mn-2, Mn-3, and Mn-4, respectively. Figure 19.3(a) shows (001) layers of Th octahedra. Each layer consists of large Th octahedra, centered by a Mn-1 atom, and smaller uncentered Th octahedra. The large and small octahedra share vertices along [001] directions. One can think of the octahedron centers as being similar to an NaCl structure, with the smaller octahedra occupying the face centers and the unit cell corners, and the larger ones occupying the edge centers. The lattice constant is twice the sum of the center-to-vertex distances of each of the octahedra. Figure 19.3(b) shows that Mn-1 atoms are coordinated by Mn-3 atom cubes that also decorate the

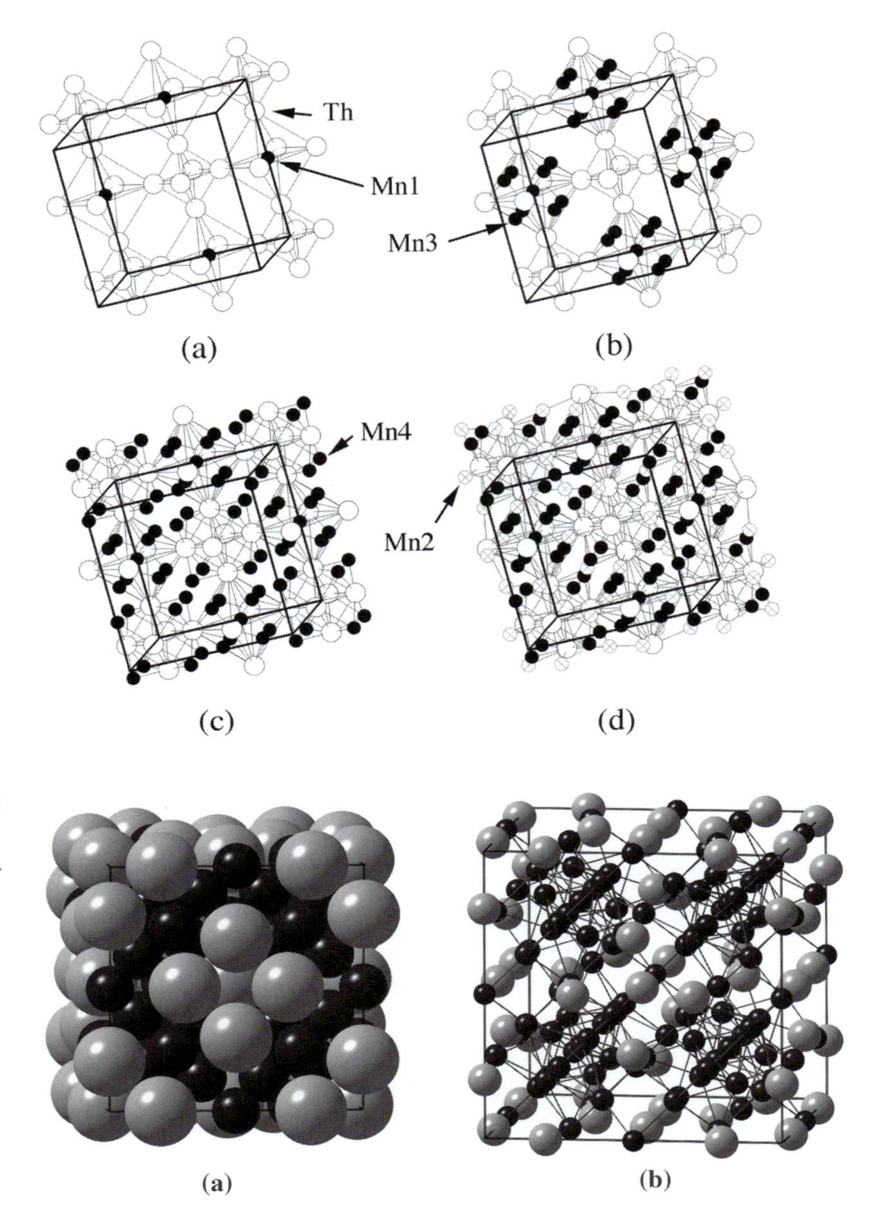

Fig. 19.3. Decomposition of the Th_6Mn_{23} structure illustrating the connectivity of Th octahedra (a) and successive decorations with inequivalent Mn atoms in (b), (c), and (d).

Fig. 19.4. Space filling (a) and (b) ball-and-stick depiction of the cubic unit cell of Th_6Mn_{23}.

eight faces of the larger Th octahedra. The Mn-1 to Mn-3 distance is similar to that observed for *bcc* late transition metals. Figure 19.3(c) shows that the empty centers of the smaller Th octahedra are coordinated by Mn-4 atom cubes which also decorate the faces of the smaller Th octahedra. Finally, Fig. 19.3(d) shows Mn-2 atoms to form cuboctahedrons centered by the same voids in the smaller octahedra. Figure 19.4 shows the complete unit cell in the space filling and ball-and-stick representations.

The Th_6Mn_{23} structure is observed for many early transition metal – late transition metal systems. There is also a B_6Cr_{23} structural prototype that represents a structure commonly adopted by borides and carbides. This structure possesses the same cubic **Fm$\bar{3}$m** (O_h^5) space group with four formula units in the cubic cell, but with different site occupations. The decomposition of the B_6Cr_{23} structure is discussed in detail in Chapter 21.

19.3.3 The $LaCo_{13}$ phase

The $LaCo_{13}$ is a cubic phase rich in Co. It has space group **Fm$\bar{3}$c** (O_h^6), and the special positions are shown as **Structure 45** in the on-line structures appendix. The prototype structure is the $NaZn_{13}$ structure. This structure type is not common among *RT* systems. Figure 19.5(a) and (b) show space filling and ball-and-stick depictions of the $LaCo_{13}$ structure, which can be decomposed into a $2 \times 2 \times 2$ *bcc* supercell, where the body center is occupied by a Co-centered Co_{13} icosahedron and the cube vertex sites are occupied by La (Fig. 19.5(c)). Substitutions into the RT_2 Laves phase structure can be represented as:

$$RT_2 - T + T_{12} = RT_{13}, \tag{19.2}$$

where a 12-atom icosahedron is centered by a *T*-atom from the parent structure.

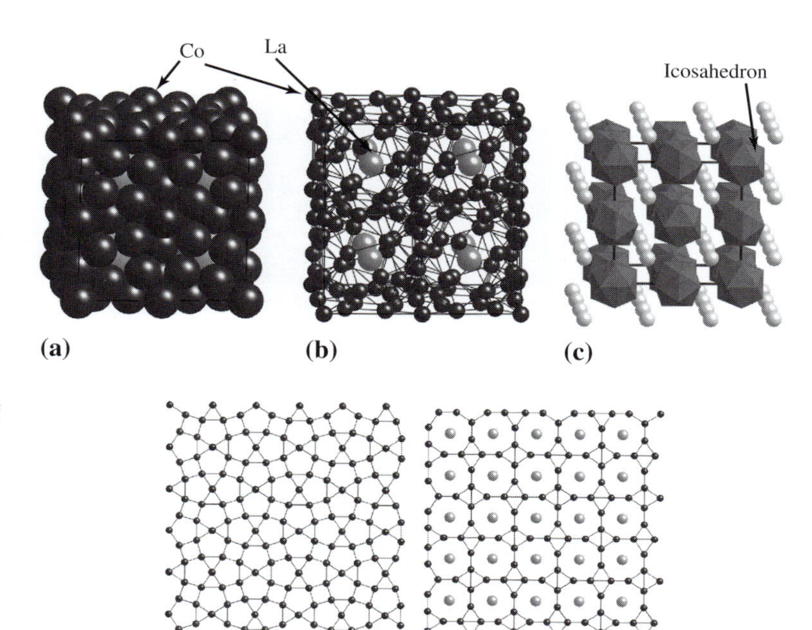

Fig. 19.5. The cubic unit cell of the $LaCo_{13}$ phase in (a) space filling and (b) ball-and-stick depictions; (c) decomposition into a $2 \times 2 \times 2$ *bcc* supercell, (d) the tiling of Co atoms in the $z = 0$ plane and (e) the additional two Co layers and La layer between $z = 0.01$ and $z = 0.25$.

Figure 19.5(d) and (e) illustrate interesting tilings in the $LaCo_{13}$ phase structure. The first shows Co(1) and Co(2) atoms in the $z = 0$ plane. This tile is decorated with nearly regular pentagons, squares, and triangles. The second shows the projection of all additional atoms in the first quarter of the cell along the c-axis. Puckered layers of Co(2) atoms form pseudo-octagonal "nests" in which the larger La atoms reside.

An example of the $NaZn_{13}$ structure occurs in the self-assembled nanostructures illustrated in Box 19.1. This lattice is decorated with nanoparticles of two different sizes. Crystallization of large PbSe and small Pd nanoparticles into $NaZn_{13}$ and $MgZn_2$ (Chapter 18) superstructures is achieved by evaporating (under a reduced pressure of about 3.2 kPa at 45–50 °C) dispersions in mixtures of toluene and tetrachloroethylene containing a ~ 20-fold and ~ 5-fold, respectively, excess of Pd particles.

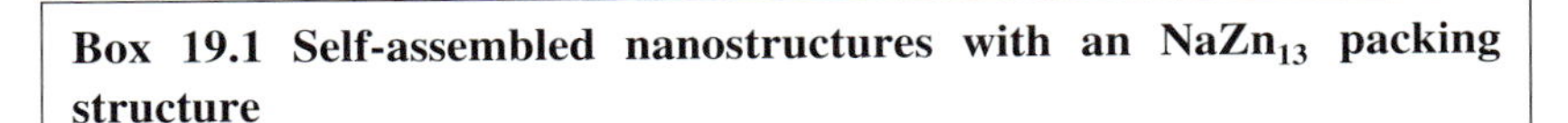

Box 19.1 Self-assembled nanostructures with an $NaZn_{13}$ packing structure

Another beautiful example of a self-assembled structure is illustrated below for a system of 5.8 nm PbSe and 3.0 nm Pd nanoparticles.

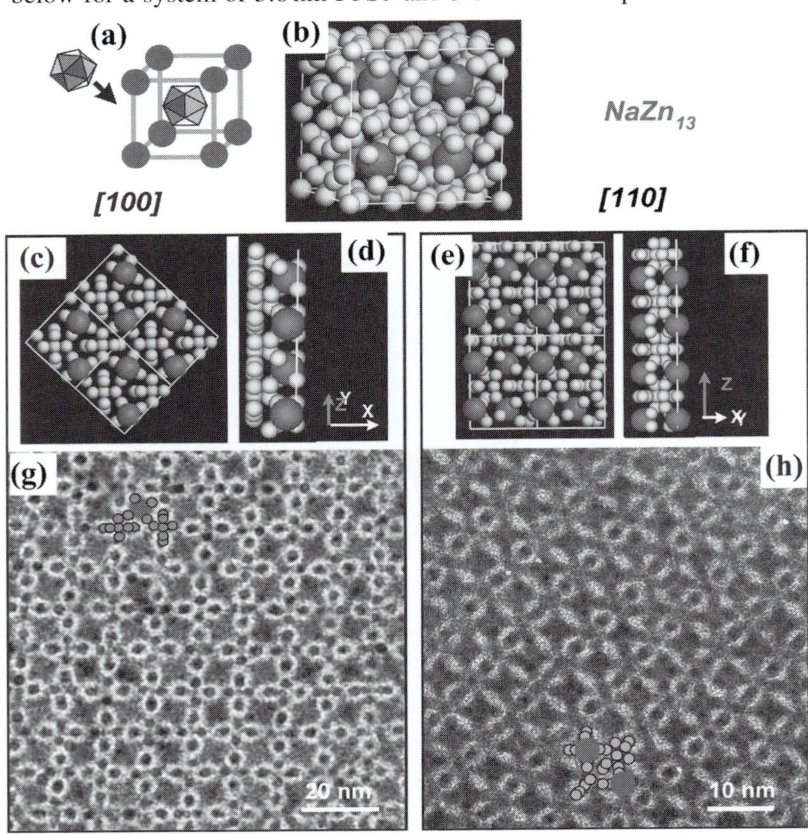

The figure (courtesy of E. Shevshenko and Chris Murray, IBM) shows (a) a subcell of $NaZn_{13}$: the 12 small T spheres are arranged around the central T atom at the vertices of a slightly distorted icosahedron; (b) the unit cell of $NaZn_{13}$ consisting of 112 atoms (8 subcells). (c) depicts the [100] projection and (e) the [110] projection of the unit cell; (d) and (f) represent the repeat units of these projections. (g) and (h) are experimental high resolution TEM micrographs of the two projections, indicating that the nano-particles do arrange themselves in this particular structure (Shevchenko *et al.*, 2005).

19.4 *Non-cubic phases

Table 19.2 provides a summary of the structural information for several non-cubic RT phases that are introduced in the following sections. Additional structural details can be found in the on-line structures appendix.

19.4.1 $SmCo_3$ and $SmCo_5$ phases

In this section, we describe an example of a rhombohedral and a hexagonal RT compound. The first, $SmCo_3$, is too rich in R to be a useful permanent magnet but again has an interesting structure. The second, $SmCo_5$, is a premiere permanent magnet that is structurally related to the permanent magnets discussed in later sections.

19.4.1.1 The $SmCo_3$ phase

The rhombohedral $SmCo_3$ has space group **R$\bar{3}$m** (D^5_{3d}) (**Structure 46**), with Sm in the special positions $3a$ and $6c$ ($z = 0.140$), and Co in the special positions $3b$, $6c$ ($z = 0.334$), and $18h$ ($z = 0.504, y = 0.496, z = 0.082$) with lattice constants $a = 0.50584$ nm and $c = 2.4618$ nm, respectively. The prototype for this structure is the intermetallic compound Be_3Nb. The unit cell contains nine formula units for a total of 36 atoms (Fig. 19.6).

Figure 19.6(a) and (b) illustrates the structure (three unit cells in a hexagonal prismatic representation) in space filling and ball-and-stick depictions. Notice the long c-axis for this structure. Figure 19.6(c) shows the projection of the three cells into a (110) plane. Figure 19.6(d), (e), and (f) show projections of the structure the (001) plane. Figure 19.6(d) shows all atoms within one c-axis lattice constant (from the three rhombohedral cells) projected along the [001] direction. Note the similarity between this projection and the one for the hexagonal $MgNi_2$ Laves phase structure. The $MgNi_2$ Laves phase has a hexagonal space group whereas $SmCo_3$ has a rhombohedral space group. This similarity results from the three-fold roto-inversion operation that is a symmetry operation of the **R$\bar{3}$m** (D^5_{3d}) space group. To see this, we project

Table 19.2. Summary of structural information for several non-cubic *RT* phases.

Compound	Space group	Z	Lattice parameter (nm)	Structures appendix
$SmCo_3$	**R$\bar{3}$m** (D_{3d}^5)	9	$a = 0.50584$ $c = 2.4618$	**Structure 46**
$SmCo_5$	**P6/mmm** (D_{6h}^1)	1	$a = 0.50002$ $c = 0.3964$	**Structure 47**
β-Sm_2Co_{17}	**P6$_3$/mmc** (D_{6h}^4)	2	$a = 0.8384$ $c = 0.8159$	**Structure 48**
α-Sm_2Co_{17}	**R$\bar{3}$m** (D_{3d}^5)	3	$a = 0.8420$ $c = 1.2210$	**Structure 49**
$SmZn_{12}$	**I4/mmm** (D_{4h}^{17})	2	$a = 0.8927$ $c = 0.5215$	**Structure 50**
$Nd_2Fe_{14}B$	**P4$_2$/mnm** (D_{4h}^{14})	4	$a = 0.8804$ $c = 1.2205$	**Structure 51**
Nd_3Fe_{29}	**A2/m** (C_{2h}^3)	2	$a = 1.06382$ $b = 0.85892$ $c = 0.97456$ $\beta = 96.93°$	**Structure 52**
α-$Sm_2Fe_{17}N_3$	**R$\bar{3}$m** (D_{3d}^5)	3	$a = 0.843$ $c = 1.222$	**Structure 53**
$Sm(Fe,Ti)_{12}N$	**I4/mmm** (D_{4h}^{17})	2	$a = 0.893$ $c = 0.522$	**Structure 54**

only the top and bottom halves of the three unit cells along the [001] direction in Figs. 19.6(e) and (f).

19.4.1.2 The $SmCo_5$ phase

The hexagonal material $SmCo_5$ (Fig. 19.7) is the REPM with the largest observed value for magnetocrystalline anisotropy. It was first reported in 1967, by Strnat *et al.* (1967) after non-magnetic RT_5 compounds were first synthesized by the group of Wallace (1960). The phase has 17 at% *R*, so in many applications it was thought that other REPM materials, notably Sm_2Co_{17}, with larger *T* fractions and consequently larger inductions, offer greater promise. However, the Sm_2Co_{17} phases do not achieve levels of magnetocrystalline anisotropy comparable to $SmCo_5$.

In state of the art Sm–Co permanent magnets, a two-phase microstructure combines the large magnetic anisotropy of the $SmCo_5$ phase with the high magnetic moment of the Sm_2Co_{17} phase. Alloys of composition near $SmCo_{7.7}$ have achieved the largest magnetic energy products (highest stored magnetic energy) in this system. In these non-stoichiometric two-phase materials, some Co may be substituted by Fe, Cu, and/or Zr (Buschow, 1988).

The $SmCo_5$ phase is a structural template for many important permanent magnet phases. The hexagonal **P6/mmm** (D_{6h}^1) structure of $CaCu_5$ is the prototype for $SmCo_5$ (Wernick and Geller, 1959). Figure 19.7 shows a single

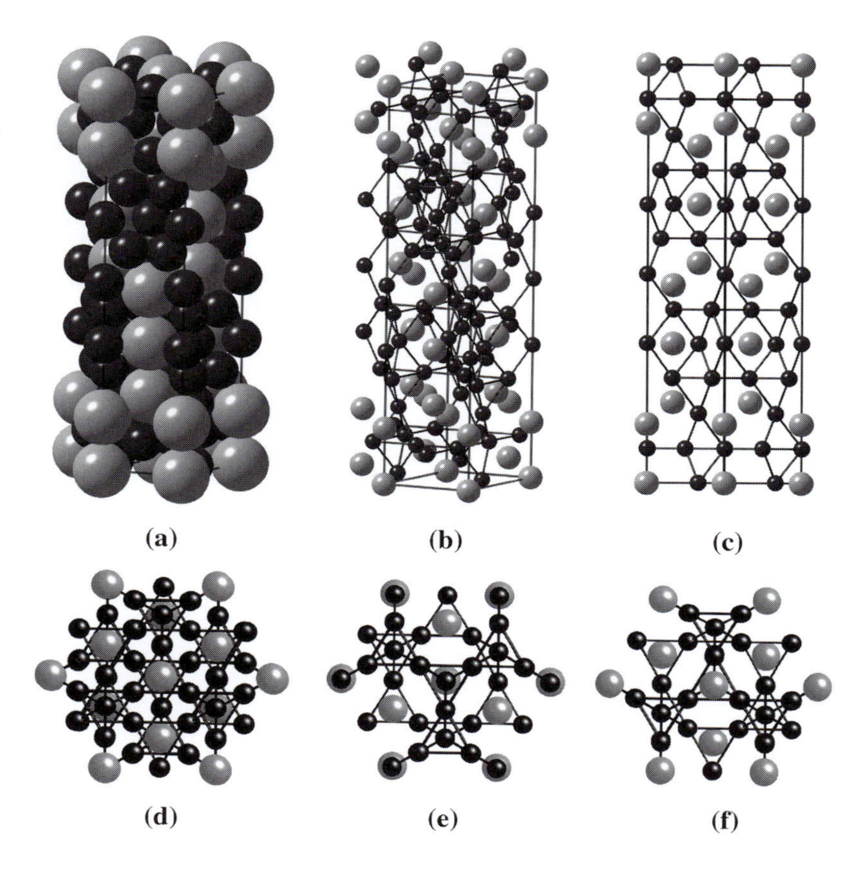

Fig. 19.6. 3 rhombohedral cells in a hexagonal prismatic representation for $SmCo_3$ (a) ball-and-stick and (b) space filling models; (c) projection of the three cells into a (110) plane. Projections into a (001) plane (d) entire cell; (e) and (f) top and bottom halves of cells.

formula unit of $SmCo_5$ in the hexagonal unit cell. In each cell, there is one Ca atom (eight Ca atoms on the cell corners, each shared with eight other cells). The single Cu atom joining the two Cu double tetrahedra belongs to that cell only, but the other eight Cu atoms are shared with other cells. Thus, there are $1 + 8 \times (\frac{1}{2}) = 5$ Cu atoms per unit cell.

Figure 19.7(a) and (b) show ball-and-stick and space filling depictions, respectively, of the $CaCu_5$ structure. Note that the double Cu tetrahedra (sharing faces) are reminiscent of the Laves phase structures. Planes of transition metal atoms are arranged in a Kagome network, as illustrated in Fig. 19.7(d). Stacked planes of transition metal atoms alternate with planes consisting of T atoms at interstices in an *hcp* (3^6) arrangement of R atoms (the transition metal atoms form a hexagonal (6^3) net) as shown in Fig. 19.7(c). The plane in Fig. 19.7(c) is described with the Pearson notation as $[Aa]$ and the Kagome net is α. The Pearson notation for stacking in the $CaCu_5$ structure is thus given by $[Aa]\alpha$.

Figure 19.8(a) shows a ball-and-stick model of atoms in the hexagonal unit cell for the intermetallic compound $SmCo_5$ with the Co atom tetrahedra emphasized by shading. Figure 19.8(b) shows the projection of the structure

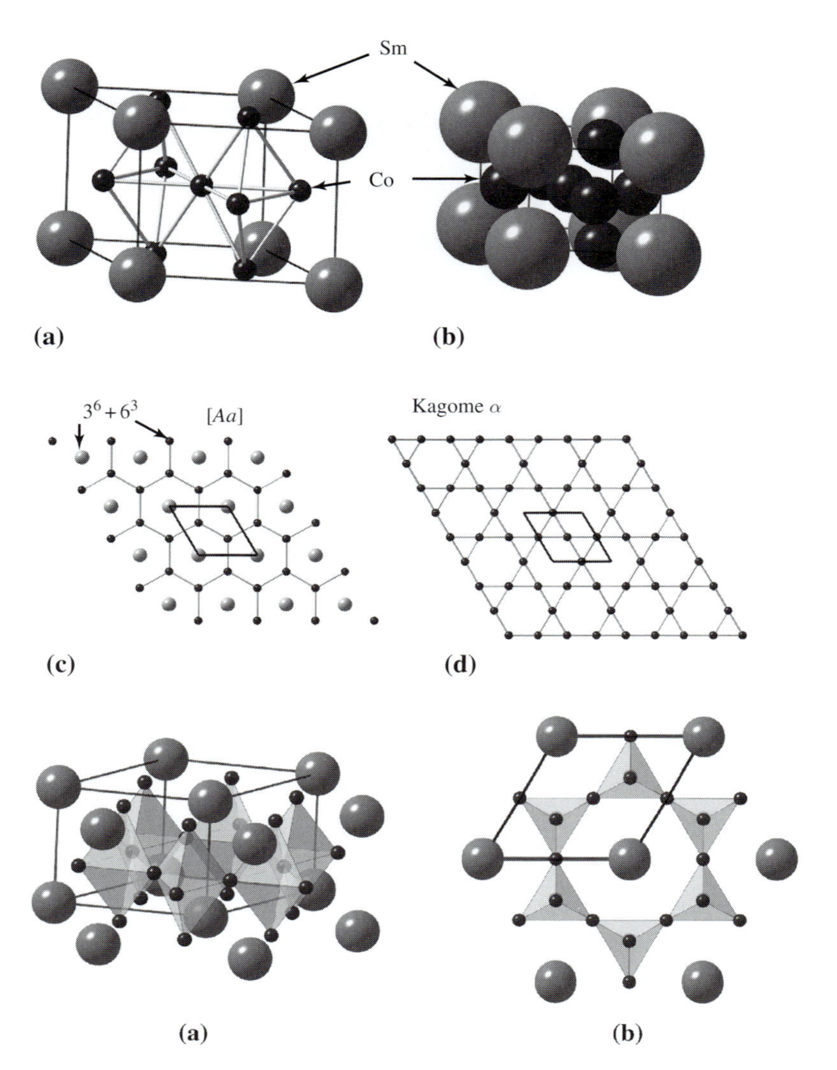

Fig. 19.7. Hexagonal unit cell of $CaCu_5$ (a) ball-and-stick and (b) space filling depictions; (c) Ca and Cu atoms in the $z = 0$ (001) plane and (d) $z = 0.5$ Cu Kagome net.

Fig. 19.8. (a) Ball-and-stick model of the hexagonal unit cell of $SmCo_5$ with the Co tetrahedra shaded; (b) projection into the (001) plane, showing connected double tetrahedral units.

into the (001) plane, emphasizing the connectivity of the double tetrahedral units. In ionic solids, tetrahedral coordination is also common for small cations, but for electrostatic reasons face sharing, as seen in the double tetrahedra, is improbable.

Note that the tetrahedra in $SmCo_5$ are empty, but this does not have to be the case. Rare earth intermetallic materials are among the most important hydrogen storage materials and H is incorporated into the tetrahedral interstices in these structures. Figure 19.8(b) shows that the double tetrahedra link by sharing vertices. In this respect, they have projected structures that are reminiscent of silicate structures, which will be discussed in Chapter 24.

The $MgZn_2$ Laves phase and the hexagonal $CaCu_5$ structure are related through a more complicated transformation than for the $MgCu_2$ Laves phase

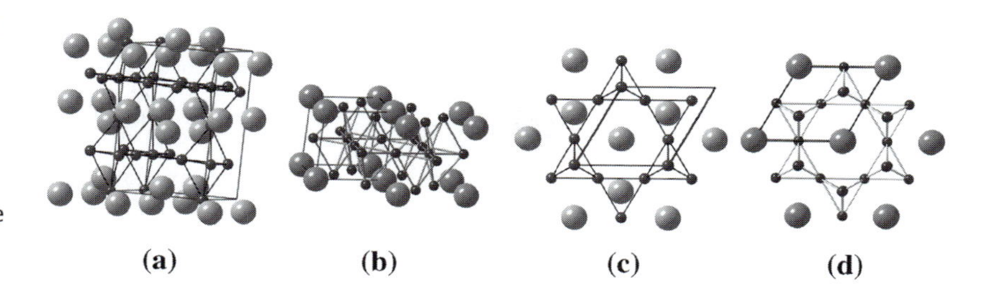

Fig. 19.9. Ball-and-stick model of three rhombohedral cells for (a) $MgZn_2$ and (b) $CaCu_5$ in a hexagonal prismatic representation; projection of the top half the three cells of $MgZn_2$ (c) and all atoms in the three cells for the $CaCu_5$ structure (d) into the (001) plane.

and the cubic UNi_5 compound. Figure 19.9(a) and (b) show the $MgZn_2$ Laves phase and the $CaCu_5$ structures in a hexagonal prismatic representation with three rhombi in the basal plane. It can be seen that the $MgZn_2$ structure has double the period of the $CaCu_5$ structure and has a B (Zn) atom at its origin as opposed to an A (Ca) atom in the former. This relationship between the two can be explained in two steps: (1) The origin of the $MgZn_2$ Laves phase is shifted from a B atom to an A atom for comparison; and (2) considering only the upper (or lower) half of this structure, substitution of B atoms for A atoms in three sites on (002) and (001) planes and removal of puckering of B-atom atomic positions in these planes yields the $CaCu_5$ structure.

Figure 19.9(c) and (d) show the projection of the top half of the three cells in the case of $MgZn_2$ and all the atoms in the three cells for the $CaCu_5$ structure into the (001) plane. Note that the projection of the $CaCu_5$ structure represents superimposing the two planes of Fig. 19.7(c) and (d). Furthermore, note that the beautiful centered Zn hexagons of Fig. 19.9(c) result from the projecton of both halves of the cells along the c-axis for this structure. This indicates that the double tetrahedra in the upper and lower halves of the $MgZn_2$ Laves phase structures are in fact oriented anti-prismatically. As a final note, in the $CaCu_5$ structure, the effective (touching sphere) size of the B element, d_B, is $\sqrt{(c^2/4 + a^2/12)}$.

It is of interest also to compare XRD patterns for the $CaCu_2$ and $CaCu_5$ phases. Figure 19.10(a) and (b) show simulated XRD patterns for a hypothetical $CaCu_2$ hexagonal Laves phase structure and the hexagonal $CaCu_5$

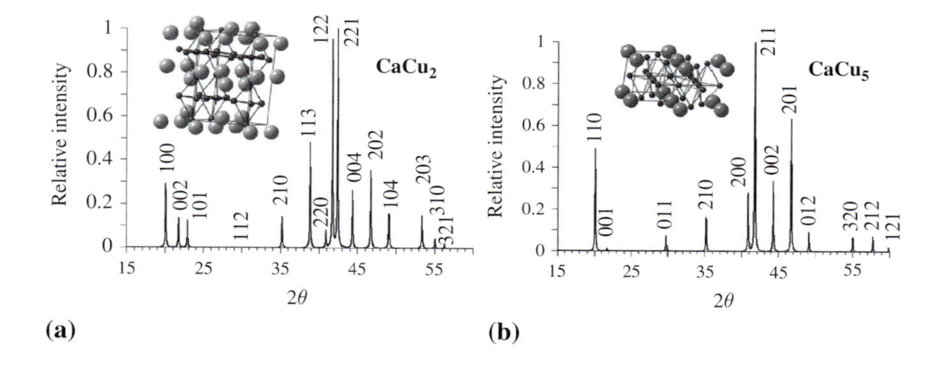

Fig. 19.10. Simulated XRD patterns assuming CuK_α radiation and the $CaCu_5$ equilibrium a lattice constant for (a) a hypothetical $CaCu_2$ hexagonal Laves phase structure and (b) the hexagonal $CaCu_5$ structure, respectively.

structure, respectively. The hypothetical $CaCu_2$ compound is chosen to have an $MgZn_2$ Laves phase structure, with an a lattice constant identical to that of the $CaCu_5$ structure. The composition and lattice constant are chosen for ease of comparison of the diffraction pattern for the hypothetical Laves phase structure and that of the $CaCu_5$ structure. It is instructive to compare the diffraction patterns that have common fundamental peaks and the superlattice reflections characteristic of its 2×1 Laves phase structure.

19.4.2 Dumbbell substitutions: α-Sm_2Co_{17} and β-Sm_2Co_{17} phases

Many of the important REPMs can be related to the $SmCo_5$ structure through so-called *dumbbell substitutions*. These substitutions replace an R atom with a pair of transition metal atoms in a dumbbell arrangement. This is another mechanism by which the stoichiometry can be changed, so as to enrich the T content at the expense of the R content. Stadelmeier (1984) proposed to use the following formula to describe various RT compounds with dumbbell transformations:

$$R_{m-n}T_{5m+2n},$$

where m and n are integers. For $m = 1$ and $n = 0$ we have the parent RT_5 structure. For other combinations, m represents the number of RT_5 formula units being considered and n the number of dumbbell substitutions within those m units. For example, a stoichiometry of R_2T_{17} is obtained for $m = 3$ and $n = 1$. Phases of this stoichiometry will be discussed further in this section. If $m = 2$ and $n = 1$, a stoichiometry of RT_{12} is obtained. For $m = 5$ and $n = 3$, a stoichiometry of R_3T_{29} is obtained. These will be discussed in later sections. The transformation by which $\frac{1}{3}$ of the R atoms in the $CaCu_5$ structure of $SmCo_5$ are replaced with pairs of transition metal atom *dumbbells*, is represented as:

$$3RT_5 - R + 2T = R_2T_{17}, \tag{19.3}$$

where a single rare earth atom (Sm) is removed from three units of the $SmCo_5$ starting structure and replaced by a transition metal (Co) dumbbell, $2T$, to yield the final Sm_2Co_{17} stoichiometry. Phases with this stoichiometry exist in both hexagonal and rhombohedral variants. Pairs of transition metal atoms are arranged along the c-axis and these are commonly referred to as "dumbbell sites" (Fig. 19.11). The substitutions require the use of three $SmCo_5$ unit cells and thus certain of the planes will be designated with a prime $'$ symbol to indicate the $3\times$ larger basal plane of the unit cell.

The Th_2Ni_{17} prototype structure (**P6$_3$/mmc** (D^4_{6h}), **Structure 48**), is shown in Fig. 19.11. It is the structure of the β-Sm_2Co_{17} phase. β-Sm_2Co_{17} has

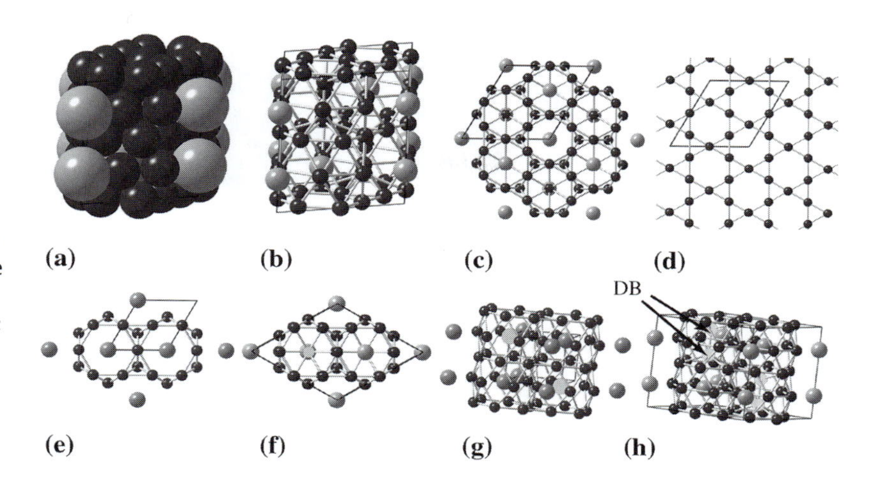

Fig. 19.11. Unit cell of the β-Sm_2Co_{17} phase, depicted in (a) space filling and (b) ball-and-stick formats. The hexagonal prismatic representation with three cells is shown projected along the [001] direction (c) with the Co Kagome net in the $z = 0$ plane (d); (e) and (f) show the cells projected into the (001) plane; (g) and (h) show $a\sqrt{3} \times a\sqrt{3} \times 2c$ $SmCo_5$ phase structure and a single unit cell of the β-Sm_2Co_{17} phase, respectively.

lattice constants $a = 0.8384$ and $c = 0.8159$ nm respectively; Sm atoms sit on the $2b$ and $2d$ special sites, both with $\bar{\mathbf{6}}\mathbf{m2}$ (D_{3h}) symmetry. Co atoms sit at the positions $4f$ ($z = 0.11$), $6g$, $12j$ ($x = \frac{1}{3}, y = 0$), and $12k$ ($x = \frac{1}{3}, z = 0$), respectively. Figure 19.11 shows the structure of β-Sm_2Co_{17}, as a single unit cell depicted in (a) space filling and (b) ball-and-stick formats. The hexagonal setting with three cells is shown projected into the (001) plane in Fig. 19.11(c) while (d) shows the Co atom Kagome net in the $z = 0$ plane.

Since the β-Sm_2Co_{17} phase contains two formula units per unit cell, we can double Equation 19.3 to determine that in six unit cells of the $SmCo_5$ phase structure, we replace two rare earth atoms with two pairs of Co dumbbells to arrive at the two formula units of the β-Sm_2Co_{17} phase in a single unit cell. To choose six convenient cells of the $SmCo_5$ phase structure, we double the c-axis length (centering on a Co plane) and consider an area $a\sqrt{3} \times a\sqrt{3}$ in the ab plane. This volume projected into the (001) plane is shown in Fig. 19.11(e). Figure 19.11(f) shows a similar projection of a single β-Sm_2Co_{17} phase unit cell with one of the dumbbell Co atoms highlighted. These dumbbell Co atoms sit in the $4f$ sites of the structure. Figure 19.11(e) and (f) show parallel views of equivalent regions of the two structures (i.e., $3\ a\sqrt{3} \times a\sqrt{3} \times 2c$) in the $SmCo_5$ phase structure and a single unit cell in the β-Sm_2Co_{17} phase. In Fig. 19.11(g) the two Sm atoms to be removed in the dumbbell transformation are highlighted. In Fig. 19.11(h) the two dumbbell pairs substituted are highlighted in the β-Sm_2Co_{17} phase structure.

If instead of following an $ABABAB\ldots$ stacking of the rare earth atoms, the stacking sequence is $ABCABCABC\ldots$ the resulting structure is rhombohedral with the Th_2Zn_{17} ($\mathbf{R\bar{3}m}$ (D^5_{3d}), **Structure 49**) prototype structure. The 2:17H (hexagonal) phase has a lattice parameter c about twice that of the 1:5H. A range of values has been reported in the crystallographic tables for the lattice parameters, averaging to $a = 0.843$ nm, and $c = 1.222$ nm for the

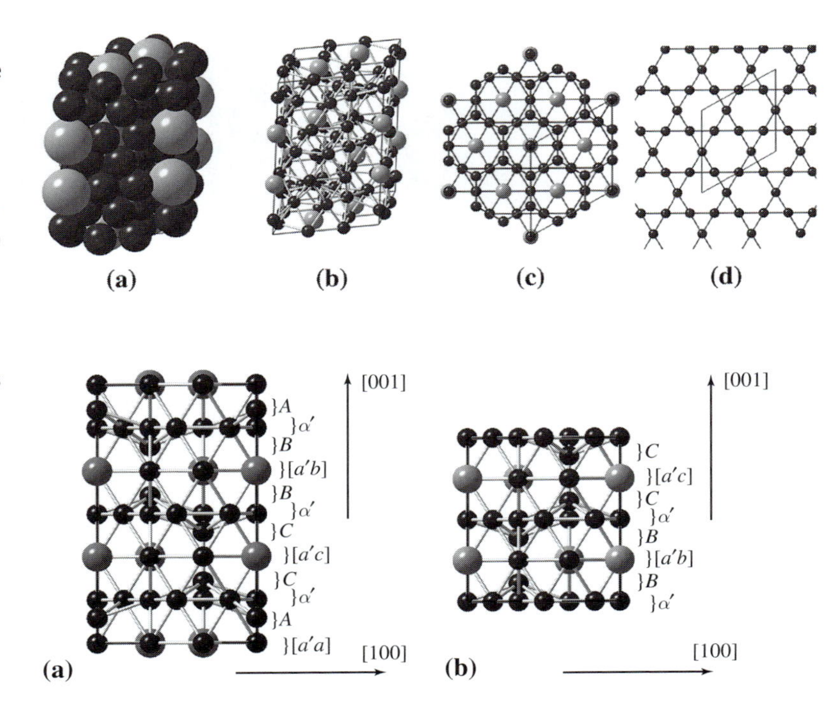

Fig. 19.12. The structure of the α-Sm_2Co_{17} phase, a single rhombohedral unit cell depicted in (a) space filling and (b) ball-and-stick formats. The hexagonal prismatic representation projected along [001] is shown in (c) while (d) shows the Co atom Kagome net in the $z = \frac{1}{6}$ plane.

Fig. 19.13. Stacking sequences in the α-Sm_2Co_{17} (a) and β-Sm_2Co_{17} (b) phases.

rhombohedral α-Sm_2Co_{17} phase. In this structure, Sm atoms reside at the $6c_2$ position and Co atoms are found on the $6c_1$ ($z = 0.097$), $9d$, $18f$ ($x = 1/3$), and $18h$ ($x = 0.5$, $z = 1/6$) positions. Twins form on the basal plane of the 2:17R (rhombohedral) phase. Figure 19.12 shows the structure of α-Sm_2Co_{17}, a single rhombohedral unit cell depicted in (a) space filling and (b) ball-and-stick formats. The hexagonal prismatic representation projected along [001] in Fig. 19.12 (c) and (d) shows the Co atom Kagome net near the $z = \frac{1}{6}$ plane, containing Co atoms in $18h$ and $9d$ special positions.

While the stacking sequence of the major R ligands is instructive, a more complete description is obtained using the Pearson notation. Figure 19.13(a) and (b) compare stacking sequences of the α-Sm_2Co_{17} and β-Sm_2Co_{17} phases for planes stacked along the [001] direction. For the α-Sm_2Co_{17} phase, the structure is idealized to ignore slight puckering of some planes. The Pearson notation for the stacking sequence of the α-Sm_2Co_{17} phase is $[a'a]A\alpha'C[a'c]C\alpha'B[a'b]B\alpha'A$. For the β-Sm_2Co_{17} phase, it is $\alpha'B[a'b]B\alpha'C[a'c]C$. Recall that prime symbols refer to the larger basal plane stacking unit.

Typical microstructures of the so-called 2-17 phase magnets are, in fact, multiphase microstructures, and each phase has an important function in developing premiere hard magnetic properties. The $SmCo_5$ phase develops on the six equivalent pyramidal planes of the 2:17R phase. A small amount of Zr added to the alloy stabilizes the 2:17 phase with respect to Fe substitution (partial substitution of Fe for Co increases the magnetization of the material,

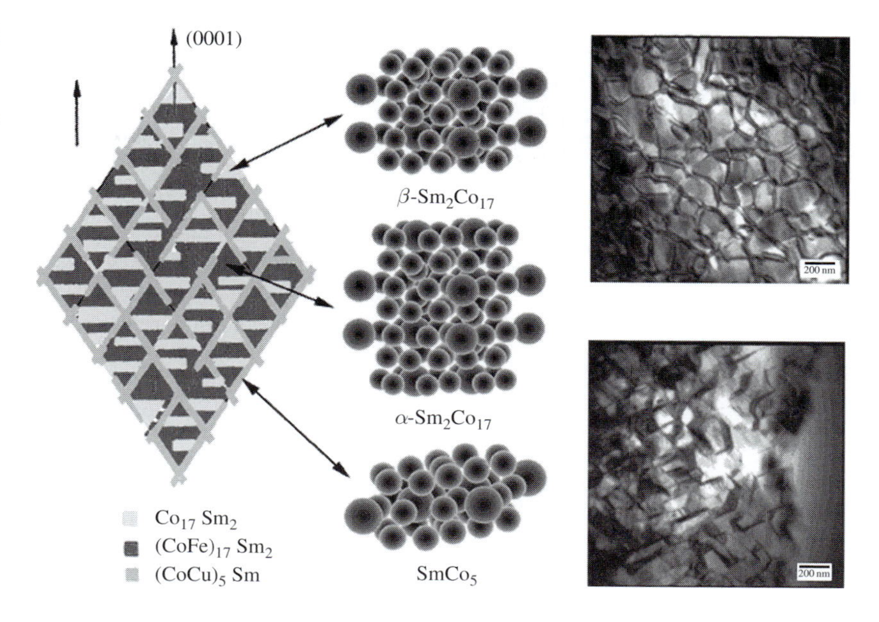

Fig. 19.14. Cartoon and typical microstructures of Sm_2Co_{17} magnetic materials with a cellular structure along with an actual TEM micrograph showing the cell structure.

but the corresponding Sm_2Fe_{17} does not possess a high Curie temperature). The mechanism for this may be the replacement of Fe–Fe pairs on the $6c$ dumbbell sites with a Zr–vacancy pair. The presence of Fe may also facilitate the formation of a *cellular structure* illustrated in Fig. 19.14. $SmCo_5$ is metastable at room temperature, but can generally be retained. Sm_2Co_{17} takes the 2:17R at room temperature, but the 2:17H phase can be retained by rapid quenching.

As for most of the structures discussed in this chapter, the specific site occupancy is critical to determining the magnetic properties of interest. It has often been the course of study to investigate a binary phase and its structure systematically with different choices for the rare earth and transition metal species. Structural variations in ternary systems are also studied. While it is beyond the scope of this text to give an exhaustive review of the conclusions of such studies, some examples will be noted. In magnetic systems in general, it is often of interest to look at mixtures of Fe and Co as the transition metal species. This is because of the fact that, in the binary Fe–Co systems, a magnetization larger than that of either pure Fe or pure Co is attainable. It is also the case that Fe is cubic whereas Co has an anisotropic *hcp* structure, and their alloys often show preferences for low or high symmetry phases, which can, of course, influence the resulting magnetocrystalline anisotropy. In phases with the Sm_2Co_{17} structure such studies have been performed. For example, Deportes *et al.* (1976) have studied the system $Y_2(Co_{1-x}Fe_x)_{17}$ and Herbst *et al.* (1982) studied preferences for T site selection in the $Nd_2(Co_{1-x}Fe_x)_{17}$ series. The interested reader is referred to the original literature for details of these studies.

19.4.3 Tetragonal phases: RT_{12} and $Nd_2Fe_{14}B$

19.4.3.1 The RT_{12} phases

If one half of the rare earth atoms in the $CaCu_5$ structure of $SmCo_5$ are replaced by pairs of transition metal atoms (*dumbbells*), the stoichiometry becomes 1:12. The transformation is represented as:

$$2RT_5 - R + 2T = RT_{12}, \qquad (19.4)$$

where a single rare earth (Sm) atom is removed from two units of the $SmCo_5$ starting structure and replaced by a transition metal (Co) dumbbell, $2T$, to yield the final RT_{12} stoichiometry. An RT_{12} phase with space group **I4/mmm** (D_{4h}^{17}) exists with structural prototype $ThMn_{12}$, where again Th is a lanthanide group element (Fig. 19.15).

In general, phases of the stoichiometry RT_{12} are not stable in binary alloys ($SmZn_{12}$ is an exception but is not strongly magnetic), but they can be stabilized by ternary alloying additions. In the systems $R(T_{12-x}M_x)$ (Coey, 1996) with R a light rare earth element and M = Ti, V, Cr, Mo, W, or Al, the 1 : 12 phase can be stabilized with x as small as 1 to 2. In the prototype $ThMn_{12}$ structure (**I4/mmm** (D_{4h}^{17}), **Structure 50**), the R (Th) atoms occupy the $2a$ sites and T (Mn) atoms are found on the $8f$, $8i$, and $8j$ sites. For the example of $SmZn_{12}$, illustrated in Fig. 19.15, we have $x = 0.353$ for the $8i$ position and $x = 0.293$ for the $8j$ position with lattice constants $a = 0.8927$ nm and $c = 0.5215$ nm.

Figure 19.15 shows the structure of the $SmZn_{12}$ phase, a single tetragonal unit cell depicted in (a) space filling and (b) ball-and-stick formats. The

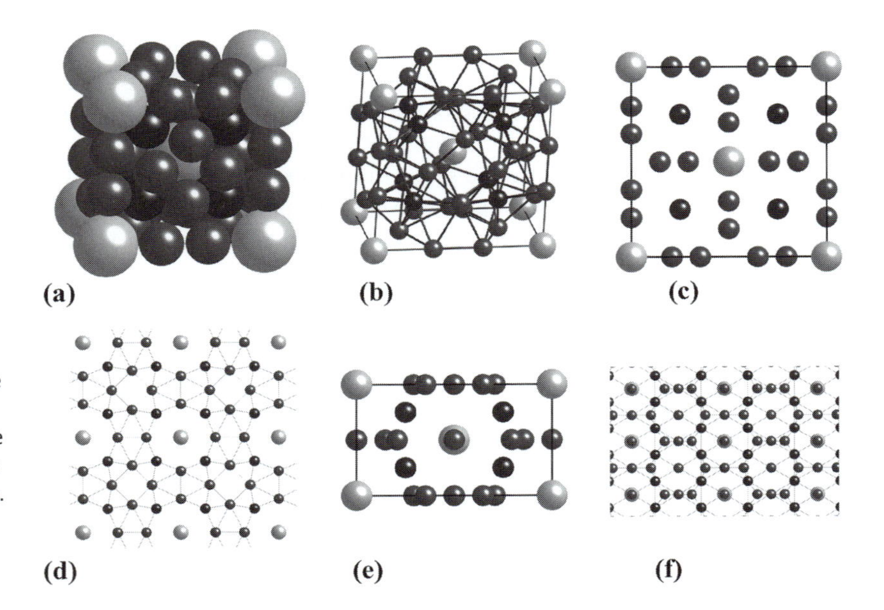

Fig. 19.15. $SmZn_{12}$ phase structure depicted in (a) space filling and (b) ball-and-stick formats, (c) projection into the (001) plane, and (d) half cells projected into the (001) plane. (010) plane projections in (e) and (f) shows tilings from a half cell projection.

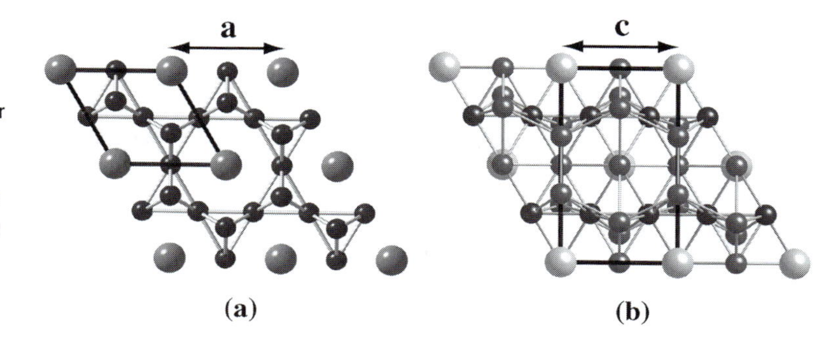

Fig. 19.16. Dumbbell transition taking the RT_5 structure into the RT_{12} structure. (a) projection of four unit cells of the RT_5 structure along the [001] direction and (b) projection of two unit cells of the RT_{12} structure along the tetragonal [100] direction.

structure is shown projected along the (001) direction in Fig. 19.15(c) and into the (010) plane in Fig. 19.15(e). Figure 19.15(d) shows half cells projected along the [001] direction yielding an interesting tiling which is nearly k-uniform with octagonal, square, and triangular tiles. Figure 19.15(f) shows half cells projected along the [001] direction yielding an interesting tiling.

Equation 19.4 can be multiplied by 2 to show that two R atoms can be removed from four RT_5 cells and replaced by two T dumbbells to yield two cells of the RT_{12} structure. The choice of the RT_5 and RT_{12} cells is illustrated in Fig. 19.16. Note that the a lattice constant for the RT_5 structure is similar to the c lattice constant for the RT_{12} structure. Figure 19.16(a) shows a projection of 4 unit cells of the RT_5 structure into an (001) plane with the unit cell emphasized. Figure 19.16(b) shows a projection of two unit cells of the RT_{12} structure along the tetragonal [100] direction with the unit cell emphasized. Note that the second cell is actually two half cells chosen so that the two projections in (a) and (b) have a nearly 1:1 correspondence. The fact that the T double tetrahedra are tilted in the 1:12 phase results from the fact that the atomic positions (using the coordinates for $SmZn_{12}$) for the T atom in the $8i$ and $8j$ sites differ slightly from $\frac{1}{3}$. This gives rise to the slight bending of the T atom chains along the horizontal axis. It can be seen that two R atoms have been replaced by T atoms (dumbbells in projection).

It is now worthwhile to summarize structural relationships between the 1:5, 2:17R, 2:17H, and 1:12 phases. Figure 19.17 shows a projection into the basal plane, of the structures showing hexagonal antiprism networks of T atoms. The smaller open and shaded circles depict T atoms from different planes forming the hexagonal antiprisms. These form "nests" which are occupied by the R atoms. The R atom planes are 3^6 tilings. The various decorations of the nests determine the 1:5, 2:17R, 2:17H, and 1:12 phase structures. The legend refers to these structures. Structures with R atom type 1 (afdg) correspond to the 1:5 unit cell. The cell with R atom type 3 (cefg) corresponds to the 2:17 unit cells. Structures with R atom type 2 (bdge) correspond to the 1:12 unit

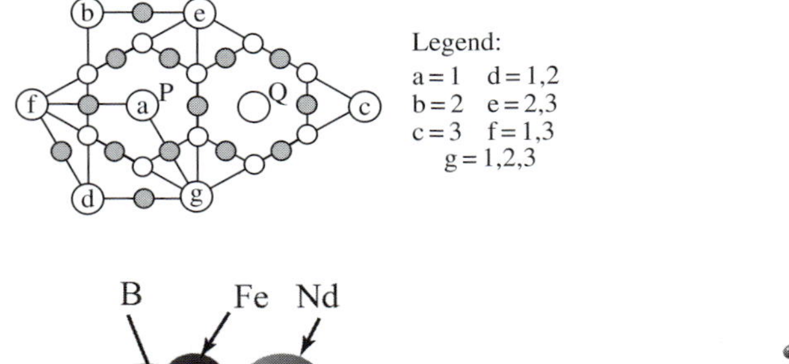

Fig. 19.17. Comparison of the atomic positions in the 1:5, 2:17R, 2:17H, and 1:12 phases (Pearson, 1972).

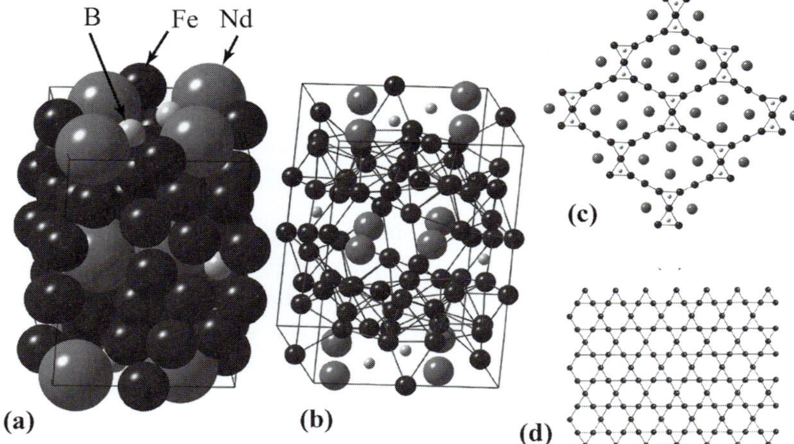

Fig. 19.18. The structure of $Nd_2Fe_{14}B$ phase, a single unit cell depicted in (a) space filling and (b) ball-and-stick formats. In (c), the structure is shown projected into the (001) plane with the [110] direction up; (d) shows a strip of T atoms projected into the (120) plane, illustrating the familiar Kagome network.

cell. The sites P and Q represent the sites at which dumbbell substitutions are made.

19.4.3.2 The $Nd_2Fe_{14}B$ phase

The most important tetragonal permanent magnet material is the $Nd_2Fe_{14}B$ (2:14:1) phase because of its large magnetocrystalline anisotropy and magnetic induction (Herbst *et al.*, 1984, Givord *et al.*, 1984). The $Nd_2Fe_{14}B$ phase has a tetragonal structure with space group **P4$_2$/mnm** (D^{14}_{4h}) (**Structure 51**) and lattice parameters $a = 0.8804$ nm and $c = 1.2205$ nm. With four formula units per unit cell, the $Nd_2Fe_{14}B$ cell contains 68 atoms. Nd atoms sit in the special positions $4f$ ($x = 0.1428$) and $4g$ ($x = 0.7302$). Fe atoms sit in the special positions $4c$, $4e$ ($z = 0.3852$), $8j_1$($x = 0.4021$; $z = 0.2955$), $8j_2$ ($x = 0.1824$; $z = 0.2543$), $16k_1$ ($x = 0.4627$; $y = 0.1404$; $z = 0.3237$) and $16k_2$($x = 0.7242$; $y = 0.0676$; $z = 0.3725$), while B occupies the $4f$ ($x = 0.3774$) site (Shoemaker *et al.*, 1984). The magnetic structure (i.e., the orientation of the magnetic moments) was solved by *neutron diffraction* by Herbst *et al.* (1984).

Figure 19.18 shows the structure of the $Nd_2Fe_{14}B$ phase, a single tetragonal unit cell depicted in (a) space filling and (b) ball-and-stick formats. This structure is quite involved and there are several projections that are helpful in understanding the structure. Figure 19.18(c) shows a projection into the (001) plane with the [110] direction pointing up of all atoms from $z = 0$ to $z = 0.128$ (just above $z = \frac{1}{4}$). The R atoms sit in groups of four in a rather open nest coordinated by Fe atoms. These oblong units are connected by B-centered trigonal prisms. Hexagonal rings sit above the rare earths but these are inclined. Figure 19.18(d) shows a projection into the (120) plane of T atoms near $y = 0$, illustrating the familiar Kagome net. For a review of 2:14:1 material properties, the reader is referred to Herbst (1991).

19.4.4 The monoclinic $R_3(Fe,Co)_{29}$ phases

The monoclinic $R_3(\mathrm{Fe}, \mathrm{Co})_{29}$ phase (**Structure 52**) is an example of a lower symmetry RT magnet for which the hard axis is not orthogonal to a basal plane. R_3T_{29} magnets were first reported in 1994 (Cadogan *et al.*, 1994). These structures can be obtained by the following dumbbell substitution:

$$5RT_5 - 2R + 4T = R_3T_{29}. \tag{19.5}$$

The prototype R_3T_{29} structure has an **A2/m** (C_{2h}^3) space group with two inequivalent R sites and 11 inequivalent T sites, respectively.[2] Relationships between the a_m, b_m, and c_m lattice constants of the monoclinic unit cell and a, b, and c lattice constants in the unit cell of the similar 1:5 derivative structure exist:

$$b_m = 3^{\frac{1}{2}}a; \qquad a_m = (4a + c)^{\frac{1}{2}}; \qquad c_m = (a + 4c)^{\frac{1}{2}}. \tag{19.6}$$

We will make use of these relationships in the analysis of the dumbbell substitutions that take the 1 : 5 into the 3 : 29 phase.

The two crystallographically inequivalent sites for the rare earth elements are the $2a$ and $4i$ sites (Hu *et al.*, 1996). The $2a$ site has a local 1:12-like environment and the $4i$ site has a 2 : 17-like environment (Li *et al.*, 1996). R ions reside in planes normal to the b-direction. The stacking sequence of the close-packed planes is $ABAB\ldots 2T$ dumbbells are positioned in the close-packed planes. A single cell of the R_3T_{29} phase has two formula units (or 64 atoms). This structure is formed by alternate stacking of 1:12 and 2:17 segments. The R_3T_{29} structures form another example of a phase that is stabilized by ternary additions. In general, the composition of the phases reported is of the form $R_3(\mathrm{Fe}, M)_{29}$, with M being a larger early transition metal. Substitution of other magnetic transition metals for Fe (notably Co) to alter magnetic properties is also possible, but does not seem to have an impact on the stability of the phase.

Figure 19.19 illustrates the structure of the $Nd_3Fe_{27.5}Ti_{1.5}$ phase with a monoclinic unit cell depicted in (a) space filling and (b) ball-and-stick formats. Fe atoms occupy 11 inequivalent sites; details of the atom positions can be

[2] Note that the official space group symbol is **C2/m** (C_{2h}^3); it is common practice to describe this crystal structure in an alternative setting of this space group.

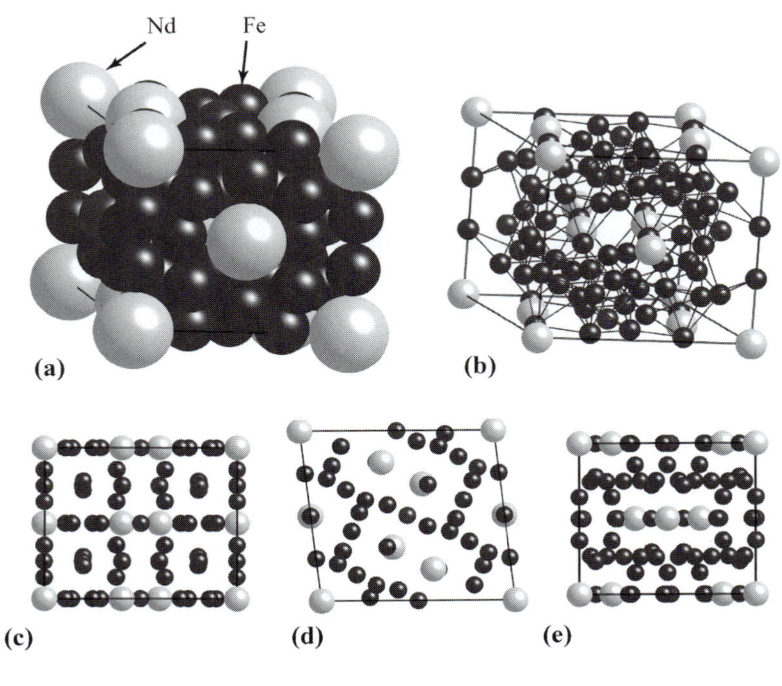

Fig. 19.19. The structure of $Nd_3(Fe,Ti)_{29}$, a single unit cell depicted in (a) space filling and (b) ball-and-stick formats. The structure is shown projected along (c) [001], (d) [010], and (e) [100].

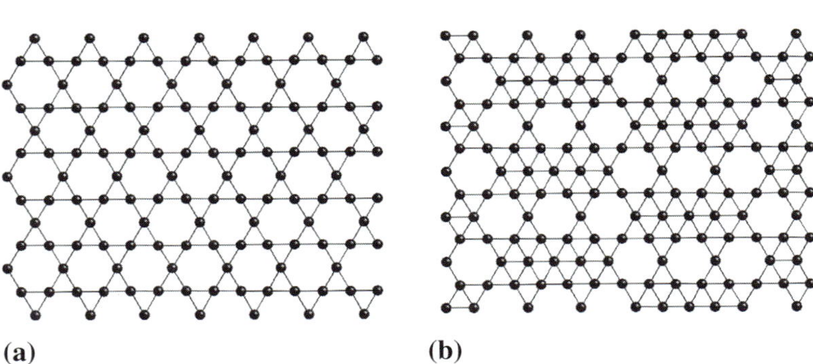

Fig. 19.20. A portion of the structure of Nd_3Fe_{29} showing a single (102) plane containing only Fe atoms and (b) projection from two adjacent (102) planes containing only Fe atoms.

found in the on-line structures appendix as **Structure 52**. The lattice constants vary with composition and typically obey *Vegard's law*. The monoclinic tilting angle is given by $\beta = \arctan(2a/c)$, which, for the structure in Fig. 19.19, amounts to 96.93°. Figure 19.19(c), (d), and (e) show the atoms from one unit cell of the structure as projected along the [001], [010], and [100] directions, respectively.

While the previous projections are interesting, they are not particularly instructive. More insight can be gained by projections along the [102] direction. Figure 19.20(a) shows a single (102) plane containing only Fe atoms where, once again, we find the Kagome tile. Figure 19.20(b) shows the projection of atoms in two successive (102) planes containing only Fe atoms, illustrating the $3^4 \cdot 6$ Archimedean tiling! Further consideration of the structure reveals that the large Nd atoms sit nested in the large hexagonal voids in these layers.

To understand the dumbbell transformation that takes the RT_5 phase into the R_3T_{29} phase beyond the stoichiometric description given by Equation 19.5 is not a trivial undertaking. Equation 19.5 can be multiplied by 2 to show that four R atoms can be removed from ten RT_5 cells and replaced by eight T dumbbells (four pairs) to yield two formula units of the R_3T_{29} phase. Table 19.2 shows that two formula units of the R_3T_{29} phase are contained in a single monoclinic unit cell. Thus the first task is to decide *which* ten RT_5 phase cells will be combined to form the single unit cell of the R_3T_{29} phase. Figure 19.21(a) shows a projection of five unit cells of the RT_5 structure along [001]. One would be tempted to just look at replicating these cells two unit cells deep along the c-axis and then making an appropriate dumbbell substitution for four R atoms to arrive at the 3 : 29 phase structure. Unfortunately, this does not work.

Figure 19.21(b) shows the actual relationship between the 3:29 phase and 1 : 5 phase structures. Here, the five cells of RT_5 are replicated three times along the c-axis. The monoclinic unit cell is then constructed with the basis vectors [201] and [$\bar{1}$02], as illustrated in the figure. The vectors $\mathbf{A}'$ and $\mathbf{C}'$ represent vectors in the directions of the previously defined $\mathbf{a}_m$ and $\mathbf{b}_m$ lattice vectors for the 3:29 phase. Their relationship with the 1:5 lattice is easily discerned from the figure. Four R atoms at lattice points in the ac plane of the monoclinic Bravais lattice are also highlighted for emphasis. From this figure, it is possible to rationalize the previous relationships between the a_m and c_m monoclinic lattice constants and the a and c lattice constants of the 1:5 unit cell.

We are now in a position to understand the dumbbell transformation that takes the 1:5 phase into the 3:29 phase stoichiometry. From the definition of the monoclinic unit cell, we can see that it will be profitable to compare (001)-type planes in the 1:5 structure with (010)-type planes in the 3:29 structure. Figure 19.22(a) illustrates the projection of nine cells of the RT_5 structure into the monoclinic (010) basal plane. Only the atoms inside the

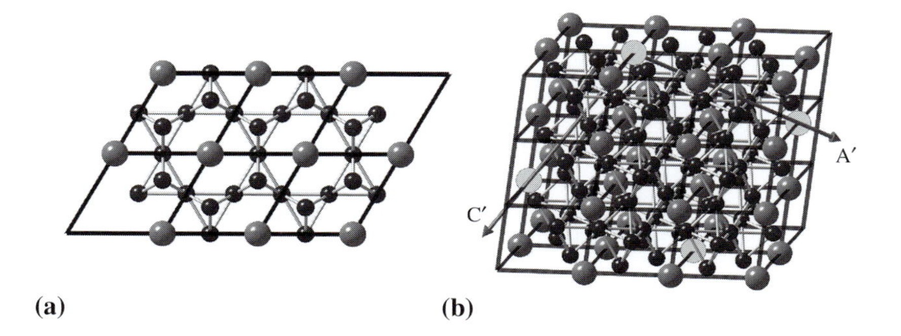

Fig. 19.21. Relationship between the RT_5 phase and the R_3T_{29} structures; (a) shows a projection of five unit cells (note the incomplete cells at lower left and upper right) of the RT_5 structure along [001], (b) shows the actual relationship between the 3:29 phase and 1:5 phase structures.

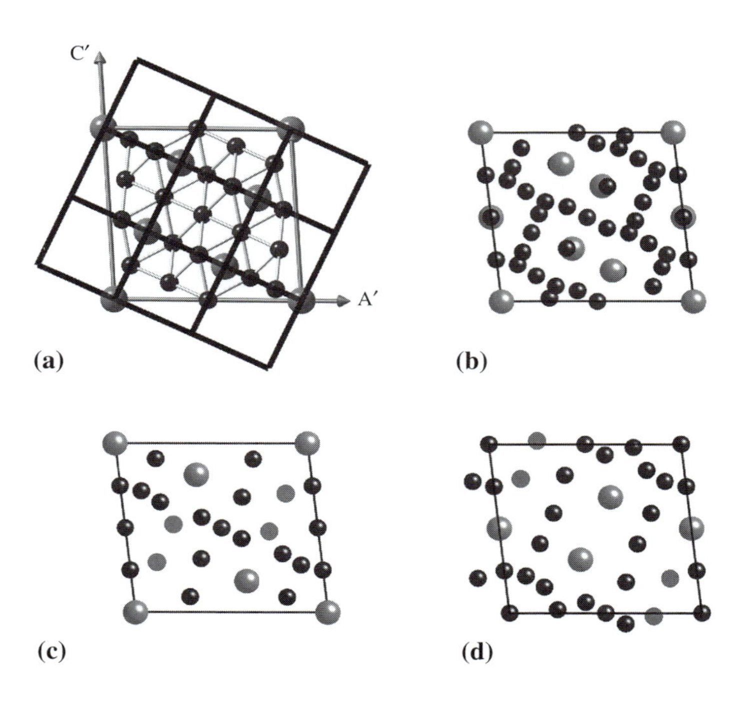

Fig. 19.22. (a) Projection of nine cells of the RT_5 structure into the monoclinic (010) basal plane; (b) comparison with the projection of a single cell into the (010) plane of the R_3T_{29} phase structure. (c) and (d) show the projections of the R_3T_{29} structure from the bottom ($y = -0.25$ to $y = 0.25$) and top ($y = 0.25$ to $y = 0.75$) halves of the unit cell into a (010) plane, respectively.

monoclinic AC cell are retained. With a bit of thought, it can be seen that this contains the atoms from five basal cells! Since the projection will not change if we add another five cells in the direction normal to the plane of the drawing, we can also think of this as the projection of ten RT_5 cells into an (010) plane. Now we have the number of cells consistent with the dumbbell transformation equation to yield the two formula units in the 3 : 29 phase unit cell.

Figure 19.22(b) shows for comparison the projection of a single cell of the R_3T_{29} structure into the (010) plane. While the similarities between these two projections are apparent, it is still difficult to discern the positions of the dumbbell substitutions. As a final step, we recognize that the two formula units included in the R_3T_{29} cell, are translated with respect to one another (the amount of the translation is left as an exercise). This can be seen by plotting the projection of the R_3T_{29} structure from the bottom half ($y = -0.25$ to $y = 0.25$) and top half ($y = 0.25$ to $y = 0.75$) of the unit cell into an (010) plane, as illustrated in Fig. 19.22(c) and (d), respectively. Here the dumbbell atoms are highlighted for clarity. In the bottom half projection, a bit more than a unit cell is illustrated so as to emphasize one additional feature. That is that the dumbbell substitutions place atoms in neighboring cells. By translational symmetry arguments, these atoms must be replicated to an equivalent position in the original cell. The combination of the translation shift (c) to (d) and the placement of atoms in neighboring cells by the dumbbell substitutions explain the total cell projection shown in (b).

The $R_3(Fe,TE)_{29}$ (3 : 29) (TE = early transition metal, such as Ti, etc.) compounds, which have been reported as potential candidates for high temperature permanent magnet applications (Shah *et al.*, 1998, 1999), also crystallize in the monoclinic structure with **A2/m** (C_{2h}^3) space group. Compounds of the type $Pr_3(Fe_{1-x}Co_x)_{27.5}Ti_{1.5}$ with $x = 0, 0.1, 0.2, 0.3, 0.4$, and 0.5 (up to 50% Co substitution for Fe) have been studied. Co substitutions in these alloys have been shown to significantly increase the Curie temperature, induction, and anisotropy field. Of fundamental importance is the fact that, with a larger T content, the magnetic exchange interaction and consequent Curie temperatures of these magnets can be increased. Some of the attractiveness of the high $T : R$ ratio in this phase is mitigated by the fact that 1.5 of the 29 T atoms are replaced by Ti to stabilize this metastable phase. The site selection in the substitution of Ti for Fe in these systems was studied using combined EXAFS and neutron diffraction observations (Harris *et al.*, 1999). Ti was found to substitute on the Fe sites in the $4g$ and $4i$ special positions, consistent with observations in $Pr_3Fe_{27.5}Ti_{1.5}$ (Yelon and Hu, 1996, Hu *et al.*, 1996). Furthermore, it was concluded that Co substituted without preference on the sites not occupied by Ti.

19.5 Interstitial modifications

Another technique for altering and, often, improving the magnetic properties of rare earth permanent magnet materials is through interstitial modifications with small light elements. These elements occupy interstitial sites (often octahedral sites) in the REPM lattice and, therefore, should have covalent radii less than about 0.1 nm (Skomski, 1996). While B, with a covalent radius of 0.088 nm, fits this description, it has a strong preference for trigonal prismatic coordinations which often dictates the structure, as we saw was the case for the 2:14:1 magnets. On the other hand, C and N with covalent radii of 0.077 nm and 0.070 nm, respectively, are excellent interstitial atoms for REPM structures. While these atoms are small, they still have noticeable effects on the structure, promoting volume expansions of the unit cell by as much as 8%. In some cases, the volume expansion is anisotropic and the interstitial modification can increase the magnetocrystalline anisotropy. More dramatic, however, are the effects on the magnetization and Curie temperatures of interstitially modified REPM materials, especially those containing Fe as the T species. Even slightly more spacing between the Fe atoms causes narrowing of the Fe electronic d-bands and closing of the majority spin d-band. In layman's terms, this increases iron's magnetic dipole moment, further increasing the magnetization. The increased separation of Fe atoms also favorably influences the magnetic exchange, so as to increase the Curie temperature of these alloys. While again not providing an exhaustive survey, we will review some of the structures of interstitially modified REPMs in this section.

Nitrogen interstitial modification of the Sm_2Fe_{17} phase structures has been investigated by Coey and Sun (1990). Typically, this modification is performed by gas-phase reaction with fine particles. The nitrogenation reaction can be written as:

$$2Sm_2Fe_{17} + (3-\delta)N_2 = 2Sm_2Fe_{17}N_{3-\delta}.$$

At temperatures above 720 K, the nitride disproportionates by the reaction:

$$2Sm_2Fe_{17}N_3 \ = \ 2SmN + Fe_4N + 13Fe.$$

In the nitrided materials, N occupies large octahedral interstitial sites. The structural features of the modification in the rhombohedral Sm_2Fe_{17} phase can be explained by considering a volume expansion of the parent α-Sm_2Co_{17} phase lattice with the occupation of originally empty octahedral interstitial sites. Figure 19.23(a) shows a single unit cell, containing three formula units of the rhombohedral Sm_2Co_{17} phase into which nine N atoms (highlighted) have been incorporated in octahedral interstices to result in the compound $Sm_2Co_{17}N_3$ (**Structure 53**). These N interstitials sit at the 9*e* special positions of the **R$\bar{3}$m** (D^5_{3d}) space group with the sites of the *R* and *T* atoms the same as in the parent phase. Figure 19.23(b) shows the same in a polyhedral representation, illustrating the N octahedral coordination polyhedra. Note that the vertex sharing polyhedra are connected along the [100] and [010] directions. Interstitial modification with C to form $Sm_2Fe_{17}C_x$ has also been demonstrated. In this case, it has been shown that solid-state diffusion can be used as an interstitial modification method (Skomski and Coey, 1993).

Nitrogen interstitial modification of the $Sm(Fe,Ti)_{12}$ phase structures has been investigated as well (Yang *et al.*, 1993). Here, the N also occupies large octahedral interstices. The $Sm(Fe,Ti)_{12}$ phase requires one of the 12 *T* sites to be occupied by Ti in order to stabilize this Fe-containing compound. The structural features of the modification in the tetragonal $Sm(Fe,Ti)_{12}$ phase can also be explained by considering a volume expansion of the parent phase lattice with the occupation of originally empty octahedral interstitial sites. Figure 19.24(a) shows a single unit cell, containing two formula units of the tetragonal $Sm(Fe,Ti)_{12}$ phase, for which two (per cell) N atoms (highlighted) have been incorporated in octahedral interstices to result in the compound $Sm(Fe,Ti)_{12}N$ (**Structure 54**). Figure 19.24(b) shows the same in a polyhedral setting, illustrating the N octahedral coordination polyhedra. Note that the polyhedra share vertices along the [001] direction.

Hydrogen, nitrogen, and carbon interstitial modifications of the $Sm_3(Fe,Ti)_{29}$ phase structures have been investigated by Cadogan *et al.* (1994). Alloys of composition $Sm_3(Fe,Ti)_{29}N_5$ have been synthesized with

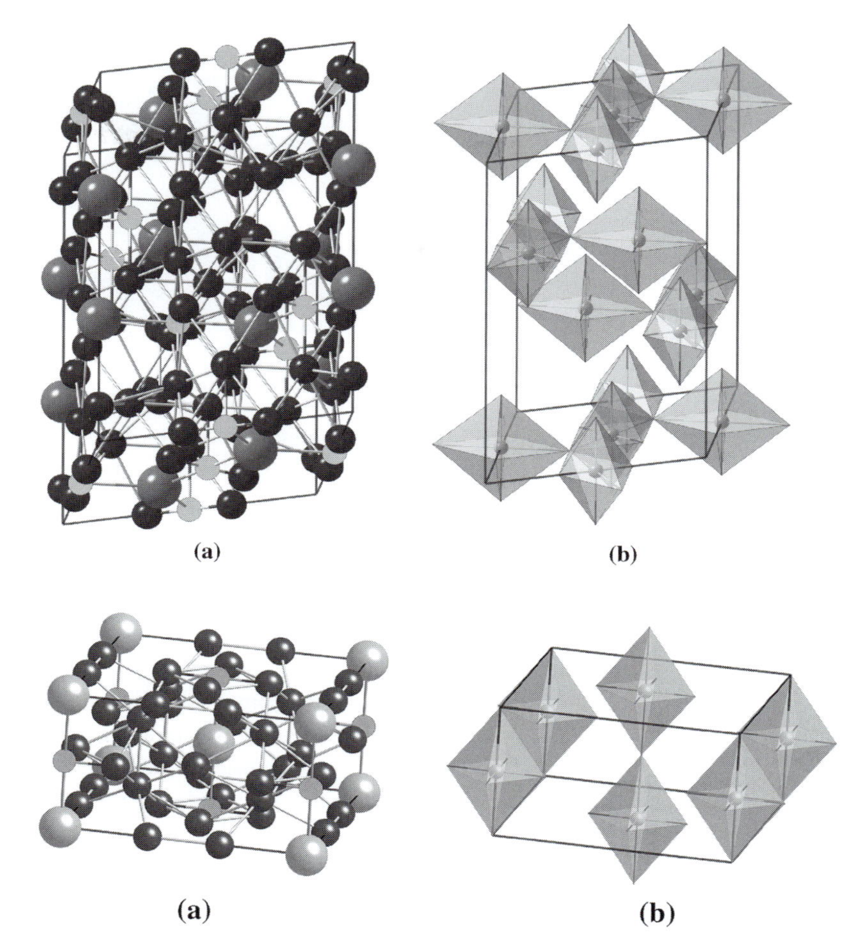

Fig. 19.23. The structure of $Sm_2Co_{17}N_3$: (a) a single unit cell, containing three formula units of the rhombohedral Sm_2Co_{17} phase into which nine N atoms (highlighted) have been incorporated in octahedral interstices and (b) the same in a polyhedral setting, illustrating the N octahedral coordination polyhedra.

Fig. 19.24. The structure of $Sm(Fe,Ti)_{12}N$: (a) a single unit cell, containing two formula units of the tetragonal $Sm(Fe,Ti)_{12}$ phase into which two (per cell) N atoms (highlighted) have been incorporated in octahedral interstices and (b) the same in a polyhedral setting, illustrating the N octahedral coordination polyhedra.

a notable increase in the anisotropy fields and increases in both saturation magnetization and Curie temperature.

19.6 Historical notes

Victor Moritz Goldschmidt (1853–1933) was born January 27, 1888, in Zürich, Switzerland. He was named after Victor Mayer, a famous chemist (Mason, 1992). His family moved to Kristiania (now Oslo), Norway in 1900. Goldschmidt was a pupil of the Norwegian geologist Waldemar C. Brogger at the University of Oslo, where he completed his Ph.D. thesis in 1911. He was appointed professor and subsequently made Director of the Mineralogic Institute of the University of Oslo in 1914. In 1917 he was appointed Chair and Laboratory Director of the Norwegian government "Commission on Raw

Fig. 19.25. (a) Victor Moritz Goldschmidt (1853–1933) (picture courtesy of J. Lima-de-Faria and originally from Spencer (Spencer, 1936)) and (b) Clara Brink Shoemaker and her husband David P. Shoemaker (picture courtesy of Clara Shoemaker).

Materials." He held appointments in Göttingen (1929–35) and in Oslo again (1935–42), and he served as the Director of the Geological Museum in Oslo.

Goldschmidt was a prominent geochemist considered the father of the field of Earth Sciences (geochemistry) (Goldschmidt, 1954). He was influential in estimating the chemical composition of the Earth's crust. He compared the chemistry of meteorites and the silicates in the Earth's crust. In his thesis work, he developed a *mineralogical phase rule* which was an adaptation of the *Gibbs phase rule*.

Goldschmidt was instrumental in laying the foundation for the field of crystal chemistry. He estimated the ionic sizes of the rare earth elements. In doing so, he observed a decrease in their ionic size with increasing atomic number, a phenomenon that he called the *lanthanide contraction*. Goldschmidt also developed rules governing elemental distributions in rocks and minerals. He calculated the ionic size of many elements from oxide structures. He developed a set of ionic and atomic radii (Chapter 17) and crystal chemistry rules that were later extended by Linus Pauling. These are discussed in Chapter 22. He noted that an increase in coordination number is accompanied by an increase in interatomic distances in ionic solids. Goldschmidt also trained Fritz Laves. He worked with Laves to develop a topological approach to structure derivation describing the structure of intermetallic compounds.[3] Goldschmidt is remembered today through the Goldschmidt conference and the Goldschmidt medal, the highest honour in the field of geochemistry.

Clara Brink Shoemaker (1921–) was born in Rolde, Province of Drente, in the Netherlands. Clara majored in chemistry with a minor in physics at the University of Leiden beginning in 1938. After the Nazis closed Leiden during World War II, Clara continued work towards her doctoral degree at The University of Utrecht. In 1943, Clara moved back to Leiden to avoid

[3] More details on the life of Goldschmidt (written by Eric Reardon) can be found at the University of Waterloo website: http://www.science.uwaterloo.ca/earth/waton/gold.html. The book "Victor Moritz Goldschmidt: Father of Modern Geochemistry" (Mason, 1992) also is an excellent biographical source.

having to sign a German loyalty oath. She continued her studies informally until after the war when she was able to take a final doctoral examination. She received her doctorate in chemistry in 1950 from Leiden. Her major professor for the Ph.D. was Anton Eduard van Arkel.

From 1950–51 Clara worked at Dorothy Hodgkin's laboratory in Oxford, England, where she contributed to the structural determination of vitamin B_{12} (Brink *et al.*, 1964, Brink-Shoemaker *et al.*, 1964, Abrahams, 1983). After returning to Leiden for a period, Clara moved again to join the group of **David P. Shoemaker** at MIT. She married David Shoemaker in 1955 while working as a research fellow. Clara moved to Oregon when David became Chemistry Department Chair at Oregon State University in 1970. Clara Shoemaker became a senior research professor at Oregon State in 1982 and retired as Professor Emeritus.

Clara's work on transition metal phases coined the name "tetrahedrally close-packed" as synonymous with topologically close-packed. Out of these studies came discoveries that materials could be used for hydrogen storage. Some of the topologically close-packed phases which she and David Shoemaker studied were precursors to alloys later discovered to be quasicrystalline. She contributed to the field of permanent magnets in 1984, solving the structure of the $Nd_2Fe_{14}B$ (2:14:1) phase (Shoemaker *et al.*, 1984). A more complete biography has been written by Mary F. Singleton.[4]

David P. Shoemaker (1920–95) was born in Kooskia, Idaho. He received a B.A. from Reed College in Portland in 1942 and a Ph.D. from Caltech in 1947. He was awarded a Guggenheim Fellowship and studied at the Institute for Theoretical Physics in Copenhagen. During this time, he also worked as a crystallographer in Oxford, while Linus Pauling was visiting as the Eastman Professor. From 1948–51, he was a senior research fellow at Caltech. He moved to MIT in 1951, where he stayed for 19 years.

19.7 Problems

(i) *Rare earth atomic size*: Elemental Gd crystallizes into an *hcp* lattice with lattice constants $a = 0.3631$ nm and $c = 0.5728$ nm.

- Compare the $\frac{c}{a}$ ratio for Gd to that for an ideal *hcp* material.
- Calculate the metallic radius for Gd.
- Calculate the density of Gd.

(ii) UNi_5 *structure*: Examine the stacking of (111) planes in the UNi_5 structure. Show that the stacking sequence is the same as for the

[4] See http://osulibrary.oregonstate.edu/specialcollections/coll/shoemaker/index.html.

Cu_2Mg-type (**C15**) Laves phase, $A)\alpha(ACB)\beta(BAC)\gamma(CB$. Identify the atoms occupying each plane in both structures.

(iii) $LaCo_{13}$ *structure-tiling I*: Fig. 19.5 (c) illustrates a tiling of Co(1) and Co(2) atoms in the $z = 0$ plane of the $LaCo_{13}$ phase structure. This tile has nearly regular pentagons, squares, and triangles decorating the plane. Give the Schläfli symbol for this tiling if it were ideal.

(iv) $LaCo_{13}$ *structure-tiling II*: Figure 19.5(d) illustrates two puckered layers of Co(2) atoms of the $LaCo_{13}$ phase structure. These combine to form pseudo-octagonal "nests" that the larger La atoms sit in. Give the Schläfli symbol for the idealized (perfect octagons) tiling of the Co(2) atoms. Using a 0.125 nm metallic radius for Co, and considering touching Co atoms, determine the edge length of a perfect octagon. If the metallic radius of La is ~ 0.18 nm, determine the elevation of the La atom above the nest for touching Co and La atoms.

(v) Th_6Mn_{23} *structure*: Th_6Mn_{23} is an important *RT* structural prototype with space group **Fm$\bar{3}$m** (O_h^5). The compound has $a = 1.2523$ nm, four inequivalent Mn atoms and a Th atom in the sites tabulated below. The atomic weights of Mn and Th are: $AW_{Mn} = 54.93$ and $AW_{Th} = 232.04$. Making reference to the ITC, determine the following:

Atom	Site	x	y	z
Mn 1	$4b$	0.500	0.500	0.500
Mn 2	$24d$	0.000	0.250	0.250
Mn 3	$32f$	0.378	0.378	0.378
Mn 4	$32f$	0.178	0.178	0.178
Th 1	$24e$	0.203	0.000	0.000

(a) The Bravais lattice, Pearson symbol, and the point group symmetry at each special position.

(b) The composition, number of formula units in a unit cell, and density.

(c) The shortest distance between Mn atoms on the $4b$ and $32f$ sites.

(d) The extinctions that occur as a result of the Bravais lattice. What are the first eight peaks that you would predict to occur for this Bravais lattice?

(e) Additional extinctions occur as a result of the Fe atom at the $24d$ special position. What symmetry operation causes these extinctions? Which of the previous eight peaks will be extinct because of this?

(vi) $SmCo_3$ *structure*: Examine the stacking of (111) planes in the $SmCo_3$ structure. Determine the stacking sequence in the Pearson notation.

(vii) $CaCu_5$ *structure*: In the $CaCu_5$ structure, the touching sphere size of the B element, d_B, is $\sqrt{(c^2/4 + a^2/12)}$. Determine how good the Sm–Sm touching sphere approximation is for the $SmCo_5$ phase.

(viii) $SmCo_5$ *I*: Using tabulated lattice constants, calculate the density of $SmCo_5$.

(ix) $SmCo_5$ *II*: Using tabulated lattice constants, calculate the H capacity (atoms per unit volume) of $SmCo_5$ assuming occupancy of all of the tetrahedral interstices.

(x) $SmZn_{12}$ *structure*: Figure 19.15(d) shows half cells projected along the [001] direction yielding a tiling that is nearly k-uniform with octagonal, square, and triangular tiles. Give the Schläfli symbol for this tiling.

(xi) $SmZn_{12}$ *structure – Dumbbell substitution*: Figure 19.16 compares the RT_5 and RT_{12} structures. Explain the symmetry changes from the hexagonal $SmCo_5$ structure to the tetragonal $SmCo_{12}$ structure that result from dumbbell substitutions.

(xii) *RT structure lattice constants*: Beginning with the parent RT_5 structure and using Fig. 19.17, suggest relationships between the a, b, and c lattice constants of the RT_{12} and RT_{17} structures and the RT_5 lattice constants.

(xiii) Th_6Ni_{17} *structure*: An important RT structural prototype is that of Th_6Ni_{17} with space group **P6$_3$/mmc** (D^4_{6h}). The compound has $a =$ 0.843 nm, $c = 0.804$ nm, four inequivalent Ni atoms and two inequivalent Th atoms in sites tabulated below. The atomic weights of Ni and Th are: $AW_{Ni} = 58.69$ and $AW_{Th} = 232.04$. Making reference to the ITC, determine the following:

Atom	Site	x	y	z
Ni 1	$4f$	0.333	0.667	0.110
Ni 2	$6g$	0.500	0.000	0.000
Ni 3	$12j$	0.333	0.000	0.250
Ni 4	$12k$	0.167	0.333	0.000
Th 1	$24e$	0.203	0.000	0.250
Th 2	$2d$	0.333	0.667	0.750

(a) The Bravais lattice, Pearson symbol, and the point group symmetry at each special position.

(b) The composition, number of formula units in a unit cell, and density.

(xiv) *Site preferences in ternary alloys*: Construct a model for the unit cell of Nd_2Co_{17} (prototype Th_2Zn_{17}) using crystal structure software such as Crystalmaker. Consult Pearson's tables for the lattice constants and special positions. Enumerate the positions of the transition metal sites in this structure. Consult the article by Herbst *et al.* (1982). Describe the preference of Co and Fe for these sites in the ternary alloy $Nd_2(Co_{1-x}Fe_x)_{17}$.

(xv) $Nd_2Fe_{14}B$ *structure – B coordination*: Using a 0.124 nm metallic radius for Fe and a 0.088 nm covalent radius for B, calculate the dimensions of a trigonal prism in which Fe atoms touch in the triangular faces and Fe and the central B atom also touch.

(xvi) Nd_3Fe_{29} *structure*: Consider $Nd_3Fe_{27.5}Ti_{1.5}$ as having a typical Nd_3Fe_{29} structure with special positions noted in the text. Determine the angle between the *a* and *c*, and *b* and *c*, axes in this structure. How close is this structure to being uniaxial, i.e., having an axis orthogonal to a basal plane?

(xvii) $Sm_2Co_{17}N_3$ *structure – N coordination*: Using a 0.125 nm radius for Co, determine the largest interstitial atom that can fit in an octahedral site surrounded by touching Co atoms. Will N fit in this site or does the octahedron need to expand?

CHAPTER

20 Metallic structures IV: quasicrystals

> *"A man has one pair of rabbits at a certain place entirely surrounded by a wall. We wish to know how many pairs will be bred from it in one year, if the nature of these rabbits is such that they breed every month one other pair and begin to breed in the second month after."*
>
> Fibonacci, *Liber Abaci* (1202)

20.1 Introduction

Icosahedral orientational order in a sharply peaked diffraction pattern was first observed for a rapidly solidified Al–14% Mn alloy (Shechtman *et al.*, 1984, Shechtman and Blech, 1984). These materials were called *quasicrystals* and the Al–14% Mn alloy phase was named *Shechtmanite*. Quasicrystals have long-range orientational order but no 3-D translational periodicity. The discovery of quasicrystals caused a reexamination of the basic tenets of crystallography. Observations of icosahedral symmetry also spurred inquiries into its implications on electronic structure and magnetism (McHenry *et al.*, 1986, McHenry, 1988).

Quasi-periodic structures were introduced in Chapter 17. If a lattice is modulated with a periodicity that is an irrational fraction or multiple of the underlying periodicity, then an *incommensurate superlattice* results. An incommensurate superlattice gives rise to *quasi-periodicity*. A mathematical function is said to be quasi-periodic, if its *Fourier transform* gives rise to a set of delta functions that is not periodic, but can be described by a finite (countable) set of lengths (Cahn *et al.*, 1986).[1]

[1] If the Fourier transform results in delta functions but an infinite (uncountable) set of lengths is needed to index them, then the function is called *almost periodic*.

Quasicrystals possess quasi-periodicity and orientational order with a symmetry that does not belong to the 32 crystallographic point groups. Quasicrystal diffraction patterns, therefore, also exhibit non-crystallographic symmetry. The mathematical principles needed to describe the unique diffraction patterns for quasicrystals include:

(i) the use of *quasi-periodic functions* for the description of atomic densities, and their Fourier transforms for the description of diffraction in reciprocal space;
(ii) the *embedding* of a *non-crystallographic point group* into a higher dimensional space, followed by an (irrational) projection into lower dimensions to preserve the orientational order;
(iii) the determination of atomic configurations consistent with the quasi-periodic diffraction patterns observed in scattering experiments.

Diffraction conditions for quasicrystals are described using a basis in a higher dimensional space, in which the non-crystallographic point group is embedded. In general, a D-dimensional quasi-periodic function, requiring a basis with n basis vectors, can be constructed from a periodic n-dimensional function projected into a D-dimensional space (Bohr, 1932).

Quasicrystals have led the International Union of Crystallographers to redefine the term *crystal* as *any solid having an essentially discrete diffraction pattern*. This shifts the definition of crystallinity from direct to reciprocal space. Projection from higher to lower dimensions (e.g., from 6-D into 3-D) results in more basis vectors than are necessary to span the space. The irrational projection implies that (1) through repeated addition of the vectors it is possible to reach any point in the lower dimensional reciprocal space and (2) discrete diffraction spots will densely fill reciprocal space.[2]

This chapter presents examples of 1-D, 2-D, and 3-D quasi-periodic structures. For $D = 1$, projection from $n = 2$ results in the *Fibonacci lattice*. For $D = 2$, the *Penrose tile* and other polygonal tilings result from projection from $n = 4$. For $D = 3$, the projection from $n = 6$ results in a *3-D Penrose tile* with $\bar{\mathbf{5}}\mathbf{m}$ (D_{5d}) symmetry. After considering quasicrystals with five-fold symmetry, we generalize to other non-crystallographic symmetries. *Dodecagonal quasicrystals* have been reported in Ni–Cr alloys (Ishimasa *et al.*, 1985); *octagonal quasicrystals* were reported in V–Ni–Si and Cr–Ni–Si alloys (Wang *et al.*, 1987). Yamamoto (1996) has summarized four methods to obtain quasi-periodic tilings. These are (1) the *inflation–deflation method*, (2) the *dual method*, (3) the *projection method*, and (4) the *section method*. We will illustrate these methods in our discussion of quasicrystalline structures.

[2] Although, in practice, only the most intense reflections will be observed.

20.2 The golden mean and pentagonal symmetry

The irrational number, τ, the *golden mean* or *golden ratio*, is significant both in number theory and pentagonal geometry. This number is defined as:

$$\boxed{\tau = 1.618034\ldots = \tfrac{1}{2}(1+\sqrt{5}).} \tag{20.1}$$

and is important in quasicrystals with pentagonal, decagonal, or icosahedral orientational order. The golden mean, τ, and its inverse, $-1/\tau$, are the two solutions to the polynomial equation $x^2 - x - 1 = 0$. Restated as $x = 1 + 1/x$, the equation represents the set of *regular continued fractions* that converges most slowly.

The recursion relation for generating the set of continued fractions, called *Fibonacci numbers*, was developed by Leonardo Pisano Fibonacci. Fibonacci was interested in describing the growth of rabbit populations with the sequence of these regular continued fractions. Box 20.1 summarizes useful trigonometric relationships and algebraic identities involving τ.

Box 20.1 Trigonometric relationships, algebraic identities, and inflationary properties of the golden mean

Trigonometric relationships involving τ follow from pentagonal geometry. These are:

$$\tau = 1 + 2\cos(\tfrac{2\pi}{5}) = 1 + 2\cos(\tfrac{8\pi}{5}) = -2\cos(\tfrac{4\pi}{5}) = -2\cos(\tfrac{6\pi}{5}).$$

The golden mean also satisfies the algebraic identities:

$$\tau - \tau^{-1} = 1 \quad \text{and} \quad \tau^3 - 2\tau^2 = -1.$$

Finally, the golden mean also has the following inflationary properties:

$$\tau^2 = \tau + 1 \quad \text{and} \quad \tau^3 = \tau^2 + \tau \ldots \quad \text{and} \quad \tau^{n+2} = \tau^{n+1} + \tau^n;$$

the last expression is true for any integer n. Examples of integer *Lucas numbers* are:

$$\tau^2 + \frac{1}{\tau^2} = 3 \quad \text{and} \quad \tau^3 - \frac{1}{\tau^3} = 4 \quad \text{and} \quad \tau^4 + \frac{1}{\tau^4} = 7 \quad \text{and} \quad \tau^5 - \frac{1}{\tau^5} = 11.$$

The recursion relationship for the Lucas numbers, will be explored in a reader exercise.

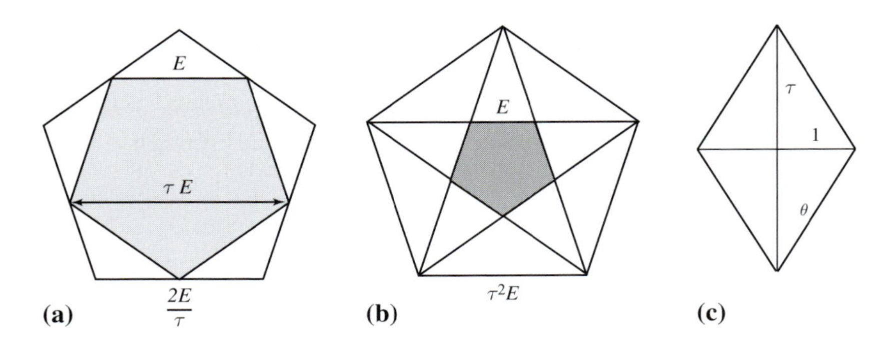

Fig. 20.1. (a) Relationship between a pentagon and a circumscribed pentagon, (b) construction of a pentagram by expansion, and (c) the golden rhomb.

Figure 20.1(a) shows that τ is the ratio of the distance between two non-adjacent vertices of a pentagon to the length of a side. A pentagon circumscribed about another pentagon is scaled in size by $2/\tau = 2(\tau - 1)$. Figure 20.1(b) shows the *star extension* or *stellation* of the pentagon to yield a *pentagram*. The ratio of the outer to inner pentagon edge lengths is $\tau^2 = \tau + 1$. These constructions are examples of *inflations* of the pentagon and yield structures (the larger pentagons) which are related through a rotation and scaling to the original structure (the small pentagon).[3]

An important attribute of *inflation* operations is the scaling by powers of the golden mean. The golden mean obeys the following useful power law:

$$\tau^n = m\tau + p \qquad (20.2)$$

where m, n, and p are integers.[4] The coordinates of any inflated pentagon can be determined from the vertex coordinates of the original pentagon, by inversion followed by a linear transformation determined by the two indices m and p.

The ancient Greeks recognized the importance of a particular rhombus, the *golden rhomb*, for which τ is the ratio of the long to the short diagonal. This rhomb forms the 30 faces of the *rhombic triacontahedron*, which is one of the shells used in describing the $Mg_{32}(Al,Zn)_{49}$ crystal structure in Chapter 18.

Coordinates of polyhedra with icosahedral symmetry also have τ as a parameter. We summarize these in Box 20.2. We use the shorthand I, for icosahedron; PD, for pentagonal dodecahedron; RT for rhombic triacontahedron; ID, for icosiododecahedron; and TI for truncated icosahedron. A normalization constant ensures that the vertices are at unit distance from the origin. Since powers of τ can be expressed as $m\tau + p$, each of the coordinates of these polyhedra can be expressed in terms of two indices, one which scales an integer, the other scales τ.

[3] In this particular example, the pentagonal symmetry occurs as a singularity in that only the the central site has five-fold symmetry (Kramer, 1982).

[4] The reader is encouraged to prove this relation as an exercise.

Box 20.2 Vertex coordinates of polyhedra with icosahedral symmetry

The vertex coordinates for polyhedra, with all vertices at unit length from the origin and with two-fold axes oriented along the Cartesian coordinate axes, are (as before, cp stands for cyclic permutations):

$$I = \frac{1}{\sqrt{1+\tau^2}}(\pm 1, 0, \pm\tau) + \text{cp}$$

$$PD = \frac{1}{\sqrt{3}}(0, \pm\tau, \pm\frac{1}{\tau}), \frac{1}{\sqrt{3}}(\pm 1, \pm 1, \pm 1) + \text{cp}$$

$$= \frac{1}{\sqrt{3}}(0, \pm\tau, \pm(\tau - 1)), \frac{1}{\sqrt{3}}(\pm 1, \pm 1, \pm 1) + \text{cp}$$

$$RT = \frac{1}{\sqrt{3}}(0, \pm\tau, \pm\frac{1}{\tau}), \frac{1}{\sqrt{3}}(\pm 1, \pm 1, \pm 1), \frac{1}{\sqrt{1+\tau^2}}(\pm 1, 0, \pm\tau) + \text{cp}$$

$$= \frac{1}{\sqrt{3}}(0, \pm\tau, \pm(\tau - 1)), \frac{1}{\sqrt{3}}(\pm 1, \pm 1, \pm 1), \frac{1}{\sqrt{1+\tau^2}}(\pm 1, 0, \pm\tau) + \text{cp}$$

$$ID = \frac{1}{2}(\pm 2, 0, 0), \frac{1}{2}(\pm\tau, \pm\frac{1}{\tau}, \pm 1) + \text{cp}$$

$$= \frac{1}{2}(\pm 2, 0, 0), \frac{1}{2}(\pm\tau, \pm(\tau - 1), \pm 1) + \text{cp}$$

$$TI = \frac{1}{\sqrt{1+9\tau^2}}(0, \pm 1, \pm 3\tau), \frac{1}{\sqrt{1+9\tau^2}}(\pm 2, \pm\tau^3, \pm\tau),$$
$$\frac{1}{\sqrt{1+9\tau^2}}(\pm 1, \pm 2\tau, \pm(\tau^2 + 1)) + \text{cp}$$

$$= \frac{1}{\sqrt{1+9\tau^2}}(0, \pm 1, \pm 3\tau), \frac{1}{\sqrt{1+9\tau^2}}(\pm 2, \pm(2\tau + 1), \pm\tau),$$
$$\frac{1}{\sqrt{1+9\tau^2}}(\pm 1, \pm 2\tau, \pm(\tau + 2)) + \text{cp}$$

20.3 One-dimensional quasicrystals

20.3.1 The Fibonacci sequence and Fibonacci lattice derived by recursion

A one-dimensional (1-D) quasi-periodic lattice is known as a *Fibonacci lattice* and is related to the *Fibonacci numbers*, $(F_0, F_1, F_2, \ldots, F_n)$ defined by the recursion relation in Box 20.3. Of particular importance is the limit:

$$\lim_{k\to\infty} \frac{F_{k+1}}{F_k} = \tau; \tag{20.3}$$

Box 20.3 Recursion relationship for the Fibonacci numbers

The following recursion relationship defines the Fibonacci numbers:

$$F_0 = 1; \quad F_1 = 1; \quad \ldots \quad (F_k + F_{k+1}) = F_{k+2}.$$

Thus $F_2 = F_1 + F_0 = 2$, $F_3 = F_2 + F_1 = 3$, $F_4 = F_3 + F_2 = 5, \ldots$

i.e., the ratio of consecutive Fibonacci numbers is a convergent series of *continued fractions* which converges to the golden mean τ.

The famous *Fibonacci series* of number theory is summarized in Table 20.1. The previously defined power law relationship for the golden mean can be written as:

$$\tau^n = F_{n-1}\tau + F_{n-2}, \tag{20.4}$$

stating that the integer indices previously called m and p are Fibonacci numbers! The reader can prove this as an exercise.

The *quasi-periodic Fibonacci lattice* is constructed by extending long period commensurate superlattices with successively longer periods to the limit of an (∞ period) quasi-periodic lattice (Venkataraman *et al.*, 1989). This succession yields *periodic approximants* to the 1-D quasi-periodic Fibonacci lattice, in analogy with Table 20.1 where the ratio of successive Fibonacci numbers gives successively better approximations to τ.

To construct the periodic approximants we use two 1-D "lattice constants", a_S and a_L, to define a short step, S, and a long step, L, respectively. An index, k, numbers the particular sequence. We begin ($k = 0$) with a 1-D periodic lattice:

$$x_n = na_S, \tag{20.5}$$

i.e., with only short steps. We use the following substitutional rules to build subsequent periodic approximants, as long period superlattices:

(i) All short steps, a_S, are replaced with a long step, a_L, $S \to L$;
(ii) All long steps, a_L, are replaced with a long and a short step, $L \to L + S$.

Starting with the 1-D periodic lattice ($k = 0$), $x_n = na_S$, we use the first rule (there are no long steps): $S \to L$ to produce the lattice $x_n = na_L$ as the $k = 1$ periodic approximant. To produce the $k = 2$ approximant we use the second rule ($L \to L + S$) to produce a lattice with period $2(a_L + a_S)$. In subsequent approximants, we use both rules to produce lattices with periods $5, 8, 13, 21\ldots$ i.e., the periods are Fibonacci numbers. The repeat sequence,

Table 20.1. Fibonacci numbers and the ratio of successive Fibonacci numbers (Venkataraman *et al.*, 1989).

k	0	1	2	3	4	5	6	7		...
F_k	1	1	2	3	5	8	13	21		...
F_{k+1}/F_k	$\frac{1}{1}$	$\frac{2}{1}$	$\frac{3}{2}$	$\frac{5}{3}$	$\frac{8}{5}$	$\frac{13}{8}$	$\frac{21}{13}$	$\frac{34}{21}$	...	1.6182...

Table 20.2. Periodic Fibonacci lattice approximants to 1-D Fibonacci Lattice.

k	Sequence, **S(k)**	Period
0	*S*	1
1	*L*	1
2	*LS*	2
3	*LSL*	3
4	*LSLLS*	5
5	*LSLLSLSL*	8
6	*LSLLSLSLLSLLS*	13
7	*LSLLSLSLLSLLSLSLLSLSL*	21
8	*LSLLSLSLLSLLSLSLLSLSLLSLLSLSLLSLLS*	34
9	*LSLLSLSLLSLLSLSLLSLSLLSLLSLSLLSLLSLSLLSLSLLSLLSLSLLSLLS*	55

S(k), of the superlattice is summarized in Table 20.2 in terms of the sequence of L and S steps. The period is the sum of the number of short and long steps in the sequence.

In Table 20.2, the ratio of the number of long to short elements in the sequence ***S(k)*** is given by the ratio F_{k+1}/F_k of the two Fibonacci numbers. These ratios become better approximations to the golden mean as the period becomes longer. The infinite period tiling is called the Fibonacci tiling or *Fibonacci quasilattice*. It is an infinite period, incommensurate, and quasi-periodic lattice! The 1-D Fibonacci quasilattice has long to short elements in the ratio τ to 1. The substitution rule can be expressed in matrix notation as:

$$I\begin{pmatrix} L \\ S \end{pmatrix} = \begin{pmatrix} L+S \\ L \end{pmatrix}, \tag{20.6}$$

where the matrix operation I is called an *inflation*. The inverse matrix operation, I^{-1}, is a *deflation*. The sequences generated by this operation are self-similar and the numbers of S and L segments increase under inflation, I, and decrease under deflation, I^{-1}. The *inflation* process can be expressed in matrix notation as:

$$\begin{pmatrix} 1 & 1 \\ 1 & 0 \end{pmatrix}\begin{pmatrix} L \\ S \end{pmatrix} = \begin{pmatrix} L+S \\ L \end{pmatrix}, \tag{20.7}$$

identifying the inflation operator as:

$$I = \begin{pmatrix} 1 & 1 \\ 1 & 0 \end{pmatrix}. \tag{20.8}$$

Any of the sequences in Table 20.2 can be generated by the successive operation of the inflation operator. If we associate a column vector with two entries representing the number of L and S segments in the sequence, then the recursive relationship for counting the number of long and short elements in the sequence (beginning with $k = 1$, i.e., a single L segment) can be written in the compact notation (Giacovazzo, 2002b):

$$\mathcal{S}(k) = I^k = \begin{pmatrix} 1 & 1 \\ 1 & 0 \end{pmatrix}^k. \tag{20.9}$$

The first few of these successive operations are shown in Box 20.4. A closed form for the recursion relationship can be written in terms of the Fibonacci numbers as:

$$\begin{pmatrix} F_{k+1} & F_k \\ F_k & F_{k-1} \end{pmatrix} \begin{pmatrix} 1 \\ 0 \end{pmatrix} = \begin{pmatrix} F_{k+1} \\ F_k \end{pmatrix}. \tag{20.10}$$

Decorating the long and short segments with different atoms, the ratio of the number of long to short tiles represents the *stoichiometry* of the quasicrystal, which also approaches τ. The infinite period Fibonacci sequence can also be written as: $\ldots LSLLS \ldots$ or $\ldots SLSLLSL \ldots$ where in the second case the sequence starts at a short segment. The existence of two tiles with occurrences in the ratio of the golden mean is necessary but not sufficient for constructing the quasi-periodic Fibonacci tiling. The sequence of tiles must also be prescribed.

20.3.2 Lattice positions in the Fibonacci lattice (following Venkataraman *et al.*, 1989)

Choosing the sequence $LSLLS\ldots$ with $S = a$ and $L = \tau a$, the position of the n-th lattice point from the origin at the beginning of the chain can be expressed as:

$$x_n = na + \left(\frac{a}{\tau}\right) \text{Int}^{<} \left(\frac{n+1}{\tau}\right) = na + a(\tau - 1)\text{Int}^{<}((n+1)(\tau - 1)), \tag{20.11}$$

where the function $\text{Int}^{<}(y)$ chooses the greatest integer less than or equal to y. For an arbitrary origin and sequencing of tiles, x_n is expressed more generally as:

$$x_n = (na + \alpha) + \left(\frac{a}{\tau}\right) \text{Int}^{<} \left(\frac{n}{\tau} + \beta\right); \tag{20.12}$$

Box 20.4 First few $\mathcal{S}(k)$ in the recursion relationship for the Fibonacci chain

Beginning with a single long segment at $k = 1$:

$$\begin{pmatrix} 1 & 1 \\ 1 & 0 \end{pmatrix}\begin{pmatrix} 1 \\ 0 \end{pmatrix} = \begin{pmatrix} 1 \\ 1 \end{pmatrix} = \begin{pmatrix} 1 & 1 \\ 1 & 0 \end{pmatrix}\begin{pmatrix} 1 \\ 0 \end{pmatrix},$$

yielding a long and short segment. For $k = 2$:

$$\begin{pmatrix} 1 & 1 \\ 1 & 0 \end{pmatrix}\begin{pmatrix} 1 \\ 1 \end{pmatrix} = \begin{pmatrix} 2 \\ 1 \end{pmatrix} = \begin{pmatrix} 2 & 1 \\ 1 & 2 \end{pmatrix}\begin{pmatrix} 1 \\ 0 \end{pmatrix},$$

yielding a two long and short segment. For $k = 3$:

$$\begin{pmatrix} 1 & 1 \\ 1 & 0 \end{pmatrix}\begin{pmatrix} 2 \\ 1 \end{pmatrix} = \begin{pmatrix} 3 \\ 2 \end{pmatrix} = \begin{pmatrix} 3 & 2 \\ 2 & 1 \end{pmatrix}\begin{pmatrix} 1 \\ 0 \end{pmatrix},$$

yielding a three long and two short segments. For $k = 4$:

$$\begin{pmatrix} 1 & 1 \\ 1 & 0 \end{pmatrix}\begin{pmatrix} 3 \\ 2 \end{pmatrix} = \begin{pmatrix} 5 \\ 3 \end{pmatrix} = \begin{pmatrix} 5 & 3 \\ 3 & 2 \end{pmatrix}\begin{pmatrix} 1 \\ 0 \end{pmatrix},$$

where in all cases the left-hand equation represents the next step in the recursion and the right-hand side represents I^k operating on the original long segment.

α and β are arbitrary real numbers, with α determining the origin and β the sequencing of the tiles. Choosing $\alpha = 0$, $\beta = \frac{1}{\tau}$, Box 20.5 shows the first few x_n and compares them with distances calculated from the sequences in Table 20.2. We define a *density function* for the lattice as:

$$\rho(x) = \Sigma_n \delta(x - x_n), \tag{20.13}$$

i.e., with δ-functions located at each tile position $x = x_n$.

20.3.3 Construction of the Fibonacci lattice by the projection method

Projection from higher-dimensional spaces is a general method to construct quasi-periodic structures. A 1-D quasi-periodic structure can also be generated by projecting proximate points of a 2-D square lattice onto a line with an irrational slope. The steps involved in constructing the 1-D Fibonacci tiling by the *cut and project method* (shown schematically in Fig. 20.2) are:

Box 20.5 First few x_n lengths in the Fibonacci series

Beginning with a single long segment at $n = 1$, we have:

$$2(\tau - 1) = 1.236 \rightarrow \text{Int}^{<}(2(\tau - 1)) = 1 \rightarrow x_1 = 1a + a(\tau - 1)1 = 1\tau a + 0,$$

for the distance to the end of a single L segment. For $n = 2$, we have:

$$3(\tau - 1) = 1.854 \rightarrow \text{Int}^{<}(3(\tau - 1)) = 1 \rightarrow x_2 = 2a + a(\tau - 1)1 = 1\tau a + a,$$

for the distance to the end of the LS segment. For $n = 3$, we find:

$$4(\tau - 1) = 2.472 \rightarrow \text{Int}^{<}(4(\tau - 1)) = 2 \rightarrow x_3 = 3a + a(\tau - 1)2 = 2\tau a + a,$$

for the distance to the end of the LSL segment. For $n = 4$, we have:

$$5(\tau - 1) = 3.090 \rightarrow \text{Int}^{<}(5(\tau - 1)) = 3 \rightarrow x_4 = 4a + a(\tau - 1)3 = 3\tau a + a,$$

for the distance to the end of the $LSLL$ segment. For $n = 5, 6, 7, 8, \ldots$ the distances correspond to $LSLLS$, $LSLLSL$, $LSLLSLS$, $LSLLSLSL, \ldots$ Notice that when n is a Fibonacci number, x_n is the length of one of the approximants of Table 20.2.

(i) construct a 2-D square lattice (for convenience the lattice constant is taken as unity);
(ii) select an arbitrary lattice point as the origin, O. The basis directions defining the space of this lattice are labeled x and y, respectively.
(iii) draw a line (1) through the origin with an irrational slope, τ, i.e., if θ is the angle between the line and the x-axis, then $\tan\theta = \tau$. The direction parallel to this line is labeled as $x_{\|}$ and the direction orthogonal to this line as $x_{\perp}$.
(iv) draw another line (2) parallel to the first, intersecting another lattice point. For convenience, intersect another lattice point along the diagonal of a square containing the origin.
(v) project all atoms in the strip between line (1) and line (2) onto line (1), using an orthogonal projection.

Figure 20.2 illustrates the use of the *cut and project method* to produce a quasi-periodic structure (Elser, 1985). Figure 20.2(a) shows the construction of a 2-D square lattice. The shaded strip between lines (1) and (2) represents the projecting space; $x_{\|}$ is the 1-D space (line (1)) into which the lattice is projected; $x_{\perp}$ is a pseudo-space orthogonal to the projected space. An orthogonal pseudo-space is always present in the cut and project method

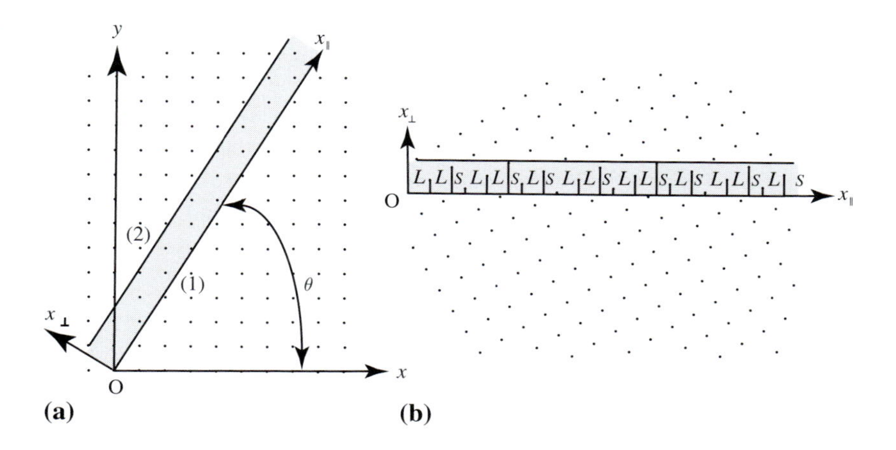

Fig. 20.2. 2-D cut and project method for generating the Fibonacci lattice.

because of the subduction of the higher dimensional space. The dimension of the projected space added to the dimension of the perpendicular pseudo-space is equal to the dimension of the original projecting space (in this case $2 = 1 + 1$).

If all the lattice points of the 2-D space are projected onto $x_{||}$, then the projected structure will densely fill the $x_{||}$ line. In order to get a discrete pattern, we use the further restriction that the only points in the 2-D lattice that are projected are those for which the magnitude of the perpendicular projected vectors $|x_{\perp} - x_0|$ is less than some constant, i.e., the projection strip must be of finite width. In Fig. 20.2(a), the grid points to be projected are chosen only within the shaded strip.

Figure 20.2(b) shows the lattice of Fig. 20.2(a) rotated clockwise by θ. Lines parallel to $x_{\perp}$ are drawn from each lattice point within the cut strip to where they meet $x_{||}$. It can be seen that the projection subdivides $x_{||}$ into a set of long, L, and short, S, segments in the sequence, $LLSLLSLSLLSLLSLSLLS\ldots$ i.e., the Fibonacci lattice! The relative lengths are in the ratio of τ, but the scale depends on the width of the projecting strip, as does the beginning segment for the sequence.

To discuss diffraction from this quasi-periodic structure it is necessary to construct a reciprocal lattice. This is accomplished by considering a dual square 2-D reciprocal lattice from which an analogous projection into one dimension yields a set of reciprocal quasilattice vectors $g_{||}$. In this case, we do not restrict the lattice points projected using analogous proximity arguments, i.e., we project *all* points. This leaves the problem that the irrational projection results in a dense filling of the projected line. However, it appears logical to associate with $g_{||}$ an intensity that is related to its distance from the projected surface (line) $g_{\perp}$. Therefore, an irregular distribution of intensities is expected as a function of $g_{\perp}$ and the density of the projected reciprocal quasilattice points is determined by the resolution of the spectrometer.

20.3.4 *The Fourier transform of the Fibonacci lattice (following Venkataraman *et al.*, 1989)

We choose $a = 1/\sqrt{(1+\tau^2)}$ as the lattice spacing, i.e., $a = \sin(\arctan(\tau))$; this is a normalization constant that represents the natural length scale obtained by projecting points from a 2-D square lattice with unit spacings along a line of slope $\tan\tau$, as used in the cut and project method of the previous section. For numerical convenience, we choose $\alpha = 0$ and $\beta = \frac{1}{2}$, so that:

$$x_n = \frac{1}{\sqrt{(1+\tau^2)}}\left(n + \frac{1}{\tau}\mathrm{Int}^{<}\left(\frac{n+1}{\tau} + \frac{1}{2}\right)\right). \tag{20.14}$$

The density of lattice points in the unit square lattice of the projecting space is:

$$L(x, y) = \frac{1}{4\pi^2}\Sigma_{i,j}\delta(x-i)\delta(y-j). \tag{20.15}$$

This can be expressed in the rotated coordinate system as:

$$L(x_\perp, x_{||}) = L(x_\perp \sin\theta + x_{||}\cos\theta, x_\perp\cos\theta - x_{||}\sin\theta). \tag{20.16}$$

Finally the projected *density function*, $\rho(x_{||})$, for the lattice is:

$$\rho(x_{||}) = \int_{-w/2}^{w/2} S(x_\perp)L(x_\perp, x_{||})\mathrm{d}x_\perp, \tag{20.17}$$

where $S(x_\perp)$ is the strip of width w from which the 2-D lattice points are being projected, chosen for convenience to be symmetric about the origin, i.e., between $-w/2$ and $w/2$.

The *Fourier transform*, ρ(q) of ρ(x) is:

$$\rho(q) = \frac{1}{2\pi}\int \tilde{S}(-p)\tilde{L}(p, q)\mathrm{d}p, \tag{20.18}$$

where $\tilde{S}(-p)$ and $\tilde{L}(p, q)$ are the Fourier transforms of the shape and lattice densities, respectively:

$$\tilde{S}(p) = \frac{w\sin\frac{pw}{2}}{\frac{pw}{2}}, \tag{20.19}$$

and:

$$\tilde{L}(p, q) = \delta(p - 2\pi(n\cos\theta + m\sin\theta))\delta(q - 2\pi(-n\sin\theta + m\cos\theta)); \tag{20.20}$$

where m and n are integers. This leads to:

$$\rho(q) = \frac{1}{2\pi}\tilde{S}(n\sin\theta - m\cos\theta)\delta(q - 2\pi(-n\sin\theta + m\cos\theta)). \tag{20.21}$$

From the Fourier coefficients and the fact that:

$$\sin\theta = \frac{1}{\sqrt{1+\tau^2}} \quad \text{and} \quad \cos\theta = \frac{\tau}{\sqrt{1+\tau^2}},$$

we conclude that the diffraction condition is:

$$q_{mn} = \frac{2\pi(m+n\tau)}{\sqrt{(1+\tau^2)}}. \tag{20.22}$$

This implies that the positions x_n can be described in terms of superimposed waves with wave vectors that are incommensurate, i.e., $n\tau$ is irrational. The diffraction spots require two Miller indices (m, n), whereas a periodic 1-D lattice would require only one. Because of the strip function, reflections with high Miller indices (m, n) are less intense than those with lower indices. Finally, the decoration of the two tiles with atoms determines the stoichiometry.

20.4 *Two-dimensional quasicrystals

In higher-dimensional analogues to the Fibonacci lattice, the projection can be chosen to preserve symmetries of the higher dimensional space. An important example of a two-dimensional (2-D) quasicrystal is the *Penrose tile*, Fig. 20.3, where an irrational projection preserves non-crystallographic five-fold symmetry. The irrational cut could equally well be chosen to preserve other symmetries. We will consider analogous tilings which preserve 8-, 10-, and 12-fold symmetries, for example.

20.4.1 2-D quasicrystals: Penrose tilings

In 1974, Roger Penrose discovered a set of prototiles which, following matching rules, tiled a plane quasi-periodically while preserving five-fold rotational

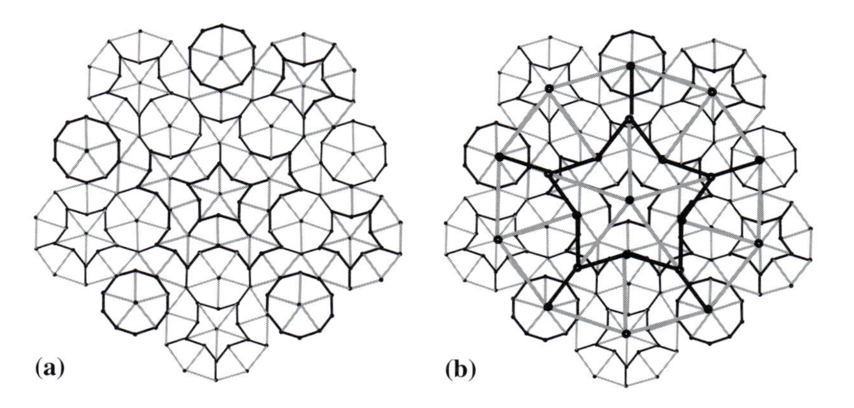

Fig. 20.3. (a) Penrose tiling with dart and kite prototiles and (b) the same tile with its self-similar central patch inflated by τ^2 and superimposed.

symmetry. The *Penrose tile* (Penrose, 1974) has since been the motivation for much of the theoretical work on quasicrystals. The Penrose tiling can be understood in terms of a rule for connecting two *Penrose rhombs*, which have areas in the ratio of the golden mean, τ. The Penrose tile can be constructed from its prototiles through a similar inflationary process as that used to construct the 1-D Fibonacci lattice. It can also be understood in terms of basis vectors projected from a higher dimensional space.

20.4.1.1 The Penrose tile derived by the inflation–deflation method

Figure 20.3(a) illustrates a Penrose tile where the two prototiles are called a *dart* and *kite*, respectively (Gardner, 1977). Figure 20.3(b) shows the central patch of the Penrose tile of Fig. 20.3(a) scaled by τ^2 and superimposed on the original tile. The process of combining prototiles to form new tiles that are larger versions of the originals is called *inflation*; the inverse mapping is called *deflation*. The tilings (before and after inflation or deflation) are said to be self-similar, in that similar structures are observed on the two scales. *Self-similarity* means that an equivalent set of tiles may be constructed from the original set such that the new tiles are related to the old ones through the same scaling.

Figure 20.4(a) illustrates the dart (top) and kite (bottom) prototiles of the Penrose tile. The vertices of these tiles are marked with black and white dots to describe the matching rules for constructing the Penrose tiles. It is only possible to tile the plane quasi-periodically if the shared edges between prototiles have the same coloring of the dots. Figure 20.4(b) shows that the dart and kite can be decorated with two more basic prototiles in the form of acute and obtuse rhombi. Figure 20.4(c) shows these basic units to be an acute and obtuse rhombus with pentagonal (decagonal) angles. The edge lengths of both rhombi are equal and the angles ($\theta = \pi/5$) are illustrated in Fig. 20.4(c).

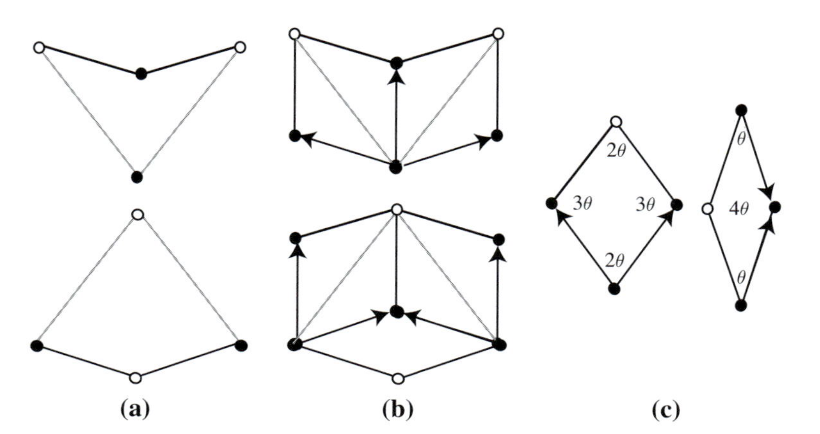

Fig. 20.4. (a) Dart (top) and kite prototiles, (b) their decoration by Penrose rhombs, and (c) angular relationships for the rhombic tiles.

Arrow directions on the edges are another method of enforcing the matching rules, to insure that the tiling is quasi-periodic.[5]

It is instructive to further examine the Penrose tiling of Fig. 20.3, to understand local environments in the tiling and how a 2-D quasicrystal would grow. There are only seven configurations of kites and darts around any vertex in this tiling (Grünbaum and Shepard, 1987). Figure 20.5 illustrates these side by side with their decoration with Penrose rhombs. These vertex configurations have been given affectionate names (a) sun, (b) star, (c) deuce, (d) ace, (e) king, (f) queen, and (g) jack, respectively. Socolar (1991) has enumerated deterministic rules by which a tiling can grow, using only local information at the vertex to determine which tiles to add next. We will not detail these rules but it is comforting to know that these growth models do in fact exist.

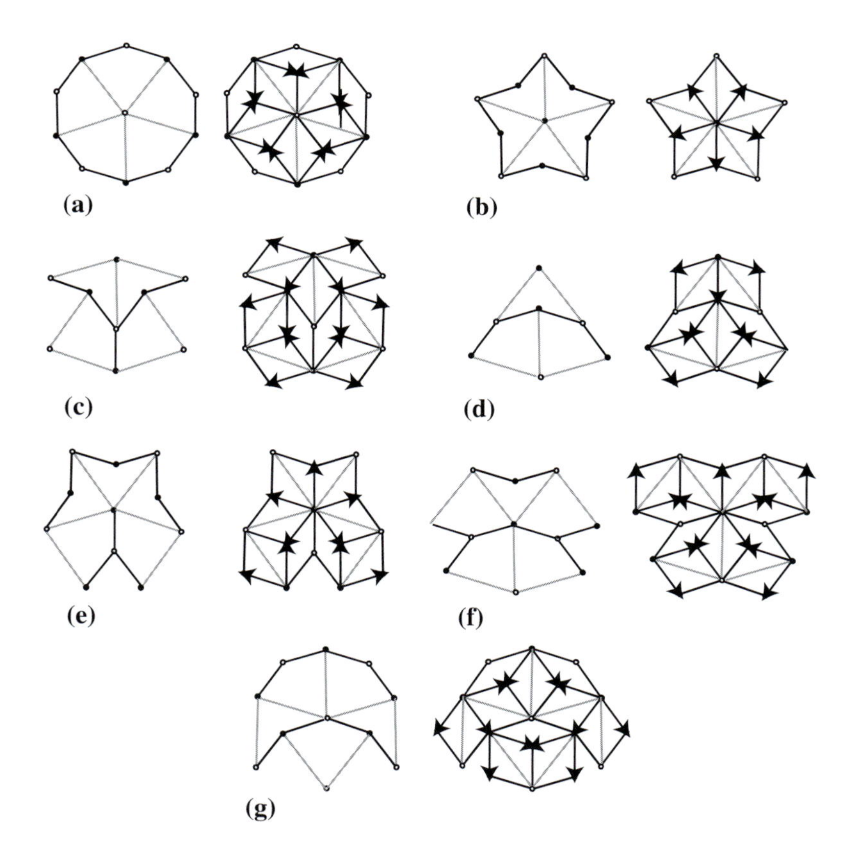

Fig. 20.5. The seven vertex configurations of kites and darts in the Penrose tile and decoration by Penrose rhombs: (a) sun, (b) star, (c) deuce, (d) ace, (e) king, (f) queen, and (g) jack.

[5] The colored dots (poles) and arrowed edges are two equivalent ways of enforcing Penrose matching rules commonly used in the literature.

20.4.1.2 The Penrose tiles: tilings with Penrose rhombs

Figure 20.6 shows two Penrose tilings that are tiled by Penrose rhombs. These tilings are related by an inflation (deflation) factor of τ. Notice that an inflation by τ does not yield a tiling with coincident sites with the first, but is rotated by π (i.e., pentagons that point up (down) in the original tiling point down (up) in the inflated tiling). It is only through a τ^2 inflation (deflation) that the pentagons are self-similar.

The quasi-periodicity of the Penrose tilings can be demonstrated by taking a diffraction pattern of the vertex sites (Levine and Steinhardt, 1985). The long range orientational order is also demonstrated by observing that the orientation of the many decagons in the structure is preserved. The Penrose lattice illustrated in Fig. 20.6 shows the composition of the lattice in terms of the two tiles. The frequencies of occurrence of the rhombs in an infinite Penrose tile are in the ratio τ to 1. The tiling rules admit an infinite number of different tilings.

Figure 20.7 further examines the inflation/deflation process considering a (decagon) patch from the Penrose tiling. The recursion rule for replacement of acute and obtuse rhombi has been illustrated by Mackay (1982). In deflating to the next generation, acute rhombi subdivide according to the rule: [Acute $\rightarrow$ 2Acute$'$ + Obtuse$'$] and obtuse rhombi follow the rule [Obtuse $\rightarrow$ Acute$'$ + Obtuse$'$].

Mackay's decoration of the Penrose rhombs is illustrated in Fig. 20.7(a). In the decagon, we begin with equal numbers of acute and obtuse rhombi. After the first deflation, Fig. 20.7(b), the ratio of acute to obtuse rhombi is 3 to 2 (within the area defined by the first decagon). The deflated tiling has the central pentagonal star rotated by π. After the second deflation, Fig. 20.7(c), the ratio of acute to obtuse rhombi is now 8 to 5. The central pentagonal star is rotated by π so that it is again parallel to the original star of Fig. 20.7(a). Decoration of the nodes of the τ^2 deflated tile with white and black circles is shown in (d). The thick lines illustrate seven of eight possible different vertex

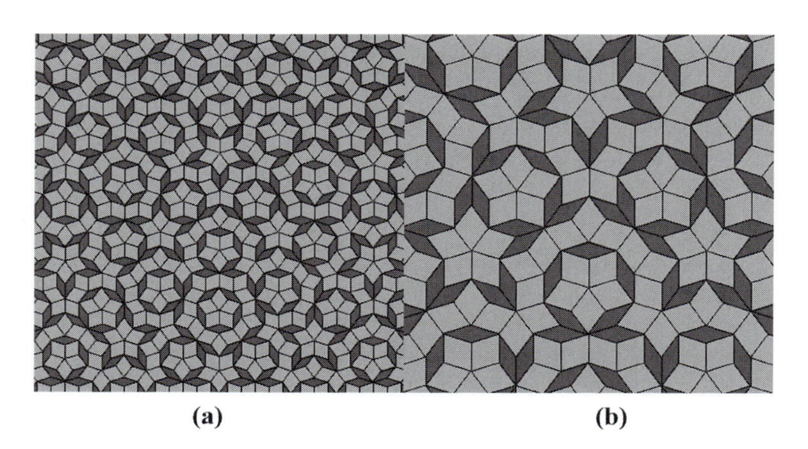

Fig. 20.6. Penrose tilings (a) and (b) are related by *inflation* (*deflation*).

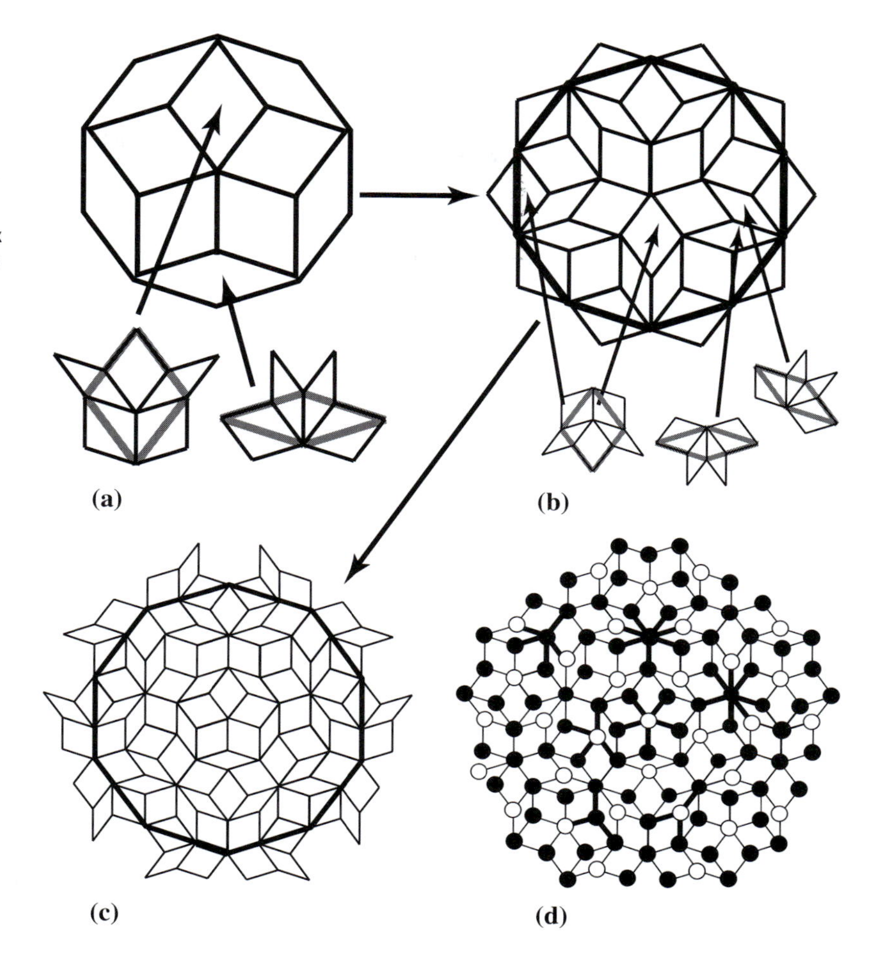

Fig. 20.7. (a) Penrose tiling and recursive substitutions yielding a *deflation* by τ (b) and τ^2 (c). Decoration of the nodes of the τ^2 deflation with white and black circles is shown in (d), thick lines illustrate seven different vertex configurations (figure courtesy J. L. Woods).

configurations in the Penrose tile. The last one occurs out of the field of view of the patch of the tiling shown.

Mackay's recursion relation is analogous to that of the Fibonacci sequence. The frequency of obtuse and acute tiles in successive tilings are Fibonacci numbers. It is left as a reader exercise to show that the ratio of acute to obtuse rhombi contained in the region of the original obtuse rhombus is: F_{2k}/F_{2k-1} where k is the number of deflations. Similarly, the ratio of acute to obtuse rhombi contained in the region of the original acute rhombus is: $F_{2k-1}/F_{2(k-1)}$. Finally, the ratio of acute to obtuse rhombi in total is: F_{2k+1}/F_{2k}.

In analogy with Fig. 20.5, we now consider the vertex configurations for the Penrose rhombs. The decoration of the dart and kite vertex arrangements can yield more than one Penrose rhomb vertex site. A systematic method for determining the arrangement of vertex sites with Penrose rhombs relies on the fact that the sum of the angles of rhombs meeting at a single vertex must

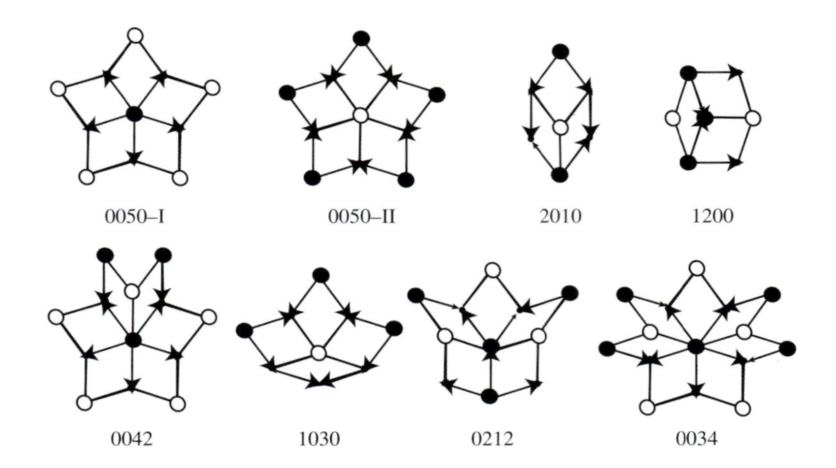

Fig. 20.8. The eight configurations of Penrose rhombs around any vertex in the Penrose tiling.

equal 2π. From this we conclude that:

$$n_1\left(\frac{4\pi}{5}\right)+n_2\left(\frac{3\pi}{5}\right)+n_3\left(\frac{2\pi}{5}\right)+n_4\left(\frac{\pi}{5}\right)=2\pi \tag{20.23}$$

where n_1, n_2, n_3, and n_4 are integers. We will use the symbol $(n_1n_2n_3n_4)$ to designate an allowed vertex configuration. The eight possible configurations are illustrated in Fig. 20.8. The integers have upper bounds and restrictions posed by the Penrose matching rules. Obviously, n_1 and n_2 cannot exceed 2, n_3 cannot exceed 5 and n_4 cannot exceed 10 in any configuration.

Figure 20.8 shows that the only two possible configurations with $n_5=5$ satisfying the matching rules are (0050)−I and (0050)−II. One is centered by a black dot and the other by a white dot. The only ways of surrounding a vertex with three edges gives rise to the configurations (2010) and (1200). For the (1200) configuration, the central vertex is black and we can replace the obtuse vertex to yield the (0212) configuration shown. This accounts for all configurations where n_1 or n_2 equals 2.

Penrose matching rules require that for the (2010) configuration the vertex is white. This is the only way to pair two obtuse triangles. There is no way for two obtuse rhombus pairs to be adjacent at a vertex, nor can they occur as three adjacent obtuse rhombs. This means that $n_4\neq 10, 9, 8, 7, 6, 5$. Further, when the odd angles $3\pi/5$ or $\pi/5$ meet at a vertex, symmetry requires that they occur in pairs and cannot pair with each other. This implies that $n_4\neq 3, 1$ and $n_2\neq 1$. In all remaining configurations $n_1=0, 1$, $n_2=0$, $n_3=0, 1, 2, 3, 4$, and $n_4\neq 0, 2, 4$.

Figure 20.8 shows that the three remaining possibilities with $n_1\neq 0$ are (1030), (1022), and (1014). Configuration (1030) satisfies the matching rule at a vertex with a white dot. Pairs of obtuse rhombi with black dot vertices cannot be substituted for acute rhombi, so that (1022) and (1014) are not possible. The final possible configurations are (0042) and (0034); (0026) and (0018) are not possible for the reasons mentioned above.

20.4.2 The Penrose tiling derived by projection

In analogy to the cut and projection from a 2-D space to yield the Fibonacci lattice, it is possible to construct the Penrose lattice considering a 5-D space projected onto 2-D. The projection is chosen to preserve the orientational relationship between the basis vectors which, in 2-D, point to the vertices of a regular pentagon (Socolar and Steinhart, 1986) as shown in Fig. 20.9. These basis vectors are expressed in terms of the five vertices of the pentagon as:

$$\mathbf{e}_n = \left[\cos\left(\frac{2\pi n}{5}\right), \sin\left(\frac{2\pi n}{5}\right)\right], \tag{20.24}$$

where $n = 1, 2, \ldots 5$.

The choice of the projection from 5-D onto the vertex vectors of the regular pentagon in Fig. 20.9(a) is sufficient to preserve five-fold symmetry as a long range orientational order throughout the structure. The reciprocal lattice basis vectors of the pentagon describe a rotated dual pentagon:

$$\mathbf{e}_n^* = \left[\sin\left(\frac{\pi(2n+1)}{5}\right), \cos\left(\frac{\pi(2n+1)}{5}\right)\right], \tag{20.25}$$

indicating that diffraction spots can be indexed in terms of five Miller indices $(h_1, h_2, h_3, h_4, h_5)$. In practice, only four Miller indices are necessary because:

$$\sum_{n=1}^{5} \mathbf{e}_n^* = 0, \tag{20.26}$$

so that one of the reciprocal lattice vectors is not independent. It is thus possible to construct a general scattering vector, $\mathbf{Q}$, in terms of the four independent vectors:

$$\mathbf{Q} = \sum_{i=1}^{4} h_i \mathbf{e}_i^* \tag{20.27}$$

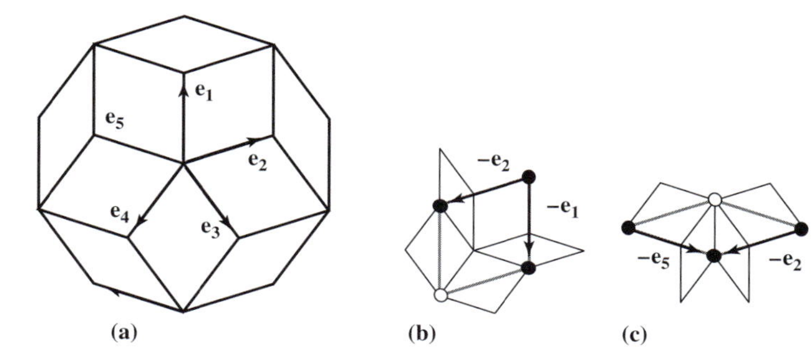

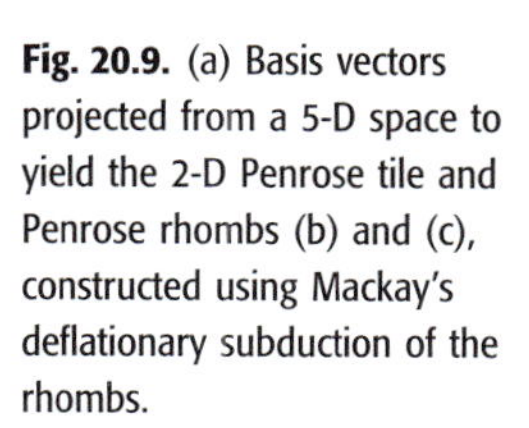
Fig. 20.9. (a) Basis vectors projected from a 5-D space to yield the 2-D Penrose tile and Penrose rhombs (b) and (c), constructed using Mackay's deflationary subduction of the rhombs.

An analogous projection from a 10-D space is possible where the basis vectors are related to the vertex vectors of the decagon. In the next section, we will illustrate the projection of the polygonal unit vectors of the octagonal, decagonal, and dodecagonal lattices.

The projection from a 5-D space which gives rise to the Penrose tile is only one possible projection from a higher dimensional space that gives rise to a quasi-periodic 2-D tiling. Again, these are constructed by taking a rational or irrational cut of a higher dimensional space and projecting the lattice points into two dimensions. The Penrose tiles in Fig. 20.6 were generated using QuasiTiler.[6] QuasiTiler implements a generalization of the above ideas of the Penrose tiling. It can generate different tilings by choosing the dimension of the higher dimensional space, the generating plane, and the translation of the projection of the unit cell in the orthogonal space.

20.4.3 2-D quasicrystals: other polygonal quasicrystals

After the description of the Penrose tiling (Penrose, 1974), its interpretation as a projection of a 5-dimensional lattice structure into a 2-D subspace (de Bruijn, 1981), and the discovery of 3-D quasicrystalline alloys (Shechtman *et al.*, 1984), several new 2-D quasicrystalline alloys have been discovered. These alloys included octagonal Cr–Ni quasicrystals (Wang *et al.*, 1987), decagonal Al–Mn quasicrystals (Bendersky, 1985), and dodecagonal Ni–Cr quasicrystals (Ishimasa *et al.*, 1985). These are called *polygonal quasicrystals* because they possess non-crystallographic (8-, 10-, and 12-fold) symmetries within stacked planes, but no rotational symmetries orthogonal to these planes. It is possible to construct 2-D tilings with these symmetries using the previously discussed projection and inflation methods. Figure 20.10 (top) illustrates examples of tilings with (a) 8-fold, (b) 10-fold, and (c) 12-fold symmetries. Note that these tiling examples are constructed using four, five, and six tiles (rhombs), respectively.

All three tilings in Fig. 20.10 give rise to diffraction patterns with the symmetries of the tiles and discrete diffraction spots. Like the case of the Penrose tiling, the reciprocal space lattice vectors can be obtained by projection from higher dimensional spaces. A set of basis vectors pointing to the vertices of an n-gon is again over-determined because one or more of the projected unit vectors of the *polygonal reciprocal lattices* can be described as a linear combination of the others. In each of these cases, the unique vectors can be described in terms of a 4-D space and, therefore, four *Miller indices* are required to index the diffraction patterns. Figure 20.10 (middle row) shows a choice of these reciprocal lattice vectors for the (a) octagonal, (b) decagonal, and (c) dodecagonal quasicrystalline tilings, respectively.

[6] http://www.geom.uiuc.edu/apps/quasitiler/.

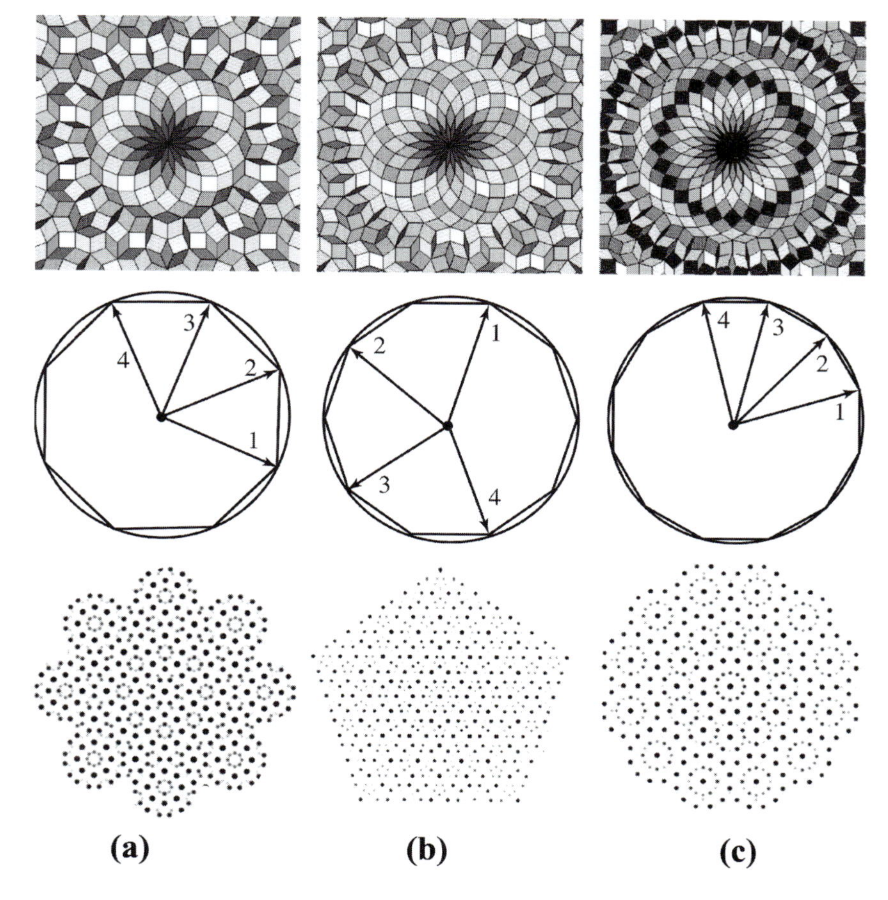

Fig. 20.10. 2-D tilings (top), projection of the unit vectors of the polygonal reciprocal lattices in a 4-D space (middle) and simulated diffraction patterns, all along the highest symmetry zone axis for (a) octagonal, (b) decagonal, and (c) dodecagonal quasicrystals. Tilings were produced using QuasiTiler from the Geometry Center at the University of Minnesota.

20.4.3.1 The law of rational indices revisited

In Chapter 8 it was shown that rotational symmetries consistent with translational periodicity can be determined by considering two lattice points separated by a unit translation vector, **t**. Rotation of both lattice points by the angle $\pm\alpha$ about an axis normal to the plane, shows that the rotations are consistent with the translational symmetry of a periodic lattice, if the distance t' (the length of $\mathbf{t}'$) between the new positions can be written as:

$$t' = t + 2t\cos\alpha = t(1 + 2\cos\alpha) = mt. \tag{20.28}$$

where m is an integer. Note that the distance t is *scaled* by $1 + 2\cos\alpha$ to yield the distance t'. The quantity $1 + 2\cos\alpha$ is the trace of the matrix representing the rotation by the angle α and is an invariant quantity for the representation of that rotation.

We now consider the quantity:

$$T = 1 + 2\cos\left(\frac{2\pi}{n}\right), \tag{20.29}$$

which is the trace of a matrix representing a rotation by an angle $\alpha = 2\pi/n$ about an n-fold rotation axis. We will explore below *quasilattices* associated with the values $n = 5, 8, 10$, and 12. Note that for $n = 1, 2, 3, 4$, and 6, corresponding to the identity, two-fold, three-fold, four-fold, and six-fold rotation operations, $T = 3, -1, 0, 1$ *and* 2, respectively. For the rotations compatible with translational symmetry, the traces of the rotation matrices are integers. For $n = 5, 8, 10$, and 12, corresponding to five-fold, eight-fold, ten-fold, and twelve-fold rotation operations, $T = \tau$, $1 + \sqrt{2}$, $1 + \tau$, and $1 + \sqrt{3}$, respectively.

The consequence of these irrational traces for diffraction is profound. Consider $n = 8$, 10, and 12: if a unit distance in the reciprocal lattice occurs every 45°, 36°, or 30° (i.e., a unit vector replicated by the rotation operations), then adding two adjacent vectors defines a vector ending on a ring of radius τ, $\sqrt{2}$ or, $\sqrt{3}$, respectively. Subsequent rings will also be found at integral powers of these irrational numbers.

20.4.3.2 Rotations in higher dimensional spaces

Point group symmetries can be combined with the translations of the higher dimensional space to construct higher dimensional space groups. For octagonal, decagonal, and dodecagonal quasicrystals, the rotational symmetries can be described in terms of their action on the 4 unique projected basis vectors of the reciprocal lattice (or the direct space tilings, since the symmetries are the same) of Fig. 20.10. The 4-D rotation matrices $D(n)$ for the **8** (C_8), **10** (C_{10}), and **12** (C_{12}) rotation operations (Yamamoto, 1996), are:

$$D(8) = \begin{pmatrix} 0 & 1 & 0 & 0 \\ 0 & 0 & 1 & 0 \\ 0 & 0 & 0 & 1 \\ -1 & 0 & 0 & 0 \end{pmatrix}; \quad D(10) = \begin{pmatrix} 0 & 1 & 0 & 0 \\ 0 & 0 & 1 & 0 \\ 0 & 0 & 0 & 1 \\ -1 & -1 & -1 & -1 \end{pmatrix};$$

$$D(12) = \begin{pmatrix} 0 & 1 & 0 & 0 \\ 0 & 0 & 1 & 0 \\ 0 & 0 & 0 & 1 \\ -1 & 0 & 1 & 0 \end{pmatrix}. \tag{20.30}$$

Note that in all cases these are n-fold rotation axes and repeated operation reveals the orders of the **8** (C_8), **10** (C_{10}), and **12** (C_{12}) point groups to be 8, 10, and 12, respectively. Dihedral groups of orders 16, 20, and 24, respectively, can also be generated by combining the set of pure rotation operations with the operation for a horizontal mirror plane given by:

$$D(\sigma) = \begin{pmatrix} 0 & 0 & 0 & 1 \\ 0 & 0 & 1 & 0 \\ 0 & 1 & 0 & 0 \\ 1 & 0 & 0 & 0 \end{pmatrix}. \tag{20.31}$$

The higher dimensional representations offer the simplicity of matrix elements that are all 0s and 1s.

As with the Penrose tiling, the choice of the set of unit vectors is not unique. Another set can be constructed by an inflation or deflation (i.e., a scaling by τ^n for dodecagonal quasicrystals). For a more detailed discussion the reader is referred to Yamamoto (1996). The projection of the unit vectors of the polygonal reciprocal lattices leads to a set of reciprocal lattice basis vectors which can be used in an inflationary process to simulate diffraction patterns for the polygonal lattices. The positions of rings of spots occur at integral powers of the particular irrational number related to the rotational symmetry. Pseudo-intensities can also be represented by taking the size of the diffraction spots as proportional to the sum of the lengths of the reciprocal lattice vectors used to locate a position in reciprocal space. Figure 20.10 (bottom row) shows simulated diffraction patterns for (a) the octagonal, (b) the decagonal, and (c) the dodecagonal polygonal quasicrystals along the highest symmetry zone axis.[7]

20.5 *Three-dimensional quasicrystals

The study of quasicrystallography began with the challenge of explaining the five-fold electron diffraction patterns observed in Al–Mn alloys as well as sharply peaked X-ray diffraction patterns. The challenge in explaining the original experimental observations was to devise a scheme where the space group of the quasicrystal could preserve the icosahedral or decagonal symmetry observed for this quasicrystal. Such a model should also be able to index XRD peaks of the quasicrystal. It should predict peak intensities and variations in the peak widths if the decoration of the quasilattice is known. The model should also be able to describe the various symmetries of zone axes observed in electron diffraction. To achieve this goal, it is necessary to know the atomic positions in the quasicrystal; this is still an active area of research.

Two early models were proposed to explain quasicrystalline diffraction patterns. The first, proposed by Elser and Henley (1985), Elser (1985, 1986), considered the diffraction patterns of projected structures. Quasi-periodic structures in a lower-dimensional space were generated through projection from a higher dimensional space. An irrational cut of the higher dimensional space was chosen so that projected vectors in the lower dimensional space constituted a representation of the icosahedral group. This cut and project method was discussed in previous sections for simpler lower-dimensional quasilattices, where projection from a 2-D square lattice into 1-D space yielded

[7] Simulated diffraction patterns courtesy of Steffen Weber.

the 1-D Fibonacci lattice and projection from a 5-D space into 2-D space yielded the 2-D Penrose lattice. Extension of these ideas to consider a projection from a 12-D (or 6-D) space into three dimensions (3-D) which preserves icosahedral symmetry was shown to yield a 3-D Penrose lattice that is one of the generally accepted models of quasicrystalline structure.

The second early model of quasicrystalline structure was called the icosahedral glass model (Stephens and Goldman, 1986). This model considered the quasicrystal as constructed from the random packing of icosahedral units (clusters). The packing of these units is specified by vertex, edge, and face sharing constraints. These steric constraints are shown to yield atomic positions which can be described by almost periodic functions (Hendricks and Teller, 1942). The Fourier transforms of these almost periodic functions yield sharply peaked diffraction patterns (with some finite broadening), consistent with those observed for quasicrystals.

20.5.1 3-D Penrose tilings

Like the preceding example of the 2-D Penrose tiling, the 3-D Penrose tiling may be described as an aperiodic packing of two tiles whose volumes and frequency of occurrence both scale in the ratio of the golden mean. The first of these Penrose bricks is shown in Fig. 20.11. Figure 20.11(a) illustrates an atom-centered icosahedron with atoms at the vertices labeled C1–C12 and the central atom labeled C13. Figure 20.11(b) shows a choice of basis vectors (with atoms attached to each end) that can be used to construct one of the Penrose bricks, the *oblate rhombohedron*. These basis vectors point from the central C13 atom to the vertices C5, C11, and C13, respectively, where these atoms are now colored a darker gray to distinguish them from the other icosahedral vertices. Figure 20.11(c) illustrates the addition of these basis vectors to define the entire parallelogram. Figure 20.11(d) show the resulting Penrose *oblate rhombohedron*.

Using the symmetry operations of the icosahedral group, it is possible to determine all possible orientations of this brick. The derivation of the second Penrose brick is shown in Fig. 20.12. The icosahedral vertex positions

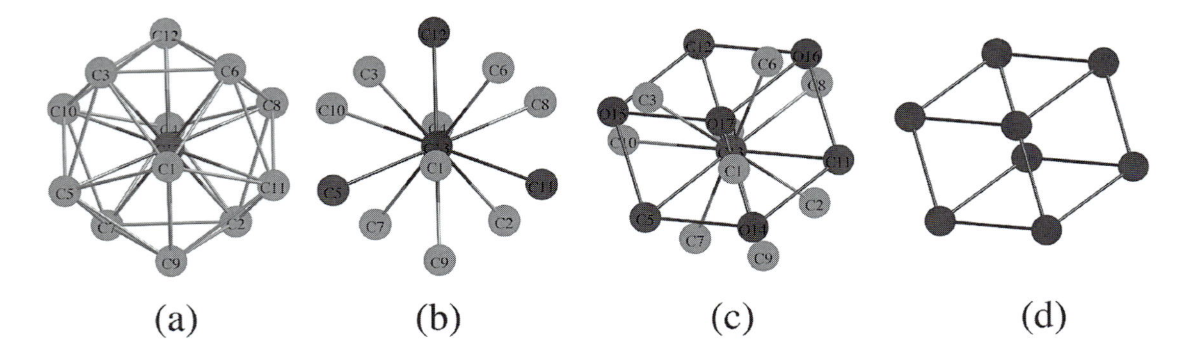

Fig. 20.11. The definition of the *oblate rhombohedron* Penrose brick: (a) reference icosahedron, (b) basis vectors for the brick, (c) extension to form a parallelogram, and (d) the free standing *oblate rhombohedron*.

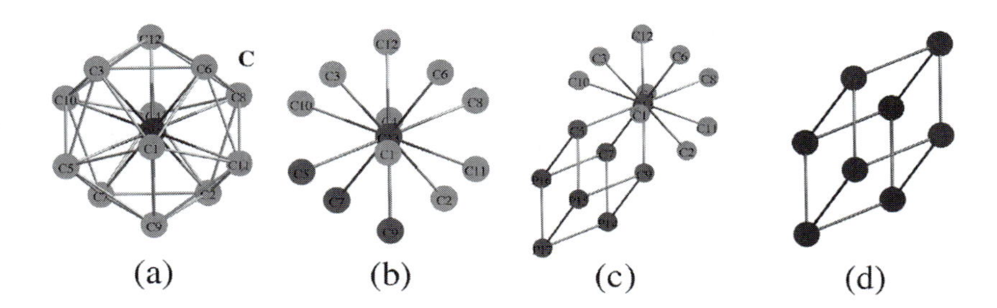

Fig. 20.12. The definition of the *prolate rhombohedron* Penrose brick: (a) reference icosahedron, (b) basis vectors for the brick, (c) extension to form a parallelogram, and (d) the free standing *prolate rhombohedron*.

are shown in Fig. 20.12(a), the definition of the basis vectors is shown in Fig. 20.12(b), and the addition of the basis vectors to define the *prolate rhombohedron* in Fig. 20.12(c). Figure 20.12(d) illustrates a free standing *prolate rhombohedron* Penrose brick.

The edge vectors of each of the Penrose bricks can be normalized by $(1+\tau^2)^{-1/2}$, so that they correspond to basis vectors of the projecting space. Only the six unique basis vectors of the 6-D space are required to produce these bricks in all possible orientations found in the Penrose tile. In the bricks illustrated, only five of a set of six spanning basis vectors have been defined. The remaining basis vector could be inferred but it is useful to have a more general definition of the basis vectors of the 6-D space.

The basis set for this 6-D space can be described with reference to a coordinate system having a five-fold icosahedral axis parallel to the z-axis of the coordinate system as follows:

$$\mathbf{e}_n = \left[\cos\left(\frac{2\pi n}{5}\right), \sin\left(\frac{2\pi n}{5}\right), \frac{1}{\sqrt{5}}\right] \tag{20.32}$$

for $n = 2, 3, \ldots, 6$ and $\mathbf{e}_1 = (1, 0, 0)$. Alternatively, in the literature it is also common to orient the two-fold symmetry axes of the icosahedron along the coordinate axes, so that the basis vectors may be expressed as the cyclic permutations of $(1, \tau, 0)$ with appropriate normalization (Elser, 1985). Note that there are $2^6 \times 6 = 384$ unique possibilities for labeling these vectors.

The star of these vectors defines a polyhedron called the rhombic triacontahedron illustrated in Fig. 20.15(d) on page 608 which can be constructed from ten of each type of Penrose brick shown in Fig. 20.11 and Fig. 20.12, respectively. Because of the unique duality relationships of the icosahedral groups, this polyhedron is also the star of the reciprocal lattice vector, i.e., the vectors of 20.15(d) also span reciprocal space.

20.5.2 Indexing icosahedral quasicrystal diffraction patterns

Cahn *et al.* (1986) summarized a method for indexing quasicrystalline diffraction patterns using a variation of the basis vectors suggested by Elser. This first involved a description of direct and reciprocal space vectors and planes. In 3-D space, a vector is denoted $[UVW]$ and a plane by (HKL) where the capital letter designations distinguish a quasilattice from a Bravais lattice. As described in Chapter 15, the most convenient coordinate system for describing icosahedral symmetry operations is one for which three orthogonal two-fold axes are chosen to coincide with the Cartesian coordinate system axes. In such a system, the actions of the symmetry operations of the icosahedral point groups $\mathbf{m\bar{3}\bar{5}}$ (I_h) and **532** (I), on a position $[UVW]$ can be summarized as follows:

(i) The two-fold axes give pairs of sign changes (e.g., $[\bar{U}\bar{V}W]$, $[\bar{U}V\bar{W}]$, ...) and mirrors give individual sign changes (e.g., $[\bar{U}VW]$, $[U\bar{V}W]$ $[UV\bar{W}]$, ...) and inversion changes all signs (only for $\mathbf{m\bar{3}\bar{5}}$ (I_h)).
(ii) The three-fold axes give cyclic permutations (e.g., $[WUV]$, and $[VWU]$).
(iii) The five-fold axes change the magnitudes of U, V, and W and yield linear combinations involving the golden mean.

We use the choice made by Cahn *et al.* (1986) for the reciprocal lattice basis vectors:

$$\begin{aligned} \mathbf{e}_1^* &= \mu[1, \tau, 0]; & \mathbf{e}_2^* &= \mu[\tau, 1, 0]; \\ \mathbf{e}_3^* &= \mu[0, 1, \tau]; & \mathbf{e}_4^* &= \mu[-1, \tau, 0]; \\ \mathbf{e}_5^* &= \mu[\tau, 0, -1]; & \mathbf{e}_6^* &= \mu[0, -1, \tau], \end{aligned} \tag{20.33}$$

where $\mu = \surd(1+\tau^2)$ is a normalization constant. Again, this choice of basis vectors is one of $2^6 \times 6 = 384$ unique possibilities for choosing these vectors. These vectors satisfy the orthogonality relationship:

$$\mathbf{e}_i^* \cdot \mathbf{e}_j^* = \surd 5 \, (i \neq j) \qquad \text{and} \qquad \mathbf{e}_i^* \cdot \mathbf{e}_i^* = 1. \tag{20.34}$$

Given a unit distance in the reciprocal lattice, adding two adjacent vectors defines a vector ending on a ring at a distance τ. Subsequent rings will also be found at linear combinations of integral powers of irrational numbers which, as proved above, can always be written as $m\tau + p$. As a result, the Miller indices H, K, and L for a plane can be expressed as:

$$\begin{aligned} H &= h + h'\tau \\ K &= k + k'\tau \\ L &= l + l'\tau \end{aligned} \tag{20.35}$$

where h, h', k, k', l, l' are all integers and constitute six Miller indices of an icosahedral quasicrystal reflection. The six index scheme for labeling icosahedral reflections is, therefore, of the form:

$$(h/h' \, k/k' \, l/l'). \tag{20.36}$$

As was the case with the polygonal quasicrystals, it is also possible to discuss point group symmetries within the context of the 6-D space basis of the icosahedral quasicrystals. Here, it is sufficient to construct a rotation matrix for a 5 (C_5), and 3 (C_3) rotation operation (Yamamoto, 1996), and generate the rest of the icosahedral group elements using the generating relationship of Chapter 15. The generators are expressed as follows:

$$D(5) = \begin{pmatrix} 1 & 0 & 0 & 0 & 0 & 0 \\ 0 & 0 & 1 & 0 & 0 & 0 \\ 0 & 0 & 0 & 1 & 0 & 0 \\ 0 & 0 & 0 & 0 & 1 & 0 \\ 0 & 0 & 0 & 0 & 0 & 1 \\ 0 & 1 & 0 & 0 & 0 & 0 \end{pmatrix}; D(3) = \begin{pmatrix} 0 & 0 & 0 & 0 & 0 & 1 \\ 1 & 0 & 0 & 0 & 0 & 0 \\ 0 & 0 & 0 & 0 & 1 & 0 \\ 0 & 0 & -1 & 0 & 0 & 0 \\ 0 & 0 & 0 & -1 & 0 & 0 \\ 0 & 1 & 0 & 0 & 0 & 0 \end{pmatrix}; \quad (20.37)$$

where, once again, the representation in the higher dimensional space has the convenience of only 1s and 0s occurring as entries in the matrix representation. It is possible to construct a generalized Bragg scattering vector in terms of the six basis vectors:

$$b\mathbf{Q}_{\|} = \tau \sum_{i=1}^{6} h_i \mathbf{e}_i^*, \quad (20.38)$$

where b is a scale factor depending on the interatomic spacing. It is thus seen that this construction has left us with an indexing scheme in which there are six Miller indices (as discussed further in the section on indexing icosahedral diffraction patterns). The preservation of the icosahedral vertices as a scattering space basis insures that the icosahedral symmetry operations are maintained in scattering. The Bragg scattering condition is satisfied for the wave vector k $(= 2\sin\theta/\lambda)$ when:

$$k = |Q_{\|}| \quad (20.39)$$

which can be specified in units of the scale factor, b, for each different peak. To determine the value of the scale factor, b, the experimentally determined values of k for various peaks can be plotted as a function of $|Q_{\|}|$, with the slope determining the scale factor, b. This scale factor can be related to the so-called quasilattice constant, as discussed in more detail below.

Figure 20.13 shows the other common choice of reciprocal lattice basis vectors for the 6-D projection representation (note that these point to the vertices of an icosahedron). These vectors span the previously described 6-D space. The basis set for this 6-D space is described with reference to a coordinate system having a five-fold icosahedral axis parallel to the z-axis of the coordinate system as follows:

$$\mathbf{e}_n^* = \left[\cos\left(\frac{2\pi n}{5}\right), \sin\left(\frac{2\pi n}{5}\right), \frac{1}{\sqrt{5}}\right] \quad (20.40)$$

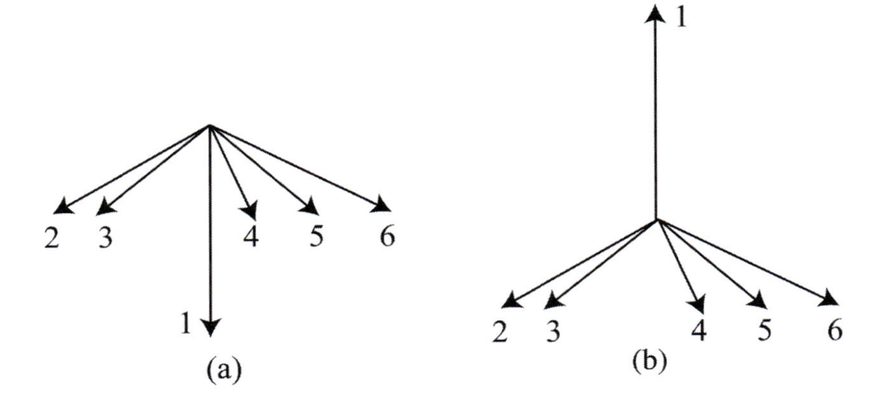

Fig. 20.13. The basis vectors for the six-dimensional projection representation of a 3-D icosahedral quasicrystal.

for $n = 2, 3, \ldots, 6$ and $\mathbf{e}_1^* = (1, 0, 0)$. In this representation, there are $6!/5! = 6$ possibilities for choosing one of the six indices of a 6-D vector, resulting in vectors (with inverses) passing through the 12 vertices of an icosahedron. There are $6!/4!2! = 15$ possibilities for choosing pairs of the six indices of a 6-D vector, resulting in the two-fold axes; there are $\frac{6!}{3!3!} = 20$ possibilities for choosing three of the six indices of a 6-D vector, corresponding to the three-fold axes.

Icosahedral quasicrystals can, therefore, be thought of as being indexed in terms of basis vectors which point to the vertices of a simple 6-D hypercube. Just as it is possible in 3-D to have *bcc* and *fcc* lattices in addition to the primitive lattice, so too can the hypercubic lattice be *bcc* or *fcc*. These original hypercubic lattices carry with them implications for the extinction rules for the icosahedral reciprocal lattice vectors. These rules are summarized in Table 20.3.

20.5.3 Icosahedral quasicrystal diffraction patterns and quasilattice constants

Figure 20.14(a) shows an X-ray diffraction pattern for a $Ti_{56}Ni_{23}Fe_5Si_{16}$ quasicrystalline alloy. The peaks are indexed using the six basis vectors defined above and labeled with the six Miller indices defining the Bragg scattering vector. Figure 20.14(b) shows experimentally determined values of k for various peaks plotted as a function of $|Q_\parallel|$. The slope of this plot gives a value of $b = 0.1119$ nm for the scale factor. This scale factor can be used to assign a quasilattice constant a_R which has been defined by Elser (1985) as:

$$a_R = b\sqrt{5(1+\tau^2)}. \tag{20.41}$$

For the $Ti_{56}Ni_{23}Fe_5Si_{16}$ quasicrystalline alloy, the quasilattice constant is then determined to be 0.4761 nm. The significance of the quasilattice constant will be discussed in the next section. It can be used to decide between possible atomic arrangements of atoms decorating the Penrose bricks in the quasicrystalline structure.

Table 20.3. Extinction rules for icosahedral reciprocal quasilattices.

	P(a*)	F(2a*)	I(2a*)
HKL	$h+k'=2n$ $k+l'=2n$ $l+h'=2n$	same as P plus $h+k+l=2n$ $h'+k'+l'=2n$	all even plus $h+l+h'+k'=4n$ $h+k+l'+k'=4n$ $l+k+h'+l'=4n$
n_i	all integers	$\sum n_i=2n$	all even or all odd

20.5.4 3-D Penrose tiles: stacking, decoration and quasilattice constants

Figure 20.15(a) illustrates the two Penrose bricks described above and shows the edge-sharing relationship between the two bricks that were chosen. Figure 20.15(b) shows a relative translation of the two bricks (by one of the basis vectors) so as to yield a face-sharing arrangement between the two bricks. As was the case with the 2-D Penrose tile, the mere definition of the bricks is not sufficient to insure that their replication results in a quasicrystalline packing. Figure 20.15(c) illustrates a periodic stacking of the two bricks which if extended would give rise to a crystalline structure. Figure 20.15(d) illustrates the projected arrangement of the bricks in the *rhombic triacontahedron*.

Notice that the two 3-D Penrose tiles have the same edge length but have complementary angles. The tiles can be considered as analogous to unit cells in conventional crystals and the common edge length can be identified as a *quasilattice constant*. Just as in the case of a crystalline lattice, the quasilattice constant can be determined by indexing a diffraction pattern. This is not without some ambiguity, however, because inflated and deflated structures can give rise to the same diffraction pattern with quasilattice constants reflecting the scaling of these operations. Furthermore, the notion of projecting from a higher dimensional space does not allow for a direct means of determining atomic positions, as was the case for conventional crystals whose diffraction patterns aided by translational periodicity can be used to infer the lattice and

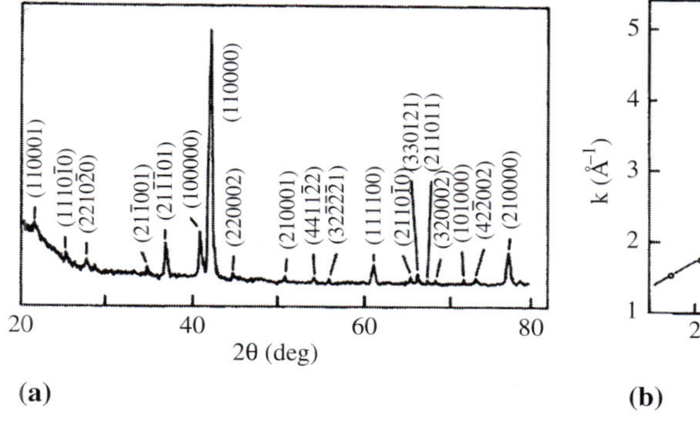

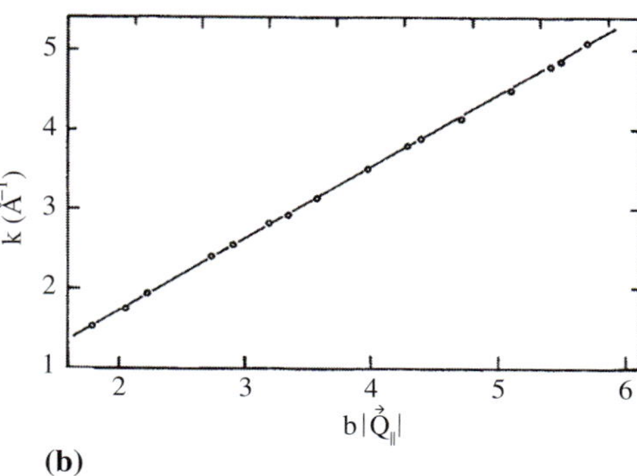

Fig. 20.14. (a) Cu-Kα X-ray diffraction pattern for a $Ti_{56}Ni_{23}Fe_5Si_{16}$ quasicrystalline alloy indexed with the Miller indices for the generalized Bragg scattering vector and (b) experimentally determined values of k for various peaks plotted as a function of $|Q_{\parallel}|$ with the slope determining the scale factor, b. (R. A. Dunlap, M. E. McHenry, R. Chaterjee and R. C. O'Handley, Physical Review B **37**, 8484–7, 1988; Copyright (1988) by the American Physical Society).

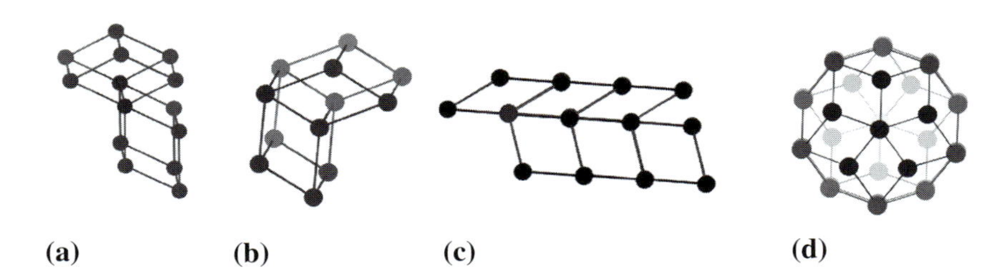

Fig. 20.15. (a) edge-sharing Penrose bricks, (b) face-sharing Penrose bricks, (c) beginning of a periodic stacking of the Penrose bricks, and (d) the rhombic triacontahedron.

the lattice decoration. Nonetheless, certain geometric reasoning can be used to infer reasonable atomic positions within the two Penrose tiles. This involves determination of a quasilattice constant from a diffraction experiment and constructing trial decorations of the Penrose bricks using atomic arrangements similar to those observed in crystalline phases of similar compositions, for example quasicrystal approximant phases. Of particular interest is the quasicrystalline phase of stoichiometry $(Al,Zn)_{49}Mg_{32}$ (Ramachandrarao and Sastry, 1985). This metastable icosahedral quasicrystalline phase, produced by rapid solidification has the same stoichiometry as the *bcc* quasicrystal approximant whose structure was discussed in Chapter 18 (Bergman *et al.*, 1952). It is, therefore, reasonable to assume that structural similarities exist in the atomic arrangements between the two phases. Elser and Henley (1985) and Henley and Elser (1986) have determined relationships between the cubic lattice constants, a, in α-Al–Mn–Si (Cooper and Robinson, 1966) and the *bcc* $(Al,Zn)_{49}Mg_{32}$ phases, respectively, and their quasilattice constant, a_R. This geometric relationship is given by:

$$a = \sqrt{4+8\sqrt{5}}a_R. \tag{20.42}$$

In the case of α-AlMnSi, a quasilattice constant $a_R = 0.4604$ nm is inferred, while for the *bcc* $(Al,Zn)_{49}Mg_{32}$ phase $a_R = 0.514$ nm. From these quasicrystal lattice constants, and the knowledge of the atomic arrangements in the crystalline approximants, it is possible to infer reasonable decorations of the two 3-D Penrose tiles for each of these structures (Elser and Henley, 1985, Henley and Elser, 1986). Figure 20.16 shows the decoration of the Penrose bricks inferred by Henley and Elser (1986) for the $(Al,Zn)_{49}Mg_{32}$ phase. Note that, in both cases, the decoration involves Al(Zn) atoms at the vertices and mid-edges of the bricks. This leaves enough room in the prolate rhombohedron for two Mg atoms along the body diagonal.

The first quasicrystals to be discovered in which the majority species were transition metal atoms were in the Ti–Ni–V system (Zhang *et al.*, 1985). These quasicrystalline alloys were found through an investigation of several transition metal systems in which Frank–Kasper phases were found. The structure of this phase has typically been characterized as microquasicrystalline and has typically coexisted with the stable Ti_2Ni phase as well as

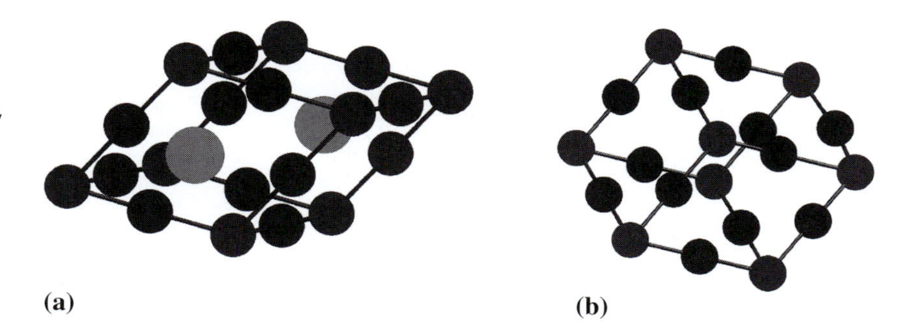

Fig. 20.16. Decoration of the (a) prolate rhombohedron Penrose brick and (b) oblate rhombohedron Penrose bricks, with Al(Zn) atoms (dark gray) and Mg atoms (light gray).

with an amorphous phase with approximately the same composition. It was subsequently shown that the addition of Si to Ti–Ni promotes the growth of larger quasicrystallites during solidification. The work by Dunlap *et al.* (1988b,a), described quenching techniques whereby a single phase icosahedral alloy was produced. A series of alloys of composition $Ti_{56}Ni_{28-x}Fe_xSi_{16}$ were produced and magnetic properties characterized and compared with data for crystalline Ti_2Ni. For the single composition $Ti_{56}Ni_{23}Fe_5Si_{16}$, a detailed X-ray study was performed and Mössbauer measurements were carried out (a short discussion of Mössbauer spectroscopy is presented in Chapter 21).

As shown in Table 20.4, the quasilattice constant for $Ti_{56}Ni_{23}Fe_5Si_{16}$ has been calculated to be 0.476 nm. As discussed above, Henley and Elser (1986) have described two different decorations of these rhombohedral units for the Al–Mn–Si and Al–Zn–Mg quasicrystals, respectively. These two decorations have been distinguished empirically through the relationship between the quasilattice constant a_R and the average interatomic spacing d in an analogous crystalline alloy of the same composition. When a/d is large, the Al–Zn–Mg decoration is favored; when it is small, the Al–Mn–Si decoration is favored. Table 20.4 shows this relationship for four systems in which quasicrystals were observed soon after the intial discovery. It can be seen that both for Ti–Ni–Si (Dunlap *et al.*, 1988b,a) and U–Pd–Si (Poon *et al.*, 1985) the ratio d/a_R is intermediate between the values for Al–Zn–Mg and Al–Mn–Si.

20.5.5 3-D Penrose tiles: projection method

The construction of a 3-D Penrose lattice is analogous to the construction previously described for the 2-D Penrose tiling. Here, the projecting space can be taken as a 12-D (or, alternatively, 6-D) space. The 12-D space has basis vectors which are projected onto the vertex vectors of a regular icosahedron. The point group that describes the cubic 12-D space from which the projection takes place is called the *hyperoctahedral group* of twelve dimensions (Kramer and Neri, 1984) or $\Omega(12)$, which, like the octahedral group in 3-D, has a symmetric group as a subgroup. This symmetric group $S(12)$ further

Table 20.4. Quasilattice constant and ratio of quasi lattice and crystalline spacing of the quasicrystal approximant phases in four quasicrystalline systems.

Alloy system	a_R (nm)	Citation	Phase (struct.)	d (nm)	Citation	$\frac{d}{a_R}$
Al–Mn–Si	0.46	Elser and Henley (1985)	α-Al-Mn-Si SC	0.279	Cooper and Robinson (1966)	1.65
Al–Zn–Mg	0.514	Henley and Elser (1986)	$(Al,Zn)_{49}Mg_{32}$ *bcc*	0.257	Bergman *et al.* (1952)	2.00
U–Pd–Si	0.54	Poon *et al.* (1985)	UPd_3 HEX	0.290	Heal and Williams (1955)	1.77
Ti–Ni(Fe)–Si	0.476	Dunlap *et al.* (1988b), Dunlap *et al.* (1988a)	Ti_2Ni *fcc*	0.261	Yurko *et al.* (1959)	1.82

has the alternating symmetry group $A(5)$, which, of course, is isomorphic to the icosahedral group, as a subgroup. Thus, projections from 12 to 3 dimensions are desired in which the elements of the $A(5)$ subgroup are preserved. Kramer and Neri (1984) addressed this problem by first using the inherent inversion symmetry of the icosahedral group to reduce the basis vectors or projecting space to six dimensions. They described a method for projecting a cubic 12-grid into a cubic 6-grid. Taking the basis functions $\mathbf{b}_1 \ldots \mathbf{b}_{12}$, corresponding to the 12 vertices of the icosahedron, it is possible to construct two six-dimensional subspaces $E6_1$ (with basis vectors $\mathbf{c}_i$) and $E6_2$ (with basis vectors $\mathbf{c}_{i+6}$) which are duals of one another. This reduction is accomplished as follows:

$$\mathbf{c}_i = \frac{\mathbf{b}_i - \mathbf{b}_{13-i}}{\sqrt{2}} \quad \text{and} \quad \mathbf{c}_{i+6} = \frac{\mathbf{b}_i + \mathbf{b}_{13-i}}{\sqrt{2}}. \tag{20.43}$$

The details of the projection are complex and require a thorough understanding of *irreducible representation group theory*; the interested reader is referred to the following citations for more information: Kramer and Neri (1984), and Elser and Henley (1985). Ultimately, when all is said and done, the projections from the 12-D space through two dual 6-D spaces yield the 3-D Penrose tiling.

20.6 ∗Multiple twinning and icosahedral glass models

In the early days after the discovery of quasicrystals, other explanations were given for the sharp diffraction peaks. These explanations often involved considerations of crystalline structures in which icosahedral coordination was preserved out to many shells (as in the Bergman polyhedron discussed in Chapter 18). In particular, it was the contention of Linus Pauling that the apparent icosahedral symmetry was due to directed *multiple twinning* of cubic crystals (Pauling, 1985). In the case of the Ti–Ni(Fe)–Si quasicrystals (Dunlap *et al.,* 1988b,a), a plausible explanation for a diffraction pattern was given in terms of a multiply twinned *bcc* structure. Analysis of both weak and strong peaks in the powder X-ray diffraction pattern for the icosahedral quasicrystal $Ti_{56}Ni_{23}Fe_{5}Si_{16}$ showed it to be consistent with a primitive 1012-atom unit cube, which, by icosahedral twinning, produces aggregates with icosahedral point group symmetry. The icosahedral symmetry is preserved in the diffraction patterns because of the multiple twinning which masks the crystalline periodicity.

Stephens and Goldman (1986) offered another alternative means of understanding the sharp diffraction patterns associated with quasicrystals. They considered densely packed groupings of icosahedra, such that bond orientational order is enforced throughout the sample. The bond orientational order can be preserved by imposing local matching conditions on the icosahedra. These matching conditions require that, wherever possible, the icosahedra share either vertices, edges, or faces, giving rise to three different models: the *vertex model*, the *edge model*, and the *face model*, respectively. This construction suggests that a local short-range order is present leading to a structure of densely packed units of local icosahedral symmetry, but that the resulting structure can be called glassy in the sense that no long-range positional order exists within it. Simulations of such structures reveal that, indeed, they yield diffraction patterns remarkably similar to those of real quasicrystals.

Stephens and Goldman went on to further demonstrate that, although a sharply peaked diffraction pattern was obtained, the peaks were not Bragg peaks in the sense that they do not result from coherent contributions from each site in the sample. Instead, they pointed out how the diffraction pattern could be obtained by considering random sequences of several characteristic distances, similar to a model proposed by Hendricks and Teller (1942); this model of scattering from partially ordered structures considers a random sequence of sites, separated by two or more characteristic distances $\mathbf{d}_1, \mathbf{d}_2, \ldots$ Relatively sharp interference maxima were found to occur at scattering vectors $\mathbf{Q} = 2\pi m/d_1, 2\pi n/d_2, \ldots$ for integral $m, n, \ldots$ The scattered intensity per site may be expressed in closed form as $(1+C)/(1-C)$ where C is given by:

$$C = \sum_{k=1}^{n} f_k \cos(\mathbf{Q}\cdot\mathbf{d}_k), \tag{20.44}$$

where the sum is over all possible site separations $\pm\mathbf{d}_k$ and f_k is the probability of occurrence of each characteristic distance. In the case of quasicrystals, the simplest notion in this light is to consider the sum as running over the six independent vectors pointing to the vertices of a regular icosahedron. In this case, the scattering function can be generalized as:

$$C(\mathbf{Q}) = \frac{1}{6}\sum_{k=1}^{n}\cos(\mathbf{Q}\cdot\mathbf{d}_k). \tag{20.45}$$

This simple scattering function has been shown to give remarkably good agreement with experimental observations for quasicrystals. It gives the further insight that, since the icosahedral vertex vectors are incommensurate, there are no Qs for which all of the cosines are simultaneously unity. Thus, this proves that each peak is of finite intensity and has a finite maximum width and, therefore, is not a Bragg peak (which is a delta function). It should also be noted that in the Stephens and Goldman (1986) work, only the edge and vertex models gave good agreement with quasicrystalline diffraction patterns while the face model did not.

20.7 *Microscopic observations of quasicrystal morphologies

Selected area electron diffraction (SAED) studies of quasicrystalline alloys have been used to confirm the presence of icosahedral symmetry elements. In the first Al–Mn quasicrystals, the typical sizes of the quasicrystals were 0.5–2 μm. According to Cahn (1986), there was no congruent metastable melting point found for the icosahedral phase in the early samples, and, therefore, some small portion of *fcc* Al was always evident (although this was subsequently minimized through the addition of Si). Figure 20.17(a) shows a bright field electron micrograph of an icosahedral phase single crystal exhibiting rhombic triacontahedral faceting. Figure 20.17(b) shows an electron diffraction pattern from the icosahedral phase in a melt-spun $Al_{74}Mn_{20}Si_6$ alloy oriented along a five-fold zone axis (McHenry, 1988). This ten-fold diffraction

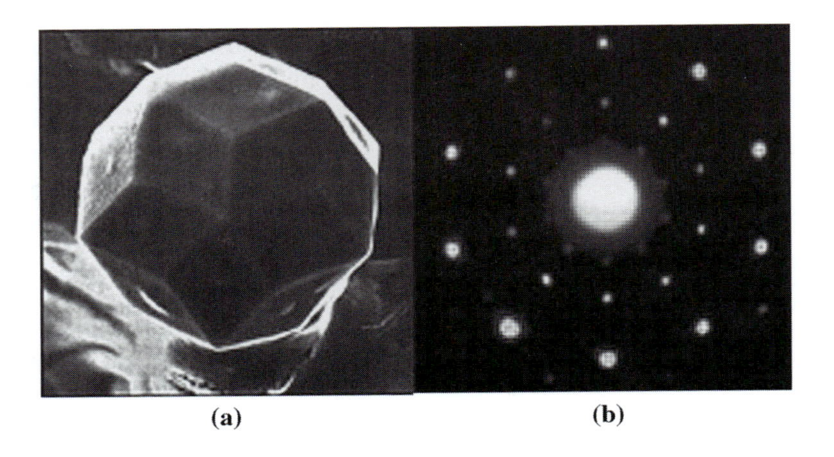

Fig. 20.17. (a) Bright field electron micrograph of an icosahedral phase single crystal exhibiting rhombic triacontahedral faceting (courtesy of F. Gayle, NIST Gaithersburg, Gayle (1987)) and (b) electron diffraction pattern from the icosahedral phase in a melt-spun $Al_{74}Mn_{20}Si_6$ alloy along a five-fold zone axis (McHenry, 1988).

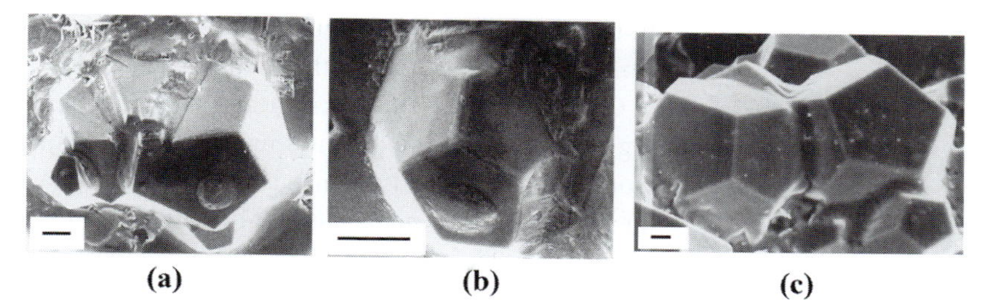

Fig. 20.18. Scanning electron micrographs of melt-grown Ga Mg Zn icosahedral phase single crystals exhibiting pentagonal dodecahedral faceting (courtesy of W. Ohashi and F. Spaepen). The micron bar in the lower left corner of each image corresponds to a distance of 10 μm. (a) and (b) were originally published in Ohashi and Spaepen (1987) and (c) appears in the Harvard Ph.D. thesis of W. Ohashi.

pattern was used for subsequent dark-field microscopy, in order to demonstrate that this ten-fold pattern was not the result of twinning. This diffraction pattern illustrates some of the important features of the quasicrystalline structure. As pointed out by Cahn, the pentagonal geometry means that when we add two proximate reciprocal lattice vectors, each of which is one unit in length, we obtain another reciprocal lattice vector, τ units long. This is, therefore, a simple manifestation of an irrational length scale and, thus, demonstrates quasi-periodicity (Cahn, 1986).

Rhombic triacontahedral faceting is not the only morphology observed for icosahedral quasicrystals. Ohashi and Spaepen (1987) at Harvard University discovered stable GaMgZn quasicrystals. These crystals were formed by a peritectic reaction from a liquid melt. Figure 20.18 illustrates several examples of these large (tens of microns) single quasicrystals grown from the melt. They are seen to possess pentagonal dodecahedral solidification morphologies with exclusively pentagonal faces.

20.8 Historical notes

John Werner Cahn received his B.S. in Chemistry from the University of Michigan in 1949. He did his Ph.D. research in Physical Chemistry under the direction of Professor Richard E. Powell at the University of California, Berkeley. In 1953, he was awarded a Ph.D. for his thesis entitled "The Oxidation of Isotopically Labelled Hydrazine." From 1952 until 1954, Cahn served as Instructor at the Institute for the Study of Metals at the University of Chicago. This was followed by 10 years as a research associate in the Metallurgy and Ceramics Department Research Laboratory of General Electric in Schenectady, NY. From 1964 until 1978, he was Professor of Materials Science at the Massachusetts Institute of Technology. He joined the National Bureau of Standards in 1977 as Center Scientist in the Center for Materials Science. From 1984 until present he has been Senior NIST Fellow at the

Materials Science and Engineering Laboratory at the National institute of Standards and Technology in Gaithersburg, MD.

Cahn has made seminal contributions to the field of thermodynamics including the important theory of spinodal decomposition. Cahn is known for his work on the thermodynamics and kinetics of phase transitions and diffusion, on interface phenomena. In 1984, working in the group of John Cahn, **Dani Shechtman** and co-workers discovered a new class of structures, icosahedral quasicrystals (QCs), in $Al_{86}Mn_{14}$. This discovery led to the reexamination of the basic tenets of crystallography and to the redefining of what is meant by the word *crystal*.

John Cahn is the recipient of many awards and honors. Among these are the 1985 Von Hippel Award of the Materials Research Society and the 1993 Hume-Rothery Award of The Materials Society (TMS). Cahn was awarded the National Medal of Congress in 1998.[8]

Roger Penrose (1931–), was born August 8, 1931 in Colchester, Essex, England. Roger's family moved to London, Ontario, by way of the United States in 1939. After the end of World War II in 1945, the Penrose family returned to England. Roger Penrose attended University College London where his father was a Professor of Human Genetics. He received his B.Sc. degree with first class honours in Mathematics. Penrose received a Ph.D. in 1957 for work in algebra and geometry from Cambridge University. Penrose had appointments as assistant lecturer in Pure Mathematics at Bedford College, London, research fellow at St. John's College, Cambridge, a NATO fellow at Princeton and Syracuse Universities, a research associate at King's College, London and Visiting Associate Professor at University of Texas, Austin. In 1964 Penrose accepted a Readership at Birkbeck College, London

Fig. 20.19. (a) John Cahn (picture courtesy of the Materials Research Society, Warrendale, PA), and (b) Roger Penrose (1931–) (picture courtesy of J. Lima-de-Faria).

[8] A full biography of John Cahn can be found at http://www.ctcms.nist.gov/~cahn/bio.html.

followed by promotion to Professor of Applied Mathematics in 1966. In 1973 Penrose was made Rouse Ball Professor of Mathematics at the University of Oxford, a position he maintained until becoming emeritus in 1998. In 1998 he was made Gresham Professor of Geometry at Gresham College, London.

Penrose contributed significantly to the fields of both mathematics and physics. He contributed to the field of cosmology and the mathematics of general relativity. He worked in efforts to unify quantum theory and general relativity including the introduction of twistor theory. He is the inventor of the Penrose tiling, a quasi-periodic tiling of two- or three-dimensional space which does not require traditional crystal periodicity as discussed in this chapter. His work on non-periodic tilings was begun while he was a graduate student at Cambridge. The Dutch mathematician *Nicholas G. de Bruijn* discovered the connection between the Penrose tiles and lattices in a five-dimensional space. Penrose tilings in three dimensions have been used to postulate the structures of icosahedral quasicrystals. Penrose is also the author of popular books on physics, computer science, the philosophy of science and consciousness.[9]

20.9 Problems

(i) *The golden mean, τ*:

(a) Show that τ, $-1/\tau$ are both solutions to the algebraic equation:

$$x^2 - x - 1 = 0$$

(b) Consider rotation matrices through the angles $2\pi/5$, $4\pi/5$, $6\pi/5$ and $8\pi/5$. Show that the traces of these matrices are, respectively, τ and $-1/\tau$.

(c) Draw a regular pentagon and then circumscribe the smallest regular pentagon containing the vertices of the first and to which the vertex vectors of the first are normal to the edge vectors of the second. Show that the ratio of the length of the edges of the two pentagons is τ.

(ii) *The golden mean, Fibonacci numbers and continued fractions*:

(a) Show that τ can be evaluated using the following expression involving infinite square roots:

$$\tau = \sqrt{1 + \sqrt{1 + \sqrt{1 + \sqrt{1 + \ldots}}}}$$

[9] For a more complete biography of Penrose, see the article by J. J. O'Connor and E. F. Robertson School of Mathematics and Statistics, University of St Andrews, Scotland at http://www-groups.dcs.st-and.ac.uk/~history/Mathematicians/Penrose.html.

(b) Show that τ can be evaluated using the following expression involving infinite continued fractions:

$$\tau = \frac{1}{1+\frac{1}{1+\frac{1}{1+\frac{1}{1+\dots}}}}.$$

(c) Calculate *approximants* to the continued fractions of the last exercise, truncating the fraction at 0, 1, 2, terms, etc. Show that the numerator and denominator of these fractions are Fibonacci numbers.

(iii) *The golden mean, identity*: Prove the following identity involving the golden mean:

$$\tau^4 + 1 = 3\tau^2.$$

(iv) *The golden rhomb*: Calculate the interior angles for the golden rhomb. What is its area?

(v) *Fibonacci series*:

(a) Express τ, τ^2, τ^3, τ^4, as $m\tau + p$ where m and p are integers.
(b) Show that $\tau^n = F_{n-1}\tau + F_{n-2}$.
(c) Show that $\tau^n(\pm)^n\tau^{-n} = F_{n+1}\tau + F_{n-1}$.

(vi) *Lucas numbers*: The integer solutions for n to: $L_n = \tau^n + (-1/\tau)^n$ are called *Lucas numbers*, L_n. They satisfy the recurrence relationship: $L_n = L_{n-1} + L_{n-2}$ with $L_1 = 1$ and $L_2 = 3$.

(a) For the first five Lucas numbers with $n \geq 2$ show that:

$$L_n = \text{Int}^{>}(\tau^n)$$

where the function $\text{Int}^{>}(y)$ chooses the smallest integer greater than or equal to y.

(b) For the first five Lucas numbers with $n \geq 4$ show that:

$$L_n = \text{Int}^{<}\left(\frac{L_n(1+\sqrt{5})+1}{2}\right).$$

(c) Show that $L_n = F_{n+1}\tau + F_{n-1}$.
(d) Show that $L_{2n} = \frac{1}{2}(5F_n^2 + L_n^2)$.
(e) Show that $L_n^2 - L_{n-1}L_{n+1} = 5(-1)^n$.

(vii) *Inflation operator for a 1-D quasicrystal*:

(a) Express the inverse of the inflation operator I^{-1}.
(b) Determine the matrix representation for I^5 and I^6.

(c) Write the matrix representation for I^5 and I^6 in terms of Fibonacci numbers F_k. Show that these agree with the generalization of the recursion relationship.

(viii) *Composition of a 1-D quasicrystal*: Consider a 1-D quasicrystal with a Fibonacci series of L and S cells:

(a) Suppose S cells are centered by A atoms and L cells are centered by B atoms. Calculate the composition of the quasicrystal.
(b) Suppose that the S cells are each decorated by two A atoms each shared with adjacent cells and the L cells also have two A atoms shared with adjacent cells and a B atom centering the cell. Calculate the composition of the quasicrystal.

(ix) *Lattice positions in the Fibonacci lattice*: Determine the position of the n-th lattice point, x_n, for $n = 8$ and $n = 13$ in the Fibonacci series. How does this compare with the length of the repeat distance in the periodic approximants to the Fibonacci lattice?

(x) *Cut and project method*: Beginning with a 2-D square lattice, repeat the cut and project procedure used to generate the Fibonacci lattice, but this time pick a line of slope, $\sqrt{3}$ rather than τ for your projection. Project the lattice points within one lattice constant distance from the line. Describe the result.

(xi) *Dart and kite*: From the matching rules for the oblate and prolate 2-D Penrose rhombs and their matching rules, rationalize the matching rules for the dart and kite tiles in producing a Penrose tiling.

(xii) *Deflation of the 2-D Penrose tiling*: Consider a decagonal patch in a 2-D Penrose tiling and a deflation process given by the rules: [Acute $\rightarrow$ 2 Acute$'$ + 1 Obtuse$'$] and [Obtuse $\rightarrow$ 1 Acute$'$ + 1 Obtuse$'$]. Show the following:

(a) That the ratio of acute to obtuse rhombi contained in the region of the original obtuse rhombus is given by F_{2k}/F_{2k-1} where k is the number of deflations.
(b) That the ratio of acute to obtuse rhombi contained in the region of the original acute rhombus is given by $F_{2k-1}/F_{2(k-1)}$.
(c) That the ratio of acute to obtuse rhombi in total is F_{2k+1}/F_{2k}.

(xiii) *Vertex configuration of the 2-D Penrose tiling-I*: Show that the (2010) vertex configuration illustrated in Fig. 20.8 is the only such possibility. Show that the pairing in this configuration is the only possible edge-sharing pair of the obtuse rhombs that obeys the Penrose matching rules.

(xiv) *Vertex configuration of the 2-D Penrose tiling-I*: Determine the local symmetry about each vertex site in the 2-D Penrose tile.

(xv) *Composition of a 2-D quasicrystal*: Consider a 2-D quasicrystal tiled with oblate and prolate Penrose bricks:

(a) Suppose prolate cells are centered by A atoms and oblate cells are centered by B atoms. Calculate the composition of the quasicrystal.

(b) Suppose that the prolate cells are decorated by A atoms at the vertice and the oblate cells also have A atoms at the vertices and a B atom centering the cell. Calculate the composition of the quasicrystal.

(c) Suppose that the prolate cells are decorated by A atoms at the vertice and the oblate cells also have A atoms at the vertices and two B atoms along the long diagonal of the cell. Calculate the composition of the quasicrystal. What are the the restrictions on the A–A, A–B, and B–B bond distances?

(xvi) *Rhombic triacontahedron*: Show that 3-D Penrose bricks can be used to construct the rhombic tricontahedron. Compare this shape with the facets of icosahedral single crystals.

CHAPTER

21 Metallic structures V: amorphous metals

"The methods of the scientist would be of little avail if he had not at his disposal an immense stock of previous knowledge and experience. None of it probably is quite correct, but it is sufficiently so for the active scientist to have advanced points of departure for the work of the future. Science is an ever-growing body of knowledge built of sequences of the reflections and ideas, but even more of the experience and actions, of a great stream of thinkers and workers."

J. D. Bernal (1901–71), *Science in History*

21.1 Introduction

The word *amorphous* means without shape or structure. In *amorphous solids*, atomic positions lack crystalline (periodic) or quasicrystalline order but they do have short-range order. Amorphous metals are usually structurally and chemically homogeneous, which gives them isotropic properties attractive for many applications. Chemical and structural homogeneity can lead to corrosion resistance while isotropic magnetic properties are important in materials for power transformation and inductive components. The absence of crystallinity alters the traditional micromechanisms for deformation of the solid, giving many amorphous metals attractive mechanical properties.

Amorphous metal can be synthesized by rapid solidification processing in alloy systems with phase diagrams that exhibit deep eutectics. In such systems, the liquid phase remains stable to low temperatures, i.e., a large *undercooling* is required to begin solidification.[1] Undercooling provides a sufficiently large free energy difference between the liquid and solid phase, so that the surface energy barrier to the *nucleation* of stable crystalline phase(s) can be overcome.

Amorphous metals were first synthesized by *piston and anvil splat cooling* by Pol Duwez at the California Institute of Technology (Duwez *et al.*, 1960). These metastable amorphous phases had a structure postulated to be that of a frozen liquid. If a liquid is undercooled rapidly to a temperature below the *glass transition temperature*, T_g, the reduced atomic mobility can provide a kinetic barrier to reaching equilibrium by *heterogeneous nucleation* of the crystalline phase. On heating the frozen amorphous phase above T_g, where the atomic mobility increases, structural relaxation of the glass is possible (Davies, 1985). Nucleation of the stable crystalline phase subsequently occurs at a higher *primary crystallization temperature*, T_{x1}. For amorphous alloys of different composition, there can be different reaction pathways to reach equilibrium through primary and *secondary crystallization* at successively higher temperatures.

The structure of glassy metals differs from that of window glass which has random covalent networks of $(SiO_4)^{4-}$ tetrahedra. Amorphous metal structures lack directional bonding, can be nearly topologically close-packed, and/or possess coordination polyhedra similar to those of stable crystalline phases of similar composition (O'Handley, 1987). Icosahedral coordination polyhedra are common in the amorphous phase.

21.2 Order in amorphous and nanocrystalline alloys

Table 21.1 classifies amorphous and nanocrystalline alloys in terms of *short-range order*, *long-range order*, and the length scales of the ordering (O'Handley, 1987). Crystalline alloys classified by crystallite size are designated macrocrystalline, microcrystalline, or nanocrystalline. Amorphous alloys with local order similar to crystalline counterparts are known as Amorphous I alloys, whereas amorphous alloys with non-crystalline local order belong to the Amorphous II type.

In amorphous metals, the atomic correlations do not extend past the first few coordination shells (0.1 to 0.5 nm out from the central atom), resulting in significant broadening of peaks and fewer features in X-ray diffraction patterns (recall the powder pattern for a regular glass microscope slide in Fig. 1.7). In *nanocrystalline alloys*, finite size effects give rise to *Scherrer broadening* of the XRD peaks. Figure 21.1 shows simulated XRD patterns

[1] Undercooling is the amount of cooling below the normal melting (freezing) temperature (Turnbull, 1950).

Table 21.1. Classification of materials by range and type of atomic order.

SRO	Range of SRO	LRO	Range of LRO	Material classification
Crystalline	$\geq 10\mu$m	Crystalline	$\geq 10\mu$m	Macrocrystalline
Crystalline	100 nm–10μm	Crystalline	$\geq$ 100 nm	Microcrystalline
Crystalline	$\leq$100 nm	Crystalline	$\leq$ 100 nm	Nanocrystalline
Crystalline	$\sim$1 nm	No LRO		Amorphous I
Non cryst.	$\sim$1 nm	No LRO		Amorphous II
Non cryst.		Quasiperiodic	$\sim$1 μm–0.1 m	Quasicrystalline

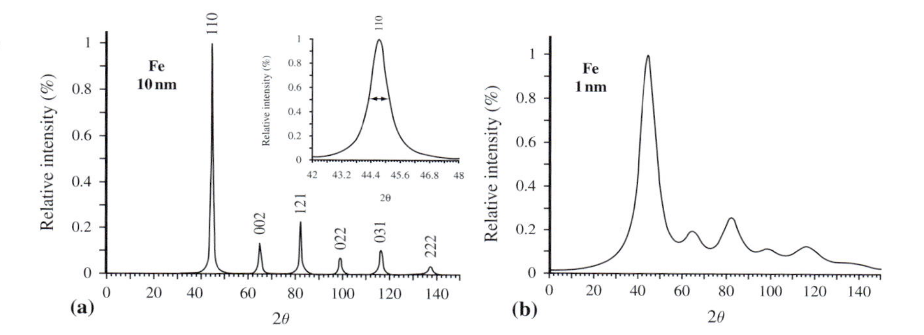

Fig. 21.1. Fine particle broadening in XRD patterns for (a) 10 nm and (b) 1 nm *bcc* Fe particles.

for (a) 10 nm and (b) 1 nm *bcc* Fe particles, showing fine particle broadening (Scherrer, 1918). The inset of Fig. 21.1(a) shows that for 10 nm particle size, the *full width at half maximum (FWHM)* of the fundamental *bcc* (110) reflection is nearly 1 degree. For a particle size of 1 nm (i.e., about four unit cells) the peaks have broadened so much that they overlap and the high angle peaks are no longer resolved. The peak broadening is a signature of the *nanocrystalline structure*.

Consider a finite crystal of thickness, $t = md$, where m is an integer, d is the distance between crystalline planes, and t is the crystal thickness. Figure 21.2(a) shows a δ-function Bragg diffraction peak for an infinite crystal and, in Fig. 21.2(b), the same peak in a nanocrystal is shown. If the broadened Bragg peak in Fig. 21.2(b) begins at an angle $2\theta_2$ and ends at $2\theta_1$, then the peak's FWHM is:

$$w = \frac{1}{2}(2\theta_1 - 2\theta_2).$$

Following the argument in Cullity (1978), we consider the path length differences for each of the two angles θ_1 and θ_2, for X-rays traveling through the thickness of the crystal:

$$(m+1)\lambda = 2t \sin\theta_1 \quad \text{and} \quad (m-1)\lambda = 2t \sin\theta_2. \tag{21.1}$$

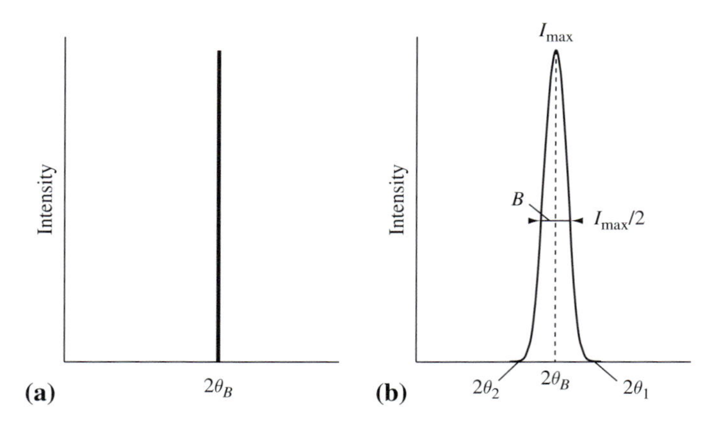

Fig. 21.2. In a crystalline material, XRD peaks will be narrow, with a peak width that reflects only instrumental broadening (a). In a fine-grained material, peak broadening will occur; the quantities involved in the derivation of the Scherrer formula are indicated in (b).

After subtraction:

$$\lambda = t(\sin\theta_1 - \sin\theta_2) = 2t\cos\left(\frac{\theta_1+\theta_2}{2}\right)\sin\left(\frac{\theta_1-\theta_2}{2}\right). \tag{21.2}$$

Since $\theta_1+\theta_2 \sim 2\theta_B$ and $\sin((\theta_1-\theta_2)/2) \sim (\theta_1-\theta_2)/2$ we obtain:

$$t = \frac{\lambda}{w\cos\theta_B}. \tag{21.3}$$

A more exact treatment yields *Scherrer's formula*:

$$t = \frac{0.9\lambda}{w\cos\theta_B}. \tag{21.4}$$

For $t = 50$ nm diameter particles, a broadening of about 0.2° is predicted. In order to use this equation, one must first correct the measured values of w for the *instrumental broadening*; instrumental broadening combines a variety of broadening sources in a single parameter, w_{ib}. Those sources include, among others, the slit width, imperfect focusing, and sample size. The instrumental broadening is θ-dependent and can be measured from the FWHM values for a sample with known large grain size, so that the grain size broadening is minimal. Polycrystalline Si with large grains is a commonly used sample to determine w_{ib}. Since the FWHM parameter occurs in a squared form in the pseudo-Voigt peak shape function, we can find the real FWHM, w_{real}, from

$$w_{real}^2(\theta) = w^2(\theta) - w_{ib}^2(\theta).$$

Combining this with Sherrer's formula results in:

$$t = \frac{0.9\lambda}{\cos\theta\sqrt{w^2(\theta) - w_{ib}^2(\theta)}}.$$

X-ray diffraction patterns taken from amorphous materials exhibit one or more broad peaks in the scattered intensity, often at angles near where diffraction peaks occur in a similar crystalline alloy. A d-spacing naively calculated for this peak (using Bragg's law) corresponds to a distance that can be rationalized in terms of models such as the *dense random packing of hard spheres (DRPHS)*. It has been shown (Guinier and Dexter, 1963) that $d \sim 0.815D$, where d is the calculated spacing and D is the atomic diameter (in a single component system). A Scherrer analysis (Scherrer, 1918) of the breadth of such X-ray scattering peaks indicates that the "crystallite size" is in the order of atomic dimensions, consistent only with short-range atomic correlations.

Structural information in amorphous solids is contained in the *radial distribution function* (RDF). The RDF is determined from a Fourier transform of a normalized X-ray scattered intensity (expressed as a function of the wave vector, $\mathbf{k}$), as explained in more detail in Section 21.8. The RDF is used to test structural models of the short-range order in amorphous alloys. These include the *dense random packing of hard spheres (DRPHS)*, the *microcrystalline short-range order*, and the *icosahedral short-range order* models.

21.3 Atomic positions in amorphous alloys

Atomic distances in an amorphous solid can be described by the *pair correlation function,* $g(r)$ (Fig. 21.3). The pair correlation function is defined as the probability that a pair of atoms are separated by a distance, r. We consider N atoms in a volume Ω; let $\mathbf{r}_1, \mathbf{r}_2, \ldots \mathbf{r}_N$ represent the positions of these atoms with respect to an arbitrary origin. The distance $r = |\mathbf{r}_i - \mathbf{r}_j|$ is the length of the vector connecting atoms i and j. Related atomic distribution functions are the spatially dependent atomic density, $\rho_{\text{atom}}(r)$, defined as:

$$\rho_{\text{atom}}(r) = \frac{N}{\Omega} g(r), \tag{21.5}$$

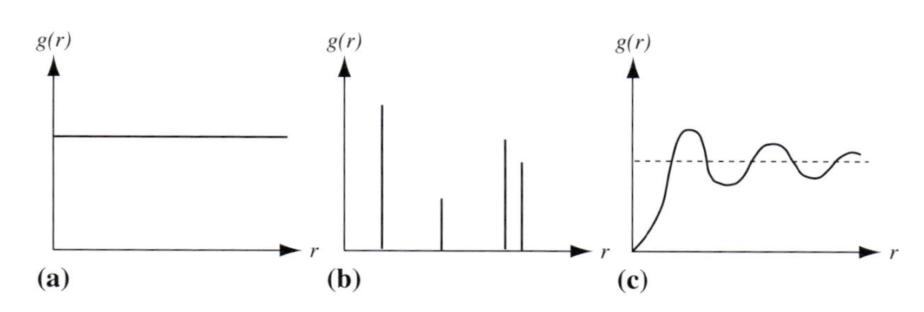

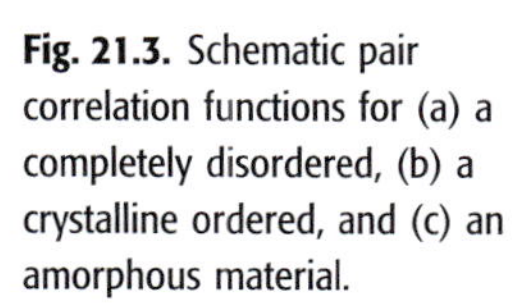
Fig. 21.3. Schematic pair correlation functions for (a) a completely disordered, (b) a crystalline ordered, and (c) an amorphous material.

and the *radial distribution function*, RDF(r), defined in terms of the atomic density:

$$\boxed{\mathrm{RDF}(r) = 4\pi r^2 \rho_{\mathrm{atom}}(r)\mathrm{d}r,} \tag{21.6}$$

The radial distribution function, RDF(r), represents the number of atoms between the distances r and $r + \mathrm{d}r$.

The functions $\rho_{\mathrm{atom}}(r)$ and $g(r)$ are determined from scattering experiments using wavelengths of the order of the atomic distances. Figure 21.3 shows schematically $g(r)$ for a completely disordered state, a (crystalline) completely ordered state and an (amorphous) short range ordered state. A completely structurally disordered material (e.g., a gas) has a uniform probability of finding neighboring atoms at all possible distances (larger than twice the atomic radius), leading to a featureless $g(r)$. In a crystalline solid, $g(r)$ is represented by a set of *delta*-functions related to the discrete distances between pairs of atoms (i.e., a diffraction pattern). In amorphous alloys, broad peaks in $g(r)$ reflect the presence of short range order.

21.4 Atomic volume, packing, and bonding in amorphous solids

We have defined a crystal structure to consist of a *lattice* and a *basis* decorating the lattice. As described in Chapter 3, in a *Bravais lattice*, with the origin on any one of an infinite set of equivalent lattice points, the positions of all lattice points can be described by *translation vectors*, **t**, consisting of three integers multiplied by the *primitive basis vectors* of the lattice: $\mathbf{t} = u\mathbf{a} + v\mathbf{b} + w\mathbf{c}$. The *primitive unit cell* is then a parallelepiped bounded by these primitive basis vectors. For a monatomic basis, the primitive unit cell contains the volume of one atom. It is the smallest volume that completely fills the space when translated using all possible translation vectors.

An alternative unique primitive unit cell is the *Wigner–Seitz primitive cell*. The Wigner–Seitz (WS) cell is defined in 3-D (2-D) as the volume (area) that is closer to a lattice point than to any other. The WS cell is bounded by the set of planes bisecting the shortest lattice vectors connecting a central atom with its neighboring atoms. We also know that with each Bravais lattice we can associate a reciprocal lattice. The Wigner–Seitz cell of the reciprocal lattice is commonly known as the *first Brillouin zone*. The first Brillouin zone contains all the points that are closer to the origin than to any other reciprocal lattice point.

Figure 21.4 illustrates the first Brillouin zones for the *fcc* and *bcc* lattices. Recall that the *bcc* lattice has a dual *fcc* reciprocal lattice (taking into account the systematic absences due to the lattice centering) and vice versa. The *fcc* first Brillouin zone is a *rhombic dodecahedron* and the *bcc* first Brillouin

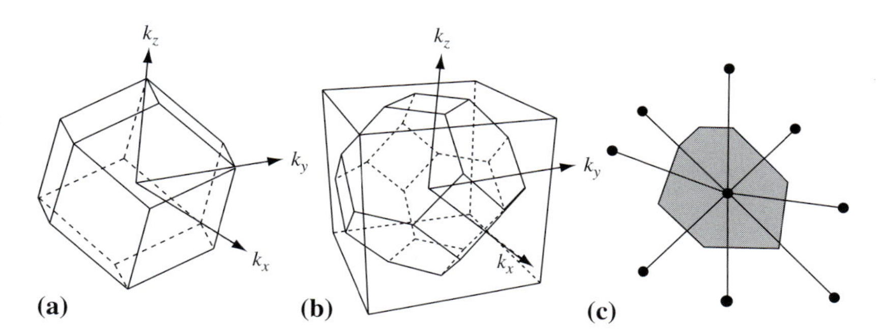

Fig. 21.4. First Brillouin zones for (a) *fcc* (rhombic dodecahedron), (b) *bcc* (truncated octahedron) lattices, and (c) a Voronoi polyhedron.

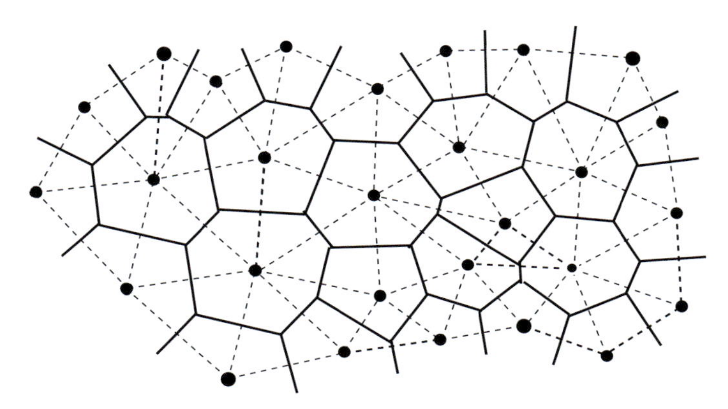

Fig. 21.5. Atoms in an amorphous solid showing Voronoi polyhedra (solid lines) and the Delaunay network (dashed lines). Note that the Voronoi cell edges are all perpendicular to the Delaunay line segments.

zone is a *truncated octahedron*. The coordinate axes in reciprocal space are labeled k_x, k_y, and k_z. Note that, in physics, the reciprocal lattice is often defined with a prefactor of 2π, i.e., the reciprocal lattice vector $\mathbf{a}^*$ would be defined as $2\pi\mathbf{b} \times \mathbf{c}/V$. The reader should be aware of this scaling factor when consulting the literature.

For amorphous materials, a construction equivalent to that of the WS cell can be carried out; it is known as the *Voronoi construction*, and the equivalent of the WS cell in an amorphous material is the *Voronoi polyhedron* (Fig. 21.4(c)). The Voronoi construction partitions space in atomic arrangements lacking traditional periodicity. A *Delaunay network* (Delaunay, 1937) illustrated in Fig. 21.5 is a network analogous to the crystalline lattice. The *Voronoi polyhedron* is defined by the bisector planes of the lines connecting neighboring points and represents the volume (in 3-D) that is closer to an atomic site than to any other. The *Delaunay network* is constructed by joining atom pairs if their Voronoi polyhedra share a common face (edge in 2-D).

The Voronoi tessellation (Voronoi, 1908) and the Delaunay network are duals that uniquely describe the atomic positions in the amorphous solid. A Voronoi tessellation is constructed by assigning all points in space to their nearest atom. For N atoms, this divides space into N Voronoi polyhedra, each containing a single atom.

21.4.1 DRPHS model

Bernal (Bernal, 1964) modeled a liquid as a set of spherical atoms interacting with a potential function with no angular dependence, i.e., a radial potential. He postulated that the structure of a liquid is determined by *volume exclusion* (Finney, 1983). Since metal liquid densities are typically only a few per cent smaller than those of the solid, it was conjectured that the high coordination numbers of the solid (8–12) are replicated in the liquid. This notion can be used to model atom positions in amorphous metals according to the dense random packing of hard spheres (DRPHS) model (Bernal and Mason, 1960, Bernal, 1965).

Figure 21.6 depicts possible atomic arrangements in the DRPHS model. The DRPHS model does not allow for relaxation of atomic positions in the local interatomic potential. The DRPHS configurations have been investigated experimentally, using a large set of ball bearings, and theoretically, using Monte Carlo simulations of atoms interacting through a pair potential. Commonly recurring Voronoi polyhedra for various atomic sites are known as *Bernal polyhedra*. These coordination polyhedra include the Platonic solids, trigonal prisms, and square anti-prisms. Figure 21.6 compares configurations in the DRPHS model with atomic arrangements in a crystalline solid and near grain boundaries. Certain incoherent grain boundaries are modeled as having structures similar to the DRPHS. In *bulk amorphous alloys*, which are discussed further below, the DRPHS of metals of different relative radii gives rise to more efficient packing and increased stability of the amorphous metal.

Table 21.2 compares packing densities in cubic crystals with predictions of the DRPHS. Packing densities are listed with type of structure, the number of atoms per unit cell (N), the coordination number (CN), and the ratio of the lattice constant and the touching atom radius. The DRPHS model gives a packing fraction of 63.8%.

From analysis of computer simulations, the average coordination number for DRPHS has been determined to be about 8.5. Both the predicted packing fraction and the average coordination number for DRPHS are lower than the experimentally observed values. The discrepancies lie in part with the

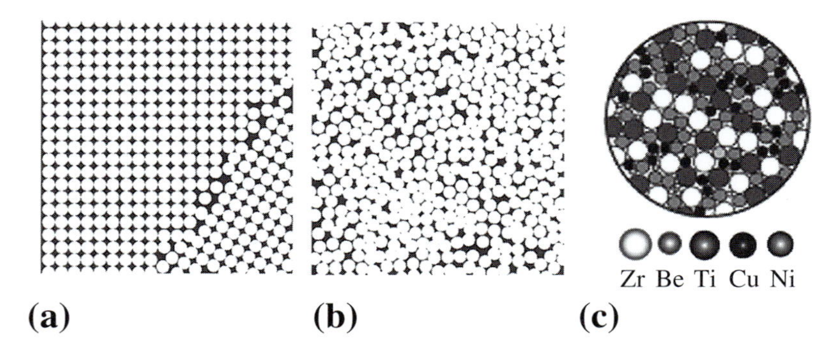

Fig. 21.6. Two-dimensional atomic models: (a) a crystalline solid with grain boundaries; (b) an amorphous solid in a DRPHS model (Bernal and Mason, 1960, Bernal, 1965) and (c) DRPHS of metals of different relative radii in Liquidmetal™bulk amorphous alloys ((a) and (b) courtesy of J. Hess and (c) N. Hayward).

Table 21.2. Packing densities in representative crystals and by the DRPHS model.

Structure	N	CN	$\frac{a}{r}$	Packing fraction (%)
SC	1	6	2	52
BCC	2	8	$4\sqrt{3}$	68
FCC	4	12	$2\sqrt{2}$	74
HCP	6	12		74
DRPHS		8.5		63.8

assumptions of the model. In particular, we have seen (e.g., in Frank–Kasper phases) that atomic relaxations can result in different atomic sizes in different environments, calling into question the use of a hard-sphere potential. Also, bonding preferences call into question the assumption of a potential with no angular dependence. Nonetheless, the model gives a good qualitative description of the structure of the liquid and amorphous states.

21.4.2 Binding in clusters: crystalline and icosahedral short range order

The stability of the icosahedron and its postulated importance in liquid metals (Frank, 1952) lead to the suggestion of icosahedral short range order in amorphous metals. The lower energy of icosahedra as compared to crystalline close-packed clusters has been used to explain the size of the nucleation energy barrier to crystallization. Recently, Kelton *et al.* (2003) have investigated nucleation from the liquid using containerless solidification of melts and studied the amount of *undercooling* possible.

Much evidence of local icosahedral order in amorphous phases is inferred from molecular dynamic simulations. Briant and Burton (1978) proposed a model of the amorphous state as containing a large density of icosahedral units. Before this model was proposed, the two models advanced to explain the structure of the amorphous state were the DRPHS model discussed above and the microcrystalline model of Wagner (1969).

In Chapter 17, we used a Lennard-Jones (Lennard-Jones, 1924) pair potential in the extended solid. For clusters (transitory entities in the liquid state and frozen in the amorphous solid), simple pair potentials are also useful to discuss stability. Recall that the *Lennard-Jones potential* of an atom pair as a function of separation, r, is given by:

$$V(r) = 4\epsilon[(\frac{\sigma}{r})^{12} - (\frac{\sigma}{r})^{6}], \tag{21.7}$$

with ϵ and σ setting the scale of the potential energy and interatomic spacing, respectively. The force between two atoms is given by $-\mathrm{d}V/\mathrm{d}r$. The total cluster energy can be calculated by summing over all N atomic pairs:

$$V_{\mathrm{TOT}} = \frac{N}{2}(4\epsilon)\left[\left(\frac{\sigma}{R}\right)^{12}\sum_{ij}\frac{1}{p_{ij}^{12}} - \left(\frac{\sigma}{R}\right)^{6}\sum_{ij}\frac{1}{p_{ij}^{6}}\right], \quad (21.8)$$

where the sum excludes atoms pairing with themselves; $p_{ij}R$ is the distance between an atom i and an atom j and R is the nearest neighbor distance. The factor $N/2$ avoids double counting of pair interactions. Summations of the p_{ij} factors can be performed for the 12-fold coordinated *hcp*, cuboctahedral, and icosahedral coordinations. Minimizing the resulting potentials shows that the icosahedral coordination gives rise to a smaller equilibrium interatomic spacing and a larger *cohesive energy* (the energy per atom required to break all of the bonds in the cluster). It is precisely this argument that Frank (1952) used to predict icosahedral stability in the liquid. It is equally valid for the frozen liquid and for an amorphous solid.

21.4.3 Icosahedral short range order models

The icosahedral cluster model is rooted both in Frank's observations (Frank, 1952) based on the Lennard-Jones potential as well as in the similarity of calculated scattering intensities for icosahedral micro-clusters and those experimentally observed in many amorphous metals. The *reduced intensity function*, $i(k)$, or *interference function*, for a cluster is expressed as:

$$i(k) = 1 + \frac{2}{N}\sum_{i,j}\frac{\sin(kr_{ij})}{kr_{ij}}, \quad (21.9)$$

where N is the number of atoms in the cluster and r_{ij} the interatomic pair separation between atom i and atom j, respectively. Although the *microcrystalline model* often gives excellent agreement with experimental interference functions, for other amorphous metals the icosahedral cluster model gives better agreement with the structure observed in the amorphous state.

The molecular dynamic simulations of Hoare (1978) revealed large icosahedral clusters that he called *amorphons*. These were often observed to be superstructures or substructures of the 55-atom *Mackay icosahedron* (Mackay, 1962). Molecular dynamics simulations of supercooled liquids (Steinhardt and Levine, 1984) revealed long range icosahedral orientational order. Haymet (1983) suggested that glass formation may be due to supercooling of the liquid to temperatures where the diffusion constant is so small that thermal fluctuations do not disturb regions with a stable local icosahedral symmetry. Haymet also showed that for any purely repulsive potential of the form:

$$V(r) = \epsilon\left(\frac{\sigma}{r}\right)^{n}, \quad (21.10)$$

where r is the interatomic distance and ϵ and σ are empirical parameters, the icosahedral 13-atom cluster always has the lowest total energy when compared to the 13-atom *fcc* and *hcp* clusters, extending Frank's hypothesis.

While it is not possible to fill 3-D space with regular tetrahedra, Sadoc and Mosseri (1984) showed that a 4-D space can be tiled by regular tetrahedra while maintaining icosahedral symmetry. He proposed schemes for projections into 3-D that yield icosahedral clusters. The projected structures have densities similar to those observed in metallic glasses as well as similar features in the interference function. Sachdev and Nelson (1985) also considered a description of short-range icosahedral order in metallic glasses based on projection from an ideal curved space icosahedral crystal. The projection introduces frustration in the resulting 3-D structure due to the incompatibility of the curved space crystal with the flat space into which it is projected. The disorder is described in terms of tangled arrays of Frank and Kasper's *disclination lines* (Widom, 1985).

21.5 Amorphous metal synthesis

Since the pioneering work of Duwez *et al.* (1960), a variety of metallic glasses have been produced through rapid solidification processing. Amorphous alloy synthesis typically requires cooling rates in excess of 10^4 K/s (McHenry *et al.*, 1999). Examples of rapid solidification techniques include *splat quenching*, *melt spinning*, and *planar flow casting*. In melt spinning, melts at temperatures typically greater than 1300 K are cooled to room temperature in about 1 ms, at cooling rates around 10^6 K/s. Figure 21.7 (a) is an illustration of the melt spinning process, where an alloy charge is placed in a crucible with a small hole at one end. The alloy is typically induction melted. Surface tension keeps the melt in the crucible until an inert gas over-pressure pushes the melt through the hole onto a rotating Cu wheel below. The stream rapidly solidifies into about 20 μm thick amorphous ribbons. Good reviews of the

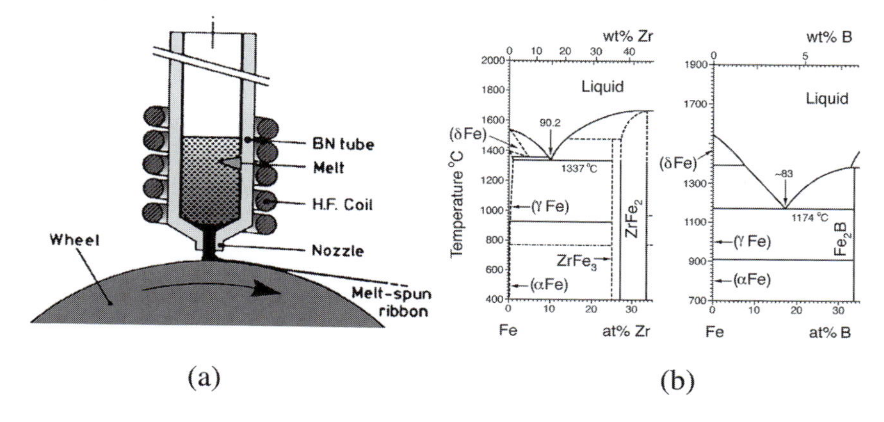

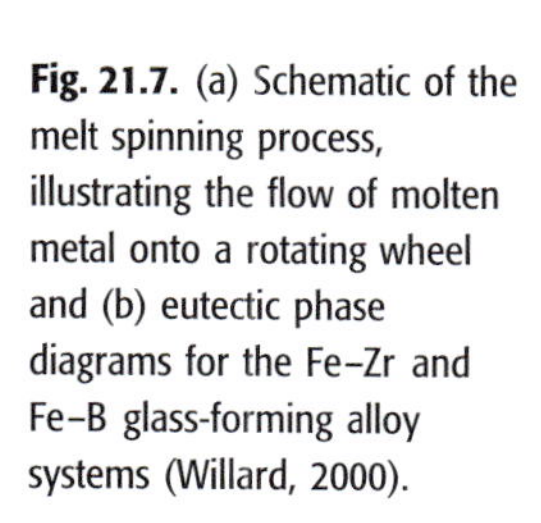
Fig. 21.7. (a) Schematic of the melt spinning process, illustrating the flow of molten metal onto a rotating wheel and (b) eutectic phase diagrams for the Fe–Zr and Fe–B glass-forming alloy systems (Willard, 2000).

melt spinning process include those by Liebermann (1983), Davies (1985), and Boettinger and Perepezko (1985).

Many other techniques have been used in amorphous metal synthesis including:

- *solidification processing of bulk amorphous alloys: bulk amorphous alloys* are formed by conventional solidification routes with slower cooling rates. The large glass forming abilities of these alloys allow for the production of amorphous materials with dimensions (up to several cm) that are much larger than those obtainable by melt spinning.
- *powder synthesis techniques:* amorphous metals may be synthesized as free standing powders or compacted to form bulk alloys with an amorphous structure. Examples of these techniques include *plasma torch synthesis* (Turgut, 1999), *gas atomization*, and *mechanical milling*. Rapid solidification in the gas phase (e.g., ultrasonic gas atomization) or splatting onto a substrate can lead to the formation of nanometer-sized glassy droplets or *nano-glasses* (Gleiter, 1989), as opposed to nanocrystals.
- *solid state mechanical processing: mechanical alloying* is a processing route that has been used to produce amorphous alloys. The energy of the milling process and the thermodynamic properties of the constituents determine whether amorphization will occur. Mechanical alloying is also a means of synthesizing amorphous alloys by solid state reaction of two crystalline elemental metals in multilayer systems with a fine interlayer thickness (Johnson *et al.*, 1985).
- *amorphization by irradiation:* crystalline alloys can be made amorphous by irradiation by energetic particle beams (Matteson and Nicolet, 1982). Amorphization is an effect that is often observed at high particle fluences in radiation damaged materials (Sutton, 1994).
- *thin film processing: thin film deposition* techniques were shown as early as 1963 (Mader *et al.*, 1963) to be capable of producing amorphous alloys.

21.6 Thermodynamic and kinetic criteria for glass formation

The *glass forming ability* (GFA) of a material involves the suppression of crystallization by preventing nucleation and growth of the stable crystalline phase. The solidification of a eutectic liquid involves partitioning of the constituents so as to form the stable crystalline phase. Glass Forming Ability can be correlated with the *reduced glass forming temperature,* T_{rg}, defined as:

$$T_{rg} = \frac{T_g}{T_L}, \tag{21.11}$$

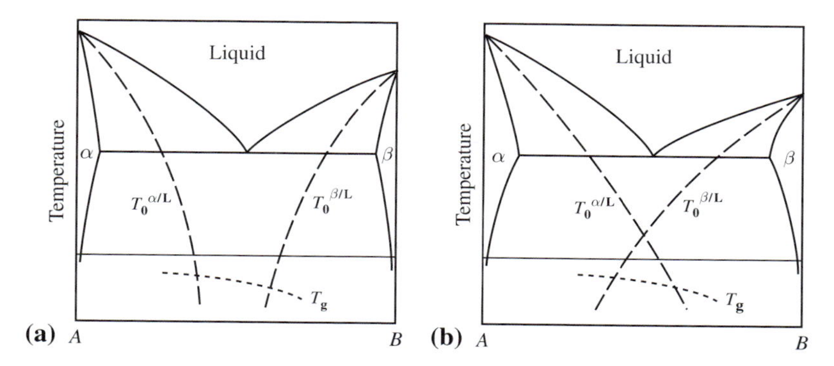

Fig. 21.8. (a) T_0 construction for an *AB* binary alloy with a deep eutectic. (b) In an alloy for which the T_0 curves intersect above T_g, partitionless solidification is not possible (Willard, 2000).

where T_L and T_g are the liquidus and glass transition temperatures, respectively. Below the glass transition temperature, T_g, the atomic mobility is too small for diffusional partitioning of alloy constituents.

The thermodynamic condition for glass formation is described by the T_0 construction, illustrated in Fig. 21.8 for an $A-B$ eutectic system. The T_0 curve is defined as the locus of temperature – composition points where the liquid and solid phase free energies are equal. For compositions between the T_0 curves, the liquid phase can lower its free energy only through the partitioning of the chemical components, nucleating an A-rich or B-rich region that expels the other (B or A) constituent as it grows. This nucleation and growth process requires long range diffusion to continue. If an alloy can be quenched below its glass transition temperature, T_g, in the region of compositions between the two T_0 curves, then the atomic motion necessary for this partitioning will not be possible and the material will retain the configuration of the liquid. The glass forming ability is increased for materials where the reduced glass forming temperature, T_{rg}, is large. Figure 21.7(b) illustrates examples of two alloy systems that exhibit relatively deep eutectics and can be rapidly solidified to form a metallic glass. The Fe–Zr system is an example of a eutectic in a late transition metal/early transition metal system. The Fe–B system is an example of a eutectic in a (transition) metal/metalloid system.

Massalski (1981) presented thermodynamic and kinetic considerations for the synthesis of amorphous metals. The criteria suggested for *partitionless freezing* (no compositional change) of a liquid to form a metallic glass are:

(i) *quenching to below the T_0 curve:* the T_0 curve represents the temperature below which there is no thermodynamic driving force for partitioning and the liquid freezes into a solid of the same composition.

(ii) *morphological stability:* this depends on the comparison of imposed heat flow and the velocity of the interface between the amorphous and liquid phases.

(iii) *heat flow:* to prevent segregation, the supercooling of the liquid phase must exceed L/C, where L is the latent heat of solidification and C is the specific heat of the liquid.

(iv) *kinetic criteria:* a critical cooling rate, R_c, for quenching of the liquid is empirically known to depend on the reduced glass forming temperature, T_{rg}.
(v) *structural*: differences between atomic radii exceeding about 13% (consistent with the *Hume-Rothery rules*) retard the diffusion necessary for partitioning.

Criterion (i) is a condition on the supercooling of the liquid. Criterion (iii) requires that heat must be transported quickly enough from the moving solidification front. This is determined by heat transfer between the amorphous solid phase and the melt spinning wheel and depends on the wheel conductivity, speed, and the degree of wetting by the liquid. Criterion (iv) defines a critical cooling rate required to prevent the *nucleation and growth* of the crystalline phase. Deep eutectics occur in systems with large positive heats of mixing and consequent atomic size differences, motivating criterion (v).

Crystallization is a solid state phase transformation often controlled by *nucleation and growth* kinetics. Empirical rules for the stability of the glassy phase have been developed with consideration for the crystallization kinetics. The progress of an isothermal phase transformation is represented by plotting the transformed volume fraction (of the primary crystalline phase), $X(t, T)$, as a function of temperature, T, and time, t, in a *TTT diagram*. Crystallization products can be combinations of other metastable amorphous and metastable or stable crystalline phases. Crystallization reactions provide a structural reaction path from the amorphous phase to their end stable crystalline products. Thermally activated crystallization reactions are characterized by an *activation energy* for the transformation. Activation energies for crystallization are typically several eV/atom reflecting the energy barriers to the nucleation and growth of the new phase.

21.7 Examples of amorphous metal alloy systems

Common glass forming systems include Group IIa-transition metal, actinide-transition metal, and early transition metal (TE)-Group IIa alloy systems (Elliot, 1983). There are several classes of alloy systems that have been shown to have commercial potential. The first of these, *metal-metalloid amorphous systems* include simple metal, early transition metal and late transition metal metalloid systems. The second class are the *rare earth transition metal (RETM) amorphous systems*, and the third class are the *late transition metal (TL) – early transition metal (TE) systems*. In this section, we discuss each of these classes as well as a few interesting multicomponent systems. Finally, we discuss systems that can be used to develop interesting nanocomposites.

21.7.1 Metal–metalloid systems

Metal–metalloid systems include early (TE) or late (TL) transition metals along with *metalloids* (M = C, B, P, Si, etc.). Eutectic compositions are found near 20–30 at% M in typical TL–M systems. The eutectic alloy composition is often bracketed by a solid solution and an intermetallic alloy with composition richer in M. There may also be other M-rich high temperature phases and/or metastable intermetallic phases. It is instructive to understand the M coordination polyhedra in the TE or TL rich crystalline phases to gain insight about coordination preferences in the amorphous phase.

Figure 21.9(a) shows the trigonal prismatic coordination of Fe around C in Fe_3C or *cementite* (Meinhardt and Krisement, 1962). This coordination is also observed in an isostructural metastable Fe_3B phase (Khan *et al.*, 1982). Figure 21.9(b) shows the square antiprismatic coordination of Fe around B in the Fe_2B phase, discussed in Chapter 18. In Fe-rich alloys, B is incorporated interstitially in the *bcc* Fe structure (in the larger octahedral interstices). The competition between the simple interstitial coordination in B-poor alloys and the more complicated polyhedral environments in the B-rich phases contributes to the competing interactions in the eutectic phase diagram.

Another important terminal structure in transition metal borides and carbides is the $T_{23}M_6$ phase, for which C_6Cr_{23} is a prototype. The C_6Cr_{23} phase has the cubic $\mathbf{Fm\bar{3}m}$ (O_h^5) space group, with 4 formula units per cell and a cubic lattice parameter of 1.0659 nm. The space group and stoichiometry are the same as for the Th_6Mn_{23} structure discussed in Chapter 19, but the assignment of atoms to special positions is different. C sits in the 24*e* special position with $x = 0.275$. There are four inequivalent Cr atoms at the 4*a*, 8*c*, 32*f* ($x = 0.385$), and 48*h* ($x = 0.165$) special positions. The C_6Cr_{23} structural prototype is also adopted by several B_6T_{23} phases.

The structure of B_6Cr_{23} is best understood by considering Fig. 21.10. Figure 21.10(a) shows (001) layers of B octahedra. Each layer consists of large B octahedra, centered by a Cr 1 atom, and smaller uncentered B octahedra. The large and small octahedra share vertices along [001] directions, with the lattice constant equal to twice the sum of the center vertex distance of each of the octahedra. Figure 21.10(b) shows that Cr 1 atoms are coordinated by Cr 4 cuboctahedra that also decorate the edges of the larger B octahedra.

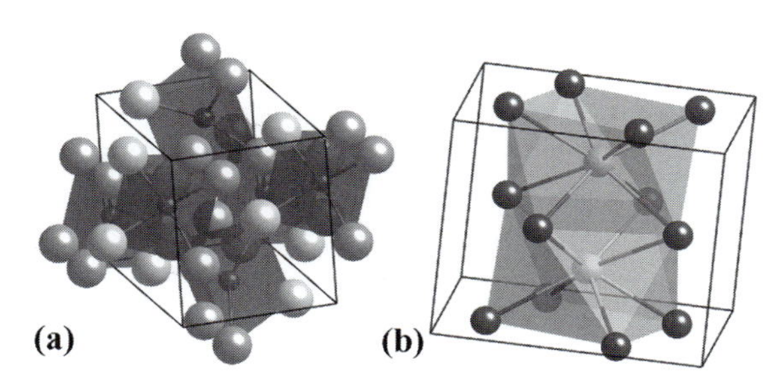

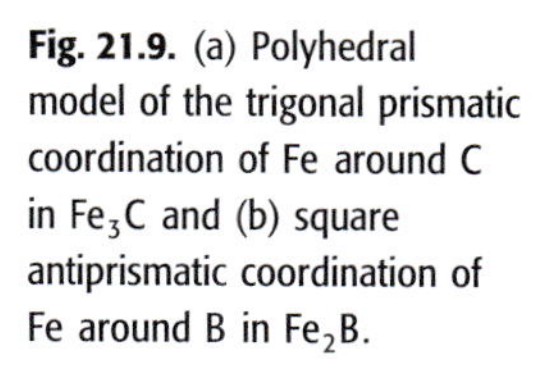

Fig. 21.9. (a) Polyhedral model of the trigonal prismatic coordination of Fe around C in Fe_3C and (b) square antiprismatic coordination of Fe around B in Fe_2B.

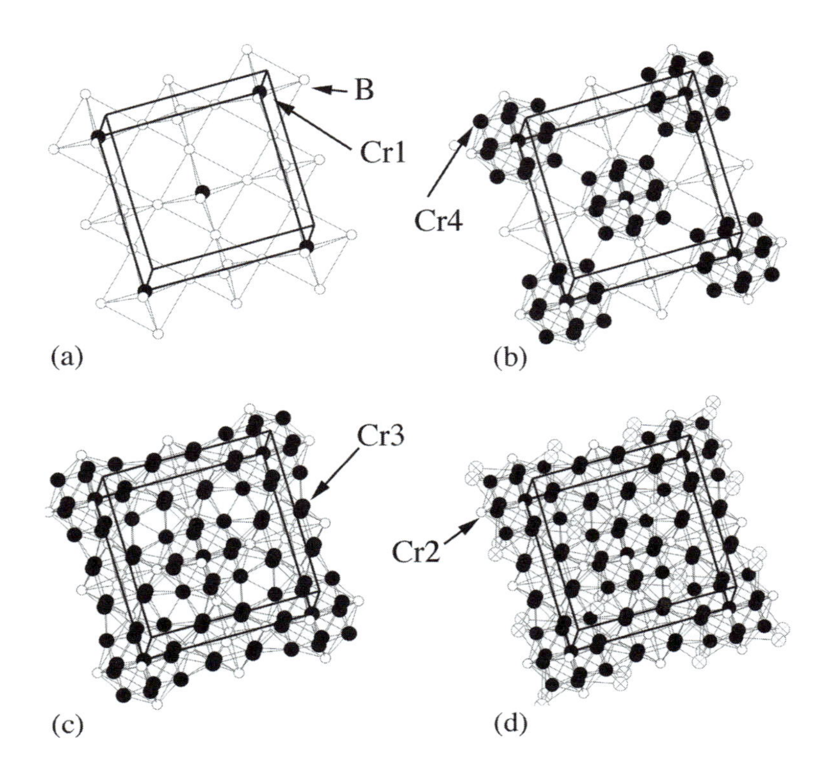

Fig. 21.10. Decomposition of the B_6Cr_{23} structure, illustrating connectivity of B octahedra (a) and successive decoration with inequivalent Cr atoms (b), (c), and (d).

The Cr1–Cr4 distance is similar to that observed for *bcc* late transition metals. Figure 21.10(c) shows that the empty centers of the smaller B octahedra are coordinated by Cr4 atom cubes, which decorate the faces of these octahedra.

Figure 21.10(d) shows Cr2 atoms to form cubes centered by the same voids in the smaller octahedra. These atoms occupy the fluorite (diamond) sites in the *fcc* lattice. These 8*c* sites provide a larger cubic framework to the structure. Widom and Mihalkovik (2005) have recently suggested that alloying with large atoms that disrupt the 8*c* sites can lead to destabilization of this crystalline phase, which may help to stabilize bulk amorphous materials in ternary systems by decreasing the likelihood of crystallization into this structure.

The TL–M systems are among the important amorphous systems for magnetic applications. In these alloys, considerations of glass forming abilities and the desire to maintain high early transition metal (typically Fe or Co and sometimes Ni) concentrations are paramount. A series of eutectics that form near the Fe- and Co-rich edges of TL–M binary phase diagrams are summarized in Table 21.3.

Table 21.3. Glass forming ability parameters (eutectic composition and temperature, and solubility) in binary TL–M systems (McHenry *et al.*, 1999).

Binary alloy	x_e (at%)	x_e (wt%)	T_e (°C)	Solubility of X at 600 (°C) (at%)	Terminal phases
Fe–B	17	3.8	1174	0	Fe, Fe_2B
Co–B	18.5	4.0	1110	0	Co, Co_2B, (Co_3B)
Cr–C	14.0	3.6	1530	0	Co, C_6Cr_{23}
Fe–P	17	10.2	1048	1	Fe, Fe_3P
Co–P	19.9	11.5	1023	0	Co, Co_2P
Fe–Si	33	20	1200	10	Fe, β-Fe_2Si, (Fe_3Si)
Co–Si	23.1	12.8	1204	8	Co, α-Co_2Si, (Co_3Si)

21.7.2 Rare earth–transition metal systems

Amorphous *rare earth–transition metal (RETM) systems* have been studied widely, in part because of their importance as *magneto-optic materials*. $Co_{80}Gd_{20}$ amorphous alloys were the first materials considered for magneto-optic recording. Chaudhari *et al.* (1973) discovered the phenomenon of perpendicular magnetic anisotropy (PMA) in amorphous GdCo films. In a presumably isotropic amorphous material, this anisotropy was puzzling. Atomic structure anisotropy (ASA) was proposed as a source of the large PMA. This anisotropy results from preferential ordering of atomic pairs in the amorphous materials. Another model (Gambino and Cuomo, 1978) proposed selective resputtering to explain the ASA. In 1992, (Harris and Sachidanandam, 1985) employed the polarization properties of EXAFS (see Section 21.9) to measure and describe the anisotropic atomic structure and relate it to the amplitude of the PMA in amorphous TbFe. They reported a direct measure of the ASA in a series of amorphous TbFe films and correlated it with the growth conditions and the magnetic anisotropy energy (Harris and Pokhil, 2001).

21.7.3 Early transition metal – late transition metal systems

Early transition metal – late transition metal binary alloy systems can have eutectics on both the TE- and TL-rich sides of the phase diagram. Of technological importance for magnetic applications are the TL-rich eutectics, since Fe, Co, and Ni, the ferromagnetic transition metals, are TL species. The TL-rich eutectics are of interest in that they typically occur at 8–20 at% of the TE species. These alloys do not have as deep a eutectic, making them harder to synthesize, but they do have larger T_{x1} temperatures, making the resulting amorphous alloys more stable.

Eutectics forming near the Fe- and Co-rich edges of TL–TE binary phase diagrams are summarized in Table 21.4. Terminal alloy compounds and other

Table 21.4. Glass forming ability parameters (eutectic composition and temperature, and solubility) in binary TL–TE systems (McHenry *et al.*, 1999).

Binary alloy	x_e (at%)	x_e (wt%)	T_e (°C)	Solubility of X at 600 (°C) (at%)	Terminal phases
Fe–Zr	9.8	15.1	1337	0	Fe, Fe_3Zr
Co–Zr	9.5	14	1232	0	Co, γ-Co_5Zr, (δ-Co_4Zr)
Fe–Hf	7.9	21.9	1390	0	Fe, λ-Fe_7Hf_3
Co–Hf	11	27.2	1230	0.5	Co, Co_7Hf_2, $Co_{23}Hf_6$
Fe–Nb	12.1	18.6	1373	0	Fe, Fe_2Nb
Co–Nb	13.9	20.3	1237	0.5	Co, Co_3Nb
Fe–Ta	7.9	21.7	1442	0	Fe, Fe_2Ta
Co–Ta	13.5	32.4	1276	3	Co, Co_2Ta

phases in proximity to the eutectic in these systems include Fe_2Zr, Fe_2Hf, Co_2Zr, and Co_2Hf phases, which all have the cubic $MgCu_2$ Laves phase structure illustrated in Chapter 18. The Fe_2Ta, Co_2Ta, Fe_2Nb, and λ-Fe_7Hf_3 phases have the hexagonal $MgZn_2$ Laves phase structures. Other compounds include Fe_3Zr, δ-Co_4Zr, and $Co_{23}Hf_6$, which have the cubic Th_6Mn_{23} structure that was illustrated in Chapter 19. This structure is analyzed again in Fig. 21.11 in terms of the connectivity of Fe2 cuboctahedra. For Fe_3Zr, some substitution of Fe on the Zr sites is required to preserve the Fe_3Zr stoichiometry, as opposed to the $Fe_{23}Zr_6$ stoichiometry of the prototype structure.

Figure 21.11(a) shows an (001) layer in the structure; a checkerboard arrangement of polyhedral environments around vacant cube vertex sites and Fe 1-centered cube edge sites is apparent. Figure 21.11(b) illustrates the decomposition of the polyhedral environments about these two sites.

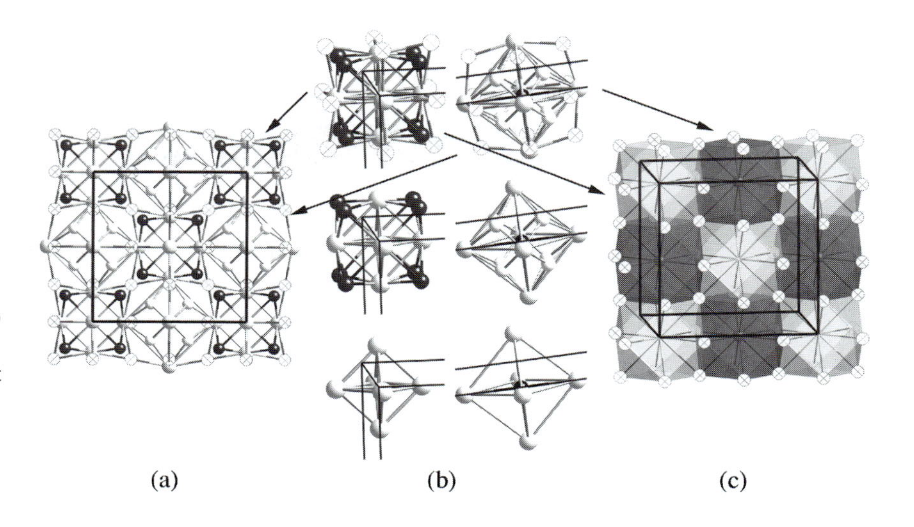

Fig. 21.11. $Fe_{23}Zr_6$ structure showing (a) an (001) layer, (b) the decomposition of the polyhedral environments about the (0, 0, 0) and (1/2, 0, 0) positions in the structure, and (c) the connectivity of Fe2 cuboctahedra.

Figure 21.11(c) illustrates the connectivity of the Fe2 cuboctahedra in this structure. These cuboctahedra fill space, with half of them centered by Fe1 and half centered by a vacant site.

21.7.4 Multicomponent systems for magnetic applications

Since magnetic properties stem from chemical and structural variations on a nanoscale, magnetic applications are among those most significantly impacted by amorphous metals. Recent research has focused on *nanocomposite* and *bulk amorphous alloys* for many of these applications. A *metal/amorphous nanocomposite* is produced by the controlled crystallization of a multicomponent amorphous precursor, to yield a nanocrystalline phase embedded in an amorphous matrix. Important nanocomposite magnets have nanocrystalline grains of a (*bcc*, $D0_3$ or CsCl) Fe(Co)X phase, consuming 20 to 90% of the volume in an amorphous matrix. An example of a metal/amorphous nanocomposite is illustrated in Box 21.1.

Table 21.5 lists examples of magnetic nanocomposite and bulk amorphous systems. Magnetic applications also exist for *bulk amorphous alloys*, which can often be synthesized at cooling rates as low as 1 K/s.

Box 21.1 A metal/amorphous nanocomposite: HITPERM

The figure below shows a typical nanocomposite microstructure with metal nanocrystals embedded in an amorphous metal matrix (Willard, 2000).

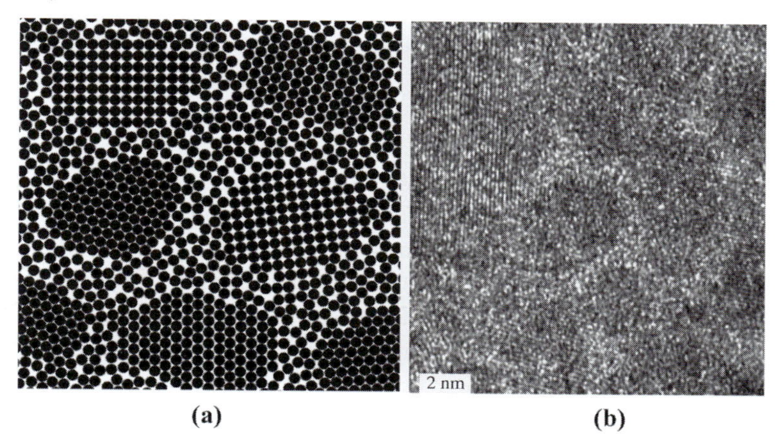

(a) (b)

(a) shows a cartoon of a nanocomposite with nanocrystals embedded in an amorphous matrix. A high resolution TEM image of a nanocrystallized HITPERM material (Willard *et al.*, 1998) is shown in (b). HITPERM is a high induction, high temperature soft magnetic material.

Table 21.5. Examples of nanocrystalline and bulk amorphous alloy systems along with relevant citations.

Alloy composition	Citation
Fe–Si–B–Nb–Cu	Yoshizawa *et al.* (1988)
Fe–Si–B–Nb–Au	Kataoka *et al.* (1989)
Fe–Si–B–V–Cu	Sawa and Takahashi (1990)
Fe–(Zr,Hf)–B	Suzuki *et al.* (1990)
Fe–(Ti,Zr,Hf,Nb,Ta)–B–Cu	Suzuki *et al.* (1991)
Fe–Si–B–(Nb,Ta,Mo,W,Cr,V)–Cu	Yoshizawa and Yamauchi (1991)
Fe–P–C–(Mo,Ge)–Cu	Fujii *et al.* (1991)
Fe–Ge–B–Nb–Cu	Yoshizawa *et al.* (1992)
Fe–Si–B–(Al,P,Ga,Ge)–Nb–Cu	Yoshizawa *et al.* (1992)
Fe–Zr–B–Ag	Kim *et al.* (1993)
Fe–Al–Si–Nb–B	Watanabe *et al.* (1993)
Fe–Al–Si–Ni–Zr–B	Chou *et al.* (1993)
Fe–Si–B–Nb–Ga	Tomida (1994)
Fe–Si–B–U–Cu	Sovák *et al.* (1995)
Fe–Si–B–Nd–Cu	Müller *et al.* (1996)
Fe–Si–P–C–Mo–Cu	Liu *et al.* (1996)
Fe–Zr–B–(Al,Si)	Inoue and Gook (1996)
Fe–Ni–Zr–B	Kim *et al.* (1996)
Fe–Co–Nb–B	Kraus *et al.* (1997)
Fe–Ni–Co–Zr–B	Inoue *et al.* (1997)
Fe–Co–Zr–B–Cu	Willard *et al.* (1998)

Understanding *primary nanocrystallization* is important in the development of nanocomposites (McHenry *et al.*, 2003). Fe-based metallic glass crystallization is the most widely studied process (Luborsky, 1977). Typical commercial Fe-based metallic glass alloys are *hypoeutectic* (Fe-rich), and have been observed to crystallize in a two-step process. The primary crystallization reaction (Am $\rightarrow$ Am$'$ + α-Fe) is followed by secondary crystallization of the glass former enriched amorphous phase, Am$'$ (Koster and Herold, 1981). Luborsky and Lieberman (1978) studied the crystallization kinetics of Fe_xB_{1-x} alloys using *differential scanning calorimetry (DSC)*. Activation energies for crystallization were determined to be largest for the eutectic compositions.

Ramanan and Fish (1982) observed that replacement of B by Si increases the activation energy barriers in Fe-based metallic glasses. Donald and Davies (1978) studied the primary crystallization temperature, T_{x1}, as a function of transition metal substitution, X, in $M_{78-x}X_xSi_{10}B_{12}$ compounds. They explained variations in T_{x1} using the *Hume-Rothery rules*, correlating T_{x1} with the cohesive energies of the pure X species and the atomic size. Two important observations were that: (1) Cu additions, which promote the nucleation of the primary nanocrystals by clustering, resulted in significant reductions in T_{x1}, with additions as small as 0.5–1.0 at%; and (2) early transition metal (e.g.,

Zr, Hf, Mo) additions impede the growth and result in the largest primary crystallization temperatures.

Amorphous alloy precursors to nanocomposites are typically based on ternary (or higher order) systems. These are often variants of TL/TE/M systems. In many cases, a small amount of a fourth element (making the alloys *quaternary*) such as Cu, Ag, or Au can be added to promote nucleation of the nanocrystalline phase. Five- and six-component systems are also commonplace if more than one metalloid and/or early transition metal species are used as glass formers. A matrix of typical elements in many amorphous phases is given by (Willard, 2000):

$$\begin{bmatrix} \mathrm{Fe} \\ \mathrm{Co} \\ \mathrm{Ni} \end{bmatrix} \begin{bmatrix} \mathrm{Ti} & \mathrm{V} & \mathrm{Cr} \\ \mathrm{Zr} & \mathrm{Nb} & \mathrm{Mo} \\ \mathrm{Hf} & \mathrm{Ta} & \mathrm{W} \end{bmatrix} \begin{bmatrix} \mathrm{B} & \mathrm{C} & \\ \mathrm{Al} & \mathrm{Si} & P \\ \mathrm{Ga} & \mathrm{Ge} & \end{bmatrix} \begin{bmatrix} \mathrm{Cu} \\ \mathrm{Ag} \\ \mathrm{Au} \end{bmatrix}$$

Magnetic metal/amorphous nanocomposites (McHenry *et al.*, 1999) have excellent soft magnetic properties as measured by the figures of merit of combined magnetic induction and permeability, high frequency magnetic response, and retention of magnetic softness at elevated temperatures. Applications have been identified for the patented Fe–Si–B–Nb–Cu alloys (trade-name FINEMET) (Yoshizawa *et al.*, 1988) and Fe*M*BCu alloys (trade-name NANOPERM) (Suzuki *et al.*, 1990). Another nanocomposite $(\mathrm{Fe}_{1-x}\mathrm{Co}_x)_{88}\mathrm{M}_7\mathrm{B}_4\mathrm{Cu}$ (M = Nb, Zr, Hf) soft magnetic material is known as HITPERM (Willard *et al.*, 1998); HITPERM has a superior high temperature magnetic induction.

21.7.5 Multicomponent systems for non-magnetic applications

Although magnetic systems have been emphasized in the previous section, many amorphous systems containing noble metals or without TL species have been developed for non-magnetic structural applications. In fact, Pd–Si alloys were the first demonstrated metallic glass systems (Klement *et al.*, 1960). The main feature of the Pd–Si phase diagram of interest for glass forming ability is the deep eutectic at about 15–20 at.% Si. The eutectic temperature is 835 °C. The Cu–Si phase diagram also has a similar deep eutectic. Pd–Cu–Si amorphous alloys have been investigated for hydrogen separation membrane materials (Kircheim *et al.*, 1982) and, recently, Pd–Cu–Si–P bulk amorphous alloys with excellent mechanical properties have been reported (Liu *et al.*, 2005).

Lanthanides are also components in other interesting amorphous alloy systems; Loffler (2003) gives a history of the development of these alloys. In synthesizing bulk amorphous materials, the considerations for glass formation include the *confusion principle*, which suggests that having a multicomponent systems is crucial for successful synthesis. The more components present in a system, the more "confused" the system will become during cooling, hence promoting the formation of an amorphous compound. These multi-component systems typically involve variations based on a known binary glass forming system, with additions consistent with the confusion principle, and other physical or electronic properties considerations.

Inoue and Gook (1996) have proposed empirical rules for the synthesis of bulk amorphous alloys with large glass forming abilities. These are multicomponent systems with (1) three or more constituent elements, (2) significantly different atomic size ratios, typically with differences exceeding about 13%, and (3) negative heats of mixing among the constituents. Important alloys for mechanical properties are based on the La–Al–Ni(Cu) (Inoue *et al.*, 1989) and Zr(Ti)–Ni(Cu)–Be systems (Peker and Johnson, 1993). Commercial alloys with the trade-name *Liquidmetal* have been developed in the Zr(Ti)–Ni(Cu)–Be system.[2]

The structure of bulk amorphous alloys is of considerable interest. Figure 21.6(c) illustrates the relative sizes of the metallic radii of the Zr, Be, Ti, Cu, and Ni atomic components of *Liquidmetal* bulk amorphous alloys. Figure 21.6(c) also illustrates the dense random packing of hard spheres of different sizes that serves as a model for the structure of a typical bulk amorphous alloy. If this is compared with Fig. 21.6(b), it can be seen that having significantly different atomic size ratios increases the packing density. This packing makes it difficult for atoms to move past one another, as would be required for a mechanical deformation process to proceed (Um, 2006).

The low atomic mobility in bulk amorphous solids makes difficult the non-conservative conversion of kinetic energy into heat or permanent plastic deformation upon impact with another object. The wide range of atomic sizes reduces the internal friction associated with the atomic displacements on impact. As a result, "frictional losses" are significantly reduced and such collisions can be considered as nearly perfectly elastic. Elastic momentum transfer makes these materials ideal for applications such as golf club heads. Here, the increased momentum transfer between the head and a golf ball promotes longer drives. A particularly elegant demonstration of this principle can be observed in devices known as "atomic trampolines."[3]

21.8 *X-ray scattering in amorphous materials

To understand X-ray scattering from amorphous materials it is useful to consider first the scattering from a very small piece of crystal; we will follow Warren (1990) to derive an expression for the scattered intensity from an

[2] Additional information can be found at www.liquidmetaltechnologies.com.

[3] See http://www.mrsec.wisc.edu/edetc/amorphous for laboratory demonstration kits and a demonstration video.

amorphous collection of atoms. We know from our discussions in Chapter 12 that an individual atom scatters X-rays according to the *atomic scattering factor*, $f(s)$, where $s = |\mathbf{s}| = \sin\theta/\lambda$ with θ the scattering angle, and

$$\mathbf{s} = \frac{1}{2}(\mathbf{k}' - \mathbf{k});$$

$\mathbf{k}$ is the incident wave vector, and $\mathbf{k}'$ the scattered wave vector. As before we will only consider elastic scattering events, for which the length of the two vectors is equal to the inverse of the wave length, λ. Scattering from a group of atoms is expressed by means of the structure factor:

$$F(\mathbf{s}) = \sum_{n=1}^{N} f_n(s) \mathrm{e}^{4\pi \mathrm{i}\mathbf{s}\cdot\mathbf{r}_n}.$$

Since the scattered intensity is proportional to the modulus squared of the structure factor, we have:

$$I(\mathbf{s}) = F(\mathbf{s})F^*(\mathbf{s}) = \sum_{n=1}^{N}\sum_{m=1}^{N} f_n(s) f_m(s) \mathrm{e}^{4\pi \mathrm{i}\mathbf{s}\cdot\mathbf{r}_{nm}},$$

with $\mathbf{r}_{nm} = \mathbf{r}_n - \mathbf{r}_m$. We find that the diffracted intensity depends only on the relative vectors connecting each pair of atom sites. All the terms with $n = m$ (and therefore $\mathbf{r}_{nm} = \mathbf{0}$) can be collected and put up front, so that we have (dropping the argument s on the atomic scattering factors):

$$I(\mathbf{s}) = Nf^2 + \sum_{n=1}^{N}\sum_{m=1}^{N}{}' f_n f_m \mathrm{e}^{4\pi \mathrm{i}\mathbf{s}\cdot\mathbf{r}_{nm}},$$

where the prime indicates that only terms with $n \neq m$ are counted.

In an amorphous material, there is no preferential direction, so that we can average the intensity over all orientations. If we represent by ϕ the angle between $\mathbf{r}_{nm}$ and $\mathbf{s}$, then the argument of the exponential becomes $4\pi \mathrm{i} s r_{nm}\cos\phi$. The end-point of the vector $\mathbf{r}_{nm}$ lies anywhere on a spherical surface with radius r_{nm}, so that the orientational average can be written as:

$$\langle \mathrm{e}^{4\pi \mathrm{i}\mathbf{s}\cdot\mathbf{r}_{nm}} \rangle = \frac{1}{4\pi r_{nm}^2}\int_0^{\pi} \mathrm{d}\phi\, \mathrm{e}^{4\pi \mathrm{i} s r_{nm}\cos\phi} 2\pi r_{nm}^2 \sin\phi = \frac{\sin k r_{nm}}{k r_{nm}},$$

where $k = 4\pi s$. This leads to the *Debye scattering equation*:

$$\boxed{I(k) = Nf^2 + \sum_{n=1}^{N}\sum_{m=1}^{N}{}' f_n f_m \frac{\sin k r_{nm}}{k r_{nm}}.} \tag{21.12}$$

Before we discuss amorphous materials, let us first try to understand what this equation means. Consider a small primitive cubic crystallite, consisting of only eight identical atoms at the corners of a cube. To compute the scattered intensity from an assembly of randomly oriented cubes we must first determine all possible interatomic vectors. If the cube has edge length a, then it is not too hard to see that there are 24 interatomic vectors of length a (note that we distinguish between $\mathbf{r}_{nm}$ and $\mathbf{r}_{mn}$, so that the 12 edges of the cube must be double counted); there are also 24 diagonal interatomic vectors of length $a\sqrt{2}$, and 8 interatomic vectors of length $a\sqrt{3}$. The Debye scattering equation for this randomly oriented simple cubic "material" is then written as:

$$\bar{I}_{\text{sc}}(k) \equiv \frac{I_{\text{sc}}(k)}{Nf^2} = 1 + 3\frac{\sin(ka)}{ka} + 3\frac{\sin(ka\sqrt{2})}{ka\sqrt{2}} + \frac{\sin(ka\sqrt{3})}{ka\sqrt{3}}. \qquad (21.13)$$

We leave it as an exercise for the reader to show that for a nine-atom *bcc* cell, we have

$$\bar{I}_{\text{bcc}}(k) = 1 + \frac{16}{9}\frac{\sin(ka\sqrt{3/2})}{ka\sqrt{3/2}} + \frac{24}{9}\frac{\sin(ka)}{ka} + \frac{24}{9}\frac{\sin(ka\sqrt{2})}{ka\sqrt{2}} + \frac{8}{9}\frac{\sin(ka\sqrt{3})}{ka\sqrt{3}}, \qquad (21.14)$$

and for a 14-atom *fcc* crystallite:

$$\bar{I}_{\text{fcc}}(k) = 1 + \frac{36}{7}\frac{\sin(ka/\sqrt{2})}{ka/\sqrt{2}} + \frac{15}{7}\frac{\sin(ka)}{ka} + \frac{24}{7}\frac{\sin(ka\sqrt{3/2})}{ka\sqrt{3/2}} + \frac{12}{7}\frac{\sin(ka\sqrt{2})}{ka\sqrt{2}} + \frac{4}{7}\frac{\sin(ka\sqrt{3})}{ka\sqrt{3}}. \qquad (21.15)$$

Figure 21.12(a) shows these three curves as a function of the variable ka; each curve is normalized by the number of atoms in the crystallite, $N = 8$ for simple cubic, $N = 9$ for *bcc*, and $N = 14$ for *fcc*. The vertical dashed lines indicate the positions of the Bragg peaks for an infinite crystal with lattice parameter a. It is clear that, even for such small crystallites, there is already quite a bit of "structure" in the scattered intensity, and the three cells give rise to very different scattered intensities. If we increase the number of unit cells from 1 to 2, 4, and 6, then there are more terms in the summation of the Debye scattering equation. For the simple cubic system, we have $N = 8$, 27, 125, and 343, respectively. Figure 21.12(b) shows how the scattered intensity changes from the slowly oscillating curve for $N = 8$ to a curve that peaks at each dashed line. In the limit of $N \to \infty$, the scattered intensity would consist of infinitely narrow peaks at each of the reciprocal lattice spacings, i.e., we would recover the reciprocal lattice of the infinite crystal and its corresponding diffraction pattern.

Figure 21.12(c) and (d) show the same progression for the *bcc* and *fcc* cells, respectively. For the *bcc* case, we have $N = 9$, 35, 189, and 559. We can observe clearly how the lattice planes with $h+k+l$ odd have a vanishing intensity, corresponding to the systematic absences of the infinite crystal. Similarly, for the *fcc* case, we have $N = 14$, 63, 365, and 1099; the systematic absences of the *fcc* lattice clearly appear, even for rather small cells.

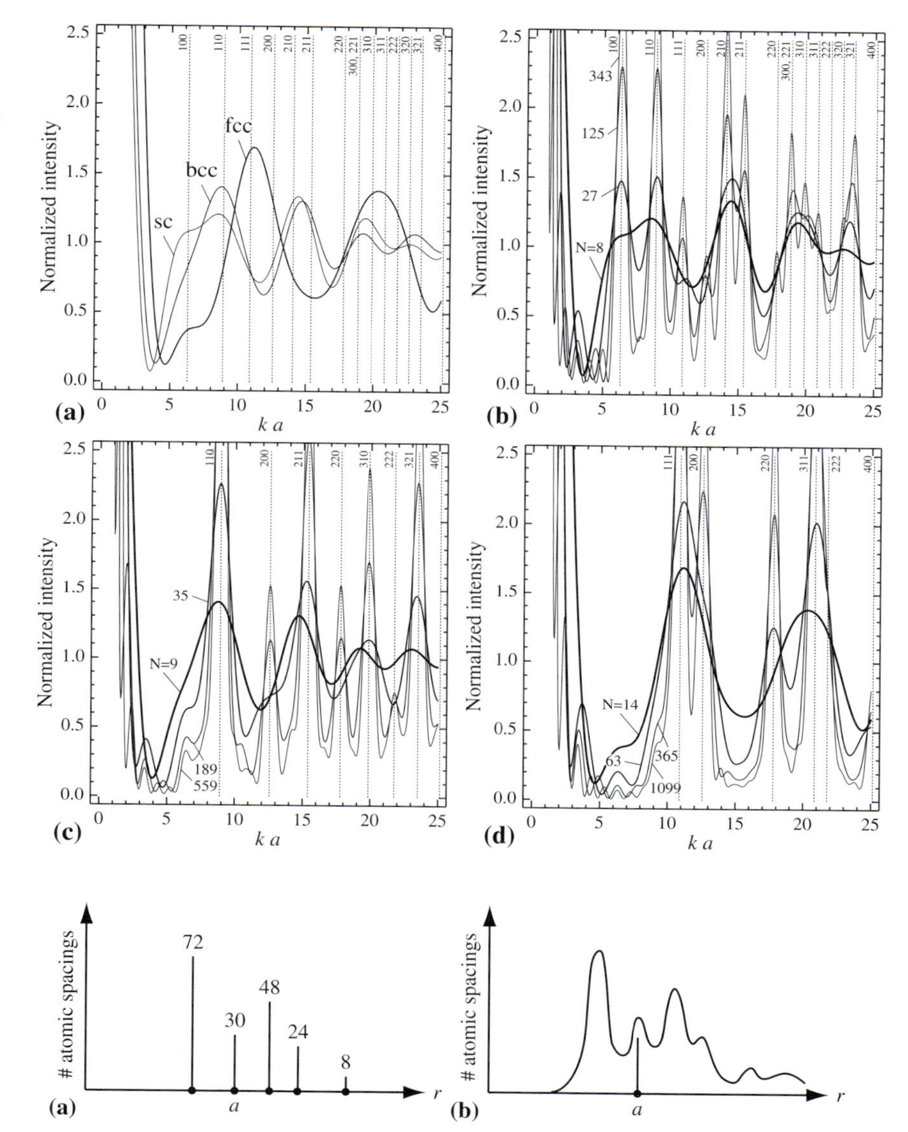

Fig. 21.12. Normalized scattered intensity for simple cubic, *bcc*, and *fcc* crystallites (equations 21.13–21.15) (a); in (b), (c), and (d), the evolution of the scattered intensity is shown for the simple cubic, *bcc*, and *fcc* cases, respectively, as a function of increasing number of atoms.

Fig. 21.13. Number of interatomic spacings versus distance (distance pair correlation function) for the 14-atom *fcc* cluster (a), and after randomly moving atoms away from their lattice sites (b).

Now we are ready to describe the scattered intensity from an amorphous material. Figure 21.13(a) shows the distribution of interatomic spacings in the face-centered cubic crystallite with 14 atoms; this is a discrete spectrum of distances because the atoms are arranged in a highly regular fashion. Assume now that we randomly move each atom away from its lattice site. If we do this in all crystallites that make up our "sample," then some interatomic spacings will shorten, others will lengthen, and the resulting distribution function, the *distance pair correlation function*, will become continuous rather than discrete, as shown schematically in Fig. 21.13(b).

We can re-derive the Debye scattering equation by introducing the atomic density function $\rho_{\text{atom}}(\mathbf{r})$ (Warren, 1990):

$$
\begin{aligned}
I(\mathbf{s}) =& Nf^2 + f^2 \sum_m \sum_{n\neq m} \mathrm{e}^{4\pi i \mathbf{s}\cdot\mathbf{r}_{nm}};\\
=& Nf^2 + f^2 \sum_m \int \mathrm{d}V_n\, \rho_{\text{atom}}(\mathbf{r}_{nm})\mathrm{e}^{4\pi i \mathbf{s}\cdot\mathbf{r}_{nm}},
\end{aligned} \tag{21.16}
$$

where $\rho_{\text{atom}}(\mathbf{r}_{nm})\mathrm{d}V_n$ is the number of atom centers in the volume element $\mathrm{d}V_n$ at position $\mathbf{r}_{nm}$ relative to the atom at $\mathbf{r}_m$. We can then write the atomic density function as $\rho_{\text{atom}} = (\rho_{\text{atom}} - \rho_a) + \rho_a$, where ρ_a is the average atomic density. After substitution in the integral above, it can be shown that the last integral, containing only ρ_a, is usually negligible for scattering directions away from the forward direction (Warren, 1990). Carrying out the orientational average as before, we find for the Debye scattering equation of a monatomic amorphous solid:

$$
I(k) = Nf^2 + Nf^2 \int_0^\infty \mathrm{d}r\, 4\pi^2[\rho_{\text{atom}} - \rho_a]\frac{\sin kr}{kr}. \tag{21.17}
$$

It is customary to introduce the *reduced intensity function*, $i(k)$, as:

$$
i(k) \equiv \frac{I(k)/N - f^2}{f^2} = \int_{r=0}^\infty \mathrm{d}r\, 4\pi r^2[\rho_{\text{atom}}(r) - \rho_a]\frac{\sin(kr)}{kr}. \tag{21.18}
$$

The reduced intensity function is the experimental observable in an X-ray scattering experiment. The function $\phi(k) \equiv ki(k)$ is often used in the analysis of scattering data from amorphous materials:

$$
\phi(k) = 4\pi \int_{r=0}^\infty \mathrm{d}r\, r[\rho(r) - \rho_a]\sin(kr). \tag{21.19}
$$

The function $\phi(k)$ is recognized to be the Fourier transform of a function $f(r)$:

$$
\phi(k) = 4\pi \int_{r=0}^\infty \mathrm{d}r\, f(r)\sin(kr)
$$

and $f(r) = r[\rho(r) - \rho_a]$ can be expressed as:

$$
f(r) = \frac{1}{2\pi^2}\int_{k=0}^\infty \mathrm{d}k\, \phi(k)\sin(kr).
$$

This allows us to express the radial distribution function as:

$$
\mathrm{RDF}(r) = 4\pi r^2\rho(r) = 4\pi r^2\rho_a + \frac{2r}{\pi}\int_{k=0}^\infty \mathrm{d}k\, ki(k)\sin(kr). \tag{21.20}
$$

If we measure $ki(k)$ from a scattering experiment, then we can use this relation to convert the measured data directly into the radial distribution function, which shows the distribution of interatomic spacings in the amorphous material. Much of the structural information obtainable in the form of atomic configurations in amorphous solids comes from this determination of the radial distribution function. The treatment of X-ray scattering from an amorphous solid can be generalized to a material containing two or more atomic species (Warren, 1990). In that case, the information contained in the RDF reflects *all* of the atomic species. This can be complemented by atom-specific radial distribution functions (the RDF about a specific atom type) attainable from an extended X-ray absorption fine structure (EXAFS) measurement, near an X-ray absorption edge of the atomic species of interest.

21.9 *Extended X-ray absorption fine structure (EXAFS)

X-ray absorption, *X-ray fluorescence*, and *Auger electron spectroscopy* are all experimental techniques that rely on X-ray absorption and/or emission by the *photoelectric effect*. Figure 21.14 illustrates the electronic transitions associated with each of these phenomena. Keeping in mind that the energy scale for X-rays is between 1 and 500 keV, corresponding to wavelengths of 0.003 to 1.2 nm, it is possible to predict which electronic transitions can be associated with the absorption or emission of X-rays. Figure 21.14(a) illustrates a typical process of *X-ray absorption*. In this process, an X-ray photon is absorbed by an atom, if its energy corresponds to the energy needed to release a *photoelectron* from a core shell into the continuum. This process could involve a K, L, or M shell electron, resulting in different characteristic energies for the absorption process. The atom is left in an excited state with a core level hole (an empty level) into which another higher energy electron could fall in a subsequent process.

X-ray absorption fine structure (XAFS) is the modulation of the X-ray absorption coefficient at energies near or just above an X-ray absorption edge. *X-ray absorption spectroscopy* (XAS) refers to the spectroscopic techniques used to probe the XAFS. Two techniques are prevalent: (1) *X-ray absorption*

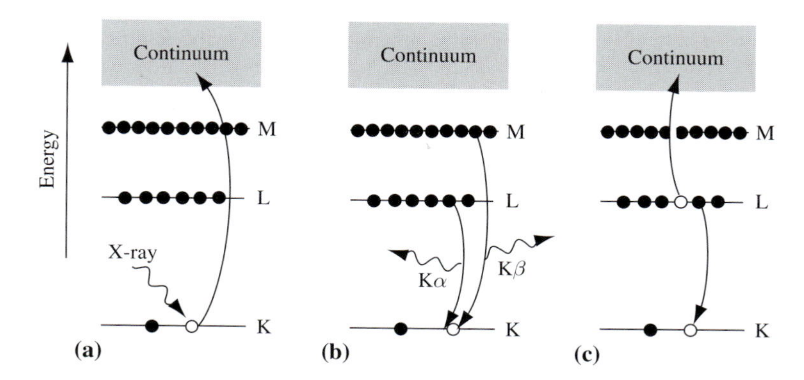

Fig. 21.14. Electronic transitions associated with the photoelectric effect for (a) X-ray absorption, (b) X-ray fluorescence, and (c) Auger effect.

near-edge spectroscopy (XANES) and (2) *extended X-ray absorption fine structure (EXAFS)*. Each provides information about the absorbing atom's local environment, including local atomic coordination numbers, types of coordinating atoms, chemical oxidation state, etc.; EXAFS is more widely used for studying amorphous solids and will be emphasized here.

Figure 21.14(b) shows processes associated with *X-ray fluorescence*. These involve relaxation of the excited state by higher energy electrons falling into the core holes. An X-ray photon of energy equal to the difference between the core energy levels is emitted. For an L to K transition, the emitted x-radiation is called Kα radiation, for an M to K transition it is known as Kβ radiation, etc. The energy is characteristic of the absorbing atom. Figure 21.14(c) illustrates the process associated with the emission of an *Auger electron*. Here, an electron from the L shell decays to the K shell, and the energy released in this event is used to promote another electron to the continuum.

X-ray absorption in solids results from the *X-ray photoelectric effect*, in which the incident X-ray photon is responsible for an electronic excitation on the absorber atom. Absorption results in a loss of X-ray intensity, which depends on the amount of material traversed, and is expressed by *Beer's law*:

$$\frac{I}{I_0} = \mathrm{e}^{-(\frac{\mu}{\rho})\rho z}, \tag{21.21}$$

where I is the transmitted intensity (through a thickness z) and I_0 is the incident intensity of the X-rays; μ/ρ is the *mass absorption coefficient* for X-rays.

Figure 21.15(a) shows a block diagram of an X-ray absorption experiment. It includes an X-ray source, which could be a standard cathode, a rotating anode or a synchrotron source. The X-ray photons from these sources typically have a wide spectrum of energies. A monochromatic source is obtained by scattering off a *single crystal monochromator*; by rotating the monochromator, the photon wave length can be selected. The incident intensity of monochromatic X-rays is measured at a detector before reaching the sample. A portion of the transmitted intensity is measured at another detector. The portion of the absorbed X-ray intensity that is reemitted at energies associated with fluorescence events is collected and counted using a *fluorescence*

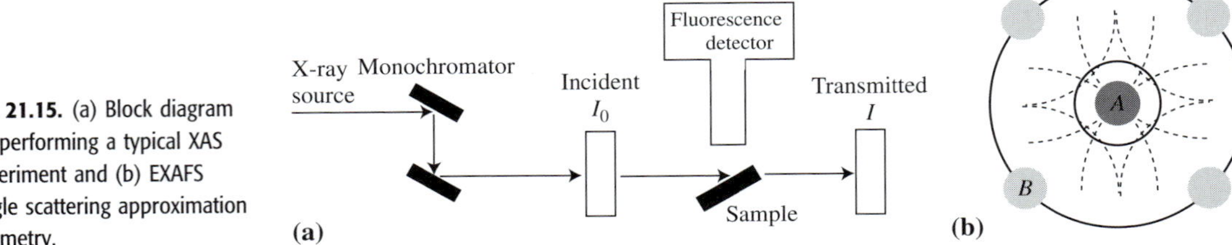

Fig. 21.15. (a) Block diagram for performing a typical XAS experiment and (b) EXAFS single scattering approximation geometry.

detector. The absorption spectrum is typically inferred from the transmitted intensity.

The sequence of events leading to the EXAFS fine structure is (Fig. 21.15(b)):

(i) X-ray photons of appropriate energy are absorbed at an atomic site, A, giving rise to a core level transition (e.g., Kα, Kβ, etc.). The energy at which this transition occurs is atom specific.
(ii) The excited photoelectron leaves site A as a spherical wavefront. At site B, however, it can be treated as a plane wave.
(iii) The photoelectron plane wave is backscattered from the nearest neighbor B sites and travels back to the A site.
(iv) Constructive and destructive interference between the outgoing photoelectron and the backscattered wavefronts modulates the transition probability of the absorption event. This backscattered wavefront carries information about the local environment around the absorbing atom.

As a result of interference between the forward scattered and backscattered waves, the amount of radiation absorbed is modulated. This manifests itself in an oscillation in the X-ray absorption spectrum for energies exceeding the absorption threshold. Analysis of the EXAFS fine structure (Fig. 21.16) is described in Box 21.2. The RDF is used in the *EXAFS single scattering approximation* to determine structural information. A single back-scattering event is assumed and a plane wave returns from the backscatterer site. The *absorption function* $\chi(k)$ is defined by:

$$k\chi(k) = \frac{1}{k}\sum_j \frac{N_j}{r_j} f_j(\pi) e^{-2\sigma_j^2 k^2} \sin(2kr_j - 2\delta + \psi_j) \tag{21.22}$$

where k is the X-ray wave number, σ is the *Debye-Waller factor*, $f_j(\pi)$ is the *backscattering phase shift*, δ is the phase shift due to the exciter atomic potential and ψ is the phase shift due to the backscatterer atomic potential. Analyses of this type provide information on the local environment of atoms, regardless of whether or not the atom resides in a crystalline environment.

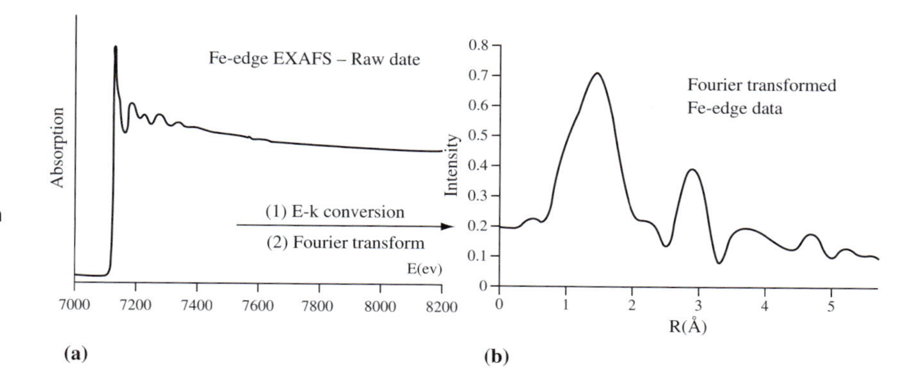

Fig. 21.16. EXAFS analysis involves processing raw data (a) to yield a Fourier transformed radial distribution function (b) shown for an Fe-edge in Ni Zn ferrite nanoparticles (Swaminathan, 2005) (figure courtesy of R. Swaminathan).

Box 21.2 Reduction of EXAFS data to obtain the RDF

The procedure for converting EXAFS data to a radial distribution function is as follows:

(i) The *absorption coefficient*, $\mu(k)$, is determined as a function of the magnitude of the wave vector, k, using the relationship:

$$\mu(k) = \ln\left[\frac{I}{I_0}\right].$$

(ii) The monotonically decreasing absorption coefficient background is subtracted to yield the part that carries the information as to the type, number, and distance of nearest neighbors. The function $k\chi(k)$ is defined as:

$$k\chi(k) = k\left[\mu(k) - \mu_0(k)\right], \tag{21.23}$$

where $\mu_0(k)$ is the non-oscillatory part of the absorption coefficient.

(iii) The Fourier transform of the function $k\chi(k)$ results in a site specific radial distribution function about the absorbing atom.

21.10 Mössbauer spectroscopy

Mössbauer spectroscopy is another powerful technique for studying the local structure of amorphous solids. The *Mössbauer effect* refers to recoil-free γ-ray resonance absorption. Mössbauer spectroscopy is used as a probe of local environments and magnetic structure in solids. Since Mössbauer spectroscopy determines the local symmetries and chemical environments of atoms, it is particularly useful in probing amorphous materials. ^{57}Fe Mössbauer spectroscopy is also useful to understand the structural and magnetic properties of nanocrystalline magnetic materals (Sorescu *et al.*, 2003); the technique provides information on Fe-coordination, particle size, and surface effects.

The Mössbauer effect involves the decay of an atomic nucleus from an excited state to its ground state, with the emission of a γ-ray. In a solid, the entire lattice takes up the recoil energy required for momentum conservation. The large mass of the lattice means that the recoil energy ($p^2/2m$) is negligible and the γ-ray energy is precisely the difference in energy between the excited and ground states of the emitting nucleus. An emitted γ-ray photon can, in turn, be absorbed, to pump a nucleus into an exited state; after a mean lifetime τ, this excited nucleus will fall back into the ground state, re-emitting the γ-ray isotropically. This γ-ray resonance fluorescence is used as a spectroscopic tool by modulating the emitted γ-ray energy by a Doppler

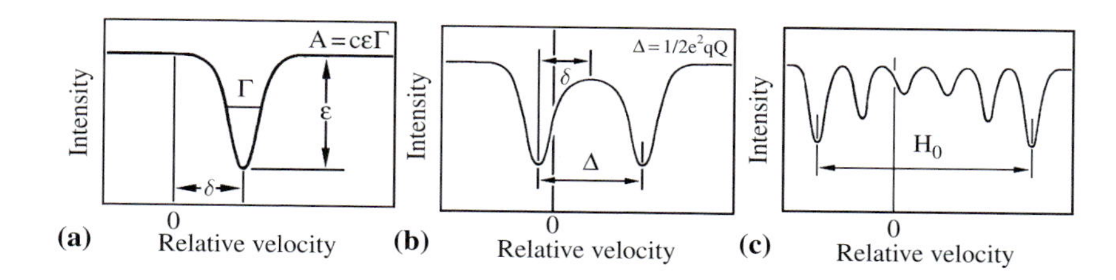

Fig. 21.17. Mössbauer spectroscopy phenomena: (a) resonance peak, (b) quadrupole splitting, and (c) hyperfine field splitting.

shift, associated with the relative velocity of the γ-ray source and the emitting nuclei. The Doppler energy shift is given by $E = (v/c)E_\gamma$, where v is the source velocity, c is the speed of light and E_γ is the γ-ray energy. The most common Mössbauer experiment involves a radioactive source containing the Mössbauer isotope in an excited state (^{57}Fe) and an absorber, consisting of the material under investigation, that contains the same isotope in the ground state.

A Mössbauer spectrum (Herber, 1982) consists of the γ-ray intensity measured at a γ-ray detector and collected as a function of the source/absorber relative velocity (and, therefore, the modulation energy, E). A resonance occurs for $E = \Gamma = h/2\pi\tau$. For a ^{57}Fe Mössbauer active nucleus, Γ is 4.6×10^{-12} keV. Figure 21.17(a) shows a typical unsplit Mössbauer resonance line, characterized by a position, δ, a line width, Γ, and area $A = c\epsilon\Gamma$. Mössbauer spectra splittings are useful to obtain chemical and field signatures. A Mössbauer active nucleus in a non-cubic environment is subject to a quadrupole splitting, Δ, which is proportional to the electric field gradient at the nucleus resulting from the chemical environment of the surrounding nuclei (Fig. 21.17(b)). For a magnetic environment, the hyperfine field, H_0, splits the Mössbauer signal into a hextet (Fig. 21.17(c)).

21.11 Historical notes

George Feodosevich Voronoi (1868–1908) was a Russian mathematician who made contributions in the theory of tiling of N-dimensional space. Voronoi was born in Zhuravka (a small village 160 km east of Kiev), Russia (now the Ukraine). He attended the University of St. Petersburg in physics and mathematics, where he graduated in 1889. He continued at St Petersburg where in 1894 he received a Master's Degree. Voronoi became Professor of Pure Mathematics at Warsaw University. While there, he was awarded a doctorate from the University of St. Petersburg for work on algorithms for continued fractions. Voronoi's masters and doctoral research were awarded the Bunyakovsky Prize by the St. Petersburg Academy of Sciences.[4]

Voronoi contributed to the theory of numbers, algebraic numbers, and the geometry of numbers. He published 12 papers, most of which were influential. The Voronoi polyhedron and diagram are named after him. He published this geometric construction in 1908 (Voronoi, 1908). This construction was also discovered previously by the mathematician, **Johann Peter Gustav Lejeune Dirichlet**, (1805–59) (Dirichlet, 1850) and, therefore, the Voronoi diagram is sometimes called the *Dirichlet tessellation*. The Voronoi diagram is one of the most fundamental constructs used to describe non-periodic lattices. It has also been applied in such diverse areas as city planning (optimizing the locations of service facilities), astrophysical mapping of galaxy clusters, mapping of tumor cells, cell and crystal growth, and protein molecular volume analysis, for example.

John Desmond Bernal (1901–71) was an Irish physicist and X-ray crystallographer. He was born in 1901 in Nenagh County, Tipperary, Ireland. He was a student at Cambridge after which he worked from 1923–27 with W. H. Bragg at the Royal Institution in London. He returned to Cambridge and stayed from 1927–37. In 1937, he was appointed Chair of Physics at Birbeck College at the University of London and was elected as a Fellow of the Royal Society. From 1938–63 he remained as Professor of Physics at the University of London. He was awarded a Royal Medal of the Royal Society in 1945. From 1963–68 he was Professor of Crystallography at the University of London.

Bernal made many contributions to X-ray crystallography. His work included studying the structure of solids and liquids. Bernal also conducted

(a)

(b)

Fig. 21.18. (a) Georgii Feodosevich Voronoi (1868–1908) (picture courtesy of J. Lima-de-Faria (Lima-de-Faria, 1990)) and (b) John Desmond Bernal (1901–71) (picture courtesy of J. Lima-de-Faria (Lima-de-Faria, 1990) and originally from his obituary in Acta Cryst. A28, 359, 1972, photograph by Henry Grant).

[4] More biographical information on Voronoi can be found in the article by J. J. O'Connor and E. F. Robertson at http://www-maths.mcs.st-andrews.ac.uk/Mathematicians/Voronoy.html

research in molecular biology. During the 1930s, his group made the first X-ray studies of the amino acids, several steroids, proteins, and viruses (the tobacco mosaic virus). Bernal and Dorothy Crowfoot studied crystals of the protein pepsin. This was the first use of the X-ray technique to study biological molecules at the Cavendish Laboratory. He also studied the structure and properties of water. Two of his students, D. Crowfoot-Hodgkin and M. Perutz went on to win Nobel prizes for their work. Bernal made important contributions to the chemical origin of life and used them to speculate on general principles guiding life in the universe. The dense random packing of hard spheres model of amorphous solids was proposed by him (Bernal, 1965), and he was a major contributor to the theory of amorphous metals. He made important contributions to the structure of liquids, liquid crystals, and hydrogen bonding. Bernal also published *The Social Functions of Science* (1939) and *Science in History* (1954).

21.12 Problems

(i) *Packing density*: Calculate the area packing density of close-packed circles (i.e., the fraction of the total area covered by circles). Calculate the area packing density of a square array of touching circles.

(ii) *DRPHS*: Figure 21.6 depicts the 2-D atomic arrangements in (a) an amorphous solid using the dense random packing of hard spheres (DRPHS) model and (b) a crystalline solid with grain boundaries. Describe how the 2-D packing density differs in an amorphous solid or at a grain boundary from that of a crystalline solid (Hint: Consider void sizes).

(iii) *Scherrer broadening*: Consider the (110) and (220) reflections for *bcc* Fe:

(a) Calculate the Bragg angles for each reflection for Cu-Kα radiation.
(b) Calculate the Scherrer broadening for a 50, 10, and 1 nm particle for each peak.
(c) If the integrated intensity under the peaks is conserved and doubling the peak breadth roughly halves the maximum intensity of the peak, predict how the relative intensity of the two peaks will change with particle size.

(iv) *bcc direct and reciprocal lattice and Brillouin zones I*: Determine the following for the *bcc* lattice:

(a) Construct real space basis vectors and reciprocal lattice vectors for the *bcc* lattice.
(b) Determine the shortest reciprocal lattice vectors for the *bcc* lattice; how many of these vectors are there?

(c) Construct the first Brillouin zone for the *bcc* lattice. Show that it is a rhombic dodecahedron.
(d) Determine the shortest vectors to the surface of the first Brillouin zone.

(v) *bcc direct and reciprocal lattice and Brillouin zones II*: Consider the conventional 2-D (monatomic) rectangular unit cell with orthogonal edges $a = 2b$ and with sites occupied by atoms of species A.

(a) Identify a set of primitive unit cell vectors, describe them in a Cartesian coordinate system and calculate the area of the primitive unit cell.
(b) Express the reciprocal unit cell vectors and calculate the k-space area of the primitive reciprocal unit cell.
(c) Carefully sketch the shape of the first and second Brillouin zones.

(vi) *Pair potentials*: Calculate the total energy for the 13-atom icosahedral, cuboctahedral, and twinned cuboctahedral (*hcp*) configurations of atoms using a standard Lennard-Jones pair potential.

(vii) *Hard sphere liquid*: Consider the following simple model for a monatomic liquid. Each atom has n nearest neighbors at a distance, d, and a uniform density ρ_a of neighboring atomic centers starting at a distance R, such that $(4\pi/3)R^3\rho_a = n + 1$. Let N be the number of atoms in the sample.

(a) Derive an expression for I/Nf^2 as a function of n, kd, and $\Phi(kR)$.
(b) Plot I/Nf^2 versus k for the range $k = 0 - 4$, choosing $n = 10$, d = 0.4 nm and $R = 0.5$ nm.
(c) What would be the significance of negative values of I/Nf^2?

(viii) *Multicomponent DRPHS*: Figure 21.6(c) shows a 2-D arrangement of atoms of different sizes for an amorphous multicomponent amorphous alloy. Comment on the 2-D packing density for a multicomponent system. If crystallization of an amorphous alloy requires atoms sliding past one another, postulate whether crystallization will be easier in a single or multicomponent amorphous alloy.

(ix) *Secondary crystallization product*: An important secondary crystallization product of certain metallic glasses is a transition metal boride that has space group **Fm$\bar{3}$m** (O_h^5) (Consult the International Tables for space group information). The compound has $a = 1.059$ nm, four inequivalent Fe atoms and one B atom on the sites tabulated below. The atomic weights of Fe and B are: $\text{AW}_{\text{Fe}} = 55.85$, $\text{AW}_{\text{B}} = 10.8$. Determine the following:

(a) The Bravais lattice, Pearson symbol, and point group symmetry at each special position.

(b) The composition, number of formula units in a unit cell, and density.
(c) The shortest distance between Fe atoms on the $4a$ and $32f$ sites.
(d) The extinctions that occur as a result of the Bravais lattice. What are the first eight peaks that you would predict to occur for this Bravais lattice?
(e) The additional extinctions that occur as a result of the Fe atom at the $8c$ special position. What symmetry operation causes these extinctions? Which of the previous eight peaks will be extinct because of this?

Atom	site	x	y	z
Fe1	4a	0.00000	0.00000	0.00000
Fe2	8c	0.25000	0.25000	0.25000
Fe3	32f	0.38103	0.38104	0.38104
Fe4	48h	0.00000	0.17115	0.17115
B1	24e	0.27644	0.00000	0.00000

(x) *Bulk amorphous alloys*: Zr is the largest and Be is the smallest atom in the Zr(Ti)–Ni(Cu)–Be bulk amorphous system. The high temperature crystalline β-phase of Zr has a *bcc* crystal structure with $a = 0.361$ nm. The high temperature crystalline β-phase of Be also has a *bcc* crystal structure with $a = 0.255$ nm.

(a) From this data, calculate the metallic radii for Zr and Be.
(b) Calculate a percentage difference between the metallic radii and comment on whether this satisfies the empirical rules for synthesis of bulk amorphous alloys with large glass forming abilities.

CHAPTER

22 Ceramic structures I

> *"The best way to have a good idea is to have lots of ideas."*
>
> Linus Pauling, (1901–1994)

22.1 Introduction

Webster's dictionary defines *ceramic* as: (1) *of or relating to pottery, earthenware, tiles, porcelain, etc.;* (2) *of ceramics – the art or work of making objects of baked clay, as pottery, earthenware, etc.; an object made of such materials.* Materials scientists identify ceramics with the following rather broad characteristics: (a) ceramics are typically hard but brittle materials. These mechanical properties arise from their unique ionic or covalent bonding, which is also the reason for their high melting temperatures; (b) ceramics are often (but not exclusively) electrical insulators or semiconductors; (c) ceramics can be crystalline or amorphous (glasses).

Ionic ceramics are compounds of an electropositive *cation*, M, and an electronegative *anion*, denoted by X; examples include Al_2O_3 ($Al_2^{3+}O_3^{2-}$), MgO ($Mg^{2+}O^{2-}$), ZrO_2 ($Zr^{4+}O_2^{2-}$), NaCl (Na^+Cl^-), ... From a structural point of view, we will consider ionic radii, bond lengths, and coordination numbers as important descriptive factors in ceramic structures. The cohesive energy in ceramics results predominantly from Coulomb interactions between anions and cations; often, these interactions involve *charge transfer* from the cation to the anion. *Bond lengths* in ionic crystals are well described by adding together the *ionic radii*. The ionic radii, in turn, depend on the ionic charge and the coordination number. The cation *coordination polyhedra* are understood by considering the sizes of cations and the interstitial sites in the (often close-packed) anion sublattice.

Covalent ceramics are compounds, such as GaAs, SiC, ZnO, SiO_2, . . . , in which *covalent bonding* is dominant. This type of bonding involves the build-up of electric charge density *between* atomic sites, leading to directional bonds. We can accurately calculate bond lengths in covalent solids by adding *covalent radii*. These radii may depend on the coordination number.

The distinction between ionic and covalent ceramics is often blurred because *electronegativity differences* between the elemental species in a ceramic material can have a wide range. Strongly ionic compounds include the *alkali-halides*, with a large electronegativity difference between the alkali and halide species. Semiconducting compounds with exclusively group IV (e.g., SiC) or group III and group V species (e.g., GaAs) have small electronegativity differences and are examples of more covalent ceramics. We discussed the structures of the most covalent materials in Chapter 17. In this chapter, we analyze structures on the basis of their ionic radii. While structures based on cubic close packing tend to be more ionic, those based on hexagonal close packing tend to be more covalent.

We begin this chapter with a compilation of ionic radii. Because ionic species are not charge neutral, understanding their relative sizes is complicated. Sizes also depend on the polyhedral coordination of the ions. Next, we will present *Pauling's rules*, which will help us to understand ionic structures based on the anion packing, the occupation of interstices by cations, and the connectivity of cation coordination polyhedra. We will discuss the radius ratio rules for the occurrence of important coordination polyhedra. Finally, we will illustrate representative ionic structures and the connectivity of cation coordination polyhedra. We will particularly note adherence or exceptions to Pauling's rules.

22.2 Ionic radii

Atomic, metallic, covalent, and ionic radii are distinguished by the types of interactions between atomic species: *atomic radii* characterize free atoms, while *metallic radii* reflect a compressible electron gas. Directional bonding characterizes *covalent radii* and *ionic radii* reflect charge transfer. While a cation (e.g., Mn^{2+}, Fe^{2+}) is smaller than its respective neutral counterpart, an anion (e.g., Cl^-, O^{2-}) is larger. Anions are larger due to the increased electron–electron Coulomb repulsion, whereas cations are smaller due to the larger electron–proton (nucleus) attraction. The higher the ionization state, the smaller is the cation and the larger the anion radius, because more electrons have been transferred. Isoelectronic anions in the same series of the periodic table have approximately the same ionic radii. For example, Table 22.1 shows

Table 22.1. Anion radii (in nm) for six-fold coordination around cations

X — R	X — R	X — R	X — R	X — R
Br^- — 0.196	Cl^- — 0.181	F^- — 0.133	I^- — 0.22	O^{2-} — 0.140
OH^- — 0.137	S^{2-} — 0.184	Se^{2-} — 0.198	Te^{2-} — 0.221	

Table 22.2. Cation (M) radii (in nm) for four-fold coordination (the superscript *s* indicates square planar coordination).

	M — R	M — R	M — R	M — R	M — R
+	Ag^+ — 0.100	Cu^+ — 0.060	K^+ — 0.137	Li^+ — 0.059	Na^+ — 0.099
2+	Ag^{2+s} — 0.079	Be^{2+} — 0.027	Cd^{2+} — 0.078	Co^{2+} — 0.056	Cu^{2+s} — 0.057
	Hg^{2+} — 0.096	Fe^{2+} — 0.063	Mg^{2+} — 0.057	Mn^{2+} — 0.066	Ni^{2+s} — 0.049
	Pd^{2+s} — 0.064	Pt^{2+s} — 0.060	Zn^{2+} — 0.074		
3+	Au^{3+s} — 0.064	Al^{3+} — 0.039	Fe^{3+} — 0.049	Ga^{3+} — 0.047	Tl^{3+} — 0.075
4+	Cr^{4+} — 0.041	Ge^{4+} — 0.039	Mn^{4+} — 0.039	Pb^{4+} — 0.065	Si^{4+} — 0.041
	Sn^{4+} — 0.055	Te^{4+} — 0.066	Ti^{4+} — 0.042	Zr^{4+} — 0.059	
5+	Mn^{5+} — 0.033	Nb^{5+} — 0.048	Mo^{5+} — 0.046	V^{5+} — 0.036	
6+	Cr^{6+} — 0.026	Mn^{6+} — 0.026	Mo^{6+} — 0.041	Se^{6+} — 0.028	Te^{6+} — 0.043
	W^{6+} — 0.042				

that the ionic radius for O^{-2} (for six-fold coordination about a cation) is 0.140 nm while that of F^{-1} is 0.133 nm.

Ionic radii and their classification were considered extensively in the early works of Goldschmidt (1926), Pauling (1927), Zachariessen (1931), and Ahrens (1952). More modern and widely cited ionic radii are tabulated in Shannon and Prewitt (1969), Shannon (1976); Shannon's numbers are used in the tables in this chapter. Ionic radii, which vary with coordination number (CN), are reported for typical charge states of anions and cations. As CN increases, so does the radius. It is common to compare radii at other CNs to those observed for CN = 6 (i.e., octahedral coordination). Radii for CN = 4 (tetrahedral coordination) are typically 93–96% of those for CN = 6; radii for CN = 8 (cubic coordination) are typically about 104% of those for CN = 6. Anion radii are summarized in Table 22.1 for six-fold coordination. Selected cation radii are summarized in Table 22.2 for four-fold coordination, Table 22.3 for six-fold coordination, and Table 22.4 for eight-fold coordination, respectively.

In a given row (period) of the periodic table, divalent anions are typically larger than monovalent anions. For either divalent or monovalent anions, the radii are larger in lower rows. The effect of coordination on ionic radii is apparent for O^{-2}, which has radii of 0.121 nm, 0.14 nm, and 0.142 nm, respectively for two-, six-, and eight-fold coordination. Fumi and Tosi (1964)

Table 22.3. Cation (M) radii (in nm) for six-fold coordination.

	M — R	*M — R*	*M — R*	*M — R*	*M — R*
+	Ag^{+} — 0.115	Au^{+} — 0.137	Cs^{+} — 0.167	Cu^{+} — 0.077	Hg^{+} — 0.119
	K^{+} — 0.138	Li^{+} — 0.076	Na^{+} — 0.102	Rb^{+} — 0.158	Tl^{+} — 0.150
2+	Ag^{2+} — 0.094	Ba^{2+} — 0.135	Be^{2+} — 0.045	Ca^{2+} — 0.100	Cd^{2+} — 0.095
	Co^{2+} — 0.065	Cr^{2+} — 0.073	Cu^{2+} — 0.073	Dy^{2+} — 0.107	Eu^{2+} — 0.117
	Fe^{2+} — 0.061	Ge^{2+} — 0.073	Hg^{2+} — 0.102	Mg^{2+} — 0.072	Mn^{2+} — 0.083
	Ni^{2+} — 0.069	Pb^{2+} — 0.119	Pd^{2+} — 0.086	Pt^{2+} — 0.080	Sm^{2+} — 0.119
	Sr^{2+} — 0.118	Ti^{2+} — 0.086	Tm^{2+} — 0.101	V^{2+} — 0.079	
3+	Al^{3+} — 0.054	Au^{3+} — 0.085	Bi^{3+} — 0.103	Ce^{3+} — 0.101	Co^{3+} — 0.055
	Cr^{3+} — 0.062	Dy^{3+} — 0.091	Er^{3+} — 0.089	Eu^{3+} — 0.095	Fe^{3+} — 0.055
	Ga^{3+} — 0.062	Gd^{3+} — 0.094	La^{3+} — 0.103	Mn^{3+} — 0.058	Mo^{3+} — 0.069
	Nb^{3+} — 0.072	Nd^{3+} — 0.098	Ni^{3+} — 0.056	Pd^{3+} — 0.076	Pr^{3+} — 0.099
	Rh^{3+} — 0.067	Ru^{3+} — 0.068	Sb^{3+} — 0.076	Sm^{3+} — 0.096	Ta^{3+} — 0.072
	Tb^{3+} — 0.092	Ti^{3+} — 0.067	Tl^{3+} — 0.089	V^{3+} — 0.064	Y^{3+} — 0.090
4+	Ce^{4+} — 0.087	Cr^{4+} — 0.055	Ge^{4+} — 0.053	Ir^{4+} — 0.063	Mn^{4+} — 0.053
	Mo^{4+} — 0.065	Nb^{4+} — 0.068	Pb^{4+} — 0.078	Pd^{4+} — 0.062	Pr^{4+} — 0.096
	Pt^{4+} — 0.063	Re^{4+} — 0.063	Rh^{4+} — 0.060	Ru^{4+} — 0.062	Se^{4+} — 0.050
	Sn^{4+} — 0.069	Ta^{4+} — 0.068	Tb^{4+} — 0.076	Te^{4+} — 0.097	Th^{4+} — 0.094
	Ti^{4+} — 0.061	V^{4+} — 0.058	W^{4+} — 0.066	Zr^{4+} — 0.072	
5+	Bi^{5+} — 0.076	Ir^{5+} — 0.057	Mo^{5+} — 0.061	Nb^{5+} — 0.064	Os^{5+} — 0.058
	Re^{5+} — 0.058	Rh^{5+} — 0.055	Ru^{5+} — 0.057	Sb^{5+} — 0.060	Ta^{5+} — 0.064
	V^{5+} — 0.054	W^{5+} — 0.062			
6+	Cr^{6+} — 0.044	Mo^{6+} — 0.059	Os^{6+} — 0.055	Re^{6+} — 0.055	Se^{6+} — 0.042
	Te^{6+} — 0.056	W^{6+} — 0.060			

Table 22.4. Cation (M) radii (in nm) for eight-fold coordination.

	M — R	*M — R*	*M — R*	*M — R*	*M — R*
+	Ag^{+} — 0.128	Cs^{+} — 0.174	K^{+} — 0.151	Li^{+} — 0.092	Na^{+} — 0.118
	Rb^{+} — 0.161	Tl^{+} — 0.159			
2+	Ba^{2+} — 0.142	Ca^{2+} — 0.112	Cd^{2+} — 0.110	Co^{2+} — 0.090	Cs^{2+} — 0.174
	Dy^{2+} — 0.119	Eu^{2+} — 0.125	Fe^{2+} — 0.092	Hg^{2+} — 0.114	Mg^{2+} — 0.089
	Mn^{2+} — 0.096	Pb^{2+} — 0.129	Sm^{2+} — 0.127	Sr^{2+} — 0.126	V^{2+} — 0.101
3+	Dy^{3+} — 0.103	Er^{3+} — 0.100	Eu^{3+} — 0.107	Fe^{3+} — 0.078	Gd^{3+} — 0.105
	La^{3+} — 0.116	Nd^{3+} — 0.112	Pr^{3+} — 0.113	Tb^{3+} — 0.104	Tl^{3+} — 0.098
	Y^{3+} — 0.102	Yb^{3+} — 0.099	Nb^{3+} — 0.079		
4+	Ce^{4+} — 0.097	Pb^{4+} — 0.094	Sn^{4+} — 0.081	Tb^{4+} — 0.088	Th^{4+} — 0.105
	Ti^{4+} — 0.074	V^{4+} — 0.072	Zr^{4+} — 0.084		

tabulated the *crystal radii* for many atoms for which the ionic radii are listed in the tables in this section. Crystal radii, R_C, are determined from *neutral* close packing considerations, while the ionic radii, R_I, account for the bonds that are shortened by cation/anion interactions. As a rule of thumb, we have $R_I = R_C + 0.0014$ nm (for anions) and $R_I = R_C - 0.0015$ nm (for cations).

Cation size decreases from monovalent to divalent to trivalent cations due to the increased positive charge in the nucleus. Other cation oxidation states, such as M^{7+} and M^{8+}, are known but are not listed here because they are rare. The multiple coordination numbers and oxidation states of transition metals and rare earth atoms make their crystallography and properties correspondingly rich. These oxidation states are also complicated because the d and f orbitals of these elements are prone to asymmetric charge distributions. The quantum states chosen by these electrons are influenced by the symmetry of the so-called *crystal field*, the electrostatic field caused by the coordinating ions. Often, the energy of these orbitals will be lower in low symmetry coordination environments, which is the driving force for symmetry lowering distortions through the *Jahn–Teller effect*.

The cation radii in Table 22.2 are typical for tetrahedral coordination. Planar three- and four-fold coordination numbers are also occasionally observed. In Table 22.2, four-fold square planar coordination is denoted by a superscript s. Square planar coordination is observed for Cu^{2+} in many of the cuprate superconductors (Chapter 23). Selected radii are summarized in Table 22.3 for six-fold coordinated cations.[1] Heavy transition metal and rare earth cations are observed more frequently in six-fold coordination because the octahedral interstices are larger than the tetrahedral interstices. Selected radii are summarized in Table 22.4 for eight-fold coordinated cations. The four-fold, six-fold, and eight-fold coordinations are the most commonly observed coordinations, although coordination numbers of 5 (e.g., an octahedron with one missing vertex), 9, 10, and 12 (dodecahedral coordination, e.g., in garnet structures) are sometimes observed.

22.3 Bonding energetics in ionic structures

As discussed in Chapter 17, crystalline bond energies can be treated empirically through the consideration of a simple pair potential, such as the *Lennard-Jones potential*. The form of the Lennard-Jones potential applicable to inert gas solids is:

$$V(r) = 4\epsilon \left[\left(\frac{\sigma}{r} \right)^{12} - \left(\frac{\sigma}{r} \right)^{6} \right] \tag{22.1}$$

[1] This is typically for octahedral coordination, although sometimes six-fold trigonal coordination can be observed, usually in covalent solids.

where we recall that the parameters ϵ and σ set the scale of the potential energy, V, and interatomic spacing, r, respectively.

In a simple AX ionic solid, this pair potential must be modified in two ways:

(i) The energy scale, with reference to the neutral M and X atoms, must be shifted, to reflect the energy required to ionize the M cation and transfer the electron(s) to ionize the X anions. This results in an additional energy, $I_M - A_X$, where I_M is the ionization energy of the M cation and A_X is the electron affinity of the X anion.

(ii) A Coulombic attraction energy must be added to the pair potential to reflect the interaction between charged, rather than neutral, atoms in the pair.

A modified pair potential can be written to include the ionization energy shift and the Coulomb potential, $V_c(r)$, as:

$$V(r) = I_M - A_X + V_c(r) + 4\epsilon\left[\left(\frac{\sigma}{r}\right)^{12} - \left(\frac{\sigma}{r}\right)^{6}\right]. \tag{22.2}$$

The number of electrons moved for the formation of ion i, Z_i, is taken to be positive if the electrons were added to the ion and negative if they were removed. In a crystalline solid with N ions, the total energy can then be calculated by summing over all ion pairs in the crystal:

$$V_{\text{tot}} = \frac{N}{2}(I_M - A_X) + \frac{4N\epsilon}{2}\left[\sum_{ij}{}' \left(\frac{\sigma}{p_{ij}R}\right)^{12} - \sum_{ij}{}' \left(\frac{\sigma}{p_{ij}R}\right)^{6}\right] + \frac{N}{2}\sum_{ij}{}' \frac{\pm Z_i Z_j e^2}{p_{ij}R}, \tag{22.3}$$

where the prime on the summations signs indicates that i must be different from j. This expression is more complicated than the one encountered in Chapter 17, because now we need to distinguish between at least two types of ions. The first term is trivial. The second term is the same as the previously defined (scaled) lattice sum for the Lennard-Jones potential, where the summation depended on the type of lattice (Chapter 17 considers the *fcc* and *bcc* lattices). The last term is the *Madelung sum* (Madelung, 1909) over lattice sites of the Coulombic part of the ionic pair potential, where $\pm$ refers to pairs with like and opposite charge respectively. It is parameterized in terms of a number called the *Madelung constant* α. The relationship between the Madelung sum and Madelung constant is expressed as:

$$\alpha = \sum_{ij}{}' \left(\frac{\pm}{p_{ij}}\right); \quad \text{or} \quad \frac{\alpha}{R} = \sum_{ij}{}' \left(\frac{\pm}{r_{ij}}\right). \tag{22.4}$$

The Madelung sum is specific to a given lattice. In 1-D, this is a converging series that can be expressed analytically. In 3-D, we must calculate the summations using numerical methods. Once we calculate the Madelung constant for a given structure, we can express the Madelung sum as $N\alpha(\pm Z_i Z_j)e^2/R$. We can determine the equilibrium lattice constant and cohesive energy as detailed in Chapter 17. Again, due to the approximate and parameterized form of the potential, we can obtain more accurate results by employing first principles quantum mechanical calculations.[2] We note that inclusion of the Madelung energy term leads, once again, to the discovery of a universal binding potential, as demonstrated in Smith *et al.* (1991).

22.4 Rules for packing and connectivity in ionic crystals

22.4.1 Pauling's rules for ionic structures

Many ionic crystal prototypes have structures that can be analyzed in terms of *Pauling's rules*, summarized in Box 22.1. Typical ionic structures have a close packed anion lattice, and cations that occupy a fraction of the octahedral

Box 22.1 Pauling's rules for ionic structures

Pauling's rules for ionic structures are:

(i) A coordination polyhedron of anions is formed about each cation. The cation–anion distance is determined by the sum of their radii. The *coordination number* is determined by the radius ratio.

(ii) The *bond strength, s*, of a cation–anion bond is defined as the ratio of the cation charge to the cation coordination number. In a stable structure, the total strength of the bonds that reach an anion from all neighboring cations in the coordination polyhedron is equal to the charge of the anion.

(iii) The linkage of coordination polyhedra is such that edges, and especially faces, tend not to be shared. If edges are shared, they tend to be shortened.

(iv) Because edge sharing decreases the stability of a structure, cations with high charge and low coordination number especially do not share edges.

(v) The number of crystallographically or chemically distinct kinds of atoms in a structure tends to be small.

[2] An explanation of these techniques is beyond the scope of this book.

and tetrahedral interstices. *fcc* anion lattices are described in terms of an $ABCABC\ldots$ stacking sequence; *hcp* anion lattices are described in terms of an $ABAB\ldots$ stacking of close-packed planes. In both structures, each anion has two tetrahedral and one octahedral interstices nearby. The composition of an ionic material can be determined by the cation fractional occupation on these interstitial sites. This occupancy is further dictated by size considerations, charge balance, and electrostatic interactions. From previous arguments, we can conclude that typically $R_X > R_M$. We can determine the lattice constant for a crystal structure by summing neighboring anion–cation ionic radii and multiplying this sum by an appropriate geometric constant.

22.4.2 Radius ratio rules for ionic compounds

To understand coordination in ionic solids, we make use of simple geometrical considerations. We will determine the restrictions on the radii of A and B ions when A must be coordinated by a certain number of B ions. These restrictions define the smallest possible radius of an A ion as that for which the B ions just touch. The B ions will not be able to approach each other any closer because of electrostatic repulsion between ions of like charge. For CN $= 1$ or 2, the restriction on the ratio R_A/R_B is trivial, as any value between 0 and ∞ is possible without having the B ions touch each other. Figure 22.1 illustrates the geometrical constraints for a single A atom coordinated by B atoms for CN $= 3$, 4, 6, 8, and 12, respectively.

While CN $= 3$ is a planar triangular coordination, CN $= 4$ is tetrahedral, CN $= 6$ is octahedral, CN $= 8$ is cubic, and CN $= 12$ is *cuboctahedral* (*fcc*) or hexagonal close packed (*hcp*). In each case, a triangle with sides

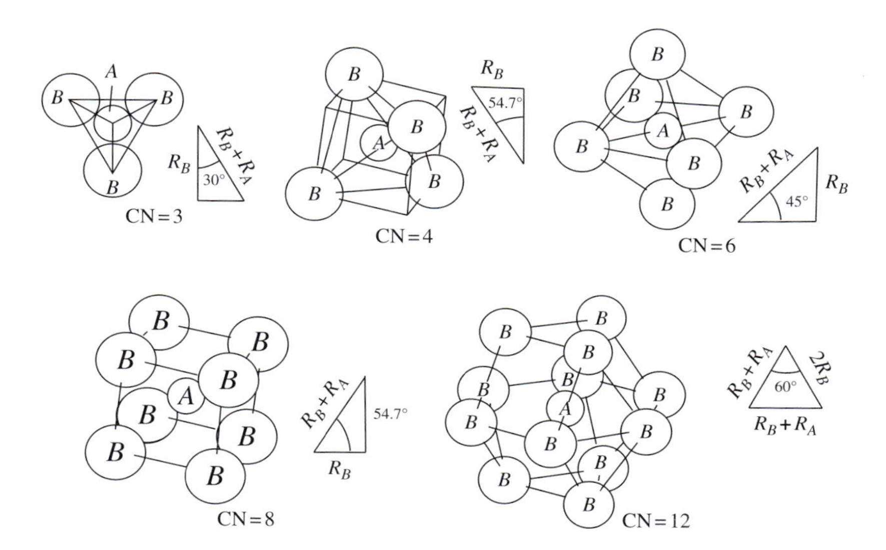

Fig. 22.1. Ideal geometries for a single A ion coordinated by B ions for CN $= 3$, 4, 6, 8, and 12.

corresponding to R_B and $R_A + R_B$ as well as an angle determined by the geometry of this triangle are shown for the case where the B ions touch along the direction of closest approach. From this geometry, the limiting value of the radius ratio R_A/R_B, called the *critical radius ratio*, ρ_C, can be computed. An example of such a calculation is shown Box 22.2 for triangular coordination.

Four-fold coordination can occur in two forms: square planar or tetrahedral. For tetrahedral coordination (Figure 22.1), the critical geometry occurs when B atoms touch along the face diagonal of a cube (circumscribed about the tetrahedron) and the A and B atoms touch along the body diagonal of this cube. The critical radius radio for this coordination is calculated in Box 22.3.

Six-fold coordination can be hexagonal, *trigonal prismatic*, or octahedral. For octahedral coordination (Figure 22.1), the critical geometry occurs when the B atoms touch along the octahedron's edges and A and B atoms touch along the base diagonal. The critical radius ratio for this coordination is derived in Box 22.4.

Box 22.2 Calculation of ρ_C for triangular coordination

For triangular coordination (Figure 22.1), the larger B atoms touch along the sides of an equilateral triangle. A right-angled triangle where A and B atoms touch along the hypotenuse and the base length is R_B has a 30° ($\pi/6$) angle between R_B and $R_A + R_B$, so that:

$$\frac{R_A + R_B}{R_B} = \frac{2}{\sqrt{3}}, \quad \text{and} \quad \rho_C = \frac{R_A}{R_B} = \frac{2}{\sqrt{3}} - 1 = 0.155.$$

Box 22.3 Calculation of ρ_C for tetrahedral coordination.

For tetrahedral coordination (Figure 22.1), the tetrahedron has vertices coinciding with four of the eight vertices of a cube with edge length a_0. The B atoms touch along the cube face diagonal, whereas A and B atoms touch along the body diagonal, so that:

$$2R_B = a_0\sqrt{2} \text{ and } R_B + R_A = a_0\sqrt{3}.$$

$R_B + R_A$ forms the hypotenuse of a right-angled triangle with R_B as the shorter side and tetrahedral angle, 54.7°, opposite R_B. The critical radius ratio is then computed from:

$$\frac{R_A + R_B}{R_B} = \sqrt{\frac{3}{2}} \text{ so that } \rho_C = \frac{R_A}{R_B} = \sqrt{\frac{3}{2}} - 1 = 0.225.$$

Box 22.4 Calculation of ρ_C for octahedral coordination.

For octahedral coordination (Figure 22.1), an octahedron is inscribed in a cube. Atoms on the octahedral vertices coincide with the six face centers of the cube with edge length a_0. For this geometry:

$$2(R_B + R_A) = a_0 \text{ and } 4R_B = a_0\sqrt{2}.$$

The critical radius ratio then follows from:

$$\frac{(R_A + R_B)}{R_B} = \sqrt{2}, \text{ so that } \rho_C = \frac{R_A}{R_B} = \sqrt{2} - 1 = 0.414.$$

The octahedron, viewed along a three-fold axis, can also be seen to be an *anti-prismatic coordination polyhedron*, i.e., the upper and lower triangles are *staggered*. The *trigonal prism* is another CN $= 6$ polyhedron, with a *prismatic* coordination. Although prismatic coordination is sometimes observed in crystal structures, it is usually not the lowest energy configuration. To find the coordination scheme that produces the lowest energy, we could consider the Coulomb repulsion terms in the interatomic potentials. It is left to the reader to calculate the critical radius ratio, ρ_C, for the prismatic coordination.

Coordination numbers 7, 8, and 9 can be obtained by decorating 1, 2, or 3 of the rectangular faces of the trigonal prism. However, these coordinations are not commonly observed. For cubic coordination, with CN $= 8$, the critical geometry occurs when the B atoms touch along the edge of the cube and the A and B atoms along the body diagonal (Figure 22.1). Calculation of ρ_C for the cubic coordination is also left as a reader exercise, as is the computation of ρ_C for the twelve-fold coordination of the *fcc* and *hcp* packings. For CN $= 12$, the cubic close-packed *cuboctahedron* has a *staggered structure* whereas the *hcp* structure is known as an *eclipsed structure*.

The critical radius ratio calculations give the lower bounds on the radius ratio based on the coordination of B atoms about A atoms. We can derive the upper bounds by considering the coordination of A atoms about B atoms. These give results identical to those above, but instead the restrictions will be in terms of the inverse ratio, R_B/R_A. Table 22.5 summarizes the restrictions on radius ratios for different coordination numbers in terms of the radius ratio R_A/R_B.

The results of Table 22.5 can be generalized to determine restrictions on radius ratios for non-equiatomic compounds, as summarized in Table 22.6. For a compound A_nB_m, the coordination numbers must satisfy the following relation, in order to preserve the stoichiometry:

$$\frac{\mathrm{CN}_A}{\mathrm{CN}_B} = \frac{m}{n} \tag{22.5}$$

Table 22.5. Radius ratios for different coordination numbers for A and B atoms.

CN_A	$\frac{R_A}{R_B}$	CN_B	$\frac{R_B}{R_A}$	$\frac{R_A}{R_B}$
1	0–∞	1	∞–0	0–∞
1	0–∞	1	∞–0	0–∞
3	0.155–∞	3	∞–0.155	0–6.45
4	0.225–∞	4	∞–0.225	0–4.44
6	0.414–∞	6	∞–0.414	0–2.41
8	0.732–∞	8	∞–0.732	0–1.37
12	1.0–∞	12	∞–0.225	0–1.0

Table 22.6. Radius ratio ranges for different coordination numbers in A_nB_m compounds.

AB		AB_2		A_2B		A_2B_3	
$\frac{CN_A}{CN_B}$	$\frac{R_A}{R_B}$	$\frac{CN_A}{CN_B}$	$\frac{R_A}{R_B}$	$\frac{CN_A}{CN_B}$	$\frac{R_A}{R_B}$	$\frac{CN_A}{CN_B}$	$\frac{R_A}{R_B}$
2:2	0–∞	2:1	0–∞	1:2	0–∞	3:2	0.155–∞
3:3	0.155–6.45	4:2	0.225–∞	2:4	0–4.44	6:4	0.414–4.44
4:4	0.225–4.44	6:3	0.414–6.45	3:6	0.155–2.41	12:8	1–1.37
6:6	0.414–2.41	8:4	0.732–4.44	4:8	0.225–1.37		
8:8	0.732–1.37	12:6	1–2.41	6:12	0.414–1.0		
12:12	1–1						

In addition, we must consider the charge balance: for example, an A_2B_3 compound would have trivalent A atoms and divalent B atoms to maintain charge balance.

In the following sections, we illustrate structures within the framework of Pauling's rules. As a result of geometrical restrictions on coordination numbers as well as electrostatic considerations, ionic structures are commonly described in terms of cation–anion distances, coordination polyhedra, and their linkages. We will consider prototypical examples of ionic crystal structures and discuss to what extent they adhere to Pauling's rules. While we will introduce many different ceramic structures in this chapter, we postpone the discussion of structures based on the $(SiO_4)^{4-}$ tetrahedral unit until Chapter 24; there are so many different ways in which these tetrahedral units can be arranged in space that we will dedicate an entire chapter to the discussion of mineral compounds and gemstones.

22.5 Halide salt structures: CsCl, NaCl, and CaF_2

The most basic ionic structures are those of the simple salts CsCl (B2, **Structure 10** in the on-line structure appendix) and NaCl (rocksalt, **Structure 7**).

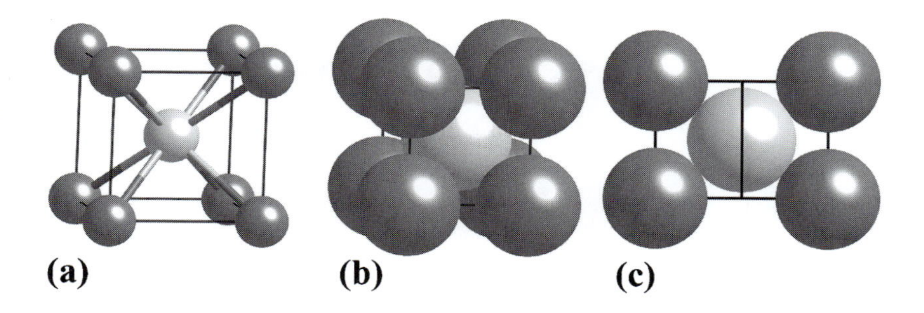

Fig. 22.2. CsCl structure: (a) ball-and-stick model, (b) close-packed model, and (c) a (110) plane in which Cs and Cl touch along the [111] direction.

These *AB* compounds have monovalent anions and cations. As discussed in Chapter 17, the CsCl structure is a *bcc* derivative where Cs^+ cations occupy one of the interpenetrating simple cubic lattices and Cl^- anions occupy the other. CsCl (Fig. 22.2) does not fully adhere to Pauling's rules; in particular, the anions are not close packed. The next simple ionic structure is NaCl, which does have close packed anions and better adherence to Pauling's rules.

CsCl has a diatomic basis with Cs at the $(0, 0, 0)$ position and Cl at $(1/2, 1/2, 1/2)$. Although the structure is a *bcc* derivative, the $[1/2, 1/2, 1/2]$ translation is not a symmetry operation, because this would move a Cs^+ cation into a Cl^{-1} anion site or vice versa. Similarly, there are no atoms at face-centered sites; thus, by elimination, the Bravais lattice of CsCl is simple cubic.[3] Figure 22.2 shows a ball-and-stick model (a) and a close-packed model (b) of the CsCl structure. In (b), the atomic spheres have the correct relative sizes, allowing for a direct examination of the ionic packing. Figure 22.2(c) displays a close-packed (110) plane of the CsCl structure. It is clear that the anions and cations touch along the diagonal [111] direction.

In the XRD pattern for a metallic alloy with the **B2** structure, namely equiatomic FeCo in Chapter 17, superlattice reflections were difficult to detect because of the similar scattering factors of Fe and Co. Figure 22.3(a) illustrates the XRD pattern (for Cu-Kα radiation) for CsCl, the prototype **B2** structure. Here, superlattice reflections are evident because of the differences in scattering factors for the Cs^+ and Cl^- ions. Many more peaks are also evident in the same range of 2θ, because CsCl has a larger lattice constant than FeCo.

Let us now analyze this structure in terms of Pauling's rules. The Cs^+ cation sits in an eight-fold cubic coordination. The anion coordination polyhedron is also a cube, and neighboring cubes share faces. In eight-fold coordination, Cs^+ has an ionic radius of 0.174 nm. The ionic radius for Cl^- is only reported for the more common six-fold coordination, for which it is 0.181 nm. This

[3] It is a common mistake to say that CsCl is a body-centered structure. While there *is* an atom at the center of the cube, this atom is *not* equivalent to the one in the origin, and, therefore, this is not a body-centered cell.

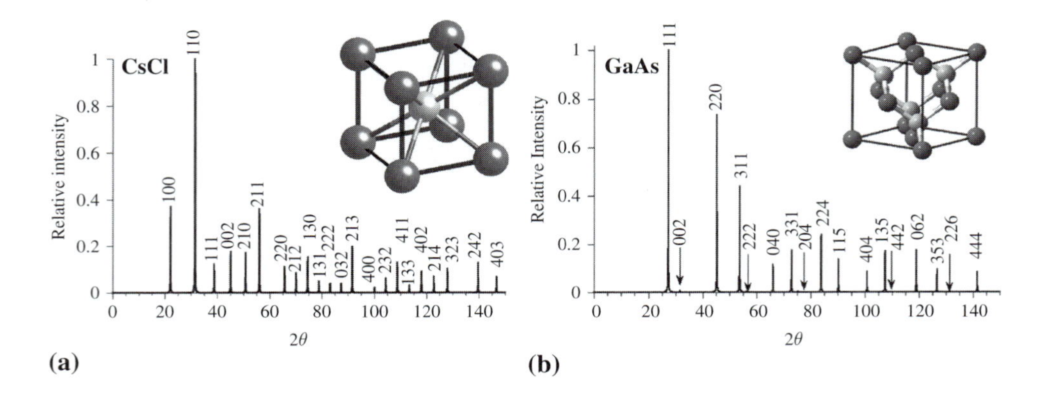

Fig. 22.3. Simulated XRD patterns assuming Cu-Kα radiation and equilibrium lattice constants for (a) the CsCl structure, and (b) GaAs with the zinc-blende structure.

radius is likely to be larger for CN = 8. The anion and cation radii are hence very nearly the same, which is not typical for ionic structures. The Cl^- anion is the second of the group VIIA elements (with a smaller radius than later group VIIA elements), while the Cs^+ cation is the sixth of the group IA elements (with a larger radius than earlier group IA elements).

The Cs^+ cation cannot fit in either the octahedral or tetrahedral interstices of a hypothetical close packed Cl^- anion sublattice; instead, it prefers the more open structure associated with CN = 8. The structure is an exception to one of Pauling's rules, namely the one stating that face sharing is not energetically favorable. As cubic coordination polyhedra share faces, the CsCl structure tends to occur with monovalent anions for which the Coulombic repulsion is low. Anions and cations touch along the [111] directions, so that the lattice constant, a, equals, $2(R_A + R_B)/\sqrt{3}$. Taking the Cl^- ionic radius as 0.181 nm and the Cs^+ ionic radius of 0.174 nm, we predict that the cubic lattice constant is 0.412 nm. This prediction is in excellent agreement with the experimentally observed CsCl lattice constant of 0.4123 nm.

To determine the CsCl size limitations, we refer to Table 22.6 for an AB compound, where both ions have CN = 8. The CsCl structure is predicted to form with radius ratios in the range:

$$0.732 < \frac{R_A}{R_B} < 1.37.$$

The ratio for CsCl is close to 1 (0.961), which falls near the middle of this range.

The NaCl structure, shown in Figure 22.4, is a *fcc* derivative structure with Na^+ on one of the interpenetrating *fcc* lattices and Cl^- on the other. Referring to the tables of ionic radii, we see that Na^+ has an ionic radius of 0.102 nm; Cl^- has a radius of 0.181 nm for the six-fold octahedral coordination. This large anion to cation size ratio is typical of many common ionic structures. This size difference is large enough so that Na^+ cations fit comfortably in all the octahedral interstices of the close packed Cl^- sublattice. Figure 22.4(c)

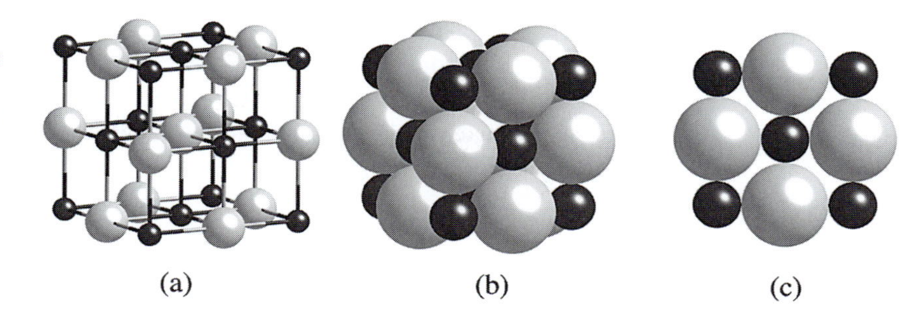

Fig. 22.4. NaCl structure: (a) ball-and-stick and (b) close packed representations of the interpenetrating *fcc* Na and Cl lattices; (c) a (100) plane in which the anions and cations touch.

shows a (100) plane of the NaCl structure with the ions drawn to size. The anions touch cations along the $\langle 100 \rangle$ directions. The lattice constant for this structure is predicted to be $a = 2(R_{\text{Na}} + R_{\text{Cl}}) = 0.566\,\text{nm}$, which compares favorably with the experimental value of 0.564 nm.

The only coordination polyhedron present in the NaCl structure is the octahedron. Figure 22.5 shows the arrangement of Cl octahedra. It is apparent that all octahedra share all 12 edges. The edges are oriented along $\langle 110 \rangle$-type directions. The larger coordination number (6) helps to explain deviations from Pauling's rule as to edge-rather than vertex-sharing polyhedra. The strength of the anion–cation bond and the bond sum are:

$$s = \frac{+1}{6} \text{ and } \Sigma s = 6\left(\frac{+1}{6}\right) = 1.$$

Because both Na and Cl have CN = 6, we expect the rocksalt structure to form with cation/anion radius ratios between:

$$0.414 < \frac{R_A}{R_B} < 2.41.$$

For NaCl we have an anion-to-cation radius ratio of 1.775, which falls in the stated range. Among oxides with the rocksalt structure, we note MgO, in which the O^{2-} anions are close packed in an *fcc* lattice and the Mg^{2+} cations occupy the octahedral interstices. We leave it to the reader to compute the radius ratio for this compound.

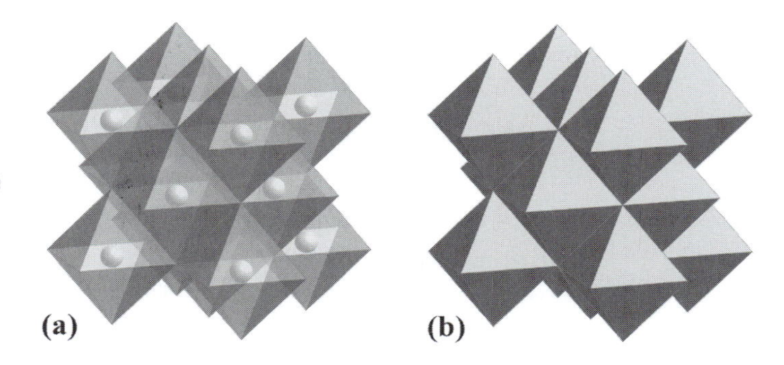

Fig. 22.5. Polyhedral representations of the NaCl structure: (a) shows the Na atoms with the Cl coordination octahedra. In (b), the octahedra are rendered as solid polyhedra and the edge-sharing arrangement is highlighted.

The occupation of all of the tetrahedral sites in an *fcc* anion sublattice results in the fluorite structure of CaF_2. The CaF_2 structure was discussed in Chapter 17. This occupancy gives rise to an AB_2 stoichiometry. We will leave it to the reader to compute the radius ratio for this structure, and to verify that the ratio falls within the predicted range.

22.6 Close packed sulfide and oxide structures: ZnS and Al_2O_3

Another simple structure based on anions in an *fcc* lattice is the previously introduced *zinc-blende* structure (also known as *sphalerite*, **Structure 11**), of which ZnS is the prototype. This structure is illustrated again in Figure 22.6, where coordination polyhedra and their connectivity are emphasized. In this structure, the S^{2-} anions are close packed with Zn^{2+} occupying *half* of the tetrahedral interstices to form an interpenetrating *fcc* cation sub-lattice. The atomic positions are identical to those of the diamond cubic lattice, but they are ordered such that anions occupy one and cations the other of the two sublattices. We choose the ZnS structure to illustrate a structure factor computation in Box 22.5.

Next, we consider the coordination polyhedra and connectivity in the zinc-blende structure within the framework of Pauling's rules. The Zn^{2+} cation is four-fold tetrahedrally coordinated and has a radius of 0.074 nm. Combining this with the S^{2-} radius of 0.184 nm, we find a radius ratio of 0.402. Because the low coordination number favors smaller cations, the prototype structure has a divalent cation as opposed to the monovalent cations in the alkali halide salts. The anion coordination polyhedron is a vertex-sharing tetrahedron, which is compatible with the larger ionic charges. The strength of the anion–cation bond and the bond sum, respectively, are given by:

$$s = \frac{+2}{4}, \qquad \text{so that} \qquad \Sigma s = 2\left(\frac{+2}{4}\right) = 1.$$

Because two tetrahedra meet at any shared vertex, the bond sum is 1. As shown in Fig. 22.6(c) and (d), the anions and cations touch along [111] directions passing through the cube vertices and the tetrahedral sites occupied by the Zn^{2+} cations.

Box 22.5 Calculation of the structure factor for the zinc-blende structure

The ZnS structure has an *fcc* Bravais lattice with a 2-atom basis with S on $(0, 0, 0)$ and Zn on $(1/4, 1/4, 1/4)$. The six additional atoms in the unit cell are obtained by adding face centering translations. Since this structure is face centered, we know that the structure factor will vanish for mixed parity Miller indices; when the Miller indices have the same parity, the structure factor for the atomic basis is multiplied by a factor of 4. We find:

$$F_{hkl} = 4\sum_{j=1}^{\frac{N}{4}} f_j e^{2\pi i(hx_j+ky_j+lz_j)} = 4\left[f_S + f_{Zn}\, e^{\frac{\pi i}{2}(h+k+l)}\right].$$

This structure factor is complex, so we take its modulus ($|F_{hkl}|^2 = F_{hkl}F^*_{hkl}$):

$$F^2_{hkl} = 16\left[f_S + f_{Zn}\, e^{\frac{\pi i}{2}(h+k+l)}\right] \times \left[f_S + f_{Zn}\, e^{-\frac{\pi i}{2}(h+k+l)}\right],$$

to yield:

$$F^2_{hkl} = 16\left[f_S^2 + f_{Zn}^2 + 2f_S f_{Zn} \cos\frac{\pi}{2}(h+k+l)\right].$$

We find three special cases among the allowed reflections:

$$|F_{hkl}|^2 = 16(f_S + f_{Zn})^2 \qquad \text{for } h+k+l = 4n;$$
$$|F_{hkl}|^2 = 16(f_S - f_{Zn})^2 \qquad \text{for } h+k+l = 2(2n+1);$$
$$|F_{hkl}|^2 = 16(f_S^2 + f_{Zn}^2) \qquad \text{for } h,k,l \text{ all odd.}$$

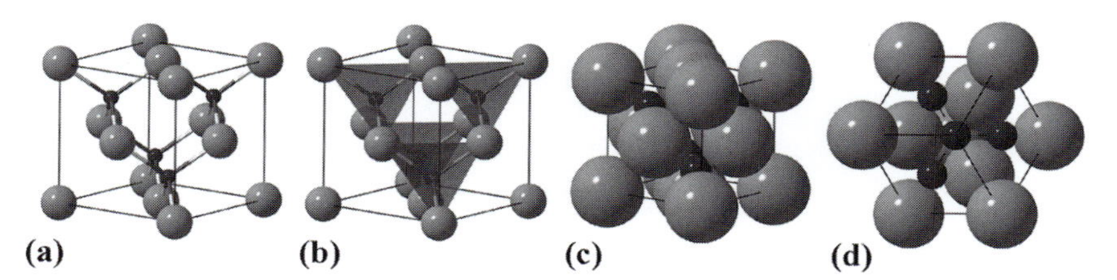

Fig. 22.6. (a) Zinc-blende structure with interpenetrating *fcc* Zn^{2+} and S^{2-} sublattices; (b) connectivity of the S^{2-} tetrahedra; (c) close packed setting; and (d) projection along the [111] direction showing touching anions and cations.

Zinc-blende is an important prototype structure for many AB semiconducting compounds. As discussed in Chapter 17, this is a superlattice derivative of the diamond cubic structure, the structure of the elemental Si semiconductor. In many AB compound semiconductors with group III and group V atoms (or, in the case of ZnS, group II and group VI), the group III (or II) atoms occupy one of the interpenetrating *fcc* lattices, while the group V (or VI) atoms occupy the other. In these compounds, a total of eight valence electrons is maintained, which is the number required to completely fill the sp^3 bonding states and to leave the corresponding anti-bonding states unfilled.

For III–V and II–VI semiconductors, the bonding becomes increasingly more ionic (less covalent) as the difference in the number of valence electrons between the A and B species increases. GaAs is an important example of a III–V semiconductor.

Figure 22.3(b) on page 666 illustrates the XRD pattern (Cu-Kα radiation) for the III–V semiconductor GaAs. It can be seen that the (002), (222), (204), (442), and (226) reflections (arrowed) have very small intensities. These are reflections for which $h+k+l=2(2n+1)$; according to the structure factor computation in Box 22.5, these reflections have intensities proportional to the difference between the atomic scattering factors. Because of their close proximity in the periodic table, Ga and As have nearly identical scattering factors, resulting in a low intensity for these reflections.

A simple structure based on hexagonal close packing of anions is the wurtzite or B4 structure (**Structure 14**), for which another allotrope of ZnS is the prototype. This structure is illustrated in Figure 22.7. The S^{2-} anions are hexagonal close-packed with the Zn^{2+} cations again occupying half of the tetrahedral interstices. The four-fold, tetrahedrally coordinated, Zn^{2+} cation coordination polyhedra share vertices, as in the zinc-blende structure. This structure was introduced in Chapter 17, and in Fig. 22.7(b), we illustrate the connectivity of the tetrahedral S^{2+} coordination polyhedra.

Alumina is the chemical name for the compound Al_2O_3, while *corundum* is the mineral name for the gemstone *ruby* (depending on the impurity content, we also use the names *sapphire* and *emerald*). The corundum structure (**Structure 58**) is based on a nearly hexagonally close packed array of O^{2-} anions. The Al^{3+} cations occupy two-thirds of the octahedral interstices in this lattice. Al_2O_3 has the rhombohedral space group $\mathbf{R\bar{3}c}$ (D^6_{3d}) with $a=0.47617$ nm and $c=1.29947$ nm (Finger and Hazen, 1978), leading to a structure with a large c/a ratio of 2.73. Al^{3+} cations occupy the $12c$ $(0,0,z)$ positions with $z=0.3521$, and the O^{2-} anions occupy the $18e$ $(x,0,1/4)$

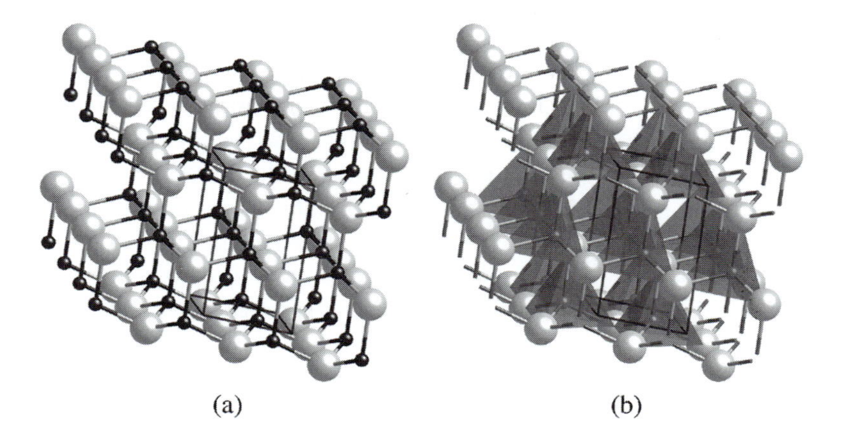

Fig. 22.7. (a) Wurtzite structure with an *hcp* S^{2-} sublattice and Zn^{2+} occupying half of the tetrahedral interstices; (b) connectivity of the tetrahedral S^{2+} coordination polyhedra.

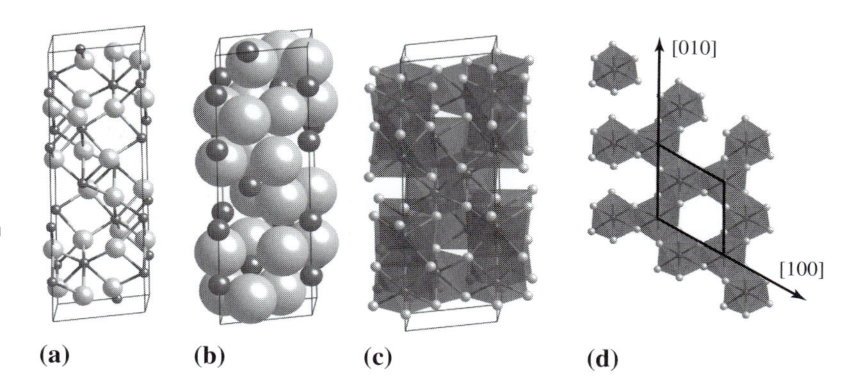

Fig. 22.8. Al_2O_3 corundum structure in (a) ball-and-stick and (b) space filling depiction of the unit cell; (c) illustration of the Al^{3+} octahedral coordination polyhedra stacked in (001) plane sheets along the [001] direction; (d) a single (001) plane sheet of octahedral (several unit cells).

positions with $x = 0.3065$. Figure 22.8(a) and (b) show a ball-and-stick and space filling depiction, respectively, of the hexagonal unit cell; as usual, we employ a hexagonal unit cell for this rhombohedral structure. The polyhedral representation of Fig. 22.8(c) shows the connectivity of the Al^{3+} octahedra in a single unit cell. The octahedra are oriented such that they have anti-prismatic triangular faces normal to the [001] direction.

The connectivity of the Al^{3+} octahedra in a single (001) sheet is illustrated in Fig. 22.8(d). The trigonally aligned octahedra form simple hexagonal networks. These sheets of octahedra then stack along the [001] direction, in such a way that the next layer lies above (below) the hexagonal void in the layer below (above) it. The same type of stacking can be found in the stacking of hexagonal C-sheets in graphite. Figure 22.8(d) shows the octahedra (trigonal anti-prisms) to be edge sharing.

22.7 Perovskite and spinel structures

In this section, we consider two important structure types with *fcc* arrangements of O^{2-} anions: compounds with ABO_3 (perovskite) and AB_2O_4 (spinel) stoichiometries. These structures have two (or more) different types of cations. In perovskites, one cation shares the close-packed sites with the $3O^{2-}$ anions, and the other cation occupies exclusively octahedral sites coordinated by O. In the spinel structure, the *fcc* array is exclusively made from O^{2-} anions and A and B cations occupy a fraction of the octahedral and tetrahedral interstices. Both structures are of significant technological importance.

22.7.1 Perovskites: ABO_3

The name *perovskite* refers to a large family of crystalline ceramics with structures based on that of the natural mineral $CaTiO_3$, also known as perovskite. The mineral name derives from the last name of the Russian mineralogist Count Lev Aleksevich von Perovski (1792–1856). The perovskite

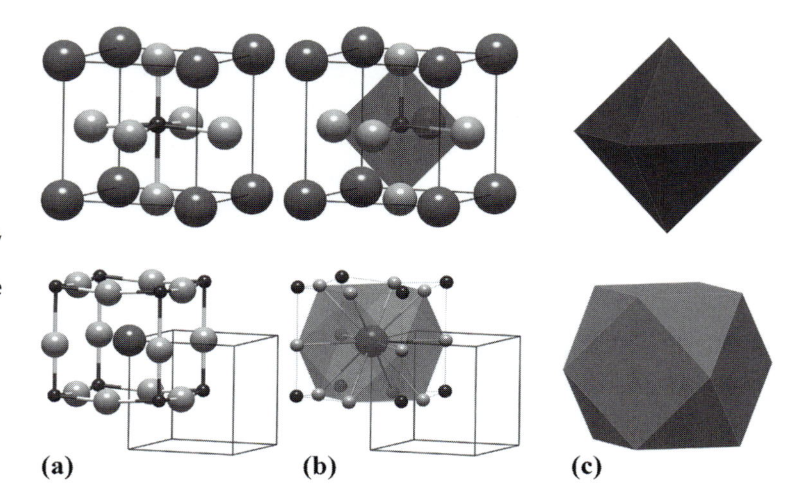

Fig. 22.9. $BaTiO_3$ perovskite crystal structure in a *B*-setting centered on Ti (top) and *B*-setting centered on Ba (bottom); (a) ball-and-stick models; (b) shows the TiO_6 octahedron (top) and the BaO_{12} cuboctahedron (bottom, the *A*-setting unit cell is also shown); and (c) outlines of the two coordination polyhedra.

structure (**Structure 59**) is shown in Fig. 22.9. Perovskite has been given the *StrukturBericht symbol* $E2_1$. Many oxides with an ABO_3 stoichiometry adopt the perovskite structure with $CaTiO_3$ as the prototype. This structure is based on mixed *fcc* packing of three O^{2-} anions and one *A* cation. This partitions the *fcc* array to yield 12-fold cuboctahedral coordination of O^{2-} anions about the larger *A* cation. The *B* cations are octahedrally coordinated by O^{2-} anions. The structure is usually illustrated as an idealized cubic structure but is more commonly tetragonal or orthorhombic because of tilting of octahedra or Jahn–Teller distortions. $MgSiO_3$ is an example of a true cubic perovskite. The perovskite structure is a substructure in many of the oxide superconductors that we will examine in Chapter 23.

In the *B*-setting of the perovskite structure of $BaTiO_3$, Ti (*B*) cations occupy the body center and Ba (*A*) cations occupy the cube vertices; O^{2-} anions decorate the face-centers. Figure 22.9(a) shows (top) the perovskite structure in the *B*-setting and (bottom) in the *A*-setting. Figure 22.9(b) shows the octahedral coordination around the Ti^{4+} cations (top) and the 12-fold cuboctahedral coordination about the Ba^{2+} cations (bottom). The original unit cell of the *B*-setting is also shown for reference. The *A*-setting center and the origin of the *B*-centered unit cell are related by a $[1/2, 1/2, 1/2]$ translation. Figure 22.9(c) shows the two coordination polyhedra for emphasis. The main reason for the presence of two very different coordination environments is the fact that the Ba^{2+} cation has approximately the same radius as the O^{2-} anion, favoring a close-packed configuration, whereas the radius of Ti^{2+} is significantly smaller than that of the anion, favoring an octahedral coordination.

As illustrated in Fig. 22.10(a), the 12-fold Ba coordination polyhedra share square faces, whereas Fig. 22.10(b) shows that the octahedrally coordinated Ti cations are connected in a vertex linked octahedral structure. The bond

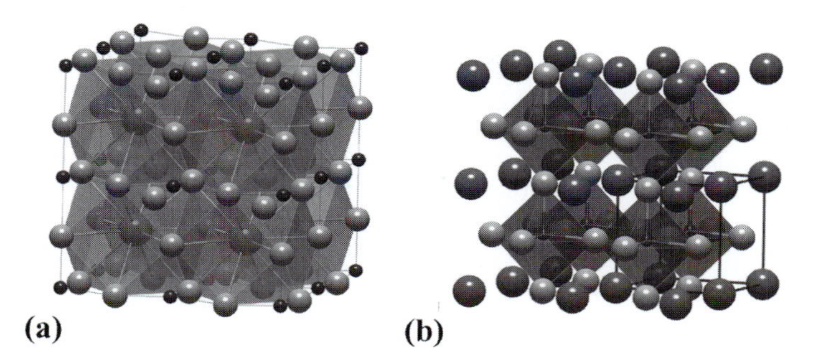

Fig. 22.10. 2 x 2 x 2 cells showing the connectivity of (a) the BaO_{12} cuboctahedra and (b) the TiO_6 octahedra.

strength for Ba^{2+} cations is, therefore, $s_{Ba} = +2/12$, and that for the Ti^{4+} cations is $s_{Ti} = +4/6$. At any O site, two octahedra and four cuboctahedra are shared. The sum of the bond strengths is then $\sum s = 4s_{Ba} + 2s_{Ti} = 2$, which balances the O^{2-} charge.

The room temperature $BaTiO_3$ crystal structure is not precisely cubic, as can be seen from the slightly puckered Ti–O bonds in Fig. 22.9. In fact, the Ti atoms sit slightly off-center, causing the structure to be tetragonal. However, in many perovskites the structure is idealized as cubic because the bond distortion is typically small. We will calculate the structure factor for the $BaTiO_3$ perovskite structure in Chapter 23. In Box 22.6, we use the metric tensor formalism introduced in the first part of this book to compute, among others, the distance between the Ti atom position and the center of the unit cell for the tetragonal $BaTiO_3$ structure. At room temperature, the Ti^{4+} cations are permanently displaced from the cell center, and, as a result, the center-of-charge for the positive charges no longer coincides with that for the negative charges; this gives rise to a *permanent dipole moment* in the unit cell. Since the original perovskite structure is cubic, there are six equivalent directions for the shift of the Ti^{4+} cation, namely along the six cubic directions of the type $\langle 100 \rangle$.

Box 22.6 Lattice computations in tetragonal $BaTiO_3$

Tetragonal $BaTiO_3$ has space group **P4mm** (C^1_{4v}) with lattice parameters $a = 0.399\,98$ nm and $c = 0.4018$ nm. The Ti atom is located at the $1b$ $(1/2, 1/2, z)$ position with $z = 0.482$; Ba is located in the origin, and there are two types of O sites: $1b$ $(1/2, 1/2, z)$ with $z = 0.016$, and $2c$ $(1/2, 0, z)$ with $z = 0.515$. The Ti atom is located slightly below the center of the unit cell. We can use the metric tensor formalism to compute the distance between the cell center and the Ti position. The metric tensor for this tetragonal structure is given by:

$$g_{ij} = \begin{bmatrix} a^2 & 0 & 0 \\ 0 & a^2 & 0 \\ 0 & 0 & c^2 \end{bmatrix} = \begin{bmatrix} 0.159\,98 & 0 & 0 \\ 0 & 0.159\,98 & 0 \\ 0 & 0 & 0.161\,44 \end{bmatrix}.$$

The distance, ℓ, to the center is the length of the vector $(1/2, 1/2, 1/2) - (1/2, 1/2, 0.482) = (0, 0, 0.018)$. Using the metric tensor formalism we find:

$$\ell^2 = \begin{bmatrix} 0 & 0 & 0.018 \end{bmatrix} \begin{bmatrix} 0.15998 & 0 & 0 \\ 0 & 0.15998 & 0 \\ 0 & 0 & 0.16144 \end{bmatrix} \begin{bmatrix} 0 \\ 0 \\ 0.018 \end{bmatrix} = 5.23 \times 10^{-5},$$

so that $\ell = 7.23$ pm (pico meters).

In the cubic perovskite structure, the Ti atom lies on a line connecting two opposite O anions with coordinates $(0, 1/2, 1/2)$ and $(1, 1/2, 1/2)$; in the tetragonal structure, this is no longer a straight line, and we can compute the O–Ti–O bond angle ω using the formalism introduced in Chapter 4. The position vectors of the O anions with respect to the Ti atom are $(\pm 1/2, 0, 0.033)$. Using the alternative method to compute bond angles (Section 4.3.5 on page 90) we find the 2×2 matrix:

$$\begin{pmatrix} -\frac{1}{2} & 0 & 0.033 \\ \frac{1}{2} & 0 & 0.033 \end{pmatrix} \begin{bmatrix} 0.15998 & 0 & 0 \\ 0 & 0.15998 & 0 \\ 0 & 0 & 0.16144 \end{bmatrix} \begin{pmatrix} -\frac{1}{2} & \frac{1}{2} \\ 0 & 0 \\ 0.033 & 0.033 \end{pmatrix}$$

$$= \begin{pmatrix} 0.04017 & -0.03982 \\ -0.03982 & 0.04017 \end{pmatrix},$$

from which we derive that

$$\cos\omega = \frac{-0.03982}{\sqrt{0.04017 \times 0.04017}} = -0.99125 \rightarrow \omega = 172.41^\circ.$$

If the direction of the permanent dipole moment can be changed by means of an externally applied electric field, then we say that the material is a *ferroelectric*. The perovskite structure is the parent structure for many important *ferroelectric materials*. $BaTiO_3$ is an important *electroceramic*, used in capacitors and piezoelectric or electromechanical devices. Some ceramics display a property called *electrostriction*, which is analogous to the previously discussed magnetostriction (Chapter 19).

Single crystals of $BaTiO_3$ generally do not have a net polarization because a crystal contains many *domains*. Each domain corresponds to a different direction of the Ti displacement and, hence, a different direction of the spontaneous polarization. Because the vector sum of the different domain polarizations adds up to zero, the macroscopic polarization can be negligible in the absence of an electric field. Applying an electric field causes the growth of those domains for which the polarization is favorably oriented with respect to the electric field direction. This growth occurs at the expense of the shrinking domains that have unfavorably oriented polarization. As a result, the net polarization grows as the magnitude of the applied electric field increases. When the entire crystal has its dipole moments aligned along the same direction, the material is said to be *saturated*.

22.7.2 Spinels: AB_2O_4

The prototype of the spinel structure is $MgAl_2O_4$; the general formula for a spinel compound is AB_2O_4, where A and B refer to tetrahedral and octahedral interstitial sites, respectively, in the *fcc* sublattice of O anions. In this structure (Fig. 22.11) a fraction of the octahedral and tetrahedral interstices are occupied by cations (Bragg, 1915b). The cations are mixed valent, e.g., divalent Mg^{2+} and trivalent Al^{3+}. $MgAl_2O_4$ is an example of a *normal spinel* for which Mg^{2+} occupies tetrahedral sites, A-sites, and Al^{3+} occupies octahedral sites, B-sites, respectively.

Figure 22.11 shows (a) a ball-and-stick depiction of the unit cell for the $MgAl_2O_4$ cubic spinel structure (space group **Fd$\bar{3}$m** (O_h^7), **Structure 60**) and (b) the connectivity of the Mg^{2+} coordination tetrahedra (T^+ and T^-) and Al^{+3} coordination octahedra (O) in the structure. The T^+ and T^- designations for the tetrahedral sites refer to the fact that, for any given close-packed plane, half of the tetrahedra that can be formed by placing neighboring close-packed planes on both sides of the original plane point in one direction, whereas the other half point in the opposite direction (normal to the planes). The directions are of the type $\langle 111 \rangle$ for the *fcc* lattice. Although only a fraction of the tetrahedral interstices are occupied by cations in the spinel structure, the T^+ and T^- tetrahedra have equal occupation.

The spinel structure can be understood further by taking it apart plane by plane, as will be illustrated in the section on ferrites below. We should note that alternating (001) planes have exclusively Mg tetrahedra or Al octahedra. The Al octahedra form chains along the $\langle 110 \rangle$-type directions. The structure can be understood in terms of alternating chains of edge-sharing octahedra that also share vertices with Mg tetrahedra in intervening planes. The planes in which the chains reside do not possess four-fold symmetry even though

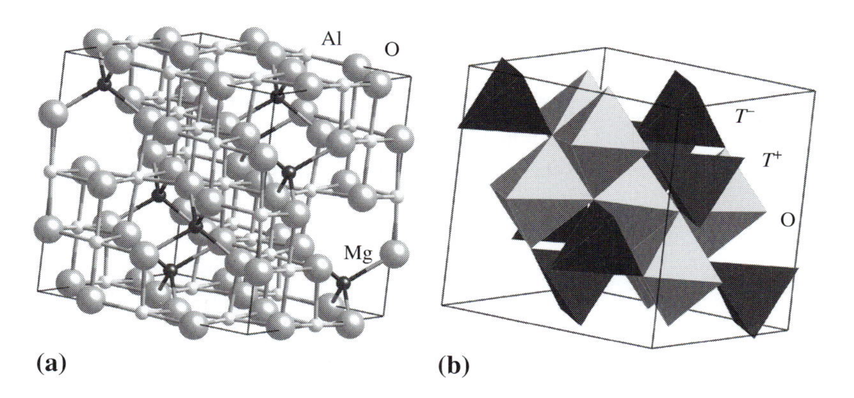

Fig. 22.11. (a) Ball-and-stick depiction of the unit cell for the $MgAl_2O_4$ cubic spinel structure and (b) the connectivity of the Mg^{2+} tetrahedra and Al^{3+} octahedra.

the material is cubic. This is due to the 4_1 screw axes that result in alternating chains of octahedra in planes normal to the cubic axes.

There are many magnetic oxides that adopt the spinel structure. In these materials, the A and B cations are magnetic ions that give rise to technologically important magnetic properties. In the following sub-sections, we will discuss these oxides in more detail.

22.7.2.1 Ferrites

Magnetic oxides with a spinel structure include the important *ferrites*, defined by the chemical formula: $M\mathrm{Fe_2O_4}$, where M is a divalent metal and Fe is nominally trivalent. When M is Fe^{2+}, the compound is known as *magnetite*, $\mathrm{Fe_3O_4}$, which is one of the important oxides of iron. Magnetic oxides are important for high frequency devices because their large electrical resistivities limit eddy current losses.[4] For example, Ni–Zn ferrites are the pre-eminent magnetic materials for microwave applications (in the 100 MHz to 1 GHz range) such as power supply cores and sensors because they have the highest resistivity among the ferrites.

Figure 22.11 shows a spinel unit cell (i.e., a $2 \times 2 \times 2$ stacking of *fcc* units) with $32\mathrm{O}^{2-}$ anions, $16M^{3+}$ cations, and $8M^{2+}$ cations in a cell with lattice parameter around 0.85 nm. In the normal spinel structure, the trivalent atoms occupy $\frac{1}{2}$ of the available octahedral B sites and the divalent atoms occupy $\frac{1}{8}$ of the tetrahedral A sites in the unit cell. As the cation distribution can deviate from the normal spinel structure, it is instructive to express the spinel formula more generally as: $(M_{\delta}^{2+}M_{1-\delta}^{3+})[M_{1-\delta}^{2+}M_{1+\delta}^{3+}]\mathrm{O}_4$ where the cations in parentheses, (), are on A-sites and those in brackets, [], are on B-sites. The case of $\delta = 1$ corresponds to the *normal spinel structure*; $\delta = 0$ corresponds to the *inverse spinel structure*; and $\delta = \frac{1}{3}$ corresponds to a *random distribution* of cations.

A non-equilibrium distribution of cations in the spinel structure can have a strong influence on the magnetic properties of these materials. Such a distribution can result from the valence state preferences of transition metal species to the A- and B-sites. For example, Mn^{2+} has a magnetic dipole moment of $5\mu_B$ whereas Mn^{3+} has a moment of $4\mu_B$ (see Table 13.3 on page 335); Mn^{2+} prefers to occupy the tetrahedral sites, and Mn^{3+} prefers the octahedral sites in the spinel structure. In so-called 4-2 spinels, it is possible to have Mn^{4+} ions with a dipole moment of $3\mu_B$. In simple magnetic spinels, the magnetic moment per formula unit is evaluated by adding or subtracting the individual cation moments (depending on the sign of the

[4] A key attribute of a spinel ferrite as a magnetic material is the anti-parallel coupling of the two Fe atomic *magnetic dipole moments*. This coupling results from a phenomenon called *superexchange*, a magnetic interaction, that is mediated by the intervening O^{2-} anions between the two Fe sites. For a comprehensive review of spinel ferrites and their magnetic properties, see Gorter (1954).

exchange interactions). In Mn_3O_4 and Zn-rich ferrites (and others), non-collinear magnetic dipole moment arrangements complicate such an analysis; this complication adds to the richness of these interesting magnetic structures.

22.7.2.2 Spinel surfaces

Finite size and surface effects in ferrite nanoparticles can include changes in the degree of inversion (cation preference for *A* and *B* sites), preferential breaking of like (*B–B*) or unlike (*A–B*) bonds at terminating surfaces, symmetry breaking, and chemistry changes at the surface. Chemical and structural changes influence the manifestation of such magnetic properties as surface superexchange interactions, dipole moments, and magnetic anisotropy. Oxygen mediated *B–B* and *A–B* superexchange bonds determine the temperature dependence of the magnetization and whether the alignment of cation magnetic dipole moments is collinear or non-collinear (Swaminathan *et al.*, 2006).

We can consider the stacking sequence in the spinel structure in terms of the cation coordination polyhedra to predict the structures of terminating surfaces in spinel crystals, as illustrated in Fig. 22.12. The crystallography of terminating faces determines the symmetry of the atomic environments

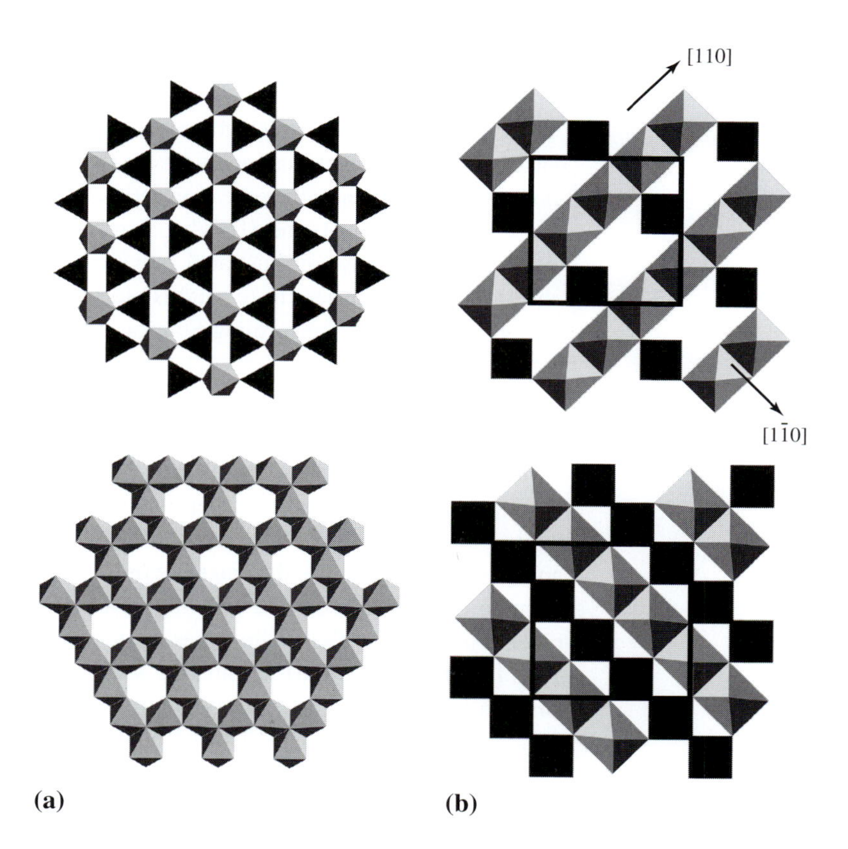

Fig. 22.12. Polyhedral representation showing tetrahedral *A*-sites and octahedral *B*-sites in proposed terminating (a) [111] mixed and octahedral and (b) [100] (octahedral chain) surfaces.

Box 22.7 Facetted spinel ferrite nanoparticles

An example of facetted spinel surfaces is shown below for $Ni_{0.5}Zn_{0.5}Fe_2O_4$ nanoparticles.

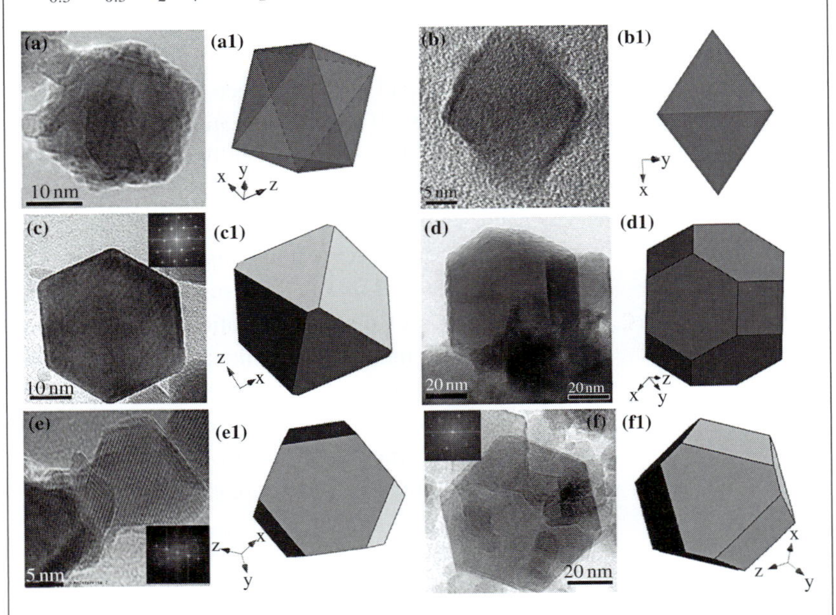

These are high-resolution TEM images (courtesy of R. Swaminathan) of small octahedral ferrite nanoparticles (a) and (b); larger truncated octahedral nanoparticles with truncation ratios 0.43, 0.36, 0.36, and 0.43 are shown in (c), (d), (e), and (f). The corresponding polyhedral models in different projections are illustrated as well. The inset in (c) correspond to the [110] and the insets in (e) and (f) to the [111] zone axis diffraction patterns obtained by performing Fourier transforms of the images.

at the surface. For oxides, we find it more profitable to consider cation coordination polyhedral units, rather than atoms, as basic structural units. The changes in polyhedral environments at the surface, compared with the bulk, can significantly influence the magnetic properties (Swaminathan *et al.*, 2006).

Figure 22.12 considers the stacking of planes of polyhedral units in the spinel structure. The figure illustrates the terminating coordination polyhedra in (111) (a) and (100) (b) terminating surfaces (Swaminathan *et al.*, 2006). For each surface, there are two possibilities; these can be thought of as the surfaces on either side of a cleavage plane of a crystal. The (100) surface exhibits chains of octahedra along [110] and [1$\bar{1}$0] directions. The broken four-fold symmetry at these surfaces reflects the fact that the **Fd$\bar{3}$m** (O_h^7)

space group does not have an ordinary four-fold rotation axis. Instead, it has a 4_1 screw axis, which can have important implications for magnetic surface anisotropy in (100) terminated ferrites. The (111) terminations can be called octahedral and mixed. The first of these has pseudo-hexagonal arrays of octahedra. The mixed surface has octahedra coordinated by six tetrahedra. We can see that the symmetry is trigonal as the apices of the T^+ and T^- tetrahedra alternately point into and out of the surface plane.

Examples of facetted spinel ferrite nanoparticles are shown in Box 22.7 (Swaminathan *et al.*, 2006). The electron micrographs show small (≤ 20 nm) and large (>20 nm) $Ni_{0.5}Zn_{0.5}Fe_2O_4$ ferrite nanoparticles, that have either perfectly octahedral or truncated octahedral morphologies.

22.8 Non-cubic close-packed structures: NiAs, CdI_2, and TiO_2

The *hcp* anion lattice is not as frequently observed in ionic structures because the tetrahedra alternately share vertices and faces along the *c*-axis in the *hcp* structure; octahedra share all edges normal to the *c*-axis and faces parallel to the *c*-axis. Edge- and particularly face-sharing polyhedra violate Pauling's rules. The likelihood of the occurrence of an *hcp* anion lattice is increased in compounds with more covalent bonding, such as NiAs (Fig. 22.13). The likelihood of an *hcp* anion lattice is also higher in compounds such as CdI_2, where the cation Coulomb repulsion by face-sharing octahedra is eliminated only by the cations occupying alternating layers. Similarly, the occupation of alternating rows of tetrahedra in the lower symmetry TiO_2 rutile structure lessens the Coulomb repulsion. All three of these structures are illustrated in this section.

The NiAs structure (Strukturbericht symbol $B8_1$, **Structure 15**), illustrated in Fig. 22.13, is based on an *hcp* anion lattice. This structure has space group **P6$_3$/mmc** (D^4_{6h}). Arsenic anions are arranged in an *hcp* array and Ni cations occupy octahedral interstices. The As anion coordination polyhedra have six Ni atoms arranged at the vertices of a trigonal prism. The As trigonal prisms

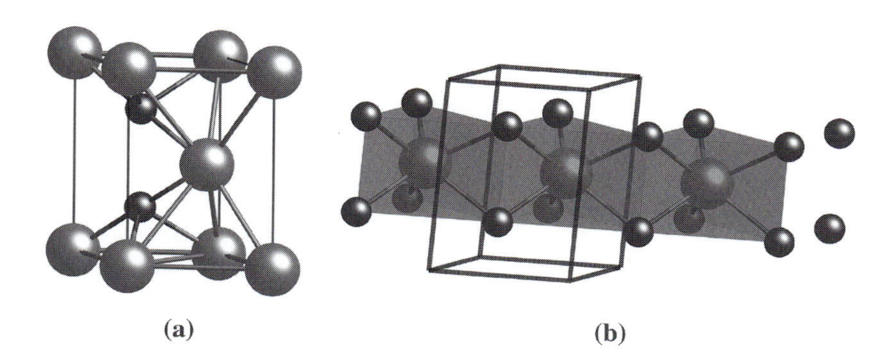

Fig. 22.13. (a) NiAs structure in the hexagonal setting and (b) the connectivity of the anion trigonal prismatic coordination polyhedra.

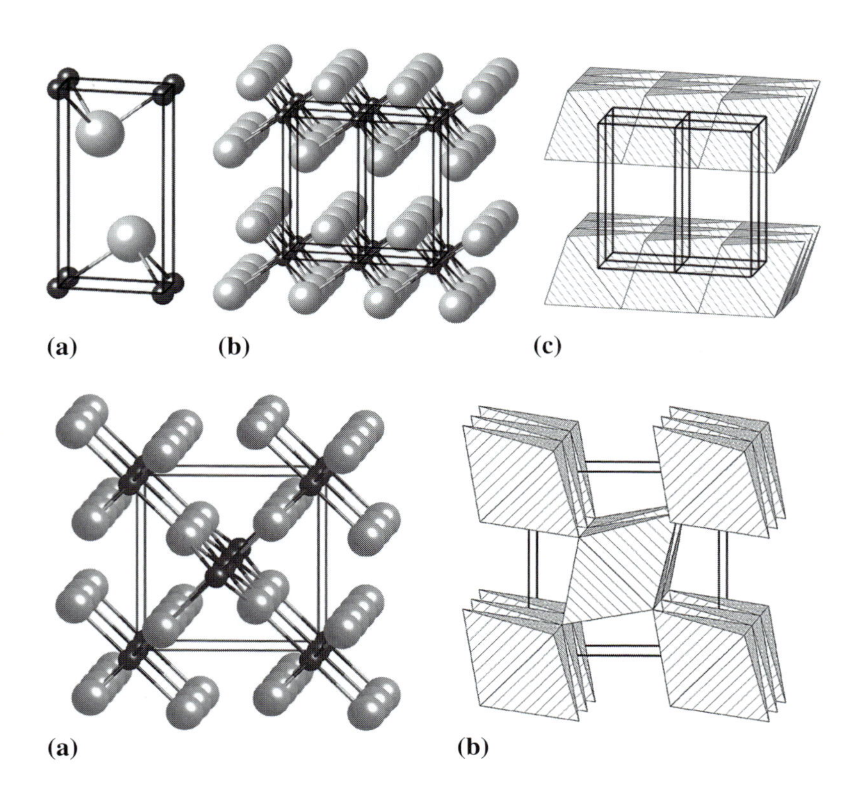

Fig. 22.14. (a) CdI_2 crystal structure, (b) several cells, showing the hexagonal close-packed nature of the anion lattice, and (c) connectivity of the Cd^{2+} edge-sharing octahedra.

Fig. 22.15. a) TiO_2 rutile crystal structure viewed along the [001] direction, and (b) connectivity of the Ti^{4+} edge-sharing octahedra in [001] chains.

share faces, violating Pauling's third rule. The NiAs structure type is generally observed for *MX* compounds in which the bonding is appreciably covalent or metallic. Another interesting attribute of this structure is the edge sharing between the anion coordination polyhedra. When we first introduced the NiAs structure in Chapter 17, we illustrated the crystal structure in the hexagonal setting and showed the connectivity of the edge-sharing cation octahedra. Here, we illustrate the connectivity of the anion coordination polyhedra in Fig. 22.13(b).

Another structure based on an *hcp* anion lattice is the CdI_2 structure illustrated in Fig. 22.14 (C6, **Structure 61**). The Cd^{2+} cations are octahedrally coordinated by I^- anions. The octahedra share edges, but are filled only in alternate layers. This leads to strong two-dimensional bonding within alternate planes, but weak bonding between planes. Structures such as these with weak intraplanar bonding are easily cleaved. Mica is an example of a mineral that has weak intraplanar bonding and is notable for its ease in cleaving; such structures are referred to as *micaceous*.

To explore an oxide structure with lower symmetry, we consider the structure of a tetragonal form of TiO_2, known as *rutile* (C4, **Structure 62**, illustrated in Fig. 22.15). Here, the Ti^{4+} cations are octahedrally coordinated by O^{2-} anions. The Ti ions occupy half of the octahedral sites along alternate [100] rows. The structure consists of edge-sharing octahedra arranged in chains parallel to the [001] direction. Octahedra of neighboring chains share corners. The filled rows are staggered between alternate layers, maintaining isotropy in the bonding (not observed in micaceous structures such as CdI_2). Because the octahedra are vertex-sharing, they maintain large distances between the tetravalent Ti^{4+} cations.

22.9 *Layered structures

In this section, we illustrate examples of layered oxide and carbosulfide structures. The layered high temperature superconductors (HTSCs) are the subject of Chapter 23.

22.9.1 Magnetoplumbite phases

Hexa-aluminates and hexaferrites are important layered oxide materials. Hexaferrites are compounds with stoichiometry $(MO)(Fe_2O_3)_6$; hexa-aluminates have general formula $(MO)(Al_2O_3)_6$. In both compounds, M is a divalent cation like Ca^{2+}, Ba^{2+}, Sr^{2+}, ... The prefix *hexa* refers to the six Fe_2O_3 or Al_2O_3 units in each compound, but can also refer to the hexagonal β-alumina or magnetoplumbite structure. Compounds with these structures exhibit a wide variety of technically important properties, including ionic conductivity, and optical and magnetic properties.

Barium hexaferrite, $BaFe_{12}O_{19}$, is an important hexagonal magnetic ferrite used as a permanent magnet material. This is an example of a material that adopts the *magnetoplumbite structure* (**Structure 63**). Figure 22.16 shows ball-and-stick (a), space filling (b), and polyhedral representations (c) of the hexagonal crystal structure of barium hexaferrite. The connectivity of the

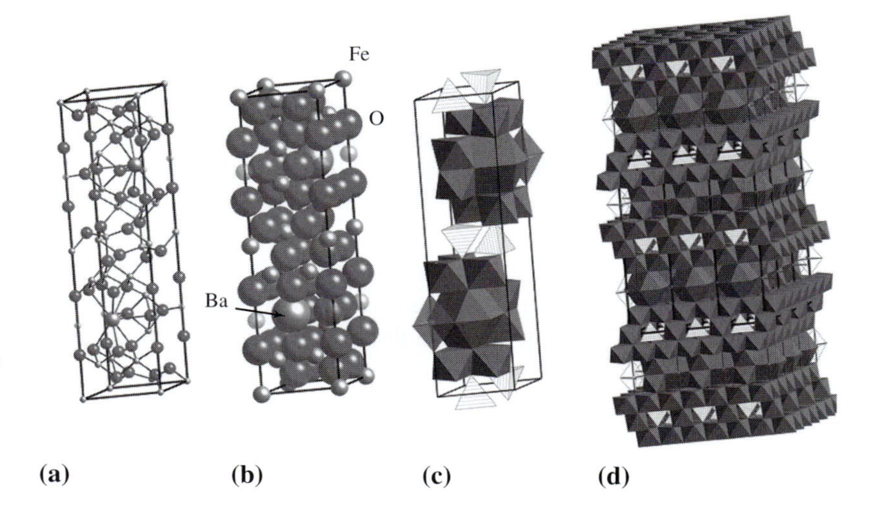

Fig. 22.16. Ball-and-stick (a), space filling (b), and polyhedral representations of the hexagonal crystal structure of barium hexaferrite, $BaFe_{12}O_{19}$. The connectivity of the cation coordination polyhedra is illustrated in (d), which shows the polyhedra from several neighboring cells.

cation coordination polyhedra is illustrated in (d) which shows the polyhedra from several neighboring cells. In this structure, the large Ba atoms are 12-fold coordinated by O. There are five different Fe sites, three of which are octahedrally coordinated and one is tetrahedrally coordinated. The last Fe site has the rather uncommon five-fold triangular prismatic coordination. The structure has layers containing BaO and the trigonal prismatically coordinated Fe, separated by layers containing the Fe^{3+} octahedra and tetrahedra.

22.9.2 Aurivillius phases

A large class of multilayered interstitial oxide compounds was discovered by Bengt Aurivillius (Aurivillius, 1950, 1951, 1952). The *Aurivillius phases* are described by the chemical formula: $(Bi_2O_2)^{2+}(A_{n-1}B_nO_{3n+1})^{2-}$. In these phases, the A cation (Ca, Sr, Ba, ...) is cuboctahedrally (12-fold) coordinated, and the smaller B cation (Ti, Nb, Ta, ...) is octahedrally (6-fold) coordinated. Bismuth has 8-fold square anti-prismatic coordination. The integer n can be 1, 2, 3, 4, or 5, determining the stacking of the layers. An example of an $n = 1$ phase is the compound Bi_2MoO_6, with alternating $(Bi_2O_2)^{2+}$ and $(MoO_4)^{2-}$ layers. An $n = 2$ compound, $Bi_2O_2AB_2O_7$, would consist of stacking of $(Bi_2O_2)^{2+}$ layers that alternate with double perovskite-like $(AB_2O_7)^{2-}$ layers.

Important ferroelectric materials exist among the Aurivillius phases. Ferroelectricity was first observed in the compound $PbBi_2Nb_2O_9$ (Bi_2O_2-$PbNb_2O_7$) (Smolenski and Agranovskaya, 1959; Smolenski *et al.*, 1961). A single unit cell of the orthorhombic crystal structure of the $PbBi_2Nb_2O_9$ phase is illustrated in Fig. 22.17(a) in a ball-and-stick representation (**Structure 64**). This structure has lattice constants $a = 0.550$ nm, $b = 0.549$ nm, and $c = 2.551$ nm. Early work by Aurivillius (1950) assigned a face-centered orthorhombic space group **F222** (D_2^7), but recent x-ray studies have re-assigned the space group as **Cmc2₁** (C_{2v}^{12})[5] (Rae *et al.*, 1990).

$PbBi_2Nb_2O_9$ consists of vertex-linked perovskite-like sheets that are separated by $(Bi_2O_2)^{2+}$ layers. Figure 22.17(b) shows the Nb-centered octahedral coordination polyhedra. In Fig. 22.17(c), several cells are shown with octahedra linked along [110]. Note the double layers of octahedra (outlined by a dashed rectangle) and the separating $(Bi_2O_2)^{2+}$ layers. The term *octahedral sheets* refers to a grouping of more than a single layer of octahedra. The plate-like crystal structure of $PbBi_2Nb_2O_9$ gives rise to highly anisotropic ferroelectric properties. The variations in the ferroelectric properties are likely the result of cation disorder, as indicated by recent structural evidence (Srikanth

[5] The space group is reported as **A2₁am** in Rae *et al.* (1990), but this is simply a different setting of the **Cmc2₁** (C_{2v}^{12}) space group.

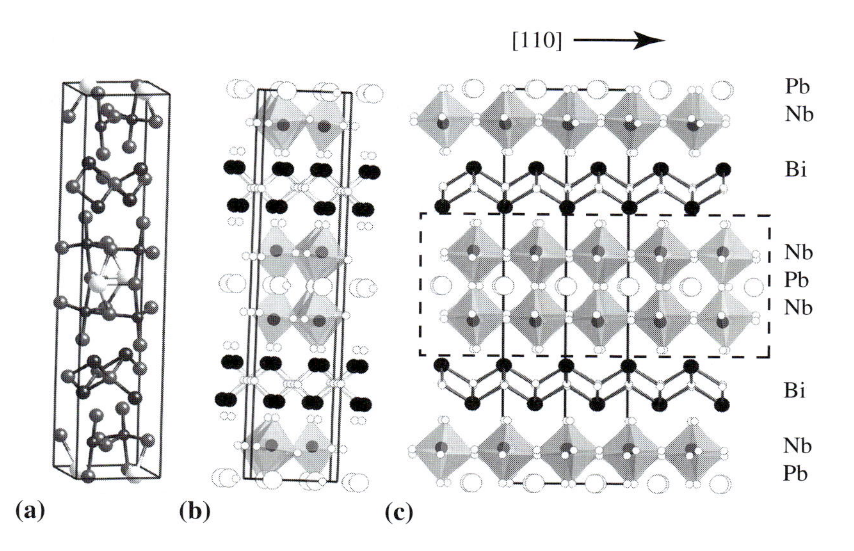

Fig. 22.17. (a) Orthorhombic crystal structure of the $PbBi_2Nb_2O_9$, (b) a single (extended) cell, and (c) several cells in a polyhedral depiction, showing octahedra linked along [110].

et al., 1996). The similar ionic sizes of Pb and Bi make their interchange probable.

Following the discovery of $PbBi_2Nb_2O_9$, Subbarao (1962) discovered several new ferroelectrics. Of current interest among these materials are $SrBi_2Ta_2O_9$ (strontium bismuth tantalate) and related phases, which are candidates for nonvolatile ferroelectric memories. The orthorhombic crystal structure of $SrBi_2Ta_2O_9$ was first determined by Newnham *et al.* (1973). It is a more complex orthorhombic derivative structure of the original $n = 2$ Aurivillius phase.

22.9.3 Ruddelson–Popper phases

Ruddelson–Popper phases (RP) (Ruddelson and Popper, 1958) have layered oxide structures consisting of the repeated stacking of simpler building block units. A prototype set of RP phases is described by the chemical formula: $A_2(A'_{n-1}B_nX_{3n+1})$, where X is an anion, B is an octahedrally coordinated cation, and A and A' are two other cations. When A and A' are the same cation, this formula becomes $A_{n+1}B_nX_{3n+1}$. These structures are characterized by *layers* or *sheets* of BO_6 perovskite-like octahedra in 2-D networks that are separated by A cations. As in the *Aurivillius phases*, a sheet consists of two or more layers. Figure 22.18 shows two typical $A_{n+1}B_nX_{3n+1}$ RP structures, which have a parent tetragonal **I4/mmm** (D^{17}_{4h}) space group. Figure 22.18(a) shows a polyhedral representation of the unit cell for $n = 1$, an A_2BX_4 compound with layers of BO_6 octahedra at $z = 0$ and $z = \frac{1}{2}$. Figure 22.18(b) shows a polyhedral representation of the unit cell for $n = 2$, an $A_3B_2X_7$ compound with sheets of double-layer BO_6 octahedra. The sheets can be observed clearly in the multiple unit cell stack illustrated in

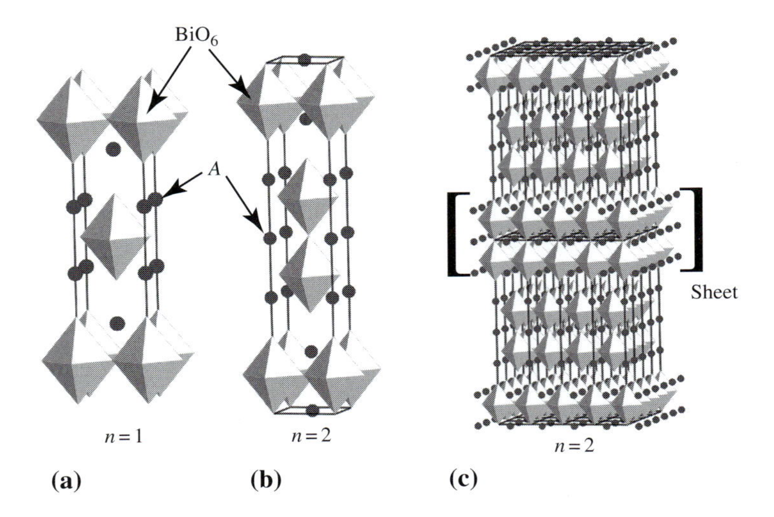

Fig. 22.18. Polyhedral representations of idealized unit cells of the (a) $n = 1$ (A_2BX_4) RP phase; (b) $n = 2$ ($A_3B_2X_7$) compound, and (c) multiple unit cells of the $n = 2$ compound.

Fig. 22.18(c). The body centering operation of the space group gives rise to the back-and-forth shifting of the sheets along the c-axis of the tetragonal unit cell.

Compounds described by the chemical formula $La_{n+1}Cu_nO_{3n+1}$ also form phases with the RP crystal structure. These structures have perovskite layers in A- and B-settings with CuO_6 octahedra forming layers or sheets separated by LaO layers. The $n = 1$ phase, La_2CuO_4, is isostructural with the *high temperature superconductor (HTSC)* phases $(La,Ba)_2CuO_4$ and $(La,Sr)_2CuO_4$. The La_2CuO_4 phase is an insulator. All three phases can be described in terms of alternating perovskite layers in A- and B-settings, as illustrated for $(La,Sr)_2CuO_4$ in Fig. 22.19(a). The $n = 2$ phase,

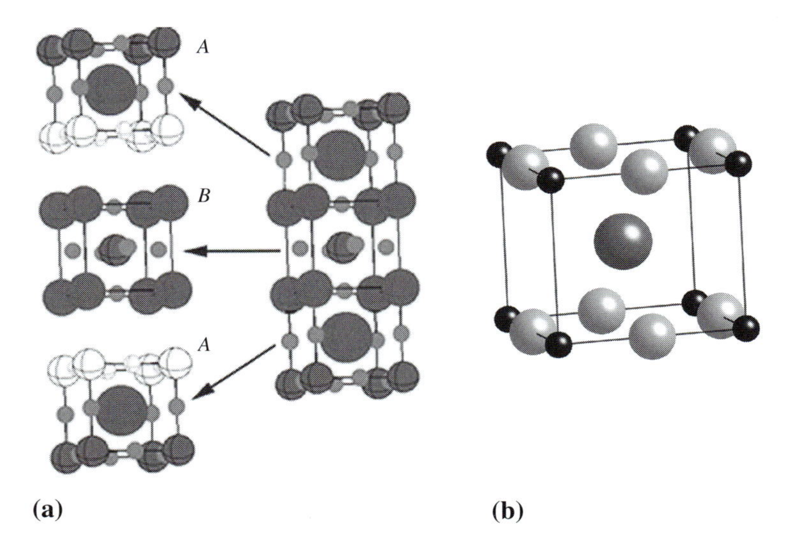

Fig. 22.19. (a) The crystal structure of the $(La,Sr)_2CuO_7$ Ruddelson–Popper phase and its decomposition into stacked structural units and (b) the ∞-layer perovskite structure of $LaCuO_3$.

$La_3Cu_2O_7$, has two perovskite layers and one rocksalt layer, respectively. The $n = 3$ phase, $La_4Cu_3O_{10}$, has three perovskite layers and one rocksalt layer, and so on. In the limit of $n \to \infty$, the stoichiometry is $LaCuO_3$ and the structure is the simple perovskite structure; this end-member structure could be called the *infinite-layer structure*. The insulating $LaCuO_3$ structure is shown in Fig. 22.19(b). The *high temperature superconductors* in the Bi–Sr–Ca–Cu–O and Tl–Ba–Ca–Cu–O systems are examples of RP phases; they are discussed further in Chapter 23.

22.9.4 Tungsten bronzes

Tungsten bronze is a term used to describe alkali metal tungstates, vanadates, molybdates, titanates, and niobates. The prototype was named tungsten bronze because of its color; its chemical formula is Na_xWO_3 ($x \sim 0.1$). This oxide is unique because W has an anomalously large cation charge of 5+. The compound $NaWO_3$ adopts a *perovskite structure* with octahedral W^{5+} and cuboctahedral Na^+ coordination polyhedra. In the Na-deficient tungsten bronze, both Na^+ and W^{5+} cations are octahedrally coordinated.

Alkali additions to WO_3 remarkably change the electronic structure of this compound and, consequently, its electrical, magnetic, and optical properties. As a result, these materials are interesting for optical devices (electrochromic windows) and ion-exchange batteries (Rohrer, 2001). Recently, tungsten bronze high temperature superconductors have been reported. Surface-doped Na–W bronzes have been reported to have superconducting transition temperatures as high as 91 K. These tungsten bronzes are significant as examples of non-Cu-containing high temperature superconductors (Reich and Tsabba, 1999).

To explore an example of a tungsten bronze-related oxide structure, we consider the structure of *sanmartinite*, $ZnWO_4$, shown in Fig. 22.20 (Filipenko *et al.*, 1968). This oxide is also unique because W has an anomalously large 6+ charge. In this structure, both the Zn^{2+} and the W^{6+} cations have octahedral anion coordination polyhedra. Figure 22.20(b) depicts a ball-and-stick model of the unit cell of the $ZnWO_4$ monoclinic crystal structure with space group **P2/c** (C_{2h}^4) and lattice parameters $a = 0.472\,\text{nm}$, $b = 0.57\,\text{nm}$,

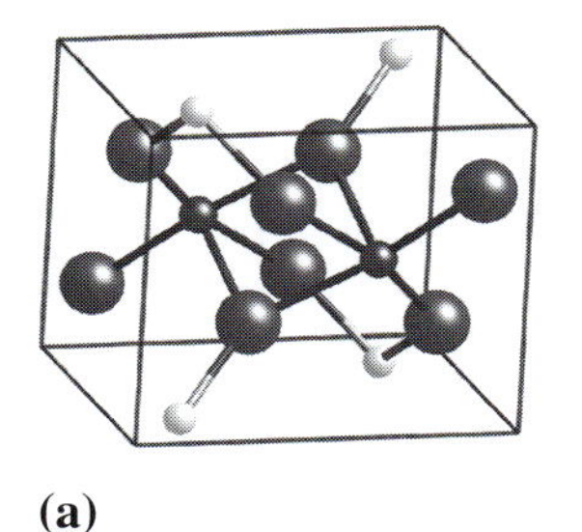
(a)

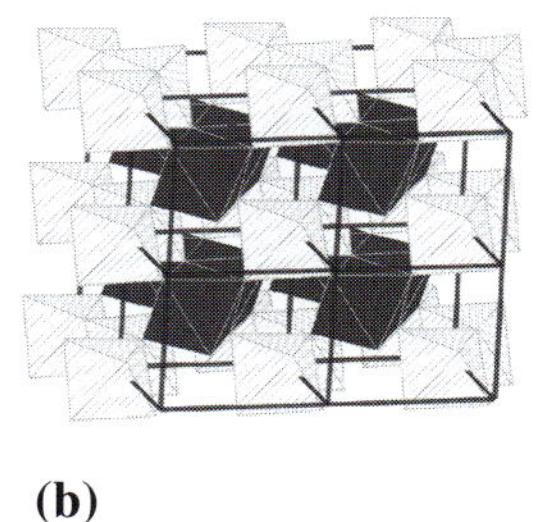
(b)

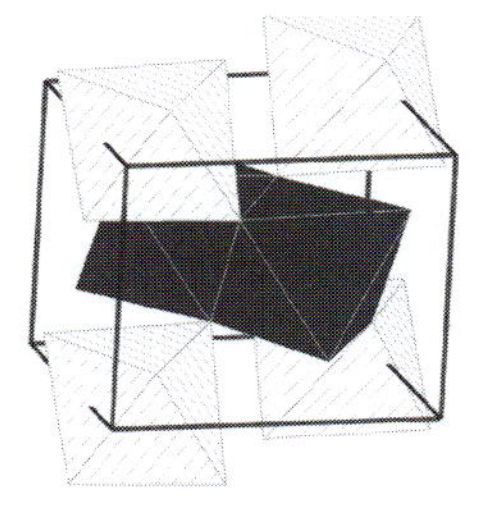
(c)

Fig. 22.20. (a) A unit cell for the cubic $ZnWO_4$ structure; (b) the connectivity of the Zn^{2+} and W^{6+} octahedra, and (c) connectivity of W^{6+} and Zn^{2+} octahedra in a single cell.

and $c = 0.495$ nm ($\beta = 90.15°$). There are two inequivalent O^{2-} anions at the special positions $(0.22, 0.11, 0.95)$ and $(0.26, 0.38, 0.39)$, W^{6+} cations are located at $(0.0, 0.179, 0.25)$, and Zn^{2+} cations at $(0.5, 0.674, 0.25)$.

Figure 22.20(b) illustrates the connectivity of these octahedral coordination polyhedra. The Zn^{2+} octahedra share edges with each other and vertices with the W^{6+} octahedra; this sharing can be seen more clearly in Fig. 22.20(c). The W^{6+} octahedra also share edges. This connectivity yields a layered structure with alternating sheets of edge-sharing Zn^{2+} and W^{6+} octahedra. The layers are linked by the vertices that are shared between the two types of cation-centered octahedra.

22.9.5 Titanium carbosulfide

While the previous structure types involved layering of perovskite octahedra, interesting layer structures can exist that exhibit both covalent and ionic bonding. Ti_2CS is an example of such a material. In this compound, Ti–C bonds are covalent whereas Ti–S bonds are ionic. Ti_2CS is an important inclusion in many steel alloys (Ramalingum, 1998) because of its potential to embrittle the steel. It has a hexagonal structure with space group $\mathbf{P6_3/mmc}$ (D^4_{6h}); structure data can be found in the on-line structures appendix as **Structure 66**. This structure can be regarded as a layered structure, with TiC and TiS layers that share Ti planes, as illustrated in Fig. 22.21. Because of the mix of ionic and covalent bonding, this structure is not well described by the rules for ionic structures discussed thus far in this chapter. Instead, the structure consists of hexagonal networks of Ti atoms in parallel (001) planes. This leads to the less commonly observed hexagonal prismatic coordination of both the C and S atoms in the structure. These *drums* share vertices, and their size reflects the relative sizes of the S and C atoms.

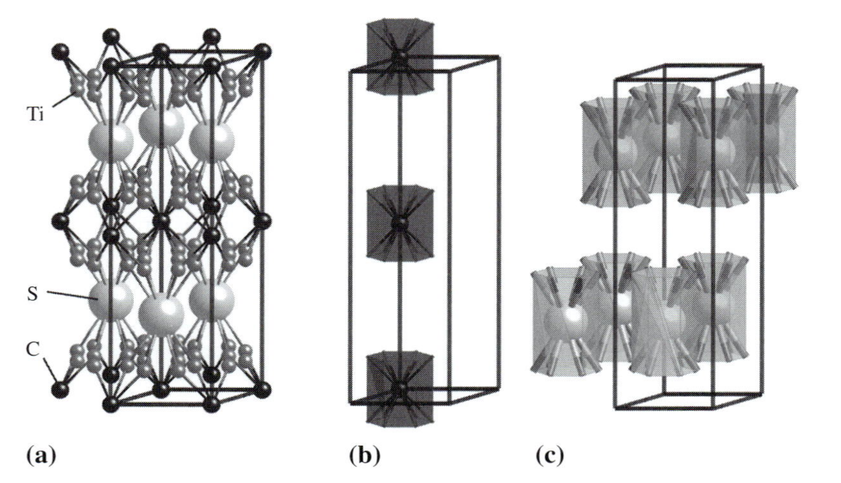

Fig. 22.21. (a) Ti_2CS structure in a hexagonal setting showing the content of three neighboring unit cells; (b) C, and (c) S hexagonal prismatic coordination polyhedra.

Table 22.7. Comparison of some of the structures described in this and earlier chapter in terms of the filling of tetrahedral and octahedral sites.

Anion lattice	T^+	T^-	O	Example
fcc	–	–	1	NaCl (rocksalt)
	1	–	–	ZnS (sphalerite)
	–	–	$\frac{1}{2}$	$CdCl_2$
	1	1	–	CaF_2 (fluorite)
	$\frac{1}{8}$	$\frac{1}{8}$	$\frac{1}{2}$	Al_2MgO_4 (spinel)
hcp	–	–	1	NiAs
	1	–	–	ZnS (wurtzite)
	–	–	$\frac{1}{2}$	CdI_2
	–	–	$\frac{1}{2}$	TiO_2 (rutile)
	–	–	$\frac{2}{3}$	Al_2O_3 (corundum)

22.10 Additional remarks

Before we conclude this chapter with a short section on defects in ceramics, it is perhaps useful to summarize some of the structures that we have introduced in this and earlier chapters in terms of the occupation of octahedral and tetrahedral sites in the close-packed anion lattices. Table 22.7 lists for both *fcc* and *hcp* anion lattices, what fraction of the interstitial sites is occupied for a few structure prototypes. We leave it up to the reader to add more structures to this table.

22.11 *Point defects in ceramics

Most of the material in this book deals with perfect crystal structures, i.e., all the atoms are located at their proper positions. What makes *real* materials particularly interesting is the fact that the atoms are not always where they are supposed to be. Real materials have *defects* in them, and these defects are often responsible for physical behavior that is quite different from that of the perfect or defect-free crystal. Defects are commonly categorized according to their dimensionality: single atom defects (missing atoms, known as *vacancies*, or incorrectly positioned atoms) are point-like defects and are, hence, labeled as zero-dimensional defects. Linear or one-dimensional defects are most often identified with edge and screw *dislocations*, although dislocations are only one member of a larger family of line defects, also encompassing *disclinations*, *dispirations*, and *disconnections*. There are many two-dimensional defects, such as *stacking faults*, where the normal stacking sequence of the crystal is interrupted; *anti-phase boundaries*, where there is a sudden shift in the occupation of sub-lattices (this mostly occurs in ordered alloy structures); *domain boundaries*, where there is a sudden change in the orientation of the crystal lattice and the associated magnetization or polarization; *inversion boundaries*, which occur in non-centrosymmetric crystals; and, last but not least, the *external surfaces* of crystals, of which we have seen an example in the section on spinel surfaces. Three-dimensional defects include grains and grain boundaries, which separate regions of the material where the unit cell orientation is different.

In ceramic materials, point defects are charged because the atoms all carry a charge; thus, point defects can dramatically affect properties such as diffusion, electrical and ionic conduction, and charge transfer. In this section, we will introduce briefly the most important defects in ceramic (ionic) materials, along with the notation used to describe them.

The *Kroger–Vink notation* is used to describe charge neutrality in ionic solids. The notation consists of three components: the species (cation, anion, or vacancy), the charge state, and the crystallographic position. A superscript is used to identify the charge state with respect to the ideal crystal (a prime $'$ is used to label a negative charge; a bullet $\bullet$ indicates a positive charge, and x denotes a neutral species with respect to a perfect, charge-balanced lattice). A subscript is used to designate the crystallographic position. Square brackets around the defect symbol indicate the concentration of that defect.

Let us consider a few examples for an MO (metal oxide) crystal:

$V_M'' \rightarrow$ Vacancy species, V, on a metal site, M, with a charge of $2-$;

$V_O^{\bullet\bullet} \rightarrow$ Vacancy species, V, on the oxygen site, O, with a charge of $2+$.

We can write defect generation in terms of chemical reactions. For example, a *Schottky defect* is a pair of nearby charged vacancies on each of the anion and cation sub-lattices. It is shown in Fig. 22.22(a) and is described by the reaction:

$$0 \leftrightarrow V_M'' + V_O^{\bullet\bullet};$$

here, 0 denotes the perfect crystal state without either vacancy. The atoms that formerly occupied the vacancy positions are thought to have migrated to the surface of the crystal. A second, less common form of Schottky defect involves occupation of two interstitial sites with oppositely charged ions:

$$0 \leftrightarrow O_I'' + M_I^{\bullet\bullet},$$

where the subscript I denotes the interstitial site.

A *Frenkel defect* is a charged vacancy and an interstitial of the same sub-lattice. It is shown in Fig. 22.22(b) and is described by the reaction:

$$M_M \leftrightarrow V_M'' + M_I^{\bullet\bullet}.$$

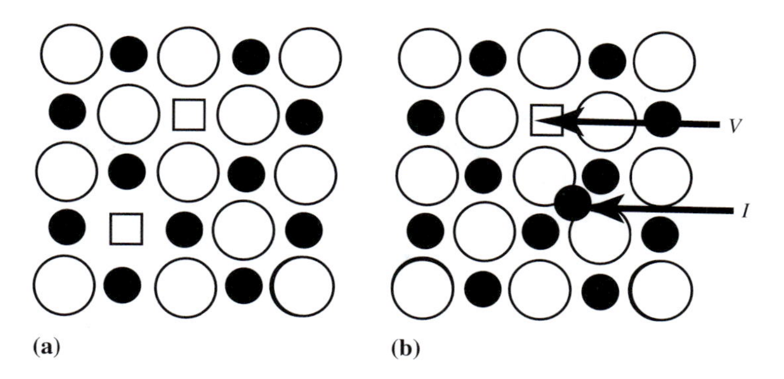

Fig. 22.22. (a) Schottky defect and (b) Frenkel pair.

A less common form of a Frenkel defect involves an oxygen anion vacancy and an interstitial:

$$O_O \leftrightarrow V_O^{\bullet\bullet} + O_I''.$$

The concentrations of defects can be calculated using chemical equilibrium theory, which considers the energy of formation of the defects. We can write equilibrium constants (per atom) for the (more common) reactions:

$$K_S = \exp\left(\frac{-\Delta G_S^f}{k_B T}\right) = \left[V_M''\right] \cdot \left[V_O^{\bullet\bullet}\right], \tag{22.6}$$

and

$$K_F = \exp\left(\frac{-\Delta G_F^f}{k_B T}\right) = \left[V_M''\right] \cdot \left[M_I^{\bullet\bullet}\right], \tag{22.7}$$

where the subscript S indicates Schottky and F Frenkel defects. ΔG_S^f is the excess (Gibbs) free energy required to form the Schottky defect, and ΔG_F^f is the excess (Gibbs) free energy required to form the Frenkel defect. Here, the square brackets are used to denote mole or site fractions of the vacancies or interstitial species.

Chemical reactions that lead to non-stoichiometric concentrations of the chemical species can also be written in the Kroger–Vink notation. For example, consider a chemical reaction generating oxygen non-stoichiometry:

$$\frac{1}{2}O_2 \leftrightarrow O_O^x + V_M^x.$$

It is also possible to have anti-site defects: $M_X^{\bullet\bullet}$ and X_M''. These are not common in ionic crystals because a large electrostatic repulsion energy gives rise to a large formation energy. In the oxides, this energy is often so large that these defects can safely be ignored. However, in covalent solids, such as the semiconductor GaAs, anti-site defects can be important.

22.12 Historical notes

Helen D. Megaw (1907–2002) was born in Northern Ireland. She studied at Queen's University, Belfast, (1925–6) and Girton College, Cambridge, where she received a B.S. in 1930. She received a Ph.D. in mineralogy and petrology at the University of Cambridge in 1934. Her doctoral studies were performed under the direction of J. D. Bernal. Megaw spent 1935 in Vienna followed by a stint at Clarendon Laboratory, Oxford. She then spent several years teaching, after which she took a position at Philips Lamps Ltd. in 1943. It was at Philips Lamps that she determined the crystal structure of barium titanate (Megaw, 1945). In 1945, Megaw moved to Birkbeck College, London, where she again worked with Bernal as Assistant Director of Research in Crystallography. From 1946–72, she was Fellow, Lecturer, and Director of Studies in Physical Sciences, at Girton College, Cambridge. At Cambridge, she interacted with notable crystallographers under the direction of W. L. Bragg. She continued to study minerals, making important contributions to the crystal structures of feldspars.

Megaw's accomplishments include comparing the structures of ordinary and *heavy* ice (this was later recognized by the naming of Megaw Island in the Antarctic). With Bernal, she studied hydrogen bonding in metal hydroxides. Her determination of the crystal structure of barium titanate, $BaTiO_3$ (Megaw, 1945), was important to the field of ferroelectric materials. She continued to contribute to ferroelectricity theory through research on titanates and perovskites (Megaw, 1952), and wrote the first book on the subject, "Ferroelectricity in Crystals" (Megaw, 1957). In 1989, Megaw was the first woman to win the Roebling Medal of the Mineralogical Society of America.

Fig. 22.23. (a) Helen D. Megaw (1907–2002) (picture courtesy of J. Lima-de-Faria and originally from *The American Mineralogist* **75**, 715-717, (1990), and (b) Linus Carl Pauling (1901–94) (picture courtesy of the Nobel e-Museum).

Later, in 2000, she received an honorary degree at Queen's University, Belfast.[6]

Linus Carl Pauling (1901–94) was born in Portland, Oregon. He received a B.Sc. degree in Chemical Engineering in 1922 from Oregon State College. He earned a Ph.D. (summa cum laude) in chemistry with minors in physics and mathematics from the California Institute of Technology in 1925. His thesis work, directed by Roscoe G. Dickinson, was on crystal structure determination by x-ray diffraction. Pauling wrote an early paper with Dickinson that detailed the determination of the crystal structure of MoS_2, molybdenite (Dickinson and Pauling, 1923), by employing Laue photographic techniques taught to Dickinson by W. G. Wyckoff. Pauling was awarded a prestigous Guggenheim fellowship in 1926 to study at the Arnold Sommerfeld Institute of Theoretical Physics in Munich. While in Europe, he also studied with Niels Bohr and Erwin Schrödinger, among others. His studies of quantum mechanics would influence and shape his future ideas on the nature of the chemical bond and lead him to become one of the first scientists to apply quantum theory to calculations of molecular structures.

Pauling returned to Caltech as an assistant professor in 1927 and became a Professor of Chemistry in 1931, a position he held until the early 1960s. Pauling was an influential scientist who contributed to experimental XRD determination of crystal structures and the interpretation of these structures in terms of atomic or ionic radii. His contributions included the development of ionic radii and Pauling's rules for ionic structures which are discussed in this chapter. He applied quantum mechanics to the calculation of electronic states in atoms, the structures of metals and intermetallic compounds, and the theory of ferromagnetism. He also contributed to the understanding of the structure of proteins and many other subjects, including quasicrystals, on which he worked late in his life (Pauling, 1985, 1989).

Pauling was the recipient of the 1954 Nobel Prize in Chemistry "for research into the nature of the chemical bond and its application to the elucidation of the structure of complex substances." In 1962, when awarded the Nobel Peace prize, he became the first scientist to win two unshared Nobel Prizes. Pauling also competed with Watson and Crick in the race to solve the structure of DNA. A detailed biography of Pauling appears in *Nobel Lectures*, Chemistry 1942–62, Elsevier Publishing, Amsterdam.

[6] An electronic archive, Contributions to 20th Century Physics, maintained at the UCLA website: http://www.physics.ucla.edu/ cwp, has details on the life and accomplishments of Megaw. Professor M. Glazer, Clarendon Laboratory, Oxford, wrote an obituary for Megaw, for the British Crystallography Association. This can be found at the URL: http://bca.cryst .bbk. ac.uk/BCA/CNews/2002/P81Ob.htm.

22.13 Problems

(i) *Ionic radii*: Compare the ionic radii for S^{2-} (CN = 6) and Cl^-.

(ii) *Cubic coordination*: Calculate the critical radius ratio, ρ_c, for eight-fold, cubic, coordination.

(iii) *CN = 12 coordination*: Calculate the critical radius ratio, ρ_c, for 12-fold, *fcc* and *hcp* coordination.

(iv) *Trigonal prismatic coordination*: Calculate the following for an AB_6 coordination polyhedron:

(a) The interbond angles for an ion in octahedral (six-fold antiprismatic) coordination.
(b) The pyramidal bond angles for the octahedron.
(c) The interbond angles in six-fold prismatic coordination if the anion–anion separations are equal.
(d) The critical radius ratio for this polyhedron.

(v) *Trigonal dipyramidal coordination of anions*: Derive the permissible range of radius ratios for five-fold *trigonal dipyramidal coordination* (apical neighbors above and below a triangular group) of anions, B, around a central cation, A.

(a) Are all anion–anion contacts the same length?
(b) What is the critical separation?
(c) Explain why this coordination would be rare in comparison to octahedral or tetrahedral coordination.

(vi) *Permissible radius ratios*: Consider an ionic A_3B_4 compound. What different pairs of coordination numbers are consistent with the stoichiometry? Using values of critical radius ratios derived in this chapter, determine the permissible ranges for radius ratios for an A_3B_4 compound if A atoms assume tetrahedral coordination.

(vii) *Pauling's rules*: As a function of the sum of the cation and anion radii, calculate the maximum cation to cation separation that occurs for coordination polyhedra sharing vertices, edges, or faces for (a) cubic, (b) octahedral, and (c) tetrahedral coordinations.

(viii) *Madelung constant*: Calculate the Madelung constant for an octahedral arrangement of doubly charged anions about a doubly charged cation. Use r_0 for the anion–cation distance. (Hint: Remember that the Madelung constant considers the sum of Coulomb interactions over *all* atom pairs.)

(ix) CsCl *structure*: Given $R_{Cs^+} = 0.175$ nm and $R_{Cl^-} = 0.181$ nm, predict the CsCl lattice constant.

(x) NaCl *structure I*: MgO is an oxide material with the NaCl structure. Given the ionic radii $R_{Mg^{2+}} = 0.089$ nm and $R_{O^{2-}} = 0.140$ nm, calculate the lattice constant for MgO.

(xi) NaCl *structure II*: The compounds KBr, KCl, RbBr, and RbCl all crystallize with the rocksalt structure. Their lattice constants are, respectively, 0.65966, 0.62931, 0.6889, and 0.65810 nm. Use this data to estimate each of the ionic radii, R_{K^+}, R_{Rb^+}, R_{Br^-}, and R_{Cl^-}.

(xii) NaCl *structure III*: Determine the packing density as a function of the ratio of the radii of the two spheres in the NaCl structure; then determine the radius ratio which gives either a maximum or a minimum packing density.

(xiii) *Alkali halide crystal structures*: Experimentally, all of the alkali halides are found to assume the rocksalt structure, with the exception of CsCl, CsBr, and CsI. Using tabulated values of the ionic radii, predict structures for all of the alkali halides involving Li^+, Na^+, K^+, Rb^+, Cs^+, F^-, Cl^-, Br^-, and I^-.

(xiv) *Fluorite – critical radius ratio*: Fluorite is *fcc* with four CaF_2 units per face-centered unit cell; Ca sits in the special (0, 0, 0) and (1/4, 1/4, 1/4) positions and F is found on (3/4, 3/4, 3/4). The space group is **Fm$\bar{3}$m** (O_h^5).

(a) Draw the structure showing all the atoms in a cubic unit cell. You should try and do this without consulting the figures in the text!
(b) Determine the critical radius ratio, ρ_c, for cation coordination in this structure.
(c) Discuss the linkage of the coordination polyhedra and analyze the structure in terms of Pauling's rules.
(d) Compare the structure to the CsCl and diamond cubic structures.

(xv) *Fluorite – structure factor*: Compute a simple expression for the structure factor of fluorite; determine the extinction conditions.

(xvi) KBr *density*: KBr crystallizes in an NaCl structure with a lattice constant $a = 0.659\,66$ nm. Determine the density of KBr, and the distance between a K atom and a touching Br atom in the structure. Rationalize your result with the sum of appropriate ionic radii for K and Br.

(xvii) *Coordination numbers in* TiO_2: Using the appropriate ionic radii, predict the coordination numbers that you would anticipate finding for Ti^{4+} and O^{2-} in TiO_2.

(xviii) AgI *bond lengths*: AgI has a wurtzite structure at room temperature with lattice constants $a = 0.4598$ nm and $c = 0.7514$ nm. Compute the bond length between Ag and I in the wurtzite phase. How does this compare to that which you would predict on the basis of ionic radii for Ag and I?

(xix) AgI *phase transformation*: At 147 °C, this compound transforms from a wurtzite structure to a structure in which the I anions assume a *bcc* lattice and the Ag cations are disordered over all available tetrahedral interstices. Compute the lattice constant of the *bcc* phase if the bond length between Ag and I remains unchanged in the phase transformation.

(xx) *Sphalerite*: Calculate the radius of the largest ion that could fit in the octahedral interstice in the sphalerite form of ZnS.

(xxi) *Molybdenite*: Look up Dickinson and Pauling's original work on molybdenite (Dickinson and Pauling, 1923) (alternatively find a more recent reference for this structure) and draw the structure (or plot using crystal structure software).

(xxii) *Ferrite structure*: Consider the spinel ferrite structure.

(a) Determine the cation coordination about the O ions in the structure.
(b) Determine the strength of the anion–cation bonds and the bond sum for this structure.
(c) *Ferrimagnetic superexchange* coupling between magnetic moments requires a bridging O ion between the magnetic cations; enumerate all types of such interactions in AB_2O_4 spinels.

(xxiii) $MnFe_2O_4$ *properties*: $MnFe_2O_4$ is a spinel with $a = 0.85$ nm.

(a) How many formula units of $MnFe_2O_4$ are there per unit cell?
(b) Calculate the density of $MnFe_2O_4$.

(xxiv) $ZnFe_2O_4$ *net dipole moment*: $ZnFe_2O_4$ has an inverse spinel structure.

(a) Determine the magnetic dipole moments for Zn^{2+} and Fe^{3+}.
(b) Compute the net dipole moment per formula unit.

(xxv) TiO_2 *rutile structure factor*: The rutile structure is tetragonal and described as **Structure 62** in the structures appendix.

(a) Derive a simplified form for the structure factor.
(b) Determine the extinction conditions independent of the parameter, x, that is part of the $4f$ site description.
(c) Is the Bravais lattice primitive or body-centered?

(xxvi) *Defects I*: Consider a monovalent ionic solid MX. Write a defect reaction for a cation Schottky defect in the Kroger–Vink notation.

(xxvii) *Defects II*: Consider a monovalent ionic solid MX. Write a defect reaction for a cation Frenkel defect in the Kroger–Vink notation.

(xxviii) *Defect concentrations*: Consider a crystal for which the Frenkel and Schottky defect formation energies are $\Delta G_F^f = 200$ kJ/mol and $\Delta G_S^f = 100$ kJ/mol. At $T = 100$ K, calculate the metal and oxygen vacancy and interstitial metal defect concentrations.

CHAPTER

23 Ceramic structures II: high temperature superconductors

> *"Mercury has passed into a new state, which on account of its extraordinary electrical properties may be called the superconductive state."*
>
> Heike Kamerlingh Onnes, (1853–1926)

23.1 Introduction: superconductivity

The discovery of *high temperature superconductors* (HTSC) was a major scientific achievement at the end of the twentieth century. *Superconductivity* is the phenomenon by which free electrons form so-called *Cooper pairs*, which move cooperatively below a temperature called the *superconducting transition temperature,* T_c. This *electron pairing* represents an *electronic* phase transition of the superconductor (i.e., the atomic positions do not change upon crossing T_c). Paired electrons are able to correlate their motion to avoid scattering off of the vibrating crystalline lattice (Bardeen *et al.*, 1957). This phase transition from a *normal conductor* ($> T_c$) to a superconductor ($\leq T_c$) is accompanied by the abrupt loss of electrical resistivity, *perfect conductivity*, and the exclusion of magnetic flux, the *Meissner effect*.

Superconductivity was discovered by Heike Kamerlingh Onnes (Onnes, 1911) in 1911 during a study of the temperature dependence of the electrical resistance of metals. Liquid helium (LHe) had only recently been produced by Onnes and Clay (1908) and this led to the ability to cool materials to temperatures less than 4.2 K. In his initial experiments with platinum, Onnes found that the metal's resistance decreased to a low level which varied with purity. As mercury was the purest metal available at the time, he measured the resistance of Hg as a function of temperature and found that the resistance fell sharply at 4 K. Below this temperature, Hg exhibited no resistance at all.

The value of T_c is different for each superconductor; it is 4.1 K for Hg, 18.1 K for Nb_3Sn, and 23.9 K for thin film Nb_3Ge.[1] Between 1964 and 1975, oxide superconductors were discovered with T_c values in excess of 15 K. Of these, the $BaPb_{1-x}Bi_xO_3$ (BPBO) system is notable because of its *perovskite* crystal structure, which is the building block of the HTSC materials. The *BPBO superconductors* (Sleight *et al.*, 1975), have T_c values less than that for Nb_3Ge (Gavaler, 1973). A decade later, the HTSCs, Cu-based perovskites with higher T_c values, were discovered by Bednorz and Müller (1986) ($La_{1-x}Ba_xCuO_4$ or LBCO, and $La_{1-x}Sr_xCuO_4$ or LSCO) and Wu *et al.* (1987) ($YBa_2Cu_3O_{7-x}$ or YBCO).

In 1986, a T_c value of 36 K was observed for $La_{1-x}Ba_xCuO_4$ (Bednorz and Müller, 1986) by J. Georg Bednorz and K. Alex Müller of the IBM Zürich research facility. This was the first time that an oxide superconductor had a T_c value surpassing that of metals. The history of this discovery, for which Bednorz and Müller won the 1987 Nobel Prize in Physics, is detailed in Müller and Bednorz (1987). Values of T_c above the liquid nitrogen (LN_2) boiling point (77 K) were first observed by the group of Paul Chu at the University of Houston (90 K for $YBa_2Cu_3O_{7-x}$, (Wu *et al.*, 1987)). A T_c value exceeding 77 K has important economic implications, as LN_2 is an order of magnitude cheaper than LHe, the coolant for conventional superconductors. The $YBa_2Cu_3O_{7-x}$ structure led the way to more complex perovskite-based HTSC materials containing Bi, Tl, and Hg, with even higher T_c values.

In 1988, Maeda *et al.* (1988) discovered superconductors with T_c reaching 115 K in the $(Bi,Pb)_2Sr_2Ca_2Cu_3O_{10+x}$ (Bi-2223, *BSCCO superconductor*) system. Compounds isostructural with the BSCCO superconductors were discovered in the Tl–Ba–Ca–Cu–O (*TBCCO superconductor*) system by Sheng and Hermann (1988). These compounds have T_c values near 125 K in the $Tl_2Ba_2Ca_2Cu_3O_{10+x}$ (Tl-2223) material. Schilling *et al.* (1993) discovered an Hg-containing superconductor, $HgBa_2Ca_2Cu_3O_{8+x}$ (*HBCCO superconductor*), with a T_c value of about 133 K. Even more recently, an $Ag_{1-x}Cu_xBa_2Ca_{n-1}Cu_nO_{2n+3-x}$ (*ACBCCO superconductor*) family of superconductors was synthesized by a high pressure synthesis route (Ihara *et al.*, 1994). The correlation between T_c and crystal structure is the subject of active scientific discussion.

[1] Prior to the discovery of HTSC materials, thin film Nb_3Ge was the highest temperature superconductor. It was discovered at the Westinghouse Science and Technology Center, in Pittsburgh (Gavaler, 1973).

23.2 High temperature superconductors: nomenclature

In this chapter, we illustrate several important HTSC structures. Each section will provide a table summarizing structural information for each of the HTSC phases discussed in that section. These include the type of compound, the formula unit, the crystal system and space group, the number of formula units per unit cell, the lattice constants, and the special positions occupied by cations, C, and oxygen, O, atoms, respectively. Atom coordinates for all special positions are provided in the on-line crystal structures appendix when they are known.

Often, HTSC superconductors are referred to in a shorthand notation with numbers that reflect the cation stoichiometry. For instance, $\mathrm{La}_{2-x}\mathrm{Sr}_x\mathrm{CuO}_4$ is referred to as the *214 phase* and $\mathrm{YBa_2Cu_3O_{7-x}}$ as the *123 phase*; alternatively, short character sequences, such as LSCO or YBCO, are used as a shorthand indicator of the chemical species occurring in the compound. With few exceptions, the HTSC materials are cuprates with superconducting $\mathrm{CuO_2}$ sheets or blocks, called *conducting blocks*, sandwiched between insulating layers or blocks.

An alternative, widely used, nomenclature for describing layered HTSC structures uses a four-number scheme (Shaked *et al.*, 1994). The four numbers describe the number of *insulating layers*, *spacing layers*, *separating layers*, and *cuprate conducting blocks* in the structure. The first number designates the number of insulating layers between $\mathrm{CuO_2}$ blocks. These typically include heavy elements such as Hg, Bi, Tl, or Pb. The second number designates the number of spacing layers. Spacing layers lie between identical $\mathrm{CuO_2}$ blocks. There are always two spacing layers per $\mathrm{CuO_2}$ block. Cations found in these layers include La and alkaline earths, such as Sr, Ba, etc. (these are also known as *rocksalt blocks*). The third number refers to the number of separating layers that reside between $\mathrm{CuO_2}$ layers within an individual conducting block. These layers are typically occupied by cations such as Ca, Sr, or a rare earth ion, typically with non-coplanar O ions. The fourth number refers to the number of $\mathrm{CuO_2}$ planes within a *conducting block*. This nomenclature will be illustrated in the structural examples discussed in the following sections.

23.3 *Perovskite-based high temperature superconductors

23.3.1 Single layer perovskite high temperature superconductors

Table 23.1 summarizes structural information for the HTSC phases discussed in this section. The $\mathrm{BaPb}_{1-x}\mathrm{Bi}_x\mathrm{O}_3$ (*BPBO superconductor*) compound has a single layer perovskite structure (**Structure 67** in the on-line structures appendix). In $\mathrm{BaPb}_{1-x}\mathrm{Bi}_x\mathrm{O}_3$, the single perovskite unit is not cubic but tetragonal and, unlike $\mathrm{BaTiO_3}$ (discussed in Chapter 22), it has a large c/a ratio. This is to be distinguished from other tetragonal HTSCs (to be discussed

Table 23.1. Structural information for BKBO and BPBO HTSC phases.

Type	Space group *(System)*	Z	a, c (nm)	O sites	C sites
$Ba_{1-x}K_xBiO_3$ $(0.37 \leq x \leq 0.5)$	**Pm$\bar{3}$m** (O_h^1) Cubic	1	0.429	3d	1a (Bi), 1b (Ba)
$BaPb_{1-x}Bi_xO_3$ $(0.05 \leq x \leq 0.30)$	**I4/mcm** (D_{4h}^{18}) Tetragonal	4	0.605, 0.8621	4d, 8e	4c (Ba) 4b (Pb,Bi)

in later sections) where tetragonality occurs because of the unidirectional stacking of cubic perovskite units.

BPBO has a maximum T_c of about 13 K for $x = 0.30$, although superconductivity is observed in the entire Pb-rich composition range $(0.05 \leq x \leq 0.3)$. Combined Pb and K doping reduces T_c to about 12 K in BPBO. Substitution of alkali metals for Ba in $BaBiO_3$ was also predicted to result in superconductivity. In 1988, superconductivity above 30 K was observed in the $Ba_{1-x}K_xBiO_3$ system, BKBO (Fig. 23.1) (Mattheiss and Hamann, 1988). In the BPBO system, Pb substitutes on the Bi site; in BKBO, K substitutes on Ba sites. The cubic phase has space group **Pm$\bar{3}$m** (O_h^1) (Cava *et al.*, 1988).

As discussed in Chapter 22, the perovskite unit cell can be represented in an A-type (A centered) or B-type (B centered) setting. Figure 23.1 shows the cubic perovskite structure of the superconducting compound $Ba_{1-x}K_xBiO_3$

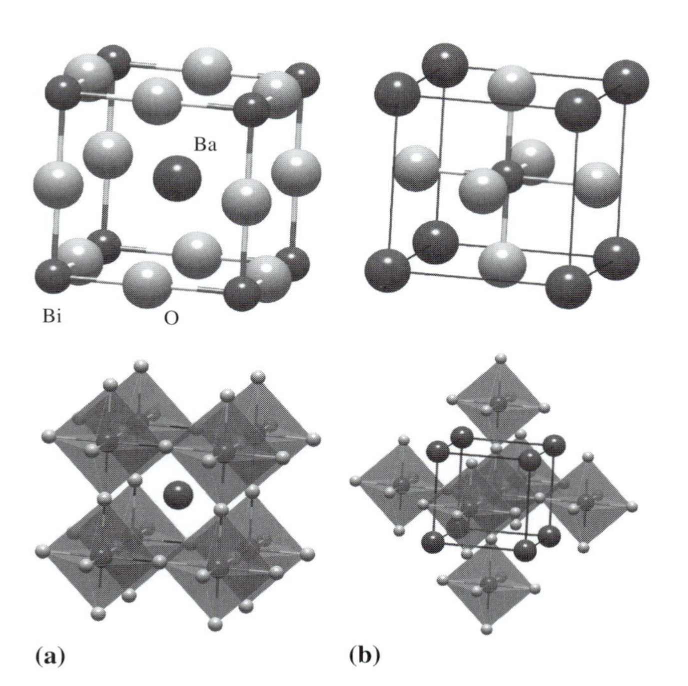

Fig. 23.1. Ball-and-stick models of the cubic perovskite superconductor $Ba_{1-x}K_xBiO_3$ shown in the A- (a) and the B- (b) settings. At the bottom, the BiO_6 coordination octahedra are highlighted.

(**Structure 68**). The top of Fig. 23.1(a) shows the A setting, with an A (Ba) atom at the body center, B (Bi) atoms at the cube vertices and O atoms at the center of all cube edges; K substitutes on Ba sites. In the B setting, shown in Fig. 23.1(b), a B (Bi) atom sits at the body center, A (Ba) atoms at the cube vertices and O atoms at the center of all cube faces. Also depicted in the bottom half of the figure is the connectivity of the BiO_6 coordination octahedra. These two settings are important for the description of the triple perovskites in the next section. The BKBO superconductor does not have a symbol in the four-number scheme because it does not have CuO_2 layers.

$Ba_{1-x}K_xBiO_3$ superconductors received much attention because (1) they are non-Cu containing HTSCs; and (2) they have isotropic 3-D structures without the 2-D copper oxide planes of the layered structures discussed in later sections. Instead, they have the 3-D network of BiO_6 octahedra illustrated in Fig. 23.1(a) and (b). Interest in the role of symmetry on superconducting properties fueled studies in a number of areas:

(i) the role of symmetry lowering tilting of the BiO_6 octahedra (Hinks *et al.*, 1988);
(ii) the low temperature structure of $Ba_{0.6}K_{0.4}BiO_3$ (Kwei *et al.*, 1989);
(iii) the structural phase diagram of the $Ba_{1-x}K_xBiO_3$ system illustrated in Fig. 23.2 (Pei *et al.*, 1990); and
(iv) the role of isotropic superconductivity in intrinsic and extrinsic properties such as flux pinning (McHenry *et al.*, 1989) and pair tunneling (Baumert, 1994) through superconductor–insulator–superconductor (SIS) junctions.

Since the formal cationic charge states are Ba^{2+} and K^+, BKBO is an example of a *mixed valence compound*.

Figure 23.2 (Pei *et al.*, 1990) shows a schematic phase diagram for the $Ba_{1-x}K_xBiO_{3-\delta}$ system. Superconductivity in $Ba_{1-x}K_xBiO_{3-\delta}$ occurs at low temperatures over the range of stoichiometries $0.37 \geq x \leq 0.5$. In this range,

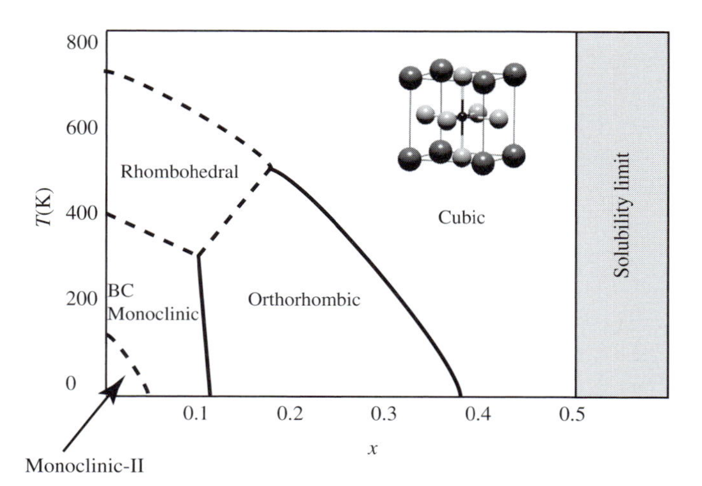

Fig. 23.2. $T-x$ phase diagram for $Ba_{1-x}K_xBiO_3$ showing the K-rich cubic phase, as well as orthorhombic, rhombohedral and two monoclinic variants.

the lattice constant varies approximately linearly from 0.4289 to 0.4270 nm. At $x = 0.5$, the solubility limit for K in the cubic phase is reached. The low temperature cubic to orthorhombic phase transformation at $x = 0.37$ coincides with the onset of superconductivity in this system.[2] At high temperature, a cubic to rhombohedral phase transition occurs due to tilting of the BiO_6 octahedra. A rhombohedral to monoclinic transition involves a further tilt of the octahedra coupled by O displacements. Since the cubic perovskite structure is a basic building block for most HTSC structures, we provide, as an example, the structure factor calculation for the prototype $BaTiO_3$ perovskite (in the A-setting) in Box 23.1.

The volatility of K renders it difficult to maintain proper stoichiometry during the fabrication of the BKBO superconductors. The cubic lattice constant is quite sensitive to both the O stoichiometry and the amount of K, so that

Box 23.1 Computation of the structure factor for the $BaTiO_3$ structure

The $BaTiO_3$ perovskite has a simple cubic Bravais lattice decorated with a three-atom basis with Ba at (0,0,0), Ti at (1/2, 1/2, 1/2), and O at the (1/2, 1/2, 0)-type positions. It is instructive to consider the individual contributions of each atom type to the structure factor:

$$F^{\mathrm{Ba}}_{hkl} = \sum_{j=1}^{N} f_{\mathrm{Ba}}\, \mathrm{e}^{2\pi \mathrm{i}(hx_j+ky_j+lz_j)} = f_{\mathrm{Ba}};$$

$$F^{\mathrm{Ti}}_{hkl} = \sum_{j=1}^{N} f_{\mathrm{Ti}}\, \mathrm{e}^{2\pi \mathrm{i}(hx_j+ky_j+lz_j)} = f_{\mathrm{Ti}}\, \mathrm{e}^{\pi \mathrm{i}(h+k+l)};$$

$$F^{\mathrm{O}}_{hkl} = \sum_{j=1}^{N} f_{\mathrm{O}}\, \mathrm{e}^{2\pi \mathrm{i}(hx_j+ky_j+lz_j)} = f_{\mathrm{O}} \left[\mathrm{e}^{\pi \mathrm{i}(h+k)} + \mathrm{cp}\right]$$

where cp stands for cyclic permutation (i.e., $h \to k$, $k \to l$, and $l \to h$). The total structure factor is then the sum of the above three contributions:

$$F_{hkl} = f_{\mathrm{Ba}} + f_{\mathrm{Ti}}\, \mathrm{e}^{\pi \mathrm{i}(h+k+l)} + f_{\mathrm{O}} \left[\mathrm{e}^{\pi \mathrm{i}(h+k)} + \mathrm{cp}\right].$$

Since the Bravais lattice is primitive, there are no systematic absences for this structure.

[2] The schematic phase diagram implies that the two-phase fields are infinitely narrow and are represented by single lines. This may or may not be correct.

Table 23.2. Structural information for triple-layer perovskite-based HTSC phases.

Type [various notations]	Space group crystal system	Z	a b (nm)	c (nm)	O sites	C sites
$La_{2-x}Ba_xCuO_4$ ($x = 0.15$) [0201, LBCO-214]	**I4/mmm** (D_{4h}^{17}) tetragonal	2	0.378	1.329	4c, 4e	4e (La), 2a (Cu)
$Ln_{2-x}Ce_xCuO_{4-y}$ ($x = 0.15$; $y = 0.08$) [0201 T', 214 T']	**I4/mmm** (D_{4h}^{17}) tetragonal	2	0.395	1.207	4c, 4d	4e (Ln), 2a (Cu)
$(Nd,Sr,Ce)_2CuO_{4-y}$ ($y = 0.08$) [0201 T*, 214 T*]	**P4/nmm** (D_{4h}^{7}) tetragonal	2	0.386	1.25	2a, 4f 8j (23% occ.)	2c (Nd,Sr,Ce) 2c (Cu)
$YBa_2Cu_3O_{7-x}$ [1212C, YBCO-123]	**Pmmm** (D_{2h}^{1}) orthorhombic	1	0.38198 0.38849	1.16762	1e, 2s, 2r	1h (Y), 2t (Ba), 1a, 2t (Cu)

care must be exercised in the interpretation of these competing effects. For example, a precise O stoichiometry is imperative for the production of superconducting materials by thin film deposition for the fabrication of tunnel junctions. This requires adequate post-oxygenation annealing to achieve the correct O stoichiometry (Baumert, 1994). The crystal chemistry of BKBO and BPBO superconductors is discussed extensively in a review article by Norton (Norton, 1992).

23.3.2 Triple-layer perovskite-based high temperature superconductors

Triple layer perovskite materials are some of the first and, ultimately, most important HTSC materials. Examples include the La(Ba,Sr)–Cu–O, Ln–(Ba,Sr)–Cu–O and Y–Ba–Cu–O systems, which will be illustrated in this section. Table 23.2 summarizes the relevant structural information.

23.3.2.1 $La_{2-x}M_xCuO_4$ with $x < 0.5$ (M = Ba, Sr, or Ca)

The layered perovskites $La_{2-x}M_xCuO_4$ with $x < 0.5$ (M = Ba, Sr, or Ca) crystallize into a triple perovskite cell (Cava *et al.*, 1987b, Jorgensen *et al.*, 1987). These A_2BO_4 compounds possess the K_2NiF_4 prototype structure, illustrated in Fig. 23.3 as $La_{2-x}Ba_xCuO_4$. The structure has space group **I4/mmm** (D_{4h}^{17}) (**Structure 69**).

Figure 23.3 shows a space filling (a), a ball-and-stick (b), and a polyhedral representation (c), of this structure. Figure 23.3(c) shows that the Cu cations are octahedrally coordinated, with octahedra sharing vertices. The Cu cations can be Cu^{3+} or Cu^{2+}, with the amount of each required to balance charge determined by the composition variable x, which determines the relative fraction of the La^{3+} and Ba^{2+} cations present in the superconductor. This is

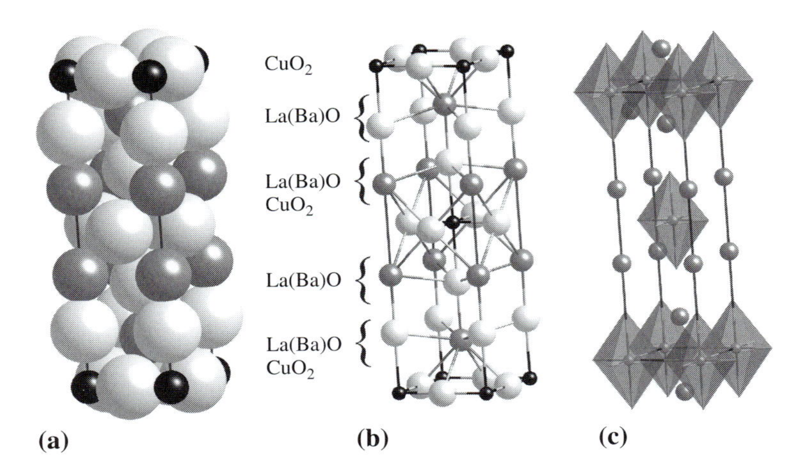

Fig. 23.3. Structure of $La_{2-x}M_xCuO_4$ (M = Ba, Sr, or Ca): (a) space filling model, (b) ball-and-stick model, and (c) polyhedral model. The octahedral coordination of O anions about Cu^{2+} cations and the cubic coordination about La^{3+} cations are highlighted.

another example of a *mixed valence compound*. The mixed valence of the Cu cations is an important prerequisite for superconductivity.

Figure 23.3 illustrates the *LBCO superconductor* structure (known as the *214 phase*). It can be viewed as a perovskite layer in a B-setting sandwiched between two perovskite layers in an A-setting. The triple perovskite unit cell is tetragonal with $c \sim 3a = 3b$. Substitution of the M atoms constitutes p-type doping with respect to the neutral ($x = 0$) compound, with superconductivity depending on the M concentration, x. In the LBCO system with $x \sim 0.15$, superconductivity with a $T_c > 30$ K was first observed. For the $La_{2-x}Sr_xCuO_4$ (*LSCO*) compound, even higher T_c values around 39 K (for $x = 0.15$) were observed. Superconductivity with T_c in excess of 30 K has also been observed in La_2CuO_4 in which holes are generated by excess O interstitials produced by electrochemical oxidation. In the four-number scheme, we can describe the $La_{2-x}M_xCuO_4$ structure as 0201, as illustrated in Box 23.2.

Box 23.2 Four-number symbol for the $La_{2-x}M_xCuO_4$ structure

The $La_{2-x}M_xCuO_4$ (M = Ba, Sr, or Ca) structure is designated as the 0201 structure in the four-number scheme. The analysis is as follows:

(i) It can be seen that the first number is 0 because there are no *insulating layers* between CuO_2 blocks (single planes).
(ii) The second number is 2 because there are two La(Ba)O (La(Sr)O, etc.) *spacing layers* between each of the CuO_2 planes.
(iii) The third number is again 0 because there can be no *separating layers* if there is only a single CuO_2 layer in a conducting block.
(iv) The fourth number is 1 since there is a single CuO_2 plane in each conducting block.

Care must be taken in analyzing the lattice constant for these structures in terms of ionic radii. For example, using the O^{2-} radius of 0.14 nm and the Cu^{2+} ionic radius of 0.073 nm (for six-fold coordination), we predict a lattice constant of $2(0.14 + 0.073) = 0.426$ nm, assuming touching Cu and O ions. This must be revised by allowing for the presence of Cu^{3+}, which has a smaller ionic radius, yielding a value closer to the observed lattice constant of 0.38 nm. The apical oxygens on the octahedra are further away from the Cu cation than the in-plane oxygen anions, as is clearly seen in Fig. 23.3(b). In many of the oxide superconductors described in later sections, a smaller Cu cation radius must be used because of the square planar coordination.

23.3.2.2 $Ln_{2-x}Ce_xCuO_{4-y}$ and $(Nd,Sr,Ce)_2CuO_{4-y}$

The octahedral coordination of Cu is an important feature of the $La_{2-x}M_xCuO_4$ superconducting compounds. In other Ln_2CuO_4 materials ($Ln \neq La$), the Cu coordination is square planar and superconductivity does not occur. There are many other systems for which square planar coordination and superconductivity are observed. For example, superconductivity has been observed in Nd_2CuO_4 (Fig. 23.4) and in the system $Ln_{2-x}Ce_xCuO_{4-y}$ (Tokura *et al.*, 1989) with $Ln =$ Nd, Pr, Eu, or Sm; these systems have layered perovskite crystal structures similar to LBCO, but with square planar Cu coordination. For $Ln =$ Nd and $x = 0.15$, a T_c of 24 K is observed. This structure has the space group **I4/mmm** (D_{4h}^{17}) (**Structure 70**). Interestingly, since Ce exists in a 4+ state, these materials are *electron-doped* while LBCO and LSCO are *hole doped*.

Figure 23.4 shows the structure of the Nd_2CuO_4 compound which, when doped with Ce on the Nd sites, becomes superconducting at low temperature. A close packed depiction is shown in Fig. 23.4(a) and a ball-and-stick model in Fig. 23.4(b). The structure consists of CuO_2 planes and Nd_2O_3 regions

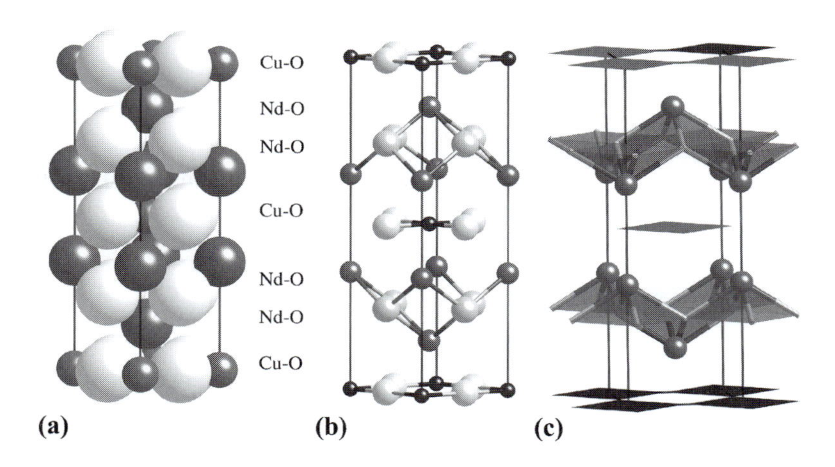

Fig. 23.4. Structure of Nd_2CuO_4: (a) Space filling model of a single unit cell, (b) ball-and-stick model, and (c) model illustrating cation coordination polyhedra and showing the square planar coordination of O^{2-} anions about Cu^{2+} cations.

(Nd and O planes) as illustrated. Even though the Nd and O are not co-planar, they are identified as a layer, i.e., a spacing layer. The Nd^{3+}–O^{2-} bonds along with the square planar coordination about the Cu^{2+} cations are illustrated in Fig. 23.4(c). In the four-number designation, this is an 0201 structure, called 0201 T′, since it is distinct from 0201 by virtue of the differences in atom positions.

Another example of an 0201 structure, $(Nd, Sr, Ce)_2CuO_{4-y}$ (Izumi *et al.*, 1989), also has space group **I4/mmm** (D_{4h}^{17}), but with two lanthanide and three O special positions. It is given the designation 0201 T*.

23.3.2.3 $YBa_2Cu_3O_7$ and related structures

Since the discovery of $YBa_2Cu_3O_{7-x}$ (Y-123, YBCO), a number of phases have been identified in the Y_2O_3–BaO–CuO system. The $YBa_2Cu_3O_{7-x}$ compound has both orthorhombic ($0.0 < x < 0.2$) and tetragonal ($x > 0.2$) structures. The former is the superconducting phase, while the latter is non-superconductive and often referred to as the high-temperature phase. The superconducting phase, $YBa_2Cu_3O_{7-x}$ ($0.0 < x < 0.2$), has space group **Pmmm** (D_{2h}^{1}) (**Structure 71**). Figure 23.5 shows the crystal structure of this O-deficient perovskite with ordered vacancies. The layer stacking sequence can be represented as:

$$YBa_2Cu_3O_{7-x} = \ldots - \overset{][}{\underset{\uparrow}{CuO}} - BaO - CuO_2 - \underset{\uparrow}{Y} - CuO_2 - BaO - \overset{][}{\underset{\uparrow}{CuO}} - \ldots$$

where the layers in brackets, [ ], represent one unit cell, and the vertical arrows indicate the positions of mirror planes. Ba^{2+} is larger than the Y^{3+} cation that centers the cell.

Figure 23.5(a) shows a space filling, (b) a ball-and-stick (Cu–O bonds only), and (c) a polyhedral model for the YBCO structure. The polyhedral

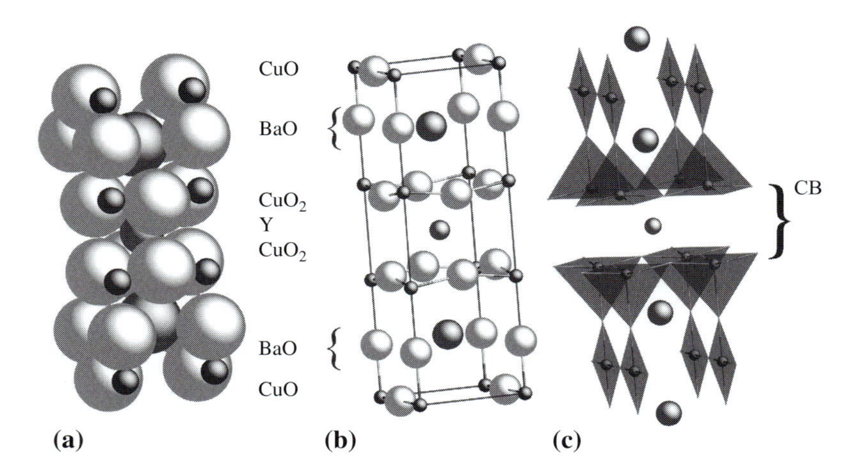

Fig. 23.5. The triple perovskite cell of the $YBa_2Cu_3O_{7-x}$ superconductor in (a) space filling, (b) ball-and-stick, and (c) polyhedral models. The planar and chain coordination polyhedra of O anions about the Cu cations are highlighted.

model shows the square planar coordination polyhedra of O anions about Cu cations in the CuO planes and the half-octahedral (five-fold) coordination of Cu in the CuO_2 planes. In the CuO layers, each Cu atom is coordinated by four O atoms, which is different from the five O atoms surrounding the Cu atom in the CuO_2 layer.

It is generally thought that the Cu–O layer is insulating and the CuO_2 planes are conducting; in the figure, the conducting block is denoted CB. It is easy to determine the composition of a plane through consideration of how the atoms are shared among cells. The slight difference between the a and b lattice constants in the orthorhombic superconducting phase is caused by O vacancy ordering in the CuO chain layers that are sandwiched between BaO layers. The b/a ratio varies as a function of the O stoichiometry. The non-superconducting phase ($x > 0.2$), with a disordered tetragonal structure, can be formed if the O $1e$ sites and the vacancy sites are equally occupied, resulting in a structure with space group **P4/mmm** (D^1_{4h}). The superconductor is designated as the 123 phase using the cation ratio nomenclature. The four-number scheme for $YBa_2Cu_3O_{7-x}$ ($0.0 < x < 0.2$) structure is 1212C, as explained in Box 23.3.

The *critical current density*, J_c, defined as the current density at which the superconducting state ceases to exist, is an important engineering property of the HTSC materials. High values of J_c can be obtained by introducing a fine dispersion of a second, non-superconducting phase; these second-phase

Box 23.3 Four-number scheme analysis of the $YBa_2Cu_3O_{7-x}$ structure

In the four-number scheme, we rewrite the compound formula as $CuBa_2YCu_2O_{7-x}$, recognizing both the stacking sequence and the fact that the Cu–O layer is the insulating layer. We analyze the structure as follows:

(i) The first number is 1 because there is a Cu–O *insulating layer* between each CuO_2 conducting block.
(ii) The second number is 2 because there are two BaO *spacing layers* between each CuO_2 block. This is easier to see if the cell is centered at $z = \frac{1}{2}$ rather than at $z = 0$.
(iii) The third number is 1 because there is a Y *separating layer* between CuO_2 planes.
(iv) The fourth number is 2 since there are two CuO_2 planes in a conducting block.

This structure is designated as the 1212C structure in the four-number scheme, where the additional C reflects the fact that the insulating layer contains CuO *chains*.

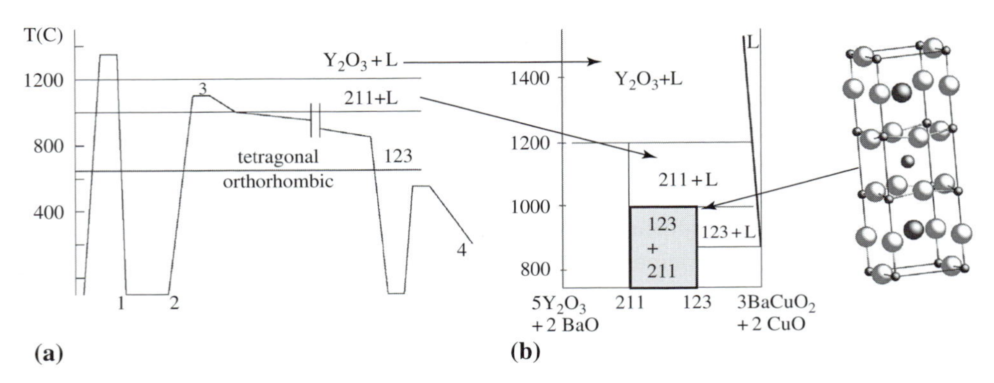

Fig. 23.6. (a) MPMG thermal processing route (Kung, 1993) and (b) pseudo-binary phase diagram along the tie line between the 211 and 123 stoichiometries in the YBCO system.

particles act as *flux pinning sites*. In the YBCO system, high J_c values have been reported in dense, large-grained, textured Y-123, with fine precipitates of the non-superconducting Y_2BaCuO_5 compound, known as the 211 phase. Such two-phase materials are produced by the *melt-powder-melt-growth (MPMG)* method (see Fig. 23.6).

Figure 23.6(a) (Kung, 1993) shows a schematic drawing of the MPMG thermal processing route (Murakami *et al.*, 1989, 1991). The principle behind the MPMG method can be explained with reference to a *pseudo-binary phase diagram*, shown in Fig. 23.6(b). The superconducting YBCO phase can be produced by the following *peritectic reaction*:

$$Y_2BaCuO_5 + L(BaCuO_2, CuO) \rightarrow YBa_2Cu_3O_{7-x}, \qquad (23.1)$$

whereas the 211 phase is formed from the peritectic reaction:

$$Y_2O_3 + L \rightarrow Y_2BaCuO_5. \qquad (23.2)$$

In the peritectic reaction, Y_2O_3 particles act as *nucleation sites* for the 211 phase, allowing for the possibility of producing a fine dispersion of this phase under appropriate processing conditions. Processing in a two-phase, liquid+solid region of the phase diagram can also be used to promote *extraordinary grain growth*. Both the large grains and the dispersed pinning centers increase J_c and, hence, the ability of the superconductor to levitate a magnet; *magnetic levitation* is an important application area for superconductors.

YBCO thin films grow by a *spiral growth* mechanism, illustrated in Box 23.4. Spiral growth occurs by attachment of atoms at the cores of *screw dislocations*. A screw dislocation is a defect in a perfect crystal. The spiral growth results in nanostructures called *ledges* that can have significant implications on the current carrying capacity of the thin films. This is because the screw dislocations are good flux pinning sites.

Box 23.4 Spiral growth nanostructures in YBCO thin films

The figure below shows a *screw dislocation* leaving the surface of a crystalline solid (a). Also shown (b) is the surface of a 450 nm thick YBCO film grown on an MgO single crystal substrate by off-axis radio-frequency (RF) sputter deposition at a rate of 1–2 nm/min (image courtesy of M. Hawley, Los Alamos National Lab). The surface is imaged using *scanning tunneling microscopy* (STM) (Hawley *et al.*, 1991).

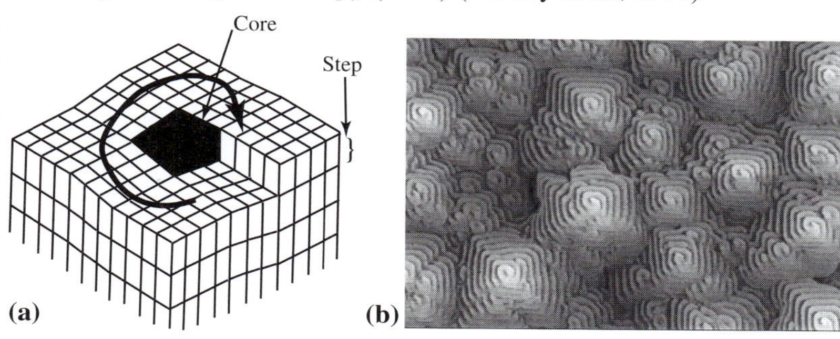

The screw displacement on the crystal surface allows the crystal to grow clockwise by atom attachment to the ledge, resulting in a growth spiral. For each revolution around the core, the crystal grows in height by a single atomic step. A stepped surface is observed, with spirals emanating from a number of randomly positioned mounds corresponding to the exit points of screw dislocations. The spiral growth mechanism has been shown to be important in HTSC thin films. In addition, screw dislocations act as flux pinning sites in HTSC films.

23.4 *BSCCO, TBCCO, HBCCO, and ACBCCO HTSC layered structures

Applications of the HTSC oxide materials have been plagued by a variety of challenging materials problems, many of which are rooted in their anisotropic crystal structures, and in chemistries that include volatile and/or toxic elements. As a result, only a few of these materials have found their way into bulk current carrying applications. Most prominent among the materials that have been pursued for these applications are the Bi(Pb)–Sr–Ca–Cu–O (BSCCO) compounds which have been developed into wires. BSCCO superconductors belong to the class of layered HTSC material that will be discussed in this section.

The BSCCO, Tl–Ba(Sr)–Ca–Cu–O (TBCCO), Hg–Ba–Ca–Cu–O (HBCCO) and Ag–Cu–Ba–Ca–Cu–O (ABCCO) compounds are Ruddleson–Popper phases (Matheis and Snyder, 1990). Each of these compounds consists of a stacking of rocksalt-like blocks, a Ca (O) plane, a CuO_2 plane and

Table 23.3. Structural information for some BSCCO HTSC phases.

Type Symbol	Space group (System)	Z	a b (nm)	c (nm)	O-sites	C-sites
$Bi_2Sr_2CuO_{6+x}$ **Structure 72** BSCCO-2201	**Ama2** (C_{2v}^{16}) Orthorhombic	2	0.5362 0.5374	2.4622	8g, 8l, 8l	8l (Bi), 8l (Sr), 4e (Cu)
$Bi_2Sr_2CaCu_2O_{8+x}$ **Structure 73** BSCCO-2212	**Fmmm** (D_{2h}^{23}) Orthorhombic	2	0.5414 0.5418	3.089	16j, 8i, 8i	8i (Bi),4a (Bi), 8i (Sr),4a (Ca), 8i (Cu)
$Bi_2Sr_2Ca_2Cu_3O_{10+x}$ **Structure 74** BSCCO-2223	**I4/mmm** (D_{4h}^{17}) Tetragonal	2	0.3814	3.700	4c, 8g, 4e, 4e	4e (Bi),4e (Sr) 2b (Cu),4e (Cu) 4e (Ca)
$Bi_2Sr_2Ca_3Cu_4O_{12+x}$ **Structure 75** BSCCO-2234	**I4/mmm** (D_{4h}^{17}) Tetragonal	2	0.3828	4.434	8g, 8g, 4e, 4e	4e (Bi),4e (Ca) 2a (Ca),4e (Sr) 4e (Cu),4e (Cu)

perovskite-like regions. Both one-layer and two-layer conducting blocks are observed in certain of these systems; for instance, the BSCCO compounds have only double layer structures, while to date only single layer systems have been investigated in HBCCO compounds.

23.4.1 The BSCCO double-layer high temperature superconductors

In the BSCCO system, Pb is typically substituted for Bi in small amounts, to promote higher T_c values. BSCCO is used in several applications due to its unique properties. While BSCCO compounds exhibit strongly anisotropic (directional) behavior, processing routes which enable alignment of the favorable directions for current flow have been developed. Table 23.3 summarizes structural information for the BSCCO HTSC phases discussed in this section.

The 2212 and 2223 phases have found the most widespread use because of their high T_c values. The 2223 phase $Bi(Pb)_2Sr_2Ca_2Cu_3O_x$ (BSCCO-2223) has the highest transition temperature, $T_c = 110\,K$. The 2223 phase is also the most difficult to synthesize as a single phase. It was not until 1994 that Pb-free single crystals of the 2223 phase were first grown (Balestrino *et al.*, 1994) in molten KCl. Magnetic properties and flux pinning were measured in collections of oriented Pb-free single crystals (Chu and McHenry, 1998b,a, 2000). More recently, traveling solvent float zone techniques have been successful in growing larger single crystals (for a review, see Giannini *et al.* (2004)). The successful production of larger crystals has been made possible by the Pb-additions.

As stated in the introduction, the BSCCO compounds are Ruddleson–Popper phases. The compounds that have been observed (or suggested) in the

BSCCO system are all two-layer structures with chemical compositions of the type:

$$Bi_2Sr_2Ca_{n-1}Cu_nO_{2(2+n)+x},$$

(where $n = 1, \ldots, 4$) and x measures a small O non-stoichiometry. The superconducting transition temperatures, T_c, for the $n = 1, \ldots, 4$ phases are 5–20, 85, 110, and 90 K, respectively. All of these compounds have tetragonal or pseudo-tetragonal layered structures. These structures have similar lattice parameters along the a-axis (about 0.54 nm), where $a = a_p\sqrt{2}$ and $a_p \sim 0.38$ nm is the lattice parameter of a cubic perovskite.

The 2201 (Torardi *et al.*, 1988) and 2212 (Sunshine *et al.*, 1988) phases are both orthorhombic with only a slight difference in the a and b lattice constants. The 2223 (Subramanian *et al.*, 1988b) and hypothetical 2234 phases (Matheis and Snyder, 1990) have a tetragonal structure. The 2201 phase, $Bi_2Sr_2CuO_{6+x}$, is the lowest T_c phase and has no CaO separating layer. With the exception of the slight orthorhombic distortion this compound is isostructural with the compound $Tl_2Ba_2CuO_{6+x}$. The orthorhombic 2201 phase with space group **Ama2** (C_{2v}^{16}) is illustrated for $Tl_2Ba_2CuO_{6+x}$ in the next section.

Figure 23.7 illustrates the BSCCO-2212 crystal structure in (a) a space filling, (b) a ball-and-stick, and (c) polyhedral representation, with BiO, SrO,

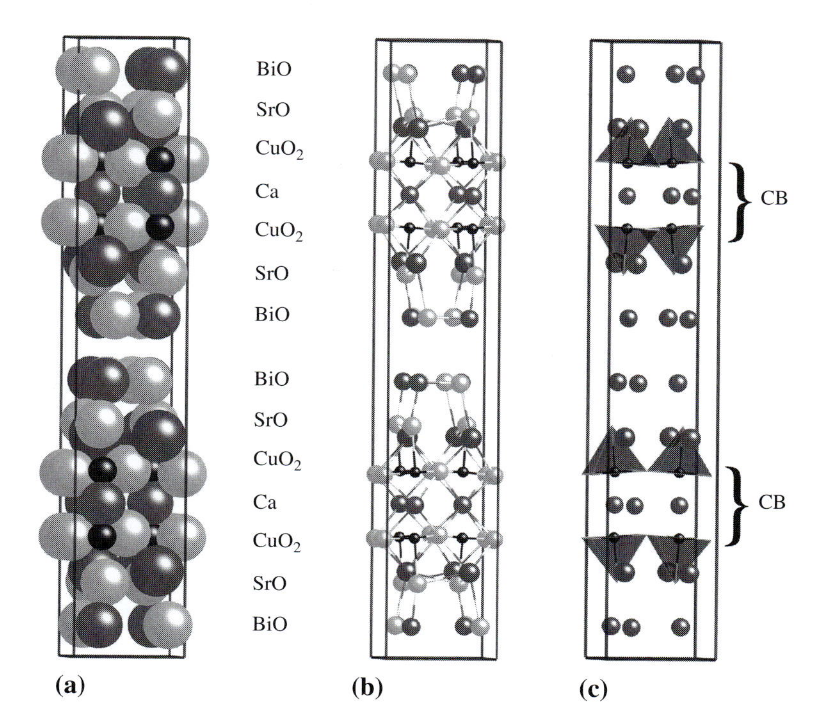

Fig. 23.7. BSCCO-2212 crystal structure (a) space filling, (b) ball-and-stick, with BiO, SrO, CuO_2, and Ca planes labeled, and (c) polyhedral representations.

CuO_2, and Ca planes labeled; CB denotes the conducting block. The stacking sequence is seen to be:

$$Bi_2Sr_2CaCu_2O_{9+x} = \ldots - \overset{\displaystyle][}{\underset{\uparrow}{}} \overbrace{BiO - SrO - CuO_2 - \underset{\uparrow}{Ca} - CuO_2 - SrO - BiO}^{2\times} \overset{\displaystyle][}{\underset{\uparrow}{}} - \ldots$$

where the block between square brackets is repeated twice within a single unit cell, and the vertical arrows indicate the positions of mirror planes normal to the c-axis. The bonding is very weak between the BiO insulating planes resulting in a micaceous structure. The structure is designated as the 2212 phase using the cation ratio nomenclature, as explained in Box 23.5.

The $Bi_2Sr_2Ca_2Cu_3O_{10+x}$, 2223, phase has a tetragonal structure (Calestani *et al.*, 1989). The stacking sequence can be represented as:

$$Bi_2Sr_2Ca_2Cu_3O_{10+x} =$$

$$\ldots - \overset{\displaystyle][}{\underset{\uparrow}{}} \overbrace{BiO - SrO - CuO_2 - Ca - \underset{\uparrow}{CuO_2} - Ca - CuO_2 - SrO - BiO}^{2\times} \overset{\displaystyle][}{\underset{\uparrow}{}} - \ldots$$

In the four-number scheme, the 2223 phase differs from the 2212 phase in that there are *two* Ca *separating layers* between three CuO_2 layers in the *conducting block*. $Bi_2Sr_2Ca_2Cu_3O_{10+x}$ is isostructural with the $Tl_2Ba_2Ca_2Cu_3O_{10+x}$ phase, which will be illustrated in the next section.

Box 23.5 Four-number scheme analysis of the $Bi_2Sr_2CaCu_2O_{9+x}$ structure

The four-number notation, 2212, can be obtained as follows:

(i) The first number is 2 because there are two BiO *insulating layers* between each conducting block.
(ii) The second number is 2 because there are two SrO *spacing layers* between the CuO_2 blocks, one on either side of the BiO insulating layers.
(iii) The third number is 1 for a single CaO *separating layer* between CuO_2 layers.
(iv) The fourth number is 2 for two CuO_2 planes within a conducting block.

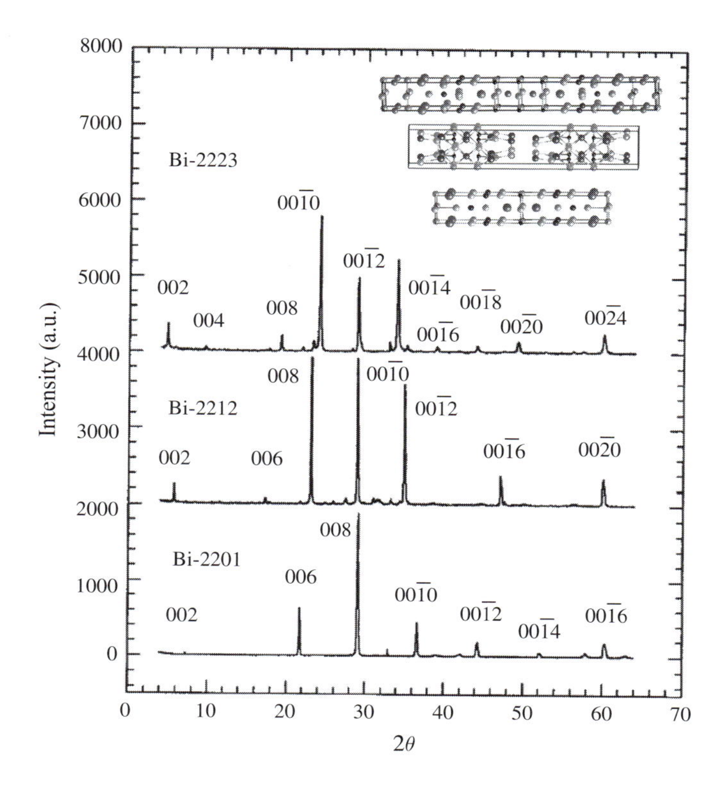

Fig. 23.8. XRD patterns of c-axis oriented BSCCO-2223, 2212, and 2201 single crystals taken using Cu K_α radiation (Chu and McHenry, 1998a) (figure courtesy of S. Chu).

Figure 23.8 shows XRD patterns of a c-axis oriented BSCCO-2223 single crystal, compared with the same for Bi-2212, and Bi-2201 single crystals taken with Cu K_α radiation (Chu and McHenry, 1998a). These are different from powder patterns because of the oriented crystals and, therefore, they do not exhibit all reflections. These patterns consist of only the superlattice reflections characteristic of the respective c-axis modulations of the three phases. These are $(00\,2l)$ superlattice reflections, which are consistent with the different long c-axis lattice constants (and mirror symmetries). It is left as a reader exercise to verify, using the space group tables, that these are indeed the only allowed reflections of the type $(00l)$.

23.4.2 The TBCCO double-layer high temperature superconductors

Unlike the BSCCO superconductors, Tl–Ba–Ca–Cu–O compounds form both double and single layer structures. In this section, we illustrate some of the double layer structures. In the next section, we will discuss the single layer structures. Table 23.4 summarizes structural information for the double layer Tl–Ba–Ca–Cu–O HTSC phases; these are all tetragonal structures with space group **I4/mmm** (D_{4h}^{17}).

Table 23.4. Structural information for the double-layer TBCCO HTSC phases.

Type Symbol	Space group (System)	Z	a b (nm)	c (nm)	O-sites	C-sites
$Tl_2Ba_2CuO_{6+x}$ **Structure 76** TBCCO-2201	**I4/mmm** (D_{4h}^{17}) Tetragonal	2	0.387	2.324	4e, 4e 16n	4e (Tl), 4e (Ba) 2b (Cu)
$Tl_2Ba_2CaCu_2O_{8+x}$ **Structure 77** TBCCO-2212	**I4/mmm** (D_{4h}^{17}) Tetragonal	2	0.386	2.932	8g, 4e 16n	4e (Tl), 4e (Ba) 4e (Cu), 4e (Ca)
$Tl_2Ba_2Ca_2Cu_3O_{10+x}$ **Structure 78** TBCCO-2223	**I4/mmm** (D_{4h}^{17}) Tetragonal	2	0.385	3.588	4c, 8g 4e, 16n	4e (Tl), 4e (Ba) 2b (Cu), 4e (Cu) 4e (Ca)
$Tl_2Ba_2Ca_3Cu_4O_{12+x}$ **Structure 79** TBCCO-2234	**I4/mmm** (D_{4h}^{17}) Tetragonal	2	0.385	4.226	8g, 8g 4e, 4e	4e (Tl), 4e (Ba) 4a (Cu), 4e (Ca) 4e (Cu), 4e (Cu)

The two-layer Ruddleson–Popper phases observed (or predicted) in the Tl(Pb)–Sr–Ca–Cu–O system, have the chemical composition:

$$Tl_2Ba_2Ca_{n-1}Cu_nO_{2(2+n)+x},$$

with $n = 1, \ldots, 4$ and x again reflects a small O non-stoichiometry. Important superconducting phases in the TBCCO system include $Tl_2Ba_2CuO_{6+x}$ (2201) (Subramanian *et al.*, 1988b), $Tl_2Ba_2CaCu_2O_{8+x}$ (2212) (Subramanian *et al.*, 1988a), $Tl_2Ba_2Ca_2Cu_3O_{10+x}$ (2223) (Torardi *et al.*, 1988) and $Tl_2Ba_2Ca_3Cu_4O_{12+x}$ (2234), with T_c values of 90, 110, 125, and 119 K, respectively. These TBCCO phases are isostructural with similar BSCCO phases, with the exception of slight orthorhombic distortions in the first two ($n = 1, 2$) of the BSCCO phases.

Figure 23.9 illustrates the TBCCO-2201 crystal structure in (a) space filling, (b) ball-and-stick, with TlO, BaO, and CuO planes labeled, and (c) polyhedral representations. In the four-number scheme, the 2201 phase differs from the 2212 phase in that there are no CaO *separating layers* and only one CuO_2 layer in the *conducting block*, thus the designation 2201. $Bi_2Sr_2CuO_{6+x}$, is isostructural with the $Tl_2Ba_2CuO_{6+x}$ phase, except for a small orthorhombic distortion.

$Tl_2Ba_2Ca_2Cu_3O_{10+x}$ is also a double layer phase. Figure 23.10 illustrates the TBCCO-2223 crystal structure in (a) space filling, (b) ball-and-stick, with TlO double layers, and BaO, CuO and SrO planes labeled, and (c) polyhedral representations. In the four-number scheme, the 2223 phase differs from the 2212 phase in that there are two CaO *separating layers* between three CuO_2 layer in the *conducting block*.

The crystal structures of TBCCO and BSCCO are closely related. In the Tl-based compounds, a and b lattice constants are both equal to 0.385 nm,

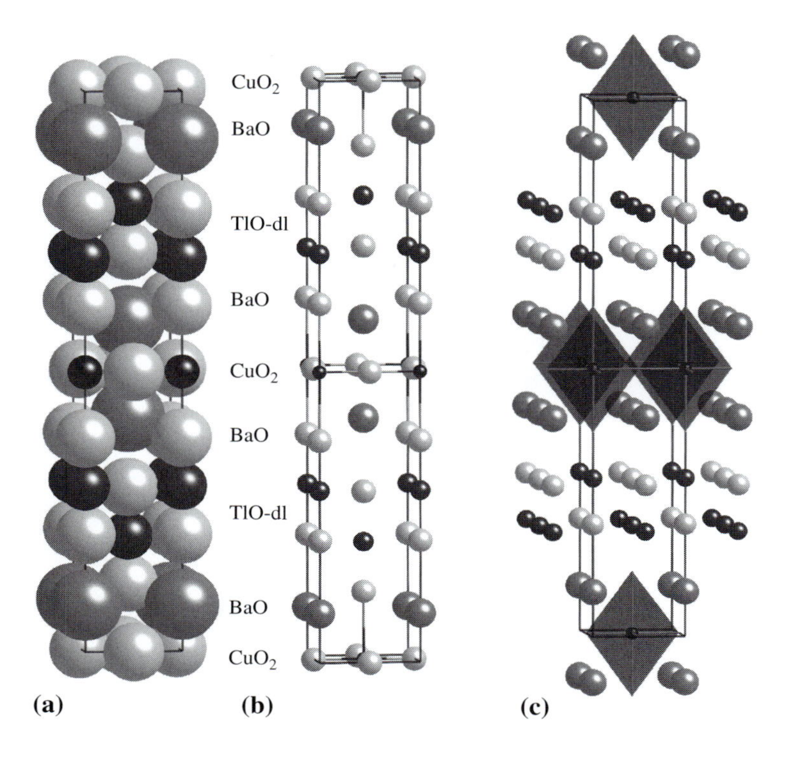

Fig. 23.9. TBCCO-2201 crystal structure in (a) a space filling, (b) ball-and-stick, and (c) polyhedral representations.

and c = 2.324, 2.932, and 3.588 nm for 2201, 2212, and 2223, respectively (Nabatame *et al.*, 1990). The space group for these three compounds is **I4/mmm** (D_{4h}^{17}). Because the CuO_2 sheets are less puckered in the Tl-compound than in the analogous Bi-compound, it is easier to synthesize Tl-2223 than (Bi,Pb)-2223 (Subramanian *et al.*, 1988b). Transmission electron microscopy observations show that the Tl-phases lack the obvious cleavage which is common in the layered Bi-phases. The TlO layers in the Tl-system, and the BiO layers in the Bi-system provide conducting holes to the CuO_2 sheets, similar to the CuO chains in the $YBa_2Cu_3O_{6+x}$ compound (Hybertsen and Mattheiss, 1988).

Figure 23.11 compares the (a) $Tl_2Ba_2CuO_{6+x}$ (2201), (b) $Tl_2Ba_2CaCu_2O_{8+x}$ (2212), and (c) $Tl_2Ba_2Ca_2Cu_3O_{10+x}$ (2223) crystal structures. Except for the 2201 phase, the CuO_2 planes and accompanying Ca planes are located at the center of the cell ($z = \frac{1}{2}$) followed to the right and left by BaO layers, TlO double layers (dl), etc.

23.4.3 The TBCCO single-layer high temperature superconductors

In addition to compounds with TlO double-layers, superconductors with a single TlO layer have also been synthesized. Table 23.5 summarizes the structural information for the single layer Tl–Ba–Ca–Cu–O phases discussed

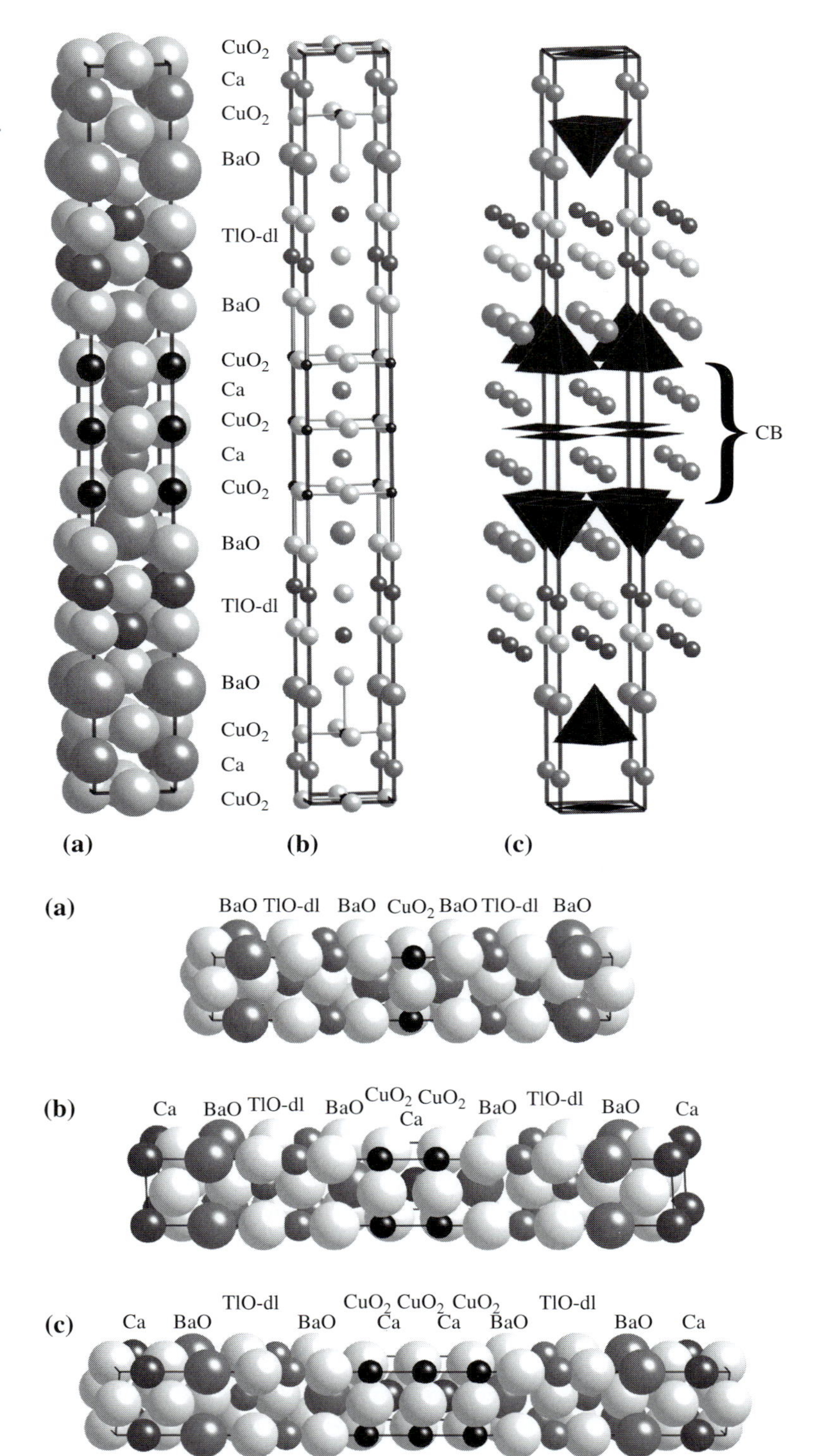

Fig. 23.10. TBCCO-2223 structure in (a) space filling, (b) ball-and-stick, and (c) polyhedral representations.

Fig. 23.11. Space filling models of (a) $Tl_2Ba_2CuO_{6+x}$ (2201), (b) $Tl_2Ba_2CaCu_2O_{8+x}$ (2212), and (c) $Tl_2Ba_2Ca_2Cu_3O_{10+x}$ (2223) crystal structures.

Table 23.5. Structural information for some single-layer TBCCO HTSC phases.

Type Symbol	Space group (System)	Z	a b (nm)	c (nm)	O-sites	C-sites
$TlBa_2CuO_5$ **Structure 80** TBCCO-1201	**P4/mmm** (D^1_{4h}) Tetragonal	1	0.385	0.954	1c, 2g 2e	1a (Tl), 2h (Ba) 1b (Cu)
$TlBa_2CaCu_2O_7$ **Structure 81** TBCCO-1212	**P4/mmm** (D^1_{4h}) Tetragonal	1	0.387	1.274	2g, 4i 1d	1b (Tl), 4m (Tl) 1c (Tl), 1c (Ca), 2h (Ba), 2h (Ba)
$TlBa_2Ca_2Cu_3O_9$ **Structure 82** TBCCO-1223	**P4/mmm** (D^1_{4h}) Tetragonal	1	0.384	1.587	2e, 4i 2g, 1d	1a (Tl), 2h (Ba) 2h (Ca), 1b (Cu) 2g (Cu)
$TlBa_2Ca_3Cu_4O_{11}$ **Structure 83** TBCCO-1234	**P4/mmm** (D^1_{4h}) Tetragonal	1	0.385	1.91	1c, 2g 4i, 4i	1a (Tl), 2h (Ba) 1d (Ca), 2h (Ca), 2g (Cu), 2g (Cu)

in this section; all these compounds are tetragonal with space group **P4/mmm** (D^1_{4h}).

The single-layer Ruddleson–Popper phases that have been observed (or suggested) in the Tl(Pb)–Sr–Ca–Cu–O system have the chemical composition:

$$TlBa_2Ca_{n-1}Cu_nO_{2n+3},$$

with $n = 1, \ldots, 4$. These phases are then $TlBa_2CuO_5$ (Matheis and Snyder, 1990), $TlBa_2CaCu_2O_7$ (Morosin *et al.*, 1988), $TlBa_2Ca_2Cu_3O_9$ (Torardi *et al.*, 1988), and $TlBa_2Ca_3Cu_4O_{11}$ (Ihara *et al.*, 1988). The general formula for this TlO single layered system was proposed independently by Gao *et al.* (1988), and Beyers *et al.* (1988). The value of T_c increases with the number of CuO layers up to $T_c = 122$ K at $n = 4$. By extending sintering times to evaporate TlO, the double-layer 2223 compound can be transformed to the single-layer compound 1223, and subsequently to 1234 (Sugise *et al.*, 1988).

Surprisingly, single layer structures analogous to the single TlO layer-like structure in Tl-1223 (or Tl-1212), have not been observed in the Bi-based cuprates. The difference in the crystal structure between 2223 and 1223 phases involves replacement of the BaO–TlO–TlO–BaO slab in the 2223 phase by the slab BaO–TlO–BaO in the 1223 phase (Parkin *et al.*, 1988). This decreases the distance between the triple CuO_2 layers and leads to a weaker magnetic field dependence (with the field parallel to the c-axis) of J_c in Tl-1223 due to a stronger coupling between the conducting blocks.

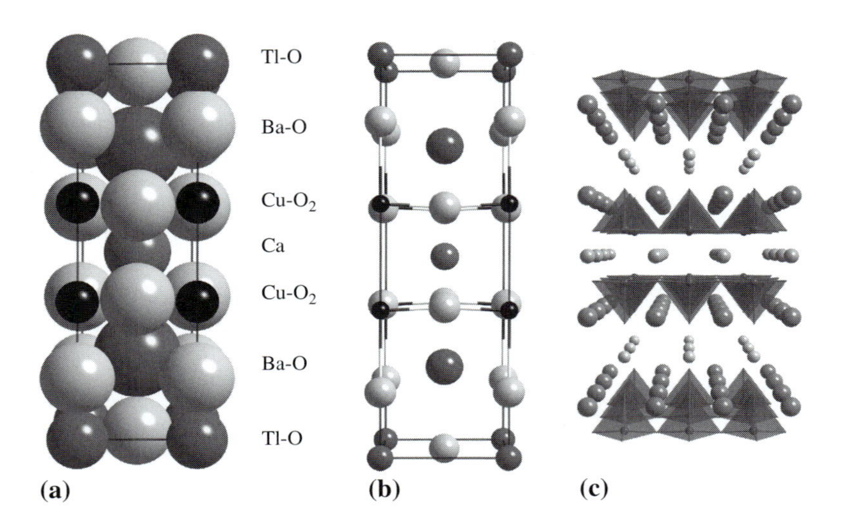

Fig. 23.12. The $TlBa_2CaCu_2O_{7-x}$, 1212 phase crystal structure in (a) space filling, (b) ball-and-stick, and (c) polyhedral representations.

For double TlO layered systems (Tl-2212, Tl-2223, etc.), the unit cell contains two perovskite units that are shifted by the lattice vector [110]/2. If we replace the block of triple CuO_2 layers in Tl-1222 by a block of double CuO_2 layers, we obtain $TlBa_2CaCu_2O_7$ (1212). Fig. 23.12 (Labbe *et al.*, 1995) illustrates the $TlBa_2CaCu_2O_7$ structure in (a) space filling, (b) ball-and-stick, and (c) polyhedral representations. In the four-number scheme, the 1212 phase differs from the 2212 phase in that there is only one TlO *insulating layer* between each CuO_2 conducting block. The chemistry of this phase is more complicated if Sr substitutes for Ba or Tl is substituted for Ca. A more general formula is $Tl_{1+x}(Ba,Sr)Ca_{1-x}Cu_2O_{7-x}$.

23.4.4 The HBCCO high temperature superconductors

We do not tabulate structural data for the Hg–Ba–Ca–Cu–O HTSCs as they are isostructural with single layer TBCCO systems. Superconducting phases in the HBCCO system are multilayer compounds in a homologous series with composition:

$$HgBa_2Ca_{n-1}Cu_nO_{2n+2+x}.$$

The $n = 1$ compound, $HgBa_2CuO_{4+x}$ (Putilin *et al.*, 1993a), known as HBCCO-1201, was observed to have a T_c of about 95 K. This discovery was followed by the observation of superconductivity at temperatures $\geq$ 130 K in HBCCO samples containing a mixture of the $n = 2$ $HgBa_2CaCu_2O_{6+x}$ (1212) and $n = 3$ $HgBa_2Ca_2Cu_3O_{8+x}$ (1223) phases (Schilling *et al.*, 1993). T_c values near 140 K were observed in a multi-phase HBCCO sample (Gao *et al.*, 1993). High pressure synthesis resulted in a nearly single phase 1212 compound with a T_c of about 120 K (Putilin *et al.*, 1993b). A solid state

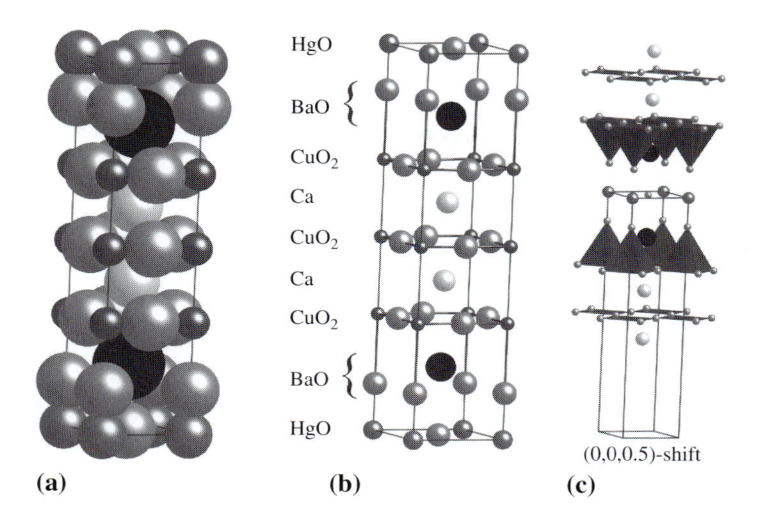

Fig. 23.13. The $HgBa_2Ca_2Cu_3O_8$ (1223) structure: (a) space filling, (b) ball-and-stick, and (c) coordination polyhedra in a cell shifted by half the c lattice constant.

synthesis route (Welp *et al.*, 1993) yielded a 1212 phase with a T_c of 128 K (Radaelli *et al.*, 1993).

Neutron diffraction studies of the $HgBa_2Ca_2Cu_3O_{8+x}$ superconductor with Tl substitutions (Dai *et al.*, 1995) revealed its structure to be tetragonal (space group **P4/mmm** (D^1_{4h})) with lattice parameters $a = 0.385$ nm and $c = 1.5816$ nm at room temperature. This 1223 phase is isostructural with the TBCCO-1223 phase. This compound has the highest known T_c of 138 K. Figure 23.13 shows the structure and stacking sequence in space filling (a) and ball-and-stick (b) representations. Figure 23.13 (c) shows a cell shifted by half the c-axis lattice constant, indicating half octahedral and square planar coordination on two different Cu sites.

In the Hg system, T_c values in the $n = 1, 2$, and 3 compounds exceed those of similar 1, 2, and 3 CuO_2 layer compounds in Bi- and Tl-based systems. Excess O is the primary doping mechanism in $HgBa_2CuO_{4+x}$ (Wagner *et al.*, 1993) with a more complicated mechanism apparent in $HgBa_2CaCu_2O_{6+x}$ (Radaelli *et al.*, 1993). Charge transfer from Hg-related bands to CuO layers is a possible self-doping mechanism (Radaelli *et al.*, 1993). The 1212 compound has been identified to have the **P4/mmm** (D^1_{4h}) space group with lattice parameters $a = 0.385$ nm and $c = 1.266$ nm), isostructural with TBCCO-1212 (Fig. 23.12).

23.4.5 The ACBCCO high temperature superconductors

T_c values exceeding 117 K have been observed in a homologous series of compounds:

$$Ag_{1-x}Cu_xBa_2Ca_{n-1}Cu_nO_{2n+3-x}.$$

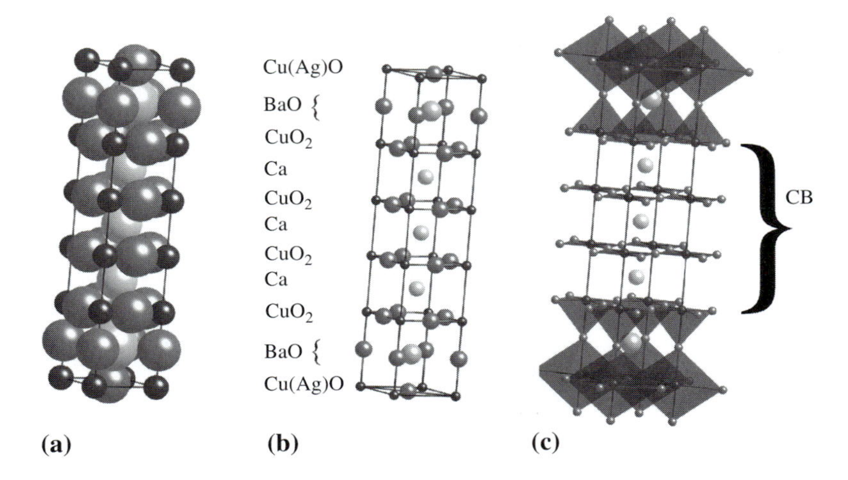

Fig. 23.14. The $AgBa_2Ca_3Cu_4O_{10}$ (1234) structure in a (a) space filling, and (b) ball-and-stick representation; (c) displays coordination polyhdedra in $\frac{1}{8}$ of the ordered bct cell.

This family of compounds, in particular a sample with nominal composition $(Ag_{0.25}\,Cu_{0.75})Ba_2Ca_3Cu_4O_{11-x}$ ($n = 4$), was produced by a high pressure solid state synthesis technique (Ihara *et al.*, 1994). The HTSC compounds in the $Ag_{1-x}Cu_xBa_2Ca_{n-1}Cu_n\,O_{2n+3-x}$ family are of particular interest because they do not contain the toxic components (Bi, Tl, Pb, Hg) found in other HTSC materials with $T_c \geq 100$ K. The $n = 4$ structure is illustrated in Fig. 23.14.

The structure of $AgBa_2Ca_3Cu_4O_{10}$ (Ag-1234, **Structure 84**) has been identified as body-centered tetragonal with a probable **I4/mmm** (D_{4h}^{17}) space group. Its primitive unit cell is similar to that of Tl-1234 but with Ag_2O layers. If Ag randomly occupies Cu sites, then the crystal structure becomes primitive tetragonal with space group **P4/mmm** (D_{4h}^{1}) ($a = 0.38635$ nm and $c = 1.8111$ nm). A proposed primitive unit cell of the disordered structure is illustrated in Fig. 23.14. In the ordered structure, there are eight of these cells centered by an Ag ion in a bct cell. The interested reader is referred to the original literature (Ihara *et al.*, 1994) for further discussion. These materials have a lower anisotropy than Bi-, Hg- and Tl-based compounds, as is evident from a short, 0.85 nm CuO_2 interlayer spacing.

23.4.6 Rutheno-cuprate high temperature superconductors

The low temperature ruthenate superconductor Sr_2RuO_4 was studied to explore superconductivity in layered perovskites without Cu (Maeno *et al.*, 1994). Subsequently, interesting *rutheno-cuprate superconductors* were synthesized with the coexistence of superconductivity and ferromagnetism (Fig. 23.15). These properties were thought to be mutually exclusive. The rutheno-cuprates are isostructural with layered cuprate superconductors

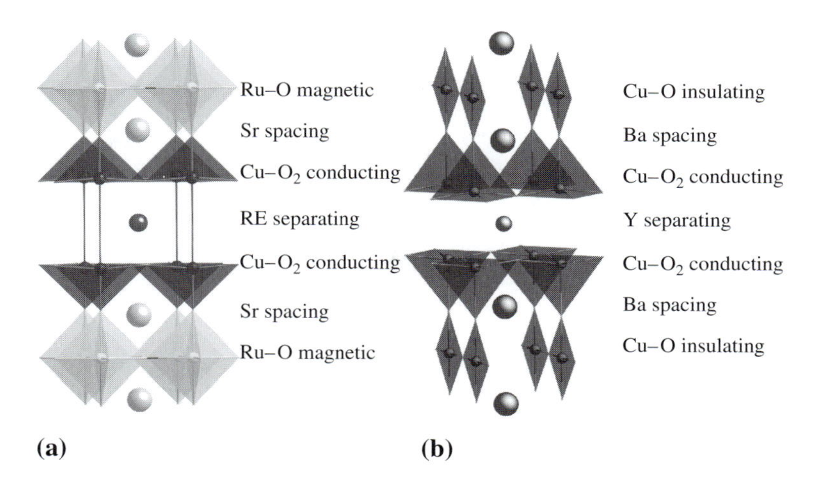

Fig. 23.15. Comparison of the structures of the (a) $RuSr_2GdCu_2O_8$ and (b) $YBa_2Cu_3O_7$.

but with the substitution of RuO_2 magnetic planes for the insulating planes.

The $RuSr_2GdCu_2O_8$ superconductor (Bernhard *et al.*, 1999) is a triple perovskite 1212 phase, isostructural with $YBa_2Cu_3O_7$. This superconductor exhibits a T_c of about 30 K and a ferromagnetic Curie temperature of 133 K. Figure 23.15 compares the $RuSr_2GdCu_2O_8$ and $YBa_2Cu_3O_7$ structures. It can be seen that the Gd rare earth ion occupies the Y site (Gd, Sm, Sm, Eu, and Pr have all been substituted on this site (Awana *et al.*, 2001)) and Sr occupies the Ba site of the Y-123 superconductor. Ru–O octahedra replace the Cu–O chains, with the higher coordination of Ru reflecting the Ru^{4+} charge state.

23.4.7 Infinite-layer high temperature superconductors

Several HTSC structures have *no insulating layer* between each CuO_2 conducting block. These include the previously illustrated 0201 (T, T′, and T*) phases as well as the 0212, 0223, and 0234 structures. The four-number scheme for these structures is of the form $02\,(n-1)\,n$ (with $n = 1, 2, 3, 4, \ldots$). In the limit of $n \to \infty$ a structure called the *infinite layer structure* results. This structure is that of the superconductor $(Ba,Sr)CuO_2$ (Takano *et al.*, 1991) (**Structure 85**), illustrated in Fig. 23.16. It is described in the four-number scheme as the $02\,(\infty - 1)\,\infty$ phase. This structure is also called the *checkerboard structure*, because of the square planar Cu coordination.

The infinite layer phase can be either electron doped or hole doped. $(La,Sr)CuO_2$ is an example of an electron doped infinite layer structure with a T_c of 42 K, whereas $(Ca,Sr)CuO_2$ is a hole doped infinite layer structure with a T_c of 110 K.

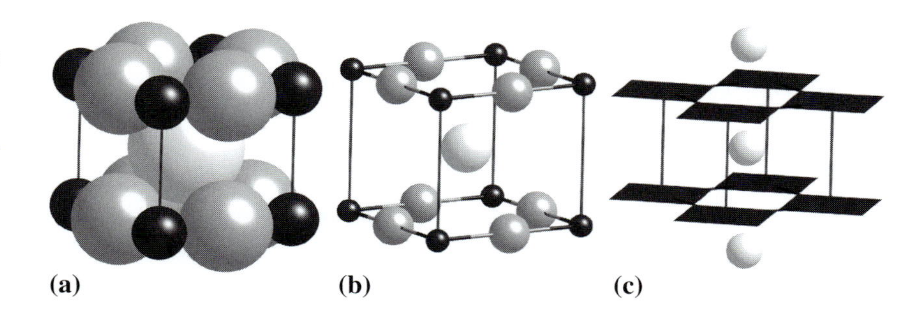

Fig. 23.16. The infinite layer superconductor $(Ba,Sr)CuO_2$: (a) space filling, (b) ball-and-stick and (c) polyhedral representation.

23.5 ∗Structure–properties relationships in HTSC superconductors

23.5.1 Type I and Type II superconductors

Superconductors are characterized by zero electrical resistance below the transition temperature T_c. In addition to zero resistance, a second experimental manifestation of superconductivity is the *Meissner effect*, the exclusion of magnetic flux from a superconductor (Meissner and Ochsenfeld, 1933). The Meissner effect identifies the superconducting state as a true thermodynamic state. For a *Type I superconductor*, this flux exclusion is complete up to a *thermodynamic critical field,* H_c (an externally applied magnetic field), above which the superconductive state is destroyed. Below H_c, flux exclusion is perfect, so that: $B = H + 4\pi M = 0$ inside the superconductor. The Meissner effect is illustrated in Fig. 23.17(a) where a typical magnetization curve for a Type I superconductor (top left) satisfies perfect flux exclusion up to H_c.

In a Type I superconductor, the entire material is either in the superconducting state, or in the normal state. Another type of superconductor exists, for which both superconducting and normal states can coexist. This *Type II superconductor* has two temperature dependent critical fields, the *lower critical field*, $H_{c1}(T)$, and the *upper critical field*, $H_{c2}(T)$, as illustrated in Fig. 23.17(b). The persistence of superconductivity above $H_{c1}(T)$ and below $H_{c2}(T)$ has been explained in terms of a mixed state in which the superconductor coexists with quantized units of magnetic flux called *Abrikosov vortices* or *fluxons* (which can be thought of as tubes of magnetic flux, Fig. 23.18).

The *field–temperature phase diagram* (Fig. 23.17(b), bottom) for a Type II superconductor exhibits a Meissner phase for $H < H_{c1}(T)$, a mixed state, $H_{c1}(T) < H < H_{c2}(T)$, in which the superconducting and normal states coexist, and a non-superconducting normal state for $H > H_{c2}(T)$. Type II superconductors are technically important because of their high current-carrying capacity. The previously discussed A15 superconductors (Chapter 18) and all high temperature superconductors are Type II. Type II superconductors were recognized as a separate class of superconductors with the work of Alexei A. Abrikosov (1927–) (Abrikosov, 1957).

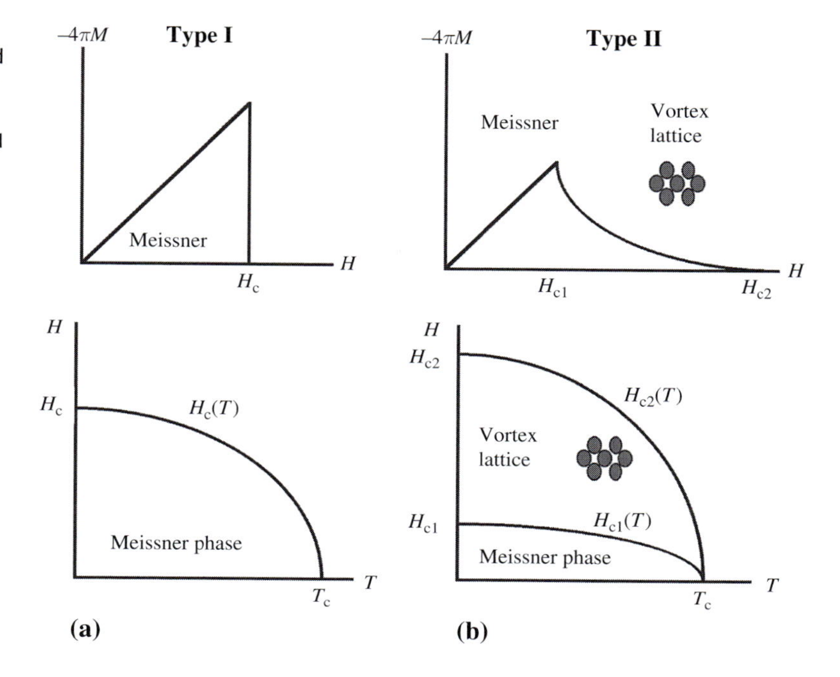

Fig. 23.17. Isothermal magnetization curves (top) and T-dependence of critical fields (bottom) for a (a) Type I superconductor and (b) Type II superconductor.

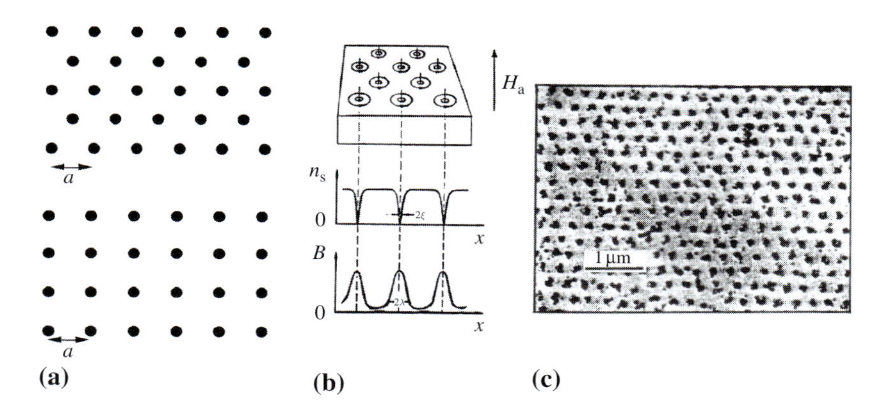

Fig. 23.18. (a) Triangular and square Abrikosov vortex lattices; (b) a periodic arrangement of vortices in a Type II superconductor, the spatial variation in the supercurrent electron pair density and the magnetic flux density near the vortices, and (c) an example of a vortex lattice in an A15 superconductor that has been decorated with magnetic nanoparticles.

23.5.2 The flux lattice and flux pinning in Type II superconductors

The coexistence of superconducting and normal states (the *mixed state*) is accomplished by (quantized) magnetic flux lines entering the superconductor for $H > H_{c1}(T)$. Abrikosov showed that the lowest energy configurations of magnetic flux vortices in the mixed state is a periodic 2-D lattice. At first, Abrikosov suggested a square vortex lattice but subsequent calculations showed that, for an isotropic superconductor, the triangular lattice is the lower energy configuration. Figure 23.18(a) depicts both triangular and square

Abrikosov vortex lattices. Abrikosov won the 2003 Nobel Prize in Physics for his contributions to the field of superconductivity.

Figure 23.18(b) shows a periodic arrangement of vortices in a Type II superconductor (top), the spatial variation in the supercurrent electron pair density (center) and the magnetic flux density near the vortices (bottom). It can be seen that flux is concentrated in the cores of the vortices. Figure 23.18(c) shows an example of a vortex lattice in an **A15** superconductor that has been decorated with magnetic nanoparticles. Flux lattice decoration offers experimental proof of Abrikosov's theory. Magnetic nanoparticles are attracted to the flux concentrated in the cores of vortices in the same way magnetic filings will follow magnetic field lines in the gap between the poles of a horseshoe magnet. Because the vortex cores are magnetic, *neutron diffraction* can be a powerful tool for determining the periodicity of the flux lattice.

In practical superconductors, it is important to pin the magnetic flux lines. *Flux pinning* prevents energy dissipation by vortex motion resulting from a Lorentz force interaction with the supercurrent density. Pinning is one of the primary determinants of the *critical current density*, J_c, that destroys the perfect conductivity of the superconductor. In practice, J_c represents the current density that causes magnetic flux lines to overcome pinning. There has been a significant amount of research concerning the nature of the pinning sites. Lattice dislocations can provide potent pinning sites in metallic superconductors (Kramer and Bauer, 1967). 211-phase precipitates are effective pinning sites in YBCO-type superconductors (Kung, 1993). Often, pinning sites can be introduced by neutron irradiation (Lessure *et al.*, 1991, Sutton, 1994, Sutton *et al.*, 1997). Differences in the dissipative magnetic response of HTSCs with and without strong pinning sites were also studied (Silva, 1994). For a review of pinning and dissipation in superconductors see (McHenry and Sutton, 1994).

Research after the discovery of HTSCs made clear the crucial role played by the crystalline anisotropy in determining the properties of HTSCs. The limited attainable J_c has been of concern for HTSC applications. Thermally activated dissipation, called *flux creep*, was identified as an important limiting factor for J_c. The time dependent decay of the magnetization is used as a method for studying the distribution of pinning energies in HTSCs (Maley *et al.*, 1990). The problems with the large anisotropies in the layered superconductors lead to new physical models of the $H-T$ phase diagram in HTSCs (Nelson, 1988). This includes the notions of *vortex liquid* and *vortex glass* phases (Fisher, 1989).

Proposed $H-T$ phase diagrams with such features are described in Box 23.6. Due to anisotropy, the nature of pinning in HTSC materials is different from that in conventional superconductors. The underlying $H-T$ phase diagram and the density and strength of flux pinning sites determine the magnitude, field, and temperature dependence of the technologically

Box 23.6 Magnetic phase diagram for high temperature superconductors

Examples of proposed $H-T$ phase diagrams for HTSC materials are illustrated below.

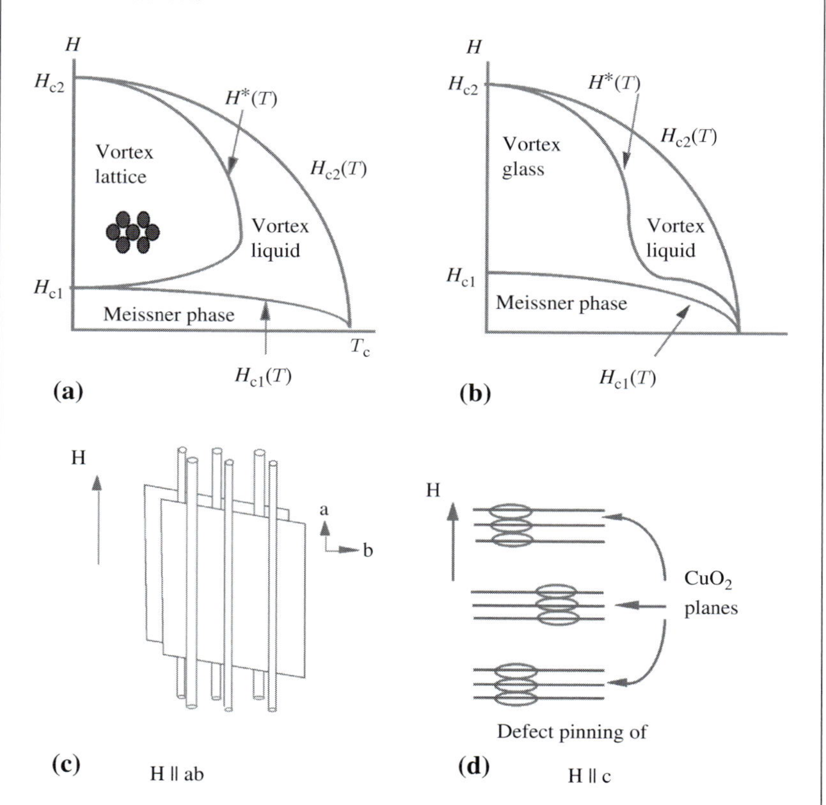

Vortex lattices exhibit a melting transition for an anisotropic material with weak random pinning (a) and a material with strong random pinning (b). The *vortex melting line* of (a) and (b) is intimately related to the state of disorder as provided by flux pinning sites. The bottom row shows schematics of intrinsic pinning (c) and pinning of pancake vortices (d) in an anisotropic superconductor.

important J_c. In anisotropic materials, flux pinning and J_c values are different for fields aligned parallel and perpendicular to a crystal's c-axis. The pinning of vortex pancakes is different than for Abrikosov vortices. The concept of *intrinsic pinning* has been used to describe low energy positions for fluxons in regions between Cu–O planes (for field parallel to the (001) plane in anisotropic materials). Models based on weakly coupled pancake vortices have been proposed for fields parallel to the c-axis for layered oxide materials.

23.6 Historical notes

Bernard Raveau (1940–) received a degree in science in 1961 and a Ph.D. in Physics in 1966 from Caen University. He became Professor in 1970, Full Professor in 1978 and Exceptional Class Professor in 1990 at the University of Caen. From 1986–2004 he served as Director of the Laboratoire de Cristallographie et Sciences des Matériaux (CRISMAT) Laboratory and Research Centre for Superconductivity at Caen. He has been Director of the National Research Centre of Technology, CNRT "Materiaux" since 2001. His many research awards include most recently the Bernd Matthias Prize (1994), and the Chevalier de l'Ordre National de la Légion d'Honneur (2001). In 2002, he was made a member of the French Academy of Sciences.

Raveau's research has concentrated on structure and physical properties relationships in oxides. He has performed work on non-stoichiometric oxides. He has studied zeolites and other structures for applications in cation exchange and ionic conduction. These include the newly discovered *fullerenoid oxides* with cage structures, discussed in Chapter 24. He contributed to the study of transition metal phosphates with large cationic charge states and mixed valence systems for applications in catalysis. He has studied anisotropic conductivity in the tungsten bronzes, and radiation damage in oxides by heavy ion bombardment, including the barium hexaferrites and garnets and high

Fig. 23.19. (a) Bernard Raveau (1940–) (picture courtesy of Bernard Raveau) and (b) K. Alex Müller (1928–) (picture courtesy of the Nobel e-Museum)

temperature superconductors. He has also contributed to the field of conventional and microwave sintering of ceramics.

Raveau's most influential work has been in mixed-valent and oxygen deficient perovskite materials for applications as sensors, high temperature superconductors, and colossal magnetoresistance (CMR). Michel, Er-Rahko, and Raveau first synthesized lanthanum barium copper oxide for applications as oxygen sensors (Michel and Raveau, 1982). In 1986, Müller and Bednorz heard of the synthesis of $La_{2-x}Ba_xCuO_4$ by Raveau's group and recognized it to be a potential oxide superconductor. Raveau has continued his work on perovskites in the study of the manganates, an important class of oxides which exhibit colossal magnetoresistance (CMR).

K. Alex Müller (1927–) studied in the Physics and Mathematics Department of the Swiss Federal Institute of Technology (ETH) in Zürich. He studied nuclear physics with Paul Scherrer and took courses from Wolfgang Pauli. His diploma research under Professor G. Busch was on the Hall effect in gray tin. After receiving his diploma in 1955, he worked for one year in the Department of Industrial Research (AFIF) of the ETH and then rejoined Busch's group to begin a thesis on electron paramagnetic resonance lines (EPR) due to Fe^{3+} in $SrTiO_3$ materials.

After receiving his Ph.D. in 1958 from ETH, Müller joined the Battelle Memorial Institute (Geneva). He became lecturer and Professor in 1990 at the University of Zürich. He joined the IBM Zürich Research Laboratory, Ruschlikon, in 1963 where, except for a two year stint at IBM's Watson Research Center in Yorktown Heights, N.Y., has been ever since. He managed the Physics group at Ruschlikon from 1972–85 and was involved in the hiring of **Gerd Binnig** who started the Scanning Tunneling Microscope (STM) project and won the Nobel Prize in Physics in 1986.

J. Georg Bednorz (1950–) performed a Ph.D. thesis on the crystal growth of perovskite-type solid solution, at the Laboratory of Solid State Physics at ETH Zürich under the supervision of Professors Heini Gränicher and Alex Müller. After joining IBM in 1982, he was made an IBM fellow and in 1985 he stepped down as manager to devote more time to his personal research. In 1983, Bednorz began a collaboration with Müller to search for high-T_c superconducting oxides culminating in the discovery in 1986 of HTSC materials with transition temperatures of 35 K in $La_{2-x}Ba_xCuO_4$ (Bednorz and Müller, 1986). Bednorz and Müller won the Nobel prize in Physics in 1987 "*for their important break-through in the discovery of superconductivity in ceramic materials.*"[3]

[3] A complete biography can be found at /http://nobelprize.org/nobel (Nobel Lectures, Physics 1981–90, Tore Frängsmyr, and Gösta Ekspång, editors, World Scientific Publishing Co., Singapore, 1993)

23.7 Problems

(i) *Perovskite unit cell I*: Draw the $BaTiO_3$ perovskite crystal structure. Consider this structure to answer the following questions:

(a) Given the ionic radii $R_{Ba^{2+}} = 0.136$ nm, $R_{Ti^{4+}} = 0.060$ nm, and $R_{O^{2-}} = 0.14$ nm, calculate the lattice constant for cubic $BaTiO_3$.
(b) The superconductor $Ba_{1-x}K_xBiO_3$ has the perovskite crystal structure; what conclusions can be made about the Bi valence in this structure if K is monovalent (i.e., K^+)?
(c) How does this structure differ from the other high T_c superconductors?

(ii) *Perovskite unit cell II*: Calculate the structure factor for the prototype $BaTiO_3$ perovskite using the *A*-setting. Does it agree with that calculated using the *B*-setting?
(iii) *Perovskite structure factor*: Determine the structure factor F_{hkl} as a function of f_{Ba}, f_{Ti} and f_O for the following $[hkl]$ reflections for the $BaTiO_3$ perovskite: (100), (110), (111), (200), (210), (211), (221), and (222).
(iv) $La_{2-x}Ba_xCuO_4$ *coordination polyhedra*: Determine the coordination polyhedron for La(Ba) in the $La_{2-x}Ba_xCuO_4$ structure. How does it differ from that of Nd in the Nd_2CuO_4 structure?
(v) $La_{2-x}Ba_xCuO_4$ *a lattice constant*: Using values of the ionic radii from Chapter 22, predict the *a* lattice constant for the $La_{2-x}Ba_xCuO_4$ phase. How does it depend on using the radii for Cu^{3+} and Cu^{2+}?
(vi) $La_{2-x}Ba_xCuO_4$ *unit cell dimensions*: Using values of the ionic radii from Chapter 22 predict the lattice constants for the $La_{2-x}Ba_xCuO_4$ phase. How do they compare with those tabulated in this chapter?
(vii) $YBa_2Cu_3O_7$ *reciprocal metric tensor*: $YBa_2Cu_3O_7$ is an important high temperature superconductor with an orthorhombic unit cell.

(a) Write an expression for the magnitude and directions of the reciprocal lattice vectors for this orthorhombic crystal in terms of the direct lattice vectors **a**, **b**, and **c**.
(b) Express the orthorhombic reciprocal metric tensor and derive a formula for the spacing between parallel planes with Miller indices (hkl).
(c) Given X-ray diffraction determined values for the spacing between the (005), (110), and (013) planes of 0.234, 0.273, and 0.275 nm, respectively, determine the *a*, *b*, and *c* lattice constants.
(d) The unit cell is primitive with one formula unit per cell. Calculate the density of $YBa_2Cu_3O_7$.

(e) What is the angle between the (005) and (110) plane normals?
Data: $N_A = 6.023 \times 10^{23}$. The atomic weights for Y, Ba, Cu, and O are 88.91, 137.33, 63.55, and 16 g/cm^3, respectively.

(viii) $YBa_2Cu_3O_7$ *unit cell dimensions*: Using values of the ionic radii from Chapter 22 predict the lattice constants for the $YBa_2Cu_3O_7$ phase. How do they compare to those tabulated in this chapter?

(ix) $YBa_2Cu_3O_7$ CuO *planes*: Explain how there are two symmetrically inequivalent CuO planes in the $YBa_2Cu_3O_7$ structure. Explicitly determine the composition of these planes by considering the atoms in the plane and how they are shared among unit cells.

(x) $YBa_2Cu_3O_7$ *polyhedral environments*: Consider cation coordination polyhedra in $YBa_2Cu_3O_7$:

(a) What are the coordination polyhedra about the Y^{3+} and Ba^{2+} cations?
(b) What is the coordination polyhedron about Ba^{2+} in the tetragonal $YBa_2Cu_3O_7$ (insulating) phase?
(c) What is the linkage between the cation coordination polyhedra?
(d) How many inequivalent O sites are there in this structure?

(xi) $YBa_2Cu_3O_7$ *bond valence*: Perform a bond valence calculation (as discussed in Chapter 22) at each of the O vertices in the orthorhombic YBCO structure. What do you conclude about the charge state for Cu cations in the chains and conducting planes in this structure?

(xii) $YBa_2Cu_3O_7$ *MPMG materials*: In the MPMG process, it is desired to produce a two-phase mixture of the YBCO-123 phase and a non-superconducting 211 flux pinning phase. Determine the amounts of Y_2O_3, BaO, and CuO needed to produce a two-phase mixture containing 90 at% of the stoichiometric $YBa_2Cu_3O_7$ (123) high temperature superconductor phase and 10 at% of the Y_2BaCuO_5 (211) second phase and total weight 20 grams.

(xiii) $YBa_2Cu_3O_7$ *crystal growth*: It is a general observation that crystals grow most slowly in the crystallographic directions for which the structure is more complicated. Use this concept to predict the shapes of YBCO crystals.

(xiv) Bi(Pb)–Sr–Ca–Cu–O *phase*: Predict the composition, stacking sequence and c lattice constant for the $n = 5$ phase in the BSCCO double-layer system.

(xv) Bi(Pb)–Sr–Ca–Cu–O *crystal growth*: Explain why crystalline grains of Bi(Pb)–Sr–Ca–Cu–O superconductors would be predicted to have plate-like morphologies. Which would be expected to have the larger aspect ratio, the 2212 phase or the 2223 phase?

(xvi) Tl(Pb)–Sr–Ca–Cu–O *phase*: Predict the composition, stacking sequence, and c lattice constant of the $n = 5$ phase in the TBCCO single-layer system.

(xvii) Tl(Pb)–Sr–Ca–Cu–O *Superlatice reflections*: Predict the diffraction angles at which the first four $(00(2l))$ superlattice reflections will occur for the TBCCO-2201, -2212, and -2223 phases in an XRD experiment using Cu K_α radiation.

(xviii) Tl(Pb)–Sr–Ca–Cu–O *pseudo-binary phase diagram*: The double-layer structures in the Tl(Pb)–Sr–Ca–Cu–O system have the chemical composition: $Tl_2Ba_2Ca_{n-1}Cu_nO_{2(2+n)+x}$. Draw a pseudo-binary phase diagram in which one end member is the $n = 1$ phase. Locate the $n = 2, 3, 4$, and 5 phases on this diagram. What is the composition of the other end member? Comment on the proximity of the phases as n is increased.

(xix) Tl(Pb)–Sr–Ca–Cu–O *layer transformation*: Two-layer structures in the Tl(Pb)–Sr–Ca–Cu–O system have compositions: $Tl_2Ba_2Ca_{n-1}Cu_nO_{2(2+n)+x}$, and the single-layer compound has a composition given by: $TlBa_2Ca_{n-1}Cu_nO_{2n+3}$. Express a chemical reaction that takes a single-layer compound into a double-layer compound for $n = 1, 2, 3$, and 4. Given that TlO is volatile, is the forward or reverse reaction more likely to occur?

(xx) *Double-layer superconductors*: Using the data tabulated in this chapter, plot the value of the c lattice constant as a function of n in the BSCCO and TBCCO. Explain the value of the slope in each case. Do the curves coincide? Why or why not?

(xxi) Hg–Ba–Ca–Cu–O *phase diagram*: Single-layer structures in the Hg–Ba–Ca–Cu–O system have chemical compositions: $HgBa_2Ca_{n-1}Cu_nO_{2n+2+x}$. Draw a pseudo-binary phase diagram in which one end member is the $n = 1$ phase. Locate the $n = 2, 3, 4$, and 5 phases on this diagram. What is the composition of the other end member of this phase diagram?

(xxii) *ABCCO superconductors*: Consider the ordered $n = 4$ ABCCO superconductor with composition: $Ag_{1-x}Cu_xBa_2Ca_{n-1}Cu_nO_{2n+3-x}$, for which 1/8 of the ordered body-centered tetragonal cell is shown in Fig. 23.14.

(a) Show that the c-lattice constant satisfies $c/2 = c_1 + c_2 \times n$. Determine c_1 and c_2.
(b) Predict the c-lattice parameter for the $n = 3$ and $n = 5$ phases.

(xxiii) $RuSr_2GdCu_2O_8$ *polyhedral environments*: Consider the $RuSr_2GdCu_2O_8$ superconductor. What are the coordination polyhedra about the Gd^{3+} and Sr^{2+} cations? What is the linkage between the cation coordination polyhedra in this structure? How many inequivalent O sites are there in this structure?

(xxiv) $RuSr_2GdCu_2O_8$ *bond valence*: Perform a bond valence calculation (see Chapter 22) at each of the O vertices in the $RuSr_2GdCu_2O_8$ structure. What do you conclude about the charge state for Cu cations in the chains and conducting planes in this structure? Does this agree with what you would predict based on O stoichiometry?

(xxv) *Meissner effect*: Show that the Meissner effect implies that the slope of a magnetization versus field curve (in Gaussian units) is $-1/4\pi$.

(xxvi) *Flux pinning I*: Flux pinning occurs when all or part of a magnetic flux line resides in a non-superconducting region. What might you conclude as to the optimum geometries for pinning in an Abrikosov vortex and a pancake vortex?

(xxvii) *Flux pinning II*: Discuss the relative merits of flux pinning by precipitates and by screw dislocations, respectively.

CHAPTER

24 Ceramic structures III: silicates and aluminates

> *"The meek shall inherit the earth, but not the mineral rights."*
>
> J. Paul Getty

24.1 Introduction

Klein and Hurlbut (1985) define a *mineral* as *a naturally occurring homogeneous solid with a definite (but generally not fixed) chemical composition and a highly ordered atomic arrangement. It is usually formed by inorganic processes.* This is a very broad definition that includes a huge variety of compounds. Intuitively, we think of minerals as *gem stones*, but not every gem stone is a mineral, since coral, opal, and pearl, for instance, are formed by organic processes. There are more than 3000 recognized minerals, and in this chapter we will introduce mostly members of the silicate class.[1] Before we do so, we must first consider briefly the classification of minerals into classes.

There are twelve major classes of minerals; they are listed in Table 24.1. Each class may have sub-classes based on the types of structures associated with the minerals. Classes consist of *families* (similar chemical type), families consist of *groups* (structural similarity), groups consist of *species* and species can have several *varieties*. The examples in the table are by no means exhaustive. Several of the mineral classes actually have more than

[1] In this chapter, we shall rely heavily on the text book *Manual of Mineralogy* by Klein and Hurlbut (1985), since this is an authoritative and well known text.

Table 24.1. The twelve classes of minerals, along with sub-classes, a few families and groups, and examples.

Mineral class	Sub-classes/Families	Groups	Examples
Native elements	native metals	gold group	Au, Ag, Cu
		platinum group	Pt
		iron group	Fe, Ni
	semi-metals	arsenic group	As, Bi
	non-metals		S, C
Sulfides		sulfides	Ag_2S, PbS, ZnS
		sulfarsenides	FeAsS
		arsenides	NiAs
		tellurides	$AuTe_2$
Sulfosalts			Ag_3SbS_3, $Cu_{12}As_4S_{13}$
Oxides	simple	XO, X_2O	ZnO, Cu_2O
		hematite group	Al_2O_3, Fe_2O_3
		rutile group	TiO_2, MnO_2
	multiple	spinel group	$MgAlO_4$, Fe_3O_4
	hydroxides	goethite group	αFeO·OH
Halides			NaCl, KCl, CaF_2
Carbonates		calcite group	$CaCO_3$
		aragonite group	$BaCO_3$
		dolomite group	$CaMg(CO_3)_2$
Nitrates			$NaNO_3$, KNO_3
Borates			$Na_2B_4O_5(OH)_4 \cdot 8H_2O$ (borax)
Phosphates			Li(Mn, Fe) PO_4
		apatite group	$Ca_5(PO_4)_3(F,Cl,OH)$
Sulfates		barite group	$BaSO_4$
Tungstates			$CaWO_4$, $PbMoO_4$
Silicates	see Table 24.2		

one type of chemistry; for instance, the tungstates are usually combined with the molybdates, and the phosphates with arsenates and vanadates. The last mineral class, silicates, forms the topic of this chapter. It is a very large class, and its sub-classes are listed in Table 24.2. Note that many of the mineral classes have been considered in other chapters, usually without reference to the term "mineral."

A large portion of the Earth's crust is formed from silicates. As a mineral class, silicates have the largest number of identified species and some of the most commonly occurring mineral species. They tend to have high hardness, good resistance to corrosion and they are often transparent or translucent. With few exceptions, the basic unit of silicate structures is the $(SiO_4)^{4-}$

Table 24.2. The sub-classes of the silicate class of minerals, with a few examples.

Sub-class	Anionic unit	Examples
Orthosilicates (Nesosilicates)	$(SiO_4)^{4-}$	$(Mg,Fe)_2SiO_4$, $Y_3Al_2Si_3O_{12}$, $ZrSiO_4$
Pyrosilicates (Sorosilicates)	$(Si_2O_7)^{6-}$	$Ca_2MgSi_2O_7$, $Zn_4Si_2O(OH)_2 \cdot 2H_2O$
Metasilicates (Inosilicates)	$(SiO_3)^{2-}$	$CaMg(SiO_3)_2$, $MgSiO_3$
Metasilicates (Inosilicates)	$(Si_4O_{11})^{6-}$	$Mg_7Si_8O_{22}(OH)_2$
Cyclosilicates	$(Si_6O_{18})^{12-}$	$Be_3Al_2Si_6O_{18}$
Phyllosilicates	$(Si_2O_5)^{2-}$	$Al_2Si_4O_{10}(OH)_2$, $KAl_3Si_3O_{10}(OH)_2$
Tectosilicates	$(SiO_2)^0$	SiO_2, $KAlSi_2O_6$, $Na_4Al_3Si_3O_{12}CO$

tetrahedron, with a cationic valence state of Si^{4+}. Many *aluminates* and *aluminosilicates* also belong to the class of minerals. Al^{3+} is often octahedrally coordinated and, therefore, many structures have linkages between octahedra and tetrahedra.

Silicates are classified by considering the linkage of $(SiO_4)^{4-}$ tetrahedra. In general, the tetrahedra link by sharing vertices. Each tetrahedron can be viewed as a charged radical unit which is combined with other cationic species to form compounds whose structures are based on the placement of these cations in the "network" of linked $(SiO_4)^{4-}$ tetrahedral groups. The ways in which the vertices are linked together leads to the classification of silicates in terms of *structural groups* (Klein and Hurlbut, 1985). Figure 24.1 shows structural units that are common to the silicates, including tetrahedra,

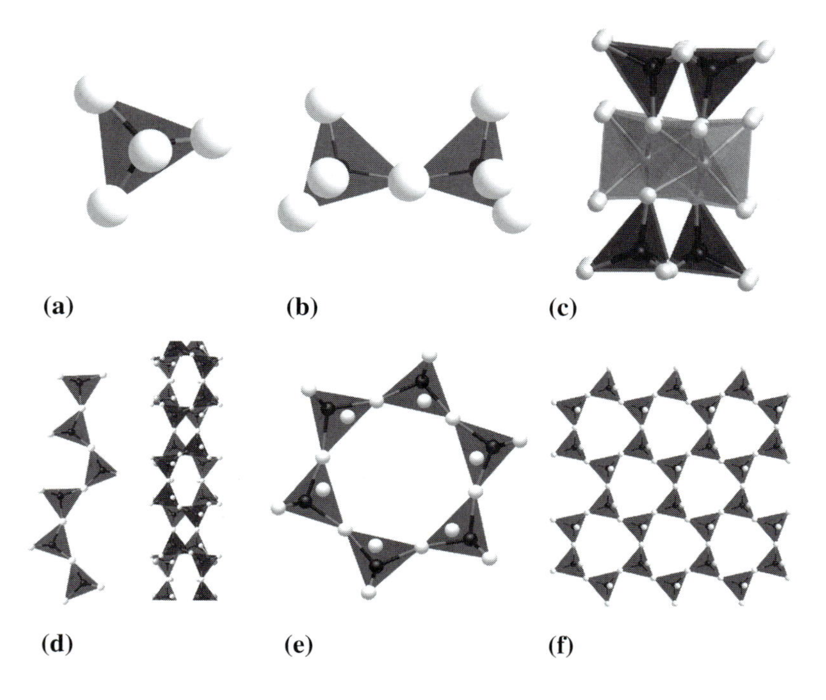

Fig. 24.1. Structural units commonly found in silicate structures, (a) isolated tetrahedra, (b) double tetrahedra, (c) double tetrahedral *I-beam*, (d) chains and double chains, (e) rings, and (f) sheets of tetrahedra.

double tetrahedra, double tetrahedral *I-beam*, chains and double chains, rings, and sheets of tetrahedra. The structural classification of silicates is based on the presence of these structural groups. Table 24.2 lists the sub-classes of the silicate mineral class, along with the chemistry and charge state of the common anionic building unit and a few examples. In the following paragraphs, we discuss briefly some of the main characteristics of the sub-classes; the remainder of this chapter describes more detailed examples for all sub-classes.

(i) *Orthosilicates* (*nesosilicates*) contain isolated $(SiO_4)^{4-}$ tetrahedra, and all linkages are with other (often octahedral) coordination polyhedra. Orthosilicates include the *olivine series*, *humites*, *garnets*, and *epidotes*. Examples are $(Mg,Fe)_2SiO_4$ (olivine); $(M_3^{2+}M_2^{3+})(SiO_4)_3$ (garnet) with M^{2+} = Ca, Mg, Fe, etc. and M^{3+} = Al, Cr, Fe, etc.; $CaMgSiO_4$ (*monticellite*); Ca_2SiO_4; and $ZrSiO_4$ (*zircon*).

Subsilicates are related to the orthosilicates. These have independent SiO_4 tetrahedra but also additional O, not included in the tetrahedra, i.e., SiO_5 units. Examples include *andalusite* (Al_2SiO_5), *sillimanite* (Al_2SiO_5), *kyanite* (Al_2SiO_5), *malayite* ($CaSnSiO_5$), *titanite* ($CaTiSiO_5$), etc.

(ii) *Pyrosilicates* (*sorosilicates*) contain isolated pairs of vertex-sharing tetrahedra which make up $(Si_2O_7)^{6-}$ units. Pyrosilicates include $Ca_2MgSi_2O_7$; *hemimorphite* ($Zn_4Si_2O_7(OH)_2 \cdot 2H_2O$); *bertrandite* ($Be_4Si_2O_7(OH)_2$); *danburite* ($CoB_2Si_2O_8$), etc.; Al is usually not present in these compounds.

(iii) Single chain *metasilicates* (*inosilicates*) contain chains of SiO_4 tetrahedra, each sharing two vertices with other tetrahedra. The basic unit is $(SiO_3)^{2-}$, in which two O ions in the $(SiO_4)^{4-}$ unit are shared with other tetrahedra. The length of the chains defines the metasilicates as pyroxenes, pyroxenoids, or ring structures. Vertex-sharing $(SiO_3)^{2-}$ tetrahedra chain together to form longer structures, e.g., Si_2O_5, Si_3O_8, Si_4O_{11}, etc., or ring structures $[Si_nO_{3n}]$. The pyroxene group includes *diopside* ($CaMg(SiO_3)_2$) and *enstatite* ($MgSiO_3$). Silicate chains lie parallel to one another and are linked together by cations. *Jadeite* ($NaAlSi_2O_6$) is a metasilicate with an *I-beam structure*.

(iv) Double chain *metasilicates* (*inosilicates*) have structures in which two types of tetrahedra exist, with two and three vertices shared, respectively. In the *amphiboles*, double chains occur in which alternate tetrahedra share two (edges) and three (faces) O atoms, respectively; the repeat unit is $(Si_4O_{11})^{6-}$. The sharing of a third vertex in alternating tetrahedra allows for the formation of a double chain. *Tremolite* ($Ca_2Mg_5(OH)_2(Si_4O_{11})_2$) is a prominent example of an amphibole mineral.

(v) *Cyclosilicates* have chains of tetrahedra that are closed into a ring. The basic building unit is of the type $(Si_6O_{18})^{12-}$, although there are minerals with a structure based on the $(Si_3O_9)^{6-}$ and $(Si_4O_{12})^{8-}$ rings. The mineral *beryl* ($Be_3Al_2Si_6O_{18}$) is a primary example, along with its gem varieties *emerald* (green), *aquamarine* (greenish-blue), and *morganite* (pink).

(vi) *Phyllosilicates* have SiO_4 tetrahedra grouped in planar close-packed sheets with hexagonal or pseudo-hexagonal symmetry. Examples include *pyrophyllite* ($Al_2Si_4O_{10}(OH)_2$); *kaolin* ($Al_3Si_2O_5(OH)_4$); *talc* ($Mg_3Si_4O_{10}(OH)_2$); *serpentine* ($Mg_3Si_2O_5(OH)_2$); etc. The *mica group* includes *muscovite* ($KAl_3Si_3O_{10}(OH)_2$); *biotite* ($K(Mg,Fe)_3AlSi_3O_{10}(OH)_2$), etc. *Micaceous* structures have weak interlayer bonding and cleave easily.

(vii) *Tectosilicates*, also known as *framework silicates*, have structures in which SiO_4 tetrahedra form networks with all four vertices shared. *Quartz*, a polymorph of SiO_2, has a network structure and is one of the most common minerals on earth. When Si is replaced by Al, the $(SiO_4)^{4-}$ building block becomes an $(AlO_4)^{5-}$ unit, which requires other cations to balance the charge. Among these *aluminosilicates* we find *feldspars* ($M(Al,Si)_4O_8$), rock-forming minerals); *feldspathoids* such as *leucite* ($KAlSi_2O_6$) and *nepheline* $(Na,K)(Al,Si)_2O_4$, minerals forming in magma); the *sodalite group*, such as *sodalite* ($Na_4Al_3Si_3O_{12}CO$) and *lazurite* ($Na_{4-5}Al_3Si_3O_{12}S$) which form in alkaline igneous and plutonic rocks and metamorphosed limestones, respectively. The *scapolites* include *marialite* and *meionite*. The *zeolite* structures are based on a secondary unit with 24 silica or alumina tetrahedra linked in a *sodalite cage*. Zeolites include *natrolite* ($Na_2Al_2Si_3O_{10} \cdot 2H_2O$) and *stilbite* ($NaCa_2Al_5Si_{13}O_{36} \cdot 14H_2O$).

In the following sections, we will describe some important examples of each of the silicate sub-classes. As before, structural information for many of the compounds described in this chapter can be found in the on-line structures appendix.

24.2 Orthosilicates (nesosilicates)

> The *orthosilicates* include all silicates with $(SiO_4)^{4-}$ tetrahedra that are unbonded to other tetrahedra.

Orthosilicates are the simplest of the silicate subclasses and are similar to the sulfates and phosphates which have SO_4 and PO_4 tetrahedra, respectively. The orthosilicates have structures with strong bonding, high density and high hardness. More gemstones belong to the orthosilicate subclass than to any other silicate subclass. In this section, we illustrate a few examples of orthosilicates: the olivine series, garnets, and other orthosilicate minerals, which include the *subsilicates*.

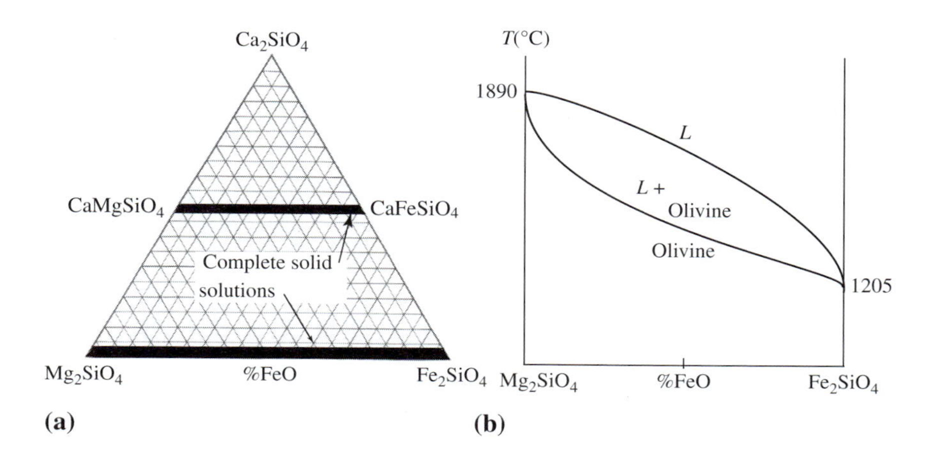

Fig. 24.2. (a) Ternary Ca_2SiO_4–Mg_2SiO_4–Fe_2SiO_4 phase diagram and pseudo-binary cuts with complete solid solutions; (b) pseudo-binary Mg_2SiO_4–Fe_2SiO_4 phase diagram.

24.2.1 Olivine minerals and gemstones

Most *olivine minerals* belong to the *quaternary system* CaO–MgO–FeO–SiO_2 (Klein and Hurlbut, 1985). They are often represented in a ternary phase diagram, with as corner compounds, Ca_2SiO_4, Mg_2SiO_4, and Fe_2SiO_4, as illustrated in Fig. 24.2(a). Important olivines occur in the *pseudo-binary* cuts of the ternary diagram. The line between $CaMgSiO_4$ (*monticellite*), and $CaFeSiO_4$ (*kirschteinite*), represents compositions for which there is a complete solid solution series with general formula $CaMg_{1-x}Fe_xSiO_4$. A complete solid solution series also exists between Mg_2SiO_4 (*forsterite*, Fig. 24.3), and Fe_2SiO_4 (*fayalite*); those solid solutions have compositions of the type $(Mg_{1-x}Fe_x)SiO_4$. In order to have complete solid solutions, the end members must be isostructural.

Figure 24.2(a) provides us with an occasion to describe the representation of a phase diagram in a ternary system. Constant temperature cuts of a ternary phase diagram are represented with the aid of the Gibbs triangle shown in Fig. 24.2(a). As a shorthand notation, we designate the terminal compositions as $A = Ca_2SiO_4$, $B = Mg_2SiO_4$, and $C = Fe_2SiO_4$. Compositional information is then conveyed as follows:

(i) At any point in the triangle, the distances to A, B, and C must add up to 100 per cent (atomic or by weight) or 1 (mole fraction).
(ii) Along the line connecting A and C, B must be zero. Similarly along the line connecting A and B, C is zero and along the line connecting C and B, A is zero.
(iii) At any apex, one composition equals 1 (mole fraction) while both others are zero.
(iv) Any line parallel to an edge of the triangle reflects a constant value of the concentration of the component in the opposite corner.

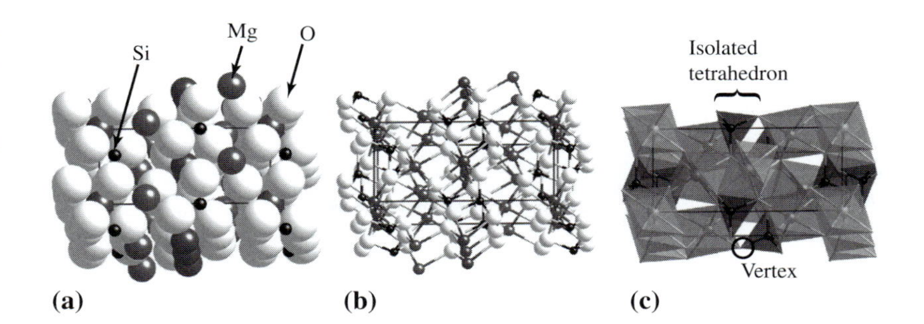

Fig. 24.3. Mg_2SiO_4 forsterite structure in (a) close-packed, and (b) ball-and-stick depictions, and (c) showing the linkage of the tetrahedral and octahedral coordination polyhedra.

(v) The intersection of two lines is a point in composition space. A line parallel to A at composition X_A intersecting a line parallel to B at composition X_B determines a point in composition space: $(X_A, X_B, 1 - X_A - X_B)$

Figure 24.2(b) shows an example of a system with a separated liquidus and solidus curve. While complete liquid solutions exist above the liquidus and complete solid solutions below the solidus, only the pure components melt or freeze congruently, i.e., without a change in composition. This means that the solid solutions will typically have Mg- and Fe-rich regions, due to chemical partitioning that occurs in the liquid + olivine region of the phase diagram.

Olivines are one of the minerals found in the igneous rock, *basalt*. Olivine's gemstone variety is known as *peridot*; peridot is the August birth stone. Figure 24.3 illustrates the forsterite structure in (a) space filling and (b) ball-and-stick representations. Figure 24.3 (c) shows the linkage of the tetrahedral and octahedral coordination polyhedra. The forsterite structure (**Structure 86**) has a quasi-hexagonal close-packed arrangement of O^{2-} anions. The Mg^{2+} cations occupy two crystallographically inequivalent octahedral interstices, for a total of half of the octahedral sites in the close-packed anion sublattice. The Si^{4+} cations occupy $\frac{1}{8}$ of the tetrahedral interstices in the anion sublattice. The polyhedral representation shows twisted chains of vertex sharing Mg-coordinating octahedra. Chains of octahedra are connected to each other by sharing edges with the isolated $(SiO_4)^{4-}$ tetrahedra.

The structure of forsterite can be compared with that of spinel. The spinel structure, also discussed in Chapter 22, has a cubic-close-packed arrangement of O anions, also with octahedral and tetrahedral cation environments. In the spinel structure, however, the tetrahedra do not share edges with the octahedra.

24.2.2 Garnets

Garnet is another example of an orthosilicate for which the SiO_4 tetrahedra are isolated. It is a gem stone mineral and the January birth stone. It is often

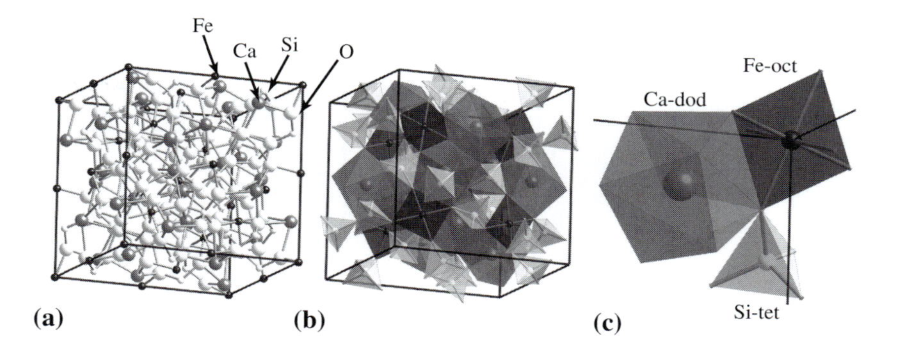

Fig. 24.4. Unit cell of the andradite garnet, (a) ball-and-stick depiction, (b) cation coordination polyhedra, and (c) connectivity of octahedra, tetrahedra, and dodecahedra.

found in metamorphic rocks and is a major component of the Earth's upper mantle and transition zone. The chemical formula for garnets is:

$$C_3A_2D_3O_{12}, \tag{24.1}$$

where the C cations occupy dodecahedral sites, the A cations occupy octahedral sites, and the D cations occupy tetrahedral sites in the structure. An example of a garnet is the mineral *Andradite* ($Ca_3Fe_2Si_3O_{12}$, Fig. 24.4, **Structure 87**), which has Fe^{3+} cations in octahedral, Si^{4+} cations in tetrahedral, and large Ca^{2+} cations in dodecahedral sites. This structure has space group **Ia$\bar{3}$d** (O_h^{10}), with the dodecahedral cation at the $(1/8, 0, 1/4)$ sites, the octahedral cation at $(0, 0, 0)$, and the tetrahedral cation at $(3/8, 0, 1/4)$, respectively. Figure 24.4(a) shows all atoms in a single cubic unit cell in a ball-and-stick depiction; Fig. 24.4(b) shows the cation coordination polyhedral. The connectivity of the three types of cation coordination polyhedra is shown in Fig. 24.4(c). The cubic unit cell of garnet contains eight formula units for a total of 160 atoms.

Figure 24.5 shows a sequential view of the building of the garnet unit cell, starting with an Fe^{3+} octahedron (a), attachment of the Si^{4+} tetrahedra (b), attachment of the Ca^{2+} dodecahedra (c), and completion of the cell (d). The garnet structure is seen to have isolated SiO_4 tetrahedra; Fe^{3+} octahedra connect the Si^{4+} tetrahedra and Ca^{2+} cations occupy the large dodecahedral sites formed within the iron-silicate framework.

Magnetic garnets form an important class of engineering materials. The compound $Gd_3Fe_5O_{12}$ has Gd^{3+} occupying the dodecahedral sites and Fe^{3+} on both octahedral and tetrahedral sites. A wide variety of cations in different valence states can reside on these different sites (especially on the large dodecahedral sites). Ionic size is the primary consideration for site occupancy. The similarity in the ionic radii of the rare earth ions causes rare earth iron garnets to readily form solid solutions. Y (ionic radius = 0.090 nm) and Gd (ionic radius = 0.0938 nm) form a complete family of solid solutions with general formula $Y_xGd_{3-x}Fe_5O_{12}$ ($0 \leq x \leq 3$); substitution of Y for Gd occurs on the dodecahedral site. Figure 24.6 shows the unit cell of the

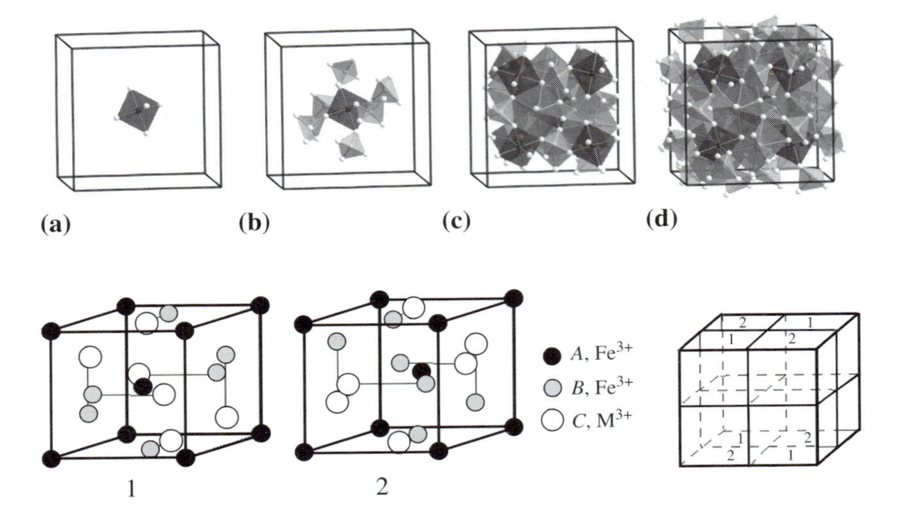

Fig. 24.5. Sequential view of the building of the garnet unit cell starting with an Fe^{3+} octahedron (a), attachment of the Si^{4+} tetrahedra (b), attachment of the Ca^{2+} dodecahedra, and additional octahedra (c) and completion of the cell (d).

Fig. 24.6. Cations in an octant (1) of the garnet structure of $M_3Fe_5O_{12}$ (left), a second octant (2) with mirror symmetry (middle), and the stacking of these units in a 2 x 2 x 2 supercell (right).

$M_3Fe_5O_{12}$ garnet as a stacking of two different building blocks, 1 and 2. All of the metal ions in an octant cell are drawn (O ions have been omitted for clarity). The second octant is the inverse of the first (inversion with respect to the octant center). Both octants are used to decorate a $2 \times 2 \times 2$ supercell, which forms the unit cell of the garnet structure.

24.2.3 Other orthosilicate minerals

Monticellite ($CaMgSiO_4$), *kirschteinite* ($CaFeSiO_4$), and *zircon* ($ZrSiO_4$) are other examples of orthosilicates. The structure of zircon (**Structure 88**) has Zr in dodecahedral polyhedra linked by isolated SiO_4 tetrahedra. Figure 24.7 illustrates the structure of the gem stone *zircon* in (a) a ball-and-stick depiction and (b) showing the linkage of the tetrahedral and dodecahedral coordination polyhedra. Figure 24.7(c) illustrates another example of isolated $(SiO_4)^{4-}$ tetrahedra in the structure of the gem stone *topaz* ($[Al(F,OH)]_3SiO_4$). Colorless zircon can easily be mistaken for diamond (the April birth stone), but, unlike diamond, it is an affordable gem stone. Zircon, the December birth stone, is not the same as the artificial gem stone *cubic zirconia*, which is not a silicate but an oxide. Zircon is commonly found in brown and green colors, but can be heated to obtain blue or gold colors. Topaz (the November birth stone) is another commonly occurring gem stone. It is usually yellow or brown, but blue varieties do exist.

Isolated tetrahedra are also observed in the *subsilicates*. The three polymorphs of Al_2SiO_5, *andalusite*, *sillimanite*, and *kyanite*, are examples of compounds with subsilicate structures. Figure 24.8 illustrates two examples of isolated SiO_4 tetrahedra in the subsilicate structures of kyanite (a) and sillimanite (b). These illustrate two different features observed in the aluminosilicate, Al_2SiO_5.

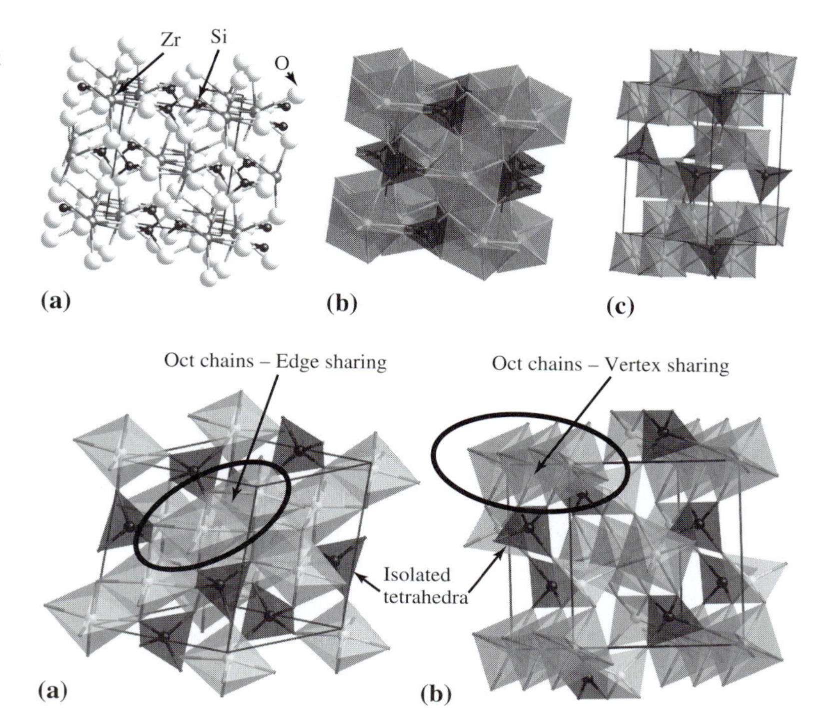

Fig. 24.7. Zircon $ZrSiO_4$ structure in (a) a ball-and-stick depiction and (b) showing the linkage of the (isolated) tetrahedra and dodecahedra. (c) Polyhedral linkages in the gemstone topaz ($[Al(F,OH)]_3SiO_4$).

Fig. 24.8. Examples of subsilicate structures and polymorphs of Al_2SiO_5, illustrating isolated SiO_4 tetrahedra in kyanite (a) and sillimanite (b).

The kyanite structure (**Structure 89**), Fig. 24.8(a), is triclinic. It has all Al^{3+} ions octahedrally coordinated. In all of the polymorphs, Si is tetrahedrally coordinated. The structure has edge-linked AlO_6 octahedra running parallel to the *c*-axis; these are cross-linked by the SiO_4 tetrahedra. The andalusite structure (not shown) is orthorhombic with Al^{3+} in five-fold coordination. In the sillimanite structure, Fig. 24.8(b), Al^{3+} has four-fold tetrahedral coordination. Sillimanite (**Structure 90**) is orthorhombic, with edge-linked AlO_6 octahedra in chains parallel to the *c*-axis. These chains are connected with alternating AlO_4 and SiO_4 tetrahedra. The SiO_4 tetrahedra that link only with AlO_4 tetrahedra are isolated from each other.

24.3 Pyrosilicates (sorosilicates)

The *pyrosilicates* (*sorosilicates*) have isolated pairs of vertex sharing tetrahedra which contain $(Si_2O_7)^{6-}$ structural units.

The tetrahedra pairs share an O atom at the vertex of 2 $(SiO_4)^{4-}$ tetrahedra, reducing the number of O atoms by one. These compounds are also known as *disilicates*. The mineral *epidote* ($Ca_2(Al,Fe)Al_2Si_3O_{13}H$, **Structure 91**) is

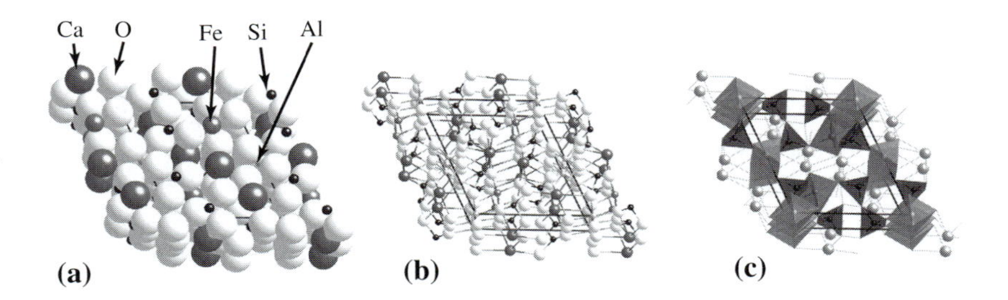

Fig. 24.9. Epidote structure in a (a) space filling and (b) ball-and-stick depictions. (c) Linkages and connectivity of the coordination polyhedra.

an example of a sorosilicate. Minerals of the epidote group can be represented by the formula:

$$A_2M_3Si_3O_{13}H, \qquad (24.2)$$

where the A sites are occupied by large cations such as Ca, Sr, etc., and the M sites are occupied by octahedrally coordinated cations (e.g., Al^{3+}, Fe^{3+}, Mg^{2+}, etc.). Two of the M atoms (usually Al) form parallel chains of MO_6 and $MO_4(OH)_2$ octahedra that define the epidote structure. Epidote has both single and double silicate tetrahedra.

Figure 24.9(a) shows a ball-and-stick depiction of the epidote structure and (b) illustrates the connectivity of the coordination polyhedra. This shows the structure to consist of chains of edge-sharing Al^{3+}-centered octahedra stacked in the direction normal to the page. These octahedra are linked by single SiO_4 tetrahedra and double (Si_2O_7) tetrahedral groups, as illustrated in Fig. 24.9(c). The structure thus contains both isolated tetrahedra and the isolated pairs of vertex-sharing tetrahedra which define the sub-class of pyrosilicates. Large cavities in the framework contain Ca cations which are observed in both 9-fold and 10-fold coordination; their unconventional coordination polyhedra are not shown (Dollase, 1971).

24.4 Chains of tetrahedra, metasilicates (inosilicates)

> The *metasilicates* have isolated, infinite chains of SiO_4 tetrahedra or chains that close to form rings (*cyclosilicates*).

The mineral *wollastonite* ($CaSiO_3$) is a metasilicate with a chain structure and is a member of the *pyroxenoid group*. Wollastonite forms in hot magmas from reactions of *calcite* ($CaCO_3$) in limestone with silica; the reaction is given by:

$$CaCO_3 + SiO_2 \rightarrow CaSiO_3 + CO_2.$$

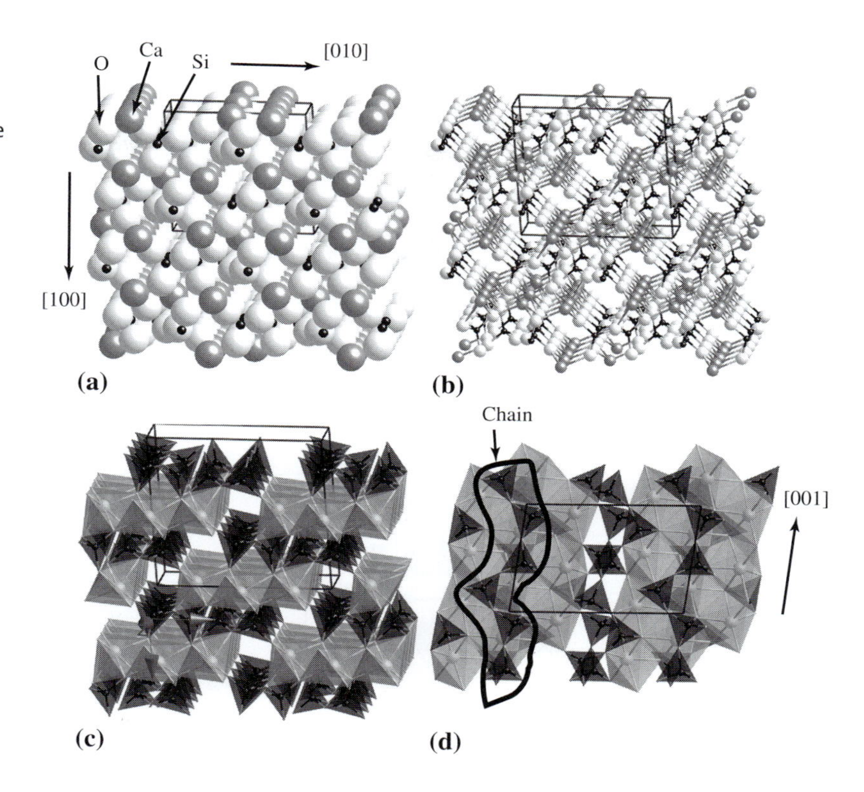

Fig. 24.10. Wollastonite structure in a space filling (a) and ball-and-stick (b) depiction (several unit cells are shown); linkages and connectivity of the coordination polyhedra are shown in (c), and chains of tetrahedra in (d) (as viewed along a [100] direction).

Wollastonite (Fig. 24.10) is named after William Hyde Wollaston (1766–1828), a mineralogist who, in the 1800s, made contributions to physical crystallography, including the design of a polarizing prism and the invention of the reflecting goniometer. Wollastonite is a relatively common mineral, important in refractory ceramics.

There are two low-temperature polymorphs of $CaSiO_3$: wollastonite-1T (triclinic) (**Structure 92**), and wollastonite-2M (monoclinic), which is also known as *parawollastonite*. Other rare and obscure polymorphs of $CaSiO_3$ with the names of wollastonite-3T, -4T, -5T, and -7T have been proposed. Figure 24.10 shows space filling (a) and ball-and-stick (b) models of the triclinic wollastonite-1T structure. The ball-and-stick (b) and polyhedral (c) models show the structure viewed with the *c*-axis approximately normal to the plane of the paper. It can be seen that infinite chains of SiO_4 tetrahedra connect parallel to the *c*-axis. The tetrahedral chains are linked to one another through distorted Ca–O octahedra (Ohashi and Finger, 1978). The structure of the chains can be further understood by viewing them along a [100] direction. Figure 24.10(d) shows chains of tetrahedra, with two vertices shared. The chains are, in fact, "kinked," with a three-tetrahedron repeat distance. In general, the chain structures in metasilicates can be complex and depend on the cations in the structure, with different chain repeat distances possible.

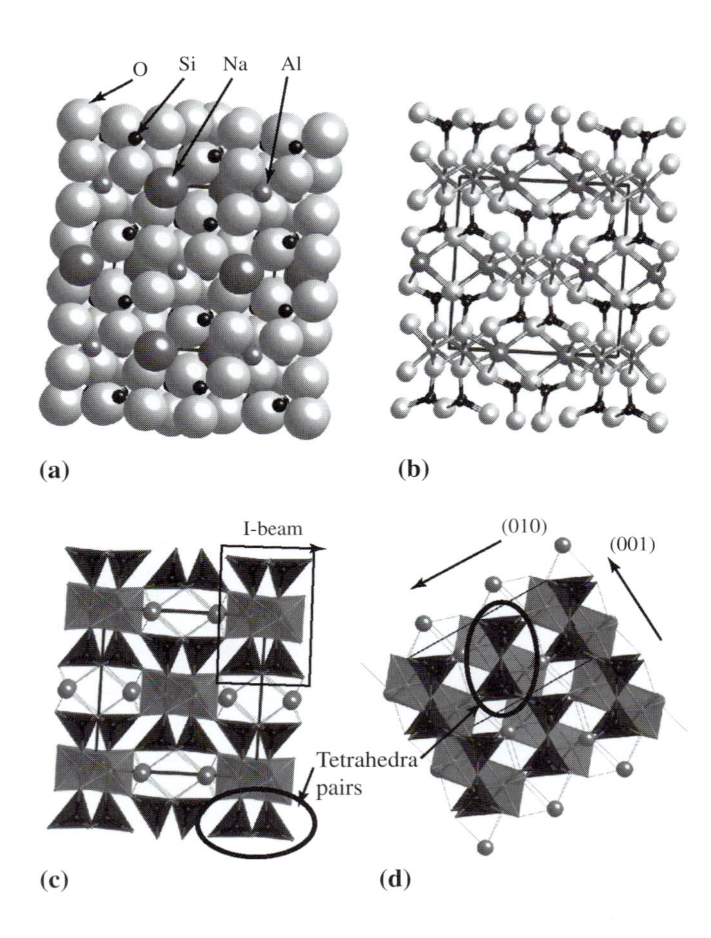

Fig. 24.11. Jadeite in a space filling (a) and ball-and-stick (b) depiction; (c) shows coordination polyhedra, in an (001) plane and chains of tetrahedra are visible in (d) (viewed along a [100] direction).

Jadeite ($NaAlSi_2O_6$, **Structure 93**) is another example of a metasilicate mineral. Jadeite is a *pyroxene mineral* with infinite chains of SiO_4 tetrahedra with a two tetrahedron repeat distance, shown in Fig. 24.11. *Nephrite* and jadeite are two minerals that are both known as *jade*. Jade has long been used as an ornamental gem stone in China.

Figure 24.11(a) and (b) show space filling and ball-and-stick depictions of jadeite. Figure 24.11(c) shows the connection of the cation coordination polyhedra in the jadeite structure in an (001) plane. Again SiO_4 tetrahedral chains are linked through Al cation octahedra. The Na cations sit in large dodecahedral coordination polyhedra (not shown) in the voids in the structure. The SiO_4 tetrahedral chains pack in pairs above and below a block of Al cation octahedra, forming a characteristic *I-beam structure*; these are, in turn, interlinked, as illustrated in Fig. 24.11(c). Figure 24.11(d) shows jadeite coordination polyhedra viewed along a [100] direction in a (100) plane.

Metasilicates can also have chain structures that remain finite by closing on themselves. *Beitoite*, $BaTi(SiO_3)_3$, is an example of a structure where SiO_4 tetrahedra close to form three-member rings. A stunning example of

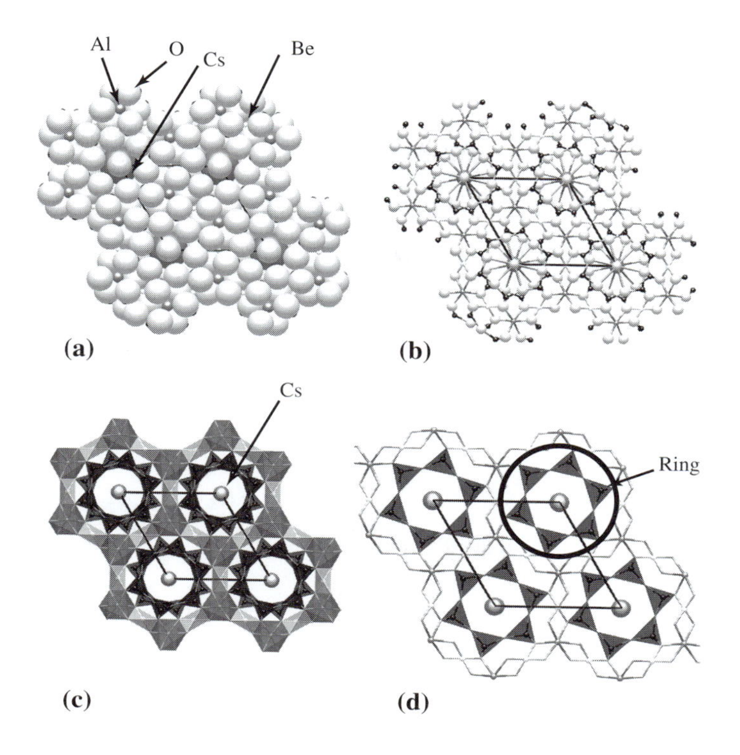

Fig. 24.12. Beryl $Be_3Al_2Si_6O_{18}$ structure in a space filling (a) and ball-and-stick (b) depiction, linkages and connectivity of the coordination polyhedra (c), and tetrahedra in the half cell (d).

a structure where SiO_4 tetrahedral chains close to form six-member rings, that act as $[Si_6O_{18}]^{12-}$ anions, is that of the gem stone *beryl* (Fig. 24.12, **Structure 94**).

Figure 24.12 shows space filling (a) and ball-and-stick representations of beryl, which has the idealized composition $Be_3Al_2Si_6O_{18}$. The structure depicted is a Cs–Li beryl, with composition $Be_3Al_2Si_6Li_{0.45}Na_{0.3}(H_2O)_{0.7}Cs_{0.14}O_x$, with Li and Na occupying Be sites, and Cs and H_2O occupying the same sites (Hawthorne and Cerný, 1977). Figure 24.12(c) shows the linkage of the cation coordination polyhedra. The 12-fold Cs coordination is not shown. The large cations (and H_2O groups) reside in the large holes in the center of two six-membered SiO_4 tetrahedra rings. The rings are connected through Be–O tetrahedra which also share edges with Al cation octahedra. Figure 24.12 (d) shows SiO_4 tetrahedra in half a unit cell (along [001]), with two vertices shared, closed to form six-member rings.

The deep green gem stone version of beryl is known as *emerald*, while a sea-green variety is known as *aquamarine*; emerald is the May birth stone, whereas aquamarine is the March birth stone. There are also multi-colored *borosilicate* gem stones with six-fold $[Si_6O_{18}]^{12-}$ anion rings. These are known as *tourmalines*; tourmaline is the October birth stone.

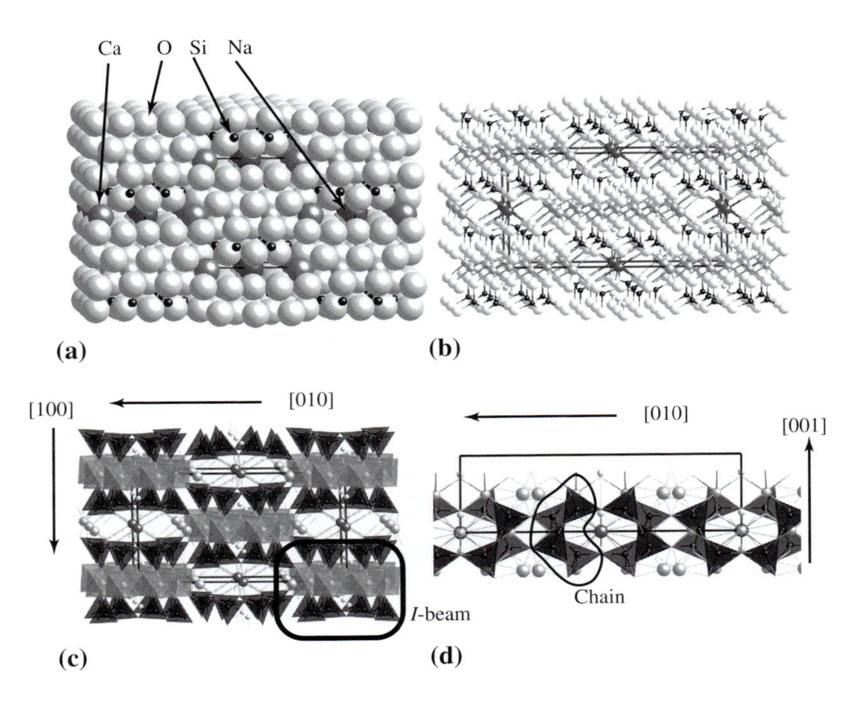

Fig. 24.13. Tremolite structure in (a) space filling, (b) ball-and-stick depiction, (c) showing the linkages and connectivity of the coordination polyhedra, and (d) showing double chains of tetrahedra in the structure (as viewed along a [100] and snaking along the [0$\bar{1}$0] direction).

24.5 Double chains of tetrahedra

Silicates can also possess infinite tetrahedral chains that are linked together to form double chains. *Tremolite* ($Ca_2Mg_5(Si_8O_{22})(OH)_3$, Fig. 24.13) is an example of an amphibole mineral with a double chain silicate structure (**Structure 95**).

Figure 24.13 shows the tremolite structure in space filling (a) and ball-and-stick (b) depictions. Figure 24.13(c) shows the linkages and connectivity of the coordination polyhedra. The SiO_4 tetrahedra connect parallel to the [010] axis and the chains are linked to double chains below through the Mg^{2+} octahedra. The large Ca^{2+} cations (and Na substitutions) fill the large voids in the structure. Figure 24.13(d) shows the double chains of tetrahedra in the structure as viewed along a [100]; the chains "snake" along the [0$\bar{1}$0] direction. The chains can be seen to continue by connecting alternately above and below the plane of the page. This structure consists of staggered *I*-beams.

24.6 Sheets of tetrahedra, phyllosilicates

The *phyllosilicates* have SiO_4 tetrahedra which share three of their four vertices to form infinite layers.

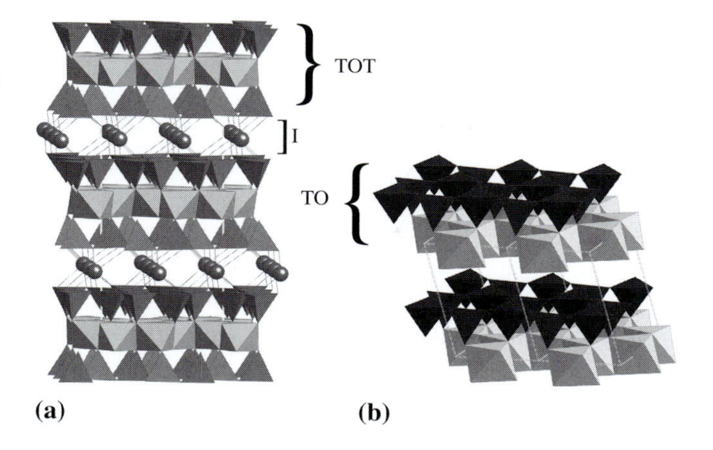

Fig. 24.14. Examples of trioctahedral (mica) (a) and dioctahedral (kaolinite) layered structures.

Figure 24.14 illustrates *trioctahedral* and *dioctahedral* layered *phyllosilicates*. In the phyllosilicates, the sheet can be thought of as a $(Si_2O_4)^{2-}$ anion of infinite spatial extent. This tetrahedral (T) sheet has the triangular faces of the tetrahedra forming a hexagonal basal network with the apical oxygens pointing upward. These T sheets are stacked with octahedral (O) sheets to form 3-D structures. Both TO (1:1) and TOT (2:1) structures are observed. The structures are further classified by the filling of the octahedral interstices in the O sheets. They are called *trioctahedral* if all of the octahedral sites are filled and *dioctahedral* if $\frac{1}{3}$ of the octahedral sites are empty.

24.6.1 Mica

Micas are layered structures with weak interlayer bonding, allowing for easy cleavage. The mica group includes *muscovite* ($KAl_3Si_3O_{10}(OH)_2$). The mica structure has one O sheet between two T sheets. These sheets form a layer that is separated by planes of non-hydrated interlayer cations I (Rieder *et al.*, 1998). It is, therefore, a TOT (2:1) structure, with stacking sequence $ITOT\ldots$ A simplified mica formula is given by:

$$IM_{2-3}\square_{1-0}T_4O_{10}A_2, \tag{24.3}$$

where $\square$ represents a vacancy in the octahedral layer. Muscovite is denoted ($KAl_2\square\ AlSi_3O_{10}(OH)_2$). The 12-fold coordinated, interlayer cation, I, can be Cs, K, Na, NH_4, Rb, Ba, Ca, etc. If fewer than half of the I-cations are monovalent, then the compound is a true mica; if more than half of the I-cations are monovalent, then it is a *brittle mica*. Octahedrally coordinated M cations are commonly Li, Fe^{3+}, Fe^{2+}, Mg, Mn^{3+}, Mn^{2+}, Zn, Al, Cr, V, or Ti, and are coordinated by A anions (Cl, F, OH, O, or S) and apical O anions of the tetrahedral sheets. Tetrahedral sheets have composition T_2O_5 where the

T cations are commonly Be, Al, B, Fe^{3+}, or Si. The tetrahedra share basal plane vertices with apices pointing out of the sheet. Micas are *dioctahedral* if they contain less than 2.5 M cations per formula unit and *trioctahedral* for more. Muscovite, a common mica that forms in nearly transparent flakes, is used in applications as capacitors and thermal insulators.

24.6.2 Kaolinite

Kaolinite ($Al_2Si_2O_5(OH)_4$) is a clay mineral example of a *phyllosilicate*. It has a structure with SiO_4 tetrahedra sharing three vertices to form sheets with large hexagonal voids (**Structure 96**). Figure 24.15 shows the structure in space filling (a) and ball-and-stick (b) depictions. Figure 24.15(c) shows linkages and connectivity of the coordination polyhedra. In kaolinite, sheets of tetrahedra sit above sheets of edge-sharing Al^{3+} octahedra. $(OH)^-$ units lie between two double sheets (tetrahedra, octahedra) in this TO (1:1) structure. Because of the weak bonding between the double sheets coupled by the $(OH)^-$ units, the structures are micaceous and cleave easily along these planes. Figure 24.15(d) shows sheets of tetrahedra in the (001) plane of the structure. The hexagonal network of tetrahedra is different from the isolated hexagonal rings in beryl.

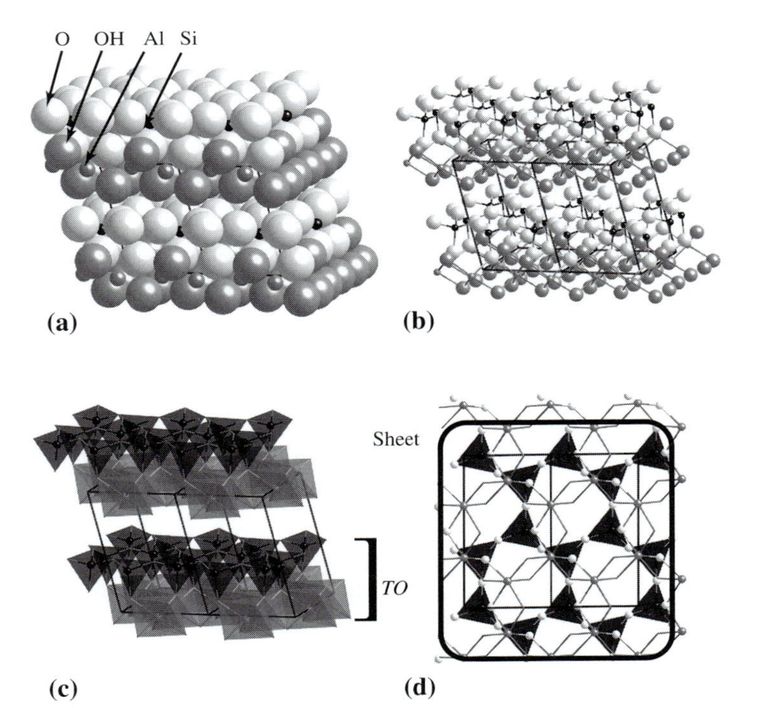

Fig. 24.15. Kaolinite structure in space filling (a) and ball-and-stick (b) depictions; linkages and connectivity of the coordination polyhedra (c), and a sheet of tetrahedra in the (001) plane (d).

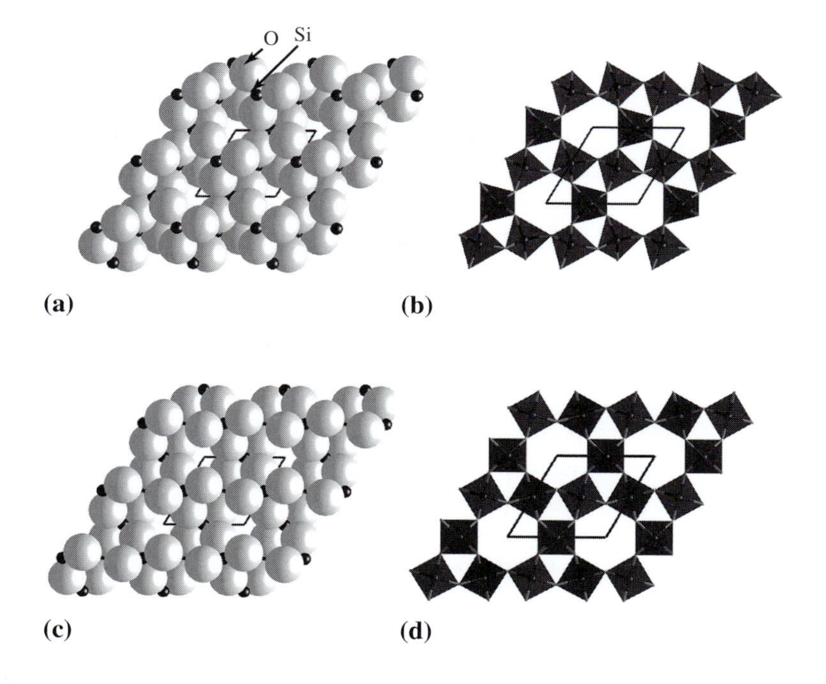

Fig. 24.16. α- and β-quartz structure in space filling settings (a) and (c); (b) and (d) show polyhedral settings projected into (001) planes.

24.7 Networks of tetrahedra, tectosilicates

The *tectosilicates* contain networks where all four vertices of the $(SiO_4)^{4-}$ tetrahedra are shared with other tetrahedra.

These vertices may be shared exclusively with other SiO_4 tetrahedra, leading to the charge-neutral composition SiO_2, as is the case for quartz (Fig. 24.16). Alternative tectosilicate structures have some of the tetrahedral vertices shared with cation–oxygen tetrahedra for which the formal charge is not 4−, such as $(AlO_4)^{5-}$ tetrahedra, which results in a tetrahedral network that is not charge neutral. Instead, the charge neutrality is achieved through the introduction of additional cations within the network structure. These include such mineral families as the *feldspars*, the *sodalite group*, the *scapolites* and the caged structures of the *zeolite family*. In this section, we illustrate examples of such network structures.

24.7.1 Quartz

A prototypical network structure is that of SiO_2, which has many polymorphs that are stable at different ranges of temperature and pressure, including α- and β-quartz. Figure 24.16 shows the structures of α- and β-quartz: α-quartz (Kihara (1990), **Structure 97**) has a trigonal space group $\mathbf{P3_221}$

(D_3^6). The lattice constants are $a = 0.491\,37$ nm and $c = 0.540\,47$ nm at room temperature. β-Quartz (Kihara (1990), **Structure 98**) has a hexagonal space group **P6$_2$** (C_6^4). The lattice constants are $a = 0.499\,65$ nm and $c = 0.545\,46$ nm at 1078 K.

Figure 24.16(b) and (d) show the structures of α- and β-quartz in polyhedral depictions where the networks of the tetrahedral coordination polyhedra can be distinguished. The structures are similar, differing only in the tilting of the tetrahedra, which leaves the α-phase trigonal and the β-phase hexagonal, respectively. The linkage of SiO_4 tetrahedra in (001) planes, shows that the tetrahedra share three of their vertices in the basal plane. The fourth is shared with tetrahedra in adjacent (001) planes.

Quartz gem stones also exist: *amethyst*, the February birth stone, is the purple variety of quartz and is a popular gem stone. Another tectosilicate, sodium calcium aluminum silicate, is known as *oligoclase*, and is used in semi-precious gem stones (*sunstone* and *moonstone*; moonstone is the June birth stone).

Quartz is considered to be the prototype mineral for the silicate class because it has the maximal linkage of the $(SiO_4)^{4-}$ tetrahedral groups. Quartz was introduced as a ceramic structure in Chapter 22. In addition to its α and β forms, other allotropic forms include *trydimite* and α- and β-*cristobalite*. The α and β allotropes of quartz and cristobalite are also called low and high forms, respectively. Some of the properties of quartz are discussed in Box 24.1.

Box 24.1 Quartz and Pauling's rules

Quartz is a good structure for the illustration of Pauling's rules:

(i) A tetrahedral coordination polyhedron of O^{2-} anions is formed around the Si^{4+} cation. At room temperature, α-quartz has two tetrahedral Si–O bonds of length 0.1605 nm and two of length 0.1613 nm (Kihara, 1990). Given the 0.138 nm ionic radius for the O^{2-} anion in four-fold coordination, the Si^{4+} cation radius is thus inferred to have the small value of about 0.023 nm. The cation to anion radius ratio is thus:

$$\frac{R_{Si^{4+}}}{R_{O^{2-}}} = 0.167,$$

which is somewhat less than the critical radius ratio (CRR) of 0.225 for tetrahedral coordination. An ideal CRR would have the Si^{4+} cation a bit larger and the O^{2-} anion a bit smaller.

(ii) The strength, s, of the anion–cation bond and the bond sum, Σs, are given by:

$$s = \frac{+4}{4}; \qquad \Sigma s = 2(\frac{+4}{4}) = 2.$$

The structure is stable since the total strength of the bonds connecting two tetrahedra meeting at an O^{2-} vertex is 2+, exactly balancing the 2− anion charge.

(iii) The tetrahedra are linked sharing vertices, ensuring that the distance between the electropositive Si^{4+} cations is large. For α-quartz, at room temperature, the Si^{4+} cations are separated by 0.44 nm.

(iv) There is no edge sharing of the tetrahedra and the electropositive Si^{4+} cations sit in high symmetry sites.

(v) There is only one crystallographically distinct cation in the structure.

Low quartz has a room temperature molar volume of density of 22.67 cm^3/mol and a density of 2.65 g/cm^3, which can be calculated from the lattice parameters as a reader exercise.

24.7.2 Cage structures in the tectosilicates

In analogy with the large icosahedral cluster structures discussed in Chapter 18, we now consider large oxide structures (silicate, aluminosilicate, etc.) in which caged molecules are packed in the structure. Materials such as *sodalite* have large isolated cavities in their structures. The *zeolites* have separated or interconnected large cavities that make these materials interesting for such applications as ion-exchange, molecular sieves, catalysts, or for their absorptive properties. In this section, we illustrate sodalite, the structure of *chabazite* (a zeolite example), and the *fullerenoid oxide* structure (Hervieu *et al.*, 2004) in which Al_{84} cages resemble the structure of one of the isomers of the fullerene C_{84}. These are *tetrahedral oxides* which are reminiscent of the large cluster topologically close-packed metallic phases.

24.7.2.1 Sodalite

Sodalite (Fig. 24.17, **Structure 99**) is an example of a *feldspathoid* mineral, a low density aluminosilicate, typically with a Si:Al ratio near 1:1, and large openings in the crystal structure. These openings are isolated so that they are not useful for the movement of ions and molecules. The openings are occupied by large ions, including Cl, CO_3, and SO_4.

The sodalites are a group of minerals with similar structures and chemistries, within the feldspathoid group. The name derives from the presence of Na, and the entire group is named for the mineral sodalite. Minerals in this group have the chemical composition:

$$(\mathrm{Na},\mathrm{Ca})_8(\mathrm{SiO}_4)_6(\mathrm{SO}_4,\mathrm{OH},\mathrm{S},\mathrm{Cl})_2. \tag{24.4}$$

Figure 24.17 illustrates the sodalite structure for the mineral with chemical composition $Na_4Al_3Si_3O_{12}Cl$ and two formula units per cell. This is a cubic

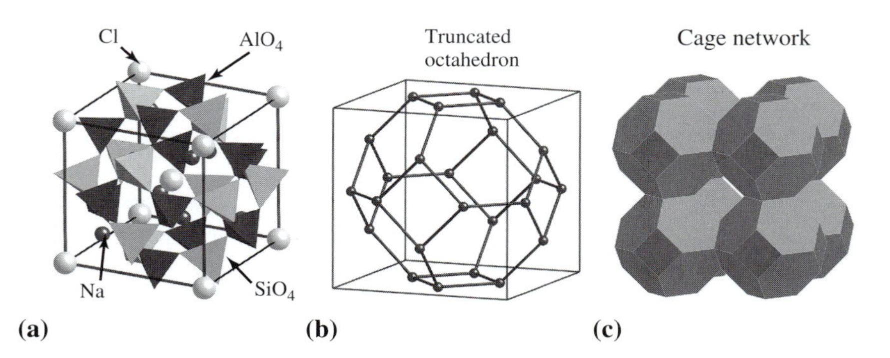

Fig. 24.17. (a) Structure of sodalite, $Na_4Al_3Si_3O_{12}Cl$: (a) vertex-sharing Si and Al tetrahedra, (b) the sodalite cage, and (c) network of cuboctahedral cages.

structure with space group **P$\bar{4}$3n** (T_d^4), and a lattice constant of 0.891 nm. In this structure, both Si and Al are tetrahedrally coordinated, with the Si tetrahedra sharing vertices with Al tetrahedra, as illustrated in Fig. 24.17(a). These form a 24-tetrahedron complex that forms the cage. Figure 24.17(b) shows the positions of the centers of the tetrahedra that form a truncated octahedron, known as the *sodalite cage*. Sodium and the large anions sit in the voids left in the cage network, illustrated in Fig. 24.17(c). The reader may show as an exercise that the sodalite cage consists of an interpenetrating Si and Al icosahedron.

24.7.2.2 Zeolites

The *zeolites* are framework silicates with connected SiO_4 and AlO_4 tetrahedra and two O-s per one (Si + Al).

The alumino-silicate portion of the zeolite structure is negatively charged and charge balance is achieved with positively charged cations located inside the cages. The zeolites have large vacant spaces or channels in their structures, making them interesting for a variety of applications. If the voids or channels are interconnected, they can be used as filters or chemical reaction sites. Some molecules can pass through the pores in zeolites, and others will not or will react in the pores. The zeolites can thus be used as molecular sieves. The large voids result in low zeolite densities.

A common use for zeolites is in water softeners, where Na ions charged in the zeolites are exchanged with Ca in hard water, allowed to pass through the channels. Zeolites can also absorb unwanted ions and molecules to act as toxin absorbers. Water in zeolite structures can be removed by heating, with the crystal structure remaining intact. Municipal water supplies are often processed through zeolite filters. Zeolites have basically three different structural variations. These are chain structures, of which *natrolite* is an example, sheet structures, such as *heulandite*, and framework structures like *chabazite* (Fig. 24.18, **Structure 100**).

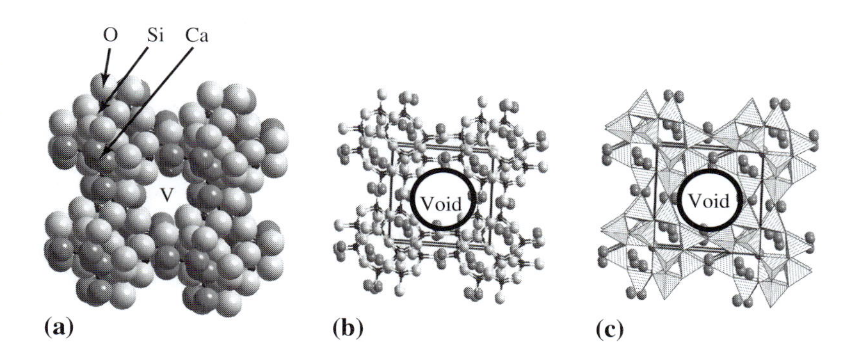

Fig. 24.18. Unit cell of the chabazite structure in a space filling (a), ball-and-stick (b), and polyhedral representation (c).

Figure 24.18 shows a unit cell of the chabazite structure, $CaAl_2Si_4O_{12} - 6H_2O$. Figure 24.18(a) and (b) show space filling and ball-and-stick representations. Note the large pore in the center of the structure. Figure 24.18(c) shows a polyhedral representation. The pores in this structure form an interconnected network of channels. Chabazite is used commercially for cation removal and as a desiccant. For example, chabazite is used to remove Cs and Sr isotopes from radioactive effluents. Chabazite is also of industrial importance in the conversion of methanol to olefins.

Zeolites extracted from the mineral *lazurite* (found in a rock called *lapis lazuli*) contain *ultramarine*. Ultramarine is a pigment with the chemical composition $Na_{8-10}\,Al_6Si_6O_{24}S_{2-4}$. Lapis lazuli (and sapphire) are September's birth stones.

24.7.2.3 Fullerenoid oxides

In 2004, poly-tetrahedral bismuth aluminum *fullerenoid oxides* were fabricated (Hervieu *et al.*, 2004). Their structure was determined to consist of a 3-D framework of AlO_4 tetrahedra, where the central Al ions form an 84-atom cage with the same geodesic structure as a C_{84} molecule. This led to the name *fullerenoid oxide* for this new class of materials (Fig. 24.19).

The fullerenoid cage structure has exclusively pentagonal and hexagonal faces, the same as the D_{2d} isomer of the C_{84} molecule. These Al_{84} pseudo-spheres can be packed in an *fcc* structure. We can decompose the structure, which has around 1200 atoms in the unit cell, into a truncated tetrahedral O_{12} unit surrounded by a Bi_{16} unit, a $Sr_{32}(Bi_{8.25}\square_{3.75})$ unit and an O_{126} shell which sit inside each Al_{84} cage (which is 1.85 nm in diameter). Figure 24.19(a) shows a single unit cell of the $Sr_{32}Bi_{24+\delta}Al_{48}O_{141+\frac{3\delta}{2}}$ fullerenoid oxide. The structure (**Structure 101**) has space group $\mathbf{F\bar{4}3m}$ (T_d^2) and a 2.509 nm cubic lattice constant. Figure 24.19(b) shows the connectivity of the AlO_4 tetrahedra in the cell. Figure 24.19(c) shows the framework of the Al_{84} pseudo-spheres in an *fcc* arrangement. Notice that the *fcc* "cell" for the Al_{84} packing is larger than that of the crystal unit cell. Figure 24.19(d) shows a chain of Al_{84} units along a [111] direction. Note that the Al_{84} units are centered on (1/4, 3/4, 1/4)-type positions in the unit cell. While this structure is technically not a silicate,

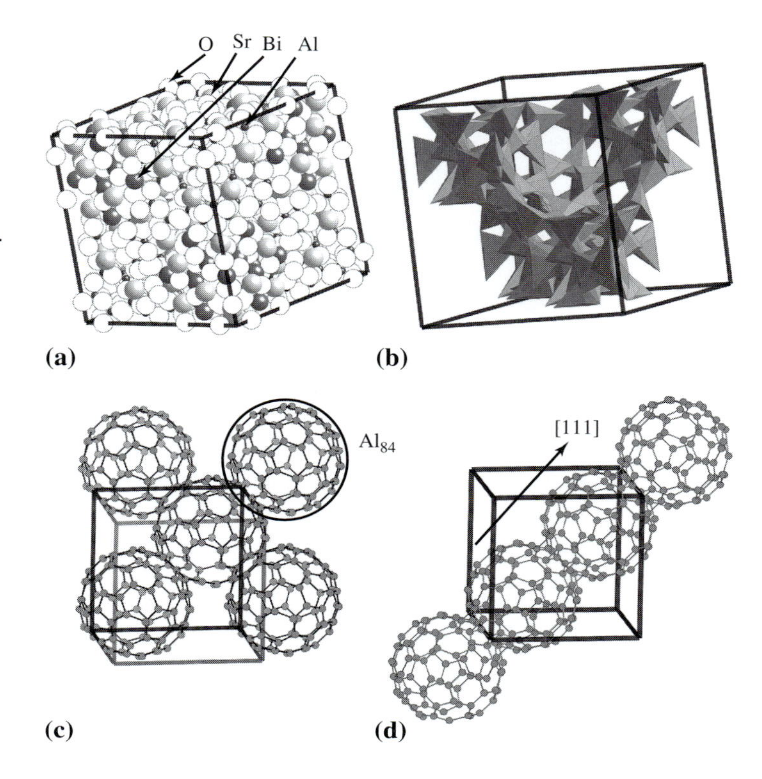

Fig. 24.19. $Sr_{32}Bi_{24+\delta}Al_{48}O_{141+3\delta/2}$: (a) single unit cell, (b) connectivity of the AlO_4 tetrahedra, (c) *fcc* arrangement of Al_{84} pseudo-spheres, and (d) Al_{84} chains along a [111] direction.

it was included in this section on cage structures because of its complex arrangement of 84-atom cages.

24.8 Random networks of tetrahedra: *silicate glasses*

Amorphous silica, a-SiO_2, is an example of a *network glass*. Network glasses are among the most important amorphous materials. Zachariessen (1932) proposed the *continuous random network (CRN) model* for the structure of network glasses. In the CRN model (Fig. 24.20), the glass has a well-defined local structure determined by preferred cation coordination numbers. For silicate glasses, the preferred coordination is tetrahedral: each Si atom has four O neighbors and each O has two Si neighbors. The structure does not, however, possess long-range translational periodicity. The tetrahedra share vertices but the randomness of the network imposes fewer restrictions on the SiO_2 bond angles. Defects in such a structure include dangling bonds, where the Si coordination is 3, or O coordination is 1.

It is common and instructive to illustrate a continuous random network in 2-D for a Si_2O_3 glass, to convey the concept. Figure 24.20(a) illustrates the structure of a hypothetical 2-D crystalline quartz (Si_2O_3) structure as a

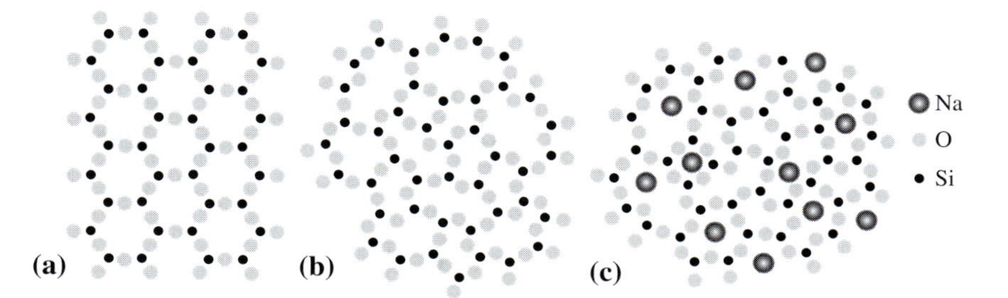

Fig. 24.20. (a) Hypothetical structure of 2-D crystalline quartz (Si_2O_3) as a hexagonal network; (b) A continuous random network of 2-D amorphous quartz (Si_2O_3); (c) 2-D CRN model of a sodium silicate glass (Na is a network modifier).

hexagonal network. Figure 24.20(b) shows a continuous random network of 2-D amorphous quartz. Figure 24.20(c) shows a proposed 2-D CRN model of a sodium silicate glass, where Na ions occupy larger voids in the CRN and eliminate some of the dangling bonds in the structure.

Zachariessen also proposed a set of empirical rules, similar to *Pauling's rules*, to predict which oxides would be good glass formers. *Zachariessen's rules* are:

(i) Networks in oxide glasses are determined by O coordination polyhedra.
(ii) O acts as a bridge between cations. Each O has no more than two cation neighbors, i.e., O is two-fold coordinated.
(iii) The cation coordination number should be 3 or 4, leading to tetrahedral $(SiO_4)^{4-}$ structures in silicate glasses and triangular $(BO_3)^{3-}$ structures in borate glasses.
(iv) Each cation coordination polyhedron shares at least three vertices with neighboring polyhedra, but does not share any faces or edges.

Oxides obeying Zachariessen's rules form 3-D network structures and are referred to as *network formers*. Oxides of alkali, alkaline earths, and others do not form network structures. When added to a network-forming glass, they are called *network modifiers*. Glasses with network modifiers will flow at a lower temperature, making them easier to process. They are also more likely to lose their shape.

24.9 Mesoporous silicates

A new and active area of nanoparticle research in which silicates figure prominently is the study of *mesoporous silicates*. The well-ordered hexagonal pore structure of a mesoporous ceramic known as MCM-41 has been used as a template for the growth of many interesting nanostructures, including *magnetic nanowires*. MCM-41, first synthesized in 1992 by the Mobil Corporation (Beck *et al.*, 1992, Kresge *et al.*, 1992), is composed of porous silica (SiO_2). It contains cylindrical pores in the mesoscale size range (2 – 50 nm in diameter), which arrange themselves in a hexagonal lattice. These self-assembled arrays have a high porosity, both in terms of the specific pore

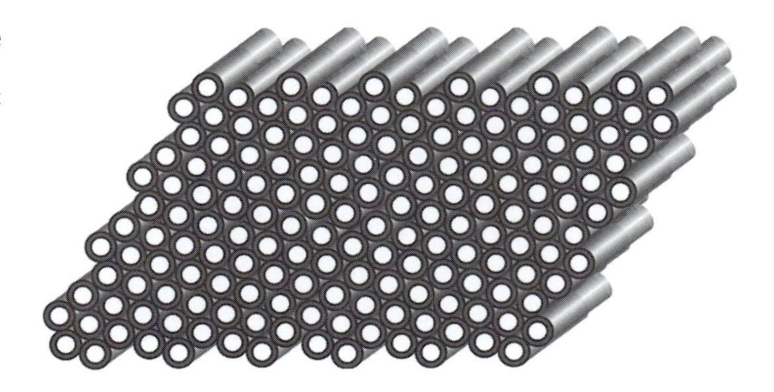

Fig. 24.21. Schematic structure of the 2-D hexagonal network of hollow silicate cylinders that form the mesoporous ceramic MCM-41 (Beck *et al.*, 1992, Kresge *et al.*, 1992).

volume and the surface area. Figure 24.21 illustrates the schematic structure of the mesoporous ceramic MCM-41.

MCM-41 is synthesized by the catalyzed hydrolysis of TEOS (tetraethylorthosilicate, tetraethoxysilane, $Si(OCH_2CH_3)_4$) in a partially to wholly aqueous surfactant solution near room temperature. *Micelles* (in aqueous solution) consist of collections of surfactant molecules in which all of the hydrocarbon chains try to avoid contacting the solution by packing tightly together, which results in a polar outer surface and a non-polar interior. Because of this, three major shapes are seen: *spherical micelles*, *columnar micelles*, and *lamellar micelles*. The shape is determined by the surfactant to water ratio, and during MCM-41 formation this ratio is tailored so as to give columnar micelles. Due to intermicellular interactions, these columnar micelles tend to self-assemble to form a hexagonal liquid-crystal template. Adding an acid or base to a TEOS solution with the liquid-crystal template acts to catalyze the decomposition to silica. Micelles and the process of *self-assembly* in polymers are discussed in more detail in the next chapter.

Like the zeolites, mesoporous silicates are investigated for applications as molecular sieves, catalysts, and *nanotesttubes*. The term nanotesttubes denotes the fact that they can serve as hosts for other chemical reactions that take place within the pores.

24.10 Sol-gel synthesis of silicate nanostructures

Another active area in the production of nano-structured materials is the study of *sol-gel synthesis*. Among products produced by sol-gel synthesis are *sols*, *wet gels*, *xerogels*, *aerogels*, *xerogel films*, and monodisperse ceramic powders. These can be further processed to produce dense ceramics, dense films, ceramic fibers, and other useful engineered structures. The silicates have figured prominently as materials produced by sol-gel synthesis, and their applications include silica-gel sensors. Figure 24.22 illustrates examples of processing steps and products made by sol-gel technologies.

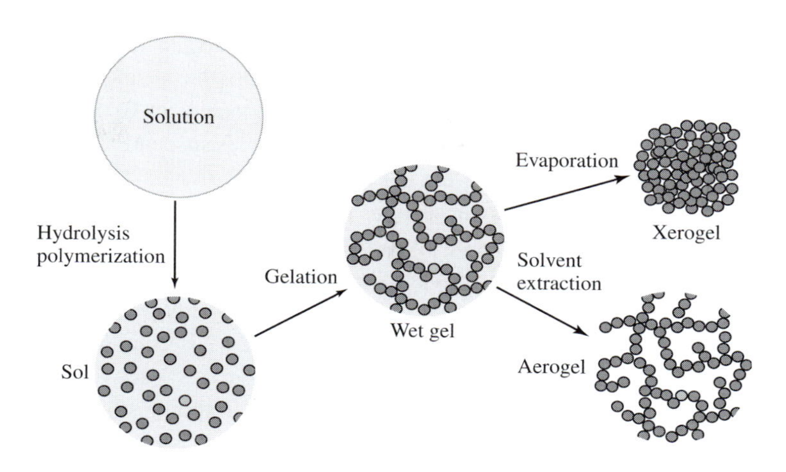

Fig. 24.22. Examples of processing steps and products made by sol-gel technologies.

Sol-gel synthesis begins with a solution, typically of inorganic metal salts or metal organic compounds, in a solvent, often ethanol. Metal *alkoxides*, common *organo-metallic compounds*, are used as precursors in sol-gel processing. The first processing step produces a *sol* by hydrolysis and polymerization reactions. A sol is a dispersion of solid particles in a liquid. These particles are typically small, in the range 1–100 nm diameter. Paint would be a prominent example of a sol. A *gel* is the opposite of a sol, since it has liquid droplets inside a solid; examples include gelatin and opal.

The discussion in Box 24.2 describes the basic steps in the synthesis of a silica sol. A condensation reaction connects Si atoms through a bridging oxygen atom. We represent these groups as linear siloxanes, Si–O–Si, but the reaction can also give rise to tetrahedral linkages. The repeated reactions are similar to the polymerization reactions that will be discussed in the next

Box 24.2 Propagation reaction to form polyethylene

The first step is a hydrolysis step in which water is added to an alkoxide solution:

$$\mathrm{Si-O}R+\mathrm{H-O-H} = \mathrm{Si-OH} + R\mathrm{-O-H},$$

where R represents the alcohol group. This step produces a *silanol*, Si–OH. The second step involves a condensation reaction to attach two silanols to form a *siloxane* group:

$$\mathrm{Si-OH} + \mathrm{Si-OH} = \mathrm{Si-O-Si} + \mathrm{H-O-H}$$

chapter. The second step illustrated in Fig. 24.22, *gelation*, is a water condensation step, but it could also be an alcohol condensation step. When a sufficient number of interconnected Si–O–Si bonds form in a region, they interact to form colloidal particles suspended in solution. The gelation process transforms the sol into a gel.

A sol can serve as the precursor for producing a variety of other forms. Thin films can be produced by *spin-coating* or *dip-coating*. By varying the sol viscosity, ceramic fibers can be drawn. Chemically uniform ceramic nanopowders can be formed by *precipitation*, *spray pyrolysis*, or *emulsion* methods. If the condensation reaction of Box 24.2 is continued for a long time, the colloidal particles can link together to form a 3-D network. At the gelation point, the network becomes a solid skeleton, trapping solvent in its pores; this is known as a *wet gel*. If the liquid in a wet gel is extracted by supercritical (freeze) drying, a porous and very low density *aerogel* is formed. If the liquid in a wet gel is further evaporated, the skeleton collapses and a fine *xerogel* is formed.

24.11 Historical notes

Gerolamo Cardano (1501–76) (also spelled Girolamo Cardanus and in English Jerome Cardan) was an Italian astrologer, inventor, philosopher, and mathematician. He began his studies in medicine at Pavia University but, after the university closed, he moved to the University of Padua. Cardano earned a doctorate in medicine in 1525. He was appointed as mathematics chair in Milan in 1532. He published Tartaglia's solution of the cubic equation and Ferrari's solution of the quartic equation. He was one of the first mathematicians to consider square roots of negative numbers, later called imaginary numbers. His book *Ars magna* (Great Art) is an influential text in the history of algebra. His work on probability theory provided the foundation on which the laws of probability were constructed by later mathematicians.

Fig. 24.23. (a) G. Cardano (1501–76) (Cardano, 1550) (picture courtesy of J. Lima-de-Faria and originally courtesy of Professor S. Menchetti of Florence University, Italy) and (b) Jose Lima-de-Faria (1925–) (picture courtesy of J. Lima-de-Faria).

Cardano studied tuberculosis, asthma, and venereal disease. With notable achievements in mathematics and medicine, he was a true renaissance man. He also contributed in more obscure areas with lasting impact. For example, he invented the cardan shaft, a device used in automobiles today.

Cardano wrote the book *De Subtilate* ("Transcendental Philosophy" Cardano, 1550) in which he attempted to interpret the hexagonal form of quartz in terms of close packing of spherical particles. His was possibly the first attempt at describing close packing and may have been motivated by observations of the stacking of cannon balls. Cardano expressed ideas as to the healing effects of quartz crystals and on the optical properties of glass.

Cardano died by committing suicide in 1576, fulfilling his prophecy that he would live to the age of seventy five (Wykes, 1969). An extended biography of Cardano can be found at the history website of the School of Mathematics and Statistics at the University of St. Andrews, Scotland[2] and the World Research Foundation website.[2,3]

Jose Lima-de-Faria (1925–) is the Director and Head of the Department of Earth Sciences of the Centro de Cristalografia e Mineralogia, Instituto de Investiga Cientifica Tropical, in Lisbon, Portugal. Lima-de-Faria and Figueiredo extended the structural classification of silicates to all inorganic structures. He developed a chart of inorganic structural units and a method for tabulating inorganic structure types (Lima-de-Faria and Figueiredo, 1976, 1978). Lima-de-Faria edited the book *Historical Atlas of Crystallography*, published for the International Union of Crystallographers (Lima-de-Faria, 1990), which has been a source of much information used in compiling the historical sections of this book. This book has time maps, geographical information and portraits of crystallographers and scientists influential in the development of crystallography. It is highly recommended as a reference on the history of crystallography. The book has chapters on geometrical crystallography, by Marjorie Senechal; physical crystallography, by W. A. Wooster; the chemical crystallography of inorganic compounds, by P. B. Moore, the chemical crystallography of organic compounds, by Jenny P. Glusker; crystal structure determination, by Martin J. Buerger, title pages of important works on crystallography, by J. Lima-de-Faria; and the domain of crystallography, by Helen D. Megaw.

24.12 Problems

(i) *Quartz*: α-quartz (low quartz) has space group **P3$_2$21** (D_3^6). The silica oxide compound has lattice parameters $a = 0.491\,37$ nm and $c = 0.540\,47$ nm, with atoms located in the sites tabulated below. Using the

[2] See http://www-history.mcs.st-andrews.ac.uk/history/Mathematicians/Cardan.html.
[3] See http://www.wrf.org/news/news0002.htm.

International Tables for Crystallography as a reference, determine the following:

(a) The Bravais lattice, the Pearson symbol, and the point group symmetry at each special position.
(b) The composition, number of formula units in a unit cell and density.
(c) The distance between Si^{4+} cations and O^{2-} anions in the cation coordination polyhedron.

Atom	Site	x	y	z
Si	3a	0.4697	0.0000	0.0000
O	6c	0.4133	0.2672	0.1188

(ii) *Forsterite*: Compare the structure of forsterite and that of the $MgAlO_4$ spinel (discussed in Chapter 22).
(iii) *Garnet bond valence*: Perform a bond valence calculation (as discussed in Chapter 22) at the O vertex connecting an octahedron, tetrahedron, and dodecahedron in the garnet structure.
(iv) *Garnet* O *position*: Consider the $Gd_3Fe_5O_{12}$ garnet. Draw a triangle in a plane showing the positions of the centers of the octahedron, tetrahedron, and dodecahedron. Considering the ionic radii of O and the cation species, estimate the position of the O atom shared at the vertex of the polyhedra in the garnet cell. Is it coplanar with the cations?
(v) *Zircon*: Zircon has four formula units per tetragonal unit cell. At 1823 K it has lattice constants $a = 0.6649$ nm and $c = 0.604$ nm. At 298 K it has lattice constants $a = 0.661$ nm and $c = 0.6001$ nm. Calculate the density of Zircon at 298 K and estimate the *coefficient of thermal expansion*.
(vi) *Beryl*: Determine the radius of the ring in the Beryl structure.
(vii) *Tremolite*: Determine the dimensions of the *I*-beams in the tremolite structure.
(viii) *Mica*: Estimate the thickness of the *TOT* layers in the mica structure. What is the thickness of the *I* layer coupling the *TO* layers? Rationalize the easy cleavage of mica.
(ix) *Kaolinite*: Estimate the thickness of the *TO* layers in the kaolinite structure. What is the thickness of the OH layer coupling the *TOT* layers?
(x) *Sodalite cage I*: Show that the sodalite cage consists of an interpenetrating Si icosahedron and an Al icosahedron. What is the orientation of each?
(xi) *Sodalite cage II*: Calculate the size of the largest sphere that touches the atoms on a sodalite cage.
(xii) *Sodalite density*: Given a 0.891 nm cubic lattice constant for sodalite, calculate the density of this mineral.

(xiii) *Chabazite*: Estimate the radius of the void in the center of the Chabazite structure.

(xiv) *Fullerenoid oxide*: Determine the size of the *fcc* cell referring to the packing of Al_{84} pseudo-spheres with reference to the actual crystal unit cell size for $Sr_{32}Bi_{24+\delta}Al_{48}O_{141+\frac{3\delta}{2}}$.

(xv) *Glass I*: Use Zachariessen's rules to determine whether ZnO would be a candidate network-forming glass.

(xvi) *Glass II*: Consider the bond density in a network-forming glass. Does this increase or decrease with the addition of network modifiers? Explain. What influence will this have on the temperature at which the glass flows?

(xvii) *Mesoporous Silicate I*: Consider a simple model for a mesoporous silicate with SiO_2 rods of 11 nm outer diameter and 10 nm inner diameter. Estimate the specific surface area of the pores in units of m^2/g.

(xviii) *Mesoporous Silicate II*: Again consider SiO_2 rods of 11 nm outer diameter and 10 nm inner diameter that organize in a 2-D hexagonal lattice. Calculate the d-spacings for the (10), (11), and (21) reflections. Predict the angles of XRD peaks for these reflections assuming Cu-$K\alpha$ radiation.

(xix) *Mesoporous Silicate III*: Calculate the density of a mesoporous silicate crystal for which SiO_2 rods of 11 nm outer diameter and 10 nm inner diameter organize in a 2-D hexagonal lattice.

CHAPTER

25 Molecular solids

> *"We've discovered the secret of life."*
>
> Francis Harry Compton Crick, 1953

25.1 Introduction

Molecular solids are defined loosely as solids for which the building blocks are conveniently described in terms of molecular, rather than individual atomic constituents. We have already seen that it is useful to represent some ceramic and silicate structures in terms of molecular units. This chapter emphasizes structures based on low atomic number constituents, such as C, H, O, N, ... *Organic chemistry* is defined as the chemistry of carbon compounds; this encompasses all molecules that occur in living organisms and in materials important for life. An older definition of *organic* as "compounds derived from living organisms" is broadened here to include synthetic materials, which are important in man-made compounds such as polymers and fullerene-based solids.

Molecular crystals often have strong bonding within the molecular units, with weaker *intermolecular* interactions, that give rise to a weak solid cohesion. In many instances, the solid is held together by *van der Waals forces* or *hydrogen bonding*. Van der Waals forces (also known as *London dispersion forces*) are attractive forces between instantaneous atomic or molecular dipole moments. As the dipole moment interaction energy falls off rapidly with distance between the dipoles, this accounts for the weak cohesion in the molecular crystals. A hydrogen bond is a bond between two or more (usually electronegative) atoms or molecules, mediated by electropositive hydrogen atoms. If two electronegative ions, A_1, A_2, (e.g., F, O) form a short ionic or covalent bond, then a longer and weaker hydrogen bond can result from $(A_1H) - A_2$ dipolar interactions. While the *weak forces* are deemed weak on an individual basis, they can be quite strong when taken in aggregate. Hydrogen bonds, in particular, can be very important in determining material properties.

Examples of molecular crystals are abundant. They tend to be characterized by small cohesive energies, resulting in low melting and boiling temperatures, and large coefficients of thermal expansion; many of these materials are mechanically soft in the crystalline state.[1] Examples include *ice*, *solid* CO_2, *solid benzene*, and *simple sugars*. Many organic structures share structural features with inorganic materials; an example of an organic caged structure is found in the *clathrates*, which are similar to the zeolite structures discussed in the previous chapter. One of the most important attributes of multiphase molecular systems is their ability to *self-assemble* on different length scales. We illustrate some of the structures that result from such self-assembly processes in this chapter.

Polymers, molecules with repeated chemical units called *mers*, are important *macromolecular solids*. The word "polymer" is rooted in the Greek language, and means "many membered." Polymers and other macromolecular solids can be crystalline, semicrystalline, or amorphous. Amorphous molecular and macromolecular solids discussed in the chapter include allotropic forms of ice and various polymers. Other molecular crystals are classified as *biological macromolecular crystals*. These include *proteins*, *deoxyribonucleic acid* (*DNA*) and *ribonucleic acid* (*RNA*), *polysaccharides*, etc. Many important viruses can also be considered as macromolecules, coated with other protecting molecules that tile the surface of a sphere. In this sense, they are 2-D crystals in curved space. Fullerenes and carbon nanotubes are examples where a C-network tiles a curved space. There are now many examples of solids (e.g., *fullerites* and *fullerides*) that can be constructed from the fullerene molecules; some of these solids are also illustrated in this chapter.

25.2 Simple molecular crystals: ice, dry ice, benzene, the clathrates, and self-assembled structures

25.2.1 Solid H_2O: ice

Water, an essential molecule for the existence of life on Earth, is perhaps one of the most extensively studied molecules, in gaseous, liquid, and solid

[1] The mechanical softness varies from one material to another. Single crystal ice has a typical Young's modulus of about 10 GPa, whereas polymers have typical moduli between 0.1 and 10 GPa. The upper end of this range approaches values observed for soft metals or glasses.

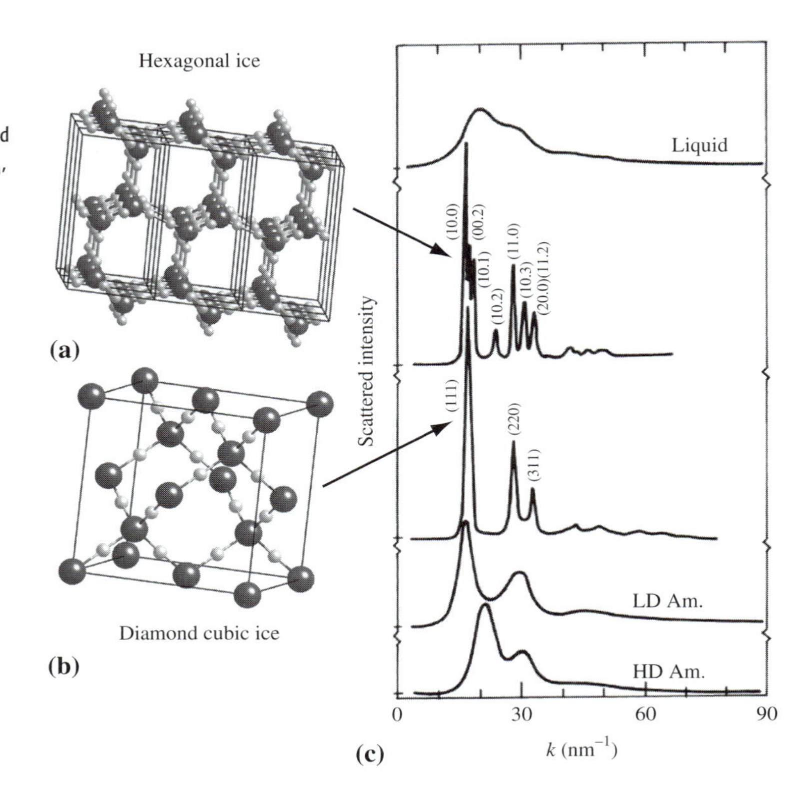

Fig. 25.1. The hexagonal structure (a) of ice I_h and the (metastable) cubic structure (b) of ice I_c; (c) X-ray scattered intensity of liquid water, ice I_h, ice I_c, and low and high density amorphous ice. (L. Bosio, G. P. Johari and J. Teixeira, Phys. Rev. Lett., **56**:460–463, 1986; Copyright (1986) by the American Physical Society).

forms. The various arrangements of the H_2O molecule as a function of the thermodynamic state variables (temperature and pressure) have been studied widely. High pressure studies of ice have identified at least nine crystalline *allotropic* forms (Gaskell, 1993); a tenth was discovered in the mid 1980s (Mishima *et al.*, 1985). We illustrate examples of the structure of ice in Fig. 25.1.

The structural determination of compounds or crystals containing hydrogen can be very difficult, because, as a light element, H is not a strong scatterer. Often, structural determination can be made easier by isotopic substitution of deuterium, D, for H. Isotopic effects on properties can be large in D-substituted materials because the mass of the atom is doubled. Although deuterium is chemically similar to hydrogen, it has a larger cross section for neutron scattering. This cross section helps to refine the determination of H (D) positions in D_2O crystals. A comparison of the cell dimensions of ordinary and "heavy ice" was first made by Helen D. Megaw.

At standard temperature and pressure, ice is hexagonal (ice I_h). Hexagonal ice (**Structure 102**) has space group **P6$_3$/mmc** (D_{6h}^4), and an open, low density structure. This allotrope has a lower density than liquid water at atmospheric pressure, which has important ramifications; one vital effect is that solid ice floats on its liquid. Although the hydrogen bonding in ice structures is often depicted as static, it actually changes over time. However, on average, two H atoms do reside near each O atom, locally preserving the H_2O stoichiometry and one H atom bridges each O–O bond. In general, the H–O–H bond angle in the crystal structure is similar to that of the isolated molecule. Figure 25.1(a) shows several unit cells of the hexagonal crystal structure of ice I_h and (b) a single unit cell of the (metastable) cubic crystal structure of ice I_c (**Structure 103**).

Figure 25.1(c) shows the X-ray scattered intensity of liquid water, ice I_h, ice I_c, and low density and high density amorphous ice (Bosio *et al.*, 1986). The I_h structure has unit cell dimensions of $a = 0.4511$ nm and $c = 0.7351$ nm (Goto *et al.*, 1990), with O at the $(1/3, 2/3, 0.0618)$ position and H at $(1/3, 2/3, 0.173)$ and $(0.437, 0.873.0.024)$. Hydrogen atoms are disordered over these two sites, with four H_2O molecules in the unit cell. Ice I_c has a diamond cubic structure with unit cell dimensions of $a = 0.635$ nm (Dowell and Rinfret, 1960), with O in the $(0, 0, 0)$, and H in the $(1/8, 1/8, 1/8)$ special positions. The unit cell contains eight H_2O molecules. This is a low temperature (high pressure) allotrope of ice that has an open structure and nearly the same density as ice I_h. Other allotropes of crystalline ice include ice II, which can be produced from ice I_h by cooling about 25 °C below its freezing point, and then pressurizing to 2200 times atmospheric pressure. At very high pressures, the molecules in ice crystals form atomic chain structures.

Two allotropic forms of *amorphous ice* have been identified. When hexagonal ice I_h is compressed at or above pressures of 1 GPa at 77 K, it transforms into a high density (1.31 g/cm^3 at 1 GPa) amorphous solid (Mishima *et al.*, 1984). Upon reducing the pressure, we observe the amorphous ice phase to elastically expand to a density of 1.19 g/cm^3 at atmospheric pressure. Subsequently, scientists found that high density amorphous ice will transform to a new amorphous phase upon heating to 117 K in zero pressure (Mishima *et al.*, 1985). This allotrope has a density of 0.94 g/cm^3 at atmospheric pressure.

Low density amorphous ice can be "quenched" to 77 K, where it is metastable. If this allotrope is subsequently compressed at 77 K, it transforms to the high density allotrope at 0.6 GPa. The pressure dependence of this transformation is very sharp. The amorphous-I to amorphous-II transition has been postulated to be important in planets that grow large enough to achieve these pressures. Figure 25.1(c) compares the scattering from liquid water, and the amorphous ice allotropes (Bosio *et al.*, 1986).

25.2.2 Solid CO_2: dry ice

Carbon dioxide, CO_2, is another molecule for which the crystal structure (Fig. 25.2) has been widely studied. Carbon dioxide is a linear chain molecule with C centered between two O atoms and a molecular length of about 0.24 nm at room temperature. Each O atom is bonded to C by a covalent

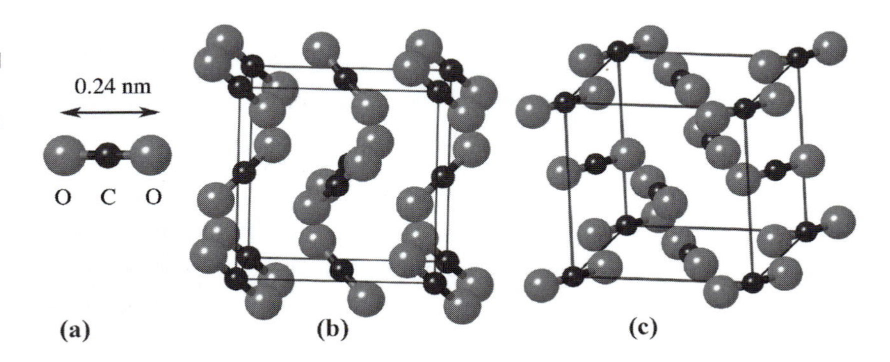

Fig. 25.2. The CO_2 molecule (a), the cubic structure of solid CO_2, dry ice (b), and orthorhombic structure at high pressure and room temperature (c).

double bond, so that all three atoms acquire a filled outer shell. The C=O double bond length in CO_2 is 0.12 nm. The remaining four electrons on each O reside in Lewis pairs above and below the axis of the molecule on the side of the O atom, opposite the bond. These unbound electrons give rise to a *molecular quadrupole moment*, which is important for the molecular properties and bonding in CO_2 crystals.

Solid CO_2, more commonly known as *dry ice*, forms at a pressure of 1.5 GPa at room temperature. The crystal structure of solid CO_2 is primitive cubic (**Structure 104**) with the CO_2 bond axis oriented along the [111] crystallographic axes (Keesom and Kohler, 1934). Figure 25.2(a) shows a ball-and-stick representation of the cubic crystal structure of CO_2. Aligning the linear molecule in this way results in the space group **Pa$\bar{3}$** (T_h^6). The symmetry of this structure will be reduced if the molecules become tilted with respect to the high symmetry cubic axes. In fact, we observe this reduction in the high pressure solid phase of CO_2 at room temperature. Recent high pressure structural data for solid CO_2 shows that a cubic to orthorhombic phase transition occurs at 11.8 GPa (Aoki *et al.*, 1994). The orthorhombic phase has lattice parameters $a = 0.433$ nm, $b = 0.4657$ nm, and $c = 0.5963$ nm. The linear CO_2 molecules lie in the $y - z$ plane, inclined at an angle $\phi = 52°$ with respect to the c-axis (Aoki *et al.*, 1994). Figure 25.2(c) shows the orthorhombic structure of solid CO_2.

25.2.3 Hydrocarbon crystals

Hydrocarbon crystals can have structural units in the form of linear chains, rings (as in *aromatic hydrocarbons*), and so on. The C–C bond determines the properties of many of these organic materials. Carbon forms four bonds using the s and p electrons in its outer shell. The strongest hybrid sp^3 *bonding* occurs in diamond. Graphite forms planar sp^2 *bonding* with the fourth valence electron in delocalized π bonds. The bond length of C–C bonds varies from 0.154 nm for single bonds, to 0.134 nm for C=C double bonds, and 0.12 nm for C≡C triple bonds. Planar aromatic hydrocarbon bond lengths

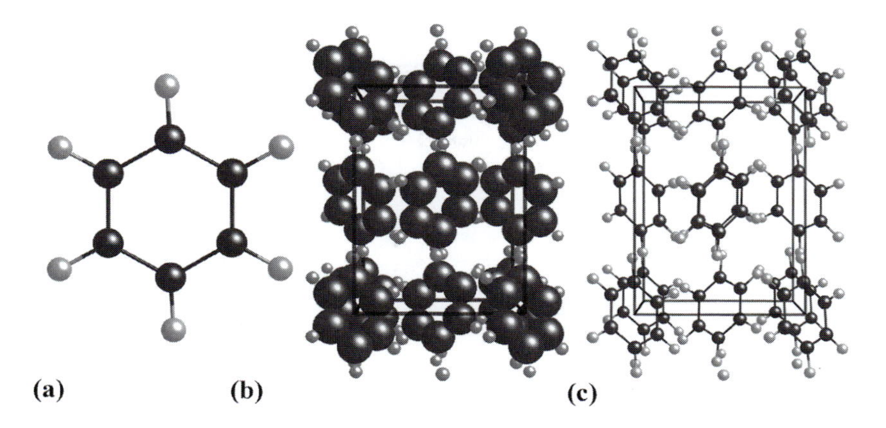

Fig. 25.3. (a) Benzene molecule. Space filling (b) and ball-and-stick (c) models of solid benzene.

vary between 0.13 and 0.15 nm, depending on the chemical environment. A model of the hydrocarbon molecule *benzene*, C_6H_6, is shown in Fig. 25.3(a).

Benzene is a liquid at room temperature and atmospheric pressure. It has at least three crystalline allotropes as a function of pressure. The first high pressure measurements were performed by Bridgman (1911). The lowest pressure allotrope crystallizes at 0.7 kbar at room temperature and has an orthorhombic space group **Pbca** (D_{2h}^{15}) with lattice constants $a = 0.744$ nm, $b = 0.955$ nm, and $c = 0.692$ nm (Piermarini *et al.*, 1969) (**Structure 105**). Figure 25.3(a) illustrates a single benzene molecule. It has six C atoms and six localized C–H single bonds radiating from a hexagonal ring. Each C atom forms a localized single bond with its two neighbors, defining the ring circumference. The remaining electron on each C resides in delocalized *π-bonds* around the ring.

Kathleen Yardley Lonsdale (1903–71) used XRD to study experimentally the structure of the benzene ring, showing conclusively that the C–C bonds were all of the same length and all C–C–C bond angles were 120° (Lonsdale, 1929). She confirmed the ring structure proposed by Friedrich August Kekule in 1865. Figure 25.3 shows the structure of the low pressure allotrope of solid benzene in space filling (b) and ball-and-stick (c) representations. The structure has strongly bonded hexagonal rings, but the crystal binding is rooted in the hydrogen bonds between molecules. The high pressure allotropic forms are monoclinic (Piermarini *et al.*, 1969). Delocalized π-electrons give rise to diamagnetic ring currents in response to a magnetic field, in accordance with *Lenz's law*.

25.2.4 Clathrates

Clathrate is the generic term for a compound that is formed by the incorporation of atoms or molecules into a crystalline lattice formed by other

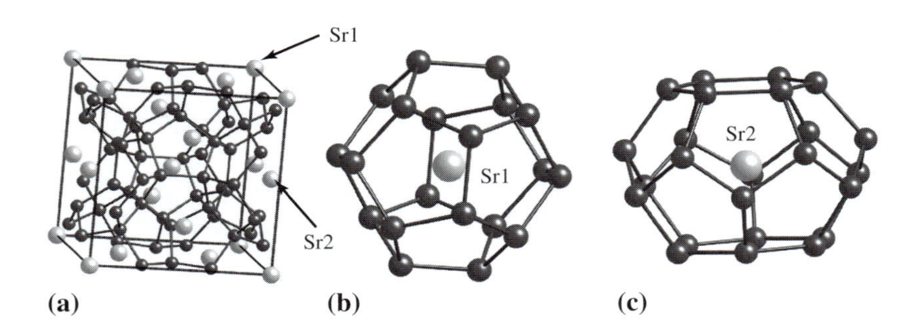

Fig. 25.4. (a) Unit cell of the Type I ice clathrate, $Sr_8Ga_{16}Ge_{30}$, (b) pentagonal dodecahedral coordination polyhedra about Sr1, and (c) tetrakaidecahedral coordination about Sr2.

molecules. More specifically, clathrates are polyatomic compounds in which one component forms a cage structure surrounding the other (Nolas *et al.*, 2001). Important clathrates include molecular complexes of water with simple molecules, such as Cl_2.[2] These are also called *gas hydrates*. They consist of crystalline H_2O cages with trapped gas molecules that can include noble gases, hydrocarbons, SO_2, etc. Hydrate structures have been studied by several groups, including Pauling's (Pauling and Marsh, 1952).

Ice clathrates consist of water molecules tetrahedrally bonded to four neighboring H_2O molecules, as in ice I_h and ice I_c. However, they have a more open structure with cavities that can accommodate atoms or molecules of different sizes. Two common forms of ice clathrates are the Type I and Type II clathrates. The Type I clathrate is illustrated in Fig. 25.4 for the semiconducting compound $Sr_8Ga_{16}Ge_{30}$.

The Type I ice clathrate is cubic with a lattice constant a of about 1.2 nm and contains 46 water molecules. The cavities in this structure include two *pentagonal dodecahedra* with 20 H_2O molecules arranged at the vertices, and six *truncated octahedra* (*tetrakaidecahedra*) with 24 H_2O molecules arranged at the vertices. Filling all eight interstices with a species, M, gives rise to the general formula $8M - 46H_2O$. Type II ice clathrates have cubic cells with a lattice constant a of about 1.7 nm and contain 136 water molecules. Water molecules form 16 pentagonal dodecahedral cages and 28 H_2O molecules form 8 *hexakaidecahedra*, resulting in the general formula $24M - 136H_2O$.

Semiconducting clathrates are the subject of much active research (Nolas *et al.*, 2001). Kasper *et al.* (1965) first determined the structure of Na_8Si_{46} to be isomorphic with the Cl_2 hydrate clathrates. Alkali metals in Si, Ge, or Sn hosts form both Type I and Type II clathrate structures. Scientists have observed semiconducting clathrates to have thermal conductivities with temperature dependences similar to those of amorphous materials, spurring interest in their *thermoelectric properties*.

[2] Michael Faraday (Faraday, 1823) proposed the clathrate compound Cl_2–$10H_2O$ in 1823.

25.2.5 Amphiphiles and micelles

When a molecule repels water, it is said to be *hydrophobic*, from the Greek words for "water hating." A molecule that is attracted to water is known as a *hydrophilic* molecule. Molecules that have one hydrophobic end and one hydrophilic end are known as *amphiphiles*. The names amphiphile and *surfactant* are often used interchangeably (Hamley, 2000). Surfactant is short for *surface active agent*. Amphiphiles have a tendency to phase separate from the solvent on a microscopic length scale. Solutions of amphipiles minimize the total energy associated with the attractive and repulsive interactions by arranging the molecules in a variety of ordered structures on length scales much larger than the atomic dimensions. This organization, known as *self-assembly*, is important both in solutions of amphiphilic molecules and solutions of amphiphilic *block copolymers*, as discussed below.

Amphipiles are polar hydrocarbon molecules that tend to phase separate from aqueous solutions as oil does from water. Oils are hydrocarbons that have hydrophobic interactions with water; oils tend to disrupt the hydrogen bonding in water. Because of this disruption, water molecules rearrange their hydrogen bonds near the hydrocarbon molecules, creating a large interfacial energy between oil and water. To minimize this interfacial energy, the hydrocarbon molecules segregate to interact with other hydrocarbon molecules and water with other water molecules. This is the underlying reason for the phase separation.

Amphiphiles differ from oils because amphiphiles have polar character: The *hydrophilic* ends of the molecules are attracted to water while the hydrophobic ends tend to join to one another, either at surfaces or at two-phase interfaces. This tendency is the underlying cause for the self-assembly of amphiphiles and the rich variety of structural morphologies that can be observed in two-phase water–amphipile systems (Fig. 25.5).

Unlike oils, which separate from water on a macroscopic scale, amphiphiles undergo *microphase separation*, characterized by microscopic and thermodynamically stable aggregates called *micelles*. Figure 25.5 illustrates common micellular morphologies, including spheres, cylinders, and bilayers. Each morphology has a different optimum aggregation number of amphiphilic

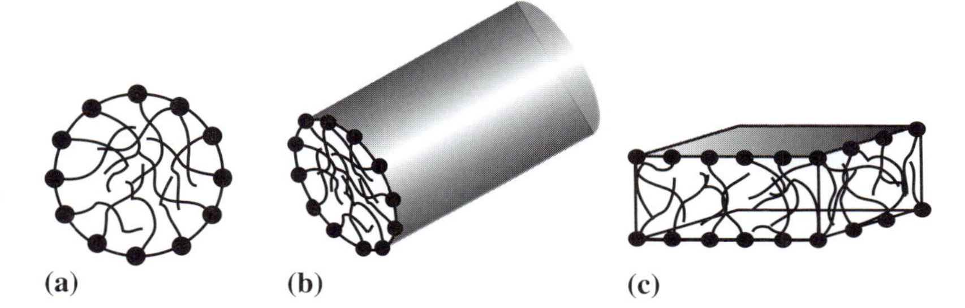

Fig. 25.5. Micelles observed in amphiphile solutions: (a) sphere, (b) cylinder, and (c) bilayers.

molecules and consequent *critical micelle concentration* in the solution. The optimum aggregation number is governed by an entropic desire to limit the interaction between water and the hydrophobic ends on the interior of the micelle. To do so, hydrophilic ends are forced into the hydrophobic interior.

25.3 Polymers

Polymers consist of repeated chemical units called *mers* or *monomers*; a monomer is a single molecular unit. An *oligomer* typically has only a few molecular units repeated as a group (usually less than about 50), whereas a polymer has many more monomers linked together in a long chain. Figure 25.6(a) shows an ethylene mer embedded in polyethylene. The term *polymer* is usually reserved for chains with more than about 50 mers. A *polyethylene* molecule (Fig. 25.6(a)) is a long chain of C atoms, with two H atoms attached to each C atom normal to the chain length. The chain can be thousands of atoms long. The polymerization reaction for polyethylene is shown schematically in Fig. 25.6(b); such a reaction is usually aided by a catalyst or high pressure, etc.

Because polymers like the polyethylene molecule may contain many thousands of monomer units, we may also call them *macromolecules*. The molecular weight of a polymer depends on when the polymerization reaction is terminated. The *degree of polymerization* is a measure of the average size or total length of a stretched polymer chain. The degree of polymerization, $\overline{X_n}$, is defined in terms of the average (over chains) molecular weight of the polymer, $\overline{M}$, and the molecular weight, M_W, of the mer as:

$$\overline{X_n} = \frac{\overline{M}}{M_W}. \tag{25.1}$$

For a polymer with a single mer, the molecular weight of the repeat unit is simply that of the mer. For polymers with many mers, the *average molecular weight* of the polymer can be defined as a *number average molecular weight* $\overline{M_\mathrm{n}}$:

$$\overline{M_\mathrm{n}} = \sum x_i M_i = \frac{\sum N_i M_i}{\sum N_i} \tag{25.2}$$

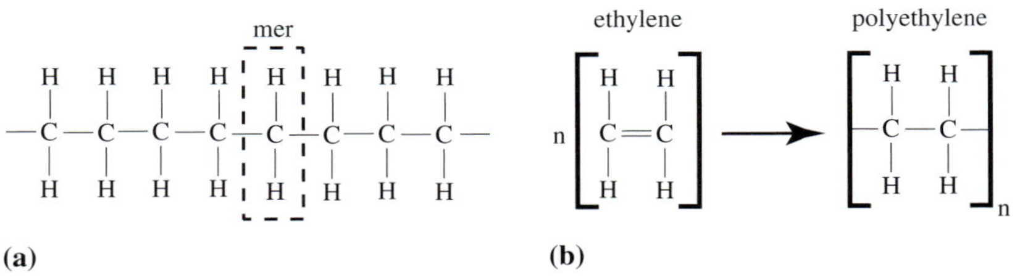

mer

(a)

ethylene

polyethylene

(b)

Fig. 25.6. (a) Repeating ethylene monomer units in polyethylene and (b) reaction of *n* ethylene monomers to form polyethylene.

or a *weight average molecular weight*, $\overline{M_w}$:

$$\overline{M_w} = \sum f_i M_i = \frac{\sum N_i (M_i)^2}{\sum N_i M_i} = \frac{\sum x_i (M_i)^2}{\sum x_i M_i} \qquad (25.3)$$

where f_i is the weight fraction of chains having a molecular weight M_i, x_i is the number fraction of chains having a molecular weight M_i, and N_i is the number of chains having a molecular weight M_i. For samples with one mer, $\overline{M_n} = \overline{M_w}$. The quantity $\overline{M_n}/\overline{M_w}$ is a measure of the mass dispersion.

25.3.1 Polymer classification

Polyethylene is classified by molecular weight as *low-density polyethylene* (LDPE), or *high-density polyethylene* (HDPE). In LDPE, some C atoms may have one or both of their H atoms replaced by another polyethylene chain. These C atoms and polyethylene chains form a branched structure. Polyethylene with molecular weights of three to six million is referred to as *ultra-high molecular weight polyethylene* (UHMWPE). UHMWPE can be used to make fibers that are stronger than the Kevlar fibers used in bullet proof vests.

The distinction between HDPE and LDPE leads to a morphological way of classifying polymers. This classification specifies whether the polymer is a *linear polymer*, a *branched polymer*, or a *network polymer*. Figure 25.7(a) illustrates a schematic of a linear polymer, a polymer that has one backbone with no branches. Note that the term linear does not imply a straight line; in fact, the polymer chain may be quite convoluted. A branched polymer (Fig. 25.7(b)) has smaller secondary chains emanating from the primary chain.

A branched structure tends to lower the degree of crystallinity and density of a polymer. High-density polyethylene is an example of a polymer that has no branching, and the chain is linear polyethylene. It is much stronger than branched polyethylene, but branched polyethylene is easier to produce. *Cross-linking* in polymers occurs when primary valence bonds are formed between separate polymer chain molecules. A third polymer morphological type is the *network polymer*, illustrated in Fig. 25.7(c).

In polymeric solids, like plastics, the chain positions are rigidly fixed in space. More common mers include *ethylene*, *styrene*, *ester*, and *acrylamide*;

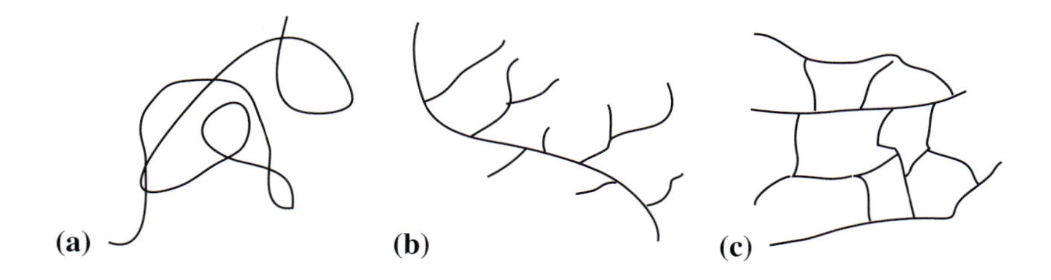

Fig. 25.7. (a) Linear unbranched, (b) linear branched with secondary offshoot chains, and (c) interconnected network polymers.

these are polymerized to produce *polyethylene*, *polystyrene*, *polyester*, and *polyacrylamide*. Polyethylene is a soft, clear plastic used in plastic bags, bottles, and toys. Polystyrene is a milky-white plastic used in covers for soft drink cups. Polyester is a fabric material, and polyacrylamide is a hard clear plastic used in such applications as compact disks. These solids contain densely packed, entangled polymer chains.

An alternative classification of polymers is based on thermal and mechanical properties. In this scheme, the three major classes of polymers are *thermoplastic polymers*, *thermosetting polymers*, and *elastomers*. Thermoplastic polymers consist of long, linear, or branched chains that are not cross-linked. The individual linear chains are typically intertwined and bonded by weak van der Waals interactions. Thermoplastic polymers can be crystalline or amorphous. By stretching a thermoplastic, we can straighten its chains; and by heating a thermoplastic, we can shape it. Thermosetting polymers consist of long linear or branched chains that are strongly cross-linked. The cross-linking reaction leads to a 3-D network (Fig. 25.7(c)) of connections between intertwined chains. Because of these linkages, thermosetting polymers are generally stronger, but more brittle, than thermoplastics. The irreversibility of the cross-linking reaction makes reprocessing of these polymers more difficult. Elastomers (or *rubbers*) may be thermoplastics or weakly cross-linked thermosetting polymers, and are distinguished by the large elastic elongation they exhibit.

25.3.2 Polymerization reactions and products

The *polymerization* process produces high molecular mass molecules from monomers. Two important polymerization processes are: *chain-reaction (addition) polymerization* and *step-reaction (condensation) polymerization*. Chain-reaction polymerization proceeds in three steps: (1) *initiation*, (2) *propagation*, and (3) *termination*. In the initiation step, a monomer (e.g., ethylene, $CH_2 = CH_2$) reacts with a free radical, R–O$^\bullet$, breaks the $C = C$ double bond, and transfers the free radical to form a complex, R–O–CH_2–$CH_2^\bullet$. A common *radical initiator* is benzoyl peroxide. Propagation is a repetitive process in which the double bonds of successive monomers are opened and the monomers react with the polymer chain. During the propagation process, the free electron from the original radical is passed down the line of the chain to the final carbon atom. The propagation reaction is described in Box 25.1.

The termination step occurs when another free radical (R–O$^\bullet$), meets the end of the growing chain. The free radical terminates the chain by linking with the last CH_2 in the polymer chain. Termination can also occur when two unfinished chains bond together. This type of termination is called *combination*. In Box 25.2 we describe both types of termination reactions.

Other polymers can be formed in a manner analogous to that illustrated for polyethylene. Four such polymerization reactions (Dieter, 1976) are illustrated

Box 25.1 Propagation reaction to form polyethylene

The first propagation reaction in the formation of polyethylene can be written as:

$$R\text{–O–CH}_2\text{–CH}_2^\bullet + (\text{CH}_2\text{=CH}_2) \rightarrow R\text{–O–CH}_2\text{–CH}_2\text{–CH}_2\text{–CH}_2^\bullet.$$

The n-th step in this propagation reaction can be written as:

$$R\text{–O–(CH}_2)_{2n-1}\text{–CH}_2^\bullet + (\text{CH}_2\text{=CH}_2) \rightarrow R\text{–O–(CH}_2)_{2n+1}\text{–CH}_2\text{–CH}_2\text{–CH}_2^\bullet.$$

The driving force for this reaction is the formation of stable single bonds in the polymer chain; these bonds are more stable than the double bonds of the monomer.

Box 25.2 Termination reactions ending the polyethylene propagation reaction

Termination by linking with the final CH_2 molecule can be written as:

$$R\text{–O–(CH}_2)_{2n-1}\text{–CH}_2^\bullet + R\text{–O}^\bullet \rightarrow R\text{–O–(CH}_2)_{2n}\text{–O–}R.$$

Termination by the bonding of two unfinished chains can be written as:

$$R\text{–O–(CH}_2)_{2n-1}\text{–CH}_2^\bullet + R\text{–O–(CH}_2)_{2m-1}\text{–CH}_2^\bullet \rightarrow R\text{–O–(CH}_2)_{2(n+m)}\text{–O–}R.$$

in Fig. 25.8: Figure 25.8(a) shows a vinyl monomer that has one H atom in ethylene replaced by a side group R_1; the polymerization reaction is analogous to that of polyethylene.

If the side group R_1 is a Cl atom, the resulting polymer is named *polyvinyl chloride* (PVC), a common polymer used in plastic pipes. If the side group is a propylene (CH_3) molecule, the resulting polymer is named *polypropylene*, a polymer that can be molded into parts. The clear polymer *polystyrene* results from polymerization of the monomer where R_1 is a benzene ring (C_6H_6). The polymer for which R_1 is a nitrile molecule ($C \equiv N$) is called *acrylonitrile*, which is produced under the trade-name Orlon and used as a fiber. If the acetate molecule ($Ac = CH_3 - CO_2$) substitutes for the side group R_1, the resulting polymer is *polyvinyl acetate*, which is used in adhesives.

Figure 25.8(b) shows a vinylidene monomer where two H atoms in ethylene are replaced by side groups R_2 and R_3. The polymerization reaction is analogous to that of polyethylene. If both R_2 and R_3 are Cl atoms, the polymer *polyvinylidene chloride* results. This is used as a film wrap with the

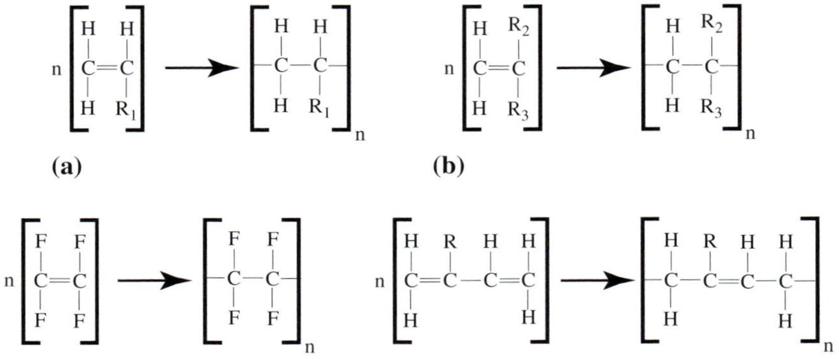

Fig. 25.8. Polymerization of (a) vinyl, (b) vinylidene, (c) tetrafluoroethylene, and (d) diene.

trade-name Saran. If R_2 and R_3 are CH_3 and acetate molecules, respectively, the polymer *polymethyl methacrylate* (PMMA) is formed, a common glazing material. Figure 25.8(c) shows the tetrafluoroethylene polymerization reaction in which F atoms replace all of the H atoms in the ethylene monomer. Figure 25.8(d) shows the longer diene monomer with a single side group R. If the side group is an H atom, the *polybutadiene* polymer results; if it is a Cl atom, the *polychloroprene* polymer (neoprene) results. Finally, if the side group is a CH_3 molecule, the *polyisoprene* polymer (natural rubber) results.

The other important polymer reaction *step-reaction polymerization* (*condensation polymerization*) involves two different types of di-functional monomers or end groups. These groups react with one another to form a chain. Condensation polymerization also produces a small molecular by-product (water, HCl, etc.). In Box 25.3 an example is shown of the formation of Nylon 66, which involves one each of two monomers hexamethylene diamine, $(NH_2)-(CH_2)_6-(NH_2)$, and adipic acid, $(C{=}OOH)-(CH_2)_4-(C{=}OOH)$, reacting to form a dimer of Nylon 66 (Askeland and Phule, 2003). The polymer can continue to grow (in either direction) by bonding to another molecule of hexamethylene diamine, adipic acid, or another dimer.

Box 25.3 Reaction to form a dimer of Nylon 66

The Nylon 66 reaction can be written as:

$$(NH_2)-(CH_2)_6-(NH_2)+(C{=}OOH)-(CH_2)_4-(C{=}OOH)\rightarrow$$
$$(NH_2)-(CH_2)_6-(NH_2)-(C{=}O)-(CH_2)_4-(C{=}OOH)+H_2O$$

The name refers to the fact that both the diamine and the acid each provide six carbon atoms to the polymer chain.

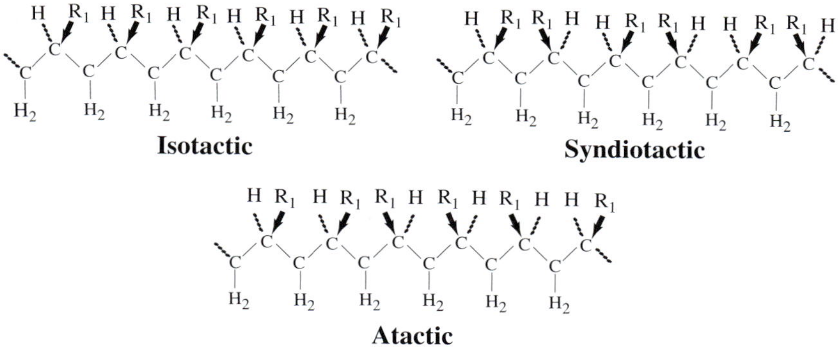

Fig. 25.9. Configurational isomers: isotactic, syndiotactic and atactic.

Polymers can also be classified in terms of their *stereo-chemistry*. While polyethylene and other symmetric monomers can be joined in only one way, mono-substituted polymers can have their mers arranged in more than one configuration. Figure 25.9 shows three possible configurations for an R_1 substituted ethylene mer. The polymer chain is drawn in a zig-zag fashion; each of the R_1 groups is located either above or below the plane of the C chain. If all R_1 groups lie on the same side of the C chain, the chain is said to be *isotactic*. If they alternate from side to side, it is called *syndiotactic*, and if they are arranged randomly, the chain is *atactic*.

25.3.3 Polymer chains: spatial configurations

Many useful properties of polymers, such as their electrical and mechanical properties, are attributed to their long molecular chains. Unlike crystals, where the exact location of atoms can be determined by precise rules (i.e., lattice sites are periodically arranged and decorated with the same atomic basis), polymer chains can have many configurations, despite the simple reaction rules for their growth. Although we can predict the sequence of atoms in a polymer chain, we cannot predict the precise location of an individual atom because the chain itself is not rigid. Figure 25.7(a) shows this to be the case for a linear polymer, where the random winding of the chain causes the distance between the ends of the polymer strand to be much less than the actual length of the chain.

The average distance between the two ends of the polymer strand can be calculated using the tool of *statistical mechanics*, which considers the randomness of the twists and turns along the chain length. Here, we present a simple model that employs the statistics of a *random walk*. We begin by assuming that the polymer chain has N mers, each of length ℓ.[3] The statistical aspect of the argument comes from the assumption that the direction in space of any

[3] To use statistical mechanics, N must be large.

Box 25.4 Calculation of the root-mean-square end-to-end distance of a polymer

Each mer of length ℓ has equal probability of being oriented along any direction in 3-D. Statistical mechanics then expresses the probability, $P(r)$, that after N steps the chain end is a distance, r, from the initial point as a Gaussian distribution:

$$P(r) = \left(\frac{3}{2\pi N\ell^2}\right)^{\frac{3}{2}} e^{\frac{-3r^2}{2N\ell^2}}.$$

The *mean square end-to-end distance* (*radius of gyration*), $\overline{r^2}$, for such a free chain is:

$$\overline{r^2} = \int_{r=0}^{\infty} P(r) 4\pi r^2 \mathrm{d}r = N\ell^2.$$

The *root-mean-square end-to-end distance* for the polymer chain is then:

$$(\overline{r^2})^{\frac{1}{2}} = \sqrt{N}\ell.$$

subsequent link with respect to its predecessor is completely random.[4] Box 25.4 considers a random walk process beginning at an origin located at one end of a polymer chain, to calculate the root-mean-square end-to-end distance of this chain. The result of Box 25.4 states that the length of the tangled strand grows as $\sqrt{N}$ and not linearly as would be the case for a rigid 1-D chain.

25.3.4 Copolymers and self-assembly

Many of the previously discussed polymers are examples of *homopolymers*, constructed from a single repeating unit. Many functionally useful polymers are, however, produced by polymerization of two or more monomers together. Such polymers are known as *copolymers*; in a sense, copolymers can be thought of as "alloys" of the monomers. As such, the monomers may order randomly, in a periodic sequence, or in blocks along the polymer chain. For example, a polymer formed from equal amounts of A and B monomers may have a random sequence of A and B monomers, an ordered (ABABAB . . .) sequence, or a "blocked sequence" (*block copolymer*) of

[4] This is oversimplified in that (1) certain bond angles between mers are more probable than others and (2) the reverse direction has zero probability because the next mer cannot occupy the same space as the first. Consideration of each of these facts suggests corrections to the results of the random walk calculation.

the type (AAA... ABBB...B). A copolymer with a random arrangement of mers is called a *random copolymer*; its properties are often intermediate between those of the two homopolymers that would form from the individual monomer components.

The simplest block copolymers are *linear diblock copolymers* which are comprised of two distinct polymer chains covalently bonded at their endpoints to form a chain. Since block copolymers are single-component systems, they cannot macrophase separate in the melt like a pair of linear homopolymers (Jones, 2002). The interactions *between* the blocks, however, can cause copolymer solutions to microphase separate, choosing a morphology that emphasizes the favorable interactions between the blocks. This leads to *self-assembly*, which is an important phenomenon used to tailor polymeric microstructures. The simplest AB diblock copolymers segregate on a local scale to form lamellar, cylindrical, cubic spherical, or interconnected morphologies. Polymer phase diagrams can be used to illustrate phases with distinct morphologies. Figure 25.10 illustrates an example of a prototypical phase diagram for an amphiphilic block copolymer in water (Alexandris *et al.*, 1996).

Microdomain structures, such as the ones found in block copolymers, can act as a host for sequestering nanoscopic inclusions of appropriate chemical

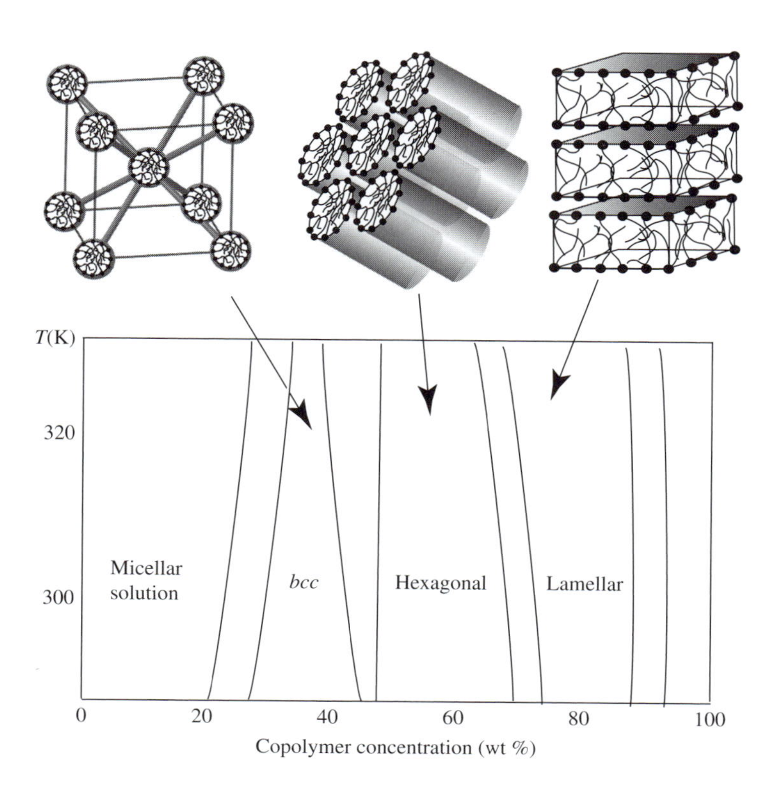

Fig. 25.10. Typical phase diagram for an amphiphilic block copolymer in water. The polymer exhibits *bcc* cylindrical *hcp* and lamellar ordered phases.

affinity and geometry. Combining host materials of a particular geometry with inclusions opens up several possibilities:

- maximum enhancement of effective physical properties of the composite material;
- applications that arise from the long-range order of the nanoscopic inclusions; and
- applications that capitalize on the physical properties of the host material, as well as the nano-specific characteristics of the sequestered inclusions.

An example of a lamellar block copolymer nanostructure is illustrated in Box 25.5. Scientists are actively exploring synthesis and structure–property relationships in nanocomposite structures such as these. Special attention is devoted to block copolymers and particle characteristics, since control of the location of nanoscopic inclusions within a host material provides opportunities to both maximize property enhancement and induce novel properties that are not inherent to either of the constituent materials.

Box 25.5 Lamellar block copolymer nanocomposites

The figure below (courtesy of M. Bockstaller) shows cross-sectional transmission electron micrographs of a lamellar block copolymer sample blended with aliphatic surface modified gold and silica nanoparticles (Bockstaller *et al.*, 2003, Bockstaller and Thomas, 2004, Bockstaller *et al.*, 2005).

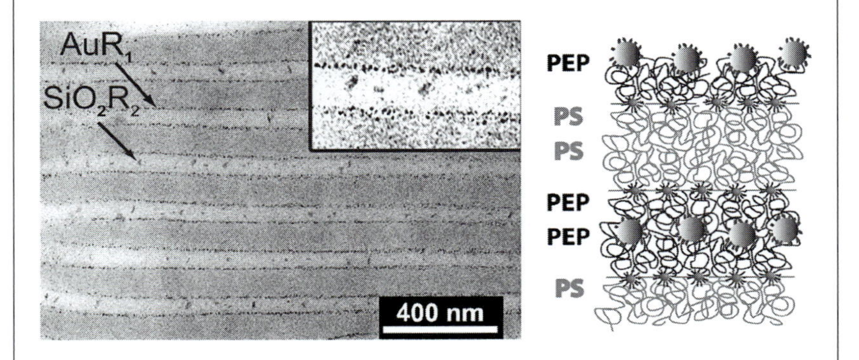

Elemental analysis reveals that gold nanoparticles (small dark spots in the micrograph) segregate to the inter-material dividing surface whereas the larger silica particles concentrate at the central regions of the chemically compatible polymer domain.

25.3.5 Conducting and superconducting polymers

A polymer chain and its constituent periodically repeated mer units serve as an example of a 1-D crystal. The conductivity of 1-D crystals is an active area of recent research. In general, most organic polymers are electrical insulators because their valence electrons reside in relatively localized bonds. To construct a conducting polymeric solid, we must have a system in which the electrons are delocalized along the length of the polymer chain. Delocalization can be accomplished in polymers with π-bonds along the length of the chain. This notion is analogous to the ring currents in benzene; the diamagnetic ring currents can be associated with the delocalized π-electrons that move freely around the circumference of the ring. In this section, we consider examples of 1-D conducting and superconducting polymers.

25.3.5.1 Conducting polymers: polyacetylene, etc.

The polymer *polyacetylene* is an example of a long-chain polymer with a 1-D network of conjugated *π-bonds*. Polyacetylene is formed by polymerizing the acetylene (ethyne) monomer, as illustrated in Fig. 25.11. Polyacetylene is an example of a polymer which has two polymorphous forms. These are the *cis*- and *trans*- forms, illustrated as the polymerization reaction products in Fig. 25.11.

The *trans*- form of polyacetylene is an example of a 1-D chain with evenly spaced C units. There is a clear sawtooth arrangement of the C atoms that is accompanied by the localization of the π-electrons in the double bonds. In the *cis*- case, the localization is even more pronounced. This localization prevents the *cis*- or *trans*-forms of polyacetylene from conducting.[5] The

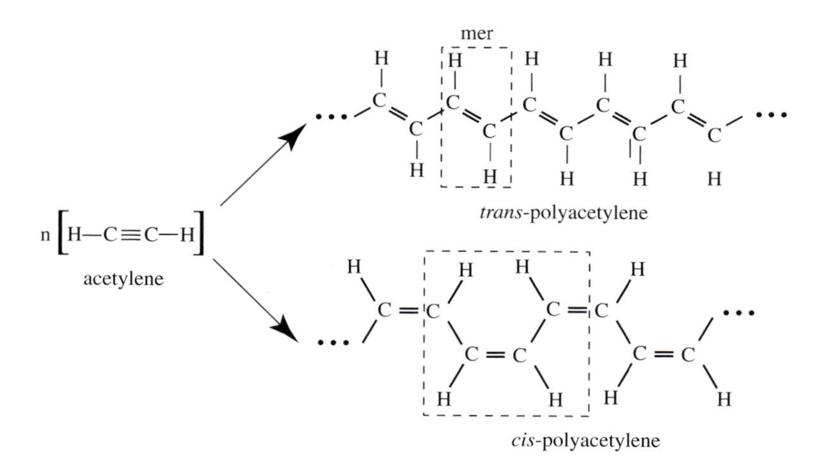

Fig. 25.11. Reaction of *n* acetylene monomers to form *cis*- and *trans*- polyacetylene.

[5] This is consistent with the famous *Peierl's theorem* stating that a 1-D metal is always unstable with respect to a symmetry lowering distortion into a non-metallic state.

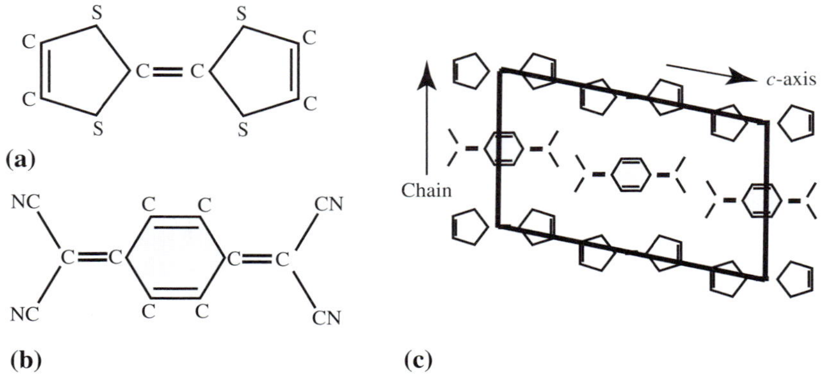

Fig. 25.12. Molecular structures of (a) tetrathiafulvalene (TTF) and (b) tetracyanoquinonedimethane (TCNQ) (H atoms not shown) and crystalline TTF-TCNQ(c).

energy gap between the localized π-bonding and anti-bonding states is small in the *cis*- and even smaller in the *trans*-forms of polyacetylene. Therefore, they behave as semiconductors. Dopants that are electron donors or acceptors can greatly increase the conductivity of polyacetylene by providing either holes or electrons as carriers. Other examples of conducting polymers with conjugated π-electrons include polypyrrole, polythiophene, polyaniline, and polyphenylenevinylene.

25.3.5.2 Superconducting polymers: TTF-TCNQ

TTF-TCNQ was the first organic superconductor, studied by the group of Alan J. Heeger (Coleman *et al.*, 1973), winner of the 2000 Nobel Prize in Chemistry. One-dimensional polymer chains in TTF-TCNQ are built of alternately arranged cyclic molecules of *tetrathiafulvalene* (TTF) and tetracyanoquinonedimethane (TCNQ). Figure 25.12(a) and (b) illustrate the structures of these two mers (absent the H atoms). Crystals of this polymer have chains aligned orthogonal to the *c*-axis, Fig. 25.12(c).

In solid form with aligned chains, TTF-TCNQ exhibits metallic conductivity in the direction parallel to the chains at room temperature. TTF-TCNQ is an example of a *charge transfer compound* (salt) in which the TTF serves as the cation and TCNQ the anion. TTF-TCNQ was of considerable importance as the first studied organic superconductor. It has a low superconducting transition temperature of less than 2 K, but opened new research into organic superconductors. Organic high temperature superconductors, the fulleride salts, will be discussed below.

25.3.6 Polymeric derivatives of fullerenes

Recently, researchers have made significant advances in synthesizing polymer derivatives of fullerenes (Chiang and Wang, 2000). Because fullerene molecules are not particularly reactive, it took time to develop ways of performing arbitrary functionalizations. As the number of organic reagents

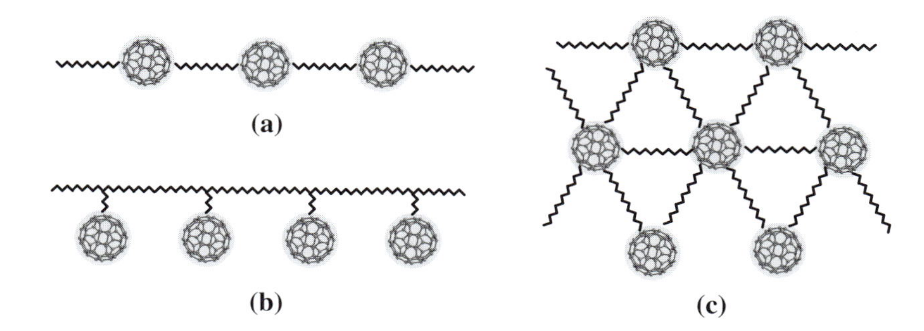

Fig. 25.13. Fullerene-containing macromolecules: (a) simple linear chain copolymer, (b) fullerene grafted side-chain polymer, and (c) fullerene-containing polymer network structure.

known to react increased, researchers naturally extended their work to design *fullerene-containing macromolecules* (Fig. 25.13). The "Buckyballs" serve as mers in polymerization reactions and provide new functionality to the polymers. Researchers have postulated that applications for these polymers include virus inhibitors and conducting polymers.

Figure 25.13 illustrates a few examples of fullerene-containing polymers following the review article by Chiang and Wang (2000). Figure 25.13 (a) depicts a simple linear chain copolymer where we idealize C_{60} molecules to be connected by other mer units. Figure 25.13(b) shows an example of a *fullerene grafted side-chain polymer* with C_{60} units grafted onto a linear polymer. Figure 25.13(c) illustrates a fullerene-containing polymer network structure. Other structures have also been studied; such structures include dendritic and "starburst" polymers, which are not illustrated here.

25.4 Biological macromolecules

25.4.1 DNA and RNA

DNA (deoxyribonucleic acid) and proteins are the essential molecules of the cell nucleus. Francis Harry Compton Crick solved the structure of DNA using XRD data provided by Rosalind Franklin. In 1953, he and James Dewey Watson proposed a mathematical theory of the diffraction from the DNA double helix. Watson and Crick's double helix structure was instrumental in the further understanding of nucleic acid replication and transcription. Watson, Crick, and Maurice Hugh Frederick Wilkins shared the Nobel Prize in Physiology and Medicine in 1962. The field of DNA sequencing and the mapping of the human genome have burgeoned in the recent past.

DNA is a polymer with monomer units called *nucleotides*. The polymer is, therefore, called a *polynucleotide*. Each nucleotide includes *deoxyribose*, a five-carbon sugar, a nitrogen-containing base attached to the sugar, and a phosphate group. There are four different types of nucleotides found in DNA (Fig. 25.14) which differ only in the base that is bound to the sugar and phosphate groups shown on the left.

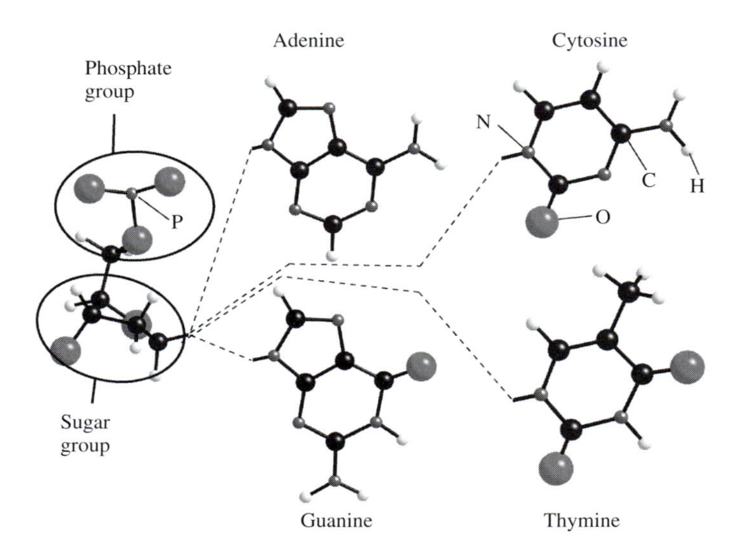

Fig. 25.14. The four nitrogen bases found in DNA: adenine, cytosine, guanine, and thymine. Structural parameters have been translated from Protein Data Bank files.

The four nucleotides are given one-letter abbreviations as shorthand for the four nitrogen bases: *adenine* (A), *thymine* (T), *guanine* (G), and *cytosine* (C). Adenine and guanine are *purines*, the larger of the two types of bases found in DNA. Purines consist of a six-membered and a five-membered N-containing ring, fused together. The bases cytosine and thymine are *pyrimidines*. Pyrimidines have only a six-membered N-containing ring. The four N bases found in the DNA structure are illustrated in Fig. 25.14; each base is connected to the deoxyribose sugar backbone at the left-most nitrogen atom (dashed lines).

The attachment of a sugar (ribose in RNA or 2-deoxyribose in DNA) to a nitrogen base results in a compound that is called a *nucleoside*. A C atom of the sugar attached to an N is a purine base or a pyrimidine base. The names of purine nucleosides end in -osine (*adenosine*, *guanosine*) while the names of pyrimidine nucleosides end in -idine (*cytidine*, *thymidine*). To indicate that the sugar is 2′-deoxyribose, we place a d before the name. Thus, the nucleosides for DNA are: d-Adenosine, d-Cytidine, d-Guanosine, and d-Thymidine. Figure 25.15 shows the adenine base (a) followed by the attachment to the deoxyribose sugar to form d-Adenosine (b) and finally the attachment of a phosphate group to form the A nucleotide (c) (adding one or more phosphates to the sugar portion of a nucleoside results in a *nucleotide*).

Nucleotides are joined by *phosphodiester bonds* to form *polynucleotides*. A single helix half of the DNA molecule is formed by linking the deoxyribose sugars of sequenced nucleotides to form a backbone. These form a helical structure with base groups radiating from the backbone. Polymerization of ribonucleotides will produce an RNA, while polymerization of deoxyribonucleotides leads to DNA.

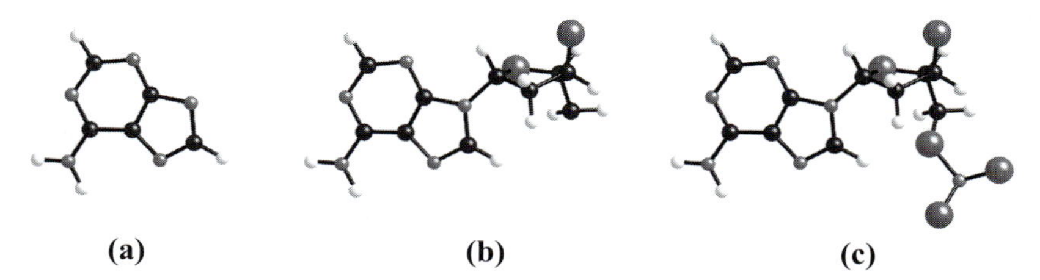

Fig. 25.15. Building the A nucleotide: (a) the adenine base, (b) the attachment to the deoxyribose sugar to form d-Adenosine, and (c) attachment of a phosphate group to form the A nucleotide. Structural parameters have been translated from Protein Data Bank files.

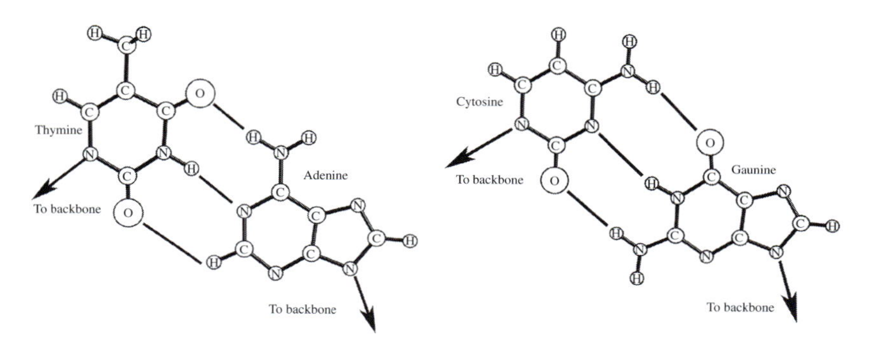

Fig. 25.16. Complementary base pairs (a) A and T, (b) G and C and their connections through hydrogen bonds shown as solid lines (after Silbey *et al.* (2005)). Structural parameters have been translated from Protein Data Bank files.

Early experiments showed that the ratios of nitrogen bases A to T and G to C are constant in the DNA of all life forms. Figure 25.16 shows the complementary base pairs ((a) A and T, (b) G and C) and their connections to one another through hydrogen bonds. The double helix in DNA is constructed by two chains of the same helicity separated by ladder rungs, i.e., a spiral staircase of base pairs (A with T and G with C).

A base and its complement are each attached to different strands of the double-stranded sugar–phosphate backbone. The two helices have the same chirality and twist around one another, coupled together by the hydrogen bonds between the complementary base pairs. Figure 25.17 shows the double helix structure of a portion of a DNA molecule in a ball-and-stick (a) and space filling representation (c). Figure 25.17 (b) shows a schematic of the structure adapted from the original Nature paper of Watson and Crick (Watson and Crick, 1953). This shows the sugar backbones of the two helices with base-pair rungs connecting the two backbone ribbons.

Many crystals can be grown by heating and supersaturating a solution. On cooling, solute precipitates out and crystallizes. When DNA is heated or placed in an aggressive chemical environment, the double strand dissociates into two single strands. This process is called *denaturation*. Denaturation occurs at relatively low temperatures for many proteins, which increases the potential for the solution growth of protein crystals. An alternative method for growing protein crystals is through vapor growth.

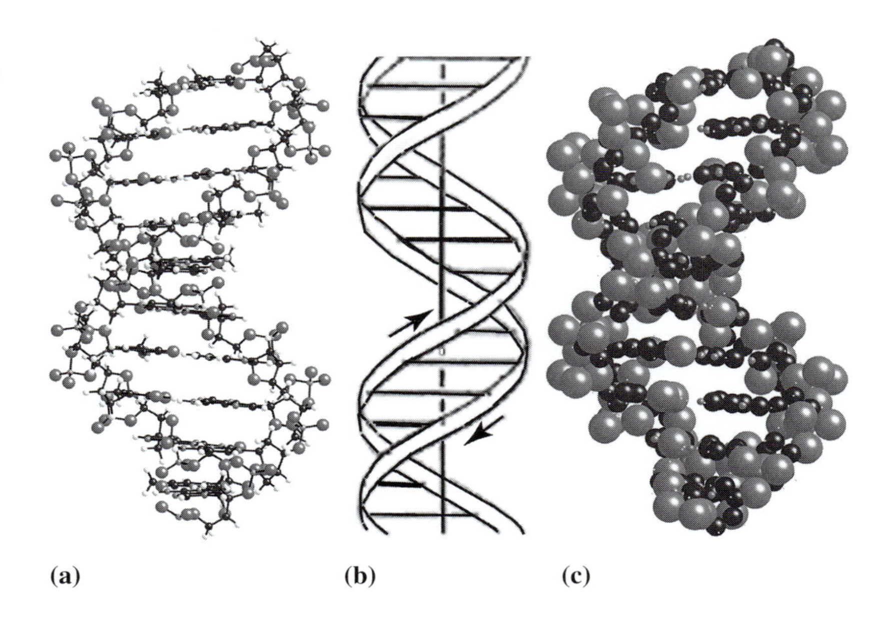

Fig. 25.17. Portion of a DNA molecule showing the double helix structure (a) ball-and-stick representation, (b) simplified drawing showing the sugar backbones of the helices and the base-pair rungs (Watson and Crick, 1953), and (c) space filling representation.

One of the true triumphs of structure determination is the detailed 3-D structure of DNA, solved by X-ray crystallography. In 1953, James Watson and Francis Crick, at the Cavendish Laboratory in Cambridge, England showed that the DNA molecule is a double helix with the complementary A-T and G-C pairs forming the rungs of the twisted DNA molecule ladder structure.[6] Because of the A-T and G-C pairing, half of the DNA ladder acts as a template for recreating the other half in DNA replication. In their landmark paper (Watson and Crick, 1953), Watson and Crick recognized the significance of the base pairing: *"It has not escaped our notice that the specific pairing [of bases in the double helical structure] we have postulated immediately suggests a possible copying mechanism for the genetic material."*

25.4.2 Virus structures

A *virus* is a parasitic entity, which depends on a live host cell to multiply. Viruses contain a nucleic acid (DNA or RNA) and the proteins encoded by the nucleic acid. Some viruses have a membrane envelope. Viruses can range in size from ≤ 100 nm to several hundred nm in spherical, helical, or complex morphologies.[7] A virus can have a lipid *envelope* acquired from a host cell. Some viruses encode a few structural proteins that are included

[6] The discovery of DNA is discussed in a more historical context in the History section at the end of this chapter.
[7] For a gallery of virus morphologies see http://www.med.sc.edu:85/mhunt/intro-vir.htm.

in the *virion* (fully grown virus particle). These enzymes aid in the gene replication process. Others encode proteins that end up in the virion, but they also encode many others that are used in viral replication and then discarded.

The nucleic acid genome (RNA or DNA) in a virus is protected by a protein coating or *capsid*. The capsid surrounds the *nucleocapsid*. The capsid is the protein coating that surrounds the nucleic acid in a virus. *Capsomers* are polypetide chain structural units of the virus. Virion nucleocapsid structures exhibit three major symmetry classes:

(i) *Icosahedral viruses* have capsid shells in which protein units decorate the faces of an icosahedron to form geodesic structures. Examples of icosahedral viruses include the human adenovirus and the herpes virus.
(ii) *Helical viruses* can have rod-like morphologies (for instance, the tobacco mosaic virus) or flexible coiled morphologies (such as the flu and rabies viruses).
(iii) *Complex symmetry viruses* have lower symmetries. Examples include Molluscum contagiosum and the pox viruses.

We will discuss examples of the first two symmetry classes in the next sections.

25.4.2.1 Icosahedral virus structures

While explaining virus structures, Donald Caspar and Aaron Klug (Caspar and Klug, 1963) presented a *theory of quasi-equivalence*. They argued that viruses form by *self-assembly* of chemically identical subunits that bond identically with neighboring subunits. Viruses grow by self-assembly into symmetric subunits through specific chemical bonds and they are surrounded by a shell that protects the genetic information. The high symmetry of the icosahedron allows for efficient self-assembly. For a subunit that is not located on a symmetry axis of the icosahedron, the symmetry operations of the icosahedral rotational group serve to replicate it 60 times with identical bonds. When more than 60 subunits form a shell, the subunits cannot have equivalent bonds. However, because more than one type of quasi-equivalent bonding environment for the subunits is allowed, replication by the icosahedral group operations could give rise to multiples of 60 subunits in a shell. If some subunits sit on high-symmetry axes, then different multiples will be produced, some of which were described in Chapter 16.

Caspar and Klug (1963) used the concept of the *triangulation numbers*, introduced in Chapter 16, to discuss the geometry of virus structures. Here, we discuss triangulation numbers with reference to a honeycomb, 6^3 lattice. Figure 25.18 illustrates the 6^3 lattice with coordinate axes h and k. To describe various virus structures, we will cut equilateral triangles from this lattice and thereby decorate the faces of an icosahedron. In describing the size of the equilateral triangles, we will draw an edge from the origin to another

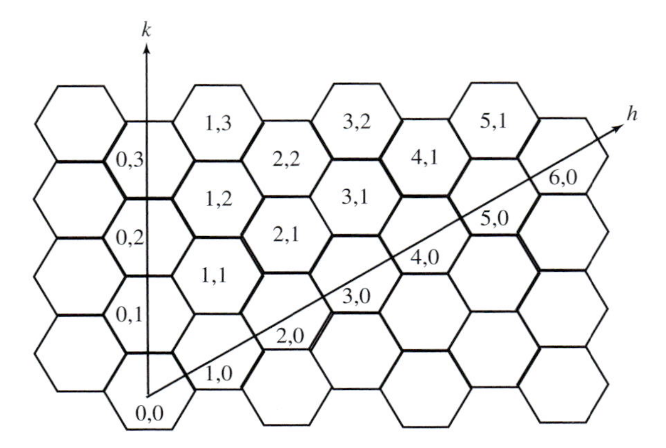

Fig. 25.18. 6^3 lattice with coordinate axes h and k. Lattice points (hexagon centers) are labeled by the pairs (h, k).

(h, k) lattice point in the honeycomb lattice. Figure 25.18 labels lattice points, located at the hexagon centers, as (h, k).

Figure 25.19 shows the decoration of icosahedral virus structures where the triangular faces are replaced with sections of the honeycomb lattice. In each case, the edge of the equilateral triangle is the distance between the origin and the lattice point (h, k). The other two edges are determined by symmetry. Figure 25.19(a) shows construction of a $(1, 1)$ structure, (b) a $(2, 0)$ structure, and (c) the two *enantiomers* of a $(2, 1)$ structure. For lattice points of the form $(h, 0)$, $(0, k)$, or (h, h), only one structure results. Others have right-and left-handed enantiomers denoted *dextro*, for right-handed or *levo*, for left-handed.

The number of lattice points contained in an equilateral triangle is given by the *triangulation number*, $T = h^2 + hk + k^2$. The tomato bushy stunt virus (TBV) is an example of a $(1, 1)$, $T = 3$ virus structure. The Sindbis virus has a $(2, 0)$, $T = 4$ structure and the polyoma virus has a $(1, 2)$ structure with $T = 7$. The herpes virus is an example of a $T = 16$ virus that will be explored in an exercise at the end of this chapter.

25.4.2.2 Helical virus structures

The *tobacco mosaic virus* (TMV) was one of the early virus structures solved; it is an example of a helical virus. It has a single type of protein molecular subunit, with three nucleotides per molecule, that grows out from a helical strand of RNA. A general picture of the molecule shown in Fig. 25.20(a) was completed by Rosalind Franklin prior to her death in 1958 (Klug, 1993). The virus has a rod-shaped morphology that results from its helical growth. There is a 4 nm cylindrical hole in the center of the virus. The single strand RNA backbone threads the exterior of this hole, surrounded by the protein coating. There are $16\frac{1}{3}$ protein subunits per turn. Each protein subunit is identical and has identical surroundings. The TMV can be *self-assembled* from its constituent protein and nucleic acid components.

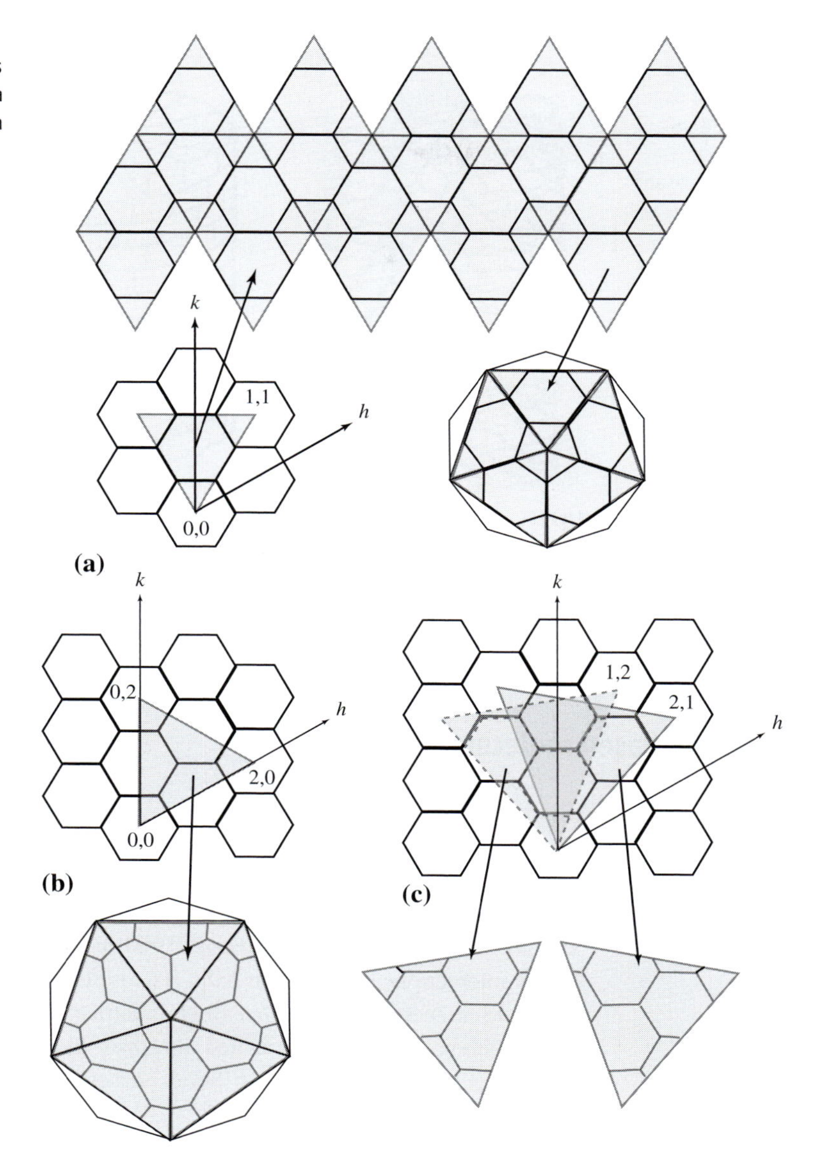

Fig. 25.19. Icosahedral virus structures with triangular faces are replaced with sections of a 6^3 tile. A $(1, 1)$ structure (a), a $(2, 0)$ structure (b), and the two enantiomers of a $(2, 1)$ structure (c), along with their origins in the 6^3 tiling.

Aaron Klug, a one-time post-doctoral associate with Rosalind Franklin, won the 1982 Nobel Prize in Chemistry *"for his development of crystallographic electron microscopy and his structural elucidation of biologically important nucleic acid–protein complexes."* Klug solved the intricate details of the TMV virus structure by isolating protein subunits, solving the protein structure by XRD, and then comparing the results to morphological observations of the virus. Obtaining protein crystals and understanding how the proteins interacted to form ordered aggregates was a major accomplishment, detailed in Klug's 1982 Nobel Prize lecture. Figure 25.20(b) shows a schematic of a single helical

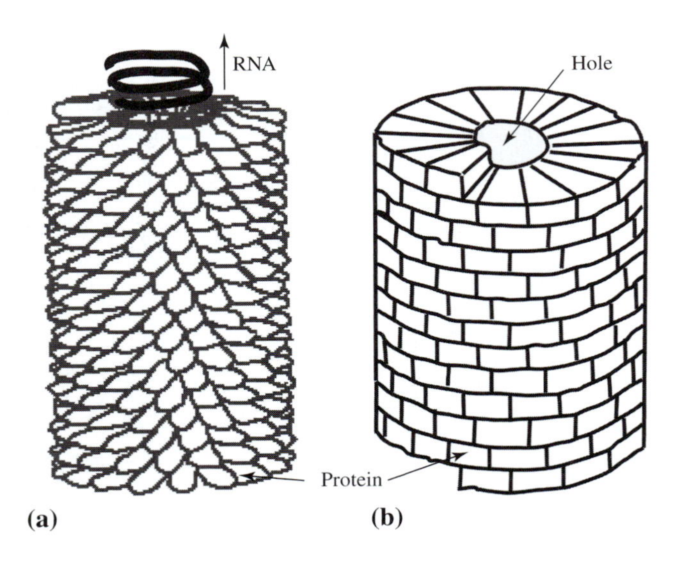

Fig. 25.20. (a) Schematic of a portion of TMV showing a helical arrangement of protein subunits emanating from a single RNA strand and surrounding a 4 nm hole and (b) single helical structure with wedge-shaped protein units grown from low pH solutions (Klug, 1993).

structure with wedge-shaped protein units grown from low-pH solutions; this schematic is instrumental in furthering the understanding of TMV.

25.5 Fullerene-based molecular solids

Improved fullerene synthesis methods have resulted in more economical means of producing these materials; the new procedures have increased research efforts to exploit these molecules as precursors for synthesizing other exotic materials. A notable early discovery is the crystallization of C_{60} into a molecular solid (Kratschmer *et al.*, 1990) held together weakly by Van der Waals interactions between the balls. The molecular solid has a spinning C_{60} molecule at each site of the *fcc* Bravais lattice. The solids constructed from pure fullerene molecules are called *fullerites*. In pristine fullerites, weak Van der Waals bonding is observed. Solid phases of fullerites with the C_{60} and C_{70} molecules have been observed.

The structures of several fullerite solids as well as phase transitions between these structures have been studied in detail. Researchers have also determined how the rotational degrees of freedom of the fullerene molecules on their lattice sites influence the symmetry of the crystalline phase. The fullerites have been produced in both single crystal and thin film forms. Figure 25.21 shows an SEM micrograph of a vapor-grown C_{60} single crystal. This crystal has an *fcc* lattice with spinning C_{60} molecules decorating its sites. The "hard sphere" radius for this structure is the radius of the Buckyball.

After the discovery of the fullerites, scientists alloyed the C_{60} solids with alkali and alkaline earth metals to produce intercalated solids (Haddon *et al.*, 1991). In these solids, the metal atoms occupy tetrahedral and octahedral

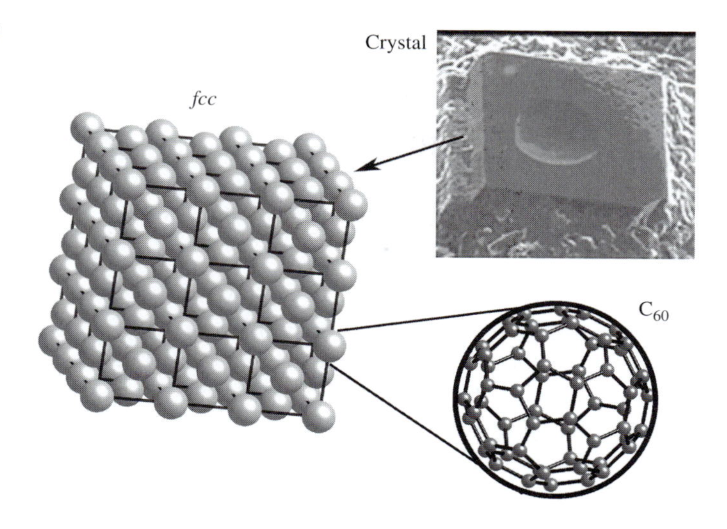

Fig. 25.21. SEM micrograph of a vapor-grown C_{60} single crystal (courtesy of S.-Y. Chu) having an *fcc* crystal structure with the Bravais lattice decorated with C_{60} molecules.

interstices in the cubic lattice of the original fullerite solid. These electropositive atoms donate their valence electron(s) to the electronegative C_{60} clusters, forming *fullerides* or *buckide salts*. The fullerides have an interesting temperature – composition phase diagram in M_xC_{60} $(0 < x < 6)$ solids, where M denotes an alkali element. This phase diagram is of interest because fulleride salts display several solid phases and the metal atom concentration dramatically influences the electronic structure within the system. These systems exhibit intrinsic semiconducting, extrinsic semiconducting, metallic conducting, superconducting and insulating behavior, depending on the concentration and type of M.

The field of fullerene science and technology has evolved to where scientists can construct *fullerene cluster assembled materials*. One novel property of fullerene-based solids is the presence of superconductivity in the buckide salts. Buckide salts are notable because, with a T_c of 36 K, the Rb_3C_{60} superconductor has become the highest temperature organic superconductor observed to date. Since several fundamental superconducting properties compare favorably with those of high temperature oxide superconductors, including the manifestation of a desirable isotopic superconducting energy gap, microelectronic applications of fullerene-based solids may exist in the future.

The first reported observation of superconductivity in fullerene-based materials was by Hebard *et al.* (1991). They observed a superconducting transition temperature, T_c, of 18 K in potassium-doped fulleride salts. This transition temperature constituted a record for organic superconductors and stimulated one of the most active areas of fullerene research. Scientists soon learned that the superconducting transition temperature depends on the K concentration in K_xC_{60} compounds. It was also found that the optimum stoichiometry is K_3C_{60}. Figure 25.24(b,c) illustrate the fulleride solid K_3C_{60} in both space filling and ball-and-stick depictions. Given the initial success with K, scientists soon substituted other alkali metals. Transition temperatures for Rb_xC_{60} were observed to be 28 K by Rosseinsky *et al.* (1991), and 30 K by Holczer *et al.* (1991).

25.5.1 Fullerites

Fullerites are molecular crystals of the fullerenes. In pristine C_{60} or C_{70} crystals and films, molecules pack as pseudo-spheres in close-packed arrangements. The C_{60} molecule, with its spherical shape, prefers cubic close-packing (Liu *et al.*, 1991). The C_{70} molecule, with its oblate (egg-shaped) geometry, is more stable in an *hcp* structure. Pristine C_{60} solids have C_{60} molecules decorating an *fcc* lattice. To assign the space group precisely, one must determine the alignment of the symmetry axes of the C_{60} molecule and those of the *fcc* lattice. At room temperature, the balls have been observed to spin, maintaining all of the symmetries of the *fcc* structure with the probable space group **Fm$\bar{3}$m** (O_h^5). At lower temperatures, an orientational ordering transition to a (simple) cubic structure is observed. In this transition, the balls have a symmetry axis in common with the cube axes, making the cube vertex and face sites no longer equivalent. Figure 25.22(a) illustrates the high temperature *fcc* C_{60} solid in space filling and ball-and-stick depictions, respectively.

The original observations of a crystalline state in fullerene based solids (Kratschmer *et al.*, 1990), performed on crystals formed from a fullerene-saturated benzene solution, revealed the solids to have an *hcp* structure. It was suggested that individual buckyballs packed as nearly hard spheres and, therefore, preferred close-packed structures. Soon after, it was discovered that the significant concentration of higher fullerenes in the original benzene solution as well as residual solvent in the crystals contributed to the

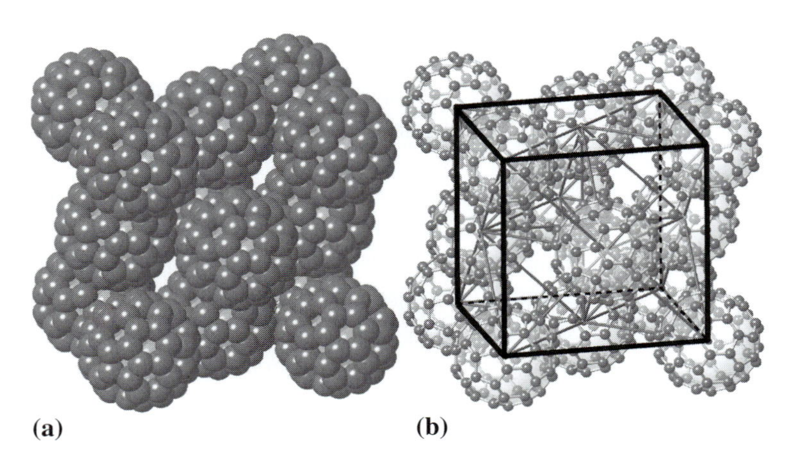

Fig. 25.22. *fcc* C_{60} solid, space filling (a) and ball-and-stick (b) models.

stability of the *hcp* phase. After developing techniques for separation of the various fullerenes and for removing residual solvents, scientists found that in pristine C_{60} fullerite, the *fcc* phase was actually more stable, confirming the original premise that the C_{60} molecules prefer a close packing as hard spheres.

The history of the structure determination of the solid fullerites provides a nice example of the fact that it is not always easy to figure out the structure of a molecular solid, even when the building blocks are well known. The early structural characterization of C_{60} fullerite solids has included crystal structure models based on the $\mathbf{Fm\bar{3}}$ (T_h^3) (Fleming *et al.*, 1991) and $\mathbf{Fm\bar{3}m}$ (O_h^5) space groups (Dravid *et al.*, 1991), each with a 1.42 nm lattice constant for the cubic unit cell at room temperature. *fcc* single crystals of the C_{60} fullerite exhibited a first-order phase transition to a primitive cubic cell at 249 K, as evident from synchrotron X-ray powder diffraction studies (Heiney *et al.*, 1991). This observation differed from those made on the basis of nuclear magnetic resonance (NMR) data (Yannoni *et al.*, 1991, Tycko *et al.*, 1991); the time scale of NMR measurements is much shorter than that for X-rays. From the NMR experiments, the possibility of a rapid rotation of the individual C_{60} molecules below 249 K was suggested. In light of the unambiguous ordering transition at 249 K, as evident from X-ray data, the samples for which NMR data were reported were subject to thermally activated rotational modes below 249 K.

Heiney *et al.* (1991) measured the ordered C_{60} phase to have a lattice constant of 1.404(1) nm at 11 K. The synchrotron X-ray data of Heiney *et al.* (1991) were used by Sachidanandam and Harris (1985) to assign a $\mathbf{Pa\bar{3}}$ (T_h^6) space group for the ordered phase. These observations agreed with neutron diffraction data on powders (Copley *et al.*, 1992). Subsequent X-ray data at 110 K on twinned crystals were interpreted in terms of an ordered structure model (Liu *et al.*, 1991). Systematic noncrystallographic extinctions of zonal reflections with two odd indices were explained by the twinning, which occurs as a result of cooling through the ordering temperature. Twinning induced extinctions allowed for the assignment of the space group as $\mathbf{Pa\bar{3}}$ (T_h^6), in agreement with the best fits to synchrotron X-ray and neutron powder data. In this space group, the individual molecules possess the crystallographically imposed $\mathbf{\bar{3}}$ (C_{3i}) point group symmetry. The crystallographic constraints require that 8 of the 20 hexagonal faces of the truncated icosahedron must be aligned.[8] The $\mathbf{Pa\bar{3}}$ (T_h^6) space group is also that of the solid CO_2 cubic phase discussed above.

In C_{70} solids an *hcp* structure is observed. Figure 25.23(a) illustrates the *hcp* C_{70} solid in space filling and ball-and-stick depictions, respectively.

[8] This same constraint is also imposed on the truncated icosahedra in the $Mg_{32}(Al,Zn)_{49}$ phase (Bergman *et al.*, 1957).

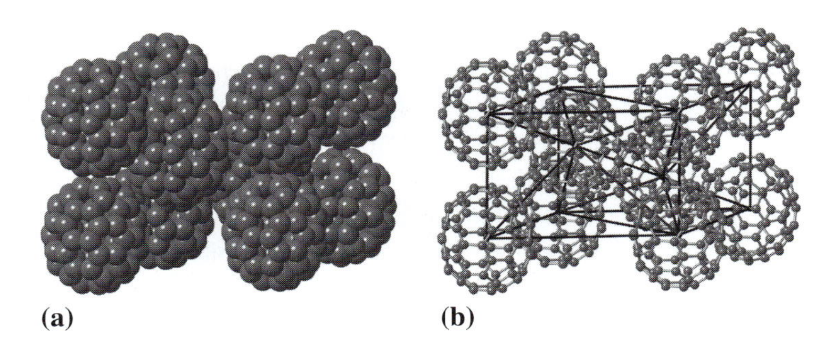

Fig. 25.23. The HCP C_{70} solid in (a) space filling depiction and (b) ball-and-stick representation.

25.5.2 Fullerides

Scientists have shown that doping C_{60} *solids* with alkali metals dramatically influences the electronic structure, crystalline phase formation, and electronic properties of these solids. Alkali metal additions in the range of 0–6 atoms per C_{60} molecule have been reported. It is now clear that, at low concentrations, these alkali metal atoms are intercalated into the octahedral and tetrahedral interstitial sites in the close packed solids. However, when the alkali metal content exceeds three per C_{60} molecule (at the composition M_3C_{60}), there are too many alkali metal atoms for the three interstitial sites per ball in the *fcc* structure. Consequently, a body-centered tetragonal (BCT) and eventually a *bcc* structure become stable. These phases are known as *fullerides*. Fullerides are intermetallic compounds formed with C_{60} as a constituent. The BCT phase occurs (at low temperatures) as a line compound of composition M_4C_{60} whereas the *bcc* phase has composition M_6C_{60}. For M_xC_{60}, the *fcc*-I phase occurs for $x \leq 3$, the immiscible *fcc*-II and BCT phases coexist for $3 \leq x < 4$, and the BCT and *bcc* phases coexist for $4 \leq x \leq 6$. The $x = 3$ composition is notable because it corresponds to the highest temperature superconductivity in the *fcc*-II phase. The *fcc*-I, BCT, and *bcc* phases do not superconduct.

Figure 25.21 shows a binary phase diagram for the M_xC_{60} $(0 < x < 6)$ system. The α phase refers to the *fcc* structure of C_{60} (fullerite). The β phase is also *fcc*, with M atoms at all of the the octahedral and tetrahedral interstices. This phase is a line compound with composition M_3C_{60} (Fig. 25.24(b) and (c)). The γ phase possesses a tetragonal crystal structure, illustrated in a body centered setting. The γ phase has a composition of M_4C_{60} at low temperatures but exhibits a stability range at elevated temperatures. Finally, the δ phase is *bcc*, with M atoms sharing octahedral interstices in the structure. The terminal δ phase composition is M_6C_{60}.

25.5.3 Carbon nanotubes

Carbon nanotubes (CNT) (*Bucky tubes*) were first discovered by electron microscopist Sumio Iijima in 1991 (Iijima, 1991). Similar structures were also discovered at the Institute of Chemical Physics in Moscow, where researchers

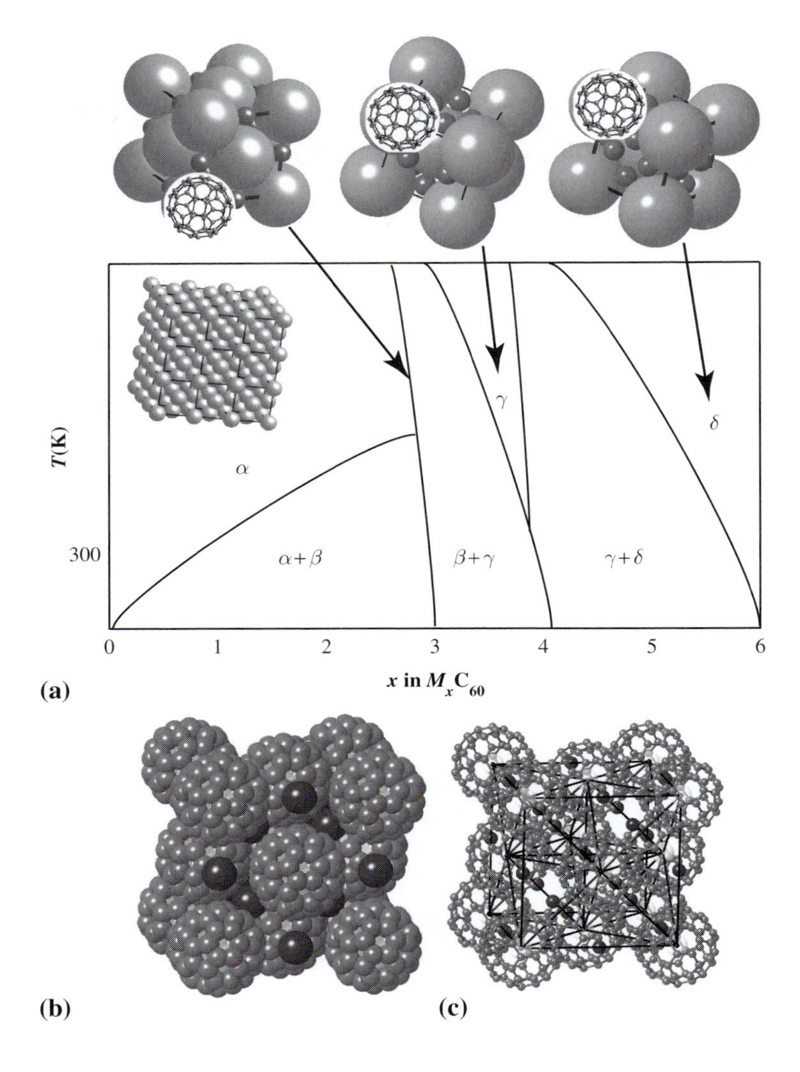

Fig. 25.24. (a) Schematic temperature–composition (x) phase diagram for the binary M_xC_{60} system and crystal structures of the various solid solutions and compounds that occur in this system. The fulleride solid K_3C_{60} is shown in space filling (b) and ball-and-stick (b) representations.

called them *barrelenes* (Dresselhaus *et al.*, 1998). These carbon fibers consist of graphitic sheets wrapped into a cylinder and capped at each end by half of a fullerene molecule (i.e., a molecule containing precisely six pentagonal rings). The length of the *graphene sheets* (height of the cylinder) is arbitrary, allowing for tubes of arbitrary length, but their diameter is geometrically constrained. Typical nanotubes can be grown with diameters of several nanometers and lengths of several microns. Carbon nanotubes have remarkable mechanical properties, including a Young's modulus in excess of 1 TPa and a tensile strength of 200 GPa. This has lead to an investigation of their use as reinforcements in polymer composites. Two such nanotubes can be capped with halves of the C_{60} fullerene molecule. These are designated the zigzag (9, 0) structure and the armchair (5, 5) structures,

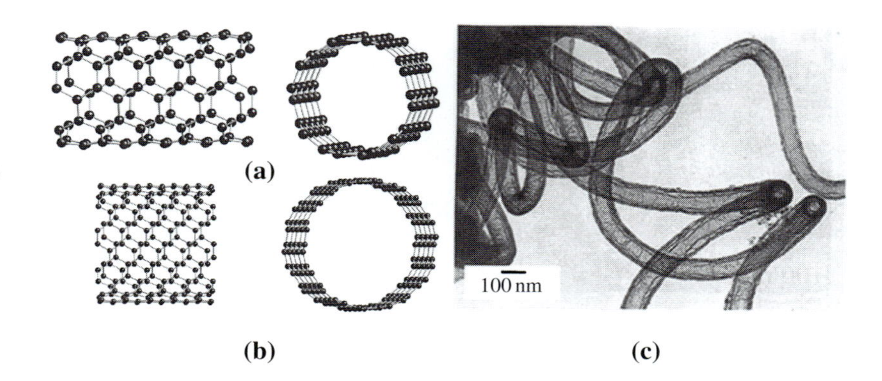

Fig. 25.25. The two single-walled nanotubes (SWNTs) that can be capped with halves of the C_{60} fullerene molecule, (a) the zigzag (9,0) structure, (b) the armchair (5,5) structure, and (c) the TEM image of a single-walled nanotube.

respectively, illustrated in Fig. 25.25(a) and (b). The nomenclature (n, m) is used to describe the method by which a portion of a planar hexagonal graphene network is cut and wrapped to produce the carbon tube that defines the middle of the nanotube. These indices are explained below.

The nanotubes first produced by Iijima in 1991 usually contained two or more concentric tubular shells and are presently known as *multi-walled nanotubes (MWNT)*. In subsequent work, Iijima and Ichihashi (1993) at NEC reported the synthesis of *single-walled nanotubes (SWNT)*. These were discovered independently by Don Bethune and coworkers at the IBM Almaden Research Center (Bethune *et al.*, 1993); they are illustrated in Fig. 25.25. Initial interest in carbon nanotubes stemmed from the premise that they could be used as prototypes for 1-D quantum wires (Mintmire *et al.*, 1992, Hamada *et al.*, 1992). The electronic structure of graphene tubules has been calculated by Saito *et al.* (1992). The conductivity of carbon nanotubes and subsequently B-containing carbon nanotubes (Hsu *et al.*, 2000) has remained an active area of research. As with graphite, the C π-electrons account for interesting diamagnetic properties. A variety of other applications of carbon nanotubes have been suggested and investigated. Notable among these has been the study of field emission from carbon nanotubes (Rinzler *et al.*, 1995) and use of the electric breakdown of nanotubes in nanotube circuits (Collins *et al.*, 2001). In many applications, the controlled introduction of impurities, either exohedrally or endohedrally, determines the nanotube properties (Clougherty, 2003). The field of carbon nanotubes has been growing rapidly since their discovery in the early 1990s.[9]

To understand the structure of carbon nanotubes, we begin with a review of the structure of graphite, previously discussed in Chapter 17. Figure 25.26(a)

[9] Excellent resources exist such as the book *Carbon Nanotubes and Related Structures: New Materials for the 21st Century*, Harris (1999).

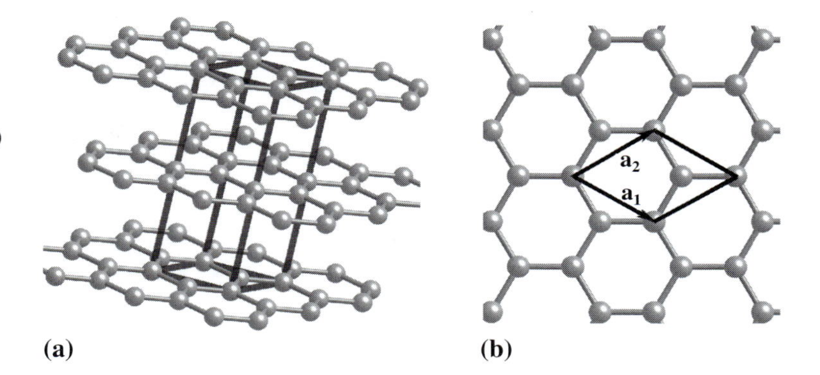

Fig. 25.26. (a) Ball-and-stick depiction of the crystal structure of graphite showing the staggered stacking of hexagonal carbon nets and (b) a single 2-D graphene sheet existing as a (00.1) plane in the graphite structure.

shows a ball-and-stick depiction of the crystal structure of graphite. The graphite structure can be regarded as a staggered stacking of hexagonal (6^3) carbon nets. Figure 25.26(b) shows a single graphene sheet (a (00.1) plane in the graphite structure) with the two-dimensional basis vectors $\mathbf{a}_1$ and $\mathbf{a}_2$. It is this graphene sheet that provides the basis for the construction of carbon nanotubes. We restrict our discussion to SWNTs. The classification of nanotubes in terms of the integers n and m of the chiral vector has been described by Dresselhaus *et al.* (1998). A general *chiral vector*, $\mathbf{C}_h$, can be defined using the graphene basis as:

$$\mathbf{C}_h = n\mathbf{a}_1 + m\mathbf{a}_2, \tag{25.4}$$

where n and m are integers. The chiral vector $\mathbf{C}_h$, also defines a chiral angle, ϕ, between 0° and 30°, which is the angle between $\mathbf{C}_h$ and the $\mathbf{a}_1$ basis vector. As an exercise, the reader may show that the chiral angle can be expressed simply as a function of the integers n and m:

$$\phi = \arctan \frac{n\sqrt{3}}{2m+n}. \tag{25.5}$$

To construct a CNT, the graphene sheet is rolled up like a rug in the direction of the chiral vector. A single-walled nanotube can be constructed by cutting the graphene sheet along a line normal to the chiral vector and passing through the origin and along a second, parallel line beginning at the end point of the chiral vector. Rolling up the rectangular section of graphene sheet so that the two cuts meet then results in a cylindrical tube. We can see that the chiral vector has now been curved into a circle, defining the circumference of the SWNT. As an exercise, The reader may show that the diameter of an SWNT, d_{SWNT}, can be expressed simply as a function of the integers n and m and the length of a C–C bond, $r_{\mathrm{C–C}}$, as:

$$d_{\mathrm{SWNT}} = r_{\mathrm{C–C}} \frac{\sqrt{3}}{\pi} \sqrt{n^2 + nm + m^2}. \tag{25.6}$$

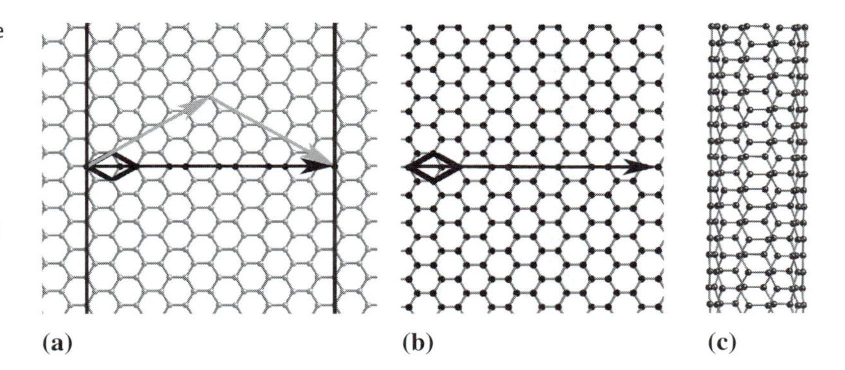

Fig. 25.27. (a) Graphene plane showing a (5, 5) chiral vector. Parallel lines normal to the chiral vector pass through the origin and the end of the chiral vector. (b) Atoms between the two parallel lines of (a). (c) The (5, 5) nanotube constructed after rolling the atoms in (b) along the chiral vector.

To completely close each end of the tube, each of the end caps must contain six pentagons. Minimal caps will consist of right and left halves of fullerenes.

Three types of nanotubes can be constructed in this way: they are called *armchair CNT*, *zigzag CNT*, and *chiral CNT*. These designations refer to the arrangement of C atoms in circumferential rings in the nanotube. For an $m = 0$ CNT, the chiral angle $\phi = 0°$ and a zigzag structure results. For $m = n$, the chiral angle $\phi = 30°$ and an armchair structure results. For arbitrary values of m and n that do not satisfy the previous conditions, a chiral structure results.

Nanometer sized clusters and tubes have properties that can differ significantly from those of bulk crystalline solids of similar chemistry, in particular the low-dimensional transport properties and magnetism. From an understanding of these structures, one can design novel materials with special properties. Examples include filling of the hollow cores of carbon nanotubes with selected metals. This is accomplished by using metal-packed anodes, by sucking liquid droplets into opened carbon nanotubes, by using iodide-catalysis reactions, and by employing other wet chemical techniques. Since the nanotube acts as a nano-scale container, the term *nanotesttube* has been introduced. The encapsulation of other chemical species in C buckyballs and buckytubes continues to lead to interesting and exciting findings, and provides for an exciting area of new science and technology (McHenry and Subramoney, 2000).

25.6 Historical notes

Francis Harry Compton Crick (1928–2004) obtained a B.Sc. in physics from University College, London, in 1937. During World War II, Crick worked as a scientist for the British Admiralty. In 1947, with support from the Medical Research Council (MRC), he began studies in biology at Cambridge. In 1949, he joined the MRC Unit at Cavendish Laboratory, Cambridge. In 1950, he was accepted as a member of Caius College, Cambridge, and obtained a Ph.D. in 1954 on a thesis entitled *X-ray diffraction: polypeptides and proteins*. Crick's friendship with **James Dewey Watson** (1928–) began

Fig. 25.28. (a) Francis Harry Compton Crick (1928–2004) and (b) Dorothy Crowfoot Hodgkin (1910–94) (pictures courtesy of the Nobel e-Museum).

in 1951. In 1952, Crick's theory of helical diffraction allowed for the interpretation of XRD patterns of DNA obtained by **Rosalind Franklin** (1920–58) and provided by **Maurice Hugh Frederick Wilkins** (1916–). In 1953, they proposed the double-helical structure for DNA and its replication scheme. In 1962, Watson and Crick shared the Nobel Prize for Medicine with Wilkins (Watson and Crick, 1953).

Watson entered the University of Chicago at age 15. He received a Ph.D. in 1950 under the direction of Salvador Luria at Indiana University. After spending time in Copenhagen, he moved to the Cavendish Laboratory, where he shared an office with Crick, who was a Ph.D. student.[10] During this time, there was an exciting race with Linus Pauling to solve the structure of DNA. Pauling's model of the α-helix (Pauling *et al.*, 1951) was instrumental in motivating Cochran, Crick, and Vand to work out the theory of helical diffraction (Cochran *et al.*, 1952). Watson was a member of the faculty of the Biology Department at Harvard University between 1956 and 1976. From 1988 to 1992, Watson directed the Human Genome Project at the National Institutes of Health.

Maurice Wilkins performed XRD studies on DNA at King's College. He received a degree in physics in 1938, and a Ph.D. in 1940 from Cambridge. He researched the luminescence of solids and worked on mass spectrograph separation of U isotopes and the Manhattan Project in California. After the

[10] Watson wrote the entertaining book *The Double Helix: A Personal Account of the Discovery of the Structure of DNA* (Watson, 1968).

war, he lectured in physics at St. Andrews' University, Scotland, and moved to King's College in 1946.

Rosalind Franklin received a B.A. in 1941 and a Ph.D. in 1945 from Cambridge. She worked at the British Coal Utilization Research Association (CURA) from 1942–46 and from 1947–50 as chercheur, Laboratoire Central des Services Chimiques de l'Etat, Paris. From 1951–58, she was Turner-Newall Research Fellow at the University of London, where her XRD patterns provided evidence for the proposal of the double helix structure of DNA. She significantly contributed to the field of crystallography, including the structure of viruses and crystals in graphitizing and non-graphitizing carbon. Franklin died in 1958 at the age of 37. Because the Nobel Prize is not awarded posthumously, Franklin could not share the award, even though her contribution to the discovery of DNA was comparable to that of those receiving the award.

Dorothy Crowfoot Hodgkin (1910–94) graduated from Oxford with a degree in chemistry in 1932. She joined the group of J. D. Bernal at Cambridge where she began research using XRD to investigate biological molecules. In 1934, she returned to Oxford, where she determined the crystal structure of cholesterol iodide. She received her doctorate from Cambridge in 1937. Between 1948 and 1956, she held positions at Cambridge and Oxford. She was named a Fellow and Chemistry Tutor at Somerville at Oxford. During that period, Margaret Thatcher worked as a student in her lab.

Hodgkin is known as a founder of the science of protein crystallography. She continued to contribute to determining the structures of cholesterol, lactoglobulin, ferritin, tobacco mosaic virus, penicillin, vitamin B-12, and insulin. She also developed methods for XRD analysis. Hodgkin was made a Fellow of the Royal Society (1947) for determining the structure of penicillin. In 1964, she was awarded the Nobel Prize in Chemistry for determining the structure of vitamin B-12. At that time, she was only the third woman to win the Nobel Prize in Chemistry. [11]

25.7 Problems

(i) *Density of solid ice*: Using the crystal structure data for solid Ice I_h, calculate its density. How does it compare with literature reported values for liquid water?

(ii) *Density of dry ice*: Using the crystal structure data for solid CO_2 calculate its density. How does it compare with literature reported values for liquid water? Will dry ice float on water?

[11] A biography of Hodgkin is available at http://www.physics.ucla.edu/~cwp and another by Linda J. Cohen appears at http://www.almaz.com/nobel/chemistry/dch.html

(iii) *Clathrate*: Calculate the touching sphere radius of an atom at the center of a pentagonal dodecahedron formed by touching spheres of equal radii, r. Do the same for the tetrakaidecahedron.

(iv) *Degree of polymerization*: Consider a sample of polyethylene containing 2 000 chains with molecular weights ranging from 0 to 4 000 g/mol; 4 000 chains with molecular weights ranging from 4 000 to 8 000 g/mol; 3 000 chains with molecular weights ranging from 8000 to 12 000 g/mol; and 1 500 chains with molecular weights ranging from 12 000 to 16 000 g/mol. Calculate the following:

(a) the weight average molecular weight;
(b) the number average molecular weight;
(c) the ratio $\overline{M_n}/\overline{M_w}$ (the polydispersity);
(d) the degree of polymerization.

(v) *Random walk I*: Choose the length of an individual mer in a long chain polymer to be $\ell = 1$. Plot the Gaussian probability distribution, $P(r)$, for $N = 100$, 1000, and 10000 steps. Calculate the root-mean-square end-to-end distance for each.

(vi) *Random walk II*: Describe how you would correct the random walk analysis to reflect the fact that certain bond angles between mers are more probable than others.

(vii) *Polyethylene*: Express the first and n-th steps in a propagation reaction to form the polymer polyethylene.

(viii) *Polyethylene crystal*: Polyethylene crystallizes in an orthorhombic cell with lattice constants, $a = 0.493$ nm, $b = 0.253$ nm, and $c = 0.74$ nm, where the polymer chains extend along the b-axis. If each cell contains two monomers, calculate the density of crystalline polyethylene and the distance between parallel chains.

(ix) *Insulin*: One of the molecular crystal structures solved by Dorothy Crowfoot Hodgkin was that of the protein insulin. Insulin crystals adopt an orthorhombic unit cell with $a = 13.0$ nm, $b = 7.48$ nm, $c = 3.09$ nm, and six insulin molecules per unit cell. The density of insulin is 1.315 g/cm^3. What is the molar mass of the insulin molecule?

(x) *Triangulation numbers I*: Draw the two triangular face configurations for enantiomers with triangulation number $T = 13$.

(xi) *Triangulation numbers II*: Determine the area per lattice point in the honeycomb, 6^3, lattice. Determine the area of an equilateral triangle with an edge beginning at $(0, 0)$ and ending at (h, k). Determine the number of lattice points contained in the triangle.

(xii) *Triangulation numbers III*: Identify the seven nodes in the $T = 7$ icosahedral decoration.

(xiii) *Triangulation numbers IV*: Describe the structure of the herpes virus which has triangulation number $T = 16$.

(xiv) *Triangulation numbers IV*: Show that the number of protein subunits in an icosahedral virus is $20T$ if they sit on general positions and $20(T-1)+12$ if there are subunits at the vertices of the icosahedron.

(xv) *Virus symmetry*: Identify the symmetry operations and point group for a $T=7$ virus.

(xvi) C_{60} *molecule*: Calculate the touching sphere radius of an entity at the center of a 60-atom truncated icosahedron formed by touching spheres of equal radii, r. What is this size for a C_{60} molecule?

(xvii) *Fullerites I*: Determine the space group for solid C_{60} if the molecules decorate the sites of an *fcc* Bravais lattice and are either spinning or randomly oriented. What is the space group if the three-fold axes of the molecule are aligned with the [111] directions?

(xviii) *Fullerites II*: Using the reported low temperature lattice constant, $a = 1.404(1)$ nm for the *fcc* C_{60} fullerite solid, calculate the following:

(a) the apparent radius of the C_{60} sphere;
(b) the maximum atomic size for an atom filling an octahedral interstice in this structure;
(c) the maximum atomic size for an atom filling a tetrahedral interstice in this structure.

(xix) *Fullerites III*: Using the radius ratio rules of Chapter 22, determine the maximum atomic size for an atom sitting in the following coordination, surrounded by C_{60} spheres:

(a) planar, triangular three-fold coordination;
(b) cubic eight-fold coordination;
(c) twelve-fold close-packed coordination.

(xx) *Fullerides*: Compare the densities of solid C_{60} and K_3C_{60}.

(xxi) *Chiral angle*: Show that the chiral angle, ϕ, can be expressed simply as a function of the integers n and m as:

$$\phi = \arctan \frac{n\sqrt{3}}{2m+n}.$$

(xxii) *CNT diameter*: Show that the diameter of a SWNT, d_{SWNT}, can be expressed simply as a function of the integers n and m and the length of a C–C bond, $r_{\mathrm{C–C}}$ as:

$$d_{\mathrm{SWNT}} = r_{\mathrm{C–C}} \frac{\sqrt{3}}{\pi} \sqrt{n^2 + nm + m^2}.$$

(xxiii) *CNT chiral angles and diameters*: Calculate the chiral angles, ϕ, and diameters, d_{SWNT}, of the following SWNTs: $(5,5)$, $(9,0)$, and $(9,3)$. How do these diameters compare with the diameter of a C_{60} buckyball?

References

S. C. Abrahams. (1983). In *Crystallography in North America. eds.* D. McLachlan Jr. and J. P. Glusker. New York, American Crystallographic Association.

A. A. Abrikosov. (1957). On the magnetic properties of superconductors of the second group. *Sov. Phys. JETP-USSR*, **5**:1174–83.

C. Adenis, V. Langer, and O. Lindqvist. (1989). Reinvestigation of the structure of tellurium. *Acta Cryst. C*, **45**:941–2.

Georgius Agricola. (1546). *De Natura Fossilium.* Basel, Per H. Frobenivm et N. Episcopivm.

Georgius Agricola. (1556). *De Re Metallica.* Basel, Per H. Frobenivm et N. Episcopivm.

T. J. Ahrens. (1952). The use of ionization potentials. 1. Ionic radii of the elements. *Geochim. Cosmochim. Acta Cryst.*, **2**:155–69.

P. Alexandris, D. Zhou, and A. Khan. (1996). Lyotropic liquid crystallinity in amphiphilic block copolymers: Temperature effects on phase behavior and structure for poly(ethylene oxide)-b-poly(propylene oxide)-b-poly(ethylene oxide) copolymers of different composition. *Langmuir*, **12**:2690–700.

G. B. Amici. (1844). Note sur un appareil de polarisation. *Ann. Chim. Phys.*, **12**: 114–17.

K. Aoki, H. Yamawaki, M. Sakashita, Y. Gotoh, and K. Takemura. (1994). Crystal structure of the high pressure phase of solid CO_2. *Science*, **263**:356–8.

D. F. J. Arago. (1811). Mémoire sur une modification remarquable quéprovent les rayons lumineux dans leur passage a travers certain corps diaphanes d'optique. *Mem. Inst. Paris*, **12**:93–134.

U. W. Arndt and A. J. Wonacott, eds. (1977). *The rotation method in crystallography.* Amsterdam, North Holland.

H. Arnfelt. Crystal structure of Fe_7W_6. *Jernkontorets Annaler*, 119:185–7, 1935.

J. E. Askeland and P. Phule. (2003). *The Science and Engineering of Materials.* Pacific Grove, USA, Brooks/Cole-Thomson Learning.

B. Aurivillius. (1950). Mixed oxides with layer lattices.1. The structure type of $CaNb_2Bi_2O_9$. *Arkiv. Kemi.*, **1**:463–80.

B. Aurivillius. (1951). Mixed oxides with layer lattices. 2. Structure of $BaBi_4Tl_4O_{15}$. *Arkiv. Kemi.*, **2**:519–27.

B. Aurivillius. (1952). Mixed oxides with layer lattices. 3. *Arkiv. Kemi.*, **5**:39–47.

V. P. S. Awana, J. Nakamura, M. Karppinen, H. Yamauchi, S. K. Malik, and W. B. Yelon. (2001). Synthesis and magnetism of Pr-based rutheno-cuprate compound $RuSr_2PrCu_2O_{8-\delta}$. *Physica C*, **357–360**:121–5.

J. E. Baggott. (1994). *Perfect Symmetry: The Accidental Discovery of the Fullerenes*. Oxford, Oxford University Press.

G. Balestrino, E. Milani, A. Paoletti, A. Tebano, Y. H. Wang, A. Ruosi, R. Vaglio, M. Valentino, and P. Paroli. (1994). Fast growth of $Bi_2Sr_2Ca_2Cu_3o_{10+x}$ and $Bi_2Sr_2CaCu_2o_{8+x}$ thin crystals at the surface of KCl fluxes. *Appl. Rev. Lett*, **64**: 1735–7.

J. Bardeen, L. N. Cooper, and J. R. Schreiffer. (1957). Theory of superconductivity. *Phys. Rev.*, **108**:1175–1204.

William Barlow. (1894). Über die geometrischen Eigenschaften homogener starrer Strukturen. *Z. Kristallogr. Mineral.*, **23**:1–63.

K. Barmak, C. Michaelsen, S. Vivekanand, and F. Ma. (1998). Formation of the first phase in sputter-deposited Nb/Al multilayer thin films. *Phil. Mag. A*, **77**:167–85.

C. Barrett and T. B. Massalski. (1980). *Structure of Metals, Crystallographic Methods, Principles and Data*, 3rd revised edition. International Series on Materials Science and Technology 35. Oxford, Pergamon.

B. A. Baumert. (1994). *Structural, Chemical, and Electronic Properties of Superconducting Thin Films and Junctions*. Ph.D. thesis, Carnegie Mellon University, Pittsburgh, USA.

W. H. Beamer and C. R. Maxwell. (1949). Physical properties of Polonium. 2. X-ray studies and crystal structure. *J. Chem. Phys.*, **17**:1293–8.

J. S. Beck, J. C. Vartuli, W. J. Roth, M. E. Leonwicz, C. T. Kresge, K. D. Schmitt, C. T-W. Chu, D. H. Olson, E. W. Sheppard, S. B. McCullen, J. B. Higgins, and J. L. Schlenker. (1992). A new family of mesoporous molecular-sieves prepared with liquid crystal templates. *J. Am. Chem.Soc.*, **114**:10834–43.

J. G. Bednorz and K. A. Müller. (1986). Possible high T_c superconductivity in the Ba–La–Cu–O system. *Z. Phys. B*, **64**:189–92.

C. A. Beevers, T. R. R. McDonald, J. H. Robertson, and F. Stern. (1952). The crystal structure of sucrose. *Acta Cryst.*, **5**:689–90.

V. K. Bel'skii, B. M. Bulychev, and A. V. Golubeva. (1983). A repeated determination of the $NaAlH_4$ structure. *Zh. Neorg. Khim.*, **28**:2694–6.

L. Bendersky. (1985). Quasicrystal with one-dimensional translational symmetry and a tenfold rotation axis. *Phys. Rev. Lett.*, **55**:1461–3.

G. Bergman and D. P. Shoemaker. (1954). The determination of the crystal structure of the sigma-phase in the iron chromium and iron molybdenum systems. *Acta Cryst.*, **7**:857–65.

G. Bergman, L. T. Waugh, and L. Pauling. (1952). The crystal structure of the metallic phase $Mg_{32}(Al,Zn)_{49}$ and related structures. *Nature*, **169**:1057–8.

G. Bergman, L. T. Waugh, and L. Pauling. (1957). The crystal structure of the metallic phase $Mg_{32}(Al,Zn)_{49}$. *Acta Cryst.*, **10**:254–9.

T. O. Bergman. (1773). *Variae Crystallorum Formae a Spatho Orthae*, volume 2. Upsala, Johannes Edman.

T. O. Bergman. (1784). De formis crystallorum praesertim e spatho ortis. In English trans. by E. Cullen, ed., In *T. Bergman, Physical and Chemical Essays* (3 Vols.). London, J. Murray.

J. D. Bernal. (1964). Bakerian lecture 1962 – Structure of liquids. *Proc. Roy. Soc. London*, **A280**:299–322.

J. D. Bernal. (1965). *Liquids: Structure Properties, Solid Interactions*. New York, Elsevier.

J. D. Bernal and J. Mason. (1960). Co-ordination of randomly packed spheres. *Nature*, **188**:910–911.

C. Bernhard, J. L. Tallon, Ch. Niedermayer, Th. Blasius, A. Golnik, E. Brucher, R. K. Kremer, D. R. Noakes, C. E. Stronack, and E. J. Asnaldo. (1999). Coexistence of ferromagnetism and superconductivity in the hybrid ruthenate-cuprate compound $RuSr_2GdCu_2O_8$. *Phys. Rev. B*, **59**:14099.

H. A. Bethe. (1928). Theorie der Beugung von Elektronen an Kristallen. *Ann. Physik*, **87**:55–129.

D. S. Bethune, C. H. Kiang, M. S. DeVries, G. Gorman, R. Savoy, and R. Beyers. (1993). Cobalt-catalysed growth of carbon nanotubes with single-atomic-layer walls. *Nature*, **363**:605–7.

R. S. Beyers, S. P. Parkin, V. Y. Lee, A. I. Nazzal, R. Savoy, G. Gorman, T. C. Huang, and S. La Placa. (1988). Crystallography and microstructure of Tl-Ca-Ba-Cu-O superconducting oxides. *Appl. Phys. Lett.*, **53**:432–4.

D. M. Bishop. (1973). *Group Theory and Chemistry*. New York, Dover.

M. R. Bockstaller and E. L. Thomas. (2004). Proximity effects in self-organized binary particle-block copolymer blends. *Phys. Rev. Lett.*, **93**:166106.

M. R. Bockstaller, Y. Lapetnikov, and E. L. Thomas. (2003). Size-selective organization of enthalpic compatibilized nanocrystals in ternary block copolymer/particle mixtures. *J. Am. Chem. Soc.*, **125**:5276–7.

M. R. Bockstaller, R. Mickiewicz, and E. L. Thomas. (2005). Block copolymer nanocomposites — new perspectives for functional materials. *Adv. Mater.*, **17**: 1331–49.

W. J. Boettinger and J. H. Perepezko. (1985). Fundamentals of rapid solidification. In *Rapidly Solidified Crystalline Alloys*. Proceedings of a TMS-AIME North East Regional Meeting, May 1–3 1985.

B. Bogdanovic and M. Schwickhardi. (1997). Ti-doped alkali metal aluminum hydrides as potential hydrogen storage materials. *J. Alloys Compd.*, **253**:1–9.

H. A. Bohr. (1932). *Almost Periodic Functions*. London, Cambridge University Press.

N. Bohr. (1913a). On the constitution of atoms and molecules. Part i. *Phil. Mag.*, **26**: 1–25.

N. Bohr. (1913b). On the constitution of atoms and molecules. Part ii. *Phil. Mag.*, **26**:476–502.

N. Bohr. (1913c). On the constitution of atoms and molecules. Part iii. *Phil. Mag.*, **26**:857–75.

R. Borchardt-Ott. (1995). *Crystallography*, 2nd edition. New York, Springer.

B. Boren. (1933). Röntgenuntersuchung der Legierungen von Silizium mit Chrom, Mangan, Kobalt und Nickel. *Arkiv Kemi. Mineral. Geol.*, **11A**:10.

R. G. Boscovich. (1758). Philosophiae Naturalis Theoria Redacta ad Unicam Legem Virium in Natura Existentium. In *Officina Libraria Kaliwodiana*, Vienna, 2nd edition.

L. Bosio, G. P. Johari, and J. Teixeira. (1986). X-ray study of high-density amorphous water. *Phys. Rev. Lett.*, **56**:460–3.

A. J. Bradley. (1935). The crystal structure of gallium. *Z. Krist.*, **A91**:302–16.

W. H. Bragg. (1915a). IX Bakerian lecture: X-rays and crystal structure. *Phil. Trans. A*, **215**:253–74.

W. H. Bragg. (1915b). The structure of magnetite and the spinels. *Nature*, **95**:561.

W. H. Bragg and W. L. Bragg. (1913). The intensity of reflection of X-rays by crystals. *Proc. R. Soc. London Ser. A*, **88**:428–38.

W. L. Bragg. (1912). The specular reflexion of X-rays. *Nature*, **90**:410.

W. L. Bragg. (1914). The analysis of crystals by the X-ray spectrometer. *Proc. R. Soc. London Ser. A*, **89**:468–89.

W. L. Bragg. (1920). The arrangement of atoms in crystals. *Phil. Mag.*, **40**:169–89.

W. L. Bragg. (1929). The determination of parameters of crystal structures by means of Fourier series. *Proc. Roy. Soc. London Ser. A*, **123**:537–59.

W. L. Bragg. (1930). The structure of silicates. *Z. Kristall.*, **74**:237–305.

A. Bravais. (1850). Memoire sur les systèmes formé par des points distribués régulièrement sur un plan ou dans l'espace. *J. Ec. Polytech.*, **19**:1–128.

D. Brewster. (1830a). *Double Refraction and Polarization of Light. (in Optics, Part 3).* Edinburgh, Edinburgh encyclopedia.

D. Brewster. (1830b). *On the Production of regular Double Refraction in the Molecules of Bodies by Simple Pressure; with Observations on the Origin of the Doubly Refracting Structure.* William Blackwood ; Edinburgh, T. Cadell.

D. Brewster. (1830c). Ueber die Hervorbringung einer regelmessigen Doppelbrechung in Körperteilchen durch blossen Druck, nebst Betrachtungen über den Ursprung des doppelt-brechenden Geffes. *Poggendorff Ann. Phys.*, **19**:527–38.

C.L. Briant and J.J. Burton. (1978). Icosahedral micro-clusters – possible structural unit in amorphous metals. *Phys. Status Solidi*, **85**:393–402.

P.W. Bridgman. (1911). Change of phase under pressure. I. The phase diagram of eleven substances with especial reference to the melting curve. *Phys. Rev.*, **3**: 153–203.

C. Brink, D.C. Hodgkin, J. Lindsey, J. Pickworth, J.H. Robertson, and J.G. White. (1964). X-ray crystallographic evidence on the structure of vitamin-B_{12}. *Nature*, **174**:1169–71.

C. Brink-Shoemaker, D.W.C. Cruickshank, D.C. Hodgkin, M.J Kamper, and D. Pilling. (1964). Structure of vitamin-B_{12}. 6. structure of crystals of vitamin-B_{12} grown from and immersed in water. *Proc. Roy. Soc. London A*, **A278**:1–26.

M. Buerger. (1956). *Elementary Crystallography.* New York, Wiley.

H.J. Bunge. (1983). *Texture Analysis in Materials Science: Mathematical Methods*, 2nd edn. London, Butterworth-Heinemann.

H.B. Burgi, E. Blanc, D. Schwarzenbach, S. Liu, Y.J. Lu, M.M. Kappes, and J.A. Ibers. (1994). The structure of C_{60} – orientational disorder in the low-temperature modification of C_{60}. *Agnew. Chem. Int. Edit.*, **31**:640–3.

G. Burns and A.M. Glazer. (1990). *Space Groups for Solid State Scientists*, 2nd edn. Boston, Academic Press Inc.

H. Busch. (1926). Calculation of the paths of cathode rays in axial symmetric electromagnetic fields. *Ann. Phys.*, **81**:974–93.

K.H.J. Buschow. (1988). "Rare Earth Compounds" in *Ferromagnetic Materials*. E.P. Wohlfarth and K.H.J. Buschow, eds. Amsterdam, Elsevier Science Publishers.

F. Buta, M.D. Sumption, and E.W. Collings. (2003). Phase stability at high temperatures in the Nb–Al system. *IEEE Trans. Appl. Super.*, **13**:3462–5.

B.F. Buxton, J.A. Eades, J.W. Steeds, and G.M. Rackham. (1976). The symmetry of electron diffraction zone axis patterns. *Phil. Trans. R. Soc.*, **281**:171–94.

M. Cadogan, H.S. Li, A. Margarian, J.B. Dunlop, D.H. Ryan, S.J. Collocott, and R.L. Davis. (1994). New rare-earth intermetallic phases $R_3(Fe,M)_{29}X_n$ – (R = Ce, Pr, Nd, Sm, Gd, M ¯Ti, V, Cr, Mn, and X ¯H, N, C). *J. Appl. Phys.*, **76**:6138–43.

J.W. Cahn. (1986). Quasiperiodic crystals: A revolution in crystallography. *MRS Bulletin*, **11**:9–11.

J.W. Cahn, Shechtman D., and D. Gratias. (1986). Indexing of icosahedral quasiperiodic crystals. *J. Mat. Res.*, **1**:13–26.

G. Calestani, C. Rozzoli, G.D. Andreetti, E. Buluggiu, D.C. Giori, A. Valenti, A. Ver, and G. Amoretti. (1989). Composition effects on the formation and superconducting character of $c \approx 31$ Å and $c \approx 37$ Å phases in the Bi-Sr-Ca-Cu-o and Bi-Pb-Sr-Ca-Cu-o systems – X-ray and ESR analysis. *Physica C*, **158**:217–24.

A. Carangeot. (1783). Goniomètre ou mesure-angle. *Obs. Phys. Hist. Nat. Arts*, **22**: 193–7.

G. Cardano. (1550). *De Subtilitate.* Nuremburg, John Petreius.

D. L. D. Caspar and A. Klug. (1963). Physical principles in the construction of regular viruses. *Symp. Quant. Biol.*, **27**:1–24.

M. Caspar. (1993). *Kepler.* (translated by C. D. Hellman, introduction, references and bibliographical citations by O. Gingerich and A. Segonds). New York, Dover.

R. J. Cava, B. Battlogg, C. H. Chen, E. A. Reitman, S. M. Zahurak, and D. Werder. (1987a). Single-phase 60 K bulk superconductor in annealed $Ba_2YCu_3o_{7-\delta}$ ($0.3 \leq \delta \leq 0.4$) with correlated oxygen vacancies in the Cu-O chains. *Phys. Rev. B*, **36**: 5719–22.

R. J. Cava, R. B. Van Dover, B. Battlogg, and E. A. Reitman. (1987b). Bulk superconductivity at 36 K in $La_{1.8}Sr_{0.2}CuO_4$. *Phys. Rev. Lett.*, **58**:408–410.

R. J. Cava, B. Batlogg, J. J. Krajewski, R. Farrow, L. W. Pupp Jr., A. E White, K. Short, W. F. Peck, and T. Kometani. (1988). Superconductivity near 30 K without copper: The $Ba_{0.6}K_{0.4}Bio_3$ perovskite. *Nature*, **332**:814–16.

P. Chaudhari, J. J. Cuomo, and R. J. Gambino. (1973). Amorphous films for magnetic bubble and magneto-optic applications. *Appl. Phys. Lett.*, **22**:337–9.

L. Y. Chiang and L. Y. Wang. (2000). Polymer derivatives of fullerenes. In K. M. Kadish and R. S. Ruoff, eds. *Fullerenes: Chemistry, Physics and Technology*. New York, John Wiley.

T. Chou, M. Igarashi, and Y. Narumiya. (1993). Soft magnetic properties of microcrystalline Fe-Al-Si-Ni-Zr-B alloys. *J. Magn. Soc. Jpn.*, **17**:197–200.

J. W. Christian. (1997). Golden years at Oxford. *Materials World*, **5**:219–220.

S. Y. Chu and M. E. McHenry. (1998a). Growth and characterization of $(Bi,Pb)_2 Sr_2Ca_2 Cu_3o_x$ single crystals. *J. Mat. Res.*, **13**:589–95.

S. Y. Chu and M. E. McHenry. (2000). Critical current density in high-T_c Bi-2223 single crystals using AC and DC magnetic measurements. *Physica C*, **337**:229–337.

S. Y. Chu and M. E. McHenry. (1998b). Irreversibility lines and pinning force density of aligned $(Bi,Pb)_2Sr_2Ca_2Cu_3O_x$ single crystals. *IEEE Trans. Appl. Supercond.*, **13**:589–95.

G. F Clark, B. K. Tanner, and H. T. Savage. (2003). Synchrotron X-ray topography studies of the magnetization process in $Tb_{0.27}Dy_{0.73}Fe_2$. *Phil. Mag. B*, **46**: 331–42.

D. P. Clougherty. (2003). Endohedral impurities in carbon nanotubes. *Phys. Rev. Lett.*, **90**:035507.

W. Cochran, F. H. C. Crick, and V. Vand. (1952). The structure of synthetic polypetides I: The transform of atoms on a helix. *Acta. Cryst.*, **5**:581–6.

J. M. D. Coey. (1996). *Rare Earth Permanent Magnets*. Oxford, Oxford Science Publications, Clarendon Press.

J. M. D. Coey and H. Sun. (1990). Improved magnetic properties by treatment of iron-based rare earth intermetallic compounds in ammonia. *J. Magn. Magn. Mater.*, **87**:L251–4.

L. B. Coleman, M. J. Cohen, D. J. Sandman, F. G. Yamagishi, A. F. Garito, and A. J. Heeger. (1973). Superconducting fluctuations and peierls instability in an organic solid. *Sol. State Comm.*, **12**:1125–32.

P. C. Collins, M. S. Arnold, and P. Avouris. (2001). Engineering carbon nanotubes and nanotube circuits using electrical breakdown. *Science*, **292**:706–9.

M. Cooper and K. Robinson. (1966). Crystal structure of ternary alloy α-(AlMnSi). *Acta. Cryst.*, **20**:614–17.

J. R. D. Copley, D. A. Neumann, R. L. Cappeletti, W. A. Kamitakahara, E. Prince, N. Coustel, J. P. McCauley Jr., N. C. Maliszewskyj, J. E. Fischer, A. B. III Smith, K. M. Creegan, and D. M. Cox. (1992). Structure and low-energy dynamics of solid C_{60}. *Physica B*, **180**:706–8.

H. S. M. Coxeter. (1973). *Regular Polytopes*. New York, Dover.

P. Cucka and C. S. Barrett. (1962). Crystal structure of Bi and of solid solutions of Pb, Sn, Sb, and Te in Bi. *Acta Cryst.*, **15**:865–72.

B. D. Cullity. (1978). *Elements of X-ray Diffraction*. Reading, MA, Addison-Wesley Publishing Company Inc.

B. D. Cullity and S. R. Stock. (2001). *Elements of X-ray Diffraction, Third Edition*. Upper Saddle River, NJ, Prentice Hall, Inc.

H. M. Cundy and A. P. Rollet. (1952). *Mathematical Models*. Oxford, Clarendon Press.

R. F. Curl and R. E. Smalley. (1991). Fullerenes. *Sci. Am.*, **256**:32–41.

J. C. L. Daams, P. Villars, and J. H. N. van Vucht eds. (1991). *Atlas of Crystal Structure Types for Intermetallic Phases*. Materials Park, Ohio, ASM International.

P. Dai, B. C. Chakoumakos, G. F. Sun, K. W. Wong, Y. Xin, and D. F. Lu. (1995). Synthesis and neutron powder diffraction study of the superconductor $HgBa_2Ca_2Cu_3o_{8+\delta}$ by Tl substitution. *Physica C*, **243**:201–6.

H. A. Davies. (1985). The magnetic and mechanical properties of rapidly quenched iron-silicon alloys. In S. Steeb and H. Warlimont, eds. *Rapidly Quenched Metals*, volume 2. Amsterdam, North Holland, pp. 1639–42.

C. J. Davisson and L. H. Germer. (1927a). The scattering of electrons by a single crystal of nickel. *Nature (London)*, **119**:558–60.

C. J. Davisson and L. H. Germer. (1927b). Diffraction of electrons by a nickel crystal. *Phys. Rev.*, **30**:705–40.

N. G. de Bruijn. (1981). Algebraic theory of Penrose nonperiodic tilings of the plane. *Proc. K. Ned. Akad. Wet. Ser. A*, **84**:39–66.

M. De Graef. (1998). A novel way to represent the 32 crystallographic point groups. *J. Mater. Educ.*, **20**:31–42.

M. De Graef. (2003). *Introduction to Conventional Transmission Electron Microscopy*. Cambridge, UK, Cambridge University Press.

P. M. De Wolff, T. Janssen, and A. Janner. (1981). The superspace groups for incommensurate crystal structures with a one-dimensional modulation. *Acta Crystallographica, Section A*, **37**:625–36.

L. de Broglie. (1923). Radiations – ondes et quant. *Comptes Rendus de l'Academie des Sciences.*, **177**:507–10.

B. N. Delaunay. (1937). The geometry of positive quadratic forms. *Uspekhi Mat. Nauk*, **3**:16–62.

J. Deportes, D. Givord, R. Lemaire, H. Nagai, and Y. J. Yang. (1976). Influence of substitutional pairs of cobalt atoms on magnetocrystalline anisotropy of cobalt-rich rare-earth compounds. *J. Less-Common Met.*, **44**:273–9.

G. J. Dickins, A. M. B. Douglas, and W. H. Taylor. (1951). Sigma-phase in the Co-Cr and Fe-Cr systems. *J. Iron Steel*, **167**:27.

G. J. Dickins, A. M. B. Douglas, and W. H. Taylor. (1956). The crystal structure of the Co-Cr σ phase. *Acta Cryst.*, **9**:297–303.

R. G. Dickinson and Linus Pauling. (1923). The crystal structure of molybdenite. *J. Am. Chem. Soc.*, **45**:1466–71.

George E. Dieter. (1976). *Mechanical Metallurgy*. New York, McGraw Hill.

P. A. M. Dirac. (1928). The quantum theory of the electron. *Proc. Roy. Soc. A*, **117**: 610–24.

G. L. Dirichlet. (1850). Über die Reduktion der positiven quadratischen Formen mit drei unbestimmten ganzen Zahlen. *J. Reine Agnew. Math.*, **40**:209–27.

W. Dollase. (1971). Refinement of the crystal structures of epidote, allanite and hancockite. *Am. Mineral.*, **56**:447–64.

I. W. Donald and H. A. Davies. (1978). Prediction of glass forming ability for metallic systems. *J. Non-Cryst. Solids*, **30**:77–85.

L. G. Dowell and A. P. Rinfret. (1960). Low-temperature forms of ice as studied by X-ray diffraction. *Nature*, **188**:1144–8.

P. A. Doyle and P. S. Turner. (1968). Relativistic Hartree-Fock X-ray and electron scattering factors. *Acta Cryst.*, **A24**:390–7.

V. P. Dravid, S. Z. Liu, and Kappes M. M. (1991). Transmission electron-microscopy of chromatographically purified solid-state C_{60} and C_{70}. *Chem. Phys. Lett.*, **185**: 75–81.

J. Drenth. (2002). *Principles of Protein X-ray Crystallography*, 2nd edn. New York, Springer.

M. S. Dresselhaus, G. Dresselhaus, Ecklund P. C., and R. Saito. (1998). Carbon nanotubes. *Physics World*, **11**:33–8.

R. A. Dunlap, M. E. McHenry, R. Chatterjee, and R. C. O'Handley. (1988a). Quasicrystal structure of rapidly solidified Ti-Ni-based alloys. *Phys. Rev. B*, **37**:8484–7.

R. A. Dunlap, M. E. McHenry, R. C. O'Handley, D. Bahadur, and V. Srinivas. (1988b). High-symmetry transition metal sites in $Ti_{56}Ni_{28-x}Fe_xSi_{16}$ quasicrystals. *J. Appl. Phys.*, **64**:5956–8.

P. Duwez, R. H. Willens, and Klement W. (1960). Continuous series of metastable solid solutions in silver-copper alloys. *J. Appl. Phys.*, **31**:1136–7.

D. D. Ebbing. (1984). *General Chemistry*. Boston, Houghton-Mifflin.

A. Einstein. (1905). Zur Elektrodynamik bewegter Körper. *Ann. Phys.*, **17**:891–921.

R. Elliot. (1983). *Eutectic Solidification Processing, Crystalline and Glassy Alloys*. London, Butterworths.

V. Elser. (1985). Indexing problem in quasicrystal diffraction. *Phys. Rev. B*, **32**: 4892–8.

V. Elser. (1986). The diffraction pattern of projected structures. *Acta Cryst. A*, **42**:36–43.

V. Elser and C. L. Henley. (1985). Crystal and quasicrystal structures in Al-Mn-Si alloys. *Phys. Rev. Lett.*, **55**:2883–6.

L. Esaki and R. Tsu. (1970). Superlattice and negative differential conductivity in semiconductors. *IBM J. Res. Dev.*, **14**:61–5.

R. C. Evans. (1966). *An Introduction to Crystal Chemistry*. Cambridge, Cambridge University Press.

Paul Ewald. (1913). Zur Theorie der Interferenzen der Röntgenstrahlen in Kristallen. *Phys. Z.*, **14**:465–72.

Paul Ewald, ed. (1962). *Fifty Years of X-ray Diffraction*. Utrecht, Oosthoek.

M. Faraday. (1823). On hydrate chlorin. *Quart. J. Sci. Lit. Arts*, **15**:71.

E. S. Federov. (1885). The elements of the study of configurations (in Russian). *Transactions of the St. Petersburg Mineralogical Society*, Part 21.

E. S. Federov. (1891). The symmetry of regular systems of figures (in Russian). *Not. Imp. St. Petersburg Mineral. Soc. Ser. 2*, **28**:1–146.

J. Ferrante, H. Schlosser, and J. R. Smith. (1991). Global expression for representing diatomic potential-energy curves. *Phys. Rev. A*, **28**:3487–94.

O. S. Filipenko, E. A. Pobedimskaya, and N. V. Belov. (1968). Crystal structure of $ZnWO_4$. *Soviet Physics Crystallography*, **13**:127–9.

L. W. Finger and R. M. Hazen. (1978). Crystal structure and compression of ruby to 46 kbar. *J. Appl. Phys.*, **49**:5823–6.

J. L. Finney. (1983). Modelling the atomic structure [amorphous alloys]. In F. E. Luborsky, ed. *Amorphous Metallic Alloys*. London, Butterworths, pp. 42–57.

M. P. A. Fisher. (1989). Vortex-glass superconductivity: A possible new phase in bulk high-T_c oxides. *Phys. Rev. Lett.*, **62**:1415–18.

R. M. Fleming, A. Kortan, B. Hessen, T. Siegrist, Thiel F. A., P. Marsh, R. C. Haddon, R. Tycko, G. Dabbagh, M. L. Kaplan, and A. M. Mujsce. (1991). Pseudotenfold symmetry in pentane-solvated C_{60} and C_{70}. *Phys. Rev. B*, **44**:888–91.

P. W. Fowler and J. I. Steert. (1987). The leapfrog principle – a rule for electron counts of carbon clusters. *J Chem. Soc. Chem. Commun.*, **18**:1403–5.

F. C. Frank. (1949). On the equations of motion of crystal dislocations. *Proc. Phys. Soc. London Sect. A*, **62**:131–5.

F. C. Frank. (1951). The growth of carborundum: Dislocations and polytypism. *Philos. Mag.*, **42**:1014–21.

F. C. Frank. (1952). Supercooling of liquids. *Proc. Roy. Soc. London*, **A215**:43–6.

F. C. Frank and J. S. Kasper. (1958). Complex alloy structures regarded as sphere packings: 1. Definitions and basic principles. *Acta Cryst.*, **11**:184–90.

F. C. Frank and J. S. Kasper. (1959). Complex alloy structures regarded as sphere packings: 2. Analysis and classification of representative structures. *Acta Cryst.*, **12**:483–99.

J. B. Friauf. (1927a). The crystal structure of magnesium di-zincide. *Phys. Rev.*, **29**: 34–40.

J. B. Friauf. (1927b). Crystal structure of two intermetallic compounds. *J. Am. Chem. Soc.*, **49**:3107–14.

W. Friedrich, P. Knipping, and M. von Laue. (1912). Interferenz-Erschetigungen bei Röntgenstrahlen. *Sitzungber. Math. Phys. Kl. K. Bayer Akad. Wiss. München*, pp. 303–22.

Y. Fujii, H. Fujita, A. Seki, and T. Tomida. (1991). Magnetic properties of fine crystalline Fe-P-C-Cu-X alloys. *J. Appl. Phys.*, **70**:6241–3.

B. T. Fultz and J. M. Howe. (2002). *Diffractometry and Transmission Electron Microscopy in Materials Science*, 2nd. ed. Berlin, Springer-Verlag.

F. G. Fumi and M. P. Tosi. (1964). Ionic sizes and born repulsive parameters in NaCl-type alakali halides, 2. the generalized Huggins-Mayer form. *J. Phys. Chem. Solids*, **25**:31–43.

R. V. Galiulin. (2003). To the 150th anniversary of the birth of Evgraf Stepanovich Fedorov (1853-1919); irregularities in the fate of the theory of regularity. *Crystallography Reports*, **48**:899–913.

R. J. Gambino and J. J. Cuomo. (1978). Selective resputtering induced anisotropy in thin films. *J. Vac. Sci. Tech.*, **15**:296–301.

H. X. Gao, L.-M. Peng, and J. M. Zuo. (1999). Lattice dynamics and Debye-Waller factors of some compounds with the sodium chloride structure. *Acta Cryst.*, **A55**:1014–25.

L. Gao, Z. J. Huang, R. L. Meng, P. H. Hor, J. Bechtold, Y. Y. Sun, C. W. Wu, Z. Z. Sheng, and A. M. Hermann. (1988). Bulk superconductivity in $Tl_2CaBa_2Cu_2O_{8+\delta}$ up to 120 K. *Nature*, **332**:623–4.

L. Gao, Z. J. Huang, R. L. Meng, G. Lin, F. Chen, L. Beuvais, Y. Y. Sun, Y. Y. Xue, and C. W. Chu. (1993). Study of superconductivity in the Hg-Ba-Ca-Cu-O system. *Physica C*, **213**:261–5.

M. Gardner. (1977). Mathematical games: Extraordinary nonperiodic tiling that enriches the theory of tiles. *Scientific American*, **237**:110–21.

D. R. Gaskell. (1993). *Introduction to Metallurgical Thermodynamics*, volume 237. Oxford, Pergamon Press.

J. R. Gavaler. (1973). Superconductivity in Nb-Ge films above 22 K. *Appl. Phys. Lett.*, **23**:480–2.

F. Gayle. (1987). Free surface solidification habit and point group symmetry of a faceted, icosahedral Al-Li-Cu phase. *J. Materials Research*, **2**:1–4.

J. C. Gehler. (1757). *De Characteribus Fossilum Externis*. Leipzig, House of Langenheim.

C. Giacovazzo. (2002a). Crystallographic computing. In C. Giacovazzo, ed. *Fundamentals of Crystallography*, chapter 2. Oxford, International Union of Crystallography, Oxford Science Publications.

C. Giacovazzo, ed. (2002b). *Fundamentals of Crystallography*, 2nd edn. Oxford, International Union of Crystallography, Oxford Science Publications.

E. Giannini, V. Garnier, R. Gladyshevskii, and R. Flukiger. (2004). Growth and characterization of $Bi_2Sr_2Ca_2Cu_3O_{10}$ and $(Bi,Pb)_2Sr_2Ca_2Cu_3O_{10-\delta}$ single crystals. *Supercond. Sci. Technol.*, **17**:220–6.

D. Givord, H. S. Li, and J. M. Moreau. (1984). Magnetic properties and crystal structure of $Nd_2Fe_{14}B$. *Solid State Commun.*, **50**:497–9.

H. Gleiter. (1989). Nanocrystalline materials. *Prog. Mat. Sci.*, **33**:223–315.

J. P. Glusker and K. N. Trueblood. (1985). *Crystal Structure Analysis: a primer*, 2nd edn. Oxford, Oxford University Press.

M. Goldberg. (1934). The isoperimetric problem for polyhedra. *Tohoku Math. J.*, **40**: 226–36.

M. Goldberg. (1937). A class of multi-symmetric polyhedra. *Tohoku Math. J.*, **43**:104–8.

V. M. Goldschmidt. (1926). Geochemische Verteilungsgesetze der Elemente. *R. Norske Vide Nsk. Acad.*, **2**.

V. M. Goldschmidt. (1928). Über Atomabstände in Metallen. *Z. Phys. Chem. Leipzig*, **133**:397–419.

V. M. Goldschmidt. (1954). *Geochemistry*, A. Muir, ed. London, Oxford University Press.

E. Goldstein. (1876). Vorläufige Mittheilungen über elektrische Entladungen in verdünnten Gasen. *Monatsber. Ak. der Wiss. Berlin*, pp. 279–96.

E. W. Gorter. (1954). Saturation magnetization and crystal chemistry of ferrimagnetic oxides. *Philips Res. Rep.*, **9**:295–320.

A. Goto, T. Hondoh, and S. Mae. (1990). The electron density distribution in ice Ih determined by single-crystal X-ray diffractometry. *J. Chem. Phys.*, **93**: 1412–17.

K. J. Gross, S. Guthrie, Takara S., and G. Thomas. (2000). In-situ X-ray diffraction study of the decomposition of $NaAlH_4$. *Alloys and Compounds*, **297**:270–81.

B. Grünbaum and G. S. Shepard. (1987). *Tilings and Patterns*. New York, W. H. Freeman.

D. Guglielmini. (1688). *Riflessioni Filosofiche Dedotte Dalle Figure de Sali*. Bologna, Antonio Pisarri.

D. Guglielmini. (1705). *De Salibus*. Venice, Dissertatio Epistolaris Physico-Medico-Mechanica.

A. Guinier and D. L. Dexter. (1963). *X-ray Studies of Materials*. New York, Interscience Publishers.

B. C. Guo, K. P. Kerns, and A. W. Castleman. (1992). Ti_8C_{12} metallo-carbohedrenes – a new class of molecular clusters. *Science*, **255**:1411–13.

R. C. Haddon, A. F. Hebard, M. J. Rosseinsky, D. W. Murphy, S. J. Duclos, K. B. Lyons, B. Miller, J. M. Rosamilia, R. M. Fleming, A. R. Kurtan, S. H. Glarum, A. V. Makhija, A. J. Muller, R. H. Eick, S. M. Zahurak, R. Tycko, G. Dabbagh, and F. A. Theil. (1991). Conducting films of C_{60} and C_{70} by alkali-metal doping. *Nature*, **350**:320–2.

T. Hahn, ed. (1989). *The International Tables for Crystallography Vol A: Space-Group Symmetry*. Dordrecht, Kluwer Academic Publishers.

T. Hahn, ed. (1996). *The International Tables for Crystallography Vol A: Space-Group Symmetry*. Dordrecht, Kluwer Academic Publishers.

N. Hamada, S. Sawada, T. A. Rabedeau, and A. Oshiyama. (1992). New one-dimensional conductors – graphitic microtubules. *Phys. Rev. Lett.*, **68**: 1679–82.

I. W. Hamley. (2000). *Introduction to Soft Matter:Polymers, Colloids, Amphiphiles and Liquid Crystals*. Chichester, Wiley.

G. R. Harp, D. Weller, T. A. Rabedeau, R. F. C. Farrow, and M. F. Toney. (1993). Magnetooptical Kerr spectroscopy of a new chemically ordered alloy – Co_3Pt. *Phys. Rev. Lett.*, **71**:2493–6.

A. B. Harris and R. Sachidanandam. (1985). Orientational ordering of icosahedra in solid C_{60}. *Phys. Rev. B*, **46**:4944–57.

P. J. F. Harris. (1999). *Carbon Nanotubes and Related Structures: New Materials for the 21st Century*. Cambridge, Cambridge University Press.

V. G. Harris and T. Pokhil. (2001). Selective-resputtering-induced perpendicular magnetic anisotropy in amorphous TbFe films. *Phys. Rev. Lett.*, **87**:067207.

V. G. Harris, Q. Huang, V. R. Shah, G. Markandeyulu, K. V. S. Rama Rao, M. Q. Huang, K. Sirisha, and M. E. McHenry. (1999). Neutron diffraction and extended X-ray absorption fine structure studies of $Pr_3(Fe_{1-x}Co_x)_{27.5}Ti_{1.5}$ permanent magnet compounds. *IEEE Trans. Magn.*, **35**:3286–8.

R.-J. Haüy. (1784). *Essai d'une Théorie sur la Structure des Cristaux*. Paris, Gogu and Né de la Rochelle.

R.-J. Haüy. (1801). *Traité de Mineralogie, 5 Volumes*. Paris, Delance.

R.-J. Haüy. (1822). *Traité de Crystallographie, 3 Volumes*. Paris, Delance.

M. Hawley, Raistrick K., J. Beery, and R. Houlton. (1991). Growth mechanism of sputtered films of $YBa_2Cu_3O_7$ studied by scanning tunneling microscopy. *Science*, **251**:1587–9.

F. C. Hawthorne and P. Cerný. (1977). The alkali-metal positions in Cs-Li beryl. *Canadian Mineralogist*, **15**:414P421.

A. D. J. Haymet. (1983). Orientational freezing in 3 dimensions – mean field theory. *Phys. Rev. B*, **27**:1725–31.

T. J. Heal and G. I. Williams. (1955). Compounds of uranium with transition metals of the 2nd and 3rd long periods. *Acta Cryst.*, **8**:494–8.

J. R. Heath, S. C. O'Brien, Q. Zhang, Y. Liu, R. F. Curl, H. W. Kroto, F. K. Tittel, and R. E. Smalley. (1985). Lanthanum complexes of spheroidal carbon shells. *J. Am. Chem. Soc.*, **107**:7779–80.

A. F. Hebard, M. J. Rosseinsky, R. C. Haddon, D. W. Murphy, Glarum S. H., T. T. M. Palstra, A. P. Ramirez, and A. R. Kortan. (1991). Superconductivity at 18 K in potassium-doped C_{60}. *Nature*, **350**:600–1.

P. A. Heiney, J. E. Fischer, A. R. McGhie, W. J. Romanow, A. M. Denenstein, J. P. McCauley Jr., A. B. Smith III, and D. E. Cox. (1991). Orientational ordering transition in solid C_{60}. *Phys. Rev. Lett.*, **66**:2911–14.

W. C. Heisenberg. (1925). Über den anschaulichen Inhalt der quantumtheoretischen Kinematik und Mechanik. *Zeits. f. Phys.*, **33**:879–93.

W. Heitler and F. London. (1927). Wechselwirkung neutraler Atome und homöpolare Binding nach der Quantenmechanik. *Zeit. Phys.*, **44**:455–72.

S. B. Hendricks and E. Teller. (1942). X-ray interference in partially ordered layer lattices. *J. Chem. Phys.*, **10**:147–58.

C. L. Henley and V. Elser. (1986). Quasi-crystal structure of $(Al,Zn)_{49}Mg_{32}$. *Phil. Mag.*, **B53**:L59–66.

R. H. Herber. (1982). *Mössbauer Effect, in the McGraw Hill Encyclopedia of Physics*. New York, McGraw Hill.

J. F. Herbst. (1991). $R_2Fe_{14}B$ materials: Intrinsic properties and technological applications. *Rev. Mod. Phys.*, **63**:819–98.

J. F. Herbst, J. J. Croat, and R. W. Lee. (1982). Neutron diffraction studies of $Nd_2(Co_xFe_{1-x})_{17}$ alloys: Preferential site occupation and magnetic structure. *J. Appl. Phys.*, **53**:250–6.

J. F. Herbst, J. J. Croat, and F. E. Pinkerton. (1984). Relationships between crystal structures and magnetic properties in $Nd_2Fe_{14}B$. *Phys. Rev. B*, **29**:4176–8.

M. Hervieu, B. Mellene, R. Retoux, S. Boudin, and B. Raveau. (2004). Route to fullernoid oxides. *Nature Materials*, **3**:269–74.

R. J. Hill and C. J. Howard. (1986). A computer program for Rietveld analysis of fixed wavelength X-ray and neutron diffraction patterns. Technical report, Australian Atomic Energy Commission Research Report M112.

D. G. Hinks, B. Dabrowski, J. D. Jorgenson, A. W. Mitchell, D. R. Richards, S. Pei, and D. Shi. (1988). Synthesis structure and superconductivity in the $Ba_{1-x}K_xBiO_{3-y}$ system. *Nature*, **333**:836–8.

M. R. Hoare. (1978). Packing models and structural specificity. *J. Non-Cryst. Sol.*, **31**:157–79.

P. E. Hodgson, E. Gadioli, and E. Gadioli Erba. (1999). *Introductory Nuclear Physics*. Oxford, Oxford Science Publications.

K. Holczer, O. Klein, S. M. Huang, R. B. Kaner, K. J. Fu, R. L. Whetten, and F. Diederich. (1991). Alkali-fulleride superconductors – synthesis, composition, and diamagnetic shielding. *Science*, **252**:1154–7.

R. Hooke. (1665). *Micrographia*. London, Jo. Martin & Js. Allestry.

E. Hovestreydt. (1983). FINAX: A computer program for correcting diffraction angles, refining cell parameters and calculating powder patterns. *J. Appl. Cryst.*, **16**:651–3.

A. Hsiao. (2001). *The Crystallization Kinetics of* Fe-Zr *Based Amorphous and Nanocrystalline Soft Magnetic Alloys*. Ph.D. thesis, Carnegie Mellon University, Pittsburgh, PA.

A. Hsiao, M. E. McHenry, D. E. Laughlin, M. J. Kramer, C Ashe, and T. Okubo. (2002). The thermal, magnetic and structural characterization of the crystallization kinetics of amorphous and soft magnetic materials. *IEEE Trans. Mag.*, **38**:2946–8.

W. K. Hsu, S. Y. Chu, E. Munoz-Picone, J. L. Boldu, S. Firth, S. Franchi, B. P. Roberts, A. Schilder, H. Terrones, N. Grobert, Y. Q. Zhu, M. Terrones, M. E. McHenry, H. W. Kroto, and D. R. M. Walton. (2000). Metallic behaviour of boron-containing carbon nanotubes. *Chem. Phys. Lett.*, **323**:572–9.

Z. Hu, W. B. Yelon, O Kaligirou, and V. Psycharis. (1996). Site occupancy and lattice changes on nitrogenation in $Nd_3Fe_{29-x}Ti_xN_y$. *J. Appl. Phys.*, **80**:2955–9.

W. Hume-Rothery. (1926). Researches on the nature, properties, and conditions of formation of intermetallic compounds with special reference to certain compounds of tin. I-V. *J. Inst. Met.*, **35**:295–361.

W. Hume-Rothery. (1952). *Atomic Theory for Students of Metallurgy*. London, Institute of Metals Monograph and Report Series, No. 3.

W. Hume-Rothery and G. V. Raynor. (1954). *The Structure of Metals and Alloys*, 3rd edition. London, Institute of Metals Monograph and Report Series, No. 1.

C. S. Hurlbut and C. Klein. (1977). *Manual of Mineralogy 9th edition*. New York, John Wiley and Sons.

D. E. Hutton. (2004). Biography: Prof. Ted Massalski in "A Festschrift Issue in Honor of T. B. Massalski". *Prog. Mat. Sci.*, **49**:215–16.

Christian Huyghens. (1690). *Traité de la Lumière*. Leiden, Pierre van der Aa.

M. S. Hybertsen and L. F. Mattheiss. (1988). Electronic band-structure of $CaBi_2Sr_2Cu_2O_8$. *Phys. Rev. Lett.*, **60**:1661–4.

J. E. Iglesias. (2006). Zhdanov's rules work both ways. *Acta Cryst. A*, **62**:195–200.

H. Ihara, R. Sugise, K. Hayashi, N. Terada, M. Jo, M. Hirabayashi, A. Negishi, N. Atado, H. Oyanagi, T. Shimomura, and S. Ohashi. (1988). Crystal structure of a new high-T_c $TlBa_2Ca_3Cu_4O_{11}$ superconductor by high resolution electron microscopy. *Phys. Rev. B*, **38**:11952–4.

H. Ihara, K. Tokiwa, H. Ozawa, M. Hirabayashi, H. Matuhata, A. Negishi, and Y. S. Song. (1994). New high tc superconductor $Ag_{1-x}Cu_xBa_2Ca_{n-1}Cu_nO_{2n+3-\delta}$ family with $T_c \geq 117$ K. *Jpn. J. App. Phys.*, **33**:300–3.

S. Iijima. (1991). Helical microtubules of graphitic carbon. *Nature*, **354**:56–8.

S. Iijima and T. Ichihashi. (1993). Single-shell carbon nanotubes of 1-nm diamete. *Nature*, **363**:603–15.

A. Inoue and J. S. Gook. (1996). Effect of additional elements on the thermal stability of supercooled liquid in $Fe_{72-x}Al_5Ga_2P_{11}C_6B_4M_x$ glassy alloys. *Mat. Trans. JIM.*, **37**:32–8.

A. Inoue, T. Zhang, and T. Masumoto. (1989). Al-La-Ni amorphous alloys with a wide supercooled liquid region. *Mat. Trans. JIM.*, **30**:965–72.

A. Inoue, T. Zhang, T. Itoi, and A. Takeuchi. (1997). New Fe-Co-Ni-Zr-B amorphous alloys with wide supercooled liquid regions and good soft magnetic properties. *JIM*, **38**:359–62.

T. Ishimasa, H. U. Nissen, and Y. Fukano. (1985). New ordered state between crystalline and amorphous in Ni-Cr particles. *Phys. Rev. Lett.*, **55**:511–13.

Y. F. Izumi, E. Takayama-Muromachi, A. Fujimori, T. Kamiyama, H. Asano, J. Akimitsu, and H. Sawa. (1989). Metal ordering and oxygen displacements in $(Nd,Sr,Ce)_2CuO_{4-y}$. *Physica C*, **158**:440–8.

H. Jagodzinski. (1949). Eindimensionale Fehlordnung in Kristallen und ihr Einfluss auf die Röntgeninterferenzen. I. Berechnung des Fehlordnungsgrades aus den Röntgenintensitäten. *Acta Cryst.*, **2**:201–7.

H. Jagodzinski. (1971). Vorwort to Fritz-Laves-Festband. *Zeit. Krist.*, **133**:2–6.

S. Jeong. (1994). *Structure and Magnetic Properties of Polycrystalline* FePt *and* CoPt *Thin Films for High Density Magnetic Recording Media.* Ph.D. thesis, Carnegie Mellon University, Pittsburgh, PA.

K. H. Johnson, M. E. McHenry, and D. P. Clougherty. (1991). High T_c superconductivity in potassium-doped fullerene, K_xC_{60}, via coupled $C_{60}(p\pi)$ cluster molecular orbitals and dynamic Jahn-Teller coupling. *Physica C*, **182**:319–23.

W. L. Johnson, M. Atzmon, M. Van Rossum, B. P. Dolgin, and X. L. Yeh. (1985). Metallic glass formation by solid state diffusion reactions – relationship to rapid quenching. In S. Steeb and H. Warlimont, eds. *Rapidly Quenched Metals*, volume 2. New York, Elsevier, North Holland, pp. 1515–20.

R. A. L. Jones. (2002). *Soft Condensed Matter.* New York, Oxford University Press.

J. D. Jorgensen, H. B. Schüttller, D. G. Hinks, D. W. Capone II, K. Zhang, M. B. Brodsky, and D. J. Scalapino. (1987). Lattice instability and high-T_c superconductivity in $La_{2-x}Ba_xCuO_4$. *Phys. Rev. Lett.*, **58**:1024–7.

J. S. Kasper, B. F. Decker, and J. R. Belanger. (1951). The crystal structure of the sigma-phase in the Co-Cr system. *J. Appl. Phys.*, **22**:361–2.

J. S. Kasper, P. Hagenmuller, M. Pouchard, and C. Cros. (1965). Clathrate structure of silicon Na_8Si_{46} and Na_xSi_{136} ($x < 11$). *Science*, **22**:361–2.

N. Kataoka, T. Matsunaga, A. Inoue, and T. Masumoto. (1989). Soft magnetic properties of bcc Fe-Au-X-Si-B (X = early transition metal) alloys with fine grain structure. *Mat. Trans. JIM*, **30**:947–50.

W. Kaufmann. (1897). Magnetischen Ablenkbarkeit der Kathodenstrahlen und ihre Abhängigkeit von Enladungspotential. *Ann. der Phys. und Chem.*, **61**:544–68.

W. H. Keesom and J. W. L. Kohler. (1934). New determination of the lattice constant of carbon dioxide. *Physica (Utrecht)*, **1**:167–74.

K. F. Kelton, G. W. Lee, A. K. Gangopadhyay, R. W. Hyers, T. J. Rathz, J. R. Rogers, M. B. Robinson, and D. S. Robinson. (2003). First X-ray scattering studies on electrostatically levitated metallic liquids: Demonstrated influence of local icosahedral order on the nucleation barrier. *Phys. Rev. Lett.*, **90**:195504.

Johannnes Kepler. (1611). *Strena Seu de Nive Sexangul.* Frankfurt, Gottfied Tampach.

Johannnes Kepler. (1966). *The Six-Cornered Snowflake.* Oxford University Press, Oxford.

Y. Khan, E. Kneller, and M. Sostarich. (1982). The phase Fe_3B. *Z. Metallk.*, **73**: 624–6.

K. Kihara. (1990). An X-ray study of the temperature dependence of the quartz structure sample: at $T = 298$ K. *Eur. J. Mineral.*, **2**:63–77.

K.-S. Kim, S.-C. Yu, K.-Y. Kim, T.-H. Noh, and I.-K. Kang. (1993). Low temperature magnetization in nanocrystalline $Fe_{88}Zr_7B_4Cu_1$ alloy. *IEEE Trans. Magn.*, **29**: 2679–81.

K. S. Kim, L. Driouch, V. Strom, B. J. Jonsson, K. V. Rao, and S. C. Yu. (1996). Magnetic properties of glassy $Fe_{91-x}Zr_7B_2Ni_x$. *IEEE*, **32**:5148–50.

I. R. Kircheim, F. Sommer, and G. Schluckebier. (1982). Hydrogen in amorphous metals – I. *Acta Met.*, **30**:1059–62.

C. Kittel. (1990). *Introduction to Solid State Physics*. New York, John Wiley and Sons.

C. Klein and C. S. Hurlbut, Jr. (1985). *Manual of Mineralogy* 20th edn. *(after James D. Dana)*. New York, J. Wiley and Sons.

F. Klein. (1876). Lectures on the icosahedron. *Math. Annalen.*, **9**:183.

F. Klein. (1956). *Lectures on the Icosahedron and the Solution of Equations of the Fifth Degree*, 2nd edn. New York, Dover.

W. Klement, R. H. Willens, and P. Duwez. (1960). Non-crystalline structure in solidified gold-silicon alloys. *Nature*, **187**:869.

T. Klemmer, D. Hoydick, H. Okumura, B. Zhang, and W. A. Soffa. (1995). Magnetic hardening and coercivity mechanisms in $L1_0$ ordered FePd ferromagnets. *Scripta Met. Mater.*, **33**:1793–1805.

A. Klug. (1993). From macromolecules to biological assemblies. In B. G. Malström, ed. *Nobel Lectures in Chemistry 1981–1990.* World Scientific Publishing Co., pp. 77–112.

M. Knoll and E. Ruska. (1932a). The Electron Microscope. *Z. Physik*, **78**:318–39.

M. Knoll and E. Ruska. (1932b). Geometric Electron Optics. *Ann. Physik*, **12**:607–40.

U. F. Kocks, C. N. Tomé, and H.-R. Wenk. (2001). *Texture and Anisotropy*. Cambridge, Cambridge University Press.

U. Koster and U. Herold. (1981). *Glassy Metals I. Topics in Physics 46.* Berlin, Springer-Verlag.

E. J. Kramer and C. L. Bauer. (1967). Internal-friction and Young's-modulus variations in the superconducting, mixed, and normal states of niobium. *Phys. Rev.*, **163**:407–19.

M. J. Kramer, L. Margulies, and R. W. McCallum. (1978). New high temperature furnace for structure refinement by powder diffraction in controlled atmospheres using synchrotron radiation. *Rev. Sci. Ins.*, **70**:3554–61.

P. Kramer. (1982). Non-periodic central space filling with icosahedral symmetry using copies of 7 elementary cells. *Acta Cryst.*, **38**:257–64.

P. Kramer and R. Neri. (1984). On periodic and non-periodic space fillings of EM obtained by projection. *Acta Crystallogr.*, **40**:580–7.

W. Kratschmer, L. D. Lamb, K. Fostiropoulos, and D. R. Huffman. (1990). Solid C_{60} – a new form of carbon. *Nature*, **347**:354–8.

I. Kraus, V. Haslar, P. Duhaj, Svec P., and V. Studnicka. (1997). The structure and magnetic properties of nanocrystalline $Co_{21}Fe_{64-x}Nb_xB_{15}$ alloys. *Mat. Sci. Eng. A*, **226**:626–30.

C. T. Kresge, M. E. Leonwicz, W. J. Roth, J. C. Vartuli, and J. S. Beck. (1992). Ordered mesoporous molecular-sieves synthesized by a liquid crystal template mechanism. *Nature*, **359**:710–12.

P. Krishna and D. Pandey. (2001). Close-packed structures. In C. A. Taylor, ed. *Commission on Crystallographic Teaching of the International Union of Crystallography Pamphlet Series.* University College Cardiff Press for the International Union of Crystallography.

H. W. Kroto, J. R. Heath, S. C. O'Brien, R. F. Curl, and R. E. Smalley. (1985). C_{60}: Buckminsterfullerene. *Nature*, **318**:162–3.

P-J. Kung. (1993). *The Combined Roles of Anisotropy and Microstructural Inhomgeneity in Determing the Effective Pinning Energy in High Temperature Superconductors.* Ph.D. thesis, Carnegie Mellon University, Pittsburgh, PA.

G. H. Kwei, J. A. Goldstone, A. C. Lawson, J. D. Thompson, and A. Williams. (1989). Low temperature structure of $Ba_{0.6}K_{0.4}BiO_3$. *Phys. Rev. B*, **39**:7378–80.

P. Labbe, M. Ledesert, and A. Maignan. (1995). Single-crystal study of the BaSr-1212 superconductor $Tl_{1+x}BaSrCa_{1+x}Cu_2O_{7-\delta}$. *Acta Cryst. B*, **51**:18–22.

D. E. Laughlin. (2004). Preface to "A Festschrift Issue in Honor of T.B. Massalski". *Prog. Mat. Sci.*, **49**:211–12.

F. Laves and H. Witte. (1935). Die Kristallstruktur des $MgNi_2$ und seine Beziehungen zu den Typen des $MgCu_2$ und $MgZn_2$. *Metallwirtsch. Metallwiss. Metalltech.*, **14**: 645–9.

F. Laves, K. Löhberg, and P. Rahlfs. (1934). Über die Isomorphei von Mg_3Al_2 und α-Mangan. *Nachr. Ges. Wiss. Göttingen Fachgr. IV*, **1**:67–71.

F. Lennard-Jones. (1924). On the determination of molecular fields. *Proc. Roy. Soc.*, **106**:463–77.

H. S. Lessure, S. Simizu, B. A. Baumert, S. G. Sankar, M. E. McHenry, M. P. Maley, J. R. Cost, and J. O. Willis. (1991). Flux pinning in neutron irradiated Y123. *IEEE Trans. Mag.*, **27**:1043–6.

D. Levine and P. J. Steinhardt. (1985). Quasicrystals. *J. Non-Cryst Solids*, **75**:85–9.

H. S. Li, J. M. Xu, D. Courtois, J. M. Cadogan, H. K. Liu, and S. X. Dou. (1996). Structure and magnetic properties of the ternary compound $Gd_3(Fe,Ti)_{29}$. *J. Phys.-Cond. Mat.*, **8**:2881–6.

H. H. Liebermann. (1983). Chapter 2. In F. E. Luborsky, ed. *Amorphous Metallic Alloys*, London, Butterworths.

J. Lima-de-Faria, ed. (1990) *Historical Atlas of Crystallography (published for the International Union of Crystallographers).* Dordrecht, Kluwer Academic Publishers.

J. Lima-de-Faria and M. O. Figueiredo. (1976). Classification, notation and ordering on a table of inorganic structure types. *J. Solid State Chem.*, **16**:7–20.

J. Lima-de-Faria and M. O. Figueiredo. (1978). General chart of inorganic structural units and building units. *Garcia de Orta Ser. Geol.*, **2**:69–76.

C. Linnaeus. (1768). *Systema Naturae, Vol.* Stockholm, Laurenti Salvii.

J. Q. Liu, M. Skowronski, C. Hallin, R. Soderholm, and H. Lendenmann. (2002). Structure of recombination-induced stacking faults in high-voltage SiC p-n junctions. *Appl. Phys. Lett.*, **80**:749–51.

L. Liu, A Inoue, and T. Zhang. (2005). Formation of bulk Pd-Cu-Si-P glass with good mechanical properties. *Mat. Trans. JIM*, **46**:376–8.

S. Liu, Y.-J. Lu, M. M. Kappes, and J. A. Ibers. (1991). The structure of the C_{60} molecule – X-ray crystal-structure determination of a twin at 110 K. *Science*, **254**: 408–10.

T. Liu, Y. F. Gao, Z. X. Xu, Z. T. Zhao, and R. Z. Ma. (1996). Compositional evolution and magnetic properties of nanocrystalline $Fe_{81.5}Cu_{0.5}Mo_{0.5}P_{12}C_3Si_{2.5}$. *J. Appl. Phys.*, **80**:3972–6.

M. Livio. (2005). *The equation that couldn't be solved.* New York, Simon & Schuster.

J. F. Loffler. (2003). Bulk metallic glasses. *Intermetallics*, **11**:529–40.

K. Y. Lonsdale. (1929). The structure of the benzene ring in $C_6(CH_3)_6$. *Proc. Roy. Soc. A*, **123**:494–515.

F. E. Luborsky. (1977). *Amorphous Magnetism II*, chapter Perspective on Application of Amorphous Alloys in Magnetic Devices. New York, Plenum Press, pp. 345–68.

F. E. Luborsky and H. Lieberman. (1978). Crystallization kinetics of Fe-B amorphous alloys. *Appl. Phys. Lett.*, **33**:233–6.

A. L. Mackay. (1962). A dense non-crystallographic packing of equal spheres. *Acta Cryst.*, **15**:916–18.

A. L. Mackay. (1975). Generalized crystallography. *Izvj. Jugosl. centr. krist. (Zagreb)*, **10**:15–36.

A. L. Mackay. (1981). De nive quinquangula: On the pentagonal snowflake. *Sov. Phys. Crystallogr.*, **26**:517–22.

A. L. Mackay. (1982). Crystallography and the Penrose pattern. *Physica A*, **114**:609–13.

J. M. MacLaren, M. E. McHenry, M. E. Eberhart, and S. Crampin. (1990). Magnetic and electronic properties of Fe/Au multilayers and interfaces. *J. Appl. Phys.*, **67**: 5406–8.

E. Madelung. (1909). Molekulare Eigenschwingungen. *Nachr. K. Ges. Wiss. Gottingen*, 304:100–106.

S. Mader, H. Widmer, F. M. d'Huerle, and A. S. Nowick. (1963). Metastable alloys of Cu-Co and Cu-Ag thin films deposited in vacuum. *Appl. Phys. Lett.*, **3**:201–3.

H. Maeda, Y. Tanaka, M. Fukutomi, and T. Asano. (1988). A new high-T_c oxide superconductor without a rare-earth element. *Jpn. J. Appl. Phys. 2*, **7**:L209–10.

Y. Maeno, H. Hashimoto, K. Yoshida, S. Nishizaka, T. Fujita, J. G. Bednorz, and F. Lichtenberg. (1994). Superconductivity in a layered perovskite without copper. *Nature*, **372**:532.

M. P. Maley, J. O. Willis, H. Lessure, and M. E. McHenry. (1990). Dependence of flux creep activation energy upon current density in grain-aligned $YBa_2Cu_3O_{7-x}$. *Phys. Rev. B*, **42**:2639–42.

M. Maret, M. C. Cadeville, W. Staiger, E. Beaurepaire, R. Poinsot, and A. Herr. (1996). Perpendicular magnetic anisotropy in Co_xPt_{1-x} alloy films. *Thin Solid Films*, **275**:224–7.

C. Marín and E. Diéguez. (1999). *Orientation of single crystals by back-reflection Laue pattern simulation*. Singapore, World Scientific.

B. Mason. (1992). *Victor Moritz Goldschmidt: Father of Modern Geochemistry*. San Antonio, Geochemical Society.

T. B. Massalski. (1981). Relationships between metallic glass formation diagrams and phase diagrams. In T. Masumoto and K. Suzuki, eds. *Proc. 4th Int. Conf. on Rapidly Quenched Metals*, ps. 203–8. The Japan Insitute of Metals.

D. P. Matheis and R. L. Snyder. (1990). The crystal structures and powder diffraction patterns of the bismuth and thallium ruddelson-popper copper oxide superconductors. *Powder Diffraction*, **5**:8–25.

S. Matteson and M. A. Nicolet. (1982). in *Metastable Materials Formation by Ion Implantation*. S. T. Picraux and W. J. Choyke. eds. New York, Elsevier.

L. F. Mattheiss and P. R. Hamann. (1988). Electronic structure of the high-T_c superconductor $Ba_{1-x}K_xBiO_3$. *Phys. Rev. Lett.*, **60**:2681–4.

M. E. McHenry. (1988). *Electronic Structure and Magnetism in Metallic Alloys Exhibiting Local Icosahedral Order*. Ph.D. thesis, Massachusetts Institute of Technology.

M. E. McHenry and S. Subramoney. (2000). Synthesis, structure, and properties of carbon encapsulated metal nanoparticles. In K. M. Kadish and R. S. Ruoff, eds. *Fullerenes: Chemistry, Physics and Technology*. New York, John Wiley.

M. E. McHenry and R. A. Sutton. (1994). Flux pinning and dissipation in high temperature oxide superconductors. *Prog. Mat. Sci.*, **38**:159–310.

M. E. McHenry, M. E. Eberhart, R. C. O'Handley, and K. H. Johnson. (1986). Calculated electronic structure of icosahedral Al and Al-Mn alloys. *Phys. Rev. Lett.*, **56**: 81–5.

M. E. McHenry, R. C. O'Handley, and K. H. Johnson. (1987). Physical significance of local density functional eigenvalues. *Phys. Rev. B*, **35**:3555–9.

M. E. McHenry, M. P. Maley, G. H. Kwei, and J. D. Thompson. (1989). Flux creep in a polycrystalline of $Ba_{0.6}K_{0.4}BiO_3$ superconductor. *Phys. Rev. B*, **39**:7339–42.

M. E. McHenry, J. M. MacLaren, M. E. Eberhart, and S. Crampin. (1990). Electronic and magnetic properties of Fe/Au multilayers and interfaces. *J. Mag. Mag. Mat.*, **88**:134–50.

M. E. McHenry, J. M. MacLaren, and D. P. Clougherty. (1991). Monolayer magnetism of 3d transition metals in Ag, Au, Pd, and Pt hosts: Systematics of local moment variation. *J. Appl. Phys.*, **70**:5902–34.

M. E. McHenry, M. A. Willard, and D. E. Laughlin. (1999). Amorphous and nanocrystalline materials for applications as soft magnets. *Prog. Mat. Sci.*, **44**:291–433.

M. E. McHenry, F. Johnson, H. Okumura, T. Ohkubo, V. R. V. Ramanan, and D. E. Laughlin. (2003). The kinetics of nanocrystallization and microstructural observations in FINEMET, NANOPERM and HITPERM nanocomposite magnetic materials. *Scripta Mater.*, **48**:881–7.

D. McKie and C. McKie. (1986). *Essentials of Crystallography*. Boston, Blackwell Scientific.

A. G. McLellan. (1961). Eigenfunctions for integer and half-odd integer values of J symmetrized according to the icosahedral group and the group c_{3v}. *J. Chem. Phys.*, **34**:1350–9.

H. D. Megaw. (1945). Crystal structure of barium titanate. *Nature*, **155**:484–5.

H. D. Megaw. (1952). Origin of ferroelectricity in barium titanate and other perovskite-type crystal. *Acta Cryst.*, **5**:739–49.

H. D. Megaw. (1957). *Ferroelectricity in Crystals*. London, Methuen.

D. Meinhardt and O. Krisement. (1962). Strukturuntersuchungen an Karbiden des Eisens, Wolframs und Chroms mit Thermischen Neutronen. *Archiv fuer das Eisenhuettenwesen*, **33**:493–9.

W. Meissner and R. Ochsenfeld. (1933). Ein neuer Effekt bei Eintritt der Supraleitfähigkeit. *Die Naturwissenschaften*, **21**:787–8.

C. Michel and C. Raveau. (1982). Les oxydes A_2BaCuO_5 (A = Y, Sm, Eu, Gd, Dy, Ho, Er, Yb). *J. Sol. St. Chem.*, **43**.

William Hallowes Miller. (1839). *A Treatise on Crystallography*. Cambridge, Deighton.

J. W. Mintmire, B. I. Dunlap, and C. T. White. (1992). Are fullerene tubules metallic? *Phys. Rev. Lett.*, **68**:631–4.

O. Mishima, L. D. Calvert, and E. Whalley. (1984). Melting ice at 77 K and 10 kbar – a new method of making amorphous solids. *Nature*, **310**:393–5.

O. Mishima, L. D. Calvert, and E. Whalley. (1985). An apparently fiirst order transition between two amorphous phases of ice induced by pressure. *Nature*, **314**:76–8.

Friedrich Mohs. (1822). *Grundriss der Mineralogie*. Dresden, Arnold.

C. E. Moore. (1970). *Ionization Potentials and Ionization Limits Derived from the Analysis of Optical Spectra*, volume 34. National Bureau of Standards.

B. Morosin, D. S. Ginley, P. F. Hlava, M. J. Carr, R. J. Baughman, J. E. Schirber, E. L. Venturini, and J. F. Kwak. (1988). Structural and compositional characterization of polycrystals and single-crystals in the Bi-superconductor and Tl-superconductor systems – crystal structure of $TlCaBa_2Cu_2O_7$. *Physica C*, **152**:413–23.

K. H. Müller and J. G. Bednorz. (1987). The discovery of a class of high temperature superconductors. *Science*, **237**:1133–9.

M. Müller, N. Mattern, and U. Kühn. (1996). Correlation between magnetic and structural properties of nanocrystalline soft magnetic alloys. *J. Mag. Magn. Mat.*, **157/158**:209–10.

P. Müller. (1977). Supraleitung in Quasibinären Legierungsreihen vom Typ A_3B-Nb_3Si mit A15 Struktur. *Z. Metallk.*, **68**:421–7.

R. S. Mulliken. (1949). Quelques aspects de la théorie des orbitales moleculaires. *J. Chem. Phys.*, **46**:497–542.

M. Murakami, M. Morita, and N. Koyama. (1989). Magnetization of a $YBa_2Cu_3O_7$ crystal prepared by the quench and melt growth process. *Jpn. J. Appl. Phys.*, **28**: L1125–7.

M. Murakami, H. Fujimoto, S. Gotoh, K. Yamaguchi, N. Koshizuka, and S. Tanaka. (1991). Flux pinning due to nonsuperconducting particles in melt processed YBaCuO superconductors. *Physica C*, **185**:321–6.

T. Nabatame, Y. Saito, K. Aihara, T. Kamo, and S. P. Matsuda. (1990). Properties of $Tl_2Ba_2Ca_2Cu_3O_x$ thin-films with a critical temperature of 122 K prepared by excimer laser ablation. *Jpn. J. Appl. Phys.*, **29**:L1813–15.

D. R. Nelson. (1988). Vortex entanglement in high-T_c superconductors. *Phys. Rev. Lett.*, **60**:1973–6.

R. E. Newnham, R. W. Wolfe, R. S. Horsey, F. A. Diaz-Colon, and M. I. Kay. (1973). Crystal-structure of $(Sr,Ba)Bi_2Ta_2O_9$. *Mater. Res. Bull.*, **8** (10):1183–95.

R. E. Newnham. (2004). *Properties Of Materials : Anisotropy, Symmetry, Structure*. Oxford, Oxford University Press.

G. S. Nolas, G. A. Slick, and S. B. Schujman. (2001). Semiconductor clathrates: A phonon glass electron crystal material with potential for thermoelectric applications. *Semiconductors and Semimetals*, **69**:255–300.

M. Norton. (1992). *Chemistry of Superconducting Materials*, page 347. Park Ridge, N. J., Noyes Publications.

A. S. Nowick. (1995). *Crystal Properties via Group Theory*. Cambridge, Cambridge University Press.

J. F. Nye. (1957). *Physical Properties of Crystals, their Representation by Tensors and Matrices*. Oxford, Clarendon Press.

R. C. O'Handley. (1987). Physics of ferromagnetic amorphous alloys. *J. Appl. Phys.*, **62**:R15–49.

W. Ohashi and F. Spaepen. (1987). Stable GaMgZn quasi-periodic crystals with pentagonal dodecahedral solidification morphology. *Nature*, **330**:555–6.

Y. Ohashi and L. W. Finger. (1978). The role of octahedral cations in pyroxenoid crystal chemistry. I. Bustamite, Wollastonite, and the Pectolite-Schizolite-Serandite series. *Amer. Min.*, **63**:274–88.

P. R. Okamoto and G. Thomas. (1968). On 4-axis hexagonal reciprocal lattice and its use in indexing of transmission electron diffraction patterns. *Phys. Stat. Solidi*, **25**: 81–5.

H. K. Onnes. (1911). The superconductivity of mercury. *Commun. Phys. Lab. Univ. Leiden*, **122** and **124**.

H. K. Onnes and J. Clay. (1908). On the change of the resistance of the metals at very low temperatures and the influence on it by small amounts of admixtures. *Commun. Phys. Lab. Univ. Leiden*, **99c**:17–26.

S. M. Opalka and D. L. Anton. (2003). First principles study of sodium-aluminum-hydrogen phases. *J. Alloys and Compounds*, **357**:486–9.

A. Pais. (1986). *Inward Bound: of Matter and Forces in the Physical World*. Oxford, Clarendon Press.

C. S. Pandey. (2000). Microstructural aspects of high and low T_c superconductors. *Mater. Phys. Mech.*, **2**:1–10.

S. S. P. Parkin, V. Y. Lee, A. I. Nazzal, R. Savoy, R. Beyers, and S. J. La Placa. (1988). $TlCa_{n-1}Ba_2Cu_nO_{2n+3}$ (n=1,2,3) – a new class of crystal-structures exhibiting volume superconductivity at up to congruent to 11 k. *Phys. Rev. Lett.*, **61**:750–3.

A. L. Patterson. (1934). A fourier series method for the determination of the components of interatomic distances in crystals. *Phys. Rev.*, **46**:372–6.

W. Pauli. (1925). Relation between the closing in of electron groups in the atom and the structure of complexes in the spectrum. *Zeits. f. Phys.*, **31**:765–83.

L. Pauling. (1923). The crystal structure of magnesium stannide. *J. Am. Chem. Soc.*, **45**:2777–80.

L. Pauling. (1927). The sizes of ions and the structure of ionic crystals. *J. Am. Chem. Soc.*, **49**:765–90.

L. Pauling. (1946). *Nature of the Chemical Bond, and subsequent editions*. Ithaca, New York, Cornell University.

L. Pauling. (1985). Apparent icosahedral symmetry is due to directed multiple twinning of cubic crystals. *Nature*, **317**:512–14.

L. Pauling. (1989). Quasicrystal structure of rapidly solidified Ti-Ni-based alloys – comment. *Phys. Rev. B*, **39**:1964–5.

L. Pauling. (1932). The nature of the chemical bond IV. The energy of single bonds and the relative electronegativity of atoms. *J. Am. Chem. Soc.*, **54**:3570–82.

L. Pauling and R. E. Marsh. (1952). The structure of chlorine hydrate. *Proc. Natl. Acad. Sci. USA*, **38**:106–11.

L. Pauling, R. B. Corey, and H. R. Branson. (1951). The structure of proteins: Two hydrogen-bonded helical configurations of the polypeptide chain. *Proc. Natl. Acad. Sci. USA*, **37**:205–11.

W. B. Pearson, ed. (1972). *The Crystal Chemistry and Physics of Metals and Alloys*. New York, Wiley-Interscience.

W. B. Pearson, J. W. Christian, and W. Hume-Rothery. (1951). New sigma-phases in binary alloys of the transition elements of the 1st long period. *Nature*, **167**:110.

S. Pei, J. D. Jorgenson, B. Dabrowski, D. G. Hinks, D. R. Richards, A. W. Mitchell, J. M. Newsam, S. K. Sinha, D. Vaknin, and A. J. Jacobson. (1990). Structural phase diagram of the $Ba_{1-x}K_xBiO_3$ system. *Phys. Rev. B*, **333**:4126–41.

A. Peker and W. L. Johnson. (1993). A highly processible metallic glass $Zr_{41.2}Ti_{13.8}Cu_{12.5}Ni_{10.0}Be_{22.5}$. *Appl. Phys. Lett.*, **63**:2342–4.

L.-M. Peng, G. Ren, S. L. Dudarev, and M. J. Whelan. (1996). Debye-Waller factors and absorptive scattering factors of elemental crystals. *Acta Crystall. A*, **52**:456–70.

R. Penrose. (1974). The role of aesthetics in pure and applied mathematical research. *Bulletin of the Institute of Mathematics and Its Applications*, **10**:266–71.

R. Penrose. (1978). Pentaplexity: A class of nonperiodic tilings of the plane. *Eureka*, **39**:468–89.

G. J. Piermarini, A. D. Mighell, C. E. Weir, and S. Block. (1969). Crystal structure of benzene-2 at 25-kilobars. *Science*, **165**:1250–5.

S. J. Poon, A. J. Drehman, and K. R. Lawless. (1985). Glassy to icosahedral phase-transformation in Pd-U-Si alloys. *Phys. Rev. Lett.*, **55**:2324–7.

S. N. Putilin, E. V. Antipov, O. Chmaissem, and M. Marezio. (1993a). Superconductivity at 94 K in $HgBa_2CuO_{4+\delta}$. *Nature (London)*, **362**:226–8.

S. N. Putilin, E. V. Antipov, and M. Marezio. (1993b). Superconductivity above 120 K in $HgBa_2Cu_2O_{6+\delta}$. *Physica C*, **212**:266–70.

P. G. Radaelli, J. L. Wagner, B. A. Hunter, M. A. Beno, G. S. Knapp, J. D. Jorgensen, and D. G. Hinks. (1993). Structure, doping and superconductivity in $HgBa_2Cu_2O_{6+\delta}$ ($T_c \leq 128$ K). *Physica C*, **216**:29–39.

A. A. Radzig and B. M. Smirnov. (1985). Reference data on atoms, molecules, and ions. In Vol. 31 of *Chemical Physics*. Berlin, Springer-Verlag.

A. D. Rae, J. G. Thompson, R. L. Withers, and A. C. Willis. (1990). Structure refinement of commensurately modulated bismuth titanate, $Bi_4Ti_3O_{12}$. *Acta Cryst.*, **B46**: 474–87.

P. Ramachandrarao and G. V. S. Sastry. (1985). A basis for the synthesis of quasicrystals. *Pramana*, **25**:L225–30.

B. Ramalingum. (1998). *Studies of Bonding and Void Nucleation at Inclusion-Matrix Interfaces in Steel*. Ph.D. thesis, Carnegie Mellon University.

V. R. V. Ramanan and G. Fish. (1982). Crystallization kinetics in Fe-B-Si metallic glasses. *J. Appl. Phys.*, **53**:2273–5.

L. S. Ramsdell. (1947). Studies on silicon carbide. *Am. Mineralogist*, **32**:64–82.

V. Randle. (2000). *Introduction to Texture Analysis: Macrotexture, Microtexture and Orientation Mapping*. CRC Press.

Rayshade. (1997). Rayshade 4.0 ray tracing system (public domain software). URL: http://graphics. stanford.edu/˜cek/rayshade/

S. Reich and Y. Tsabba. (1999). Possible nucleation of a 2D superconducting phase on WO_3 single crystal surfaces doped with Na^+. *European Physical Journal*, **B9**:1–4.

G. Rhodes. (2000). *Crystallography Made Crystal Clear* 2nd edn. New York, Academic Press.

D. Richter and J. M. Rowe. (2003). New frontiers in the application of neutron scattering to materials science. *MRS Bulletin*, **28**:903–6.

M. Rieder, G. Cavazzini, Y. S. Yakonov, V. A. Frank-Kamenetskii, G. Gottardi, S. Guggenheim, P. V. Koval, G. Müller, A. M. R. Neiva, E. W. Radoslovich, J.-L. Robert, F. P. Sassi, Z. Takeda, H. Weiss, and D. R. Wones. (1998). Nomenclature of the micas. *Can. Min.*, **36**:905–12.

W. Rieger, H. Nowotny, and F. Benesovsky. (1966). Die Kristallstruktur von W_2CoB_2 und Isotopen Phasen. *Monatshwfte Fuer Chemie*, **97**:378–82.

H. M. Rietveld. (1967). Line profiles of neutron powder-diffraction peaks for structure refinement. *Acta Cryst.*, **22**:151–2.

H. M. Rietveld. (1969). A profile refinement method for nuclear and magnetic structures. *J. Appl. Phys.*, **2**:65–71.

A. G. Rinzler, J. H. Hafner, P. Nikolaev, L. Lou, S. G. Kim, D. Tomanek, P. Nordlander, D. T. Colbert, and R. E. Smalley. (1995). Unraveling nanotubes: Field emission from an atomic wire. *Science*, **269**:1550–3.

A. F. Rogers. (1935). A tabulation of crystal forms and discussion of form-names. *Amer. Mineral.*, **20**:838–51.

E. A. Rohlfing, D. M. Cox, and A. Kaldor. (1984). Production and characterization of supersonic carbon cluster beams. *J. Phys. Chem.*, **81**:3322–30.

G. S. Rohrer. (2001). *Structure and Bonding in Crystalline Material*. Cambridge, Cambridge University Press.

J. B. L. Romé de L'Isle. (1772). *Essai de Cristallographie*. Paris, Didot Jeune.

J. B. L. Romé de L'Isle. (1783). *Essai de Cristallographie*. Paris, De limprimériés de Monsieur.

W. K. Röntgen. (1896). Über eine neue Art von Strahlen (on a new kind of rays). *Situngeber. Wurzburger Phys. Med. Ges.*, **28**:132–41.

J. H. Rose, J. R. Smith, F. Guinea, and J. Ferrante. (1984). Universal features of the equation of state of metals. *Phys. Rev. B*, **29**:2963–9.

M. J. Rosseinsky, A. P. Ramirez, S. H. Glarum, D. W. Murphy, R. C. Haddon, A. F. Hebard, T. T. M. Palstra, A. R. Kortan, S. M. Zahurak, and A. V. Makhija. (1991). Superconductivity at 28 K in Rb_xC_{60}. *Phys. Rev. Lett.*, **66**:2830–2.

S. N. Ruddelson and P. Popper. (1958). The compound Sr_2TiO_7 and its structure. *Acta Cryst.*, **11**:54–5.

P. S. Rudman. (1965a). Atomic volume of metallic elements. *Trans. AIME*, **233**:864–72.

P. S. Rudman. (1965b). Atomic volume in laves phases: A hemisubstitutional solid-solution elastic model. *Transactions of the Metallurgical Society of AIME*, **233**: 872–8.

E. Ruska. (1934a). Advances in building and performance of the magnetic electron microscope. *Z. Physik*, **87**:580–602.

E. Ruska. (1934b). Magnetic objective for the electron microscope. *Z. Physik*, **89**: 90–128.

S. Sachdev and D.R. Nelson. (1985). Order in metallic glasses and icosahedral crystals. *Phys. Rev. B*, **32**:4592–606.

R. Sachidanandam and A. B. Harris. (1985). Orientational ordering transition in solid C_{60} – comment. *Phys. Rev. Lett.*, **67**:1467.

J. Sadoc and R. Mosseri. (1984). Modeling of the structure of glasses. *J. Non-Crystalline Solids*, **61–2**:487–98.

R. Saito, M. Fujita, G. Dresselhaus, and M. S. Dresselhaus. (1992). Electronic structure of graphene tubules based on C_{60}. *Phys. Rev. B*, **46**:1804–11.

R. Salmon and M. Slater. (1987). *Computer Graphics, Systems and Concepts*. Wokingham, England, Addison-Wesley Publishing Company.

S. Samson. (1965). The crystal structure of the phase beta Mg_2Al_3. *Acta Cryst.*, **19**: 401–13.

S. Samson. (1967). The crystal structure of the intermetallic compound Cu_4Cd_3. *Acta Cryst.*, **23**:586–600.

S. Samson. (1969). Structural principles of giant cells. In B. C. Giessen, ed. *Developments in the Structural Chemistry of Alloy Phases*, pp. 65–106. Plenum Press.

T. Sawa and Y. Takahashi. (1990). Magnetic properties of Fe-Cu-(3d transition-metals)-Si-B alloys with fine-grain structure. *J. Appl. Phys.*, **67**:5565–7.

P. Scherrer. (1918). Estimation of the size and internal structure of colloid particles by measured Röntgen rays. *Nachr. Gesell. Wiss. Göttingen*, **2**:98–100.

A. Schilling, M. Cantoni, J. D. Guo, and H. R. Oh. (1993). Superconductivity above 130 K in the Hg-Ba-Ca-Cu-O system. *Nature*, **363**:56–8.

Arthur Moritz Schönflies. (1891). *Kristallsysteme und Krystallstruktur*. Leipzig, Teubner.

E. Schrödinger. (1926). Quantisierung als Eigenwertproblem. *Ann. Phys.*, **79**: 361–76.

I. K. Schüller. (1980). A new class of layered materials. *Phys. Rev. Lett.*, **44**:1597–1600.

M. Senechal. (1995). *Quasicrystals and Geometry*. Cambridge, Cambridge University Press.

V. R. Shah, G. Markandeyulu, K. V. S. Rama Rao, M. Q. Huang, K. Sirisha, and M. E. McHenry. (1998). Structural and magnetic properties of $Pr_3(Fe_{1-x}Co_x)_{27.5}Ti_{1.5}$ ($x = 0.0, 0.1, 0.2, 0.3$). *J. Magn. Magn. Mat.*, **190**:233–9.

V. R. Shah, G. Markandeyulu, K. V. S. Rama Rao, M. Q. Huang, K. Sirisha, and M. E. McHenry. (1999). Effects of co substitution on magnetic properties of $Pr_3(Fe_{1-x}Co_x)_{27.5}Ti_{1.5}$ ($x = 0$–0.3). *J. Appl. Phys.*, **85**:4678–80.

H. Shaked, P. M. Keane, J. C. Rodriguez, F. F. Owen, R. L. Hitterman, and J. D. Jorgensen. (1994). *Crystal Structures of the High-T_c Superconducting Copper-Oxides*. Amsterdam, Elsevier Science.

R. D. Shannon and C. T. Prewitt. (1969). Effective ionic radii in oxides and fluorides. *Acta. Cryst.*, **B25**:925–46.

R. T. Shannon. (1976). Revised effective ionic radii in halides and chalcogenides. *Acta. Cryst.*, **A32**:751–67.

D. Shechtman and I. Blech. (1984). The microstructure of rapidly solidified Al_6Mn. *Met. Trans.*, **16A**:1005–12.

D. Shechtman, I. Blech, D. Gratias, and J. W. Cahn. (1984). Metallic phase with long-range orientational order and no translational symmetry. *Phys. Rev. Lett.*, **53**: 1951–3.

Z. Z. Sheng and A. M. Hermann. (1988). Bulk superconductivity at 120 K in the Tl-Ca-Ba-Cu-O system. *Nature*, **332**:138–9.

E. V. Shevchenko, D. V. Talapin, S. O'Brien, and C. B. Murray. (2005). Polymorphism in AB_{13} nanoparticle superlattices: An example of semiconductor-metal metamaterials. *J. Am. Chem. Soc.*, **127**:8741–7.

C. B. Shoemaker. (1978). Refinement of the structure of beta-manganese and of a related phase in the Mn-Ni-Si system. *Acta Cryst.*, **B34**:3573–6.

C. B. Shoemaker and D. P. Shoemaker. (1969). Structural properties of some σ-phase related phases. In B. C. Giessen, ed. *Developments in the Structural Chemistry of Alloy Phases*, pp. 107–39. Plenum Press.

C. B. Shoemaker, D. P. Shoemaker, and R. Fruchart. (1984). The structure of a new magnetic phase related to the sigma phase iron-neodymium boride $Nd_2Fe_{14}B$. *Acta Cryst.*, **C40**:1665–8.

C. B. Shoemaker, D. A. Keszler, and D. P. Shoemaker. (1989). Structure of μ-$MnAl_4$ with composition close to that of quasicrystal phases. *Acta Cryst.*, **B45**:13–20.

D. P. Shoemaker. (1950). The crystal structure of a sigma phase, FeCr. *J. Am. Chem. Soc.*, **72**:5793.

A. V. Shubnikov and N. V. Belov. (1964). *Colored Symmetry*. Oxford, Oxford Pergamon Press.

R. J. Silbey, R. A. Alberty, and M. G. Bawendi. (2005). *Physical Chemistry*. Hoboken, NJ, John Wiley and Sons, Inc.

C. Silva. (1994). *Measurement and Interpretation of AC Susceptibility and Loss in High T_c Superconductors*. Ph.D. thesis, Carnegie Mellon University.

A. K. Sinha. (1972). Topologically close-packed structures of transition metal alloys. *Progress in Materials Science*, **15**:79–185.

R. Skomski. (1996). Interstitial modification, chapter 4. In *Rare Earth Permanent Magnets*. Oxford, Oxford Science Publications, Clarendon Press.

R. Skomski and J. M. D. Coey. (1993). Nitrogen diffusion in Sm_2Fe_{17} and local elastic and magnetic-properties. *J. Appl. Phys.*, **73**:7602–11.

A. W. Sleight, J. L. Gillson, and P. E. Bierstedt. (1975). High-temperature superconductivity in $BaPb_{1-x}Bi_xO_3$ system. *Sol. State Comm.*, **17**:27–8.

G. H. Smith and E. J. Burge. (1962). Analytical representation of atomic scattering amplitudes for electrons. *Acta Cryst.*, **A15**:182–6.

J. R. Smith, H. Schlosser, W. Leaf, J. Ferrante, and J. H. Rose. (1991). Global expression for representing cohesive-energy curves. *Physical Review B*, **44**:9696–9.

G. A. Smolenski and A. I. Agranovskaya. (1959). A new group of ferroelectrics (with layered structure). *Sov. Phys. Solid State*, **1**:149–50.

G. A. Smolenski, V. A. Isupov, and A. I. Agranovskaya. (1961). Ferroelectrics of the oxygen-octahedral type with layered structure. *Sov. Phys. Solid State*, **3**:651–5.

J. E. S. Socolar. (1991). Growth rules for quasicrystals. In D. P. DiVincenzo and P. J. Steinhardt, eds. *Quasicrystals: The State of the Art, volume 11 of Directions in Condensed Matter Physics*. Singapore, World Scientific.

J. E. S. Socolar and P. J. Steinhart. (1986). Quasicrystals II: unit-cell configurations. *Phys. Rev. B*, **34**:617–47.

M. Sorescu, A. Grabias, D. Tarabasnu-Mihaila, and L. Diamandescu. (2003). Bulk versus surface effects in magnetic thin films obtained by pulsed laser deposition. *Applied Surface Science*, **217**:233–8.

P. Sovák, P. Petrovič, P. Kollár, M. Zatroch, and M. Knoč. (1995). Structure and magnetic properties of an $Fe_{73.5}Cu_1U_3Si_{13.5}B_9$ nanocrystalline alloy. *J. Mag. Magn. Mat.*, **140–4**:427–8.

F. H. Spedding, A. H. Daane, and K. W. Hermann. (1956). The crystal structure and lattice parameters of high-purity, scandium, yttrium and rare earth metals. *Acta Cryst.*, **9**:559–63.

A. Speiser. (1937). *Die Theorie der Gruppen von Endlicher Ordnung*. New York, Dover.

J. C. H. Spence. (1988). *Experimental High-Resolution Electron Microscopy*. Oxford, Oxford University Press.

L. J. Spencer. (1936). Biographical notices of mineralogists recently deceased. *Miner. Mag.*, **24**:287–9.

M. Springford, ed. (1997). *Electron: A Centenary Volume*. Cambridge, Cambridge University Press.

V. Srikanth, H. Idink, W. B. White, E. C. Subbarao, H. Rajogopal, and A. Sequeira. (1996). Cation disorder in ferroelectric $PbBi_2Nb_2O_9$. *Acta Cryst.*, **B52**:432–9.

H. H. Stadelmeier. (1984). Structural classification of transition metal rare earth boride permanent compounds between T and T_5R. *Z. Metallkde*, **75**:227–30.

P. J. Steinhardt and D. Levine. (1984). Quasicrystals – a new class of ordered structures. *Phys. Rev. Lett.*, **53**:2477–80.

N. Steno. (1669). *De Solido Intra Solidum Naturaliter Contento*. Florence, Star.

P. W. Stephens and A. I. Goldman. (1986). Sharp diffraction maxima from an icosahedral glass. *Phys. Rev. Lett.*, **56**:1168–71.

G. J. Stoney. (1891). Cause of double lines and of equidistant satellites in the spectra of gases. *Transactions of the Royal Dublin Society*, **4**:563–9.

K. Strnat, G. Hoffer, J. C. Olson, W. Ostertag, and J. J. Becker. (1967). A family of new cobalt-base permanent magnet materials. *J. Appl. Phys.*, **38**:1001–2.

E. C. Subbarao. (1962). A family of ferroelectric bismuth compounds. *J. Phys. Chem. Solids*, **23**:665–76.

M. A. Subramanian, Calabrese, C. C. Torardi, J. Gopalakrishnan, T. R. Askew, R. B. Flippen, K. J. Morrissey, U. Chowdhry, and A. W. Sleight. (1988a). Crystal structure of the high-temperature superconductor $Tl_2Ba_2CaCuO_8$. *Nature*, **332**:420–2.

M. A. Subramanian, C. C. Torardi, J. Gopalakrishnan, P. L. Gai, J. C. Calabrese, T. R. Askew, R. B. Flippen, and A. W. Sleight. (1988b). Bulk superconductivity up to 122 K in the Tl-Pb-Sr-Ca-Cu-o system. *Science*, **242**:249–52.

M. J. Sugise, T. M. Hirabayahi, N. Terada, T. Shimomura, and H. Ihara. (1988). The formation process of new high-t_c superconductors with single-layer thallium-oxide, $TlBa_2Ca_3Cu_3O_y$ and $TlBa_2Ca_3Cu_4O_y$. *Jpn. J. App. Phys.*, **27**:L1709–11.

S. A. Sunshine, T. Siegrist, L. F. Schneemeyer, D. W. Murphy, R. J. Cava, B. Batlogg, R. B. van Dover, R. M. Fleming, S. H. Glarum, S. Nakahara, R. Farrow, J. J. Krajewski, S. M. Zahurak, J. V. Waszczak, J. H. Marshall, P. Marsh, L. W. Rupp Jr., and W. F. Peck. (1988). Structure and physical properties of single crystals of the 84 K superconductor $Bi_{2.2}Sr_2Ca_{0.8}Cu_2O_{8+\delta}$. *Phys. Rev. B*, **38**:893–6.

R. A. Sutton. (1994). *Irradiation Defects in Oxide Superconductors: Their Role in Flux Pinning*. Ph.D. thesis, Carnegie Mellon University, Pittsburgh, PA.

R. A. Sutton, M. E. McHenry, and K. E. Sickafus. (1997). Annealing of neutron irradiation induced defects in LSCO crystals. *IEEE Trans. Appl. Superconductivity.*, **7**:2001–4.

K. Suzuki, N. Kataoka, A. Inoue, A. Makino, and T. Masumoto. (1990). High saturation magnetization and soft magnetic properties of bcc Fe-Zr-B alloys with ultrafine grain structure. *Mat. Trans. JIM*, **31**:743–6.

K. Suzuki, A. Makino, N. Kataoka, A. Inoue, and T. Masumoto. (1991). High saturation magnetization and soft magnetic properties of bcc Fe-Zr-B and Fe-Zr-B-m (m = transition metal) alloys with nanoscale grain size. *Mat. Trans. JIM*, **32**: 93–102.

R. Swaminathan. (2005). *Influence of Surface Structure on the Magnetic Properties of RF Plasma Synthesized* NiZn *Ferrite Nanoparticles*. Ph.D. thesis, Carnegie Mellon University, Pittsburgh, PA.

R. Swaminathan, Willard M. A., and M. E. McHenry. (2006). Experimental observations and nucleation and growth theory of polyhedral magnetic ferrite nanoparticles synthesized using an RF plasma torch. *Acta Materialia*, **54**:807–16.

H. E. Swanson and E. Tatge. (1953). Standard X-ray diffraction patterns. *National Bureau of Standards (U.S.)*, Circular 539, Vol. I.

M. Takano, M. Azuma, Z. Hiroi, Y. Bando, and Y. Takeda. (1991). Superconductivity in the Ba-Sr-Cu-O system. *Physica C*, **176**:441–4.

M. Tanaka, H. Sekii, and T. Nagasawa. (1983). Space-group determination by dynamic extinction in convergent-beam electron diffraction. *Acta Crystall. A*, **39**:825–37.

R. Taton, ed. (1969). *Histoire Generale des Sciences. Tome II, (p.288)*. Paris, Presses Universitaires de France.

J. J. Thomson. (1897). Cathode rays. *Phil. Mag.*, **44**:311–35.

J. J. Thomson. (1899). The masses of the ions in gases at low pressures. *Phil. Mag.*, **48**:547–67.

Y. Tokura, H. Takagi, and S. Uchida. (1989). A superconducting copper-oxide compound with electrons as the charge-carriers. *Nature*, **337**:345–7.

T. Tomida. (1994). Crystallization of an Fe-Si-B-Ga-Nb amorphous alloy. *Mat. Sci. Eng.*, **A179/180**:521–5.

C. C. Torardi, M. A. Subramanian, J. C. Calabrese, J. Gopalakrishnan, E. M. McCarron, K. J. Morrissey, T. R. Askew, R. B. Flippen, U. Chowdhry, and A. W. Sleight. (1988). Structures of the superconducting oxides $Tl_2Ba_2CuO_6$ and $Bi_2Sr_2CuO_6$. *Phys. Rev. B*, **38**:225–31.

C. W. Tucker. (1950). An approximate crystal structure for the beta-phase of Uranium. *Science*, **112**:448.

Z. Turgut. (1999). *Thermal Plasma Synthesis of Coated* FeCo-FeCoV *Nanoparticles as Precursors for Compacted Nanocrystalline Bulk Magnets*. Ph.D. thesis, Carnegie Mellon University, Pittsburgh, PA.

D. Turnbull. (1950). Formation of crystal nuclei in liquid metals. *J. Appl. Phys.*, **21**: 1022–8.

R. Tycko, R. C. Haddon, G. Dabbagh, S. H. Glarum, D. C. Douglass, and A. M. Mujsce. (1991). Solid-state magnetic-resonance spectroscopy of fullerenes. *J. Phys. Chem.*, **95**:518–20.

C.-Y. Um. (2006). *Fe-based Amorphous and Nanocrystalline Nanocomposite Soft Ferromagnetic Materials*. Ph.D. thesis, Carnegie Mellon University, Pittsburgh, PA.

Anton Van Leewenhoek. (1685a). *Arcana Natura Detecta*. Delft, Henrik Van Krooneveld.

Anton Van Leewenhoek. (1685b). Concerning the various figures of the salts contained in the several substances. *pht*, **15**:1073–90.

G. Van Tendeloo and S. Amelinckx. (1974). Group-theoretical considerations concerning domain formation in ordered alloys. *Acta Cyst. A*, **30**:431–9.

C. F. Varley. (1871). Experiment on the discharge of electricity through rarefied media and the atmosphere. *Proc. Roy. Soc.*, **19**:236–42.

G. Venkataraman, D. Sahoo, and V. Balakrishnan. (1989). *Beyond the Crystalline State: An Emerging Perspective*. Springer Series in Solid State Sciences 84.

P. Villars, ed. (1997). *Pearson's Handbook Desk Edition*. Materials Park, Ohio, ASM International.

P. Villars and L. D. Calvert, eds. (1991). *Pearson's Handbook of Crystallographic Data for Intermetallic Phases*. Materials Park, Ohio, ASM International.

Paul Heinrich Von Groth. (1895). *Physikalische Krystallographie und Einleitung in die krystallographische Kenntniss der wichtigeren Substanzen*. Leipzig, Engelmann.

M. G. Voronoi. (1908). Nouvelles applications des paramètres continus a la théorie des formes quadratiques. *J. Reine u. Angew. Math.*, **134**:198–287.

C. N. J. Wagner. (1969). Structure of amorphous alloy films. *J. Vac. Sci. Technol.*, **6**: 650–4.

J. L. Wagner, P. G. Radaelli, B. A. Hunter, M. A. Beno, G. S. Knapp, J. D. Jorgensen, and D. G. Hinks. (1993). Structure and superconductivity of $HgBa_2CuO_{4+\delta}$. *Physica C*, **210**:447–54.

B. H. Walker. (1995). *Modern Optical Engineering*. New York, McGraw-Hill.

M. G. Wallace. (1960). Intermetallic compounds between lanthanons and transition metals of the first long period 1. Preparation, existence and structural studies. *J. Phys. Chem. Sol.*, **16**:123–30.

N. Wang, H. Chen, and K. H. Kuo. (1987). Two-dimensional quasi-crystal with eightfold rotational symmetry. *Phys. Rev. Lett.*, **59**:1010–13.

B. E. Warren. (1990). *X-ray Diffraction*. New York, Dover.

H. Watanabe, H. Saito, and M. Takahashi. (1993). Soft magnetic properties and structures of nanocrystalline Fe-Al-Si-Nb-B alloy ribbons. *Trans. Mag. Soc. Jpn*, **8**:888–94.

J. D. Watson. (1968). *The Double Helix: A Personal Account of the Discovery of the Structure of DNA*. New York, Penguin Books.

J. D. Watson and F. H. C. Crick. (1953). A structure for deoxy ribose nucleic acid. *Nature (London)*, **171**:737–8.

D. Weller, A. Moser, L. Folks, M. E. Best, W. Lee, M. F. Toney, M. Schwickert, J-U. Thiele, and M. F. Doerner. (2002). High K_u materials approach to 100 Gbits/in^2. *IEEE Trans. Mag.*, **36**:10–15.

U. Welp, G. W. Crabtree, J. L. Wagner, D. G. Hinks, P. G. Radaelli, J. D. Jorgensen, and J. F. Mitchell. (1993). The irreversibility line of $HgBa_2CuO_{4+\delta}$. *Appl. Phys. Lett.*, **63**:693–5.

M. J. Wenninger. (1971). *Polyhedron Models*. Cambridge, Cambridge University Press.

A. G. Werner. (1774). *Von den usserlichen Kennzeichen der Fossilien*. Leipzig, Seigfried Lebrecht Crusius.

Ab. G. Werner. (1775). *Course on Mineralogy*. Sec Amoros.

J. H. Wernick and S. Geller. (1959). Transition element – rare earth compounds with the $CaCu_5$ structure. *Acta Cryst.*, **12**:662–5.

M. Widom. (1985). Icosahedral order in glass: Electronic properties. *Phys. Rev. B*, **31**:6456–68.

M. Widom and M. Mihalkovik. (2005). Stability of Fe-based alloys with structure type C_6Cr_{23}. *J. Mater. Res.*, **20**:237–42.

J. E. Wiechert. (1897). Wesen der Elektric. und Exper. über Kathodenstrahlen. *Schriften der Phys.-Ökonomischen Ges. zu Königsberg (Abh)*, **38**:3–19.

M. A. Willard. (2000). *Structural and Magnetic Characterization of HITPERM Soft Magnetic Materials for High Temperature Applications*. Ph.D. thesis, Carnegie Mellon University, Pittsburgh, PA.

M. A. Willard, D. E. Laughlin, M. E. McHenry, K. Sickafus, J. O. Cross, V. G. Harris, and D. Thoma. (1998). Structure and magnetic properties of $(Fe_{0.5}Co_{0.5})_{88}Zr_7B_4Cu_1$ nanocrystalline alloys. *J. Apply. Phys.*, **84**:6773–7.

D. B. Williams and C. B. Carter. (1996). *Transmission Electron Microscopy, a Textbook for Materials Science*. New York, Plenum Press.

S. D. Willoughby. (2002). *Theoretical Investigations of* Co-Pt *Alloys and Supercells*. Ph.D. thesis, Tulane University, New Orleans, LA.

S. D. Willoughby, J. M. MacLaren, T. Ohkubo, S. Jeong, M. E. McHenry, D. E. Laughlin, and S-J. Choi. (2002). Electronic, magnetic, and structural properties of $L1_0$ $FePt_xPd_{1-x}$ alloys. *J. Appl. Phys.*, **91**:8822–4.

S. D. Willoughby, R. A. Stern, R. Duplessis, J. M. MacLaren, D. E. Laughlin, and M. E. McHenry. (2003). Electronic structure calculations of hexagonal and cubic phases of Co_3Pt. *J. Appl. Phys.*, **93**:7145–7.

C. G. Wilson and F. J. Spooner. (1960). The crystal structure of Zr_3Al_2. *Acta Cryst.*, **13**:358–9.

C. G. Wilson, D. K. Thomas, and F. J. Spooner. (1960). The crystal structure of Zr_4Al_3. *Acta Cryst.*, **13**:56–7.

W. H. Wollaston. (1809). Description of a reflective goniometer. *Phil. Trans. Royal Soc.*, **99**:253–6.

W. H. Wollaston. (1813). On the elementary particles of certain crystals. *Phil. Trans. Royal Soc.*, **103**:51–63.

M. K. Wu, J. R. Ashburn, C. J. Torng, P. H. Hor, R. L. Meng, L. Gao, Z. J. Huang, Y. A. Wang, and C. W. Chu. (1987). Superconductivity at 93 K in a new mixed-phase Y-Ba-Cu-O compound system at ambient pressure. *Phys. Rev. Lett.*, **58**:908–10.

Y. V. Wulff. (1902). Untersuchungen im Gebiete der optischen Eigenschaften isomorpher Krystalle. *Z. Kristallogr.*, **36**:1–28.

Y. V. Wulff. (1908). Zur Theorie der Krystallhabitus. *Z. Kristallogr.*, **45**:433–72.

R. W. G. Wyckoff. (1963). *Crystal Structures, Volume 1, and subsequent volumes.* New York, John Wiley.

A. Wykes. (1969). *Doctor Cardano; Physician Extraordinaire.* London, Frederick Muller Ltd.

A. Yamamoto. (1996). Crystallography of quasiperiodic crystals. *Acta Cryst.*, **A52**: 509–60.

Y. C. Yang, Q. Pan, X. D. Zhang, M. H. Zhang, C. L. Yang, Y. Li, S. L. Ge, and B. F. Zhang. (1993). Structural and magnetic properties of $RMo_{1.5}Fe_{10.5}N_x$. *J. Appl. Phys.*, **74**:4066–71.

C. S. Yannoni, R. D. Johnson, G. Meijer, *et al.* (1991). C^{13} NMR-study of the C_{60} cluster in the solid-state – molecular-motion and carbon chemical-shift anisotropy. *J. Phys. Chem.*, **95**:9–10.

W. B. Yelon and Z. Hu. (1996). Neutron diffraction study of lattice changes in $Nd_2Fe_{17-x}\,Si_x(C_y)$. *J. Appl. Phys.*, **78**:7196–201.

Y. Yoshizawa and K. Yamauchi. (1991). Magnetic properties of Fe-Cu-Cr-Si-B, Fe-Cu-V-Si-B, and Fe-Cu-Mo-Si-B alloys. *Mat. Sci. Eng. A*, **133**:176–9.

Y. Yoshizawa, S. Oguma, and K. Yamauchi. (1988). New Fe-based soft magnetic alloys composed of ultrafine grain structure. *J. Appl. Phys.*, **64**:6044–6.

Y. Yoshizawa, Y. Bizen, K. Yamauchi, and H. Sugihara. (1992). Improvement of magnetic properties in Fe-based nanocrystalline alloys by addition of Si, Ge, C, Ga, P, Al elements and their applications. *Trans. IEE of Japan*, **112A**:553–8.

R. A. Young. (2000). *The Rietveld method.* Oxford, Oxford University Press.

G. A. Yurko, J. W. Burton, and J. G. Parr. (1959). The crystal structure of Ti_2Ni. *Acta Cryst.*, **12**:909–11.

W. H. Zachariessen. (1931). A set of empirical crystal radii for ions with inert gas configuration. *Z. Kristall.*, **80**:137–53.

W. H. Zachariessen. (1932). The atomic arrangement in glass. *J. Am. Chem. Soc.*, **54**: 3841–51.

Z. Zhang, H. Q. Ye, and K. H. Kuo. (1985). A new icosahedral phase with m35 symmetry. *Phil. Mag.*, **A52**:L49–52.

Y. Zuxiang. (1986). Two new minerals Gupeiite and Xifengite in cosmic dusts from Yanshan. *Am. Min.*, **71**:228.

Index